3.4 FORMULAS AND LITERAL EQUATIONS

Figure	Area	Figure	Volume
Square	$A = s^2$	Cube	$V = s^3$
Rectangle	$A = lw$	Rectangular solid	$V = lwh$
Circle	$A = \pi r^2$	Sphere	$V = \frac{4}{3}\pi r^3$
Triangle	$A = \frac{1}{2}bh$	Cylinder	$V = Bh$*
Trapezoid	$A = \frac{1}{2}h(b_1 + b_2)$	Cone	$V = \frac{1}{3}Bh$*
		Pyramid	$V = \frac{1}{3}Bh$*

*B is the area of the base.

3.5 ABSOLUTE VALUE EQUATIONS

If $k \geq 0$, then

$|x| = k$ is equivalent to $x = k$ or $x = -k$.
$|a| = |b|$ is equivalent to $a = b$ or $a = -b$.

3.6 LINEAR INEQUALITIES

If a, b, and c are real numbers and $a < b$, then

$a + c < b + c \qquad a - c < b - c$

$ac < bc \quad (c > 0) \qquad ac > bc \quad (c < 0)$

$\frac{a}{c} < \frac{b}{c} \quad (c > 0) \qquad \frac{a}{c} > \frac{b}{c} \quad (c < 0)$

$c < x < d$ is equivalent to $c < x$ and $x < d$.

3.7 INEQUALITIES WITH ABSOLUTE VALUES

If $k > 0$, then

$|x| < k$ is equivalent to $-k < x < k$.
$|x| > k$ is equivalent to $x < -k$ or $x > k$.

4.1–4.4 FACTORING POLYNOMIALS

$ax + bx = x(a + b)$

$(a + b)x + (a + b)y = (a + b)(x + y)$

$ax + ay + cx + cy = a(x + y) + c(x + y)$
$\qquad = (x + y)(a + c)$

$x^2 - y^2 = (x + y)(x - y)$

$x^3 + y^3 = (x + y)(x^2 - xy + y^2)$

$x^3 - y^3 = (x - y)(x^2 + xy + y^2)$

$x^2 + 2xy + y^2 = (x + y)(x + y)$

$x^2 - 2xy + y^2 = (x - y)(x - y)$

4.5 SOLVING EQUATIONS BY FACTORING

Zero-factor theorem: Let a and b be real numbers, then

If $ab = 0$, then $a = 0$ or $b = 0$.

6.1 THE RECTANGULAR COORDINATE SYSTEM

The graph of $x = a$ is a vertical line with x-intercept at $(a, 0)$.

The graph of $y = b$ is a horizontal line with y-intercept at $(0, b)$.

Midpoint formula: If $P(x_1, y_1)$ and $Q(x_2, y_2)$ are endpoints of segment PQ, the midpoint of PQ is

$$\left(\frac{x_1 + x_2}{2}, \frac{y_1 + y_2}{2}\right)$$

6.2 SLOPE OF A NONVERTICAL LINE

The **slope** of the line passing through $P(x_1, y_1)$ and $Q(x_2, y_2)$ is

$$m = \frac{\Delta y}{\Delta x} = \frac{y_2 - y_1}{x_2 - x_1} \quad (x_2 \neq x_1)$$

Horizontal lines have a slope of 0.

Vertical lines have no defined slope.

Lines with the same slope are parallel.

Lines with slopes that are negative reciprocals are perpendicular.

6.3 EQUATIONS OF LINES

Point–slope form of the equation of a line:

$y - y_1 = m(x - x_1)$

Slope–intercept form of the equation of a line:

$y = mx + b$

General form of the equation of a line:

$Ax + By = C$

Equation of a horizontal line: $x = a$

Equation of a vertical line: $y = b$

6.6 INTRODUCTION TO FUNCTIONS

$f(a)$ is the value of $y = f(x)$ when $x = a$.

6.7 PROPORTION AND VARIATION

Direct variation: $y = kx$

Inverse variation: $y = \frac{k}{x}$

Joint variation: $y = kxz$

Combined variation: $y = \frac{kx}{z}$

Intermediate Algebra

Books in the Gustafson/Frisk Series

Essential Mathematics with Geometry, Second Edition
Beginning Algebra, Fourth Edition
Intermediate Algebra, Fourth Edition
Algebra for College Students, Fourth Edition
College Algebra, Fifth Edition
Plane Trigonometry, Fourth Edition
College Algebra and Trigonometry, Fourth Edition
Functions and Graphs, Second Edition

4 Intermediate Algebra

R. David Gustafson
Peter D. Frisk

Rock Valley College

Brooks/Cole Publishing Company

I(T)P™ An International Thomson Publishing Company

Pacific Grove • Albany • Bonn • Boston • Cincinnati • Detroit • London • Madrid • Melbourne
Mexico City • New York • Paris • San Francisco • Singapore • Tokyo • Toronto • Washington

GWO A Gary W. Ostedt Book

Editor: *Gary W. Ostedt*
Editorial Associate: *Carol Ann Benedict*
Production Editor: *Ellen Brownstein*
Production Service: *Hoyt Publishing Services*
Manuscript Editor: *David Hoyt*
Permissions Editor: *May Clark*
Interior and Cover Design: *E. Kelly Shoemaker*
Marketing Team: *Patrick Farrant* and *Jean Vevers Thompson*
Interior Illustration: *Lori Heckelman*
Cover Photo: *COMSTOCK, INC.*
Photo Editor: *Kathleen Olson*
Typesetting: *The Clarinda Company*
Cover Printing: *Color Dot*
Printing and Binding: *R. R. Donnelley & Sons, Crawfordsville*

For more information, contact:

BROOKS/COLE PUBLISHING COMPANY
511 Forest Lodge Road
Pacific Grove, CA 93950
USA

International Thomson Publishing
Berkshire House 168-173
High Holborn
London WC1V 7AA
England

Thomas Nelson Australia
102 Dodds Street
South Melbourne, 3205
Victoria, Australia

Nelson Canada
1120 Birchmount Road
Scarborough, Ontario
Canada M1K 5G4

International Thomson Editores
Campos Eliseos 385, Piso 7
Col. Polanco
11560 México D.F., México

International Thomson Publishing Gmbh
Königwinterer Strasse 418
53227 Bonn
Germany

International Thomson Publishing Asia
221 Henderson Road #05-10
Henderson Building
Singapore 0315

International Thomson Publishing—Japan
Hirakawacho-cho Kyowa Building, 3F
2-2-1 Hirakawacho-cho
Chiyoda-ku, Tokyo 102
Japan

Printed in the United States of America.

10 9 8 7 6 5 4 3 2 1

Library of Congress Cataloging-in-Publication Data

Gustafson, R. David (Roy David), [date]
Intermediate algebra / R. David Gustafson, Peter D. Frisk.—4th ed.
p. cm.
Includes index.
ISBN 0-534-23022-9
1. Algebra. I. Frisk, Peter D., [date]. II. Title.
QA154.2.G874 1994
512.9—dc20 94-29955
CIP

Photo credits: P. 3, The British Museum; **p. 67 (both),** Frank Rossotto/The Stock Market; **p. 311,** Courtesy of Texas Instruments; **p. 410,** Kathleen Olson; and **p. 641,** Archaeological Consulting/Gary Breschini & Trudy Haverstat.

To
Harold and Monie,
Harder and Evelyn,
with love and affection

PREFACE FOR THE INSTRUCTOR

Intermediate Algebra, Fourth Edition, is the second of a two-volume series designed to prepare students for college mathematics. It presents all of the topics associated with a second course in algebra. We believe that it will hold student attrition to a minimum while preparing students to succeed at the next stage—whether that is college algebra, trigonometry, statistics, finite mathematics, liberal arts mathematics, or everyday life.

Our goal in preparing this new edition was to write a book that

1. is enjoyable to read,
2. is easy to understand,
3. is relevant, and
4. will develop the necessary skills for success in future academic courses or on the job.

The Fourth Edition retains the basic philosophy of the highly successful previous edition. However, we have made several improvements in line with the NCTM standards and the current trends in mathematics reform. For example, much more emphasis has been placed on problem solving.

Changes in the Fourth Edition

To make the book more enjoyable to read, we have:

- used a new and more open four-color design.
- enlarged section heads to make them easier to find. Sections are now divided into subsections.

- redrawn all art, most with added color. Art that accompanies application problems is now much more representational.
- placed application exercises in a two-column format. Each application now has a title.
- added a Perspective to each chapter: brief, interesting stories that pertain to the material in the chapter.
- included pictures of famous mathematicians to provide a taste of mathematics history.

To make the book easier to understand, we have:

- revised the explanations and simplified the language.
- added many more author's notes to explain more steps in the problem-solving process.
- added **Warning!** notices to warn students of common errors and misconceptions.
- added Oral Exercises before each exercise set. These problems enable the instructor to check student understanding before assigning homework.

To make the book more relevant, we have:

- strengthened the problem-solving emphasis in the book by adding many more application problems and distributing them throughout the book.
- ended each chapter with a discussion of a career that uses mathematics. The necessary qualifications for the job and the job outlook are discussed, and there is a sample application using the material taught in the chapter.

To develop the necessary skills for success in future academic courses or on the job, we have:

- added **Writing Exercises** to each exercise set. These will help students clarify ideas.
- added **Something to Think About** exercises to each exercise set. These require extra thought and insight.
- added **graphing calculator material.** Although this material is integrated throughout the book, it can be omitted with no loss in continuity.
- added **Cumulative Review Exercises** after every third chapter. These are in addition to the Review Exercises in each exercise set, the Chapter Review Exercises, the Sample Chapter Tests, and the Sample Final Examination.
- added a **Project** to each chapter. The Projects can be used for extended assignments or for cooperative learning.

In addition to these changes, we have made the book more accessible to students by reordering some topics. For example,

- The chapter on equations and inequalities has been placed before factoring. This gets the student to problem solving more quickly, which is in line with the recommendations of many mathematics organizations. Application problems are more plentiful and more realistic.

The chapter is now one section shorter, because the section on solving equations by factoring has been moved to the factoring chapter.

- Chapter 4, on factoring, has been streamlined. Factoring by grouping has been integrated into each section. By including the section on solving equations by factoring in Chapter 4, we can give applications of factoring sooner and use factoring to solve problems.
- We have placed the chapter on graphing linear equations before the chapter on radicals. The chapter is now shorter and easier to teach because the material on the algebra of functions, composition of functions, and inverse functions has been moved to Chapter 9.

  Because the distance formula cannot be discussed before radicals, the distance formula now appears as an application of radical equations in Chapter 7.
- The material on graphing and functions is now placed in three shorter chapters rather than two long chapters.
- We have included graphing calculator material throughout the book. At first, the graphing calculator is used as a tool for graphing. However, as the book progresses, the calculator is used to enhance understanding of some topics.

Organization

The book is designed primarily for the chapters to be taught in order. However, for instructors who would like to teach graphing early, the chapters can be taught in the following sequence without loss of continuity:

1, 2, 3, 6, 10, 4, 5, 7, 8, 9, 11, 12

If graphing is taught early, a small number of Review Exercises will have to be omitted. For easy reference, these Review Exercises are noted in the margin of the Annotated Instructor's Edition.

Calculators

The use of calculators is assumed throughout the book. We believe that students should learn calculator skills in the mathematics classroom. They then will be prepared to use calculators in science and business classes and for nonacademic purposes.

We have also included material on the graphing calculator. We recommend that instructors try it. However, it can be skipped without interrupting the flow of ideas.

The directions within each exercise set indicate which exercises require calculators. There are two symbols, for scientific calculators and for graphing calculators.

Accuracy

To guarantee the accuracy of the answers, each exercise has been worked by both authors and two other problem checkers. The manuscript has been read by many reviewers, and the printed pages have been read by many proofreaders.

Writing Exercises ■ *Write a paragraph using your own words.*

1. Explain the meanings of the terms *coefficient* and *exponent*.
2. Explain the rules that determine the order in which operations are performed.

Something to Think About ■

1. In the definition of x^{-1}, x cannot equal 0. Why not?
2. Find the sum: $2^{-1} + 3^{-1} - 4^{-1}$ $\frac{7}{12}$

Review Exercises ■ *Graph each interval on a number line.*

1. $-2 < x \le 4$
2. $-5 \le x \le 5$
3. $[-2, 8]$
4. $(-5, -3)$

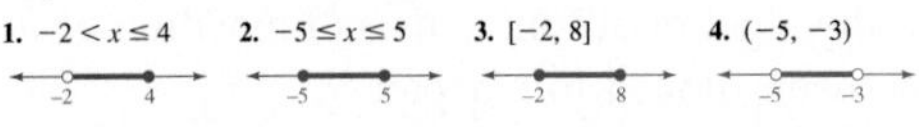

If $a = 4$, $b = -2$, and $c = 5$, find each value.

5. $\dfrac{ab + 2c}{a + b}$ 1
6. $\dfrac{ac - bc}{6ab + b}$ $-\frac{3}{5}$

◀ Writing Exercises are now included.

◀ Something to Think About exercises are more challenging.

2.2 Scientific Notation

■ Using Scientific Notation to Simplify Computations ■ Calculators

◀ Section heads are easier to find and sections are divided into subsections.

Scientists often use very large and very small numbers. For example, the speed of light is approximately 29,980,000,000 centimeters per second, and the mass of a hydrogen atom is 0.000000000000000000000001673 gram. The large number of zeros in these numbers makes them difficult to read and hard to remember. However, exponents make it possible for scientists to write these numbers compactly in a form called

Scientific Notation A number is ... number whose ... power of 10.

(a, c) **EXAMPLE 1** Change a. 2 ... c. −0.0013

Solution a. The numb... point in 2... multiplying... 29,980...

Amalie Noether (1882–1935)
Albert Einstein described Noether as the most creative female mathematical genius since the beginning of higher education for women. Her work was in the area of abstract algebra. Although she received a doctoral degree in mathematics, she was denied a mathematics position in Germany because she was a woman.

The graph of this inequality is shown in Figure 3-17. The solution set is $(-\infty, -3] \cup [8, \infty)$.

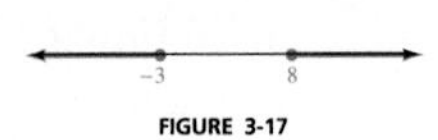

FIGURE 3-17

The word *or* in the statement $x \le -3$ or $x \ge 8$ indicates that only one of the inequalities needs to be true to make the statement true. ■

WARNING! In the statement $x \le -3$ or $x \ge 8$, it is incorrect to string the equalities together as $8 \le x \le -3$, because that would imply that $8 \le -3$, which is impossible.

ORALS *Solve each inequality.*

1. $2x < 4$ $x < 2$
2. $3x + 1 \ge 10$ $x \ge 3$
3. $-3x > 12$ $x < -4$
4. $-\dfrac{x}{2} \le 4$ $x \ge -8$
5. $-2 < 2x < 8$ $-1 < x < 4$
6. $3 \le \dfrac{x}{3} \le 4$ $9 \le x \le 12$

▲ Examples on video tape are marked with a special symbol.

Students are warned about common errors. ▶

Preparatory Oral Exercises precede each exercise set. ▶

69. Integer problem The sum of three consecutive even integers is 384. Find the integers. 126, 128, 130

70. Integer problem The sum of the first and third integers of three consecutive odd integers is 58. Find the middle integer. 29

71. Cutting a board The carpenter in Illustration 1 saws a board into two pieces. He wants one piece to be 1 foot longer than twice the length of the shorter piece. Find the length of each piece. 7 ft, 15 ft

ILLUSTRATION 1

72. Cutting a beam A 30-foot steel beam is to be into two pieces. The longer piece is to be 2 feet than 3 times as long as the shorter piece. Find th length of each piece. 7 ft, 23 ft

73. Cutting a rope A 60-foot rope is cut into four pieces, with each successive piece being twice as long as the previous one. Find the length of the longest piece. 32 ft

74. Cutting a cable A 185-foot cable is cut into t pieces of equal length and a third piece that is to 5 feet longer than the sum of the equal pieces. F the length of the longer piece. 95 ft

75. Buying a TV and a VCR See Illustration 2. If TV costs $55 more than the VCR, how much do the TV cost? $355

76. Buying a washer and dryer The total cost of a washer and dryer is $590. If the washer costs $40 more than the dryer, find the cost of the dryer. $275

77. Finding dimensions The rectangular garden shown in Illustration 3 is twice as long as it is wide. Find its dimensions. 12 m by 24 m

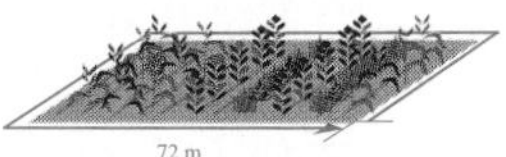

ILLUSTRATION 3

78. Finding dimensions The width of a rectangular swimming pool is one-third its length. If its perimeter is 96 meters, find the dimensions of the pool. 12 m by 36 m

◀ Application problems are in a two-column format.

◀ Art has been redrawn and made more colorful.

◀ All applications have titles.

Teaching Tip: You may want to remind students of the following divisibility rules:

1. An even integer is divisible by 2.
2. If the sum of the digits in an integer is divisible by 3, the integer is divisible by 3.
3. An integer ending with 0 or 5 is divisible by 5.
4. If a number ends with 0, it is divisible by 10.

There is no natural number greater than 12 that divides 60, 84, and 180.

Algebraic monomials also have greatest common factors. Consider three monomials with their prime factorizations:

$$6a^2b^3c = 3 \cdot 2 \cdot a \cdot a \cdot b \cdot b \cdot b \cdot c$$
$$9a^3b^2c = 3 \cdot 3 \cdot a \cdot a \cdot a \cdot b \cdot b \cdot c$$
$$18a^4c^3 = 2 \cdot 3 \cdot 3 \cdot a \cdot a \cdot a \cdot a \cdot c \cdot c \cdot c$$

Since each monomial has at least one factor of 3, two factors of a, and one factor of c in common, their greatest common factor is

$$3^1 \cdot a^2 \cdot c^1 = 3a^2c$$

To find the greatest common factor of several monomials, we follow these steps.

Steps for Finding the Greatest Common Factor

1. Find the prime factorization of each monomial.
2. List the prime factors that are common to each monomial.
3. Find the product of the factors found in Step 2, with each factor raised to the smallest power that occurs in any one monomial.

Strategy boxes ▶ are shaded in green.

PERSPECTIVE

No End of Primes

A composite number always has prime factors that are smaller than the number itself. The composite number 30, for example, has the prime factors 2, 3, and 5–all less than 30. We might think that the larger a number is, the more likely it is to be composite, because there are so many smaller prime numbers that could be possible factors. Maybe all numbers, if large enough, would be composite, and therefore there would be a largest prime number.

That is not what happens, however. In about 300 B.C. the Greek mathematician Euclid proved that the number of primes is unlimited–that there are infinitely many prime numbers. This is an important fact of a branch of mathematics called **number theory.** Although Euclid is best known for his study of geometry, many of his writings deal with number theory.

Who could possibly care that prime numbers can be very large? Cryptographers working for the federal government care, because several codes that are very difficult to crack rely on the properties of large primes.

Euclid (about 300 B.C.)

Perspectives are found ▶ in each chapter. They tell interesting stories about mathematics.

CHAPTER SUMMARY 221

Review Exercises

1. List the prime numbers less than 10.
   2, 3, 5, 7
2. List the composite numbers between 7 and 17.
   8, 9, 10, 12, 14, 15, 16
3. The formula for the volume of a sphere is $V = \frac{4}{3}\pi r^3$. Find the volume when $r = 21$ centimeters. Give the answer to the nearest hundredth. 38,792.39 cm³
4. The formula for the volume of a cone is $V = \frac{1}{3}\pi r^2 h$. Find the volume when $r = 12$ meters and $h = 14$ meters. 2111.15 m³

◀ Review Exercises are abundant throughout the book.

PROJECT Give a Little, Take a Little

The trinomial $x^4 + x^2 + 1$ does not factor as a perfect square, because $(x^2 + 1)^2$ is $x^4 + 2x^2 + 1$, which has a middle term of $2x^2$ instead of $1x^2$. We can change the middle term of the trinomial into $2x^2$ by adding x^2 to the trinomial. To ensure that adding x^2 does not change the value of the trinomial, we also subtract x^2.

$x^4 + x^2$

Because the
$(x^2 + 1)^2$, we

$x^4 + x^2$

- Use this te
- Factor x^6 (*Hint:* Use
- Factor x^6 –
- Factor x^6 – two square

◀ Projects are included for extended assignments or cooperative learning.

Chapte

KEY WORDS

difference o
difference o
factor (4.1)

6.4 GRAPHS OF OTHER EQUATIONS 311

WARNING! The graph of a nonlinear equation is not a line. To graph nonlinear equations, we must usually plot many points to recognize the shape of the graph.

EXAMPLE 2 Graph the equation $y = -|x| + 3$.

Solution We make a table of values by substituting numbers for x and finding the corresponding values of y. For example, if we substitute -2 for x, we get

$$y = -|x| + 3$$
$$y = -|-2| + 3$$
$$y = -(2) + 3$$
$$y = 1$$

After plotting the points listed in the table shown in Figure 6-25, we obtain the graph of the equation.

$y = -|x| + 3$

x	y
−3	0
−2	1
−1	2
0	3
1	2
2	1
3	0

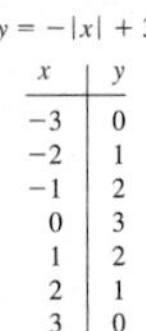

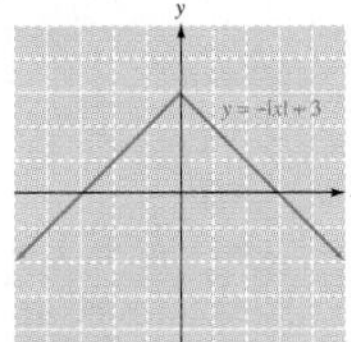

FIGURE 6-25

Many pieces of mathematics art are placed on grids. ▶

Graphing calculator material is included. ▶

FIGURE 6-26

Graphing Calculators

So far, we have graphed equations by making a table of ordered pairs and plotting points. This method is often tedious and time-consuming. Graphing is much easier when we use a graphing calculator.

All graphing calculators have a **viewing window** that is used to display graphs (see Figure 6-26.) To see the proper picture of a graph, we must often set the minimum and maximum values for the x- and y-coordinates. The standard RANGE settings of

Xmin = −10 Xmax = 10 Ymin = −10 Ymax = 10

indicate that −10 is the minimum x- and y-coordinate that will be used in the graph, and that 10 is the maximum x- and y-coordinate that will be used. If these settings do not appear when you turn on your calculator and select the RANGE option, move the cursor to the desired position by pressing the cursor keys ◀, ▶,

Student Support

We have included many features that make the book very accessible to students.

Worked Examples The text contains more than 500 worked examples. Extensive use of explanatory notes makes them easy to follow.

Chapter Summaries Each chapter concludes with a summary of key words, key ideas, and formulas used in the chapter.

Author's Notes Author's notes explain the steps in the solutions of most examples.

Functional Use of Color For easy reference, definitions and theorems are boxed in blue. Strategy boxes are green. In addition, the book uses color to highlight terms and expressions that you would point to in a classroom discussion.

End Papers Key formulas and ideas are listed inside the front and back covers for easy reference.

Exercises The book includes more than 5000 carefully graded exercises, with answers to the odd-numbered exercises provided in an appendix in the Student Edition.

Systematic Review Each exercise set is followed by Review Exercises. In addition, there are Chapter Review Exercises, Chapter Tests, Cumulative Review Exercises, and a Sample Final Examination.

Video tapes Many examples in the book are taught on videotape. These examples are marked with the symbol [video symbol] in the book.

Computer Software Students can get additional practice with BCX software. BCX drills students on problems similar to those in the book and will give hints and show complete solutions when necessary.

Ancillaries for the Instructor

Annotated Instructor's Edition The Annotated Instructor's Edition has the answer to every exercise printed next to that exercise. Annotations in the margin give alternate methods of presenting material, as well as teaching hints and strategies.

Test Manual
Teresa Bittner The *Test Manual* contains four ready-to-use forms of every chapter test. Two of the tests are free-response and two are multiple-choice.

Computer Testing Software
Teresa Bittner Available with the book are two extensive electronic question banks, one free-response and one multiple-choice. Each bank contains approximately 1700 questions and is available for either IBM-compatible or Macintosh computers. The testing program gives all of the features of a state-of-the-art word processor and more,

including the ability to see all technical symbols, fonts, and formatting on the screen just the way they will appear when printed. The question banks can be edited.

EXPTEST™ runs on IBM and compatible computers.

ExamBuilder™ runs on Macintosh computers.

Transparencies

Color transparencies of key graphics from the book are available to assist the instructor in the classroom.

Videotapes

A set of 24 book-specific videotapes is available without charge for adoptions of 100 books or more. The videos include the solutions of all examples in the book that are marked with the symbol [OO]. The instructors appearing on the videotapes are David Gustafson and Peter Frisk (the authors) and Diane Koenig.

Computer-Aided Instruction
Teresa Bittner

BCX is book-specific tutorial software that drills students on problems similar to those found in the book. There is a set of questions for each section in the book. BCX provides hints to students, and if they cannot answer a question correctly, the complete solution will be displayed. BCX monitors student progress and includes a reporting system.

Ancillaries for the Student

Study Guide
George Grisham

The *Study Guide* provides more explanation, worked examples, practice problems, and practice tests. It is available for sale at your college bookstore.

Student Solutions Manual
Michael Welden

The *Student Solutions Manual* gives complete solutions for the odd-numbered exercises in the book. It is available for sale at your college bookstore.

TO THE STUDENT

Congratulations. You now own a state-of-the-art textbook that has been written especially for you. To use the book properly, read it carefully, do the exercises, and check your progress with the Review Exercises and the Chapter Tests. Be sure to read and use the hints on studying algebra listed below.

A *Student Solutions Manual,* available for sale at your college bookstore, contains solutions to the odd-numbered exercises. A *Study Guide* that contains additional explanations, worked examples, and practice problems is also available for sale.

When you finish this course, consider keeping your book. It is the single reference source that will keep at your fingertips the information that you have learned. You may need this reference material in future mathematics, science, or business courses.

We wish you well.

Hints on Studying Algebra

The phrase "Practice makes perfect" is not quite true. It is *perfect* practice that makes perfect. For this reason, it is important that you learn how to study algebra to get the most out of this course.

Although we all learn differently, there are some hints on how to study algebra that most students find useful. Here is a list of some things you should consider.

Plan a Strategy for Success

To get where you want to be, you need a goal and a plan. Set a goal of passing this course with a grade of A or B. To meet this goal, you must have a good plan. Your plan should include several points:

- getting ready for class,
- attending class,

- doing homework,
- arranging for special help when you need it, and
- having a strategy for taking tests.

Getting Ready for Class To get the most out of every class period, you will need to prepare. One of the best things that you can do is to read the material in the book before your instructor discusses it. You may not understand all of what you read, but you will be better able to understand it when your instructor presents the material in class.

Be sure to do your work every day. If you get behind and attend class without understanding prior material, you will be lost, and your classroom time will be wasted. Even worse, you will become frustrated and discouraged. Promise yourself that you will always prepare for class, and then keep your promise.

Attending Class The classroom experience is your opportunity to learn from your instructor. Make the most of it by attending every class. Sit near the front of the room where you can see and hear well and where you won't be distracted. It is your responsibility to follow the instructor's discussion, even though that might be hard work.

Pay attention to your instructor and jot down the important things that he or she says. However, do not spend so much time taking notes that you fail to concentrate on what your instructor is explaining. It is much better to listen and understand the *big picture* than it is merely to copy solutions to problems.

Don't be afraid to ask questions. If something is unclear to you, it is probably unclear to other students as well. They will appreciate your willingness to ask. Besides, asking questions will make you an active participant in class. This will help you pay attention and keep you alert and involved.

Doing Homework Everyone knows that it requires practice to excel at tennis, master a musical instrument, or learn a foreign language. It also requires practice to learn mathematics. Since *practice* in mathematics is the homework, homework is your opportunity to practice skills and experiment with ideas.

It is very important to pick a definite time to study and do homework. Set a formal schedule and stick to it. Try to study in a place that is comfortable and quiet. If you can, do some homework shortly after class, or at least before you forget what was discussed in class. This quick follow-up will help you remember the skills and concepts your instructor taught that day.

Study Sessions

Each formal study session should include three parts:

1. Begin every study session with a review period. Look over previous chapters and see if you can do a few problems from previous sections. Keeping old skills alive will greatly reduce the amount of time that you will need to cram for tests.

2. After reviewing, read the assigned material. Resist the temptation of diving into the problems without reading and understanding the examples. Instead, work the examples with pencil and paper. Only after you completely understand the underlying principles behind them should you try to work the problems.

Once you begin to work the problems, check your answers with the printed answers in the back of the book. If one of your answers differs from the printed answer, see if you can reconcile the two. Sometimes answers can have more than one form. If you still believe that your answer is incorrect, compare your work to the example in the book that most closely resembles the problem and try to find your mistake. If you cannot find an error, consult the *Student Solutions Manual*. If nothing works, mark the problem and ask about it during your next class meeting.

3. After you complete the written assignment, read the next section. That preview will be helpful when you hear that material discussed during the next class period.

You probably know that the rule of thumb for doing homework is two hours of homework for every hour spent in class. If mathematics is hard for you, plan on spending even more time on homework.

To make homework more enjoyable, study with one or more friends. The interaction will clarify ideas and help you remember them. If you must study alone, try talking to yourself. A good study technique is to explain the material to yourself.

Arranging for Special Help

Take advantage of any special help available from your instructor. Often, your instructor can clear up difficulties in a very short time.

Find out if your college has a free tutoring program. Peer tutors also can be of great help. Be sure to use the videotapes and BCX software.

Taking Tests

Students often get nervous before taking a test, because they are afraid that they will not do well. The most common reason for this fear is that students are not confident that they know the material.

To build confidence in your ability to work tests, rework many of the problems in the exercise sets, work the Review Exercises at the end of each chapter, and work the Chapter Tests. Check all your answers with the answers printed at the back of the book.

Then guess what the instructor will ask and make up your own tests and work them. Once you know your instructor, you will be surprised at how good you can get at picking test questions. With this preparation, you will have some idea of what will be on the test and will have more confidence in your ability to do well.

When you take a test, work slowly and deliberately. Scan the test and first work the easy problems that you know you can do. This will build confidence. Tackle the hardest problems last.

ACKNOWLEDGMENTS

We are grateful to the following people who have reviewed the book at various stages of its development:

Lee R. Clancy
Golden West College

Linda Crabtree
Longview Community College

Elias Deeba
University of Houston-Downtown

Mary Catherine Dooley
University of New Orleans

Robert B. Eicken
Illinois Central College

Paul Finster
El Paso Community College

Ruth Flourney
University of Alaska-Anchorage

Lenore Frank
SUNY-Stony Brook

Margaret J. Greene
Florida Community College-Jacksonville

George Grisham
Bradley University

Charlotte Grossbeck
SUNY-Cobleskill

David W. Hansen
Monterey Peninsula College

Steven Hatfield
Marshall University

Rose Ann Haw
Mesa College

Denise Hennicke
Collin County College

Ingrid Holzner
University of Wisconsin

John Hooker
Southern Illinois University

William A. Hutchings
Diablo Valley College

Herbert Kasube
Bradley University

John Robert Kennedy II
Santa Monica College

Diane Koenig
Rock Valley College

Ralph A. Liguori
University of Texas

Thomas McCready
California State University-Chico

James W. Newsom
Tidewater Community College

Christine Panoff
University of Michigan-Flint

Kenneth Shabell
Riverside Community College

Pat Stone
Tom Ball College

Salli Takenaka
Santa Monica College

Ray Tebbetts
San Antonio College

Jerry Wilkerson
Missouri Western State College

We wish to thank Diane Koenig and Robert Hessel, who read the entire manuscript and worked every problem. We also wish to thank Bill Hinrichs, Jerry Frang, George Mader, Michael Welden, Jennifer Dollar, and Rob Clark for their helpful suggestions. We give special thanks to Gary Ostedt, Ellen Brownstein, David Hoyt, Kelly Shoemaker, Lori Heckelman, Kathleen Olson, Audra Silverie, and Carol Benedict for their assistance in the production process.

R. David Gustafson
Peter D. Frisk

CONTENTS

3 Equations and Inequalities 108

4 Factoring Polynomials 180

5 Rational Expressions 228

1 The Real Number System

Algebra is the result of contributions from many cultures over thousands of years. The ancient Babylonians may have known some algebra, but its first recorded traces are found in the writings of an Egyptian priest named Ahmes, who lived before 1700 B.C.

Diophantus, a Greek who lived around A.D. 300, was the first to use special symbols to represent unknown quantities. The first known woman mathematician, a Greek scholar named Hypatia, studied and explained the work of Diophantus.

In A.D. 415, the Arabians were adding, subtracting, and working with fractions. The symbols we now use to write numbers were developed by the Hindu and Arabic cultures. They were introduced to the west by the 13th-century Italian merchant Leonardo of Pisa, also known as Fibonacci.

The word *algebra* comes from the title of a book written by the Arabian mathematician Al-Khowarazmi around A.D. 800. Its title, *Ihm al-jabr wa'l muqabalah*, means restoration and reduction, a process then used to solve equations.

During the middle ages, there was little further development of algebra. Not until the 16th century did scholars again become interested in mathematics. At that time, France and Spain were at war. François Viête, a French lawyer with an interest in mathematics, devised a system to break the codes the Spaniards used to send secret messages. His system used the same algebraic notation we use today.

Because the concept of number is basic to all of mathematics, we begin by discussing the various sets of real numbers.

1.1 Sets and the Real Number System

■ Subsets of Sets ■ Union and Intersection of Two Sets ■ Sets of Numbers

A **set** is a collection of objects. To denote a set, we often use braces to enclose a list of its **members** or **elements**. For example, the notation

$$\{a, b, c\}$$

denotes the set with elements a, b, and c. To show that b is an element of this set, we write

$b \in \{a, b, c\}$ Read $\in$ as "is an element of."

To show that d is not an element of $\{a, b, c\}$, we write

$d \notin \{a, b, c\}$ Read $\notin$ as "is not an element of."

Capital letters are used to name sets. For example, the expression

$$A = \{a, e, i, o, u\}$$

means that A is the set containing the vowels a, e, i, o, and u.

In **set-builder notation**, a rule is given that establishes membership in a set. The set of vowels in the English alphabet, for example, can be denoted as

$$B = \{x \mid x \text{ is a vowel of the English alphabet.}\}$$

The statement above is read as "B is the set of all letters x such that x represents a vowel of the English alphabet." Because x can represent many different elements of the set, x is called a **variable**.

Equality of Sets Two sets such as A and B are **equal** when they have exactly the same elements. If sets A and B are equal, we write $A = B$.

If two sets A and B do not have the same elements, they are not equal, and we write $A \neq B$.

Subsets of Sets

If $B = \{a, c, e\}$ and $A = \{a, b, c, d, e\}$, each element of B is also an element of A. When this is so, we say that B is a **subset** of A, and we write

$B \subseteq A$ Read as "B is a subset of A."

PERSPECTIVE

All Dark Things

The ancient Egyptians developed two systems of writing. Each symbol in one system, hieroglyphics, was a picture of an object. Because hieroglyphic writing was usually inscribed on stone, many examples still survive. For daily purposes, Egyptians used hieratic writing. Simpler than hieroglyphics, hieratic writings were done with ink on papyrus sheets. Papyrus, made from plants, is very delicate and quickly dries and crumbles. Few Egyptian papyri survive.

Those that do survive provide important clues to the content of ancient mathematics. One, the Rhind Papyrus, was discovered in 1858 by a British archeologist, Henry Rhind. Also known as the Ahmes Papyrus after its ancient author, it begins with a description of its contents: *Directions for Obtaining the Knowledge of All Dark Things.*

The Ahmes Papyrus and another, the Moscow Papyrus, together contain 110 mathematical problems and their solutions. Many of these were probably for education, because they represented situations that the scribes, priests, and other workers in government and temple administration were expected to be able to solve.

The Ahmes Papyrus (the British Museum)

We can illustrate this relationship graphically as in Figure 1-1. If the interior of circle B represents the elements in set B, and the interior of circle A represents the elements of set A, the figure shows that every element in set B is also an element of set A.

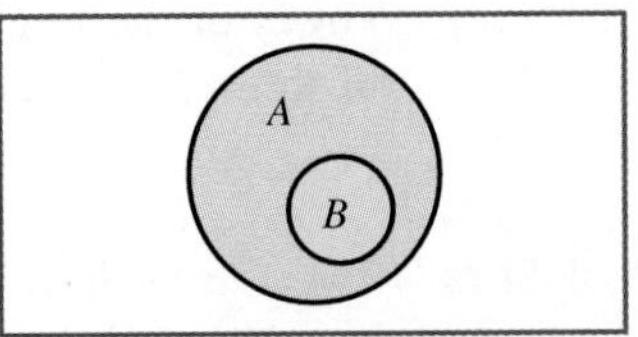

FIGURE 1-1

François Vieta (Viête)
(1540–1603)
By using letters in place of unknown numbers, Vieta simplified the subject of algebra and brought its notation closer to the notation that we use today. One symbol he didn't use was the equal sign.

Since set A contains elements such as b and d that are not included in set B, set A is not a subset of set B:

$A \not\subseteq B$ Read as "A is not a subset of B."

Since every element in set A is an element in set A, set A is a subset of itself. In symbols,

$A \subseteq A$

In general, any set is a subset of itself.

A set with no elements is called the **empty set**, or the **null set**, and is denoted as $\emptyset$. Thus,

$\emptyset = \{ \quad \}$

The empty set is considered to be a subset of every set.

Union and Intersection of Two Sets

If the elements of some set A are joined with the elements of some set B, the **union** of set A and set B is formed. The union of set A and set B is denoted as

$A \cup B$ Read as "the union of set A and set B."

The elements of $A \cup B$ are *either* elements of set A, *or* elements of set B, *or* both, as shown in Figure 1-2(a).

The set of elements that are common to set A and set B is called the **intersection** of set A and set B. The intersection of set A and set B is denoted as

$A \cap B$ Read as "the intersection of set A and set B."

The elements of $A \cap B$ are elements that are in *both* set A *and* set B, as shown in Figure 1-2(b).

If A and B have no elements in common, then $A \cap B = \emptyset$, and the sets A and B are said to be **disjoint**, or **mutually exclusive**.

$A \cup B$

A B

(a)

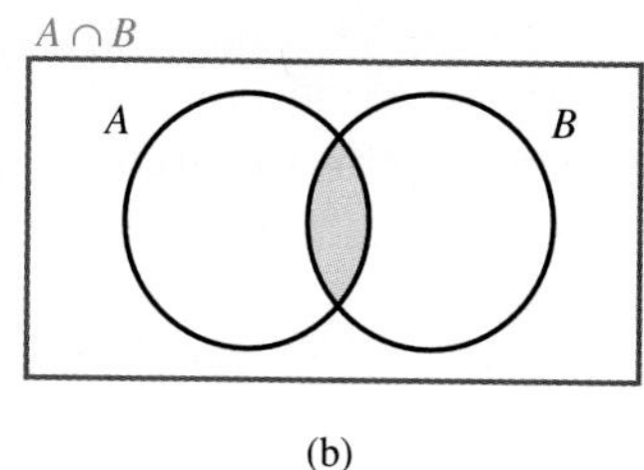

(b)

FIGURE 1-2

Because $\emptyset$ is the empty set, it follows that

$A \cup \emptyset = A$ The elements in set A joined with no elements is set A.

and

$A \cap \emptyset = \emptyset$ Since $\emptyset$ has no elements, sets A and $\emptyset$ have no elements in common.

EXAMPLE 1 If $A = \{a, e, i, o, u\}$ and $B = \{a, b, c, d, e\}$, find **a.** $A \cup B$, **b.** $A \cap B$, and **c.** $(A \cap \emptyset) \cup B$.

Solution **a.** $A \cup B = \{a, b, c, d, e, i, o, u\}$ **b.** $A \cap B = \{a, e\}$

c. $(A \cap \emptyset) \cup B = \emptyset \cup B$ Do the work in parentheses first.

$= B$

$= \{a, b, c, d, e\}$ ■

Sets of Numbers

The most basic set of numbers is called the set of **natural numbers**, the numbers we use for counting.

Natural Numbers The set of **natural numbers** is the set

$\{1, 2, 3, 4, 5, 6, 7, 8, 9, . . .\}$

The three dots in the previous definition, called the **ellipsis**, indicate that the list of natural numbers continues endlessly. Since the set of natural numbers has an unlimited number of elements, it is an **infinite set**. A set with a limited number of elements is called a **finite set**.

Two important subsets of the natural numbers are the **prime numbers** and the **composite numbers**.

Prime and Composite Numbers A **prime number** is a natural number greater than 1 that is divisible only by itself and 1.

A **composite number** is a natural number greater than 1 that is not a prime number.

Historical Note: The largest known prime was discovered in January of 1994. It is

$2^{859433} - 1$

It is 258,716 digits long and would fill eight newspaper pages.

Teaching Tip: 1 is not a prime. If it were, every prime would be the product of primes:

$5 = 5 \cdot 1$

The prime numbers less than 20 are

2, 3, 5, 7, 11, 13, 17, and 19

and the composite numbers less than 20 are

4, 6, 8, 9, 10, 12, 14, 15, 16, and 18

WARNING! Note that the number 1 is the only natural number that is neither prime nor composite.

If we join 0 to the set of natural numbers, we have the set of **whole numbers**.

Whole Numbers The set of **whole numbers** is the set

$\{0, 1, 2, 3, 4, 5, 6, 7, 8, 9, \ldots\}$

We often use numbers to indicate direction as well as quantity—for example,

- profit or loss
- temperatures above or below 0
- gains or losses in the stock market

To do so, we must extend the set of whole numbers to include the negatives of the natural numbers. These negatives are denoted with $-$ signs. For example, the negative of 7 is written as -7, and is read as "negative 7."

The union of the set of whole numbers and the set of negatives of the natural numbers forms the set of **integers**.

Integers The set of **integers** is the set

$\{\ldots, -7, -6, -5, -4, -3, -2, -1, 0, 1, 2, 3, 4, 5, 6, 7, \ldots\}$

Integers that are divisible by 2 are called **even integers**, and integers that are not divisible by 2 are called **odd integers**. The even integers from -10 to 10 are

$-10, -8, -6, -4, -2, 0, 2, 4, 6, 8, 10$

Teaching Tip: Point out that if the last digit is even, the integer is even. If the last digit is odd, the integer is odd.

and the odd integers from -10 to 10 are

$-9, -7, -5, -3, -1, 1, 3, 5, 7, 9$

When 8 is divided by 5, we do not obtain another integer. Instead, we obtain the fraction $\frac{8}{5}$. Fractions such as $\frac{8}{5}$ that have integers in both their numerator and their denominator are called **rational numbers**.

Rational Numbers A **rational number** is any number that can be written in the form $\frac{a}{b}$, where a and b are integers and $b \neq 0$.

A number that is written in the form $-\frac{a}{b}$, where a and b are integers and $b \neq 0$, is also a rational number: It is $\frac{-a}{b}$. Thus, the negative of the rational number $\frac{a}{b}$ is a rational number.

Some examples of rational numbers are

$$\frac{2}{3}, \quad -\frac{44}{23}, \quad \frac{-315}{476}, \quad 16, \quad \text{and} \quad -0.25$$

The number 16 is a rational number because it can be written in fractional form as $\frac{16}{1}$, $\frac{32}{2}$, $\frac{48}{3}$, and so on. The number -0.25 is a rational number because it can be written as $-\frac{1}{4}$.

WARNING! We note that $\frac{24}{8} = 3$ because $8 \cdot 3 = 24$, and $\frac{0}{9} = 0$ because $9 \cdot 0 = 0$. However, the fraction $\frac{5}{0}$ is undefined, because there is no number that when multiplied by 0 gives 5.

The fraction $\frac{0}{0}$ is indeterminate, because *all* numbers when multiplied by 0 give 0. Thus, the denominator of a fraction can never be 0.

EXAMPLE 2

a. The fraction $\frac{5}{3}$ is a rational number, because it is the quotient of two integers and the denominator is not 0.

b. -7 is a rational number, because it can be written in the form $\frac{-7}{1}$.

c. 0.125 is a rational number, because it can be written in the form $\frac{1}{8}$.

d. $-0.666\ldots$ is a rational number, because it can be written in the form $-\frac{2}{3}$. ■

Every rational number written in fractional form can be written in decimal form. For example, to change $\frac{3}{4}$ to a decimal fraction, we divide 3 by 4 to obtain 0.75.

$$\begin{array}{r} .75 \\ 4\overline{)3.00} \\ \underline{2\,8} \\ 20 \\ \underline{20} \\ 0 \end{array}$$

Since the division leaves a remainder of 0, the division stops. The quotient 0.75 is called a **terminating decimal**.

If we change a fraction such as $\frac{421}{990}$ to a decimal fraction, we obtain 0.4252525. . . , called a **repeating decimal**, in which the block of digits 25 repeats forever.

$$\begin{array}{r} 0.42525\ldots \\ 990\overline{)421.0000} \\ \underline{396\ 0} \\ 25\ 00 \\ \underline{19\ 80} \\ 5\ 200 \\ \underline{4\ 950} \\ 250 \end{array}$$

The repeating decimal 0.4252525 . . . is often written as $0.4\overline{25}$, where the overbar indicates the repeating block of digits.

It can be shown that all decimal forms of rational numbers are either terminating or repeating decimals.

It is easy to write a terminating decimal as a rational number in fractional form. For example, to write 0.25 as a fraction, we note that 0.25 means $\frac{25}{100}$ and that $\frac{25}{100}$ can be simplified to obtain $\frac{1}{4}$.

$$0.25 = \frac{25}{100} = \frac{25 \cdot 1}{25 \cdot 4} = \frac{\overset{1}{\cancel{25}} \cdot 1}{\underset{1}{\cancel{25}} \cdot 4} = \frac{1}{4}$$

Later, we will show that any repeating decimal can be written as a rational number in fractional form.

Numbers whose decimal representations neither terminate nor repeat are called **irrational numbers**. The decimal

0.31 331 3331 . . .

follows a pattern, but it will never have a repeating block of digits. Thus, it is an irrational number.

Irrational Numbers An **irrational number** is a decimal that cannot be written in the form $\frac{a}{b}$, where a and b are integers and $b \neq 0$.

Other examples of irrational numbers are

$\sqrt{3} = 1.732050808\ldots$ and $\pi = 3.141592653\ldots$

The union of the set of rational numbers and the set of irrational numbers is the set of **real numbers**, the set of all decimals. The set of real numbers is denoted by the symbol $\mathfrak{R}$.

The relationships between the sets of numbers developed so far are shown in Figure 1-3.

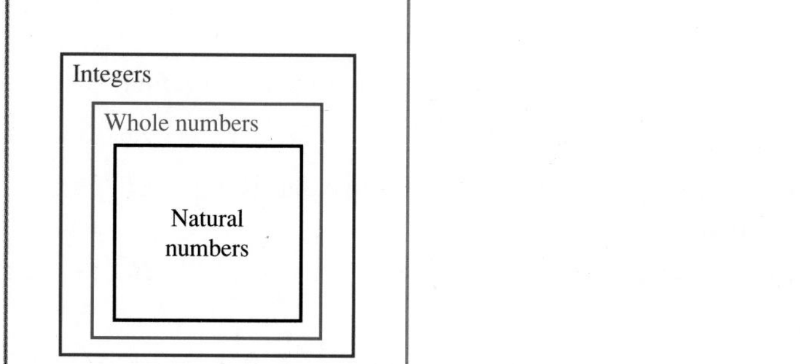

FIGURE 1-3

ORALS *Give an example of . . .*

1. a whole number.
2. a rational number that is not an integer.
3. a rational number that is also an integer.
4. a rational number that is represented by a terminating decimal.
5. a rational number that is represented by a repeating decimal.
6. a prime number greater than 20.

EXERCISE 1.1

In Exercises 1–8, $A = \{1, 3, 5, 9\}$, $B = \{3, 5, 9, 11\}$, and $C = \{x \mid x$ is an odd integer$\}$. Insert either a $\in$ or a $\subseteq$ symbol to make a true statement.

1. $A \subseteq C$ **2.** $3 \in A$ **3.** $9 \in B$ **4.** $\{3, 5\} \subseteq C$

5. $\emptyset \subseteq B$ **6.** $9 \in C$ **7.** $A \subseteq A$ **8.** $B \subseteq B$

In Exercises 9–12, list the elements in each set, if possible.

9. $\{x \mid x$ is a prime number less than 10.$\}$
2, 3, 5, 7

10. $\{x \mid x$ is a composite number between 3 and 7.$\}$
4, 6

11. $\{x \mid x$ is a number that is both rational and irrational.$\}$ no numbers

12. $\{x \mid x$ is a state in the United States beginning with the letter I.$\}$ Idaho, Illinois, Indiana, Iowa

In Exercises 13–20, $A = \{2, 3, 5, 7\}$, $B = \{2, 4, 6, 8, 10\}$, and $C = \{1, 3, 5, 7, 9\}$. Find each set.

13. $A \cup B$ $\{2, 3, 4, 5, 6, 7, 8, 10\}$

14. $A \cap C$ $\{3, 5, 7\}$

15. $B \cap C$ $\emptyset$

16. $A \cup C$ $\{1, 2, 3, 5, 7, 9\}$

17. $(C \cap \emptyset) \cap B$ $\emptyset$

18. $B \cup (\emptyset \cap C)$ $\{2, 4, 6, 8, 10\}$

19. $(A \cup B) \cap C$ $\{3, 5, 7\}$

20. $(A \cap B) \cup C$ $\{1, 2, 3, 5, 7, 9\}$

In Exercises 21–26, tell whether each set is a finite or an infinite set.

21. $\{x \mid x \text{ is a prime number.}\}$ infinite

22. $\{1, 2, 3, 4, 5, 6, 7, 8\}$ finite

23. $\{x \mid x \text{ is a natural number that is not prime and is not composite.}\}$ finite

24. $\{x \mid x \text{ is an even integer.}\}$ infinite

25. $\{x \mid x \text{ is a rational number.}\}$ infinite

26. $\{x \mid x \text{ is a state of the United States.}\}$ finite

In Exercises 27–38, list the elements in the set $\{-3, 0, 1, 2, 9, \frac{2}{3}, \sqrt{3}\}$ that satisfy the given condition.

27. natural number 1, 2, 9

28. whole number 0, 1, 2, 9

29. integer $-3, 0, 1, 2, 9$

30. rational number $-3, 0, 1, 2, 9, \frac{2}{3}$

31. irrational number $\sqrt{3}$

32. real number $-3, 0, 1, 2, 9, \frac{2}{3}, \sqrt{3}$

33. even natural number 2

34. odd integer $-3, 1, 9$

35. prime number 2

36. composite number 9

37. odd composite number 9

38. even prime number 2

In Exercises 39–50, simplify each expression, if necessary, and classify each result as an even integer, an odd integer, a prime number, and a composite number.

39. 8 even, composite

40. -5 odd

41. -9 odd

42. 4 even, composite

43. $\frac{12}{6}$ even, prime

44. $8 - 5$ odd, prime

45. 0 even

46. 1 odd

47. $3(5)$ odd, composite

48. -16 even

49. $\frac{0}{4}$ even

50. 2 even, prime

In Exercises 51–70, tell whether each statement is true. If a statement is false, give an example to show that it is false.

51. The product of two natural numbers is a natural number. true

52. The sum of two natural numbers is a natural number. true

53. Every whole number is a natural number. false; 0 is a whole number, but not a natural number

54. Every natural number is an integer. true

55. The sum of two prime numbers is a prime number. false; $3 + 5 = 8$, and 8 is composite

56. The product of two prime numbers is a prime number. false; $3 \cdot 5 = 15$, and 15 is not prime

57. The sum of two even integers is even. true

58. The product of two even integers is even. true

59. The sum of two odd integers is odd. false; $5 + 9 = 14$, and 14 is even

60. The product of two odd integers is odd. true

61. The sum of two composite numbers is a composite number. false; $4 + 9 = 13$, and 13 is prime

62. The product of two composite numbers is composite. true

63. The product of two prime numbers is composite.
true

64. The sum of two prime numbers is composite.
false; $2 + 3 = 5$, and 5 is not composite

65. The sum of 0 and any natural number is a natural number. true

66. The product of 0 and any natural number is a natural number. false; $0 \cdot 3 = 0$, and 0 is not natural

67. The only even prime numbers are 2 and 4.
false; 4 is not prime

68. The negative of a prime number is also prime.
false; $-(3) = -3$, and -3 is not prime

69. Every integer is a rational number.
true

70. No rational number is an integer.
false; $\frac{8}{4} = 2$, and 2 is rational and an integer

In Exercises 71–74, change each rational number to a decimal and classify the decimal as a terminating or a repeating decimal.

71. $\frac{7}{8}$ 0.875, terminating

72. $\frac{7}{3}$ $2.\overline{3}$, repeating

73. $-\frac{11}{15}$ $-0.7\overline{3}$, repeating

74. $-\frac{19}{16}$ -1.1875, terminating

In Exercises 75–78, write each terminating decimal as a fraction in simplest form.

75. 0.5 $\frac{1}{2}$

76. 0.2 $\frac{1}{5}$

77. 0.75 $\frac{3}{4}$

78. 0.125 $\frac{1}{8}$

Writing Exercises ■ *Write a paragraph using your own words.*

1. Distinguish between "is an element of" and "is a subset of."

2. Explain why every integer is a rational number, but not every rational number is an integer.

Something to Think About ■

1. List all the subsets of {1, 2}. Don't forget to include ∅ and the set itself. How many subsets are there?
4

2. List all the subsets of {1, 2, 3}. How many are there? Can you predict how many subsets {1, 2, 3, 4} has? 8, 16

Review Exercises ■

1. Add: $\begin{array}{r} 3456 \\ \underline{2947} \end{array}$
6403

2. Subtract: $\begin{array}{r} 4529 \\ \underline{2637} \end{array}$
1892

3. Multiply: $\begin{array}{r} 2375 \\ \underline{459} \end{array}$
1,090,125

4. Divide: $45\overline{)6300}$ 140

5. A man buys a sport coat for $185 and slacks for $45. If sales tax is 5%, find the total cost.
$241.50

6. A woman invests $4500 at an annual rate of 6% and $5400 at an annual rate of 7%. How much interest does she receive each year?
$648

1.2 Equality and Properties of Real Numbers

■ Properties of Real Numbers ■ The Identity Elements ■ Additive and Multiplicative Inverses

Historical Note: John Widmann (1460–?) was the first to use the + and − symbols.
Christopher Clavius (1537–1612) was the first to use the · symbol for multiplication.
William Oughtred (1574–1660) was the first to use the × symbol for multiplication.
Johann Heinrich Rahn (1622–1676) was the first to use the ÷ symbol for division.

To show that two expressions represent the same number, we use the = sign. For example,

$3 + 8 = 11$	Read as "The sum of 3 and 8 is 11."
$12 - 4 = 8$	Read as "The difference obtained when 4 is subtracted from 12 is 8."
$4 \cdot 5 = 20$	Read as "The product of 4 and 5 is 20."
$36 \div 9 = 4$	Read as "The quotient obtained when 36 is divided by 9 is 4."

The quotient $36 \div 9$ can also be written as $\frac{36}{9}$. Thus, $\frac{36}{9} = 4$.

If two numbers a and b are not equal, we write

$$a \neq b$$

■ Properties of Real Numbers

There are several properties of real numbers that we will use throughout the book.

Properties of Real Numbers If a, b, and c are real numbers, then

$a = a$	The reflexive property of equality
If $a = b$, then $b = a$.	The symmetric property of equality
If $a = b$ and $b = c$, then $a = c$.	The transitive property of equality

The reflexive property of equality states that any number is equal to itself. For example,

$$5 = 5$$

The symmetric property of equality states that if one number is equal to a second, then the second number is equal to the first. For example,

If $4 = 3 + 1$, then $3 + 1 = 4$.

The transitive property of equality states that if one number is equal to a second and the second number is equal to a third, then the first number is equal to the third. For example,

If $10 = 5 + 5$ and $5 + 5 = 2 \cdot 5$, then $10 = 2 \cdot 5$.

Another property of equality enables us to substitute a quantity for its equal in any mathematical expression without changing the meaning of the expression.

The Substitution Property If a and b are real numbers and $a = b$, then b can be substituted for a in any mathematical expression to obtain an equivalent expression.

(b,d) **EXAMPLE 1** Each statement is true because of the given reason.

a. $x - 4 = x - 4$ — The reflexive property

b. If $5 + x = 3 + y$, then $3 + y = 5 + x$. — The symmetric property

c. If $3 - x = 8$ and $8 = 2 + y$, then $3 - x = 2 + y$. — The transitive property

d. If $x + 3 = x \cdot y$ and $x = 9$, then $9 + 3 = 9 \cdot y$. — The substitution property

■

There are many properties involving operations on the real numbers.

The Closure Properties If a and b are real numbers, then

$a + b$ is a real number.

$a - b$ is a real number.

$a \cdot b$ is a real number.

$\frac{a}{b}$ is a real number, provided that $b \neq 0$.

The product of a and b is often written as $a(b)$, $(a)(b)$, or just ab.

Because of the closure properties, the sum, difference, product, and quotient of any two real numbers is, again, a real number (provided there are no divisions by 0).

The Associative Properties If a, b, and c are real numbers, then

$(a + b) + c = a + (b + c)$ — The associative property of addition

$(ab)c = a(bc)$ — The associative property of multiplication

The associative properties enable us to group the numbers in a sum, or the numbers in a product, in any way we wish and still get the same answer. For example,

$$(2 + 3) + 4 = 5 + 4 = 9 \quad \text{and} \quad 2 + (3 + 4) = 2 + 7 = 9$$

Likewise,

$$(2 \cdot 3) \cdot 4 = 6 \cdot 4 = 24 \quad \text{and} \quad 2 \cdot (3 \cdot 4) = 2 \cdot 12 = 24$$

WARNING! The operations of subtraction and division are not associative, because different groupings give different answers. For example,

$$(8 - 4) - 2 = 4 - 2 \qquad \text{but} \qquad 8 - (4 - 2) = 8 - 2$$
$$= 2 \qquad\qquad\qquad\qquad = 6$$

$$(8 \div 4) \div 2 = 2 \div 2 \qquad \text{but} \qquad 8 \div (4 \div 2) = 8 \div 2$$
$$= 1 \qquad\qquad\qquad\qquad = 4$$

EXAMPLE 2 Use an associative property to simplify $3 + (2 + a)$.

Solution

$$\begin{aligned} 3 + (2 + a) &= (3 + 2) + a && \text{Use the associative property of addition.} \\ &= 5 + a && 3 + 2 = 5. \end{aligned}$$

■

The Commutative Properties If a and b are real numbers, then

$a + b = b + a$ The commutative property of addition

$ab = ba$ The commutative property of multiplication

The commutative properties enable us to add or multiply two numbers in either order. We get the same sum whether we add the first number to the second or the second number to the first. Likewise, we get the same product whether we multiply the first number by the second or the second number by the first. For example,

$2 + 3$ and $3 + 2$ are both 5 and $7 \cdot 9$ and $9 \cdot 7$ are both 63

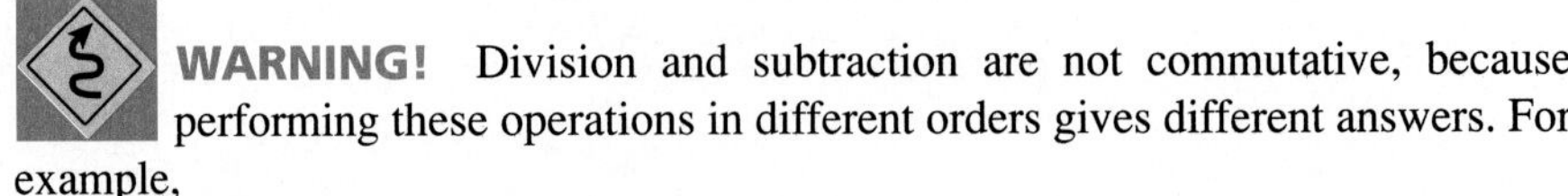

WARNING! Division and subtraction are not commutative, because performing these operations in different orders gives different answers. For example,

$$8 \div 4 = 2 \qquad \text{but} \qquad 4 \div 8 = \frac{1}{2}$$

$$8 - 4 = 4 \qquad \text{but, as we will see,} \qquad 4 - 8 = -4$$

EXAMPLE 3 Use a commutative and an associative property to simplify $(3a)4$.

Solution

$$\begin{aligned} (3a)4 &= 4(3a) && \text{Use the commutative property of multiplication.} \\ &= (4 \cdot 3)a && \text{Use the associative property of multiplication.} \\ &= 12a && 4 \cdot 3 = 12. \end{aligned}$$

■

The Distributive Property If a, b, and c are real numbers, then

$$a(b + c) = ab + ac$$

Because of the distributive property, we can evaluate many expressions that involve both a multiplication and an addition. We can either add first and then multiply, or multiply first and then add. For example, $2(3 + 7)$ can be found in two ways:

$$\begin{aligned} 2(3 + 7) &= 2 \cdot 10 \\ &= 20 \end{aligned} \qquad \Bigg| \qquad \begin{aligned} 2(3 + 7) &= 2 \cdot 3 + 2 \cdot 7 \\ &= 6 + 14 \\ &= 20 \end{aligned}$$

Either way, the answer is 20.

A more general form of the distributive property is called the **extended distributive property:**

$$a(b + c + d + e + \cdots) = ab + ac + ad + ae + \cdots$$

EXAMPLE 4 Simplify $2(x + y + 7)$.

Solution

$$\begin{aligned} 2(x + y + 7) &= 2x + 2y + 2(7) && \text{Use the extended distributive property.} \\ &= 2x + 2y + 14 \end{aligned}$$

■

The Identity Elements

Because adding 0 to a number leaves that number identically the same, 0 is called the **identity for addition.**

The Identity Element for Addition There is a single number 0, called the **additive identity**, such that

$$0 + a = a + 0 = a$$

Another property of 0, called the **multiplication property of zero**, states that the product of any number and 0 is 0.

$$a \cdot 0 = 0 \cdot a = 0$$

Because multiplying a number by 1 leaves that number identically the same, 1 is called the **identity for multiplication.**

The Identity Element for Multiplication There is a single number 1, called the **multiplicative identity**, such that

$$1 \cdot a = a \cdot 1 = a$$

Additive and Multiplicative Inverses

If the sum of two numbers is 0, those numbers are called **additive inverses**, **negatives**, or **opposites** of each other.

The Additive Inverse Elements For each real number a, there is a single number $-a$ such that

$$a + (-a) = -a + a = 0$$

The number $-a$ is called the **additive inverse**, the **negative**, or the **opposite** of a. Also, a is called the additive inverse, the negative, or the opposite of $-a$.

Because of the previous definition, the sum of a number and its negative is 0. For example,

$$5 + (-5) = 0 \qquad \text{and} \qquad -7 + 7 = 0$$

The symbol $-(-6)$ is read as "the negative of negative 6." Because the sum of two numbers that are negatives is 0, we have

$$-6 + [-(-6)] = 0 \qquad \text{but} \qquad -6 + 6 = 0$$

Because -6 has only one additive inverse, it follows that

$$-(-6) = 6$$

In general, we have the following rule.

The Double Negative Rule If a represents any number, then

$$-(-a) = a$$

If the product of two numbers is 1, the numbers are called **multiplicative inverses** or **reciprocals** of each other.

The Multiplicative Inverse Elements For every nonzero real number a, there exists a single real number $\frac{1}{a}$ such that

$$a \cdot \frac{1}{a} = \frac{1}{a} \cdot a = 1$$

The number $\frac{1}{a}$ is called the **multiplicative inverse**, or the **reciprocal**, of a. Also, a is called the multiplicative inverse, or the reciprocal, of $\frac{1}{a}$.

The existence of multiplicative inverse elements guarantees that whatever nonzero real number we start with, another real number can be found such that their product is 1. For example, the reciprocal, or multiplicative inverse, of 5 is $\frac{1}{5}$ because $5 \cdot \frac{1}{5} = 1$. The reciprocal of $\frac{3}{2}$ is $\frac{2}{3}$ because

$$\left(\frac{3}{2}\right)\left(\frac{2}{3}\right) = 1$$

The reciprocal of 0 does not exist, because $\frac{1}{0}$ does not exist.

EXAMPLE 5 The statements on the left are true because of the properties on the right.

$2 + 7$ is a real number	The closure property for addition
$2(7)$ is a real number	The closure property for multiplication
$9 + 3 = 3 + 9$	The commutative property for addition
$8 \cdot 3 = 3 \cdot 8$	The commutative property for multiplication
$9 + (2 + 3) = (9 + 2) + 3$	The associative property for addition
$2(xy) = (2x)y$	The associative property for multiplication
$2(x + 3) = 2x + 2 \cdot 3$	The distributive property
$(a + b) + c = c + (a + b)$	The commutative property for addition
$37 + 0 = 37$	The identity property for addition
$17 \cdot 1 = 17$	The identity property for multiplication
$\frac{3}{7} + \left(-\frac{3}{7}\right) = 0$	The additive inverse property
$\frac{4}{5} \cdot \frac{5}{4} = 1$	The multiplicative inverse property

■

ORALS *State each property.*

1. Commutative property of addition
2. Associative property of multiplication
3. Transitive property of equality
4. Double negative property
5. Reflexive property of equality
6. Substitution property

EXERCISE 1.2

In Exercises 1–4, insert either an = or an ≠ symbol to make a true statement.

1. $3 \; = \; 2 + 1$ **2.** $2 \cdot 3 \; = \; 3 \cdot 2$ **3.** $\pi \; \neq \; 3$ **4.** $0.375 \; = \; \frac{3}{8}$

In Exercises 5–12, tell which property of equality justifies each statement.

5. If $a = b + c$, then $b + c = a$.
symmetric property

6. If $x = y + z$ and $z = 3$, then $x = y + 3$.
substitution property

7. $a + b + c = a + b + c$
reflexive property

8. If $a = 37$ and $37 = b$, then $a = b$.
transitive property

9. If $x = y + z$ and $y + z = 10$, then $x = 10$.
transitive property

10. If $x + y = c + d$, then $c + d = x + y$.
symmetric property

11. If $3x = 3y$ and $y = 4$, then $3x = 3(4)$.
substitution property

12. $(a + b) + c = (a + b) + c$
reflexive property

In Exercises 13–16, use an associative property to help simplify each expression.

13. $5 + (2 + x)$ $\quad 5 + (2 + x) = (5 + 2) + x = 7 + x$

14. $(a + 3) + 4$ $\quad (a + 3) + 4 = a + (3 + 4) = a + 7$

15. $5(3b)$ $\quad 5(3b) = (5 \cdot 3)b = 15b$

16. $3(2x)$ $\quad 3(2x) = (3 \cdot 2)x = 6x$

In Exercises 17–20, use a commutative property and then an associative property to help simplify each expression.

17. $(3 + b) + 7$
$(3 + b) + 7 = 7 + (3 + b) = (7 + 3) + b = 10 + b$

18. $7 + (a + 3)$
$7 + (a + 3) = (a + 3) + 7 = a + (3 + 7) = a + 10$

19. $(3y)2$ $\quad (3y)2 = 2(3y) = (2 \cdot 3)y = 6y$

20. $(5z)3$ $\quad (5z)3 = 3(5z) = (3 \cdot 5)z = 15z$

In Exercises 21–26, use the distributive property to remove parentheses and then simplify, if possible.

21. $3(x + 2)$ $\quad 3x + 6$

22. $2(3y + 4)$ $\quad 6y + 8$

23. $4(2a - 1)$ $\quad 8a - 4$

24. $5(6z - 4)$ $\quad 30z - 20$

25. $5(x + y + 4)$ $\quad 5x + 5y + 20$

26. $9(3 + a + b)$ $\quad 27 + 9a + 9b$

In Exercises 27–42, tell which property of the real numbers justifies each statement.

27. 3(4) is a real number. clos. prop. of ×

28. 5 + 5 is a real number. clos. prop. of +

29. $3 + 7 = 7 + 3$ commut. prop. of +

30. $2(9 \cdot 13) = (2 \cdot 9)13$ assoc. prop. of ×

31. $3(2 + 5) = 3 \cdot 2 + 3 \cdot 5$ dist. prop.

32. $1 \cdot 3 = 3 \cdot 1$ commut. prop. of ×

33. $81 + 0 = 0 + 81$ commut. prop. of +

34. $3(9 + 2) = 3 \cdot 9 + 3 \cdot 2$ dist. prop.

35. $81 + 0 = 81$ additive identity prop.

36. $3 + (9 + 0) = (9 + 0) + 3$ commut. prop. of +

37. $5 \cdot \frac{1}{5} = 1$ mult. inverse prop.

38. $a + (3 + y) = (a + 3) + y$ assoc. prop. of +

39. $2 + (7 + 8) = (2 + 7) + 8$ assoc. prop. of +

40. $1 \cdot 3 = 3$ mult. identity prop.

41. $(2 \cdot 3) \cdot 4 = 4 \cdot (2 \cdot 3)$ commut. prop. of ×

42. $8 + (-8) = 0$ additive inverse prop.

In Exercises 43–54, find the additive inverse of each number. When necessary, simplify the expression first, and then find the additive inverse of the result.

43. 1 $\quad -1$

44. 3 $\quad -3$

45. -8 $\quad 8$

46. -7 $\quad 7$

47. 0 $\quad 0$

48. $\frac{1}{2}$ $\quad -\frac{1}{2}$

49. π $\quad -\pi$

50. a $\quad -a$

51. $2 + 8$ $\quad -10$

52. $5 - 3$ $\quad -2$

53. $-(7 - 4)$ $\quad 3$

54. $-\frac{8}{2}$ $\quad 4$

In Exercises 55–62, find the multiplicative inverse (the reciprocal) of each number, provided one exists.

55. 1 1

56. 3 $\frac{1}{3}$

57. $\frac{1}{2}$ 2

58. $-\frac{7}{5}$ $-\frac{5}{7}$

59. 0.25 4

60. 0.5 2

61. 0 none

62. 1.25 $\frac{4}{5}$

In Exercises 63–66, use a calculator to verify each statement. Identify the property of real numbers that is being illustrated.

63. $(37.9 + 25.2) + 14.3 = 37.9 + (25.2 + 14.3)$ assoc. prop. of +

64. $7.1(3.9 + 8.8) = 7.1 \cdot 3.9 + 7.1 \cdot 8.8$ distrib. prop.

65. $2.73(4.534 + 57.12) = 2.73 \cdot 4.534 + 2.73 \cdot 57.12$ distrib. prop.

66. $(6.789 + 345.1) + 27.347 = (345.1 + 6.789) + 27.347$ commut. prop. of +

In Exercises 67–72, use a calculator to verify each statement. To find the negative of a number, enter the number and press the +/− *key. To find a reciprocal of a number, enter the number and press the* 1/x *key.*

67. $3568.45 + (-3568.45) = 0$

68. $-0.00573 + (0.00573) = 0$

69. $-(-32.432) = 32.432$

70. The reciprocal of 0.03125 is 32.

71. $13.9 \cdot \frac{1}{13.9} = 1$

72. $\frac{3.7}{2.9} = \frac{1}{\frac{2.9}{3.7}}$

In Exercises 73–76, give the property of real numbers that justifies each step in the following proofs.

73. Prove that $(a + b) + c = a + (c + b)$.

$(a + b) + c = a + (b + c)$ assoc. prop. of +
$= a + (c + b)$ commut. prop. of +

74. Prove that $a(b + c) = ca + ba$.

$a(b + c) = ab + ac$ distrib. prop.
$= ac + ab$ commut. prop. of +
$= ca + ab$ commut. prop. of ×
$= ca + ba$ commut. prop. of ×

75. Prove that $(b + c)a = ba + ca$.

$(b + c)a = a(b + c)$ commut. prop. of ×
$= ab + ac$ distrib. prop.
$= ba + ac$ commut. prop. of ×
$= ba + ca$ commut. prop. of ×

76. Prove that $(ab)(cd) = (ad)(bc)$.

$(ab)(cd) = (cd)(ab)$ commut. prop. of ×
$= c[d(ab)]$ assoc. prop. of ×
$= c[(da)b]$ assoc. prop. of ×
$= c[(ad)b]$ commut. prop. of ×
$= [(ad)b]c$ commut. prop. of ×
$= (ad)(bc)$ assoc. prop. of ×

Writing Exercises *Write a paragraph using your own words.*

1. The symmetric property of equality is often confused with the commutative properties. Explain each one.

2. Explain the extended distributive property.

Something to Think About

1. Of the four basic operations (addition, subtraction, multiplication, and division), which are commutative? Associative?

2. Is there an identity for subtraction? For division? no, no

Review Exercises ■ *Let* $A = \{1, 2, 3, 4, 5\}$, $B = \{1, 3, 5, 7\}$, *and* $C = \{3, 4, 5, 6, 7\}$. *Find each set.*

1. $A \cup B$
$\{1, 2, 3, 4, 5, 7\}$

2. $B \cap C$
$\{3, 5, 7\}$

3. $(A \cup B) \cap C$
$\{3, 4, 5, 7\}$

4. $A \cap (B \cup C)$
$\{1, 3, 4, 5\}$

5. A man bought 32 gallons of gasoline at $1.29 per gallon and 3 quarts of oil at $1.35 per quart. The sales tax was included in the price of the gasoline, but 5% sales tax was added to the cost of the oil. Find the total cost. $45.53

6. On an adjusted income of $57,760, a woman must pay taxes according to the schedule shown in Table 1. Find the woman's tax bill. $14,169.85

Over	But not over		Of the amount over
$0	$18,450	. . . . . 15%	$0
18,450	44,575	$2767.50 + 28%	18,450
44,575	70,000	10,082.50 + 31%	44,575

TABLE 1

1.3 Inequalities and Graphs of Sets of Real Numbers

■ Ordering of the Real Numbers ■ Intervals ■ Absolute Value of a Number

Sets of numbers can be graphed on a number line. To construct a number line, we choose some point on a line (called the **origin**) and give it a number name (a **coordinate**) of 0. We then locate points that are equal distances to the right and to the left of 0 and label them with coordinates, as shown in Figure 1-4.

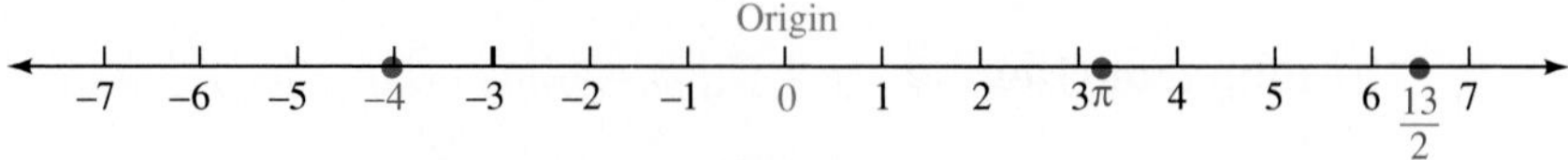

FIGURE 1-4

The **graph** of the real number -4 in Figure 1-4 is the point on the number line with a coordinate of -4. The graph of the real number $\frac{13}{2}$ (or $6\frac{1}{2}$) is the point midway between the points with coordinates 6 and 7. The graph of π is the point with coordinate 3.14159. . . . To every real number there corresponds exactly one point on the number line, its graph. And to every point on the number line there corresponds exactly one real number, its coordinate.

Real numbers such as 5 and 25.347 that are to the right of 0 are **positive numbers**. Sometimes these numbers are preceded by + signs. For example,

$$+5 = 5, \qquad +25.347 = 25.347, \qquad \text{and} \qquad +\pi = \pi$$

Real numbers such as -4 and $-\frac{17}{2}$ that are to the left of 0 are **negative numbers**.

WARNING! The number 0 is neither positive nor negative.

EXAMPLE 1 Graph the set of even integers between -5 and 5.

Solution The graph of the set of even integers between -5 and 5 includes the points with coordinates of -4, -2, 0, 2, and 4. The graph is shown in Figure 1-5.

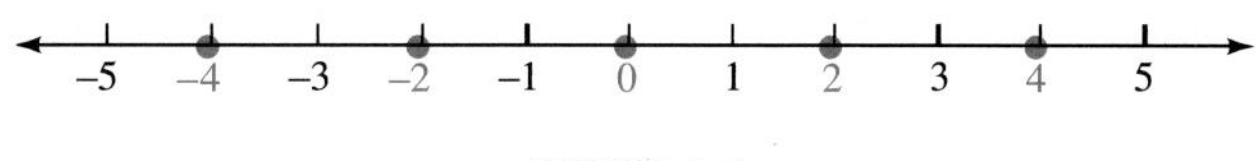

FIGURE 1-5 ■

■ Ordering of the Real Numbers

As we move from left to right on the number line, the coordinates get larger. Thus, if a point lies to the right of a second point on the number line, its coordinate is the greater. For example, on the number line, the point with coordinate 4 lies to the right of the point with coordinate -2. Thus,

$4 > -2$ Read as "4 is greater than negative 2."

If a point on the number line is to the left of another, its coordinate is the smaller. The point with coordinate -5, for example, lies to the left of the point with coordinate -1. Thus,

$-5 < -1$ Read as "Negative 5 is less than negative 1."

Two other common inequality symbols are

$\leq$ Read as "is less than or equal to."

$\geq$ Read as "is greater than or equal to."

EXAMPLE 2 **a.** $-7 > -10$ because on the number line, -7 is to the right of -10.

b. $5 < 8$ because on the number line, 5 is to the left of 8.

c. $15 \leq 15$ because $15 = 15$.

d. $20 \geq -19$ because $20 > -19$. ■

Inequality statements can be written so that the inequality symbol points in the opposite direction. For example, the inequality

$-3 \leq 9$ can be written as $9 \geq -3$

To say that a number is not less than 0, we write $x \not< 0$. This is equivalent to saying that $x \geq 0$. Likewise, we have

$x \not> 0$ is equivalent to $x \leq 0$.
$x \not\geq 0$ is equivalent to $x < 0$.
$x \not\leq 0$ is equivalent to $x > 0$.
$x \neq 0$ is equivalent to $x < 0$ or $x > 0$.

If a and b are real numbers, either a and b are equal, or they are not. And if they are not, one or the other must be the larger. The possibilities are summed up in the following property.

The Trichotomy Property If a and b are real numbers, then one of the following statements is true:

$a < b$ or $a = b$ or $a > b$

The inequality relationships $<$, $\leq$, $>$, and $\geq$ have a transitive property.

The Transitive Property of Inequality If a, b, and c are real numbers, then

If $a < b$ and $b < c$, then $a < c$.

A similar statement is true for the $>$, $\leq$, and $\geq$ symbols.

EXAMPLE 3 **a.** By the trichotomy property, if x is a real number, then

$x < 5$ or $x = 5$ or $x > 5$

b. By the transitive property,

If $x > 12$ and $12 > 5$, then $x > 5$. ■

Intervals

Graphs of sets of real numbers are often portions of the number line called **intervals**. For example, Figure 1-6 shows the graph of all real numbers greater than 3. The graph is shown as a solid arrow that includes all points with coordinates greater than 3. Thus, points with coordinates such as 5, $37\frac{2}{3}$, and $99\frac{4}{5}$ are included in the graph. The open circle at 3 indicates that the point with a coordinate of 3 is not

Alternate Notation: You may want to use a parenthesis instead of an open circle and a bracket instead of a closed circle. Then Figures 1–6 and 1–7 would look like

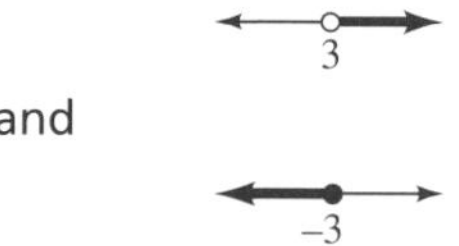

included. This interval, which has no endpoints, is called an **open interval** and is denoted by the inequality

$$x > 3$$

In **interval notation**, this interval is denoted as $(3, \infty)$, where the symbol ∞ is read as "infinity." The parentheses indicate that no endpoints are included.

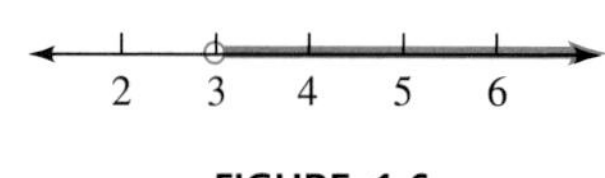

FIGURE 1-6

EXAMPLE 4 Graph all numbers x on the number line such that $x \leq -3$.

Solution The interval shown in Figure 1-7 includes all real numbers x such that $x \leq -3$. The solid circle at -3 indicates that the point with coordinate -3 is included in the graph. Because exactly one endpoint is included in the graph, this interval is called a **half-open** interval. In interval notation, it is denoted as $(-\infty, -3]$. The bracket indicates that the endpoint with coordinate -3 is included.

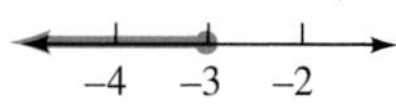

FIGURE 1-7 ■

To graph the set of real numbers between -4 and 2 (see Figure 1-8), we graph the open interval denoted by the inequalities

$$x > -4 \quad \text{and} \quad x < 2$$

or more briefly by the double inequality

$$-4 < x < 2$$ Read as "x is between -4 and 2," or as "-4 is less than x, which is, in turn, less than 2."

In interval notation, this open interval is expressed as $(-4, 2)$.

FIGURE 1-8

EXAMPLE 5 Graph all numbers x on the number line such that $-3 \leq x \leq 1$.

Solution All of the points on the number line from -3 to 1 form the interval shown in Figure 1-9. This interval is called a **closed interval** because it includes two endpoints. In interval notation, it is denoted by $[-3, 1]$. The brackets indicate that both endpoints are included.

FIGURE 1-9 ■

EXAMPLE 6 Graph all numbers x on the number line such that $0 \leq x < 5$.

Solution All of the points on the number line from 0 to 5, not including 5, form the interval shown in Figure 1-10. This half-open interval is denoted by $[0, 5)$.

FIGURE 1-10

Figure 1-11 shows several different intervals and their corresponding graphs.

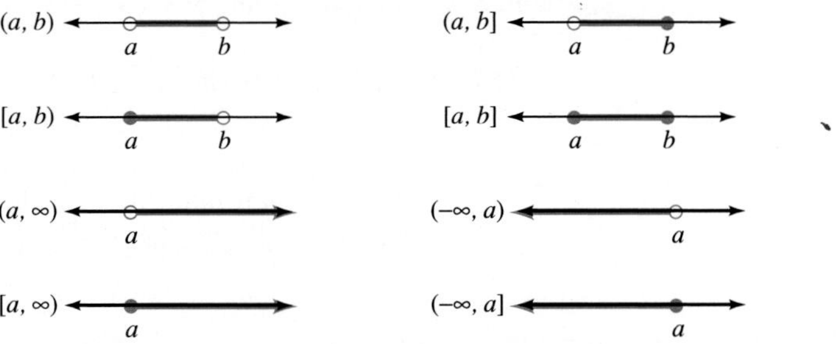

FIGURE 1-11

EXAMPLE 7 If $A = (-2, 4)$ and $B = [1, 5)$, find the graph of **a.** $A \cup B$ and **b.** $A \cap B$.

Solution **a.** The union of intervals A and B is the set of all real numbers that are elements of either set A or set B or both. Numbers between -2 and 4 are in set A, and numbers between 1 and 5 (including 1) are in set B. Numbers between -2 and 5 are in at least one of these sets. Refer to Figure 1-12(a) to see that

$$A \cup B = (-2, 4) \cup [1, 5) = (-2, 5)$$

b. The intersection of intervals A and B is the set of all real numbers that are elements of both set A and set B. The numbers that are in both of these sets are the numbers between 1 and 4 (including 1). Refer to Figure 1-12(b) to see that

$$A \cap B = (-2, 4) \cap [1, 5) = [1, 4)$$

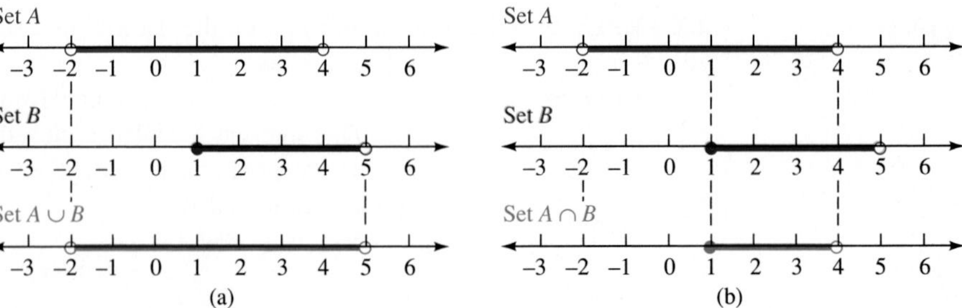

FIGURE 1-12

Absolute Value of a Number

The **absolute value** of any real number a, denoted as $|a|$, is the distance on a number line between 0 and the point with coordinate a. Because the points shown in Figure 1-13 with coordinates of 3 and -3 both lie 3 units from 0,

$$|3| = |-3| = 3$$

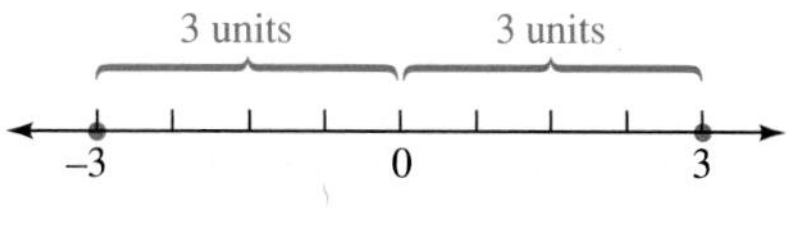

FIGURE 1-13

In general, for any real number a,

$$|a| = |-a|$$

The absolute value of a number can be defined more formally.

Absolute Value If $x \geq 0$, then $|x| = x$.

If $x < 0$, then $|x| = -x$.

The previous definition indicates that if x is a positive number or 0, then x is its own absolute value. However, if x is a negative number, then $-x$ (which is a positive number) is the absolute value of x. Thus, $|x|$ always represents a nonnegative number. In general, for all x,

$$|x| \geq 0$$

WARNING! The variable x is not always positive, nor is $-x$ always negative. If $x = -4$, for example, then

$x = -4$ and x is negative

$-x = -(-4) = 4$ and $-x$ is positive

EXAMPLE 8

a. $|3| = 3$ **b.** $|-4| = -(-4) = 4$ **c.** $|0| = 0$

d. $-|7| = -7$ **e.** $-|-7| = -(7)$
$= -7$

f. $|7 - 2| - (-3) = |5| + 3$
$= 5 + 3$
$= 8$ ■

ORALS *Evaluate each expression.*

1. $|-30|$ 30
2. $-|30|$ -30
3. $-|-30|$ -30
4. $|-(-30)|$ 30
5. $|-5| - |5|$ 0
6. $|0| + |-5|$ 5

EXERCISE 1.3

In Exercises 1–4, graph each set on the number line.

1. The set of prime numbers less than 8.

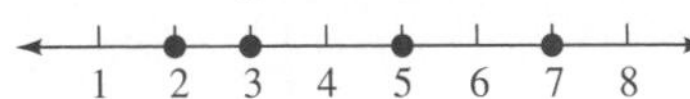

2. The set of integers between -9 and 0.

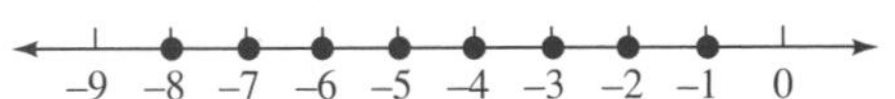

3. The set of odd integers between 10 and 20.

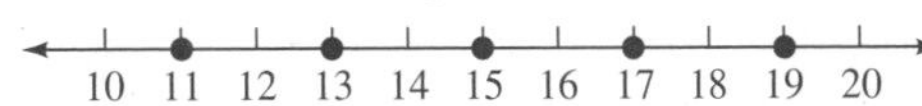

4. The set of composite numbers less than 10.

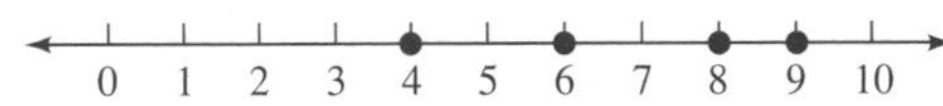

In Exercises 5–12, insert either a $<$ or a $>$ symbol to make a true statement.

5. $5 < 9$ **6.** $9 > 0$ **7.** $-5 > -10$ **8.** $-3 < 10$

9. $-7 < 7$ **10.** $0 > -5$ **11.** $6 > -6$ **12.** $-6 < -2$

In Exercises 13–20, write each statement with the inequality symbol pointing in the opposite direction.

13. $19 > 12$ $12 < 19$ **14.** $-3 \geq -5$ $-5 \leq -3$ **15.** $-6 \leq -5$ $-5 \geq -6$ **16.** $-10 < 13$ $13 > -10$

17. $5 \geq -3$ $-3 \leq 5$ **18.** $0 \leq 12$ $12 \geq 0$ **19.** $-10 < 0$ $0 > -10$ **20.** $-4 > -8$ $-8 < -4$

In Exercises 21–24, rewrite each statement using a $<$ or $>$ symbol.

21. $x \nleq 3$ $x > 3$ **22.** $y \ngeq 4$ $y < 4$ **23.** $z \ngeq -4$ $z < -4$ **24.** $t \nleq -2$ $t > -2$

In Exercises 25–28, rewrite each expression using a $\leq$ or a $\geq$ symbol.

25. $x \nless 7$ $x \geq 7$ **26.** $x \ngtr 3$ $x \leq 3$ **27.** $x \ngtr -3$ $x \leq -3$ **28.** $x \nless -7$ $x \geq -7$

In Exercises 29–40, graph each interval on the number line.

29. $x > 3$

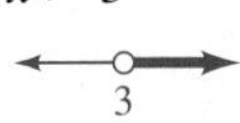

30. $x < 0$

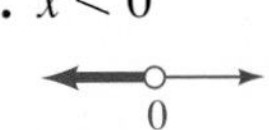

31. $-3 < x < 2$

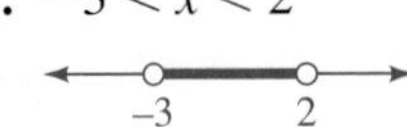

32. $-5 \leq x < 2$

33. $0 < x \leq 5$

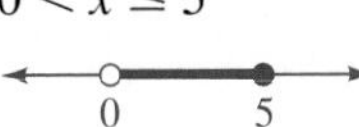

34. $-4 \leq x \leq -2$

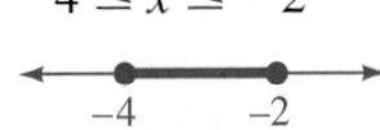

35. $(-2, \infty)$

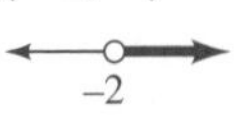

36. $(-\infty, 4]$

37. $[-6, 9]$

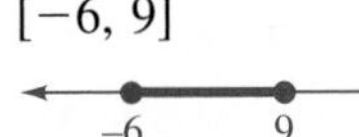

38. $(-1, 3)$

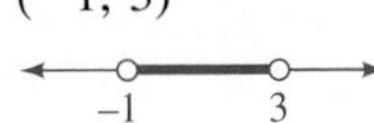

39. $(-2, 4]$

40. $[-5, 2]$

In Exercises 41–48, A, B, C, and D are intervals, with $A = [-4, 4]$, $B = (0, 6)$, $C = [2, 8)$, *and* $D = (-3, 4]$. *Graph each set.*

41. $A \cap C$

42. $B \cup C$

0 8

43. $A \cup C$

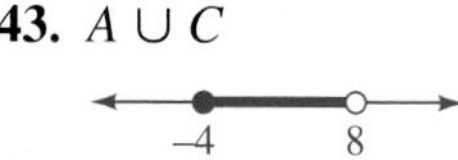

44. $A \cap B$

0 4

45. $A \cap B \cap C$

2 4

46. $A \cup B \cup C$

−4 8

47. $A \cup B \cup D$

−4 6

48. $B \cap C \cap D$

2 4

In Exercises 49–60, write each expression without using absolute value symbols and simplify the result when possible.

49. $|20|$ 20

50. $|-20|$ 20

51. $-|-6|$ −6

52. $-|8|$ −8

53. $|-5| + |0|$ 5

54. $-|0| + |-4|$ 4

55. $|-5| \cdot |4|$ 20

56. $|-6| \cdot |-3|$ 18

57. $|-15| - |10| - (-2)$ 7

58. $|-4| - |-4| - (-4)$ 4

59. $|5 - 3| - (-|-1|)$ 3

60. $|18 - 5| - [-|-18|]$ 31

61. How many integers have an absolute value that is less than 50? 99

62. How many odd integers have an absolute value between 20 and 40? 20

63. What numbers are equal to their own absolute values? all positives and 0

64. What numbers when added to their own absolute values give a sum of 0? all negatives and 0

65. Find x and y if $|x| + |y| = 0$. both are 0

66. Find the relationship between x and y if $|x + y| = 0$. $y = -x$

67. Does the absolute value of the product of two numbers equal the product of their absolute values? Explain. yes

68. Does the absolute value of the sum of two numbers equal the sum of their absolute values? Explain. no

69. If $|x| = 3$, what numbers could x be? 3 or −3

70. If $|x| = 7$, what numbers could x be? 7 or −7

In Exercises 71–74, x is a positive number. Tell whether each expression represents a positive or a negative number.

71. $-x$ negative

72. $|-x|$ positive

73. $|x|$ positive

74. $-(-x)$ positive

In Exercises 75–78, x is a negative number. Tell whether each expression represents a positive or a negative number.

75. $-x$ positive

76. $-|x|$ negative

77. $|x|$ positive

78. $-(-x)$ negative

In Exercises 79–82, x is a positive number. Place a $<$ *or* $>$ *symbol between each pair of expressions to make a true statement.*

79. x $>$ $-x$

80. 0 $<$ x

81. $-x$ $<$ 0

82. x $>$ $-|x|$

Writing Exercises ■ *Write a paragraph using your own words.*

1. The negative of a number is not necessarily a negative number. Explain.

2. Is the absolute value of a number always positive? Explain.

Something to Think About

1. If $(1, 5) \cap (a, b) = (a, b)$, what is known about a and b?
2. If $(1, 5) \cup (a, b) = (a, b)$, what is known about a and b?
3. When is $|x + y| = |x| + |y|$?
4. When is $|xy| = |x||y|$?

Review Exercises

If Q is the set of rational numbers and H is the set of irrational numbers, find each set.

1. $Q \cap H$ $\emptyset$
2. $Q \cap \emptyset$ $\emptyset$
3. $H \cup \emptyset$ H
4. $H \cap (\emptyset \cup Q)$ $\emptyset$

1.4 Arithmetic of Real Numbers

Adding Real Numbers ■ Subtracting Real Numbers ■ Multiplying Real Numbers ■ Dividing Real Numbers ■ Order of Operations ■ Properties of Fractions

If x and y are added, the result $(x + y)$ is called their **sum**. Each of the numbers x and y is called a **term** of that sum.

Adding Real Numbers

Suppose we wish to add the numbers $+2$ and $+3$. Since the positive direction on the number line is to the right, we can represent $+2$ with an arrow of length 2, pointing to the right. We can represent $+3$ with an arrow of length 3, also pointing to the right.

To find the sum $(+2) + (+3)$, we start at the origin and place the arrows end-to-end as in Figure 1-14. The endpoint of the second arrow is the point with coordinate $+5$. Thus,

$$+2 + (+3) = +5$$

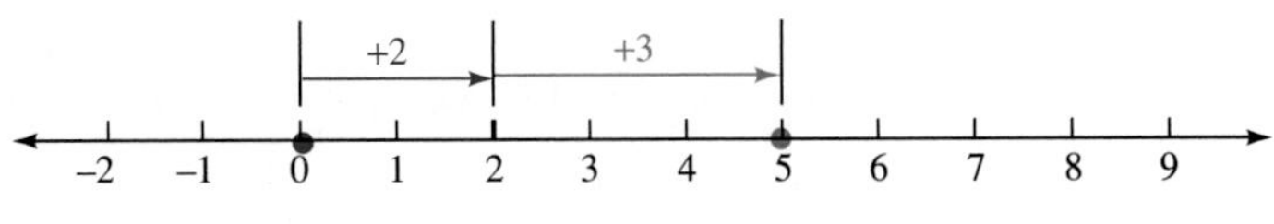

FIGURE 1-14

We can represent the numbers in the addition problem

$$(-2) + (-3)$$

with arrows as in Figure 1-15. We represent -2 with an arrow of length 2 that begins at the origin and points to the left. Then using -2 as a starting point, we represent -3 with an arrow of length 3 that also points to the left. Because the endpoint of the final arrow is -5,

$$(-2) + (-3) = -5$$

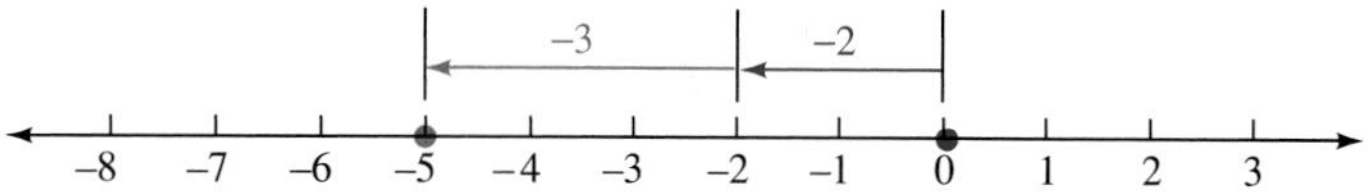

FIGURE 1-15

Since two numbers with like signs are represented by arrows pointing in the same direction, we have the following rule.

Adding Real Numbers with Like Signs To add two real numbers with the same sign, add their absolute values and keep their common sign.

EXAMPLE 1

a. $+4 + (+6) = +(|+4| + |+6|)$
$= +(4 + 6)$
$= +10$

b. $-4 + (-6) = -(|-4| + |-6|)$
$= -(4 + 6)$
$= -10$ ■

Two real numbers with unlike signs can be represented by arrows that point in opposite directions. For example, to add -6 and $+2$, we refer to the number line in Figure 1-16, where the number -6 is represented by an arrow of length 6 that begins at the origin and points to the left. The arrow representing $+2$ begins at -6, has length 2, and points to the right. The endpoint of this final arrow is the point -4. Thus,

$$(-6) + (+2) = -4$$

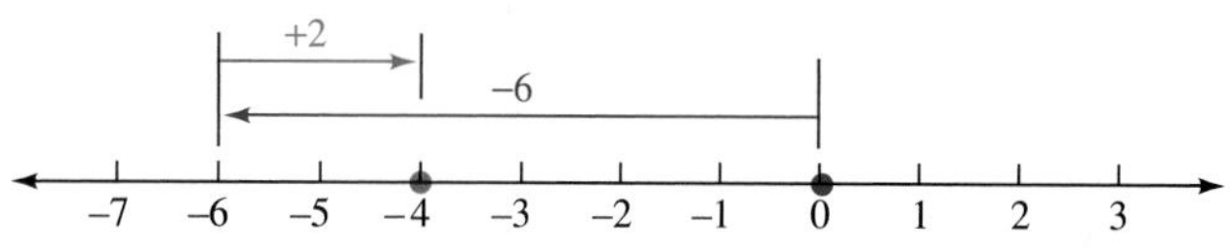

FIGURE 1-16

The arrows in Figure 1-17 represent the numbers in the addition problem

$$(+7) + (-4)$$

The first arrow begins at the origin, has length 7, and points to the right. The second arrow begins at point 7, has length 4, and points to the left. The endpoint of the final arrow is the point $+3$. Thus,

$$(+7) + (-4) = +3$$

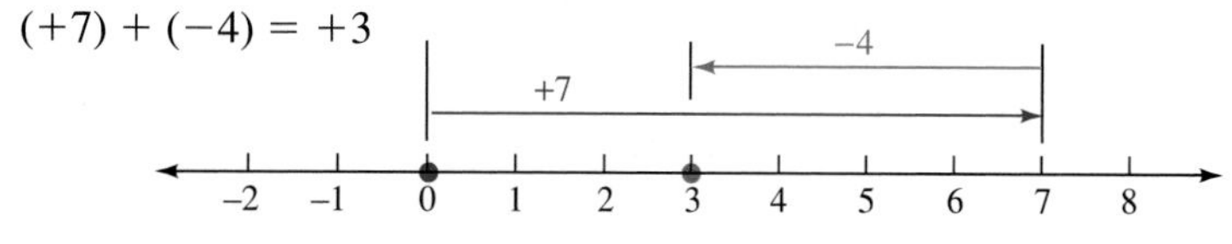

FIGURE 1-17

Because two real numbers with unlike signs are represented by arrows pointing in opposite directions, we have the following rule.

Adding Real Numbers with Unlike Signs To add two real numbers with unlike signs, subtract their absolute values (the smaller from the larger) and keep the sign of the number with the larger absolute value.

(a,d) **EXAMPLE 2**

a. $+6 + (-5) = +(|+6| - |-5|)$
$= +(6 - 5)$
$= +1$

b. $-2 + (+5) = +(|+5| - |-2|)$
$= +(5 - 2)$
$= +3$

c. $6 + (-9) = -(|-9| - |6|)$
$= -(9 - 6)$
$= -3$

d. $-10 + (4) = -(|-10| - |4|)$
$= -(10 - 4)$
$= -6$ ■

Subtracting Real Numbers

We can always express a subtraction problem as an equivalent addition problem. For example, the subtraction problem

$7 - 4$ is equivalent to the addition problem $7 + (-4)$

because they have the same answer:

$7 - 4 = 3$ and $7 + (-4) = 3$

We use this idea to define the **difference** when b is to be subtracted from a.

Subtracting Real Numbers If a and b are real numbers, then

$$a - b = a + (-b)$$

EXAMPLE 3 Evaluate **a.** $12 - 4$, **b.** $-13 - 5$, and **c.** $-14 - (-6)$.

Solution We use the rule for finding the difference of two real numbers.

a. $12 - 4 = 12 + (-4)$
$= 8$

b. $-13 - 5 = -13 + (-5)$
$= -18$

c. $-14 - (-6) = -14 + [-(-6)]$
$= -14 + 6$ Use the double negative rule.
$= -8$ ■

Multiplying Real Numbers

The result of multiplying two numbers x and y is called the **product** of x and y. Each of the numbers x and y is called a **factor** of that product.

Multiplication by a positive number can be thought of as repeated addition. The expression 5(4), for example, means that 4 is to be used as a term in an indicated sum five times:

$$\begin{aligned} 5(4) &= 4 + 4 + 4 + 4 + 4 \\ &= 20 \end{aligned}$$

Likewise, the expression $5(-4)$ means that -4 is to be used as a term in a sum five times:

$$\begin{aligned} 5(-4) &= (-4) + (-4) + (-4) + (-4) + (-4) \\ &= -20 \end{aligned}$$

If multiplication by a positive number means repeated addition, then it is reasonable to assume that multiplication by a negative number means repeated subtraction. For example, the expression $-5(4)$ means that 4 is to be used as a term in a repeated subtraction five times:

$$\begin{aligned} -5(4) &= -4 - 4 - 4 - 4 - 4 \\ &= -4 + (-4) + (-4) + (-4) + (-4) \\ &= -20 \end{aligned}$$

The expression $-5(-4)$ means that -4 is to be used as a term in a repeated subtraction five times:

$$\begin{aligned} -5(-4) &= -(-4) - (-4) - (-4) - (-4) - (-4) \\ &= 4 + [-(-4)] + [-(-4)] + [-(-4)] + [-(-4)] \\ &= 4 + 4 + 4 + 4 + 4 \\ &= 20 \end{aligned}$$

The products 5(4) and $-5(-4)$ both equal $+20$, and the products $5(-4)$ and $-5(4)$ both equal -20. These results suggest the following rule.

Multiplying Real Numbers

1. To multiply two real numbers with like signs, multiply their absolute values. The sign of the product will be positive.
2. To multiply two real numbers with unlike signs, multiply their absolute values. The sign of the product will be negative.
3. If x is any real number, then $x \cdot 0 = 0 \cdot x = 0$.

EXAMPLE 4 Find the products **a.** $4(-7)$, **b.** $-5(-6)$, **c.** $-7(6)$, and **d.** $8(6)$.

Solution **a.** $4(-7) = -28$ **b.** $-5(-6) = +30$

c. $-7(6) = -42$ **d.** $8(6) = +48$ ■

Dividing Real Numbers

When x is divided by a nonzero number y, the result is called the **quotient** of x and y, and we can write

$$\frac{x}{y} = q$$

The quotient q is that number which, when multiplied by y, gives x:

$$yq = x$$

This special relationship can be used to develop rules for dividing one real number by a nonzero real number. We consider four divisions:

$$\frac{+10}{+2} = +5 \quad \text{because } +2(+5) = +10$$

$$\frac{-10}{-2} = +5 \quad \text{because } -2(+5) = -10$$

$$\frac{+10}{-2} = -5 \quad \text{because } -2(-5) = +10$$

$$\frac{-10}{+2} = -5 \quad \text{because } +2(-5) = -10$$

These results suggest that the rules for dividing real numbers are similar to the rules for multiplying real numbers.

Dividing Real Numbers

1. To divide two real numbers with like signs, divide their absolute values. The sign of the quotient is positive.
2. To divide two real numbers with unlike signs, divide their absolute values. The sign of the quotient is negative.
3. Division by 0 is undefined.

WARNING! If x is any nonzero real number, then $\frac{0}{x} = 0$. However, $\frac{x}{0}$ is undefined.

EXAMPLE 5 Find each quotient: **a.** $\frac{36}{18}$, **b.** $\frac{-44}{11}$, **c.** $\frac{27}{-9}$, and **d.** $\frac{-64}{-8}$.

Solution **a.** $\frac{36}{18} = +2$ **b.** $\frac{-44}{11} = -4$

c. $\frac{27}{-9} = -3$ **d.** $\frac{-64}{-8} = +8$ ■

Order of Operations

The expression $4 + 2 \cdot 3$ appears to have two answers, depending on whether we add or multiply first. If we add first, the answer is

$$\begin{aligned}(4 + 2)3 &= 6 \cdot 3 \\ &= 18\end{aligned}$$

However, if we multiply first, the answer is

$$\begin{aligned}4 + (2 \cdot 3) &= 4 + 6 \\ &= 10\end{aligned}$$

Unless there is agreement on which operation is to be done first, there is no way of knowing which of these answers is correct.

The expression $50 \div 10 \div 5$ also appears to have two answers, depending on whether we divide 50 by 10 first or divide 10 by 5 first:

$$\begin{aligned}(50 \div 10) \div 5 &= 5 \div 5 \\ &= 1\end{aligned}$$

$$\begin{aligned}50 \div (10 \div 5) &= 50 \div 2 \\ &= 25\end{aligned}$$

To guarantee that every calculation has a single answer, we will agree to do operations in the following order.

Order of Operations

1. Unless grouping symbols indicate otherwise, perform all multiplications and/or divisions, in order from left to right.
2. Then perform all additions and/or subtractions, in order from left to right.

Thus, the correct calculations of the above examples are

$$\begin{aligned}4 + 2 \cdot 3 &= 4 + (2 \cdot 3) \\ &= 4 + 6 \\ &= 10\end{aligned}$$

Do the multiplication first and then the addition.

and

$$\begin{aligned}50 \div 10 \div 5 &= (50 \div 10) \div 5 \\ &= 5 \div 5 \\ &= 1\end{aligned}$$

Do the divisions in order from left to right.

EXAMPLE 6 Find **a.** $2 \cdot 3 + 4$, **b.** $2(3 + 4)$, **c.** $10 \div 5 \cdot 2$, and **d.** $10 \div (5 \cdot 2)$.

Solution **a.** Because parentheses do not indicate otherwise, we do the multiplication first.

$$\begin{aligned}2 \cdot 3 + 4 &= 2 \cdot 3 + 4 \\ &= 6 + 4 \\ &= 10\end{aligned}$$

b. Here the parentheses indicate that we must do the addition first.

$$\begin{aligned}2(3 + 4) &= 2(3 + 4) \\ &= 2 \cdot 7 \\ &= 14\end{aligned}$$

c. Since there are no grouping symbols to indicate otherwise, we do the multiplications and divisions in order from left to right.

$$\begin{aligned} 10 \div 5 \cdot 2 &= (10 \div 5) \cdot 2 \\ &= 2 \cdot 2 \\ &= 4 \end{aligned}$$

d. Here the parentheses indicate that we must do the multiplication first.

$$\begin{aligned} 10 \div (5 \cdot 2) &= 10 \div (5 \cdot 2) \\ &= 10 \div 10 \\ &= 1 \end{aligned}$$

■

EXAMPLE 7 Simplify **a.** $5 \cdot 3 - 6 \div 3 + 1$, **b.** $5(3 - 6) \div 3 + 1$, and **c.** $5(3 - 6 \div 3) + 1$.

Solution **a.** We do the multiplication and division first. Then we do the subtraction and addition.

$$\begin{aligned} 5 \cdot 3 - 6 \div 3 + 1 &= 15 - 2 + 1 \\ &= 13 + 1 \\ &= 14 \end{aligned}$$

b. Because of the parentheses, we do the subtraction first, then the multiplication and division, then the addition.

$$\begin{aligned} 5(3 - 6) \div 3 + 1 &= 5(-3) \div 3 + 1 \\ &= -15 \div 3 + 1 \\ &= -5 + 1 \\ &= -4 \end{aligned}$$

c. The operations within the parentheses must be done first, and within the parentheses, the division has priority over the subtraction.

$$\begin{aligned} 5(3 - 6 \div 3) + 1 &= 5(3 - 2) + 1 \\ &= 5 \cdot 1 + 1 \\ &= 5 + 1 \\ &= 6 \end{aligned}$$

■

A division such as $(7 + 8) \div (7 - 2)$ can be written in fractional form as

$$\frac{7 + 8}{7 - 2}$$

In this fraction, the bar is a grouping symbol that separates the numerator, $7 + 8$, from the denominator, $7 - 2$. To simplify this fraction, we must first do all indicated operations in the numerator and the denominator separately. Then we can simplify the fraction. Thus,

$$\frac{7 + 8}{7 - 2} = \frac{15}{5} = 3$$

EXAMPLE 8 If $a = 2$, $b = -3$, and $c = -5$, evaluate $\dfrac{ab + 3c}{b(c - a)}$.

Solution We substitute 2 for a, -3 for b, -5 for c, do the work in the numerator and denominator separately, and simplify.

$$\begin{aligned}\frac{ab + 3c}{b(c - a)} &= \frac{2(-3) + 3(-5)}{-3(-5 - 2)}\\ &= \frac{-6 + (-15)}{-3(-7)}\\ &= \frac{-21}{21}\\ &= -1\end{aligned}$$

■

Properties of Fractions

We summarize the properties of arithmetic fractions here. Algebraic fractions will be discussed in detail in Chapter 5.

Properties of Fractions If there are no divisions by 0, then

1. $\dfrac{a}{b} = \dfrac{c}{d}$ if and only if $ad = bc$
2. $\dfrac{a}{1} = a$, $\dfrac{a}{a} = 1 \quad (a \neq 0)$
3. $\dfrac{ak}{bk} = \dfrac{a}{b}$
4. $\dfrac{a}{b} \cdot \dfrac{c}{d} = \dfrac{ac}{bd}$ and $\dfrac{a}{b} \div \dfrac{c}{d} = \dfrac{a}{b} \cdot \dfrac{d}{c} = \dfrac{ad}{bc}$
5. $\dfrac{a}{b} + \dfrac{c}{b} = \dfrac{a + c}{b}$ and $\dfrac{a}{b} - \dfrac{c}{b} = \dfrac{a - c}{b}$
6. $-\dfrac{a}{b} = \dfrac{-a}{b} = \dfrac{a}{-b}$

Property 1 of fractions indicates that fractions are equal when the same product occurs when multiplying crosswise. For example,

$$\frac{2}{3} = \frac{8}{12} \quad \text{because} \quad 2(12) \text{ and } 3(8) \text{ are both equal to 24.}$$

Property 2 indicates that any number divided by 1 is left unchanged and that any nonzero number divided by itself is 1. For example,

$$\frac{5}{1} = 5 \qquad \text{and} \qquad \frac{5}{5} = 1$$

EXAMPLE 9 **a.** Simplify $\frac{27}{90}$ and **b.** write $\frac{3}{4}$ as a fraction with a denominator of 20.

Solution **a.** We can use Property 3 to divide out common factors that appear in both the numerator and the denominator of a fraction. To simplify $\frac{27}{90}$, we proceed as follows:

Note: We do not use the word *reduce* because many teachers find it misleading. To reduce means "to make smaller."

$$\begin{aligned} \frac{27}{90} &= \frac{9 \cdot 3}{9 \cdot 10} \\ &= \frac{\overset{1}{\cancel{9}} \cdot 3}{\underset{1}{\cancel{9}} \cdot 10} \\ &= \frac{3}{10} \end{aligned}$$

b. To write the fraction $\frac{3}{4}$ as an equivalent fraction with a denominator of 20, we use Property 3 and multiply both the numerator and the denominator of $\frac{3}{4}$ by 5:

$$\begin{aligned} \frac{3}{4} &= \frac{3 \cdot 5}{4 \cdot 5} \\ &= \frac{15}{20} \end{aligned}$$

The fraction $\frac{3}{4}$ is equal to the fraction $\frac{15}{20}$ because $3(20) = 4(15)$. ■

To justify the rule for multiplying two fractions given in Property 4, we refer to the square shown in Figure 1-18, where each side is 1 unit in length. Because the area of the square is the product of its length and width, the square in Figure 1-18 has an area of 1 square unit.

If this square is divided into 3 equal parts vertically and 5 equal parts horizontally, it is divided into 15 equal parts, and each part represents $\frac{1}{15}$ of the total area. The area of the shaded rectangle in the square is $\frac{8}{15}$, because it consists of 8 of the 15 equal parts.

Since the shaded rectangle has a width of $\frac{2}{3}$ and a length of $\frac{4}{5}$ and its area is the product of its length and width, we have

$$\begin{aligned} A &= lw \\ \frac{8}{15} &= \frac{2}{3} \cdot \frac{4}{5} \\ &= \frac{2 \cdot 4}{3 \cdot 5} \end{aligned}$$

Thus, we can find the product of $\frac{2}{3}$ and $\frac{4}{5}$ by multiplying the numerators and multiplying the denominators.

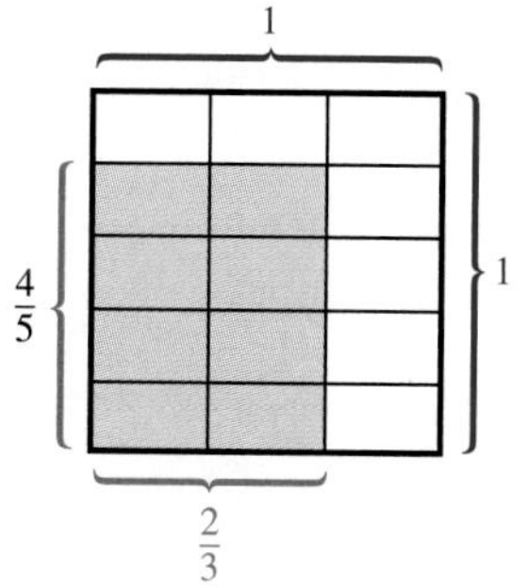

FIGURE 1-18

EXAMPLE 10 Find $\frac{3}{7} \cdot \frac{21}{5}$.

Solution By Property 4, we multiply the fractions by multiplying their numerators and multiplying their denominators. We then simplify.

$$\frac{3}{7} \cdot \frac{21}{5} = \frac{3 \cdot 21}{7 \cdot 5}$$

$$= \frac{3 \cdot 3 \cdot \overset{1}{\cancel{7}}}{\underset{1}{\cancel{7}} \cdot 5} \qquad \text{Factor 21 and divide out the common factor of 7.}$$

$$= \frac{9}{5}$$

■

To justify the rule for dividing two fractions given in Property 4, we divide $\frac{3}{7}$ by $\frac{5}{4}$, as follows:

$$\frac{3}{7} \div \frac{5}{4} = \frac{\frac{3}{7}}{\frac{5}{4}}$$

$$= \frac{\frac{3}{7} \cdot \frac{4}{5}}{\frac{5}{4} \cdot \frac{4}{5}}$$

$$= \frac{\frac{3}{7} \cdot \frac{4}{5}}{1}$$

$$= \frac{3}{7} \cdot \frac{4}{5}$$

Thus, the division $\frac{3}{7} \div \frac{5}{4}$ is equivalent to the multiplication $\frac{3}{7} \cdot \frac{4}{5}$.

EXAMPLE 11 Divide: $\frac{3}{7} \div \frac{5}{4}$.

Solution By Property 4, we invert the **divisor** (the fraction following the ÷ sign) and multiply.

$$\frac{3}{7} \div \frac{5}{4} = \frac{3}{7} \cdot \frac{4}{5}$$
$$= \frac{12}{35}$$

This result does not simplify. ■

Property 5 enables us to add or subtract fractions with common denominators. We simply add or subtract the numerators and keep the denominators.

EXAMPLE 12

a. $\frac{5}{9} + \frac{4}{9} = \frac{5+4}{9}$
$= \frac{9}{9}$
$= 1$

b. $\frac{6}{7} - \frac{2}{7} = \frac{6-2}{7}$
$= \frac{6+(-2)}{7}$
$= \frac{4}{7}$ ■

(b) **EXAMPLE 13** Find **a.** $\frac{7}{10} + \frac{11}{28}$ and **b.** $\frac{5}{6} - \frac{3}{4}$.

Solution **a.** To add fractions with unlike denominators, we use Property 3 to build the fractions into fractions with a common denominator. To find the smallest common denominator, called the **lowest** or **least common denominator (LCD)**, we write each denominator as the product of prime factors and use each prime factor the greatest number of times it appears in either product.

Teaching Tip: Use a factoring tree:

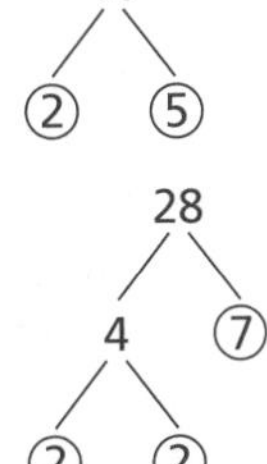

$$\left.\begin{aligned} 10 &= 2 \cdot 5 \\ 28 &= 2 \cdot 2 \cdot 7 \end{aligned}\right\} \quad \text{LCD} = 2 \cdot 2 \cdot 5 \cdot 7 = 140$$

We now write both fractions as fractions with a denominator of 140 and add.

$$\frac{7}{10} + \frac{11}{28} = \frac{7 \cdot 14}{10 \cdot 14} + \frac{11 \cdot 5}{28 \cdot 5}$$
$$= \frac{98}{140} + \frac{55}{140}$$
$$= \frac{153}{140}$$

We do not need to write the fraction $\frac{153}{140}$ as the mixed number $1\frac{13}{140}$. In algebra, **improper fractions** (fractions with numerators greater than their denominators) are acceptable.

b. To subtract the fractions, we first find the LCD to be 12.

$$\left.\begin{aligned}4 &= 2 \cdot 2\\ 6 &= 2 \cdot 3\end{aligned}\right\} \quad \text{LCD} = 2 \cdot 2 \cdot 3 = 12$$

We then write both fractions as fractions with a denominator 12 and subtract.

$$\begin{aligned}\frac{5}{6} - \frac{3}{4} &= \frac{5 \cdot 2}{6 \cdot 2} - \frac{3 \cdot 3}{4 \cdot 3}\\ &= \frac{10}{12} - \frac{9}{12}\\ &= \frac{1}{12}\end{aligned}$$

■

Property 6 points out that a − sign placed in front of a fraction can be placed either in the numerator or the denominator of the fraction. For example,

$$-\frac{2}{3} = \frac{-2}{3} = \frac{2}{-3}$$

ORALS *Evaluate each expression.*

1. $5 + (-3)$ 2
2. $5 - (-3)$ 8
3. $-5 + 3$ −2
4. $-5 - 3$ −8
5. $-5 - (-3)$ −2
6. $-5 + (-3)$ −8

EXERCISE 1.4

In Exercises 1–52, do the operations.

1. $-3 + (-5)$ −8
2. $2 + (+8)$ 10
3. $-7 + (-2)$ −9
4. $3 + (-5)$ −2
5. $2 + (-8)$ −6
6. $-5 + 3$ −2
7. $-7 + 2$ −5
8. $3 + (-7)$ −4
9. $8 + (-2)$ 6
10. $-17 + (-8)$ −25
11. $0 - (-17)$ 17
12. $0 - 21$ −21
13. $0 - 93$ −93
14. $0 - (-57)$ 57
15. $3 - 5$ −2
16. $3 - (-4)$ 7
17. $-3 - 4$ −7
18. $-11 - (-17)$ 6
19. $-33 - (-33)$ 0
20. $14 - (-13)$ 27
21. $-2(6)$ −12
22. $3(-5)$ −15
23. $-3(-7)$ 21
24. $-2(-5)$ 10
25. $\frac{-8}{4}$ −2
26. $\frac{25}{-5}$ −5
27. $\frac{-16}{-4}$ 4
28. $\frac{-5}{-25}$ $\frac{1}{5}$
29. $\frac{9-5}{5-9}$ −1
30. $\frac{8+2}{3-8}$ −2
31. $3 - 2 - 1$ 0
32. $5 - 3 - 1$ 1
33. $3 - (2 - 1)$ 2
34. $5 - (3 - 1)$ 3
35. $2 - 3 \cdot 5$ −13
36. $6 + 4 \cdot 7$ 34
37. $8 \div 4 \div 2$ 1
38. $100 \div 10 \div 5$ 2
39. $8 \div (4 \div 2)$ 4
40. $100 \div (10 \div 5)$ 50
41. $2 + 6 \div 3 - 5$ −1
42. $6 - 8 \div 4 - 2$ 2
43. $(2 + 6) \div (3 - 5)$ −4
44. $(6 - 8) \div (4 - 2)$ −1

45. $\dfrac{3(8 + 4)}{2 \cdot 3 - 9}$ -12

46. $\dfrac{5(4 - 1)}{3 \cdot 2 + 5 \cdot 3}$ $\frac{5}{7}$

47. $\dfrac{100(2 - 4)}{1000 \div 10 \div 10}$ -20

48. $\dfrac{8(3) - 4(6)}{5(3) + 3(-7)}$ 0

49. $\dfrac{5 \div (2 - 3)}{5(2 - 3)}$ 1

50. $\dfrac{4(3 - 7)}{4(3) - 8}$ -4

51. $\dfrac{60 \div (-10) \div 3}{10 - 3 - 5}$ -1

52. $\dfrac{80 \div [40 \div (-5)]}{3 - (3 - 5)}$ -2

In Exercises 53–60, $a = 3$, $b = -2$, $c = -1$, and $d = 2$. Evaluate each expression.

53. $ab + cd$ -8

54. $ad + bc$ 8

55. $a(b + c)$ -9

56. $d(b + a)$ 2

57. $\dfrac{ad + c}{cd + b}$ $-\frac{5}{4}$

58. $\dfrac{ab + d}{bd + a}$ 4

59. $\dfrac{ac - bd}{cd - ad}$ $-\frac{1}{8}$

60. $\dfrac{bc - ad}{bd + ac}$ $\frac{4}{7}$

In Exercises 61–66, $a = -1$, $b = 3$, $c = -3$, and $d = -2$. Evaluate each expression.

61. $ad - d$ 4

62. $db - d$ -4

63. $\dfrac{ad + dc - bd}{(a - d)(b + d)}$ 14

64. $\dfrac{ab - bc + cd}{(a - c)(b - d)}$ $\frac{6}{5}$

65. $\dfrac{(a + c)(b - d)}{(a + bd)(-c)}$ $\frac{20}{21}$

66. $\dfrac{(cd - b + a)(a - bc + d)}{(c - dc)(-b + c)(c - d)}$ $-\frac{2}{9}$

In Exercises 67–78, use signed numbers to solve each problem.

67. **Earning money** One day, Scott earned \$22.25 tutoring mathematics and \$39.75 tutoring physics. How much did he earn? \$62

68. **Losing weight** During an illness, Wendy lost 13 pounds. She then dieted and lost another 11 pounds. How much did she lose? 24 lb

69. **Changing temperatures** The temperature rose 17° in 1 hour and then dropped 13° in the next hour. Find the net change in temperature. +4°

70. **Displaying the flag** Before the American flag is displayed at half-mast, it should first be raised to the top of the flagpole. How far has the flag in Illustration 1 traveled? 57 ft

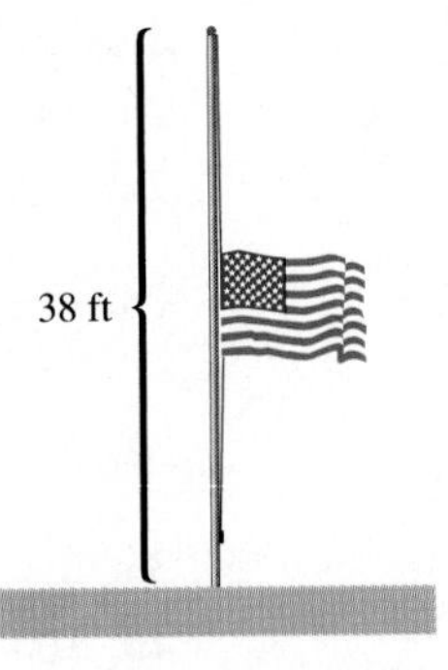

ILLUSTRATION 1

71. **Military science** An army retreated 2300 meters. After regrouping, it moved forward 1750 meters. The next day, it gained another 1875 meters. Find the army's net gain (or loss). +1325 m

72. 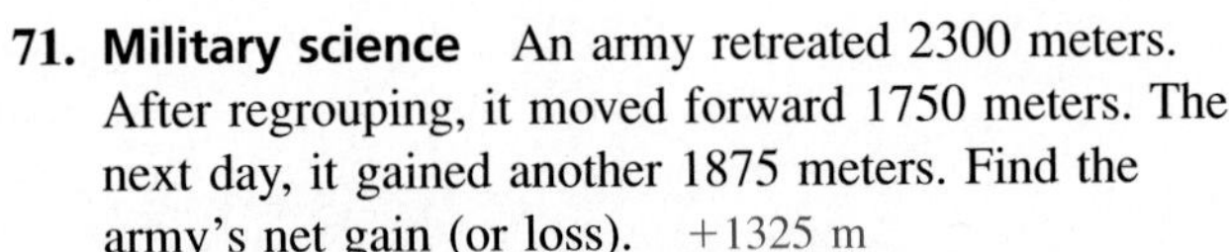 **Managing a checkbook** Sally started with \$437.37 in a checking account. One month, she had deposits of \$125.18, \$137.26, and \$145.56. That same month, she had withdrawals of \$117.11, \$183.49, and \$122.89. Find her ending balance. \$421.88

73. **Stock averages** Illustration 2 shows the daily advances and declines of the Dow Jones averages for one week. Find the total gain or loss for the week. +46

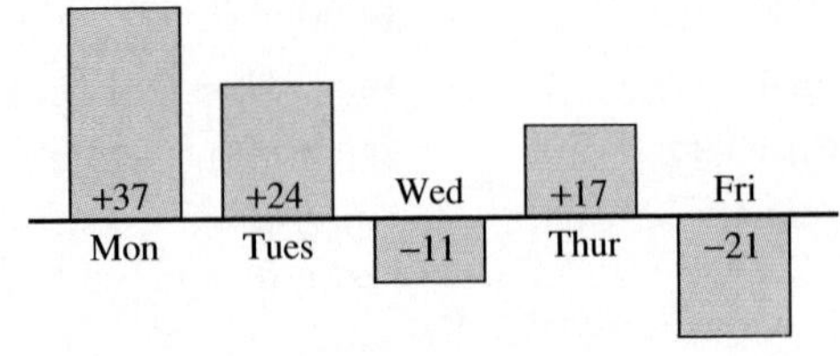

ILLUSTRATION 2

74. **Changing temperatures** If the temperature has been dropping 4° each hour, how much warmer was it 3 hours ago? 12°

75. **Playing slot machines** In Las Vegas, Harry lost \$30 per hour playing the slot machines. How much did he lose after gambling for 15 hours? \$450

76. **Filling a pool** The flow of water from a pipe fills a pool at the rate of 23 gallons per minute. How much less water was in the pool 5 hours ago? 6900 gal

77. **Draining a pool** If a drain is emptying a pool at the rate of 12 gallons per minute, how much more water was in the pool 2 hours ago? 1440 gal

78. **Grooming horses** John was paid \$8 per hour for grooming horses. If he had \$94 at the end of an 8-hour day, how much money did he have before he started work? \$30

In Exercises 79–82, tell whether the fractions are equal.

79. $\frac{14}{15}, \frac{42}{45}$ equal

80. $\frac{13}{17}, \frac{17}{21}$ not equal

81. $\frac{11}{13}, \frac{133}{169}$ not equal

82. $\frac{17}{19}, \frac{85}{135}$ not equal

In Exercises 83–86, simplify each fraction.

83. $\frac{12}{15}$ $\frac{4}{5}$

84. $-\frac{18}{36}$ $-\frac{1}{2}$

85. $-\frac{48}{72}$ $-\frac{2}{3}$

86. $\frac{539}{637}$ $\frac{11}{13}$

In Exercises 87–98, do the indicated operations and simplify the answer, if possible.

87. $\frac{1}{4} \cdot \frac{3}{5}$ $\frac{3}{20}$

88. $-\frac{3}{5}\left(\frac{20}{27}\right)$ $-\frac{4}{9}$

89. $-\frac{2}{3} \div \left(-\frac{3}{7}\right)$ $\frac{14}{9}$

90. $\frac{1}{4} \div \frac{3}{5}$ $\frac{5}{12}$

91. $-\frac{3}{5} \div \frac{9}{15}\left(-\frac{5}{27}\right)$ $\frac{5}{27}$

92. $\frac{1}{3} \div \frac{3}{4} \cdot \frac{5}{3} \div \frac{1}{2}$ $\frac{40}{27}$

93. $\frac{2}{3} + \frac{4}{3}$ 2

94. $\frac{9}{14} - \frac{3}{14}$ $\frac{3}{7}$

95. $\frac{2}{5} - \frac{7}{9}$ $-\frac{17}{45}$

96. $\frac{2}{3} + \frac{4}{5}$ $\frac{22}{15}$

97. $-\frac{3}{5}\left(\frac{1}{7} + \frac{2}{3}\right)$ $-\frac{17}{35}$

98. $\frac{2}{3}\left(-\frac{3}{4} + \frac{1}{3}\right)$ $-\frac{5}{18}$

99. **Bake sale** Jim baked 4 dozen cookies and took two-thirds of them to a bake sale. How many were left for his family to eat? 16

100. **Baking recipes** One dessert recipe calls for 8 eggs, and another recipe requires only three-fourths of that number. How many eggs are needed to make both desserts? 14

101. **Dieting** Sandy lost 15 pounds on a diet. Following the same diet, Jane lost one-third as much as Sandy, and Sue lost two-fifths as much as Sandy. How many pounds did the women lose altogether? 26

102. **Working on commission** In January, a sales representative earned \$3000 in commissions. In February, her commissions dropped to five-sixths of her January earnings. Her March commissions were two-fifths of her January commissions. How much did she earn in commissions during the first quarter of the year? \$6700

103. **Averaging test scores** On a test, Carlos scored half as many points as Maria, and Juanita scored three-fourths as many points as Maria. If Maria had a score of 80, find the average of their three scores. 60

104. **Watching the stock market** At the beginning of the year, the Dow Jones stock average was 3804. After a serious January slump, the Dow retained five-sixths of its value. After a February rally, the Dow recovered one-twelfth of its original value. Find the Dow average at the end of February. 3487

Writing Exercises *Write a paragraph using your own words.*

1. Explain how to add numbers that have opposite signs.
2. Explain how to simplify a fraction.

Something to Think About

1. If the absolute values of two different numbers are equal, what is the sum of those numbers? 0
2. Pick two numbers a and b and graph them on the number line. What distances on the number line are represented by $|a|$, $|b|$, and $|a - b|$?
3. If x and y are both negative, which of the following is largest?
   a. $x + y$ **b.** $|x|y$ **c.** $x|y|$ **d.** xy
   xy
4. Discuss whether each quantity is positive or negative when $x < 0$.
   a. $4x$ **b.** $|x|$ **c.** $-4x$ **d.** $x \cdot x$ **e.** $x \cdot x \cdot x$

Review Exercises *Graph each interval on a number line.*

1. $x > 4$

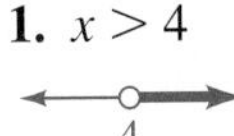

2. $x < -5$

-5

3. $(2, 10]$

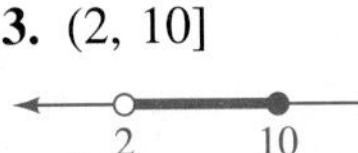

4. $[-4, 4]$

-4 4

PROJECT The Four Fours

Many of the mathematical exercises found in both modern books and ancient scrolls are not practical problems, but puzzles designed to sharpen the mind. One such puzzle, the Problem of the Four Fours, also provides practice with the rules that define the order in which operations are performed. Using these rules, the expression $4 - 4 + 4 \cdot 4$ is 16; the expression $\frac{44 + 4}{4}$ is 12. Each expression uses exactly four fours and no other numbers.

- Using the rules for order of operations; grouping symbols; any of the four operations of addition, subtraction, multiplication, and division; and exactly four fours, write each integer from 0 to 10.
- If, in a previous course, you have encountered the concepts of square root $(\sqrt{4} = 2)$ and exponents $(4^4 = 4 \cdot 4 \cdot 4 \cdot 4 = 256)$, you may succeed in going beyond 10. See how many integers between 11 and 100 you can write using only four fours. For example,

$$\frac{44 - \sqrt{4}}{\sqrt{4}} = 21 \qquad \text{and} \qquad \frac{4^4}{4 + 4} = 32$$

$\frac{4}{4} - \frac{4}{4}$, $\frac{44}{44}$, $\frac{4}{4} + \frac{4}{4}$, $\frac{4+4+4}{4}$, $4 + 4(4 - 4)$, $\frac{4 \cdot 4 + 4}{4}$, $4 + \frac{4+4}{4}$, $\frac{44}{4} - 4$, $\frac{4 \cdot 4}{4} + 4$, $4 + 4 + \frac{4}{4}$, $\frac{44 - 4}{4}$

Chapter Summary

KEY WORDS

absolute value (1.3)
closed interval (1.3)
coordinate (1.3)
difference (1.4)
disjoint sets (1.1)
element of a set (1.1)
ellipsis (1.1)
empty set (1.1)
factor (1.4)
finite set (1.1)
graph (1.3)
half-open interval (1.3)
improper fraction (1.4)
infinite set (1.1)
intersection of two sets (1.1)
interval (1.3)
least common denominator (1.4)
negative numbers (1.3)
null set (1.1)
open interval (1.3)
origin (1.3)
positive numbers (1.3)
product (1.4)
quotient (1.4)
reciprocal (1.3)
repeating decimal (1.1)
set (1.1)
set-builder notation (1.1)
subset (1.1)
sum (1.4)
term (1.4)
terminating decimal (1.1)
union of two sets (1.1)
variable (1.1)

KEY IDEAS

(1.1) Two sets are equal if they have the same elements.

The **natural numbers**: 1, 2, 3, 4, 5, . . .

A **prime number** is a natural number greater than 1 that is divisible without remainder only by itself and 1.

A **composite number** is a natural number greater than 1 that is not a prime number.

The **whole numbers**: 0, 1, 2, 3, 4, 5, . . .

The **integers**: . . . $-4, -3, -2, -1, 0, 1, 2, 3, 4,$ . . .

If an integer n is divisible by 2, then n is an **even integer**.

If an integer n is not divisible by 2, then n is an **odd integer**.

A **rational number** is any number that can be written as a fraction with an integer numerator and a nonzero integer denominator. Any rational number can be written in decimal form as either a terminating or a repeating decimal.

An **irrational number** is any number that can be written in decimal form as a nonterminating, nonrepeating decimal.

A **real number** is any number that can be written in decimal form.

(1.2) **Properties of equality**:

$a = a$ (the reflexive property)
If $a = b$, then $b = a$. (the symmetric property)
If $a = b$ and $b = c$, then $a = c$. (the transitive property)

The **substitution property**: In any expression, a quantity can be substituted for its equal without changing the meaning of the expression.

The **closure properties**:

$a + b$ is a real number.
$a - b$ is a real number.
ab is a real number.
$\frac{a}{b}$ is a real number, provided that $b \neq 0$.

The **associative properties**:
$(a + b) + c = a + (b + c)$
$(ab)c = a(bc)$

The **commutative properties**:
$a + b = b + a$
$ab = ba$

The **distributive property**:
$a(b + c) = ab + ac$

0 is the **additive identity**:
$a + 0 = 0 + a = a$

1 is the **multiplicative identity**:
$a \cdot 1 = 1 \cdot a = a$

$-a$ is the **negative** (or **additive inverse**) of a:
$a + (-a) = -a + a = 0$

The double negative rule:
$-(-a) = a$

If $a \neq 0$, then $\frac{1}{a}$ is the **multiplicative inverse** (or **reciprocal**) of a:
$a\left(\frac{1}{a}\right) = \left(\frac{1}{a}\right)a = 1$

(1.3) The **trichotomy property**:
$a < b$ or $a = b$ or $a > b$

The **transitive property** for $<$:
If $a < b$ and $b < c$, then $a < c$.

A similar statement is true for $>$, $\leq$, and $\geq$.

$|x| = x$ if $x \geq 0$ and $|x| = -x$ if $x < 0$

$|x| \geq 0$

(1.4) To add two numbers with like signs, add their absolute values and keep the same sign.

To add two numbers with unlike signs, subtract their absolute values and keep the sign of the number with the greater absolute value.

$x - y$ is equivalent to $x + (-y)$.

To multiply (or divide) two numbers with like signs, multiply (or divide) their absolute values. The sign is positive.

To multiply (or divide) two numbers with unlike signs, multiply (or divide) their absolute values. The sign is negative.

If $a \neq 0$, $\frac{a}{0}$ is undefined. $\frac{0}{0}$ is indeterminate.

Order of operations:

Unless parentheses indicate otherwise, do multiplications and/or divisions first, in order from left to right. Then do the additions and/or subtractions, from left to right. In a fraction, simplify the numerator and denominator separately. Then simplify the fraction, if possible.

Properties of fractions: If $b \neq 0$ and $d \neq 0$, then

$$\frac{a}{b} = \frac{c}{d} \quad \text{if and only if} \quad ad = bc$$

$$\frac{a}{1} = a \quad \text{and if } a \neq 0, \quad \frac{a}{a} = 1$$

If $k \neq 0$, then $\frac{ak}{bk} = \frac{a}{b}$.

$$\frac{a}{b} \cdot \frac{c}{d} = \frac{ac}{bd} \quad \text{and} \quad \frac{a}{b} \div \frac{c}{d} = \frac{a}{b} \cdot \frac{d}{c} = \frac{ad}{bc}$$

$$\frac{a}{b} + \frac{c}{b} = \frac{a + c}{b} \quad \text{and} \quad \frac{a}{b} - \frac{c}{b} = \frac{a - c}{b}$$

$$-\frac{a}{b} = \frac{-a}{b} = \frac{a}{-b}$$

Chapter 1 Review Exercises

In Review Exercises 1–8, $A = \{1, 2, 4, 6, 8, 9\}$, $B = \{1, 2, 4, 9\}$, and $C = \{3, 5, 7, 9\}$. Tell whether each statement is true. If a statement is false, change the symbol between the letters to make it true.

1. $4 \in A$ true

2. $B \subseteq C$ false; $B \not\subseteq C$

3. $B \in A$ false; $B \subseteq A$

4. $\{3, 7\} \subseteq C$ true

5. $\emptyset \subseteq B$ true

6. $\emptyset \in C$ false; $\emptyset \subseteq C$

7. $\{2, 8\} \subseteq A$ true

8. $\{4, 8\} \in A$ false; $\{4, 8\} \subseteq A$

In Review Exercises 9–16, list all the elements in the set $\left\{-4, 0, 1, -\frac{2}{3}, \pi\right\}$ that satisfy the given condition.

9. whole number
0, 1

10. natural number
1

11. rational number
$-4, 0, 1, -\frac{2}{3}$

12. integer
$-4, 0, 1$

13. irrational number
π

14. real number
$-4, 0, 1, -\frac{2}{3}, \pi$

15. negative number
$-4, -\frac{2}{3}$

16. positive number
$1, \pi$

In Review Exercises 17–24, classify each integer as an even, odd, prime, or composite number.

17. 10 even, composite

18. -8 even

19. -11 odd

20. 7 odd, prime

21. 1 odd

22. 0 even

23. -1 odd

24. 2 even, prime

In Review Exercises 25–36, state the property of equality or the property of real numbers that justifies each statement.

25. $3(4 + 2) = 3 \cdot 4 + 3 \cdot 2$ distrib. prop.

26. If $3 = 2 + 1$, then $2 + 1 = 3$. symm. prop.

27. $3 + (x + 7) = (x + 7) + 3$ commut. prop. of +

28. $3 + (x + 7) = (3 + x) + 7$ assoc. prop. of +

29. $3 + 0 = 3$ add. identity prop.

30. $3 + (-3) = 0$ add. inverse prop.

31. $xy = xy$ reflexive prop.

32. $5(3) = 3(5)$ commut. prop. of $\times$

33. $3(xy) = (3x)y$ assoc. prop. of $\times$

34. $3x \cdot 1 = 3x$ mult. identity prop.

35. $a\left(\frac{1}{a}\right) = 1 \; (a \neq 0)$ mult. inverse prop.

36. If $x = 7$ and $7 = y$, then $x = y$. trans. prop.

In Review Exercises 37–40, find the additive inverse and the multiplicative inverse (the reciprocal), if any, of each number.

37. 1 $-1, 1$

38. -3 $3, -\frac{1}{3}$

39. 0 0; no mult. inverse

40. $\frac{1}{3}$ $-\frac{1}{3}, 3$

In Review Exercises 41–44, write each expression using the $<$ and/or $>$ symbols.

41. $a \ngtr 4$
$a < 4$

42. $a \nleq b$
$a > b$

43. $a \neq b$
$a < b$ or $a > b$

44. $a \nless b$ and $a \neq b$
$a > b$

In Review Exercises 45–48, write each expression so that the inequality symbol points in the opposite direction.

45. $3 \leq 10$ $10 \geq 3$

46. $12 \geq -3$ $-3 \leq 12$

47. $-4 > -8$ $-8 < -4$

48. $-5 < 6$ $6 > -5$

49. Graph the set of prime numbers between 20 and 30.

20 21 22 23 24 25 26 27 28 29 30

50. Graph the set of composite numbers between 5 and 13.

5 6 7 8 9 10 11 12 13

In Review Exercises 51–58, graph each interval.

51. $x \geq -4$

-4

52. $-2 < x \leq 6$

-2 6

53. $(-2, 3)$

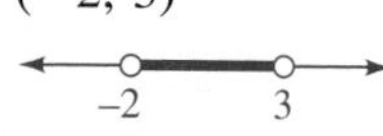

54. $[2, 6]$

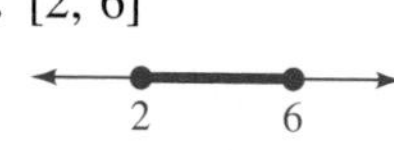

55. $(2, \infty)$

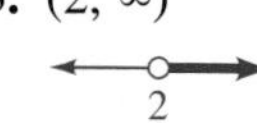

56. $(-\infty, -1)$

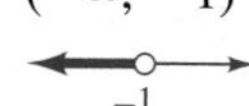

57. $(-2, 4] \cup (0, 6)$

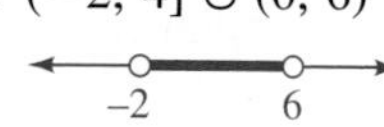

58. $(-2, 4] \cap (0, 6)$

0 4

In Review Exercises 59–62, A, B, and C are intervals, with $A = (-2, 3]$, $B = (-1, 4)$, and $C = [0, 5]$. Graph each set.

59. $A \cap B$

60. $A \cup B$

-2 4

61. $A \cup C$

-2 5

62. $B \cap C$

In Review Exercises 63–66, write each expression without using absolute value symbols.

63. $|0|$ 0

64. $|-1|$ 1

65. $-|-8|$ −8

66. $|3-8|$ 5

In Review Exercises 67–82, do the operations and simplify when possible.

67. $3+(-5)$ −2

68. $5-3$ 2

69. $-2+5$ 3

70. $-3-5$ −8

71. $-8-(-3)$ −5

72. $7-(-9)$ 16

73. $4(-3)$ −12

74. $-3(8)$ −24

75. $-4(3-6)$ 12

76. $3[8-(-1)]$ 27

77. $\frac{-8}{2}$ −4

78. $\frac{8}{-4}$ −2

79. $\frac{-16}{-4}$ 4

80. $\frac{-25}{-5}$ 5

81. $\frac{3-8}{10-5}$ −1

82. $\frac{|-32-8|}{6-16}$ −4

In Review Exercises 83–94, $a=5$, $b=-2$, $c=-3$, and $d=2$. Simplify each expression.

83. $\frac{3a-2b}{cd}$ $-\frac{19}{6}$

84. $\frac{3b+2d}{ac}$ $\frac{2}{15}$

85. $\frac{ab+cd}{c(b-d)}$ $-\frac{4}{3}$

86. $\frac{ac-bd}{a(d+c)}$ $\frac{11}{5}$

87. $\frac{a(b+d)+c}{ad-bc}$ $-\frac{3}{4}$

88. $\frac{b(c-d)-a}{a+c(b-d)}$ $\frac{5}{17}$

89. $\frac{ac}{-4b}+\frac{bc}{-2d}$ $-\frac{27}{8}$

90. $\frac{ab}{c}-\frac{-5b}{a-d}$ 0

91. $\frac{ab-c}{a-b}\cdot\frac{3c+d}{2d-c}$ 1

92. $\frac{3b-2c}{a(b-2)}\cdot(3a+bc)$ 0

93. $\frac{b(a+c)}{bc}\div\frac{d-b}{a+b}$ $-\frac{1}{2}$

94. $\frac{2a-c}{c-2a}\div\frac{a-bc}{ab-c}$ −7

Chapter 1 Test

In Problems 1–4, $A=\{1,2,3,4\}$, $B=\{3,4,5,6\}$, and $C=\{5,6,7,8\}$. Insert either a $\in$ or a $\subseteq$ symbol to make a true statement.

1. 3 $\in$ A

2. $\emptyset$ $\subseteq$ B

3. $A\cap B$ $\subseteq$ B

4. $\{7,8\}$ $\subseteq$ C

In Problems 5–8, $A=\{1,2,3,4\}$, $B=\{3,4,5,6\}$, and $C=\{5,6,7,8\}$. Find each set.

5. $A\cup B$ $\{1, 2, 3, 4, 5, 6\}$

6. $A\cap B\cap C$ $\emptyset$

7. $(A\cup\emptyset)\cap C$ $\emptyset$

8. $A\cup B\cup C$ $\{1, 2, 3, 4, 5, 6, 7, 8\}$

In Problems 9–12, let $A=\left\{-2, 0, 1, 2, \frac{6}{5}, 5, \sqrt{7}\right\}$.

9. List the elements in A that are natural numbers.
1, 2, 5

10. List the elements in A that are integers.
−2, 0, 1, 2, 5

11. List the elements in A that are rational numbers.
$-2, 0, 1, 2, \frac{6}{5}, 5$

12. List the elements in A that are irrational numbers.
$\sqrt{7}$

13. Express the fraction $\frac{7}{9}$ as a decimal. $0.\overline{7}$

14. Write 0.75 as a common fraction. $\frac{3}{4}$

In Problems 15–20, tell which property of equality or property of real numbers justifies each statement.

15. $3 = 3$ reflexive prop. of =

16. If $3 = 2 + 1$, then $2 + 1 = 3$. symm. prop.

17. $3 + 5 = 5 + 3$ comm. prop of +

18. $a(b + c) = ab + ac$ distrib. prop.

19. $7 + (4 + 3) = (7 + 4) + 3$ assoc. prop. of +

20. $(4 \cdot 3) \cdot 5 = 5 \cdot (4 \cdot 3)$ comm. prop. of ×

In Problems 21–22, graph each set on the number line.

21. The set of odd integers from -4 to 6.

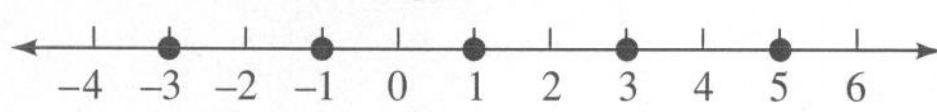

22. The set of prime numbers less than 12.

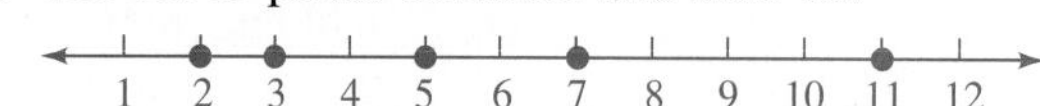

In Problems 23–24, graph each interval on the number line.

23. $-2 \leq x < 4$

24. $[-1, 3)$

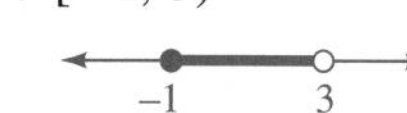

In Problems 25–26, A and B are intervals: $A = (-2, 2]$ and $B = [-2, 3)$. Graph each set.

25. $A \cap B$

26. $A \cup B$

In Problems 27–28, write each expression in simplest form without using absolute value symbols.

27. $-|8 - 5|$ -3

28. $|-5| - |2| + |0|$ 3

In Problems 29–32, do the operations.

29. $7 + (-5)$ 2

30. $-5(-4)$ 20

31. $\dfrac{12}{-3}$ -4

32. $-4 - \dfrac{-15}{3}$ 1

In Problems 33–40, $a = 2$, $b = -3$, and $c = 4$. Evaluate each expression.

33. ab -6

34. $a + bc$ -10

35. $ab - bc$ 6

36. $b(a + c)$ -18

37. $\dfrac{-3b + a}{ac - b}$ 1

38. $\dfrac{(4a + b)(b + 2c)}{-4c - ab}$ $-\frac{5}{2}$

39. $\dfrac{a - 3c}{ab} \cdot \dfrac{2c + 1}{5b}$ -1

40. $\dfrac{a + b - c}{a - b + c} \div \dfrac{a - b - c}{a + b + c}$ $-\frac{5}{3}$

CAREERS & MATHEMATICS

Computer Systems Analyst

Computer systems analysts help businesses and scientific research organizations develop analytical systems to process and interpret data. Using such techniques as cost accounting, sampling, and mathematical model building, they analyze information and often present results graphically by charts and diagrams. They may also prepare cost–benefit analyses to help management decide whether proposed solutions are satisfactory.

Once the system is accepted, systems analysts adapt its logical requirements to the capabilities of computer machinery. They work closely with programmers to rid the system of errors.

Systems analysts solve a wide range of problems in many different industries. Because the work is so varied and complex, analysts usually specialize in either business or scientific and engineering applications.

Qualifications

Businesses prefer a college degree in accounting, business management, or economics. Scientific organizations prefer a college degree in the physical sciences, mathematics, or engineering. Advanced degrees and degrees in computer science and information science are becoming more important in both employment areas.

Job Outlook

Through the year 2005, the demand for systems analysts is expected to grow more rapidly than the average for all other occupations. As competition increases worldwide, systems analysts will help organizations take advantage of technological advances in automation, telecommunications, and scientific research. Because personal computers are now found on almost every desktop, systems analysts will be needed to design computer networks that facilitate the sharing of data, while providing adequate security for sensitive information.

Example Application

The process of arranging records into a sequential order, called **sorting,** is a common and important task in electronic data processing. In any sorting operation, records must be compared with other records to determine which one should precede the other. One sorting technique, called a **selection sort,** requires C comparisons to sort N records into their proper order, where C and N are related by the following formula:

$$C = \frac{N(N-1)}{2}$$

How many comparisons are required to

a. sort 20 records?

b. sort 10,000 records?

Solution **a.** Substitute 20 for N in the formula

$$C = \frac{N(N-1)}{2}$$

and calculate C:

$$C = \frac{20(20-1)}{2}$$

$$C = \frac{20 \cdot 19}{2}$$

$$C = 190$$

Sorting 20 records requires 190 comparisons.

b. Substitute 10,000 for N in the formula

$$C = \frac{N(N-1)}{2}$$

and calculate C:

$$C = \frac{10{,}000(10{,}000-1)}{2}$$

$$C = 49{,}995{,}000$$

Sorting 10,000 records requires almost 50 million comparisons. A selection sort is not efficient for large lists.

EXERCISES

1. How many comparisons are required to sort 500 records?
2. If the computer time required to sort 20 records were 0.5 second, how long would it take to sort 10,000 records?
3. Another important data-processing task is that of finding a particular entry in a large list. In a **sequential search,** an average of C comparisons is required to find an entry in a list of N items. C and N are related by the formula

$$C = \frac{(N+1)}{2}$$

How many comparisons are needed (on the average) to search a list of 25 items?
4. How many comparisons are needed (on the average) to search 10,000 items?

(*Answers:* **1.** 124,750 **2.** 36.5 hr **3.** 13 **4.** 5000)

2 Exponents and Polynomials

A common expression in algebra is the polynomial. Because polynomials often involve exponents, we begin this chapter by discussing the properties of exponents.

2.1 Exponents

■ Properties of Exponents ■ The Zero Exponent ■ Negative Exponents
■ Calculators ■ Order of Operations

Multiplication is a shortcut for repeated addition. For example, $4x$ means $x + x + x + x$, and $3y$ means $y + y + y$. The number 4, called the **numerical coefficient** or just the **coefficient** in the expression $4x$, indicates that x is to be used as a term in an addition four times. The coefficient 3 in the expression $3y$ indicates that y is to be used as a term three times.

Exponents are used to indicate repeated multiplication. For example,

$y^2 = y \cdot y$ Read y^2 as "y to the second power" or "y squared."

$z^3 = z \cdot z \cdot z$ Read z^3 as "z to the third power" or "z cubed."

$x^4 = x \cdot x \cdot x \cdot x$ Read x^4 as "x to the fourth power."

These examples suggest the following definition.

Natural-Number Exponents If n is a natural number, then

$$x^n = \overbrace{x \cdot x \cdot x \cdot \cdots \cdot x}^{n \text{ factors of } x}$$

The exponential expression x^n is called a **power of x,** and we read the expression x^n as "x to the nth power." In this expression, x is called the **base** and n is called the **exponent.**

A natural-number exponent tells how many times the base of an exponential expression is to be used as a factor in a product. An exponent of 1 indicates that its base is to be used one time as a factor; an exponent of 2 indicates that the base is to be used two times as a factor; and so on.

$$a^1 = a \qquad b^2 = b \cdot b \qquad c^3 = c \cdot c \cdot c \qquad d^4 = d \cdot d \cdot d \cdot d$$

If the base is more than a single variable, we have

$$(-b)^1 = -b \qquad (c^3)^2 = c^3 \cdot c^3 \qquad (-5d)^3 = (-5d)(-5d)(-5d)$$

(c,d,f) **EXAMPLE 1**

a. $2^5 = 2 \cdot 2 \cdot 2 \cdot 2 \cdot 2 = 32$

b. $(-2)^5 = (-2)(-2)(-2)(-2)(-2) = -32$

c. $-4^4 = -(4^4) = -(4 \cdot 4 \cdot 4 \cdot 4) = -256$

d. $(-4)^4 = (-4)(-4)(-4)(-4) = 256$

e. $(8a)^2 = (8a)(8a) = 64a^2$

f. $(-2b)^2 = (-2b)(-2b) = 4b^2$ ■

Teaching Tip: Point out that -4^4 means "the negative of 4^4." Thus,

$-4^4 = -(4 \cdot 4 \cdot 4 \cdot 4)$

Then point out that $(-4)^4$ means "find the fourth power of -4."

WARNING! It is important to note the difference between $-x^n$ and $(-x)^n$, and between ax^n and $(ax)^n$.

$$-x^n = -(\overbrace{x \cdot x \cdot x \cdot \cdots \cdot x}^{n \text{ factors of } x}) \quad \text{and} \quad (-x)^n = \overbrace{(-x)(-x)(-x) \cdot \cdots \cdot (-x)}^{n \text{ factors of } -x}$$

$$ax^n = a \cdot \overbrace{x \cdot x \cdot x \cdot \cdots \cdot x}^{n \text{ factors of } x} \quad \text{and} \quad (ax)^n = \overbrace{(ax)(ax)(ax) \cdot \cdots \cdot (ax)}^{n \text{ factors of } ax}$$

Properties of Exponents

Many properties of exponents follow from the definition of exponent. Because x^5 means that x is to be used as a factor five times, and x^3 means that x is to be used as a factor three times, the product $x^5 \cdot x^3$ means that x will be used as a factor eight times.

$$x^5x^3 = \overbrace{x \cdot x \cdot x \cdot x \cdot x}^{5 \text{ factors of } x} \cdot \overbrace{x \cdot x \cdot x}^{3 \text{ factors of } x} = \overbrace{x \cdot x \cdot x \cdot x \cdot x \cdot x \cdot x \cdot x}^{8 \text{ factors of } x}$$

In general,

$$x^mx^n = \overbrace{x \cdot x \cdot x \cdot \cdots \cdot x}^{m \text{ factors of } x} \cdot \overbrace{x \cdot x \cdot \cdots \cdot x}^{n \text{ factors of } x} = \overbrace{x \cdot x \cdot x \cdot x \cdot \cdots \cdot x}^{m+n \text{ factors of } x}$$

In words, we have the following rule: *To multiply exponential expressions with the same base, we keep the same base and add the exponents.*

Product Rule of Exponents If m and n are natural numbers, then

$$x^mx^n = x^{m+n}$$

WARNING! The product rule of exponents applies only to exponential expressions with the same base. The expression x^5y^3, for example, cannot be simplified, because the bases of the exponential expressions are different.

(a, c) **EXAMPLE 2**

a. $x^{11}x^5 = x^{11+5} = x^{16}$

b. $a^5a^4a^3 = (a^5a^4)a^3 = a^9a^3 = a^{12}$

c. $a^2b^3a^3b^2 = a^2a^3b^3b^2 = a^5b^5$

d. $8x^4x^4 = 8x^{4+4} = 8x^8$ ■

To find a second property of exponents, we simplify $(x^4)^3$. This expression means x^4 cubed or $x^4 \cdot x^4 \cdot x^4$, which is

$$\overbrace{x \cdot x \cdot x \cdot x}^{x^4} \cdot \overbrace{x \cdot x \cdot x \cdot x}^{x^4} \cdot \overbrace{x \cdot x \cdot x \cdot x}^{x^4}$$

This long string of x's can be written as x^{12}. Thus,

$$(x^4)^3 = x^{4 \cdot 3} = x^{12}$$

In general, we have

$$(x^m)^n = \overbrace{x^m \cdot x^m \cdot x^m \cdot \cdots \cdot x^m}^{n \text{ factors of } x^m} = \overbrace{x \cdot x \cdot x \cdot x \cdot x \cdot \cdots \cdot x}^{mn \text{ factors of } x} = x^{mn}$$

In words, we have the following rule: *To raise an exponential expression to a power, we keep the same base and multiply the exponents.*

To find a third property, we square $3x$ to get

$$(3x)^2 = 3x \cdot 3x = 3 \cdot 3 \cdot x \cdot x = 3^2x^2 = 9x^2$$

In general, we have

$$(xy)^n = \overbrace{(xy)(xy)(xy) \cdot \cdots \cdot (xy)}^{n \text{ factors of } xy}$$
$$= \overbrace{xxx \cdot \cdots \cdot x}^{n \text{ factors of } x} \cdot \overbrace{yyy \cdot \cdots \cdot y}^{n \text{ factors of } y}$$
$$= x^ny^n$$

To find a fourth property, we cube $\frac{x}{3}$ to get

$$\left(\frac{x}{3}\right)^3 = \frac{x}{3} \cdot \frac{x}{3} \cdot \frac{x}{3} = \frac{x \cdot x \cdot x}{3 \cdot 3 \cdot 3} = \frac{x^3}{3^3} = \frac{x^3}{27}$$

In general, we have

$$\left(\frac{x}{y}\right)^n = \overbrace{\left(\frac{x}{y}\right)\left(\frac{x}{y}\right)\left(\frac{x}{y}\right)\cdot\dots\cdot\left(\frac{x}{y}\right)}^{n \text{ factors of } \frac{x}{y}}$$

$$= \frac{\overbrace{xxx\cdot\dots\cdot x}^{n \text{ factors of } x}}{\underbrace{yyy\cdot\dots\cdot y}_{n \text{ factors of } y}}$$

We multiply fractions by multiplying the numerators and multiplying the denominators.

$$= \frac{x^n}{y^n}$$

The previous three results are called the **power rules of exponents.**

Power Rules of Exponents If m and n are natural numbers, then

$$(x^m)^n = x^{mn} \qquad (xy)^n = x^n y^n \qquad \left(\frac{x}{y}\right)^n = \frac{x^n}{y^n} \quad (y \neq 0)$$

(b, c) **EXAMPLE 3**

a. $(3^2)^3 = 3^{2\cdot 3} = 3^6 = 729$

b. $(x^{11})^5 = x^{11\cdot 5} = x^{55}$

c. $(x^2x^3)^6 = (x^5)^6 = x^{30}$

d. $(x^2)^4(x^3)^2 = x^8x^6 = x^{14}$ ■

(b, d) **EXAMPLE 4**

Teaching Tip: You may want to name each Power Rule
a. $(x^m)^n$: Power to a power
b. $(xy)^n$: Product to a power
c. $\left(\frac{x}{y}\right)^n$: Quotient to a power

a. $(x^2y)^3 = (x^2)^3y^3 = x^6y^3$

b. $(x^3y^4)^4 = (x^3)^4(y^4)^4 = x^{12}y^{16}$

c. $\left(\frac{x}{y^2}\right)^4 = \frac{x^4}{(y^2)^4} \quad (y \neq 0)$
$= \frac{x^4}{y^8}$

d. $\left(\frac{x^3}{y^4}\right)^2 = \frac{(x^3)^2}{(y^4)^2} \quad (y \neq 0)$
$= \frac{x^6}{y^8}$ ■

The Zero Exponent

We can define an exponent of zero if we assume that the rules for natural-number exponents hold when the exponent is 0.

$$x^0x^n = x^{0+n} = x^n = 1x^n$$

Because $x^0x^n = 1x^n$, it follows that if $x \neq 0$, then $x^0 = 1$.

Zero Exponent If $x \neq 0$, then

$$x^0 = 1$$

Gottfried Wilhelm Leibniz (1646–1716)
Leibniz, a German philosopher and logician, is principally known as one of the inventors of calculus, along with Newton. He also developed the binary numeration system, which is basic to modern computers.

This definition of the zero exponent guarantees that any nonzero number raised to the 0th power is 1. For example,

$$5^0 = 1 \qquad (-7)^0 = 1 \qquad (3ax^3)^0 = 1 \qquad \left(\frac{1}{2}x^5y^7z^9\right)^0 = 1$$

WARNING! Because of the definition of a zero exponent, 0^0 is undefined.

Negative Exponents

If we assume that the rules for natural-number exponents are true for negative integer exponents and if $x \neq 0$, then we can write

$$x^{-n}x^n = x^{-n+n} = x^0 = 1$$

Because

$$\frac{1}{x^n} \cdot x^n = 1$$

we can make the following definition.

Negative Exponents If n is an integer and $x \neq 0$, then

$$x^{-n} = \frac{1}{x^n} \quad \text{and} \quad \frac{1}{x^{-n}} = x^n$$

Because of the previous definition, x^{-n} represents the reciprocal of x^n. For example,

$$5^{-2} = \frac{1}{5^2} = \frac{1}{25} \qquad 10^{-3} = \frac{1}{10^3} = \frac{1}{1000}$$

$$(2x)^{-3} = \frac{1}{(2x)^3} = \frac{1}{8x^3} \qquad 3x^{-1} = 3 \cdot \frac{1}{x} = \frac{3}{x}$$

WARNING! By the definition, $x \neq 0$. Thus, an expression such as 0^{-5} is undefined. Note that a positive number raised to a negative power is positive. For example, $5^{-2} = \frac{1}{25}$ and not -25.

With the previous two definitions, all of the rules for natural-number exponents are also true for integer exponents.

(a, b, e) **EXAMPLE 5**

a. $(3x)^0 = 1$

b. $3x^0 = 3(x^0) = 3(1) = 3$

c. $x^{-5} = \frac{1}{x^5}$

d. $\frac{1}{x^{-6}} = x^6$

e. $x^{-5}x^3 = x^{-5+3} = x^{-2} = \frac{1}{x^2}$

f. $(x^{-3})^{-2} = x^{(-3)(-2)} = x^6$

■

To develop a rule for dividing exponential expressions, we proceed as follows:

$$\frac{x^m}{x^n} = x^m\left(\frac{1}{x^n}\right) = x^m x^{-n} = x^{m+(-n)} = x^{m-n}$$

In words, we have the following rule: *To divide two exponential expressions with the same nonzero base, we keep the base and subtract the exponent in the denominator from the exponent in the numerator.*

Quotient Rule If m and n are integers and $x \neq 0$, then

$$\frac{x^m}{x^n} = x^{m-n}$$

(b, e) **EXAMPLE 6**

a. $\frac{a^5}{a^3} = a^{5-3} = a^2$

b. $\frac{x^{-5}}{x^{11}} = x^{-5-11} = x^{-16} = \frac{1}{x^{16}}$

c. $\frac{x^4x^3}{x^{-5}} = \frac{x^7}{x^{-5}} = x^{7-(-5)} = x^{12}$

d. $\dfrac{(x^2)^3}{(x^3)^2} = \dfrac{x^6}{x^6}$
$= x^{6-6}$
$= x^0$
$= 1$

e. $\dfrac{x^2y^3}{xy^4} = x^{2-1}y^{3-4}$
$= xy^{-1}$
$= \dfrac{x}{y}$

f. $\left(\dfrac{a^{-2}b^3}{a^2a^3b^4}\right)^3 = \left(\dfrac{a^{-2}b^3}{a^5b^4}\right)^3$
$= (a^{-2-5}b^{3-4})^3$
$= (a^{-7}b^{-1})^3$
$= \left(\dfrac{1}{a^7b}\right)^3$
$= \dfrac{1}{a^{21}b^3}$ ■

Part **a** of Example 6 also can be simplified by factoring the numerator and denominator of the fraction and dividing out all common factors.

$$\frac{a^5}{a^3} = \frac{a \cdot a \cdot a \cdot a \cdot a}{a \cdot a \cdot a} = \frac{\not{a} \cdot \not{a} \cdot \not{a} \cdot a \cdot a}{\not{a} \cdot \not{a} \cdot \not{a}} = a^2$$

One final property often is used to simplify exponential expressions.

Theorem If n is an integer, $x \neq 0$, and $y \neq 0$, then

$$\left(\frac{x}{y}\right)^{-n} = \left(\frac{y}{x}\right)^n$$

Proof To prove this theorem, we proceed as follows:

$$\left(\frac{x}{y}\right)^{-n} = \frac{x^{-n}}{y^{-n}}$$

$$= \frac{x^{-n}\mathbf{x^n y^n}}{y^{-n}\mathbf{x^n y^n}}$$ Multiply both numerator and denominator by x^ny^n.

$$= \frac{x^0y^n}{y^0x^n}$$

$$= \frac{y^n}{x^n}$$

$$= \left(\frac{y}{x}\right)^n$$ □

(c) **EXAMPLE 7**

a. $\left(\dfrac{2}{3}\right)^{-4} = \left(\dfrac{3}{2}\right)^4$
$= \dfrac{81}{16}$

b. $\left(\dfrac{y^2}{x^3}\right)^{-3} = \left(\dfrac{x^3}{y^2}\right)^3$
$= \dfrac{x^9}{y^6}$

c. $$\left(\frac{2x^{-2}}{3y^3}\right)^{-4} = \left(\frac{3y^3}{2x^{-2}}\right)^4 = \frac{81y^{12}}{16x^{-8}} = \frac{81y^{12}x^8}{16x^{-8}x^8} = \frac{81y^{12}x^8}{16x^0} = \frac{81y^{12}x^8}{16}$$

d. $$\left(\frac{a^{-2}b^3}{a^2a^3b^4}\right)^{-3} = \left(\frac{a^2a^3b^4}{a^{-2}b^3}\right)^3 = \left(\frac{a^5b^4}{a^{-2}b^3}\right)^3 = (a^{5-(-2)}b^{4-3})^3 = (a^7b)^3 = a^{21}b^3$$

■

We summarize the properties of exponents as follows.

Properties of Exponents If there are no divisions by 0, then for all integers m and n,

1. $x^mx^n = x^{m+n}$ **2.** $(x^m)^n = x^{mn}$ **3.** $(xy)^n = x^ny^n$ **4.** $\left(\frac{x}{y}\right)^n = \frac{x^n}{y^n}$

5. $x^0 = 1$ $(x \neq 0)$ **6.** $x^{-n} = \frac{1}{x^n}$ **7.** $\frac{x^m}{x^n} = x^{m-n}$ **8.** $\left(\frac{x}{y}\right)^{-n} = \left(\frac{y}{x}\right)^n$

Classroom Activity: Find examples to show that
$(x + y)^n \neq x^n + y^n$
$(x - y)^n \neq x^n - y^n$

Calculators

To use a scientific calculator to find powers of numbers, we use the y^x key. For example, to find 5.37^4 we enter these numbers and press these keys:

5.37 y^x 4 = Some calculators have an x^y key.

The display will read 831.5668016. To use a graphing calculator, we enter 5.37 and press these keys:

5.37 ∧ 4 ENTER Some calculators have an x^y or a y^x key.

We will obtain the same result. If neither of these methods works on your calculator, consult your owner's manual.

Order of Operations

We have seen that multiplications and/or divisions take priority over additions and/or subtractions. For example, $3 \cdot 4 + 5$ means $(3 \cdot 4) + 5$, rather than $3(4 + 5)$.

When simplifying expressions containing exponents, we will follow this convention:

Unless parentheses indicate otherwise, we find the power of each base first, and then follow the priority rules given in Section 1.4.

For example, $5 \cdot 2^3$ means $5(2^3)$ or $5(8) = 40$. However, $(5 \cdot 2)^3$ means 10^3 or 1000, since parentheses indicate that the multiplication is to be done first.

EXAMPLE 8 If $x = 2$, $y = -1$, and $z = 3$, find the value of **a.** $3x^2y^3$, **b.** $(3x)^2y^3$, **c.** $-z^4$, **d.** $(-z)^4$, and **e.** $(xyz)^0$.

Solution **a.** $3x^2y^3 = 3(2)^2(-1)^3$
$= 3(4)(-1)$
$= -12$

b. $(3x)^2y^3 = (3 \cdot 2)^2(-1)^3$
$= 6^2(-1)$
$= -36$

c. $-z^4 = -(3)^4$
$= -81$

d. $(-z)^4 = (-3)^4$
$= 81$

e. Since $xyz \neq 0$, $(xyz)^0 = 1$. Any nonzero number raised to the 0 power is 1.

■

Variable exponents follow the same rules as integer exponents.

(c, d) **EXAMPLE 9** Simplify each expression. Assume that $x \neq 0$ and $a \neq 0$.

a. $\dfrac{a^n a}{a^2} = a^{n+1-2}$
$= a^{n-1}$

b. $\dfrac{x^3x^2}{x^n} = x^{3+2-n}$
$= x^{5-n}$

c. $\left(\dfrac{x^n}{x^2}\right)^2 = \dfrac{x^{2n}}{x^4}$
$= x^{2n-4}$

d. $\dfrac{a^n a^{-3}}{a^{-1}} = a^{n+(-3)-(-1)}$
$= a^{n-3+1}$
$= a^{n-2}$

■

ORALS *Simplify each expression.*

1. 4^2 16

2. 3^3 27

3. x^2x^3 x^5

4. y^3y^4 y^7

5. 17^0 1

6. $(x^2)^3$ x^6

7. $(a^2b)^3$ a^6b^3

8. $\left(\dfrac{x}{y^2}\right)^2$ $\dfrac{x^2}{y^4}$

9. 5^{-2} $\frac{1}{25}$

10. $(x^{-2})^{-1}$ x^2

11. $\dfrac{x^5}{x^2}$ x^3

12. $\dfrac{x^2}{x^5}$ $\dfrac{1}{x^3}$

EXERCISE 2.1

In Exercises 1–8, identify the base and the exponent in each exponential expression.

1. 5^3 base is 5, exponent is 3

2. -7^2 base is 7, exponent is 2

3. $-x^5$ base is x, exponent is 5

4. $(-t)^4$ base is $-t$, exponent is 4

5. $2b^6$ base is b, exponent is 6

6. $(3xy)^5$ base is $3xy$, exponent is 5

7. $(-mn^2)^3$ base is $-mn^2$, exponent is 3

8. $(-p^2q)^2$ base is $-p^2q$, exponent is 2

In Exercises 9–94, simplify each expression and write all answers without using negative exponents. Assume that no denominators are zero.

9. 3^2 9

10. 3^4 81

11. -3^2 -9

12. -3^4 -81

13. $(-3)^2$ 9

14. $(-3)^3$ -27

15. $(-2x)^5$ $-32x^5$

16. $(-3a)^3$ $-27a^3$

17. $-(2x)^7$ $-128x^7$

18. $(-2a)^4$ $16a^4$

19. $(-2x)^6$ $64x^6$

20. $(-3y)^5$ $-243y^5$

21. 5^{-2} $\frac{1}{25}$

22. 5^{-4} $\frac{1}{625}$

23. -5^{-2} $-\frac{1}{25}$

24. -5^{-4} $-\frac{1}{625}$

25. $(-5)^{-2}$ $\frac{1}{25}$

26. $(-5)^{-4}$ $\frac{1}{625}$

27. 8^0 1

28. 9^0 1

29. -8^0 -1

30. -9^0 -1

31. $(-8)^0$ 1

32. $(-9)^0$ 1

33. x^2x^3 x^5

34. y^3y^4 y^7

35. k^0k^7 k^7

36. x^8x^{11} x^{19}

37. $x^2x^3x^5$ x^{10}

38. $y^3y^7y^2$ y^{12}

39. p^9pp^0 p^{10}

40. z^7z^0z z^8

41. aba^3b^4 a^4b^5

42. $x^2y^3x^3y^2$ x^5y^5

43. $(-x)^2y^4x^3$ x^5y^4

44. $-x^2y^7y^3x^{-2}$ $-y^{10}$

45. $(x^4)^7$ x^{28}

46. $(y^7)^5$ y^{35}

47. $(b^{-8})^9$ $\frac{1}{b^{72}}$

48. $(z^{12})^2$ z^{24}

49. $(x^3y^2)^4$ $x^{12}y^8$

50. $(x^2y^5)^2$ x^4y^{10}

51. $(r^{-3}s)^3$ $\frac{s^3}{r^9}$

52. $(m^5n^2)^{-3}$ $\frac{1}{m^{15}n^6}$

53. $(a^2a^3)^4$ a^{20}

54. $(bb^2b^3)^4$ b^{24}

55. $(-d^2)^3(d^{-3})^3$ $-\frac{1}{d^3}$

56. $(c^3)^2(c^4)^{-2}$ $\frac{1}{c^2}$

57. $(3x^3y^4)^3$ $27x^9y^{12}$

58. $\left(\frac{1}{2}a^2b^5\right)^4$ $\frac{1}{16}a^8b^{20}$

59. $\left(-\frac{1}{3}mn^2\right)^6$ $\frac{1}{729}m^6n^{12}$

60. $(-3p^2q^3)^5$ $-243p^{10}q^{15}$

61. $(x^{-2}yx^3y^4)^2$ x^2y^{10}

62. $(-a^2b^{-4}a^3b^2)^3$ $-\frac{a^{15}}{b^6}$

63. $(a^2b)^{-2}(ab^{-3})^4$ $\frac{1}{b^{14}}$

64. $(p^2q^{-2})^3(q^2)^{-2}$ $\frac{p^6}{q^{10}}$

65. $\left(\frac{a^3}{b^2}\right)^5$ $\frac{a^{15}}{b^{10}}$

66. $\left(\frac{a^2}{b^3}\right)^4$ $\frac{a^8}{b^{12}}$

67. $\left(\frac{a^{-3}}{b^{-2}}\right)^{-2}$ $\frac{a^6}{b^4}$

68. $\left(\frac{k^{-3}}{k^{-4}}\right)^{-1}$ $\frac{1}{k}$

69. $\frac{a^4a^4}{a^3}$ a^5

70. $\frac{c^3c^4}{c^2}$ c^5

71. $\frac{c^{12}c^5}{(c^5)^2}$ c^7

72. $\frac{(a^3)^{11}}{a^2a^3}$ a^{28}

73. $\frac{m^9m^{-2}}{(m^2)^3}$ m

74. $\frac{a^{10}a^{-3}}{a^5a^{-2}}$ a^4

75. $\frac{1}{a^{-4}}$ a^4

76. $\frac{3}{b^{-5}}$ $3b^5$

77. $\frac{3m^5m^{-7}}{m^2m^{-5}}$ $3m$

78. $\frac{(2a^{-2})^3}{a^3a^{-4}}$ $\frac{8}{a^5}$

79. $\left(\frac{4a^{-2}b}{3ab^{-3}}\right)^3$ $\frac{64b^{12}}{27a^9}$

80. $\left(\frac{2ab^{-3}}{3a^{-2}b^2}\right)^2$ $\frac{4a^6}{9b^{10}}$

81. $\left(\frac{3a^{-2}b^2}{17a^2b^3}\right)^0$ 1

82. $\frac{a^0+b^0}{2(a+b)^0}$ 1

83. $\left(\frac{-2a^4b}{a^{-3}b^2}\right)^{-3}$ $\frac{-b^3}{8a^{21}}$

84. $\left(\frac{-3x^{-5}y^2}{-9x^5y^{-2}}\right)^{-2}$ $\frac{9x^{20}}{y^8}$

85. $\left(\dfrac{2a^3b^2}{3a^{-3}b^2}\right)^{-3}$ $\dfrac{27}{8a^{18}}$

86. $\left(\dfrac{3x^5y^{-2}}{2x^5y^{-2}}\right)^{-4}$ $\frac{16}{81}$

87. $\dfrac{(3x^2)^{-2}}{x^3x^{-4}x^0}$ $\dfrac{1}{9x^3}$

88. $\dfrac{y^{-3}y^{-4}y^0}{(2y^{-2})^3}$ $\frac{1}{8y}$

89. $\dfrac{-3x^{-2}y^2}{(-2x^{-3})^0}$ $\dfrac{-3y^2}{x^2}$

90. $\dfrac{-4x^{-2}x^2(y^0)^2}{(-4x^2y^{-4})^0}$ -4

91. $\dfrac{(4m^{-2}n^{-3})^{-2}(m^{-4}n^{-3})^2}{-3m^{-4}n^{-3}}$ $\dfrac{-n^3}{48}$

92. $\dfrac{(3t^2s^{-2})^{-1}(2^{-3}ts^{-4})^{-2}}{(6t^2s^{-3})^{-2}}$ $768s^4$

93. $\dfrac{(-3p^4q^{-5})^{-3}(2p^{-4}q^3)^{-2}}{4p^5q^{-2}}$ $-\dfrac{q^{11}}{432p^9}$

94. $\dfrac{(6x^{-3}y^2)^3(-4x^{-3}y^4)^{-3}}{(-12x^{-3}y^3)^{-2}}$ $-\dfrac{486}{x^6}$

In Exercises 95–106, evaluate each expression when $x = -2$ and $y = 3$.

95. x^2y^3 108

96. x^3y^2 -72

97. x^y -8

98. y^x $\frac{1}{9}$

99. $\dfrac{x^{-3}}{y^3}$ $-\frac{1}{216}$

100. $\dfrac{x^2}{y^{-3}}$ 108

101. $(xy^2)^{-2}$ $\frac{1}{324}$

102. $(x^3y)^{-3}$ $-\frac{1}{13{,}824}$

103. $-y^3x^{-1}$ $\frac{27}{2}$

104. $-y^3x^{-2}$ $-\frac{27}{4}$

105. $(-yx^{-1})^3$ $\frac{27}{8}$

106. $(-y)^3x^{-2}$ $-\frac{27}{4}$

In Exercises 107–114, simplify each expression.

107. $\dfrac{a^na^3}{a^4}$ a^{n-1}

108. $\dfrac{b^9b^7}{b^n}$ b^{16-n}

109. $\left(\dfrac{b^n}{b^3}\right)^3$ b^{3n-9}

110. $\left(\dfrac{a^2}{a^n}\right)^4$ a^{8-4n}

111. $\dfrac{a^{-n}a^2}{a^3}$ $\dfrac{1}{a^{n+1}}$

112. $\dfrac{a^na^{-2}}{a^4}$ a^{n-6}

113. $\dfrac{a^{-n}a^{-2}}{a^{-4}}$ a^{2-n}

114. $\dfrac{a^n}{a^{-3}a^5}$ a^{n-2}

115. Show that $(xy)^3 = x^3y^3$.

116. Show that $\left(\dfrac{x}{y}\right)^4 = \dfrac{x^4}{y^4}$.

117. Construct an example using numbers to show that $x^m + x^n \neq x^{m+n}$.

118. Construct an example using numbers to show that $x^m + y^m \neq (x + y)^m$.

In Exercises 119–126, use a calculator to find each value.

119. 3.4^4 133.6336

120. 1.23^6 3.462825992

121. 0.0537^4 0.000008315

122. 0.2345^4 0.003023927

123. -6.25^3 -244.140625

124. -4.17^4 -302.3738432

125. $(-25.1)^5$ -9962506.263

126. $(-0.35)^4$ 0.01500625

In Exercises 127–134, use a calculator to verify that each statement is true.

127. $(3.68)^0 = 1$

128. $(2.1)^4(2.1)^3 = (2.1)^7$

129. $(7.2)^2(2.7)^2 = [(7.2)(2.7)]^2$

130. $(3.7)^2 + (4.8)^2 \neq (3.7 + 4.8)^2$

131. $(3.2)^2(3.2)^{-2} = 1$

132. $[(5.9)^3]^2 = (5.9)^6$

133. $(7.23)^{-3} = \dfrac{1}{(7.23)^3}$

134. $\left(\dfrac{5.4}{2.7}\right)^{-4} = \left(\dfrac{2.7}{5.4}\right)^4$

Writing Exercises ■ *Write a paragraph using your own words.*

1. Explain the meanings of the terms *coefficient* and *exponent*.
2. Explain the rules that determine the order in which operations are performed.

Something to Think About ■

1. In the definition of x^{-1}, x cannot equal 0. Why not?
2. Find the sum: $2^{-1} + 3^{-1} - 4^{-1}$ $\frac{7}{12}$

Review Exercises ■ *Graph each interval on a number line.*

1. $-2 < x \le 4$
2. $-5 \le x \le 5$
3. $[-2, 8]$
4. $(-5, -3)$

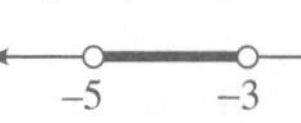

If $a = 4$, $b = -2$, and $c = 5$, find each value.

5. $\dfrac{ab + 2c}{a + b}$ 1
6. $\dfrac{ac - bc}{6ab + b}$ $-\frac{3}{5}$

2.2 Scientific Notation

■ **Using Scientific Notation to Simplify Computations** ■ **Calculators**

Scientists often use very large and very small numbers. For example, the speed of light is approximately 29,980,000,000 centimeters per second, and the mass of a hydrogen atom is 0.000000000000000000000001673 gram. The large number of zeros in these numbers makes them difficult to read and hard to remember. However, exponents make it possible for scientists to write these numbers compactly in a form called **scientific notation.**

Scientific Notation A number is written in **scientific notation** if it is written as the product of a number whose absolute value is between 1 and 10 (including 1) and an integer power of 10.

 (a, c)

EXAMPLE 1 Change **a.** 29,980,000,000, **b.** 0.000000000000000000000001673, and **c.** -0.0013 to scientific notation.

Solution **a.** The number 2.998 is between 1 and 10. To get 29,980,000,000, the decimal point in 2.998 must be moved 10 places to the right. This is accomplished by multiplying 2.998 by 10^{10}. Thus,

$$29{,}980{,}000{,}000 = 2.998 \times 10^{10}$$

b. The number 1.673 is a number between 1 and 10. To get the number 0.000000000000000000000001673, the decimal point in 1.673 must be moved 24 places to the left. This is accomplished by dividing 1.673 by 10^{24}. This is equivalent to multiplying 1.673 by 10^{-24}. Thus,

$$0.000000000000000000000001673 = 1.673 \times 10^{-24}$$

c. The number -1.3 has an absolute value between 1 and 10. To get -0.0013, the decimal point in -1.3 must be moved 3 places to the left. Thus,

$$-0.0013 = -1.3 \times 10^{-3}$$ ■

EXAMPLE 2 Change **a.** 3.7×10^5 and **b.** 1.1×10^{-3} to standard notation.

Solution **a.** Since $10^5 = 100{,}000$, it follows that $3.7 \times 10^5 = 3.7 \times 100{,}000 = 370{,}000$.

b. Since $10^{-3} = \frac{1}{1000}$, it follows that $1.1 \times 10^{-3} = \frac{1.1}{1000} = 0.0011$. ■

Each of the following numbers is written in both scientific and standard form. In each case, the exponent gives the number of places that the decimal point moves, and the sign of the exponent indicates the direction that it moves:

Teaching Tip: A number larger than 1 is associated with a positive exponent on 10.

A number smaller than 1 is associated with a negative exponent.

$5.32 \times 10^4 = 53200$
4 places to the right

$6.45 \times 10^7 = 64500000$
7 places to the right

$2.37 \times 10^{-4} = 0.000237$
4 places to the left

$9.234 \times 10^{-2} = 0.09234$
2 places to the left

$4.89 \times 10^0 = 4.89$
No movement of the decimal point

 (b)

EXAMPLE 3 Change **a.** 47.2×10^{-3} and **b.** 0.043×10^{-2} to scientific notation.

Solution Neither number is in scientific notation, because their first factors are not between 1 and 10. However, we can change these numbers to scientific notation as follows:

a.
$$\begin{aligned} 47.2 \times 10^{-3} &= (4.72 \times 10^1) \times 10^{-3} \\ &= 4.72 \times (10^1 \times 10^{-3}) \\ &= 4.72 \times 10^{-2} \end{aligned}$$

b.
$$\begin{aligned} 0.043 \times 10^{-2} &= (4.3 \times 10^{-2}) \times 10^{-2} \\ &= 4.3 \times (10^{-2} \times 10^{-2}) \\ &= 4.3 \times 10^{-4} \end{aligned}$$ ■

Using Scientific Notation to Simplify Computations

Another advantage of scientific notation becomes evident when we must multiply and divide combinations of very large and very small numbers.

EXAMPLE 4 Use scientific notation to simplify $\dfrac{(0.0000064)(24{,}000{,}000{,}000)}{0.000000048}$.

Solution After changing each number into scientific notation, we can do the arithmetic on the numbers and the exponential expressions separately.

$$\begin{aligned}\frac{(0.0000064)(24{,}000{,}000{,}000)}{0.000000048} &= \frac{(6.4 \times 10^{-6})(2.4 \times 10^{10})}{4.8 \times 10^{-8}}\\ &= \frac{(6.4)(2.4)}{4.8} \cdot \frac{10^{-6}10^{10}}{10^{-8}}\\ &= 3.2 \times 10^{12}\end{aligned}$$

In standard notation, the result is 3,200,000,000,000. ■

EXAMPLE 5 Evaluate $\dfrac{(1{,}920{,}000)(0.0015)}{(0.000032)(45{,}000)}$.

Solution We first express all numbers in scientific notation,

$$\frac{(1.92 \times 10^{6})(1.5 \times 10^{-3})}{(3.2 \times 10^{-5})(4.5 \times 10^{4})}$$

and then do the arithmetic on the numbers and the exponential expressions separately.

$$\begin{aligned}\frac{(1.92 \times 10^{6})(1.5 \times 10^{-3})}{(3.2 \times 10^{-5})(4.5 \times 10^{4})} &= 0.2 \cdot \frac{10^{3}}{10^{-1}}\\ &= \mathbf{0.2} \cdot 10^{4}\\ &= \mathbf{2.0 \cdot 10^{-1}} \cdot 10^{4} && 0.2 = 2.0 \times 10^{-1}.\\ &= 2.0 \cdot 10^{3}\\ &= 2000\end{aligned}$$

■

Calculators

Scientific and graphing calculators often give answers in scientific notation. For example, if we use a calculator to find 301.2^8, the display will read

6.77391496 19 — on a scientific calculator

or

301.2 ∧ 8
6.773914961E19 — on a graphing calculator

In either case, the answer is given in scientific notation and is to be interpreted as

$$6.77391496 \times 10^{19}$$

Numbers can also be entered into a calculator in scientific form. For example, to enter 24,000,000,000 (which is 2.4×10^{10} in scientific notation), we enter these numbers and press these keys:

2.4 EXP 10 — on a scientific calculator

or

2.4 EE 10 — on a graphing calculator

To use a calculator to simplify $\frac{(24{,}000{,}000{,}000)(0.00000006495)}{0.00000004824}$, we must enter each number in scientific notation, because each number has too many digits to be entered directly.

The three numbers written in scientific notation are

$$2.4 \times 10^{10}$$
$$6.495 \times 10^{-8}$$
$$4.824 \times 10^{-8}$$

To simplify the expression using a scientific calculator, we enter these numbers and press these keys:

2.4 EXP 10 × 6.495 EXP 8 +/− ÷ 4.824 EXP 8 +/− =

The display will read

3.231343284 10

In standard notation, the answer is 32,313,432,840.

The steps are similar on a graphing calculator.

ORALS *Write each number in scientific notation.*

1. 352 3.52×10^2
2. 5130 5.13×10^3
3. 0.002 2×10^{-3}
4. 0.00025 2.5×10^{-4}

Write each number in standard notation.

5. 3.5×10^2 350
6. 4.3×10^3 4300
7. 2.7×10^{-1} 0.27
8. 8.5×10^{-2} 0.085

EXERCISE 2.2

In Exercises 1–20, write each number in scientific notation.

1. 3900 3.9×10^3
2. 1700 1.7×10^3
3. 0.0078 7.8×10^{-3}
4. 0.068 6.8×10^{-2}
5. $-45{,}000$ -4.5×10^4
6. $-547{,}000$ -5.47×10^5
7. -0.00021 -2.1×10^{-4}
8. -0.00078 -7.8×10^{-4}
9. 17,600,000 1.76×10^7
10. 89,800,000 8.98×10^7
11. 0.0000096 9.6×10^{-6}
12. 0.000046 4.6×10^{-5}
13. 323×10^5 3.23×10^7
14. 689×10^9 6.89×10^{11}
15. 6000×10^{-7} 6.0×10^{-4}
16. 765×10^{-5} 7.65×10^{-3}
17. 0.0527×10^5 5.27×10^3
18. 0.0298×10^3 2.98×10^1
19. 0.0317×10^{-2} 3.17×10^{-4}
20. 0.0012×10^{-3} 1.2×10^{-6}

In Exercises 21–32, write each number in standard notation.

21. 2.7×10^2 270
22. 7.2×10^3 7200
23. 3.23×10^{-3} 0.00323
24. 6.48×10^{-2} 0.0648
25. 7.96×10^5 796,000
26. 9.67×10^6 9,670,000
27. 3.7×10^{-4} 0.00037
28. 4.12×10^{-5} 0.0000412
29. 5.23×10^0 5.23
30. 8.67×10^0 8.67
31. 23.65×10^6 23,650,000
32. 75.6×10^{-5} 0.000756

In Exercises 33–36, write each number in scientific notation and do the operations. Give all answers in scientific notation.

33. $\dfrac{(4000)(30{,}000)}{0.0006}$ 2×10^{11}
34. $\dfrac{(0.0006)(0.00007)}{21{,}000}$ 2×10^{-12}
35. $\dfrac{(640{,}000)(2{,}700{,}000)}{120{,}000}$ 1.44×10^7
36. $\dfrac{(0.0000013)(0.000090)}{0.00039}$ 3×10^{-7}

In Exercises 37–42, write each number in scientific notation and do the operations. Give all answers in standard notation.

37. $\dfrac{(0.006)(0.008)}{0.0012}$ 0.04
38. $\dfrac{(600)(80{,}000)}{120{,}000}$ 400
39. $\dfrac{(220{,}000)(0.000009)}{0.00033}$ 6000
40. $\dfrac{(0.00024)(96{,}000{,}000)}{640{,}000{,}000}$ 0.000036
41. $\dfrac{(320{,}000)^2(0.0009)}{12{,}000^2}$ 0.64
42. $\dfrac{(0.000012)^2(49{,}000)^2}{0.021}$ 16.464

In Exercises 43–56, use scientific notation to find each answer.

43. Speed of sound The speed of sound in air is 3.31×10^4 centimeters per second. Find the speed of sound in centimeters per hour.
1.1916×10^8 cm/hr

44. Volume of a tank Find the volume of a tank shown in Illustration 1. 8.4×10^{10} mm^3

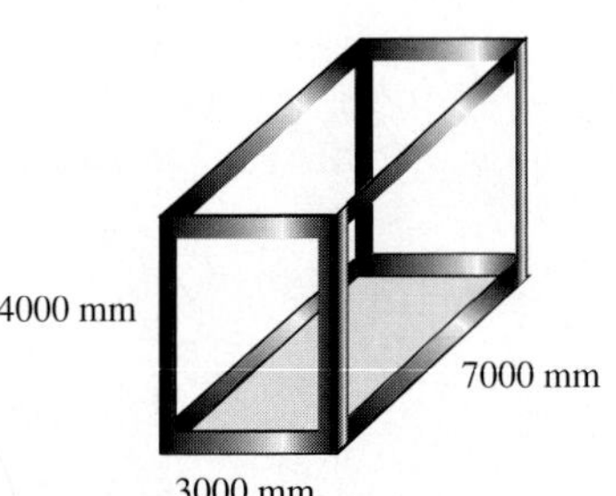

ILLUSTRATION 1

45. Mass of protons If the mass of one proton is 0.00000000000000000000000167248 gram, find the mass of 1 million protons. 1.67248×10^{-18} g

46. Speed of light The speed of light in a vacuum is about 30,000,000,000 centimeters per second. Find the speed of light in miles per hour. (*Hint:* 160,000 cm $\approx$ 1 mile. Read $\approx$ as "approximately equal to.") 6.75×10^8 mph

47. Distance to the moon The moon is about 235,000 miles from the earth. Find this distance in inches. 1.48896×10^{10} in.

48. Distance to the sun The sun is about 149,700,000 kilometers from the earth. Find this distance in miles. (*Hint:* 1 km $\approx$ 0.6214 miles.) 9.3×10^7 mi

49. Solar flares Solar flares often produce immense loops of glowing gas ejected from the sun's surface. The flare in Illustration 2 extends about 95,000 kilometers into space. Express this distance in miles. (*Hint:* 1 km $\approx$ 0.6214 mi.) 5.9×10^4 mi

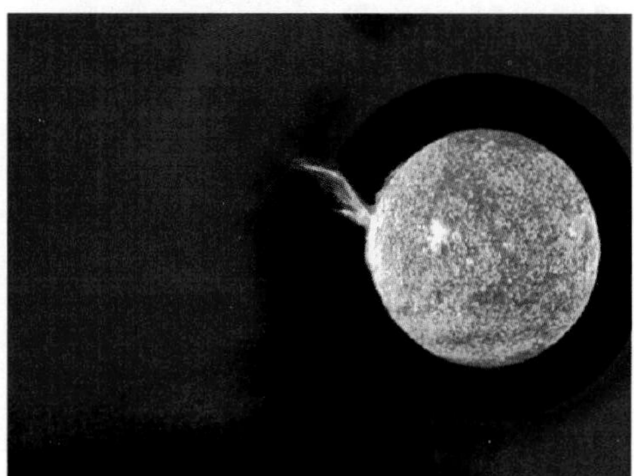

ILLUSTRATION 2

50. Earth's orbit The radius, r, of the earth's circular orbit is 93 million miles, and its circumference C (the distance around the circle) is given by $C = 2\pi r$, where $\pi \approx 3.14$. How far does the earth travel in one year? 5.8×10^8 mi

51. Angstroms per inch One **angstrom** is 0.0000001 millimeter, and one inch is 25.4 millimeters. Find the number of angstroms in one inch. 2.54×10^8

52. Range of a comet One **astronomical unit** (AU) is the distance from the earth to the sun—about 9.3×10^7 miles. Halley's Comet ranges from 0.6 to 18 AU from the sun. Express this range in miles. 5.58×10^7 mi to 1.674×10^9 mi

53. Flight to Pluto The planet Pluto is approximately 3,574,000,000 miles from earth. If a space ship can travel 18,000 miles per hour, how long will it take to reach Pluto? almost 23 years

54. Light year Light travels about 300,000,000 meters per second. A **light year** is the distance that light can travel in one year. How many meters are in one light year? 9.4608×10^{15} m

55. Distance to Alpha Centauri Light travels about 186,000 miles per second. A **parsec** is 3.26 light years. The star Alpha Centauri is 1.3 parsecs from the earth. Express this distance in miles. 2.49×10^{13} mi

56. Life of a comet The mass of a typical comet (see Illustration 3) is about 10^{16} grams. When the comet is close to the sun, matter evaporates at the rate of 10^7 grams per second. Calculate the life of a comet that appears every 50 years and spends ten days close to the sun. about 60,000 years

ILLUSTRATION 3

In Exercises 57–62, use a scientific calculator to evaluate each expression.

57. $23{,}437^3$ 1.2873779×10^{13}

58. 0.00034^4 1.336336×10^{-14}

59. $(63{,}480)(893{,}322)$ 5.6708081×10^{10}

60. $(0.0000413)(0.0000049)^2$ 9.91613×10^{-16}

61. $\dfrac{(69.4)^8(73.1)^2}{(0.0043)^3}$ 3.6166389×10^{25}

62. $\dfrac{(0.0031)^4(0.0012)^5}{(0.0456)^{-7}}$ $9.4212148 \times 10^{-35}$

Writing Exercises ■ *Write a paragraph using your own words.*

1. Explain how to change a number from standard notation to scientific notation.

2. Explain how to change a number from scientific notation to standard notation.

Something to Think About ■

1. Which is the greater: the number of heartbeats in a lifetime or the distance to the moon, in feet?
the number of heartbeats

2. How fast does your hair grow, in miles per hour?
An estimate of 1 in./mo is 3×10^{-8} mph.

Review Exercises ■ *Write each terminating decimal as a fraction.*

1. 0.2 $\frac{1}{5}$

2. 0.75 $\frac{3}{4}$

3. 0.375 $\frac{3}{8}$

4. 0.625 $\frac{5}{8}$

5. A man raises 3 to the second power, 4 to the third power, and 2 to the fourth power and finds their sum. What number does he obtain? 89

6. If $a = -2$, $b = -3$, and $c = 4$, find the value of $\frac{5ab - 4ac - 2}{3bc + abc}$. -5

2.3 Adding and Subtracting Polynomials

■ Degree of a Polynomial ■ *P*(*x*) Notation ■ Combining Like Terms ■ Adding Polynomials ■ Subtracting Polynomials ■ Adding and Subtracting Multiples of Polynomials

Expressions such as

$$5x, \quad 8y^2, \quad -12x^4y^5, \quad \text{and} \quad 52$$

that contain constants and/or variables are called **algebraic terms.** The numerical coefficients of the first three of these terms are 5, 8, and -12, respectively. Because $52 = 52x^0$, the number 52 is considered to be the numerical coefficient of the term 52.

A **polynomial** is an algebraic expression that is the sum of one or more terms containing whole number exponents on its variables.

Polynomial in One Variable A **polynomial in one variable** (say, x) is the sum of one or more terms of the form ax^n, where a is a real number and n is a whole number.

The expressions

$$3x^2 + 2x, \quad \frac{3}{2}x^5 - \frac{7}{3}x^4 - \frac{8}{3}x^3, \quad \text{and} \quad 19x^{20} + \sqrt{3}x^{14} + 4.5x^{11} - 17x^2$$

are polynomials in x. However, the expressions

$$\frac{2x}{x^2+1}, \qquad x^{1/2}-1, \qquad \text{and} \qquad x^{-3}+2x$$

are not. The first expression is the quotient of two polynomials, and the last two expressions have exponents that are not whole numbers.

Polynomial in Several Variables A **polynomial in several variables,** (say, x, y, and z) is the sum of one or more terms of the form $ax^my^nz^p$, where a is a real number and m, n, and p are whole numbers.

The expressions

$$3xy, \qquad 5x^2y+2yz^3-3xz, \qquad \text{and} \qquad u^2v^2w^2+x^3y^3+1$$

are polynomials in several variables.

A polynomial with a single term is called a **monomial;** a polynomial with two terms is called a **binomial;** and a polynomial with three terms is called a **trinomial.** Here are some examples of each:

Monomials	*Binomials*	*Trinomials*
$2x^3$	$2x^4+5$	$2x^3+4x^2+3$
a^2b	$-17t^{45}-3xy$	$3mn^3-m^2n^3+7n$
$3x^3y^5z^2$	$32x^{13}y^5+47x^3yz$	$-12x^5y^2+13x^4y^3-7x^3y^3$

Degree of a Polynomial

Degree of a Monomial If $a \neq 0$, the **degree of the monomial ax^n** is n. The degree of a monomial containing several variables is the sum of the exponents of those variables.

Because of the previous definition,

$3x^4$ is a monomial of degree 4.

$4^7x^2y^3$ is a monomial of degree 5. The sum of the exponents on x and y is 5.

$-18x^3y^2z^{12}$ is a monomial of degree 17.

3 is a monomial of degree 0. $3 = 3x^0$.

WARNING! Since $a \neq 0$ in the previous definition, 0 has no defined degree.

Degree of a Polynomial The **degree of a polynomial** is the same as the degree of the term with largest degree that is contained within the polynomial.

Using the previous definition,

$3x^5 + 4x^2 + 7$ is a trinomial of degree 5,

because the degree of the first term is 5, the degree of the second term is 2, the degree of the third term is 0, and the largest of these is 5. Likewise,

$7x^2y^8 - 3xy$ is a binomial of degree 10.

$3x + 2y - xy$ is a trinomial of degree 2.

$18x^2y^3 - 12x^7y^2 + 3x^9y^3 - 3$ is a polynomial of degree 12.

$P(x)$ Notation

Hypatia
(370 A.D.–415 A.D.)
Hypatia is the earliest known woman in the history of mathematics. She was a professor at the University of Alexandria. Because of her scientific beliefs, she was considered to be a heretic. At the age of 45, she was attacked by a mob and murdered for her beliefs.

Polynomials in one variable can be denoted with expressions such as

$P(x)$ Read $P(x)$ as "P of x."

where the letter within the parentheses represents the variable of the polynomial. As we read each of the polynomials

$$P(x) = 3x^2 + 4x + 5$$
$$Q(y) = 5y^4 + 3y^3 + 4y^2 - 3y - 4$$
$$R(z) = 2z^3 - 3z^2 + 7z + 1$$

from left to right, the degrees of successive terms get smaller, and we say that the polynomials are written in descending powers of their variables. In the polynomial

$$3 - 2x + 3x^2 + 4x^3 - 7x^4$$

the exponents on x increase, and the polynomial is written in ascending powers of x. In the polynomial

$$x^5 - 5x^4y + 10x^3y^2 - 10x^2y^3 + 5xy^4 - y^5$$

the exponents on x decrease and the exponents on y increase, and we say that the polynomial is written in descending powers of x and in ascending powers of y.

The notation $P(x)$ gives a convenient way to indicate the value of a polynomial at different values of x. For example, $P(0)$ represents the value of the polynomial $P(x)$ when $x = 0$. Likewise, $P(-5)$ represents the value of $P(x)$ when $x = -5$.

(a, c, e)

EXAMPLE 1 Consider the trinomial $P(x) = 3x^2 - 2x + 7$. Find **a.** $P(0)$, **b.** $P(4)$, **c.** $P(-5)$, **d.** $P(a)$, **e.** $P(-a)$, and **f.** $P(2t)$.

Solution

a. $P(0) = 3(0)^2 - 2(0) + 7$
$= 7$

b. $P(4) = 3(4)^2 - 2(4) + 7$
$= 47$

c. $P(-5) = 3(-5)^2 - 2(-5) + 7$
$= 92$

d. $P(a) = 3(a)^2 - 2(a) + 7$
$= 3a^2 - 2a + 7$

e. $P(-a) = 3(-a)^2 - 2(-a) + 7$
$= 3a^2 + 2a + 7$

f. $P(2t) = 3(2t)^2 - 2(2t) + 7$
$= 12t^2 - 4t + 7$ ■

EXAMPLE 2 **Height of a rocket** From the top of a building 150 feet tall, a toy rocket is launched straight up with an initial velocity of 164 feet per second. The height of the rocket above the ground is given by the polynomial

$$P(t) = -16t^2 + 164t + 150$$

where the height is $P(t)$ and t is the number of seconds after the rocket is launched. Find the height of the rocket at **a.** 0 seconds, **b.** 5 seconds, and **c.** 11 seconds.

Solution

a. To find the height after 0 seconds, we substitute 0 for t in the polynomial $P(t) = -16t^2 + 164t + 150$.

$$\begin{aligned} P(0) &= -16(0)^2 + 164(0) + 150 \\ &= 150 \end{aligned}$$

At 0 seconds, the rocket is 150 feet high, resting at the top of the building waiting to be launched.

b. To find the height after 5 seconds, we substitute 5 for t in the polynomial.

$$\begin{aligned} P(5) &= -16(5)^2 + 164(5) + 150 \\ &= -16(25) + 820 + 150 \\ &= -400 + 970 \\ &= 570 \end{aligned}$$

At 5 seconds, the rocket is 570 feet high.

c. To find the height after 11 seconds, we substitute 11 for t in the polynomial.

$$\begin{aligned} P(11) &= -16(11)^2 + 164(11) + 150 \\ &= -16(121) + 1804 + 150 \\ &= -1936 + 1954 \\ &= 18 \end{aligned}$$

At 11 seconds, the rocket is 18 feet high. It has fallen nearly back to earth. ■

EXAMPLE 3 Find the value of the polynomial $4x^2y - 5xy^3$ if $x = 3$ and $y = -2$.

Solution We substitute 3 for x and -2 for y and simplify.

$$\begin{aligned} 4x^2y - 5xy^3 &= 4(3)^2(-2) - 5(3)(-2)^3 \\ &= 4(9)(-2) - 5(3)(-8) \\ &= -72 + 120 \\ &= 48 \end{aligned}$$

■

Combining Like Terms

If two terms have the same variables with the same exponents, they are called **like** or **similar terms.**

$3x^2$, $5x^2$, and $7x^2$ are like terms. They have the same variables with the same exponents.

$5x^3y^2$, $17x^3y^2$, and $103x^3y^2$ are like terms. They have the same variables with the same exponents.

$4x^2y^2$, $12xy^5$, and $98x^7y^9$ are unlike terms. They have the same variables, but with different exponents.

$3x^4y$ and $5x^4z^2$ are unlike terms. They have different variables.

We have seen that multiplication distributes over addition and that this fact is expressed by the distributive property:

$$a(b + c) = ab + ac$$

Because multiplication is commutative, we can write the distributive property in the form

$$ba + ca = (b + c)a$$

This form of the distributive property enables us to combine like terms. For example,

$$3x + 7x = (3 + 7)x = 10x$$
$$5x^2y^3 + 22x^2y^3 = (5 + 22)x^2y^3 = 27x^2y^3$$

and since $xy^4 = 1xy^4$,

$$\begin{aligned} 9xy^4 + 6xy^4 + xy^4 &= 9xy^4 + 6xy^4 + 1xy^4 \\ &= (9 + 6 + 1)xy^4 \\ &= 16xy^4 \end{aligned}$$

The preceding examples suggest that to combine like terms, *we add their numerical coefficients and keep the same variables with the same exponents.*

WARNING! The terms in the binomials

$3x^2 - 5y^2$, $-2a^2 + 3a^3$, and $5y^2 + 17xy$

cannot be combined because they are not like terms.

EXAMPLE 4

a. $12x^2z + 13x^2z = 25x^2z$ **b.** $28x^5y^2 + 11x^5y^2 = 39x^5y^2$

c. $-22x^2y^3 + 7x^2y^3 = -15x^2y^3$ **d.** $-12r^3s^4 + 13r^3s^4 = r^3s^4$ ■

Adding Polynomials

To add polynomials, we use the distributive property to remove parentheses and combine like terms, where possible.

EXAMPLE 5

$(3x^2 - 2x + 4) + (2x^2 + 4x - 3)$

$= 1(3x^2 - 2x + 4) + 1(2x^2 + 4x - 3)$ Each polynomial has an understood coefficient of 1.

$= 3x^2 - 2x + 4 + 2x^2 + 4x - 3$ Use the distributive property.

$= 3x^2 + 2x^2 - 2x + 4x + 4 - 3$

$= 5x^2 + 2x + 1$ Combine like terms. ■

Classroom Activity:
1. Think of your age.
2. Add 10 and double the sum.
3. Subtract 30.
4. Add your age.
5. Add 10.
6. Divide by 3.
7. The result is your age.
Have the students explain why this number trick works.

The addition in Example 5 can also be done by aligning the terms vertically.

$$\begin{array}{l} 3x^2 - 2x + 4 \\ \underline{2x^2 + 4x - 3} \\ 5x^2 + 2x + 1 \end{array}$$

Subtracting Polynomials

To subtract one monomial from another, we add the negative (or opposite) of the monomial that is to be subtracted.

EXAMPLE 6

a. $8x^2 - 3x^2 = 8x^2 + (-3x^2)$
$= 5x^2$

b. $3x^2y - 9x^2y = 3x^2y + (-9x^2y)$
$= -6x^2y$

c. $-5x^5y^3z^2 - 3x^5y^3z^2 = -5x^5y^3z^2 + (-3x^5y^3z^2)$
$= -8x^5y^3z^2$ ■

To subtract polynomials, we use the distributive property to remove parentheses and combine terms, where possible.

EXAMPLE 7 **a.** $(8x^3y + 2x^2y) - (2x^3y - 3x^2y)$

$= 1(8x^3y + 2x^2y) - 1(2x^3y - 3x^2y)$ Insert the understood coefficients of 1.

$= 8x^3y + 2x^2y - 2x^3y + 3x^2y$ Use the distributive property to remove parentheses.

$= 6x^3y + 5x^2y$ Combine like terms.

b. $(3rt^2 + 4r^2t^2) - (8rt^2 - 4r^2t^2 + r^3t^2)$

$= 1(3rt^2 + 4r^2t^2) - 1(8rt^2 - 4r^2t^2 + r^3t^2)$ Insert the understood coefficients of 1.

$= 3rt^2 + 4r^2t^2 - 8rt^2 + 4r^2t^2 - r^3t^2$ Use the distributive property to remove parentheses.

$= -5rt^2 + 8r^2t^2 - r^3t^2$ Combine like terms. ■

To subtract polynomials in vertical form, we add the negative (or opposite) of the polynomial that is being subtracted.

$$\begin{array}{rr} & 8x^3y + 2x^2y \\ - & \underline{2x^3y - 3x^2y} \end{array} \Rightarrow \begin{array}{rr} & 8x^3y + 2x^2y \\ + & \underline{-2x^3y + 3x^2y} \\ & 6x^3y + 5x^2y \end{array}$$

Adding and Subtracting Multiples of Polynomials

To add multiples of one polynomial to another, or subtract multiples of one polynomial from another, we use the distributive property to remove parentheses and combine like terms.

EXAMPLE 8 Simplify $3(2x^2 + 4x - 7) - 2(3x^2 - 4x - 5)$.

Solution

$$3(2x^2 + 4x - 7) - 2(3x^2 - 4x - 5) = 6x^2 + 12x - 21 - 6x^2 + 8x + 10$$
$$= 20x - 11$$ Combine like terms. ■

ORALS *Give the degree of each polynomial.*

1. $3x^3$ 3 **2.** $-4x^3y$ 4 **3.** $4x^2y + 2x$ 3 **4.** $3x + 4xyz$ 3

If $P(x) = 2x + 1$, find each value.

5. $P(0)$ 1 **6.** $P(2)$ 5 **7.** $P(-1)$ −1 **8.** $P(-2)$ −3

Combine terms.

9. $4x^2 + 5x^2$ $9x^2$ **10.** $3y^2 - 5y^2$ $-2y^2$

EXERCISE 2.3

In Exercises 1–8, classify each polynomial as a monomial, binomial, trinomial, or none of these.

1. $3x^2$ monomial
2. $2y^3 + 4y^2$ binomial
3. $3x^2y - 2x + 3y$ trinomial
4. $a^2 + b^2$ binomial
5. $x^2 - y^2$ binomial
6. $\frac{17}{2}x^3 + 3x^2 - x - 4$ none of these
7. 5 monomial
8. $8x^3y^5$ monomial

In Exercises 9–20, find the degree of each polynomial.

9. $3x^2 + 2$ 2
10. x^{17} 17
11. $4x^8 + 3x^2y^4$ 8
12. $19x^2y^4 - y^{10}$ 10
13. $4x^2 - 5y^3z^3t^4$ 10
14. $7x$ 1
15. $121y$ 1
16. $x^2y^3z^4 + z^{12}$ 12
17. 77 0
18. 43 0
19. $x + y + z$ 1
20. $5xy - x$ 2

In Exercises 21–26, write each polynomial in descending powers of x.

21. $2x + 3x^3 - 4x^2 + 7$
$3x^3 - 4x^2 + 2x + 7$

22. $3x^3 + 5x - 3x^2 - 8$
$3x^3 - 3x^2 + 5x - 8$

23. $5xy^3 - 3x^2y^2 + 2x^3y - x^4 + y^4$
$-x^4 + 2x^3y - 3x^2y^2 + 5xy^3 + y^4$

24. $x^2y + y^2x^{10} - x^{13}y + x^5 - y$
$-x^{13}y + x^{10}y^2 + x^5 + x^2y - y$

25. $3x^3z - 4x^6y + 3 - 4x$
$-4x^6y + 3x^3z - 4x + 3$

26. $-6x^2yz + 7x^3y^2z - 8x^4yz^2 - 10$
$-8x^4yz^2 + 7x^3y^2z - 6x^2yz - 10$

In Exercises 27–30, consider the polynomial $P(x) = 2x^2 + x + 2$. Find each value.

27. $P(0)$ 2
28. $P(1)$ 5
29. $P(4)$ 38
30. $P(-3)$ 17

In Exercises 31–34, the height $P(t)$, in feet, of a ball thrown straight up with an initial velocity of 64 feet per second is given by the polynomial $P(t) = 64t - 16t^2$. Find the height of the ball after the given number of seconds.

31. 0 second 0 ft
32. 1 second 48 ft
33. 2 seconds 64 ft
34. 4 seconds 0 ft

In Exercises 35–44, consider the polynomial $P(x) = -3x^2 + 4x - 3$. Find each value.

35. $P(1)$ -2
36. $P(0)$ -3
37. $P(-2)$ -23
38. $P(3)$ -18
39. $P(t)$ $-3t^2 + 4t - 3$
40. $P(z)$ $-3z^2 + 4z - 3$
41. $P(-x)$ $-3x^2 - 4x - 3$
42. $P(-r)$ $-3r^2 - 4r - 3$
43. $P(2x)$ $-12x^2 + 8x - 3$
44. $P(3x)$ $-27x^2 + 12x - 3$

In Exercises 45–58, find each value when $x = 2$ and $y = -3$.

45. $2x^3 - 3x^2 - 4x + 2$ -2
46. $3y^3 + 4y^2 - 2y - 4$ -43
47. $x^2 + y^2$ 13
48. $x^3 + y^3$ -19
49. $x^3 - y^3$ 35
50. $x^2 - y^2$ -5
51. $3x^2y + xy^3$ -90
52. $8xy - xy^2$ -66
53. $\frac{9x^3}{y} - \frac{8y^3}{x}$ 84
54. $\frac{27x^2}{y} + \frac{16y^4}{x}$ 612
55. $\frac{3x^3 - 2y^2}{2x^2 + y^3}$ $-\frac{6}{19}$
56. $\frac{2x^4 + 3y^3}{2x^3 - y}$ $-\frac{49}{19}$
57. $\frac{3x^3 + 6y^2}{-2x - 3y^3}$ $\frac{78}{77}$
58. $\frac{4x^4 - 3y}{-2x^2 - y}$ $-\frac{73}{5}$

In Exercises 59–64, use a calculator to find each value when $x = 3.7$, $y = -2.5$, and $z = 8.9$.

59. x^2y -34.225

60. xyz^2 -732.6925

61. $\frac{x^2}{z^2}$ 0.17283171

62. $\frac{z^3}{y^2}$ 112.79504

63. $\frac{x + y + z}{xyz}$ -0.12268448

64. $\frac{x + yz}{xy + z}$ 53

In Exercises 65–72, tell whether the terms are like or unlike terms.

65. $3x, 7x$ like terms

66. $-8x, 3y$ unlike terms

67. $7x, 7y$ unlike terms

68. $3mn, 5mn$ like terms

69. $3r^2t^3, -8r^2t^3$ like terms

70. $9u^2v, 10u^2v$ like terms

71. $9x^2y^3, 3x^2y^2$ unlike terms

72. $27x^6y^4z, 8x^6y^4z^2$ unlike terms

In Exercises 73–80, simplify each expression.

73. $8x + 4x$ $12x$

74. $-2y + 16y$ $14y$

75. $5x^3y^2z - 3x^3y^2z$ $2x^3y^2z$

76. $8wxy - 12wxy$ $-4wxy$

77. $-2x^2y^3 + 3xy^4 - 5x^2y^3$ $-7x^2y^3 + 3xy^4$

78. $3ab^4 - 4a^2b^2 - 2ab^4 + 2a^2b^2$ $ab^4 - 2a^2b^2$

79. $(3x^2y)^2 + 2x^4y^2 - x^4y^2$ $10x^4y^2$

80. $(5x^2y^4)^3 - (5x^3y^6)^2$ $100x^6y^{12}$

In Exercises 81–100, do each operation.

81. $(3x^2 + 2x + 1) + (-2x^2 - 7x + 5)$
$x^2 - 5x + 6$

82. $(-2a^2 - 5a - 7) + (-3a^2 + 7a + 1)$
$-5a^2 + 2a - 6$

83. $(-a^2 + 2a + 3) - (4a^2 - 2a - 1)$
$-5a^2 + 4a + 4$

84. $(x^2 - 3x + 8) - (3x^2 + x + 3)$
$-2x^2 - 4x + 5$

85. $(7y^3 + 4y^2 + y + 3) + (-8y^3 - y + 3)$
$-y^3 + 4y^2 + 6$

86. $(6x^3 + 3x - 2) - (2x^3 + 3x^2 + 5)$
$4x^3 - 3x^2 + 3x - 7$

87. $(3x^2 + 4x - 3) + (2x^2 - 3x - 1) - (x^2 + x + 7)$ $4x^2 - 11$

88. $(-2x^2 + 6x + 5) - (-4x^2 - 7x + 2) - (4x^2 + 10x + 5)$ $-2x^2 + 3x - 2$

89. $(3x^3 - 2x + 3) + (4x^3 + 3x^2 - 2) + (-4x^3 - 3x^2 + x + 12)$ $3x^3 - x + 13$

90. $(x^4 - 3x^2 + 4) + (-2x^4 - x^3 + 3x^2) + (3x^2 + 2x + 1)$ $-x^4 - x^3 + 3x^2 + 2x + 5$

91. $(3y^2 - 2y + 4) + [(2y^2 - 3y + 2) - (y^2 + 4y + 3)]$ $4y^2 - 9y + 3$

92. $(-t^2 - t - 1) - [(t^2 + 3t - 1) - (-2t^2 + 4)]$ $-4t^2 - 4t + 4$

93. Add the polynomials:

$$\begin{array}{r} 3x^3 - 2x^2 + 4x - 3 \\ -2x^3 + 3x^2 + 3x - 2 \\ 5x^3 - 7x^2 + 7x - 12 \\ \hline 6x^3 - 6x^2 + 14x - 17 \end{array}$$

94. Add the polynomials:

$$\begin{array}{r} 7a^3 + 3a + 7 \\ -2a^3 + 4a^2 - 13 \\ 3a^3 - 3a^2 + 4a + 5 \\ \hline 8a^3 + a^2 + 7a - 1 \end{array}$$

95. Add the polynomials:

$$\begin{array}{r} -2y^4 - 2y^3 + 4y^2 - 3y + 10 \\ -3y^4 + 7y^3 - y^2 + 14y - 3 \\ - 3y^3 - 5y^2 - 5y + 7 \\ -4y^4 + y^3 - 13y^2 + 14y - 2 \\ \hline -9y^4 + 3y^3 - 15y^2 + 20y + 12 \end{array}$$

96. Add the polynomials:

$$\begin{array}{r} 17t^4 + 3t^3 - 2t^2 - 3t + 4 \\ -12t^4 - 2t^3 + 3t^2 - 5t - 17 \\ -2t^4 - 7t^3 + 4t^2 + 12t - 5 \\ 5t^4 + t^3 + 5t^2 - 13t + 12 \\ \hline 8t^4 - 5t^3 + 10t^2 - 9t - 6 \end{array}$$

97. Subtract the second polynomial from the first:
$3x^2 - 4x + 17$
$2x^2 + 4x - 5$
$x^2 - 8x + 22$

98. Subtract the second polynomial from the first:
$-2y^2 - 4y + 3$
$3y^2 + 10y - 5$
$-5y^2 - 14y + 8$

99. Subtract the second polynomial from the first:
$-5y^3 + 4y^2 - 11y + 3$
$-2y^3 - 14y^2 + 17y - 32$
$-3y^3 + 18y^2 - 28y + 35$

100. Subtract the second polynomial from the first:
$17x^4 - 3x^2 - 65x - 12$
$23x^4 + 14x^2 + 3x - 23$
$-6x^4 - 17x^2 - 68x + 11$

In Exercises 101–108, simplify each expression.

101. $2(x^3 + x^2) + 3(2x^3 - x^2)$ $8x^3 - x^2$

102. $3(y^2 + 2y) - 4(y^2 - 4)$ $-y^2 + 6y + 16$

103. $-5(2x^3 + 7x^2 + 4x) - 2(3x^3 - 4x^2 - 4x)$
$-16x^3 - 27x^2 - 12x$

104. $-3(3a^2 + 4b^3 + 7) + 4(5a^2 - 2b^3 + 3)$
$11a^2 - 20b^3 - 9$

105. $4(3z^2 - 4z + 5) + 6(-2z^2 - 3z + 4) - 2(4z^2 + 3z - 5)$ $-8z^2 - 40z + 54$

106. $-3(4x^3 - 2x^2 + 4) - 4(3x^3 + 4x^2 + 3x) + 5(3x - 4)$ $-24x^3 - 10x^2 + 3x - 32$

107. $5(2a^2 + 4a - 2) - 2(-3a^2 - a + 12) - 2(a^2 + 3a - 5)$ $14a^2 + 16a - 24$

108. $-2(2b^2 - 3b + 3) + 3(3b^2 + 2b - 8) - (3b^2 - b + 4)$ $2b^2 + 13b - 34$

109. Find the difference when $3x^2 + 4x - 3$ is subtracted from the sum of $-2x^2 - x + 7$ and $5x^2 + 3x - 1$.
$-2x + 9$

110. Find the difference when $8x^3 + 2x^2 - 1$ is subtracted from the sum of $x^2 + x + 2$ and $2x^3 - x + 9$. $-6x^3 - x^2 + 12$

111. Find the sum when $2x^2 - 4x + 3$ minus $8x^2 + 5x - 3$ is added to $-2x^2 + 7x - 4$.
$-8x^2 - 2x + 2$

112. Find the sum when $7x^3 - 4x$ minus $x^2 + 2$ is added to $5 + 3x$.
$7x^3 - x^2 - x + 3$

Writing Exercises *Write a paragraph using your own words.*

1. Explain how to find the degree of a polynomial.
2. Explain how to recognize like terms, and how to add them.

Something to Think About

1. If $P(x) = 3x^2 + 5x + c$, and $P(2) = 26$, what is c? 4
2. If $P(x) = x^3 - 3x$, is $P(2) + P(3) = P(2 + 3)$? no
3. If $P(x) = x^2 - 2x - 3$, find $P(P(0))$. 12
4. If $P(x) = 2x^2 - x - 5$, find $P(P(-1))$ 5

Review Exercises *Write each expression with a single exponent.*

1. a^3a^2 a^5
2. $\frac{b^3b^3}{b^4}$ b^2
3. $\frac{3(y^3)^{10}}{y^3y^4}$ $3y^{23}$
4. $\frac{4x^{-4}x^5}{2x^{-6}}$ $2x^7$
5. The distance from Mars to the sun is about 114,000,000 miles. Express this number in scientific notation.
1.14×10^8
6. One angstrom is about 0.0000001 millimeter. Express this number in scientific notation. 1×10^{-7}

2.4 Multiplying Polynomials

■ Multiplying a Monomial by a Monomial ■ Multiplying a Polynomial by a Monomial ■ Multiplying a Polynomial by a Polynomial ■ The FOIL Method

■ Multiplying a Monomial by a Monomial

We first consider multiplying a monomial by a monomial. In the examples, the commutative and associative properties of multiplication enable us to rearrange the factors and the parentheses.

(c) **EXAMPLE 1**

a. $(3x^2)(6x^3) = 3 \cdot x^2 \cdot 6 \cdot x^3$
$= (3 \cdot 6)(x^2 \cdot x^3)$
$= 18x^5$

b. $(-8x)(2y)(xy) = -8 \cdot x \cdot 2 \cdot y \cdot x \cdot y$
$= (-8 \cdot 2) \cdot x \cdot x \cdot y \cdot y$
$= -16x^2y^2$

c. $(2a^3b)(-7b^2c)(-12ac^4) = 2 \cdot a^3 \cdot b \cdot (-7) \cdot b^2 \cdot c \cdot (-12) \cdot a \cdot c^4$
$= 2(-7)(-12) \cdot a^3 \cdot a \cdot b \cdot b^2 \cdot c \cdot c^4$
$= 168a^4b^3c^5$ ■

The results in Example 1 suggest that to multiply monomials, *we multiply the numerical factors and then multiply the variable factors.*

■ Multiplying a Polynomial by a Monomial

To find the product of a monomial and a polynomial, we use the distributive property or the extended distributive property:

$$a(b + c + d + \cdots) = ab + ac + ad + \cdots$$

(b) **EXAMPLE 2**

a. $3x^2(6xy + 3y^2) = 3x^2 \cdot 6xy + 3x^2 \cdot 3y^2$
$= 18x^3y + 9x^2y^2$

b. $5x^3y^2(xy^3 - 2x^2y) = 5x^3y^2 \cdot xy^3 - 5x^3y^2 \cdot 2x^2y$
$= 5x^4y^5 - 10x^5y^3$

c. $-2ab^2(3bz - 2az + 4z^3) = -2ab^2 \cdot 3bz - (-2ab^2) \cdot 2az + (-2ab^2) \cdot 4z^3$
$= -6ab^3z + 4a^2b^2z - 8ab^2z^3$ ■

The results in Example 2 suggest that to multiply a polynomial by a monomial, *we multiply each term of the polynomial by the monomial.*

Multiplying a Polynomial by a Polynomial

To multiply a polynomial by a polynomial, we use the distributive property repeatedly.

(b) **EXAMPLE 3** **a.** $(3x + 2)(4x + 9) = (3x + 2) \cdot 4x + (3x + 2) \cdot 9$

$$= 12x^2 + 8x + 27x + 18$$
$$= 12x^2 + 35x + 18$$

b. $(2a - b)(3a^2 - 4ab + b^2) = (2a - b)3a^2 - (2a - b)4ab + (2a - b)b^2$

$$= 6a^3 - 3a^2b - 8a^2b + 4ab^2 + 2ab^2 - b^3$$
$$= 6a^3 - 11a^2b + 6ab^2 - b^3$$

The results of Example 3 suggest that to multiply one polynomial by another, *we multiply each term of one polynomial by each term of the other polynomial.*

It is often convenient to organize the work vertically, as in the following example.

EXAMPLE 4 **a.**

$$\begin{array}{lr} & 3x + 2 \\ & \underline{4x + 9} \\ 4x(3x+2) \longrightarrow & 12x^2 + 8x \qquad\quad \\ 9(3x+2) \longrightarrow & \underline{\qquad + 27x + 18} \\ & 12x^2 + 35x + 18 \end{array}$$

b.

$$\begin{array}{lr} & 3a^2 - 4ab + b^2 \qquad\quad \\ & \underline{2a \ - b \qquad\qquad\quad} \\ 2a(3a^2 - 4ab + b^2) \longrightarrow & 6a^3 - 8a^2b + 2ab^2 \qquad\quad \\ -b(3a^2 - 4ab + b^2) \longrightarrow & \underline{\qquad - 3a^2b + 4ab^2 - b^3} \\ & 6a^3 - 11a^2b + 6ab^2 - b^3 \end{array}$$

The FOIL Method

Multiplying one binomial by another requires that each term of one binomial be multiplied by each term of the other binomial. This fact can be emphasized by drawing curves to show the indicated products. For example, to multiply the binomials $3x + 2$ and $x + 4$, we can write

First terms   Last terms

$$(3x + 2)(x + 4) = 3x \cdot x + 3x \cdot 4 + 2 \cdot x + 2 \cdot 4$$

Inner terms
$$= 3x^2 + 12x + 2x + 8$$

Outer terms
$$= 3x^2 + 14x + 8$$

We note that

- the product of the **F**irst terms is $3x^2$,
- the product of the **O**uter terms is $12x$,
- the product of the **I**nner terms is $2x$, and
- the product of the **L**ast terms is 8.

This scheme is often called the **FOIL** method of multiplying two binomials. FOIL is an acronym for **F**irst terms, **O**uter terms, **I**nner terms, and **L**ast terms. Of course, the resulting terms of the product must be combined, if possible.

EXAMPLE 5 Find the products **a.** $(x + 3)(x + 3)$ **b.** $(y - 4)(y - 4)$, and **c.** $(a + 6)(a - 6)$.

Solution We use the FOIL method to multiply each term of one binomial by each term of the other binomial, and then simplify.

a. $(x + 3)(x + 3) = x^2 + 3x + 3x + 3^2$
$= x^2 + 6x + 9$

b. $(y - 4)(y - 4) = y^2 - 4y - 4y + 4^2$
$= y^2 - 8y + 16$

c. $(a + 6)(a - 6) = a^2 - 6a + 6a + 6^2$
$= a^2 - 36$ ■

The products discussed in Example 5 are called **special products.** Because they occur so often, it is worthwhile to learn their forms. In the exercises, you will be asked to verify the following product formulas.

Special Product Formulas

$$(x + y)^2 = (x + y)(x + y) = x^2 + 2xy + y^2$$
$$(x - y)^2 = (x - y)(x - y) = x^2 - 2xy + y^2$$
$$(x + y)(x - y) = x^2 - y^2$$

Because $x^2 + 2xy + y^2 = (x + y)^2$ and $x^2 - 2xy + y^2 = (x - y)^2$, the two trinomials are called **perfect square trinomials.**

Teaching Tip: Have students learn that to square a binomial:
1. Square the first term.
2. Multiply the terms and double the product.
3. Square the last term.

WARNING! The squares $(x + y)^2$ and $(x - y)^2$ have trinomials for their products. Don't forget to write the middle terms in these products. Remember that

$$(x + y)^2 \neq x^2 + y^2 \quad \text{and that} \quad (x - y)^2 \neq x^2 - y^2$$

Also remember that the product $(x + y)(x - y)$ is the binomial $x^2 - y^2$.

Teaching Tip: You may want to introduce the word *conjugate* here. Then say the product of two conjugate binomials is the difference of two squares.

When we use the FOIL method, it is easy to multiply binomials by sight. We find the product of the first terms, then the products of the outer terms and the inner terms and add them (when possible), and then the product of the last terms.

PERSPECTIVE

A Father of Algebra

One important figure in the history of mathematics, François Vieta (1540–1603) was one of the first mathematicians to use notation close to that which we use today. Trained as a lawyer, Vieta served in the parliament of Brittany and as the personal lawyer of Henry of Navarre. If he had continued as a successful lawyer, Vieta might now be forgotten. Fortunately, he lost his job.

When political opposition forced him out of office in 1584, Vieta had time to devote himself entirely to his hobby, mathematics. He used the time to study the writings of earlier mathematicians, and adapted and improved their ideas. Vieta was the first to use letters to represent unknown numbers, but he did not use our modern notation for exponents. To us, his notation seems awkward. For example, what we would write as

$$(x + 1)^3 = x^3 + 3x^2 + 3x + 1$$

Vieta would have written as

$\overline{x + 1}$ cubus aequalis x cubus + x quad. 3 + x in 3 + 1

 (a)

EXAMPLE 6 Find the products by sight: **a.** $(2x - 3)(3x + 2)$ and **b.** $(3x + 1)(3x + 4)$.

Solution **a.** $(2x - 3)(3x + 2) = 6x^2 - 5x - 6$

The middle term in the trinomial $6x^2 - 5x - 6$ comes from combining the outer and inner products $+4x$ and $-9x$:

$$4x + (-9x) = -5x$$

b. $(3x + 1)(3x + 4) = 9x^2 + 15x + 4$

The middle term in the trinomial $9x^2 + 15x + 4$ comes from combining the products $+12x$ and $+3x$:

$$+12x + 3x = 15x$$

■

At first glance, the expression $3[x^2 - 2(x + 3)]$ does not look like a polynomial. However, if we simplify the expression by removing the parentheses and the brackets, the expression takes the form of a polynomial.

$$\begin{aligned} 3[x^2 - 2(x + 3)] &= 3[x^2 - 2x - 6] \\ &= 3x^2 - 6x - 18 \end{aligned}$$

If an expression has one set of grouping symbols that is enclosed within another set, it is usually wise to eliminate the inner set first.

EXAMPLE 7 Find the product of $-2[y^3 + 3(y^2 - 2)]$ and $5[y^2 - 2(y + 1)]$.

Solution We change each expression into polynomial form:

$-2[y^3 + 3(y^2 - 2)]$	$5[y^2 - 2(y + 1)]$
$-2(y^3 + 3y^2 - 6)$	$5(y^2 - 2y - 2)$
$-2y^3 - 6y^2 + 12$	$5y^2 - 10y - 10$

We can now do the multiplication:

$$\begin{array}{rrrrrr} -2y^3 & - 6y^2 & + 12 & & & \\ 5y^2 & - 10y & - 10 & & & \\ \hline -10y^5 & - 30y^4 & & + 60y^2 & & \\ & + 20y^4 & + 60y^3 & & - 120y & \\ & & + 20y^3 & + 60y^2 & & - 120 \\ \hline -10y^5 & - 10y^4 & + 80y^3 & + 120y^2 & - 120y & - 120 \end{array}$$

■

The following examples show how to use the FOIL method to multiply expressions that are not polynomials.

EXAMPLE 8 Find the product of $x^{-2} + y$ and $x^2 - y^{-2}$.

Solution We use the FOIL method to multiply each term of the second expression by each term of the first expression, and then simplify.

$$\begin{aligned}(x^{-2} + y)(x^2 - y^{-2}) &= x^{-2}x^2 - x^{-2}y^{-2} + yx^2 - yy^{-2} \\ &= x^{-2+2} - x^{-2}y^{-2} + yx^2 - y^{1+(-2)} \\ &= x^0 - \frac{1}{x^2y^2} + x^2y - y^{-1} \\ &= 1 - \frac{1}{x^2y^2} + x^2y - \frac{1}{y}\end{aligned}$$

■

EXAMPLE 9 Find the product of $x^n + 2x$ and $x^n + 3x^{-n}$.

Solution We use the FOIL method to multiply each term of the second expression by each term of the first expression and simplify:

$$\begin{aligned}(x^n + 2x)(x^n + 3x^{-n}) &= x^nx^n + x^n(3x^{-n}) + 2x(x^n) + 2x(3x^{-n}) \\ &= x^{n+n} + 3x^nx^{-n} + 2x^{1+n} + 6xx^{-n} \\ &= x^{2n} + 3x^0 + 2x^{n+1} + 6x^{1+(-n)} \\ &= x^{2n} + 3 + 2x^{n+1} + 6x^{1-n}\end{aligned}$$

■

ORALS *Find each product.*

1. $(-2a^2b)(3ab^2)$ $-6a^3b^3$

2. $(4xy^2)(-2xy)$ $-8x^2y^3$

3. $3a^2(2a - 1)$ $6a^3 - 3a^2$

4. $-4n^2(4m - n)$ $-16mn^2 + 4n^3$

5. $(x + 1)(2x + 1)$ $2x^2 + 3x + 1$

6. $(3y - 2)(2y + 1)$ $6y^2 - y - 2$

EXERCISE 2.4

In Exercises 1–30, find each product.

1. $(2a^2)(-3ab)$ $-6a^3b$
2. $(-3x^2y)(3xy)$ $-9x^3y^2$
3. $(-3ab^2c)(5ac^2)$ $-15a^2b^2c^3$
4. $(-2m^2n)(-4mn^3)$ $8m^3n^4$
5. $(4a^2b)(-5a^3b^2)(6a^4)$ $-120a^9b^3$
6. $(2x^2y^3)(4xy^5)(-5y^6)$ $-40x^3y^{14}$
7. $(3x^3y^5)(2xy^2)^2$ $12x^5y^9$
8. $(a^3b^2c)^3(ab^2c^3)$ $a^{10}b^8c^6$
9. $(5x^3y^2)^4\left(\frac{1}{5}x^{-2}\right)^2$ $25x^8y^8$
10. $(4a^{-2}b^{-1})^2(2a^3b^4)^4$ $256a^8b^{14}$
11. $(-5xx^2)(-3xy)^4$ $-405x^7y^4$
12. $(-2a^2ab^2)^3(-3ab^2b^2)$ $24a^{10}b^{10}$
13. $[(-2x^3y)(5x^2y^2)]^2$ $100x^{10}y^6$
14. $[(3x^2y^3)(4xy^5)]^3$ $1728x^9y^{24}$
15. $3(x + 2)$ $3x + 6$
16. $-5(a + b)$ $-5a - 5b$
17. $-a(a - b)$ $-a^2 + ab$
18. $y^2(y - 1)$ $y^3 - y^2$
19. $3x(x^2 + 3x)$ $3x^3 + 9x^2$
20. $-2x(3x^2 - 2)$ $-6x^3 + 4x$
21. $-2x(3x^2 - 3x + 2)$ $-6x^3 + 6x^2 - 4x$
22. $3a(4a^2 + 3a - 4)$ $12a^3 + 9a^2 - 12a$
23. $5a^2b^3(2a^4b - 5a^0b^3)$ $10a^6b^4 - 25a^2b^6$
24. $-2a^3b(3a^0b^4 - 2a^2b^3)$ $-6a^3b^5 + 4a^5b^4$
25. $7rst(r^2 + s^2 - t^2)$ $7r^3st + 7rs^3t - 7rst^3$
26. $3x^2yz(x^2 - 2y + 3z^2)$ $3x^4yz - 6x^2y^2z + 9x^2yz^3$
27. $-4x^2y^3(3x^2 - 4xy + y^2)$ $-12x^4y^3 + 16x^3y^4 - 4x^2y^5$
28. $-2x^2y(3x^4y^2 - 2x^2y - 7)$ $-6x^6y^3 + 4x^4y^2 + 14x^2y$
29. $4m^2n(-3mn)(m + n)$ $-12m^4n^2 - 12m^3n^3$
30. $-3a^2b^3(2b)(3a + b)$ $-18a^3b^4 - 6a^2b^5$

In Exercises 31–66, find each product. If possible, find the product by sight.

31. $(x + 2)(x + 2)$ $x^2 + 4x + 4$
32. $(x - 3)(x - 3)$ $x^2 - 6x + 9$
33. $(a - 4)(a - 4)$ $a^2 - 8a + 16$
34. $(y + 5)(y + 5)$ $y^2 + 10y + 25$
35. $(a + b)(a + b)$ $a^2 + 2ab + b^2$
36. $(a - 2b)(a - 2b)$ $a^2 - 4ab + 4b^2$
37. $(2x - y)(2x - y)$ $4x^2 - 4xy + y^2$
38. $(3m + 4n)(3m + 4n)$ $9m^2 + 24mn + 16n^2$
39. $(x + 2)(x - 2)$ $x^2 - 4$
40. $(z + 3)(z - 3)$ $z^2 - 9$
41. $(a + b)(a - b)$ $a^2 - b^2$
42. $(2x + 3y)(2x - 3y)$ $4x^2 - 9y^2$
43. $(x + 2)(x + 3)$ $x^2 + 5x + 6$
44. $(y - 3)(y + 4)$ $y^2 + y - 12$
45. $(z - 7)(z - 2)$ $z^2 - 9z + 14$
46. $(x + 3)(x - 5)$ $x^2 - 2x - 15$
47. $(2a + 1)(a - 2)$ $2a^2 - 3a - 2$
48. $(3b - 1)(2b - 1)$ $6b^2 - 5b + 1$
49. $(3y - z)(2y - z)$ $6y^2 - 5yz + z^2$
50. $(2m + n)(3m + n)$ $6m^2 + 5mn + n^2$
51. $(2x - 3y)(x + 2y)$ $2x^2 + xy - 6y^2$
52. $(3y + 2z)(y - 3z)$ $3y^2 - 7yz - 6z^2$
53. $(3 - 2x)(3 + 2x)$ $9 - 4x^2$
54. $(2x - 5)(2x + 5)$ $4x^2 - 25$
55. $(3x + y)(3x - 3y)$ $9x^2 - 6xy - 3y^2$
56. $(2x - y)(3x + 2y)$ $6x^2 + xy - 2y^2$
57. $(4a - 3b)(2a + 5b)$ $8a^2 + 14ab - 15b^2$
58. $(3a + 2b)(2a - 7b)$ $6a^2 - 17ab - 14b^2$
59. $(u - v)^2$ $u^2 - 2uv + v^2$
60. $(u + v)^2$ $u^2 + 2uv + v^2$
61. $(2x + 1)^2$ $4x^2 + 4x + 1$
62. $(3y - 2)^2$ $9y^2 - 12y + 4$

63. $(3x + 2y)^2$ $9x^2 + 12xy + 4y^2$

64. $(3x - 2y)^2$ $9x^2 - 12xy + 4y^2$

65. $(2a - 3b)^2$ $4a^2 - 12ab + 9b^2$

66. $(4m + 3n)^2$ $16m^2 + 24mn + 9n^2$

In Exercises 67–84, find each product.

67. $(3y + 1)(2y^2 + 3y + 2)$ $6y^3 + 11y^2 + 9y + 2$

68. $(a + 2)(3a^2 + 4a - 2)$ $3a^3 + 10a^2 + 6a - 4$

69. $(2a - b)(3a^2 - 2ab + 2b^2)$ $6a^3 - 7a^2b + 6ab^2 - 2b^3$

70. $(4x - 3y)(x^2 - 2xy + y^2)$ $4x^3 - 11x^2y + 10xy^2 - 3y^3$

71. $(a + b + c)(2a - b - 2c)$ $2a^2 + ab - b^2 - 3bc - 2c^2$

72. $(x - 2y - 3z)(x + 2y + z)$ $x^2 - 2xz - 4y^2 - 8yz - 3z^2$

73. $(x + 2y + 3z)^2$ $x^2 + 4xy + 6xz + 4y^2 + 12yz + 9z^2$

74. $(3x - 2y - z)^2$ $9x^2 - 12xy - 6xz + 4yz + 4y^2 + z^2$

75. $(r + s)^2(r - s)^2$ $r^4 - 2r^2s^2 + s^4$

76. $r(r + s)(r - s)^2$ $r^4 - r^3s - r^2s^2 + rs^3$

77. $(2x - 1)[2x^2 - 3(x + 2)]$ $4x^3 - 8x^2 - 9x + 6$

78. $(x + 1)^2[x^2 - 2(x + 2)]$ $x^4 - 7x^2 - 10x - 4$

79. $[2x - 3(x^2 - x)]^3$ $-27x^6 + 135x^5 - 225x^4 + 125x^3$

80. $-[y - 2(y + y^2)]^3$ $8y^6 + 12y^5 + 6y^4 + y^3$

81. $(a + b)(a - b)(a - 3b)$ $a^3 - 3a^2b - ab^2 + 3b^3$

82. $(x - y)(x + 2y)(x - 2y)$ $x^3 - x^2y - 4xy^2 + 4y^3$

83. $[x + (2y - 1)][x - (2y - 1)]$ $x^2 - 4y^2 + 4y - 1$

84. $[x + (2a - b)]^2$ $x^2 + 4ax - 2bx + 4a^2 - 4ab + b^2$

In Exercises 85–94, find each product. Write all answers without negative exponents.

85. $x^3(2x^2 + x^{-2})$ $2x^5 + x$

86. $x^{-4}(2x^{-3} - 5x^2)$ $\frac{2}{x^7} - \frac{5}{x^2}$

87. $x^3y^{-6}z^{-2}(3x^{-2}y^2z - x^3y^{-4})$ $\frac{3x}{y^4z} - \frac{x^6}{y^{10}z^2}$

88. $ab^{-2}c^{-3}(a^{-4}bc^3 + a^{-3}b^4c^3)$ $\frac{1}{a^3b} + \frac{b^2}{a^2}$

89. $(x^{-1} + y)(x^{-1} - y)$ $\frac{1}{x^2} - y^2$

90. $(x^{-1} - y)(x^{-1} - y)$ $\frac{1}{x^2} - \frac{2y}{x} + y^2$

91. $(2x^{-3} + y^3)(2x^3 - y^{-3})$ $2x^3y^3 - \frac{2}{x^3y^3} + 3$

92. $(5x^{-4} - 4y^2)(5x^2 - 4y^{-4})$ $\frac{25}{x^2} - \frac{20}{x^4y^4} - 20x^2y^2 + \frac{16}{y^2}$

93. $(2x^2 - 3y^{-2})(2x^2 + 3y^2)^2$
$8x^6 + 24x^4y^2 + 18x^2y^4 - \frac{12x^4}{y^2} - 36x^2 - 27y^2$

94. $(-3x^{-4} + 2y^{-1})(x + y^{-2})^2$
$-\frac{3}{x^2} - \frac{6}{x^3y^2} - \frac{3}{x^4y^4} + \frac{2x^2}{y} + \frac{4x}{y^3} + \frac{2}{y^5}$

In Exercises 95–106, find each product. Consider n to be a whole number.

95. $x^n(x^{2n} - x^n)$ $x^{3n} - x^{2n}$

96. $a^{2n}(a^n + a^{2n})$ $a^{3n} + a^{4n}$

97. $(x^n + 1)(x^n - 1)$ $x^{2n} - 1$

98. $(x^n - a^n)(x^n + a^n)$ $x^{2n} - a^{2n}$

99. $(x^n - y^n)(x^n - y^{-n})$ $x^{2n} - \frac{x^n}{y^n} - x^ny^n + 1$

100. $(x^n + y^n)(x^n + y^{-n})$ $x^{2n} + \frac{x^n}{y^n} + x^ny^n + 1$

101. $(x^{2n} + y^{2n})(x^{2n} - y^{2n})$ $x^{4n} - y^{4n}$

102. $(a^{3n} - b^{3n})(a^{3n} + b^{3n})$ $a^{6n} - b^{6n}$

103. $(2x^n - y^{2n})(3x^{-n} + y^{-2n})$ $-\frac{3y^{2n}}{x^n} + \frac{2x^n}{y^{2n}} + 5$

104. $(3x^{2n} + 2x^n - 1)^2$ $9x^{4n} + 12x^{3n} - 2x^{2n} - 4x^n + 1$

105. $(x^n + y^n - 1)(x^n - y^n + 1)$ $x^{2n} - y^{2n} + 2y^n - 1$

106. $(1 - x^n)(x^{-n} - 1)$ $x^n + \frac{1}{x^n} - 2$

In Exercises 107–120, simplify each expression.

107. $3x(2x + 4) - 3x^2$ $3x^2 + 12x$

108. $2y - 3y(y^2 + 4)$ $-3y^3 - 10y$

109. $3pq - p(p - q)$ $-p^2 + 4pq$

110. $-4rs(r - 2) + 4rs$ $-4r^2s + 12rs$

111. $2m(m - n) - (m + n)(m - 2n)$ $m^2 - mn + 2n^2$

112. $-3y(2y + z) + (2y - z)(3y + 2z)$ $-2yz - 2z^2$

113. $(x + 3)(x - 3) + (2x - 1)(x + 2)$ $3x^2 + 3x - 11$

114. $(2b + 3)(b - 1) - (b + 2)(3b - 1)$ $-b^2 - 4b - 1$

115. $(3x - 4)^2 - (2x + 3)^2$ $5x^2 - 36x + 7$

116. $(3y + 1)^2 + (2y - 4)^2$ $13y^2 - 10y + 17$

117. $3(x - 3y)^2 + 2(3x + y)^2$ $21x^2 - 6xy + 29y^2$

118. $2(x - y^2)^2 - 3(y^2 + 2x)^2$ $-10x^2 - 16xy^2 - y^4$

119. $5(2y - z)^2 + 4(y + 2z)^2$ $24y^2 - 4yz + 21z^2$

120. $3(x + 2z)^2 - 2(2x - z)^2$ $-5x^2 + 20xz + 10z^2$

121. Verify that $(x + y)^2 = x^2 + 2xy + y^2$.

122. Verify that $(x - y)^2 = x^2 - 2xy + y^2$.

123. Verify that $(x + y)(x - y) = x^2 - y^2$.

124. Verify that $(x + y + z)^2 = x^2 + y^2 + z^2 + 2xy + 2xz + 2yz$.

In Exercises 125–128, use a calculator to find each product.

125. $(3.21x - 7.85)(2.87x + 4.59)$
$9.2127x^2 - 7.7956x - 36.0315$

126. $(7.44y + 56.7)(-2.1y - 67.3)$
$-15.624y^2 - 619.782y - 3815.91$

127. $(-17.3y + 4.35)^2$
$299.29y^2 - 150.51y + 18.9225$

128. $(-0.31x + 29.3)(-81x - 0.2)$
$25.11x^2 - 2373.238x - 5.86$

Writing Exercises

Write a paragraph using your own words.

1. Explain how to use the FOIL method.

2. Explain how to multiply two trinomials.

Something to Think About

1. The numbers 0.35×10^7 and 1.96×10^7 both involve the same power of 10. Find their sum.
2.31×10^7

2. Without converting to standard notation, find the sum: $1.435 \times 10^8 + 2.11 \times 10^7$. (*Hint:* The first number in the previous exercise is not in scientific notation.)
1.646×10^8

Review Exercises

Let $a = -2$ and $b = 4$ and find the value of each expression.

1. $|3a - b|$ 10 **2.** $|ab - b^2|$ 24 **3.** $-|a^2b - b^0|$ -15 **4.** $\left|\dfrac{a^3b^2 + ab}{2(ab)^2 - a^3}\right|$ 1

5. A woman owns 200 shares of ABC Company, valued at \$125 per share, and 350 shares of WD Company, valued at \$75 per share. One day ABC rose $1\frac{1}{2}$ points and WD fell $1\frac{1}{2}$ points. Find the current value of her portfolio. \$51,025

6. One light year is approximately 5,870,000,000,000 miles. Write this number in scientific notation.
5.87×10^{12}

2.5 Dividing Polynomials

■ Dividing a Monomial by a Monomial ■ Dividing a Polynomial by a Monomial ■ Dividing a Polynomial by a Polynomial ■ The Case of the Missing Terms

■ Dividing a Monomial by a Monomial

EXAMPLE 1 Simplify the expression $(3a^2b^3) \div (2a^3b)$.

Solution 1 We write the expression as a fraction and divide out all common factors:

$$\begin{aligned}\frac{3a^2b^3}{2a^3b} &= \frac{3aabbb}{2aaab}\\ &= \frac{3\cancel{a}\cancel{a}\cancel{b}bb}{2\cancel{a}\cancel{a}a\cancel{b}}\\ &= \frac{3b^2}{2a}\end{aligned}$$

Solution 2 We write the expression as a fraction and use the rules of exponents:

$$\begin{aligned}\frac{3a^2b^3}{2a^3b} &= \frac{3}{2}a^{2-3}b^{3-1}\\ &= \frac{3}{2}a^{-1}b^2\\ &= \frac{3}{2}\left(\frac{1}{a}\right)\frac{b^2}{1}\\ &= \frac{3b^2}{2a}\end{aligned}$$

■

■ Dividing a Polynomial by a Monomial

EXAMPLE 2 Divide $4x^3y^2 + 3xy^5 - 12xy$ by $3x^2y^3$.

Solution We write the expression as a fraction and then as the sum of three separate fractions:

$$\frac{4x^3y^2 + 3xy^5 - 12xy}{3x^2y^3} = \frac{4x^3y^2}{3x^2y^3} + \frac{3xy^5}{3x^2y^3} + \frac{-12xy}{3x^2y^3}$$

We then pick one of the methods used in Example 1 and simplify each of the three fractions on the right-hand side of the equal sign to get

$$\frac{4x^3y^2 + 3xy^5 - 12xy}{3x^2y^3} = \frac{4x}{3y} + \frac{y^2}{x} + \frac{-4}{xy^2}$$
$$= \frac{4x}{3y} + \frac{y^2}{x} - \frac{4}{xy^2}$$
■

Dividing a Polynomial by a Polynomial

EXAMPLE 3 Divide $x^2 + 7x + 12$ by $x + 4$.

Solution There is an **algorithm** (a repeating series of steps) to use when the divisor is not a monomial. To use the division algorithm, we write the division in long division form and proceed as follows:

$$x + 4\overline{)x^2 + 7x + 12}$$

Step 1:

$$\begin{array}{r} x \phantom{{}+7x+12} \\ x + 4\overline{)x^2 + 7x + 12} \end{array}$$

How many times does x divide x^2? $\frac{x^2}{x} = x$.

Place the x in the quotient.

Step 2:

$$\begin{array}{r} x \phantom{{}+7x+12} \\ x + 4\overline{)x^2 + 7x + 12} \\ \underline{\mathbf{x^2 + 4x}} \phantom{{}+12} \end{array}$$

Multiply each term in the divisor by x. Place the product under $x^2 + 7x$ and draw a line.

Step 3:

$$\begin{array}{r} x \phantom{{}+7x+12} \\ x + 4\overline{)x^2 + 7x + 12} \\ \underline{x^2 + 4x} \phantom{{}+12} \\ \mathbf{3x + 12} \end{array}$$

Subtract $x^2 + 4x$ from $x^2 + 7x$ and bring down the next term.

Step 4:

$$\begin{array}{r} x + 3 \phantom{{}+12} \\ x + 4\overline{)x^2 + 7x + 12} \\ \underline{x^2 + 4x} \phantom{{}+12} \\ 3x + 12 \end{array}$$

How many times does x divide $3x$? $\frac{3x}{x} = +3$. Place the $+3$ in the quotient.

Step 5:

$$\begin{array}{r} x + 3 \phantom{{}+12} \\ x + 4\overline{)x^2 + 7x + 12} \\ \underline{x^2 + 4x} \phantom{{}+12} \\ 3x + 12 \\ \underline{\mathbf{3x + 12}} \end{array}$$

Multiply each term in the divisor by 3. Place the product under $3x + 12$ and draw a line.

Step 6:

$$\begin{array}{r} x + 3 \\ x + 4 \overline{)x^2 + 7x + 12} \\ \underline{x^2 + 4x} \\ 3x + 12 \\ \underline{3x + 12} \\ \mathbf{0} \end{array}$$

Subtract $3x + 12$ from $3x + 12$ to get a remainder of 0.

The division process stops when the result of the subtraction is either a constant or a polynomial with a degree that is less than the degree of the divisor. Thus, the quotient is $x + 3$, and the remainder is 0.

We can check the quotient by multiplying the divisor by the quotient. The product should be the dividend. Because

$$\underbrace{(x + 4)}_{\text{divisor}} \cdot \underbrace{(x + 3)}_{\text{quotient}} = \underbrace{x^2 + 7x + 12}_{\text{dividend}}$$

the answer checks. ■

EXAMPLE 4 Divide $2a^3 + 9a^2 + 5a - 6$ by $2a + 3$.

Solution *Step 1:*

$$\begin{array}{r} a^2 \\ 2a + 3 \overline{)2a^3 + 9a^2 + 5a - 6} \end{array}$$

How many times does $2a$ divide $2a^3$? $\frac{2a^3}{2a} = a^2$. Place a^2 in the quotient.

Step 2:

$$\begin{array}{r} a^2 \\ 2a + 3 \overline{)2a^3 + 9a^2 + 5a - 6} \\ \underline{2a^3 + 3a^2} \end{array}$$

Multiply each term in the divisor by a^2. Place the product under $2a^3 + 9a^2$ and draw a line.

Step 3:

$$\begin{array}{r} a^2 \\ 2a + 3 \overline{)2a^3 + 9a^2 + 5a - 6} \\ \underline{2a^3 + 3a^2} \\ \mathbf{6a^2 + 5a} \end{array}$$

Subtract $2a^3 + 3a^2$ from $2a^3 + 9a^2$ and bring down the next term.

Step 4:

$$\begin{array}{r} a^2 + 3a \\ 2a + 3 \overline{)2a^3 + 9a^2 + 5a - 6} \\ \underline{2a^3 + 3a^2} \\ 6a^2 + 5a \end{array}$$

How many times does $2a$ divide $6a^2$? $\frac{6a^2}{2a} = 3a$. Place the $+ 3a$ in the quotient.

Step 5:

$$\begin{array}{r}
a^2 + 3a \phantom{{}+5a-6} \\
2a + 3\overline{)2a^3 + 9a^2 + 5a - 6} \\
\underline{2a^3 + 3a^2} \phantom{{}+5a-6} \\
6a^2 + 5a \phantom{{}-6} \\
\underline{6a^2 + 9a} \phantom{{}-6}
\end{array}$$

Multiply each term in the divisor by $3a$. Place the product under $6a^2 + 5a$ and draw a line.

Step 6:

$$\begin{array}{r}
a^2 + 3a \phantom{{}+5a-6} \\
2a + 3\overline{)2a^3 + 9a^2 + 5a - 6} \\
\underline{2a^3 + 3a^2} \phantom{{}+5a-6} \\
6a^2 + 5a \phantom{{}-6} \\
\underline{6a^2 + 9a} \phantom{{}-6} \\
-4a - 6
\end{array}$$

Subtract $6a^2 + 9a$ from $6a^2 + 5a$ and bring down the next term.

Step 7:

$$\begin{array}{r}
a^2 + 3a \;\; - 2 \phantom{{}-6} \\
2a + 3\overline{)2a^3 + 9a^2 + 5a - 6} \\
\underline{2a^3 + 3a^2} \phantom{{}+5a-6} \\
6a^2 + 5a \phantom{{}-6} \\
\underline{6a^2 + 9a} \phantom{{}-6} \\
-4a - 6
\end{array}$$

How many times does $2a$ divide $-4a$? $-\dfrac{4a}{2a} = -2$. Place the -2 in the quotient.

Step 8:

$$\begin{array}{r}
a^2 + 3a \;\; - 2 \phantom{{}-6} \\
2a + 3\overline{)2a^3 + 9a^2 + 5a - 6} \\
\underline{2a^3 + 3a^2} \phantom{{}+5a-6} \\
6a^2 + 5a \phantom{{}-6} \\
\underline{6a^2 + 9a} \phantom{{}-6} \\
-4a - 6 \\
\underline{-4a - 6}
\end{array}$$

Multiply each term in the divisor by -2. Place the product under $-4a - 6$ and draw a line.

Step 9:

$$\begin{array}{r}
a^2 + 3a \;\; - 2 \phantom{{}-6} \\
2a + 3\overline{)2a^3 + 9a^2 + 5a - 6} \\
\underline{2a^3 + 3a^2} \phantom{{}+5a-6} \\
6a^2 + 5a \phantom{{}-6} \\
\underline{6a^2 + 9a} \phantom{{}-6} \\
-4a - 6 \\
\underline{-4a - 6} \\
0
\end{array}$$

Subtract $-4a - 6$ from $-4a - 6$ to get a remainder of 0.

Since the remainder is 0, the quotient is $a^2 + 3a - 2$. We can check the quotient by verifying that

$$\underbrace{(2a + 3)}_{\text{divisor}} \cdot \underbrace{(a^2 + 3a - 2)}_{\text{quotient}} = \underbrace{2a^3 + 9a^2 + 5a - 6}_{\text{dividend}}$$

■

EXAMPLE 5 Divide $3x^3 + 2x^2 - 3x + 8$ by $x - 2$.

Solution

$$\begin{array}{r} 3x^2 + 8x + 13 \\ x - 2\overline{)3x^3 + 2x^2 - 3x + 8} \\ \underline{3x^3 - 6x^2} \qquad\qquad \\ 8x^2 - 3x \qquad \\ \underline{8x^2 - 16x} \qquad \\ 13x + 8 \\ \underline{13x - 26} \\ 34 \end{array}$$

This division gives a quotient of $3x^2 + 8x + 13$ and a remainder of 34. It is common to form a fraction with the remainder as numerator and the divisor as denominator and to write the quotient as

$$3x^2 + 8x + 13 + \frac{34}{x - 2}$$

To check, we verify that

$$(x - 2)\left(3x^2 + 8x + 13 + \frac{34}{x - 2}\right) = 3x^3 + 2x^2 - 3x + 8$$

■

EXAMPLE 6 Divide $-9x + 8x^3 + 10x^2 - 9$ by $3 + 2x$.

Solution The division algorithm works best when the polynomials in the dividend and the divisor are written in descending powers of x. We can use the commutative property of addition to rearrange the terms. Then the division is routine.

$$\begin{array}{r} 4x^2 - x - 3 \\ 2x + 3\overline{)8x^3 + 10x^2 - 9x - 9} \\ \underline{8x^3 + 12x^2} \qquad\qquad \\ -2x^2 - 9x \qquad \\ \underline{-2x^2 - 3x} \qquad \\ -6x - 9 \\ \underline{-6x - 9} \\ 0 \end{array}$$

Thus,

$$\frac{-9x + 8x^3 + 10x^2 - 9}{3 + 2x} = 4x^2 - x - 3$$

■

The Case of the Missing Terms

EXAMPLE 7 Divide $8x^3 + 1$ by $2x + 1$.

Solution When we write the terms in the dividend in descending powers of x, we see that the terms involving x^2 and x are missing. To make the division convenient, we must either include the terms $0x^2$ and $0x$ in the dividend or leave spaces for them. After this adjustment, the division is routine.

$$\begin{array}{r}
4x^2 - 2x + 1 \\
2x+1 \overline{)\, 8x^3 + 0x^2 + 0x + 1} \\
\underline{8x^3 + 4x^2 \phantom{{}+0x+1}} \\
-4x^2 + 0x \phantom{{}+1} \\
\underline{-4x^2 - 2x \phantom{{}+1}} \\
+2x + 1 \\
\underline{+2x + 1} \\
0
\end{array}$$

Hence,

$$\frac{8x^3 + 1}{2x + 1} = 4x^2 - 2x + 1$$

Check this answer. ■

EXAMPLE 8 Divide $-17x^2 + 5x + x^4 + 2$ by $x^2 - 1 + 4x$.

Solution Rewrite the problem with both the divisor and the dividend in descending powers of x. Leave space for the missing term in the dividend. Then perform the division as follows:

$$\begin{array}{r}
x^2 - 4x \phantom{{}-17x^2+5x+2} \\
x^2 + 4x - 1 \overline{)\, x^4 \phantom{{}+4x^3} - 17x^2 + 5x + 2} \\
\underline{x^4 + 4x^3 - x^2 \phantom{{}+5x+2}} \\
-4x^3 - 16x^2 + 5x \phantom{{}+2} \\
\underline{-4x^3 - 16x^2 + 4x \phantom{{}+2}} \\
x + 2
\end{array}$$

This division gives a quotient of $x^2 - 4x$ and a remainder of $x + 2$. Hence,

$$\frac{-17x^2 + 5x + x^4 + 2}{x^2 - 1 + 4x} = x^2 - 4x + \frac{x + 2}{x^2 + 4x - 1}$$ ■

EXERCISE 2.5

In Exercises 1–18, do each division. Express all answers without using negative exponents.

1. $(4x^2y^3) \div (8x^5y^2)$ $\quad \frac{y}{2x^3}$

2. $(25x^4y^7) \div (5xy^9)$ $\quad \frac{5x^3}{y^2}$

3. $\frac{33a^{-2}b^2}{44a^2b^{-2}}$ $\quad \frac{3b^4}{4a^4}$

4. $\frac{-63a^4b^{-3}}{81a^{-3}b^3}$ $\quad \frac{-7a^7}{9b^6}$

5. $\frac{45x^{-2}y^{-3}t^0}{-63x^{-1}y^4t^2}$ $\quad \frac{-5}{7xy^7t^2}$

6. $\frac{112a^0b^2c^{-3}}{48a^4b^0c^4}$ $\quad \frac{7b^2}{3a^4c^7}$

7. $\frac{-65a^{2n}b^nc^{3n}}{-15a^nb^{-n}c}$ $\quad \frac{13a^nb^{2n}c^{3n-1}}{3}$

8. $\frac{-32x^{-3n}y^{-2n}z}{40x^{-2}y^{-n}z^{n+1}}$ $\quad \frac{-4}{5x^{3n-2}y^nz^n}$

9. $(4x^2 - x^3) \div (6x)$ $\quad \frac{2x}{3} - \frac{x^2}{6}$

10. $(5y^4 + 45y^3) \div (15y^2)$ $\quad \frac{y^2}{3} + 3y$

11. $\frac{4x^2y^3 + x^3y^2}{6xy}$ $\quad \frac{2xy^2}{3} + \frac{x^2y}{6}$

12. $\frac{3a^3y^2 - 18a^4y^3}{27a^2y^2}$ $\quad \frac{a}{9} - \frac{2a^2y}{3}$

13. $\frac{24x^6y^7 - 12x^5y^{12} + 36xy}{48x^2y^3}$ $\quad \frac{x^4y^4}{2} - \frac{x^3y^9}{4} + \frac{3}{4xy^2}$

14. $\frac{9x^4y^3 + 18x^2y - 27xy^4}{9x^3y^3}$ $\quad x + \frac{2}{xy^2} - \frac{3y}{x^2}$

15. $\frac{3a^{-2}b^3 - 6a^2b^{-3} + 9a^{-2}}{12a^{-1}b}$ $\quad \frac{b^2}{4a} - \frac{a^3}{2b^4} + \frac{3}{4ab}$

16. $\frac{4x^3y^{-2} + 8x^{-2}y^2 - 12y^4}{12x^{-1}y^{-1}}$ $\quad \frac{x^4}{3y} + \frac{2y^3}{3x} - xy^5$

17. $\frac{x^ny^n - 3x^{2n}y^{2n} + 6x^{3n}y^{3n}}{x^ny^n}$ $\quad 1 - 3x^ny^n + 6x^{2n}y^{2n}$

18. $\frac{2a^n - 3a^nb^{2n} - 6b^{4n}}{a^nb^{n-1}}$ $\quad \frac{2}{b^{n-1}} - 3b^{n+1} - \frac{6b^{3n+1}}{a^n}$

In Exercises 19–54, use the division algorithm to find each quotient.

19. $\frac{x^2 + 5x + 6}{x + 3}$ $\quad x + 2$

20. $\frac{x^2 - 5x + 6}{x - 3}$ $\quad x - 2$

21. $(x^2 + 10x + 21) \div (x + 3)$ $\quad x + 7$

22. $(x^2 + 10x + 21) \div (x + 7)$ $\quad x + 3$

23. $\frac{6x^2 - x - 12}{2x + 3}$ $\quad 3x - 5 + \frac{3}{2x + 3}$

24. $\frac{6x^2 - x - 12}{2x - 3}$ $\quad 3x + 4$

25. $\frac{3x^3 - 2x^2 + x + 6}{x - 1}$ $\quad 3x^2 + x + 2 + \frac{8}{x - 1}$

26. $\frac{4a^3 + a^2 - 3a + 7}{a + 1}$ $\quad 4a^2 - 3a + \frac{7}{a + 1}$

27. $\frac{6x^3 + 11x^2 - x - 2}{3x - 2}$ $\quad 2x^2 + 5x + 3 + \frac{4}{3x - 2}$

28. $\frac{6x^3 + 11x^2 - x + 10}{2x + 3}$ $\quad 3x^2 + x - 2 + \frac{16}{2x + 3}$

29. $\frac{6x^3 - x^2 - 6x - 9}{2x - 3}$ $\quad 3x^2 + 4x + 3$

30. $\frac{16x^3 + 16x^2 - 9x - 5}{4x + 5}$ $\quad 4x^2 - x - 1$

31. $(2a + 1 + a^2) \div (a + 1)$ $\quad a + 1$

32. $(a - 15 + 6a^2) \div (2a - 3)$ $\quad 3a + 5$

33. $(6y - 4 + 10y^2) \div (5y - 2)$ $\quad 2y + 2$

34. $(-10xy + x^2 + 16y^2) \div (x - 2y)$ $\quad x - 8y$

35. $\frac{-18x + 12 + 6x^2}{x - 1}$ $\quad 6x - 12$

36. $\frac{27x + 23x^2 + 6x^3}{2x + 3}$ $\quad 3x^2 + 7x + 3 - \frac{9}{2x + 3}$

37. $\frac{-9x^2 + 8x + 9x^3 - 4}{3x - 2}$ $\quad 3x^2 - x + 2$

38. $\frac{6x^2 + 8x^3 - 13x + 3}{4x - 3}$ $\quad 2x^2 + 3x - 1$

39. $\dfrac{13x + 16x^4 + 3x^2 + 3}{4x + 3}$ $4x^3 - 3x^2 + 3x + 1$

40. $\dfrac{3x^2 + 9x^3 + 4x + 4}{3x + 2}$ $3x^2 - x + 2$

41. $(a^3 + 1) \div (a - 1)$ $a^2 + a + 1 + \frac{2}{a-1}$

42. $(27a^3 - 8b^3) \div (3a - 2b)$ $9a^2 + 6ab + 4b^2$

43. $\dfrac{15a^3 - 29a^2 + 16}{3a - 4}$ $5a^2 - 3a - 4$

44. $\dfrac{4x^3 - 12x^2 + 17x - 12}{2x - 3}$ $2x^2 - 3x + 4$

45. $y - 2\overline{)-24y + 24 + 6y^2}$ $6y - 12$

46. $3 - a\overline{)21a - a^2 - 54}$ $-18 + a$ or $a - 18$

47. $2x + y\overline{)32x^5 + y^5}$ $16x^4 - 8x^3y + 4x^2y^2 - 2xy^3 + y^4$

48. $3x - y\overline{)81x^4 - y^4}$ $27x^3 + 9x^2y + 3xy^2 + y^3$

49. $x^2 - 2\overline{)x^6 - x^4 + 2x^2 - 8}$ $x^4 + x^2 + 4$

50. $x^2 + 3\overline{)x^6 + 2x^4 - 6x^2 - 9}$ $x^4 - x^2 - 3$

51. $(x^4 + 2x^3 + 4x^2 + 3x + 2) \div (x^2 + x + 2)$
$x^2 + x + 1$

52. $(2x^4 + 3x^3 + 3x^2 - 5x - 3) \div (2x^2 - x - 1)$
$x^2 + 2x + 3$

53. $x + x^2 + 2\overline{)x^3 + 3x + 5x^2 + 6 + x^4}$ $x^2 + 3$

54. $x^3 + 1 + 2x\overline{)x^5 + 3x + 2}$ $x^2 - 2 + \dfrac{-x^2 + 7x + 4}{x^3 + 2x + 1}$

In Exercises 55–56, use a calculator to find each quotient.

55. $x - 2\overline{)9.8x^2 - 3.2x - 69.3}$
$9.8x + 16.4 + \frac{-36.5}{x-2}$

56. $2.5x - 3.7\overline{)-22.25x^2 - 38.9x - 16.65}$
$-8.9x - 28.732 + \frac{-122.9584}{2.5x - 3.7}$

Writing Exercises *Write a paragraph using your own words.*

1. Explain how to divide a monomial by a monomial.

2. Explain how to check the result of a division problem.

Something to Think About

1. Since 6 is a factor of 24, 6 divides 24 with no remainder. Decide whether $2x - 3$ is a factor of $10x^2 - x - 21$. it is

2. Is $x - 1$ a factor of $x^5 - 1$? yes

Review Exercises *Remove parentheses and simplify.*

1. $2(x^2 + 4x - 1) + 3(2x^2 - 2x + 2)$
$8x^2 + 2x + 4$

2. $3(2a^2 - 3a + 2) - 4(2a^2 + 4a - 7)$
$-2a^2 - 25a + 34$

3. $-2(3y^3 - 2y + 7) - 3(y^2 + 2y - 4) + 4(y^3 + 2y - 1)$ $-2y^3 - 3y^2 + 6y - 6$

4. $3(4y^3 + 3y - 2) + 2(3y^2 - y + 3) - 5(2y^3 - y^2 - 2)$ $2y^3 + 11y^2 + 7y + 10$

2.6 Synthetic Division (Optional)

■ The Remainder Theorem ■ The Factor Theorem

There is a shortcut method, called **synthetic division,** that we can use to divide a polynomial by a binomial of the form $x - r$. To see how this method works, we consider the division of $4x^3 - 5x^2 - 11x + 20$ by $x - 2$.

$$\begin{array}{rrrrrl}
 & 4x^2 & +\,3x & -\,5 & & \\ \cline{2-5}
x-2\,\big) & 4x^3 & -\,5x^2 & -\,11x & +\,20 & \\
 & 4x^3 & -\,8x^2 & & & \\ \cline{2-3}
 & & 3x^2 & -\,11x & & \\
 & & 3x^2 & -\,6x & & \\ \cline{3-4}
 & & & -\,5x & +\,20 & \\
 & & & -\,5x & +\,10 & \\ \cline{4-5}
 & & & & 10 & \text{(remainder)}
\end{array}
\qquad
\begin{array}{rrrrrl}
 & 4 & 3 & -\,5 & & \\ \cline{2-5}
1-2\,\big) & 4 & -\,5 & -\,11 & 20 & \\
 & 4 & -\,8 & & & \\ \cline{2-3}
 & & 3 & -\,11 & & \\
 & & 3 & -\,6 & & \\ \cline{3-4}
 & & & -5 & 20 & \\
 & & & -5 & 10 & \\ \cline{4-5}
 & & & & 10 & \text{(remainder)}
\end{array}$$

On the left is the familiar long-division process, and on the right is the skeleton form of that division. All references to the variable x have been removed. The various powers of x can be remembered without actually writing them, because the exponents of the terms in the divisor, dividend, and quotient were written in descending order.

We can further shorten the version on the right. The numbers printed in color need not be written, because they are duplicates of the numbers immediately above them. Thus, we can write the division in the following form:

$$\begin{array}{rrrrr}
 & 4 & 3 & -\,5 & \\ \cline{2-5}
1-2\,\big) & 4 & -\,5 & -\,11 & 20 \\
 & & -\,8 & & \\ \cline{2-3}
 & & 3 & & \\
 & & & -\,6 & \\ \cline{3-4}
 & & & -\,5 & \\
 & & & & 10 \\ \cline{4-5}
 & & & & 10
\end{array}$$

We can shorten the process still further by compressing the work vertically and eliminating the 1 (the coefficient of x in the divisor):

$$\begin{array}{rrrrr}
 & 4 & 3 & -\,5 & \\ \cline{2-5}
-2\,\big) & 4 & -\,5 & -\,11 & 20 \\
 & & -\,8 & -\,6 & 10 \\ \cline{2-5}
 & & 3 & -\,5 & 10
\end{array}$$

There is no reason why the quotient, represented by the numbers 4 3 -5, must appear *above* the long division. If we write the 4 on the bottom line, the bottom line gives the coefficients of the quotient, and it also gives the remainder. The entire top line can be eliminated. The division now appears as follows:

$$\begin{array}{r|rrrr}
-2 & 4 & -5 & -\,11 & 20 \\
 & & -8 & -\,6 & 10 \\ \hline
 & 4 & 3 & -\,5 & 10
\end{array}$$

The bottom line was obtained by subtracting the middle line from the top line. If we were to replace the -2 in the divisor by $+2$, the division process would reverse the signs of every entry in the middle line. Then the bottom line could be obtained by addition. Thus, we have this final form of the synthetic division.

$$\begin{array}{r|rrrr|l} +2 & 4 & -5 & -11 & 20 & \text{The coefficients of the dividend.} \\ & & 8 & 6 & -10 & \\ \hline & 4 & 3 & -5 & 10 & \text{The coefficients of the quotient, and the remainder to the right of the vertical bar.} \end{array}$$

Thus,

$$\frac{4x^3 - 5x^2 - 11x + 20}{x - 2} = 4x^2 + 3x - 5 + \frac{10}{x - 2}$$

EXAMPLE 1 Use synthetic division to divide $6x^2 + 5x - 2$ by $x - 5$.

Solution Begin by writing the coefficients of the dividend, and the 5 from the divisor, in the following form:

$$\begin{array}{r|rrr} 5 & 6 & 5 & -2 \\ & & & \\ \hline \end{array}$$

Then we follow these steps:

$$\begin{array}{r|rrr} 5 & 6 & 5 & -2 \\ & \downarrow & & \\ \hline & 6 & & \end{array}$$ Begin by bringing down the 6.

$$\begin{array}{r|rrr} 5 & 6 & 5 & -2 \\ & & 30 & \\ \hline & 6 & & \end{array}$$ Multiply 5 and 6, to get 30.

$$\begin{array}{r|rrr} 5 & 6 & 5 & -2 \\ & & 30 & \\ \hline & 6 & 35 & \end{array}$$ Add 5 and 30, to get 35.

$$\begin{array}{r|rrr} 5 & 6 & 5 & -2 \\ & & 30 & 175 \\ \hline & 6 & 35 & \end{array}$$ Multiply 5 and 35, to get 175.

$$\begin{array}{r|rrr} 5 & 6 & 5 & -2 \\ & & 30 & 175 \\ \hline & 6 & 35 & | \; 173 \end{array}$$ Add -2 and 175, to get 173.

The numbers 6 and 35 represent the quotient: $6x + 35$. The number 173 is the remainder. Thus,

$$\frac{6x^2 + 5x - 2}{x - 5} = 6x + 35 + \frac{173}{x - 5}$$

■

EXAMPLE 2 Use synthetic division to divide $5x^3 + x^2 - 3$ by $x - 2$.

Solution We begin by writing

$$\begin{array}{r|rrrr} 2 & 5 & 1 & 0 & -3 \\ \hline \end{array}$$

Write 0 for the coefficient of x, the missing term.

and complete the division as follows:

$$\begin{array}{r|rrrr} 2 & 5 & 1 & 0 & -3 \\ & & 10 & & \\ \hline & 5 & 11 & & \end{array} \qquad \begin{array}{r|rrrr} 2 & 5 & 1 & 0 & -3 \\ & & 10 & 22 & \\ \hline & 5 & 11 & 22 & \end{array} \qquad \begin{array}{r|rrr|r} 2 & 5 & 1 & 0 & -3 \\ & & 10 & 22 & 44 \\ \hline & 5 & 11 & 22 & 41 \end{array}$$

Thus,

$$\frac{5x^3 + x^2 - 3}{x - 2} = 5x^2 + 11x + 22 + \frac{41}{x - 2}$$

■

EXAMPLE 3 Use synthetic division to divide $5x^2 + 6x^3 + 2 - 4x$ by $x + 2$.

Solution First, we rewrite the dividend with the exponents in descending order:

$$6x^3 + 5x^2 - 4x + 2.$$

Then we rewrite the divisor in $x - r$ form: $x - (-2)$. Using synthetic division, we begin by writing

$$\begin{array}{r|rrrr} -2 & 6 & 5 & -4 & 2 \\ \hline \end{array}$$

and complete the division:

$$\begin{array}{r|rrr|r} -2 & 6 & 5 & -4 & 2 \\ & & -12 & 14 & -20 \\ \hline & 6 & -7 & 10 & -18 \end{array}$$

Thus,

$$\frac{5x^2 + 6x^3 + 2 - 4x}{x + 2} = 6x^2 - 7x + 10 + \frac{-18}{x + 2}$$

■

The Remainder Theorem

Synthetic division is important in mathematics because of the **remainder theorem.**

Remainder Theorem If a polynomial $P(x)$ is divided by $x - r$, then the remainder is $P(r)$.

We will illustrate the remainder theorem in Example 4.

EXAMPLE 4 If $P(x) = 2x^3 - 3x^2 - 2x + 1$, find **a.** $P(3)$ and **b.** the remainder when $P(x)$ is divided by $x - 3$.

Solution **a.** $P(3) = 2(3)^3 - 3(3)^2 - 2(3) + 1$ Substitute 3 for x.

$$= 2(27) - 3(9) - 6 + 1$$
$$= 54 - 27 - 6 + 1$$
$$= 22$$

b. Use synthetic division to find the remainder when $P(x) = 2x^3 - 3x^2 - 2x + 1$ is divided by $x - 3$.

$$\begin{array}{r|rrrr} 3 & 2 & -3 & -2 & 1 \\ & & 6 & 9 & 21 \\ \hline & 2 & 3 & 7 & \boxed{22} \end{array}$$

The remainder is 22.

The results of parts **a** and **b** show that when $P(x)$ is divided by $x - 3$, the remainder is $P(3)$. ■

It is often easier to calculate $P(r)$ by using synthetic division than by substituting r for x in $P(x)$. This is especially true if r is a number such as 2.3, 0.06, or 3.698.

The Factor Theorem

Recall that if two quantities are multiplied, each is called a **factor** of the product. Thus, $x - 2$ is one factor of $6x - 12$ because $6(x - 2) = 6x - 12$. A theorem, called the **factor theorem,** tells us how to find one factor of a polynomial if the remainder of a certain division is 0.

The Factor Theorem If $P(x)$ is a polynomial in x, then

$$P(r) = 0 \quad \text{if and only if} \quad x - r \text{ is a factor of } P(x).$$

We illustrate the factor theorem in the next example.

EXAMPLE 5 Let $P(x) = 3x^3 - 5x^2 + 3x - 10$. Show that **a.** $P(2) = 0$ and **b.** $x - 2$ is a factor of $P(x)$.

Solution **a.** Use the remainder theorem to evaluate $P(2)$ by dividing $P(x) = 3x^3 - 5x^2 + 3x - 10$ by $x - 2$. Use synthetic division.

$$\begin{array}{r|rrrr} 2 & 3 & -5 & 3 & -10 \\ & & 6 & 2 & 10 \\ \hline & 3 & 1 & 5 & 0 \end{array}$$

The remainder in this division is 0. By the remainder theorem, the remainder is $P(2)$. Thus, $P(2) = 0$.

b. Because the remainder is 0, the numbers 3, 1, and 5 in the synthetic division in part **a** represent the quotient: $3x^2 + x + 5$. Thus,

$$\underbrace{(x-2)}_{\text{divisor}} \cdot \underbrace{(3x^2 + x + 5)}_{\text{quotient}} + \underbrace{0}_{\text{remainder}} = \underbrace{3x^3 - 5x^2 + 3x - 10}_{\text{the dividend, } P(x)}$$

or

$$(x - 2)(3x^2 + x + 5) = 3x^3 - 5x^2 + 3x - 10$$

Thus, $x - 2$ is a factor of $P(x)$. ■

The result of Example 5 is true because the remainder, $P(2)$, is 0. If the remainder had not been 0, then $x - 2$ would not have been a factor of $P(x)$.

The process of finding the factors of a polynomial is called **factoring.** In Chapter 4, we will study other techniques of factoring.

EXERCISE 2.6

In Exercises 1–14, use synthetic division to do each division.

1. $(x^2 + x - 2) \div (x - 1)$ $x + 2$

2. $(x^2 + x - 6) \div (x - 2)$ $x + 3$

3. $x - 4\overline{)x^2 - 7x + 12}$ $x - 3$

4. $x - 5\overline{)x^2 - 6x + 5}$ $x - 1$

5. $(x^2 + 8 + 6x) \div (x + 4)$ $x + 2$

6. $(x^2 - 15 - 2x) \div (x + 3)$ $x - 5$

7. $x + 2\overline{)x^2 - 5x + 14}$ $x - 7 + \frac{28}{x+2}$

8. $x + 6\overline{)x^2 + 13x + 42}$ $x + 7$

9. $(3x^3 - 10x^2 + 5x - 6) \div (x - 3)$ $3x^2 - x + 2$

10. $(2x^3 - 9x^2 + 10x - 3) \div (x - 3)$ $2x^2 - 3x + 1$

11. $(2x^3 - 5x - 6) \div (x - 2)$ $2x^2 + 4x + 3$

12. $(4x^3 + 5x^2 - 1) \div (x + 2)$ $4x^2 - 3x + 6 + \frac{-13}{x+2}$

13. $x + 1\overline{)5x^2 + 6x^3 + 4}$ $6x^2 - x + 1 + \frac{3}{x+1}$

14. $x - 4\overline{)4 - 3x^2 + x}$ $-3x - 11 + \frac{-40}{x-4}$

In Exercises 15–20, use a calculator and synthetic division to do each division.

15. $x - 0.2\overline{)7.2x^2 - 2.1x + 0.5}$ $7.2x - 0.66 + \frac{0.368}{x - 0.2}$

16. $x - 0.4\overline{)8.1x^2 + 3.2x - 5.7}$ $8.1x + 6.44 + \frac{-3.124}{x - 0.4}$

17. $x + 1.7\overline{)2.7x^2 + x - 5.2}$ $2.7x - 3.59 + \frac{0.903}{x + 1.7}$

18. $x + 2.5\overline{)1.3x^2 - 0.5x - 2.3}$ $1.3x - 3.75 + \frac{7.075}{x + 2.5}$

19. $x + 57\overline{)9x^3 - 25}$ $9x^2 - 513x + 29{,}241 + \frac{-1{,}666{,}762}{x + 57}$

20. $x - 2.3\overline{)0.5x^3 + x}$ $0.5x^2 + 1.15x + 3.645 + \frac{8.3835}{x - 2.3}$

In Exercises 21–28, let $P(x) = 2x^3 - 4x^2 + 2x - 1$. *Evaluate the polynomial by substituting the given value of x into the polynomial and simplifying. Then evaluate the polynomial by using the remainder theorem and synthetic division.*

21. $P(1)$ -1

22. $P(2)$ 3

23. $P(-2)$ -37

24. $P(-1)$ -9

25. $P(3)$ 23

26. $P(-4)$ -201

27. $P(0)$ -1

28. $P(4)$ 71

In Exercises 29–36, let $Q(x) = x^4 - 3x^3 + 2x^2 + x - 3$. *Evaluate the polynomial by substituting the given value of x into the polynomial and simplifying. Then evaluate the polynomial by using the remainder theorem and synthetic division.*

29. $Q(-1)$ 2

30. $Q(1)$ -2

31. $Q(2)$ -1

32. $Q(-2)$ 43

33. $Q(3)$ 18

34. $Q(0)$ -3

35. $Q(-3)$ 174

36. $Q(-4)$ 473

In Exercises 37–44, use the remainder theorem and synthetic division to find P(r).

37. $P(x) = x^3 - 4x^2 + x - 2;\ r = 2$ -8

38. $P(x) = x^3 - 3x^2 + x + 1;\ r = 1$ 0

39. $P(x) = 2x^3 + x + 2;\ r = 3$ 59

40. $P(x) = x^3 + x^2 + 1;\ r = -2$ -3

41. $P(x) = x^4 - 2x^3 + x^2 - 3x + 2;\ r = -2$ 44

42. $P(x) = x^5 + 3x^4 - x^2 + 1;\ r = -1$ 2

43. $P(x) = 3x^5 + 1;\ r = -\frac{1}{2}$ $\frac{29}{32}$

44. $P(x) = 5x^7 - 7x^4 + x^2 + 1;\ r = 2$ 533

In Exercises 45–48, use the factor theorem and tell whether the first expression is a factor of P(x).

45. $x - 3;\ P(x) = x^3 - 3x^2 + 5x - 15$ yes

46. $x + 1;\ P(x) = x^3 + 2x^2 - 2x - 3$ yes

47. $x + 2;\ P(x) = 3x^2 - 7x + 4$ no

48. $x;\ P(x) = 7x^3 - 5x^2 - 8x$ (*Hint:* $x = x - 0$.) yes

49. Find 2^6 by using synthetic division to evaluate the polynomial $P(x) = x^6$ at $x = 2$. Then check the answer by evaluating 2^6 with a calculator. 64

50. Find $(-3)^5$ by using synthetic division to evaluate the polynomial $P(x) = x^5$ at $x = -3$. Then check the answer by evaluating $(-3)^5$ with a calculator. -243

Writing Exercises ■ *Write a paragraph using your own words.*

1. If you are given $P(x)$, explain how to use synthetic division to calculate $P(a)$.

2. Explain the factor theorem.

Something to Think About ■ *Suppose that* $P(x) = x^{100} - x^{99} + x^{98} - x^{97} + \cdots + x^2 - x + 1$.

1. Find the remainder when $P(x)$ is divided by $x - 1$. 1

2. Find the remainder when $P(x)$ is divided by $x + 1$. 101

Review Exercises *Let $P(x) = 3x^2 + 2x - 1$ and find each value.*

1. $P(1)$ 4

2. $P(-2)$ 7

3. $P(2a)$ $12a^2 + 4a - 1$

4. $P(-t)$ $3t^2 - 2t - 1$

In Review Exercises 5–6, remove the parentheses and simplify.

5. $2(x^2 + 4x - 1) + 3(2x^2 - 2x + 2)$ $8x^2 + 2x + 4$

6. $-2(3y^3 - 2y + 7) - 3(y^2 + 2y - 4) + 4(y^3 + 2y - 1)$ $-2y^3 - 3y^2 + 6y - 6$

PROJECT A Strange Polynomial

The expression $1 + x + x^2 + x^3$ is a polynomial of degree 3. The polynomial $1 + x + x^2 + x^3 + x^4$ has the same pattern, but one more term. Its degree is 4. As the pattern continues and more terms are added, the degree of the polynomial increases. If there were no end to the number of terms, the "polynomial" would have infinitely many terms, and no defined degree:

$$1 + x + x^2 + x^3 + x^4 + x^5 + x^6 + \cdots$$

Such "unending polynomials," called **power series,** are studied in calculus. However, this particular series is the result of a division of polynomials:

- Consider the division $\frac{1}{1-x}$. Find the quotient by filling in more steps of this long division:

Step 1:

$$\begin{array}{r} 1 \\ 1 - x\overline{)1 + 0x + 0x^2 +} \\ \underline{1 - x} \\ x \end{array}$$

Step 2:

$$\begin{array}{r} 1 + x \\ 1 - x\overline{)1 + 0x + 0x^2 +} \\ \underline{1 - x} \\ x + 0x^2 \\ \underline{x - x^2} \\ x^2 \end{array}$$

To determine how the fraction $\frac{1}{1-x}$ and the series $1 + x + x^2 + x^3 + x^4 + x^5 + x^6 + \cdots$ could be equal, try this experiment.

- Let $x = \frac{1}{2}$ and evaluate $\frac{1}{1-x}$.
- Again, let $x = \frac{1}{2}$ and evaluate the series. Because you cannot add infinitely many numbers, just add the first 3, or 4, or 5 terms and see if you find a pattern. Use a calculator to complete this chart:

Polynomial	Value at $x = \frac{1}{2}$
$1 + x + x^2$	
$1 + x + x^2 + x^3$	
$1 + x + x^2 + x^3 + x^4$	
$1 + x + x^2 + x^3 + x^4 + x^5$	
$1 + x + x^2 + x^3 + x^4 + x^5 + x^6$	

What number do the values in the second column seem to be approaching? That number is called the sum of the series.

- Explain why the nonterminating decimal 1.1111111 . . . represents the infinite series

$$1 + \left(\frac{1}{10}\right) + \left(\frac{1}{10}\right)^2 + \left(\frac{1}{10}\right)^3 + \left(\frac{1}{10}\right)^4 + \left(\frac{1}{10}\right)^5 + \left(\frac{1}{10}\right)^6 + \cdots$$

- Using the fraction $\frac{1}{1-x}$, explain why $1.11111 = \ldots \frac{10}{9}$.
- Verify that $\frac{10}{9} = 1.11111\ldots$ by dividing 10 by 9.

Chapter Summary

KEY WORDS

algebraic terms (2.3)
algorithm (2.5)
base of an exponential expression (2.1)
binomial (2.3)
coefficient (2.1)
degree of a polynomial (2.3)
exponent (2.1)
FOIL method for multiplying binomials (2.4)
like terms (2.3)
monomial (2.3)
polynomial (2.3)
power of x (2.1)
scientific notation (2.2)
similar terms (2.3)
synthetic division (2.6)
trinomial (2.3)

KEY IDEAS

(2.1) For any natural number n, $x^n = \underbrace{x \cdot x \cdot x \cdot \cdots \cdot x}_{n \text{ factors of } x}$.

If m and n are integers and there are no divisions by 0, then

$$x^m x^n = x^{m+n} \quad (x^m)^n = x^{mn} \quad (xy)^n = x^n y^n \quad \left(\frac{y}{x}\right)^n = \frac{y^n}{x^n}$$

$$x^0 = 1 \quad x^{-n} = \frac{1}{x^n} \quad \frac{x^m}{x^n} = x^{m-n} \quad \left(\frac{y}{x}\right)^{-n} = \left(\frac{x}{y}\right)^n$$

(2.2) A number is written in scientific notation if it is expressed as a number whose absolute value is between 1 and 10 multiplied by an integer power of ten.

(2.3) A polynomial in x is the sum of one or more terms of the form ax^n, where a is a real number and n is a whole number.

A polynomial in several variables (say, x, y, and z) is a sum of one or more terms of the form $ax^my^nz^p$, where a is a real number and m, n, and p are whole numbers.

The **degree** of the monomial ax^n is n. The degree of a monomial containing several variables is the sum of the exponents on those variables.

The **degree of a polynomial** is the degree of the term with highest degree contained within the polynomial.

If $P(x)$ is a polynomial in x, then $P(r)$ is the value of the polynomial at $x = r$.

To add like terms, add their numerical coefficients and use the same variables with the same exponents.

To add polynomials, add their like terms.

To subtract polynomials, add the negative of the polynomial that is to be subtracted.

(2.4) To multiply monomials, multiply their numerical factors and multiply their variable factors.

To multiply a polynomial by a monomial, multiply each term of the polynomial by the monomial.

To multiply polynomials, multiply each term of one polynomial by each term of the other polynomial.

If one set of grouping symbols is contained within another set, remove the inner set first.

(2. 5) To find the quotient of two monomials, express the quotient as a fraction and use the rules of exponents to simplify.

To divide a polynomial by a monomial, write the expression as a fraction. Then split the fraction into separate fractions and simplify each of the separate fractions.

To divide a polynomial by a polynomial, use long division.

(2.6) Synthetic division can be used to divide polynomials by binomials of the form $x - r$.

The remainder theorem. If a polynomial $P(x)$ is divided by $x - r$, then the remainder is $P(r)$.

The factor theorem. If a polynomial $P(x)$ is divided by $x - r$, then $P(r) = 0$, if and only if $x - r$ is a factor of $P(x)$.

Chapter 2 Review Exercises

In Review Exercises 1–28, use the rules of exponents to simplify each quantity. Write all answers without using negative exponents.

1. 3^6 729
2. -2^6 -64
3. $(-4)^3$ -64
4. $-(-5)^4$ -625
5. $(3x^4)(-2x^2)$ $-6x^6$
6. $(-x^5)(3x^3)$ $-3x^8$
7. $x^{-4}x^3$ $\frac{1}{x}$
8. $x^{-10}x^{12}$ x^2
9. $(3x^2)^3$ $27x^6$
10. $(4x^4)^4$ $256x^{16}$
11. $(-2x^2)^5$ $-32x^{10}$
12. $-(-3x^3)^5$ $243x^{15}$
13. $(x^2)^{-5}$ $\frac{1}{x^{10}}$
14. $(x^{-4})^{-5}$ x^{20}
15. $(3x^{-3})^{-2}$ $\frac{x^6}{9}$
16. $(2x^{-4})^4$ $\frac{16}{x^{16}}$
17. $\frac{x^6}{x^4}$ x^2
18. $\frac{x^{12}}{x^7}$ x^5
19. $\frac{a^7}{a^{12}}$ $\frac{1}{a^5}$
20. $\frac{a^4}{a^7}$ $\frac{1}{a^3}$
21. $\frac{y^{-3}}{y^4}$ $\frac{1}{y^7}$
22. $\frac{y^5}{y^{-4}}$ y^9
23. $\frac{x^{-5}}{x^{-4}}$ $\frac{1}{x}$
24. $\frac{x^{-6}}{x^{-9}}$ x^3
25. $(3x^2y^3)^2$ $9x^4y^6$
26. $(-4a^3b^2)^{-4}$ $\frac{1}{256a^{12}b^8}$
27. $\left(\frac{3x^2}{4y^3}\right)^{-3}$ $\frac{64y^9}{27x^6}$
28. $\left(\frac{4y^{-2}}{5y^{-3}}\right)^3$ $\frac{64y^3}{125}$

In Review Exercises 29–30, write each numeral in scientific notation.

29. 19,300,000,000 1.93×10^{10}

30. 0.0000000273 2.73×10^{-8}

In Review Exercises 31–32, write each numeral in standard notation.

31. 7.2×10^7 72,000,000

32. 8.3×10^{-9} 0.0000000083

In Review Exercises 33–36, find each value if $P(x) = -x^2 + 4x + 6$.

33. $P(0)$ 6

34. $P(1)$ 9

35. $P(-t)$ $-t^2 - 4t + 6$

36. $P(z)$ $-z^2 + 4z + 6$

37. Give the degree of $P(x) = 3x^5 + 4x^3 + 2$. 5

38. Give the degree of $9x^2y + 13x^3y^2 + 8x^4y^4$. 8

In Review Exercises 39–42, simplify each expression.

39. $(3x^2 + 4x + 9) - (2x^2 - 2x + 7) + (4x^2 - 3x - 2)$ $5x^2 + 3x$

40. $(4x^3 + 4x^2 + 7) - (-2x^3 - x - 2) + (-5x^3 - 3x^2)$ $x^3 + x^2 + x + 9$

41. $(2x^2 - 5x + 9) - (x^2 - 3) - (-3x^2 + 4x - 7)$ $4x^2 - 9x + 19$

42. $2(7x^3 - 6x^2 + 4x - 3) - 3(7x^3 + 6x^2 + 4x - 3)$ $-7x^3 - 30x^2 - 4x + 3$

In Review Exercises 43–50, find each product.

43. $(8a^2b^2)(-2abc)$ $-16a^3b^3c$

44. $(-3xy^2z)(2xz^3)$ $-6x^2y^2z^4$

45. $2xy^2(x^3y - 4xy^5)$ $2x^4y^3 - 8x^2y^7$

46. $a^2b(a^2 + 2ab + b^2)$ $a^4b + 2a^3b^2 + a^2b^3$

47. $(8x - 5)(2x + 3)$ $16x^2 + 14x - 15$

48. $(3x^2 + 2)(2x - 4)$ $6x^3 - 12x^2 + 4x - 8$

49. $(5x^2 - 4x + 5)(3x^2 - 2x + 10)$
$15x^4 - 22x^3 + 73x^2 - 50x + 50$

50. $(3x^2 + x - 2)(x^2 - x + 2)$
$3x^4 - 2x^3 + 3x^2 + 4x - 4$

In Review Exercises 51–58, do each division.

51. $\dfrac{3x^3y^2 + 6xy^3}{9xy^2}$ $\dfrac{x^2}{3} + \dfrac{2y}{3}$

52. $\dfrac{4a^3b^3 - 8ab^4 + 12a^3b^2}{2a^3b^2}$ $2b - \dfrac{4b^2}{a^2} + 6$

53. $x + 3\overline{)3x^3 - 4x^2 + 3x + 2}$ $3x^2 - 13x + 42 + \frac{-124}{x+3}$

54. $x + 2\overline{)x^5 + 32}$ $x^4 - 2x^3 + 4x^2 - 8x + 16$

55. $3x - 2\overline{)6x^3 + 5x^2 - 3x + 8}$ $2x^2 + 3x + 1 + \frac{10}{3x - 2}$

56. $x^2 - 1\overline{)x^4 + x^2 - 2}$ $x^2 + 2$

57. $x^2 + 2x + 3\overline{)x^4 - x^2 - 3}$ $x^2 - 2x + \dfrac{6x - 3}{x^2 + 2x + 3}$

58. $3x + x^3 + 1\overline{)3x^3 - x^2 + 5}$ $3 + \dfrac{-x^2 - 9x + 2}{x^3 + 3x + 1}$

In Review Exercises 59–60, use synthetic division to do each division.

59. $x - 4\overline{)x^3 - 13x - 12}$ $x^2 + 4x + 3$

60. $x + 1\overline{)x^4 + x^2 + 1}$ $x^3 - x^2 + 2x - 2 + \frac{3}{x+1}$

In Review Exercises 61–62, let $P(x) = 3x^2 - 2x + 3$. Use synthetic division and the remainder theorem to find each value.

61. $P(2)$ 11

62. $P(-1)$ 8

In Review Exercises 63–64, use the factor theorem to decide whether the first expression is a factor of P(x).

63. $x - 5$; $P(x) = x^3 - 3x^2 - 8x - 10$
yes

64. $x + 5$; $P(x) = x^3 + 4x^2 - 5x + 5$
(*Hint:* Write $x + 5$ as $x - (-5)$.) no

Chapter 2 Test

In Problems 1–6, simplify each expression. Write all answers without using negative exponents. Assume that no denominators are zero.

1. x^3x^5 x^8

2. $(x^2y^3)^3$ x^6y^9

3. $(m^{-4})^2$ $\frac{1}{m^8}$

4. $\left(\frac{a^3}{b^2}\right)^4$ $\frac{a^{12}}{b^8}$

5. $3x^0$ 3

6. $\left(\frac{m^2n^3}{m^4n^{-2}}\right)^{-2}$ $\frac{m^4}{n^{10}}$

In Problems 7–8, write each number in scientific notation.

7. 4,700,000 4.7×10^6

8. 0.00000023 2.3×10^{-7}

In Problems 9–10, write each number in standard notation.

9. 6.53×10^5 653,000

10. 24.5×10^{-3} 0.0245

11. The moon is approximately 235,000 miles from the earth. Use scientific notation to express this distance in kilometers. (*Hint:* 1 mile ≈ 1.6 kilometers.)
3.76×10^5 km

12. Find the degree of $3x^2y^3 + 4x^3y^7 - 7x^4y^5$.
10

In Problems 13–14, let $P(x) = -3x^2 + 2x - 1$ and find each value.

13. $P(2)$ -9

14. $P(-1)$ -6

In Problems 15–16, $x = 3$ and $y = -2$. Find the value of each expression.

15. $x^2 - y^2$ 5

16. $\frac{4x^2 + y^2}{-xy^2}$ $-\frac{10}{3}$

In Problems 17–28, do the operations.

17. $(2y^2 + 4y + 3) + (3y^2 - 3y - 4)$ $5y^2 + y - 1$

18. $(-3u^2 + 2u - 7) - (u^2 + 7)$ $-4u^2 + 2u - 14$

19. $3(2a^2 - 4a + 2) - 4(-a^2 - 3a - 4)$ $10a^2 + 22$

20. Add:
$$\begin{array}{r} 8x^2 + 4x - 9 \\ \underline{-2x^2 - 6x + 8} \\ 6x^2 - 2x - 1 \end{array}$$

21. $(3x^3y^2z)(-2xy^{-1}z^3)$ $-6x^4yz^4$

22. $-5a^2b(3ab^3 - 2ab^4)$ $-15a^3b^4 + 10a^3b^5$

23. $(z + 4)(z - 4)$ $z^2 - 16$

24. $(3x - 2)(4x + 3)$ $12x^2 + x - 6$

25. $(2x + 1)(x^2 - x - 3)$ $2x^3 - x^2 - 7x - 3$

26. $(x^n + y^n)(2x^n - y^n)$ $2x^{2n} + x^ny^n - y^{2n}$

27. $\dfrac{18x^2y^3 - 12x^3y^2 + 9xy}{-3xy^4}$ $\dfrac{-6x}{y} + \dfrac{4x^2}{y^2} - \dfrac{3}{y^3}$

28. $2x - 1\overline{)6x^3 + 5x^2 - 2}$ $3x^2 + 4x + 2$

29. Find the remainder: $\dfrac{x^3 - 4x^2 + 5x + 3}{x + 1}$. -7

30. **Optional** Use synthetic division to find the remainder when $4x^3 + 3x^2 + 2x - 1$ is divided by $x - 2$. 47

CAREERS & MATHEMATICS

BANKER

Practically every bank has a group of officers who make decisions affecting bank operations: the president, who directs overall operations; one or more vice presidents, who act as general managers or are in charge of bank departments, such as trust or credit; a comptroller, or cashier, who (as an executive officer) is generally responsible for all bank property; and treasurers and other senior officers as well as junior officers, who supervise sections within departments.

These officers make decisions within a framework of policy set by a board of directors and existing laws and regulations. They must have a broad knowledge of business activities to relate to the operations of their departments, since their customers will include a variety of individuals and businesses applying for loans, seeking investment advice, organizing trusts, setting up pensions, and so on. Besides supervising these financial services, officers advise individuals and businesses and participate in community projects.

Qualifications

Bank officer and management positions are filled by management trainees and by promoting outstanding bank clerks or tellers. A college degree in finance or liberal arts, including accounting, economics, commercial law, political science, and statistics, is necessary. A Master of Business Administration degree (MBA) is preferred by some banks, although people with backgrounds as diverse as nuclear physics and forestry are hired by some banks to meet the needs of the complex, high-technology industries with which they deal.

Job Outlook

Through the year 2005, employment of bank officers is expected to increase faster than the average for other occupations, due to expanding bank services, both domestic and international, and the increasing dependence on computers.

Example Application

In banking, the calculation of compound interest uses exponential expressions. At the end of each period, a savings account is credited with a dividend, calculated as a fixed percent of the savings balance. The money will grow to an amount FV, called the **future value,** according to the formula

$$FV = PV(1 + i)^n$$

where PV, called the **present value,** is the initial deposit. The percent of the current balance credited as interest is called the **periodic interest rate,** denoted by i. The **number of periods** that the money is left on deposit is n.

If \$1000 is deposited in an account that earns 6% annual interest compounded quarterly, find the amount in the account after three years.

Solution Interest paid quarterly is paid four times each year. The periodic interest rate is one-fourth of 6%, or 1.5%. Express the periodic interest rate as a decimal: $i = 0.015$. In three years, there will be $3 \cdot 4$, or 12 periods. Thus, $n = 12$. The present value is the initial deposit: $PV = \$1000$. Substitute these values into the equation and calculate FV.

$$
\begin{aligned}
FV &= PV(1 + i)^n \\
&= \mathbf{1000}(1 + \mathbf{0.015})^{\mathbf{12}} && \text{Substitute 1000 for } PV\text{, 0.015 for } i\text{, and 12 for } n. \\
&= 1000(1.015)^{12} \\
&= 1195.618171 && \text{Use a calculator.}
\end{aligned}
$$

After three years, the account will contain \$1195.62.

EXERCISES

1. If \$10,000 is deposited in an account earning 6% annual interest, compounded once a year, how much will be in the account at the end of two years?
2. If \$10,000 is deposited in an account earning 6% annual interest, compounded twice a year, how much will be in the account at the end of two years?
3. If \$10,000 is deposited in an account earning 6% annual interest, compounded four times a year, how much will be in the account at the end of two years?
4. Refer to Exercises 1–3. How much more is earned by compounding twice a year, instead of once a year? How much more is earned by compounding four times a year, instead of twice a year?

(*Answers:* **1.** \$11,236 **2.** \$11,255.09 **3.** \$11,264.93 **4.** \$19.09; \$9.84)

3 Equations and Inequalities

In this chapter, we will consider equations and inequalities, two of the most basic ideas in algebra. Equations and inequalities are used in almost every academic and vocational field, especially in science, business, economics, and electronics.

3.1 Linear Equations and Their Solutions

■ Solving Equations ■ Identities and Impossible Equations ■ Repeating Decimals ■ Problem Solving

An **equation** is a statement indicating that two mathematical expressions are equal. The equation $2 + 4 = 6$ is a true equation, and $2 + 4 = 7$ is a false equation. The equation

$$7x - 3 = 4$$

can be true or false depending on the value of x, which is called a **variable.** If $x = 1$, the equation is true, because 1 satisfies the equation:

$$\begin{aligned} 7(1) - 3 &= 4 \\ 7 - 3 &= 4 \\ 4 &= 4 \end{aligned}$$

The equation is false for all other numbers x.

The set of numbers that satisfy an equation is called its **solution set.** The elements in the solution set are called **solutions** or **roots** of the equation. Finding the solution set of an equation is called **solving the equation.**

EXAMPLE 1 Determine whether 3 is a solution of the equation $2x + 4 = 10$.

Solution We substitute 3 for x to see if it satisfies the equation.

$$\begin{aligned} 2x + 4 &= 10 \\ 2(3) + 4 &\stackrel{?}{=} 10 \qquad \text{Substitute 3 for } x. \\ 6 + 4 &\stackrel{?}{=} 10 \\ 10 &= 10 \end{aligned}$$

Since $10 = 10$, the number 3 satisfies the equation and is a solution. ■

■ Solving Equations

To find the solution set of an equation, we will replace the equation with simpler ones having the same solution set. Such equations are called **equivalent equations.**

Equivalent Equations Two or more equations are called **equivalent equations** if they have the same solution set.

The method of replacing equations with simpler but equivalent equations continues until we have isolated the variable on one side of the equation. To isolate the variable, we can use the following properties of equality, along with the properties of equality discussed in Section 1.2.

Addition and Subtraction Properties of Equality If any quantity is added to (or subtracted from) both sides of an equation, a new equation is formed that is equivalent to the original equation.

In symbols, if a, b, and c are real numbers and $a = b$, then

$$a + c = b + c \qquad \text{and} \qquad a - c = b - c$$

Multiplication and Division Properties of Equality If both sides of an equation are multiplied (or divided) by the same nonzero constant, a new equation is formed that is equivalent to the original equation.

In symbols, if a, b, and c are real numbers, $a = b$ and $c \neq 0$, then

$$ac = bc \qquad \text{and} \qquad \frac{a}{c} = \frac{b}{c}$$

The easiest equations to solve are **first-degree** or **linear equations.** Every linear equation has exactly one solution.

Linear Equation A **linear equation in one variable** (say, x) is any equation that can be written in the form

$$ax + c = 0 \qquad (a \text{ and } c \text{ are real numbers and } a \neq 0)$$

EXAMPLE 2 Solve the linear equation $2x + 8 = 0$.

Solution To find the number x that satisfies the equation, we isolate x on the left-hand side of the equation by writing a series of simplified equations, all of which are equivalent to the original equation.

Teaching Tip: Point out that dividing by 2 is like multiplying by $\frac{1}{2}$. Show that

$$\frac{2x}{2} = \frac{-8}{2}$$

is equivalent to

$$\frac{1}{2}(2x) = \frac{1}{2}(-8)$$

$2x + 8 = 0$	
$2x + 8 - 8 = 0 - 8$	Subtract 8 from both sides.
$2x = -8$	Combine terms.
$\frac{2x}{2} = \frac{-8}{2}$	Divide both sides by 2.
$x = -4$	Simplify.

Check: To show that -4 is a solution, we substitute -4 for x in the original equation and simplify:

$$2x + 8 = 0$$
$$2(-4) + 8 \stackrel{?}{=} 0$$
$$-8 + 8 \stackrel{?}{=} 0$$
$$0 = 0$$

Since -4 satisfies the original equation, it is a solution. Furthermore, -4 is the only number that satisfies the equation. Thus, the solution set is $\{-4\}$. ■

EXAMPLE 3 Solve the equation $3x + 6 = 24$.

Solution We isolate x on the left-hand side of the equation.

$3x + 6 = 24$	
$3x + 6 - 6 = 24 - 6$	Subtract 6 from both sides.
$3x = 18$	Combine terms.
$\frac{3x}{3} = \frac{18}{3}$	Divide both sides by 3.
$x = 6$	Simplify.

Check: We substitute 6 for x in the original equation and show that it satisfies the equation:

$$3x + 6 = 24$$
$$3(6) + 6 \stackrel{?}{=} 24$$
$$18 + 6 \stackrel{?}{=} 24$$
$$24 = 24$$

Since 6 satisfies the equation, it is the solution. The solution set is $\{6\}$. ■

EXAMPLE 4 Solve the equation $3(2x - 1) = 2x + 9$.

Solution We use the distributive property to remove parentheses and then isolate x on the left-hand side of the equation.

$$
\begin{aligned}
3(2x - 1) &= 2x + 9 & \\
6x - 3 &= 2x + 9 & &\text{Remove parentheses.} \\
6x - 3 + 3 &= 2x + 9 + 3 & &\text{Add 3 to both sides.} \\
6x &= 2x + 12 & &\text{Combine terms.} \\
6x - 2x &= 2x + 12 - 2x & &\text{Subtract } 2x \text{ from both sides.} \\
4x &= 12 & &\text{Combine terms.} \\
x &= 3 & &\text{Divide both sides by 4.}
\end{aligned}
$$

Check: We substitute 3 for x in the original equation to see if it satisfies the equation.

$$
\begin{aligned}
3(2x - 1) &= 2x + 9 \\
3(2 \cdot 3 - 1) &\stackrel{?}{=} 2 \cdot 3 + 9 \\
3(5) &\stackrel{?}{=} 6 + 9 \\
15 &= 15
\end{aligned}
$$

Since 3 satisfies the equation, it is the solution. The solution set is $\{3\}$. ■

To solve more complicated linear equations, we will follow these steps.

Steps for Solving Equations

1. If the equation contains fractions, multiply both sides of the equation by their least common denominator to eliminate the denominators.
2. Use the distributive property to remove all sets of parentheses, and combine like terms.
3. Use the addition and subtraction properties to get all variables on one side of the equation and all numbers on the other side. Combine terms, when possible.
4. Use the multiplication and division properties to cause the coefficient of the variable to be 1.
5. Check the result by replacing the variable with the possible solution and verifying that the number satisfies the equation.

EXAMPLE 5 Solve the equation $\frac{5}{3}(x-3) = \frac{3}{2}(x-2)+2$.

Solution *Step 1:* Since 6 is the smallest number that can be divided by both 2 and 3, we multiply both sides of the equation by 6 to eliminate the fractions:

$$\frac{5}{3}(x-3) = \frac{3}{2}(x-2)+2$$
$$6\left[\frac{5}{3}(x-3)\right] = 6\left[\frac{3}{2}(x-2)+2\right]$$
$$10(x-3) = 9(x-2)+6\cdot 2$$

Step 2: We use the distributive property to remove parentheses and then combine terms.

$$10x-30 = 9x-18+12$$
$$10x-30 = 9x-6$$

Step 3: We use the addition and subtraction properties by adding 30 to both sides and subtracting $9x$ from both sides.

$$10x-30-9x+30 = 9x-6-9x+30$$
$$x = 24 \qquad \text{Combine terms.}$$

Since the coefficient of x in the above equation is 1, Step 4 is unnecessary.

Step 5: We check by substituting 24 for x in the original equation and simplifying:

$$\frac{5}{3}(x-3) = \frac{3}{2}(x-2)+2$$
$$\frac{5}{3}(24-3) \stackrel{?}{=} \frac{3}{2}(24-2)+2$$
$$\frac{5}{3}(21) \stackrel{?}{=} \frac{3}{2}(22)+2$$
$$5(7) \stackrel{?}{=} 33+2$$
$$35 = 35$$

Since 24 satisfies the equation, it is the solution. The solution set is $\{24\}$. ■

EXAMPLE 6 Solve the equation $\dfrac{x+2}{5} - 4x = \dfrac{8}{5} - \dfrac{x+9}{2}$.

Solution

$$\frac{x+2}{5} - 4x = \frac{8}{5} - \frac{x+9}{2}$$

$$\mathbf{10}\left(\frac{x+2}{5} - 4x\right) = \mathbf{10}\left(\frac{8}{5} - \frac{x+9}{2}\right) \qquad \text{Multiply both sides by 10.}$$

$$2(x+2) - 40x = 2(8) - 5(x+9) \qquad \text{Remove parentheses and simplify.}$$

$$2x + 4 - 40x = 16 - 5x - 45 \qquad \text{Remove parentheses.}$$

$$-38x + 4 = -5x - 29 \qquad \text{Combine terms.}$$

$$-33x = -33 \qquad \text{Add } 5x - 4 \text{ to both sides.}$$

$$\frac{-33x}{\mathbf{-33}} = \frac{-33}{\mathbf{-33}} \qquad \text{Divide both sides by } -33.$$

$$x = 1 \qquad \text{Simplify.}$$

Check:

$$\frac{x+2}{5} - 4x = \frac{8}{5} - \frac{x+9}{2}$$

$$\frac{\mathbf{1}+2}{5} - 4(\mathbf{1}) \stackrel{?}{=} \frac{8}{5} - \frac{\mathbf{1}+9}{2}$$

$$\frac{3}{5} - 4 \stackrel{?}{=} \frac{8}{5} - 5$$

$$\frac{3}{5} - \frac{20}{5} \stackrel{?}{=} \frac{8}{5} - \frac{25}{5}$$

$$-\frac{17}{5} = -\frac{17}{5}$$

Since 1 satisfies the equation, it is the solution. The solution set is $\{1\}$. ■

Identities and Impossible Equations

The equations discussed so far are **conditional equations.** For these equations, some numbers x are solutions and others are not. An **identity** is an equation that is satisifed by every number x for which both sides of the equation are defined.

EXAMPLE 7 Solve the equation $2(x-1) + 4 = 4(1+x) - (2x+2)$.

Solution

$$2(x-1) + 4 = 4(1+x) - (2x+2)$$

$$2x - 2 + 4 = 4 + 4x - 2x - 2 \qquad \text{Use the distributive property to remove parentheses.}$$

$$2x + 2 = 2x + 2 \qquad \text{Combine terms.}$$

$$2 = 2 \qquad \text{Subtract } 2x \text{ from both sides.}$$

The result $2 = 2$ is a true equation, and it is a true equation for every number x. Since every number x satisfies this equation, it is an identity. ■

An **impossible equation** or a **contradiction** is an equation that has no solution.

EXAMPLE 8 Solve the equation $\dfrac{x-1}{3} + 4x = \dfrac{3}{2} + \dfrac{13x-2}{3}$.

Solution

$$\frac{x-1}{3} + 4x = \frac{3}{2} + \frac{13x-2}{3}$$

$$6\left(\frac{x-1}{3} + 4x\right) = 6\left(\frac{3}{2} + \frac{13x-2}{3}\right) \qquad \text{Multiply both sides by 6.}$$

$$2(x-1) + 6(4x) = 9 + 2(13x-2) \qquad \text{Use the distributive property to remove parentheses.}$$

$$2x - 2 + 24x = 9 + 26x - 4 \qquad \text{Remove parentheses.}$$

$$26x - 2 = 26x + 5 \qquad \text{Combine terms.}$$

$$-2 = 5 \qquad \text{Subtract } 26x \text{ from both sides.}$$

Since the result $-2 = 5$ is a false equation, no number x satisfies the equation. Its solution set is $\emptyset$. ■

Repeating Decimals

Every repeating decimal can be written as a rational number in fractional form. We can show that this is true by demonstrating a process to convert repeating decimals to fractional form.

To write the decimal $0.2\overline{54}$, for example, as a rational number in fractional form, we first note that the decimal has a repeating block of two digits. We then form an equation by setting x equal to the decimal.

1. $x = 0.2\ 54\ 54\ 54\ldots$

We can form another equation by multiplying both sides of Equation 1 by 10^2, which is 100.

2. $100x = 25.4\ 54\ 54\ 54\ldots$

We then subtract each side of Equation 1 from the corresponding side of Equation 2 to obtain

$$\begin{aligned} 100x &= 25.4\ 54\ 54\ 54\ldots \\ x &= 0.2\ 54\ 54\ 54\ldots \\ \hline 99x &= 25.2 \end{aligned}$$

Finally, we solve $99x = 25.2$ for x and simplify the fraction.

$$\begin{aligned} x &= \frac{25.2}{99} \\ &= \frac{25.2 \cdot \mathbf{10}}{99 \cdot \mathbf{10}} && \text{Multiply numerator and denominator by 10.} \\ &= \frac{252}{990} \\ &= \frac{\cancel{18} \cdot 14}{\cancel{18} \cdot 55} \\ &= \frac{14}{55} \end{aligned}$$

We can use a calculator to verify that the decimal representation of $\frac{14}{55}$ is $0.2\overline{54}$.

The key step was multiplying both sides of Equation 1 by 10^2. If there had been n digits in the repeating block of the decimal, we would have multiplied both sides of Equation 1 by 10^n.

Problem Solving

EXAMPLE 9 **Integer problem** The sum of three consecutive odd integers is 213. Find the integers.

Analysis Consecutive odd integers are odd integers like 11, 13, 15, and 17 that differ by 2. If x is an odd integer, the algebraic expressions

$$x, \quad x + 2, \quad \text{and} \quad x + 4$$

represent three consecutive odd integers. The sum of these integers can be expressed in two ways: as

$$x + (x + 2) + (x + 4) \qquad \text{or as} \qquad 213$$

Solution Let x represent the first odd integer.
Then $x + 2$ represents the second odd integer.
Then $x + 4$ represents the third odd integer.

The first odd integer	+	the second odd integer	+	the third odd integer	=	the sum of the integers.
x	+	$x + 2$	+	$x + 4$	=	213

$$\begin{aligned} 3x + 6 &= 213 && \text{Combine terms.} \\ 3x &= 207 && \text{Subtract 6 from both sides.} \\ x &= 69 && \text{Divide both sides by 3.} \\ x + 2 &= 71 \\ x + 4 &= 73 \end{aligned}$$

The three consecutive odd integers are 69, 71, and 73.

Check: The numbers 69, 71, and 73 are consecutive odd integers and their sum is 213. ■

EXAMPLE 10 **Plumbing** A plumber wants to cut a 20-foot length of pipe into two pieces. The longer piece needs to be 5 feet longer than twice the shortest piece. Where should the plumber cut the pipe?

Analysis The length of the longer piece of pipe is given in terms of the length of the shorter piece. Thus, we pick a variable to represent the length of the shorter piece. We can then express the length of the longer piece in terms of the variable.

Solution Let x represent the length of the shorter piece of pipe.
Then $2x + 5$ represents the length of the longer piece.

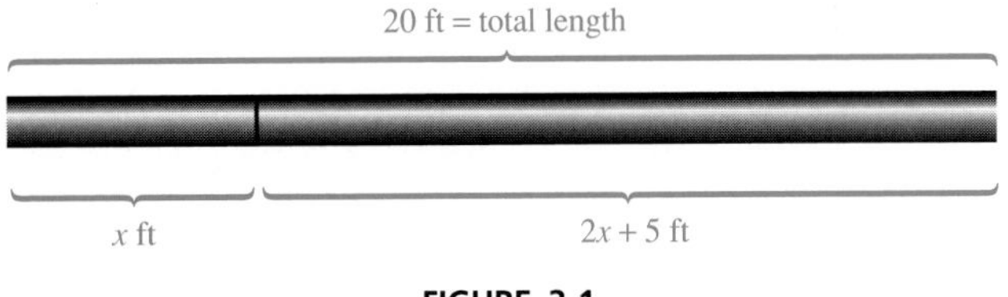

FIGURE 3-1

From Figure 3-1, we see that the sum of the individual lengths must equal the total length of the pipe.

The length of the shorter piece	+	**the length of the longer piece**	=	**the total length.**
x	+	$2x + 5$	=	20

$3x + 5 = 20$ Combine terms.

$3x = 15$ Subtract 5 from both sides.

$x = 5$ Divide both sides by 3.

The plumber should cut the pipe 5 feet from one end.

Check: Since 15 feet is 5 feet more than twice 5 feet and the sum of 5 feet and 15 feet is 20 feet, the solution checks. ■

EXAMPLE 11 **Building a dog run** A man has 28 meters of fencing to make a rectangular dog run. If he wants the dog run to be 6 feet longer than it is wide, find its dimensions.

Analysis If w is chosen to represent the width of the dog run, then $w + 6$ represents its length. (See Figure 3-2.) The perimeter, P, of a rectangle is the distance around it and can be expressed either as $2w + 2(w + 6)$ or as 28.

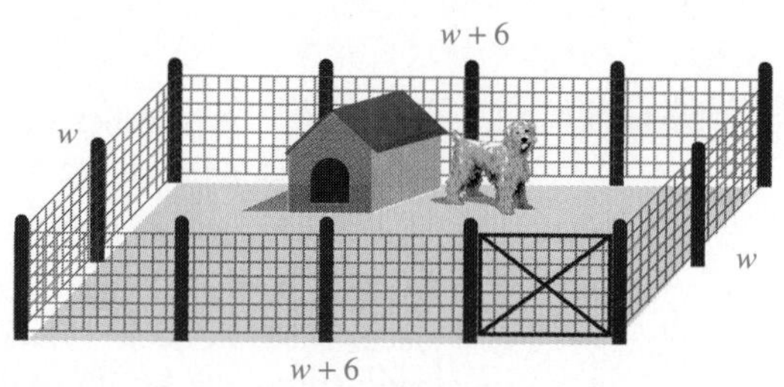

FIGURE 3-2

Solution We let w represent the width of the dog run. Then $w + 6$ represents its length.

Two widths	+	two lengths	=	the perimeter.
$2 \cdot w$	+	$2 \cdot (w + 6)$	=	28

$2w + 2w + 12 = 28$ Use the distributive property to remove parentheses.

$4w + 12 = 28$ Combine terms.

$4w = 16$ Subtract 12 from both sides.

$w = 4$ Divide both sides by 4.

$w + 6 = 10$

The dimensions of the dog run are 4 meters by 10 meters.

Check: If a dog run has a width of 4 meters and a length of 10 meters, its length is 6 meters longer than its width, and the perimeter is $2(4) + 2(10) = 28$. ■

ORALS *Tell whether each number is a solution of* $2x + 5 = 13$.

1. 3 no **2.** 4 yes **3.** 5 no **4.** 6 no

Solve each equation.

5. $3x - 2 = 7$ 3

6. $\frac{1}{2}x - 1 = 5$ 12

7. $\frac{x - 2}{3} = 1$ 5

8. $\frac{x + 3}{2} = 3$ 3

PERSPECTIVE

How to Solve It

As a young student, George Polya (1888–1985) enjoyed his studies of mathematics and understood the solutions presented by his teachers. However, Polya had questions still asked by mathematics students today: "Yes, the solution works, but how is it possible to come up with such a solution? How could *I* discover such things by myself?" These questions still concerned him years later when, as Professor of Mathematics at Stanford University, he developed an approach to teaching mathematics that was very popular with faculty and students. His book, *How to Solve It,* became a bestseller.

George Polya
(1888–1985)

Polya's problem-solving approach involves four steps.

- *Understand the problem.* What is the unknown? What information is known? What are the conditions?
- *Devise a plan.* Have you seen anything like it before? Do you know any related problems you have solved before? If you can't solve the proposed problem, can you solve a similar but easier problem?
- *Carry out the plan.* Check each step. Can you explain why each step is correct?
- *Look back.* Examine the solution. Can you check the result? Can you use the result, or the method, to solve any other problem?

EXERCISE 3.1

In Exercises 1–4, tell whether 5 is a solution of each equation.

1. $3x + 2 = 17$ yes

2. $7x - 2 = 33$ yes

3. $\frac{3}{5}x - 5 = -2$ yes

4. $\frac{2}{5}x + 12 = 8$ no

In Exercises 5–50, solve each equation.

5. $x + 6 = 8$ 2

6. $y - 7 = 3$ 10

7. $10 = z - 4$ 14

8. $12 = 7 + t$ 5

9. $2u = 6$ 3

10. $3v = 12$ 4

11. $\frac{x}{4} = 7$ 28

12. $\frac{x}{6} = 8$ 48

13. $\frac{3}{4}x = \frac{1}{2}$ $\frac{2}{3}$

14. $\frac{2}{3}x = \frac{5}{4}$ $\frac{15}{8}$

15. $2x + 1 = 13$ 6

16. $2x - 4 = 16$ 10

17. $\frac{3}{4}x - 3 = -9$ -8

18. $\frac{4}{5}x + 5 = 17$ 15

19. $2r - 5 = 1 - r$ 2

20. $5s - 13 = s - 1$ 3

21. $\frac{1}{2}x - 4 = -1 + 2x$ -2
22. $2x + 3 = \frac{2}{3}x - 1$ -3
23. $3a - 22 = -2a - 7$ 3
24. $a + 18 = 6a - 3$ $\frac{21}{5}$
25. $3(x + 1) = 15$ 4
26. $-2(x + 5) = 30$ -20
27. $3(y - 4) - 6 = 0$ 6
28. $2x + (2x - 3) = 5$ 2
29. $5(5 - a) = 37 - 2a$ -4
30. $4a + 17 = 7(a + 2)$ 1
31. $4(y + 1) = -2(4 - y)$ -6
32. $5(r + 4) = -2(r - 3)$ -2
33. $2(a - 5) - (3a + 1) = 0$ -11
34. $8(3a - 5) - 4(2a + 3) = 12$ 4
35. $3(y - 5) + 10 = 2(y + 4)$ 13
36. $2(5x + 2) = 3(3x - 2)$ -10
37. $9(x + 2) = -6(4 - x) + 18$ -8
38. $3(x + 2) - 2 = -(5 + x) + x$ -3
39. $\frac{x}{2} - \frac{x}{3} = 4$ 24
40. $\frac{x}{2} + \frac{x}{3} = 10$ 12
41. $\frac{x}{6} + 1 = \frac{x}{3}$ 6
42. $\frac{3}{2}(y + 4) = \frac{20 - y}{2}$ 2
43. $5 - \frac{x + 2}{3} = 7 - x$ 4
44. $3x - \frac{2(x + 3)}{3} = 16 - \frac{x + 2}{2}$ 6
45. $\frac{4x - 2}{2} = \frac{3x + 6}{3}$ 3
46. $\frac{t + 4}{2} = \frac{2t - 3}{3}$ 18
47. $\frac{a + 1}{3} + \frac{a - 1}{5} = \frac{2}{15}$ 0
48. $\frac{2z + 3}{3} + \frac{3z - 4}{6} = \frac{z - 2}{2}$ -2
49. $\frac{5a}{2} - 12 = \frac{a}{3} + 1$ 6
50. $\frac{5a}{6} - \frac{5}{2} = -\frac{1}{2} - \frac{a}{6}$ 2

In Exercises 51–58, solve each equation. If the equation is an identity or an impossible equation, so indicate.

51. $4(2 - 3t) + 6t = -6t + 8$ identity
52. $2x - 6 = -2x + 4(x - 2)$ impossible
53. $\frac{a + 1}{4} + \frac{2a - 3}{4} = \frac{a}{2} - 2$ -6
54. $\frac{y - 8}{5} + 2 = \frac{2}{5} - \frac{y}{3}$ 0
55. $3(x - 4) + 6 = -2(x + 4) + 5x$ impossible
56. $2(x - 3) = \frac{3}{2}(x - 4) + \frac{x}{2}$ identity
57. $y(y + 2) = (y + 1)^2 - 1$ identity
58. $x(x - 3) = (x - 1)^2 - (5 + x)$ impossible

In Exercises 59–62, solve each equation. Clear each equation of decimals by multiplying both sides by 100.

59. $0.05x + 0.04(5000 - x) = 220$ 2000
60. $0.06x + 0.08(20{,}000 - x) = 1500$ 5000
61. $0.05x + 0.1x + 0.25(100 - x) = 24.5$ 5
62. $0.09y + 0.14(10{,}000 - y) = 1275$ 2500

In Exercises 63–66, write each repeating decimal number as a fraction. Simplify the answer when possible.

63. $0.\overline{3}$ $\frac{1}{3}$
64. $0.\overline{29}$ $\frac{29}{99}$
65. $-0.34\overline{89}$ $-\frac{691}{1980}$
66. $-2.3\overline{47}$ $-\frac{1162}{495}$

In Exercises 67–82, solve each problem.

67. **Integer problem** The sum of two consecutive integers is 75. Find the integers. 37, 38

68. **Integer problem** The sum of two consecutive integers is 51. Find the integers. 25, 26

69. Integer problem The sum of three consecutive even integers is 384. Find the integers. 126, 128, 130

70. Integer problem The sum of the first and third integers of three consecutive odd integers is 58. Find the middle integer. 29

71. Cutting a board The carpenter in Illustration 1 saws a board into two pieces. He wants one piece to be 1 foot longer than twice the length of the shorter piece. Find the length of each piece. 7 ft, 15 ft

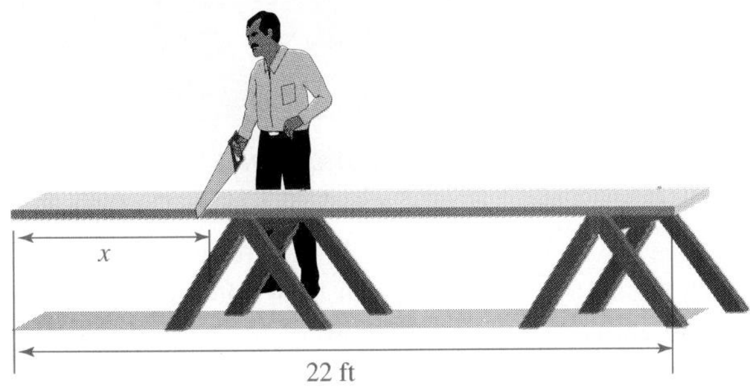

ILLUSTRATION 1

72. Cutting a beam A 30-foot steel beam is to be cut into two pieces. The longer piece is to be 2 feet more than 3 times as long as the shorter piece. Find the length of each piece. 7 ft, 23 ft

73. Cutting a rope A 60-foot rope is cut into four pieces, with each successive piece being twice as long as the previous one. Find the length of the longest piece. 32 ft

74. Cutting a cable A 185-foot cable is cut into two pieces of equal length and a third piece that is to be 5 feet longer than the sum of the equal pieces. Find the length of the longer piece. 95 ft

75. Buying a TV and a VCR See Illustration 2. If the TV costs $55 more than the VCR, how much does the TV cost? $355

ILLUSTRATION 2

76. Buying a washer and dryer The total cost of a washer and dryer is $590. If the washer costs $40 more than the dryer, find the cost of the dryer. $275

77. Finding dimensions The rectangular garden shown in Illustration 3 is twice as long as it is wide. Find its dimensions. 12 m by 24 m

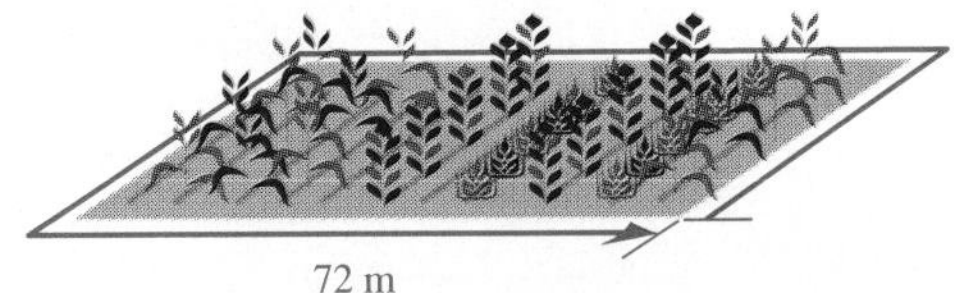

ILLUSTRATION 3

78. Finding dimensions The width of a rectangular swimming pool is one-third its length. If its perimeter is 96 meters, find the dimensions of the pool. 12 m by 36 m

79. Fencing a pasture A farmer has 624 feet of fencing to enclose the pasture shown in Illustration 4. Because a river runs along one side, fencing will be needed on only three sides. Find the dimensions of the pasture if its length is double its width. 156 ft by 312 ft

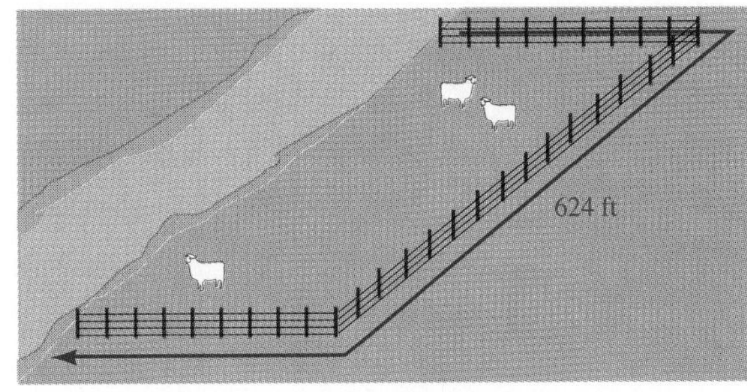

ILLUSTRATION 4

80. Fencing a pen A man has 150 feet of fencing to build the pen shown in Illustration 5. If one end is a square, find the outside dimensions of the entire pen. 20 ft by 45 ft

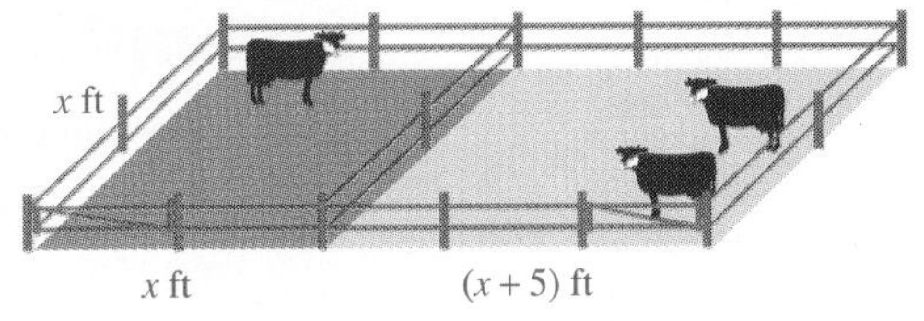

ILLUSTRATION 5

81. Enclosing a swimming pool A woman wants to enclose the swimming pool shown in Illustration 6 and have a walkway of uniform width all the way around. How wide will the walkway be if the woman uses 180 feet of fencing? 10 ft

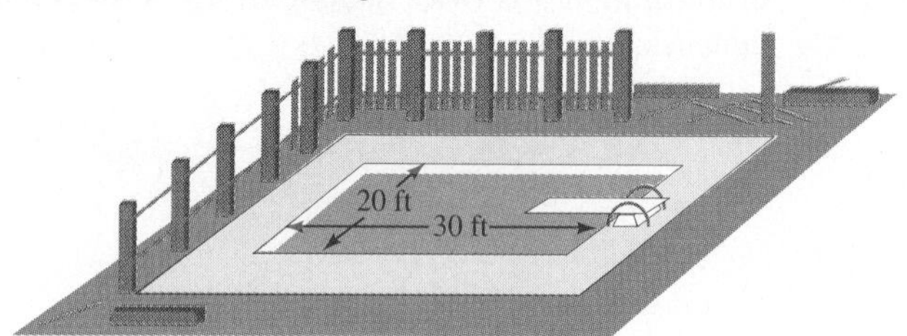

ILLUSTRATION 6

82. Framing a picture An artist wants to frame the picture shown in Illustration 7 with a frame 2 inches wide. How wide will the framed pictured be if the artist uses 70 inches of framing material? 15 in.

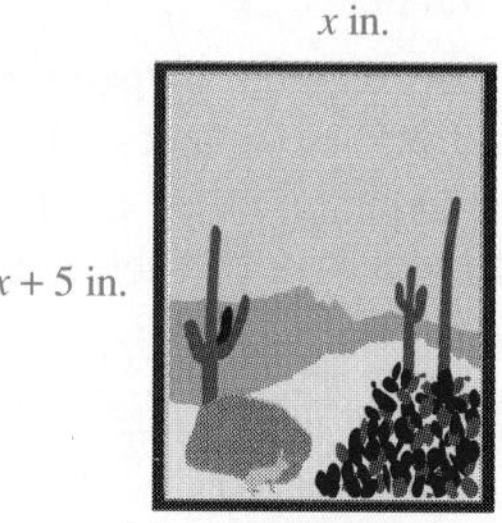

ILLUSTRATION 7

Writing Exercises *Write a paragraph using your own words.*

1. Explain the difference between a conditional equation, an identity, and an impossible equation.
2. Explain how you would solve an equation.

Something to Think About

1. Is 0.9 equal to 1? no
2. Is $0.\overline{9}$ equal to 1? yes
3. Find the error:

$$\begin{aligned} 4(x + 3) &= 16 \\ 4x + 12 &= 16 \\ 4x &= 28 \\ x &= 7 \end{aligned}$$

4. Find the error:

$$\begin{aligned} 4(x + 3) &= 16 \\ 4x + 3 &= 16 \\ 4x &= 13 \\ x &= \frac{13}{4} \end{aligned}$$

Review Exercises *Simplify each expression.*

1. $3(2x^2 - 4x + 2) + 2(3x^2 + 3x - 2)$ $12x^2 - 6x + 2$
2. $-5(2r^2 - 4r - 3) - 4(3r^2 - 3r - 3)$ $-22r^2 + 32r + 27$
3. $\frac{3}{5}(5m^3 + 10m^2 - 15m) - \frac{3}{4}(8m^2 - 12m + 4)$ $3m^3 - 3$
4. $\frac{3}{2}(3x^2 - 9x + 15) - \frac{2}{3}(8x^2 + 11x - 5)$ $-\frac{5}{6}x^2 - \frac{125}{6}x + \frac{155}{6}$

3.2 Applications

■ Business Problems ■ Lever Problems ■ Break-Point Analysis ■ Geometric Problems

In this section, we continue the discussion of problem solving. The following steps provide a strategy for solving many problems.

Strategy for Problem Solving

1. Read the problem several times and analyze the facts given. What information is given? What are you asked to find? Often a sketch or diagram will help you visualize the facts of the problem.
2. Pick a variable to represent the quantity to be found. Express all other quantities mentioned as expressions involving this single variable.
3. Find a way to express the quantity to be found in two different ways.
4. Write an equation showing that the two quantities found in Step 3 are equal.
5. Solve the equation.
6. Check the result in the words of the problem.

Business Problems

When the regular price of merchandise is reduced, the amount of reduction is called **markdown.**

Sale price = **regular price** − **markdown.**

Usually, the markdown is expressed as a percent of the regular price.

Markdown = **percent of markdown** · **regular price.**

The markdown is often called the **discount.**

EXAMPLE 1 **Furniture sale** A sofa and matching chair are on sale for \$777. If the list price was \$925, find the percent of markdown.

Analysis In this case, \$777 is the sale price, \$925 is the regular price, and the markdown is the product of \$925 and the percent of markdown.

Solution We can let r represent the percent of markdown, expressed as a decimal. We substitute \$777 for the sale price and \$925 for the regular price in the formula

Sale price = **regular price** − **markdown.**

$$777 = 925 - r \cdot 925$$

$$777 = 925 - 925r$$

$$-148 = -925r \qquad \text{Subtract 925 from both sides.}$$

$$0.16 = r \qquad \text{Divide both sides by } -925.$$

The percent of markdown is 16%.

Check: Since the markdown is 16% of \$925, or \$148, the sale price is \$925 − \$148, or \$777. ■

EXAMPLE 2 **Stock portfolio** A college foundation owns stock in IBC (selling at \$54 per share), GS (selling at \$65 per share), and ATB (selling at \$105 per share). The foundation owns equal shares of GS and IBC, but five times as many shares of ATB.

If this portfolio is worth \$450,800, how many shares of each type does the foundation own?

Analysis The value of the IBC stock plus the value of the GS stock plus the value of the ATB stock must equal \$450,800.

- If x represents the number of shares of IBC, then \$54x is the value of that stock.
- Since the foundation has equal numbers of shares of GS as IBC, x also represents the number of shares of GS. The value of this stock is \$65x.
- Since the foundation owns five times as many shares of ATB, it owns $5x$ shares of ATB. The value of this stock is \$105($5x$).

We set the sum of these values equal to \$450,800.

Solution We let x represent the number of shares of IBC.
Then x also represents the number of shares of GS.
Then $5x$ represents the number of shares of ATB.

The value of IBC stock	+	the value of GS stock	+	the value of ATB stock	=	the total value of the stock.
$54x$	+	$65x$	+	$105(5x)$	=	450,800

$$54x + 65x + 525x = 450,800 \quad 105(5x) = 525x.$$
$$644x = 450,800 \quad \text{Combine terms.}$$
$$x = 700 \quad \text{Divide both sides by 644.}$$

The foundation owns 700 shares of IBC, 700 shares of GS, and 5(700), or 3500, shares of ABT.

Check: The value of 700 shares of IBC at \$54 per share is \$37,800.
The value of 700 shares of GS at \$65 per share is \$45,500.
The value of 3500 shares of ATB at \$105 per share is \$367,500.

The sum of these values is \$450,800. ■

Lever Problems

A **lever** is a device used to increase a force and apply that force where it is needed. Crowbars and claw hammers are examples of levers used in construction; even a child's seesaw is a lever.

EXAMPLE 3 **Lever design** Design engineers must position two hydraulic cylinders as in Figure 3-3 to balance a 9500-pound force at point A. The first cylinder at the end of the lever exerts a 3500-pound force. Where should the design engineers position the second cylinder, which is capable of exerting a 5500-pound force?

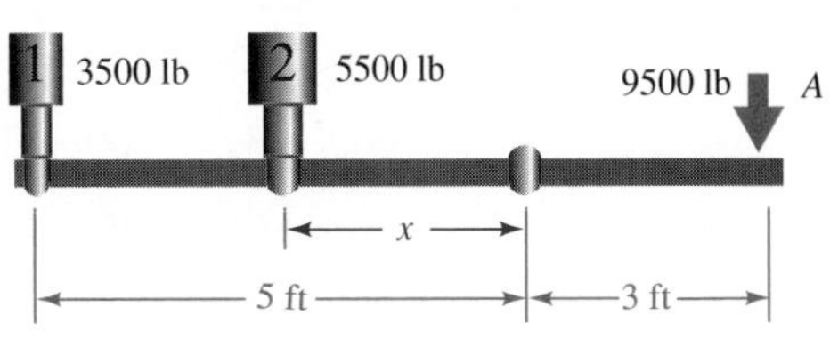

FIGURE 3-3

Analysis From physics, the lever will be in balance when the force of the first cylinder multiplied by its distance from the pivot (also called the **fulcrum**), added to the second cylinder's force multiplied by its distance from the fulcrum, is equal to the product of the 9500-pound force and its distance from the fulcrum.

Solution We let x represent the distance from the larger cylinder to the fulcrum.

Force of cylinder 1 times its distance	+	force of cylinder 2 times its distance	=	force to be balanced times its distance.
$3500 \cdot 5$	+	$5500x$	=	$9500 \cdot 3$

$$17{,}500 + 5500x = 28{,}500$$

$$5500x = 11{,}000 \quad \text{Subtract 17,500 from both sides.}$$

$$x = 2 \quad \text{Divide both sides by 5500.}$$

The design must specify that the second cylinder be positioned 2 feet from the fulcrum.

Check: $3500 \cdot 5 + 5500 \cdot 2 = 17{,}500 + 11{,}000 = 28{,}500$

$9500 \cdot 3 = 28{,}500$ ■

Break-Point Analysis

Running a machine involves two types of costs—**setup costs** and **unit costs.** Setup costs include the cost of preparing a machine to do a certain job. Unit costs depend on the number of items to be manufactured, including such costs as raw materials and labor.

Suppose that a certain machine has a setup cost of \$600 and a unit cost of \$3. If x items are manufactured using this machine, the cost will be

$$\text{Cost} = 600 + 3x$$

Furthermore, suppose that a larger and more efficient machine has a setup cost of \$800 and a unit cost of \$2. The cost of manufacturing x items using this machine is

$$\text{Cost on larger machine} = 800 + 2x$$

The **break point** is the number of units x that need to be manufactured to make the cost the same using either machine. It can be found by setting the two costs equal to each other and solving for x.

$$600 + 3x = 800 + 2x$$
$$x = 200 \quad \text{Subtract 600 and } 2x \text{ from both sides.}$$

The break point is 200 units, because the cost using either machine is \$1200:

$$\begin{aligned}\text{Cost on small machine} &= 600 + 3x\\ &= 600 + 3(200)\\ &= 600 + 600\\ &= 1200\end{aligned}$$

$$\begin{aligned}\text{Cost on larger machine} &= 800 + 2x\\ &= 800 + 2(200)\\ &= 800 + 400\\ &= 1200\end{aligned}$$

EXAMPLE 4 **Finding the break point** One machine has a setup cost of \$400 and a unit cost of \$1.50, and another machine has a setup cost of \$500 and a unit cost of \$1.25. Find the break point.

Analysis If x represents the number of items to be manufactured, the cost C_1 using machine 1 is

$$C_1 = 400 + 1.5x$$

and the cost C_2 using machine 2 is

$$C_2 = 500 + 1.25x$$

The break point is the value of x when $C_1 = C_2$.

Solution Let x represent the number of items to be manufactured. Then $400 + 1.5x$ represents the cost using machine 1, and $500 + 1.25x$ represents the cost using machine 2.

The break point occurs when these two costs are equal.

The cost of using machine 1	=	the cost of using machine 2.
$400 + 1.5x$	=	$500 + 1.25x$

$$\begin{aligned}1.5x &= 100 + 1.25x && \text{Subtract 400 from both sides.}\\ 0.25x &= 100 && \text{Subtract } 1.25x \text{ from both sides.}\\ x &= 400 && \text{Divide both sides by 0.25.}\end{aligned}$$

The break point is 400 units.

Check: The cost using machine 1 is $400 + 1.5(400) = 400 + 600 = 1000$.
The cost using machine 2 is $500 + 1.25(400) = 500 + 500 = 1000$.
Since the costs are equal, the break point is 400. ■

Geometric Problems

EXAMPLE 5 **Isosceles triangles** Find the measure of each base angle of the isosceles triangle shown in Figure 3-4.

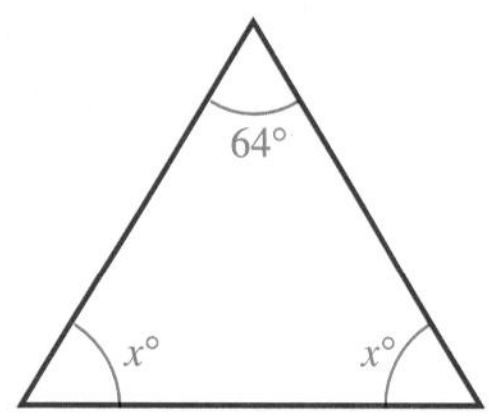

FIGURE 3-4

Analysis An **isosceles triangle** is a triangle with two equal sides that meet to form the **vertex angle.** The angles opposite the equal sides, called the **base angles,** are also equal. If we let $x°$ represent the measure of one base angle, then the measure of the other base angle is also $x°$. In any triangle, the sum of the measures of its three angles is 180°.

Solution We let x represent the measure of one base angle.
Then x also represents the measure of the other base angle.

We can form the equation

The measure of one base angle	+	the measure of the other base angle	+	the measure of the vertex angle	=	180°.
x	+	x	+	64	=	180

$$x + x + 64 = 180$$
$$2x + 64 = 180 \quad \text{Combine terms.}$$
$$2x = 116 \quad \text{Subtract 64 from both sides.}$$
$$x = 58 \quad \text{Divide both sides by 2.}$$

The measure of each base is 58°.

Check: The sum of the measures of each base angle and the vertex angle is 180°:

$$58° + 58° + 64° = 180°$$ ■

EXAMPLE 6 **Dimensions of a rectangle** The width of a rectangle is 4 centimeters less than its length. If the width is increased by 2 centimeters and the length is increased by 8 centimeters, the area is increased by 84 square centimeters. Find the dimensions of the original rectangle.

Analysis The area, A, of a rectangle is given by the formula $A = lw$, where l is the length of the rectangle and w is its width.

If the length of the original rectangle is l, then the width is $l - 4$, and the area is $l(l - 4)$, as shown in Figure 3-5. If the width is increased by 2 centimeters and the length is increased by 8 centimeters, the width and the length of the larger rectangle are $l - 2$ and $l + 8$. Thus, the area of the larger rectangle is $(l - 2)(l + 8)$.

If 84 square inches are added to the area of the original rectangle, the result is the same as the area of the larger rectangle.

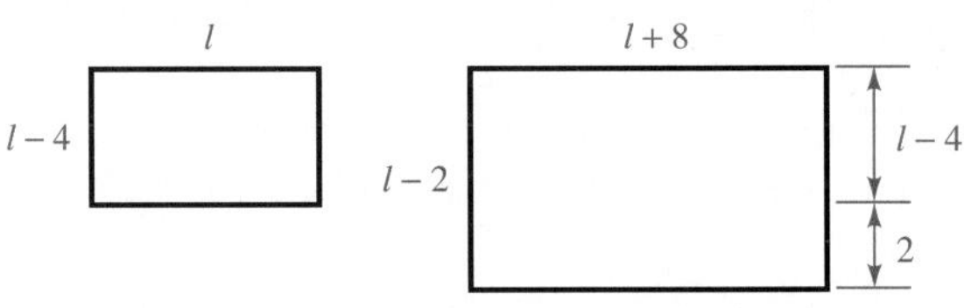

FIGURE 3-5

Solution Let l represent the length of the original rectangle.
Then $l - 4$ represents the width of the original rectangle.
Then $l - 2$ represents the width of the larger rectangle.
Then $l + 8$ represents the length of the larger rectangle.

The area of the original rectangle	+	84	=	the area of the larger rectangle.
$l(l - 4)$	+	84	=	$(l - 2)(l + 8)$

$$l^2 - 4l + 84 = l^2 + 6l - 16 \quad \text{Remove parentheses.}$$
$$-4l + 84 = 6l - 16 \quad \text{Subtract } l^2 \text{ from both sides.}$$
$$100 = 10l \quad \text{Add } 4l + 16 \text{ to both sides.}$$
$$10 = l \quad \text{Divide both sides by 10.}$$

$$l = 10$$

and $$l - 4 = 6$$

The dimensions of the original rectangle are 6 centimeters by 10 centimeters.

Check: The area of the original rectangle is $6 \cdot 10 = 60$ square centimeters. The area of the larger rectangle is $8 \cdot 18 = 144$ square centimeters. The larger rectangle is 84 square centimeters greater than the original rectangle. ■

ORALS *Find each value.*

1. 20% of 500 100

2. $33\frac{1}{3}$% of 600 200

If a stock costs $54, find the cost of

3. 5 shares $270

4. x shares $\$54x$

Find the area of the rectangle with the given dimensions.

5. 6 meters long, 4 meters wide 24 m^2

6. l meters long, $l - 5$ meters wide $l(l - 5)\ m^2$

EXERCISE 3.2

In Exercises 1–22, solve each problem.

1. Buying a washer and dryer Find the percent of markdown of the sale in Illustration 1. 20%

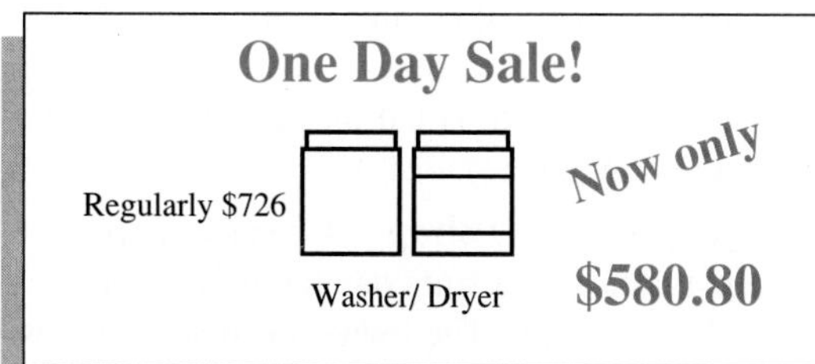

ILLUSTRATION 1

2. Buying furniture A bedroom set regularly sold for $983. If it is on sale for $737.25, what is the percent of markdown? 25%

3. Buying a calculus book A bookstore buys a used calculus book for $12 and sells it for $40. Find the percent of markup. 233%

4. Selling stuffed animals The owner of a gift shop buys stuffed animals for $18 and sells them for $30. Find the percent of markup. 67%

5. Value of an IRA In an Individual Retirement Account (IRA) valued at $53,900, a student has 500 shares of stock in Big Bank Corporation and Safe Savings and Loan. If Big Bank sells for $115 per share and Safe Savings sells for $97 per share, how many shares of each does the student own? 300 shares of BB, 200 shares of SS

6. Assets of a pension fund A pension fund owns 12,000 shares in mutual stock funds and mutual bond funds. Currently, the stock funds sell for $12 per share, and the bond fund sells for $15 per share. How many shares of each does the pension fund own if the value of the securities is $165,000? 5000 shares of stock funds, 7000 shares of bond funds

7. Selling calculators Last month, the bookstore sold 85 calculators, generating $3875 of income. (See Illustration 2.) How many of each type of calculator did the bookstore sell? 35 $15 calculators, 50 $67 calculators

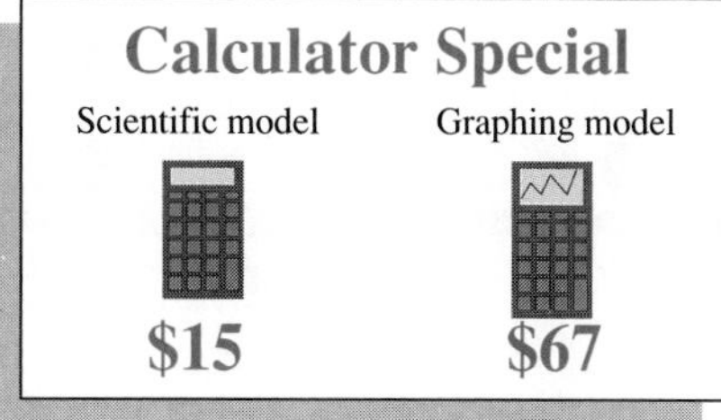

ILLUSTRATION 2

8. Selling grass seed A seed company sells two grades of grass seed. A 100-pound bag of a mixture of rye and Kentucky bluegrass sells for $245, and a 100-pound bag of bluegrass sells for $347. How many bags of each are sold if the receipts for 19 bags are $5369? 12 bags of mixture, 7 bags of bluegrass

9. **Buying roses** A man with $21.25 stops after work to buy some roses for his wife's birthday. If each rose costs $1.25 and there is a delivery charge of $5, how many roses can he buy? 13

10. **Renting a truck** To move to Wisconsin, a man can rent a truck for $29.95 per day plus 19¢ per mile. If he keeps the truck for one day, how many miles can he drive for a cost of $77.45? 250

11. **Balancing a seesaw** A seesaw is 20 feet long and the fulcrum is in the center. If an 80-pound boy sits at one end, how far will the boy's 160-pound father have to sit from the fulcrum to balance the seesaw? 5 ft

12. **Establishing equilibrium** Two forces—110 pounds and 88 pounds—are applied to opposite ends of an 18-foot lever. How far from the greater force must the fulcrum be placed so that the lever is balanced? 8 ft

13. **Moving a stone** A woman uses a 10-foot bar to lift a 210-pound stone. If she places another rock 3 feet from the stone to act as the fulcrum, how much force must she exert to move the stone? 90 lb

14. **Lifting a car** A 350-pound football player brags that he can lift a 2500-pound car. If he uses a 12-foot bar with the fulcrum placed 3 feet from the car, will he be able to lift the car? no

15. **Balancing a lever** Forces are applied to a lever as indicated in Illustration 3. Determine x, the distance of the smallest force from the fulcrum. 4 ft

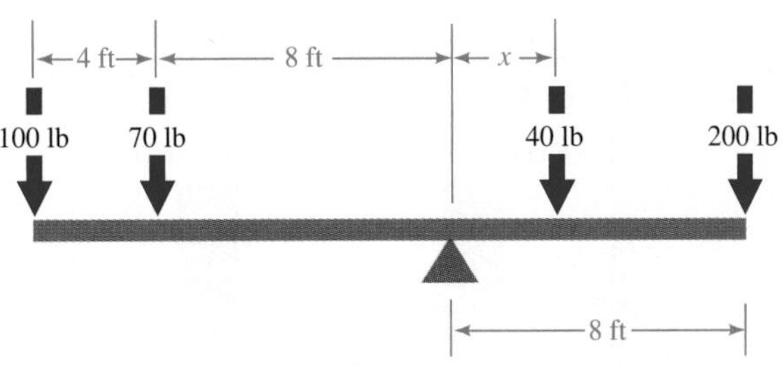

ILLUSTRATION 3

16. **Balancing a seesaw** Jim and Bob sit at opposite ends of an 18-foot seesaw, with the fulcrum at its center. Jim weighs 160 pounds, and Bob weighs 200 pounds. Kim sits 4 feet in front of Jim, and the seesaw balances. How much does Kim weigh? 72 lb

17. **Milling brass plates** Two machines can mill a brass plate. One machine has a setup cost of $300 and a cost per plate of $2. The other machine has a setup cost of $500 and a cost per plate of $1. Find the break point. Which machine should be used if 1000 plates are needed? 200 plates; the second machine

18. **Printing books** A printer has two presses. One has a setup cost of $210 and can print the pages of a certain book for $5.98. The other press has a setup cost of $350 and can print the pages of the same book for $5.95. Find the break point. Which machine should be used to print a run of 4500 copies? $4666\frac{2}{3}$; use the first press

19. **Managing a computer store** The manager of a computer store knows that his fixed costs are $8925 per month and that his unit cost is $850 for every computer sold. If he can sell all the computers he can get for $1275 each, how many computers must he sell each month to break even? (*Hint:* The **break-even point** is the number of computers that must be sold for the wholesale cost of the computers plus the fixed costs to equal the revenue from the sale of the computers.) 21

20. **Managing a beauty shop** A beauty shop specializing in permanents has fixed costs of $2101.20 per month. The owner estimates that the cost for each permanent is $23.60. This cost covers labor, chemicals, and electricity. If her shop can give as many permanents as she wants at a price of $44 each, how many must be given each month to break even? 103

21. **Running a small business** A person invests $18,375 to set up a small business, producing a piece of computer software that will sell for $29.95. If each piece can be produced for $5.45, how many pieces must be sold to break even? 750

22. **Running a record company** Three people invest $35,000 each to start a record company that will produce reissues of classic jazz. Each release will be a set of 3 CDs that will retail for $15 per disc. If each set can be produced for $18.95, how many sets must be sold for the investors to make a profit? 1344

In Exercises 23–26, a paint manufacturer can choose between two processes for manufacturing house paint, with monthly costs as shown in Table 1. Assume the paint sells for $18 per gallon.

Process	Fixed costs	Unit cost (per gallon)
A	$32,500	$13
B	$80,600	$ 5

TABLE 1

23. Find the break-even point for process A.
6500 gal per month

24. Find the break-even point for process B.
6200 gal per month

25. If expected sales are 6000 gallons per month, which process should the company use? A

26. If expected sales are 7000 gallons per month, which process should the company use? B

In Exercises 27–32, a manufacturer of automobile water pumps is considering retooling for one of two manufacturing processes, with monthly fixed costs and unit costs as indicated in Table 2. Each water pump can be sold for $50.

Process	Fixed costs	Unit cost
A	$12,390	$29
B	$20,460	$17

TABLE 2

27. Find the break-even point for process A.
590 units per month

28. Find the break-even point for process B.
620 units per month

29. If expected sales are 550 per month, which process should be used? A (smaller loss)

30. If expected sales are 600 per month, which process should be used? A

31. If expected sales are 650 per month, which process should be used? A

32. At what monthly sales level is process B better?
673 units

33. Temperature scales The Celsius and Fahrenheit temperature scales are related by the equation $C = \frac{5}{9}(F - 32)$. At what temperature will a Fahrenheit and a Celsius thermometer give the same reading?
$-40°$

34. Width of a rectangle A rectangle has a length of 9 centimeters and an area of 54 square centimeters. Find its width. 6 cm

35. Dimensions of a rectangle A rectangle is 6 feet longer than it is wide, and its area is 30 more than the square of its width. Find its dimensions.
5 ft by 11 ft

36. Dimensions of a rectangle The length of a rectangle is 2 inches longer than its width. If both the length and the width are increased by 5 inches, the area is increased by 75 square inches. Find the dimensions of the original rectangle.
6 in. by 4 in.

37. Dimensions of a rectangle The length of a rectangle is 1 foot longer than its width. If the length is increased by 2 feet and the width is decreased by 6 feet, the area is decreased by 50 square feet. Find the dimensions of the original rectangle.
8 ft by 9 ft

38. **Perimeter of a rectangle** A rectangle is 11 meters longer than it is wide. If the length is decreased by 12 meters and the width is increased by 10 meters, the area is decreased by 24 square meters. Find the perimeter of the original rectangle. 50 m

39. **Height of a triangle** The height of the larger triangle in Illustration 4 is three times the height of the smaller triangle, and the area is 60 square feet greater. Find the height of the smaller triangle. (*Hint:* $A = \frac{1}{2}bh$.) 5 ft

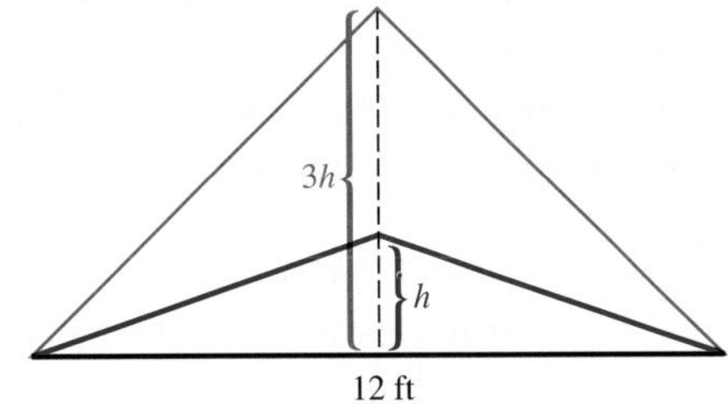

ILLUSTRATION 4

40. **Height of a triangle** If the height of a triangle with a base of 8 inches is tripled, its area is increased by 96 square inches. Find the height of the triangle. 12 in.

41. **Engineering design** The width, w, of the flange in the engineering drawing in Illustration 5 has not yet been determined. Find w so that the area of the rectangular portion is exactly one-half of the total area. 18 in.

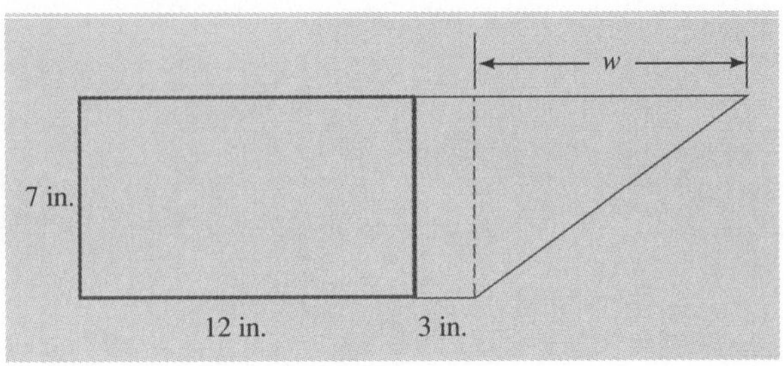

ILLUSTRATION 5

42. **Installing solar heating** One solar panel in Illustration 6 is to be 6 feet wider than the other, but to be equally efficient, they must have the same area. Find the width of each. 16 and 22 feet

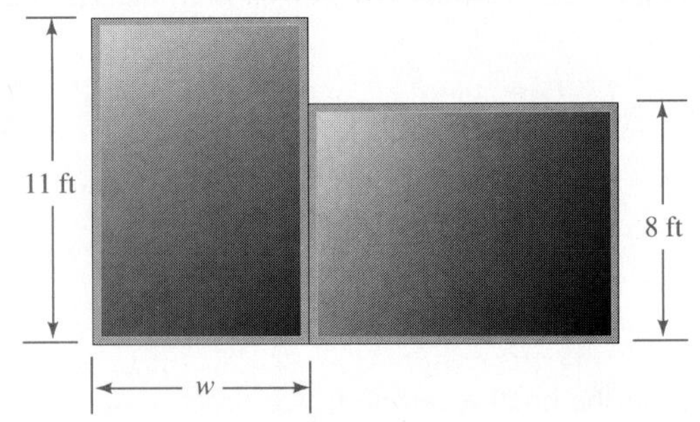

ILLUSTRATION 6

Writing Exercises ■ *Write a paragraph using your own words.*

1. Explain the steps for solving an applied problem.
2. Explain how to check the solution of an applied problem.

Something to Think About ■

1. Find the distance x required to balance the lever.
2. Interpret the answer to Question 1.

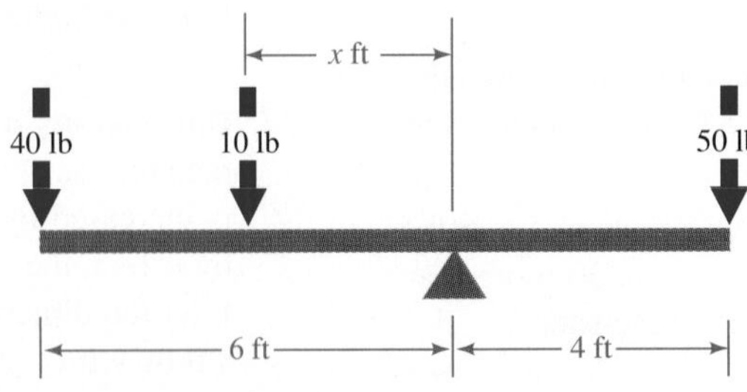

Review Exercises *Simplify each expression.*

1. $\left(\frac{3x^{-3}}{4x^2}\right)^{-4}$ $\frac{256x^{20}}{81}$

2. $\left(\frac{r^{-3}s^2}{r^2r^3s^{-4}}\right)^{-5}$ $\frac{r^{40}}{s^{30}}$

3. $\frac{a^m a^3}{a^2}$ a^{m+1}

4. $\left(\frac{b^m}{b^3}\right)^3$ b^{3m-9}

3.3 More Applications of Equations

■ Investment Problems ■ Uniform Motion Problems ■ Mixture Problems

Investment Problems

EXAMPLE 1 **Financial planning** A professor has \$15,000 to invest for one year. She invests some at 8% and the rest at 7%. If her total income from these investments is \$1110, how much did she invest at each rate?

Analysis We will add the interest from the 8% investment to the interest from the 7% investment and set the sum equal to the total interest earned.

For simple interest, the interest earned is computed by the formula $i = prt$, where p is the principal, i is the annual interest rate, and t is the length of time the principal is invested. Thus, if \$$x$ are invested at 8%, the interest earned is \$$0.08x$. If the remaining \$$(15{,}000 - x)$ is invested at 7%, the amount earned on that investment is \$$0.07(15{,}000 - x)$. The sum of these two amounts equals \$1110.

Solution We can let x represent the number of dollars invested at 8%.
Then $15{,}000 - x$ represents the number of dollars invested at 7%.

The interest earned at 8%	+	the interest earned at 7%	=	the total interest.
$0.08x$	+	$0.07(15{,}000 - x)$	=	1110

$8x + 7(15{,}000 - x) = 111{,}000$ Multiply both sides by 100 to eliminate the decimals.

$8x + 105{,}000 - 7x = 111{,}000$ Use the distributive property to remove parentheses.

$x + 105{,}000 = 111{,}000$ Combine terms.

$x = 6000$ Subtract 105,000 from both sides.

$15{,}000 - x = 9000$

She invested \$6000 at 8% and \$9000 at 7%.

Check: The interest on \$6000 is 0.08(\$6000) = \$480. The interest earned on \$9000 is 0.07(\$9000) = \$630. The total interest is \$1110. ■

Uniform Motion Problems

EXAMPLE 2 **Travel time** A car leaves Rockford traveling toward Wausau at the rate of 55 miles per hour. At the same time, another car leaves Wausau traveling toward Rockford at the rate of 50 miles per hour. How long will it take them to meet if the cities are 157.5 miles apart?

Analysis In this case, the cars are traveling toward each other as shown in Figure 3-6.

Uniform motion problems are based on the formula $d = rt$, where d is distance, r is rate, and t is time. We can organize the information given in the chart shown in Figure 3-7.

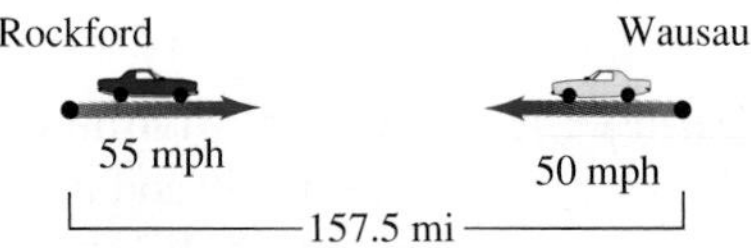

FIGURE 3-6

We know that one car is traveling at 55 miles per hour and that the slower car is traveling at 50 miles per hour. We also know that they travel for the same amount of time—say, t hours. Thus, the distance that the faster car travels is $55t$ miles, and the distance that the slower car travels is $50t$ miles. The sum of these distances equals 157.5 miles, the distance between the cities.

	d =	r ·	t
Faster car	$55t$	55	t
Slower car	$50t$	50	t

FIGURE 3-7

Solution We can let t represent the time that each car travels.
Then $55t$ represents the distance traveled by the faster car.
Then $50t$ represents the distance traveled by the slower car.

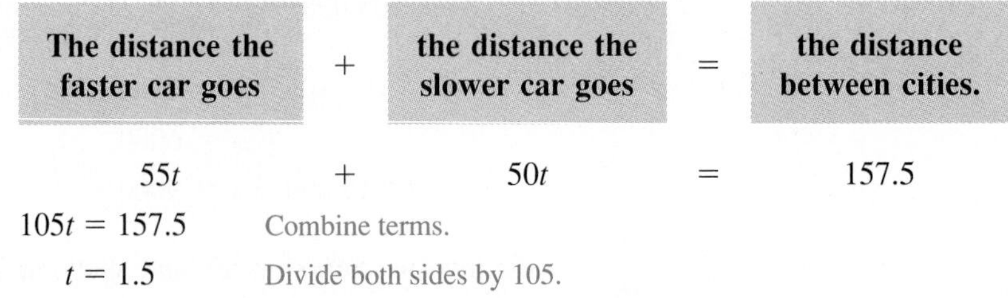

$$55t + 50t = 157.5$$

$$105t = 157.5 \quad \text{Combine terms.}$$

$$t = 1.5 \quad \text{Divide both sides by 105.}$$

The two cars will meet in $1\frac{1}{2}$ hours.

Check: The faster car travels $1.5(55) = 82.5$ miles. The slower car travels $1.5(50) = 75$ miles. The total distance traveled is 157.5 miles. ■

Mixture Problems

EXAMPLE 3 **Mixed nuts** The owner of a candy store notices that 20 pounds of cashews are getting stale. They did not sell because of their high price ($12 per pound). The store owner decides to mix peanuts with the cashews to lower the price per pound. If peanuts sell for $3 per pound, how many pounds of peanuts must be mixed with the cashews to make a mixture that could be sold for $6 per pound?

Analysis This problem is based on the formula $V = pn$, where V represents value, p represents the price per pound, and n represents the total number of pounds.

We can let x represent the total number of pounds in the mixture and enter the known information in the chart shown in Figure 3-8. The value of the cashews plus the value of the peanuts will be equal to the value of the mixture.

	V =	p ·	n
Cashews	240	12	20
Peanuts	$3x$	3	x
Mixture	$6(20 + x)$	6	$20 + x$

FIGURE 3-8

Solution We can let x represent the number of pounds of peanuts to be used. Then $20 + x$ represents the number of pounds in the mixture.

The value of the cashews	+	the value of the peanuts	=	the value of the mixture.
240	+	$3x$	=	$6(20 + x)$

$240 + 3x = 120 + 6x$ Use the distributive property to remove parentheses.

$120 = 3x$ Subtract $3x$ and 120 from both sides.

$40 = x$ Divide both sides by 3.

The store owner should mix 40 pounds of peanuts with the 20 pounds of cashews.

Check: Since

The cashews are valued at \$12(20) = \$240 and
The peanuts are valued at \$3(40) = \$120,
The mixture is valued at \$6(60) = \$360.

The value of the cashews plus the value of the peanuts equals the value of the mixture. ■

EXAMPLE 4 **Milk production** A container is partially filled with 12 liters of whole milk containing 4% butterfat. How much 1% milk must be added to get a mixture that is 2% butterfat?

Analysis If the first container shown in Figure 3-9 contains 12 liters of 4% milk, it contains 0.04(12) liters of butterfat. To this container, we will add the contents of the second container, which holds $0.01l$ liters of butterfat. The sum of these two amounts of butterfat will be the amount of butterfat in the third container, which is $0.02(12 + l)$ liters of butterfat.

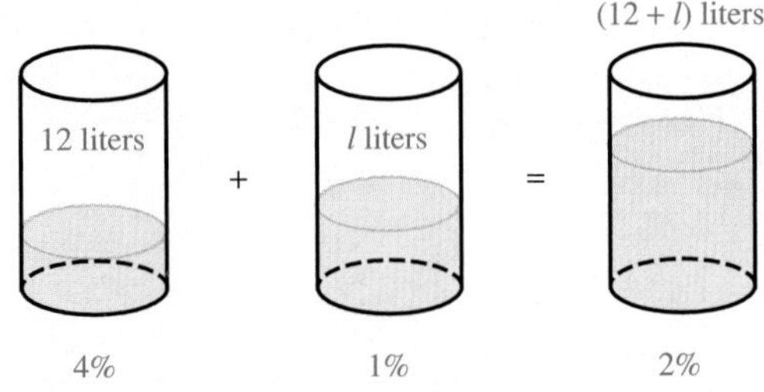

FIGURE 3-9

Solution We can let l represent the number of liters of 1% milk to be added. Then

The amount of butterfat in 12 liters of 4% milk	+	the amount of butterfat in l liters of 1% milk	=	the amount of butterfat in $(12 + l)$ liters of mixture.
$0.04(12)$	+	$0.01l$	=	$0.02(12 + l)$

$4(12) + 1l = 2(12 + l)$ Multiply both sides by 100.

$48 + l = 24 + 2l$ Use the distributive property to remove parentheses.

$24 = l$ Subtract 24 and l from both sides.

Thus, 24 liters of 1% milk should be added.

Check: 12 liters of 4% milk contains 0.48 liters of butterfat.
24 liters of 1% milk contains 0.24 liters of butterfat.

This gives a total of 36 liters of a mixture that contains 0.72 liters of butterfat. This is a 2% solution. ■

ORALS *Assume all investments are for one year.*

1. How much interest will $1500 earn if invested at 6%? $90
2. How much interest will $$x$ earn if invested at 5%? $0.05x$
3. If $$x$ of $30,000 is invested at 5%, how much is left to be invested at 6%? $$(30,000 - x)$
4. If Brazil nuts are worth $$x$ per pound, how much will 20 pounds be worth? $$20x$
5. If whole milk is 4% butterfat, how much butterfat is in 2 gallons? 0.08 gal

EXERCISE 3.3

1. **Investing money** Lured by the advertisement in Illustration 1, a woman invested $12,000, some in a money market account and the rest in a 5-year CD. How much was invested in each account if the income from both investments is $1060 per year? $2000 at 8%, $10,000 at 9%

First Republic Savings and Loan

Account	Rate
NOW	5.5%
Savings	7.5%
Money market	8.0%
Checking	4.0%
5-year CD	9.0%

ILLUSTRATION 1

2. **Investing money** A man invested $14,000, some at 7% and some at 10% annual interest. The annual income from these investments was $1280. How much did he invest at each rate? $4000 at 7%, $10,000 at 10%
3. **Supplemental income** A teacher wants to earn $1500 per year in supplemental income from a cash gift of $16,000. She puts $6000 in a credit union that pays 7% annual interest. What rate must she earn on the remainder to achieve her goal? 10.8%
4. **Inheriting money** Paul split an inheritance between two investments, one paying 7% annual interest and the other 10%. He invested twice as much in the 10% investment as he did in the 7% investment. If his combined annual income from the two investments was $4050, how much did he inherit? $45,000
5. **Investing money** Maria has some money to invest. If she could invest $3000 more, she could qualify for an 11% investment. Otherwise, she could invest the money at 7.5% annual interest. If the 11% investment would yield twice as much annual income as the 7.5% investment, how much does she have on hand to invest? $8250
6. **Supplemental income** A bus driver wants to earn $3500 per year in supplemental income from an inheritance of $40,000. If the driver invests $10,000 in a mutual fund paying 8%, what rate must he earn on the remainder to achieve his goal? 9%
7. **Concert receipts** For a jazz concert, student tickets were $2 each and adult tickets were $4 each. If 200 tickets were sold and the total receipts were $750, how many student tickets were sold? 25
8. **School play** At a school play, 140 tickets were sold, with total receipts of $290. If adult tickets cost $2.50 each and student tickets cost $1.50 each, how many adult tickets were sold? 80

9. **Computing time** One car leaves Chicago headed for Cleveland, a distance of 343 miles. At the same time, a second car leaves Cleveland headed toward Chicago. If the first car averages 50 miles per hour and the second car averages 48 miles per hour, how long will it take the cars to meet? $3\frac{1}{2}$ hr

10. **Cycling** A cyclist leaves Las Vegas riding at the rate of 18 miles per hour. One hour later, a car leaves Las Vegas going 45 miles per hour in the same direction. How long will it take the car to overtake the cyclist? $\frac{2}{3}$ hr

11. **Computing distance** At 2 P.M. two cars leave Eagle River, WI, one headed north and one headed south. If the car headed north averages 50 miles per hour and the car headed south averages 60 miles per hour, when will the cars be 165 miles apart? $1\frac{1}{2}$ hr

12. **Running a marathon race** Two marathon runners leave the starting gate, one running 12 miles per hour and one running 10 miles per hour. If they maintain the pace, how long will it take for them to be one-quarter of a mile apart? $\frac{1}{8}$ hr

13. **Riding a jet ski** A jet ski can go 12 miles per hour in still water. If a rider goes upstream for 3 hours against a current of 4 miles per hour, how long will it take the rider to return? (*Hint:* Upstream speed is $(12 - 4)$ mph; how far can the rider go in 3 hours?) $1\frac{1}{2}$ hr

14. **Taking a walk** Sarah walked north at the rate of 3 miles per hour and returned at the rate of 4 miles per hour. How many miles did she walk if the round trip took 3.5 hours? 12 mi

15. **Computing travel time** Grant traveled a distance of 400 miles in 8 hours. Part of the time, his rate of speed was 45 miles per hour, and part of the time his rate of speed was 55 miles per hour. How long did Grant travel at each rate? 4 hr at each rate

16. **Riding a motorboat** The motorboat in Illustration 2 can go 18 miles per hour in still water. If it can go 80 miles downstream in 4 hours and return in 5 hours, find the speed of the current. 2 mph

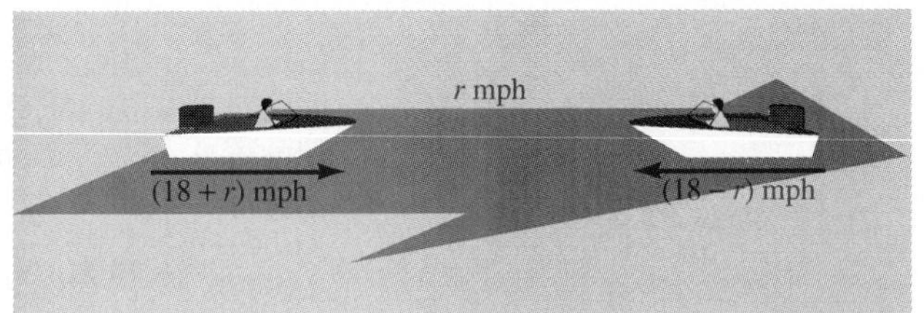

ILLUSTRATION 2

17. **Mixing candies** The owner of a candy store wants to make a 30-pound mixture of two candies to sell for \$1 per pound. If one candy sells for 95¢ per pound and the other candy sells for \$1.10 per pound, how many pounds of each should be used? 20 lb of 95¢ candy; 10 lb of \$1.10 candy

18. **Computing selling price** A mixture of candy is made to sell for 89¢ per pound. If 32 pounds of a cheaper candy, selling for 80¢ per pound, are used along with 12 pounds of a more expensive candy, find the price per pound of the better candy. \$1.13

19. **Diluting solutions** How much water should be added to 20 ounces of a 15% solution of alcohol in Illustration 3 to dilute it to a 10% solution? 10 oz

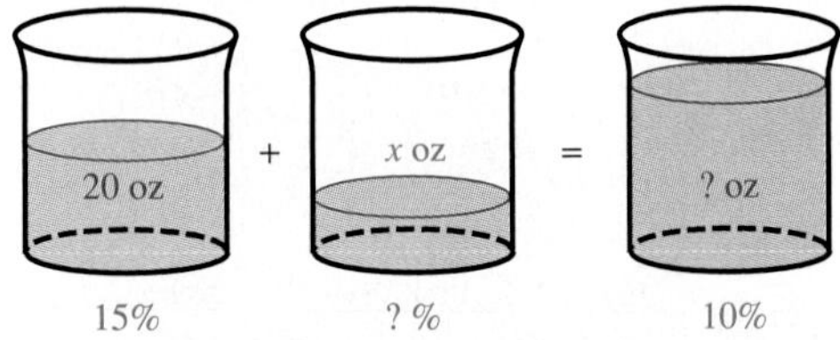

ILLUSTRATION 3

20. **Increasing concentration** How much water must be boiled away to increase the concentration of 300 gallons of a salt solution from 2% to 3%? 100 gal

21. **Making whole milk** Cream is approximately 22% butterfat. How many gallons of cream must be mixed with milk testing at 2% butterfat to give 20 gallons of milk containing 4% butterfat? 2 gal

22. **Mixing solutions** How much acid must be added to 60 grams of a solution that is 65% acid to obtain a new solution that is 75% acid? 24 g

23. **Raising grades** A student had a score of 70% on a test that contained 30 questions. To improve his score, the instructor agreed to let him work 15 additional questions. How many must he get right to raise his grade to 80%? 15

24. **Raising grades** On a second exam, the student in Exercise 23 earned a score of 60% on another 30-question test. This time, the instructor allowed him to work 20 extra problems to improve his score. How many must he get right to raise his grade to 70%? 17

25. **Computing grades** Before the final, Maria had earned a total of 375 points on four tests. To receive an A in the course, she must have 90% of a possible total of 450 points. Find the lowest number of points that she can earn on the final and still receive an A. 30

26. **Computing grades** A student has earned a total of 435 points on five algebra tests. To receive a B in the course, he must have 80% of a possible total of 600 points. Find the lowest number of points that the student can make on the final and still receive a B. 45

27. **Managing a bookstore** A bookstore sells a calculus book for \$65. If the bookstore makes a profit of 30% on each book, what does the bookstore pay the publisher for each book? (*Hint:* The retail price = the wholesale price + the markup.) \$50

28. **Managing a bookstore** A bookstore sells a textbook for \$39.20. If the bookstore makes a profit of 40% on each sale, what does the bookstore pay the publisher for each book? (*Hint:* The retail price = the wholesale price + the markup.) \$28

29. **Making furniture** A woodworker wants to put two partitions crosswise in a drawer that is 28 inches deep. (See Illustration 4.) He wants to place the partitions so that the spaces created increase by 3 inches from the front to back. If the thickness of each partition is $\frac{1}{2}$ inch, how far from the front end should he place the first partition? 6 in.

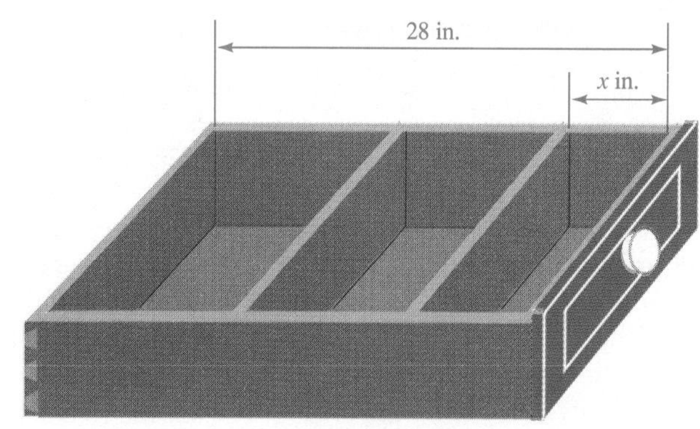

ILLUSTRATION 4

30. **Building shelves** A carpenter wants to put four shelves on an 8-foot wall so that the five spaces created decrease by 6 inches as we move up the wall. (See Illustration 5.) If the thickness of each shelf is $\frac{3}{4}$ inch, how far will the bottom shelf be from the floor? $30\frac{3}{5}$ in.

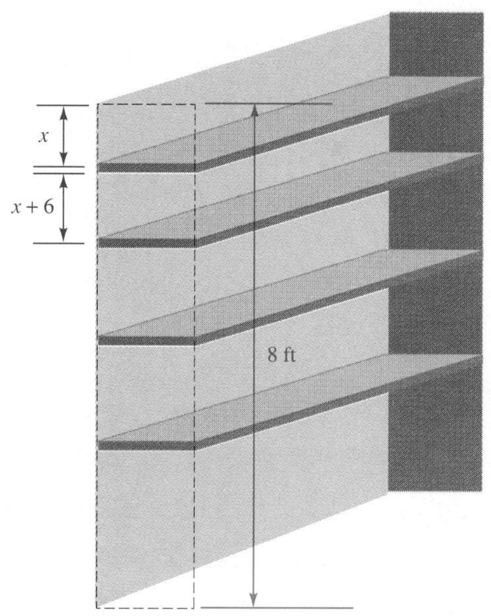

ILLUSTRATION 5

Writing Exercises ■ *Write a paragraph using your own words.*

1. What do you find most difficult in solving application problems, and why?

2. Which type of application problem do you find easiest, and why?

Something to Think About ■

1. Discuss the difficulties in solving this problem:
A man drives 100 miles at 30 miles per hour. How fast should he drive on the return trip to average 60 mph for the entire trip?

2. Discuss the difficulties in solving this problem:
Adult tickets cost \$4, and student tickets cost \$2. Sales of 71 tickets brought in \$245. How many of each were sold?

Review Exercises ■ *Multiply.*

1. $3x^2y^3(1 + 2x - 4x^2)$
$3x^2y^3 + 6x^3y^3 - 12x^4y^3$

2. $(x + 2y)(x - 2y)$
$x^2 - 4y^2$

3. $(3t + 2)(2t - 3)$
$6t^2 - 5t - 6$

4. $(2r^2 + 3b^3)(4r^4 - 6r^2b^3 + 9b^6)$
$8r^6 + 27b^9$

3.4 Formulas and Literal Equations

■ Solving Literal Equations

Equations with several variables are called **literal equations.** Often these equations are **formulas.** For example, we have seen that the formula for computing the area of a rectangle is

$$A = lw$$

where l is the length of the rectangle and w is its width. To compute the area of a rectangle with a length of 30 centimeters and a width of 15 centimeters, we use the formula and proceed as follows:

$$A = lw$$
$$A = \mathbf{30}(\mathbf{15}) \quad \text{Substitute 30 for } l \text{ and 15 for } w.$$
$$= 450$$

The area of the rectangle is 450 square centimeters.

Table 3.1 shows many formulas from geometry that we can use to compute the areas and volumes of many geometric figures.

Figure	Name	Area	Figure	Name	Volume
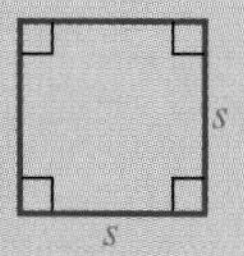	Square	$A = s^2$	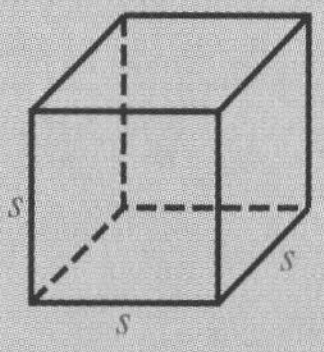	Cube	$V = s^3$
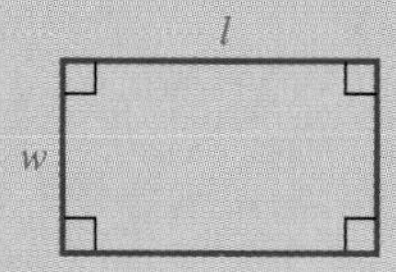	Rectangle	$A = lw$	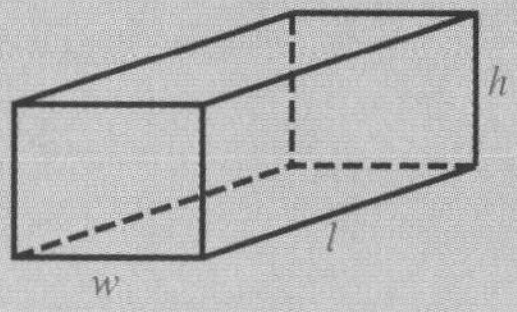	Rectangular solid	$V = lwh$
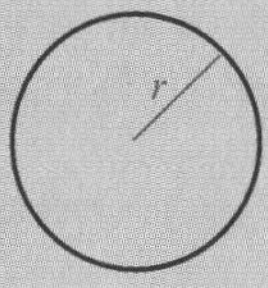	Circle	$A = \pi r^2$	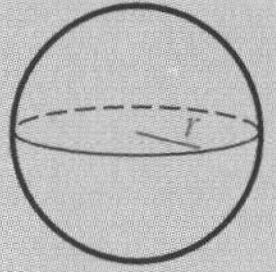	Sphere	$V = \frac{4}{3}\pi r^3$
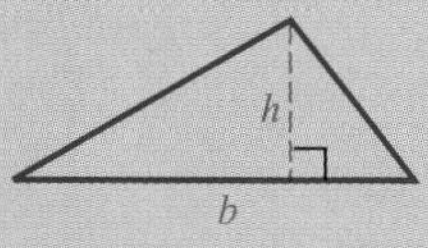	Triangle	$A = \frac{1}{2}bh$	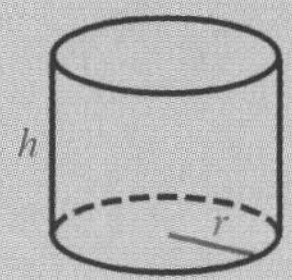	Cylinder	$V = Bh$*
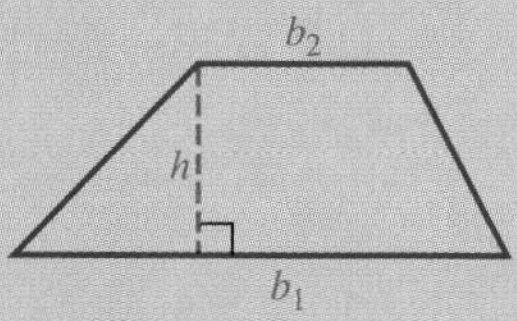	Trapezoid	$A = \frac{1}{2}h(b_1 + b_2)$	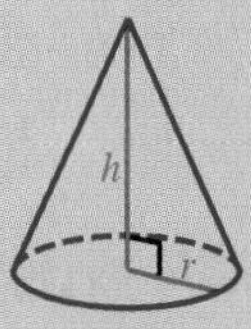	Cone	$V = \frac{1}{3}Bh$*
			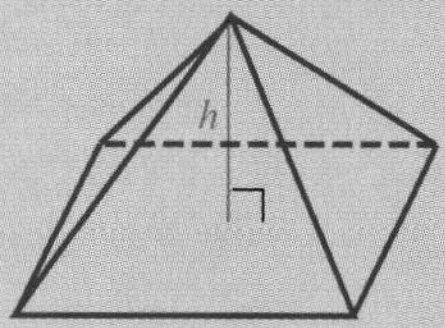	Pyramid	$V = \frac{1}{3}Bh$*

*B represents the area of the base.

TABLE 3.1

 (a)

EXAMPLE 1 **Area of a trapezoid and volume of a sphere** Find **a.** the area of a trapezoid with bases measuring 12 inches and 16 inches and a height of 8 inches and **b.** the volume of a sphere with a diameter of 20 centimeters.

Solution **a.** We substitute 12 for b_1, 16 for b_2, and 8 for h in the formula for the area of a trapezoid and simplify.

Teaching Tip: You may want to develop the formula for the area of a trapezoid:

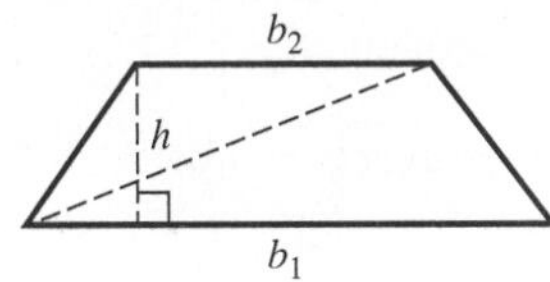

A = area of one triangle + area of other triangle

$$= \frac{1}{2}b_1h + \frac{1}{2}b_2h$$

$$= \frac{1}{2}h(b_1 + b_2)$$

$$A = \frac{1}{2}h(\boldsymbol{b_1} + \boldsymbol{b_2})$$

$$A = \frac{1}{2}(\mathbf{8})(\mathbf{12} + \mathbf{16})$$

$$= 4(28)$$

$$= 112$$

The area of the trapezoid is 112 square inches.

b. Because a radius is half as long as a diameter, the radius of the sphere is 10 centimeters. To find the volume of the sphere, we substitute 10 for r in the formula for the volume of a sphere and simplify.

$$V = \frac{4}{3}\pi \boldsymbol{r}^3$$

$$V = \frac{4}{3}\pi(\mathbf{10})^3$$

$$= \frac{4}{3}\pi(1000)$$

$$\approx 4188.790205 \qquad \text{Use a calculator.}$$

To the nearest tenth, the volume of the sphere is 4188.8 cubic centimeters. ■

Solving Literal Equations

Suppose we want to use the formula for the area of a triangle to find the heights of several triangles whose areas and bases are known. It would be very tedious to substitute values of A and b into the formula $A = \frac{1}{2}bh$ and then repeatedly solve the formula for h. It would be better to solve for h first and then substitute values for A and b and compute h directly.

To solve a formula for a variable means to isolate that variable on one side of an equation and isolate all other quantities on the other side. To isolate the desired variable on one side of the equation, we follow the same steps that we used to solve linear equations.

EXAMPLE 2 Solve the formula $A = \frac{1}{2}bh$ for h.

Solution

$$A = \frac{1}{2}bh$$

$$2A = bh \qquad \text{Multiply both sides by 2.}$$

$$\frac{2A}{b} = h \qquad \text{Divide both sides by } b.$$

$$h = \frac{2A}{b} \qquad \text{Use the symmetric property of equality.}$$

■

EXAMPLE 3 **Savings accounts** The formula $A = p + prt$ gives the amount of money in a savings account at the end of a specific time. A represents the amount, p the principal, r the rate of interest, and t the time. We can solve the formula for t as follows:

Solution

$$A = p + prt$$

$$A - p = prt \qquad \text{Subtract } p \text{ from both sides.}$$

$$\frac{A - p}{pr} = t \qquad \text{Divide both sides by } pr.$$

$$t = \frac{A - p}{pr} \qquad \text{Use the symmetric property of equality.}$$

■

EXAMPLE 4 Solve the formula $F = \frac{9}{5}C + 32$ for C.

Solution The formula $F = \frac{9}{5}C + 32$ converts degrees Celsius to degrees Fahrenheit. We can solve the formula for C as follows:

$$F = \frac{9}{5}C + 32$$

$$F - 32 = \frac{9}{5}C \qquad \text{Subtract 32 from both sides.}$$

$$\frac{5}{9}(F - 32) = \frac{5}{9}\left(\frac{9}{5}C\right) \qquad \text{Multiply both sides by } \tfrac{5}{9}.$$

$$\frac{5}{9}(F - 32) = C \qquad \tfrac{5}{9} \cdot \tfrac{9}{5} = 1.$$

or

$$C = \frac{5}{9}(F - 32)$$

■

EXAMPLE 5 **Wages and commissions** A sales clerk in a high-fashion dress shop earns \$200 per week plus a 5% commission on the value of the merchandise she sells. What dollar volume must she sell each week to earn \$250, \$300, and \$350 in three successive weeks?

Solution The weekly earnings e are computed using the formula

1. $e = 200 + 0.05v$

where v represents the value of the merchandise sold. To find v for the three values of e, we first solve Equation 1 for v.

$$e = 200 + 0.05v$$

$$e - 200 = 0.05v \qquad \text{Subtract 200 from both sides.}$$

$$\frac{e - 200}{0.05} = v \qquad \text{Divide both sides by 0.05.}$$

We can now substitute \$250, \$300, and \$350 for e and compute v.

$$v = \frac{e - 200}{0.05} \qquad v = \frac{e - 200}{0.05} \qquad v = \frac{e - 200}{0.05}$$

$$v = \frac{\mathbf{250} - 200}{0.05} \qquad v = \frac{\mathbf{300} - 200}{0.05} \qquad v = \frac{\mathbf{350} - 200}{0.05}$$

$$v = 1000 \qquad v = 2000 \qquad v = 3000$$

She must sell \$1000 worth of merchandise the first week, \$2000 worth in the second week, and \$3000 worth in the third week. ■

ORALS *Find each area.*

1. A square with a side 6 meters long. 36 m^2
2. A triangle with a base 6 meters long and a height of 6 meters. 18 m^2
3. Solve $C = \pi D$ for D. $D = \frac{C}{\pi}$
4. Solve $\frac{d}{r} = t$ for d. $d = rt$
5. Solve $y = mx + b$ for x. $x = \frac{y - b}{m}$

EXERCISE 3.4

In Exercises 1–8, find the area of each figure. Round all answers to the nearest unit.

1. 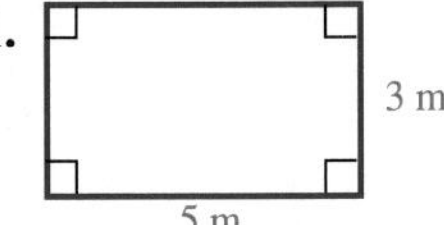

15 m^2

2. 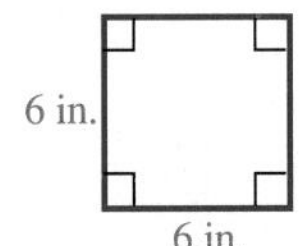

36 in.^2

3. 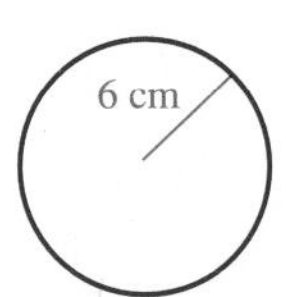

113 cm^2

4. 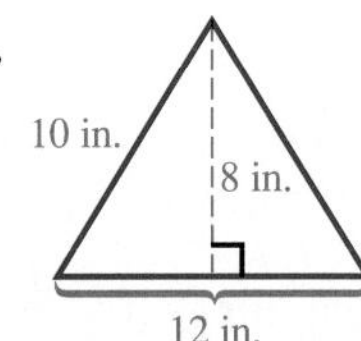

48 in.^2

5. 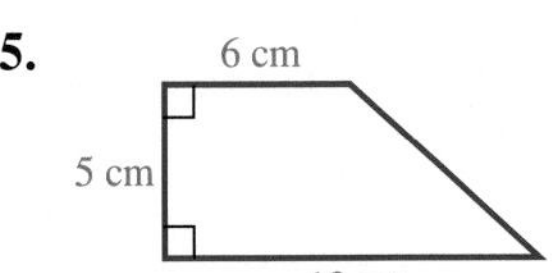

45 cm^2

6. 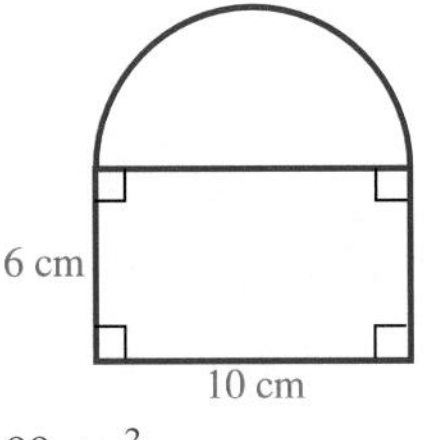

99 cm^2

7. 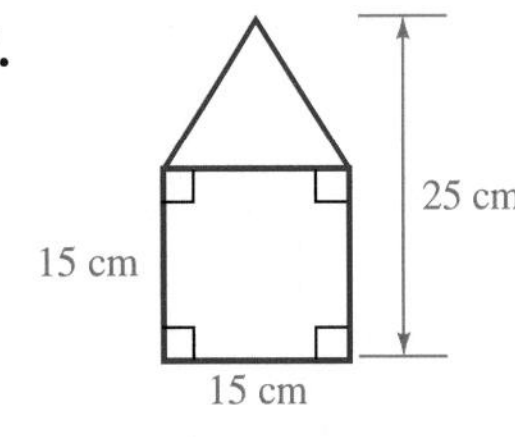

300 cm^2

8. 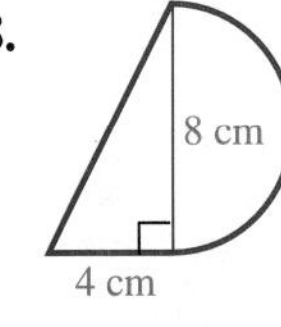

41 cm^2

In Exercises 9–16, find the volume of each figure. Round all answers to the nearest unit.

9. 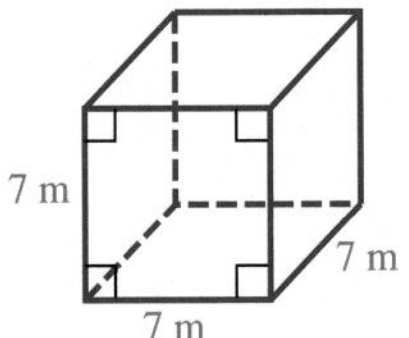

343 m^3

10. 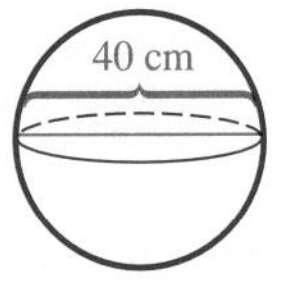

$33{,}510 \text{ cm}^3$

11. 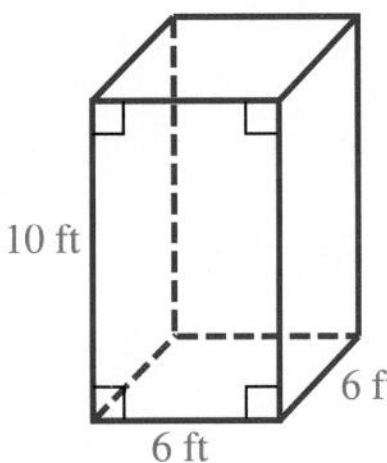

360 ft^3

12. 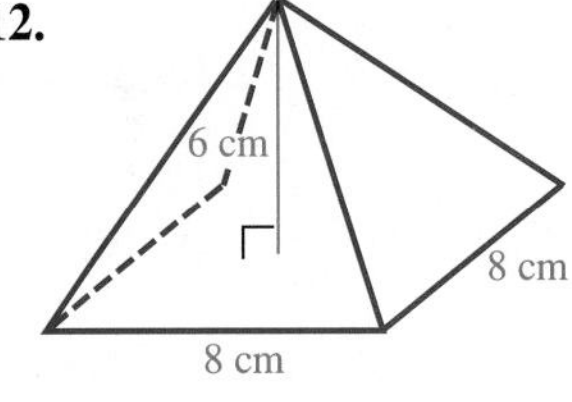

128 cm^3

13. 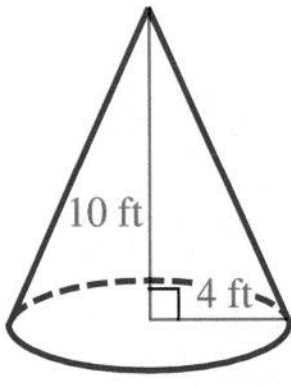

168 ft^3

14. 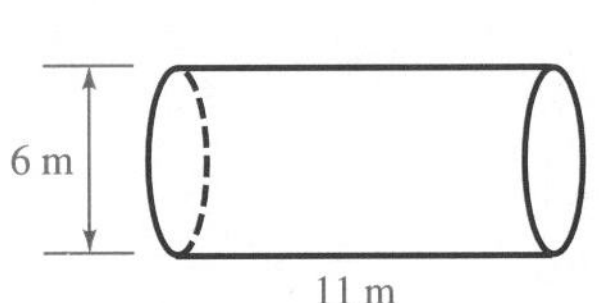

311 m^3

15. 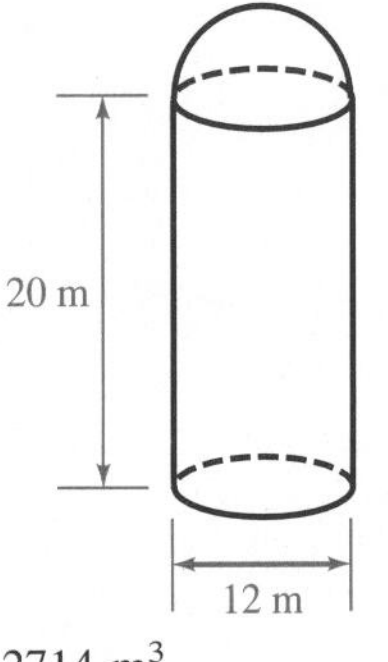

2714 m^3

16. 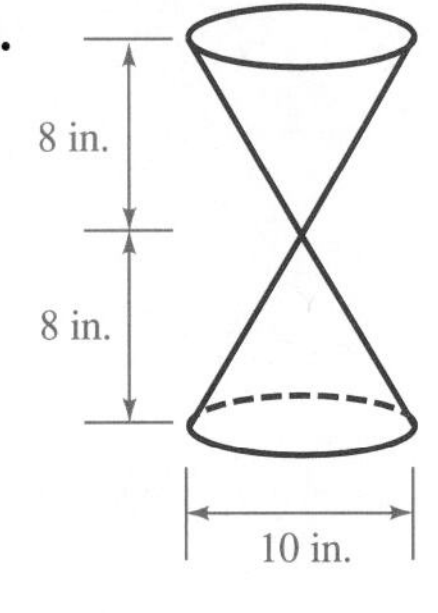

419 in.^3

In Exercises 17–50, solve each equation for the indicated variable.

17. $A = lw$ for w $\quad w = \dfrac{A}{l}$

18. $p = 4s$ for s $\quad s = \dfrac{p}{4}$

19. $A = \pi r^2$ for r^2 $\quad r^2 = \dfrac{A}{\pi}$

20. $A = \dfrac{1}{2}bh$ for b $\quad b = \dfrac{2A}{h}$

21. $V = \dfrac{1}{3}Bh$ for B $\quad B = \dfrac{3V}{h}$

22. $V = \pi r^2 h$ for h $\quad h = \dfrac{V}{\pi r^2}$

23. $I = prt$ for t $\quad t = \dfrac{I}{pr}$

24. $I = prt$ for r $\quad r = \dfrac{I}{pt}$

25. $p = 2l + 2w$ for w $\quad w = \dfrac{p - 2l}{2}$

26. $p = 2l + 2w$ for l $\quad l = \dfrac{p - 2w}{2}$

27. $A = \dfrac{1}{2}h(b_1 + b_2)$ for b_1 $\quad b_1 = \dfrac{2A}{h} - b_2$

28. $A = \dfrac{1}{2}h(b_1 + b_2)$ for b_2 $\quad b_2 = \dfrac{2A}{h} - b_1$

29. $z = \dfrac{x - \mu}{\sigma}$ for x $\quad x = z\sigma + \mu$

30. $z = \dfrac{x - \mu}{\sigma}$ for μ $\quad \mu = x - z\sigma$

31. $y = mx + b$ for x $\quad x = \dfrac{y - b}{m}$

32. $y = mx + b$ for m $\quad m = \dfrac{y - b}{x}$

33. $l = a + (n - 1)d$ for n $\quad n = \dfrac{l - a + d}{d}$

34. $l = a + (n - 1)d$ for d $\quad d = \dfrac{l - a}{n - 1}$

35. $ax + by = c$ for x $\quad x = \dfrac{c - by}{a}$

36. $ax + by = c$ for y $\quad y = \dfrac{c - ax}{b}$

37. $\sigma^2 = \dfrac{\Sigma x^2}{n} - \mu^2$ for Σx^2 $\quad \Sigma x^2 = n(\sigma^2 + \mu^2)$

38. $\sigma^2 = \dfrac{\Sigma x^2}{n} - \mu^2$ for n $\quad n = \dfrac{\Sigma x^2}{\sigma^2 + \mu^2}$

39. $S = \dfrac{a - lr}{1 - r}$ for l $\quad l = \dfrac{a - S + Sr}{r}$

40. $C = \dfrac{5}{9}(F - 32)$ for F $\quad F = \dfrac{9}{5}C + 32$

41. $P = L + \dfrac{s}{f}i$ for s $\quad s = \dfrac{f(P - L)}{i}$

42. $P = L + \dfrac{s}{f}i$ for f $\quad f = \dfrac{si}{P - L}$

43. $S = \dfrac{n(a + l)}{2}$ for l $\quad l = \dfrac{2S - na}{n}$

44. $S = \dfrac{n(a + l)}{2}$ for n $\quad n = \dfrac{2S}{a + l}$

45. $y - y_1 = m(x - x_1)$ for x $\quad x = \dfrac{y - y_1 + mx_1}{m}$

46. $y - y_1 = m(x - x_1)$ for x_1 $\quad x_1 = \dfrac{mx - y + y_1}{m}$

47. $\dfrac{x^2}{a^2} + \dfrac{y^2}{b^2} = 1$ for a^2 $\quad a^2 = \dfrac{x^2b^2}{b^2 - y^2}$

48. $\dfrac{x^2}{a^2} - \dfrac{y^2}{b^2} = 1$ for y^2 $\quad y^2 = \dfrac{x^2b^2 - a^2b^2}{a^2}$

49. $y = a(x^2 - h) + k$ for h

$h = x^2 - \dfrac{y - k}{a} = \dfrac{ax^2 - y + k}{a}$

50. $y = a(x^2 - h) + k$ for x^2

$x^2 = \dfrac{y - k}{a} + h = \dfrac{y - k + ah}{a}$

51. Stopping distance The distance a car travels before stopping depends on the driver's reaction time and the braking distance. (See Illustration 1.) After one driver decides to apply the brakes, the distance his car travels before stopping is given by $d = 0.04v^2 + 0.9v$, where v is the velocity of the car in miles per hour and d is the distance measured in feet. Find the stopping distances when the car is going 50 mph and 65 mph. 145 ft, 227.5 ft

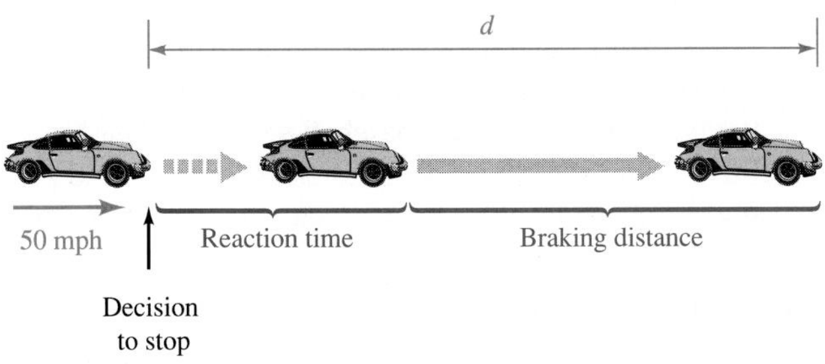

ILLUSTRATION 1

52. Pulleys The approximate length L of a belt joining two pulleys of radii r and R feet with centers D feet apart is given by the formula $L = 2D + 3.25(r + R)$. (See Illustration 2.) Find the length of belt required to join pulleys with radii 4 and 7 inches, which are 17 inches apart. 69.75 in.

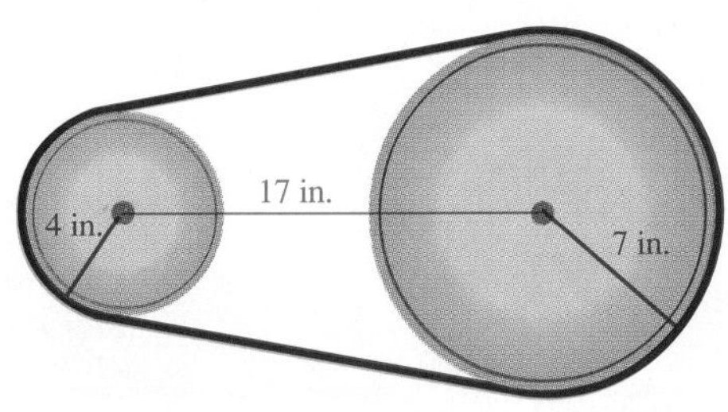

ILLUSTRATION 2

53. Force of gravity The masses of the two objects in Illustration 3 are m and M. The force of gravitation, F, between the masses is

$$F = \frac{GmM}{d^2}$$

where G is a constant and d is the distance between them. Solve for m. $m = Fd^2/(GM)$

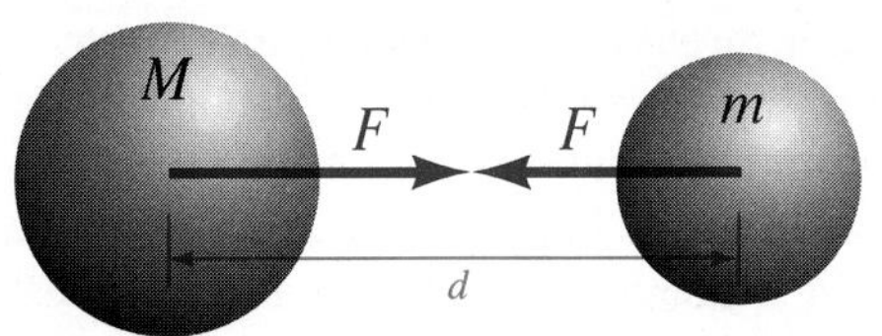

ILLUSTRATION 3

54. Thermodynamics In thermodynamics, the Gibbs free-energy function is given by

$$G = U - TS + pV$$

Solve this equation for S. $S = (U + pV - G)/T$

55. Converting temperatures Solve the formula $F = \frac{9}{5}C + 32$ for C and find the Celsius temperatures that correspond to Fahrenheit temperatures of 32°, 70°, and 212°. 0°, 21.1°, 100°

56. Doubling money A man intends to invest \$1000. Solve the formula $A = p + prt$ for t and find the number of years required to double his money at the rates of 5%, 7%, and 10%. 20, 14.29, 10

57. Cost of electricity The cost of electricity in a certain city is given by the formula

$$C = 0.07n + 6.50$$

where C is the cost and n is the number of kilowatt hours used. Solve the formula for n and find the number of kilowatt hours used for costs of \$49.97, \$76.50, and \$125. 621, 1000, 1692.86

58. Cost of water A monthly water bill in a certain city is calculated by using the formula

$$n = \frac{5000C - 17{,}500}{6}$$

where n is the number of gallons used and C is the monthly cost. Solve the formula for C and compute the bill for quantities used of 500, 1200, and 2500 gallons. \$4.10, \$4.94, \$6.50

59. Ohm's law The formula $E = IR$, called **Ohm's law,** is used in electronics. Solve for R and then calculate the resistance R if the voltage E is 56 volts and the current I is 7 amperes. (Resistance has units of **ohms.**) $R = E/I$; $R = 8$ ohms

60. **Earning interest** An amount P, invested at a simple interest rate r, will grow to an amount A in t years according to the formula $A = P(1 + rt)$. Solve the formula for P. After $t = 5$ years, a woman has an amount $A = \$6693.75$ on deposit. If the interest rate, r, is 5.5%, what amount P did she start with?
$P = A/(1 + rt)$; $P = \$5250$

61. **Angles of a polygon** A regular polygon has n equal sides and n equal angles. The measure a of an interior angle is given by the formula $a = 180°\left(1 - \frac{2}{n}\right)$. Solve this formula for n. Find the number of sides of the regular polygon in Illustration 4, if an interior angle is 135°.
$n = 360°/(180° - a)$; 8 sides

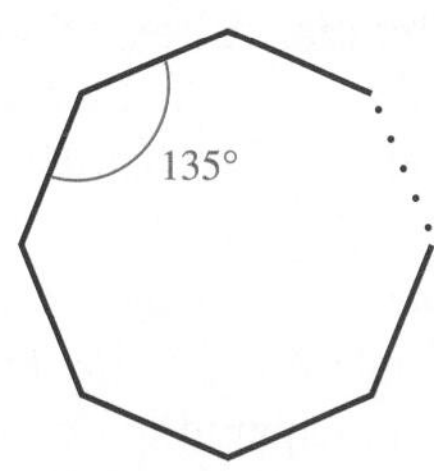

ILLUSTRATION 4

62. **Power loss** Illustration 5 is the schematic diagram of a resistor connected to a voltage source of 60 volts. As a result, the resistor dissipates power in the form of heat. The power P lost when a voltage V is placed across a resistance R is given by the formula

$$P = \frac{E^2}{R}$$

Solve for R. If P is 4.8 watts and E is 60 volts, find R. $R = E^2/P$; $R = 750$ ohms

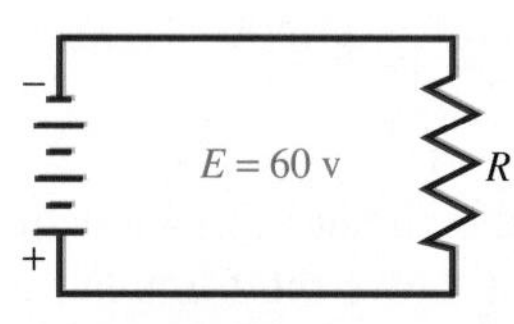

ILLUSTRATION 5

63. **Car rental** While waiting for his car to be repaired, a man rents a car for \$12 per day plus 10 cents per mile. If he keeps the car for 2 days, how many miles can he drive for a total cost of \$30? How many miles can he drive for a total cost of \$36? 60 mi, 120 mi

64. **Computing salaries** A student earns \$17 per day for delivering overnight packages. She is paid \$5 per day plus 60¢ for each package delivered. How many more deliveries must she make each day to increase her daily earnings to \$23? 10

65. **Computing partial salaries** A baseball team agreed to pay Jim \$1750 plus a scholarship for being bat boy for the 5-month baseball season. When he had to quit after 3 months, he was paid \$810 and given the scholarship. How much was the scholarship worth? \$600

66. **Computing rents** The landlord of a duplex apartment collected \$8730 rent by renting both units for one year. If one apartment rented for \$60 more per month than the other but was vacant for 3 months, find the monthly rent of the more expensive apartment. \$450

67. **Volume of a box** A square sheet of cardboard with its corners cut out as in Illustration 16 is to be folded along the dashed lines to form a box with a square base and an open top. Find a formula that relates the volume of the box to the values a and b.
$V = a(b - 2a)^2$

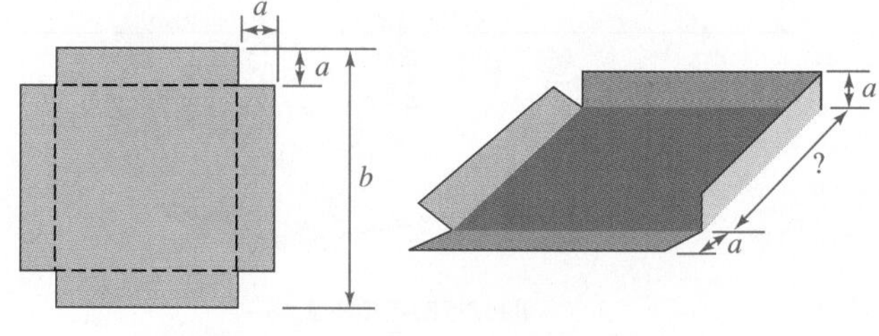

ILLUSTRATION 6

Writing Exercises ■ *Write a paragraph using your own words.*

1. Other than in a mathematics class, where have you needed to use a formula?

2. What do you think is the difference between a literal equation and a formula?

Something to Think About

1. The volume of a pyramid is given by $V = \frac{1}{3}Bh$, where B is the area of the base. Write formulas for the volume of a pyramid with
   a. a triangular base **b.** a square base **c.** a trapezoidal base
2. Find the volume of a pyramid with a circular base, and compare your equation with the equation for the volume of a cone. Explain the similarity.

Review Exercises

Do the operations.

1. $2(3x^2 + 2x - 5) + 5(2x^2 - 4x + 2)$
   $16x^2 - 16x$
2. $(2a - 5b)(3a + 2b)$
   $6a^2 - 11ab - 10b^2$
3. $(x - y)(2x^2 - xy + y^2)$
   $2x^3 - 3x^2y + 2xy^2 - y^3$
4. $3a - 4b\overline{)6a^3 - 11a^2b + 7ab^2 - 4b^3}$
   $2a^2 - ab + b^2$

3.5 Absolute Value Equations

■ Absolute Value Equations ■ Equations with Two Absolute Values

We begin this section by reviewing the definition of the absolute value of x.

Absolute Value If $x \geq 0$, then $|x| = x$.

If $x < 0$, then $|x| = -x$.

Historical Note: Karl Weierstrauss (1815–1897) was the first to use the absolute value symbol.

This definition gives a way for associating a nonnegative real number with any real number.

- If $x \geq 0$, then x is its own absolute value.
- If $x < 0$, then $-x$ (which is positive) is the absolute value.

Either way, $|x|$ is positive or 0:

$$|x| \geq 0 \qquad \text{for all real numbers } x$$

 (a, b, c)

EXAMPLE 1 Find **a.** $|9|$, **b.** $|-5|$, **c.** $|0|$, and **d.** $|2 - \pi|$.

Solution

a. Since $9 \geq 0$, 9 is its own absolute value: $|9| = 9$.

b. Since $-5 < 0$, the negative of -5 is the absolute value: $|-5| = -(-5) = 5$.

c. Since $0 \geq 0$, 0 is its own absolute value: $|0| = 0$.

d. Since $\pi \approx 3.14$, it follows that $2 - \pi < 0$. Thus,

$$|2 - \pi| = -(2 - \pi) = \pi - 2$$

■

WARNING! The placement of a − sign in an expression containing an absolute value symbol is important. For example, $|-19| = 19$, but $-|19| = -19$.

(a, c) **EXAMPLE 2** Find **a.** $-|-10|$, **b.** $-|13|$, and **c.** $-(-|-3|)$.

Solution **a.** $-|-10| = -(10) = -10$ **b.** $-|13| = -13$

c. $-(-|-3|) = -(-3) = 3$ ■

Absolute Value Equations

In the equation $|x| = 5$, x can be either 5 or −5, because

$$|5| = 5 \quad \text{and} \quad |-5| = 5$$

Thus, if $|x| = 5$, then $x = 5$ or $x = -5$. In general, the following is true.

Absolute Value Equations If $k > 0$, then

$$|x| = k \quad \text{is equivalent to} \quad x = k \quad \text{or} \quad x = -k$$

The absolute value of x can be interpreted as the distance on the number line that a point is from the origin. The solutions of $|x| = k$ are represented by the two points that lie exactly k units from the origin. (See Figure 3-10.)

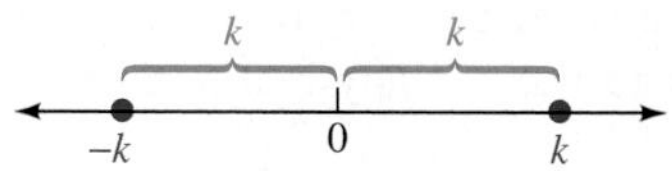

FIGURE 3-10

The equation $|x - 3| = 7$ indicates that a point on the number line with a coordinate of $x - 3$ is 7 units from the origin. Thus, $x - 3$ can be either 7 or −7.

$$x - 3 = 7 \quad \text{or} \quad x - 3 = -7$$
$$x = 10 \quad | \quad x = -4$$

The solutions of the equation $|x - 3| = 7$ are 10 and −4. (See Figure 3-11.) If either of these numbers is substituted for x in the equation $|x - 3| = 7$, the equation is satisfied:

$\|x - 3\| = 7$	$\|x - 3\| = 7$
$\|10 - 3\| \stackrel{?}{=} 7$	$\|-4 - 3\| \stackrel{?}{=} 7$
$\|7\| \stackrel{?}{=} 7$	$\|-7\| \stackrel{?}{=} 7$
$7 = 7$	$7 = 7$

−4 10

FIGURE 3-11

EXAMPLE 3 Solve the equation $|3x - 2| = 5$.

Solution We can write $|3x - 2| = 5$ as

$$3x - 2 = 5 \quad \text{or} \quad 3x - 2 = -5$$

and solve each equation for x:

$$\begin{array}{rcl|rcl} 3x - 2 &=& 5 & 3x - 2 &=& -5 \\ 3x &=& 7 & 3x &=& -3 \\ x &=& \frac{7}{3} & x &=& -1 \end{array}$$

Verify that both solutions check. ■

EXAMPLE 4 Solve the equation $\left|\frac{2}{3}x + 3\right| + 4 = 10$.

Solution We first isolate the absolute value on the left-hand side.

$$\left|\frac{2}{3}x + 3\right| + 4 = 10$$

1. $$\left|\frac{2}{3}x + 3\right| = 6 \qquad \text{Subtract 4 from both sides.}$$

We can now write Equation 1 as

$$\frac{2}{3}x + 3 = 6 \quad \text{or} \quad \frac{2}{3}x + 3 = -6$$

and solve each equation for x:

$$\begin{array}{rcl|rcl} \frac{2}{3}x + 3 &=& 6 & \frac{2}{3}x + 3 &=& -6 \\ \frac{2}{3}x &=& 3 & \frac{2}{3}x &=& -9 \\ 2x &=& 9 & 2x &=& -27 \\ x &=& \frac{9}{2} & x &=& \frac{-27}{2} \end{array}$$

Verify that both solutions check. ■

EXAMPLE 5 Solve the equation $\left|7x + \frac{1}{2}\right| = -4$.

Solution Since the absolute value of a number cannot be negative, no value of x can make $\left|7x + \frac{1}{2}\right| = -4$. Since this equation has no solutions, its solution set is $\emptyset$. ■

EXAMPLE 6 Solve the equation $\left|\frac{1}{2}x - 5\right| - 4 = -4$.

Solution We first isolate the absolute value on the left-hand side.

$$\left|\frac{1}{2}x - 5\right| - 4 = -4$$

$$\left|\frac{1}{2}x - 5\right| = 0 \qquad \text{Add 4 to both sides.}$$

Since 0 is the only number whose absolute value is 0, the binomial $\frac{1}{2}x - 5$ must be 0, and we have

$$\frac{1}{2}x - 5 = 0$$

$$\frac{1}{2}x = 5 \qquad \text{Add 5 to both sides.}$$

$$x = 10 \qquad \text{Multiply both sides by 2.}$$

Verify that 10 satisfies the original equation. ■

Equations with Two Absolute Values

The equation $|a| = |b|$ is true when $a = b$ or when $a = -b$. For example,

$$|3| = |3| \quad \text{or} \quad |3| = |-3|$$

$$3 = 3 \qquad | \qquad 3 = 3$$

Thus, we have the following result.

Equations with Two Absolute Values If a and b represent algebraic expressions, the equation $|a| = |b|$ is equivalent to the pair of equations

$$a = b \quad \text{or} \quad a = -b$$

EXAMPLE 7 Solve the equation $|5x + 3| = |3x + 25|$.

Solution This equation is true when $5x + 3 = 3x + 25$, or when $5x + 3 = -(3x + 25)$. We solve each equation for x.

$$5x + 3 = 3x + 25 \quad \text{or} \quad 5x + 3 = -(3x + 25)$$

$$2x = 22 \qquad\qquad 5x + 3 = -3x - 25$$

$$x = 11 \qquad\qquad 8x = -28$$

$$x = -\frac{28}{8}$$

$$x = -\frac{7}{2}$$

Verify that both solutions check. ■

ORALS *Find each value.*

1. $|-5|$ 5
2. $-|5|$ -5
3. $-|-6|$ -6
4. $-|4|$ -4

Solve each equation.

5. $|x| = 8$ 8 or -8
6. $|x| = -5$ no solution
7. $|x - 5| = 0$ 5
8. $|x + 1| = 1$ 0, -2

EXERCISE 3.5

In Exercises 1–12, find the value of each expression.

1. $|8|$ 8
2. $|-18|$ 18
3. $|-12|$ 12
4. $|15|$ 15
5. $-|2|$ -2
6. $-|-20|$ -20
7. $-|-30|$ -30
8. $-|25|$ -25
9. $-(-|50|)$ 50
10. $-(-|-20|)$ 20
11. $|\pi - 4|$ $4 - \pi$
12. $|2\pi - 4|$ $2\pi - 4$

In Exercises 13–24, select the smaller of the two numbers.

13. $|2|, |5|$ $|2|$
14. $|-6|, |2|$ $|2|$
15. $|5|, |-8|$ $|5|$
16. $|6|, |3|$ $|3|$
17. $|-2|, |10|$ $|-2|$
18. $|-6|, -|6|$ $-|6|$
19. $|-3|, -|-4|$ $-|-4|$
20. $|-3|, |-2|$ $|-2|$
21. $-|-5|, -|-7|$ $-|-7|$
22. $-|-8|, -|20|$ $-|20|$
23. $-x, |x + 1|$ $(x > 0)$ $-x$
24. $y, |y - 1|$ $(y > 0)$ $|y - 1|$

In Exercises 25–58, solve each equation, if possible.

25. $|x| = 8$ 8, -8
26. $|x| = 9$ 9, -9
27. $|x - 3| = 6$ 9, -3
28. $|x + 4| = 8$ 4, -12
29. $|2x - 3| = 5$ 4, -1
30. $|4x - 4| = 20$ 6, -4
31. $|3x + 2| = 16$ $\frac{14}{3}$, -6
32. $|5x - 3| = 22$ 5, $-\frac{19}{5}$
33. $\left|\frac{7}{2}x + 3\right| = -5$ no solution
34. $|2x + 10| = 0$ -5

35. $\left|\frac{x}{2} - 1\right| = 3$ 8, −4

36. $\left|\frac{4x - 64}{4}\right| = 32$ 48, −16

37. $|3 - 4x| = 5$ 2, $-\frac{1}{2}$

38. $|8 - 5x| = 18$ $\frac{26}{5}$, −2

39. $|3x + 24| = 0$ −8

40. $|x - 21| = -8$ no solution

41. $\left|\frac{3x + 48}{3}\right| = 12$ −4, −28

42. $\left|\frac{x}{2} + 2\right| = 4$ 4, −12

43. $|x + 3| + 7 = 10$ 0, −6

44. $|2 - x| + 3 = 5$ 0, 4

45. $\left|\frac{3}{5}x - 4\right| - 2 = -2$ $\frac{20}{3}$

46. $\left|\frac{3}{4}x + 2\right| + 4 = 4$ $-\frac{8}{3}$

47. $|2x + 1| = |3x + 3|$ −2, $-\frac{4}{5}$

48. $|5x - 7| = |4x + 1|$ 8, $\frac{2}{3}$

49. $|3x - 1| = |x + 5|$ 3, −1

50. $|3x + 1| = |x - 5|$ 1, −3

51. $|2 - x| = |3x + 2|$ 0, −2

52. $|4x + 3| = |9 - 2x|$ 1, −6

53. $\left|\frac{x}{2} + 2\right| = \left|\frac{x}{2} - 2\right|$ 0

54. $|7x + 12| = |x - 6|$ −3, $-\frac{3}{4}$

55. $\left|x + \frac{1}{3}\right| = |x - 3|$ $\frac{4}{3}$

56. $\left|x - \frac{1}{4}\right| = |x + 4|$ $-\frac{15}{8}$

57. $|3x + 7| = -|8x - 2|$ no solutions

58. $-|17x + 13| = |3x - 14|$ no solutions

59. Construct several examples to show that $|a \cdot b| = |a| \cdot |b|$

60. Construct several examples to show that $\left|\frac{a}{b}\right| = \frac{|a|}{|b|}$

61. Construct several examples to show that $|a + b| \neq |a| + |b|$

62. Construct several examples to show that $|a - b| \neq |a| - |b|$

Writing Exercises ■ *Write a paragraph using your own words.*

1. Explain how to find the absolute value of a given number.

2. Explain why the equation $|x| + 5 = 0$ has no solution.

Something to Think About ■

1. For what values of k does $|x| + k = 0$ have exactly two solutions?

2. For what value of k does $|x| + k = 0$ have exactly one solution?

Review Exercises ■ *Solve each equation.*

1. $3(2a - 1) = 2a$ $\frac{3}{4}$

2. $\frac{t}{6} - \frac{t}{3} = -1$ 6

3. $\frac{5x}{2} - 1 = \frac{x}{3} + 12$ 6

4. $4b - \frac{b + 9}{2} = \frac{b + 2}{5} - \frac{8}{5}$ 1

3.6 Linear Inequalities

■ Properties of Inequalities ■ Linear Inequalities ■ Double Inequalities

Inequalities are statements that indicate that two quantities are unequal. Inequalities can be recognized by the use of one or more of the following symbols.

Inequality Symbols

$a < b$ means "a is less than b."

$a > b$ means "a is greater than b."

$a \leq b$ means "a is less than or equal to b."

$a \geq b$ means "a is greater than or equal to b."

To show that 2 is less than 3, we can write the inequality $2 < 3$. To show that a number x is greater than 4 or equal to 4, we can write the inequality $x \geq 4$.

By definition, the inequality $a < b$ means that "a is less than b," but it also means that $b > a$. If a lies to the left of b on the number line, then $a < b$.

In Chapter 1, we saw that many inequalities can be graphed as intervals on a number line. Table 3-2 shows the possibilities when a and b are real numbers.

Kind of interval	Inequality	Graph	Interval
Open interval	$x > a$	a	(a, ∞)
	$x < a$	a	$(-\infty, a)$
	$a < x < b$	a b	(a, b)
Half-open interval	$x \geq a$	a	$[a, \infty)$
	$x \leq a$	a	$(-\infty, a]$
	$a \leq x < b$	a b	$[a, b)$
	$a < x \leq b$	a b	$(a, b]$
Closed interval	$a \leq x \leq b$	a b	$[a, b]$

TABLE 3.2

Inequalities like $x + 1 > x$ that are true for all numbers x are called **absolute inequalities.** Inequalities like $3x + 2 < 8$ that are true for some numbers x, but not all numbers x, are called **conditional inequalities.**

Properties of Inequalities

There are several basic properties of inequalities.

The Trichotomy Property For two real numbers a and b, exactly one of the following statements is true:

$$a < b, \qquad a = b, \qquad \text{or} \qquad a > b$$

The trichotomy property indicates that one and only one of three statements is true about any two real numbers. Either

- the first number is less than the second, or
- the first number is equal to the second, or
- the first number is greater than the second.

The Transitive Property If a, b, and c are real numbers with $a < b$ and $b < c$, then $a < c$.

If a, b, and c are real numbers with $a > b$ and $b > c$, then $a > c$.

The first part of the transitive property indicates that if we have three numbers and

- the first number is less than the second number and
- the second number is less than the third number, then
- the first number is less than the third number.

The second part of the transitive property is similar, with the words "is greater than" substituted for "is less than."

Property 1 of Inequalities Any real number can be added to (or subtracted from) both sides of an inequality to produce another inequality with the same direction.

Property 1 indicates that any number can be added to both sides of a true inequality to get another true inequality with the same direction. For example, if 4 is added to both sides of the inequality $3 < 12$, we get

$$3 + \mathbf{4} < 12 + \mathbf{4}$$
$$7 < 16$$

and the $<$ symbol remains an $<$ symbol. Thus, adding 4 to both sides does not change the direction (sometimes called the **order**) of the inequality.

Subtracting 4 from both sides of $3 < 12$ does not change the direction of the inequality either.

$$3 - 4 < 12 - 4$$
$$-1 < 8$$

Property 2 of Inequalities If both sides of an inequality are multiplied (or divided) by a positive number, another inequality results, with the same direction as the original inequality.

Property 2 indicates both sides of a true inequality can be multiplied by any positive number to get another true inequality with the same direction. For example, if both sides of the true inequality $-4 < 6$ are multiplied by $+2$, we get

$$2(-4) < 2(6)$$
$$-8 < 18$$

and the $<$ symbol remains an $<$ symbol. Thus, multiplying both sides by $+2$ does not change the direction of the inequality.

Dividing both sides by $+2$ does not change the direction of the inequality either.

$$\frac{-4}{2} < \frac{6}{2}$$
$$-2 < 3$$

Property 3 of Inequalities If both sides of an inequality are multiplied (or divided) by a negative number, another inequality results, but with the opposite direction of the original inequality.

Property 3 indicates that if both sides of a true inequality are multiplied by any negative number, another true inequality results, but with the opposite direction. For example, if both sides of the true inequality $-4 < 6$ are multiplied by -2, we get

$$-4 < 6$$
$$-2(-4) > -2(6)$$
$$8 > -12$$

and the $<$ symbol becomes an $>$ symbol. Thus, multiplying both sides by -2 reverses the direction of the inequality.

Dividing both sides by -2 also reverses the direction of the inequality.

$$-4 < 6$$
$$\frac{-4}{-2} > \frac{6}{-2}$$
$$2 > -3$$

WARNING! We must remember to reverse the inequality symbol every time we multiply or divide by a negative number.

Linear Inequalities

A **linear inequality** is any inequality that can be expressed in the form

$$ax + c < 0 \qquad ax + c > 0 \qquad ax + c \leq 0 \qquad \text{or} \qquad ax + c \geq 0 \quad (a \neq 0)$$

We can solve linear inequalities by using the same steps that we used for solving linear equations, with one exception. If we multiply or divide both sides by a *negative* number, we must reverse the direction of the inequality.

EXAMPLE 1 Solve the linear inequality $3(2x - 9) < 9$.

Solution We solve the inequality as if it were an equation:

$3(2x - 9) < 9$	
$6x - 27 < 9$	Use the distributive property to remove parentheses.
$6x < 36$	Add 27 to both sides.
$x < 6$	Divide both sides by 6.

The solution set is the interval $(-\infty, 6)$. The graph of the solution set is shown in Figure 3-12. The open circle at 6 indicates that 6 is not included in the solution set.

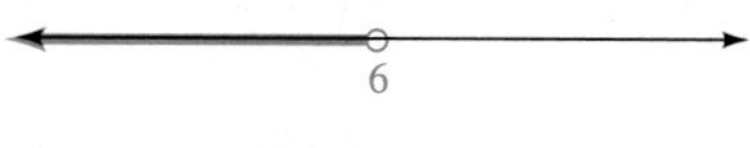

FIGURE 3-12 ■

EXAMPLE 2 Solve the linear inequality $-4(3x + 2) \leq 16$.

Solution We solve the inequality as if it were an equation:

$-4(3x + 2) \leq 16$	
$-12x - 8 \leq 16$	Use the distributive property to remove parentheses.
$-12x \leq 24$	Add 8 to both sides.
$x \geq -2$	Divide both sides by -12 and reverse the $\leq$ symbol.

The solution set is the interval $[-2, \infty)$. The graph of the solution set is shown in Figure 3-13. The solid circle at -2 indicates that -2 is included in the solution set.

FIGURE 3-13

EXAMPLE 3 Solve $\frac{2}{3}(x + 2) > \frac{4}{5}(x - 3)$.

Solution We solve the inequality as if it were an equation:

$$\frac{2}{3}(x + 2) > \frac{4}{5}(x - 3)$$

$$15 \cdot \frac{2}{3}(x + 2) > 15 \cdot \frac{4}{5}(x - 3)$$ Multiply both sides by 15.

$$10(x + 2) > 12(x - 3)$$ Simplify.

$$10x + 20 > 12x - 36$$ Use the distributive property to remove parentheses.

$$-2x + 20 > -36$$ Add -12 to both sides.

$$-2x > -56$$ Subtract 20 from both sides.

$$x < 28$$ Divide both sides by -2 and reverse the $>$ symbol.

The solution set is the interval $(-\infty, 28)$, whose graph is shown in Figure 3-14.

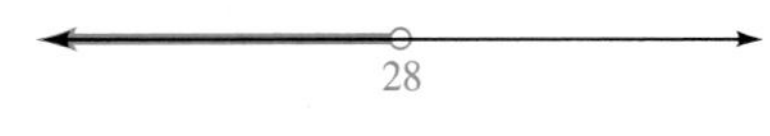

FIGURE 3-14

Double Inequalities

To say that x is between -3 and 8, we write a double inequality:

$-3 < x < 8$ Read as "-3 is less than x and x is less than 8."

This double inequality contains two different linear inequalities:

$-3 < x$ *and* $x < 8$

These two inequalities mean that $x > -3$ and $x < 8$. The word *and* indicates that these two inequalities are true at the same time.

Double Inequalities The double inequality $c < x < d$ is equivalent to $c < x$ and $x < d$.

WARNING! The inequality $c < x < d$ cannot be expressed as

$$c < x \quad \text{or} \quad x < d$$

EXAMPLE 4 Solve the inequality $-3 \le 2x + 5 < 7$.

Solution This inequality means that $2x + 5$ is between -3 and 7. We can solve it by isolating x between the inequality symbols:

$$\begin{aligned} -3 &\le 2x + 5 < 7 \\ -8 &\le 2x < 2 && \text{Subtract 5 from all three parts.} \\ -4 &\le x < 1 && \text{Divide all three parts by 2.} \end{aligned}$$

The solution set is the interval $[-4, 1)$. Its graph is shown in Figure 3-15.

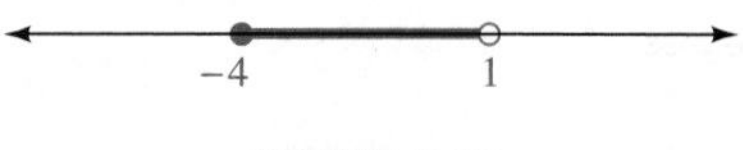

FIGURE 3-15 ■

EXAMPLE 5 Solve the inequality $x + 3 < 2x - 1 < 4x - 3$.

Solution Since it is impossible to isolate x between the inequality symbols, we solve each of its linear inequalities separately.

$$\begin{array}{ccc} x + 3 < 2x - 1 & \text{and} & 2x - 1 < 4x - 3 \\ 4 < x & \Big| & 2 < 2x \\ & & 1 < x \end{array}$$

Only those numbers x where $x > 4$ and $x > 1$ are in the solution set. Since all numbers greater than 4 are also greater than 1, the solutions are the numbers x where $x > 4$. Thus, the solution set is the interval $(4, \infty)$. The graph is shown in Figure 3-16.

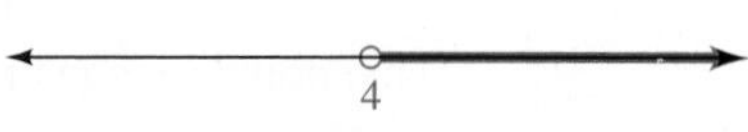

FIGURE 3-16 ■

EXAMPLE 6 Solve the double inequality $x \le -3$ or $x \ge 8$.

Solution The double inequality $x \le -3$ or $x \ge 8$ has two parts. Either

$$x \le -3 \quad \text{or} \quad x \ge 8$$

Amalie Noether
(1882–1935)
Albert Einstein described Noether as the most creative female mathematical genius since the beginning of higher education for women. Her work was in the area of abstract algebra. Although she received a doctoral degree in mathematics, she was denied a mathematics position in Germany because she was a woman.

The graph of this inequality is shown in Figure 3-17. The solution set is $(-\infty, -3] \cup [8, \infty)$.

FIGURE 3-17

The word *or* in the statement $x \le -3$ or $x \ge 8$ indicates that only one of the inequalities needs to be true to make the statement true. ■

WARNING! In the statement $x \le -3$ or $x \ge 8$, it is incorrect to string the equalities together as $8 \le x \le -3$, because that would imply that $8 \le -3$, which is impossible.

ORALS *Solve each inequality.*

1. $2x < 4$ $\quad x < 2$

2. $3x + 1 \ge 10$ $\quad x \ge 3$

3. $-3x > 12$ $\quad x < -4$

4. $-\dfrac{x}{2} \le 4$ $\quad x \ge -8$

5. $-2 < 2x < 8$ $\quad -1 < x < 4$

6. $3 \le \dfrac{x}{3} \le 4$ $\quad 9 \le x \le 12$

EXERCISE 3.6

In Exercises 1–42, solve each inequality. Give the result in interval notation and graph the solution set.

1. $x + 4 < 5$ $\quad (-\infty, 1)$; graph: open circle at 1

2. $x - 5 > 2$ $\quad (7, \infty)$; graph: open circle at 7

3. $-3x - 1 \le 5$ $\quad [-2, \infty)$; graph: closed circle at −2

4. $-2x + 6 \ge 16$ $\quad (-\infty, -5]$; graph: closed circle at −5

5. $5x - 3 > 7$ $\quad (2, \infty)$; graph: open circle at 2

6. $7x - 9 < 5$ $\quad (-\infty, 2)$; graph: open circle at 2

7. $8 - 9y \geq -y$ $(-\infty, 1]$

8. $4 - 3x \leq x$ $[1, \infty)$

9. $-3(a + 2) > 2(a + 1)$ $(-\infty, -8/5)$

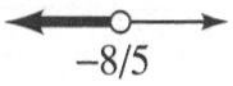

10. $-4(y - 1) < y + 8$ $(-4/5, \infty)$

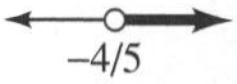

11. $3(z - 2) \leq 2(z + 7)$ $(-\infty, 20]$

12. $5(3 + z) > -3(z + 3)$ $(-3, \infty)$

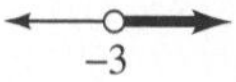

13. $-11(2 - b) < 4(2b + 2)$ $(-\infty, 10)$

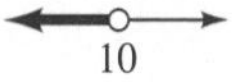

14. $-9(h - 3) + 2h \leq 8(4 - h)$ $(-\infty, 5]$

15. $\frac{1}{2}y + 2 \geq \frac{1}{3}y - 4$ $[-36, \infty)$

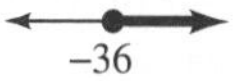

16. $\frac{1}{4}x - \frac{1}{3} \leq x + 2$ $[-28/9, \infty)$

17. $\frac{3}{4}x - 4 < \frac{4}{5}x + 1$ $(-100, \infty)$

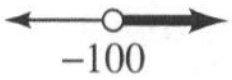

18. $\frac{7}{8}(a - 3) < \frac{3}{4}(a + 3)$ $(-\infty, 39)$

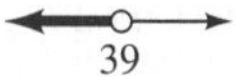

19. $\frac{2}{3}x + \frac{3}{2}(x - 5) \leq x$ $(-\infty, 45/7]$

20. $\frac{5}{9}(x + 3) - \frac{4}{3}(x - 3) \geq x - 1$ $(-\infty, 15/4]$

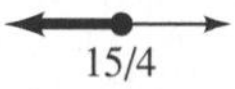

21. $-2 < -b + 3 < 5$ $(-2, 5)$

22. $4 < -t - 2 < 9$ $(-11, -6)$

23. $9 < 2x - 7 < 15$ $(8, 11)$

24. $7 < 3x - 2 < 25$ $(3, 9)$

25. $-6 < -3(x - 4) \leq 24$ $[-4, 6)$

26. $-4 \leq -2(x + 8) < 8$ $(-12, -6]$

27. $6 < \frac{1}{2}x - 4 < 0$ no solution

28. $-6 \leq \frac{1}{3}a + 1 < 0$ $[-21, -3)$

29. $0 \leq \frac{4 - x}{3} \leq 2$ $[-2, 4]$

30. $-2 \leq \frac{5 - 3x}{2} \leq 2$ $[1/3, 3]$

31. $x + 3 < 3x - 1 < 2x + 2$ $(2, 3)$

32. $x - 1 \leq 2x + 4 \leq 3x - 1$ $[5, \infty)$

33. $4x \geq -x + 5 \geq 3x - 4$ $[1, 9/4]$

34. $x + 2 < -\frac{1}{3}x < \frac{1}{2}x$ no solution

35. $5(x+1) \le 4(x+3) < 3(x-1)$ $(-\infty, -15)$ −15

36. $-5(2+x) < 4x+1 < 3x$ $(-11/9, -1)$ −11/9 −1

37. $3x+2 < 8$ or $2x-3 > 11$ $(-\infty, 2) \cup (7, \infty)$ 2 7

38. $3x+4 < -2$ or $3x+4 > 10$ $(-\infty, -2) \cup (2, \infty)$ −2 2

39. $-4(x+2) \ge 12$ or $3x+8 < 11$ $(-\infty, 1)$ 1

40. $5(x-2) \ge 0$ and $-3x < 9$ $[2, \infty)$ 2

41. $x < -3$ and $x > 3$ no solution

42. $x < 3$ or $x > -3$ $(-\infty, \infty)$

43. Is the relation $\le$ reflexive? yes

44. Is the relation $\ge$ reflexive? yes

45. Is the relation $\le$ symmetric? no

46. Is the relation $>$ symmetric? no

47. Is the relation $\ge$ transitive? yes

48. If $x < 3$, must it be true that $x^2 < 9$? no

49. If $x > -3$, must it be true that $x^2 > 9$? no

50. If $x > 2$, must it be true that $x^2 > 4$? yes

51. Finding profit The wholesale cost of a clock is \$27. A store owner knows that in order to sell, the clock must be priced under \$42. If p is the profit, express the possible profit as an inequality. $p < \$15$

52. Investing money If a woman invests \$10,000 at 8% annual interest, how much more must she invest at 9% so that her annual income will exceed \$1250? more than \$5000

53. Geometry If one side of a **regular pentagon** (a figure with five equal sides) is at least 37 inches but less than 52 inches, find the range in values of its perimeter p. 185 in. $\le p <$ 260 in.

54. Buying compact disks A student can afford to spend up to \$330 on a stereo system and some compact disks. If the stereo costs \$175 and the disks are \$8.50 each, find the greatest number of disks the student can buy. 18

55. Grade averages A student has scores of 70, 77, and 85 on three exams. What score is needed on a fourth exam to make the average 80 or better? 88 or higher

56. Geometry A rectangle is 3 feet longer than it is wide. If its perimeter must be at least 34 feet but not more than 54 feet, how wide can the rectangle be? 7 ft $\le w \le$ 12 ft

Writing Exercises

Write a paragraph using your own words.

1. The techniques for solving linear equations and linear inequalities are similar, yet different. Explain.

2. Explain the concepts of *absolute* and *conditional* inequalities.

Something to Think About

1. Which of these relations are transitive?
a. $=$ **b.** $\le$ **c.** $\not\ge$ **d.** $\ne$ $=, \le$

2. The following solution of $\frac{1}{3} > \frac{1}{x}$ is not correct. Why?

$$\frac{1}{3} > \frac{1}{x}$$

$$3x \cdot \frac{1}{3} > \frac{1}{x} \cdot 3x \quad \text{Multiply both sides by } 3x.$$

$$x > 3 \quad \text{Simplify.}$$

Review Exercises ■ *Simplify each expression.*

1. $\left(\dfrac{t^3t^5t^{-6}}{t^2t^{-4}}\right)^{-3}$ $\dfrac{1}{t^{12}}$

2. $\left(\dfrac{a^{-2}b^3a^5b^{-2}}{a^6b^{-5}}\right)^{-4}$ $\dfrac{a^{12}}{b^{24}}$

3. A man invested \$1200 in baking equipment to make pies. Each pie requires \$3.40 in ingredients. If the man can sell all the pies he can make for \$5.95 each, how many pies will he have to make to earn a profit? 471 or more

4. A woman invests \$15,000, part at 7% annual interest and the rest at 8%. If she earns \$2200 in income over a two-year period, how much did she invest at 7%? \$10,000

3.7 Inequalities with Absolute Values

■ Inequalities of the Form $|x| < k$ ■ Inequalities of the Form $|x| > k$

■ Inequalities of the Form $|x| < k$

The inequality $|x| < 5$ indicates that a point with coordinate x is less than 5 units from the origin. (See Figure 3-18.)

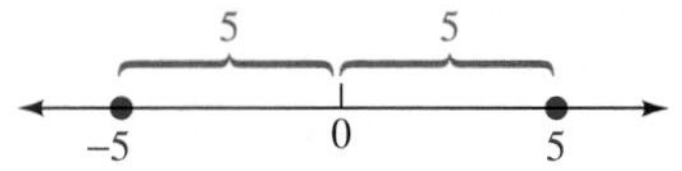

FIGURE 3-18

Thus, x is between -5 and 5, and

$$|x| < 5 \qquad \text{is equivalent to} \qquad -5 < x < 5$$

The solution to the inequality $|x| < k$ $(k > 0)$ includes the coordinates of the points on the number line that are less than k units from the origin. (See Figure 3-19.)

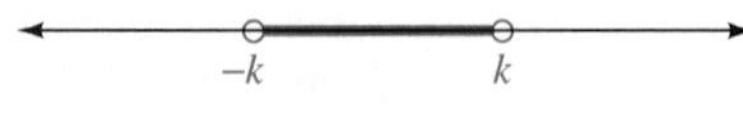

FIGURE 3-19

Thus, we have the following theorem.

Theorem If $k > 0$, then

$$|x| < k \qquad \text{is equivalent to} \qquad -k < x < k$$

$$|x| \leq k \qquad \text{is equivalent to} \qquad -k \leq x \leq k \qquad (k \geq 0)$$

EXAMPLE 1 Solve the inequality $|2x - 3| < 9$.

Solution We write the inequality as a double inequality

$$|2x - 3| < 9 \qquad \text{is equivalent to} \qquad -9 < 2x - 3 < 9$$

and solve for x:

$$-9 < 2x - 3 < 9$$

$$-6 < 2x < 12 \qquad \text{Add 3 to all three parts.}$$

$$-3 < x < 6 \qquad \text{Divide all parts by 2.}$$

Any number between -3 and 6, not including either -3 or 6, is in the solution set. This is the interval $(-3, 6)$. The graph is shown in Figure 3-20.

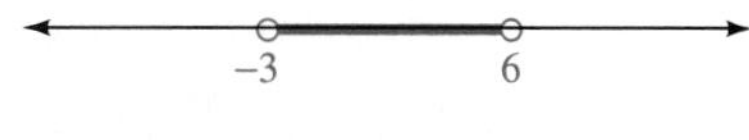

FIGURE 3-20

EXAMPLE 2 Solve the inequality $|3x + 2| \leq 5$.

Solution We write the expression as a double inequality and solve for x:

$$|3x + 2| \leq 5 \qquad \text{is equivalent to} \qquad -5 \leq 3x + 2 \leq 5$$

and solve for x:

$$-5 \leq 3x + 2 \leq 5$$

$$-7 \leq 3x \leq 3 \qquad \text{Subtract 2 from all three parts.}$$

$$-\frac{7}{3} \leq x \leq 1 \qquad \text{Divide all three parts by 3.}$$

The solution set is the interval $\left[-\frac{7}{3}, 1\right]$, whose graph is shown in Figure 3-21.

FIGURE 3-21

Inequalities of the Form $|x| > k$

The inequality $|x| > 5$ can be interpreted to mean that a point with coordinate of x is greater than 5 units from the origin. (See Figure 3-22.)

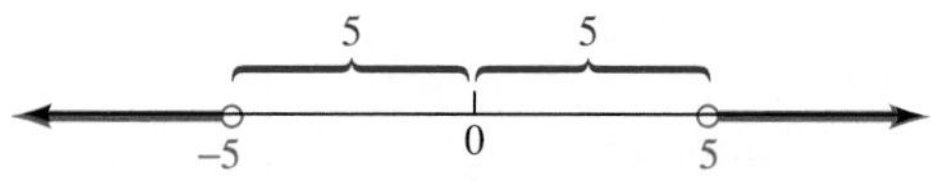

FIGURE 3-22

Thus, $x < -5$ or $x > 5$.

In general, the inequality $|x| > k$ can be interpreted to mean that a point with coordinate of x is more than k units from the origin. (See Figure 3-23.)

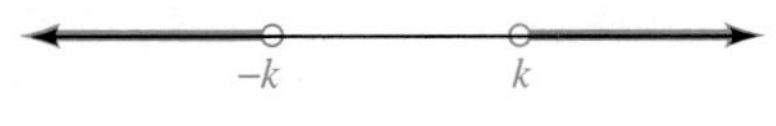

FIGURE 3-23

Thus,

$$|x| > k \quad \text{is equivalent to} \quad x < -k \quad \text{or} \quad x > k$$

The *or* indicates an either/or situation. It is necessary for x to satisfy only one of the two conditions to be in the solution set.

Theorem If k is a nonnegative constant, then

$$|x| > k \quad \text{is equivalent to} \quad x < -k \text{ or } x > k$$
$$|x| \geq k \quad \text{is equivalent to} \quad x \leq -k \text{ or } x \geq k$$

EXAMPLE 3 Solve the inequality $|5x - 10| > 20$.

Solution We write the inequality as two separate inequalities

$$|5x - 10| > 20 \quad \text{is equivalent to} \quad 5x - 10 < -20 \quad \text{or} \quad 5x - 10 > 20$$

and solve each one for x:

$5x - 10 < -20$	or	$5x - 10 > 20$	
$5x < -10$		$5x > 30$	Add 10 to both sides.
$x < -2$		$x > 6$	Divide both sides by 5.

Thus, x is either less than -2 or greater than 6:

$$x < -2 \text{ or } x > 6$$

This is the interval $(-\infty, -2) \cup (6, \infty)$. The graph appears in Figure 3-24.

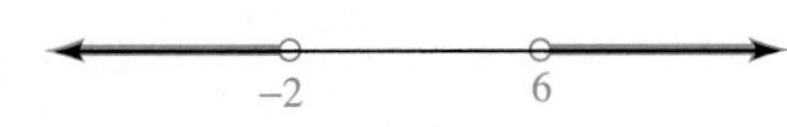

FIGURE 3-24

■

EXAMPLE 4 Solve the inequality $\left|\frac{3 - x}{5}\right| \geq 6$.

Solution We write the inequality as two separate inequalities

$$\left|\frac{3 - x}{5}\right| \geq 6 \quad \text{is equivalent to} \quad \frac{3 - x}{5} \leq -6 \quad \text{or} \quad \frac{3 - x}{5} \geq 6$$

and solve each one for x:

$$\frac{3-x}{5} \leq -6 \quad \text{or} \quad \frac{3-x}{5} \geq 6$$

$3 - x \leq -30$	$3 - x \geq 30$	Multiply both sides by 5.
$-x \leq -33$	$-x \geq 27$	Subtract 3 from both sides.
$x \geq 33$	$x \leq -27$	Divide both sides by -1 and reverse the direction of the inequality symbol.

The solution set is the interval $(-\infty, -27] \cup [33, \infty)$, whose graph appears in Figure 3-25.

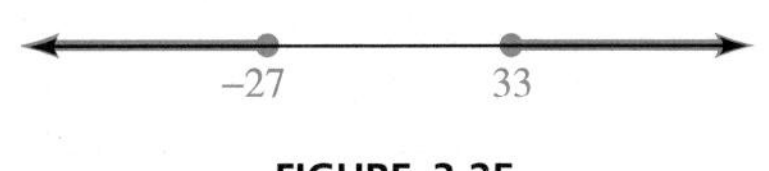

FIGURE 3-25

■

EXAMPLE 5 Solve the inequality $\left|\frac{2}{3}x - 2\right| - 3 > 6$.

Solution We begin by adding 3 to both sides to isolate the absolute value on the left-hand side. We then proceed as follows:

$$\left|\frac{2}{3}x - 2\right| - 3 > 6$$

$$\left|\frac{2}{3}x - 2\right| > 9 \qquad \text{Add 3 to both sides.}$$

$$\frac{2}{3}x - 2 < -9 \quad \text{or} \quad \frac{2}{3}x - 2 > 9$$

$\frac{2}{3}x < -7$	$\frac{2}{3}x > 11$	Add 3 to both sides.
$2x < -21$	$2x > 33$	Multiply both sides by 3.
$x < -\frac{21}{2}$	$x > \frac{33}{2}$	Divide both sides by 2.

The solution set is $\left(-\infty, -\frac{21}{2}\right) \cup \left(\frac{33}{2}, \infty\right)$, whose graph appears in Figure 3-26.

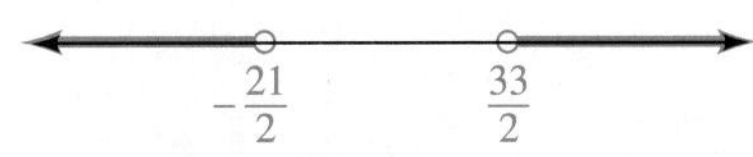

FIGURE 3-26

■

EXAMPLE 6 Solve the inequality $|3x - 5| \geq -2$.

Solution Since the absolute value of any number is nonnegative, and since any nonnegative number is larger than -2, the inequality is true for all x. The solution set is the interval $(-\infty, \infty)$, whose graph appears in Figure 3-27.

ORALS *Solve each inequality.*

1. $|x| < 8$ $\quad -8 < x < 8$
2. $|x| > 8$ $\quad x < -8$ or $x > 8$
3. $|x| \geq 4$ $\quad x \leq -4$ or $x \geq 4$
4. $|x| \leq 7$ $\quad -7 \leq x \leq 7$
5. $|x + 1| < 2$ $\quad -3 < x < 1$
6. $|x + 1| > 2$ $\quad x < -3$ or $x > 1$

EXERCISE 3.7

In Exercises 1–44, solve each inequality. Write the solution set in interval notation and graph it.

1. $|2x| < 8$ $\quad (-4, 4)$

2. $|3x| < 27$ $\quad (-9, 9)$

3. $|x + 9| \leq 12$ $\quad [-21, 3]$

4. $|x - 8| \leq 12$ $\quad [-4, 20]$

5. $|3x + 2| \leq -3$ $\quad$ no solution

6. $|3x - 2| < 10$ $\quad (-8/3, 4)$

7. $|4x - 1| \leq 7$ $\quad [-3/2, 2]$

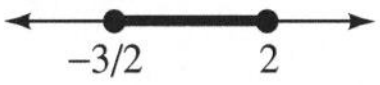

8. $|5x - 12| < -5$ $\quad$ no solution

9. $|3 - 2x| < 7$ $\quad (-2, 5)$

10. $|4 - 3x| \leq 13$ $\quad [-3, 17/3]$

11. $|5x| > 5$ $\quad (-\infty, -1) \cup (1, \infty)$

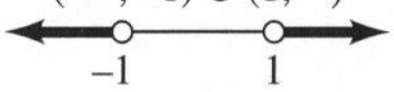

12. $|7x| > 7$ $\quad (-\infty, -1) \cup (1, \infty)$

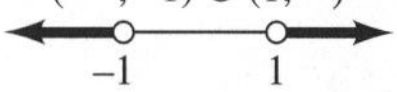

13. $|x - 12| > 24$ $\quad (-\infty, -12) \cup (36, \infty)$

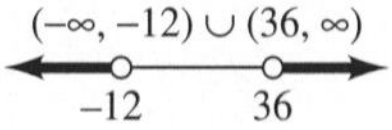

14. $|x + 5| \geq 7$ $\quad (-\infty, -12] \cup [2, \infty)$

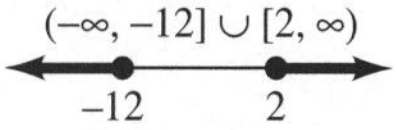

15. $|3x + 2| > 14$ $\quad (-\infty, -16/3] \cup (4, \infty)$

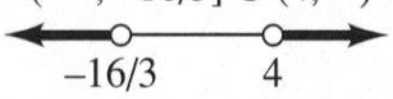

16. $|2x - 5| > 25$ $\quad (-\infty, -10) \cup (15, \infty)$

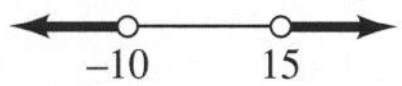

17. $|4x + 3| > -5$ $(-\infty, \infty)$

18. $|4x + 3| > 0$ $(-\infty, -3/4) \cup (-3/4, \infty)$; $-3/4$

19. $|2 - 3x| \geq 8$ $(-\infty, -2] \cup [10/3, \infty)$; -2, $10/3$

20. $|-1 - 2x| > 5$ $(-\infty, -3) \cup (2, \infty)$; -3, 2

21. $-|2x - 3| < -7$ $(-\infty, -2) \cup (5, \infty)$; -2, 5

22. $-|3x + 1| < -8$ $(-\infty, -3) \cup (7/3, \infty)$; -3, $7/3$

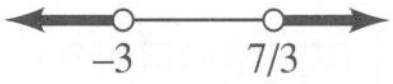

23. $|8x - 3| > 0$ $(-\infty, 3/8) \cup (3/8, \infty)$; $3/8$

24. $|7x + 2| > -8$ $(-\infty, \infty)$

25. $\left|\dfrac{x-2}{3}\right| \leq 4$ $[-10, 14]$; -10, 14

26. $\left|\dfrac{x-2}{3}\right| > 4$ $(-\infty, -10) \cup (14, \infty)$; -10, 14

27. $|3x + 1| + 2 < 6$ $(-5/3, 1)$; $-5/3$, 1

28. $|3x - 2| + 2 \geq 0$ $(-\infty, \infty)$

29. $3|2x + 5| \geq 9$ $(-\infty, -4] \cup [-1, \infty)$; -4, -1

30. $-2|3x - 4| < 16$ $(-\infty, \infty)$

31. $|5x - 1| + 4 \leq 0$ no solution

32. $-|5x - 1| + 2 < 0$ $(-\infty, -1/5) \cup (3/5, \infty)$; $-1/5$, $3/5$

33. $\left|\dfrac{1}{3}x + 7\right| + 5 > 6$ $(-\infty, -24) \cup (-18, \infty)$; -24, -18

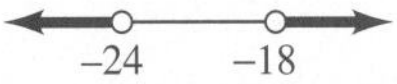

34. $\left|\dfrac{1}{2}x - 3\right| - 4 < 2$ $(-6, 18)$; -6, 18

35. $\left|\dfrac{1}{5}x - 5\right| + 4 > 4$ $(-\infty, 25) \cup (25, \infty)$; 25

36. $\left|\dfrac{1}{6}x + 6\right| + 2 < 2$ no solution

37. $\left|\dfrac{3}{5}x + \dfrac{7}{3}\right| < 2$ $(-65/9, -5/9)$; $-65/9$, $-5/9$

38. $\left|\dfrac{7}{3}x - \dfrac{3}{5}\right| \geq 1$ $(-\infty, -6/35] \cup [24/35, \infty)$; $-6/35$, $24/35$

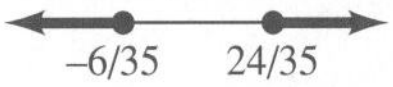

39. $\left|3\left(\dfrac{x+4}{4}\right)\right| > 0$ $(-\infty, -4) \cup (-4, \infty)$; -4

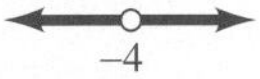

40. $3\left|\dfrac{1}{3}(x - 2)\right| + 2 \leq 3$ $[1, 3]$; 1, 3

41. $\left|\frac{1}{7}x + 1\right| \le 0$ [−7, −7] (number line: solid dot at −7)

42. $|2x + 1| + 2 \le 2$ [−1/2, −1/2] (number line: solid dot at −1/2)

43. $\left|\frac{x - 5}{10}\right| \le 0$ [5, 5] (number line: solid dot at 5)

44. $\left|\frac{3}{5}x - 2\right| + 3 \le 3$ [10/3, 10/3] (number line: solid dot at 10/3)

Writing Exercises

Write a paragraph using your own words.

1. Explain the use of open and solid circles in graphing inequalities.

2. If $k > 0$, explain the differences between the solution sets of $|x| < k$ and $|x| > k$.

Something to Think About

1. Under what conditions is $|x| + |y| > |x + y|$?
x and y must have different signs.

2. Under what conditions is $|x| + |y| = |x + y|$?
x and y must have the same sign.

Review Exercises

Solve each formula for the given variable.

1. $A = p + prt$ for t $\quad t = \frac{A - p}{pr}$

2. $A = p + prt$ for r $\quad r = \frac{A - p}{pt}$

3. $P = 2w + 2l$ for l $\quad l = \frac{P - 2w}{2}$

4. $V = \frac{1}{3}Bh$ for B $\quad B = \frac{3V}{h}$

PROJECT Guess and Adjust

The early Egyptians were clever at solving mathematical problems, but they were limited by their notation and an inadequate understanding of numbers. They did not solve equations as we do today but rather used a method called **the rule of false position.** In this method, the Egyptians would guess at a solution to an equation, check to see how far off that guess was, and then adjust the guess to find the correct answer.

For example, to solve

$$x + \frac{x}{5} = 42$$

by false position, we would guess a convenient value for x, such as $x = 5$. Then $x + \frac{x}{5}$ would be equal to 6 instead of 42. Because 6 must be multiplied by 7 to obtain 42, we would multiply our guess by 7: the correct solution is $x = 5 \cdot 7 = 35$.

1. Use the rule of false position to solve **a.** $x + \frac{x}{7} = 24$ and **b.** $2x + \frac{3x}{2} = 21$. Check your solutions.

2. Use the rule of false position to solve **a.** $\frac{x+2}{3} = 5$ and **b.** $x^2 + x = 42$. Did the method work? Explain.

Chapter Summary

KEY WORDS

absolute inequalities (3.6)
conditional equation (3.1)
conditional inequalities (3.6)
contradiction (3.1)
equation (3.1)
equivalent equations (3.1)
identity (3.1)
impossible equation (3.1)
linear equation (3.1)
linear inequality (3.6)
literal equation (3.4)
solution of an equation (3.1)
solution set (3.1)
variable (3.1)

KEY IDEAS

(3.1) An **equation** is a statement indicating that two quantities are equal.

The numbers that satisfy an equation are called **solutions** or **roots** of the equation.

If a and b are real numbers and $a = b$, then

$$a + c = b + c \qquad a - c = b - c$$

$$ac = bc \qquad \frac{a}{c} = \frac{b}{c} \quad (c \neq 0)$$

An **identity** is an equation that is true for all values of its variable.

A **conditional equation** is true for some but not all values of its variable.

An **impossible equation** is true for no values of its variable.

(3.2–3.3) To solve an application problem, express a quantity in two ways to form an equation. Then solve the equation.

(3.4) **Literal equations** are equations that contain several letters. Literal equations are often formulas.

Use the methods for solving equations to solve literal equations for an indicated variable.

(3.5) If $x \geq 0$, then $|x| = x$.
If $x < 0$, then $|x| = -x$.

If $k > 0$, $|x| = k$ is equivalent to $x = k$ or $x = -k$.

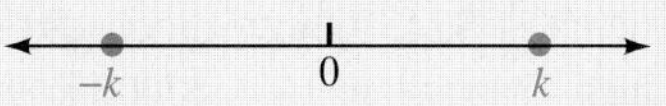

(3.6) The **trichotomy property** states that if a and b are two real numbers, then

$$a < b, \; a = b, \text{ or } a > b$$

(continued)

The relationships $<$, $>$, $\leq$, and $\geq$ obey the transitive property.

Solve a **linear inequality** in the same way as a linear equation. However, remember to change the order of the inequality when both sides are multiplied or divided by a negative number.

(3.7) If $k > 0$, then

$|x| < k$ is equivalent to $-k < x < k$

$|x| > k$ is equivalent to $x < -k$ or $x > k$

Chapter 3 Review Exercises

In Exercises 1–8, solve and check each equation.

1. $5x + 12 = 37$ 5

2. $-3x - 7 = 20$ -9

3. $4(y - 1) = 28$ 8

4. $3(x + 7) = 42$ 7

5. $13(x - 9) - 2 = 7x - 5$ 19

6. $\dfrac{8(x - 5)}{3} = 2(x - 4)$ 8

7. $\dfrac{3y}{4} - 13 = -\dfrac{y}{3}$ 12

8. $\dfrac{2y}{5} + 5 = \dfrac{14y}{10}$ 5

In Exercises 9–16, solve each problem.

9. Integer problem The sum of three consecutive even integers is 270. Find the integers. 88, 90, 92

10. Carpentry A carpenter wants to cut a 20-foot rafter so that one piece is 3 times as long as the other. Where should he cut the board? 5 ft from one end

11. Geometry A rectangle is 4 meters longer than it is wide. If the perimeter of the rectangle is 28 meters, find its area. 45 m^2

12. Balancing a seesaw Sue weighs 48 pounds, and her father weighs 180 pounds. If Sue sits on one end of a 20-foot long seesaw with the fulcrum in the middle, how far from the fulcrum should the father sit to balance the seesaw? $2\frac{2}{3}$ ft

13. Investment problem Sally has $25,000 to invest. She invests some money at 10% interest and the rest at 9%. If her total annual income from these two investments is $2430, how much did she invest at each rate? $18,000 at 10%, $7000 at 9%

14. Mixing solutions How much water must be added to 20 liters of a 12% alcohol solution to dilute it to an 8% solution? 10 liters

15. Motion problem A car and a motorcycle both leave from the same point and travel in the same direction. (See Illustration 1.) The car travels at an average rate of 55 miles per hour and the motorcycle at an average rate of 40 miles per hour. How long will it take before they are 5 miles apart? $\frac{1}{3}$ hr

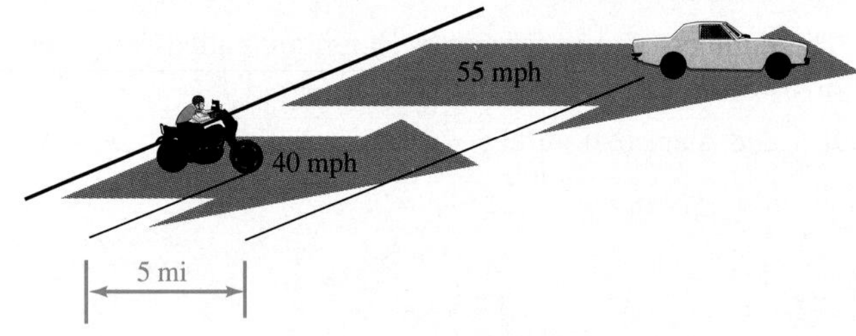

ILLUSTRATION 1

16. Finding the break point One machine has setup costs of $850 and a unit cost of $3.50 to make a circuit board. A newer machine has setup costs of $1050 and a unit cost of $1.50. Find the break point. 100 units

In Exercises 17–20, solve for the quantity indicated.

17. $V = \frac{4}{3}\pi r^3$ for r^3 $\quad r^3 = \frac{3V}{4\pi}$

18. $V = \frac{1}{3}\pi r^2 h$ for h $\quad h = \frac{3V}{\pi r^2}$

19. $V = \frac{1}{6}ab(x + y)$ for x $\quad x = \frac{6V}{ab} - y$

20. $V = \pi h^2\left(r - \frac{h}{3}\right)$ *for* r $\quad r = \frac{V}{\pi h^2} + \frac{h}{3}$

In Exercises 21–26, solve and check each equation.

21. $|3x + 1| = 10$ $\quad 3, -\frac{11}{3}$

22. $\left|\frac{3}{2}x - 4\right| = 9$ $\quad \frac{26}{3}, -\frac{10}{3}$

23. $|3x + 2| = |2x - 3|$ $\quad \frac{1}{5}, -5$

24. $|5x - 4| = |4x - 5|$ $\quad -1, 1$

25. $\left|\frac{2 - x}{3}\right| = 4$ $\quad 14, -10$

26. $\left|\frac{3 - 2x}{2}\right| = \left|\frac{3x - 2}{3}\right|$ $\quad \frac{13}{12}$

In Exercises 27–30, solve each inequality. Give each solution set in interval notation and graph it.

27. $\frac{1}{3}y - 2 \geq \frac{1}{2}y + 2$ $\quad (-\infty, -24]$

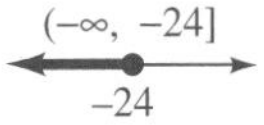

28. $\frac{7}{4}(x + 3) < \frac{3}{8}(x - 3)$ $\quad (-\infty, -51/11)$

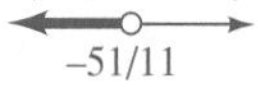

29. $3 < 3x + 4 < 10$ $\quad (-1/3, 2)$

30. $4x > 3x + 2 > x - 3$ $\quad (2, \infty)$

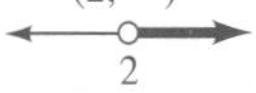

In Exercises 31–36, solve each inequality. Give each solution in interval notation and graph it.

31. $|2x + 7| < 3$ $\quad (-5, -2)$

32. $|3x - 8| \geq 4$ $\quad (-\infty, 4/3] \cup [4, \infty)$

33. $\left|\frac{3}{2}x - 14\right| \geq 0$ $\quad (-\infty, \infty)$

34. $\left|\frac{2}{3}x + 14\right| < 0$ $\quad$ no solutions

35. $|5 - 3x| \leq 14$ $\quad [-3, 19/3]$

36. $\left|\frac{1 - 5x}{3}\right| > 7$ $\quad (-\infty, -4) \cup (22/5, \infty)$

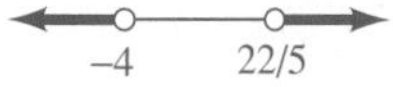

Chapter 3 Test

1. Is -12 a solution of $\frac{1}{3} + \frac{1}{x} = \frac{1}{4}$? yes

In Problems 2–3, solve each equation.

2. $9(x + 4) + 4 = 4(x - 5)$ -12

3. $\frac{y - 1}{5} + 2 = \frac{2y - 3}{3}$ 6

4. A 20-foot pipe is to be cut into three pieces. One piece is to be twice as long as another, and the third piece is to be six times as long as the shortest. Find the length of the longest piece. $13\frac{1}{3}$ ft

5. A rectangle with a perimeter of 26 centimeters is 5 centimeters longer than it is wide. Find its area. 36 m^2

6. Bob invests part of $10,000 at 9% annual interest and the rest at 8%. His annual income from these investments is $860. How much does he invest at 8%? $4000

7. Two women drove a car a total of 280 miles. One woman drove at an average speed of 50 mph and the other at 60 mph. If the trip took 5 hours, how long did the first woman drive? 2 hr

8. How many liters of water are needed to dilute 20 liters of a 5% salt solution to a 1% solution? 80 liters

9. Solve $p = L + \frac{s}{f}i$ for i. $i = \frac{f(P - L)}{s}$

10. Solve $n = \frac{360}{180 - a}$ for a. $a = \frac{180n - 360}{n}$

In Problems 11–12, write each expression without using absolute value symbols.

11. $|5 - 8|$ 3

12. $|4 - 4\pi|$ $4\pi - 4$

In Problems 13–16, solve each equation.

13. $|2x + 3| = 11$ 4, -7

14. $|4 - 3x| = 19$ -5, $\frac{23}{3}$

15. $|3x + 4| = |x + 12|$ 4, -4

16. $|3 - 2x| = |2x + 3|$ 0

In Problems 17–22, solve each inequality and graph it.

17. $-2(2x + 3) \geq 14$ $(-\infty, -5]$ (graph: −5)

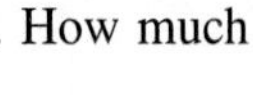

18. $-2 < \frac{x - 4}{3} < 4$ $(-2, 16)$ (graph: −2, 16)

19. $|x + 3| \leq 4$ $[-7, 1]$ (graph: −7, 1)

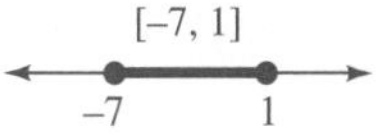

20. $|2x - 4| > 22$ $(-\infty, -5) \cup (6, \infty)$ (graph: −5, 6)

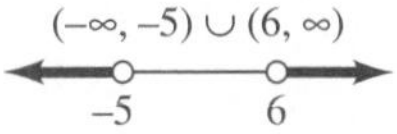

21. $|4 - 2x| > 2$ $(-\infty, 1) \cup (3, \infty)$ (graph: 1, 3)

22. $|2x - 4| \leq 2$ $[1, 3]$ (graph: 1, 3)

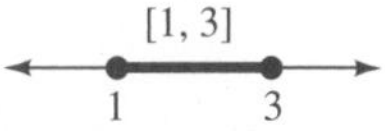

Cumulative Review Exercises

In Exercises 1–10, tell which numbers in the set $\left\{-2, 0, 1, 2, \frac{13}{12}, 6, 7, \sqrt{5}, \sqrt{-7}, \pi\right\}$ *are in each category.*

1. natural numbers 1, 2, 6, 7

2. whole numbers 0, 1, 2, 6, 7

3. rational numbers $-2, 0, 1, 2, \frac{13}{12}, 6, 7$

4. irrational numbers $\sqrt{5}, \pi$

5. imaginary numbers $\sqrt{-7}$

6. real numbers $-2, 0, 1, 2, \frac{13}{12}, 6, 7, \sqrt{5}, \pi$

7. prime numbers 2, 7

8. composite numbers 6

9. even numbers $-2, 0, 2, 6$

10. odd numbers 1, 7

In Exercises 11–12, simplify each expression.

11. $-(|5| - |3|)$ -2

12. $\dfrac{|-5| + |-3|}{-|4|}$ -2

13. Write the terminating decimal 0.875 in fractional form. $\frac{7}{8}$

14. Draw a number line and graph the prime numbers from 50 to 60.

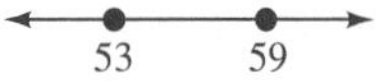

In Exercises 15–18, tell which property of real numbers justifies each statement.

15. If $3 = x$ and $x = y$, then $3 = y$. transitive prop.

16. $3(x + y) = 3x + 3y$ distrib. prop.

17. $(a + b) + c = c + (a + b)$ commut. prop. of +

18. $(ab)c = a(bc)$ assoc. prop. of ×

19. Find the additive inverse of -5. 5

20. Find the multiplicative inverse of $-\dfrac{2}{3}$. $-\frac{3}{2}$

In Exercises 21–24, do the operations.

21. $2 + 4 \cdot 5$ 22

22. $\dfrac{8 - 4}{2 - 4}$ -2

23. $20 \div (-10 \div 2)$ -4

24. $\dfrac{6 + 3(6 + 4)}{2(3 - 9)}$ -3

In Exercises 25–28, do the operations and simplify when possible.

25. $-\dfrac{5}{6} \cdot \dfrac{3}{20}$ $-\frac{1}{8}$

26. $\dfrac{2}{3}\left(-\dfrac{4}{5} \div \dfrac{28}{35}\right)$ $-\frac{2}{3}$

27. $-\left(\dfrac{1}{3} + \dfrac{3}{4}\right)\left(\dfrac{5}{3} + \dfrac{1}{2}\right)$ $-\frac{169}{72}$

28. $\dfrac{1}{2} + \left(\dfrac{2}{3} - \dfrac{3}{4}\right)$ $\frac{5}{12}$

In Exercises 29–32, simplify each expression. Assume that all variables are positive numbers and write all answers without negative exponents.

29. $(x^2y^3)^4$ x^8y^{12}

30. $\dfrac{c^4c^8}{(c^5)^2}$ c^2

31. $\left(-\dfrac{a^3b^{-2}}{ab}\right)^{-1}$ $-\dfrac{b^3}{a^2}$

32. $\left(\dfrac{-3a^3b^{-2}}{6a^{-2}b^3}\right)^0$ 1

33. Change 0.00000497 to scientific notation. 4.97×10^{-6}

34. Change 9.32×10^8 to standard notation. 932,000,000

35. Is $3 + x + 4x^2$ a monomial, a binomial, or a trinomial? trinomial

36. Find the degree of $3 + x^2y + 17x^3y^4$. 7

37. If $P(x) = -3x^3 + x - 4$, find $P(-2)$. 18

38. Evaluate $\dfrac{x^2 - y^2}{2x + y}$ when $x = 2$ and $y = -3$. -5

In Exercises 39–44, do the operations and simplify.

39. $(3x^2 - 2x + 7) + (-2x^2 + 2x + 5) + (3x^2 - 4x + 2)$ $4x^2 - 4x + 14$

40. $(-5x^2 + 3x + 4) - (-2x^2 + 3x + 7)$ $-3x^2 - 3$

41. $(3x + 4)(2x - 5)$ $6x^2 - 7x - 20$

42. $(2x^n - 1)(x^n + 2)$ $2x^{2n} + 3x^n - 2$

43. $(x^2 + 9x + 20) \div (x + 5)$ $x + 4$

44. $(2x^2 + 4x - x^3 + 3) \div (x - 1)$ $-x^2 + x + 5 + \frac{8}{x-1}$

In Exercises 45–48, solve each equation.

45. $2x - 5 = 11$ 8

46. $\dfrac{2}{3}x - 2 = x + 7$ -27

47. $4(y - 3) + 4 = -3(y + 5)$ -1

48. $2x - \dfrac{3(x - 2)}{2} = 7 - \dfrac{x - 3}{3}$ 6

49. The sum of three consecutive even integers is 90. Find the integers. 28, 30, 32

50. A rectangle is three times as long as it is wide. If its perimeter is 112 centimeters, find its dimensions. 14 cm by 42 cm

51. Solve the formula $S = \dfrac{n(a + l)}{2}$ for a. $a = \dfrac{2S}{n} - l$

52. Solve the formula $A = \dfrac{1}{2}h(b_1 + b_2)$ for h. $h = \dfrac{2A}{b_1 + b_2}$

In Exercises 53–54, solve each equation.

53. $|4x - 3| = 9$ $3, -\frac{3}{2}$

54. $|2x - 1| = |3x + 4|$ $-5, -\frac{3}{5}$

In Exercises 55–60, solve each inequality.

55. $-3(x - 4) \geq x - 32$ $x \leq 11$

56. $-8 < -3x + 1 < 10$ $-3 < x < 3$

57. $x - 2 \leq 3x + 1 \leq 5x - 4$ $x \geq \frac{5}{2}$

58. $|3x - 2| \leq 4$ $-\frac{2}{3} \leq x \leq 2$

59. $|2x + 3| > 5$ $x < -4$ or $x > 1$

60. $\left|\frac{3a}{5} - 2\right| + 1 \geq \frac{6}{5}$ $a \leq 3$ or $a \geq \frac{11}{3}$

CAREERS & MATHEMATICS

STATISTICIAN

Statisticians devise and carry out experiments and surveys and interpret the numerical results. They then apply their knowledge to subject areas such as economics, human behavior, natural science, and engineering.

Often they are able to obtain accurate information about a group of people or things by surveying a small portion, called a **sample,** rather than the whole group. This technique requires statisticians to decide where and how to get the data, how to determine the type and size of the sample group, and how to develop the survey questionnaire or reporting form.

Qualifications A bachelor's degree in statistics or mathematics is the minimum educational requirement for many beginning jobs in statistics. For other entry-level statistics jobs, however, it is preferable to have a bachelor's degree with a major in an applied field such as economics or natural science and a minor in statistics.

Job Outlook Employment opportunities for persons who combine training in statistics with knowledge of a field of application such as manufacturing, engineering, scientific research, or business are expected to be favorable through the year 2005.

Example Application A researcher wishes to estimate the mean (average) property tax paid by homeowners living in the city. To do so, he decides to select a **random sample** of homeowners and compute the mean tax paid by the homeowners in that sample. How large must the sample be for the researcher to be 95% certain that his computed sample mean will be within \$35 of the population mean—that is, within \$35 of the average tax paid by *all* homeowners in the city? Assume that the standard deviation, σ, of all tax bills is known to be \$120.

Solution From elementary statistics, the researcher has the formula

$$\frac{3.84\sigma^2}{N} < E^2$$

where σ^2, called the **population variance,** is the square of the standard deviation, E is the maximum acceptable error, and N is the sample size. He substitutes 120 for σ and 35 for E in the previous formula and solves for N.

$$\frac{(3.84)(120^2)}{N} < 35^2$$

$$\frac{55{,}296}{N} < 1225 \qquad \text{Simplify.}$$

$$55{,}296 < 1225N \qquad \text{Multiply both sides by } N.$$

$$\frac{55{,}296}{1225} < N \qquad \text{Divide both sides by 1225.}$$

$$45.1 < N \qquad \text{Simplify.}$$

To be 95% certain that the sample mean will be within \$35 of the population mean, the researcher must sample *more than* 45.1 homeowners: the sample must contain at least 46 homeowners.

EXERCISES

1. What sample size must the researcher choose to keep the error less than \$20?

2. What must the sample size be to keep the error less than \$10?

3. If the researcher cuts the acceptable error in half, by what factor is the sample size increased?

4. If the researcher decides to double the acceptable error, by what factor is the sample size decreased?

(*Answers:* **1.** 139 **2.** 553 **3.** 4 **4.** $\frac{1}{4}$)

4 Factoring Polynomials

We have discussed how to multiply polynomials by polynomials. We now reverse that procedure and discuss how to undo multiplications. The process of finding the individual factors of a known product is called **factoring.**

In Section 4.5, we will use factoring to solve equations, and in Chapter 5 we will use factoring to simplify algebraic fractions.

4.1 The Greatest Common Factor

■ Prime-Factored Form of a Natural Number ■ Factoring Out the Greatest Common Factor ■ Factoring by Grouping ■ Literal Equations

■ Prime-Factored Form of a Natural Number

When one number divides another number, the first number is called a **factor** of the other. For example, 2 is a factor of 6, because 2 divides 6. The factors of 6 are

1, 2, 3, and 6

because each of these numbers divides 6. Similarly, the factors of 24 are

1, 2, 3, 4, 6, 8, 12, and 24

because each of these numbers divides 24.

Teaching Tip: Use a factoring tree:

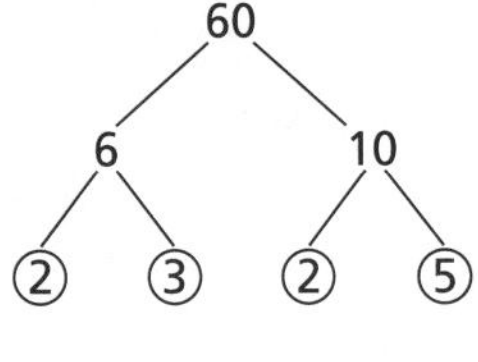

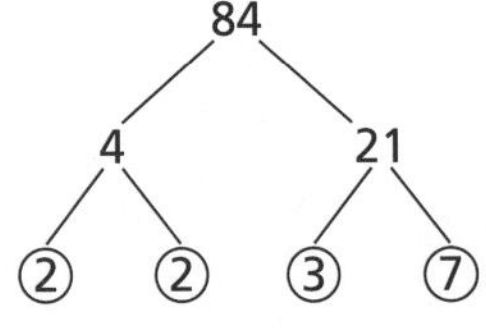

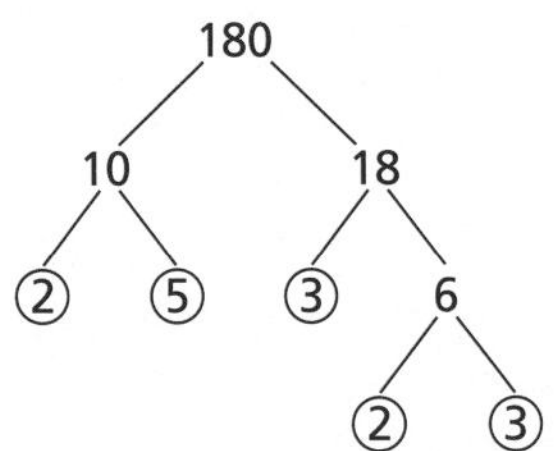

A natural number p is a **prime number** if $p > 1$ and the only natural number factors of p are 1 and p. The first ten primes are

2, 3, 5, 7, 11, 13, 17, 19, 23, 29

To factor a natural number means to write the number as a product of other natural numbers. If each of the factors in the product is prime, the number is written in **prime-factored form.** The statements

$$60 = 6 \cdot 10 = 2 \cdot 3 \cdot 2 \cdot 5 = 2^2 \cdot 3 \cdot 5$$

$$84 = 4 \cdot 21 = 2 \cdot 2 \cdot 3 \cdot 7 = 2^2 \cdot 3 \cdot 7$$

and

$$180 = 10 \cdot 18 = 2 \cdot 5 \cdot 3 \cdot 6 = 2 \cdot 5 \cdot 3 \cdot 3 \cdot 2 = 2^2 \cdot 3 \cdot 3 \cdot 5$$

show the prime-factored forms of 60, 84, and 180, respectively.

The largest natural number that divides 60, 84, and 180 is called the **greatest common factor** or **greatest common divisor** of these three numbers. Because 60, 84, and 180 all have at least two factors of 2 and one factor of 3, the greatest common factor of these three numbers is $2^2 \cdot 3 = 12$. We note that

$$\frac{60}{12} = 5, \qquad \frac{84}{12} = 7, \qquad \text{and} \qquad \frac{180}{12} = 15$$

Teaching Tip: You may want to remind students of the following divisibility rules:

1. An even integer is divisible by 2.
2. If the sum of the digits in an integer is divisible by 3, the integer is divisible by 3.
3. An integer ending with 0 or 5 is divisible by 5.
4. If a number ends with 0, it is divisible by 10.

There is no natural number greater than 12 that divides 60, 84, and 180.

Algebraic monomials also have greatest common factors. Consider three monomials with their prime factorizations:

$$6a^2b^3c = 3 \cdot 2 \cdot a \cdot a \cdot b \cdot b \cdot b \cdot c$$
$$9a^3b^2c = 3 \cdot 3 \cdot a \cdot a \cdot a \cdot b \cdot b \cdot c$$
$$18a^4c^3 = 2 \cdot 3 \cdot 3 \cdot a \cdot a \cdot a \cdot a \cdot c \cdot c \cdot c$$

Since each monomial has at least one factor of 3, two factors of a, and one factor of c in common, their greatest common factor is

$$3^1 \cdot a^2 \cdot c^1 = 3a^2c$$

To find the greatest common factor of several monomials, we follow these steps.

Steps for Finding the Greatest Common Factor

1. Find the prime factorization of each monomial.
2. List the prime factors that are common to each monomial.
3. Find the product of the factors found in Step 2, with each factor raised to the smallest power that occurs in any one monomial.

PERSPECTIVE

No End of Primes

A composite number always has prime factors that are smaller than the number itself. The composite number 30, for example, has the prime factors 2, 3, and 5—all less than 30. We might think that the larger a number is, the more likely it is to be composite, because there are so many smaller prime numbers that could be possible factors. Maybe all numbers, if large enough, would be composite, and therefore there would be a largest prime number.

That is not what happens, however. In about 300 B.C. the Greek mathematician Euclid proved that the number of primes is unlimited—that there are infinitely many prime numbers. This is an important fact of a branch of mathematics called **number theory.** Although Euclid is best known for his study of geometry, many of his writings deal with number theory.

Euclid (about 300 B.C.)

Who could possibly care that prime numbers can be very large? Cryptographers working for the federal government care, because several codes that are very difficult to crack rely on the properties of large primes.

Factoring Out the Greatest Common Factor

We have seen that the distributive property provides a method for multiplying a polynomial by a monomial. For example,

$$2x^3y^3(3x^2 - 4y^3) = 2x^3y^3 \cdot 3x^2 - 2x^3y^3 \cdot 4y^3$$
$$= 6x^5y^3 - 8x^3y^6$$

If the product of a multiplication is $6x^5y^3 - 8x^3y^6$, we can use the distributive property to find the individual factors.

$$6x^5y^3 - 8x^3y^6 = 2x^3y^3 \cdot 3x^2 - 2x^3y^3 \cdot 4y^3$$
$$= 2x^3y^3(3x^2 - 4y^3)$$

Since $2x^3y^3$ is the greatest common factor of the terms of $6x^5y^3 - 8x^3y^6$, this process is called **factoring out the greatest common factor.**

EXAMPLE 1 Factor $25a^3b + 15ab^3$.

Solution We begin by factoring each monomial:

$$25a^3b = 5 \cdot 5 \cdot a \cdot a \cdot a \cdot b$$
$$15ab^3 = 5 \cdot 3 \cdot a \cdot b \cdot b \cdot b$$

Since each term has at least one factor of 5, one factor of a, and one factor of b in common, and there are no other common factors, $5ab$ is the greatest common factor of the two terms. We can use the distributive property to factor out $5ab$.

$$25a^3b + 15ab^3 = 5ab \cdot 5a^2 + 5ab \cdot 3b^2$$
$$= 5ab(5a^2 + 3b^2)$$

■

EXAMPLE 2 Factor $3xy^2z^3 + 6xz^2 - 9xyz^4$.

Solution We begin by factoring each monomial:

$$3xy^2z^3 = 3 \cdot x \cdot y \cdot y \cdot z \cdot z \cdot z$$
$$6xz^2 = 3 \cdot 2 \cdot x \cdot z \cdot z$$
$$-9xyz^4 = -3 \cdot 3 \cdot x \cdot y \cdot z \cdot z \cdot z \cdot z$$

Since each term has one factor of 3, one factor of x, and two factors of z in common, and because there are no other common factors, $3xz^2$ is the greatest common factor of the three terms. We can use the distributive property to factor out the $3xz^2$.

$$3xy^2z^3 + 6xz^2 - 9xyz^4 = 3xz^2 \cdot y^2z + 3xz^2 \cdot 2 - 3xz^2 \cdot 3yz^2$$
$$= 3xz^2(y^2z + 2 - 3yz^2)$$

■

EXAMPLE 3 Factor $x^3y^3z^3 + xyz$.

Solution Since each term has one factor of x, one factor of y, and one factor of z in common, and there are no other common factors, the greatest common factor of the two terms is xyz, and the expression factors as

$$\begin{aligned} x^3y^3z^3 + xyz &= xyz \cdot x^2y^2z^2 + xyz \cdot 1 \\ &= xyz(x^2y^2z^2 + 1) \end{aligned}$$

WARNING! The last term xyz of the given binomial has an understood coefficient of 1. When the xyz is factored out, remember to write the 1. ■

A polynomial that cannot be factored is called a **prime polynomial** or an **irreducible polynomial.**

EXAMPLE 4 Factor $3x^2 + 4y + 7$.

Solution We factor each monomial:

$$3x^2 = 3 \cdot x \cdot x$$
$$4y = 4 \cdot y$$
$$7 = 7$$

Since there are no common factors other than 1, this polynomial cannot be factored. It is a prime polynomial. ■

EXAMPLE 5 Factor the negative of the greatest common factor from $-6u^2v^3 + 8u^3v^2$.

Solution Because the greatest common factor of the two terms is $2u^2v^2$, the negative of the greatest common factor is $-2u^2v^2$. To factor out $-2u^2v^2$, we proceed as follows:

$$\begin{aligned} -6u^2v^3 + 8u^3v^2 &= -2u^2v^2 \cdot 3v + 2u^2v^2 \cdot 4u \\ &= -2u^2v^2 \cdot 3v - (-2u^2v^2)4u \\ &= -2u^2v^2(3v - 4u) \end{aligned}$$

■

We can also factor out a common factor with a variable exponent.

EXAMPLE 6 Factor x^{2n} from $x^{4n} + x^{3n} + x^{2n}$.

Solution We can write the trinomial in the form

$$x^{2n} \cdot x^{2n} + x^{2n} \cdot x^n + x^{2n} \cdot 1$$

and factor out the x^{2n}.

$$\begin{aligned} x^{4n} + x^{3n} + x^{2n} &= x^{2n} \cdot x^{2n} + x^{2n} \cdot x^n + x^{2n} \cdot 1 \\ &= x^{2n}(x^{2n} + x^n + 1) \end{aligned}$$

■

EXAMPLE 7 Factor $a^{-2}b^{-2}$ from $a^{-2}b - a^3b^{-2}$.

Solution We rewrite the expression $a^{-2}b - a^3b^{-2}$ in the form $a^{-2}b^{-2} \cdot b^3 - a^{-2}b^{-2} \cdot a^5$ and factor out the $a^{-2}b^{-2}$.

$$\begin{aligned} a^{-2}b - a^3b^{-2} &= a^{-2}b^{-2} \cdot b^3 - a^{-2}b^{-2} \cdot a^5 \\ &= a^{-2}b^{-2}(b^3 - a^5) \end{aligned}$$ ■

A common factor can have more than one term. For example, in the expression

$$x(a + b) + y(a + b)$$

the binomial $a + b$ is a factor of both terms. We can factor it out to get

$$\begin{aligned} x(a + b) + y(a + b) &= (a + b)x + (a + b)y \quad \text{Use the commutative property of multiplication.} \\ &= (a + b)(x + y) \end{aligned}$$

EXAMPLE 8 Factor $a(x - y + z) - b(x - y + z) + 3(x - y + z)$.

Solution We can use the distributive property to factor out the greatest common factor of $(x - y + z)$.

$$\begin{aligned} a(x - y + z) - b(x - y + z) + 3(x - y + z) &= (x - y + z)a - (x - y + z)b + (x - y + z)3 \\ &= (x - y + z)(a - b + 3) \end{aligned}$$ ■

Factoring by Grouping

Suppose we wish to factor

$$ac + ad + bc + bd$$

Although there is no factor common to all four terms, there is a common factor of a in the first two terms and a common factor of b in the last two terms. We can factor out these common factors to get

$$ac + ad + bc + bd = a(c + d) + b(c + d)$$

We can now factor out the common factor of $c + d$ on the right-hand side:

$$ac + ad + bc + bd = (c + d)(a + b)$$

The grouping in this type of problem is not always unique. For example, if we write the expression $ac + ad + bc + bd$ in the form

$$ac + bc + ad + bd$$

and factor c from the first two terms and d from the last two terms, we obtain

$$\begin{aligned} ac + bc + ad + bd &= c(a + b) + d(a + b) \\ &= (a + b)(c + d) \end{aligned}$$

The method used in the previous examples is called **factoring by grouping.**

EXAMPLE 9 Factor $3ax^2 + 3bx^2 + a + 5bx + 5ax + b$.

Solution Although there is no factor common to all six terms, $3x^2$ can be factored out of the first two terms and $5x$ can be factored out of the fourth and fifth terms, to get

$$3ax^2 + 3bx^2 + a + \mathbf{5bx} + \mathbf{5ax} + b = 3x^2(a + b) + a + \mathbf{5x(b + a)} + b$$

This result can be written in the form

$$3ax^2 + 3bx^2 + a + 5bx + 5ax + b = 3x^2(a + b) + 5x(a + b) + (a + b)$$

Since $a + b$ is common to all three terms, it can be factored out to get

$$3ax^2 + 3bx^2 + a + 5bx + 5ax + b = (a + b)(3x^2 + 5x + 1)$$ ■

To factor an expression, it is often necessary to factor more than once, as the following example illustrates.

EXAMPLE 10 Factor $3x^3y - 4x^2y^2 - 6x^2y + 8xy^2$.

Solution We begin by factoring out the common factor of xy.

$$3x^3y - 4x^2y^2 - 6x^2y + 8xy^2 = xy(3x^2 - 4xy - 6x + 8y)$$

We can now factor $3x^2 - 4xy - 6x + 8y$ by grouping:

$$\begin{aligned}3x^3y - 4x^2y^2 - 6x^2y + 8xy^2 &= xy(3x^2 - 4xy - 6x + 8y)\\ &= xy[x(3x - 4y) - 2(3x - 4y)] && \text{Factor } x \text{ from } 3x^2 - 4xy \text{ and } -2 \text{ from } -6x + 8y.\\ &= xy(3x - 4y)(x - 2) && \text{Factor out } 3x - 4y.\end{aligned}$$

Because no more factoring can be done, the factorization is complete. ■

WARNING! Whenever you factor an expression, always factor it completely. Each factor of a completely factored expression will be prime.

Literal Equations

Factoring is often required to solve a literal equation for one of its variables.

EXAMPLE 11 The formula $A = p + prt$ gives the amount of money in a savings account at the end of a specific time. Solve the formula for p.

Solution We must isolate p on one side of the equation. Because p appears twice, we factor p from the right-hand side of the equation. Then we solve for p.

$$A = p + prt$$

$$A = p(1 + rt) \quad \text{Factor out } p.$$

$$\frac{A}{1 + rt} = p \quad \text{Divide both sides by } 1 + rt.$$

$$p = \frac{A}{1 + rt}$$

■

EXAMPLE 12 The formula $r_1r_2 = rr_2 + rr_1$ is used in electronics to relate the combined resistance, r, of two resistors wired in parallel. The variable r_1 represents the resistance of the first resistor, and the variable r_2 represents the resistance of the second. Solve the equation for r_2.

Solution To isolate r_2 on one side of the equation, we get all terms involving r_2 on the left-hand side of the equation and all terms not involving r_2 on the right-hand side and proceed as follows.

$$r_1r_2 = rr_2 + rr_1$$

$$r_1r_2 - rr_2 = rr_1 \quad \text{Subtract } rr_2 \text{ from both sides.}$$

$$r_2(r_1 - r) = rr_1 \quad \text{Factor out } r_2.$$

$$r_2 = \frac{rr_1}{r_1 - r} \quad \text{Divide both sides by } r_1 - r.$$

■

ORALS *Factor each expression.*

1. $3x^2 - x$ $\quad x(3x - 1)$

2. $7t^3 + 14t^2$ $\quad 7t^2(t + 2)$

3. $-3a^2 - 6a$ $\quad -3a(a + 2)$

4. $-4x^2 + 12x$ $\quad -4x(x - 3)$

5. $3(a + b) + x(a + b)$ $\quad (a + b)(3 + x)$

6. $a(m - n) - b(m - n)$ $\quad (m - n)(a - b)$

EXERCISE 4.1

In Exercises 1–8, find the prime factorization of each number.

1. 6 $\quad 2 \cdot 3$

2. 10 $\quad 2 \cdot 5$

3. 135 $\quad 3^3 \cdot 5$

4. 98 $\quad 2 \cdot 7^2$

5. 128 $\quad 2^7$

6. 357 $\quad 3 \cdot 7 \cdot 17$

7. 325 $\quad 5^2 \cdot 13$

8. 288 $\quad 2^5 \cdot 3^2$

In Exercises 9–16, find the greatest common factor of each set of monomials.

9. 36, 48 $\quad 12$

10. 45, 75 $\quad 15$

11. 42, 36, 98 $\quad 2$

12. 16, 40, 60 $\quad 4$

13. $4a^2b, 8a^3c$ $\quad 4a^2$

14. $6x^3y^2z, 9xyz^2$ $\quad 3xyz$

15. $18x^4y^3z^2, -12xy^2z^3$ $\quad 6xy^2z^2$

16. $6x^2y^3, 24xy^3, 40x^2y^2z^3$ $\quad 2xy^2$

In Exercises 17–20, complete each factorization.

17. $3a - 12 = 3(a - 4)$ **18.** $5t + 25 = 5(t + 5)$ **19.** $8z^2 + 2z = 2z(4z + 1)$ **20.** $9t^3 - 3t^2 = 3t^2(3t - 1)$

In Exercises 21–42, factor each expression, if possible.

21. $2x + 8$ $2(x + 4)$

22. $3y - 9$ $3(y - 3)$

23. $2x^2 - 6x$ $2x(x - 3)$

24. $3y^3 + 3y^2$ $3y^2(y + 1)$

25. $5xy + 12ab^2$ prime

26. $7x^2 + 14x$ $7x(x + 2)$

27. $15x^2y - 10x^2y^2$ $5x^2y(3 - 2y)$

28. $11m^3n^2 - 12x^2y$ prime

29. $63x^3y^2 + 81x^2y^4$ $9x^2y^2(7x + 9y^2)$

30. $33a^3b^4c - 16xyz$ prime

31. $14r^2s^3 + 15t^6$ prime

32. $13ab^2c^3 - 26a^3b^2c$ $13ab^2c(c^2 - 2a^2)$

33. $27z^3 + 12z^2 + 3z$ $3z(9z^2 + 4z + 1)$

34. $25t^6 - 10t^3 + 5t^2$ $5t^2(5t^4 - 2t + 1)$

35. $24s^3 - 12s^2t + 6st^2$ $6s(4s^2 - 2st + t^2)$

36. $18y^2z^2 + 12y^2z^3 - 24y^4z^3$ $6y^2z^2(3 + 2z - 4y^2z)$

37. $45x^{10}y^3 - 63x^7y^7 + 81x^{10}y^{10}$
$9x^7y^3(5x^3 - 7y^4 + 9x^3y^7)$

38. $48u^6v^6 - 16u^4v^4 - 3u^6v^3$
$u^4v^3(48u^2v^3 - 16v - 3u^2)$

39. $25x^3 - 14y^3 + 36x^3y^3$
prime

40. $9m^4n^3p^2 + 18m^2n^3p^4 - 27m^3n^4p$
$9m^2n^3p(m^2p + 2p^3 - 3mn)$

41. $24a^3b^5 + 32a^5b^3 - 64a^5b^5c^5$
$8a^3b^3(3b^2 + 4a^2 - 8a^2b^2c^5)$

42. $32a^4 + 9b^2 + 5a^4b^2$
prime

In Exercises 43–52, factor out the negative of the greatest common factor.

43. $-3a - 6$ $-3(a + 2)$

44. $-6b + 12$ $-6(b - 2)$

45. $-3x^2 - x$ $-x(3x + 1)$

46. $-4a^3 + a^2$ $-a^2(4a - 1)$

47. $-6x^2 - 3xy$ $-3x(2x + y)$

48. $-15y^3 + 25y^2$ $-5y^2(3y - 5)$

49. $-18a^2b - 12ab^2$ $-6ab(3a + 2b)$

50. $-21t^5 + 28t^3$ $-7t^3(3t^2 - 4)$

51. $-63u^3v^6z^9 + 28u^2v^7z^2 - 21u^3v^3z^4$
$-7u^2v^3z^2(9uv^3z^7 - 4v^4 + 3uz^2)$

52. $-56x^4y^3z^2 - 72x^3y^4z^5 + 80xy^2z^3$
$-8xy^2z^2(7x^3y + 9x^2y^2z^3 - 10z)$

In Exercises 53–66, factor out the designated common factor.

53. x^2 from $x^{n+2} + x^{n+3}$ $x^2(x^n + x^{n+1})$

54. y^3 from $y^{n+3} + y^{n+5}$ $y^3(y^n + y^{n+2})$

55. y^n from $2y^{n+2} - 3y^{n+3}$ $y^n(2y^2 - 3y^3)$

56. x^n from $4x^{n+3} - 5x^{n+5}$ $x^n(4x^3 - 5x^5)$

57. $2t^{2n}$ from $2t^{4n} + 4t^{3n} - 8t^{2n}$ $2t^{2n}(t^{2n} + 2t^n - 4)$

58. $9p^{3n}$ from $36p^{3n} - 27p^{6n} + 18p^{9n}$ $9p^{3n}(4 - 3p^{3n} + 2p^{6n})$

59. x^{-2} from $x^4 - 5x^6$ $x^{-2}(x^6 - 5x^8)$

60. y^{-4} from $7y^4 + y$ $y^{-4}(7y^8 + y^5)$

61. t^{-3} from $t^5 + 4t^{-6}$ $t^{-3}(t^8 + 4t^{-3})$

62. p^{-5} from $6p^3 - p^{-2}$ $p^{-5}(6p^8 - p^3)$

63. $4y^{-2n}$ from $8y^{2n} + 12 + 16y^{-2n}$ $4y^{-2n}(2y^{4n} + 3y^{2n} + 4)$

64. $7x^{-3n}$ from $21x^{6n} + 7x^{3n} + 14$ $7x^{-3n}(3x^{9n} + x^{6n} + 2x^{3n})$

65. $r^{-3n}s^2$ from $r^{9n}s^4 - 3r^{3n}s^3 - 2s^2$ $r^{-3n}s^2(r^{12n}s^2 - 3r^{6n}s - 2r^{3n})$

66. $t^{-4n}s$ from $t^{8n}s^3 - s^2 - t^{-3n}s$ $t^{-4n}s(t^{12n}s^2 - t^{4n}s - t^n)$

In Exercises 67–78, factor each expression.

67. $4(x + y) + t(x + y)$ $(x + y)(4 + t)$

68. $5(a - b) - t(a - b)$ $(a - b)(5 - t)$

69. $(a - b)r - (a - b)s$ $(a - b)(r - s)$

70. $(x + y)u + (x + y)v$ $(x + y)(u + v)$

71. $3(m + n + p) + x(m + n + p)$ $(m + n + p)(3 + x)$

72. $x(x - y - z) + y(x - y - z)$ $(x - y - z)(x + y)$

73. $(x + y)(x + y) + z(x + y)$ $(x + y)(x + y + z)$

74. $(a - b)^2 + (a - b)$ $(a - b)(a - b + 1)$

75. $(u + v)^2 - (u + v)$ $(u + v)(u + v - 1)$

76. $a(x - y) - (x - y)^2$ $(x - y)(a - x + y)$

77. $-a(x + y) + b(x + y)$ $-(x + y)(a - b)$

78. $-bx(a - b) - cx(a - b)$ $-x(a - b)(b + c)$

In Exercises 79–90, factor by grouping.

79. $ax + bx + ay + by$ $(x + y)(a + b)$

80. $ar - br + as - bs$ $(r + s)(a - b)$

81. $x^2 + yx + 2x + 2y$ $(x + 2)(x + y)$

82. $2c + 2d - cd - d^2$ $(c + d)(2 - d)$

83. $3c - cd + 3d - c^2$ $(3 - c)(c + d)$

84. $x^2 + 4y - xy - 4x$ $(x - y)(x - 4)$

85. $a^2 - 4b + ab - 4a$ $(a + b)(a - 4)$

86. $7u + v^2 - 7v - uv$ $(v - u)(v - 7)$

87. $ax + bx - a - b$ $(a + b)(x - 1)$

88. $x^2y - ax - xy + a$ $(xy - a)(x - 1)$

89. $x^2 + xy + xz + xy + y^2 + zy$ $(x + y)(x + y + z)$

90. $ab - b^2 - bc + ac - bc - c^2$ $(b + c)(a - b - c)$

In Exercises 91–96, factor by grouping. Factor out all common monomials first.

91. $mpx + mqx + npx + nqx$ $x(m + n)(p + q)$

92. $abd - abe + acd - ace$ $a(b + c)(d - e)$

93. $x^2y + xy^2 + 2xyz + xy^2 + y^3 + 2y^2z$
$y(x + y)(x + y + 2z)$

94. $a^3 - 2a^2b + a^2c - a^2b + 2ab^2 - abc$
$a(a - b)(a - 2b + c)$

95. $2n^4p - 2n^2 - n^3p^2 + np + 2mn^3p - 2mn$
$n(2n - p + 2m)(n^2p - 1)$

96. $a^2c^3 + ac^2 + a^3c^2 - 2a^2bc^2 - 2bc^2 + c^3$
$c^2(a^2 + 1)(c + a - 2b)$

If the greatest common factor of several terms is 1, the terms are called ***relatively prime.*** *In Exercises 97–104, tell whether the terms in each set are relatively prime.*

97. 14, 45 yes

98. 24, 63, 112 yes

99. 60, 28, 36 no

100. 55, 49, 78 yes

101. $12x^2y, 5ab^3, 35x^2b^3$ yes

102. $18uv, 25rs, 12rsuv$ yes

103. $9(a - b), 16(a + b), 25(a + b + c)$ yes

104. $44(x + y - z), 99(x - y + z), 121(x + y + z)$ no

In Exercises 105–112, solve for the indicated variable.

105. $r_1r_2 = rr_2 + rr_1$ for r_1 $r_1 = \dfrac{rr_2}{r_2 - r}$

106. $r_1r_2 = rr_2 + rr_1$ for r $r = \dfrac{r_1r_2}{r_2 + r_1}$

107. $d_1d_2 = fd_2 + fd_1$ for f $f = \dfrac{d_1d_2}{d_2 + d_1}$

108. $d_1d_2 = fd_2 + fd_1$ for d_1 $d_1 = \dfrac{fd_2}{d_2 - f}$

109. $S(1 - r) = a - lr$ for r $r = \dfrac{S - a}{S - l}$

110. $Sn = (n - 2)180°$ for n $n = \dfrac{360°}{180° - S}$

111. $H(a + b) = 2ab$ for a $a = \dfrac{Hb}{2b - H}$

112. $H(a + b) = 2ab$ for b $b = \dfrac{Ha}{2a - H}$

Writing Exercises

1. Explain how to find the greatest common factor of two natural numbers.
2. Explain how to recognize if a number is prime.

Something to Think About

1. Pick two natural numbers. Divide their product by their greatest common factor. The result is called the **least common multiple** of the two numbers you picked. Why?
2. The number 6 is called a **perfect number** because the sum of all the divisors of 6 is twice 6: $1 + 2 + 3 + 6 = 12$. Verify that 28 is also a perfect number.

Review Exercises

Do each multiplication.

1. $(a + 4)(a - 4)$ $a^2 - 16$
2. $(2b + 3)(2b - 3)$ $4b^2 - 9$
3. $(4r^2 + 3s)(4r^2 - 3s)$ $16r^4 - 9s^2$
4. $(5a + 2b^3)(5a - 2b^3)$ $25a^2 - 4b^6$
5. $(m + 4)(m^2 - 4m + 16)$ $m^3 + 64$
6. $(p - q)(p^2 + pq + q^2)$ $p^3 - q^3$

4.2 The Difference of Two Squares; The Sum and Difference of Two Cubes

■ The Difference of Two Squares ■ The Sum and Difference of Two Cubes ■ Factoring by Grouping

To factor the difference of two squares, we need to know the integers that are **perfect squares.** The number 400, for example, is a perfect square because $20^2 = 400$. The first twenty integers that are perfect squares are

1, 4, 9, 16, 25, 36, 49, 64, 81, 100, 121, 144, 169, 196, 225, 256, 289, 324, 361, 400

Expressions like $x^6y^4z^2$ are also perfect squares, because they can be written as the square of another quantity:

$$x^6y^4z^2 = (x^3y^2z)^2$$

All such exponential expressions with only even-number exponents are perfect squares.

The Difference of Two Squares

We begin by recalling a special product formula discussed in Section 2.4:

1. $(x + y)(x - y) = x^2 - y^2$

From Equation 1, we see that when we multiply a binomial of the form $x + y$ by a binomial of the form $x - y$, we get another binomial. The binomial $x^2 - y^2$ is called the **difference of two squares,** because x^2 represents the square of x, y^2 represents the square of y, and $x^2 - y^2$ represents the difference of these squares.

Equation 1 can be written in reverse order to give a formula for factoring the difference of two squares.

Factoring the Difference of Two Squares

$$x^2 - y^2 = (x + y)(x - y)$$

If we think of the difference of two squares as the square of a **F**irst quantity minus the square of a **L**ast quantity, we have the formula

$$F^2 - L^2 = (F + L)(F - L)$$

In words, we say

To factor the square of a **F***irst quantity minus the square of a* **L***ast quantity, we multiply the* **F***irst plus the* **L***ast by the* **F***irst minus the* **L***ast.*

To factor $49x^2 - 16$, for example, we write $49x^2 - 16$ in the form $(7x)^2 - (4)^2$ and use the formula for factoring the difference of two squares:

$$\begin{aligned} 49x^2 - 16 &= (7x)^2 - (\mathbf{4})^2 \\ &= (7x + \mathbf{4})(7x - \mathbf{4}) \end{aligned}$$

We can verify this result by multiplying $7x + 4$ and $7x - 4$.

$$\begin{aligned} (7x + 4)(7x - 4) &= 49x^2 - 28x + 28x - 16 \\ &= 49x^2 - 16 \end{aligned}$$

F L I O

We note that if $49x^2 - 16$ is divided by $7x + 4$, the quotient is $7x - 4$, and that if $49x^2 - 16$ is divided by $7x - 4$, the quotient is $7x + 4$.

WARNING! Expressions such as $(7x)^2 + (4)^2$ that represent the sum of two squares cannot be factored in the real number system. Thus, the binomial $49x^2 + 16$ is a prime binomial.

EXAMPLE 1 Factor $64x^4 - 25y^2$.

Solution Because $64x^4$ is the square of $8x^2$ and $25y^2$ is the square of $5y$, the binomial is a difference of two squares. Its two factors are the sum of $8x^2$ and $5y$, and the difference of $8x^2$ and $5y$.

$$\begin{aligned} 64x^4 - 25y^2 &= (8x^2)^2 - (\mathbf{5y})^2 \\ &= (8x^2 + \mathbf{5y})(8x^2 - \mathbf{5y}) \end{aligned}$$

Verify by multiplication that $(8x^2 + 5y)(8x^2 - 5y) = 64x^4 - 25y^2$. ■

EXAMPLE 2 Factor $a^4 - 1$.

Solution Because the binomial is the difference of the squares of a^2 and 1, it factors into the sum of a^2 and 1, and the difference of a^2 and 1.

$$\begin{aligned} a^4 - 1 &= (a^2)^2 - (1)^2 \\ &= (a^2 + 1)(a^2 - 1) \end{aligned}$$

The factor $a^2 + 1$ is the sum of two quantities and is prime. However, the factor $a^2 - 1$ is the difference of two squares and can be factored as $(a + 1)(a - 1)$. Thus,

$$\begin{aligned} a^4 - 1 &= (a^2 + 1)(\mathbf{a^2 - 1}) \\ &= (a^2 + 1)(\mathbf{a + 1})(\mathbf{a - 1}) \end{aligned}$$ ■

EXAMPLE 3 Factor $(x + y)^4 - z^4$.

Solution This expression is the difference of two squares that can be factored:

$$\begin{aligned} (x + y)^4 - z^4 &= [(x + y)^2]^2 - (z^2)^2 \\ &= [(x + y)^2 + z^2][(x + y)^2 - z^2] \end{aligned}$$

Alternate Approach: You may want to suggest substituting to make these problems easier. Let $(x + y) = u$. Then

$$\begin{aligned} u^4 - z^4 &= (u^2 + z^2)(u^2 - z^2) \\ &= (u^2 + z^2)(u + z)(u - z) \\ &= [(x + y)^2 + z^2](x + y - z) \cdot (x + y + z) \end{aligned}$$

The factor $(x + y)^2 + z^2$ is the sum of two squares and is prime. However, the factor $(x + y)^2 - z^2$ is the difference of two squares and can be factored as $(x + y + z)(x + y - z)$. Thus,

$$\begin{aligned} (x + y)^4 - z^4 &= [(x + y)^2 + z^2][\mathbf{(x + y)^2 - z^2}] \\ &= [(x + y)^2 + z^2]\mathbf{(x + y + z)(x + y - z)} \end{aligned}$$ ■

When possible, we will always factor out a common factor before factoring the difference of two squares. The factoring process is easier when all common factors are factored out first.

EXAMPLE 4 Factor $2x^4y - 32y$.

Solution

$$\begin{aligned} 2x^4y - 32y &= 2y(x^4 - 16) && \text{Factor out } 2y. \\ &= 2y(x^2 + 4)(x^2 - 4) && \text{Factor } x^4 - 16. \\ &= 2y(x^2 + 4)(x + 2)(x - 2) && \text{Factor } x^2 - 4. \end{aligned}$$ ■

The Sum and Difference of Two Cubes

The number 64 is called a perfect cube, because $4^3 = 64$. To factor the sum or difference to two cubes, it is helpful to know the first ten perfect cubes:

1, 8, 27, 64, 125, 216, 343, 512, 729, 1000

Expressions like $x^9y^6z^3$ are also perfect cubes, because they can be written as the cube of another quantity:

$$x^9y^6z^3 = (x^3y^2z)^3$$

In Section 2.4, we also discussed the following special product formulas:

2. $(x + y)(x^2 - xy + y^2) = x^3 + y^3$

3. $(x - y)(x^2 + xy + y^2) = x^3 - y^3$

To verify Equation 2, we multiply $x^2 - xy + y^2$ by $x + y$.

$$\begin{aligned}(x + y)(x^2 - xy + y^2) &= (x + y)x^2 - (x + y)xy + (x + y)y^2 \\ &= x^3 + x^2y - x^2y - xy^2 + xy^2 + y^3 \\ &= x^3 + y^3\end{aligned}$$

Equation 3 can also be verified by multiplication.

If we write the previous formulas as follows, we have formulas for factoring the **sum and difference of two cubes.**

Sum and Difference of Two Cubes

$$x^3 + y^3 = (x + y)(x^2 - xy + y^2)$$
$$x^3 - y^3 = (x - y)(x^2 + xy + y^2)$$

If we think of the sum of two cubes as the sum of the cube of a **F**irst quantity plus the cube of a **L**ast quantity, we have the formula

$$F^3 + L^3 = (F + L)(F^2 - FL + L^2)$$

In words, we say

To factor the cube of a **F***irst quantity plus the cube of a* **L***ast quantity, we multiply the sum of the* **F***irst and* **L***ast by*

- *the* **F***irst squared*
- *minus the* **F***irst times the* **L***ast*
- *plus the* **L***ast squared.*

The formula for the difference of two cubes is

$$F^3 - L^3 = (F - L)(F^2 + FL + L^2)$$

In words, we say

To factor the cube of a **F***irst quantity minus the cube of a* **L***ast quantity, we multiply the difference of the* **F***irst and* **L***ast by*

- *the* **F***irst squared*
- *plus the* **F***irst times the* **L***ast*
- *plus the* **L***ast squared.*

EXAMPLE 5 Factor $a^3 + 8$.

Solution Since $a^3 + 8$ can be written as $a^3 + 2^3$, we have the sum of two cubes, which factors as follows:

$$\begin{aligned} F^3 + L^3 &= (F + L)(F^2 - FL + L^2) \\ a^3 + 2^3 &= (a + 2)(a^2 - a2 + 2^2) \\ &= (a + 2)(a^2 - 2a + 4) \end{aligned}$$

Thus, $a^3 + 8 = (a + 2)(a^2 - 2a + 4)$. Check by multiplication. ■

EXAMPLE 6 Factor $27a^3 - 64b^3$.

Solution Since $27a^3 - 64b^3$ can be written as $(3a)^3 - (4b)^3$, we have the difference of two cubes, which factors as follows:

$$\begin{aligned} F^3 - L^3 &= (F - L)(F^2 + FL + L^2) \\ (3a)^3 - (4b)^3 &= (3a - 4b)[(3a)^2 + (3a)(4b) + (4b)^2] \\ &= (3a - 4b)(9a^2 + 12ab + 16b^2) \end{aligned}$$

Thus, $27a^3 - 64b^3 = (3a - 4b)(9a^2 + 12ab + 16b^2)$. Check by multiplication. ■

EXAMPLE 7 Factor $a^3 - (c + d)^3$.

Solution

$$\begin{aligned} a^3 - (c + d)^3 &= [a - (c + d)][a^2 + a(c + d) + (c + d)^2] \\ &= (a - c - d)(a^2 + ac + ad + c^2 + 2cd + d^2) \end{aligned}$$ ■

EXAMPLE 8 Factor $x^6 - 64$.

Solution This expression is the difference of two squares and factors into the product of a sum and a difference.

$$\begin{aligned} x^6 - 64 &= (x^3)^2 - 8^2 \\ &= (x^3 + 8)(x^3 - 8) \end{aligned}$$

Each of these factors further, however, for one is the sum of two cubes and the other is the difference of two cubes:

$$x^6 - 64 = (x + 2)(x^2 - 2x + 4)(x - 2)(x^2 + 2x + 4)$$ ■

EXAMPLE 9 Factor $2a^5 + 128a^2$.

Solution We first factor out the common monomial factor of $2a^2$ to obtain

$$2a^5 + 128a^2 = 2a^2(a^3 + 64)$$

Then we factor $a^3 + 64$ as the sum of two cubes to obtain

$$2a^5 + 128a^2 = 2a^2(a + 4)(a^2 - 4a + 16)$$ ■

EXAMPLE 10 Factor $16r^{6m} - 54t^{3n}$.

Solution

$16r^{6m} - 54t^{3n} = 2(8r^{6m} - 27t^{3n})$ — Factor out a 2.

$= 2[(2r^{2m})^3 - (3t^n)^3]$ — Write $8r^{6m}$ as $(2r^{2m})^3$ and $27t^{3n}$ as $(3t^n)^3$.

$= 2[(2r^{2m} - 3t^n)(4r^{4m} + 6r^{2m}t^n + 9t^{2n})]$ — Factor $(2r^{2m})^3 - (3t^n)^3$.

■

Factoring by Grouping

EXAMPLE 11 Factor $x^2 - y^2 + x - y$

Solution If we group the first two terms and factor the difference of two squares, we have

$x^2 - y^2 + x - y = (x + y)(x - y) + (x - y)$ — Factor $x^2 - y^2$.

$= (x - y)(x + y + 1)$ — Factor out $x - y$. ■

ORALS *Factor each expression, if possible.*

1. $x^2 - 1$ $(x + 1)(x - 1)$

2. $a^4 - 16$ $(a^2 + 4)(a + 2)(a - 2)$

3. $x^3 + 1$ $(x + 1)(x^2 - x + 1)$

4. $a^3 - 8$ $(a - 2)(a^2 + 2a + 4)$

5. $2x^2 - 8$ $2(x + 2)(x - 2)$

6. $x^4 + 25$ prime

EXERCISE 4.2

In Exercises 1–20, factor each expression, if possible.

1. $x^2 - 4$ $(x + 2)(x - 2)$

2. $y^2 - 9$ $(y + 3)(y - 3)$

3. $9y^2 - 64$ $(3y + 8)(3y - 8)$

4. $16x^4 - 81y^2$ $(4x^2 + 9y)(4x^2 - 9y)$

5. $x^2 + 25$ prime

6. $144a^2 - b^4$ $(12a + b^2)(12a - b^2)$

7. $625a^2 - 169b^4$ $(25a + 13b^2)(25a - 13b^2)$

8. $4y^2 + 9z^4$ prime

9. $81a^4 - 49b^2$ $(9a^2 + 7b)(9a^2 - 7b)$

10. $64r^6 - 121s^2$ $(8r^3 + 11s)(8r^3 - 11s)$

11. $36x^4y^2 - 49z^4$ $(6x^2y + 7z^2)(6x^2y - 7z^2)$

12. $4a^2b^4c^6 - 9d^8$ $(2ab^2c^3 + 3d^4)(2ab^2c^3 - 3d^4)$

13. $(x + y)^2 - z^2$ $(x + y + z)(x + y - z)$

14. $a^2 - (b - c)^2$ $(a + b - c)(a - b + c)$

15. $(a - b)^2 - c^2$ $(a - b + c)(a - b - c)$

16. $(m + n)^2 - p^4$ $(m + n + p^2)(m + n - p^2)$

17. $x^4 - y^4$ $(x^2 + y^2)(x + y)(x - y)$

18. $16a^4 - 81b^4$ $(4a^2 + 9b^2)(2a + 3b)(2a - 3b)$

19. $256x^4y^4 - z^8$ $(16x^2y^2 + z^4)(4xy + z^2)(4xy - z^2)$

20. $225a^4 - 16b^8c^{12}$ $(15a^2 + 4b^4c^6)(15a^2 - 4b^4c^6)$

In Exercises 21–28, factor each expression.

21. $2x^2 - 288$ $2(x + 12)(x - 12)$

22. $8x^2 - 72$ $8(x + 3)(x - 3)$

23. $2x^3 - 32x$ $2x(x + 4)(x - 4)$

24. $3x^3 - 243x$ $3x(x + 9)(x - 9)$

25. $5x^3 - 125x$ $5x(x + 5)(x - 5)$

26. $6x^4 - 216x^2$ $6x^2(x + 6)(x - 6)$

27. $r^2s^2t^2 - t^2x^4y^2$ $t^2(rs + x^2y)(rs - x^2y)$

28. $16a^4b^3c^4 - 64a^2bc^6$ $16a^2bc^4(ab + 2c)(ab - 2c)$

In Exercises 29–38, factor each expression.

29. $r^3 + s^3$ $(r + s)(r^2 - rs + s^2)$

30. $t^3 - v^3$ $(t - v)(t^2 + tv + v^2)$

31. $x^3 - 8y^3$ $(x - 2y)(x^2 + 2xy + 4y^2)$

32. $27a^3 + b^3$ $(3a + b)(9a^2 - 3ab + b^2)$

33. $64a^3 - 125b^6$ $(4a - 5b^2)(16a^2 + 20ab^2 + 25b^4)$

34. $8x^6 + 125y^3$ $(2x^2 + 5y)(4x^4 - 10x^2y + 25y^2)$

35. $125x^3y^6 + 216z^9$ $(5xy^2 + 6z^3)(25x^2y^4 - 30xy^2z^3 + 36z^6)$

36. $1000a^6 - 343b^3c^6$ $(10a^2 - 7bc^2)(100a^4 + 70a^2bc^2 + 49b^2c^4)$

37. $x^6 + y^6$ $(x^2 + y^2)(x^4 - x^2y^2 + y^4)$

38. $x^9 + y^9$ $(x + y)(x^2 - xy + y^2)(x^6 - x^3y^3 + y^6)$

In Exercises 39–48, factor each expression.

39. $5x^3 + 625$ $5(x + 5)(x^2 - 5x + 25)$

40. $2x^3 - 128$ $2(x - 4)(x^2 + 4x + 16)$

41. $4x^5 - 256x^2$ $4x^2(x - 4)(x^2 + 4x + 16)$

42. $2x^6 + 54x^3$ $2x^3(x + 3)(x^2 - 3x + 9)$

43. $128u^2v^3 - 2t^3u^2$ $2u^2(4v - t)(16v^2 + 4vt + t^2)$

44. $56rs^2t^3 + 7rs^2v^6$ $7rs^2(2t + v^2)(4t^2 - 2tv^2 + v^4)$

45. $(a + b)x^3 + 27(a + b)$ $(a + b)(x + 3)(x^2 - 3x + 9)$

46. $(c - d)r^3 - (c - d)s^3$ $(c - d)(r - s)(r^2 + rs + s^2)$

47. $6a^3b^3 - 6z^3$ $6(ab - z)(a^2b^2 + abz + z^2)$

48. $18x^3y^3 + 18c^3d^3$ $18(xy + cd)(x^2y^2 - xycd + c^2d^2)$

In Exercises 49–58, factor each expression. Assume that m and n are natural numbers.

49. $x^{2m} - y^{4n}$ $(x^m + y^{2n})(x^m - y^{2n})$

50. $a^{4m} - b^{8n}$ $(a^{2m} + b^{4n})(a^m + b^{2n})(a^m - b^{2n})$

51. $100a^{4m} - 81b^{2n}$ $(10a^{2m} + 9b^n)(10a^{2m} - 9b^n)$

52. $25x^{8m} - 36y^{4n}$ $(5x^{4m} + 6y^{2n})(5x^{4m} - 6y^{2n})$

53. $x^{3n} - 8$ $(x^n - 2)(x^{2n} + 2x^n + 4)$

54. $a^{3m} + 64$ $(a^m + 4)(a^{2m} - 4a^m + 16)$

55. $a^{3m} + b^{3n}$ $(a^m + b^n)(a^{2m} - a^mb^n + b^{2n})$

56. $x^{6m} - y^{3n}$ $(x^{2m} - y^n)(x^{4m} + x^{2m}y^n + y^{2n})$

57. $2x^{6m} + 16y^{3m}$ $2(x^{2m} + 2y^m)(x^{4m} - 2x^{2m}y^m + 4y^{2m})$

58. $24 + 3c^{3m}$ $3(2 + c^m)(4 - 2c^m + c^{2m})$

In Exercises 59–64, factor each expression by grouping.

59. $a^2 - b^2 + a + b$ $(a + b)(a - b + 1)$

60. $x^2 - y^2 - x - y$ $(x + y)(x - y - 1)$

61. $a^2 - b^2 + 2a - 2b$ $(a - b)(a + b + 2)$

62. $m^2 - n^2 + 3m + 3n$ $(m + n)(m - n + 3)$

63. $2x + y + 4x^2 - y^2$ $(2x + y)(1 + 2x - y)$

64. $m - 2n + m^2 - 4n^2$ $(m - 2n)(1 + m + 2n)$

Writing Exercises ■ *Write a paragraph using your own words.*

1. Describe the pattern used to factor the difference of two squares.
2. Describe the patterns used to factor the sum and the difference of two cubes.

Something to Think About

1. Factor $x^{32} - y^{32}$.
2. Without using a calculator, how can you tell that $459^3 - 451^3$ is divisible by 8? Is $2573^3 - 2563^3$ divisible by 10?
3. Find the error: Suppose that $x = y$. Then

$$x^2 = xy \quad \text{Multiply both sides by } x.$$
$$x^2 - y^2 = xy - y^2 \quad \text{Subtract } y^2 \text{ from both sides.}$$
$$(x + y)(x - y) = y(x - y) \quad \text{Factor.}$$
$$\frac{(x + y)(x - y)}{x - y} = \frac{y(x - y)}{x - y} \quad \text{Divide both sides by } x - y.$$
$$x + y = y \quad \text{Simplify.}$$
$$y + y = y \quad \text{Substitute } y \text{ for } x.$$
$$2y = y \quad \text{Combine terms.}$$
$$\frac{2y}{y} = \frac{y}{y} \quad \text{Divide both sides by } y.$$
$$2 = 1 \quad \text{Simplify.}$$

4. **a.** Think of any integer (11, for example).
   b. Square it (121).
   c. Find the product of the integers on either side of your chosen number (10 and 12; product, 120).
   d. The answer in part **c** is 1 less than the answer in part **b** (120 is 1 less than 121).

   Try this for several integers. Why does it always work?

Review Exercises

Do each multiplication.

1. $(x + 1)(x + 1)$ $x^2 + 2x + 1$
2. $(2m - 3)(m - 2)$ $2m^2 - 7m + 6$
3. $(2m + n)(2m + n)$ $4m^2 + 4mn + n^2$
4. $(3m - 2n)(3m - 2n)$ $9m^2 - 12mn + 4n^2$
5. $(a + 4)(a + 3)$ $a^2 + 7a + 12$
6. $(3b + 2)(2b - 5)$ $6b^2 - 11b - 10$
7. $(4r - 3s)(2r - s)$ $8r^2 - 10rs + 3s^2$
8. $(5a - 2b)(3a + 4b)$ $15a^2 + 14ab - 8b^2$

4.3 Factoring Trinomials

■ Factoring Trinomials with Lead Coefficients of 1 ■ Factoring Trinomials with Lead Coefficients Other Than 1 ■ Test for Factorability ■ Using Substitution to Factor Trinomials ■ Factoring by Grouping ■ Using Grouping to Factor Trinomials

Many trinomials can be factored by using two special product formulas discussed in Section 2.4.

1. $(x + y)(x + y) = x^2 + 2xy + y^2$
2. $(x - y)(x - y) = x^2 - 2xy + y^2$

The trinomials $x^2 + 2xy + y^2$ and $x^2 - 2xy + y^2$ are **perfect square trinomials.**

To factor the perfect square trinomial $x^2 + 6x + 9$, for example, we note that the trinomial can be written in the form $x^2 + 2(3)x + 3^2$. If $y = 3$, this form matches the right-hand side of Equation 1. Thus, $x^2 + 6x + 9$ factors as

$$\begin{aligned} x^2 + 6x + 9 &= x^2 + 2(3)x + 3^2 \\ &= (x + 3)(x + 3) \end{aligned}$$

This result can be verified by multiplication:

$$\begin{aligned} (x + 3)(x + 3) &= x^2 + 3x + 3x + 9 \\ &= x^2 + 6x + 9 \end{aligned}$$

To factor the perfect square trinomial $x^2 - 4xz + 4z^2$, we note that the trinomial can be written in the form $x^2 - 2x(2z) + (2z)^2$. If $y = 2z$, this form matches the right-hand side of Equation 2. Thus, $x^2 - 4xz + 4z^2$ factors as

$$\begin{aligned} x^2 - 4xz + 4z^2 &= x^2 - 2x(2z) + (2z)^2 \\ &= (x - 2z)(x - 2z) \end{aligned}$$

This result can also be verified by multiplication.

Many second-degree trinomials cannot be factored by using special product formulas. We begin our discussion of these **general trinomials** by considering trinomials with lead coefficients (the coefficient of the squared term) of 1.

Factoring Trinomials with Lead Coefficients of 1

We have seen that the product of two binomials is often a trinomial. Thus, we can expect that many trinomials will factor as two binomials. For example, to factor the trinomial $x^2 + 7x + 12$, we must find two binomials $x + a$ and $x + b$ such that

$$x^2 + 7x + 12 = (x + a)(x + b)$$

where $ab = 12$ and $ax + bx = 7x$

To find the numbers a and b, we list the possible factorizations of 12 and find the one where the sum of the factors is 7.

The one to choose
↓

12(1) 6(2) 4(3) −12(−1) −6(−2) −4(−3)

Thus, $a = 4$, $b = 3$, and

$$x^2 + 7x + 12 = (x + a)(x + b)$$

3. $x^2 + 7x + 12 = (x + 4)(x + 3)$

This factorization can be verified by multiplying $x + 4$ and $x + 3$ and observing that the product is $x^2 + 7x + 12$.

$$\begin{aligned} (x + 4)(x + 3) &= x^2 + 3x + 4x + 12 \\ &= x^2 + 7x + 12 \end{aligned}$$

Because of the commutative property of multiplication, the order of the factors in Equation 3 is not important.

To factor trinomials with lead coefficients of 1, we follow these steps.

Factoring Trinomials

1. Write the trinomial in descending powers of one variable.
2. List the factorizations of the third term of the trinomial.
3. Pick the factorization where the sum of the factors is the coefficient of the middle term.

EXAMPLE 1 Factor $x^2 - 6x + 8$.

Solution Since this trinomial is already written in descending powers of x, we can move to Step 2 and list the possible factorizations of the third term, which is 8.

The one to choose ↓

$8(1) \qquad 4(2) \qquad -8(-1) \qquad \mathbf{-4(-2)}$

In this trinomial, the coefficient of the middle term is -6. The only factorization where the sum of the factors is -6 is $-4(-2)$. Thus, $a = -4$, $b = -2$, and

$$\begin{aligned} x^2 - 6x + 8 &= (x + \boldsymbol{a})(x + \boldsymbol{b}) \\ &= (x - \mathbf{4})(x - \mathbf{2}) \end{aligned}$$

We can verify this result by multiplication:

$$\begin{aligned} (x - 4)(x - 2) &= x^2 - 2x - 4x + 8 \\ &= x^2 - 6x + 8 \end{aligned}$$

■

EXAMPLE 2 Factor $-x + x^2 - 12$.

Solution We begin by writing the trinomial in descending powers of x:

$$-x + x^2 - 12 = x^2 - x - 12$$

The possible factorizations of the third term are

The one to choose ↓

$12(-1) \qquad 6(-2) \qquad 4(-3) \qquad 1(-12) \qquad 2(-6) \qquad \mathbf{3(-4)}$

In this trinomial, the coefficient of the middle term is -1. The only factorization where the sum of the factors is -1 is $3(-4)$. Thus, $a = 3$, $b = -4$, and

$$\begin{aligned} -x + x^2 - 12 &= (x + \boldsymbol{a})(x + \boldsymbol{b}) \\ &= (x + \mathbf{3})(x - \mathbf{4}) \end{aligned}$$

Verify the result by multiplication. ■

EXAMPLE 3 Factor $30x - 4xy - 2xy^2$.

Solution We begin by writing the trinomial in descending powers of y:

$$30x - 4xy - 2xy^2 = -2xy^2 - 4xy + 30x$$

Each term in this trinomial has a monomial factor of $-2x$, which should be factored out.

$$30x - 4xy - 2xy^2 = -2x(y^2 + 2y - 15)$$

To factor $y^2 + 2y - 15$, we list the factors of -15 and find the pair whose sum is 2.

The one to choose ↓

$$15(-1) \quad \mathbf{5(-3)} \quad 1(-15) \quad 3(-5)$$

The only factorization where the sum of the factors is 2 (the coefficient of the middle term of $y^2 + 2y - 15$) is $5(-3)$. Thus, $a = 5$, $b = -3$, and

$$\begin{aligned} 30x - 4xy - 2xy^2 &= -2x(\mathbf{y^2 + 2y - 15}) \\ &= -2x(\mathbf{y + 5})(\mathbf{y - 3}) \end{aligned}$$

Verify this result by multiplication. ■

WARNING! In Example 3, be sure to include all factors in the final answer. It is a common error to forget to write the $-2x$.

Factoring Trinomials with Lead Coefficients Other Than 1

There are more combinations to consider when factoring trinomials with lead coefficients other than 1. To factor $5x^2 + 7x + 2$, for example, we must find two binomials of the form $ax + b$ and $cx + d$ such that

$$5x^2 + 7x + 2 = (ax + b)(cx + d)$$

Since the first term of the trinomial $5x^2 + 7x + 2$ is $5x^2$, the first terms of the binomial factors must be $5x$ and x.

$$5x^2 + 7x + 2 = (\overbrace{5x + b)(x}^{5x^2} + d)$$

Since the product of the last terms must be 2, and the sum of the products of the outer and inner terms must be $7x$, we must find two numbers whose product is 2 that will give a middle term of $7x$.

$$5x^2 + 7x + 2 = (5x + \overbrace{b)(x + d}^{2})$$

$$\text{O} + \text{I} = 7x$$

Because both 2(1) and (−2)(−1) give a product of 2, there are four possible combinations to consider:

$$(\mathbf{5x + 2})(\mathbf{x + 1}) \qquad (5x - 2)(x - 1)$$
$$(5x + 1)(x + 2) \qquad (5x - 1)(x - 2)$$

Of these possibilities, only the first one gives the proper middle term of $7x$. Thus,

1. $5x^2 + 7x + 2 = (5x + 2)(x + 1)$

We can verify this result by multiplication:

$$(5x + 2)(x + 1) = 5x^2 + 5x + 2x + 2$$
$$= 5x^2 + 7x + 2$$

Test for Factorability

If a trinomial has the form $ax^2 + bx + c$, with integer coefficients and $a \neq 0$, we can test to see if it is factorable. If the value of $b^2 - 4ac$ is a perfect square, the trinomial can be factored using only integers. If the value is not a perfect square, the trinomial cannot be factored using only integers.

For example, $5x^2 + 7x + 2$ is a trinomial in the form $ax^2 + bx + c$ with

$$a = 5, \qquad b = 7, \qquad \text{and} \qquad c = 2$$

For this trinomial, the value of $b^2 - 4ac$ is

$$\begin{aligned} \boldsymbol{b}^2 - 4\boldsymbol{ac} &= \mathbf{7}^2 - 4(\mathbf{5})(\mathbf{2}) \\ &= 49 - 40 \\ &= 9 \end{aligned}$$

Since 9 is a perfect square, the trinomial is factorable. Its factorization is shown in Equation 1.

Test for Factorability A trinomial of the form $ax^2 + bx + c$, with integer coefficients and $a \neq 0$, will factor into two binomials with integer coefficients if the value of

$$b^2 - 4ac$$

is a perfect square. If $b^2 - 4ac = 0$, the factors will be the same.

EXAMPLE 4 Factor $3p^2 - 4p - 4$.

Solution In this trinomial, $a = 3$, $b = -4$, and $c = -4$. To see if it factors, we evaluate $b^2 - 4ac$.

$$\begin{aligned} \boldsymbol{b}^2 - 4\boldsymbol{ac} &= (\mathbf{-4})^2 - 4(\mathbf{3})(\mathbf{-4}) \\ &= 16 + 48 \\ &= 64 \end{aligned}$$

Since 64 is a perfect square, the trinomial is factorable.

To factor the trinomial, we note that the first terms of the binomial factors must be $3a$ and a to give the first term of $3a^2$.

$$3a^2 - 4a - 4 = (\overset{3a^2}{\overbrace{3a \quad ?)(a}} \quad ?)$$

The product of the last terms must be -4, and the sum of the products of the outer terms and the inner terms must be $-4a$.

$$3a^2 - 4a - 4 = (3a \quad ?)(a \quad ?) \qquad \text{last terms: } -4, \quad \text{O} + \text{I} = -4a$$

Because $1(-4)$, $-1(4)$, and $-2(2)$ all give a product of -4, there are six possible combinations to consider:

$$\begin{array}{ll} (3a+1)(a-4) & (3a-4)(a+1) \\ (3a-1)(a+4) & (3a+4)(a-1) \\ (3a-2)(a+2) & \mathbf{(3a+2)(a-2)} \end{array}$$

Of these possibilities, only the last gives the required middle term of $-4a$. Thus,

$$3a^2 - 4a - 4 = (3a + 2)(a - 2)$$

Verify this result by multiplying. ■

EXAMPLE 5 Factor $4t^2 - 3t - 5$, if possible.

Solution In this trinomial, $a = 4$, $b = -3$, and $c = -5$. To see if the trinomial is factorable, we evaluate $b^2 - 4ac$ by substituting the values of a, b, and c.

$$\begin{aligned} b^2 - 4ac &= (-3)^2 - 4(4)(-5) \\ &= 9 + 80 \\ &= 89 \end{aligned}$$

Since 89 is not a perfect square, the trinomial is not factorable using only integer coefficients. ■

It is not easy to give specific rules for factoring general trinomials, because some guesswork is often necessary. However, the following hints are helpful.

Factoring a General Trinomial

1. Write the trinomial in descending powers of one variable.
2. Test the trinomial for factorability.
3. Factor out any greatest common factor (including -1 if that is necessary to make the coefficient of the first term positive).
4. When the sign of the first term of a trinomial is + and the sign of the third term is +, the signs between the terms of each binomial factor are the same as the sign of the middle term of the trinomial.
   When the sign of the first term is + and the sign of the third term is −, the signs between the terms of the binomial are opposite.
5. Try various combinations of first terms and last terms until you find the one that works.
6. Check the factorization by multiplication.

EXAMPLE 6 Factor $24y + 10xy - 6x^2y$.

Solution We write the trinomial in descending powers of x and factor out the common factor of $-2y$:

$$\begin{aligned} 24y + 10xy - 6x^2y &= -6x^2y + 10xy + 24y \\ &= -2y(3x^2 - 5x - 12) \end{aligned}$$

In the trinomial $3x^2 - 5x - 12$, $a = 3$, $b = -5$, and $c = -12$. Thus,

$$\begin{aligned} b^2 - 4ac &= (-5)^2 - 4(3)(-12) \\ &= 25 + 144 \\ &= 169 \end{aligned}$$

Since 169 is a perfect square, the trinomial will factor.

Because the sign of the third term of $3x^2 - 5x - 12$ is −, the signs between the binomial factors will be opposite. Because the first term is $3x^2$, the first terms of the binomial factors must be $3x$ and x.

$$24y + 10xy - 6x^2y = -2y(\overbrace{3x \quad)(x}^{3x^2} \quad)$$

The product of the last terms must be -12, and the sum of the outer terms and the inner terms must be $-5x$.

$$24y + 10xy - 6x^2y = -2y(3x \quad ?)(x \quad ?)$$

-12 (product of last terms); $\text{O} + \text{I} = -5x$

Because $1(-12)$, $2(-6)$, $3(-4)$, $12(-1)$, $6(-2)$, and $4(-3)$ all give a product of -12, there are 12 possible combinations to consider.

Teaching Tip: Emphasize that any combination of factors where one factor has a common factor won't work.

	$(3x + 1)(x - 12)$	$(3x - 12)(x + 1)$
	$(3x + 2)(x - 6)$	$(3x - 6)(x + 2)$
	$(3x + 3)(x - 4)$	$(3x - 4)(x + 3)$
	$(3x + 12)(x - 1)$	$(3x - 1)(x + 12)$
	$(3x + 6)(x - 2)$	$(3x - 2)(x + 6)$
The one to choose →	$(3x + 4)(x - 3)$	$(3x - 3)(x + 4)$

The six combinations marked in color cannot work, because one of the factors has a common factor. This implies that $3x^2 - 5x - 12$ would have a common factor, which it doesn't.

After mentally trying the remaining combinations, we find that only $(3x + 4)(x - 3)$ gives the proper middle term of $-5x$. Thus,

$$\begin{aligned} 24y + 10xy - 6x^2y &= -2y(3x^2 - 5x - 12) \\ &= -2y(3x + 4)(x - 3) \end{aligned}$$

Verify this result by multiplication. ■

EXAMPLE 7 Factor $6y + 13x^2y + 6x^4y$.

Solution We write the trinomial in descending powers of x and factor out the common factor of y to obtain

$$\begin{aligned} 6y + 13x^2y + 6x^4y &= 6x^4y + 13x^2y + 6y \\ &= y(6x^4 + 13x^2 + 6) \end{aligned}$$

Because the coefficients of the first and last terms of the trinomial $6x^4 + 13x^2 + 6$ are positive, the signs between the terms in each binomial will be $+$.

Since the first term of the trinomial is $6x^4$, the first terms of the binomial factors must be either $2x^2$ and $3x^2$, or perhaps x^2 and $6x^2$.

Since the product of the last terms of the binomial factors must be 6, we must find two numbers whose product is 6 that will lead to a middle term of $13x^2$. After trying some combinations, we find the one that works.

$$\begin{aligned} 6y + 13x^2y + 6x^4y &= y(6x^4 + 13x^2 + 6) \\ &= y(2x^2 + 3)(3x^2 + 2) \end{aligned}$$

Verify this result by multiplication. ■

EXAMPLE 8 Factor $x^{2n} + x^n - 2$.

Solution Since the first term is x^{2n}, the first terms of the binomial factors must be x^n and x^n.

$$x^{2n} + x^n - 2 = (\overbrace{x^n \quad)(x^n}^{x^{2n}} \quad)$$

Since the third term of the trinomial is -2, the last terms of the binomial factors must have opposite signs, have a product of -2, and lead to a middle term of x^n. The only combination that works is

$$x^{2n} + x^n - 2 = (x^n + 2)(x^n - 1)$$

Verify this result by multiplication. ■

Using Substitution to Factor Trinomials

EXAMPLE 9 Factor the trinomial $(x + y)^2 + 7(x + y) + 12$.

Solution We rewrite the trinomial $(x + y)^2 + 7(x + y) + 12$ as $z^2 + 7z + 12$, where $z = x + y$. The trinomial $z^2 + 7z + 12$ factors as $(z + 4)(z + 3)$.

To find the factorization of $(x + y)^2 + 7(x + y) + 12$, we substitute $x + y$ for z in the expression $(z + 4)(z + 3)$ to obtain

$$z^2 + 7z + 12 = (z + 4)(z + 3)$$
$$(x + y)^2 + 7(x + y) + 12 = (x + y + 4)(x + y + 3)$$ ■

Factoring by Grouping

EXAMPLE 10 Factor $x^2 + 6x + 9 - z^2$.

Solution We group the first three terms together and factor the trinomial to get

$$x^2 + 6x + 9 - z^2 = (x + 3)(x + 3) - z^2$$
$$= (x + 3)^2 - z^2$$

We can now factor the difference of two squares to get

$$x^2 + 6x + 9 - z^2 = (x + 3 + z)(x + 3 - z)$$ ■

Sonya Kovalevskaya (1850–1891) This talented young Russian woman hoped to study mathematics at the University of Berlin, but strict rules prohibited women from attending lectures. Undaunted, Sonya studied privately with the great mathematician Karl Weierstrauss and published several important papers.

Using Grouping to Factor Trinomials

The method of factoring by grouping can be used to help factor trinomials of the form $ax^2 + bx + c$. For example, to factor the trinomial $6x^2 + 7x - 3$, we proceed as follows:

1. First determine the product ac: $6(-3) = -18$. This number is called the **key number.**
2. Find two factors of the key number -18 whose sum is $b = 7$:

   $9(-2) = -18$ and $9 + (-2) = 7$

3. Use the factors 9 and -2 as coefficients of terms to be placed between $6x^2$ and -3:

   $$6x^2 + 7x - 3 = 6x^2 + 9x - 2x - 3$$

4. Factor by grouping:

$$6x^2 + 9x - 2x - 3 = 3x(2x + 3) - (2x + 3)$$
$$= (2x + 3)(3x - 1) \qquad \text{Factor out } 2x + 3.$$

We can verify this factorization by multiplication.

ORALS *Factor each expression.*

1. $x^2 + 3x + 2$ $(x + 2)(x + 1)$
2. $x^2 + 5x + 4$ $(x + 4)(x + 1)$
3. $x^2 - 5x + 6$ $(x - 3)(x - 2)$
4. $x^2 - 3x - 4$ $(x + 1)(x - 4)$
5. $2x^2 + 3x + 1$ $(2x + 1)(x + 1)$
6. $3x^2 + 4x + 1$ $(3x + 1)(x + 1)$

EXERCISE 4.3

In Exercises 1–10, use a special product formula to factor each perfect square trinomial.

1. $x^2 + 2x + 1$ $(x + 1)^2$
2. $y^2 - 2y + 1$ $(y - 1)^2$
3. $a^2 - 18a + 81$ $(a - 9)^2$
4. $b^2 + 12b + 36$ $(b + 6)^2$
5. $4y^2 + 4y + 1$ $(2y + 1)^2$
6. $9x^2 + 6x + 1$ $(3x + 1)^2$
7. $9b^2 - 12b + 4$ $(3b - 2)^2$
8. $4a^2 - 12a + 9$ $(2a - 3)^2$
9. $9z^2 + 24z + 16$ $(3z + 4)^2$
10. $16z^2 - 24z + 9$ $(4z - 3)^2$

In Exercises 11–22, test each trinomial for factorability and factor it, if possible.

11. $x^2 + 5x + 6$ $(x + 3)(x + 2)$
12. $y^2 + 7y + 6$ $(y + 1)(y + 6)$
13. $x^2 - 7x + 10$ $(x - 2)(x - 5)$
14. $c^2 - 7c + 12$ $(c - 4)(c - 3)$
15. $b^2 + 8b + 18$ prime
16. $x^2 - 12x + 35$ $(x - 7)(x - 5)$
17. $x^2 - x - 30$ $(x + 5)(x - 6)$
18. $a^2 + 4a - 45$ $(a + 9)(a - 5)$
19. $a^2 + 5a - 50$ $(a + 10)(a - 5)$
20. $b^2 + 9b - 36$ $(b + 12)(b - 3)$
21. $y^2 - 4y - 21$ $(y - 7)(y + 3)$
22. $x^2 + 4x - 28$ prime

In Exercises 23–34, factor each trinomial. If the coefficient of the first term is negative, begin by factoring out -1.

23. $3x^2 + 12x - 63$ $3(x + 7)(x - 3)$
24. $2y^2 + 4y - 48$ $2(y + 6)(y - 4)$
25. $a^2b^2 - 13ab^2 + 22b^2$ $b^2(a - 11)(a - 2)$
26. $a^2b^2x^2 - 18a^2b^2x + 81a^2b^2$ $a^2b^2(x - 9)(x - 9)$
27. $b^2x^2 - 12bx^2 + 35x^2$ $x^2(b - 7)(b - 5)$
28. $c^3x^2 + 11c^3x - 42c^3$ $c^3(x + 14)(x - 3)$
29. $-a^2 + 4a + 32$ $-(a - 8)(a + 4)$
30. $-x^2 - 2x + 15$ $-(x + 5)(x - 3)$
31. $-3x^2 + 15x - 18$ $-3(x - 3)(x - 2)$
32. $-2y^2 - 16y + 40$ $-2(y + 10)(y - 2)$
33. $-4x^2 + 4x + 80$ $-4(x - 5)(x + 4)$
34. $-5a^2 + 40a - 75$ $-5(a - 3)(a - 5)$

In Exercises 35–72, factor each trinomial. Factor out all common monomials first (including −1 if the first term is negative). If a trinomial is prime, so indicate.

35. $6y^2 + 7y + 2$ $(3y + 2)(2y + 1)$

36. $6x^2 - 11x + 3$ $(3x - 1)(2x - 3)$

37. $8a^2 + 6a - 9$ $(4a - 3)(2a + 3)$

38. $15b^2 + 4b - 4$ $(5b - 2)(3b + 2)$

39. $6x^2 - 5x - 4$ $(3x - 4)(2x + 1)$

40. $18y^2 - 3y - 10$ $(6y - 5)(3y + 2)$

41. $5x^2 + 4x + 1$ prime

42. $6z^2 + 17z + 12$ $(2z + 3)(3z + 4)$

43. $8x^2 - 10x + 3$ $(4x - 3)(2x - 1)$

44. $4a^2 + 20a + 3$ prime

45. $6z^2 + 7z - 20$ $(3z - 4)(2z + 5)$

46. $7x^2 - 23x + 6$ $(7x - 2)(x - 3)$

47. $a^2 - 3ab - 4b^2$ $(a + b)(a - 4b)$

48. $b^2 + 2bc - 80c^2$ $(b + 10c)(b - 8c)$

49. $2y^2 + yt - 6t^2$ $(2y - 3t)(y + 2t)$

50. $3x^2 - 10xy - 8y^2$ $(3x + 2y)(x - 4y)$

51. $3x^3 - 10x^2 + 3x$ $x(3x - 1)(x - 3)$

52. $6y^2 + 7y + 2$ $(2y + 1)(3y + 2)$

53. $-3a^2 + ab + 2b^2$ $-(3a + 2b)(a - b)$

54. $-2x^2 + 3xy + 5y^2$ $-(2x - 5y)(x + y)$

55. $9t^2 + 3t - 2$ $(3t + 2)(3t - 1)$

56. $3t^3 - 3t^2 + t$ $t(3t^2 - 3t + 1)$

57. $9x^2 - 12x + 4$ $(3x - 2)^2$

58. $4a^2 + 28a + 49$ $(2a + 7)^2$

59. $-4x^2 - 9 + 12x$ $-(2x - 3)^2$

60. $6x + 4 + 9x^2$ prime

61. $15x^2 + 2 - 13x$ $(5x - 1)(3x - 2)$

62. $-90x^2 + 2 - 8x$ $-2(9x - 1)(5x + 1)$

63. $5a^2 + 45b^2 - 30ab$ $5(a - 3b)^2$

64. $x^2 + 324y^2 - 36xy$ $(x - 18y)^2$

65. $8x^2z + 6xyz + 9y^2z$ $z(8x^2 + 6xy + 9y^2)$

66. $x^3 - 60xy^2 + 7x^2y$ $x(x + 12y)(x - 5y)$

67. $15x^2 + 74x - 5$ $(15x - 1)(x + 5)$

68. $15x^2 - 7x - 30$ $(3x - 5)(5x + 6)$

69. $21x^4 - 10x^3 - 16x^2$ $x^2(7x - 8)(3x + 2)$

70. $16x^3 - 50x^2 + 36x$ $2x(x - 2)(8x - 9)$

71. $6x^2y^2 - 17xyz + 12z^2$ $(3xy - 4z)(2xy - 3z)$

72. $6u^2v^2 - uvz + 12z^2$ prime

In Exercises 73–84, factor each trinomial.

73. $x^4 + 8x^2 + 15$ $(x^2 + 5)(x^2 + 3)$

74. $x^4 + 11x^2 + 24$ $(x^2 + 8)(x^2 + 3)$

75. $y^4 - 13y^2 + 30$ $(y^2 - 10)(y^2 - 3)$

76. $y^4 - 13y^2 + 42$ $(y^2 - 7)(y^2 - 6)$

77. $a^4 - 13a^2 + 36$ $(a + 3)(a - 3)(a + 2)(a - 2)$

78. $b^4 - 17b^2 + 16$ $(b + 1)(b - 1)(b + 4)(b - 4)$

79. $z^4 - z^2 - 12$ $(z^2 + 3)(z + 2)(z - 2)$

80. $c^4 - 8c^2 - 9$ $(c^2 + 1)(c + 3)(c - 3)$

81. $4x^3 + x^6 + 3$ $(x^3 + 3)(x + 1)(x^2 - x + 1)$

82. $a^6 - 2 + a^3$ $(a^3 + 2)(a - 1)(a^2 + a + 1)$

83. $8 + y^6 - 9y^3$ $(y - 2)(y^2 + 2y + 4)(y - 1)(y^2 + y + 1)$

84. $8 + x^6 + 9x^3$ $(x + 2)(x^2 - 2x + 4)(x + 1)(x^2 - x + 1)$

In Exercises 85–92, factor each expression. Assume that n is a natural number.

85. $x^{2n} + 2x^n + 1$ $(x^n + 1)^2$

86. $x^{4n} - 2x^{2n} + 1$ $(x^n + 1)^2(x^n - 1)^2$

87. $2a^{6n} - 3a^{3n} - 2$ $(2a^{3n} + 1)(a^{3n} - 2)$

88. $b^{2n} - b^n - 6$ $(b^n - 3)(b^n + 2)$

89. $x^{4n} + 2x^{2n}y^{2n} + y^{4n}$ $(x^{2n} + y^{2n})^2$

90. $y^{6n} + 2y^{3n}z + z^2$ $(y^{3n} + z)^2$

91. $6x^{2n} + 7x^n - 3$ $(3x^n - 1)(2x^n + 3)$

92. $12y^{4n} + 10y^{2n} + 2$ $2(3y^{2n} + 1)(2y^{2n} + 1)$

In Exercises 93–102, factor each expression.

93. $(x + 1)^2 + 2(x + 1) + 1$ $(x + 2)^2$

94. $(a + b)^2 - 2(a + b) + 1$ $(a + b - 1)^2$

95. $(a + b)^2 - 2(a + b) - 24$ $(a + b + 4)(a + b - 6)$

96. $(x - y)^2 + 3(x - y) - 10$ $(x - y + 5)(x - y - 2)$

97. $6(x + y)^2 - 7(x + y) - 20$ $(3x + 3y + 4)(2x + 2y - 5)$

98. $2(x - z)^2 + 9(x - z) + 4$ $(2x - 2z + 1)(x - z + 4)$

99. $5(x^2 - 4x + 4) - 4(x - 2) - 1$ $(5x - 9)(x - 3)$

100. $6(x^2 + 2x + 1) + 5(x + 1) + 1$ $(3x + 4)(2x + 3)$

101. $(4x^2 - 8x + 4) + (-5x + 5) + 1$ $(4x - 5)(x - 2)$

102. $(5x^2 + 30x + 45) + (6x + 18) + 1$ $(5x + 16)(x + 4)$

In Exercises 103–112, factor each expression.

103. $x^2 + 4x + 4 - y^2$ $(x + 2 + y)(x + 2 - y)$

104. $x^2 - 6x + 9 - 4y^2$ $(x - 3 + 2y)(x - 3 - 2y)$

105. $x^2 + 2x + 1 - 9z^2$ $(x + 1 + 3z)(x + 1 - 3z)$

106. $x^2 + 10x + 25 - 16z^2$ $(x + 5 + 4z)(x + 5 - 4z)$

107. $c^2 - 4a^2 + 4ab - b^2$ $(c + 2a - b)(c - 2a + b)$

108. $4c^2 - a^2 - 6ab - 9b^2$ $(2c + a + 3b)(2c - a - 3b)$

109. $a^2 - b^2 + 8a + 16$ $(a + 4 + b)(a + 4 - b)$

110. $a^2 + 14a - 25b^2 + 49$ $(a + 7 + 5b)(a + 7 - 5b)$

111. $4x^2 - z^2 + 4xy + y^2$ $(2x + y + z)(2x + y - z)$

112. $x^2 - 4xy - 4z^2 + 4y^2$ $(x - 2y + 2z)(x - 2y - 2z)$

In Exercises 113–120, use factoring by grouping to help factor each trinomial.

113. $a^2 - 17a + 16$ $(a - 16)(a - 10)$

114. $b^2 - 4b - 21$ $(b + 3)(b - 7)$

115. $2u^2 + 5u + 3$ $(2u + 3)(u + 1)$

116. $6y^2 + 5y - 6$ $(2y + 3)(3y - 2)$

117. $20r^2 - 7rs - 6s^2$ $(5r + 2s)(4r - 3s)$

118. $6s^2 + st - 12t^2$ $(2s + 3t)(3s - 4t)$

119. $20u^2 + 19uv + 3v^2$ $(5u + v)(4u + 3v)$

120. $12m^2 + mn - 6n^2$ $(4m + 3n)(3m - 2n)$

Writing Exercises ■ *Write a paragraph using your own words.*

1. Explain how you would factor -1 from a trinomial.

2. Explain how you would test the polynomial $ax^2 + bx + c$ for factorability.

Something to Think About ■

1. Because it is the difference of two squares, $x^2 - q^2$ always factors. Does the test for factorability predict this?

2. The polynomial $ax^2 + ax + a$ factors; a is a common factor. Does the test for factorability predict this? Is there something wrong with the test? Explain.

Review Exercises ■ *Solve each equation.*

1. $\dfrac{2 + x}{11} = 3$ 31

2. $\dfrac{3y - 12}{2} = 9$ 10

3. $\dfrac{2}{3}(5t - 3) = 38$ 12

4. $3(p + 2) = 4p$ 6

5. $11r + 6(3 - r) = 3$ -3

6. $2q^2 - 9 = q(q + 3) + q^2$ -3

4.4 Summary of Factoring Techniques

In this section, we will discuss ways to approach a randomly chosen factoring problem. For example, suppose we wish to factor the trinomial

$$x^2y^2z^3 + 7xy^2z^3 + 6y^2z^3$$

We begin by attempting to identify the problem type. The first to look for is **factoring out a common monomial.** Because the trinomial has a common monomial factor of y^2z^3, we factor it out:

$$x^2y^2z^3 + 7xy^2z^3 + 6y^2z^3 = y^2z^3(x^2 + 7x + 6)$$

We note that $x^2 + 7x + 6$ is a trinomial that can be factored as $(x + 6)(x + 1)$. Thus,

$$\begin{aligned} x^2y^2z^3 + 7xy^2z^3 + 6y^2z^3 &= y^2z^3(x^2 + 7x + 6) \\ &= y^2z^3(x + 6)(x + 1) \end{aligned}$$

To identify the type of factoring problem, we follow these steps.

Strategy for Identifying the Type of Factoring Problem

1. Factor out all common monomial factors.
2. If an expression has two terms, check to see if the problem type is
   a. **The difference of two squares:** $(x^2 - y^2) = (x + y)(x - y)$
   b. **The sum of two cubes:** $(x^3 + y^3) = (x + y)(x^2 - xy + y^2)$
   c. **The difference of two cubes:** $(x^3 - y^3) = (x - y)(x^2 + xy + y^2)$
3. If an expression has three terms, attempt to factor the trinomial as a **general trinomial.**
4. If an expression has four or more terms, try factoring by **grouping.**
5. Continue until each individual factor is prime.
6. Check the results by multiplying.

EXAMPLE 1 Factor $48a^4c^3 - 3b^4c^3$.

Solution We begin by factoring out the common monomial factor of $3c^3$:

$$48a^4c^3 - 3b^4c^3 = 3c^3(16a^4 - b^4)$$

Since the expression $16a^4 - b^4$ has two terms, we check to see if it is the difference of two squares, which it is. As the difference of two squares, it factors as $(4a^2 + b^2)(4a^2 - b^2)$. Thus,

$$\begin{aligned} 48a^4c^3 - 3b^4c^3 &= 3c^3(\mathbf{16a^4 - b^4}) \\ &= 3c^3(\mathbf{4a^2 + b^2})(\mathbf{4a^2 - b^2}) \end{aligned}$$

The binomial $4a^2 + b^2$ is the sum of two squares and is prime. However, the binomial $4a^2 - b^2$ is the difference of two squares and factors as $(2a + b)(2a - b)$. Thus,

$$\begin{aligned} 48a^4c^3 - 3b^4c^3 &= 3c^3(16a^4 - b^4) \\ &= 3c^3(4a^2 + b^2)(\mathbf{4a^2 - b^2}) \\ &= 3c^3(4a^2 + b^2)(\mathbf{2a + b})(\mathbf{2a - b}) \end{aligned}$$

Since each of the individual factors is prime, the factorization is complete. ■

EXAMPLE 2 Factor $x^5y + x^2y^4 - x^3y^3 - y^6$.

Solution We begin by factoring out the common monomial factor of y:

$$x^5y + x^2y^4 - x^3y^3 - y^6 = y(x^5 + x^2y^3 - x^3y^2 - y^5)$$

Because the expression $x^5 + x^2y^3 - x^3y^2 - y^5$ has four terms, we try factoring by grouping to obtain

$$\begin{aligned} x^5y + x^2y^4 - x^3y^3 - y^6 &= y(x^5 + x^2y^3 - x^3y^2 - y^5) \\ &= y[x^2(x^3 + y^3) - y^2(x^3 + y^3)] \\ &= y(x^3 + y^3)(x^2 - y^2) && \text{Factor out } x^3 + y^3. \end{aligned}$$

Finally, we factor $x^3 + y^3$ (the sum of two cubes) and $x^2 - y^2$ (the difference of two squares) to obtain

$$x^5y + x^2y^4 - x^3y^3 - y^6 = y(x + y)(x^2 - xy + y^2)(x + y)(x - y)$$

Because each of the individual factors is prime, the factorization is complete. ■

EXAMPLE 3 Factor $p^{-4} - p^{-2} - 6$.

Solution We factor this expression as if it were a trinomial:

$$\begin{aligned} p^{-4} - p^{-2} - 6 &= (p^{-2} - 3)(p^{-2} + 2) \\ &= \left(\frac{1}{p^2} - 3\right)\left(\frac{1}{p^2} + 2\right) \end{aligned}$$

■

EXAMPLE 4 Factor $x^3 + 5x^2 + 6x + x^2y + 5xy + 6y$.

Solution Since there are more than three terms, we try factoring by grouping.

We can factor x from the first three terms and y from the last three terms and proceed as follows:

$$\begin{aligned} x^3 + 5x^2 + 6x + x^2y + 5xy + 6y &= x(x^2 + 5x + 6) + y(x^2 + 5x + 6) \\ &= (x^2 + 5x + 6)(x + y) && \text{Factor out } x^2 + 5x + 6. \\ &= (x + 3)(x + 2)(x + y) && \text{Factor } x^2 + 5x + 6. \end{aligned}$$

■

EXAMPLE 5 Factor $x^4 + 2x^3 + x^2 + x + 1$.

Solution Since there are more than three terms, we try factoring by grouping.

We can factor x^2 from the first three terms and proceed as follows:

$$\begin{aligned} x^4 + 2x^3 + x^2 + x + 1 &= x^2(x^2 + 2x + 1) + (x + 1) \\ &= x^2(\mathbf{x + 1})(x + 1) + (\mathbf{x + 1}) && \text{Factor } x^2 + 2x + 1. \\ &= (\mathbf{x + 1})[x^2(x + 1) + 1] && \text{Factor out } x + 1. \\ &= (x + 1)(x^3 + x^2 + 1) \end{aligned}$$

■

ORALS *Factor each expression.*

1. $x^2 - y^2$ $(x + y)(x - y)$
2. $2x^3 - 4x^4$ $2x^3(1 - 2x)$
3. $x^2 + 4x + 4$ $(x + 2)^2$
4. $x^2 - 5x + 6$ $(x - 3)(x - 2)$
5. $x^3 - 8$ $(x - 2)(x^2 + 2x + 4)$
6. $x^3 + 8$ $(x + 2)(x^2 - 2x + 4)$

EXERCISE 4.4

In Exercises 1–52, factor each polynomial, if possible.

1. $x^2 + 8x + 16$ $(x + 4)^2$
2. $20 + 11x - 3x^2$ $-(3x + 4)(x - 5)$
3. $8x^3y^3 - 27$ $(2xy - 3)(4x^2y^2 + 6xy + 9)$
4. $3x^2y + 6xy^2 - 12xy$ $3xy(x + 2y - 4)$
5. $xy - ty + xs - ts$ $(x - t)(y + s)$
6. $bc + b + cd + d$ $(b + d)(c + 1)$
7. $25x^2 - 16y^2$ $(5x + 4y)(5x - 4y)$
8. $27x^9 - y^3$ $(3x^3 - y)(9x^6 + 3x^3y + y^2)$
9. $12x^2 + 52x + 35$ $(6x + 5)(2x + 7)$
10. $12x^2 + 14x - 6$ $2(3x - 1)(2x + 3)$
11. $6x^2 - 14x + 8$ $2(3x - 4)(x - 1)$
12. $12x^2 - 12$ $12(x + 1)(x - 1)$
13. $56x^2 - 15x + 1$ $(8x - 1)(7x - 1)$
14. $7x^2 - 57x + 8$ $(7x - 1)(x - 8)$
15. $4x^2y^2 + 4xy^2 + y^2$ $y^2(2x + 1)(2x + 1)$
16. $100z^2 - 81t^2$ $(10z + 9t)(10z - 9t)$
17. $x^3 + (a^2y)^3$ $(x + a^2y)(x^2 - a^2xy + a^4y^2)$
18. $4x^2y^2z^2 - 26x^2y^2z^3$ $2x^2y^2z^2(2 - 13z)$
19. $2x^3 - 54$ $2(x - 3)(x^2 + 3x + 9)$
20. $4x^3y^3 + 256$ $4(xy + 4)(x^2y^2 - 4xy + 16)$
21. $ae + bf + af + be$ $(a + b)(f + e)$
22. $a^2x^2 + b^2y^2 + b^2x^2 + a^2y^2$ $(x^2 + y^2)(a^2 + b^2)$
23. $2(x + y)^2 + (x + y) - 3$ $(2x + 2y + 3)(x + y - 1)$
24. $(x - y)^3 + 125$ $(x - y + 5)[(x - y)^2 - 5(x - y) + 25]$
25. $625x^4 - 256y^4$ $(25x^2 + 16y^2)(5x + 4y)(5x - 4y)$
26. $2(a - b)^2 + 5(a - b) + 3$ $(2a - 2b + 3)(a - b + 1)$
27. $36x^4 - 36$ $36(x^2 + 1)(x + 1)(x - 1)$
28. $6x^2 - 63 - 13x$ $(2x - 9)(3x + 7)$
29. $2x^6 + 2y^6$ $2(x^2 + y^2)(x^4 - x^2y^2 + y^4)$
30. $x^4 - x^4y^4$ $-x^4(y^2 + 1)(y + 1)(y - 1)$
31. $a^4 - 13a^2 + 36$ $(a + 3)(a - 3)(a + 2)(a - 2)$
32. $x^4 - 17x^2 + 16$ $(x + 1)(x - 1)(x + 4)(x - 4)$
33. $x^2 + 6x + 9 - y^2$ $(x + 3 + y)(x + 3 - y)$
34. $x^2 + 10x + 25 - y^8$ $(x + 5 + y^4)(x + 5 - y^4)$
35. $4x^2 + 4x + 1 - 4y^2$ $(2x + 1 + 2y)(2x + 1 - 2y)$
36. $9x^2 - 6x + 1 - 25y^2$ $(3x - 1 - 5y)(3x - 1 + 5y)$
37. $z^2 + 8z + 16 - 16y^2$ $(z + 4 + 4y)(z + 4 - 4y)$
38. $x^2 + 10x + 25 - 25y^2$ $(x + 5 + 5y)(x + 5 - 5y)$
39. $x^5 + x^2 - x^3 - 1$ $(x + 1)(x^2 - x + 1)(x + 1)(x - 1)$
40. $x^5 - x^2 - 4x^3 + 4$ $(x - 1)(x^2 + x + 1)(x + 2)(x - 2)$
41. $x^5 - 9x^3 + 8x^2 - 72$ $(x + 3)(x - 3)(x + 2)(x^2 - 2x + 4)$
42. $x^5 - 4x^3 - 8x^2 + 32$ $(x + 2)(x - 2)^2(x^2 + 2x + 4)$
43. $2x^5z - 2x^2y^3z - 2x^3y^2z + 2y^5z$ $2z(x + y)(x - y)^2(x^2 + xy + y^2)$
44. $x^2y^3 - 4x^2y - 9y^3 + 36y$ $y(x + 3)(x - 3)(y + 2)(y - 2)$

45. $x^{2m} - x^m - 6$ $(x^m - 3)(x^m + 2)$

46. $a^{2n} - b^{2n}$ $(a^n + b^n)(a^n - b^n)$

47. $a^{3n} - b^{3n}$ $(a^n - b^n)(a^{2n} + a^n b^n + b^{2n})$

48. $x^{3m} + y^{3m}$ $(x^m + y^m)(x^{2m} - x^m y^m + y^{2m})$

49. $x^{-2} + 2x^{-1} + 1$ $\left(\frac{1}{x} + 1\right)^2$

50. $4a^{-2} - 12a^{-1} + 9$ $\left(\frac{2}{a} - 3\right)^2$

51. $6x^{-2} - 5x^{-1} - 6$ $\left(\frac{3}{x} + 2\right)\left(\frac{2}{x} - 3\right)$

52. $x^{-4} - y^{-4}$ $\left(\frac{1}{x^2} + \frac{1}{y^2}\right)\left(\frac{1}{x} + \frac{1}{y}\right)\left(\frac{1}{x} - \frac{1}{y}\right)$

Writing Exercises ■ *Write a paragraph using your own words.*

1. What is your strategy for factoring a polynomial?

2. Explain how you can know that a factorization is correct.

Something to Think About ■

1. If you have the choice of factoring a polynomial as the difference of two squares or as the difference of two cubes, which would you do first? Why?

2. Can several polynomials have a greatest common factor? Find the GCF of $2x^2 + 7x + 3$ and $x^2 - 2x - 15$. $x + 3$

Review Exercises ■ *Do the operations.*

1. $(3a^2 + 4a - 2) + (4a^2 - 3a - 5)$
$7a^2 + a - 7$

2. $(-4b^2 - 3b - 2) - (3b^2 - 2b + 5)$
$-7b^2 - b - 7$

3. $5(2y^2 - 3y + 3) - 2(3y^2 - 2y + 6)$
$4y^2 - 11y + 3$

4. $4(3x^2 + 3x + 3) + 3(x^2 - 3x - 4)$
$15x^2 + 3x$

5. $(m + 4)(m - 2)$
$m^2 + 2m - 8$

6. $(3p + 4q)(2p - 3q)$
$6p^2 - pq - 12q^2$

7. $x + 2\overline{)x^3 + 6x^2 + 7x + 6}$
$x^2 + 4x - 1 + \frac{8}{x+2}$

8. $2z - 1\overline{)6z^3 - 7z^2 + 6z - 2}$
$3z^2 - 2z + 2$

4.5 Solving Equations by Factoring

■ Solving Quadratic Equations ■ Solving Higher-Degree Polynomial Equations ■ Applications

We have previously solved linear or first-degree equations. In this section, we solve certain equations of higher degree.

■ Solving Quadratic Equations

An equation such as $3x^2 + 4x - 7 = 0$ or $-5y^2 + 3y + 8 = 0$ is called a **quadratic** or **second-degree** equation.

Quadratic Equations A **quadratic equation** is any equation that can be written in the form

$$ax^2 + bx + c = 0$$

where a, b, and c are real numbers and $a \neq 0$.

Many quadratic equations can be solved by factoring and then using the **zero-factor theorem.**

Zero-Factor Theorem If a and b are real numbers, then

If $ab = 0$, then $a = 0$ or $b = 0$.

The zero-factor theorem states that *if the product of two or more numbers is 0, then at least one of the numbers must be 0.*

To solve the quadratic equation

$$x^2 + 5x + 6 = 0$$

we factor its left-hand side to obtain

$$(x + 3)(x + 2) = 0$$

Since the product of $x + 3$ and $x + 2$ is 0, then at least one of the factors is 0. Thus, we can set each factor equal to 0 and solve each resulting linear equation for x:

$$\begin{array}{lcl} x + 3 = 0 & \text{or} & x + 2 = 0 \\ \quad x = -3 & | & \quad x = -2 \end{array}$$

To check these solutions, we substitute -3 and -2 for x in the equation and verify that each number satisfies the equation.

$$\begin{array}{rcr} x^2 + 5x + 6 = 0 & \text{or} & x^2 + 5x + 6 = 0 \\ (-3)^2 + 5(-3) + 6 \stackrel{?}{=} 0 & | & (-2)^2 + 5(-2) + 6 \stackrel{?}{=} 0 \\ 9 - 15 + 6 \stackrel{?}{=} 0 & | & 4 - 10 + 6 \stackrel{?}{=} 0 \\ 0 = 0 & | & 0 = 0 \end{array}$$

Both -3 and -2 are solutions, because they both satisfy the equation.

If $c = 0$ in an equation of the form $ax^2 + bx + c = 0$, the equation can always be solved by factoring.

EXAMPLE 1 Solve the equation $3x^2 + 6x = 0$.

Solution To solve the equation, we factor the left-hand side, set each factor equal to 0, and solve each resulting equation for x.

$$3x^2 + 6x = 0$$

$$3x(x + 2) = 0 \quad \text{Factor out the common factor of } 3x.$$

$$3x = 0 \quad \text{or} \quad x + 2 = 0$$

$$x = 0 \quad | \quad x = -2$$

Verify that both solutions check. ■

WARNING! In Example 1, do not divide both sides by $3x$ or you will lose the solution $x = 0$.

EXAMPLE 2 Solve the equation $x^2 - 16 = 0$.

Solution To solve the equation, we factor the difference of two squares on the left-hand side, set each factor equal to 0, and solve each resulting equation.

$$x^2 - 16 = 0$$

$$(x + 4)(x - 4) = 0$$

$$x + 4 = 0 \quad \text{or} \quad x - 4 = 0$$

$$x = -4 \quad | \quad x = 4$$

Verify that both solutions check. ■

Many equations that do not appear to be quadratic can be put into quadratic form and then solved by factoring.

EXAMPLE 3 Solve the equation $x = \frac{6}{5} - \frac{6}{5}x^2$.

Solution We begin writing the equation in quadratic form.

$$x = \frac{6}{5} - \frac{6}{5}x^2$$

$$5x = 6 - 6x^2 \quad \text{Multiply both sides by 5.}$$

$$6x^2 + 5x - 6 = 0 \quad \text{Add } 6x^2 \text{ to both sides and subtract 6 from both sides.}$$

$$(3x - 2)(2x + 3) = 0 \quad \text{Factor the trinomial.}$$

$$3x - 2 = 0 \quad \text{or} \quad 2x + 3 = 0$$
$$3x = 2 \quad\quad 2x = -3$$
$$x = \frac{2}{3} \quad\quad x = -\frac{3}{2}$$

Verify that both solutions check. ■

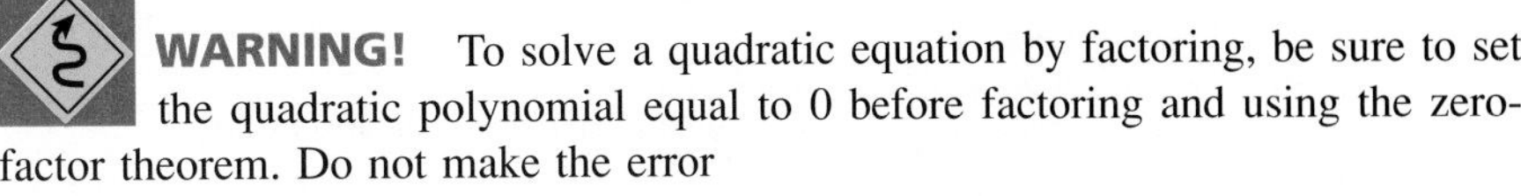

WARNING! To solve a quadratic equation by factoring, be sure to set the quadratic polynomial equal to 0 before factoring and using the zero-factor theorem. Do not make the error

$$6x^2 + 5x = 6$$
$$x(6x + 5) = 6$$
$$x = 6 \quad \text{or} \quad 6x + 5 = 6$$
$$x = \frac{1}{6}$$

Neither solution checks.

Solving Higher-Degree Polynomial Equations

We can often use factoring to solve polynomial equations with degree greater than 2.

EXAMPLE 4 Solve the equation $6x^3 - x^2 - 2x = 0$.

Solution We factor x from the third-degree polynomial on the left-hand side and proceed as follows:

$$6x^3 - x^2 - 2x = 0$$
$$x(6x^2 - x - 2) = 0 \quad \text{Factor out an } x.$$
$$x(3x - 2)(2x + 1) = 0 \quad \text{Factor } 6x^2 - x - 2.$$
$$x = 0 \quad \text{or} \quad 3x - 2 = 0 \quad \text{or} \quad 2x + 1 = 0 \quad \text{Set each factor equal to 0.}$$
$$x = \frac{2}{3} \quad\quad x = -\frac{1}{2}$$

Verify that all three solutions check. ■

EXAMPLE 5 Solve the equation $x^4 - 5x^2 + 4 = 0$.

Solution We factor the trinomial on the left-hand side and proceed as follows:

$$x^4 - 5x^2 + 4 = 0$$
$$(x^2 - 1)(x^2 - 4) = 0$$
$$(x + 1)(x - 1)(x + 2)(x - 2) = 0 \quad \text{Factor } x^2 - 1 \text{ and } x^2 - 4.$$

$x + 1 = 0$ or $x - 1 = 0$ or $x + 2 = 0$ or $x - 2 = 0$ Set each factor equal to 0.

$x = -1$ | $x = 1$ | $x = -2$ | $x = 2$

Verify that all four solutions check. ■

Applications

EXAMPLE 6 **Finding the dimensions of a truss** The width of the triangular roof truss in Figure 4-1 is 3 times its height. The area of the triangle is 96 square feet. Find its base and height.

Solution We can let x be the positive number that represents the height of the truss. Then $3x$ represents its width. We can substitute x for h, $3x$ for b, and 96 for A in the formula for the area of a triangle and solve for x.

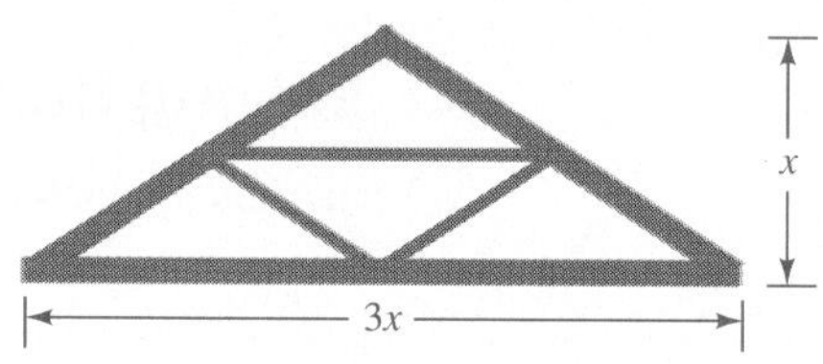

FIGURE 4-1

$$A = \frac{1}{2}bh$$
$$96 = \frac{1}{2}(3x)x$$
$$192 = 3x^2 \quad \text{Multiply both sides by 2.}$$
$$64 = x^2 \quad \text{Divide both sides by 3.}$$
$$0 = x^2 - 64 \quad \text{Subtract 64 from both sides.}$$
$$0 = (x + 8)(x - 8) \quad \text{Factor the difference of two squares.}$$

$x + 8 = 0$ or $x - 8 = 0$

$x = -8$ | $x = 8$

Since the height of a triangle cannot be negative, we must discard the negative solution. Thus, the height of the truss is 8 feet, and its width is 3(8), or 24 feet.

Check: The area of a triangle with a base of 24 feet and a height of 8 feet is 96 square feet:

$$A = \frac{1}{2}bh = \frac{1}{2}(24)(8) = 12(8) = 96$$

The solution checks. ■

EXAMPLE 7 **Ballistics** The height of an object thrown straight up into the air with an initial velocity of v feet per second is given by the formula

$$h = vt - 16t^2$$

The height h is in feet, and t represents the number of seconds since the object was released. If the initial velocity is 176 feet per second, when will the object strike the ground?

Solution When the object hits the ground, its height will be 0. Thus, we set h equal to 0, set v equal to 176, and solve for t.

$$h = vt - 16t^2$$

$$0 = 176t - 16t^2$$

$$0 = 16t(11 - t) \quad \text{Factor out } 16t.$$

$$16t = 0 \quad \text{or} \quad 11 - t = 0 \quad \text{Set each factor equal to 0.}$$

$$t = 0 \quad | \quad t = 11 \quad \text{Solve each linear equation.}$$

When $t = 0$, the object's height above the ground is 0 feet, because it has just been released. When $t = 11$, the height is again 0 feet, and the object has returned to ground. The solution is 11 seconds. ■

ORALS *Solve each equation.*

1. $(x - 2)(x - 3) = 0$ 2, 3
2. $(x + 4)(x - 2) = 0$ −4, 2
3. $(x - 2)(x - 3)(x + 1) = 0$ 2, 3, −1
4. $(x + 3)(x + 2)(x - 5)(x - 6) = 0$ −3, −2, 5, 6

EXERCISE 4.5

In Exercises 1–24, solve each equation.

1. $4x^2 + 8x = 0$ 0, −2
2. $x^2 - 9 = 0$ 3, −3
3. $y^2 - 16 = 0$ 4, −4
4. $5y^2 - 10y = 0$ 0, 2
5. $x^2 + x = 0$ 0, −1
6. $x^2 - 3x = 0$ 0, 3
7. $5y^2 - 25y = 0$ 0, 5
8. $y^2 - 36 = 0$ 6, −6
9. $z^2 + 8z + 15 = 0$ −3, −5
10. $w^2 + 7w + 12 = 0$ −3, −4

11. $y^2 - 7y + 6 = 0$ 1, 6

12. $n^2 - 5n + 6 = 0$ 2, 3

13. $y^2 - 7y + 12 = 0$ 3, 4

14. $x^2 - 3x + 2 = 0$ 1, 2

15. $x^2 + 6x + 8 = 0$ $-2, -4$

16. $x^2 + 9x + 20 = 0$ $-4, -5$

17. $3m^2 + 10m + 3 = 0$ $-\frac{1}{3}, -3$

18. $2r^2 + 5r + 3 = 0$ $-\frac{3}{2}, -1$

19. $2y^2 - 5y + 2 = 0$ $\frac{1}{2}, 2$

20. $2x^2 - 3x + 1 = 0$ $\frac{1}{2}, 1$

21. $2x^2 - x - 1 = 0$ $1, -\frac{1}{2}$

22. $2x^2 - 3x - 5 = 0$ $\frac{5}{2}, -1$

23. $3s^2 - 5s - 2 = 0$ $2, -\frac{1}{3}$

24. $8t^2 + 10t - 3 = 0$ $\frac{1}{4}, -\frac{3}{2}$

In Exercises 25–36, write each equation in quadratic form and solve it by factoring.

25. $x(x - 6) + 9 = 0$ 3, 3

26. $x^2 + 8(x + 2) = 0$ $-4, -4$

27. $8a^2 = 3 - 10a$ $\frac{1}{4}, -\frac{3}{2}$

28. $5z^2 = 6 - 13z$ $\frac{2}{5}, -3$

29. $b(6b - 7) = 10$ $2, -\frac{5}{6}$

30. $2y(4y + 3) = 9$ $\frac{3}{4}, -\frac{3}{2}$

31. $\dfrac{3a^2}{2} = \dfrac{1}{2} - a$ $\frac{1}{3}, -1$

32. $x^2 = \dfrac{1}{2}(x + 1)$ $1, -\frac{1}{2}$

33. $x^2 + 1 = \dfrac{5}{2}x$ $2, \frac{1}{2}$

34. $\dfrac{3}{5}(x^2 - 4) = -\dfrac{9}{5}x$ $1, -4$

35. $x\left(3x + \dfrac{22}{5}\right) = 1$ $\frac{1}{5}, -\frac{5}{3}$

36. $x\left(\dfrac{x}{11} - \dfrac{1}{7}\right) = \dfrac{6}{77}$ $-\frac{3}{7}, 2$

In Exercises 37–48, solve each equation.

37. $x^3 + x^2 = 0$ $0, 0, -1$

38. $2x^4 + 8x^3 = 0$ $0, 0, 0, -4$

39. $y^3 - 49y = 0$ $0, 7, -7$

40. $2z^3 - 200z = 0$ $0, 10, -10$

41. $x^3 - 4x^2 - 21x = 0$ $0, 7, -3$

42. $x^3 + 8x^2 - 9x = 0$ $1, 0, -9$

43. $z^4 - 13z^2 + 36 = 0$ $3, -3, 2, -2$

44. $y^4 - 10y^2 + 9 = 0$ $3, -3, 1, -1$

45. $3a(a^2 + 5a) = -18a$ $0, -2, -3$

46. $7t^3 = 2t\left(t + \dfrac{5}{2}\right)$ $0, 1, -\frac{5}{7}$

47. $\dfrac{x^2(6x + 37)}{35} = x$ $0, \frac{5}{6}, -7$

48. $x^2 = -\dfrac{4x^3(3x + 5)}{3}$ $0, 0, -\frac{1}{6}, -\frac{3}{2}$

In Exercises 49–54, use grouping to help solve each equation.

49. $x^3 + 3x^2 - x - 3 = 0$ $1, -1, -3$

50. $x^3 - x^2 - 4x + 4 = 0$ $1, 2, -2$

51. $2r^3 + 3r^2 - 18r - 27 = 0$ $3, -3, -\frac{3}{2}$

52. $3s^3 - 2s^2 - 3s + 2 = 0$ $1, \frac{2}{3}, -1$

53. $3y^3 + y^2 = 4(3y + 1)$ $2, -2, -\frac{1}{3}$

54. $w^3 + 16 = w(w + 16)$ $4, -4, 1$

In Exercises 55–78, solve each problem.

55. **Integer problem** The product of two consecutive even integers is 288. Find the integers.
16, 18, or $-16, -18$

56. **Integer problem** The product of two consecutive odd integers is 143. Find the integers.
11, 13 or $-11, -13$

57. **Integer problem** The sum of the squares of two consecutive positive integers is 85. Find the integers. 6, 7

58. **Integer problem** The sum of the squares of three consecutive positive integers is 77. Find the integers. 4, 5, 6

59. **Geometry** Find the perimeter of the rectangle in Illustration 1. 40 m

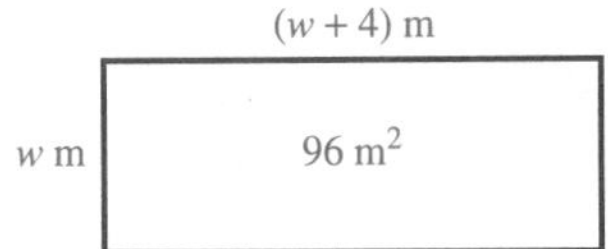

ILLUSTRATION 1

60. **Geometry** One side of a rectangle is three times longer than another. If its area is 147 square centimeters, find its dimensions. 7 cm by 21 cm

61. **Geometry** The area of a square is numerically equal to its perimeter. How long is one side of the square? 4 units

62. **Geometry** A rectangle is 2 inches longer than it is wide. Numerically, its area exceeds its perimeter by 11 inches. Find its perimeter. 24 in.

63. **Fine art** An artist intends to paint a 60-square-foot mural on the large wall shown in Illustration 2. Find the dimensions of the mural if the artist leaves a border of uniform width around it. 5 ft by 12 ft

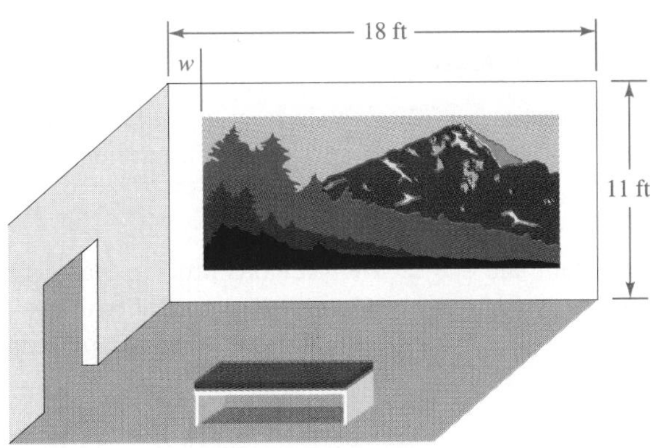

ILLUSTRATION 2

64. **Gardening** A woman plans to use one-fourth of her 48-foot-by-100-foot rectangular backyard to plant a garden. Find the perimeter of the garden if the length is to be 40 feet greater than the width. 160 ft

65. **Architecture** The rectangular room shown in Illustration 3 is twice as long as it is wide. It is divided into two rectangular parts by a partition, positioned as shown. If the larger part of the room contains 560 square feet, find the dimensions of the entire room. 20 ft by 40 ft

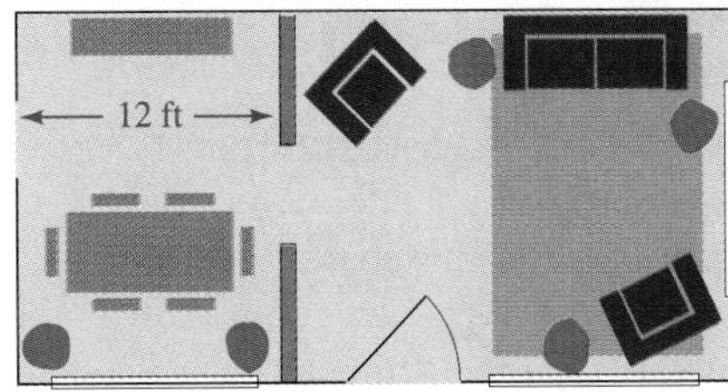

ILLUSTRATION 3

66. **Perimeter of a square** If the length of one side of a square is increased by 4 inches, the area of the square becomes 9 times greater. Find the perimeter of the original square. 8 in.

In Exercises 67–70, see Example 7.

67. **Time of flight** After how many seconds will an object hit the ground if it was thrown straight up with an initial velocity of 160 feet per second? 10 sec

68. **Time of flight** After how many seconds will an object hit the ground if it was thrown straight up with an initial velocity of 208 feet per second? 13 sec

69. **Ballistics** The muzzle velocity of a cannon is 480 feet per second. If a cannonball is fired vertically, at what times will it be at a height of 3344 feet? 11 sec and 19 sec

70. **Ballistics** A slingshot can provide an initial velocity of 128 feet per second. At what times will a stone, shot vertically upward, be 192 feet above the ground? 2 sec and 6 sec

71. Winter recreation The length of the rectangular ice skating rink in Illustration 4 is 20 meters greater than twice its width. Find the width. 50 m

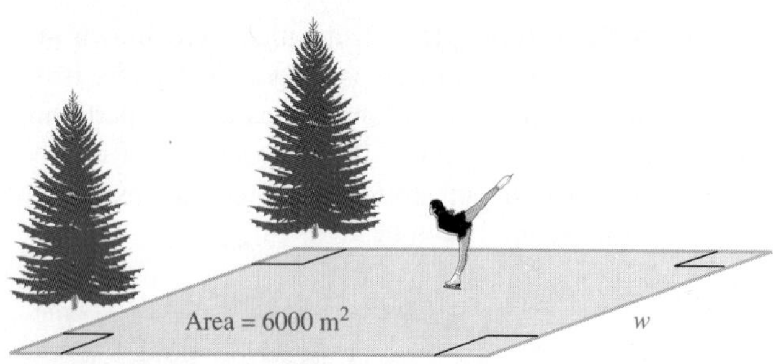

ILLUSTRATION 4

72. Carpentry A 285-square-foot room is 4 feet longer than it is wide. What length of crown molding is needed to trim the perimeter of the ceiling? 68 ft

73. Designing a swimming pool Building code requires that the rectangular swimming pool in Illustration 5 be surrounded by a uniform-width walkway of at least 516 square feet. The length of the pool is 10 feet less than twice the width. How wide should the border be? 3 ft

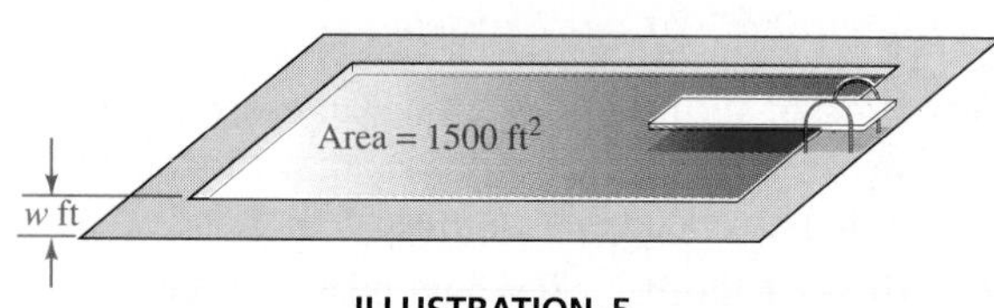

ILLUSTRATION 5

74. House construction The formula for the area of a trapezoid is $A = \frac{h(B + b)}{2}$. The area of the trapezoidal truss in Illustration 6 is 44 square feet. Find the height of the truss if the shorter base is the same as the height. 4 ft

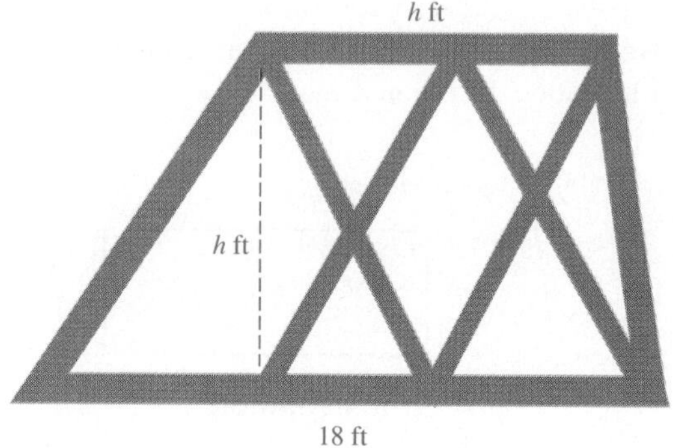

ILLUSTRATION 6

75. Algebra Find a quadratic equation whose roots are 3 and 5. $x^2 - 8x + 15 = 0$

76. Algebra Find a quadratic equation whose roots are -2 and 6. $x^2 - 4x - 12 = 0$

77. Algebra Find a quadratic equation whose roots are 0 and -5. $x^2 + 5x = 0$

78. Algebra Find a quadratic equation whose roots are $\frac{1}{2}$ and $\frac{1}{3}$. $6x^2 - 5x + 1 = 0$

Writing Exercises ■ *Write a paragraph using your own words.*

1. Describe the steps for solving an application problem.
2. Explain how to check the solution of an application problem.

Something to Think About ■

1. Attempt to solve this problem and describe the difficulties you have: The length of a rectangle is 7 feet greater than its width, and its area is 21 square feet. Find the width.
2. The previous problem has a solution. Experiment with a calculator and find an approximate solution.

Review Exercises

1. List the prime numbers less than 10.
2, 3, 5, 7

2. List the composite numbers between 7 and 17.
8, 9, 10, 12, 14, 15, 16

3. The formula for the volume of a sphere is $V = \frac{4}{3}\pi r^3$. Find the volume when $r = 21$ centimeters. Give the answer to the nearest hundredth. 38,792.39 cm^3

4. The formula for the volume of a cone is $V = \frac{1}{3}\pi r^2 h$. Find the volume when $r = 12$ meters and $h = 14$ meters. 2111.15 m^3

PROJECT Give a Little, Take a Little

The trinomial $x^4 + x^2 + 1$ does not factor as a perfect square, because $(x^2 + 1)^2$ is $x^4 + 2x^2 + 1$, which has a middle term of $2x^2$ instead of $1x^2$. We can change the middle term of the trinomial into $2x^2$ by adding x^2 to the trinomial. To ensure that adding x^2 does not change the value of the trinomial, we also subtract x^2.

$$\begin{aligned} x^4 + x^2 + 1 &= x^4 + x^2 + x^2 + 1 - x^2 \\ &= x^4 + 2x^2 + 1 - x^2 \end{aligned}$$

Because the first three terms of this expression factor are the perfect square $(x^2 + 1)^2$, we can continue:

$$\begin{aligned} x^4 + x^2 + 1 &= x^4 + 2x^2 + 1 - x^2 \\ &= (x^2 + 1)^2 - x^2 \\ &= [(x^2 + 1) + x][(x^2 + 1) - x] \quad \text{Factor the difference of two squares.} \\ &= (x^2 + x + 1)(x^2 - x + 1) \end{aligned}$$

- Use this technique to factor $x^4 + 4x^2 + 16$. (*Hint:* Add $4x^2$ and subtract $4x^2$.)
- Factor $x^6 - 64$ as the difference of two cubes. Factor it completely. (*Hint:* Use the previous result.)
- Factor $x^6 - 64$ as the difference of two squares. Do the two results agree?
- Factor $x^6 - y^6$ as the difference of two cubes, and again as the difference of two squares. Do the results agree?

Chapter Summary

KEY WORDS

difference of two cubes (4.2)
difference of two squares (4.2)
factor (4.1)
factoring by grouping (4.1)
factoring out the greatest common factor (4.1)

general trinomial (4.3)
greatest common divisor (4.1)
greatest common factor (4.1)
irreducible polynomial (4.1)
perfect squares (4.2)
perfect square trinomials (4.3)
prime-factored form (4.1)
prime polynomial (4.1)
quadratic equations (4.5)
sum of two cubes (4.2)
zero-factor theorem (4.5)

KEY IDEAS

(4.1) Always factor out common monomial factors as the first step in a factoring problem.

Use the distributive property to factor out common monomial factors.

If an expression has four or more terms, try to factor the expression by grouping.

(4.2) **Difference of two squares:**
$x^2 - y^2 = (x + y)(x - y)$

Sum of two cubes: $x^3 + y^3 = (x + y)(x^2 - xy + y^2)$

Difference of two cubes: $x^3 - y^3 = (x - y)(x^2 + xy + y^2)$

(4.3) **Special product formulas:**
$x^2 + 2xy + y^2 = (x + y)(x + y)$
$x^2 - 2xy + y^2 = (x - y)(x - y)$

Test for factorability: A trinomial of the form $ax^2 + bx + c$ (a, b, and c integers, with $a \neq 0$) will factor with integer coefficients if $b^2 - 4ac$ is a perfect square.

(4.4) Use these steps to factor a random expression:

1. Factor out all common monomial factors.

2. If an expression has two terms, check to see if the problem type is

a. The difference of two squares:
$x^2 - y^2 = (x + y)(x - y)$

b. The sum of two cubes:
$x^3 + y^3 = (x + y)(x^2 - xy + y^2)$

c. The difference of two cubes:
$x^3 - y^3 = (x - y)(x^2 + xy + y^2)$

3. If an expression has three terms, attempt to factor the trinomial as a **general trinomial.**

4. If an expression has four or more terms, try factoring by **grouping.**

5. Continue until each individual factor is prime.

6. Check the results by multiplying.

(4.5) **Zero-factor theorem:** If $xy = 0$, then $x = 0$ or $y = 0$.

Chapter 4 Review Exercises

In Exercises 1–8, factor each polynomial.

1. $4x + 8$ $\quad 4(x + 2)$

2. $3x^2 - 6x$ $\quad 3x(x - 2)$

3. $5x^2y^3 - 10xy^2$ $\quad 5xy^2(xy - 2)$

4. $7a^4b^2 + 49a^3b$ $\quad 7a^3b(ab + 7)$

5. $-8x^2y^3z^4 - 12x^4y^3z^2$ $\quad -4x^2y^3z^2(2z^2 + 3x^2)$

6. $12a^6b^4c^2 + 15a^2b^4c^6$ $\quad 3a^2b^4c^2(4a^4 + 5c^4)$

7. $27x^3y^3z^3 + 81x^4y^5z^2 - 90x^2y^3z^7$ $\quad 9x^2y^3z^2(3xz + 9x^2y^2 - 10z^5)$

8. $-36a^5b^4c^2 + 60a^7b^5c^3 - 24a^2b^3c^7$ $\quad -12a^2b^3c^2(3a^3b - 5a^5b^2c + 2c^5)$

9. Factor x^n from $x^{2n} + x^n$. $\quad x^n(x^n + 1)$

10. Factor y^{2n} from $y^{2n} - y^{4n}$. $\quad y^{2n}(1 - y^{2n})$

11. Factor x^{-2} from $x^{-4} - x^{-2}$. $\quad x^{-2}(x^{-2} - 1)$

12. Factor a^{-3} from $a^6 + 1$. $\quad a^{-3}(a^9 + a^3)$

In Exercises 13–54, factor each polynomial.

13. $5x^2(x+y)^3 - 15x^3(x+y)^4$ $5x^2(x+y)^3(1-3x^2-3xy)$

14. $-49a^3b^2(a-b)^4 + 63a^2b^4(a-b)^3$ $-7a^2b^2(a-b)^3(7a^2-7ab-9b^2)$

15. $xy + 2y + 4x + 8$ $(x+2)(y+4)$

16. $ac + bc + 3a + 3b$ $(a+b)(c+3)$

17. $x^4 + 4y + 4x^2 + x^2y$ $(x^2+4)(x^2+y)$

18. $a^5 + b^2c + a^2c + a^3b^2$ $(a^3+c)(a^2+b^2)$

19. $z^2 - 16$ $(z+4)(z-4)$

20. $y^2 - 121$ $(y+11)(y-11)$

21. $x^2y^4 - 64z^6$ $(xy^2+8z^3)(xy^2-8z^3)$

22. $a^2b^2 + c^2$ prime

23. $(x+z)^2 - t^2$ $(x+z+t)(x+z-t)$

24. $c^2 - (a+b)^2$ $(c+a+b)(c-a-b)$

25. $2x^4 - 98$ $2(x^2+7)(x^2-7)$

26. $3x^6 - 300x^2$ $3x^2(x^2+10)(x^2-10)$

27. $x^3 + 343$ $(x+7)(x^2-7x+49)$

28. $a^3 - 125$ $(a-5)(a^2+5a+25)$

29. $8y^3 - 512$ $8(y-4)(y^2+4y+16)$

30. $4x^3y + 108yz^3$ $4y(x+3z)(x^2-3xz+9z^2)$

31. $y^2 + 21y + 20$ $(y+20)(y+1)$

32. $z^2 - 11z + 30$ $(z-5)(z-6)$

33. $-x^2 - 3x + 28$ $-(x+7)(x-4)$

34. $y^2 - 5y - 24$ $(y-8)(y+3)$

35. $4a^2 - 5a + 1$ $(4a-1)(a-1)$

36. $3b^2 + 2b + 1$ prime

37. $7x^2 + x + 2$ prime

38. $-15x^2 + 14x + 8$ $-(5x+2)(3x-4)$

39. $y^3 + y^2 - 2y$ $y(y+2)(y-1)$

40. $2a^4 + 4a^3 - 6a^2$ $2a^2(a+3)(a-1)$

41. $-3x^2 - 9x - 6$ $-3(x+2)(x+1)$

42. $8x^2 - 4x - 24$ $4(2x+3)(x-2)$

43. $15x^2 - 57xy - 12y^2$ $3(5x+y)(x-4y)$

44. $30x^2 + 65xy + 10y^2$ $5(6x+y)(x+2y)$

45. $24x^2 - 23xy - 12y^2$ $(8x+3y)(3x-4y)$

46. $14x^2 + 13xy - 12y^2$ $(2x+3y)(7x-4y)$

47. $x^3 + 5x^2 - 6x$ $x(x-1)(x+6)$

48. $3x^2y - 12xy - 63y$ $3y(x+3)(x-7)$

49. $z^2 - 4 + zx - 2x$ $(z-2)(z+x+2)$

50. $x^2 + 2x + 1 - p^2$ $(x+1+p)(x+1-p)$

51. $x^2 + 4x + 4 - 4p^4$ $(x+2+2p^2)(x+2-2p^2)$

52. $y^2 + 3y + 2 + 2x + xy$ $(y+2)(y+1+x)$

53. $x^{2m} + 2x^m - 3$ $(x^m+3)(x^m-1)$

54. $x^{-2} - x^{-1} - 2$ $\left(\frac{1}{x}-2\right)\left(\frac{1}{x}+1\right)$

In Review Exercises 55–58, solve for the indicated variable.

55. $V = \frac{4}{3}\pi r^3$ for r^3 $r^3 = \frac{3V}{4\pi}$

56. $V = \frac{1}{3}\pi r^2 h$ for h $h = \frac{3V}{\pi r^2}$

57. $S = 2wh + 2wl + 2lh$ for h $h = \frac{S-2wl}{2w+2l}$

58. $S = 2wh + 2wl + 2lh$ for l $l = \frac{S-2wh}{2w+2h}$

In Review Exercises 59–64, solve each equation.

59. $4x^2 - 3x = 0$ $0, \frac{3}{4}$

60. $x^2 - 36 = 0$ $6, -6$

61. $12x^2 + 4x - 5 = 0$ $\frac{1}{2}, -\frac{5}{6}$

62. $7y^2 - 37y + 10 = 0$ $\frac{2}{7}, 5$

63. $t^2(15t - 2) = 8t$ $0, -\frac{2}{3}, \frac{4}{5}$

64. $3u^3 = u(19u + 14)$ $-\frac{2}{3}, 7, 0$

65. **Volume of a rectangular solid** The volume, V, of the rectangular solid in Illustration 1 is given by the formula $V = lwh$, where l is its length, w is its width, and h is its height. If the volume is 840 cubic centimeters, the length is 12 centimeters, and the width exceeds the height by 3 centimeters, find the height. 7 cm

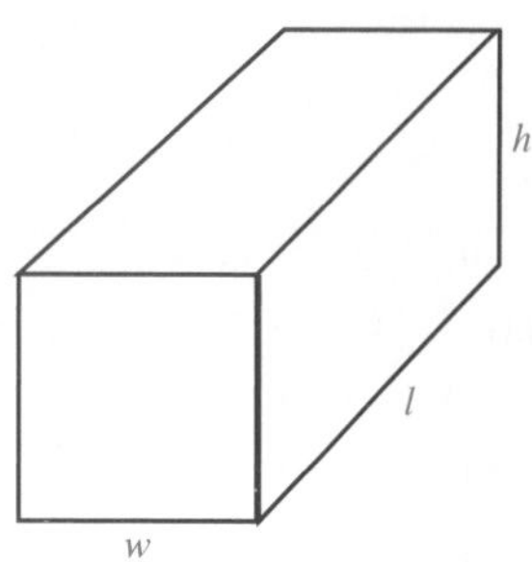

ILLUSTRATION 1

66. **Volume of a pyramid** The volume of the pyramid in Illustration 2 is given by the formula $V = \frac{Bh}{3}$, where B is the area of its base and h is its height. If the volume of the pyramid is 1020 cubic meters, find the dimensions of its rectangular base if one edge of the base is 3 meters longer than the other and the height of the pyramid is 9 meters. 17 m by 20 m

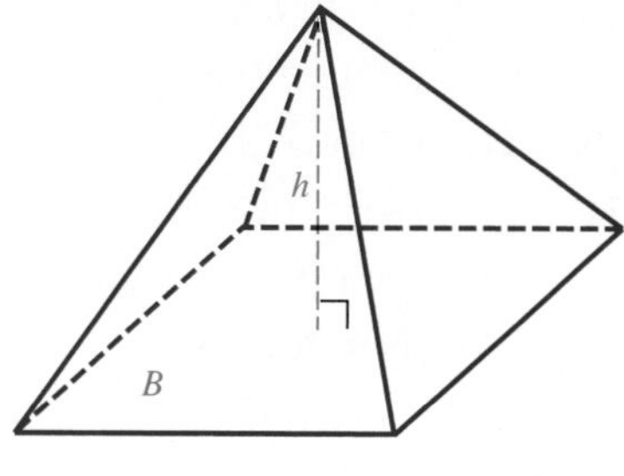

ILLUSTRATION 2

Chapter 4 Test

1. Find the prime factorization of 228. $2^2 \cdot 3 \cdot 19$

2. Which of the following polynomials is prime? $2x + 4$ $3x^2 + 2x$ $3x^2 + 2$ $3x^2 + 2$

In Problems 3–4, factor each polynomial.

3. $3xy^2 + 6x^2y$ $3xy(y + 2x)$

4. $12a^3b^2c - 3a^2b^2c^2 + 6abc^3$ $3abc(4a^2b - abc + 2c^2)$

5. Factor y^n from $x^2y^{n+2} + y^n$. $y^n(x^2y^2 + 1)$

6. Factor b^n from $a^nb^n - ab^{-n}$. $b^n(a^n - ab^{-2n})$

7. Factor y^{-2} from $x^2y^2 + y^{-2}$. $y^{-2}(x^2y^4 + 1)$

8. Factor x^{-3} from $x^3y^3 + y$. $x^{-3}(x^6y^3 + x^3y)$

In Problems 9–12, factor each expression.

9. $(u - v)r + (u - v)s$ $(u - v)(r + s)$

10. $-2x^2y(r + s) + 4xy^2(r + s)$ $-2xy(r + s)(x - 2y)$

11. $ax - xy + ay - y^2$ $(a - y)(x + y)$

12. $ax + ay + bx + by - cx - cy$ $(x + y)(a + b - c)$

In Problems 13–20, factor each polynomial.

13. $x^2 - 49$ $(x + 7)(x - 7)$

14. $2x^2 - 32$ $2(x + 4)(x - 4)$

15. $4y^4 - 64$ $4(y^2 + 4)(y + 2)(y - 2)$

16. $b^3 + 125$ $(b + 5)(b^2 - 5b + 25)$

17. $b^3 - 27$ $(b - 3)(b^2 + 3b + 9)$

18. $3u^3 - 24$ $3(u - 2)(u^2 + 2u + 4)$

19. $9z^2 - 16t^4$ $(3z + 4t^2)(3z - 4t^2)$

20. $8x^3 + 27y^3$ $(2x + 3y)(4x^2 - 6xy + 9y^2)$

In Problems 21–30, factor each trinomial.

21. $x^2 + 8x + 15$ $(x + 5)(x + 3)$

22. $x^2 - 3x - 18$ $(x + 3)(x - 6)$

23. $2a^2 - 5a - 12$ $(2a + 3)(a - 4)$

24. $6b^2 + b - 2$ $(3b + 2)(2b - 1)$

25. $6u^2 + 9u - 6$ $3(u + 2)(2u - 1)$

26. $20r^2 - 15r - 5$ $5(4r + 1)(r - 1)$

27. $6x^2 + xy - y^2$ $(2x + y)(3x - y)$

28. $5x^2 + 3xy - 2y^2$ $(x + y)(5x - 2y)$

29. $x^{2n} + 2x^n + 1$ $(x^n + 1)^2$

30. $2x^{2n} + 3x^n - 2$ $(2x^n - 1)(x^n + 2)$

In Problems 31–34, factor each expression.

31. $x^2 + 6x + 9 - y^2$ $(x + 3 + y)(x + 3 - y)$

32. $x^4 + 3x^2 + 4$ $(x^2 + x + 2)(x^2 - x + 2)$

33. $x^4 - 20x^2 + 64$ $(x + 2)(x - 2)(x + 4)(x - 4)$

34. $x^4 + 2x^3 - 2x - 1$ $(x + 1)^3(x - 1)$

35. Solve $P = L + \frac{s}{f}i$ for i. $i = \frac{f(P - L)}{s}$

36. Solve $r_1r_2 - r_2r = r_1r$ for r. $r. = \frac{r_1r_2}{r_2 + r_1}$

In Problems 37–38, solve each equation.

37. $x^2 - 5x - 6 = 0$ 6, −1

38. $t^2(1 + t) - 2t = 10t$ 0, −4, 3

39. Integer problem The product of two consecutive positive integers is 156. Find their sum. 25

40. Preformed concrete The slab of concrete in Illustration 1 is twice as long as it is wide. Its area plus the area of a 1-foot-wide border is 312 square feet. Find the dimensions of the slab.
11 ft by 22 ft

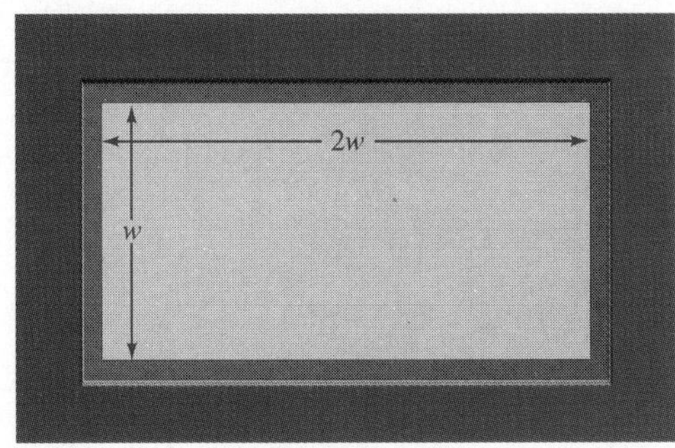

ILLUSTRATION 1

CAREERS & MATHEMATICS

COMPUTER PROGRAMMER

Computers process vast quantities of information rapidly and accurately when they are given programs to follow. Computer programmers write those programs, which logically list the steps the machine must follow to organize data, solve a problem, or do some other task.

Programmers work from descriptions (prepared by systems analysts) of the task that the computer system should perform. The programmer then writes the specific program from the description, by breaking down each step into a series of coded instructions using one of several possible computer languages.

Application programmers are usually oriented toward business, engineering, or science. System programmers maintain the **software** that controls the computer system. Because of their knowledge of operating systems, system programmers often help application programmers determine the source of problems that may occur with their programs.

Qualifications

Most programmers have taken special courses in computer programming or have degrees in computer or information science, mathematics, engineering, or the physical sciences. Graduate degrees are required for some jobs.

Some employers who use computers for business applications do not require college degrees but prefer applicants who have had college courses in data processing.

Employers look for people who can think logically, who are capable of exacting, analytical work, and who demonstrate ingenuity and imagination when solving problems.

Job Outlook

The need for computer programmers will increase as business, government, schools, and scientific organizations develop new applications and require improvements in the software they already use. System programmers will be needed to develop and maintain the complex operating systems required by new, more powerful computer languages and by networking of desktop computers. Job prospects are excellent into the next century for college graduates who have had computer-related courses—especially for those who have a major in computer science or a related field. Graduates of two-year programs in data-processing technologies also have good prospects, primarily in business applications.

Example Application The polynomial

$$3x^4 + 2x^3 + 5x^2 + 7x + 1$$

can be written as

$$3 \cdot x \cdot x \cdot x \cdot x + 2 \cdot x \cdot x \cdot x + 5 \cdot x \cdot x + 7 \cdot x + 1$$

to illustrate that it involves 10 multiplications and 4 additions. Multiplications are more time-consuming on a computer than are additions. Rewrite the polynomial to require fewer multiplications.

Solution Factor the common factor of x from the first four terms of the given polynomial as follows.

$$3x^4 + 2x^3 + 5x^2 + 7x + 1$$
$$x(3x^3 + 2x^2 + 5x + 7) + 1$$

Now factor the common x from the first three terms of the polynomial appearing within the parentheses.

$$x[x(3x^2 + 2x + 5) + 7] + 1$$

Again, factor an x from the first two terms within the set of parentheses to get

$$x\{x[x(3x + 2) + 5] + 7\} + 1$$

To emphasize the number of multiplications, we rewrite the previous expression as

$$x \cdot \{x \cdot [x \cdot (3 \cdot x + 2) + 5] + 7\} + 1$$

Each colored dot represents one multiplication. There are now only four multiplications, although there are still four additions. If the polynomial were to be evaluated for many different values of x, the savings in computer time afforded by using the revised form of the polynomial would be substantial.

EXERCISES

1. Evaluate the polynomial of the example at $x = -1$. Do this twice—once by direct substitution into the original polynomial, and then by substitution into the revised form.
2. Write $5x^7 + 3x^4 + 9x + 2$ in revised form.
3. Evaluate the polynomial of Exercise 2 at $x = 1$, by using each method.
4. Evaluate the polynomial of Exercise 2 at $x = -2$, by using each method.

(*Answers:* **1.** 0 **2.** $x \cdot \{x \cdot [x \cdot x \cdot (5 \cdot x \cdot x \cdot x + 3)] + 9\} + 2$ **3.** 19 **4.** -608)

5 Rational Expressions

Algebraic fractions such as

$$\frac{x^2 + 2}{x - 3} \quad \text{and} \quad \frac{a^3 + 2a^2 + 7}{2a^2 - 5a + 4}$$

that indicate the quotient of two polynomials are called **rational expressions.** Since division by 0 is undefined, the value of any polynomial occurring in the denominator of a rational expression cannot be 0. For example, $x \neq 7$ in the fraction

$$\frac{3x}{x - 7}$$

because if $x = 7$, the denominator would be 0.

In the fraction

$$\frac{5m + n}{8m + 16}$$

$m \neq -2$, because the denominator would be 0: $8(-2) + 16 = 0$.

The factoring skills we learned in the previous chapter will be very useful as we develop the properties of rational expressions.

5.1 Simplifying Rational Expressions

■ Simplifying Fractions ■ Simplifying Fractions by Factoring Out −1

We use the same rules to simplify, multiply, divide, add, and subtract rational expressions as we use for arithmetic fractions.

Note: We have avoided using the word *reduce.* To reduce means "to make smaller," a misleading concept. From now on, we will use the word *simplify.*

■ Simplifying Fractions

To *simplify a fraction* means to reduce it to lowest terms. To do so, we use Property 3 of fractions, first introduced in Section 1.4. This property enables us to divide out all factors that are common to both the numerator and the denominator of a fraction.

Property 3 of Fractions If $b \neq 0$ and $k \neq 0$, then

$$\frac{ak}{bk} = \frac{a}{b}$$

EXAMPLE 1 Simplify the rational expressions **a.** $\dfrac{10k}{25k^2}$ and **b.** $\dfrac{-8y^3z^5}{6y^4z^3}$.

Solution We factor each numerator and denominator and use Property 3 of fractions to divide out all common factors:

a.
$$\frac{10k}{25k^2} = \frac{5 \cdot 2 \cdot k}{5 \cdot 5 \cdot k \cdot k} = \frac{\overset{1}{\cancel{5}} \cdot 2 \cdot \overset{1}{\cancel{k}}}{\underset{1}{\cancel{5}} \cdot 5 \cdot \underset{1}{\cancel{k}} \cdot k} = \frac{2}{5k}$$

b.
$$\frac{-8y^3z^5}{6y^4z^3} = \frac{-2 \cdot 4 \cdot y \cdot y \cdot y \cdot z \cdot z \cdot z \cdot z \cdot z}{2 \cdot 3 \cdot y \cdot y \cdot y \cdot y \cdot z \cdot z \cdot z} = \frac{-\overset{1}{\cancel{2}} \cdot 4 \cdot \cancel{y} \cdot \cancel{y} \cdot \cancel{y} \cdot \cancel{z} \cdot \cancel{z} \cdot \cancel{z} \cdot z \cdot z}{\underset{1}{\cancel{2}} \cdot 3 \cdot \cancel{y} \cdot \cancel{y} \cdot \cancel{y} \cdot y \cdot \cancel{z} \cdot \cancel{z} \cdot \cancel{z}} = -\frac{4z^2}{3y}$$ ■

The fractions in Example 1 can be simplified by using the rules of exponents:

$$\frac{10k}{25k^2} = \frac{5 \cdot 2}{5 \cdot 5}k^{1-2} = \frac{2}{5} \cdot k^{-1} = \frac{2}{5k}$$

$$\frac{-8y^3z^5}{6y^4z^3} = \frac{-2 \cdot 4}{2 \cdot 3}y^{3-4}z^{5-3} = \frac{-4}{3} \cdot y^{-1}z^2 = -\frac{4}{3} \cdot \frac{1}{y} \cdot \frac{z^2}{1} = -\frac{4z^2}{3y}$$

EXAMPLE 2 Simplify the rational expression $\dfrac{x^2 - 16}{x + 4}$.

Solution We factor the numerator and divide out all common factors:

$$\frac{x^2 - 16}{x + 4} = \frac{\overset{1}{\cancel{(x + 4)}}(x - 4)}{1\underset{1}{\cancel{(x + 4)}}} = \frac{x - 4}{1} = x - 4$$ ■

EXAMPLE 3 Simplify the fraction $\dfrac{2x^2 + 11x + 12}{3x^2 + 11x - 4}$.

Solution We factor the numerator and denominator and divide out all common factors:

$$\frac{2x^2 + 11x + 12}{3x^2 + 11x - 4} = \frac{(2x + 3)\overset{1}{\cancel{(x + 4)}}}{(3x - 1)\underset{1}{\cancel{(x + 4)}}}$$

$$= \frac{2x + 3}{3x - 1}$$

WARNING! Do not divide out the x's in the fraction $\frac{2x+3}{3x-1}$. The x in the numerator is a factor of the first term only. It is not a factor of the entire numerator. Likewise, the x in the denominator is not a factor of the entire denominator. ■

Simplifying Fractions by Factoring Out −1

To simplify $\frac{b-a}{a-b}$, we factor -1 from the numerator and divide out any factors common to both the numerator and the denominator:

$$\frac{b - a}{a - b} = \frac{-a + b}{a - b}$$

$$= \frac{-\overset{1}{\cancel{(a - b)}}}{\underset{1}{\cancel{(a - b)}}}$$

$$= \frac{-1}{1}$$

$$= -1$$

In general, we have the following theorem.

Theorem The quotient of any nonzero quantity and its negative is -1.

EXAMPLE 4 Simplify the fraction $\dfrac{3x^2 - 10xy - 8y^2}{4y^2 - xy}$.

Solution We factor the numerator and denominator and apply the previous theorem:

$$\frac{3x^2 - 10xy - 8y^2}{4y^2 - xy} = \frac{(3x + 2y)\overset{-1}{\cancel{(x - 4y)}}}{y\underset{1}{\cancel{(4y - x)}}}$$

Because $x - 4y$ and $4y - x$ are negatives, their quotient is -1.

$$= \frac{-(3x + 2y)}{y}$$

$$= \frac{-3x - 2y}{y}$$

■

Many fractions we will encounter will already be in simplified form. For example, to attempt to simplify the fraction

$$\frac{x^2 + xa + 2x + 2a}{x^2 + x - 6}$$

we factor the numerator and denominator and divide out any common factors that exist:

$$\frac{x^2 + xa + 2x + 2a}{x^2 + x - 6} = \frac{x(x + a) + 2(x + a)}{(x - 2)(x + 3)} = \frac{(x + a)(x + 2)}{(x - 2)(x + 3)}$$

Since there are no common factors in the numerator and denominator, the fraction is already in lowest terms. It cannot be simplified.

EXAMPLE 5 Simplify the fraction $\dfrac{(x^2 + 2x)(x^2 + 2x - 3)}{(x^2 + x - 2)(x^2 + 3x)}$.

Solution We factor the polynomials in the numerator and denominator and divide out all common factors:

$$\frac{(x^2 + 2x)(x^2 + 2x - 3)}{(x^2 + x - 2)(x^2 + 3x)} = \frac{\overset{1}{\cancel{x}}\overset{1}{\cancel{(x+2)}}\overset{1}{\cancel{(x+3)}}\overset{1}{\cancel{(x-1)}}}{\underset{1}{\cancel{(x+2)}}\underset{1}{\cancel{(x-1)}}\underset{1}{\cancel{x}}\underset{1}{\cancel{(x+3)}}}$$

$$= 1$$

■

WARNING! Remember that only factors that are common to the entire numerator and the entire denominator can be divided out.

Terms common to both the numerator and denominator cannot be divided out. For example, consider this correct simplification:

$$\frac{3 + 7}{3} = \frac{10}{3}$$

It would be incorrect to divide out the common term of 3 in the fraction $\frac{3+7}{3}$. Doing so gives an incorrect answer:

$$\frac{3 + 7}{3} = \frac{\overset{1}{\cancel{3}} + 7}{\underset{1}{\cancel{3}}} = \frac{1 + 7}{1} = 8$$

The 3's in the fraction

$$\frac{5 + 3(2)}{3(4)}$$

cannot be divided out. The 3 in the numerator is a factor of the second term only. To be divided out, it must be a factor of the entire numerator.

It is not correct to divide out the y in the fraction

$$\frac{x^2y + 6x}{y}$$

The y is not a factor of the entire numerator.

ORALS *Simplify each fraction.*

1. $\frac{4}{6}$ $\frac{2}{3}$
2. $\frac{10}{15}$ $\frac{2}{3}$
3. $-\frac{25}{30}$ $-\frac{5}{6}$
4. $-\frac{22}{55}$ $-\frac{2}{5}$
5. $\frac{x^2}{xy}$ $\frac{x}{y}$
6. $\frac{2x - 4}{x - 2}$ 2
7. $\frac{x - 2}{2 - x}$ -1
8. $\frac{x^2 - 1}{x + 1}$ $x - 1$

EXERCISE 5.1

In Exercises 1–64, simplify each fraction when possible.

1. $\frac{12}{18}$ $\frac{2}{3}$
2. $\frac{25}{55}$ $\frac{5}{11}$
3. $-\frac{112}{36}$ $-\frac{28}{9}$
4. $-\frac{49}{21}$ $-\frac{7}{3}$
5. $\frac{288}{312}$ $\frac{12}{13}$
6. $\frac{144}{72}$ 2
7. $-\frac{244}{74}$ $-\frac{122}{37}$
8. $-\frac{512}{236}$ $-\frac{128}{59}$
9. $\frac{12x^3}{3x}$ $4x^2$
10. $-\frac{15a^2}{25a^3}$ $-\frac{3}{5a}$
11. $\frac{-24x^3y^4}{18x^4y^3}$ $-\frac{4y}{3x}$
12. $\frac{15a^5b^4}{21b^3c^2}$ $\frac{5a^5b}{7c^2}$
13. $\frac{(3x^3)^2}{9x^4}$ x^2
14. $\frac{8(x^2y^3)^3}{2(xy^2)^2}$ $4x^4y^5$
15. $-\frac{11x(x - y)}{22(x - y)}$ $-\frac{x}{2}$
16. $\frac{x(x - 2)^2}{(x - 2)^3}$ $\frac{x}{x - 2}$
17. $\frac{9y^2(y - z)}{21y(y - z)^2}$ $\frac{3y}{7(y - z)}$
18. $\frac{-3ab^2(a - b)}{9ab(b - a)}$ $\frac{b}{3}$
19. $\frac{(a - b)(b - c)(c - d)}{(c - d)(b - c)(a - b)}$ 1
20. $\frac{(p + q)(p - r)(r + s)}{(r - p)(r + s)(p + q)}$ -1
21. $\frac{x + y}{x^2 - y^2}$ $\frac{1}{x - y}$
22. $\frac{x - y}{x^2 - y^2}$ $\frac{1}{x + y}$
23. $\frac{5x - 10}{x^2 - 4x + 4}$ $\frac{5}{x - 2}$
24. $\frac{y - xy}{xy - x}$ lowest terms

25. $\dfrac{12 - 3x^2}{x^2 - x - 2}$ $\dfrac{-3(x+2)}{x+1}$

26. $\dfrac{x^2 + 2x - 15}{x^2 - 25}$ $\dfrac{x-3}{x-5}$

27. $\dfrac{3x + 6y}{x + 2y}$ 3

28. $\dfrac{x^2 + y^2}{x + y}$ lowest terms

29. $\dfrac{x^3 + 8}{x^2 - 2x + 4}$ $x + 2$

30. $\dfrac{x^2 + 3x + 9}{x^3 - 27}$ $\dfrac{1}{x-3}$

31. $\dfrac{x^2 + 2x + 1}{x^2 + 4x + 3}$ $\dfrac{x+1}{x+3}$

32. $\dfrac{6x^2 + x - 2}{8x^2 + 2x - 3}$ $\dfrac{3x+2}{4x+3}$

33. $\dfrac{3m - 6n}{3n - 6m}$ $\dfrac{m-2n}{n-2m}$

34. $\dfrac{ax + by + ay + bx}{a^2 - b^2}$ $\dfrac{x+y}{a-b}$

35. $\dfrac{4x^2 + 24x + 32}{16x^2 + 8x - 48}$ $\dfrac{x+4}{2(2x-3)}$

36. $\dfrac{a^2 - 4}{a^3 - 8}$ $\dfrac{a+2}{a^2+2a+4}$

37. $\dfrac{3x^2 - 3y^2}{x^2 + 2y + 2x + yx}$ $\dfrac{3(x-y)}{x+2}$

38. $\dfrac{x^2 + x - 30}{x^2 - x - 20}$ $\dfrac{x+6}{x+4}$

39. $\dfrac{4x^2 + 8x + 3}{6 + x - 2x^2}$ $\dfrac{2x+1}{2-x}$

40. $\dfrac{6x^2 + 13x + 6}{6 - 5x - 6x^2}$ $-\dfrac{3x+2}{3x-2}$

41. $\dfrac{a^3 + 27}{4a^2 - 36}$ $\dfrac{a^2-3a+9}{4(a-3)}$

42. $\dfrac{a - b}{b^2 - a^2}$ $\dfrac{-1}{a+b}$

43. $\dfrac{2x^2 - 3x - 9}{2x^2 + 3x - 9}$ lowest terms

44. $\dfrac{6x^2 - 7x - 5}{2x^2 + 5x + 2}$ $\dfrac{3x-5}{x+2}$

45. $\dfrac{(m+n)^3}{m^2 + 2mn + n^2}$ $m + n$

46. $\dfrac{x^3 - 27}{3x^2 - 8x - 3}$ $\dfrac{x^2+3x+9}{3x+1}$

47. $\dfrac{m^3 - mn^2}{mn^2 + m^2n - 2m^3}$ $-\dfrac{m+n}{2m+n}$

48. $\dfrac{p^3 + p^2q - 2pq^2}{pq^2 + p^2q - 2p^3}$ $-\dfrac{p+2q}{q+2p}$

49. $\dfrac{x^4 - y^4}{(x^2 + 2xy + y^2)(x^2 + y^2)}$ $\dfrac{x-y}{x+y}$

50. $\dfrac{-4x - 4 + 3x^2}{4x^2 - 2 - 7x}$ $\dfrac{3x+2}{4x+1}$

51. $\dfrac{4a^2 - 9b^2}{2a^2 - ab - 6b^2}$ $\dfrac{2a-3b}{a-2b}$

52. $\dfrac{x^2 + 2xy}{x + 2y + x^2 - 4y^2}$ $\dfrac{x}{1+x-2y}$

53. $\dfrac{x - y}{x^3 - y^3 - x + y}$ $\dfrac{1}{x^2+xy+y^2-1}$

54. $\dfrac{2x^2 + 2x - 12}{x^3 + 3x^2 - 4x - 12}$ $\dfrac{2}{x+2}$

55. $\dfrac{x^6 - y^6}{x^4 + x^2y^2 + y^4}$ $x^2 - y^2$

56. $\dfrac{6xy - 4x - 9y + 6}{6y^2 - 13y + 6}$ $\dfrac{2x-3}{2y-3}$

57. $\dfrac{x^4 - 8x^2 - 9}{x^3 + 3x^2 + x + 3}$ $x - 3$

58. $\dfrac{x^4 - 5x^2 + 4}{x^2 + 3x + 2}$ $x^2 - 3x + 2$

59. $\dfrac{(x^2 - 1)(x + 1)}{(x^2 - 2x + 1)^2}$ $\dfrac{(x+1)^2}{(x-1)^3}$

60. $\dfrac{(x^2 + 2x + 1)(x^2 - 2x + 1)}{(x^2 - 1)^2}$ 1

61. $\dfrac{(2x^2 + 3xy + y^2)(3a + b)}{(x + y)(2xy + 2bx + y^2 + by)}$ $\dfrac{3a+b}{y+b}$

62. $\dfrac{(x - 1)(6ax + 9x + 4a + 6)}{(3x + 2)(2ax - 2a + 3x - 3)}$ 1

63. $\dfrac{(a^2 + 2a + ab + 2b)(c^2 + c - dc - d)}{(a^2 - 4)(ad + bd - ac - bc)}$ $\dfrac{c+1}{2-a}$

64. $\dfrac{(mp + mq - np - nq)(pm + pn - qm - qn)}{(qm - qn - pm + pn)(pm + pn + qm + qn)}$ -1

Writing Exercises *Write a paragraph using your own words.*

1. Explain how to simplify a rational expression.
2. Explain how to recognize that a rational expression is in its simplest form.

Something to Think About

1. A student compared his answer, $\frac{a-3b}{2b-a}$, with the answer, $\frac{3b-a}{a-2b}$, in the back of the text. Is the student's answer correct? yes
2. A student shows this work:
$$\frac{3x^2+6}{3y} = \frac{\overset{1}{\cancel{3}}x^2 + \overset{2}{\cancel{6}}}{\underset{1}{\cancel{3}}y} = \frac{x^2+2}{y}$$
Is the student's work correct? yes
3. In which parts can you divide out the 4's?
   a. $\frac{4x}{4y}$ b. $\frac{4x}{x+4}$
   c. $\frac{4+x}{4+y}$ d. $\frac{4x}{4+4y}$ a, d
4. In which parts can you divide out the 3's?
   a. $\frac{3x+3y}{3z}$ b. $\frac{3(x+y)}{3x+y}$
   c. $\frac{x+3}{3y}$ d. $\frac{3x+3y}{3a-3b}$ a, d

Review Exercises *Factor each expression.*

1. $3x^2 - 9x$ $3x(x-3)$
2. $6t^2 - 5t - 6$ $(3t+2)(2t-3)$
3. $27x^6 + 64y^3$ $(3x^2+4y)(9x^4-12x^2y+16y^2)$
4. $x^2 + ax + 2x + 2a$ $(x+a)(x+2)$

5.2 Multiplying and Dividing Rational Expressions

■ Multiplying Rational Expressions ■ Dividing Rational Expressions

Property 4 of fractions, first introduced in Section 1.4, gave the rules for multiplying and dividing fractions.

Property 4 of Fractions If no denominators are 0, then

$$\frac{a}{b}\cdot\frac{c}{d} = \frac{a\cdot c}{b\cdot d} = \frac{ac}{bd} \quad \text{and} \quad \frac{a}{b}\div\frac{c}{d} = \frac{a}{b}\cdot\frac{d}{c} = \frac{ad}{bc}$$

In words, the rule for multiplying fractions says *To multiply two fractions, we multiply the numerators and multiply the denominators.* Thus,

$$\frac{3}{5}\cdot\frac{2}{7}=\frac{3\cdot 2}{5\cdot 7}=\frac{6}{35}$$

$$\frac{4}{7}\cdot\frac{5}{8}=\frac{4\cdot 5}{7\cdot 8}=\frac{\overset{1}{\cancel{2}}\cdot\overset{1}{\cancel{2}}\cdot 5}{7\cdot\underset{1}{\cancel{2}}\cdot\underset{1}{\cancel{2}}\cdot 2}=\frac{5}{14}$$

Multiplying Rational Expressions

The same rule applies to algebraic fractions.

$$\frac{x^2y}{t}\cdot\frac{xy^3}{t^3}=\frac{x^2y\cdot xy^3}{tt^3}=\frac{x^3y^4}{t^4}\qquad(t\neq 0)$$

EXAMPLE 1 Find the product of $\dfrac{x^2-6x+9}{x}$ and $\dfrac{x^2}{x-3}$.

Solution We multiply the numerators and multiply the denominators. Then we simplify the resulting fraction.

Another Approach: Some teachers prefer to factor and divide out common factors before multiplying. We prefer multiplying first, because then the problems become just like the problems in Section 5.1.

$$\frac{x^2-6x+9}{x}\cdot\frac{x^2}{x-3}=\frac{(x^2-6x+9)(x^2)}{x(x-3)}$$ Multiply the numerators and multiply the denominators.

$$=\frac{(x-3)(x-3)xx}{x(x-3)}$$ Factor the numerator.

$$=\frac{\overset{1}{\cancel{(x-3)}}(x-3)\overset{1}{\cancel{x}}x}{\underset{1}{\cancel{x}}\underset{1}{\cancel{(x-3)}}}$$ Divide out the common factors.

$$=x(x-3)$$ ■

EXAMPLE 2 Find the product of $\dfrac{x^2-x-6}{x^2-4}$ and $\dfrac{x^2+x-6}{x^2-9}$.

Solution We multiply the fractions, factor each polynomial, and simplify.

$$\frac{x^2-x-6}{x^2-4}\cdot\frac{x^2+x-6}{x^2-9}=\frac{(x^2-x-6)(x^2+x-6)}{(x^2-4)(x^2-9)}$$ Multiply the numerators and multiply the denominators.

$$= \frac{(x-3)(x+2)(x+3)(x-2)}{(x+2)(x-2)(x+3)(x-3)}$$ Factor the polynomials.

$$= \frac{\overset{1}{\cancel{(x-3)}}\overset{1}{\cancel{(x+2)}}\overset{1}{\cancel{(x+3)}}\overset{1}{\cancel{(x-2)}}}{\underset{1}{\cancel{(x+2)}}\underset{1}{\cancel{(x-2)}}\underset{1}{\cancel{(x+3)}}\underset{1}{\cancel{(x-3)}}}$$ Divide out the common factors.

$$= 1$$

■

WARNING! Note that when all factors divide out, the result is 1, not 0.

EXAMPLE 3 Multiply: $\frac{6x^2 + 5x - 4}{2x^2 + 5x + 3} \cdot \frac{8x^2 + 6x - 9}{12x^2 + 7x - 12}$.

Solution We multiply the fractions, factor each polynomial, and simplify.

$$\frac{6x^2 + 5x - 4}{2x^2 + 5x + 3} \cdot \frac{8x^2 + 6x - 9}{12x^2 + 7x - 12} = \frac{(6x^2 + 5x - 4)(8x^2 + 6x - 9)}{(2x^2 + 5x + 3)(12x^2 + 7x - 12)}$$ Multiply the numerators and multiply the denominators.

$$= \frac{(3x+4)(2x-1)(4x-3)(2x+3)}{(2x+3)(x+1)(3x+4)(4x-3)}$$ Factor the polynomials.

$$= \frac{\overset{1}{\cancel{(3x+4)}}(2x-1)\overset{1}{\cancel{(4x-3)}}\overset{1}{\cancel{(2x+3)}}}{\underset{1}{\cancel{(2x+3)}}(x+1)\underset{1}{\cancel{(3x+4)}}\underset{1}{\cancel{(4x-3)}}}$$ Divide out the common factors.

$$= \frac{2x-1}{x+1}$$

■

EXAMPLE 4 Simplify: $(2x - x^2) \cdot \frac{x}{x^2 - 5x + 6}$.

Solution We write the first factor as a fraction with a denominator of 1 and proceed as follows:

$$(2x - x^2) \cdot \frac{x}{x^2 - 5x + 6} = \frac{2x - x^2}{1} \cdot \frac{x}{x^2 - 5x + 6}$$ Write $2x - x^2$ as $\frac{2x - x^2}{1}$.

$$= \frac{(2x - x^2)x}{1(x^2 - 5x + 6)}$$ Multiply the fractions.

$$= \frac{x\overset{-1}{\cancel{(2-x)}}x}{\underset{1}{\cancel{(x-2)}}(x-3)}$$ Factor and note that the quotient of any nonzero quantity and its negative is -1.

$$= \frac{-x^2}{x-3}$$

Because $\frac{-a}{b} = -\frac{a}{b} = \frac{a}{-b}$, the negative sign can be placed in front of the fraction or in the denominator. For this reason, the answer can be written as

$$-\frac{x^2}{x-3}, \qquad \frac{x^2}{-(x-3)}, \qquad \text{or} \qquad \frac{x^2}{3-x}$$ ■

In Examples 1–4, we would obtain the same answers if we had factored first and divided out the common factors before we multiplied.

Dividing Rational Expressions

We are familiar with the following rule to divide arithmetic fractions:

$$\frac{a}{b} \div \frac{c}{d} = \frac{a}{b} \cdot \frac{d}{c}$$

In words, we say *To divide two fractions, we invert the divisor and multiply.* Thus,

$$\frac{3}{5} \div \frac{2}{7} = \frac{3}{5} \cdot \frac{7}{2} = \frac{3 \cdot 7}{5 \cdot 2} = \frac{21}{10}$$

$$\frac{4}{7} \div \frac{2}{21} = \frac{4}{7} \cdot \frac{21}{2} = \frac{4 \cdot 21}{7 \cdot 2} = \frac{\overset{1}{\cancel{2}} \cdot 2 \cdot 3 \cdot \overset{1}{\cancel{7}}}{\underset{1}{\cancel{7}} \cdot \underset{1}{\cancel{2}}} = 6$$

The same rule applies to algebraic fractions.

$$\frac{x^{-2}}{y^3z^{-2}} \div \frac{x^2}{yz^{-2}} = \frac{x^{-2}}{y^3z^{-2}} \cdot \frac{yz^{-2}}{x^2}$$ Invert the divisor and multiply.

$$= \frac{x^{-2}yz^{-2}}{x^2y^3z^{-2}}$$ Multiply the numerators and the denominators.

$$= x^{-2-2}y^{1-3}z^{-2-(-2)}$$ To divide exponential expressions with the same base, keep the base and subtract the exponents.

$$= x^{-4}y^{-2}z^0$$

$$= \frac{1}{x^4y^2}$$

EXAMPLE 5 Divide: $\frac{x^3+8}{x+1} \div \frac{x^2-2x+4}{2x^2-2}$.

Solution Using the rule for division of fractions, we invert the divisor and multiply.

$$\frac{x^3+8}{x+1} \div \frac{x^2-2x+4}{2x^2-2} = \frac{x^3+8}{x+1} \cdot \frac{2x^2-2}{x^2-2x+4}$$

$$= \frac{(x^3 + 8)(2x^2 - 2)}{(x + 1)(x^2 - 2x + 4)}$$

$$= \frac{(x + 2)\overset{1}{\cancel{(x^2 - 2x + 4)}}2\overset{1}{\cancel{(x + 1)}}(x - 1)}{\underset{1}{\cancel{(x + 1)}}\underset{1}{\cancel{(x^2 - 2x + 4)}}}$$

$2x^2 - 2 = 2(x^2 - 1)$
$= 2(x + 1)(x - 1)$

$$= 2(x + 2)(x - 1)$$ ■

EXAMPLE 6 Divide: $\frac{x^2 - 4}{x - 1} \div (x - 2)$.

Solution We proceed as follows:

$$\frac{x^2 - 4}{x - 1} \div (x - 2) = \frac{x^2 - 4}{x - 1} \div \frac{x - 2}{1}$$ Write $x - 2$ as a fraction with a denominator of 1.

$$= \frac{x^2 - 4}{x - 1} \cdot \frac{1}{x - 2}$$ Invert the divisor and multiply.

$$= \frac{x^2 - 4}{(x - 1)(x - 2)}$$ Multiply the numerators and the denominators.

$$= \frac{(x + 2)\overset{1}{\cancel{(x - 2)}}}{(x - 1)\underset{1}{\cancel{(x - 2)}}}$$ Factor the numerator.

$$= \frac{x + 2}{x - 1}$$ Divide out the common factors. ■

EXAMPLE 7 Simplify: $\frac{x^2 + 2x - 3}{6x^2 + 5x + 1} \div \frac{2x^2 - 2}{2x^2 - 5x - 3} \cdot \frac{6x^2 + 4x - 2}{x^2 - 2x - 3}$.

Solution We change the division to a multiplication. Since multiplications and divisions are performed from left to right, only the middle fraction should be inverted. Finally, we multiply the fractions, factor each polynomial, and divide out the common factors.

$$\frac{x^2 + 2x - 3}{6x^2 + 5x + 1} \div \frac{2x^2 - 2}{2x^2 - 5x - 3} \cdot \frac{6x^2 + 4x - 2}{x^2 - 2x - 3} = \frac{x^2 + 2x - 3}{6x^2 + 5x + 1} \cdot \frac{2x^2 - 5x - 3}{2x^2 - 2} \cdot \frac{6x^2 + 4x - 2}{x^2 - 2x - 3}$$

$$= \frac{(x^2 + 2x - 3)(2x^2 - 5x - 3)(6x^2 + 4x - 2)}{(6x^2 + 5x + 1)(2x^2 - 2)(x^2 - 2x - 3)}$$

$$= \frac{(x + 3)\overset{1}{\cancel{(x - 1)}}\overset{1}{\cancel{(2x + 1)}}\overset{1}{\cancel{(x - 3)}}\overset{1}{\cancel{2}}(3x - 1)\overset{1}{\cancel{(x + 1)}}}{(3x + 1)\underset{1}{\cancel{(2x + 1)}}\underset{1}{\cancel{2}}(x + 1)\underset{1}{\cancel{(x - 1)}}\underset{1}{\cancel{(x - 3)}}\underset{1}{\cancel{(x + 1)}}}$$

$$= \frac{(x + 3)(3x - 1)}{(3x + 1)(x + 1)}$$ ■

ORALS *Multiply the fractions and simplify, if possible.*

1. $\frac{3}{2} \cdot \frac{3}{4}$ $\frac{9}{8}$

2. $\frac{3x}{7} \cdot \frac{7}{6x}$ $\frac{1}{2}$

3. $\frac{x-2}{y} \cdot \frac{y}{x+2}$ $\frac{x-2}{x+2}$

Divide the fractions and simplify, if possible.

4. $\frac{3}{4} \div \frac{4}{3}$ $\frac{9}{16}$

5. $\frac{5a}{b} \div \frac{a}{b}$ 5

6. $\frac{x^2y}{ab} \div \frac{2x^2}{ba}$ $\frac{y}{2}$

EXERCISE 5.2

In Exercises 1–54, do the operations and simplify.

1. $\frac{3}{4} \cdot \frac{5}{3} \cdot \frac{8}{7}$ $\frac{10}{7}$

2. $-\frac{5}{6} \cdot \frac{3}{7} \cdot \frac{14}{25}$ $-\frac{1}{5}$

3. $-\frac{6}{11} \div \frac{36}{55}$ $-\frac{5}{6}$

4. $\frac{17}{12} \div \frac{34}{3}$ $\frac{1}{8}$

5. $\frac{x^2y^2}{cd} \cdot \frac{c^{-2}d^2}{x}$ $\frac{xy^2d}{c^3}$

6. $\frac{a^{-2}b^2}{x^{-1}y} \cdot \frac{a^4b^4}{x^2y^3}$ $\frac{a^2b^6}{xy^4}$

7. $\frac{-x^2y^{-2}}{x^{-1}y^{-3}} \div \frac{x^{-3}y^2}{x^4y^{-1}}$ $-\frac{x^{10}}{y^2}$

8. $\frac{(a^3)^2}{b^{-1}} \div \frac{(a^3)^{-2}}{b^{-1}}$ a^{12}

9. $\frac{x^2+2x+1}{x} \cdot \frac{x^2-x}{x^2-1}$ $x+1$

10. $\frac{a+6}{a^2-16} \cdot \frac{3a-12}{3a+18}$ $\frac{1}{a+4}$

11. $\frac{2x^2-x-3}{x^2-1} \cdot \frac{x^2+x-2}{2x^2+x-6}$ 1

12. $\frac{9x^2+3x-20}{3x^2-7x+4} \cdot \frac{3x^2-5x+2}{9x^2+18x+5}$ $\frac{3x-2}{3x+1}$

13. $\frac{x^2-16}{x^2-25} \div \frac{x+4}{x-5}$ $\frac{x-4}{x+5}$

14. $\frac{a^2-9}{a^2-49} \div \frac{a+3}{a+7}$ $\frac{a-3}{a-7}$

15. $\frac{a^2+2a-35}{12x} \div \frac{ax-3x}{a^2+4a-21}$ $\frac{(a+7)^2(a-5)}{12x^2}$

16. $\frac{x^2-4}{2b-bx} \div \frac{x^2+4x+4}{2b+bx}$ -1

17. $\frac{3t^2-t-2}{6t^2-5t-6} \cdot \frac{4t^2-9}{2t^2+5t+3}$ $\frac{t-1}{t+1}$

18. $\frac{2p^2-5p-3}{p^2-9} \cdot \frac{2p^2+5p-3}{2p^2+5p+2}$ $\frac{2p-1}{p+2}$

19. $\frac{3n^2+5n-2}{12n^2-13n+3} \div \frac{n^2+3n+2}{4n^2+5n-6}$ $\frac{n+2}{n+1}$

20. $\frac{8y^2-14y-15}{6y^2-11y-10} \div \frac{4y^2-9y-9}{3y^2-7y-6}$ 1

21. $(x+1) \cdot \frac{1}{x^2+2x+1}$ $\frac{1}{x+1}$

22. $\frac{x^2-4}{x} \div (x+2)$ $\frac{x-2}{x}$

23. $(x^2-x-2) \cdot \frac{x^2+3x+2}{x^2-4}$ $(x+1)^2$

24. $(2x^2-9x-5) \cdot \frac{x}{2x^2+x}$ $x-5$

25. $(2x^2-15x+25) \div \frac{2x^2-3x-5}{x+1}$ $x-5$

26. $(x^2-6x+9) \div \frac{x^2-9}{x+3}$ $x-3$

27. $\frac{x^3+y^3}{x^3-y^3} \div \frac{x^2-xy+y^2}{x^2+xy+y^2}$ $\frac{x+y}{x-y}$

28. $\frac{x^2-6x+9}{4-x^2} \div \frac{x^2-9}{x^2-8x+12}$ $\frac{-(x-3)(x-6)}{(x+2)(x+3)}$

29. $\dfrac{m^2 - n^2}{2x^2 + 3x - 2} \cdot \dfrac{2x^2 + 5x - 3}{n^2 - m^2}$ $\quad -\dfrac{x+3}{x+2}$

30. $\dfrac{x^2 - y^2}{2x^2 + 2xy + x + y} \cdot \dfrac{2x^2 - 5x - 3}{yx - 3y - x^2 + 3x}$ $\quad -1$

31. $\dfrac{ax + ay + bx + by}{x^3 - 27} \cdot \dfrac{x^2 + 3x + 9}{xc + xd + yc + yd}$ $\quad \dfrac{a+b}{(x-3)(c+d)}$

32. $\dfrac{x^2 + 3x + yx + 3y}{x^2 - 9} \cdot \dfrac{x-3}{x+3}$ $\quad \dfrac{x+y}{x+3}$

33. $\dfrac{x^2 - x - 6}{x^2 - 4} \cdot \dfrac{x^2 - x - 2}{9 - x^2}$ $\quad -\dfrac{x+1}{x+3}$

34. $\dfrac{2x^2 - 7x - 4}{20 - x - x^2} \div \dfrac{2x^2 - 9x - 5}{x^2 - 25}$ $\quad -1$

35. $\dfrac{2x^2 + 3xy + y^2}{y^2 - x^2} \div \dfrac{6x^2 + 5xy + y^2}{2x^2 - xy - y^2}$ $\quad -\dfrac{2x+y}{3x+y}$

36. $\dfrac{p^3 - q^3}{q^2 - p^2} \cdot \dfrac{q^2 + pq}{p^3 + p^2q + pq^2}$ $\quad -\dfrac{q}{p}$

37. $\dfrac{3x^2y^2}{6x^3y} \cdot \dfrac{-4x^7y^{-2}}{18x^{-2}y} \div \dfrac{36x}{18y^{-2}}$ $\quad -\dfrac{x^7}{18y^4}$

38. $\dfrac{9ab^3}{7xy} \cdot \dfrac{14xy^2}{27z^3} \div \dfrac{18a^2b^2x}{3z^2}$ $\quad \dfrac{by}{9axz}$

39. $(4x + 12) \cdot \dfrac{x^2}{2x - 6} \div \dfrac{2}{x - 3}$ $\quad x^2(x+3)$

40. $(4x^2 - 9) \div \dfrac{2x^2 + 5x + 3}{x + 2} \div (2x - 3)$ $\quad \dfrac{x+2}{x+1}$

41. $\dfrac{2x^2 - 2x - 4}{x^2 + 2x - 8} \cdot \dfrac{3x^2 + 15x}{x + 1} \div \dfrac{4x^2 - 100}{x^2 - x - 20}$ $\quad \dfrac{3x}{2}$

42. $\dfrac{6a^2 - 7a - 3}{a^2 - 1} \div \dfrac{4a^2 - 12a + 9}{a^2 - 1} \cdot \dfrac{2a^2 - a - 3}{3a^2 - 2a - 1}$ $\quad \dfrac{a+1}{a-1}$

43. $\dfrac{2t^2 + 5t + 2}{t^2 - 4t + 16} \div \dfrac{t + 2}{t^3 + 64} \div \dfrac{2t^3 + 9t^2 + 4t}{t + 1}$ $\quad \dfrac{t+1}{t}$

44. $\dfrac{a^6 - b^6}{a^4 - a^3b} \cdot \dfrac{a^3}{a^4 + a^2b^2 + b^4} \div \dfrac{1}{a}$ $\quad a(a+b)$

45. $\dfrac{x^4 - 3x^2 - 4}{x^4 - 1} \cdot \dfrac{x^2 + 3x + 2}{x^2 + 4x + 4}$ $\quad \dfrac{x-2}{x-1}$

46. $\dfrac{x^3 + 2x^2 + 4x + 8}{y^2 - 1} \cdot \dfrac{y^2 + 2y + 1}{x^4 - 16}$ $\quad \dfrac{y+1}{(y-1)(x-2)}$

47. $(x^2 - x - 6) \div (x - 3) \div (x - 2)$ $\quad \dfrac{x+2}{x-2}$

48. $(x^2 - x - 6) \div [(x - 3) \div (x - 2)]$ $\quad (x+2)(x-2)$

49. $\dfrac{3x^2 - 2x}{3x + 2} \div (3x - 2) \div \dfrac{3x}{3x - 3}$ $\quad \dfrac{x-1}{3x+2}$

50. $(2x^2 - 3x - 2) \div \dfrac{2x^2 - x - 1}{x - 2} \div (x - 1)$ $\quad \dfrac{(x-2)^2}{(x-1)^2}$

51. $\dfrac{2x^2 + 5x - 3}{x^2 + 2x - 3} \div \left(\dfrac{x^2 + 2x - 35}{x^2 - 6x + 5} \div \dfrac{x^2 - 9x + 14}{2x^2 - 5x + 2}\right)$ $\quad \dfrac{x-7}{x+7}$

52. $\dfrac{x^2 - 4}{x^2 - x - 6} \div \left(\dfrac{x^2 - x - 2}{x^2 - 8x + 15} \cdot \dfrac{x^2 - 3x - 10}{x^2 + 3x + 2}\right)$ $\quad 1$

53. $\dfrac{x^2 - x - 12}{x^2 + x - 2} \div \dfrac{x^2 - 6x + 8}{x^2 - 3x - 10} \cdot \dfrac{x^2 - 3x + 2}{x^2 - 2x - 15}$ $\quad 1$

54. $\dfrac{4x^2 - 10x + 6}{x^4 - 3x^3} \div \dfrac{2x - 3}{2x^3} \cdot \dfrac{x - 3}{2x - 2}$ $\quad 2$

Writing Exercises ■ *Write a paragraph using your own words.*

1. Explain how to multiply two rational expressions.
2. Explain how to divide one rational expression by another.

Something to Think About ■ *Insert either a multiplication or a division symbol in each box to make a true statement.*

1. $\dfrac{x^2}{y} \boxed{\div} \dfrac{x}{y^2} \boxed{\cdot} \dfrac{x^2}{y^2} = \dfrac{x^3}{y}$

2. $\dfrac{x^2}{y} \boxed{\div} \dfrac{x}{y^2} \boxed{\div} \dfrac{x^2}{y^2} = \dfrac{y^3}{x}$

Review Exercises ■ *Do each operation.*

1. $-2a^2(3a^3 - a^2)$ $\quad -6a^5 + 2a^4$
2. $(2t - 1)^2$ $\quad 4t^2 - 4t + 1$
3. $(m^n + 2)(m^n - 2)$ $\quad m^{2n} - 4$
4. $(3b^{-n} + c)(b^{-n} - c)$ $\quad \frac{3}{b^{2n}} - \frac{2c}{b^n} - c^2$

5.3 Adding and Subtracting Rational Expressions

■ Adding and Subtracting Rational Expressions with Like Denominators
■ Adding and Subtracting Rational Expressions with Unlike Denominators

Property 5 of fractions, first discussed in Section 1.4, asserts that fractions are added and subtracted according to the following rules.

Property 5 of Fractions If there are no divisions by 0, then

$$\frac{a}{b} + \frac{c}{b} = \frac{a + c}{b} \quad \text{and} \quad \frac{a}{b} - \frac{c}{b} = \frac{a - c}{b}$$

■ Adding and Subtracting Rational Expressions with Like Denominators

When we add or subtract fractions with like denominators, *we add or subtract the numerators and keep the same denominator.* Whenever possible, we should simplify the result.

(b, c) **EXAMPLE 1** Simplify: **a.** $\frac{17}{22} + \frac{13}{22}$, **b.** $\frac{3}{2x} + \frac{7}{2x}$, and **c.** $\frac{4x}{x + 2} - \frac{7x}{x + 2}$.

Solution **a.**

$$\frac{17}{22} + \frac{13}{22} = \frac{17 + 13}{22} = \frac{30}{22} = \frac{15 \cdot \cancel{2}}{11 \cdot \cancel{2}} = \frac{15}{11}$$

b.

$$\frac{3}{2x} + \frac{7}{2x} = \frac{3 + 7}{2x} = \frac{10}{2x} = \frac{\cancel{2} \cdot 5}{\cancel{2} \cdot x} = \frac{5}{x}$$

c. $\dfrac{4x}{x+2} - \dfrac{7x}{x+2} = \dfrac{4x - 7x}{x+2}$

$$= \frac{-3x}{x+2}$$

■

Adding and Subtracting Rational Expressions with Unlike Denominators

To add or subtract fractions with unlike denominators, we must convert them to fractions with the same denominator. To do so, we use Property 3 of fractions to insert whatever factors are necessary to form a common denominator.

(b, c) **EXAMPLE 2** Simplify: **a.** $\dfrac{2}{3} + \dfrac{3}{2}$, **b.** $\dfrac{3}{x} + \dfrac{4}{y}$ and **c.** $\dfrac{4x}{x+2} - \dfrac{7x}{x-2}$.

Solution **a.** $\dfrac{2}{3} + \dfrac{3}{2} = \dfrac{2 \cdot \mathbf{2}}{3 \cdot \mathbf{2}} + \dfrac{3 \cdot \mathbf{3}}{2 \cdot \mathbf{3}}$

$$= \frac{4}{6} + \frac{9}{6}$$

$$= \frac{13}{6}$$

b. $\dfrac{3}{x} + \dfrac{4}{y} = \dfrac{3\mathbf{y}}{x\mathbf{y}} + \dfrac{\mathbf{x}4}{\mathbf{x}y}$

$$= \frac{3y + 4x}{xy}$$

c. $\dfrac{4x}{x+2} - \dfrac{7x}{x-2} = \dfrac{4x\mathbf{(x-2)}}{(x+2)\mathbf{(x-2)}} - \dfrac{\mathbf{(x+2)}7x}{\mathbf{(x+2)}(x-2)}$

$$= \frac{(4x^2 - 8x) - (7x^2 + 14x)}{(x+2)(x-2)}$$

$$= \frac{4x^2 - 8x - 7x^2 - 14x}{(x+2)(x-2)}$$

$$= \frac{-3x^2 - 22x}{(x+2)(x-2)}$$

WARNING! The − sign between the fractions in Step 1 of part **c** applies to both terms of the binomial $7x^2 + 14x$. ■

To add fractions with unlike denominators, we must find a common denominator. It is easiest to use the smallest one possible, called the **least** (or lowest) **common denominator (LCD).**

Suppose the unlike denominators of three fractions are 12, 20, and 35. To find the LCD, we first find the prime factorizations of each number.

$$12 = 4 \cdot 3 = 2^2 \cdot 3$$
$$20 = 4 \cdot 5 = 2^2 \cdot 5$$
$$35 = 5 \cdot 7$$

Since the LCD is the smallest number that can be divided by 12, 20, and 35, it must contain factors of 2^2, 3, 5, and 7. Thus, LCD is

$$\begin{aligned} \text{LCD} &= 2^2 \cdot 3 \cdot 5 \cdot 7 \\ &= 420 \end{aligned}$$

Thus, 420 is the smallest number that can be divided evenly by 12, 20, and 35.

To find the least common denominator of several fractions, we follow these steps.

Finding the LCD

1. Factor the denominator of each fraction.
2. List the different factors of each denominator.
3. Write each factor found in Step 2 to the highest power that occurs in any one denominator.
4. The LCD is the product of the powers in Step 3.

EXAMPLE 3 The polynomial denominators of three fractions are $x^2 + 7x + 6$, $x^2 - 36$, and $x^2 + 12x + 36$. Find the LCD.

Solution We factor each polynomial:

$$x^2 + 7x + 6 = (x + 6)(x + 1)$$
$$x^2 - 36 = (x + 6)(x - 6)$$
$$x^2 + 12x + 36 = (x + 6)(x + 6) = (x + 6)^2$$

and list the individual factors:

$$x + 6, \; x + 1, \text{ and } x - 6$$

To find the LCD, we use the highest power of each of these factors that occurs in any one factorization.

$$\text{LCD} = (x + 6)^2(x + 1)(x - 6)$$

■

EXAMPLE 4 Add the fractions $\dfrac{x}{x^2 - 2x + 1}$ and $\dfrac{3}{x^2 - 1}$.

Solution We factor each denominator and find the LCD:

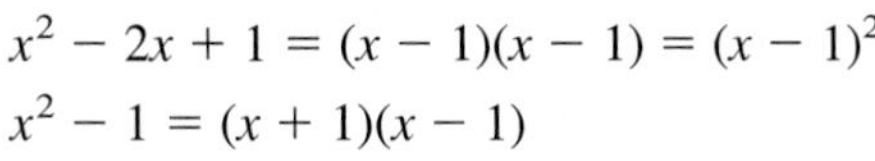

$$x^2 - 2x + 1 = (x - 1)(x - 1) = (x - 1)^2$$
$$x^2 - 1 = (x + 1)(x - 1)$$

Charles Babbage
(1792–1871)
In 1823, Babbage built a steam-powered digital calculator, which he called a difference engine. Thought to be a crackpot by his London neighbors, Babbage was a visionary. His machine embodied principles still used in modern computers.

The LCD is $(x - 1)^2(x + 1)$.

We now write each fraction with its denominator in factored form and convert the fractions to fractions with a common denominator of $(x - 1)^2(x + 1)$. Finally, we add the fractions.

$$\begin{aligned}
\frac{x}{x^2 - 2x + 1} + \frac{3}{x^2 - 1} &= \frac{x}{(x - 1)(x - 1)} + \frac{3}{(x + 1)(x - 1)} \\
&= \frac{x(x + \mathbf{1})}{(x - 1)(x - 1)(x + \mathbf{1})} + \frac{3(x - \mathbf{1})}{(x + 1)(x - 1)(x - \mathbf{1})} \\
&= \frac{x^2 + x + 3x - 3}{(x - 1)(x - 1)(x + 1)} \\
&= \frac{x^2 + 4x - 3}{(x - 1)^2(x + 1)} \qquad \text{This result does not simplify.}
\end{aligned}$$

■

EXAMPLE 5 Simplify: $\dfrac{2x}{x^2 - 4} - \dfrac{1}{x^2 - 3x + 2} + \dfrac{x + 1}{x^2 + x - 2}$.

Solution We factor each denominator to find the LCD:

$$\text{LCD} = (x + 2)(x - 2)(x - 1)$$

We then write each fraction as a fraction with the LCD as its denominator and do the subtraction and addition.

$$\begin{aligned}
&\frac{2x}{x^2 - 4} - \frac{1}{x^2 - 3x + 2} + \frac{x + 1}{x^2 + x - 2} \\
&\quad = \frac{2x}{(x - 2)(x + 2)} - \frac{1}{(x - 2)(x - 1)} + \frac{x + 1}{(x - 1)(x + 2)} \\
&\quad = \frac{2x(x - \mathbf{1})}{(x - 2)(x + 2)(x - \mathbf{1})} - \frac{1(x + \mathbf{2})}{(x - 2)(x - 1)(x + \mathbf{2})} + \frac{(x + 1)(x - \mathbf{2})}{(x - 1)(x + 2)(x - \mathbf{2})} \\
&\quad = \frac{2x(x - 1) - 1(x + 2) + (x + 1)(x - 2)}{(x + 2)(x - 2)(x - 1)} \\
&\quad = \frac{2x^2 - 2x - x - 2 + x^2 - x - 2}{(x + 2)(x - 2)(x - 1)} \\
&\quad = \frac{3x^2 - 4x - 4}{(x + 2)(x - 2)(x - 1)}
\end{aligned}$$

Here, the final result does simplify.

$$\frac{2x}{x^2-4} - \frac{1}{x^2-3x+2} + \frac{x+1}{x^2+x-2} = \frac{3x^2-4x-4}{(x+2)(x-2)(x-1)}$$

$$= \frac{(3x+2)(x-2)}{(x+2)(x-2)(x-1)} \quad \text{Factor the numerator.}$$

$$= \frac{3x+2}{(x+2)(x-1)} \quad \text{Divide out common factors.}$$

■

EXAMPLE 6 Simplify: $3 + \frac{7}{x-2}$.

Solution Since $3 = \frac{3}{1}$, we must find the sum $\frac{3}{1} + \frac{7}{x-2}$, where the fractions have an LCD of $x - 2$.

$$3 + \frac{7}{x-2} = \frac{3}{1} + \frac{7}{x-2}$$

$$= \frac{3(x-2)}{1(x-2)} + \frac{7}{x-2}$$

$$= \frac{3x-6+7}{x-2}$$

$$= \frac{3x+1}{x-2}$$

■

EXAMPLE 7 Simplify: $\frac{3x}{x-1} - \frac{2x^2+3x-2}{(x+1)(x-1)}$.

Solution We write each fraction in a form having the LCD of $(x + 1)(x - 1)$, remove the resulting parentheses in the first numerator, do the subtraction, and simplify.

1. $$\frac{3x}{x-1} - \frac{2x^2+3x-2}{(x+1)(x-1)} = \frac{(x+1)3x}{(x+1)(x-1)} - \frac{2x^2+3x-2}{(x+1)(x-1)}$$

$$= \frac{3x^2+3x}{(x+1)(x-1)} - \frac{2x^2+3x-2}{(x+1)(x-1)}$$

$$= \frac{3x^2+3x-(2x^2+3x-2)}{(x+1)(x-1)}$$

$$= \frac{3x^2+3x-2x^2-3x+2}{(x+1)(x-1)}$$

$$= \frac{x^2+2}{(x+1)(x-1)}$$

WARNING! The $-$ sign between the fractions in Equation 1 affects every term of the numerator $2x^2 + 3x - 2$. Whenever we subtract one fraction from another, we must remember to subtract each term of the numerator in the second fraction. ■

EXAMPLE 8 Simplify: $\left(\frac{x^2}{x-2} + \frac{4}{2-x}\right)^2$.

Solution We do the addition within the parentheses. Since the denominators are negatives of one another, we can write the fractions as fractions with a common denominator by multiplying both the numerator and denominator of $\frac{4}{2-x}$ by -1. We can then add the fractions, simplify, and square the result.

Teaching Tip: Point out that we can make these denominators the same by simply changing the operation, in this case, from addition to subtraction. For example,

$\frac{x^2}{x-2} + \frac{4}{2-x}$ is the same as

$\frac{x^2}{x-2} - \frac{4}{x-2}$

and $\frac{3x+1}{x-5} - \frac{2x-7}{5-x}$

is the same as $\frac{3x+1}{x-5} + \frac{2x-7}{x-5}$.

$$\begin{aligned}
\left(\frac{x^2}{x-2} + \frac{4}{2-x}\right)^2 &= \left[\frac{x^2}{x-2} + \frac{(-1)4}{-1(2-x)}\right]^2 \\
&= \left[\frac{x^2}{x-2} + \frac{-4}{x-2}\right]^2 \\
&= \left[\frac{x^2-4}{x-2}\right]^2 \\
&= \left[\frac{(x+2)(x-2)}{x-2}\right]^2 && \text{Factor } x^2-4. \\
&= (x+2)^2 && \text{Divide out } x-2. \\
&= x^2+4x+4
\end{aligned}$$
■

ORALS *Add or subtract the fractions. Simplify when possible.*

1. $\frac{x}{2} + \frac{x}{2}$ x
2. $\frac{3a}{4} - \frac{a}{4}$ $\frac{a}{2}$
3. $\frac{x}{x+2} + \frac{2}{x+2}$ 1
4. $\frac{2a}{a+4} - \frac{a-4}{a+4}$ 1
5. $\frac{2x}{3} + \frac{x}{2}$ $\frac{7x}{6}$
6. $1 + \frac{x}{3}$ $\frac{3+x}{3}$

EXERCISE 5.3

In Exercises 1–16, do the operations and simplify when possible.

1. $\frac{3}{4} + \frac{7}{4}$ $\frac{5}{2}$
2. $\frac{5}{11} + \frac{2}{11}$ $\frac{7}{11}$
3. $\frac{10}{33} - \frac{21}{33}$ $-\frac{1}{3}$
4. $\frac{8}{15} - \frac{2}{15}$ $\frac{2}{5}$

5. $\dfrac{3}{4y} + \dfrac{8}{4y}$ $\frac{11}{4y}$

6. $\dfrac{5}{3z^2} - \dfrac{6}{3z^2}$ $-\dfrac{1}{3z^2}$

7. $\dfrac{3}{a+b} - \dfrac{a}{a+b}$ $\frac{3-a}{a+b}$

8. $\dfrac{x}{x+4} + \dfrac{5}{x+4}$ $\frac{x+5}{x+4}$

9. $\dfrac{3x}{2x+2} + \dfrac{x+4}{2x+2}$ 2

10. $\dfrac{4y}{y-4} - \dfrac{16}{y-4}$ 4

11. $\dfrac{3x}{x-3} - \dfrac{9}{x-3}$ 3

12. $\dfrac{9x}{x-y} - \dfrac{9y}{x-y}$ 9

13. $\dfrac{5x}{x+1} + \dfrac{3}{x+1} - \dfrac{2x}{x+1}$ 3

14. $\dfrac{4}{a+4} - \dfrac{2a}{a+4} + \dfrac{3a}{a+4}$ 1

15. $\dfrac{3(x^2+x)}{x^2-5x+6} + \dfrac{-3(x^2-x)}{x^2-5x+6}$ $\frac{6x}{(x-3)(x-2)}$

16. $\dfrac{2x+4}{x^2+13x+12} - \dfrac{x+3}{x^2+13x+12}$ $\frac{1}{x+12}$

In Exercises 17–24, the denominators of several fractions are given. Find the LCD.

17. 8, 12, 18 72

18. 10, 15, 28 420

19. x^2+3x, x^2-9 $x(x+3)(x-3)$

20. $3y^2-6y$, $3y(y-4)$ $3y(y-2)(y-4)$

21. x^3+27, x^2+6x+9 $(x+3)^2(x^2-3x+9)$

22. x^3-8, x^2-4x+4 $(x-2)^2(x^2+2x+4)$

23. $2x^2+5x+3$, $4x^2+12x+9$, x^2+2x+1
$(2x+3)^2(x+1)^2$

24. $2x^2+5x+3$, $4x^2+12x+9$, $4x+6$
$2(2x+3)^2(x+1)$

In Exercises 25–86, do the operations and simplify when possible.

25. $\dfrac{1}{2} + \dfrac{1}{3}$ $\frac{5}{6}$

26. $\dfrac{5}{6} + \dfrac{2}{7}$ $\frac{47}{42}$

27. $\dfrac{7}{15} - \dfrac{17}{25}$ $-\frac{16}{75}$

28. $\dfrac{8}{9} - \dfrac{5}{12}$ $\frac{17}{36}$

29. $\dfrac{a}{2} + \dfrac{2a}{5}$ $\frac{9a}{10}$

30. $\dfrac{b}{6} + \dfrac{3a}{4}$ $\frac{2b+9a}{12}$

31. $\dfrac{3a}{2} - \dfrac{4b}{7}$ $\frac{21a-8b}{14}$

32. $\dfrac{2m}{3} - \dfrac{4n}{5}$ $\frac{10m-12n}{15}$

33. $\dfrac{3}{4x} + \dfrac{2}{3x}$
$\dfrac{17}{12x}$

34. $\dfrac{2}{5a} + \dfrac{3}{2b}$
$\dfrac{4b+15a}{10ab}$

35. $\dfrac{3a}{2b} - \dfrac{2b}{3a}$
$\dfrac{9a^2-4b^2}{6ab}$

36. $\dfrac{5m}{2n} - \dfrac{3n}{4m}$
$\dfrac{10m^2-3n^2}{4mn}$

37. $\dfrac{a+b}{3} + \dfrac{a-b}{7}$
$\frac{10a+4b}{21}$

38. $\dfrac{x-y}{2} + \dfrac{x+y}{3}$
$\frac{5x-y}{6}$

39. $\dfrac{3}{x+2} + \dfrac{5}{x-4}$
$\frac{8x-2}{(x+2)(x-4)}$

40. $\dfrac{2}{a+4} - \dfrac{6}{a+3}$
$\frac{-2(2a+9)}{(a+4)(a+3)}$

41. $\dfrac{x+2}{x+5} - \dfrac{x-3}{x+7}$
$\dfrac{7x+29}{(x+5)(x+7)}$

42. $\dfrac{7}{x+3} + \dfrac{4x}{x+6}$
$\dfrac{4x^2+19x+42}{(x+3)(x+6)}$

43. $x + \dfrac{1}{x}$
$\dfrac{x^2+1}{x}$

44. $2 - \dfrac{1}{x+1}$
$\dfrac{2x+1}{x+1}$

45. $\dfrac{x+8}{x-3} - \dfrac{x-14}{3-x}$
2

46. $\dfrac{3-x}{2-x} + \dfrac{x-1}{x-2}$
2

47. $\dfrac{2a+1}{3a+2} - \dfrac{a-4}{2-3a}$
$\dfrac{9a^2-11a-10}{(3a+2)(3a-2)}$

48. $\dfrac{4}{x-2} + \dfrac{5}{4-x^2}$
$\dfrac{4x+3}{(x+2)(x-2)}$

49. $\dfrac{x}{x^2+5x+6} + \dfrac{x}{x^2-4}$
$\dfrac{2x^2+x}{(x+3)(x+2)(x-2)}$

50. $\dfrac{x}{3x^2-2x-1} + \dfrac{4}{3x^2+10x+3}$
$\dfrac{x^2+7x-4}{(3x+1)(x-1)(x+3)}$

51. $\dfrac{4}{x^2 - 2x - 3} - \dfrac{x}{3x^2 - 7x - 6}$ $\quad \dfrac{-x^2 + 11x + 8}{(3x + 2)(x + 1)(x - 3)}$

52. $\dfrac{2a}{a^2 - 2a - 8} + \dfrac{3}{a^2 - 5a + 4}$ $\quad \dfrac{2a^2 + a + 6}{(a - 4)(a + 2)(a - 1)}$

53. $\dfrac{8}{x^2 - 9} + \dfrac{2}{x - 3} - \dfrac{6}{x}$ $\quad \dfrac{-4x^2 + 14x + 54}{x(x + 3)(x - 3)}$

54. $\dfrac{x}{x^2 - 4} - \dfrac{x}{x + 2} + \dfrac{2}{x}$ $\quad \dfrac{-x^3 + 5x^2 - 8}{x(x + 2)(x - 2)}$

55. $\dfrac{x}{x + 1} - \dfrac{x}{1 - x^2} + \dfrac{1}{x}$ $\quad \dfrac{x^3 + x^2 - 1}{x(x + 1)(x - 1)}$

56. $\dfrac{y}{y - 2} - \dfrac{2}{y + 2} - \dfrac{-8}{4 - y^2}$ $\quad 1$

57. $2x + 3 + \dfrac{1}{x + 1}$ $\quad \dfrac{2x^2 + 5x + 4}{x + 1}$

58. $x + 1 + \dfrac{1}{x - 1}$ $\quad \dfrac{x^2}{x - 1}$

59. $1 + x - \dfrac{x}{x - 5}$ $\quad \dfrac{x^2 - 5x - 5}{x - 5}$

60. $2 - x + \dfrac{3}{x - 9}$ $\quad \dfrac{-x^2 + 11x - 15}{x - 9}$

61. $\dfrac{3x}{x - 1} - 2x - x^2$ $\quad \dfrac{-x^3 - x^2 + 5x}{x - 1}$

62. $\dfrac{23}{x - 1} + 4x - 5x^2$ $\quad \dfrac{-5x^3 + 9x^2 - 4x + 23}{x - 1}$

63. $\dfrac{y + 4}{y^2 + 7y + 12} - \dfrac{y - 4}{y + 3} + \dfrac{47}{y + 4}$ $\quad \dfrac{-y^2 + 48y + 161}{(y + 4)(y + 3)}$

64. $\dfrac{x + 3}{2x^2 - 5x + 2} - \dfrac{3x - 1}{x^2 - x - 2}$ $\quad -\dfrac{5x + 1}{(2x - 1)(x + 1)}$

65. $\dfrac{3}{x + 1} - \dfrac{2}{x - 1} + \dfrac{x + 3}{x^2 - 1}$ $\quad \dfrac{2}{x + 1}$

66. $\dfrac{2}{x - 2} + \dfrac{3}{x + 2} - \dfrac{x - 1}{x^2 - 4}$ $\quad \dfrac{4x - 1}{(x + 2)(x - 2)}$

67. $\dfrac{x - 2}{x^2 - 3x} + \dfrac{2x - 1}{x^2 + 3x} - \dfrac{2}{x^2 - 9}$ $\quad \dfrac{3x + 1}{x(x + 3)}$

68. $\dfrac{2}{x - 1} - \dfrac{2x}{x^2 - 1} - \dfrac{x}{x^2 + 2x + 1}$ $\quad \dfrac{-x^2 + 3x + 2}{(x - 1)(x + 1)^2}$

69. $\dfrac{5}{x^2 - 25} - \dfrac{3}{2x^2 - 9x - 5} + 1$ $\quad \dfrac{2x^3 + x^2 - 43x - 35}{(x + 5)(x - 5)(2x + 1)}$

70. $\dfrac{3x}{2x - 1} + \dfrac{x + 1}{3x + 2} - \dfrac{2x}{6x^3 + x^2 - 2x}$ $\quad \dfrac{11x^2 + 7x + 1}{(2x - 1)(3x + 2)}$

71. $\dfrac{3x}{x - 3} + \dfrac{4}{x - 2} - \dfrac{5x}{x^3 - 5x^2 + 6x}$
$\dfrac{3x^2 - 2x - 17}{(x - 3)(x - 2)}$

72. $\dfrac{2x - 1}{x^2 + x - 6} - \dfrac{3x - 5}{x^2 - 2x - 15} + \dfrac{2x - 3}{x^2 - 7x + 10}$
$\dfrac{x^2 + 3x - 14}{(x - 5)(x + 3)(x - 2)}$

73. $2 + \dfrac{4a}{a^2 - 1} - \dfrac{2}{a + 1}$ $\quad \dfrac{2a}{a - 1}$

74. $\dfrac{a}{a - 1} - \dfrac{a + 1}{2a - 2} + a$ $\quad \dfrac{2a + 1}{2}$

75. $\dfrac{x + 5}{2x^2 - 2} + \dfrac{x}{2x + 2} - \dfrac{3}{x - 1}$ $\quad \dfrac{x^2 - 6x - 1}{2(x + 1)(x - 1)}$

76. $\dfrac{a}{2 - a} + \dfrac{3}{a - 2} - \dfrac{3a - 2}{a^2 - 4}$ $\quad -\dfrac{a + 4}{a + 2}$

77. $\dfrac{a}{a - b} + \dfrac{b}{a + b} + \dfrac{a^2 + b^2}{b^2 - a^2}$ $\quad \dfrac{2b}{a + b}$

78. $\dfrac{1}{x + y} - \dfrac{1}{x - y} - \dfrac{2x}{y^2 - x^2}$ $\quad \dfrac{2}{x + y}$

79. $\dfrac{7n^2}{m - n} + \dfrac{3m}{n - m} - \dfrac{3m^2 - n}{m^2 - 2mn + n^2}$
$\dfrac{7mn^2 - 7n^3 - 6m^2 + 3mn + n}{(m - n)^2}$

80. $\dfrac{3b}{2a - b} + \dfrac{2a - 1}{b - 2a} - \dfrac{3a^2 + b}{b^2 - 4ab + 4a^2}$
$\dfrac{-7a^2 + 8ab + 2a - 2b - 3b^2}{(2a - b)^2}$

81. $\dfrac{m + 1}{m^2 + 2m + 1} + \dfrac{m - 1}{m^2 - 2m + 1} + \dfrac{2}{m^2 - 1}$ (*Hint:* Think about this before finding the LCD.) $\dfrac{2}{m - 1}$

82. $\dfrac{a + 2}{a^2 + 3a + 2} + \dfrac{a - 1}{a^2 - 1} + \dfrac{3}{a + 1}$ $\quad \dfrac{5}{a + 1}$

83. $\left(\dfrac{1}{x-1}+\dfrac{1}{1-x}\right)^2$ 0

84. $\left(\dfrac{1}{a-1}-\dfrac{1}{1-a}\right)^2$ $\dfrac{4}{(a-1)^2}$

85. $\left(\dfrac{x}{x-3}+\dfrac{3}{3-x}\right)^3$ 1

86. $\left(\dfrac{2y}{y+4}+\dfrac{8}{y+4}\right)^3$ 8

87. Show that $\dfrac{a}{b}+\dfrac{c}{d}=\dfrac{ad+bc}{bd}$.

88. Show that $\dfrac{a}{b}-\dfrac{c}{d}=\dfrac{ad-bc}{bd}$.

Writing Exercises ■ *Write a paragraph using your own words.*

1. Explain how to find the least common denominator.

2. Explain how to add two fractions.

Something to Think About ■

1. Find the error:

$$\frac{8x+2}{5}-\frac{3x+8}{5}$$
$$=\frac{8x+2-3x+8}{5}$$
$$=\frac{5x+10}{5}$$
$$=x+2$$

2. Find the error:

$$\frac{(x+y)^2}{2}+\frac{(x-y)^2}{3}$$
$$=\frac{3\cdot(x+y)^2}{3\cdot 2}+\frac{2\cdot(x-y)^2}{2\cdot 3}$$
$$=\frac{3x^2+3y^2+2x^2-2y^2}{6}$$
$$=\frac{5x^2+y^2}{6}$$

Review Exercises ■ *Graph each interval on a number line.*

1. $(-2, 4] \cup [-1, 5)$

2. $(-1, 4) \cap (2, 5]$

Solve each formula for the indicated variable.

3. $P = 2l + 2w$; for w $w = \dfrac{P-2l}{2}$

4. $S = \dfrac{a-lr}{1-r}$; for a $a = S - Sr + lr$

5.4 Complex Fractions

■ **Simplifying Complex Fractions**

A **complex fraction** has a fraction in its numerator or its denominator or both. Examples of complex fractions are

$$\frac{\frac{3}{5}}{\frac{6}{7}},\qquad \frac{\frac{x+2}{3}}{x-4},\qquad \text{and}\qquad \frac{\frac{3x^2-2}{2x}}{3x-\frac{2}{y}}$$

Simplifying Complex Fractions

There are two ways to simplify the complex fraction

$$\frac{\frac{a}{b}}{\frac{c}{d}}$$

One way is to eliminate the fractions in the numerator and denominator by writing the complex fraction as a division and then using the division rule for fractions:

$$\frac{\frac{a}{b}}{\frac{c}{d}} = \frac{a}{b} \div \frac{c}{d}$$

$$= \frac{a}{b} \cdot \frac{d}{c} \qquad \text{Invert the divisor and multiply.}$$

$$= \frac{ad}{bc} \qquad \text{Multiply the numerators and multiply the denominators.}$$

The other way is to eliminate the fractions in the numerator and denominator by multiplying the numerator and denominator by bd (the LCD of $\frac{a}{b}$ and $\frac{c}{d}$).

$$\frac{(bd)\frac{a}{b}}{(bd)\frac{c}{d}} = \frac{\frac{abd}{b}}{\frac{bcd}{d}}$$

$$= \frac{ad}{bc}$$

In the numerator, divide out the common factor of b.
In the denominator, divide out the common factor of d.

With either method, the result is the same.

EXAMPLE 1 Simplify the complex fraction $\frac{\frac{3a}{b}}{\frac{6ac}{b^2}}$.

Solution 1 We write the complex fraction as a division and proceed as follows:

$$\frac{\frac{3a}{b}}{\frac{6ac}{b^2}} = \frac{3a}{b} \div \frac{6ac}{b^2}$$

$$= \frac{3a}{b} \cdot \frac{b^2}{6ac} \qquad \text{Invert the divisor and multiply.}$$

$$= \frac{b}{2c} \qquad \text{Multiply the fractions and simplify.}$$

Solution 2 We multiply the numerator and denominator by b^2, the LCD of $\frac{3a}{b}$ and $\frac{6ac}{b^2}$, and simplify:

$$\frac{\frac{3a}{b}}{\frac{6ac}{b^2}} = \frac{\frac{3a}{b} \cdot b^2}{\frac{6ac}{b^2} \cdot b^2}$$

$$= \frac{\frac{3ab^2}{b}}{\frac{6acb^2}{b^2}}$$

$$= \frac{3ab}{6ac} \quad \text{Simplify the fractions in the numerator and denominator.}$$

$$= \frac{b}{2c} \quad \text{Divide out the common factor of } 3a.$$

■

EXAMPLE 2 Simplify the complex fraction $\dfrac{\frac{1}{x} + \frac{1}{y}}{\frac{1}{x} - \frac{1}{y}}$.

Solution 1 We add the fractions in the numerator and in the denominator and proceed as follows:

$$\frac{\frac{1}{x} + \frac{1}{y}}{\frac{1}{x} - \frac{1}{y}} = \frac{\frac{1y}{xy} + \frac{x1}{xy}}{\frac{1y}{xy} - \frac{x1}{xy}}$$

$$= \frac{\frac{y+x}{xy}}{\frac{y-x}{xy}}$$

$$= \frac{y+x}{xy} \div \frac{y-x}{xy}$$

$$= \frac{y+x}{xy} \cdot \frac{xy}{y-x}$$

$$= \frac{y+x}{y-x} \quad \text{Multiply, and then divide out the common factor of } xy.$$

Solution 2 We multiply the numerator and denominator by xy (the least common denominator of the fractions appearing in the complex fraction) and simplify.

$$\frac{\frac{1}{x}+\frac{1}{y}}{\frac{1}{x}-\frac{1}{y}} = \frac{xy\left(\frac{1}{x}+\frac{1}{y}\right)}{xy\left(\frac{1}{x}-\frac{1}{y}\right)}$$

$$= \frac{\frac{xy}{x}+\frac{xy}{y}}{\frac{xy}{x}-\frac{xy}{y}}$$

$$= \frac{y+x}{y-x} \qquad \text{Simplify each fraction.}$$

■

EXAMPLE 3 Simplify $\dfrac{x^{-1}+y^{-1}}{x^{-2}-y^{-2}}$.

Solution 1 We proceed as follows:

$$\frac{x^{-1}+y^{-1}}{x^{-2}-y^{-2}} = \frac{\frac{1}{x}+\frac{1}{y}}{\frac{1}{x^2}-\frac{1}{y^2}} \qquad \text{Write the fraction without using negative exponents.}$$

$$= \frac{\frac{y}{xy}+\frac{x}{xy}}{\frac{y^2}{x^2y^2}-\frac{x^2}{x^2y^2}} \qquad \text{Get a common denominator in the numerator and denominator.}$$

$$= \frac{\frac{y+x}{xy}}{\frac{y^2-x^2}{x^2y^2}} \qquad \text{Add the fractions in the numerator and denominator.}$$

$$= \frac{y+x}{xy} \div \frac{y^2-x^2}{x^2y^2} \qquad \text{Write the fraction as a division.}$$

$$= \frac{y+x}{xy} \cdot \frac{x^2y^2}{(y-x)(y+x)} \qquad \text{Invert and multiply.}$$

$$= \frac{(y+x)x^2y^2}{xy(y-x)(y+x)} \qquad \text{Multiply the numerators and the denominators.}$$

$$= \frac{xy}{y-x} \qquad \text{Divide out the common factors of } x,\ y, \text{ and } y+x \text{ in the numerator and denominator.}$$

Solution 2 We multiply both numerator and denominator by x^2y^2, the LCD of the fractions in the problem, and proceed as follows:

$$\frac{x^{-1}+y^{-1}}{x^{-2}-y^{-2}} = \frac{\frac{1}{x}+\frac{1}{y}}{\frac{1}{x^2}-\frac{1}{y^2}}$$ Write the fraction without negative exponents.

$$= \frac{x^2y^2\left(\frac{1}{x}+\frac{1}{y}\right)}{x^2y^2\left(\frac{1}{x^2}-\frac{1}{y^2}\right)}$$ Multiply numerator and denominator by x^2y^2.

$$= \frac{xy^2+yx^2}{y^2-x^2}$$ Use the distributive property to remove parentheses.

$$= \frac{xy(y+x)}{(y+x)(y-x)}$$ Factor the numerator and denominator.

$$= \frac{xy}{y-x}$$ Divide out $y + x$. ■

WARNING! $x^{-1} + y^{-1}$ means $\frac{1}{x} + \frac{1}{y}$, and $(x + y)^{-1}$ means $\frac{1}{x+y}$. Thus,

$$(x+y)^{-1} \neq x^{-1} + y^{-1} \quad \text{and} \quad x^{-1} + y^{-1} \neq \frac{1}{x+y}$$

PERSPECTIVE

Some Complex Fractions in Architecture

Each of the complex fractions in the list

$$1+\frac{1}{2},\quad 1+\frac{1}{1+\frac{1}{2}},\quad 1+\frac{1}{1+\frac{1}{1+\frac{1}{2}}},\quad 1+\frac{1}{1+\frac{1}{1+\frac{1}{1+\frac{1}{2}}}},\quad \ldots$$

can be simplified by using the value of the expression preceding it. For example, to simplify the second expression in the list, replace $1 + \frac{1}{2}$ with $\frac{3}{2}$:

$$1+\frac{1}{1+\frac{1}{2}} = 1+\frac{1}{\frac{3}{2}} = 1+\frac{2}{3} = \frac{5}{3}.$$

To simplify the third expression, replace

$$1 + \cfrac{1}{1 + \cfrac{1}{2}} \text{ with } \frac{5}{3}:$$

$$1 + \cfrac{1}{1 + \cfrac{1}{1 + \cfrac{1}{2}}} = 1 + \cfrac{1}{\cfrac{5}{3}} = 1 + \frac{3}{5} = \frac{8}{5}.$$

Can you show that the expressions in the list simplify to the fractions $\frac{3}{2}, \frac{5}{3}, \frac{8}{5}, \frac{13}{8}, \frac{21}{13}, \frac{34}{21} \cdots$?

Do you see a pattern, and can you predict the next fraction?

Use a calculator to write each of these fractions as a decimal. The values produced get closer and closer to the irrational number 1.61803398875..., which is known as the **golden ratio.** This number often appears in the architecture of the ancient Greeks and Egyptians. The width of the stairs in front of the Greek Parthenon (Illustration 1), divided by the building's height, is the golden ratio. The height of the triangular face of the Great Pyramid of Cheops (Illustration 2), divided by the pyramid's width, is also the golden ratio.

ILLUSTRATION 1

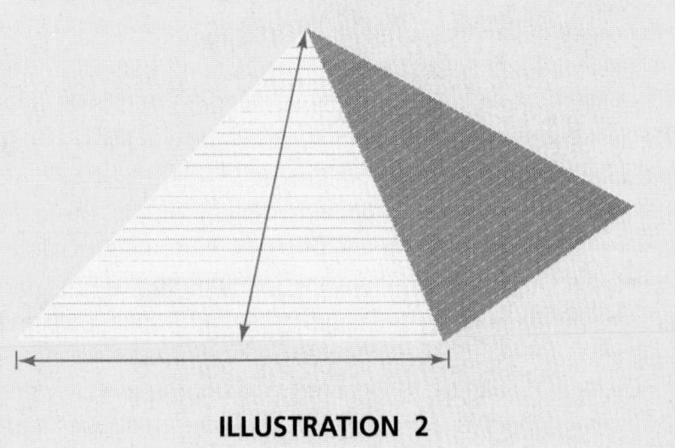

ILLUSTRATION 2

EXAMPLE 4 Simplify the fraction $\cfrac{\cfrac{2x}{1 - \cfrac{1}{x}} + 3}{3 - \cfrac{2}{x}}$.

Solution We begin by multiplying the numerator and denominator of the fraction

$$\cfrac{2x}{1 - \cfrac{1}{x}}$$

by x. This will eliminate the complex fraction in the numerator of the given fraction.

$$\frac{\dfrac{2x}{1-\dfrac{1}{x}}+3}{3-\dfrac{2}{x}} = \frac{\dfrac{x(2x)}{x\left(1-\dfrac{1}{x}\right)}+3}{3-\dfrac{2}{x}}$$

$$= \frac{\dfrac{2x^2}{x-1}+3}{3-\dfrac{2}{x}}$$

We then multiply the numerator and denominator of the previous fraction by $x(x-1)$, the LCD of $\frac{2x^2}{x-1}$, 3, and $\frac{2}{x}$, and simplify:

$$\frac{\dfrac{2x}{1-\dfrac{1}{x}}+3}{3-\dfrac{2}{x}} = \frac{x(x-1)\left(\dfrac{2x^2}{x-1}+3\right)}{x(x-1)\left(3-\dfrac{2}{x}\right)}$$

$$= \frac{2x^3+3x(x-1)}{3x(x-1)-2(x-1)}$$

$$= \frac{2x^3+3x^2-3x}{3x^2-5x+2}$$

This result does not simplify. ■

ORALS *Simplify each complex fraction.*

1. $\dfrac{\frac{3}{4}}{\frac{5}{4}}$ $\frac{3}{5}$

2. $\dfrac{\frac{a}{b}}{\frac{d}{b}}$ $\frac{a}{d}$

3. $\dfrac{\frac{3}{4}}{\frac{3}{8}}$ 2

4. $\dfrac{\frac{x+y}{x}}{\frac{x-y}{x}}$ $\frac{x+y}{x-y}$

5. $\dfrac{\frac{x}{y}-1}{\frac{x}{y}}$ $\frac{x-y}{x}$

6. $\dfrac{1+\frac{a}{b}}{\frac{a}{b}}$ $\frac{b+a}{a}$

EXERCISE 5.4

In Exercises 1–56, simplify each complex fraction.

1. $\dfrac{\frac{1}{2}}{\frac{3}{4}}$ $\quad \frac{2}{3}$

2. $-\dfrac{\frac{3}{4}}{\frac{1}{2}}$ $\quad -\frac{3}{2}$

3. $\dfrac{-\frac{2}{3}}{\frac{6}{9}}$ $\quad -1$

4. $\dfrac{\frac{11}{18}}{\frac{22}{27}}$ $\quad \frac{3}{4}$

5. $\dfrac{\frac{1}{2}+\frac{1}{3}}{\frac{1}{4}}$ $\quad \frac{10}{3}$

6. $\dfrac{\frac{1}{4}-\frac{1}{5}}{\frac{1}{3}}$ $\quad \frac{3}{20}$

7. $\dfrac{\frac{1}{6}-\frac{2}{7}}{\frac{1}{7}}$ $\quad -\frac{5}{6}$

8. $\dfrac{\frac{2}{3}+\frac{4}{5}}{\frac{1}{3}}$ $\quad \frac{22}{5}$

9. $\dfrac{\frac{4x}{y}}{\frac{6xz}{y^2}}$ $\quad \frac{2y}{3z}$

10. $\dfrac{\frac{5t^4}{9x}}{\frac{2t}{18x}}$ $\quad 5t^3$

11. $\dfrac{5ab^2}{\frac{ab}{25}}$ $\quad 125b$

12. $\dfrac{\frac{6a^2b}{4t}}{3a^2b^2}$ $\quad \frac{1}{2bt}$

13. $\dfrac{\frac{x-y}{xy}}{\frac{y-x}{x}}$ $\quad -\frac{1}{y}$

14. $\dfrac{\frac{x^2+5x+6}{3xy}}{\frac{x^2-9}{6xy}}$ $\quad \frac{2(x+2)}{x-3}$

15. $\dfrac{\frac{1}{x}-\frac{1}{y}}{xy}$ $\quad \frac{y-x}{x^2y^2}$

16. $\dfrac{xy}{\frac{1}{x}-\frac{1}{y}}$ $\quad \frac{x^2y^2}{y-x}$

17. $\dfrac{\frac{1}{a}+\frac{1}{b}}{\frac{1}{a}}$ $\quad \frac{b+a}{b}$

18. $\dfrac{\frac{1}{b}}{\frac{1}{a}-\frac{1}{b}}$ $\quad \frac{a}{b-a}$

19. $\dfrac{1+\frac{x}{y}}{1-\frac{x}{y}}$ $\quad \frac{y+x}{y-x}$

20. $\dfrac{\frac{x}{y}+1}{1-\frac{x}{y}}$ $\quad \frac{y+x}{y-x}$

21. $\dfrac{\frac{y}{x}-\frac{x}{y}}{\frac{1}{x}+\frac{1}{y}}$ $\quad y-x$

22. $\dfrac{\frac{y}{x}-\frac{x}{y}}{\frac{1}{y}-\frac{1}{x}}$ $\quad -y-x$

23. $\dfrac{\frac{1}{a}-\frac{1}{b}}{\frac{a}{b}-\frac{b}{a}}$ $\quad \frac{-1}{a+b}$

24. $\dfrac{\frac{1}{a}+\frac{1}{b}}{\frac{a}{b}-\frac{b}{a}}$ $\quad \frac{1}{a-b}$

25. $\dfrac{x+1-\frac{6}{x}}{\frac{1}{x}}$ $\quad x^2+x-6$

26. $\dfrac{x-1-\frac{2}{x}}{\frac{x}{3}}$ $\quad \frac{3(x^2-x-2)}{x^2}$

27. $\dfrac{5xy}{1+\frac{1}{xy}}$ $\quad \frac{5x^2y^2}{xy+1}$

28. $\dfrac{3a}{a+\frac{1}{a}}$ $\quad \frac{3a^2}{a^2+1}$

29. $\dfrac{a-4+\frac{1}{a}}{-\frac{1}{a}-a+4}$ $\quad -1$

30. $\dfrac{a+1+\frac{1}{a^2}}{\frac{1}{a^2}+a-1}$ $\quad \frac{a^3+a^2+1}{a^3-a^2+1}$

31. $\dfrac{1+\frac{6}{x}+\frac{8}{x^2}}{1+\frac{1}{x}-\frac{12}{x^2}}$ $\quad \frac{x+2}{x-3}$

32. $\dfrac{1-x-\frac{2}{x}}{\frac{6}{x^2}+\frac{1}{x}-1}$ $\quad \frac{x(x^2-x+2)}{(x-3)(x+2)}$

33. $\dfrac{\dfrac{1}{a+1}+1}{\dfrac{3}{a-1}+1}$ $\quad \dfrac{a-1}{a+1}$

34. $\dfrac{2+\dfrac{3}{x+1}}{\dfrac{1}{x}+x+x^2}$ $\quad \dfrac{x(2x+5)}{(x+1)(x^3+x^2+1)}$

35. $\dfrac{x^{-1}+y^{-1}}{x}$ $\quad \dfrac{y+x}{x^2y}$

36. $\dfrac{x^{-1}-y^{-1}}{y}$ $\quad \dfrac{y-x}{xy^2}$

37. $\dfrac{y}{x^{-1}-y^{-1}}$ $\quad \dfrac{xy^2}{y-x}$

38. $\dfrac{x^{-1}+y^{-1}}{(x+y)^{-1}}$ $\quad \dfrac{(x+y)^2}{xy}$

39. $\dfrac{x^{-1}+y^{-1}}{x^{-1}-y^{-1}}$ $\quad \dfrac{y+x}{y-x}$

40. $\dfrac{(x+y)^{-1}}{x^{-1}+y^{-1}}$ $\quad \dfrac{xy}{(x+y)^2}$

41. $\dfrac{x+y}{x^{-1}+y^{-1}}$ $\quad xy$

42. $\dfrac{x-y}{x^{-1}-y^{-1}}$ $\quad -xy$

43. $\dfrac{x-y^{-2}}{y-x^{-2}}$ $\quad \dfrac{x^2(xy^2-1)}{y^2(x^2y-1)}$

44. $\dfrac{x^{-2}-y^{-2}}{x^{-1}-y^{-1}}$ $\quad \dfrac{y+x}{xy}$

45. $\dfrac{1+\dfrac{a}{b}}{1-\dfrac{a}{1-\dfrac{a}{b}}}$ $\quad \dfrac{(b+a)(b-a)}{b(b-a-ab)}$

46. $\dfrac{1+\dfrac{2}{1+\dfrac{a}{b}}}{1-\dfrac{a}{b}}$ $\quad \dfrac{(3b+a)b}{b^2-a^2}$

47. $\dfrac{x-\dfrac{1}{x}}{1+\dfrac{1}{\dfrac{1}{x}}}$ $\quad \dfrac{x-1}{x}$

48. $\dfrac{\dfrac{a^2+3a+4}{ab}}{2+\dfrac{3+a}{\dfrac{2}{a}}}$ $\quad \dfrac{2}{ab}$

49. $\dfrac{b}{b+\dfrac{2}{2+\dfrac{1}{2}}}$ $\quad \dfrac{5b}{5b+4}$

50. $\dfrac{2y}{y-\dfrac{y}{3-\dfrac{1}{2}}}$ $\quad \dfrac{10}{3}$

51. $a+\dfrac{a}{1+\dfrac{a}{a+1}}$ $\quad \dfrac{3a^2+2a}{2a+1}$

52. $b+\dfrac{b}{1-\dfrac{b+1}{b}}$ $\quad b-b^2$

53. $\dfrac{x-\dfrac{1}{1-\dfrac{x}{2}}}{\dfrac{3}{x+\dfrac{2}{3}}-x}$ $\quad \dfrac{(-x^2+2x-2)(3x+2)}{(2-x)(-3x^2-2x+9)}$

54. $\dfrac{\dfrac{2x}{x-\dfrac{1}{x}}-\dfrac{1}{x}}{2x+\dfrac{2x}{1-\dfrac{1}{x}}}$ $\quad \dfrac{2x^3-x^2+1}{2x^2(x+1)(2x-1)}$

55. $\dfrac{2x+\dfrac{1}{2-\dfrac{x}{2}}}{\dfrac{4}{\dfrac{x}{2}-2}-x}$ $\quad \dfrac{2(x^2-4x-1)}{-x^2+4x+8}$

56. $\dfrac{3x-\dfrac{1}{3-\dfrac{x}{2}}}{\dfrac{3}{\dfrac{x}{2}-3}+x}$ $\quad \dfrac{3x^2-18x+2}{x^2-6x+6}$

In Exercises 57–58, factor each denominator and simplify the complex fraction.

57. $\dfrac{\dfrac{1}{x^2+3x+2}+\dfrac{1}{x^2+x-2}}{\dfrac{3x}{x^2-1}-\dfrac{x}{x+2}}$ $\quad \dfrac{-2}{x^2-3x-7}$

58. $\dfrac{\dfrac{1}{x^2-1}-\dfrac{2}{x^2+4x+3}}{\dfrac{2}{x^2+2x-3}+\dfrac{1}{x+3}}$ $\quad \dfrac{5-x}{(x+1)^2}$

Writing Exercises ■ *Write a paragraph using your own words.*

1. There are two methods used to simplify a complex fraction. Explain one of them.

2. Explain the other method of simplifying a complex fraction.

Something to Think About ■

1. Simplify: $(x^{-1}y^{-1})(x^{-1}+y^{-1})^{-1}$. $\quad \frac{1}{y+x}$

2. Simplify: $[(x^{-1}+1)^{-1}+1]^{-1}$. $\quad \frac{x+1}{2x+1}$

Review Exercises ■ *Solve each equation.*

1. $\dfrac{8(a-5)}{3}=2(a-4)$ $\quad 8$

2. $\dfrac{3t^2}{5}+\dfrac{7t}{10}=\dfrac{3t+6}{5}$ $\quad \frac{4}{3}, -\frac{3}{2}$

3. $a^4-13a^2+36=0$ $\quad 2, -2, 3, -3$

4. $|2x-1|=9$ $\quad 5, -4$

5.5 Equations Containing Rational Expressions

■ Solving Rational Equations ■ Extraneous Solutions ■ Formulas
■ Applications of Rational Equations

Equations that contain fractions with polynomial numerators and/or denominators are called **rational equations.**

■ Solving Rational Equations

To solve rational equations, we usually start by clearing the equation of fractions. To do so, we multiply both sides of the equation by the least common denominator of the fractions in the equation.

EXAMPLE 1 Solve the rational equation $\dfrac{3}{5}+\dfrac{7}{x+2}=2$.

Solution Since the denominator $x + 2$ cannot be 0, we note that x cannot be -2. We start by multiplying both sides of the equation by $5(x + 2)$ and simplifying to get

$$5(x+2)\left(\frac{3}{5} + \frac{7}{x+2}\right) = 5(x+2)2$$

$$5(x+2)\left(\frac{3}{5}\right) + 5(x+2)\left(\frac{7}{x+2}\right) = 5(x+2)2 \qquad \text{Use the distributive property on the left-hand side.}$$

$$3(x+2) + 5(7) = 10(x+2) \qquad \text{Simplify.}$$

$$3x + 6 + 35 = 10x + 20 \qquad \text{Use the distributive property and simplify.}$$

$$3x + 41 = 10x + 20 \qquad \text{Simplify.}$$

$$-7x = -21 \qquad \text{Subtract } 10x \text{ and 41 from both sides.}$$

$$x = 3 \qquad \text{Divide both sides by } -7.$$

Check: To verify that 3 satisfies the equation, we substitute 3 for x in the original equation and simplify:

$$\frac{3}{5} + \frac{7}{x+2} = 2$$

$$\frac{3}{5} + \frac{7}{3+2} \stackrel{?}{=} 2$$

$$\frac{3}{5} + \frac{7}{5} \stackrel{?}{=} 2$$

$$2 = 2$$

■

EXAMPLE 2 Solve the equation $\dfrac{-x^2 + 10}{x^2 - 1} + \dfrac{3x}{x-1} = \dfrac{2x}{x+1}$ $(x \neq 1, x \neq -1)$.

Solution We clear the equation of fractions by multiplying both sides by the LCD of the three fractions and proceed as follows:

$$\frac{-x^2 + 10}{x^2 - 1} + \frac{3x}{x-1} = \frac{2x}{x+1}$$

$$\frac{-x^2 + 10}{(x+1)(x-1)} + \frac{3x}{x-1} = \frac{2x}{x+1} \qquad \text{Factor } x^2 - 1.$$

$$\frac{(x+1)(x-1)(-x^2+10)}{(x+1)(x-1)} + \frac{3x(x+1)(x-1)}{x-1} = \frac{2x(x+1)(x-1)}{x+1} \qquad \text{Multiply both sides by } (x+1)(x-1).$$

$$-x^2 + 10 + 3x(x+1) = 2x(x-1) \qquad \text{Simplify.}$$

$$-x^2 + 10 + 3x^2 + 3x = 2x^2 - 2x \qquad \text{Remove parentheses.}$$

$$2x^2 + 10 + 3x = 2x^2 - 2x \qquad \text{Combine terms.}$$

$10 + 3x = -2x$	Add $-2x^2$ to both sides.
$10 + 5x = 0$	Add $2x$ to both sides.
$5x = -10$	Add -10 to both sides.
$x = -2$	Divide both sides by 5.

Verify that -2 is a solution to the original equation. ■

Extraneous Solutions

When we multiply both sides of an equation by a quantity that contains a variable, we sometimes get false solutions, called **extraneous solutions.** This happens when we inadvertently multiply both sides of an equation by 0 and get a solution that gives a 0 in the denominator of a fraction. We must exclude from the solution set of an equation any value that makes the denominator of a fraction 0.

The next example illustrates a rational equation with an extraneous solution.

EXAMPLE 3 Solve the equation $\dfrac{2(x+1)}{x-3} = \dfrac{x+5}{x-3}$ $(x \neq 3)$.

Solution We clear the equation of fractions by multiplying both sides by $x - 3$ and solve the resulting equation. If we obtain an apparent solution of 3, then 0 will appear in a denominator of each fraction. Thus, an apparent solution of 3 would be extraneous and should be discarded.

$$\frac{2(x+1)}{x-3} = \frac{x+5}{x-3}$$

$(x-3)\dfrac{2(x+1)}{x-3} = (x-3)\dfrac{x+5}{x-3}$	Multiply both sides by $x - 3$.
$2(x+1) = x + 5$	Simplify.
$2x + 2 = x + 5$	Remove parentheses.
$x + 2 = 5$	Add $-x$ to both sides.
$x = 3$	Add -2 to both sides.

Since x cannot be 3, it is an extraneous root and must be discarded. This equation has no solutions. Its solution set is the empty set, $\emptyset$. ■

EXAMPLE 4 Solve the equation $\dfrac{x+1}{5} - 2 = -\dfrac{4}{x}$ $(x \neq 0)$.

Solution We clear the equation of fractions by multiplying both sides by $5x$ and proceed as follows:

$$\frac{x+1}{5} - 2 = -\frac{4}{x}$$

$$5x\left(\frac{x+1}{5} - 2\right) = 5x\left(-\frac{4}{x}\right) \quad \text{Multiply both sides by } 5x.$$

$$x(x+1) - 10x = -20 \quad \text{Remove parentheses and simplify.}$$

$$x^2 + x - 10x = -20 \quad \text{Remove parentheses.}$$

$$x^2 - 9x + 20 = 0 \quad \text{Combine like terms and add 20 to both sides.}$$

$$(x-5)(x-4) = 0 \quad \text{Factor } x^2 - 9x + 20.$$

$$x - 5 = 0 \quad \text{or} \quad x - 4 = 0 \quad \text{Set each factor equal to 0.}$$

$$x = 5 \quad | \quad x = 4$$

Since 4 and 5 both satisfy the original equation, the solution set is $\{4, 5\}$. ■

Formulas

Many formulas must be cleared of fractions before we can solve them for different variables.

EXAMPLE 5 Solve the formula $\frac{1}{r} = \frac{1}{r_1} + \frac{1}{r_2}$ for r.

Solution We proceed as follows:

$$\frac{1}{r} = \frac{1}{r_1} + \frac{1}{r_2}$$

$$\frac{rr_1r_2}{r} = \frac{rr_1r_2}{r_1} + \frac{rr_1r_2}{r_2} \quad \text{Multiply both sides by } rr_1r_2.$$

$$r_1r_2 = rr_2 + rr_1 \quad \text{Simplify each fraction.}$$

$$r_1r_2 = r(r_2 + r_1) \quad \text{Factor out } r \text{ on the right-hand side.}$$

$$r = \frac{r_1r_2}{r_2 + r_1} \quad \text{Divide both sides by } r_2 + r_1 \text{ and use the symmetric property.}$$

■

Applications of Rational Equations

Many applications lead to rational equations.

EXAMPLE 6 **Drywalling a house** A contractor knows that his best crew can drywall a 2400-square-foot house in 4 days and that his second crew can drywall the same house in 5 days. One day must be allowed for the plaster coat to dry. If the contractor uses both crews, can the house be ready for installing trim in 4 days?

Analysis Because 1 day is necessary for drying, the drywallers must complete their work in 3 days.

Since the first crew can drywall the house in 4 days, it can do $\frac{1}{4}$ of the job in 1 day. Since the second crew can drywall the house in 5 days, it can do $\frac{1}{5}$ of the job in 1 day. If it takes x days for both crews to finish the house, together they can do $\frac{1}{x}$ of the job in 1 day.

We can then form the equation: The amount of work the first crew can do in 1 day plus the amount of work the second crew can do in 1 day equals the amount of work both crews can do in 1 day working together.

Solution If x represents the number of days it takes for both crews to drywall the house, we can form the equation

What crew 1 can do in one day	+	what crew 2 can do in one day	=	what they can do together in one day.
$\frac{1}{4}$	+	$\frac{1}{5}$	=	$\frac{1}{x}$

$$20x\left(\frac{1}{4} + \frac{1}{5}\right) = 20x\left(\frac{1}{x}\right) \qquad \text{Multiply both sides by } 20x.$$

$$5x + 4x = 20 \qquad \text{Remove parentheses and simplify.}$$

$$9x = 20 \qquad \text{Combine terms.}$$

$$x = \frac{20}{9} \qquad \text{Divide both sides by 9.}$$

Since it will take only $2\frac{2}{9}$ days for both crews to drywall the house and it takes one day for drying, the house will be ready for trimming in 4 days. ■

EXAMPLE 7 **Uniform motion problem** A man drove 200 miles to a convention. Because of road construction, his average speed on the return trip was 10 miles per hour less than his average speed going to the convention. If the return trip took 1 hour longer, how fast did he drive in each direction?

Analysis Because the distance traveled is given by the formula

$$d = rt \qquad (d \text{ is distance, } r \text{ is the rate of speed, and } t \text{ is time})$$

the formula for time is

$$t = \frac{d}{r} \qquad \text{Divide both sides by } r.$$

If r represents the average speed driving to the convention, then $r - 10$ is the average speed returning. We can organize the given information in the chart shown in Figure 5-1.

	d	$=$ r	$\cdot$ t
Going	200	r	$\frac{200}{r}$
Returning	200	$r - 10$	$\frac{200}{r - 10}$

FIGURE 5-1

Solution Let r represent the average rate of speed going to the meeting. Then $r - 10$ represents the average rate of speed on the return trip.

Because the return trip took 1 hour longer, we can form the following equation:

The time it took to travel to the convention	+	1 hour	=	the time it took to return.
$\frac{200}{r}$	+	1	=	$\frac{200}{r - 10}$

We can solve the equation as follows:

$$r(r - 10)\left(\frac{200}{r} + 1\right) = r(r - 10)\left(\frac{200}{r - 10}\right) \qquad \text{Multiply both sides by } r(r - 10).$$

$$200(r - 10) + r(r - 10) = 200r \qquad \text{Remove parentheses and simplify.}$$

$$200r - 2000 + r^2 - 10r = 200r \qquad \text{Remove parentheses.}$$

$$r^2 - 10r - 2000 = 0 \qquad \text{Add } -200r \text{ to both sides.}$$

$$(r - 50)(r + 40) = 0 \qquad \text{Factor } r^2 - 10r - 2000.$$

$$r - 50 = 0 \quad \text{or} \quad r + 40 = 0 \qquad \text{Set each factor equal to 0.}$$

$$r = 50 \quad | \quad r = -40$$

We must exclude the solution of -40, because a speed cannot be negative. Thus, the man averaged 50 miles per hour going to the convention, and he averaged $50 - 10$ or 40 miles per hour returning.

Check: At 50 miles per hour, the 200-mile trip took 4 hours. At 40 miles per hour, the return trip took 5 hours, which is 1 hour longer. ■

ORALS *Solve each equation.*

1. $\frac{4}{x} = 2$ 2
2. $\frac{9}{y} = 3$ 3
3. $\frac{4}{p} + \frac{5}{p} = 9$ 1
4. $\frac{5}{r} - \frac{2}{r} = 1$ 3
5. $\frac{4}{y} - \frac{1}{y} = 3$ 1
6. $\frac{8}{t} - \frac{2}{t} = 3$ 2

EXERCISE 5.5

In Exercises 1–30, solve each equation. If a solution is extraneous, so indicate.

1. $\frac{1}{4} + \frac{9}{x} = 1$ 12
2. $\frac{1}{3} - \frac{10}{x} = -3$ 3
3. $\frac{34}{x} - \frac{3}{2} = -\frac{13}{20}$ 40
4. $\frac{1}{2} + \frac{7}{x} = 2 + \frac{1}{x}$ 4
5. $\frac{3}{y} + \frac{7}{2y} = 13$ $\frac{1}{2}$
6. $\frac{2}{x} + \frac{1}{2} = \frac{7}{2x}$ 3
7. $\frac{x+1}{x} - \frac{x-1}{x} = 0$ no solution
8. $\frac{2}{x} + \frac{1}{2} = \frac{9}{4x} - \frac{1}{2x}$ $-\frac{1}{2}$
9. $\frac{7}{5x} - \frac{1}{2} = \frac{5}{6x} + \frac{1}{3}$ $\frac{17}{25}$
10. $\frac{x-3}{x-1} - \frac{2x-4}{x-1} = 0$ 1 is extraneous
11. $\frac{3-5y}{2+y} = \frac{3+5y}{2-y}$ 0
12. $\frac{x}{x-2} = 1 + \frac{1}{x-3}$ 4
13. $\frac{a+2}{a+1} = \frac{a-4}{a-3}$ 1
14. $\frac{z+2}{z+8} - \frac{z-3}{z-2} = 0$ 4
15. $\frac{x+2}{x+3} - 1 = \frac{1}{3-2x-x^2}$ 2
16. $\frac{x-3}{x-2} - \frac{1}{x} = \frac{x-3}{x}$ 4
17. $\frac{x}{x+2} = 1 - \frac{3x+2}{x^2+4x+4}$ 2
18. $\frac{3+2a}{a^2+6+5a} + \frac{2-5a}{a^2-4} = \frac{2-3a}{a^2-6+a}$ $-\frac{2}{5}$
19. $\frac{2}{x-2} + \frac{1}{x+1} = \frac{1}{x^2-x-2}$ $\frac{1}{3}$
20. $\frac{5}{y-1} + \frac{3}{y-3} = \frac{8}{y-2}$ 6
21. $\frac{a-1}{a+3} - \frac{1-2a}{3-a} = \frac{2-a}{a-3}$ 0
22. $\frac{5}{2z^2+z-3} - \frac{2}{2z+3} = \frac{z+1}{z-1} - 1$ $\frac{1}{6}$
23. $\frac{5}{x+4} + \frac{1}{x+4} = x - 1$ 2, −5
24. $\frac{2}{x-1} + \frac{x-2}{3} = \frac{4}{x-1}$ 4, −1
25. $\frac{3}{x+1} - \frac{x-2}{2} = \frac{x-2}{x+1}$ −4, 3
26. $\frac{x-4}{x-3} + \frac{x-2}{x-3} = x - 3$ 5; 3 is extraneous
27. $\frac{2}{x-3} + \frac{3}{4} = \frac{17}{2x}$ 6, $\frac{17}{3}$
28. $\frac{30}{y-2} + \frac{24}{y-5} = 13$ 8, $\frac{41}{13}$

29. $\dfrac{x+4}{x+7} - \dfrac{x}{x+3} = \dfrac{3}{8}$ 1, −11

30. $\dfrac{5}{x+4} - \dfrac{1}{3} = \dfrac{x-1}{x}$ $-\frac{3}{2}, 2$

In Exercises 31–36, solve each formula for the indicated variable.

31. $\dfrac{1}{p} + \dfrac{1}{q} = \dfrac{1}{f}$ for f $f = \frac{pq}{q+p}$

32. $\dfrac{1}{p} + \dfrac{1}{q} = \dfrac{1}{f}$ for p $p = \frac{qf}{q-f}$

33. $S = \dfrac{a - lr}{1 - r}$ for r $r = \frac{S-a}{S-l}$

34. $H = \dfrac{2ab}{a+b}$ for a $a = \frac{Hb}{2b-H}$

35. $\dfrac{1}{R} = \dfrac{1}{r_1} + \dfrac{1}{r_2} + \dfrac{1}{r_3}$ for R $R = \dfrac{r_1 r_2 r_3}{r_1 r_3 + r_1 r_2 + r_2 r_3}$

36. $\dfrac{1}{R} = \dfrac{1}{r_1} + \dfrac{1}{r_2} + \dfrac{1}{r_3}$ for r_1 $r_1 = \dfrac{-R r_2 r_3}{R r_3 + R r_2 - r_2 r_3}$

In Exercises 37–58, solve each problem.

37. Focal length The design of a camera lens uses the equation

$$\frac{1}{f} = \frac{1}{s_1} + \frac{1}{s_2}$$

which relates the focal length, f, of a lens to the image distance, s_1, and the object distance, s_2. Find the focal length of the lens in Illustration 1. (*Hint:* Convert feet to inches.) $4\frac{8}{13}$ in.

Object

Image

$s_2 = 5$ ft

$s_1 = 5$ in.

ILLUSTRATION 1

38. Lens maker's formula The focal length, f, of a lens is given by the lens maker's formula,

$$\frac{1}{f} = 0.6\left(\frac{1}{r_1} + \frac{1}{r_2}\right)$$

where f is the focal length of the lens and r_1 and r_2 are the radii of the two circular surfaces. Find the focal length of the lens in Illustration 2. $\frac{20}{3}$ cm

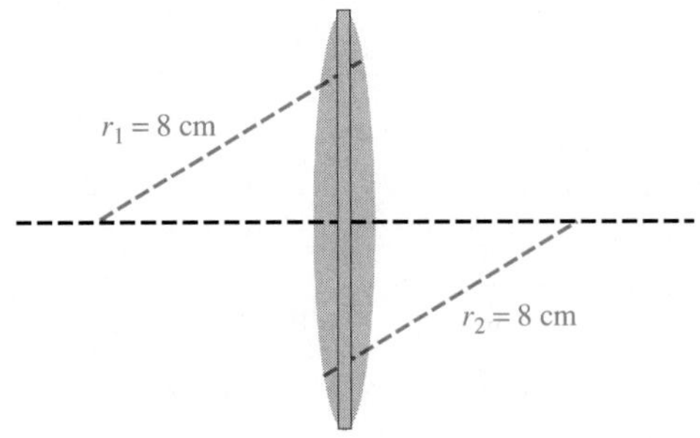

ILLUSTRATION 2

39. House painting If one painter can paint a house in 5 days and another painter can paint the same house in 3 days, how long will it take them to paint the house if they work together? $1\frac{7}{8}$ days

40. Reading proof A proofreader can read 50 pages in 3 hours, and a second proofreader can read 50 pages 1 hour. If they both work on a 250-page book, can they meet a six-hour deadline? yes

41. Storing corn In 10 minutes, a conveyor belt can move 1000 bushels of corn into the storage bin shown in Illustration 3. A smaller belt can move 1000 bushels to the storage bin in 14 minutes. If both belts are used, how long it will take to move 1000 bushels to the storage bin? $5\frac{5}{6}$ min

ILLUSTRATION 3

42. **Roofing a house** One roofing crew can finish a 2800-square-foot roof in 12 hours, and another crew can do the job in 10 hours. If they work together, can they finish before a predicted rain in 5 hours? no

43. **Draining a swimming pool** A drain can empty the swimming pool shown in Illustration 4 in 3 days. A second drain can empty the pool in 2 days. How long will it take to empty the pool if both drains are used? $1\frac{1}{5}$ days

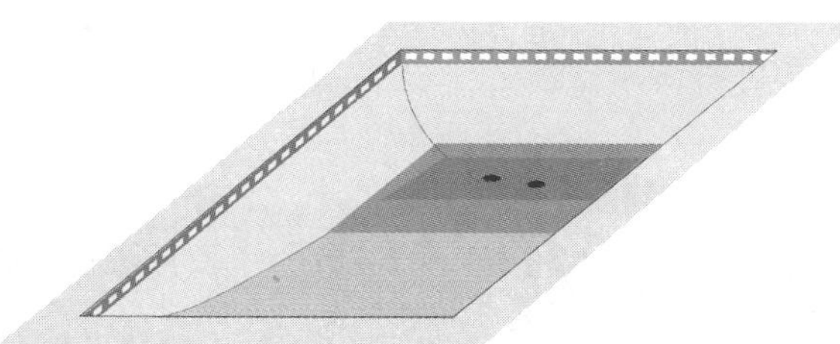

ILLUSTRATION 4

44. **Filling a pool** A pipe can fill a pool in 9 hours. If a second pipe is also used, the pool can be filled in 3 hours. How long would it take the second pipe alone to fill the pool? $4\frac{1}{2}$ hr

45. **Filling a pond** One pipe can fill a pond in 3 weeks, and a second pipe can fill the pond in 5 weeks. However, evaporation and seepage can empty the pond in 10 weeks. If both pipes are used, how long will it take to fill the pond? $2\frac{4}{13}$ weeks

46. **House cleaning** Sally can clean the house in 6 hours, and her father can clean the house in 4 hours. Sally's younger brother, Dennis, can completely mess up the house in 8 hours. If Sally and her father clean and Dennis plays, how long will it take to clean the house? $3\frac{3}{7}$ hr

47. **Making hamburgers** Sam makes hamburgers at a fast-food restaurant. He can make 600 burgers in 3 hours. If his manager helps him, together they can make 600 burgers in 2 hours. How long will it take the manager alone to make 400 burgers? 4 hr

48. **Driving a boat** A boy can drive a motorboat 45 miles down the Rock River in the same amount of time that he can drive 27 miles upstream. Find the speed of the current if the speed of the boat is 12 miles per hour in still water. 3 mph

49. **Flying a plane** If a plane that can fly 340 miles per hour in still air can travel 200 miles downwind in the same amount of time that it can travel 140 miles upwind, find the velocity of the wind. 60 mph

50. **Train travel** A train traveled 120 miles from Freeport to Chicago and returned the same distance in a total time of 5 hours. If the train traveled 20 miles per hour slower on the return trip, how fast did the train travel in each direction?
60 mph and 40 mph

51. **Rowing a boat** A woman who can row 3 miles per hour in still water rows 10 miles downstream on the Eagle River and returns upstream in a total of 12 hours. Find the speed of the current. 2 mph

52. **Touring the countryside** A man bicycles 5 miles per hour faster than he can walk. He bicycles 24 miles and walks back along the same route, completing a 11-hour tour. How fast does he walk?
3 mph

53. **Number puzzle** If three times a certain integer is added to four times its reciprocal, the result is 8. Find the integer. 2

54. **Number puzzle** If three times a number is subtracted from four times its reciprocal, the result is 11. Find the number. (There are two possibilities.)
$\frac{1}{3}, -4$

55. **Parts inventory** A repairman purchased several washing-machine motors for a total of \$224. When the unit cost decreased by \$4, he was able to buy one extra motor for the same total price. How many motors did he buy originally? (*Hint:* Cost = (unit price)(quantity).) 7 motors

56. **Price increase** An appliance store manager bought several microwave ovens for a total of \$1800. When her unit cost increased by \$25, she was able to buy one less oven for the same total price. How many ovens did she buy originally? 9 ovens

57. **Stretching a vacation** A student saved \$1200 for a trip to Europe. By cutting \$20 from her daily expenses, she was able to stay three extra days. How long had she originally planned to be gone? 12 days

58. **Planting a windscreen** A homeowner wants to plant a windscreen of several trees, spaced equally in a 120-foot row. The landscape architect can earn more money if he plants two more trees and spaces them two feet closer to each other. How many trees is the homeowner planning? (*Hint:* A row of n trees has $n - 1$ spaces between; $n + 2$ trees has $n + 2 - 1$ spaces between.) 11 trees

Writing Exercises *Write a paragraph using your own words.*

1. Why is it necessary to check the solutions of a rational equation?
2. Explain the steps you would use to solve a rational equation.

Something to Think About

1. Invent a rational equation that has an extraneous solution of 3.
2. Solve: $(x - 1)^{-1} - x^{-1} = 6^{-1}$. $-2, 3$

Review Exercises *Simplify each expression. Write each answer without using negative exponents.*

1. $(m^2n^{-3})^{-2}$ $\quad \frac{n^6}{m^4}$
2. $\frac{a^{-1}}{a^{-1} + 1}$ $\quad \frac{1}{1 + a}$
3. $\frac{a^0 + 2a^0 - 3a^0}{(a - b)^0}$ $\quad 0$
4. $(4x^{-2} + 3)(2x - 4)$ $\quad \frac{8}{x} - \frac{16}{x^2} + 6x - 12$

PROJECT Combined Operations

The expression $a(b + c)$ may be calculated in two ways. In the first method, we evaluate the expression directly by using the priority rules and performing the addition before the multiplication. In the second method, we use the distributive property of real numbers and multiply each term within the parentheses by the factor a:

$$a(b + c) = ab + ac$$

Either way, the results are equal.

The distributive property is also true for fractional expressions. For example, the expression $\left(\frac{x-1}{3}\right)\left(\frac{1}{x-1} + 1\right)$ may be simplified in either of two ways:

Using the Priority Rules

$$\left(\frac{x - 1}{3}\right)\left(\frac{1}{x - 1} + 1\right)$$

$$= \left(\frac{x-1}{3}\right)\left(\frac{1}{x-1} + \frac{x-1}{x-1}\right)$$ Find the common denominator.

$$= \left(\frac{x-1}{3}\right)\left(\frac{x}{x-1}\right)$$ Add the fractions and combine terms.

$$= \frac{(x-1)x}{3(x-1)}$$ Multiply the fractions.

$$= \frac{x}{3}$$ Simplify the fraction.

Using the Distributive Property

$$\left(\frac{x-1}{3}\right)\left(\frac{1}{x-1} + 1\right)$$

$$= \left(\frac{x-1}{3}\right)\left(\frac{1}{x-1}\right) + \left(\frac{x-1}{3}\right)(1)$$ Use the distributive property.

$$= \frac{x-1}{3(x-1)} + \frac{(x-1)}{3}$$ Multiply the fractions.

$$= \frac{1}{3} + \frac{x-1}{3}$$ Simplify the fractions.

$$= \frac{1+x-1}{3}$$ Add the fractions.

$$= \frac{x}{3}$$ Combine terms.

Either way, the results agree.

Simplify each of the following expressions in two ways: Use the priority rules and perform the addition first. Then use the distributive property. Finally, verify that the two results agree.

1. $\left(\frac{x^2-9}{3x}\right)\left(\frac{1}{x-3} - \frac{1}{x+3}\right)$ $\frac{2}{x}$ **2.** $\left(\frac{x+1}{2x+4}\right)\left(\frac{1}{x-1} + \frac{1}{x+1}\right)$ $\frac{x}{(x+2)(x-1)}$

Chapter Summary

KEY WORDS

complex fractions (5.4)
extraneous solutions (5.5)
least common denominator (5.3)
rational equations (5.5)
rational expressions (5.1)

KEY IDEAS

(5.1) Division by 0 is undefined.

$\frac{ak}{bk} = \frac{a}{b}$, provided that $b \neq 0$ and $k \neq 0$.

To simplify a fraction, factor the numerator and denominator and divide out all factors common to both the numerator and denominator.

(5.2) $\frac{a}{b} \cdot \frac{c}{d} = \frac{ac}{bd}$ $\quad (b \neq 0 \text{ and } d \neq 0)$

$\frac{a}{b} \div \frac{c}{d} = \frac{a}{b} \cdot \frac{d}{c}$ $\quad (b \neq 0, d \neq 0, \text{ and } c \neq 0)$

(5.3) $\frac{a}{b} + \frac{c}{b} = \frac{a+c}{b}$ $\quad (b \neq 0)$

$\frac{a}{b} - \frac{c}{b} = \frac{a-c}{b}$ $\quad (b \neq 0)$

Fractions must have common denominators before they can be added or subtracted.

To find the least common denominator (LCD) of two fractions, factor each denominator and use each factor the greatest number of times that it appears in any one denominator. The product of these factors is the LCD of the fractions.

(5.4) A fraction that has a fraction within its numerator or its denominator is called a **complex fraction.**

(5.5) Multiplying both sides of an equation by a quantity that contains a variable can lead to **extraneous solutions.** All possible solutions of a rational equation must be checked.

Chapter 5 Review Exercises

In Review Exercises 1–8, simplify each fraction.

1. $\frac{248x^2y}{576xy^2}$ $\quad \frac{31x}{72y}$
2. $\frac{212m^3n}{588m^2n^3}$ $\quad \frac{53m}{147n^2}$
3. $\frac{x^2-49}{x^2+14x+49}$ $\quad \frac{x-7}{x+7}$
4. $\frac{x^2+6x+36}{x^3-216}$ $\quad \frac{1}{x-6}$
5. $\frac{x^2-2x+4}{2x^3+16}$ $\quad \frac{1}{2x+4}$
6. $\frac{x-y}{y-x}$ $\quad -1$
7. $\frac{2m-2n}{n-m}$ $\quad -2$
8. $\frac{ac-ad+bc-bd}{d^2-c^2}$ $\quad \frac{-a-b}{c+d}$

In Review Exercises 9–22, do the operations and simplify.

9. $\frac{x^2+4x+4}{x^2-x-6} \cdot \frac{x^2-9}{x^2+5x+6}$ $\quad 1$
10. $\frac{x^3-64}{x^2+4x+16} \div \frac{x^2-16}{x+4}$ $\quad 1$
11. $\frac{5y}{x-y} - \frac{3}{x-y}$ $\quad \frac{5y-3}{x-y}$
12. $\frac{3x-1}{x^2+2} + \frac{3(x-2)}{x^2+2}$ $\quad \frac{6x-7}{x^2+2}$
13. $\frac{3}{x+2} + \frac{2}{x+3}$ $\quad \frac{5x+13}{(x+2)(x+3)}$
14. $\frac{4x}{x-4} - \frac{3}{x+3}$ $\quad \frac{4x^2+9x+12}{(x-4)(x+3)}$
15. $\frac{x^2+3x+2}{x^2-x-6} \cdot \frac{3x^2-3x}{x^2-3x-4} \div \frac{x^2+3x+2}{x^2-2x-8}$ $\quad \frac{3x(x-1)}{(x-3)(x+1)}$
16. $\frac{x^2-x-6}{x^2-3x-10} \div \frac{x^2-x}{x^2-5x} \cdot \frac{x^2-4x+3}{x^2-6x+9}$ $\quad 1$
17. $\frac{2x}{x+1} + \frac{3x}{x+2} + \frac{4x}{x^2+3x+2}$ $\quad \frac{5x^2+11x}{(x+1)(x+2)}$
18. $\frac{5x}{x-3} + \frac{5}{x^2-5x+6} + \frac{x+3}{x-2}$ $\quad \frac{2(3x+1)}{x-3}$

19. $\frac{3(x+2)}{x^2-1} - \frac{2}{x+1} + \frac{4(x+3)}{x^2-2x+1}$ $\frac{5x^2+23x+4}{(x+1)(x-1)(x-1)}$

20. $\frac{x}{x^2+4x+4} + \frac{2x}{x^2-4} - \frac{x^2-4}{x-2}$ $\frac{-x^4-4x^3+3x^2+18x+16}{(x-2)(x+2)^2}$

21. $\frac{x+2}{x^2-9} - \frac{x^2+6x+9}{x^2+5x+6} + \frac{3x}{x-3}$ $\frac{2x^3+13x^2+31x+31}{(x+3)(x+2)(x-3)}$

22. $\frac{-2(3+x)}{x^2+6x+9} + \frac{3(x+2)}{x^2-6x+9} - \frac{1}{x^2-9}$ $\frac{x^2+26x+3}{(x+3)(x-3)^2}$

In Review Exercises 23–34, simplify each complex fraction.

23. $\dfrac{\frac{3}{x}-\frac{2}{y}}{xy}$ $\frac{3y-2x}{x^2y^2}$

24. $\dfrac{\frac{1}{x}+\frac{2}{y}}{\frac{2}{x}-\frac{1}{y}}$ $\frac{y+2x}{2y-x}$

25. $\dfrac{2x+3+\frac{1}{x}}{x+2+\frac{1}{x}}$ $\frac{2x+1}{x+1}$

26. $\dfrac{6x+13+\frac{6}{x}}{6x+5-\frac{6}{x}}$ $\frac{3x+2}{3x-2}$

27. $\dfrac{1+\frac{3}{x}}{x+3}$ $\frac{1}{x}$

28. $\dfrac{1-\frac{1}{x}-\frac{2}{x^2}}{1+\frac{4}{x}+\frac{3}{x^2}}$ $\frac{x-2}{x+3}$

29. $\frac{(x-y)^{-2}}{x^{-2}-y^{-2}}$ $\frac{x^2y^2}{(x-y)^2(y^2-x^2)}$

30. $\frac{x^{-1}+1}{x+1}$ $\frac{1}{x}$

31. $\frac{x^{-1}+y}{x-y^{-1}}$ $\frac{y(1+xy)}{x(xy-1)}$

32. $\frac{x^{-1}+1}{x^{-1}-1}$ $\frac{1+x}{1-x}$

33. $\frac{x^{-2}+1}{x^2+1}$ $\frac{1}{x^2}$

34. $\frac{x^{-1}-y^{-1}}{x^{-1}+y^{-1}}$ $\frac{y-x}{y+x}$

In Review Exercises 35–38, solve each equation, if possible.

35. $\frac{4}{x} - \frac{1}{10} = \frac{7}{2x}$ 5

36. $\frac{2}{x+5} - \frac{1}{6} = \frac{1}{x+4}$ $-1, -2$

37. $\frac{2(x-5)}{x-2} = \frac{6x+12}{4-x^2}$ 2 and -2 are extraneous

38. $\frac{7}{x+9} - \frac{x+2}{2} = \frac{x+4}{x+9}$ $-1, -12$

In Review Exercises 39–40, solve each formula for the indicated variable.

39. $\frac{x^2}{a^2} - \frac{y^2}{b^2} = 1$ for y^2 $y^2 = \frac{x^2b^2-a^2b^2}{a^2}$

40. $H = \frac{2ab}{a+b}$ for b $b = \frac{Ha}{2a-H}$

In Review Exercises 41–44, solve each problem.

41. Trip length Traffic reduced Jim's usual speed by 10 miles per hour, which lengthened his 200-mile trip by 1 hour. Find his usual speed. 50 mph

42. Flying speed On a 600-mile trip, a pilot can save 30 minutes by increasing her usual speed by 40 miles per hour. Find her usual speed. 200 mph

43. Draining a tank If one outlet pipe can drain a tank in 24 hours and another pipe can drain the tank in 36 hours, how long will it take for both pipes to drain the tank? $14\frac{2}{5}$ hr

44. Siding a house Two men have estimated that they can side a house in 8 days. If one of them, who could have sided the house alone in 14 days, gets sick, how long will it take the other man to side the house alone? $18\frac{2}{3}$ days

Chapter 5 Test

In Problems 1–4, simplify each fraction.

1. $\dfrac{-12x^2y^3z^2}{18x^3y^4z^2}$ $\frac{-2}{3xy}$

2. $\dfrac{2x+4}{x^2-4}$ $\frac{2}{x-2}$

3. $\dfrac{3y-6z}{2z-y}$ -3

4. $\dfrac{2x^2+7x+3}{4x+12}$ $\frac{2x+1}{4}$

In Problems 5–14, do the operations and simplify, if necessary. Write all answers without negative exponents.

5. $\dfrac{x^2y^{-2}}{x^3z^2} \cdot \dfrac{x^2z^4}{y^2z}$ $\frac{xz}{y^4}$

6. $\dfrac{(x+1)(x+2)}{10} \cdot \dfrac{5}{x+2}$ $\frac{x+1}{2}$

7. $\dfrac{u^2+5u+6}{u^2-4} \cdot \dfrac{u^2-5u+6}{u^2-9}$ 1

8. $\dfrac{x^3+y^3}{4} \div \dfrac{x^2-xy+y^2}{2x+2y}$ $\frac{(x+y)^2}{2}$

9. $\dfrac{xu+2u+3x+6}{u^2-9} \cdot \dfrac{2u-6}{x^2+3x+2}$ $\frac{2}{x+1}$

10. $\dfrac{a^2+7a+12}{a+3} \div \dfrac{16-a^2}{a-4}$ -1

11. $\dfrac{3t}{t+3} + \dfrac{9}{t+3}$ 3

12. $\dfrac{3w}{w-5} + \dfrac{w+10}{5-w}$ 2

13. $\dfrac{2}{r} + \dfrac{r}{s}$ $\frac{2s+r^2}{rs}$

14. $\dfrac{x+2}{x+1} - \dfrac{x+1}{x+2}$ $\frac{2x+3}{(x+1)(x+2)}$

In Problems 15–16, simplify each complex fraction.

15. $\dfrac{\dfrac{2u^2w^3}{v^2}}{\dfrac{4uw^4}{uv}}$ $\frac{u^2}{2vw}$

16. $\dfrac{\dfrac{x}{y}+\dfrac{1}{2}}{\dfrac{x}{2}-\dfrac{1}{y}}$ $\frac{2x+y}{xy-2}$

In Problems 17–18, solve each equation.

17. $\dfrac{2}{x-1} + \dfrac{5}{x+2} = \dfrac{11}{x+2}$ $\frac{5}{2}$

18. $\dfrac{u-2}{u-3} + 3 = u + \dfrac{u-4}{3-u}$ 5; 3 is extraneous

In Problems 19–20, solve each formula for the indicated variable.

19. $\dfrac{x^2}{a^2} + \dfrac{y^2}{b^2} = 1$ for a^2 $\quad a^2 = \dfrac{x^2b^2}{b^2 - y^2}$

20. $\dfrac{1}{r} = \dfrac{1}{r_1} + \dfrac{1}{r_2}$ for r_2 $\quad r_2 = \dfrac{rr_1}{r_1 - r}$

In Problems 21–22, solve each problem.

21. **Sailing time** A boat sails a distance of 440 nautical miles. If the boat had averaged 11 nautical miles more each day, the trip would have required 2 fewer days. How long did the trip take? 10 days

22. **Investing** A student can earn $300 interest annually by investing in a bank certificate of deposit at a certain interest rate. If she were to receive an annual interest rate that is 4% higher, she could receive the same annual interest by investing $2000 less. How much would she invest at each rate? (*Hint:* 4% is 0.04.) $5000 at 6% and $3000 at 10%

CAREERS & MATHEMATICS

MECHANICAL ENGINEER

Mechanical engineers design and develop power-producing machines such as internal combustion engines, steam and gas turbines, and jet and rocket engines, as well as power-using machines such as refrigeration and air-conditioning equipment, machine tools, printing presses, and steel rolling mills. Many mechanical engineers do research, test, and design work; others work in maintenance, technical sales, and production operations. Some teach in colleges and universities or work as consultants.

Qualifications A bachelor's degree in engineering is generally required for beginning engineering jobs. College graduates with a degree in a natural science or mathematics also may qualify for some jobs.

Job Outlook Employment of mechanical engineers is expected to increase through the year 2005 due to the growing demand for complex industrial machinery and processes, as well as the need for new energy systems and solutions to environmental pollution problems.

Example Application Mechanical power is often transmitted from one location to another by means of a shaft—the drive shaft in an automobile is a good example. It is one concern of the mechanical engineer that the shaft be strong enough that it will not twist, bend, or snap. One critical value in the required calculations is the shaft's **stiffness,** denoted here by the letter k. The stiffness of a shaft is measured in units of inch pounds per radian.

If a single shaft consists of two sections (see Illustration 1), then the shaft's overall stiffness k is given by the formula

$$k = \frac{1}{\frac{1}{k_1} + \frac{1}{k_2}}$$

Section 1 Section 2

ILLUSTRATION 1

where k_1 and k_2 are the individual stiffnesses of Sections 1 and 2, respectively.

If the stiffness, k_2, of Section 2 of the shaft in Illustration 1 is 4,200,000 in. lb/rad, and the design specifies that the overall stiffness, k, of the entire shaft be 1,900,000 in. lb/rad, what must k_1, the stiffness of Section 1, be?

Solution We begin by solving the given equation for k_1.

$$k = \frac{1}{\frac{1}{k_1} + \frac{1}{k_2}}$$

$$k = \frac{1k_1k_2}{\left(\frac{1}{k_1} + \frac{1}{k_2}\right)k_1k_2}$$

Simplify the complex fraction by multiplying numerator and denominator by k_1k_2.

$$k = \frac{k_1k_2}{k_2 + k_1}$$

$$k(k_2 + k_1) = \frac{k_1k_2}{k_2 + k_1}(k_2 + k_1)$$

Clear the equation of fractions.

$$kk_2 + kk_1 = k_1k_2$$

$$kk_2 = k_1k_2 - kk_1$$

Prepare to solve for k_1 by subtracting kk_1 from both sides.

$$kk_2 = k_1(k_2 - k)$$

Factor out k_1 from the right-hand side.

$$\textbf{1.}\quad \frac{kk_2}{k_2 - k} = k_1$$

Solve for k_1 by dividing both sides by $k_2 - k$.

To find the stiffness required for Section 1, substitute 4,200,000 for k_2 and 1,900,000 for k into Equation 1, and find k_1.

$$k_1 = \frac{kk_2}{k_2 - k}$$

$$k_1 = \frac{(1{,}900{,}000)(4{,}200{,}000)}{4{,}200{,}000 - 1{,}900{,}000}$$

$$k_1 \approx 3{,}500{,}000$$

Use a calculator.

The stiffness of Section 1 must be approximately 3,500,000 in. lb/rad.

EXERCISES

1. Find k_1, the stiffness of Section 1, if $k_2 = 3{,}700{,}000$ in. lb/rad and $k = 1{,}530{,}000$ in. lb/rad.
2. What is the stiffness of a shaft whose two sections have the same stiffness, 4,000,000 in. lb/rad?
3. Show that the formula of this example may also be written as

$$k = \frac{k_1k_2}{k_1 + k_2}$$

4. The **compliance** of a shaft is the reciprocal of the stiffness. Show that the compliance of the entire shaft is the sum of the compliances of each of its sections.

(*Answers:* **1.** approximately 2,600,000 in. lb/rad **2.** 2,000,000 in. lb/rad)

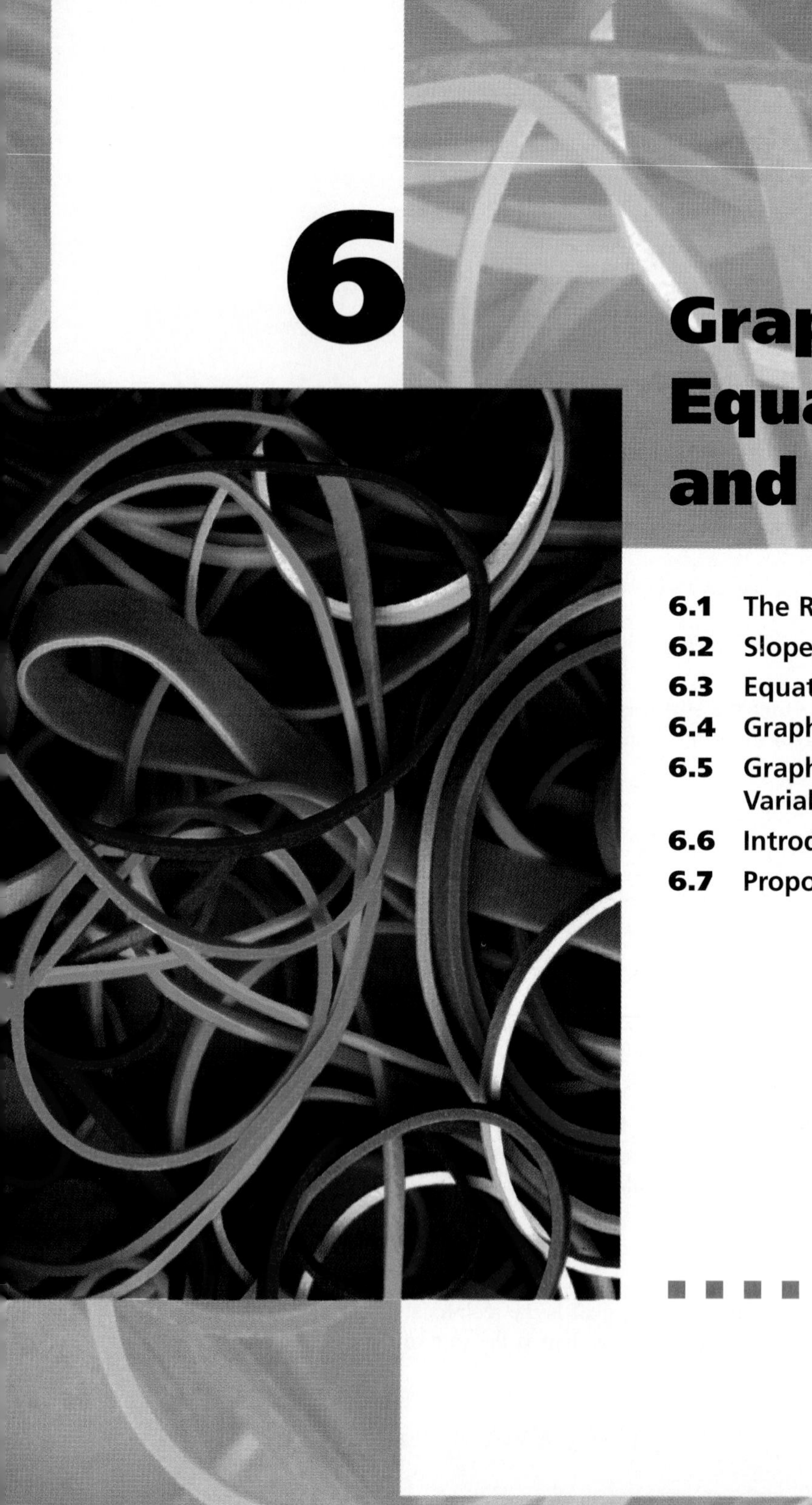

6 Graphs, Equations of Lines and Variation

Most equations we have encountered so far have involved only one variable. However, many applications of mathematics involve equations with two or more variables. In this chapter, we will discuss how to graph equations and inequalities with two variables.

6.1 The Rectangular Coordinate System

■ Graphing Equations ■ Graphing Lines Parallel to the x- and y-Axes
■ The Midpoint Formula

René Descartes (1596–1650) is credited with the idea of associating ordered pairs of real numbers with points in the geometric plane. His idea is based on two perpendicular number lines, one horizontal and one vertical, that divide the plane into four quadrants numbered as in Figure 6-1.

Historical Note: The first attempt to locate points by using coordinates appeared in the work of Nicole Oresme (1323–1382).

The horizontal number line is the ***x*-axis,** and the vertical number line is the ***y*-axis.** The point where the axes intersect, called the **origin,** is the 0 point on each number line.

The positive direction on the x-axis is to the right, the positive direction on the y-axis is upward, and the unit distance on each axis is the same. This ***xy*-plane** is called a **rectangular coordinate system** or a **Cartesian coordinate system.**

René Descartes
(1596–1650)
Descartes is famous for his work in philosophy as well as in mathematics. His philosophy is expressed in the words "I think, therefore I am." He is best known in mathematics for his invention of a coordinate system and his work with conic sections.

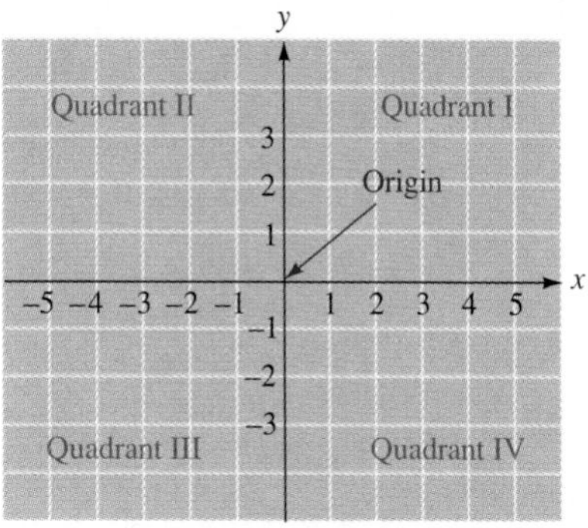

FIGURE 6-1

To plot the point associated with the pair of real numbers (2, 3), we start at the origin and count 2 units to the right and then 3 units up, as in Figure 6-2. The point P, which lies in the first quadrant, is the **graph** of the pair (2, 3). The pair (2, 3) gives the **coordinates** of point P.

To plot point Q with coordinates $(-4, 6)$, we start at the origin and count 4 units to the left and then 6 units up. Point Q lies in the second quadrant. The point R with coordinates $(6, -4)$ lies in the fourth quadrant.

WARNING! The pairs $(-4, 6)$ and $(6, -4)$ represent different points. One is in the second quadrant, and one is in the fourth quadrant. Since order is important when graphing pairs of real numbers, such pairs are called **ordered pairs.**

Historical Note: The words *coordinate, abscissa,* and *ordinate* were introduced by Gottfried Wilhelm Leibniz (1646–1716).

In the ordered pair (a, b), a is called the ***x*-coordinate** or the **abscissa,** and b is called the ***y*-coordinate** or the **ordinate.**

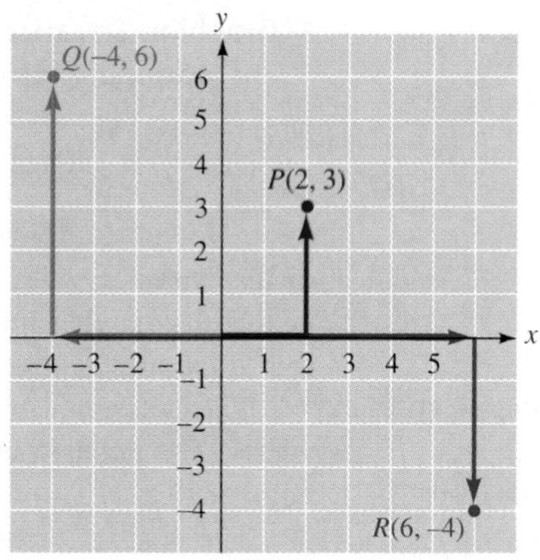

FIGURE 6-2

EXAMPLE 1 Graph the set $\{(-1, -1), (0, 0), (1, 1), (2, 2)\}$.

Solution We draw an x-axis and a y-axis that are perpendicular and plot each ordered pair. The four points shown in Figure 6-3 form the graph of the given set.

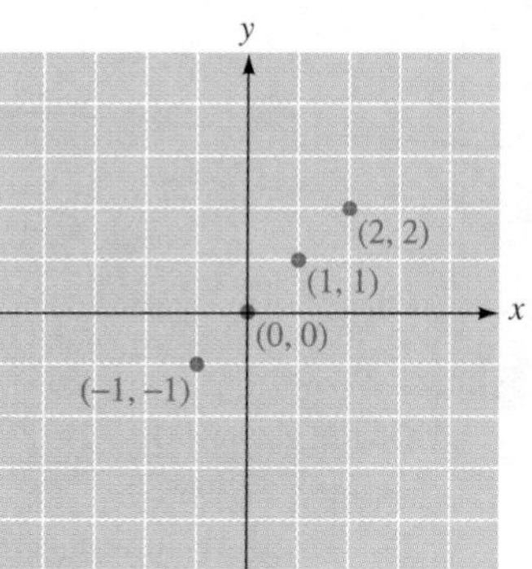

FIGURE 6-3

■

Graphing Equations

The **graph of an equation** in the variables x and y is the set of all points on a rectangular coordinate system with coordinates (x, y) that satisfy the equation.

EXAMPLE 2 Graph the equation $3x + 2y = 12$.

Solution We pick values for either x or y, substitute them in the equation, and solve for the other variable. For example, if $x = 2$, then

$$\begin{aligned} 3x + 2y &= 12 \\ 3(2) + 2y &= 12 && \text{Substitute 2 for } x. \\ 6 + 2y &= 12 && \text{Simplify.} \\ 2y &= 6 && \text{Subtract 6 from both sides.} \\ y &= 3 && \text{Divide both sides by 2.} \end{aligned}$$

One ordered pair that satisfies the equation is (2, 3).

If $y = 6$, we have

$$\begin{aligned} 3x + 2y &= 12 \\ 3x + 2(6) &= 12 && \text{Substitute 6 for } y. \\ 3x + 12 &= 12 && \text{Simplify.} \\ 3x &= 0 && \text{Subtract 12 from both sides.} \\ x &= 0 && \text{Divide both sides by 3.} \end{aligned}$$

A second ordered pair that satisfies the equation is (0, 6).

The pairs (2, 3) and (0, 6) and others that satisfy the equation are shown in the table in Figure 6-4. We plot each of these pairs on a rectangular coordinate system and join them to get the line shown in the figure. This line is the graph of the equation.

Teaching Tip: Use an overhead projector to project a grid on the chalkboard. It is surprisingly visible.

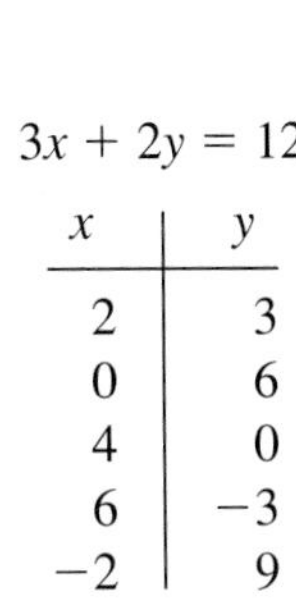
$3x + 2y = 12$

x	y
2	3
0	6
4	0
6	−3
−2	9

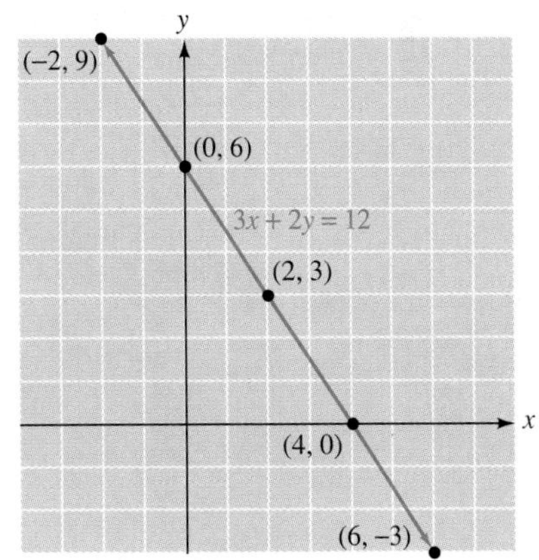

FIGURE 6-4

■

Intercepts of a Line The ***y*-intercept** of a line is the point $(0, b)$ where the line intersects the y-axis. To find b, substitute 0 for x in the equation of the line and solve for y.

The ***x*-intercept** of a line is the point $(a, 0)$ where the line intersects the x-axis. To find a, substitute 0 for y in the equation of the line and solve for x.

In Example 2, the y-intercept of the line is the point with coordinates of (0, 6), and the x-intercept is the point with coordinates of (4, 0).

EXAMPLE 3 Use the x- and y-intercepts to graph the equation $2x + 5y = 10$.

Solution To find the y-intercept, we substitute 0 for x and solve for y:

$$\begin{aligned} 2x + 5y &= 10 \\ 2(0) + 5y &= 10 && \text{Substitute 0 for } x. \\ 5y &= 10 && \text{Simplify.} \\ y &= 2 && \text{Divide both sides by 5.} \end{aligned}$$

The y-intercept is the point (0, 2).

To find the x-intercept, we substitute 0 for y and solve for x:

$$\begin{aligned} 2x + 5y &= 10 \\ 2x + 5(0) &= 10 && \text{Substitute 0 for } y. \\ 2x &= 10 && \text{Simplify.} \\ x &= 5 && \text{Divide both sides by 2.} \end{aligned}$$

The x-intercept is the point (5, 0).

Although two points are sufficient to draw the line, it is a good idea to find and plot a third point as a check. To find the coordinates of a third point, we can substitute any convenient number (such as -5) for x and solve for y:

$$\begin{aligned} 2x + 5y &= 10 \\ 2(-5) + 5y &= 10 && \text{Substitute } -5 \text{ for } x. \\ -10 + 5y &= 10 && \text{Simplify.} \\ 5y &= 20 && \text{Add 10 to both sides.} \\ y &= 4 && \text{Divide both sides by 5.} \end{aligned}$$

The line will also pass through the point $(-5, 4)$. The graph is shown in Figure 6-5.

$2x + 5y = 10$

x	y
0	2
5	0
−5	4

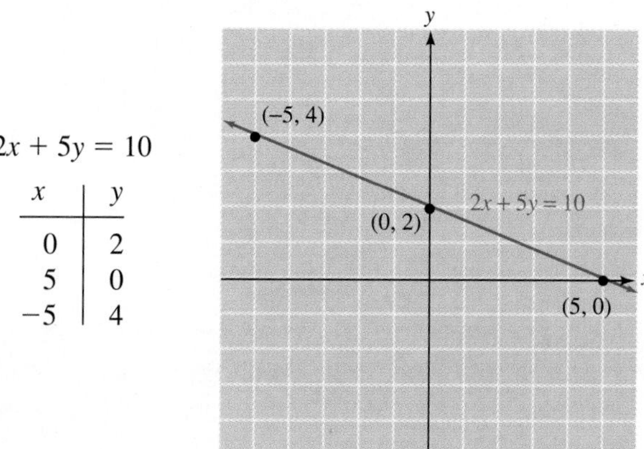

FIGURE 6-5

■

EXAMPLE 4 Graph the equation $y = 3x + 4$.

Solution We find the y- and x-intercepts: If $x = 0$, then

$$y = 3x + 4$$
$$y = 3(\mathbf{0}) + 4 \quad \text{Substitute 0 for } x.$$
$$y = 4 \quad \text{Simplify.}$$

The y-intercept is the point $(0, 4)$.

If $y = 0$, then

$$y = 3x + 4$$
$$\mathbf{0} = 3x + 4 \quad \text{Substitute 0 for } y.$$
$$-4 = 3x \quad \text{Subtract 4 from both sides.}$$
$$-\frac{4}{3} = x \quad \text{Divide both sides by 3.}$$

The x-intercept is the point $\left(-\frac{4}{3}, 0\right)$.

To find the coordinates of a third point, we can substitute 1 for x and solve for y:

$$y = 3x + 4$$
$$y = 3(\mathbf{1}) + 4 \quad \text{Substitute 1 for } x.$$
$$y = 7 \quad \text{Simplify.}$$

The point $(1, 7)$ lies on the graph. The graph of $y = 3x + 4$ is shown in Figure 6-6.

$y = 3x + 4$

x	y
0	4
$-\frac{4}{3}$	0
1	7

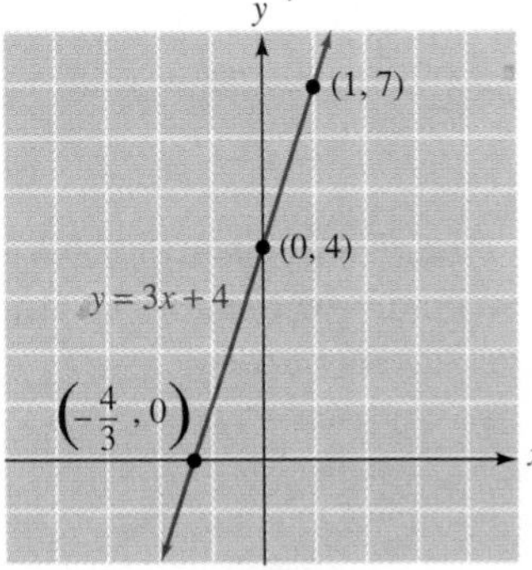

FIGURE 6-6

■

Graphing Lines Parallel to the *x*- and *y*-Axes

EXAMPLE 5 Graph the equations **a.** $y = 3$ and **b.** $x = -2$.

Solution **a.** Since the equation $y = 3$ does not contain x, the numbers chosen for x have no effect on y. The value of y is always 3.

After plotting the pairs (x, y) shown in Figure 6-7(a) and joining them with a straight line, we see that the graph is a horizontal line, parallel to the x-axis, with a y-intercept of $(0, 3)$. The line has no x-intercept.

b. Since the equation $x = -2$ does not contain y, the value of y can be any number. The value of x is always -2.

After plotting the pairs (x, y) shown in Figure 6-7(b) and joining them with a straight line, we see that the graph is a vertical line, parallel to the y-axis, with an x-intercept of $(-2, 0)$. The line has no y-intercept.

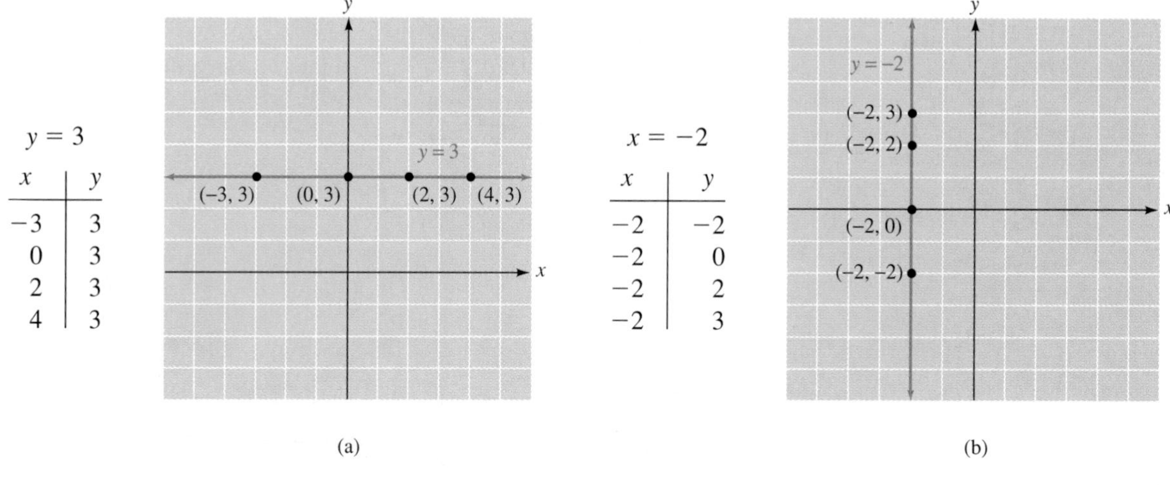

$y = 3$

x	y
−3	3
0	3
2	3
4	3

$x = -2$

x	y
−2	−2
−2	0
−2	2
−2	3

FIGURE 6-7

The results of Example 5 suggest the following facts.

Equations of Lines Parallel to the Coordinate Axes

If a and b are real numbers, then

The graph of the equation $x = a$ is a vertical line with x-intercept at $(a, 0)$. If $a = 0$, the line is the y-axis.

The graph of the equation $y = b$ is a horizontal line with y-intercept at $(0, b)$. If $b = 0$, the line is the x-axis.

The Midpoint Formula

If point M in Figure 6-8 lies midway between points $P(x_1, y_1)$ and $Q(x_2, y_2)$, point M is called the **midpoint** of segment PQ. To find the coordinates of M, we average the x-coordinates and average the y-coordinates of P and Q.

Midpoint Formula The midpoint of the line segment with endpoints at $P(x_1, y_1)$ and $Q(x_2, y_2)$ is the point M with coordinates of

$$\left(\frac{x_1 + x_2}{2}, \frac{y_1 + y_2}{2}\right)$$

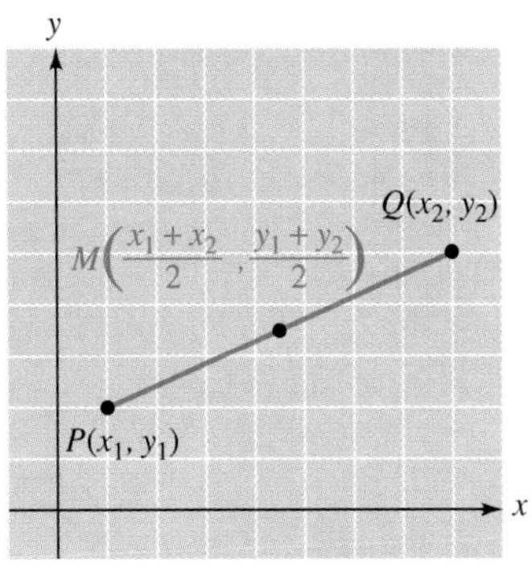

FIGURE 6-8

EXAMPLE 6 Find the midpoint of the line segment joining $P(-2, 3)$ and $Q(3, -5)$.

Solution To find the midpoint, we average the x-coordinates and the y-coordinates to get

$$\frac{x_1 + x_2}{2} = \frac{-2 + 3}{2} = \frac{1}{2} \quad \text{and} \quad \frac{y_1 + y_2}{2} = \frac{3 + (-5)}{2} = -1$$

The midpoint of segment PQ is the point $M\left(\frac{1}{2}, -1\right)$. ■

ORALS *Find the x- and y-intercepts of each line.*

1. $x + y = 3$ $(3, 0), (0, 3)$

2. $3x + y = 6$ $(2, 0), (0, 6)$

3. $x + 4y = 8$ $(8, 0), (0, 2)$

4. $3x - 4y = 12$ $(4, 0), (0, -3)$

Find the midpoint of a line segment with endpoints at

5. $(2, 4), (6, 8)$ $(4, 6)$

6. $(-4, 6), (4, -8)$ $(0, -1)$

EXERCISE 6.1

In Exercises 1–8, plot each point on the rectangular coordinate system shown in Illustration 1.

1. $A(4, 3)$ **2.** $B(-2, 1)$

3. $C(3, -2)$ **4.** $D(-2, -3)$

5. $E(0, 5)$ **6.** $F(-4, 0)$

7. $G(-3, 0)$ **8.** $H(0, 3)$

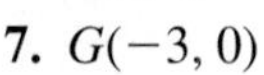

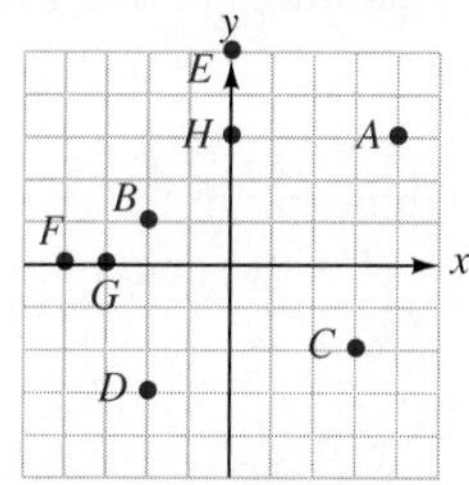

ILLUSTRATION 1

In Exercises 9–16, give the coordinates of each point shown in Illustration 2.

9. A (2, 4) **10.** B (−5, 5)

11. C (−2, −1) **12.** D (4, −4)

13. E (4, 0) **14.** F (−5, −3)

15. G (0, 0) **16.** H (0, −4)

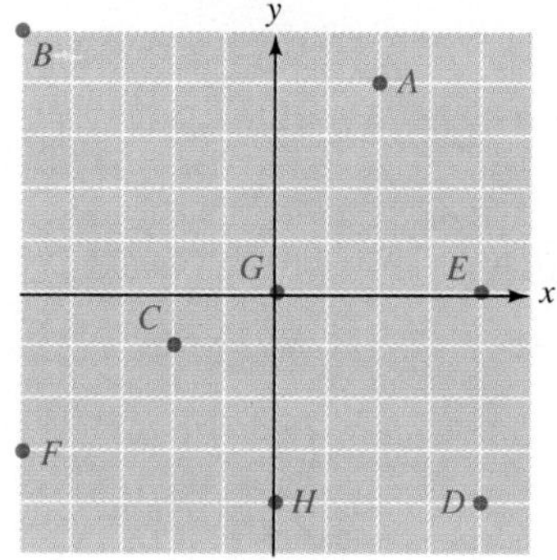

ILLUSTRATION 2

In Exercises 17–32, graph each equation.

17. $x + y = 4$

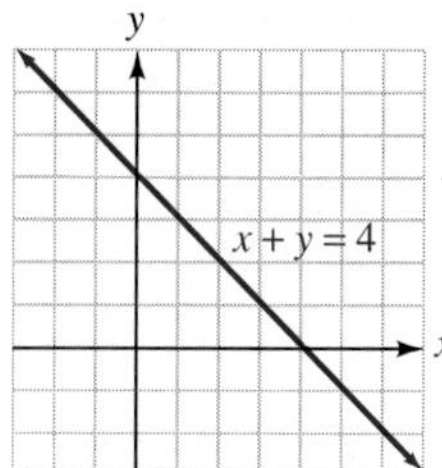

18. $x - y = 2$

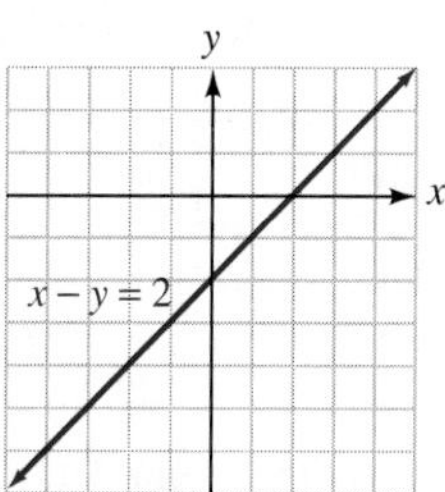

19. $2x - y = 3$

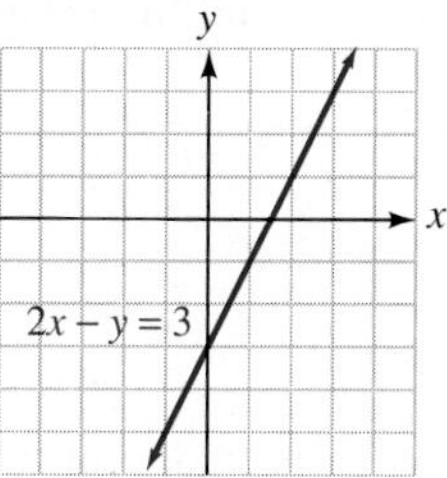

20. $x + 2y = 5$

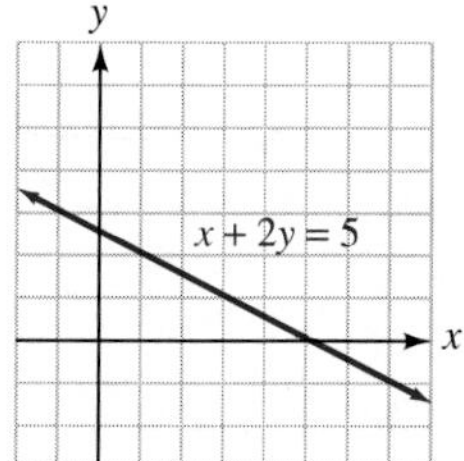

21. $3x + 4y = 12$

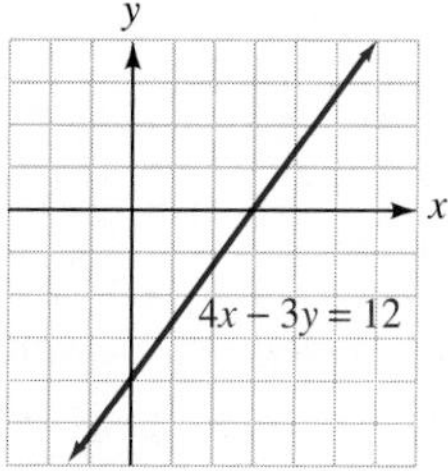

22. $4x - 3y = 12$

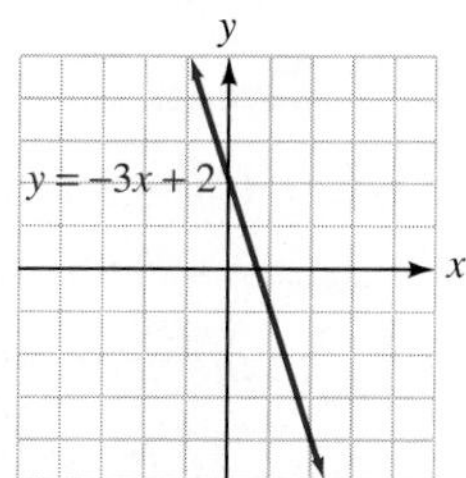

23. $y = -3x + 2$

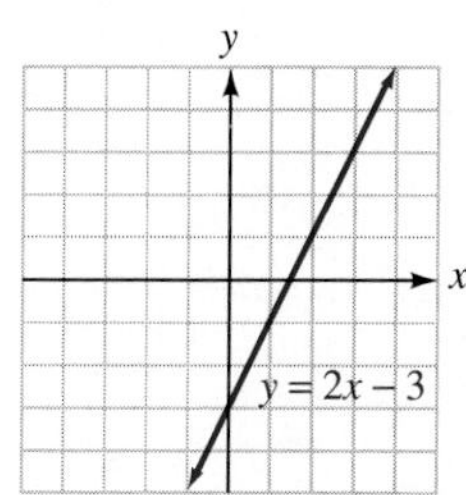

24. $y = 2x - 3$

25. $3y = 6x - 9$

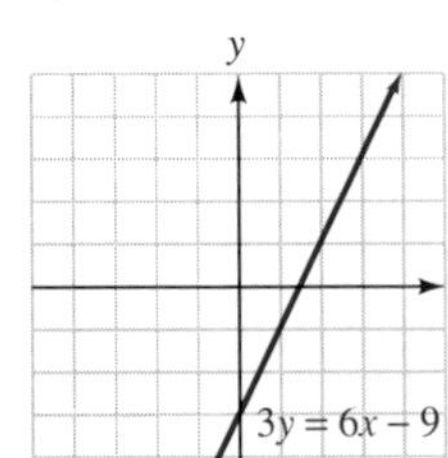

26. $2x = 4y - 10$

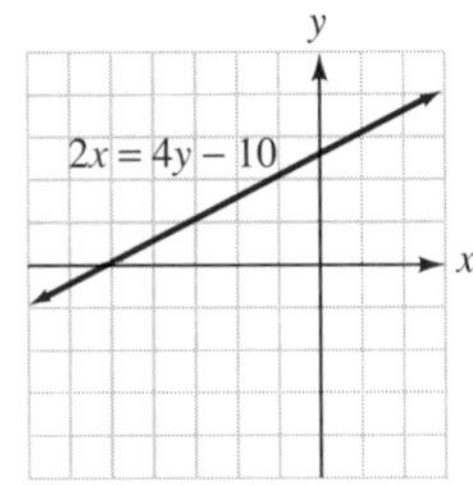

27. $3x + 4y - 8 = 0$

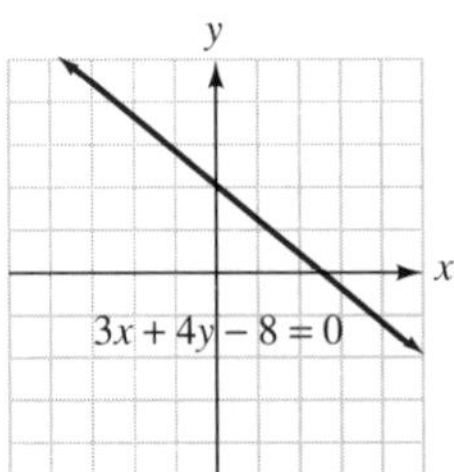

28. $-2y - 3x + 9 = 0$

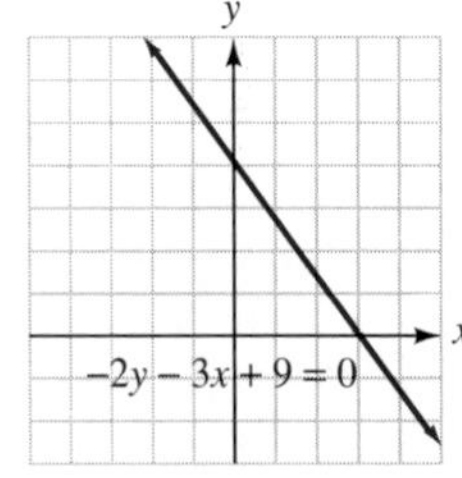

29. $x = 3$

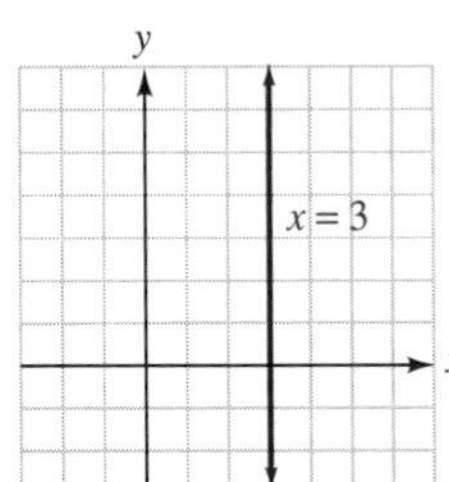

30. $y = -4$

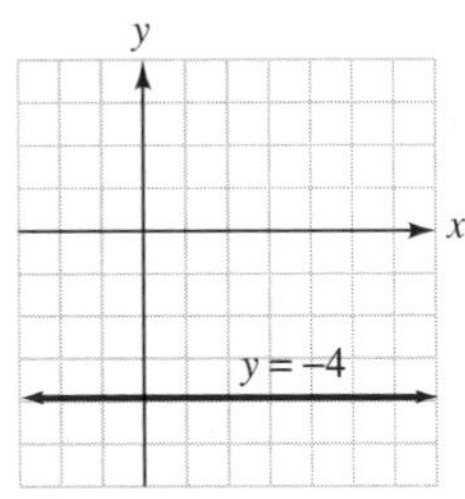

31. $-3y + 2 = 5$

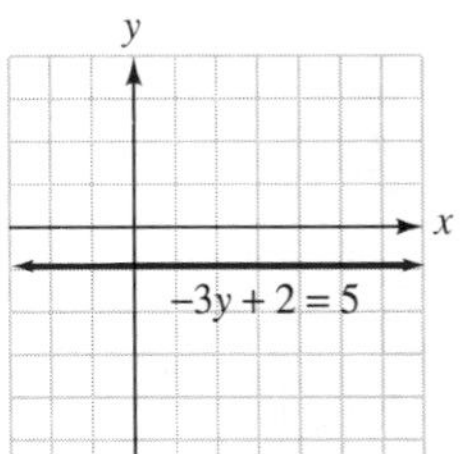

32. $-2x + 3 = 11$

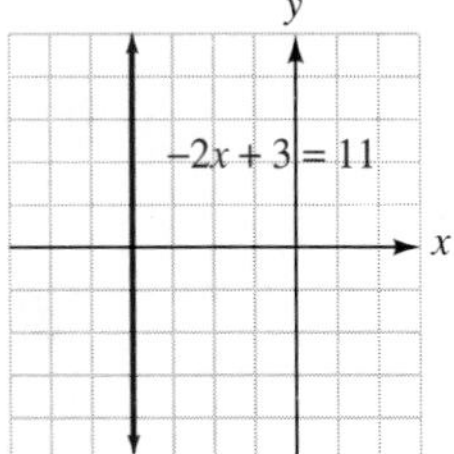

In Exercises 33–46, find the midpoint of segment PQ.

33. $P(0, 0)$, $Q(6, 8)$ $(3, 4)$

34. $P(10, 12)$, $Q(0, 0)$ $(5, 6)$

35. $P(6, 8)$, $Q(12, 16)$ $(9, 12)$

36. $P(10, 4)$, $Q(2, -2)$ $(6, 1)$

37. $P(2, 4)$, $Q(5, 8)$ $\left(\frac{7}{2}, 6\right)$

38. $P(5, 9)$, $Q(8, 13)$ $\left(\frac{13}{2}, 11\right)$

39. $P(-2, -8)$, $Q(3, 4)$ $\left(\frac{1}{2}, -2\right)$

40. $P(-5, -2)$, $Q(7, 3)$ $\left(1, \frac{1}{2}\right)$

41. $Q(-3, 5)$, $P(-5, -5)$ $(-4, 0)$

42. $Q(2, -3)$, $P(4, -8)$ $\left(3, -\frac{11}{2}\right)$

43. $Q(a, b)$, $P(4a, 3b)$ $\left(\frac{5a}{2}, 2b\right)$

44. $Q(a + b, b)$, $P(-b, -a)$ $\left(\frac{a}{2}, \frac{b-a}{2}\right)$

45. $P(a - b, b)$, $Q(a + b, 3b)$ $(a, 2b)$

46. $P(3a, a + b)$, $Q(a + 2b, a - b)$ $(2a + b, a)$

47. Finding the endpoint of a segment If $M(-2, 3)$ is the midpoint of segment PQ and the coordinates of P are $(-8, 5)$, find the coordinates of Q. $(4, 1)$

48. Finding the endpoint of a segment If $M(6, -5)$ is the midpoint of segment PQ and the coordinates of Q are $(-5, -8)$, find the coordinates of P. $(17, -2)$

49. House appreciation A house purchased for \$125,000 is expected to appreciate according to the formula $y = 7500x + 125{,}000$, where y is the value of the house after x years. Find the value of the house 5 years later and 10 years later. \$162,500, \$200,000

50. Car depreciation A car purchased for \$17,000 is expected to depreciate according to the formula $y = -1360x + 17{,}000$. When will the car be worthless? 12.5 years

51. Demand equation The number of television sets that consumers buy depends on price. The higher the price, the fewer TVs people will buy. The equation that relates price to the number of TVs sold at that price is called a **demand equation.**

If the demand equation for a 13-inch TV is $p = -\frac{1}{10}q + 170$, where p is the price and q is the number of TVs sold at that price, how many TVs will be sold at a price of \$150? 200

52. Supply equation The number of television sets that manufacturers produce depends on price. The higher the price, the more TVs manufacturers will produce. The equation that relates price to the number of TVs produced at that price is called a **supply equation.**

If the supply equation for a 13-inch TV is $p = \frac{1}{10}q + 130$, where p is the price and q is the number of TVs produced for sale at that price, how many TVs will be produced if the price is \$150? 200

53. Meshing gears The rotational speed, V, of a large gear (with N teeth) is related to the speed, v, of the smaller gear (with n teeth) by the equation $V = \frac{nv}{N}$. If the larger gear in Illustration 3 is making 60 revolutions per minute, how fast is the smaller gear spinning? 100 rpm

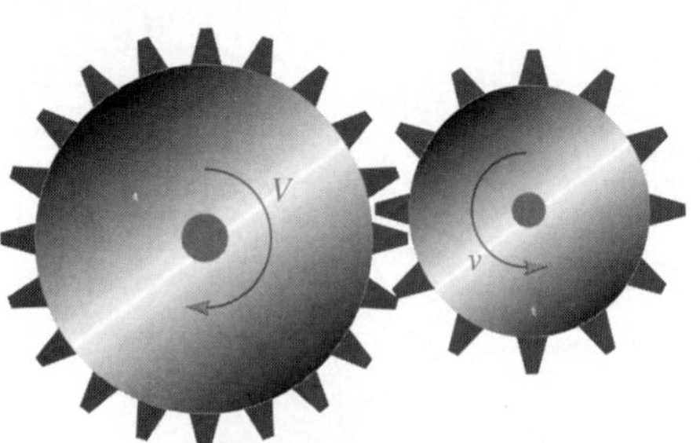

ILLUSTRATION 3

54. Crime prevention The number, n, of incidents of family violence requiring police response appears to be related to d, the money spent on crisis intervention, by the equation

$$n = 430 - 0.005d$$

What expenditure would reduce the number of incidents to 350? \$16,000

Writing Exercises ■ *Write a paragraph using your own words.*

1. Explain how to graph a line using the intercept method.
2. Explain how to determine the quadrant in which the point $P(a, b)$ lies.

Something to Think About ■

1. If the line $y = ax + b$ passes through only quadrants I and II, what can be known about the constants a and b? $a = 0, b > 0$
2. What are the coordinates of three points that divide the line segment joining $P(a, b)$ and $Q(c, d)$ into four equal parts? $\left(\frac{3a+c}{4}, \frac{3b+d}{4}\right), \left(\frac{a+c}{2}, \frac{b+d}{2}\right), \left(\frac{a+3c}{4}, \frac{b+3d}{4}\right)$

Review Exercises ■ *Graph each interval on the number line.*

1. $[-3, 2) \cup (-2, 3]$

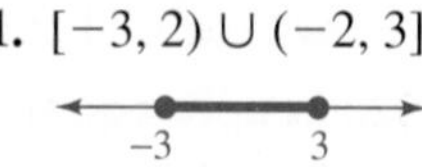

2. $(-1, 4) \cap [-2, 2]$

−1 2

3. $[-3, -2) \cap (2, 3]$ no graph; the intersection is $\emptyset$
4. $[-4, -3) \cup (2, 3]$

−4 −3 2 3

6.2 Slope of a Nonvertical Line

■ Slopes of Horizontal and Vertical Lines ■ Slopes of Parallel Lines
■ Slopes of Perpendicular Lines ■ Interpretation of Slope

The **slope of a nonvertical line** drawn in a rectangular coordinate system is a measure of its tilt or inclination. To develop a formula to find the slope of a nonvertical line, we refer to the line in Figure 6-9 that passes through points $P(2, 3)$ and $Q(6, 9)$. If line RQ is parallel to the y-axis and line PR is parallel to the x-axis, then triangle PRQ is a right triangle, and point R has coordinates of (6, 3).

The distance from point R to point Q is the change in the y-coordinates of the points, often denoted as Δy and read as "delta y." As we move from R to Q, $\Delta y = 9 - 3$ or 6 units. The distance from point P to point R is the change in the x-coordinates of the points, denoted as Δx and read as "delta x." As we move from point P to point R, $\Delta x = 6 - 2$ or 4 units.

The slope, m, of line PQ is the distance from R to Q divided by the distance from P to R.

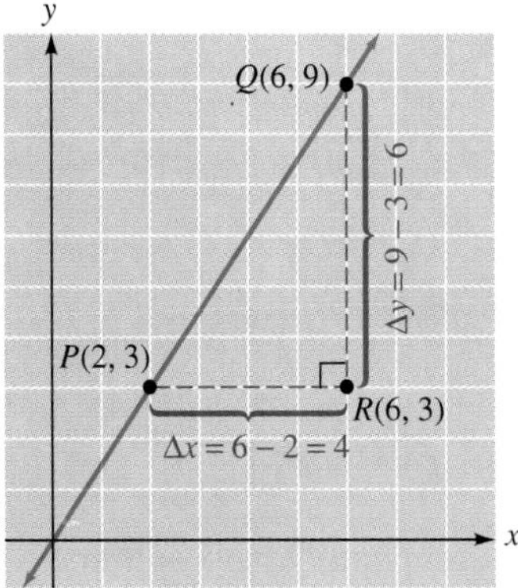

FIGURE 6-9

$$\begin{aligned} m &= \frac{\Delta y}{\Delta x} \\ &= \frac{9 - 3}{6 - 2} \\ &= \frac{6}{4} \\ &= \frac{3}{2} \end{aligned}$$

In general, we define slope as follows.

Slope The **slope of the nonvertical line** passing through the points $P(x_1, y_1)$ and $Q(x_2, y_2)$ shown in Figure 6-10(a) is

$$m = \frac{\Delta y}{\Delta x} = \frac{y_2 - y_1}{x_2 - x_1} \qquad (x_2 \neq x_1)$$

Δy is called the **rise** of the line between points P and Q, and Δx is called the **run**. The slope of the line is the ratio of the rise to the run:

$$m = \frac{\text{rise}}{\text{run}}$$

In Figure 6-10(b), point S represents an arbitrary third point on line l. Because triangles PRQ and PTS are similar, their corresponding sides are in proportion, and the ratios of the rise to the run in the two triangles are equal:

$$\begin{aligned} m &= \frac{y_2 - y_1}{x_2 - x_1} \\ &= \frac{y_3 - y_1}{x_3 - x_1} \end{aligned}$$

Thus, the slope of a nonvertical line is constant, and it can be found by using any two points on the line.

Furthermore, if point P is on a line with slope m, and the ratio of rise to run of a segment PQ is also m, then point Q is on the line.

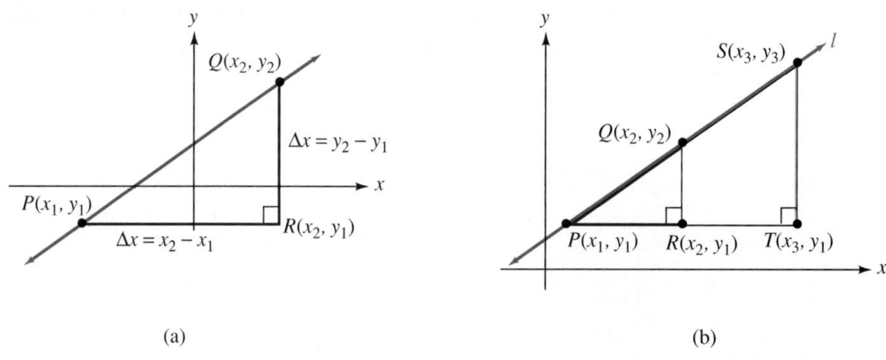

FIGURE 6-10

EXAMPLE 1 In Figure 6-11, point P has coordinates of $(-2, 5)$, and point Q has coordinates of $(6, -7)$. Find the slope of the line PQ.

Solution We can let $P(x_1, y_1) = P(-2, 5)$ and $Q(x_2, y_2) = Q(6, -7)$. Then

$$\begin{aligned} m &= \frac{\Delta y}{\Delta x} \\ &= \frac{\mathbf{y_2} - \mathbf{y_1}}{\mathbf{x_2} - x_1} \\ &= \frac{\mathbf{-7} - \mathbf{5}}{\mathbf{6} - (-2)} \quad \text{Substitute } -7 \text{ for } y_2,\ 5 \text{ for } y_1,\ 6 \text{ for } x_2, \text{ and } -2 \text{ for } x_1. \\ &= \frac{-12}{8} \\ &= -\frac{3}{2} \end{aligned}$$

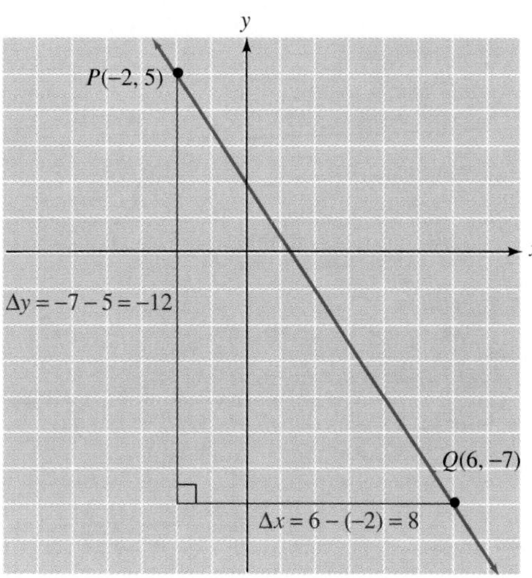

FIGURE 6-11

The slope of the line is $-\frac{3}{2}$. We would obtain the same result if we had let $P(x_1, y_1) = P(6, -7)$ and $Q(x_2, y_2) = Q(-2, 5)$. ■

WARNING! When finding the slope of a line, always subtract the y values and the x values in the same order, or your answer will have the wrong sign:

$$m = \frac{y_2 - y_1}{x_2 - x_1} \quad \text{or} \quad m = \frac{y_1 - y_2}{x_1 - x_2}$$

However,

$$m \neq \frac{y_2 - y_1}{x_1 - x_2} \quad \text{and} \quad m \neq \frac{y_1 - y_2}{x_2 - x_1}$$

EXAMPLE 2 Find the slope of the line determined by the equation $3x - 4y = 12$.

Solution We begin by finding the coordinates of two points on the line. If $x = 0$, then $y = -3$, and the point $(0, -3)$ is on the line. If $y = 0$, then $x = 4$, and the point $(4, 0)$ is on the line. (See Figure 6-12.)

We can then let $P(x_1, y_1) = P(0, -3)$, $Q(x_2, y_2) = Q(4, 0)$ and use the formula for slope to obtain

$$m = \frac{\Delta y}{\Delta x}$$

$$= \frac{y_2 - y_1}{x_2 - x_1}$$

$$= \frac{0 - (-3)}{4 - 0} \qquad \text{Substitute 0 for } y_2, -3 \text{ for } y_1, 4 \text{ for } x_2, \text{ and 0 for } x_1.$$

$$= \frac{3}{4}$$

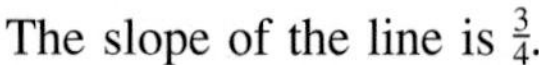

The slope of the line is $\frac{3}{4}$.

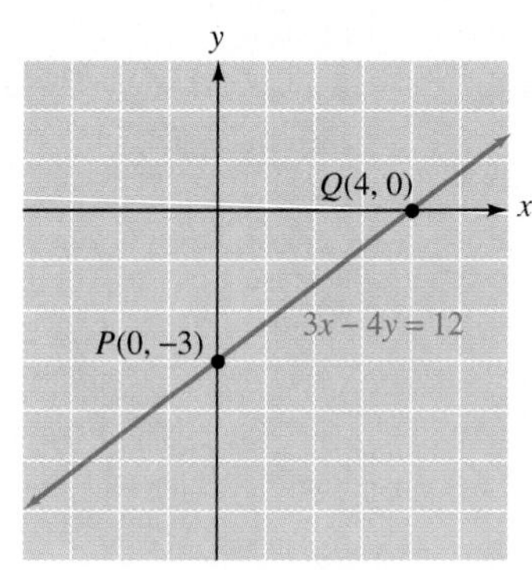

FIGURE 6-12

Slopes of Horizontal and Vertical Lines

If $P(x_1, y_1)$ and $Q(x_2, y_2)$ are points on the horizontal line shown in Figure 6-13(a), then $y_1 = y_2$, and the numerator of the fraction

$$\frac{y_2 - y_1}{x_2 - x_1} \qquad \text{On a horizontal line, } x_2 \neq x_1.$$

is 0. Thus, the value of the fraction is 0, and the slope of the horizontal line is 0.

If $P(x_1, y_1)$ and $Q(x_2, y_2)$ are two points on the vertical line shown in Figure 6-13(b), then $x_1 = x_2$, and the denominator of the fraction

$$\frac{y_2 - y_1}{x_2 - x_1} \qquad \text{On a vertical line, } y_2 \neq y_1.$$

is 0. Since the denominator of a fraction cannot be 0, a vertical line has no defined slope.

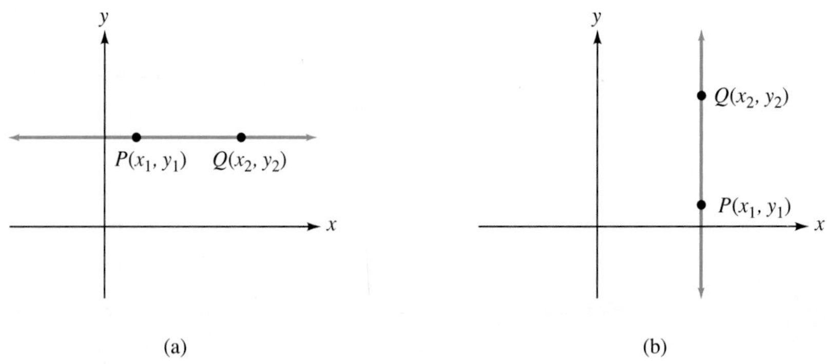

FIGURE 6-13

Slopes of Horizontal and Vertical Lines

Horizontal lines (lines with equations of the form $y = b$) have a slope of 0.

Vertical lines (lines with equations of the form $x = a$) have no defined slope.

Teaching Tip: Point out that the bigger the slope, the steeper the line. If $m = 1$, the line makes a 45° angle with the horizontal, an excellent frame of reference.

If a line rises as we follow it from left to right, as in Figure 6-14(a), its slope is positive. If a line drops as we follow it from left to right, as in Figure 6-14(b), its slope is negative. If a line is horizontal, as in Figure 6-14(c), its slope is 0. If a line is vertical, as in Figure 6-14(d), it has no defined slope.

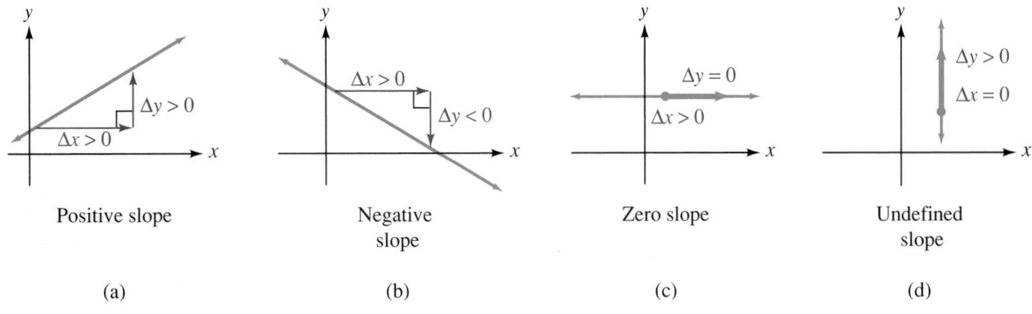

FIGURE 6-14

Slopes of Parallel Lines

To see a relationship between parallel lines and their slopes, we refer to the parallel lines l_1 and l_2 shown in Figure 6-15, with slopes of m_1 and m_2, respectively. Because right triangles ABC and DEF are similar, it follows that

$$m_1 = \frac{\Delta y \text{ of } l_1}{\Delta x \text{ of } l_1} = \frac{\Delta y \text{ of } l_2}{\Delta x \text{ of } l_2} = m_2$$

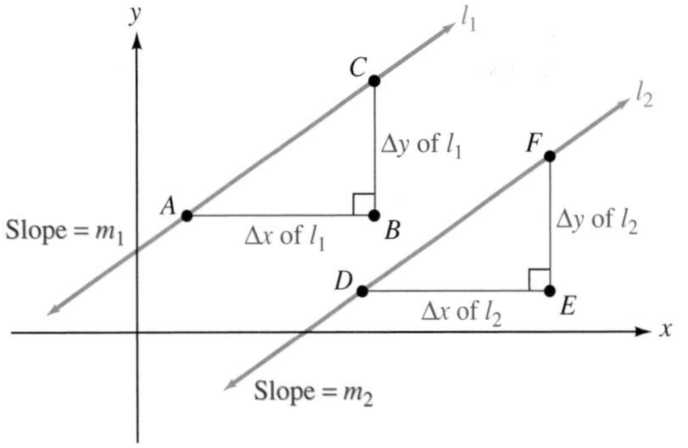

FIGURE 6-15

Thus, if two nonvertical lines are parallel, they have the same slope. It is also true that when two lines have the same slope, they are parallel.

Slopes of Parallel Lines Nonvertical parallel lines have the same slope, and lines having the same slope are parallel.

Since vertical lines are parallel, lines with no defined slope are parallel.

EXAMPLE 3 If the line passing through $P(3, -2)$ and $Q(-3, 4)$ is parallel to the line passing through $R(-2, 5)$ and $S(3, y)$, find y.

Solution Since the lines PQ and RS are parallel, they have equal slopes. (See Figure 6-16.) To find y, we find the slope of each line, set them equal, and solve the resulting equation.

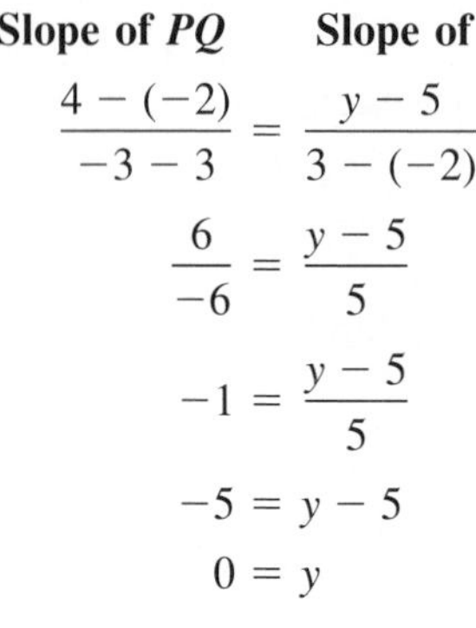

Slope of *PQ* **Slope of *RS***

$$\frac{4-(-2)}{-3-3} = \frac{y-5}{3-(-2)}$$

$$\frac{6}{-6} = \frac{y-5}{5}$$

$$-1 = \frac{y-5}{5}$$

$$-5 = y - 5$$

$$0 = y$$

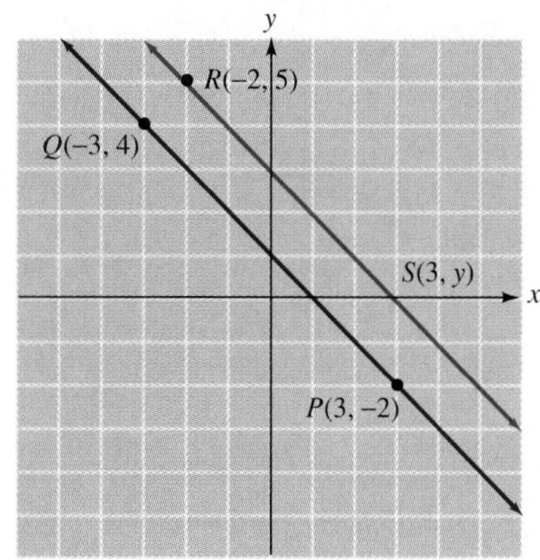

FIGURE 6-16

Thus, $y = 0$. The line passing through $P(3, -2)$ and $Q(-3, 4)$ is parallel to the line passing through $R(-2, 5)$ and $S(3, 0)$. ■

EXAMPLE 4 When graphed on a coordinate system, lines PQ and OR are distinct lines. Is the line passing through $P(-2, 5)$ and $Q(3, 9)$ parallel to the line passing through the origin O and the point $R(5, 4)$?

Solution We find the slope of each line.

$$\text{Slope of } PQ = \frac{\Delta y}{\Delta x} = \frac{y_2 - y_1}{x_2 - x_1} = \frac{9-5}{3-(-2)} = \frac{4}{5}$$

and

$$\text{Slope of } OR = \frac{\Delta y}{\Delta x} = \frac{y_2 - y_1}{x_2 - x_1} = \frac{4-0}{5-0} = \frac{4}{5}$$

Since the slopes are equal, the lines are parallel. ■

Slopes of Perpendicular Lines

Two real numbers a and b are called **negative reciprocals** if $ab = -1$. For example,

$$-\frac{4}{3} \quad \text{and} \quad \frac{3}{4}$$

are negative reciprocals because $-\frac{4}{3}\left(\frac{3}{4}\right) = -1$.

The following theorem relates perpendicular lines and their slopes.

Slopes of Perpendicular Lines

If two nonvertical lines are perpendicular, their slopes are negative reciprocals.

If the slopes of two lines are negative reciprocals, the lines are perpendicular.

Because a horizontal line is perpendicular to a vertical line, a line with a slope of 0 is perpendicular to a line with no defined slope.

EXAMPLE 5 Two lines intersect at point $P(3, -4)$. One line passes through the origin O, and the other line passes through point $Q(9, 4)$. Are the lines perpendicular?

Solution See Figure 6-17. Find the slope of lines OP and PQ.

$$\begin{aligned} \text{Slope of } OP &= \frac{\Delta y}{\Delta x} \\ &= \frac{y_2 - y_1}{x_2 - x_1} \\ &= \frac{-4 - 0}{3 - 0} \\ &= -\frac{4}{3} \end{aligned} \qquad \text{and} \qquad \begin{aligned} \text{Slope of } PQ &= \frac{\Delta y}{\Delta x} \\ &= \frac{y_2 - y_1}{x_2 - x_1} \\ &= \frac{4 - (-4)}{9 - 3} \\ &= \frac{8}{6} \\ &= \frac{4}{3} \end{aligned}$$

Since their slopes are not negative reciprocals, the lines are not perpendicular.

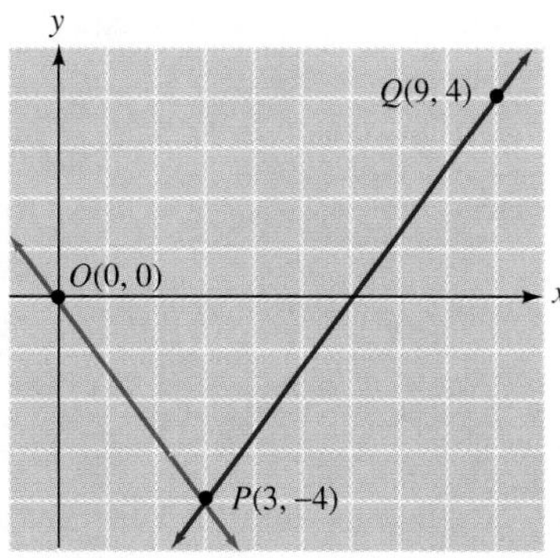

FIGURE 6-17

Interpretation of Slope

Many applications of mathematics involve equations of lines and their slopes.

EXAMPLE 6 **Cost of carpet** Assume that good-quality carpet costs \$40 per square yard. **a.** Write an equation that expresses the relationship between c (the total cost) and n (the number of square yards purchased). **b.** Graph the equation. **c.** Interpret the slope of the line.

Solution **a.** The total cost, c, of n square yards of carpet costing \$40 per square yard is the product of the number of yards purchased and the price per yard. Thus,

$$c = 40n$$

b. We can graph the equation on a rectangular coordinate system with a vertical c-axis and a horizontal n-axis. Figure 6-18 shows a table of ordered pairs and the graph.

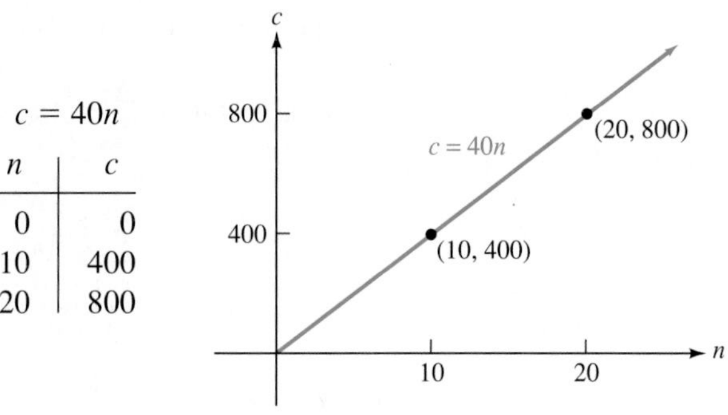

$c = 40n$

n	c
0	0
10	400
20	800

FIGURE 6-18

c. If we pick two points on the line—for example, (10, 400) and (20, 800)—we can find its slope:

$$\begin{aligned} m &= \frac{\Delta c}{\Delta n} \\ &= \frac{c_2 - c_1}{n_2 - n_1} \\ &= \frac{800 - 400}{20 - 10} && \text{Substitute 800 for } c_2\text{, 400 for } c_1\text{, 20 for } n_2\text{, and 10 for } n_1. \\ &= \frac{400}{10} \\ &= 40 \end{aligned}$$

The slope of the line is 40, which is the cost per square yard of carpet. ■

Slope often represents a **rate of change.** In Example 6, the slope gives the rate of change of the cost of the carpet in dollars per square yard.

If temperature changes with time (measured in hours), the concept of slope gives the rate of change of temperature in degrees per hour. If the amount of money in an account is increasing over the years, slope gives the rate of growth, in dollars per year.

ORALS *Find the slope of the line passing through*

1. (0, 0), (1, 3) 3
2. (0, 0), (3, 6) 2
3. Are lines with slopes of -2 and $\frac{8}{-4}$ parallel? yes
4. Find the negative reciprocal of -0.2. 5
5. Are lines with slopes of -2 and $\frac{1}{2}$ perpendicular? yes

EXERCISE 6.2

In Exercises 1–12, find the slope of the line that passes through the given points, if possible.

1. (0, 0), (3, 9) 3
2. (9, 6), (0, 0) $\frac{2}{3}$
3. $(-1, 8)$, $(6, 1)$ -1
4. $(-5, -8)$, $(3, 8)$ 2
5. $(3, -1)$, $(-6, 2)$ $-\frac{1}{3}$
6. $(0, -8)$, $(-5, 0)$ $-\frac{8}{5}$
7. $(7, 5)$, $(-9, 5)$ 0
8. $(2, -8)$, $(3, -8)$ 0
9. $(-7, -5)$, $(-7, -2)$ undefined
10. $(3, -5)$, $(3, 14)$ undefined
11. (a, b) (b, a) -1
12. (a, b), $(-b, -a)$ 1

In Exercises 13–22, find the slope of the line determined by each equation.

13. $3x + 2y = 12$ $-\frac{3}{2}$
14. $2x - y = 6$ 2
15. $3x = 4y - 2$ $\frac{3}{4}$
16. $x = y$ 1
17. $y = \frac{x - 4}{2}$ $\frac{1}{2}$
18. $x = \frac{3 - y}{4}$ -4
19. $4y = 3(y + 2)$ 0
20. $x + y = \frac{2 - 3y}{3}$ $-\frac{1}{2}$
21. $x(y + 2) = y(x - 3) + 4$ $-\frac{2}{3}$
22. $-y(x - 3) + 2 = x(4 - y)$ $\frac{4}{3}$

In Exercises 23–28, tell whether the slope of the line in each graph is positive, negative, 0, or undefined.

23.

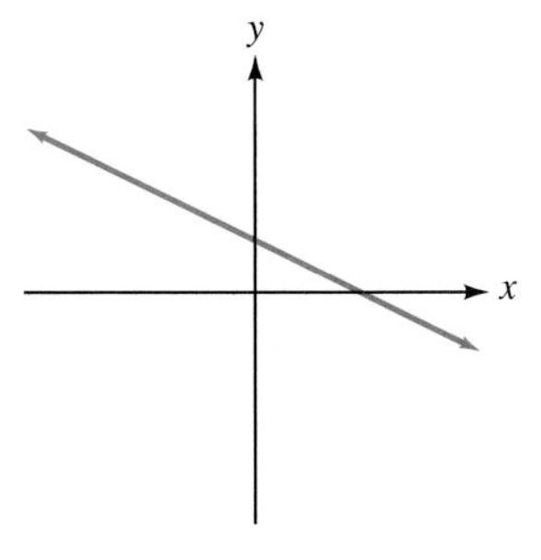

negative

24.

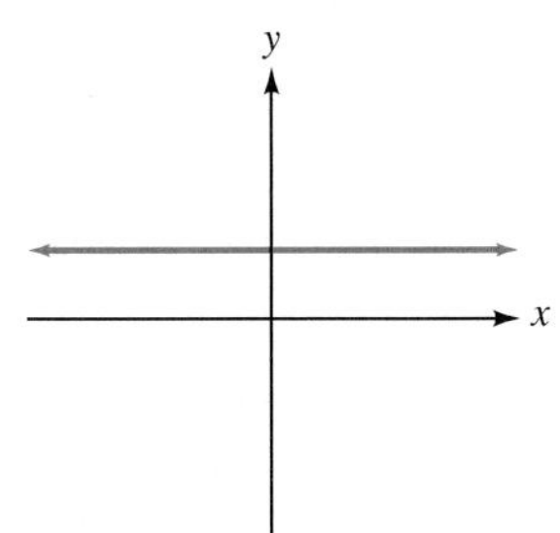

0

25.

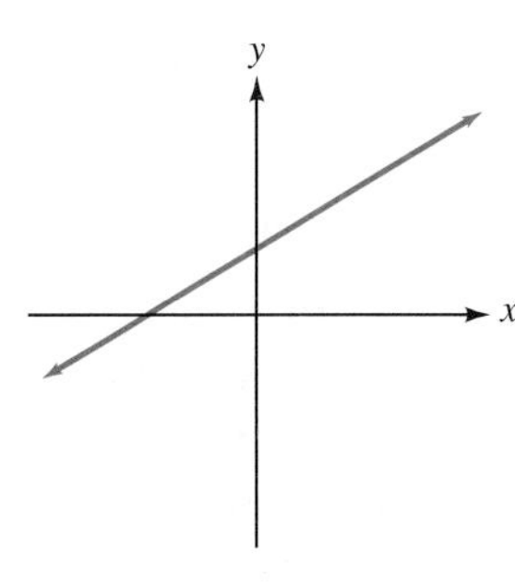

positive

26.

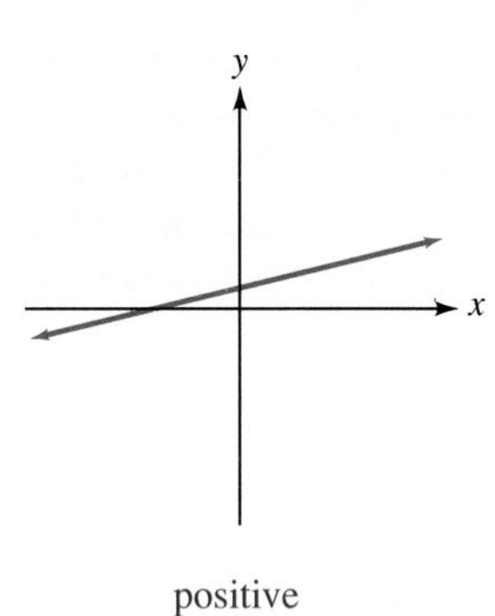

positive

27.

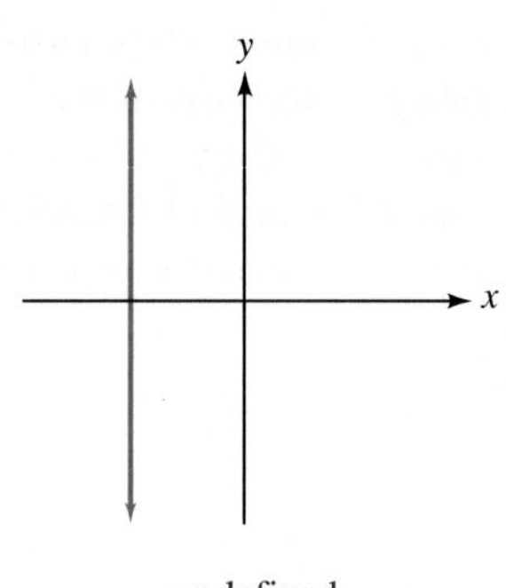

undefined

28.

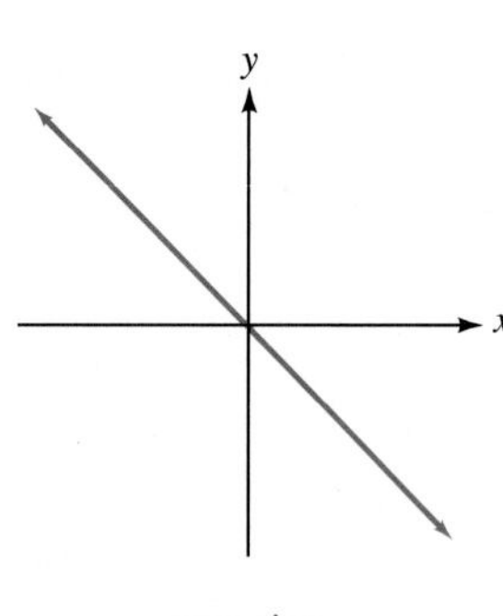

negative

In Exercises 29–36, tell whether the lines with the given slopes are parallel, perpendicular, or neither.

29. $m_1 = 3, m_2 = -\frac{1}{3}$ perpendicular

30. $m_1 = \frac{1}{4}, m_2 = 4$ neither

31. $m_1 = 4, m_2 = 0.25$ neither

32. $m_1 = -5, m_2 = \frac{1}{-0.2}$ parallel

33. $m_1 = \frac{a}{b}, m_2 = \left(\frac{b}{a}\right)^{-1}$ parallel

34. $m_1 = \frac{c}{d}, m_2 = \frac{d}{c}$ neither

35. $m_1 = \frac{a-b}{b}, m_2 = \frac{-b}{a-b}$ perpendicular

36. $m_1 = \frac{c-d}{c+d}, m_2 = -\frac{d+c}{c-d}$ perpendicular

In Exercises 37–42, tell whether the line passing through points P and Q is parallel or perpendicular to the line passing through R(2, −4) and S(−4, 8).

37. $P(3, 4), Q(4, 2)$ parallel

38. $P(6, 4), Q(8, 5)$ perpendicular

39. $P(-2, 1), Q(6, 5)$ perpendicular

40. $P(3, 4), Q(-3, -5)$ neither

41. $P(5, 4), Q(6, 6)$ neither

42. $P(-2, 3), Q(4, -9)$ parallel

In Exercises 43–48, find the slope of lines PQ and PR and tell whether the points P, Q, and R lie on the same line. (Hint: Two lines with the same slope and a point in common must be the same line.)

43. $P(-2, 4), Q(4, 8), R(10, 12)$ on same line

44. $P(6, 10), Q(0, 6), R(3, 8)$ on same line

45. $P(-4, 10), Q(-6, 0), R(-1, 5)$ not on same line

46. $P(-10, -13), Q(-8, -10), R(-12, -16)$ on same line

47. $P(-2, 4), Q(0, 8), R(2, 12)$ on same line

48. $P(8, -4), Q(0, -12), R(8, -20)$ not on same line

49. Find the equation of the x-axis and its slope.
$y = 0, m = 0$

50. Find the equation of the y-axis and its slope, if any.
$x = 0$, no defined slope

51. Rate of growth When a college started an aviation program, the administration agreed to predict enrollments using the graph of a straight line. If the enrollment during the first year was 12, and the enrollment during the fifth year was 26, find the rate of growth per year (the slope of the line). (See Illustration 1.) 3.5 students per year

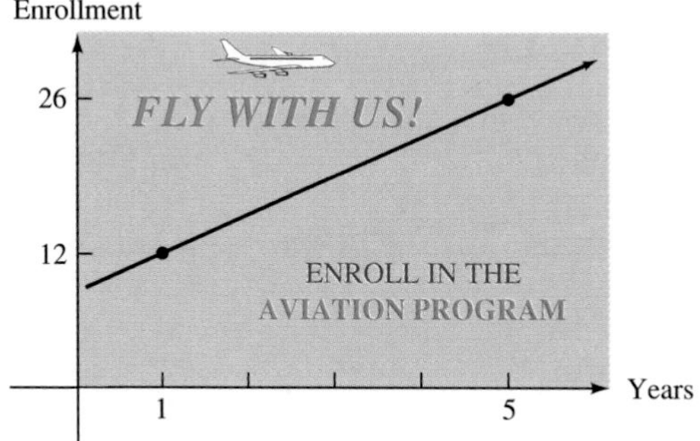

ILLUSTRATION 1

52. **Rate of growth** A small business predicts sales according to a straight-line method. If sales were \$50,000 in the first year and \$110,000 in the third year, find the rate of growth in sales per year (the slope of the line). \$30,000

53. **Rate of decrease** The price of computer technology has been dropping steadily for the past ten years. If a desktop PC cost \$6700 ten years ago, and the same computing power cost \$2200 three years ago, find the rate of decrease per year. (Assume a straight-line model). \$642.86 per year

54. **Hospital costs** Illustration 2 shows the changing average daily costs for a hospital room. Find the rates of change per year of the portion of the room cost that is absorbed by the hospital. \$26 per year

	Cost passed on to patient	Total cost to the hospital
1980	\$130	\$245
1985	214	459
1990	295	670

ILLUSTRATION 2

55. **Geometry** Show that points with coordinates of $(-3, 4)$, $(4, 1)$, and $(-1, -1)$ are vertices of a right triangle.

56. **Geometry** Show that a triangle with vertices at $(0, 0)$, $(12, 0)$, and $(14, 14)$ is not a right triangle.

57. **Geometry** A square has vertices at points $(a, 0)$, $(0, a)$, $(-a, 0)$, and $(0, -a)$. Show that its adjacent sides are perpendicular.

58. **Geometry** Show that the points $(2b, a)$, (b, b), and $(a, 0)$ are vertices of a right triangle.

59. **Geometry** Show that the points $(0, 0)$, $(0, a)$, (b, c), and $(b, a + c)$ are the vertices of a parallelogram. (*Hint:* Opposite sides of a parallelogram are parallel.)

60. **Geometry** Show that the points $(0, 0)$, $(0, b)$, $(8, b + 2)$, and $(12, 3)$ are the vertices of a trapezoid. (*Hint:* A **trapezoid** is a four-sided figure with exactly two sides parallel.)

61. The points $(3, a)$, $(5, 7)$, and $(7, 10)$ lie on a line. Find a. 4

62. The line passing through points $A(1, 3)$ and $B(-2, 7)$ is perpendicular to the line passing through points $C(4, b)$ and $D(8, -1)$. Find b. -4

Writing Exercises *Write a paragraph using your own words.*

1. Explain why a vertical line has no defined slope.
2. Explain how to determine from their slopes whether two lines are parallel, perpendicular, or neither.

Something to Think About

1. Find the slope of the line $Ax + By = C$. Follow the procedure of Example 2. $-\frac{A}{B}$
2. Follow Example 2 to find the slope of the line $y = mx + b$. m

Review Exercises ■ *Simplify each fraction.*

If you teach graphing early, skip these review exercises.

1. $\dfrac{21ab^2c^3}{14abc^4}$ $\frac{3b}{2c}$

2. $\dfrac{x^2 + xy}{x^2 - y^2}$ $\frac{x}{x-y}$

3. $\dfrac{-2x + 2y + 2z}{4x - 4y - 4z}$ $-\frac{1}{2}$

4. $\dfrac{2t^2 + t - 3}{2t^2 + 7t + 6}$ $\frac{t-1}{t+2}$

6.3 Equations of Lines

■ Point–Slope Form of the Equation of a Line ■ Slope–Intercept Form of the Equation of a Line ■ Using Slope As an Aid in Graphing ■ General Form of the Equation of a Line ■ Straight-Line Depreciation

■ Point–Slope Form of the Equation of a Line

Suppose that line l shown in Figure 6-19 has a slope of m and passes through the point $P(x_1, y_1)$. If $Q(x, y)$ is a second point on line l, we have

$$m = \frac{y - y_1}{x - x_1}$$

or

If you teach graphing early, students may need help with this step. ⟶

1. $y - y_1 = m(x - x_1)$ Multiply both sides by $x - x_1$.

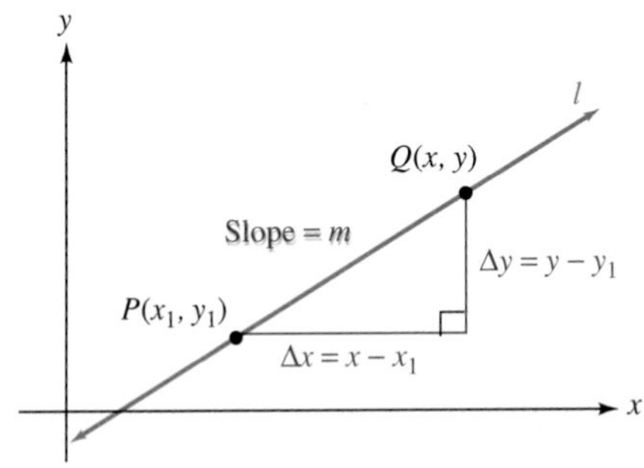

FIGURE 6-19

Because Equation 1 displays the coordinates of the point (x_1, y_1) on the line and the slope m of the line, it is called the **point–slope form** of the equation of a line.

Point–Slope Form of the Equation of a Line

The equation of the line passing through $P(x_1, y_1)$ and with slope m is

$$y - y_1 = m(x - x_1)$$

EXAMPLE 1 Use the point–slope form to write the equation of a line with a slope of $-\frac{2}{3}$ and passing through $P(-4, 5)$.

Solution We substitute $-\frac{2}{3}$ for m, -4 for x_1, and 5 for y_1 in the point–slope form and simplify.

$$y - y_1 = m(x - x_1)$$

$$y - 5 = -\frac{2}{3}[x - (-4)] \quad \text{Substitute } -\tfrac{2}{3} \text{ for } m, -4 \text{ for } x_1, \text{ and 5 for } y_1.$$

$$y - 5 = -\frac{2}{3}(x + 4) \quad -(-4) = 4.$$

$$y - 5 = -\frac{2}{3}x - \frac{8}{3} \quad \text{Use the distributive property to remove parentheses.}$$

$$y = -\frac{2}{3}x + \frac{7}{3} \quad \text{Add 5 to both sides.}$$

The equation of the line is $y = -\frac{2}{3}x + \frac{7}{3}$. ■

EXAMPLE 2 Use the point–slope form of the equation of a line to write the equation of the line passing through $P(-5, 4)$ and $Q(8, -5)$.

Solution We first find the slope of the line:

$$m = \frac{y_2 - y_1}{x_2 - x_1}$$

$$= \frac{-5 - 4}{8 - (-5)} \quad \text{Substitute } -5 \text{ for } y_2, 4 \text{ for } y_1, 8 \text{ for } x_2, \text{ and } -5 \text{ for } x_1.$$

$$= -\frac{9}{13}$$

Because the line passes through both P and Q, we can choose either one and substitute its coordinates into the point–slope form. If we choose $P(-5, 4)$, we substitute -5 for x_1, 4 for y_1, and $-\frac{9}{13}$ for m and simplify.

$$y - y_1 = m(x - x_1)$$

$$y - 4 = -\frac{9}{13}[x - (-5)] \quad \text{Substitute } -\tfrac{9}{13} \text{ for } m, -5 \text{ for } x_1, \text{ and 4 for } y_1.$$

$$y - 4 = -\frac{9}{13}(x + 5) \quad -(-5) = 5.$$

$$y - 4 = -\frac{9}{13}x - \frac{45}{13} \quad \text{Use the distributive property to remove parentheses.}$$

$$y = -\frac{9}{13}x + \frac{7}{13} \quad \text{Add 4 to both sides.}$$

The equation of the line is $y = -\frac{9}{13}x + \frac{7}{13}$. ■

Slope–Intercept Form of the Equation of a Line

Since the y-intercept of the line shown in Figure 6-20 is the point $(0, b)$, we can write the equation of the line by substituting 0 for x_1 and b for y_1 in the point–slope form and simplifying.

Teaching Tip: Some students find these words useful: Slope-intercept form is *solved for y form*. In this form, the coefficient of x is the slope. b is the y-coordinate of the y-intercept.

$$y - y_1 = m(x - x_1)$$
$$y - b = m(x - \mathbf{0})$$
$$y - b = mx$$
$$\textbf{2.} \quad y = mx + b$$

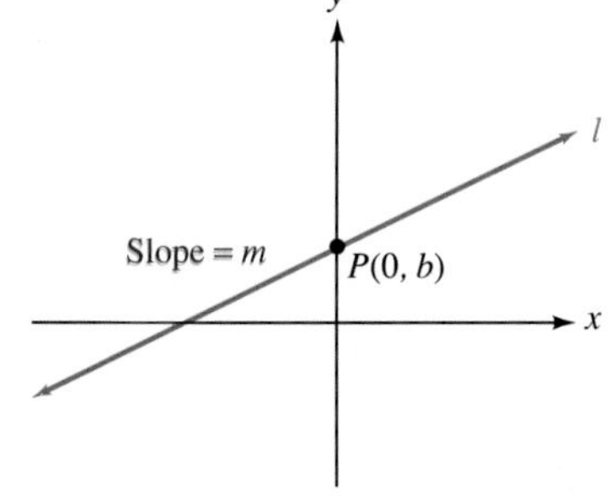

FIGURE 6-20

Because Equation 2 displays the slope m and the y-coordinate b of the y-intercept, it is called the **slope–intercept form** of the equation of a line.

Slope–Intercept Form of the Equation of a Line The equation of the line with slope m and y-intercept $(0, b)$ is

$$y = mx + b$$

EXAMPLE 3 Use the slope–intercept form to write the equation of the line with slope 4 that passes through the point $P(5, 9)$.

Solution Since we are given that $m = 4$ and that the pair $(5, 9)$ satisfies the equation, we can substitute 5 for x, 9 for y, and 4 for m in the equation $y = mx + b$ and solve for b.

$$y = \mathbf{m}\mathbf{x} + b$$
$$\mathbf{9} = \mathbf{4}(\mathbf{5}) + b$$
$$9 = 20 + b$$
$$-11 = b \qquad \text{Subtract 20 from both sides.}$$

Because $m = 4$ and $b = -11$, the equation of the line is $y = 4x - 11$. ■

Using Slope As an Aid in Graphing

It is easy to graph an equation written in slope–intercept form. For example, to graph the equation $y = \frac{4}{3}x - 2$, we first note that the y-coordinate of the y-intercept is -2. Thus, the y-intercept is $(0, -2)$. (See Figure 6-21.)

Because the slope of the line is $\frac{\Delta y}{\Delta x} = \frac{4}{3}$, we can locate another point Q on the line by starting at point P and counting 3 units to the right and 4 units up. The change in x from point P to point Q is $\Delta x = 3$, and the corresponding change in y is $\Delta y = 4$. The line joining points P and Q is the graph of the equation.

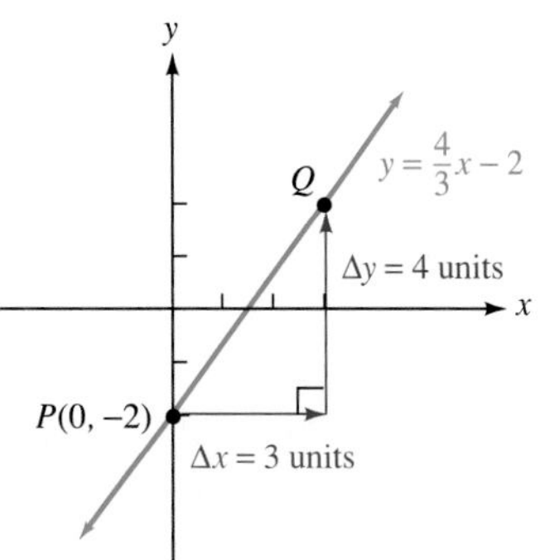

FIGURE 6-21

EXAMPLE 4 Find the slope and the y-intercept of the line with equation $2(x - 3) = -3(y + 5)$. Then graph the line.

Solution We write the equation in the form $y = mx + b$ to find the slope m and the y-intercept $(0, b)$.

$$2(x - 3) = -3(y + 5)$$

$$2x - 6 = -3y - 15 \quad \text{Use the distributive property to remove parentheses.}$$

$$3y - 6 = -2x - 15 \quad \text{Add } 3y \text{ and } -2x \text{ to both sides.}$$

$$3y = -2x - 9 \quad \text{Add 6 to both sides.}$$

$$y = -\frac{2}{3}x - 3 \quad \text{Divide both sides by 3.}$$

The slope of the line is $-\frac{2}{3}$, and the y-intercept is $(0, -3)$. To draw the graph, we plot the y-intercept $(0, -3)$ and then locate a second point on the line by moving 3 units to the right and 2 units down. We draw a line through the two points to obtain the graph shown in Figure 6-22.

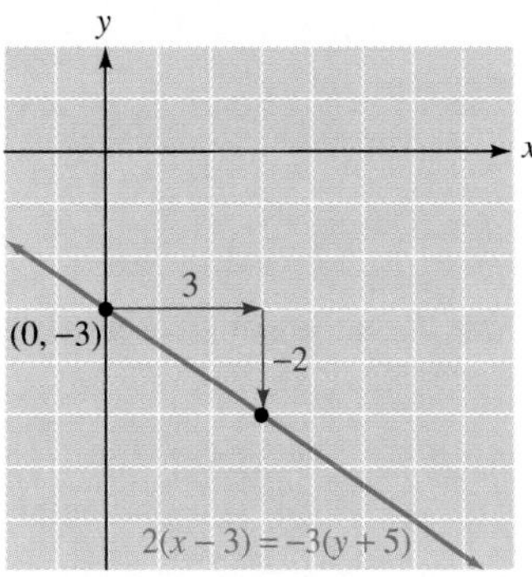

FIGURE 6-22

■

EXAMPLE 5 Show that the lines represented by $4x + 8y = 10$ and $2x = 12 - 4y$ are parallel.

Solution We solve each equation for y to see that the lines are distinct and that their slopes are equal.

$$\begin{aligned} 4x + 8y &= 10 \\ 8y &= -4x + 10 \\ y &= -\frac{1}{2}x + \frac{5}{4} \end{aligned} \qquad \begin{aligned} 2x &= 12 - 4y \\ 4y &= -2x + 12 \\ y &= -\frac{1}{2}x + 3 \end{aligned}$$

Since the values of b in these equations are different, the lines are distinct. Since the slope of each line is $-\frac{1}{2}$, they are parallel. ■

EXAMPLE 6 Show that the lines represented by $4x + 8y = 10$ and $4x - 2y = 21$ are perpendicular.

Solution We solve each equation for y to see that the slopes of their straight-line graphs are negative reciprocals.

$$\begin{aligned} 4x + 8y &= 10 \\ 8y &= -4x + 10 \\ y &= -\frac{1}{2}x + \frac{5}{4} \end{aligned} \qquad \begin{aligned} 4x - 2y &= 21 \\ -2y &= -4x + 21 \\ y &= 2x - \frac{21}{2} \end{aligned}$$

Since the slopes of the lines are $-\frac{1}{2}$ and 2 (which are negative reciprocals), the lines are perpendicular. ■

EXAMPLE 7 Use the slope–intercept form to write the equation of the line passing through the point $P(-2, 5)$ and parallel to the line $y = 8x - 2$.

Solution Since the slope of the line given by $y = 8x - 2$ is the coefficient of x, the slope of the line is 8. Thus, the line represented by the desired equation must also have a slope of 8, because it is parallel to the graph of $y = 8x - 2$.

We substitute -2 for x, 5 for y, and 8 for m in the slope–intercept form and solve for b.

$y = \boldsymbol{m}\boldsymbol{x} + b$

$5 = \mathbf{8}(\mathbf{-2}) + b$ Substitute 5 for y, 8 for m, and -2 for x.

$5 = -16 + b$ Simplify.

$21 = b$ Add 16 to both sides.

Since $m = 8$ and $b = 21$, the equation of the desired line is $y = 8x + 21$. ■

EXAMPLE 8 Use the point–slope form to write the equation of the line passing through $P(-2, 5)$ and perpendicular to the line $y = 8x - 2$.

Solution The slope of the given line is 8. Thus, the slope of the desired line must be $-\frac{1}{8}$, which is the negative reciprocal of 8.

We substitute -2 for x_1, 5 for y_1, and $-\frac{1}{8}$ for m into the point–slope form and simplify.

$$y - y_1 = m(x - x_1)$$

$$y - 5 = -\frac{1}{8}[x - (-2)] \qquad \text{Substitute 5 for } y_1,\ -\tfrac{1}{8} \text{ for } m, \text{ and } -2 \text{ for } x_1.$$

$$y - 5 = -\frac{1}{8}(x + 2) \qquad \text{Simplify.}$$

$$8y - 40 = -(x + 2) \qquad \text{Multiply both sides by 8.}$$

$$8y - 40 = -x - 2 \qquad \text{Use the distributive property to remove parentheses.}$$

$$x + 8y = 38 \qquad \text{Add } x \text{ and 40 to both sides.}$$

The equation of the line is $x + 8y = 38$. ■

General Form of the Equation of a Line

The final equation in Example 8 was written in the form

$$x + 8y = 38$$

This is an example of a form called the **general form** of the equation of a line. Any equation that is written in the form $Ax + By = C$ is said to be written in general form.

Finding the Slope and y-Intercept from the General Form

If A, B, and C are real numbers and $B \neq 0$, then the graph of the equation

$$Ax + By = C$$

is a nonvertical line with slope of $-\frac{A}{B}$ and a y-intercept of $\left(0, \frac{C}{B}\right)$.

If $B = 0$, the equation $Ax + By = C$ represents a vertical line with x-intercept of $\left(\frac{C}{A}, 0\right)$. Because the graph of any equation that can be written in the form $Ax + By = C$ is a straight line, it is called a **linear equation in *x* and *y*.**

WARNING! When writing an equation in general form, it is customary to clear the equation of fractions and make A positive.

It is also customary to eliminate common factors of A, B, and C. For example, the equation $6x + 12y = 24$ can be changed to $x + 2y = 4$ by dividing both sides by 6.

EXAMPLE 9 Show that the lines represented by $4x + 3y = 7$ and $3x - 4y = 12$ are perpendicular.

Solution To show that the lines are perpendicular, we will show that their slopes are negative reciprocals. The first equation, $4x + 3y = 7$, is written in general form, with $A = 4$, $B = 3$, and $C = 7$. By the previous theorem, the slope of the line is

$$m_1 = -\frac{A}{B} = -\frac{4}{3}$$

The second equation, $3x - 4y = 12$, is also written in general form, with $A = 3$, $B = -4$, and $C = 12$. The slope of this line is

$$m_2 = -\frac{A}{B} = -\frac{3}{-4} = \frac{3}{4}$$

Since the slopes of the two lines are negative reciprocals, they are perpendicular. ■

We summarize the various forms for the equation of a line as follows.

General Form of a Linear Equation $Ax + By = C$
A and B cannot both be 0.

Slope–Intercept Form of a Linear Equation $y = mx + b$
The slope is m, and the y-intercept is $(0, b)$.

Point–Slope Form of a Linear Equation $y - y_1 = m(x - x_1)$
The slope is m, and the line passes through (x_1, y_1).

A Horizontal Line $y = b$
The slope is 0, and the y-intercept is $(0, b)$.

A Vertical Line $x = a$
There is no defined slope, and the x-intercept is $(a, 0)$.

Straight-Line Depreciation

For tax purposes, many businesses use the equation of a line to find the declining value of aging equipment. This method is called **straight-line depreciation.**

EXAMPLE 10 **Value of a lathe** A machine shop buys a lathe for \$1970 and expects it to last for 10 years. Then the lathe can be sold as scrap for an estimated **salvage value** of \$270. If y represents the value of the lathe after x years of use and y and x are related by the equation of a line,

a. Find the equation of the line.

b. Find the value of the lathe after $2\frac{1}{2}$ years.

c. Find the economic meaning of the y-intercept of the line.

d. Find the economic meaning of the slope of the line.

Solution **a.** To find the equation of the line, we first find two points on the line and then find the line's slope. We can then use the point–slope form to find the equation of the line.

When the lathe is new, its age x is 0 and its value y is \$1970. When the lathe is 10 years old, $x = 10$ and its value is $y = \$270$. Since the line passes through the points (0, 1970) and (10, 270) as shown in Figure 6-23, the slope of the line is

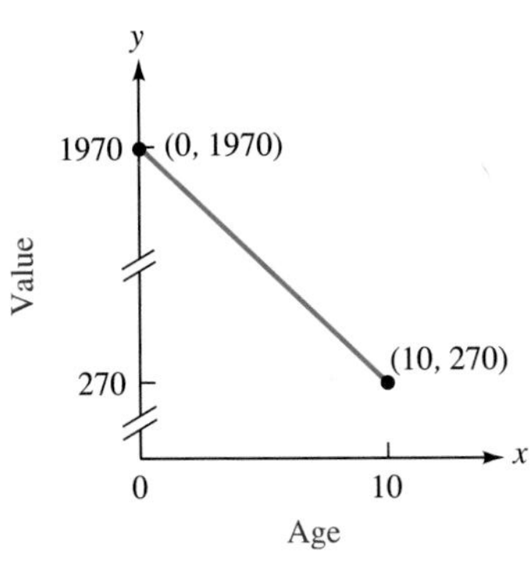

FIGURE 6-23

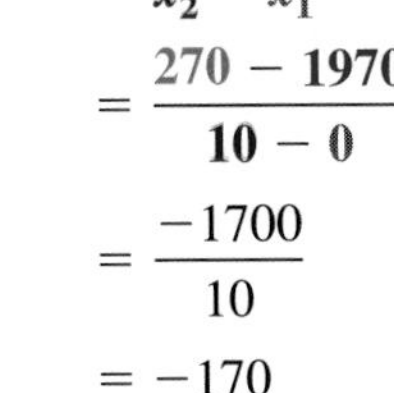

$$
\begin{aligned}
m &= \frac{y_2 - y_1}{x_2 - x_1} \\
&= \frac{270 - 1970}{10 - 0} \\
&= \frac{-1700}{10} \\
&= -170
\end{aligned}
$$

To find the equation of the line, we substitute -170 for m, 0 for x_1, and 1970 for y_1 in the point–slope form and simplify.

$$
\begin{aligned}
y - y_1 &= m(x - x_1) \\
y - 1970 &= -170(x - 0) \\
y &= -170x + 1970 \qquad (3)
\end{aligned}
$$

The current value y of the lathe is related to its age x by the equation

$$y = -170x + 1970$$

b. To find the age of the lathe after $2\frac{1}{2}$ years, we substitute 2.5 for x in Equation 3 and solve for y.

$$
\begin{aligned}
y &= -170x + 1970 \\
&= -170(2.5) + 1970 \\
&= -425 + 1970 \\
&= 1545
\end{aligned}
$$

The lathe will be worth \$1545 when it is $2\frac{1}{2}$ years old.

c. The y-intercept of the graph is $(0, b)$, where b is the value of y when $x = 0$. Thus, b is the value of a 0-year old lathe, which is the lathe's original cost.

d. Each year, the value of the lathe decreases by \$170, because the slope of the line is -170. The slope of the depreciation line is the **annual depreciation rate.** ■

ORALS *Give the point-slope form of the equation of a line with $m = 2$ and passing through the given point.*

1. $(2, 3)$ $y - 3 = 2(x - 2)$

2. $(-3, 8)$ $y - 8 = 2(x + 3)$

Give the equation of a line with $m = -3$ and y-intercept of

3. $(0, 5)$ $y = -3x + 5$

4. $(0, -7)$ $y = -3x - 7$

Tell whether the lines are parallel or perpendicular.

5. $y = 3x - 4$, $y = 3x + 5$ parallel

6. $y = -3x + 7$, $y = \frac{1}{3}x - 1$ perpendicular

EXERCISE 6.3

In Exercises 1–6, use the point–slope form to write the equation of the line with the given properties. Write each equation in general form.

1. $m = 5$, passing through $P(0, 7)$ $5x - y = -7$

2. $m = -8$, passing through $P(0, -2)$ $8x + y = -2$

3. $m = -3$, passing through $P(2, 0)$ $3x + y = 6$

4. $m = 4$, passing through $P(-5, 0)$ $4x - y = -20$

5. $m = \frac{3}{2}$, passing through $P(2, 5)$ $3x - 2y = -4$

6. $m = -\frac{2}{3}$, passing through $P(-3, 2)$ $2x + 3y = 0$

In Exercises 7–12, use the point–slope form to write the equation of the line passing through the two given points. Write each equation in slope–intercept form.

7. $P(0, 0)$, $Q(4, 4)$ $y = x$

8. $P(-5, -5)$, $Q(0, 0)$ $y = x$

9. $P(3, 4)$, $Q(0, -3)$ $y = \frac{7}{3}x - 3$

10. $P(4, 0)$, $Q(6, -8)$ $y = -4x + 16$

11. $P(-2, 4)$, $Q(3, -5)$ $y = -\frac{9}{5}x + \frac{2}{5}$

12. $P(3, -5)$, $Q(-1, 12)$ $y = -\frac{17}{4}x + \frac{31}{4}$

In Exercises 13–20, use the slope–intercept form to write the equation of the line with the given properties. Write each expression in slope–intercept form.

13. $m = 3$, $b = 17$ $y = 3x + 17$

14. $m = -2$, $b = 11$ $y = -2x + 11$

15. $m = -7$, passing through $P(7, 5)$ $y = -7x + 54$

16. $m = 3$, passing through $P(-2, -5)$ $y = 3x + 1$

17. $m = 0$, passing through $P(2, -4)$ $y = -4$

18. $m = -7$, passing through the origin $y = -7x$

19. passing through $P(6, 8)$ and $Q(2, 10)$ $y = -\frac{1}{2}x + 11$

20. passing through $P(-4, 5)$ and $Q(2, -6)$ $y = -\frac{11}{6}x - \frac{7}{3}$

In Exercises 21–26, write each equation in slope–intercept form to find the slope and the y-intercept. Then use the slope and y-intercept to draw the line.

21. $y + 1 = x$ 1, (0, −1)

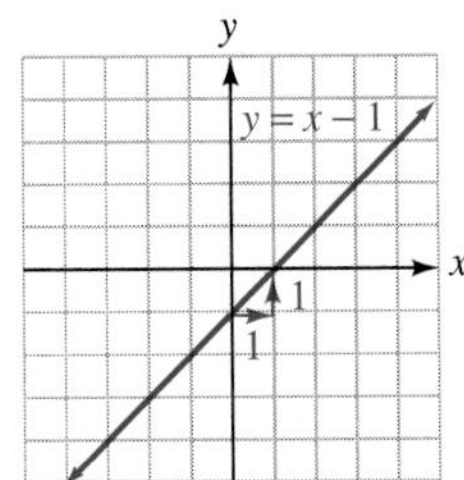

22. $x + y = 2$ −1, (0, 2)

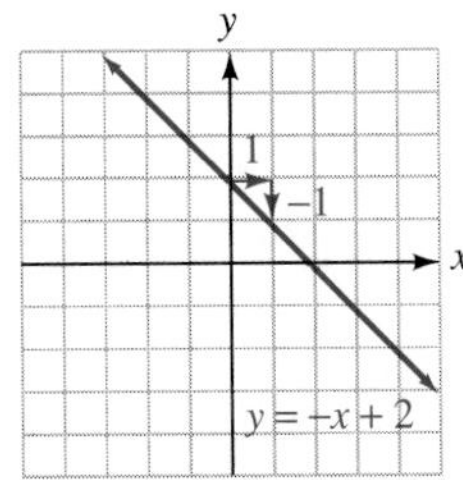

23. $x = \frac{3}{2}y - 3$ $\frac{2}{3}$, (0, 2)

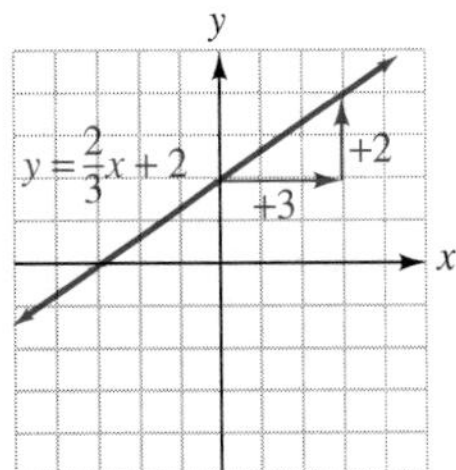

24. $x = -\frac{4}{5}y + 2$ $-\frac{5}{4}$, $\left(0, \frac{5}{2}\right)$

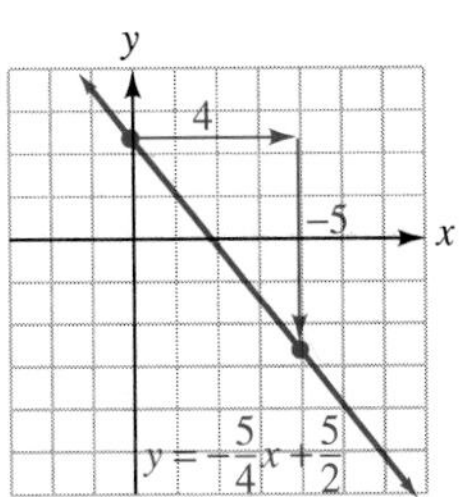

25. $3(y - 4) = -2(x - 3)$ $-\frac{2}{3}$, (0, 6)

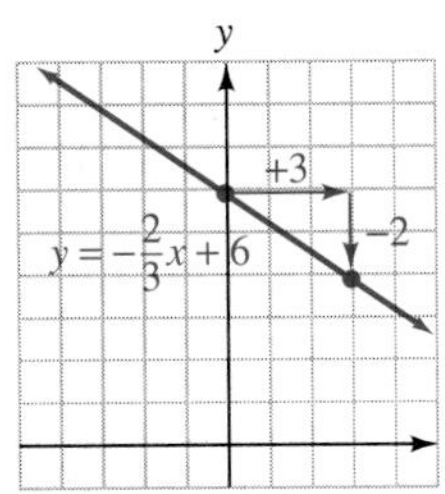

26. $-4(2x + 3) = 3(3y + 8)$ $-\frac{8}{9}$, (0, −4)

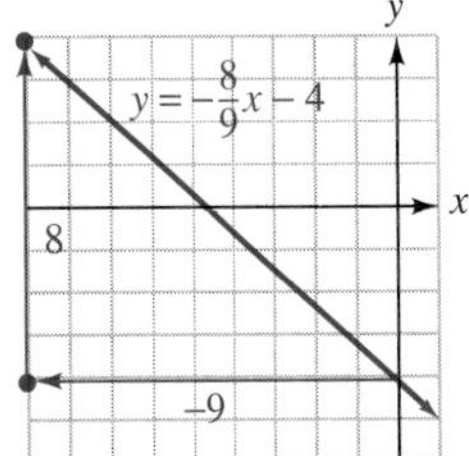

In Exercises 27–32, find the slope and the y-intercept of the line determined by the given equation.

27. $3x - 2y = 8$ $\frac{3}{2}$, (0, −4)

28. $-2x + 4y = 12$ $\frac{1}{2}$, (0, 3)

29. $-2(x + 3y) = 5$ $-\frac{1}{3}$, $\left(0, -\frac{5}{6}\right)$

30. $5(2x - 3y) = 4$ $\frac{2}{3}$, $\left(0, -\frac{4}{15}\right)$

31. $x = \dfrac{2y - 4}{7}$ $\frac{7}{2}$, (0, 2)

32. $3x + 4 = -\dfrac{2(y - 3)}{5}$ $-\frac{15}{2}$, (0, −7)

In Exercises 33–44, tell whether the graphs of each pair of equations are parallel, perpendicular, or neither.

33. $y = 3x + 4$, $y = 3x - 7$ parallel

34. $y = 4x - 13$, $y = \frac{1}{4}x + 13$ neither

35. $x + y = 2$, $y = x + 5$ perpendicular

36. $x = y + 2$, $y = x + 3$ parallel

37. $y = 3x + 7$, $2y = 6x - 9$ parallel

38. $2x + 3y = 9$, $3x - 2y = 5$ perpendicular

39. $x = 3y + 8$, $y = -3x + 7$ perpendicular

40. $3x + 6y = 1$, $y = \frac{1}{2}x$ neither

41. $y = 3$, $x = 4$ perpendicular

42. $y = -3$, $y = -7$ parallel

43. $x = \dfrac{y - 2}{3}$, $3(y - 3) + x = 0$ perpendicular

44. $2y = 8$, $3(2 + y) = 3(x + 2)$ neither

In Exercises 45–50, write the equation of the line that passes through the given point and is parallel to the given line.

45. $P(0, 0)$, $y = 4x - 7$ $y = 4x$

46. $P(0, 0)$, $x = -3y - 12$ $y = -\frac{1}{3}x$

47. $P(2, 5)$, $4x - y = 7$ $y = 4x - 3$

48. $P(-6, 3)$, $y + 3x = -12$ $y = -3x - 15$

49. $P(4, -2)$, $x = \frac{5}{4}y - 2$ $y = \frac{4}{5}x - \frac{26}{5}$

50. $P(1, -5)$, $x = -\frac{3}{4}y + 5$ $y = -\frac{4}{3}x - \frac{11}{3}$

In Exercises 51–56, write the equation of the line that passes through the given point and is perpendicular to the given line.

51. $P(0, 0)$, $y = 4x - 7$ $y = -\frac{1}{4}x$

52. $P(0, 0)$, $x = -3y - 12$ $y = 3x$

53. $P(2, 5)$, $4x - y = 7$ $y = -\frac{1}{4}x + \frac{11}{2}$

54. $P(-6, 3)$, $y + 3x = -12$ $y = \frac{1}{3}x + 5$

55. $P(4, -2)$, $x = \frac{5}{4}y - 2$ $y = -\frac{5}{4}x + 3$

56. $P(1, -5)$, $x = -\frac{3}{4}y + 5$ $y = \frac{3}{4}x - \frac{23}{4}$

In Exercises 57–60, use the method of Example 9 to find if the graphs determined by each pair of equations are parallel, perpendicular, or neither.

57. $4x + 5y = 20$, $5x - 4y = 20$ perpendicular

58. $9x - 12y = 17$, $3x - 4y = 17$ parallel

59. $2x + 3y = 12$, $6x + 9y = 32$ parallel

60. $5x + 6y = 30$, $6x + 5y = 24$ neither

61. Find the equation of the line perpendicular to the line $y = 3$ and passing through the midpoint of the segment joining $(2, 4)$ and $(-6, 10)$. $x = -2$

62. Find the equation of the line parallel to the line $y = -8$ and passing through the midpoint of the segment joining $(-4, 2)$ and $(-2, 8)$. $y = 5$

63. Find the equation of the line parallel to the line $x = 3$ and passing through the midpoint of the segment joining $(2, -4)$ and $(8, 12)$. $x = 5$

64. Find the equation of the line perpendicular to the line $x = 3$ and passing through the midpoint of the segment joining $(-2, 2)$ and $(4, -8)$. $y = -3$

65. Solve $Ax + By = C$ for y and thereby show that the slope of its graph is $-\frac{A}{B}$ and its y-intercept is $\left(0, \frac{C}{B}\right)$.
$y = -\frac{A}{B}x + \frac{C}{B}$

66. Show that the x-intercept of the graph of $Ax + By = C$ is $\left(\frac{C}{A}, 0\right)$.

In Exercises 67–77, assume straight-line depreciation or straight-line appreciation.

67. Finding a depreciation equation A taxicab was purchased for \$24,300. Its salvage value at the end of its 7-year useful life is expected to be \$1900. Find the depreciation equation.
$y = -3200x + 24{,}300$

68. Finding a depreciation equation A computer system was purchased for \$7900 and depreciated over its 4-year useful life. Its expected salvage value is \$300. Find its depreciation equation.
$y = -1900x + 7900$

69. Finding an appreciation equation An apartment building was purchased for \$475,000. The owners expect the property to double in value in 10 years. Find the appreciation equation.
$y = 47{,}500x + 475{,}000$

70. Finding an appreciation equation A house purchased for \$112,000 is expected to double in value in 12 years. Find its appreciation equation.
$y = \frac{28{,}000}{3}x + 112{,}000$

71. Finding a depreciation equation Find the depreciation equation for the TV in the want ad in Illustration 1. $y = -670x + 3900$

For Sale: 3-year-old 54-inch TV, \$3,900 new. Asking \$1,890. Call 875-5555. Ask for Mike.

ILLUSTRATION 1

72. **Depreciating a word processor** A word processor cost $555 when new and is expected to be worth $80 after 5 years. What will it be worth after 3 years? $270

73. **Finding salvage value** A copier cost $1050 when new and will be depreciated at the rate of $120 per year. If the useful life of the copier is 8 years, find its salvage value. $90

74. **Finding annual rate of depreciation** A truck that cost $27,600 when new will have no salvage value after 12 years. Find its annual rate of depreciation. $2300

75. **Finding the value of antiques** An antique table is expected to appreciate $40 each year. If the table will be worth $450 in 2 years, what will it be worth in 13 years? $890

76. **Finding the value of antiques** An antique clock is expected to be worth $350 after 2 years and $530 after 5 years. What will the clock be worth after 7 years? $650

77. **Finding the purchase price of real estate** A cottage that was purchased 3 years ago is now appraised at $47,700. If the property has been appreciating $3500 per year, find its original purchase price. $37,200

78. **Charges for computer repair** A computer repair company charges a fixed amount, plus an hourly rate, for a service call. Use the information in Illustration 2 to find the hourly rate. $17.50

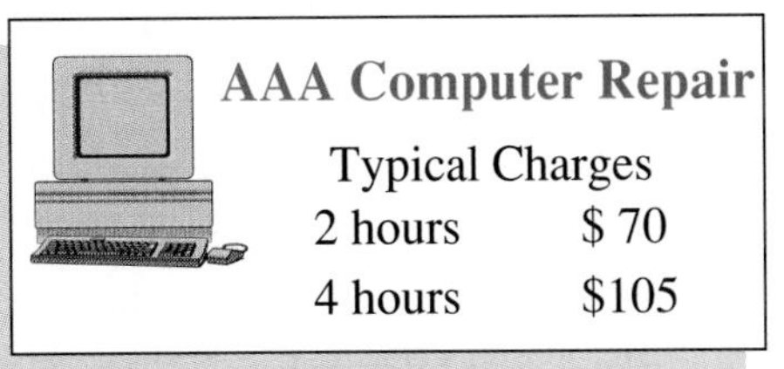

ILLUSTRATION 2

79. **Charges for automobile repair** An auto repair shop charges an hourly rate, plus the cost of parts. If the cost of labor for a $1\frac{1}{2}$-hour radiator repair is $69, find the cost of labor for a 5-hour transmission overhaul. $230

80. **Finding printer charges** A printer charges a fixed setup cost, plus $1 for every 100 copies. If 700 copies cost $52, how much will it cost to print 1000 copies? $55

81. **Estimating the cost of rain gutter** A neighbor says that an installer of rain gutter charges $60, plus a dollar amount per foot. If the neighbor paid $435 for the installation of 250 feet of gutter, how much will it cost you to have 300 feet installed? $510

82. **Predicting fires** A local fire department recognizes that city growth and the number of reported fires are related by a linear equation. City records show that 300 fires were reported in a year when the local population was 57,000 persons, and 325 fires were reported in a year when the population was 59,000 persons. How many fires can be expected when the population reaches 100,000 persons? about 838

Writing Exercises ■ *Write a paragraph using your own words.*

1. Explain how to find the equation of a line passing through two given points.
2. In straight-line depreciation, explain why the slope of the line is called the rate of depreciation.

Something to Think About ■

1. The graph of $y = ax + b$ passes through quadrants I, II, and IV. What can be known about the constants a and b? $a < 0, b > 0$
2. The graph of $Ax + By = C$ passes through the quadrants I and IV, only. What is known about the constants A, B, and C? $B = 0$; A and C have same sign

Review Exercises ■ *Solve each equation.*

If you teach graphing early, skip Review Exercises 1–4.

1. $2n^2 + n - 3 = 0$ $1, -\frac{3}{2}$
2. $6m^2 - m - 2 = 0$ $\frac{2}{3}, -\frac{1}{2}$
3. $4a^2 - 9 = 0$ $\frac{3}{2}, -\frac{3}{2}$
4. $2r^2 - r = 0$ $0, \frac{1}{2}$
5. **Mixing alloys** In 60 ounces of alloy for watch cases, there are 20 ounces of gold. How much copper must be added to the alloy so that a watch case weighing 4 ounces, made from the new alloy, will contain exactly 1 ounce of gold? 20 oz
6. **Mixing coffee** To make a mixture of 80 pounds of coffee worth \$272, a grocer mixes coffee worth \$3.25 a pound with coffee worth \$3.85 a pound. How many pounds of the cheaper coffee should the grocer use? 60 lb

6.4 Graphs of Other Equations

■ Graphing Calculators ■ Using the TRACE and ZOOM Keys
■ Solving Equations with a Graphing Calculator

Many equations have graphs that are not lines. For example, the graph of the equation $y = x^2 - 2$ is a curve, called a **parabola.**

EXAMPLE 1 Graph the equation $y = x^2 - 2$.

Solution We make a table of ordered pairs by substituting numbers for x and finding the corresponding values of y. For example, if we substitute -3 for x, we get

$$y = x^2 - 2$$
$$y = (-3)^2 - 2$$
$$y = 7$$

After plotting the points listed in the table shown in Figure 6-24 and joining them with a smooth curve, we obtain the graph of the equation.

$y = x^2 - 2$

x	y
−3	7
−2	2
−1	−1
0	−2
1	−1
2	2
3	7

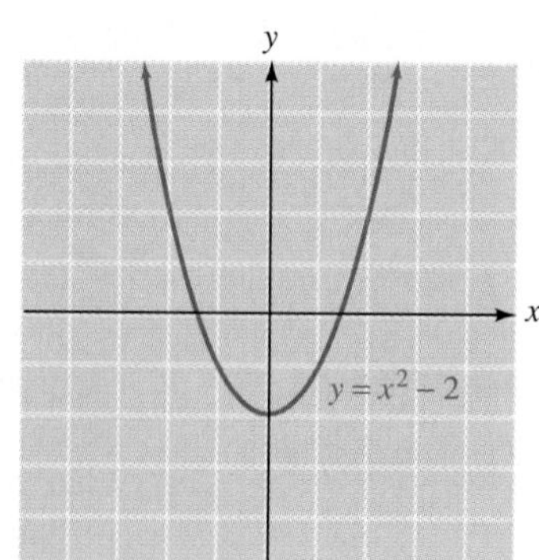

FIGURE 6-24

■

WARNING! The graph of a nonlinear equation is not a line. To graph nonlinear equations, we must usually plot many points to recognize the shape of the graph.

EXAMPLE 2 Graph the equation $y = -|x| + 3$.

Solution We make a table of values by substituting numbers for x and finding the corresponding values of y. For example, if we substitute -2 for x, we get

$$y = -|x| + 3$$
$$y = -|-2| + 3$$
$$y = -(2) + 3$$
$$y = 1$$

After plotting the points listed in the table shown in Figure 6-25, we obtain the graph of the equation.

$y = -|x| + 3$

x	y
−3	0
−2	1
−1	2
0	3
1	2
2	1
3	0

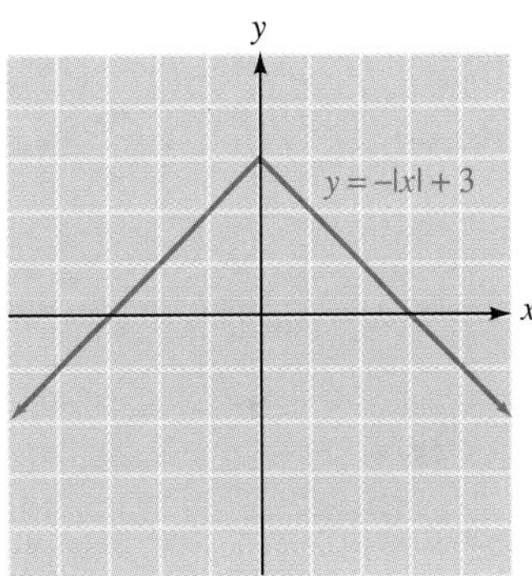

FIGURE 6-25

■

Graphing Calculators

So far, we have graphed equations by making a table of ordered pairs and plotting points. This method is often tedious and time-consuming. Graphing is much easier when we use a graphing calculator.

All graphing calculators have a **viewing window** that is used to display graphs (see Figure 6-26.) To see the proper picture of a graph, we must often set the minimum and maximum values for the x- and y-coordinates. The standard RANGE settings of

$$\text{Xmin} = -10 \qquad \text{Xmax} = 10 \qquad \text{Ymin} = -10 \qquad \text{Ymax} = 10$$

indicate that -10 is the minimum x- and y-coordinate that will be used in the graph, and that 10 is the maximum x- and y-coordinate that will be used. If these settings do not appear when you turn on your calculator and select the RANGE option, move the cursor to the desired position by pressing the cursor keys ◀, ▶,

FIGURE 6-26

▲, and ▼ and enter the standard RANGE settings. To delete any unwanted digits, press the DEL key.

To graph the equation $y = -\frac{1}{2}x^2 + 1$, we press the Y = key and enter the right-hand side of the equation after the symbol $Y_1 =$. The display will show the equation

$$Y_1 = -(1/2)X\text{^}2 + 1$$

We then press the GRAPH key to produce the graph shown in Figure 6-27(a). To show more detail, we can redraw the graph after setting other range values. The RANGE settings

Xmin = −4 Xmax = 4 Ymin = −4 Ymax = 4

produce the graph shown in Figure 6-27(b).

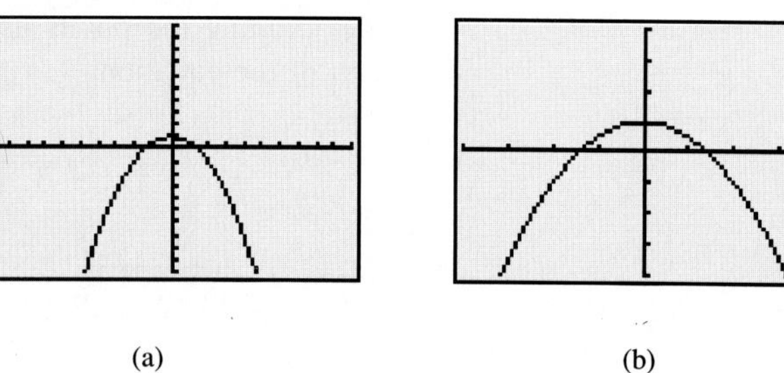

(a) (b)

FIGURE 6-27

EXAMPLE 3 Graph the equation $xy = 4$ in the following viewing windows:

a. $x = 10$ to 20 and $y = 10$ to 20

b. $x = 0$ to 5 and $y = 0$ to 5

c. $x = -6$ to 6 and $y = -6$ to 6

Solution Before we can enter the equation $xy = 4$ into the graphing calculator, we must solve the equation for y:

$$xy = 4$$

$$y = \frac{4}{x} \quad \text{Divide both sides by } x.$$

a. We enter the x and y RANGE values and enter the right-hand side of the equation. The viewing window will display

$$Y_1 = 4/X$$

When we press the GRAPH key, we obtain the blank window shown in Figure 6-28(a). The blank window shows that no portion of the graph appears on the coordinate grid seen in the viewing window.

b. We set the x and y RANGE values. There is no need to re-enter the equation. We press the GRAPH key to obtain the graph shown in Figure 6-28(b). In this case, only part of the graph appears. To see the whole graph, we need a larger viewing window.

c. We set the RANGE values and press the GRAPH key to see both branches of the curve, called a **hyperbola,** shown in Figure 6-28(c).

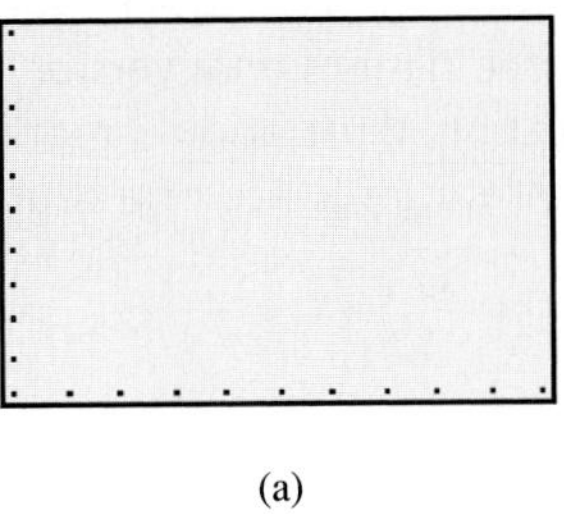
(a)

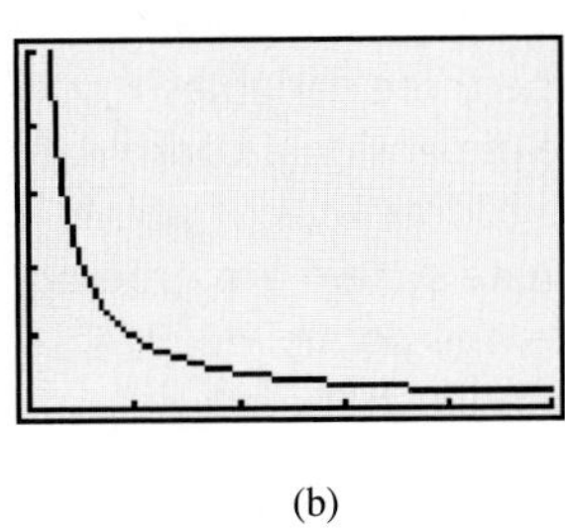
(b)

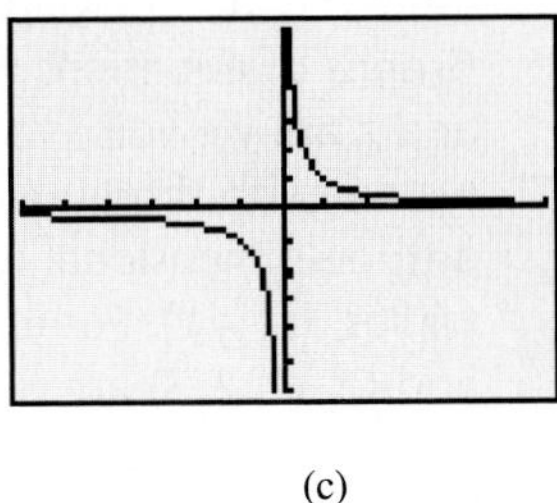
(c)

FIGURE 6-28 ■

As Example 3 shows, the choice of viewing window makes a big difference in how a graph looks. One of the challenges of using graphing calculators is finding an appropriate viewing window.

WARNING! When using a graphing calculator, be sure the viewing window does not show a misleading graph.

EXAMPLE 4 Graph the equation $y = -2x^2 + 15$ in the viewing window with RANGE values for x from 1 to 5 and for y from -5 to 5.

Solution We set the x and y RANGE values, enter the equation, and press the GRAPH key to obtain the Figure 6-29(a). Although the graph appears to be a straight line, it is not. We are seeing a part of a parabola that appears to be straight. If we pick a viewing window with x values of -5 to 5 and y values of -5 to 20, we can see the entire parabola, shown in Figure 6-29(b).

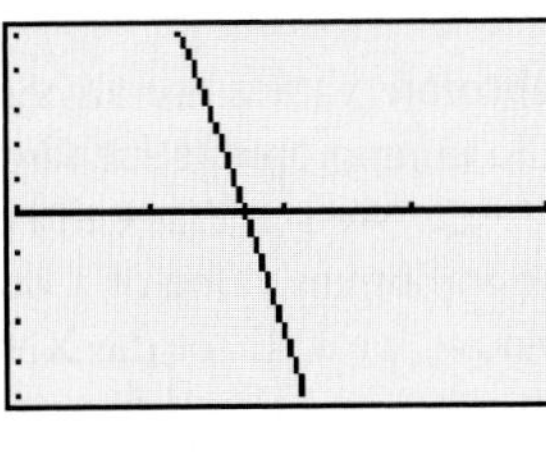
(a)

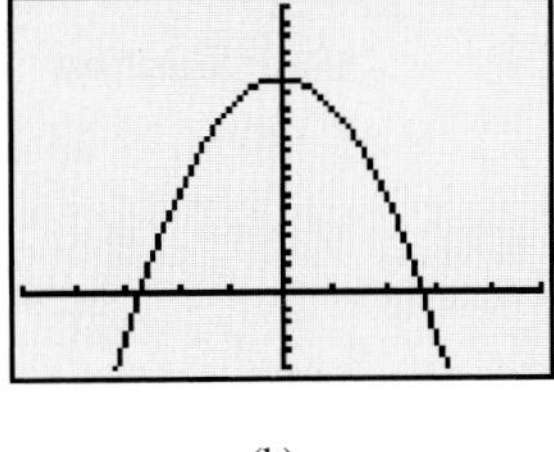
(b)

FIGURE 6-29 ■

PERSPECTIVE

Graphs in Space

In an *xy*-coordinate system, graphs of equations containing the two variables *x* and *y* are lines or curves. Other equations have more than two variables, and graphing them often requires some ingenuity and perhaps the aid of a computer. Graphs of equations with the three variables *x*, *y*, and *z* are viewed in a three-dimensional coordinate system with three axes. The coordinates of points in a three-dimensional coordinate system are ordered triples, (x, y, z). For example, the points $P(2, 3, 4)$ and $Q(-1, 2, 3)$ are plotted in Illustration 1.

Graphs of equations in three variables are not lines or curves, but flat planes or curved surfaces. Only the simplest of these equations can be conveniently graphed by hand; a computer provides the best images of others. The graph in Illustration 2 is called a paraboloid; it is the three-dimensional version of the parabolas in this section. Illustration 3 models a portion of the vibrating surface of a drum head.

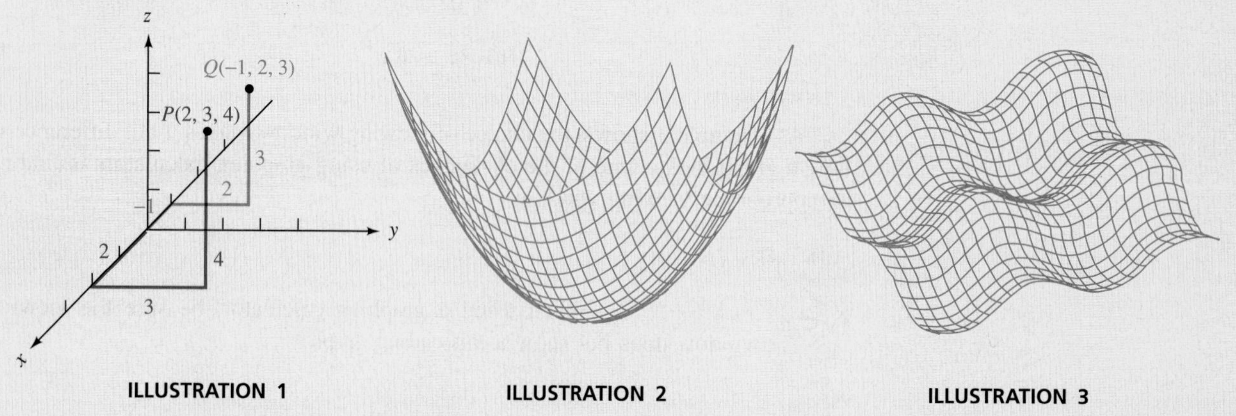

ILLUSTRATION 1 ILLUSTRATION 2 ILLUSTRATION 3

Computer programs for producing three-dimensional graphs are readily available. Some of the more powerful are Maple, Mathematica, MathCad, and Derive. Perhaps your school has such a program available for student use. With a brief introduction on the program's use, you can easily create several interesting graphs.

EXAMPLE 5 Graph the equation $y = |x - 2|$.

Solution Since absolute values are always nonnegative, the minimum value of y will be 0. To obtain an appropriate viewing window, we set the Ymin value slightly lower, at Ymin $= -2$. We set Ymax to be 10 greater than Ymin, at Ymax $= 8$.

The minimum value of y occurs when $x = 2$. To center the graph in the viewing window, we will set the Xmin and Xmax values at 5 on either side of 2:

$$\text{Xmin} = -3 \qquad \text{and} \qquad \text{Xmax} = 7$$

After entering the right-hand side of the equation, we obtain the graph shown in Figure 6-30. Consult your owner's manual to learn how to enter an absolute value.

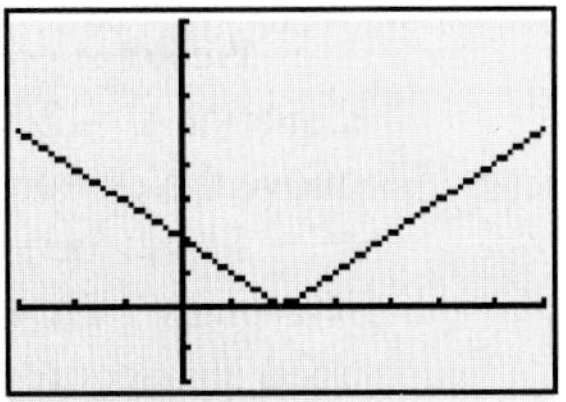

FIGURE 6-30 ■

■ Using the TRACE and ZOOM Keys

By using the TRACE key, it is possible to find the coordinates of any point on a graph. For example, if we use the standard RANGE settings, the graph of the equation $y = -\frac{1}{2}x^2 + 5$ looks like Figure 6-31(a).

To find where the parabola intersects the positive x-axis, we press the TRACE key and use the ◀ and ▶ keys to move the flashing cursor along the parabola until we approach one of the x-intercepts. The coordinates $x = 3.0526316$ and $y = .34072022$ appear at the bottom of the viewing window. This means that the cursor is at the point with coordinates (3.0526316, .34072022).

To get better results, we press the ZOOM key, zoom in, and press ENTER . We now see a magnified picture of the graph. We can press TRACE again and move the cursor to the point (3.1842105, −.0695983), as in Figure 6-31(b). Since y is nearly 0, the number 3.1842105 is nearly the x-coordinate of the x-intercept. We can achieve even better results with repeated zooms.

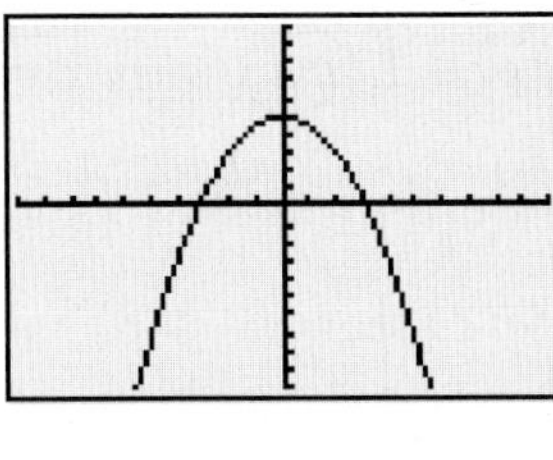

(a)

X=3.1842105 Y=-.0695983

(b)

FIGURE 6-31

■ Solving Equations with a Graphing Calculator

EXAMPLE 6 Use a graphing calculator to solve the equation $2(x - 3) + 3 = 13$.

Solution We first subtract 13 from both sides to get a 0 on the right-hand side of the equal sign.

$2(x - 3) - 10 = 0$ Subtract 13 from both sides and simplify.

The solutions of the equation $2(x - 3) - 10 = 0$ are the numbers x that will make $y = 0$ in the equation $2(x - 3) - 10 = y$. To find these numbers, we can graph the equation and read the x-intercept from the graph.

We set a graphing calculator with the standard RANGE settings and graph the equation $y = 2(x - 3) - 10$ as in Figure 6-32(a). We press the TRACE key and move the cursor close to the x-intercept, until we read an x-coordinate of 8.1052632.

To get better results, we zoom to obtain the graph shown in Figure 6-32(b). We press TRACE again and move the cursor close to the x-intercept, until we read an x-coordinate of 7.9736842.

Repeated zooms will show that the exact solution is $x = 8$. Verify that the number 8 satisfies the equation.

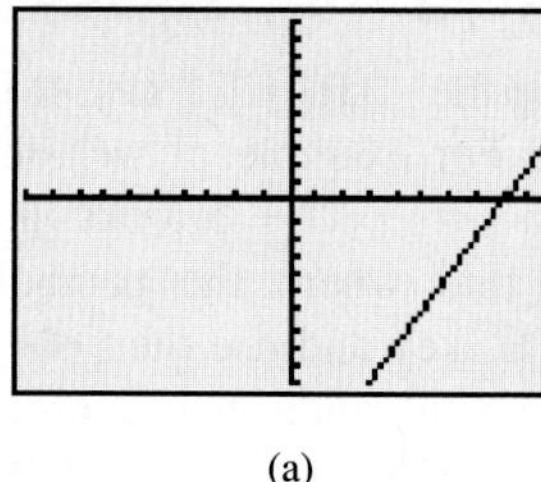

(a)

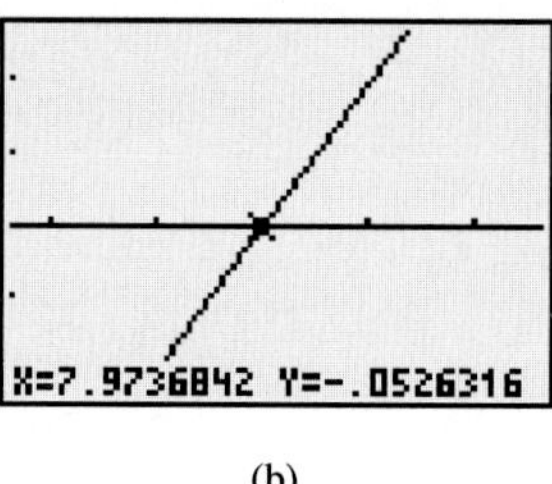

(b)

FIGURE 6-32

■

ORALS *If* $y = x^2 - 2$, *find* y *when* $x =$

1. 0 -2 **2.** 1 -1 **3.** -2 2 **4.** 5 23

If $y = |x| - 1$, *find* y *when* $x =$

5. 0 -1 **6.** 2 1 **7.** -4 3 **8.** -6 5

EXERCISE 6.4

In Exercises 1–8, graph each equation by plotting points.

1. $y = x^2 - 3$

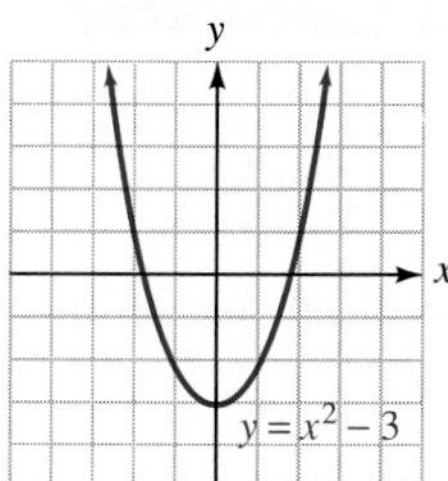

2. $y = -x^2 + 2$

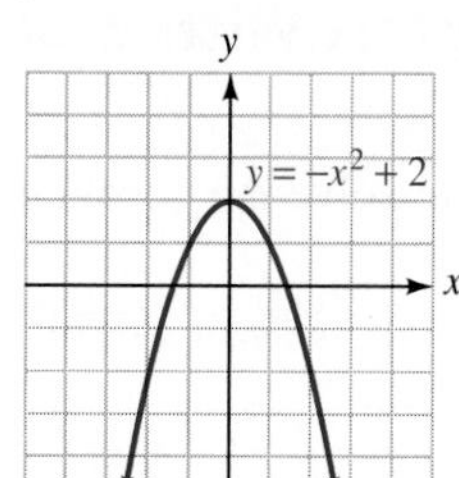

3. $y = (x - 1)^2$

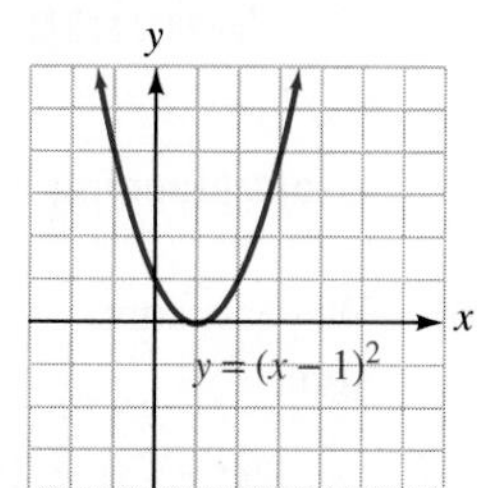

4. $y = -(x - 1)^2$

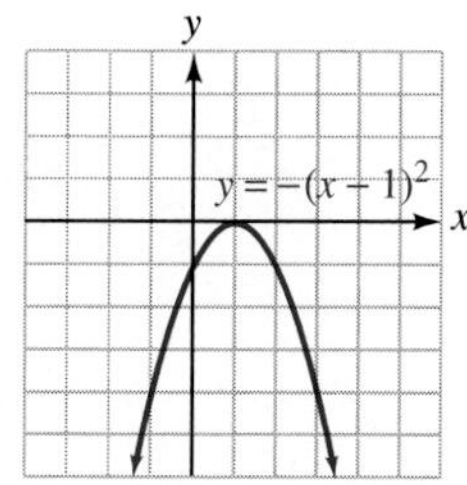

5. $y = -|x| + 2$

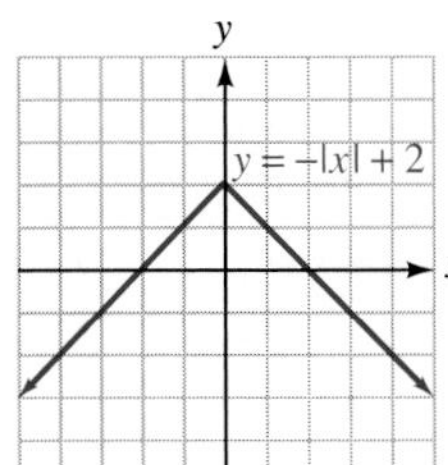

6. $y = |x + 3|$

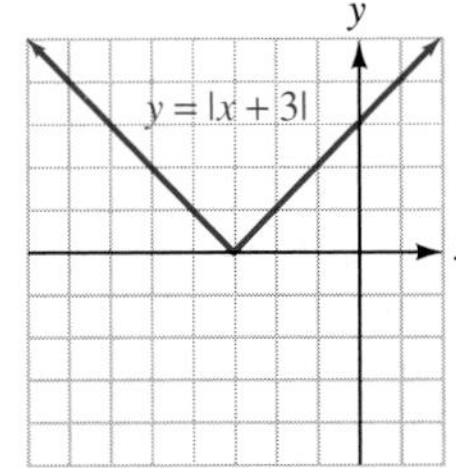

7. $y = |x - 1| - 2$

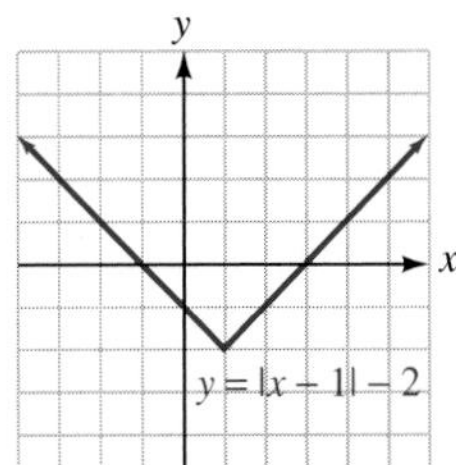

8. $y = -|x + 2| + 4$

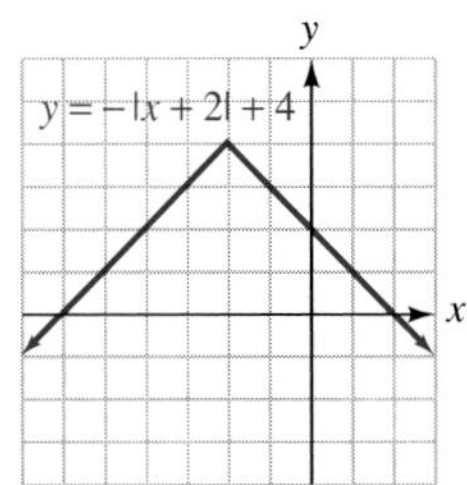

In Exercises 9–16, use a graphing calculator to graph each equation. Use a viewing window of $x = -5$ to 5 and $y = -5$ to 5. Compare the calculator graphs with your hand-made graphs in Exercises 1–8.

9. $y = x^2 - 3$ **10.** $y = -x^2 + 2$ **11.** $y = (x - 1)^2$ **12.** $y = -(x - 1)^2$

13. $y = -|x| + 2$ **14.** $y = |x + 3|$ **15.** $y = |x - 1| - 2$ **16.** $y = -|x + 2| + 4$

In Exercises 17–24, graph each equation in a viewing window of $x = -4$ to 4 and $y = -4$ to 4. The graph is not what it appears to be. Pick a better viewing window and find the true graph.

17. $y = x^2 + 8x$

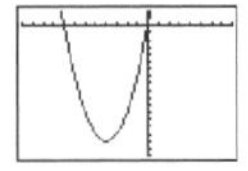

18. $y = x^2 - 8x$

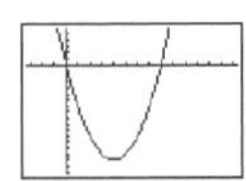

19. $y = |x + 5|$

20. $y = |x - 5|$

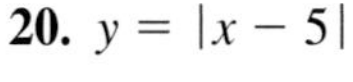

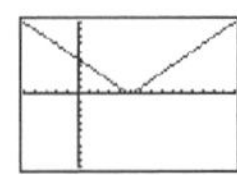

21. $y = x^2 - 9x + 14$

22. $y = x^2 + 9x + 14$

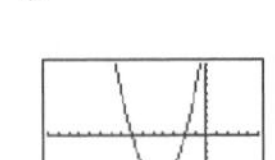

23. $y = x^3 + 8$

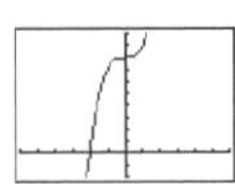

24. $y = x^3 - 12$

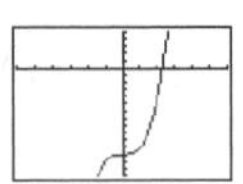

In Exercises 25–32, solve each equation for y and use a graphing calculator to graph it.

25. $2x - 5y = 10$

26. $3x + 4y = 12$

27. $x^2 + y = 6$

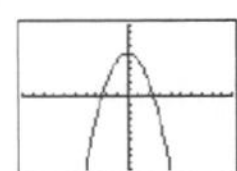

28. $x^2 - y = -4$

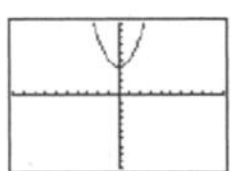

29. $|x| - 2y = 4$

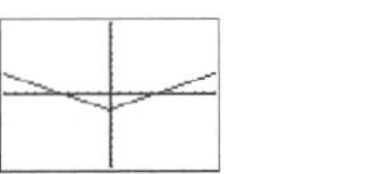

30. $|x| + \frac{1}{2}y = 2$

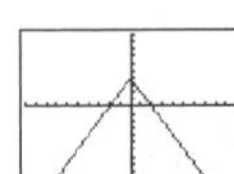

31. $xy = 12$

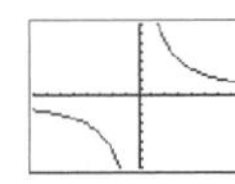

32. $xy = -6$

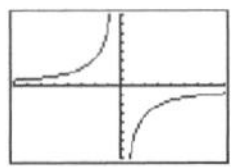

In Exercises 33–38, use a graphing calculator to solve each equation.

33. $3x + 6 = 0$ -2 **34.** $7y - 21 = 0$ 3 **35.** $4(x - 5) = 3x$ 20

36. $5(x + 7) = 6x$ 35 **37.** $11x + 6(3 - x) = 3$ -3 **38.** $3(x + 2) = 2(x - 7)$ -20

In Exercises 39–46, graph each parabola and find the coordinates of its highest or lowest point. This point is called the ***vertex*** *of the parabola.*

39. $y = x^2 + 4x$ $(-2, -4)$
40. $y = x^2 - 2x$ $(1, -1)$
41. $y = -x^2 + 6x$ $(3, 9)$
42. $y = -x^2 + 4x$ $(2, 4)$
43. $y = x^2 + 4x - 5$ $(-2, -9)$
44. $y = x^2 - 5x - 2$ $\left(\frac{5}{2}, -\frac{33}{4}\right)$
45. $y = -x^2 - 4x + 2$ $(-2, 6)$
46. $y = -x^2 - 4x + 1$ $(-2, 5)$

Writing Exercises ■ *Write a paragraph using your own words.*

1. Explain how to graph an equation by plotting points.

2. Explain why the correct choice of RANGE values is important when using a graphing calculator.

Something to Think About ■

1. *Use a graphing calculator with the standard RANGE settings to graph*
a. $y = x^2$ **b.** $y = x^2 + 1$ **c.** $y = x^2 + 2$
What do you notice?

2. *Use a graphing calculator with the standard RANGE settings to graph*
a. $y = -|x|$ **b.** $y = -|x| - 1$ **c.** $y = -|x| - 2$
What do you notice?

Review Exercises ■

1. List the prime numbers between 40 and 50. 41, 43, 47

2. State the associative property of addition. $(a + b) + c = a + (b + c)$

3. State the commutative property of multiplication. $a \cdot b = b \cdot a$

4. What is the additive identity element? 0

5. What is the multiplicative identity element? 1

6. Find the multiplicative inverse of $\frac{5}{3}$. $\frac{3}{5}$

6.5 Graphs of Linear Inequalities in Two Variables

■ Graphing Double Inequalities ■ An Application of Linear Inequalities

The **graph of a linear inequality** in x and y is the graph of all ordered pairs (x, y) that satisfy the inequality.

Linear Inequalities A **linear inequality** is any inequality that can be written in the form

$$Ax + By < C \quad \text{or} \quad Ax + By > C \quad \text{or} \quad Ax + By \leq C \quad \text{or} \quad Ax + By \geq C$$

where A, B, and C are real numbers and A and B are not both 0.

Because the inequality $y > 3x + 2$ can be written in the form $-3x + y > 2$, it is an example of a linear inequality. To graph $y > 3x + 2$, we first note that exactly one of the following statements is true:

$$y < 3x + 2, \qquad y = 3x + 2, \qquad \text{or} \qquad y > 3x + 2$$

The graph of the equation $y = 3x + 2$ is the line shown in Figure 6-33(a). The graphs of $y < 3x + 2$ and $y > 3x + 2$ are half-planes, one on each side of that line. The graph of $y = 3x + 2$ is a boundary line separating the two half-planes. It is drawn with a broken line to show that it is not part of the graph of $y > 3x + 2$.

To find which half-plane is the graph of $y > 3x + 2$, we can substitute the coordinates of the origin (0, 0) into the inequality and simplify.

$$y > 3x + 2$$
$$0 > 3(0) + 2 \qquad \text{Substitute 0 for } x \text{ and 0 for } y.$$
$$0 \not> 2$$

Since the coordinates of the origin do not satisfy the inequality, the origin is not part of the graph of $y > 3x + 2$. Thus, the graph is the half-plane on the other side of the broken line. The graph of $y > 3x + 2$ is shown in Figure 6-33(b).

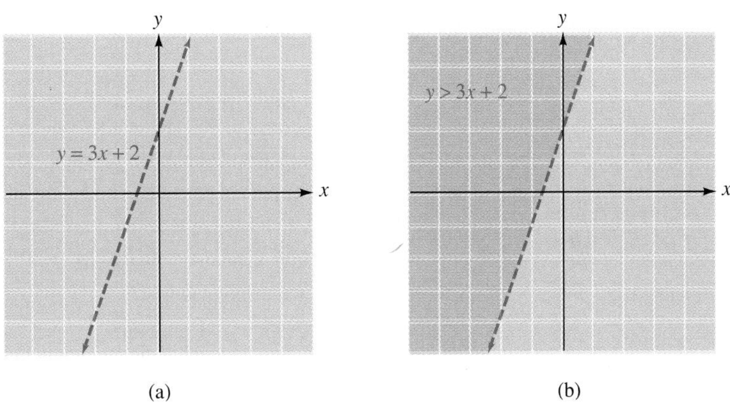

FIGURE 6-33

EXAMPLE 1 Graph the inequality $2x - 3y \leq 6$.

Solution This inequality is the combination of the inequality $2x - 3y < 6$ and the equation $2x - 3y = 6$.

We start by graphing the equation $2x - 3y = 6$ to find the boundary line that separates the two half-planes. This time, we draw the solid line shown in Figure 6-34(a), because equality is permitted. To decide which half-plane represents $2x - 3y < 6$, we check to see whether the coordinates of the origin satisfy the inequality.

$$2x - 3y < 6$$
$$2(0) - 3(0) < 6 \qquad \text{Substitute 0 for } x \text{ and 0 for } y.$$
$$0 < 6$$

Since the coordinates of the origin satisfy the inequality, the origin is in the half-plane that is the graph of $2x - 3y < 6$. The complete graph of $2x - 3y \leq 6$ is shown in Figure 6-34(b).

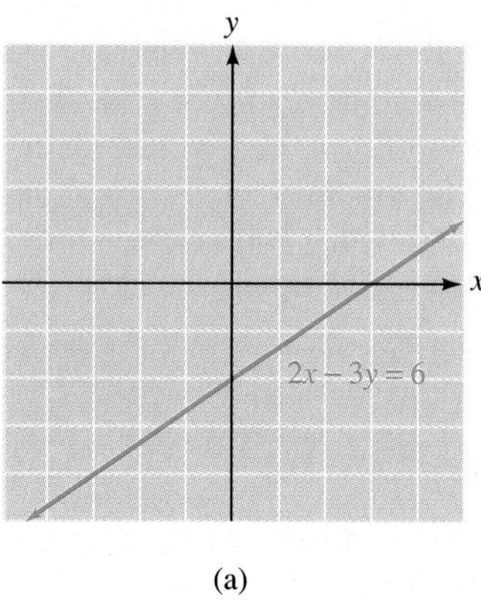

(a)

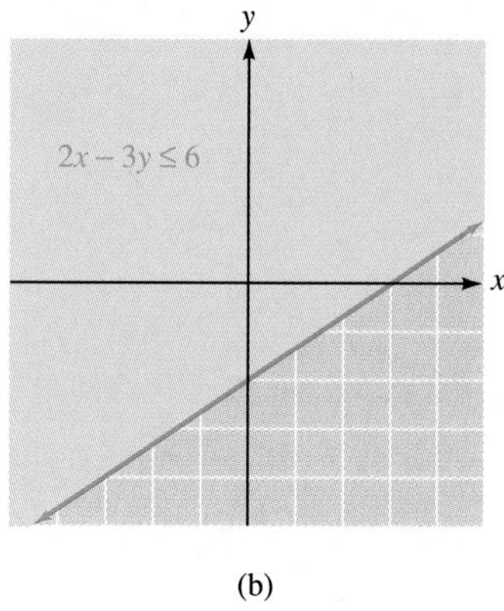

(b)

FIGURE 6-34 ■

EXAMPLE 2 Graph the inequality $y < 2x$.

Solution We graph the equation $y = 2x$, as shown in Figure 6-35(a). Because it is not part of the inequality, we draw the line as a broken line.

To decide which half-plane is the graph of $y < 2x$, we check to see whether the coordinates of some fixed point satisfy the inequality. This time we cannot use the origin as a test point, because the boundary line passes through the origin. However, we can choose a different point—say, (3, 1).

$y < 2x$

$1 < 2(3)$ Substitute 1 for y and 3 for x.

$1 < 6$

Since $1 < 6$ is a true inequality, the point (3, 1) satisfies the inequality and is in the graph of $y < 2x$, which is shown in Figure 6-35(b).

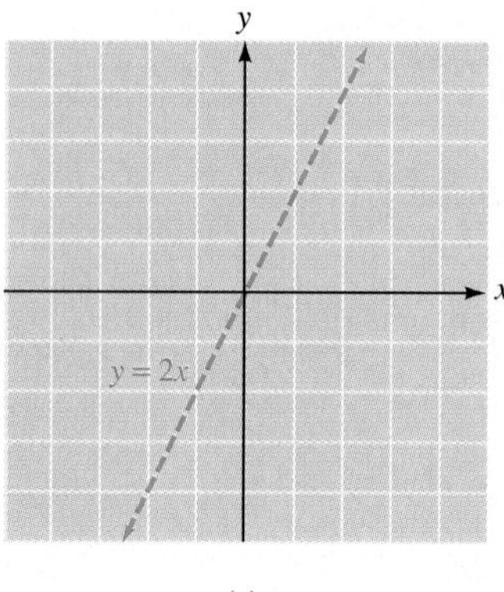

(a)

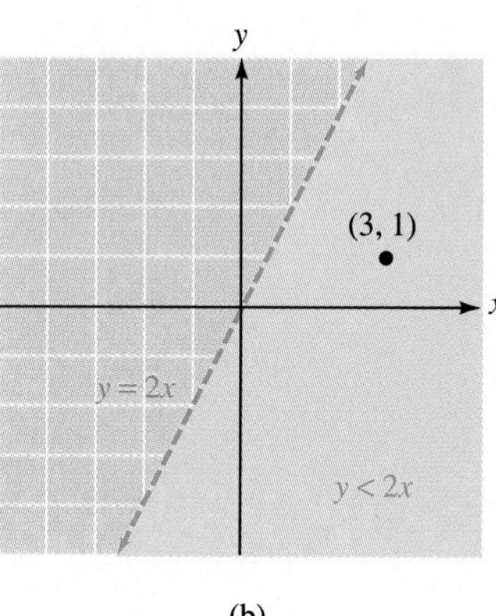

(b)

FIGURE 6-35 ■

Graphing Double Inequalities

EXAMPLE 3 Graph the inequality $2 < x \leq 5$.

Solution The double inequality $2 < x \leq 5$ is equivalent to the following pair of inequalities:

$$2 < x \quad \text{and} \quad x \leq 5$$

Thus, the graph of $2 < x \leq 5$ contains all points in the plane that satisfy the inequalities $2 < x$ and $x \leq 5$ simultaneously. These points are in the shaded region of Figure 6-36.

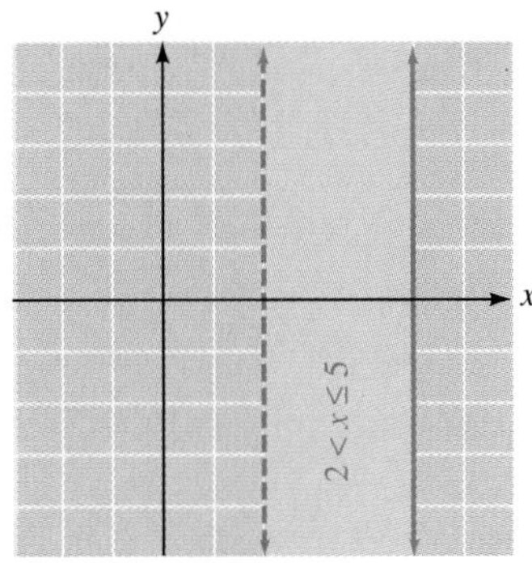

FIGURE 6-36

An Application of Linear Inequalities

EXAMPLE 4 **Earning money** Rick has two part-time jobs, one paying \$7 per hour and the other paying \$5 per hour. He must earn at least \$140 per week to pay his expenses while attending college. Write an inequality that shows the various ways he can schedule his workload to achieve his goal.

Solution If we let

x represent the number of hours he works on the first job and

y represent the number of hours he works on the second job,

we have

The hourly rate on the first job	$\cdot$	the hours worked on the first job	$+$	the hourly rate on the second job	$\cdot$	the hours worked on the second job	$\geq$	\$140.
\$7	$\cdot$	x	$+$	\$5	$\cdot$	y	$\geq$	\$140

The graph of the inequality $7x + 5y \geq 140$ is shown in Figure 6-37. Any point in the shaded region indicates a possible way he can schedule his time and earn

\$140 or more per week. For example, if he works 10 hours on the first job and 15 hours on the second job, he will earn

$$\begin{aligned} \$7(10) + \$5(15) &= \$70 + \$75 \\ &= \$145 \end{aligned}$$

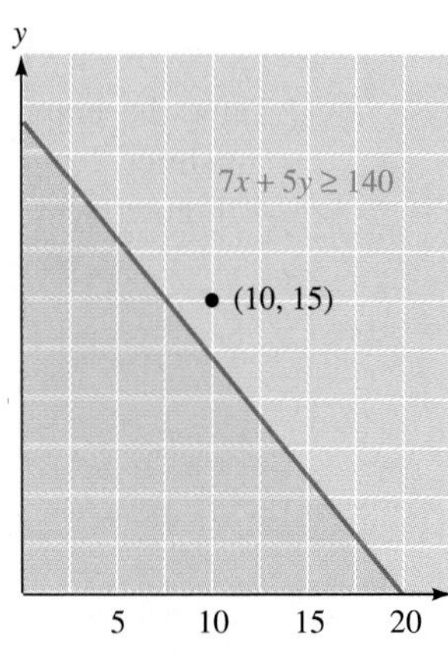

FIGURE 6-37

Since Rick cannot work a negative number of hours, the graph in the figure has no meaning when either x or y is negative. ■

ORALS *Does the given point satisfy* $2x + 3y < 12$?

1. (0, 0) yes **2.** (3, 2) no **3.** (2, 3) no **4.** (−1, 4) yes

Does the given point satisfy $3x - 2y \geq 12$?

5. (0, 0) no **6.** (3, 2) no **7.** (2, −3) yes **8.** (5, 1) yes

EXERCISE 6.5

In Exercises 1–18, graph each inequality.

1. $y > x + 1$

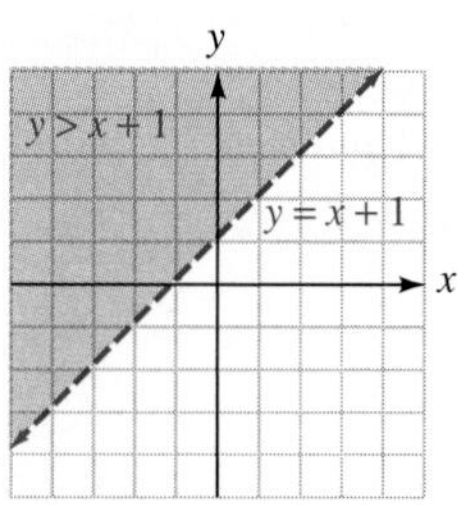

2. $y < 2x - 1$

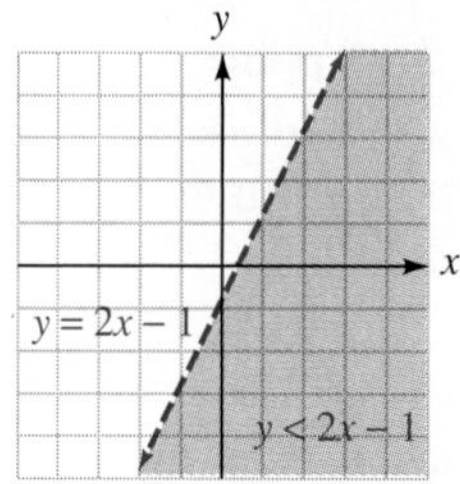

3. $y \geq x$

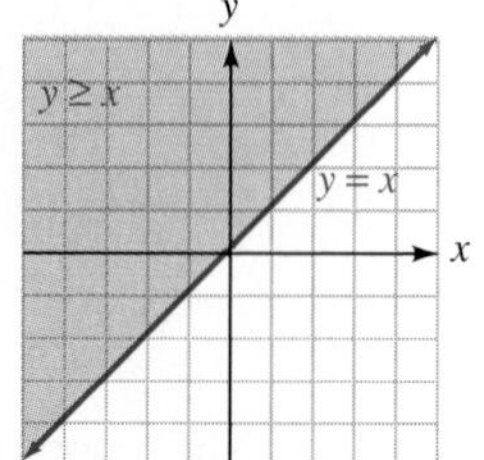

4. $y \leq 2x$

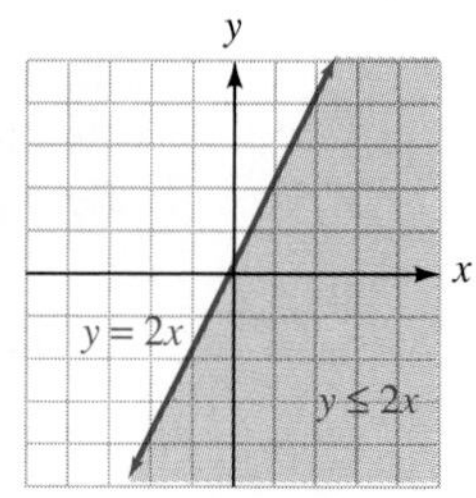

5. $2x + y \leq 6$

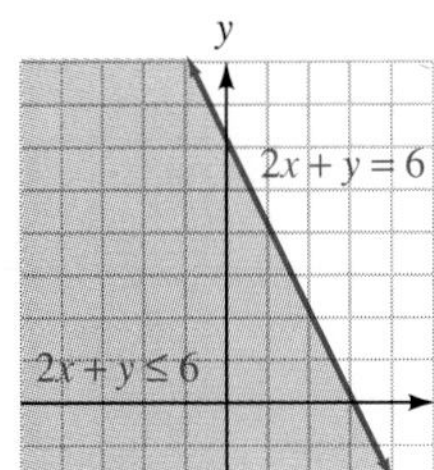

6. $x - 2y \geq 4$

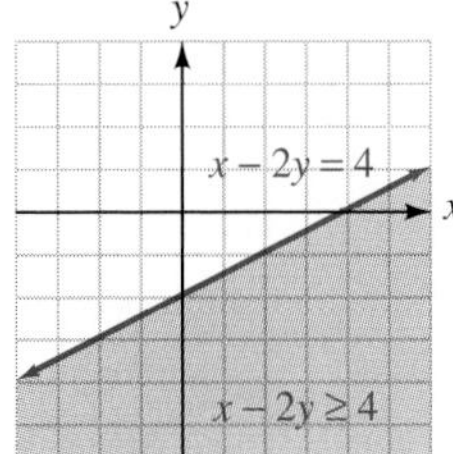

7. $3x \geq -y + 3$

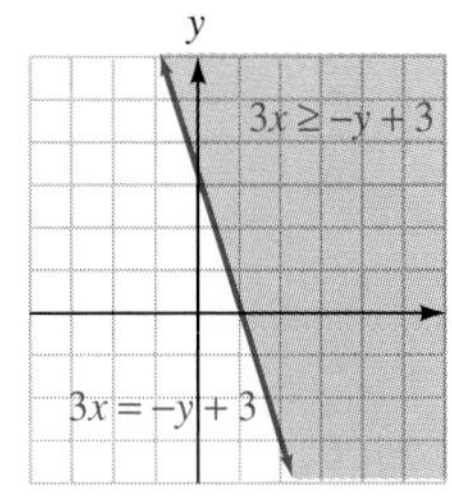

8. $2x \leq -3y - 12$

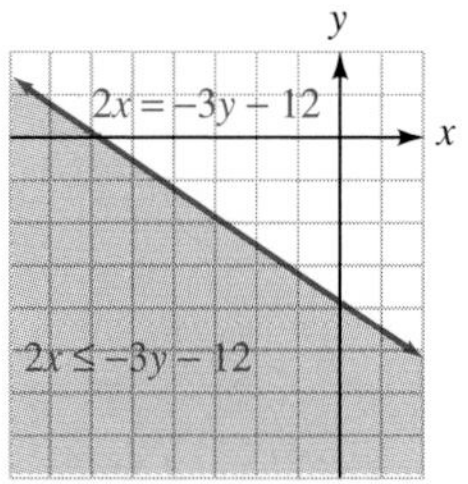

9. $y \geq 1 - \dfrac{3}{2}x$

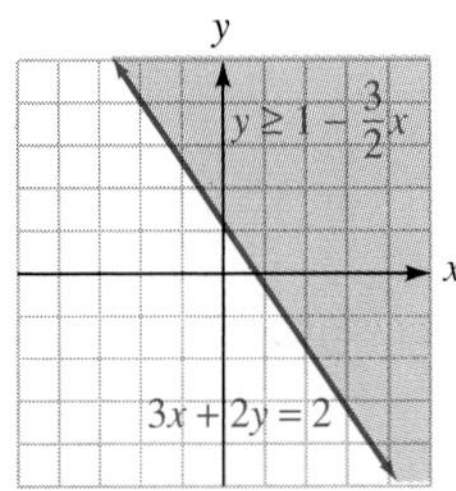

10. $y < \dfrac{1}{3}x - 1$

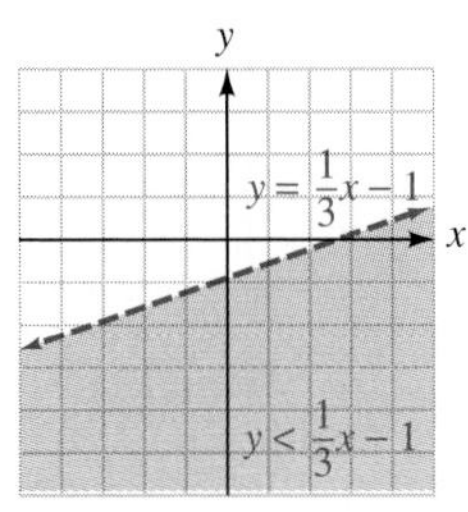

11. $x < 4$

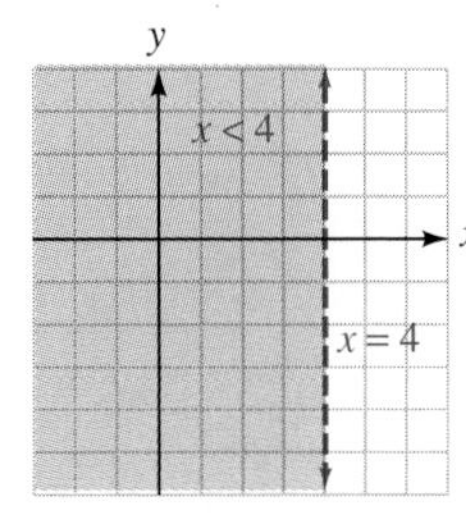

12. $y \geq -2$

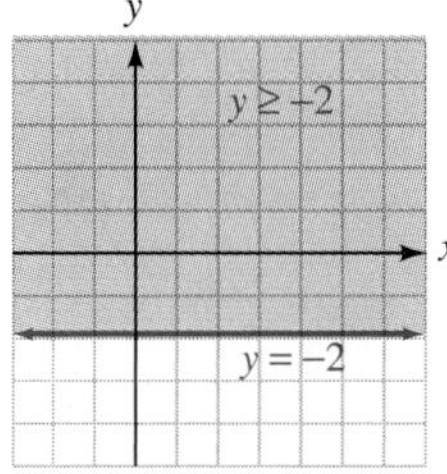

13. $-2 \leq x < 0$

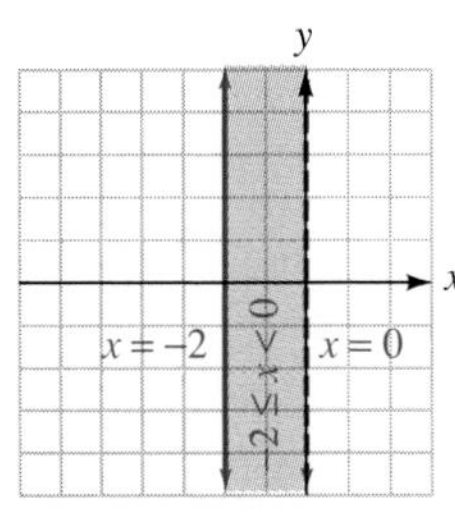

14. $0 < y \leq 5$

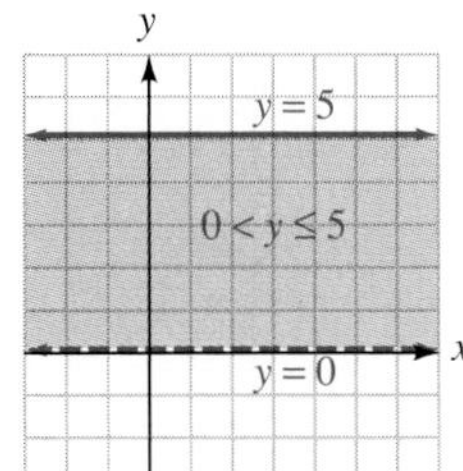

15. $y < -2$ or $y > 3$

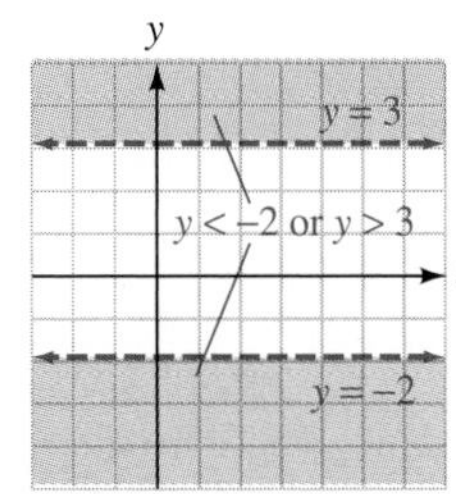

16. $-x \leq 1$ or $x \geq 2$

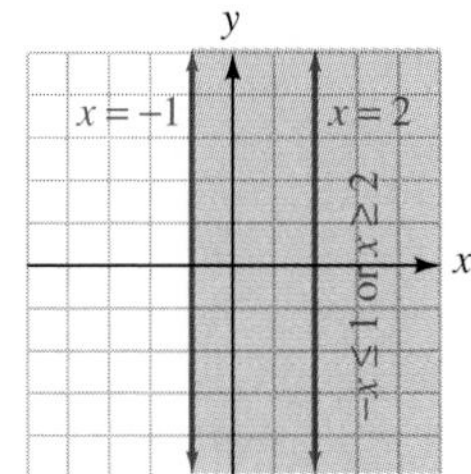

17. $-3 < y \leq -1$

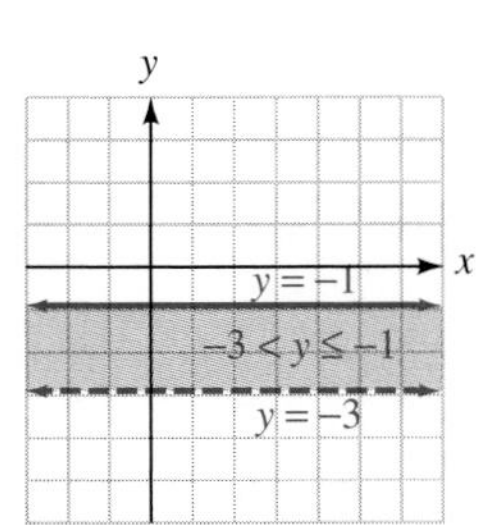

18. $-\dfrac{5}{2} \geq x > -4$

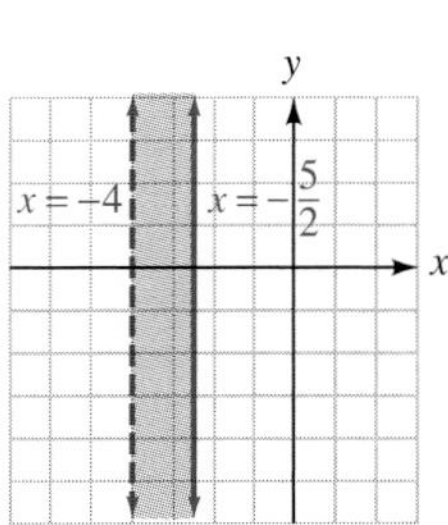

In Exercises 19–28, find the equation of the boundary line or lines. Then give the inequality whose graph is shown.

19.

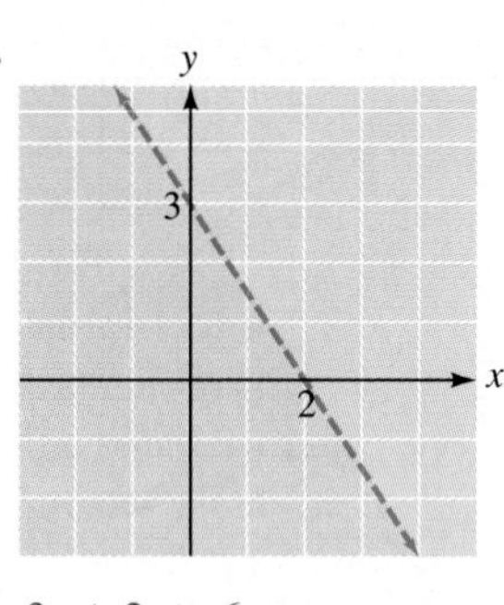

$3x + 2y > 6$

20.

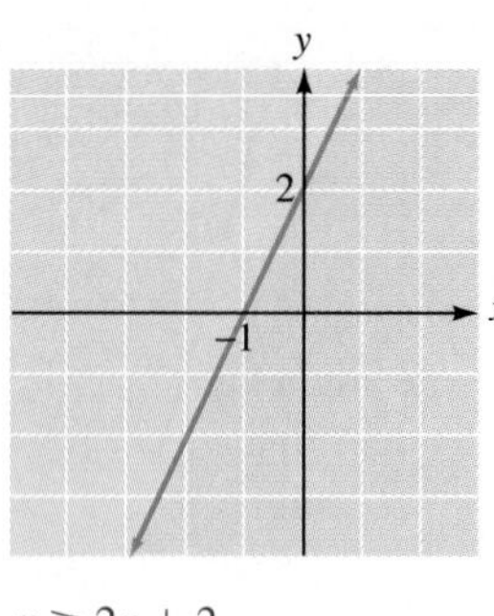

$y \geq 2x + 2$

21.

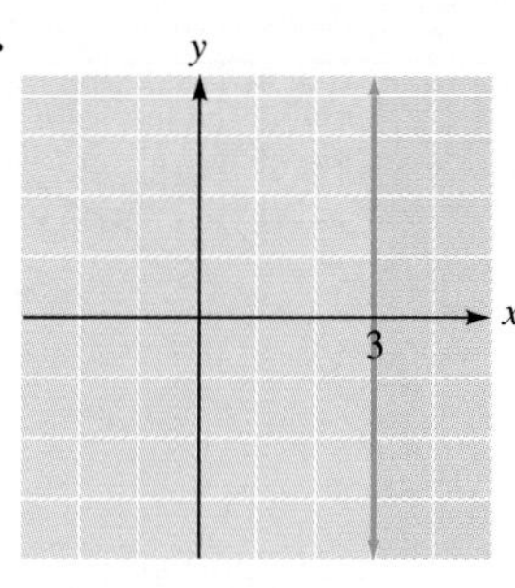

$x \leq 3$

22.

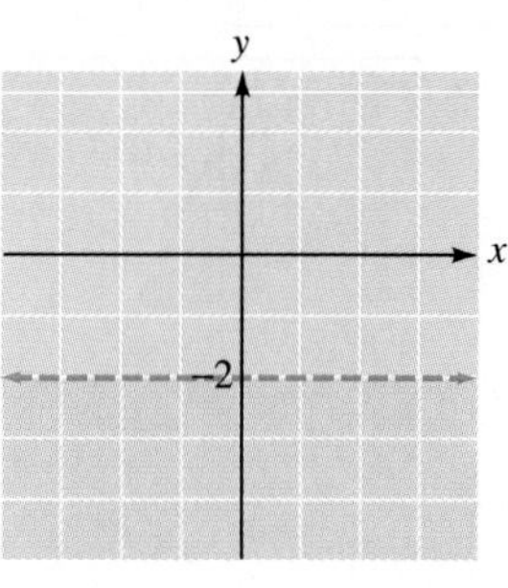

$y > -2$

23.

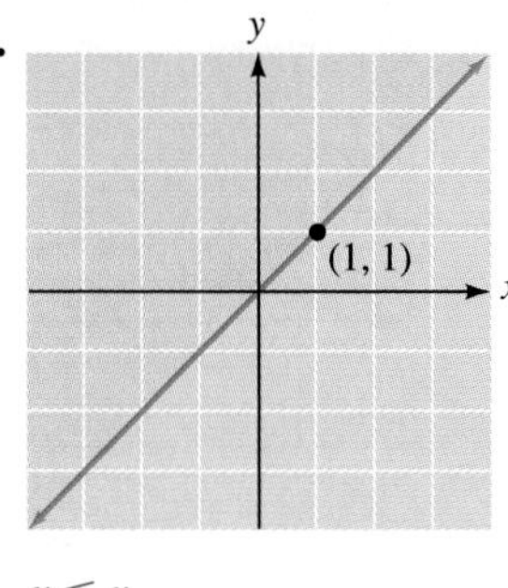

$y \leq x$

24.

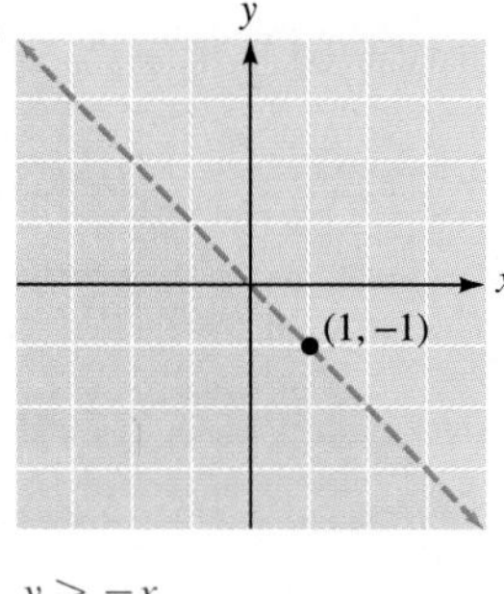

$y > -x$

25.

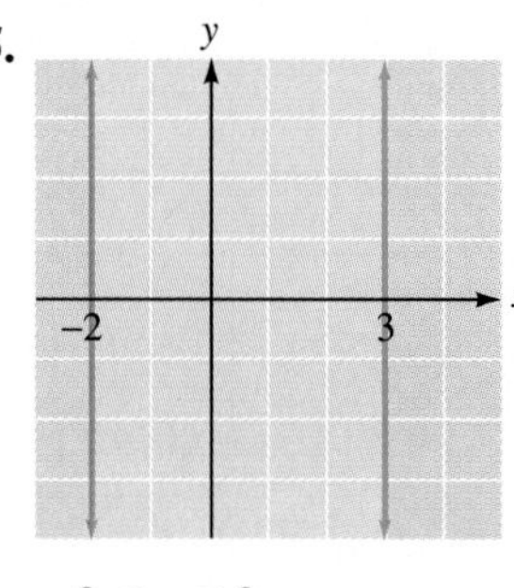

$-2 \leq x \leq 3$

26.

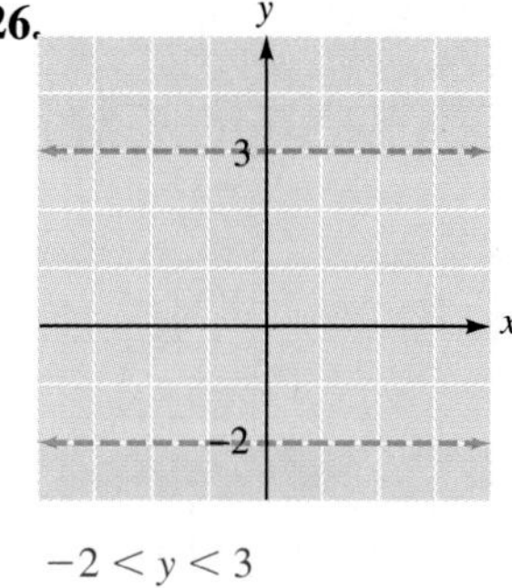

$-2 < y < 3$

27.

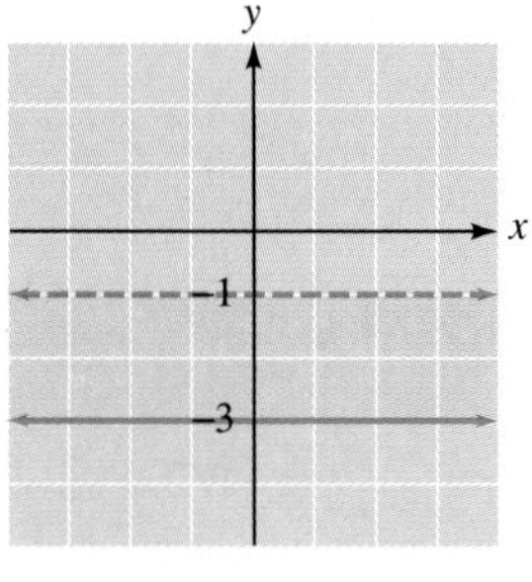

$y > -1$ or $y \leq -3$

28.

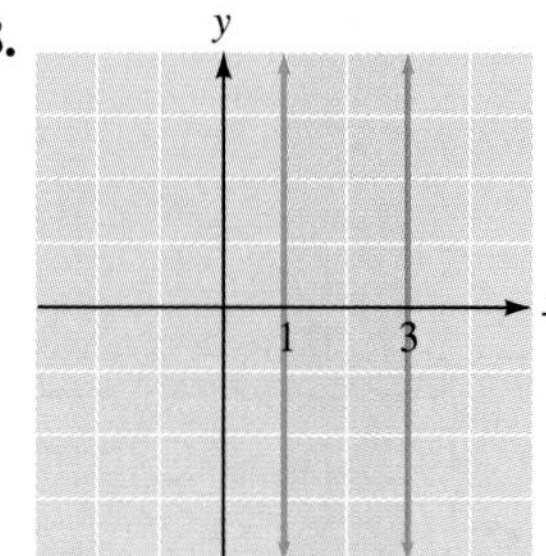

$x \leq 1$ or $x \geq 3$

In Exercises 29–34, graph each inequality for nonnegative values of x and y. Then give some ordered pairs that satisfy the inequality.

29. Figuring taxes On average, it takes an accountant 1 hour to complete a simple tax return and 3 hours to complete a complicated return. If the accountant works 9 hours per day, find an inequality that shows the possible ways that simple returns (x) and complicated returns (y) can be completed each day.
(1, 1), (2, 1), (2, 2)

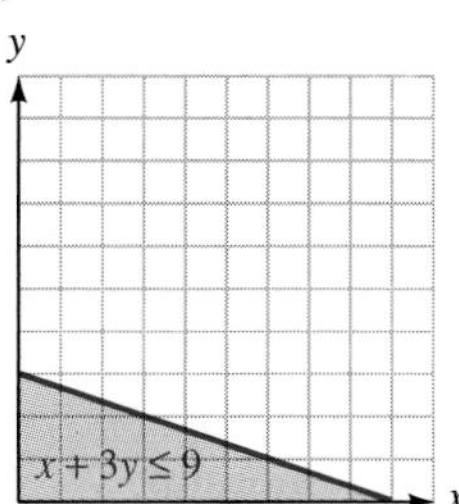

30. Selling trees During a sale, a garden store sold more than \$2000 worth of trees. If a 6-foot maple costs \$100 and a 5-foot pine costs \$125, find an inequality that shows the possible ways that maple trees (x) and pine trees (y) were sold.
(0, 17), (5, 20), (15, 10)

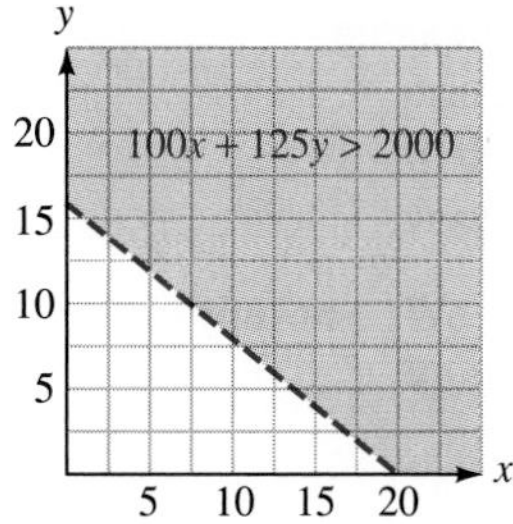

31. Choosing housekeepers One housekeeper charges \$6 per hour, and another charges \$7 per hour. If Sarah can afford no more than \$42 per week to clean her house, find an inequality that shows the possible ways that she can hire the first housekeeper (x) and the second housekeeper (y). (2, 2), (3, 3), (5, 1)

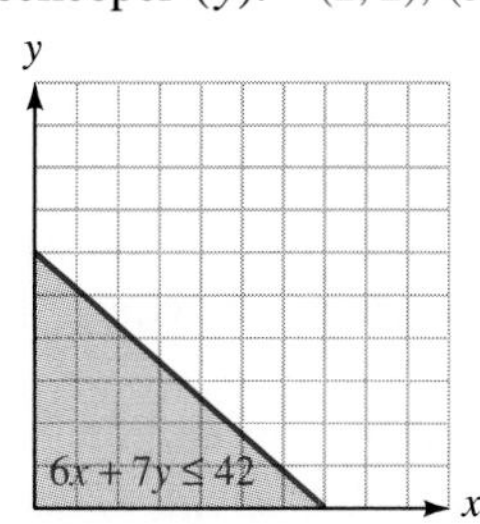

32. Making sporting goods A sporting goods manufacturer allocates at least 1200 units of time per day to make fishing rods and reels. If it takes 10 units of time to make a rod and 15 units of time to make a reel, find an inequality that shows the possible ways to schedule the time to make rods (x) and reels (y). (40, 80), (80, 80), (120, 40)

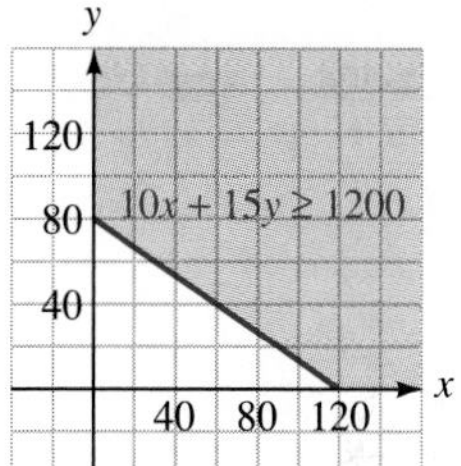

33. Investing in stocks A woman has up to \$6000 to invest. If stock in Traffico sells for \$50 per share and stock in Cleanco sells for \$60 per share, find an inequality that shows the possible ways that she can buy shares of Traffico (x) and Cleanco (y).
(40, 20), (60, 40), (80, 20)

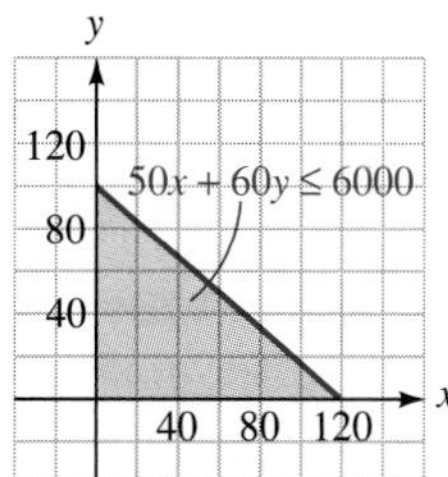

34. Buying concert tickets Tickets to a concert cost \$6 for reserved seats and \$4 for general admission. If receipts must be at least \$10,200 to meet expenses, find an inequality that shows the possible ways that the box office can sell reserved seats (x) and general admission tickets (y). (1800, 0), (1000, 1500), (2000, 2000)

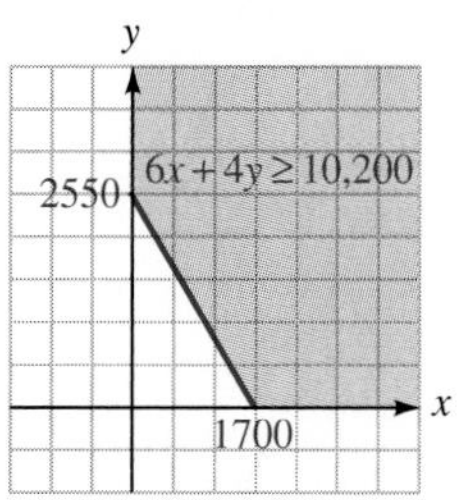

Writing Exercises *Write a paragraph using your own words.*

1. Explain how to decide where to draw the boundary of the graph of a linear inequality, and whether to draw it as a solid or a broken line.
2. Explain how to decide which side of the boundary of the graph of a linear inequality should be shaded.

Something to Think About

1. Can an inequality be an identity, one that is satisfied by all (x, y) pairs? Illustrate.
2. Can an inequality be an impossible inequality, one that has no solutions? Illustrate.

Review Exercises *Simplify each fraction.*

If you teach graphing early, skip these review exercises.

1. $\dfrac{a^3 - ab^2}{ab - a^2}$ $\quad -(a + b)$
2. $\dfrac{mp + mq + np + nq}{p^3 + q^3}$ $\quad \dfrac{m + n}{p^2 - pq + q^2}$

Do the operations and simplify.

3. $\dfrac{x + y}{x + 5y}\left(\dfrac{2}{x + y} - \dfrac{3}{x - y}\right)$ $\quad \dfrac{1}{y - x}$
4. $(2a^2 - 3a - 2) \div \dfrac{2a^2 - a - 1}{a - 2} \div (2a - 2)$ $\quad \dfrac{(a - 2)^2}{2(a - 1)^2}$

6.6 Introduction to Functions

■ Relations ■ Functions ■ Finding Domains and Ranges ■ The Vertical Line Test ■ Function Notation ■ Linear Functions

Relations

Equations and inequalities in x and y determine correspondences between numbers x and values y. For example, the equation $y = -2x + 1$ determines an infinite number of ordered pairs (x, y) where a single value of y corresponds to each number x. To find the value of y that corresponds to the number $x = 2$, we substitute 2 for x and simplify:

$$y = -2x + 1$$
$$y = -2(2) + 1 \quad \text{Substitute 2 for } x.$$
$$y = -3$$

The ordered pair $(2, -3)$ shows that a y value of -3 corresponds to the number $x = 2$. The graphs of all ordered pairs that satisfy the equation $y = -2x + 1$ form the line shown in Figure 6-38(a).

The inequality $y < -2x + 1$ also determines an infinite number of order pairs (x, y). This time, however, many values of y correspond to each number x. To find the values of y that correspond to the number $x = 2$, we substitute 2 for x and simplify:

$$y < -2x + 1$$
$$y < -2(\mathbf{2}) + 1 \quad \text{Substitute 2 for } x.$$
$$y < -3$$

Thus, if $x = 2$, then y can be any number less than -3. Ordered pairs such as

$$(2, -4), (2, -5), (2, -6), (2, -7) \quad \text{and} \quad (2, -8)$$

all satisfy the inequality $y < -2x + 1$, whose graph appears in Figure 6-38(b).

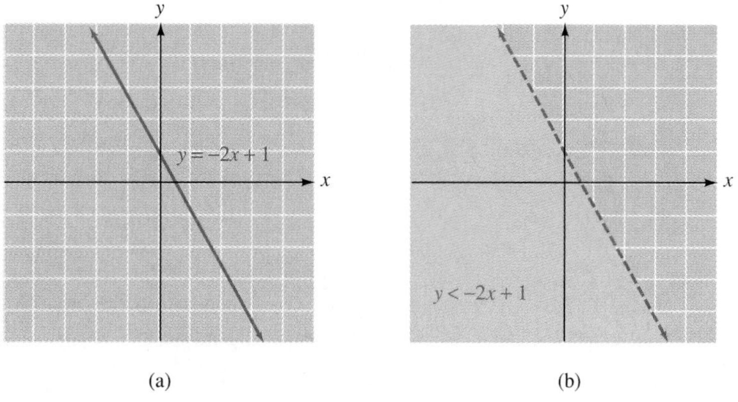

FIGURE 6-38

A set of ordered pairs that sets up a correspondence between numbers x and values y is called a **relation.** The set of ordered pairs listed in either part of Figure 6-38 is a relation.

Relation A **relation** is a correspondence that assigns to each number x one or more values of y.

The set of all numbers x for which the relation is defined is called the **domain of the relation.** The set of all possible values of y is called the **range.**

The domain and the range of both relations shown in Figure 6-38 are the set of real numbers.

Functions

The relations shown in Figure 6-38 differ in an important way. Several ordered pairs in part (b) have the same x, but different values of y. In part (a), however, a single value of y corresponds to each number x. When a single value of y corresponds to each number x, we call the relation a **function.**

In the function in part (a), determined by the equation $y = -2x + 1$, the value of y depends on a number x. For this reason, y is called the **dependent variable,** and x is called the **independent variable.**

Function A **function** is a relation that assigns to each number x exactly one value y.

The **domain** of a function is the set of all numbers x (the independent variable) for which the function is defined. The **range** is the set of all possible y values (the dependent variable).

WARNING! A function is always a relation, but a relation is not necessarily a function.

We can think of a function as a machine that takes numbers x and turns them into values y. For example, Figure 6-39 represents a function machine for the function $y = -2x + 1$. If we put numbers x into the machine, the values of y come out. For example, the machine turns the number 2 into the value -3. The set of numbers that we can put into the machine is the domain, and the set of numbers that comes out is the range.

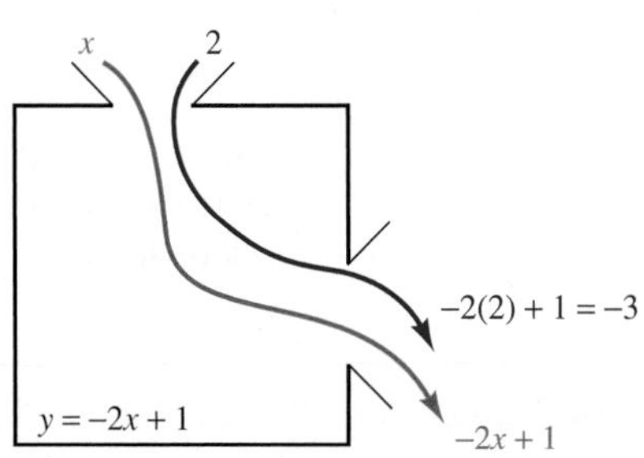

FIGURE 6-39

EXAMPLE 1 Do the equations **a.** $y = 4x - 3$ and **b.** $y^2 = x$ define y to be a function of x?

Solution

a. For a function to exist, *each* number x must determine a *single* value y. To find y in the equation $y = 4x - 3$, we multiply x by 4 and then subtract 3. Since this arithmetic always gives a single result, each choice of x determines a single value y. Thus, the equation $y = 4x - 3$ does define a function.

b. For a function to exist, *each* number x must determine a *single* value y. If we let $x = 16$, for example, y could be 4 or -4, because $4^2 = 16$ and $(-4)^2 = 16$. Since more than one value of y is determined when $x = 16$, the equation does not represent a function. ■

Finding Domains and Ranges

EXAMPLE 2 Find the domain and range of the function defined by **a.** the ordered pairs $(-2, 4)$, $(0, 6)$, $(2, 8)$ and **b.** the equation $y = \frac{1}{x-2}$.

Solution **a.** The domain is the set of numbers x: $\{-2, 0, 2\}$. The range is the set of values y: $\{4, 6, 8\}$.

b. Since any number except 2 can be substituted for x in the equation $y = \frac{1}{x-2}$, the domain is $\{x \mid x \text{ is any real number, except } 2.\}$. Since a fraction with a numerator of 1 cannot be 0, the range is $\{y \mid y \text{ is any real number, except } 0.\}$. ■

The **graph** of a function is the graph of all the ordered pairs that define the function. Since the graph gives a picture of the function, we can find its domain and range by looking at its graph.

EXAMPLE 3 Graph the function defined by $y = x^2 - 1$ and find its domain and range.

Solution The graph of $y = x^2 - 1$ is shown in Figure 6-40. Since each number x determines a corresponding y, the domain is the set of real numbers. Since the values of y are never less than -1, the range is the set of numbers greater than or equal to -1.

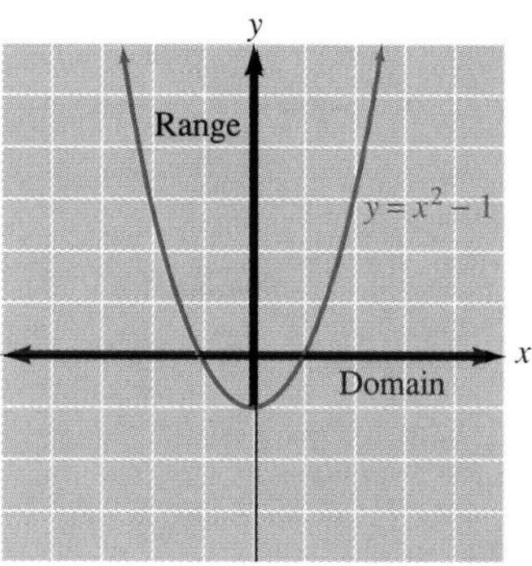

FIGURE 6-40 ■

The Vertical Line Test

A **vertical line test** can be used to determine whether the graph of an equation represents a function. If any vertical line intersects a graph more than once, the graph cannot represent a function, because to one number x there would correspond more than one value of y.

The graph in Figure 6-41(a) represents a function, because every vertical line that intersects the graph does so exactly once. However, the graph in Figure 6-41(b) does not represent a function, because some vertical lines intersect the graph more than once.

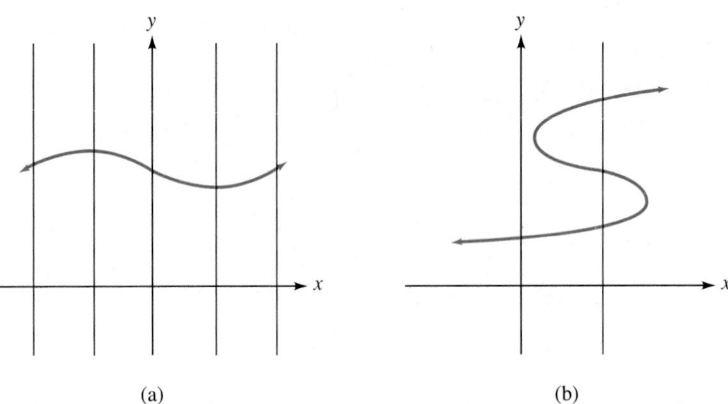

FIGURE 6-41

Function Notation

Function Notation The notation $y = f(x)$ denotes that the variable y is a function of x.

The notation $f(x)$ is read as "f of x."

WARNING! The notation $f(x)$ does not mean "f times x."

The notation $y = f(x)$ provides a way of denoting the values of y that correspond to individual numbers x. If $y = f(x)$, the value of y that is determined by $x = 3$ is denoted by $f(3)$. Similarly, $f(-1)$ represents the value of y that corresponds to $x = -1$.

EXAMPLE 4 Let $f(x) = 4x + 3$. Find **a.** $f(3)$, **b.** $f(-1)$, **c.** $f(0)$, and **d.** $f(r)$.

Solution **a.** We replace x with 3:

$$\begin{aligned} f(x) &= 4x + 3 \\ f(\mathbf{3}) &= 4(\mathbf{3}) + 3 \\ &= 12 + 3 \\ &= 15 \end{aligned}$$

b. We replace x with -1:

$$\begin{aligned} f(x) &= 4x + 3 \\ f(\mathbf{-1}) &= 4(\mathbf{-1}) + 3 \\ &= -4 + 3 \\ &= -1 \end{aligned}$$

c. We replace x with 0:

$$\begin{aligned} f(x) &= 4x + 3 \\ f(\mathbf{0}) &= 4(\mathbf{0}) + 3 \\ &= 0 + 3 \\ &= 3 \end{aligned}$$

d. We replace x with r:

$$\begin{aligned} f(x) &= 4x + 3 \\ f(r) &= 4r + 3 \end{aligned}$$

■

The letter f used in the notation $y = f(x)$ represents the word *function*. However, other letters can be used to represent functions. The notations $y = g(x)$ and $y = h(x)$ also denote functions involving the variable x.

EXAMPLE 5 Let $g(x) = x^2 - 2x$. Find **a.** $g\left(\frac{2}{5}\right)$, **b.** $g(s)$, **c.** $g(s - 1)$, and **d.** $g(-t)$.

Solution **a.** We replace x with $\frac{2}{5}$:

$$\begin{aligned} g(x) &= x^2 - 2x \\ g\left(\frac{2}{5}\right) &= \left(\frac{2}{5}\right)^2 - 2\left(\frac{2}{5}\right) \\ &= \frac{4}{25} - \frac{4}{5} \\ &= -\frac{16}{25} \end{aligned}$$

b. We replace x with s:

$$\begin{aligned} g(x) &= x^2 - 2x \\ g(s) &= s^2 - 2s \end{aligned}$$

c. We replace x with $s - 1$:

$$\begin{aligned} g(x) &= x^2 - 2x \\ g(s - 1) &= (s - 1)^2 - 2(s - 1) \\ &= (s^2 - 2s + 1) - 2s + 2 \\ &= s^2 - 4s + 3 \end{aligned}$$

d. We replace x with $-t$:

$$\begin{aligned} g(x) &= x^2 - 2x \\ g(-t) &= (-t)^2 - 2(-t) \\ &= t^2 + 2t \end{aligned}$$

■

EXAMPLE 6 Let $f(x) = 4x - 1$. Find **a.** $f(3) + f(2)$ and **b.** $f(a) - f(b)$.

Solution **a.** We find $f(3)$ and $f(2)$ separately.

$$\begin{aligned} f(x) &= 4x - 1 \\ f(3) &= 4(3) - 1 \\ &= 12 - 1 \\ &= 11 \end{aligned} \qquad \begin{aligned} f(x) &= 4x - 1 \\ f(2) &= 4(2) - 1 \\ &= 8 - 1 \\ &= 7 \end{aligned}$$

We then add the results to obtain

$$\begin{aligned} f(3) + f(2) &= 11 + 7 \\ &= 18 \end{aligned}$$

b. We find $f(a)$ and $f(b)$ separately.

$$\begin{aligned} f(x) &= 4x - 1 \\ f(a) &= 4a - 1 \end{aligned} \qquad \begin{aligned} f(x) &= 4x - 1 \\ f(b) &= 4b - 1 \end{aligned}$$

We then subtract the results to obtain

$$\begin{aligned} f(a) - f(b) &= (4a - 1) - (4b - 1) \\ &= 4a - 1 - 4b + 1 && \text{Remove parentheses.} \\ &= 4a - 4b \end{aligned}$$

■

Linear Functions

We began this chapter by graphing equations whose graphs were straight lines. These equations define basic functions, called **linear functions.**

Linear Functions A **linear function** is a function defined by an equation that can be written in the form

$$y = mx + b \qquad \text{or} \qquad f(x) = mx + b$$

where m is the slope of its line graph and $(0, b)$ is the y-intercept.

EXAMPLE 7 Solve the equation $3x + 2y = 10$ for y to show that it defines a linear function. Then graph it to find its domain and range.

Solution We solve the equation for y as follows:

$$3x + 2y = 10$$

$$2y = -3x + 10 \qquad \text{Subtract } 3x \text{ from both sides.}$$

$$y = -\frac{3}{2}x + 5 \qquad \text{Divide both sides by 2.}$$

Because the given equation can be written in the form $y = mx + b$, it defines a linear function. The slope of its line graph is $-\frac{3}{2}$, and the y-intercept is $(0, 5)$. The graph appears in Figure 6-42. From the graph, we can see that both the domain and the range are the set of real numbers.

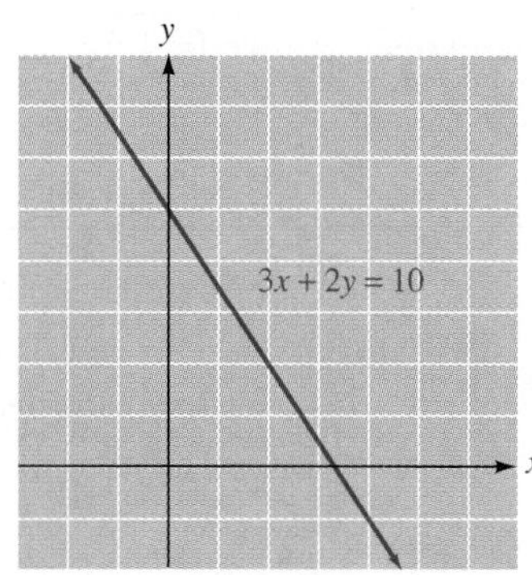

FIGURE 6-42

ORALS *Tell whether each equation or inequality determines y to be a function of x.*

1. $y = 2x + 1$ yes **2.** $y \geq 2x$ no **3.** $y^2 = x$ no

If $f(x) = 2x + 1$, *find*

4. $f(0)$ 1 **5.** $f(1)$ 3 **6.** $f(-2)$ −3

EXERCISE 6.6

In Exercises 1–12, tell whether each equation or inequality determines y to be a function of x.

1. $y = 2x + 3$ yes
2. $y = 4x - 1$ yes
3. $y = 3x^2$ yes
4. $y^2 = x + 1$ no
5. $y = 3 + 7x^2$ yes
6. $y^2 = 3 - 2x$ no
7. $y \leq x$ no
8. $y > x$ no
9. $y = |x|$ yes
10. $x = |y|$ no
11. $x = -|y|$ no
12. $y = -|x|$ yes

 *In Exercises 13–16, find the domain and range of each function.*

13. $y = -x^2 + 2$
domain = all reals; range = all reals ≤ 2

14. $y = (x - 2)^2$
domain = all reals; range = all reals ≥ 0

15. $y = |x - 2| - 2$
domain = all reals; range = all reals ≥ -2

16. $y = -|x + 1|$
domain = all reals; range = all reals ≤ 0

In Exercises 17–20, each graph represents a relation. Give the domain and range of each relation and tell whether it is a function.

17.

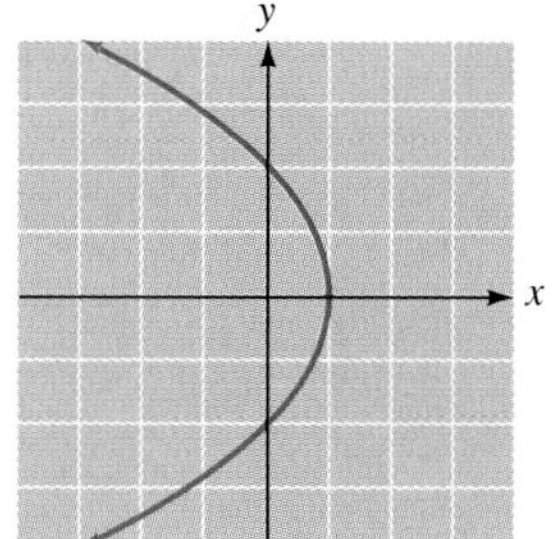

domain = all reals ≤ 1; range = all reals; not a function

18.

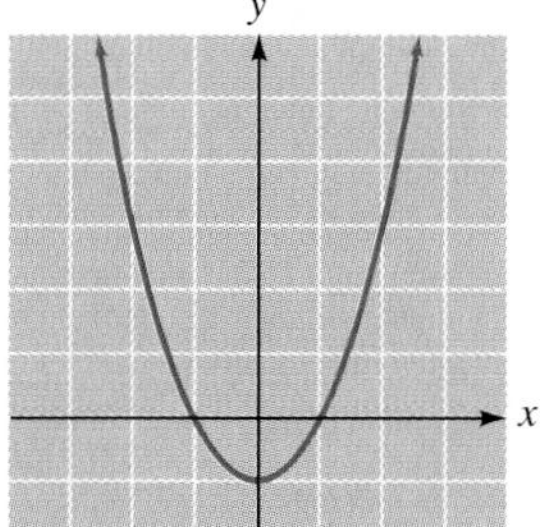

domain = all reals; range = all reals ≥ -1; a function

19.

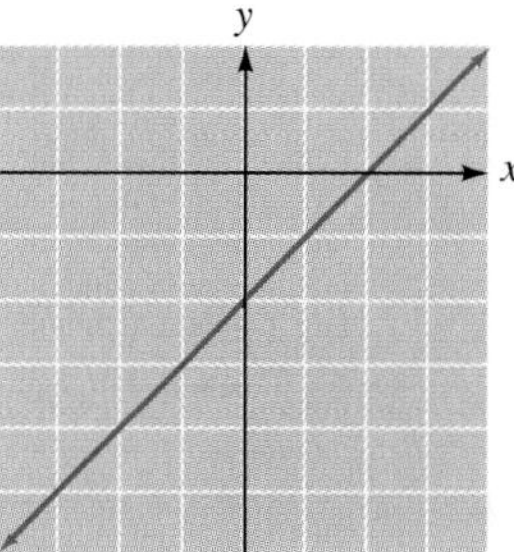

domain = all reals; range = all reals; a function

20.

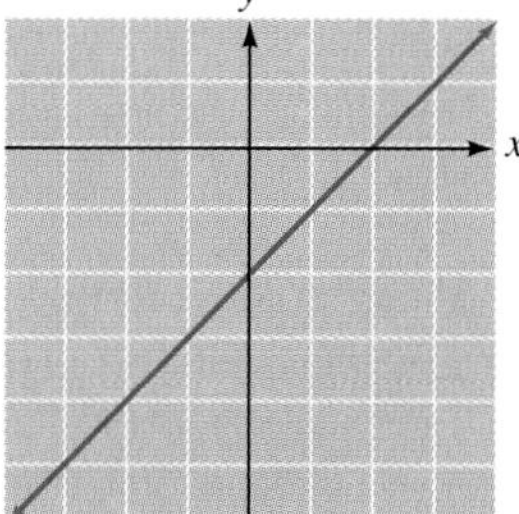

domain = all reals; range = all reals; not a function

In Exercises 21–28, find **a.** $f(3)$, **b.** $f(0)$, *and* **c.** $f(-1)$.

21. $f(x) = 3x$ 9, 0, −3
22. $f(x) = -4x$ −12, 0, 4
23. $f(x) = 2x - 3$ 3, −3, −5
24. $f(x) = 3x - 5$ 4, −5, −8
25. $f(x) = 7 + 5x$ 22, 7, 2
26. $f(x) = 3 + 3x$ 12, 3, 0
27. $f(x) = 9 - 2x$ 3, 9, 11
28. $f(x) = 12 + 3x$ 21, 12, 9

In Exercises 29–36, find **a.** $f(1)$, **b.** $f(-2)$, *and* **c.** $f(3)$.

29. $f(x) = x^2$ 1, 4, 9
30. $f(x) = x^2 - 2$ −1, 2, 7
31. $f(x) = x^3 - 1$ 0, −9, 26
32. $f(x) = x^3$ 1, −8, 27
33. $f(x) = (x + 1)^2$ 4, 1, 16
34. $f(x) = (x - 3)^2$ 4, 25, 0
35. $f(x) = 2x^2 - x$ 1, 10, 15
36. $f(x) = 5x^2 + 2x - 1$ 6, 15, 50

In Exercises 37–44, find **a.** $f(2)$, **b.** $f(1)$, *and* **c.** $f(-2)$.

37. $f(x) = |x| + 2$ 4, 3, 4
38. $f(x) = |x| - 5$ −3, −4, −3
39. $f(x) = x^2 - 2$ 2, −1, 2
40. $f(x) = x^2 + 3$ 7, 4, 7
41. $f(x) = \dfrac{1}{x + 3}$ $\frac{1}{5}, \frac{1}{4}, 1$
42. $f(x) = \dfrac{3}{x - 4}$ $-\frac{3}{2}, -1, -\frac{1}{2}$
43. $f(x) = \dfrac{x}{x - 3}$ $-2, -\frac{1}{2}, \frac{2}{5}$
44. $f(x) = \dfrac{x}{x^2 + 2}$ $\frac{1}{3}, \frac{1}{3}, -\frac{1}{3}$

In Exercises 45–52, find **a.** $g(w)$ *and* **b.** $g(w + 1)$.

45. $g(x) = 2x$ $2w, 2w + 2$
46. $g(x) = -3x$ $-3w, -3w - 3$
47. $g(x) = 3x - 5$ $3w - 5, 3w - 2$
48. $g(x) = 2x - 7$ $2w - 7, 2w - 5$
49. $g(x) = x^2 + x$ $w^2 + w, w^2 + 3w + 2$
50. $g(x) = x^2 - 2x$ $w^2 - 2w, w^2 - 1$
51. $g(x) = x^2 - 1$ $w^2 - 1, w^2 + 2w$
52. $g(x) = |x - 1|$ $|w - 1|, |w|$

In Exercises 53–60, $f(x) = 2x + 1$. Find each value.

53. $f(3) + f(2)$ 12
54. $f(1) - f(-1)$ 4
55. $f(b) - f(a)$ $2b - 2a$
56. $f(b) + f(a)$ $2b + 2a + 2$
57. $f(b) - 1$ $2b$
58. $f(b) - f(1)$ $2b - 2$
59. $f(0) + f\left(-\frac{1}{2}\right)$ 1
60. $f(a) + f(2a)$ $6a + 2$

In Exercises 61–64, draw the graph of each linear function. Give the domain and range.

61. $f(x) = 2x - 1$

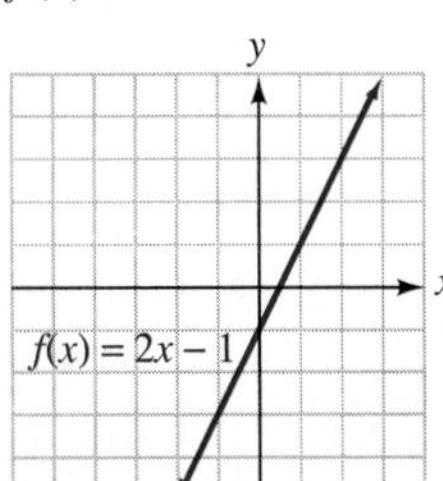

domain = all reals;
range = all reals

62. $f(x) = -x + 2$

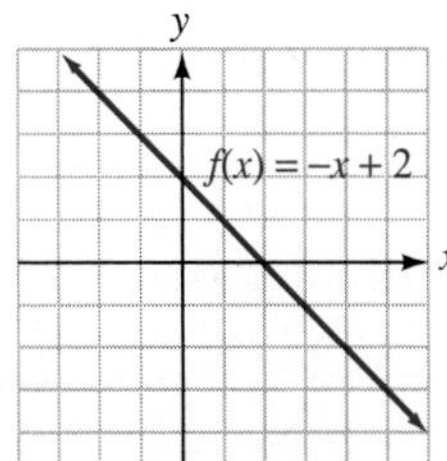

domain = all reals;
range = all reals

63. $2x - 3y = 6$

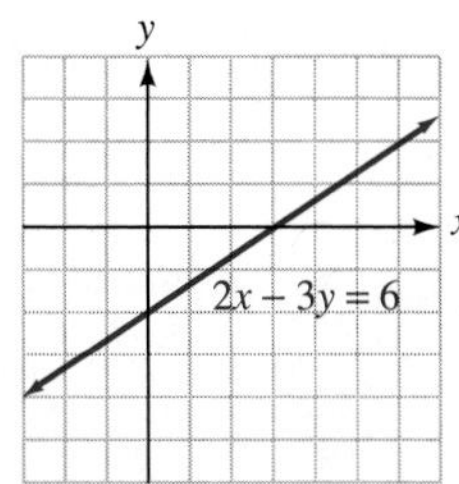

domain = all reals;
range = all reals

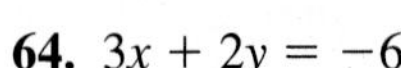

64. $3x + 2y = -6$

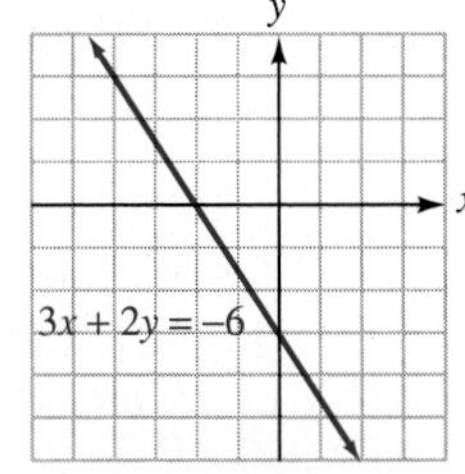

domain = all reals;
range = all reals

In Exercises 65–68, tell whether each equation defines a linear function.

65. $y = 3x^2 + 2$ no
66. $y = \dfrac{x - 3}{2}$ yes
67. $x = 3y - 4$ yes
68. $x = \dfrac{8}{y}$ no

Writing Exercises ■ *Write a paragraph using your own words.*

1. Explain why a relation is not always a function.
2. Explain the concepts of *range* and *domain.*

Something to Think About ■ *Let $f(x) = 2x + 1$ and $g(x) = x^2$ ($f(x) \neq 0$ and $g(x) \neq 0$).*

1. Is $f(x) + g(x) = g(x) + f(x)$? yes
2. Is $f(x) - g(x) = g(x) - f(x)$? no
3. Is $f(x) \cdot g(x) = g(x) \cdot f(x)$? yes
4. Is $\dfrac{f(x)}{g(x)} = \dfrac{g(x)}{f(x)}$? no

Review Exercises ■ *In Review Exercises 1–4, solve each equation.*

If you teach graphing early, skip Review Exercises 3–4.

1. $\dfrac{y + 2}{2} = 4(y + 2)$ -2
2. $\dfrac{3z - 1}{6} - \dfrac{3z + 4}{3} = \dfrac{z + 3}{2}$ -3
3. $\dfrac{2}{x - 3} - 1 = -\dfrac{1}{3}$ 6
4. $\dfrac{5}{x} + \dfrac{6}{x^2 + 2x} = \dfrac{-3}{x + 2}$ -2 is extraneous

6.7 Proportion and Variation

■ Direct Variation ■ Inverse Variation ■ Joint Variation
■ Combined Variation

The quotient of two numbers is often called a **ratio.** For example, the fraction $\frac{2}{3}$ can be read as "the ratio of 2 to 3." An equation indicating that two ratios are equal is called a **proportion.** Two examples of proportions are

$$\frac{1}{4} = \frac{2}{8} \quad \text{and} \quad \frac{4}{7} = \frac{12}{21}$$

Teaching Tip: If the proportion $\frac{a}{b} = \frac{c}{d}$ is written as
$a:b = c:d$
it is easier to recognize the extremes and the means.

In the proportion $\frac{a}{b} = \frac{c}{d}$, the terms a and d are called the **extremes** of the proportion, and the terms b and c are called the **means.**

To develop a fundamental property of proportions, we suppose that

$$\frac{a}{b} = \frac{c}{d}$$

is a proportion and multiply both sides by bd to obtain

$$bd\left(\frac{a}{b}\right) = bd\left(\frac{c}{d}\right)$$

$$\frac{\not{b}da}{\not{b}} = \frac{b\not{d}c}{\not{d}}$$

$$ad = bc$$

Thus, if $\frac{a}{b} = \frac{c}{d}$, then $ad = bc$. In a proportion, the *product of the extremes equals the product of the means.*

EXAMPLE 1 Solve the proportion $\frac{x+1}{x} = \frac{x}{x+2}$ for x.

Solution

$$\frac{x+1}{x} = \frac{x}{x+2}$$

$$(x+1)(x+2) = x \cdot x \qquad \text{The product of the extremes equals the product of the means.}$$

$$x^2 + 3x + 2 = x^2$$

$$3x + 2 = 0 \qquad \text{Subtract } x^2 \text{ from both sides.}$$

$$x = -\frac{2}{3} \qquad \text{Subtract 2 from both sides and divide by 3.}$$

■

■ Direct Variation

We consider the formula

$$C = \pi D$$

for the circumference of a circle where C is the circumference, D is the diameter, and $\pi \approx 3.14159$. If we double the diameter of a circle, we determine another circle with a larger circumference C_1 such that

$$\begin{aligned} C_1 &= \pi(2D) \\ &= 2\pi D \\ &= 2C \end{aligned}$$

Thus, doubling the diameter results in doubling the circumference. Likewise, if we triple the diameter, we triple the circumference.

In this formula, we say that the variables C and D **vary directly**, or that they are **directly proportional**. This is because as one variable gets larger, so does the other, and in a predictable way. In this example, the constant π is called the **constant of variation** or the **constant of proportionality.**

Direct Variation The words "**y varies directly with x,**" or "**y is directly proportional to x,**" mean that $y = kx$ for some constant k.

k is called the **constant of variation** or the **constant of proportionality.**

Another Approach: You may want to point out that if the quotient of two variables is constant, the variables vary directly:

$$\frac{y}{x} = k$$

An example of direct variation is Hooke's law from physics. Hooke's law states that the distance a spring will stretch varies directly with the force that is applied to it.

If d represents a distance and f represents a force, Hooke's law is expressed mathematically as

$$d = kf$$

where k is the constant of variation. If the spring stretches 10 inches when a weight of 6 pounds is attached, we can find k as follows:

$$\begin{aligned} d &= kf \\ 10 &= k(6) && \text{Substitute 10 for } d \text{ and 6 for } f. \\ \frac{5}{3} &= k && \text{Divide both sides by 6 and simplify.} \end{aligned}$$

To find the force required to stretch the spring a distance of 35 inches, we can solve the equation $d = kf$ for f, with $d = 35$ and $k = \frac{5}{3}$.

$$\begin{aligned} d &= kf \\ 35 &= \frac{5}{3}f && \text{Substitute 35 for } d \text{ and } \tfrac{5}{3} \text{ for } k. \\ 105 &= 5f && \text{Multiply both sides by 3.} \\ 21 &= f && \text{Divide both sides by 5.} \end{aligned}$$

Thus, the force required to stretch the spring a distance of 35 inches is 21 pounds.

EXAMPLE 2 **Travel distance** The distance traveled in a given time is directly proportional to speed. If a car travels 70 miles at 30 miles per hour, how far will it travel in the same time at 45 miles per hour?

Solution The words *distance is directly proportional to speed* can be expressed by the equation

1. $d = ks$

where d is distance, k is the constant of variation, and s is the speed. To find k, we substitute 70 for d, 30 for s, and solve for k.

$$d = ks$$
$$70 = k(30)$$
$$k = \frac{7}{3} \quad \text{Divide both sides by 30 and simplify.}$$

To find the distance traveled at 45 miles per hour, we substitute $\frac{7}{3}$ for k and 45 for s in Equation 1 and simplify.

$$d = ks$$
$$d = \frac{7}{3}(45)$$
$$= 105$$

In the time it took to go 70 miles at 30 miles per hour, the car could travel 105 miles at 45 miles per hour. ■

Inverse Variation

In the formula $w = \frac{12}{l}$, w gets smaller as l gets larger, and w gets larger as l gets smaller. Since these variables vary in opposite directions in a predictable way, we say that the variables **vary inversely**, or that they are **inversely proportional**. The constant 12 is the **constant of variation.**

Inverse Variation The words "**y varies inversely with x**," or "**y is inversely proportional to x**," mean that $y = \frac{k}{x}$ for some constant k.

k is called the **constant of variation**.

Another Approach: You may want to point out that if the product of two numbers is constant, the variables vary inversely:
$xy = k$

Because of gravity, an object in space is attracted to the earth. The force of this attraction varies inversely with the square of the distance that the object is from the center of the earth.

If f represents the force and d represents the distance, this information can be expressed by the equation

$$f = \frac{k}{d^2}$$

If we know that an object 4000 miles from the center of the earth is attracted to the earth with a force of 90 pounds, we can find k.

$$f = \frac{k}{d^2}$$

$$90 = \frac{k}{4000^2} \quad \text{Substitute 90 for } f \text{ and 4000 for } d.$$

$$k = 90(4000^2) \quad \text{Multiply both sides by } 4000^2.$$

$$= 1.44 \times 10^9$$

To find the force of attraction when the object is 5000 miles from the center of the earth, we proceed as follows:

$$f = \frac{k}{d^2}$$

$$f = \frac{1.44 \times 10^9}{5000^2} \quad \text{Substitute } 1.44 \times 10^9 \text{ for } k \text{ and 5000 for } d.$$

$$= 57.6$$

The object will be attracted to earth with a force of 57.6 pounds when it is 5000 miles from the earth's center.

EXAMPLE 3 **Illumination** The intensity I of light received from a light source varies inversely with the square of the distance from the light source. If the intensity from a light source 4 feet from an object is 8 candelas, find the intensity at a distance of 2 feet.

Solution The words "*intensity varies inversely with the square of the distance d*" can be expressed by the equation

$$I = \frac{k}{d^2}$$

To find k, we substitute 8 for I and 4 for d, and solve for k.

$$I = \frac{k}{d^2}$$

$$8 = \frac{k}{4^2}$$

$$128 = k \quad \text{Multiply both sides by } 4^2.$$

To find the intensity when the object is 2 feet from the light source, we substitute 2 for d and 128 for k and simplify.

$$I = \frac{k}{d^2}$$

$$I = \frac{128}{2^2}$$

$$= 32$$

The intensity at 2 feet is 32 candelas. ■

Joint Variation

There are times when one variable varies with the product of several variables. For example, the area of a triangle varies directly with the product of its base and height:

$$A = \frac{1}{2}bh$$

Such variation is called **joint variation.**

Joint Variation If one variable varies directly with the product of two or more variables, the relationship is called **joint variation**. If y varies jointly with x and z, then $y = kxz$.

The constant k is called the **constant of variation.**

EXAMPLE 4 The volume V of a cone varies jointly with its height h and the area of its base B. Express this relationship as an equation.

Solution The words *V varies jointly with h and B* mean that *V varies directly as the product of h and B*. Thus,

$$V = khB$$

The relationship can also be read as *V is directly proportional to the product of h and B*. ■

Combined Variation

Many problems involve a combination of direct and inverse variation. Such variation is called **combined variation**.

EXAMPLE 5 **Highway construction** The time it takes to build a highway varies directly with the length of the road, but inversely with the number of workers. If it takes 100 workers 4 weeks to build 2 miles of highway, how long will it take 80 workers to build 10 miles of highway?

Solution We can let t represent the time in weeks, l represent the length in miles, and w represent the number of workers. The relationship between these variables can be expressed by the equation

$$t = \frac{kl}{w}$$

We substitute 4 for t, 100 for w, and 2 for l to find k:

$$4 = \frac{k(2)}{100}$$

$400 = 2k$ Multiply both sides by 100.

$200 = k$ Divide both sides by 2.

We now substitute 80 for w, 10 for l, and 200 for k in the equation $t = \frac{kl}{w}$ and simplify:

$$t = \frac{kl}{w}$$

$$t = \frac{200(10)}{80}$$

$$= 25$$

It will take 25 weeks for 80 workers to build 10 miles of highway. ■

ORALS *Solve each proportion.*

1. $\frac{x}{2} = \frac{3}{6}$ 1
2. $\frac{3}{x} = \frac{4}{12}$ 9
3. $\frac{5}{7} = \frac{2}{x}$ $\frac{14}{5}$

Express each sentence with a formula.

4. a varies directly with b. $a = kb$
5. a varies inversely with b. $a = \frac{k}{b}$
6. a varies jointly with b and c. $a = kbc$
7. a varies directly with b but inversely with c. $a = \frac{kb}{c}$

EXERCISE 6.7

In Exercises 1–12, solve each proportion for the variable, if possible.

1. $\frac{x}{5} = \frac{15}{25}$ 3
2. $\frac{4}{y} = \frac{6}{27}$ 18
3. $\frac{r-2}{3} = \frac{r}{5}$ 5
4. $\frac{2}{c} = \frac{c-3}{2}$ 4, −1
5. $\frac{y}{4} = \frac{4}{y}$ 4, −4
6. $\frac{2}{3x} = \frac{6x}{36}$ 2, −2
7. $\frac{3}{n} = \frac{2}{n+1}$ −3
8. $\frac{4}{x+3} = \frac{3}{5}$ $\frac{11}{3}$
9. $\frac{x+1}{x-1} = \frac{6}{4}$ 5
10. $\frac{5}{5z+3} = \frac{2z}{2z^2+6}$ 5
11. $\frac{9t+6}{t(t+3)} = \frac{7}{t+3}$ −3 is extraneous
12. $\frac{(x-7)(x+2)}{2} = \frac{(x+3)(x+2)}{4}$ −2, 17

In Exercises 13–22, express each sentence as a formula.

13. A varies directly with the square of p. $A = kp^2$
14. z varies inversely with the cube of t. $z = k/t^3$
15. v varies inversely with the cube of r. $v = k/r^3$
16. r varies directly with the square of s. $r = ks^2$
17. B varies jointly with m and n. $B = kmn$
18. C varies jointly with x, y, and z. $C = kxyz$
19. P varies directly with the square of a, and inversely with the cube of j. $P = ka^2/j^3$
20. M varies inversely with the cube of n, and jointly with x and the square of z. $M = kxz^2/n^3$

21. The force of attraction between two masses m_1 and m_2 varies directly with the product of m_1 and m_2 and inversely with the square of the distance between them. $F = km_1m_2/d^2$

22. The force of wind on a vertical surface varies jointly with the area of the surface and the square of the velocity of the wind. $F = kAv^2$

In Exercises 23–30, express each formula in words. In each formula, k is the constant of variation.

23. $L = kmn$
L varies jointly with m and n.

24. $P = \frac{km}{n}$
P varies directly with m and inversely with n.

25. $E = kab^2$
E varies jointly with a and the square of b.

26. $U = krs^2t$
U varies jointly with r, the square of s, and t.

27. $X = \frac{kx^2}{y^2}$
X varies directly with x^2 and inversely with y^2.

28. $Z = \frac{kw}{xy}$
Z varies directly with w and inversely with the product of x and y.

29. $R = \frac{kL}{d^2}$
R varies directly with L and inversely with d^2.

30. $e = \frac{kPL}{A}$
e varies jointly with P and L and inversely with A.

31. Area of a circle The area of a circle varies directly with the square of its radius, and the constant of variation is π. Find the area of a circle with a radius of 6 inches. 36π in.^{2}

32. Falling objects An object in free fall travels a distance s that is directly proportional to the square of the time t. If an object falls 1024 feet in 8 seconds, how far will it fall in 10 seconds? 1600 ft

33. Finding distance The distance that a car can travel is directly proportional to the number of gallons of gasoline it consumes. If a car can go 288 miles on 12 gallons of gasoline, how far can it go on a full tank of 18 gallons? 432 miles

34. Farming A farmer's harvest in bushels varies directly with the number of acres planted. If 8 acres can produce 144 bushels, how many acres are required to produce 1152 bushels? 64

35. Farming The length of time that a given number of bushels of corn will last when feeding cattle varies inversely with the number of animals. If x bushels will feed 25 cows for 10 days, how long will the feed last for 10 cows? 25 days

36. Geometry For a fixed area, the length of a rectangle is inversely proportional to its width. A rectangle has a width of 18 feet and a length of 12 feet. If the length is increased to 16 feet and the area remains the same, find the width. 13.5 ft

37. Gas pressure Under a constant temperature, the volume occupied by a gas is inversely proportional to the pressure applied. If the gas occupies a volume of 20 cubic inches under a pressure of 6 pounds per square inch, find the volume when the gas is subjected to a pressure of 10 pounds per square inch. 12 in.^{3}

38. Value of a car The value of a car usually varies inversely with its age. If a car is worth $7000 when it is 3 years old, how much will it be worth when it is 7 years old? $3000

39. Organ pipes The frequency of vibration of air in an organ pipe is inversely proportional to the length of the pipe. (See Illustration 1.) If a pipe 2 feet long vibrates 256 times per second, how many times per second will a 6-foot pipe vibrate? 85.3

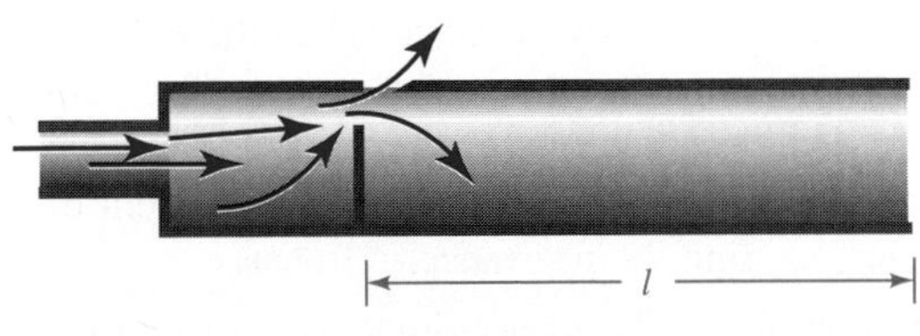

ILLUSTRATION 1

40. Geometry The area of a rectangle varies jointly with its length and width. If both the length and width are tripled, by what factor is the area multiplied? 9

41. Geometry The volume of a rectangular solid varies jointly with its length, width, and height. If the length is doubled, the width is tripled, and the height is doubled, by what factor is the volume multiplied? 12

42. **Costs of a trucking company** The costs incurred by a trucking company vary with the number of trucks in service and the number of hours they are used. When 4 trucks are used for 6 hours each, the costs are \$1800. Find the costs of using 10 trucks, each for 12 hours. \$9000

43. **Storing oil** The number of gallons of oil that can be stored in a cylindrical tank varies jointly with the height of the tank and the square of the radius of its base. The constant of proportionality is 23.5. Find the number of gallons that can be stored in the cylindrical tank in Illustration 2. 26,437.5 gal

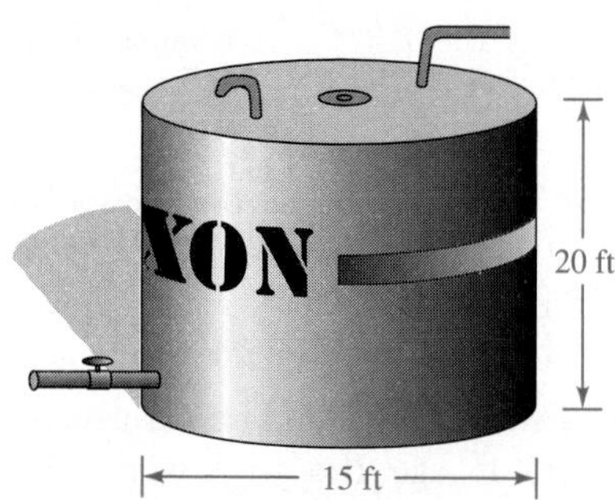

ILLUSTRATION 2

44. **Finding the constant of variation** A quantity l varies jointly with x and y and inversely with z. If the value of l is 30 when $x = 15$, $y = 5$, and $z = 10$, find k. 4

45. **Electronics** The voltage (in volts) measured across a resistor is directly proportional to the current (in amperes) flowing through the resistor. The constant of variation is the **resistance** (in ohms). If 6 volts is measured across a resistor carrying a current of 2 amperes, find the resistance. 3 ohms

46. **Electronics** The power (in watts) lost in a resistor in the form of heat is directly proportional to the square of the current (in amperes) passing through it. The constant of proportionality is the resistance (in ohms). What power is lost in a 5-ohm resistor carrying a 3-ampere current? 45 watts

47. **Building construction** The deflection of a beam is inversely proportional to its width and the cube of its depth. If the deflection of a 4-inch-by-4-inch beam is 1.1 inches, find the deflection of a 2-inch-by-8-inch beam positioned as in Illustration 3. 0.275 in.

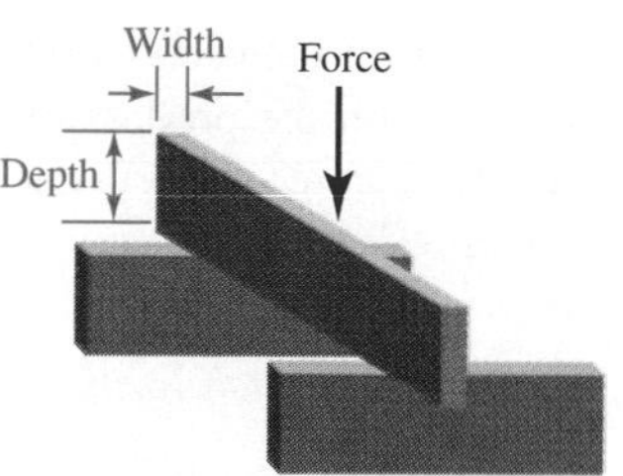

ILLUSTRATION 3

48. **Building construction** Find the deflection of the beam in Exercise 47 when the beam is positioned as in Illustration 4. 4.4 in.

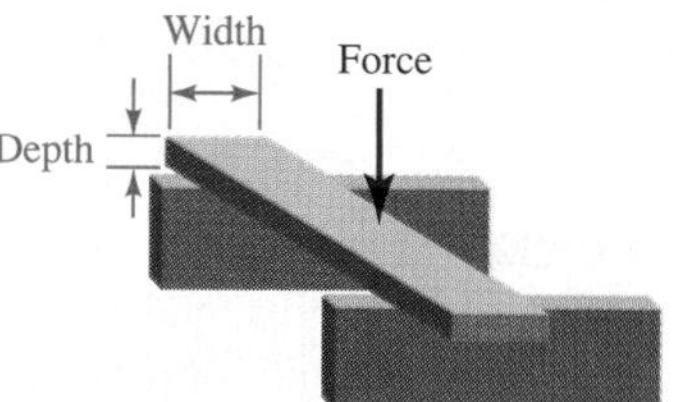

ILLUSTRATION 4

49. **Gas pressure** The pressure of a certain amount of gas is directly proportional to the temperature (measured in degrees Kelvin) and inversely proportional to the volume. A sample of gas at a pressure of 1 atmosphere occupies a volume of 1 cubic meter at a temperature of 273 Kelvin. When heated, the gas expands to twice its volume, but the pressure remains constant. To what temperature is it heated? 546 Kelvin

50. **Tension** A stone twirled at the end of a string is kept in its circular path by the tension of the string. The tension, T, is directly proportional to the square of the speed, s, and inversely proportional to the radius, r, of the circle. In Illustration 5, the tension is 32 pounds when the speed is 8 feet/second and the radius is 6 feet. Find the tension when the speed is 4 feet/second and the radius is 3 feet. 16 lb

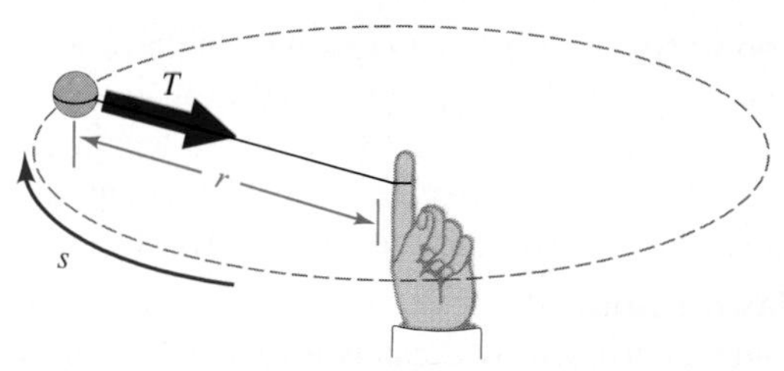

ILLUSTRATION 5

Writing Exercises *Write a paragraph using your own words.*

1. Explain the term *joint variation*.
2. Explain why the equation $\frac{y}{x} = k$ indicates that y varies directly with x.

Something to Think About

1. As temperature increases on the Fahrenheit scale, it also increases on the Celsius scale. Is this direct variation? Explain.
2. As the cost of a purchase (less than \$5) increases, the amount of change received from a five-dollar bill decreases. Is this inverse variation? Explain.

Review Exercises *Simplify each expression.*

1. $(x^2x^3)^2$ x^{10}
2. $\left(\frac{a^3a^5}{a^{-2}}\right)^3$ a^{30}
3. $\frac{b^0 - 2b^0}{b^0}$ -1
4. $\left(\frac{2r^{-2}r^{-3}}{4r^{-5}}\right)^{-3}$ 8
5. Write 357,000 in scientific notation. 3.57×10^5
6. Write 0.000357 in scientific notation. 3.57×10^{-4}
7. Write 2.5×10^{-3} in standard notation. 0.0025
8. Write 2.5×10^4 in standard notation. 25,000

PROJECT Another Equation of a Line

The graph of a line is determined by two pieces of information. If we know the line's slope and its y-intercept, we would use the *slope–intercept* form to find the equation of the line. If we know the slope of the line and the coordinates of some point on that line, we would use the *point–slope* form. We have studied several standard forms of the equation of a line. Here is one more standard form that is useful when we know a line's x- and y-intercepts.

The Intercept Form of the Equation of a Line

The equation of a line with x-intercept $(a, 0)$ and y-intercept $(0, b)$ is

$$\frac{x}{a} + \frac{y}{b} = 1$$

- Derive the intercept form of the equation of a line. *Hint:* You know two points on the line.
- Find the x- and y-intercepts of the line $\frac{x}{5} + \frac{y}{9} = 1$.
- Find the equation of the line with x-intercept $(3, 0)$ and y-intercept $(0, 7)$.
- Find the x- and y-intercepts of the line $4x + 5y = 20$ by writing the equation in intercept form.
- Graph the line $\frac{x}{k} + \frac{y}{k} = 1$ for five different values of k (your choice.) What do these lines have in common?

(continued)

PROJECT ■ **Another Equation of a Line** *(continued)*

- Graph the line $\frac{x}{3} + \frac{y}{k} = 1$ for five different values of k. What do these lines have in common?
- Can the equation of *every* line be written in intercept form? Discuss which lines can and which ones can't.

Chapter Summary

KEY WORDS

annual depreciation rate (6.3)
abscissa (6.1)
Cartesian coordinate system (6.1)
combined variation (6.7)
constant of variation (6.7)
coordinates (6.1)
dependent variable (6.6)
direct variation (6.7)
domain (6.6)
extremes of a proportion (6.7)
function (6.6)
general form of a line (6.3)
graph of an equation (6.1)
graph of an inequality (6.5)
hyperbola (6.4)
independent variable (6.6)
inverse variation (6.7)
joint variation (6.7)
linear equation in *x* and *y* (6.3)
linear function (6.6)
linear inequality (6.5)
means of a proportion (6.7)
midpoint (6.1)
negative reciprocals (6.2)
ordered pair (6.1)
ordinate (6.1)
origin (6.1)
parabola (6.4)
point–slope form of a line (6.3)
proportion (6.7)
range (6.6)
rate of change (6.2)
ratio (6.7)
rectangular coordinate system (6.1)
relation (6.6)
salvage value (6.3)
slope–intercept form of a line (6.3)
slope of a nonvertical line (6.2)
straight-line depreciation (6.3)
vertical line test (6.6)
viewing window (6.4)
x-axis (6.1)
x-coordinate (6.1)
x-intercept (6.1)
xy-plane (6.1)
y-axis (6.1)
y-coordinate (6.1)
y-intercept (6.1)

KEY IDEAS

(6.1) **The midpoint formula**: If $P(x_1, y_1)$ and $Q(x_2, y_2)$ are two points on a line, then the midpoint of the segment PQ is point M where the coordinates of M are

$$\left(\frac{x_1 + x_2}{2}, \frac{y_1 + y_2}{2}\right)$$

(6.2) **The slope of a nonvertical line:**

$$m = \frac{\Delta y}{\Delta x} = \frac{y_2 - y_1}{x_2 - x_1} \qquad (x_2 \neq x_1)$$

Nonvertical parallel lines have the same slope, and lines with the same slope are parallel.

If two nonvertical lines are perpendicular, their slopes are negative reciprocals.

If the slopes of two lines are negative reciprocals, the lines are perpendicular.

(6.3) **Point–slope form of a linear equation:**

$$y - y_1 = m(x - x_1)$$

Slope–intercept form of a linear equation: $y = mx + b$

General form of a linear equation: $Ax + By = C$

(6.4) Graphs of nonlinear equations are not straight lines.

(6.5) To graph an inequality such as $y > ax + b$, first graph the linear equation $y = ax + b$. The line graph will determine two half-planes, one on each side of the line. Then determine which half-plane represents the graph of $y > ax + b$.

(6.6) The **vertical line test** can be used to determine whether a graph represents a function.

(6.7) In a proportion, the product of the extremes is equal to the product of the means.

If $y = kx$ and k is constant, then x and y **vary directly.**

If $y = \frac{k}{x}$ and k is a constant, then x and y **vary inversely.**

If $y = kxz$ and k is a constant, then y **varies jointly** with x and z.

The expression $y = \frac{kx}{z}$ (k is a constant) indicates **combined variation**, with y and x varying directly and y and z varying inversely.

Chapter 6 Review Exercises

In Review Exercises 1–8, graph each equation.

1. $x + y = 4$

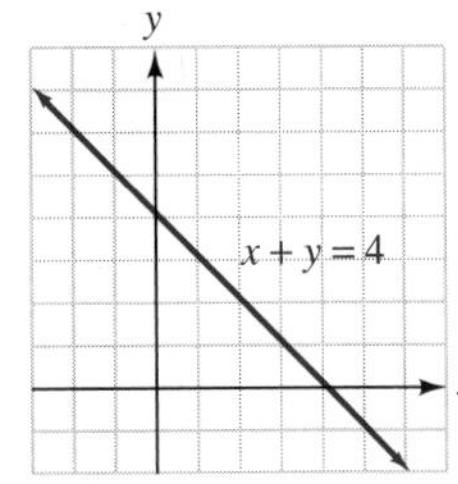

2. $2x - y = 8$

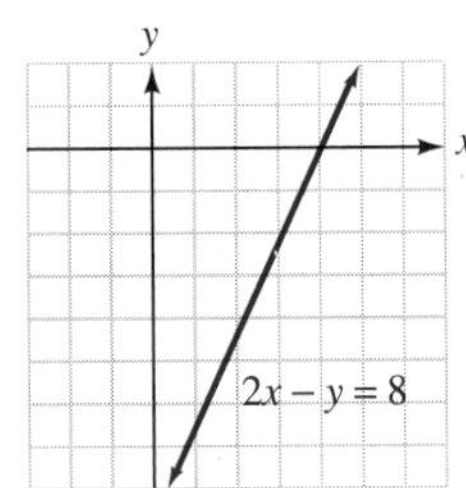

3. $y = 3x + 4$

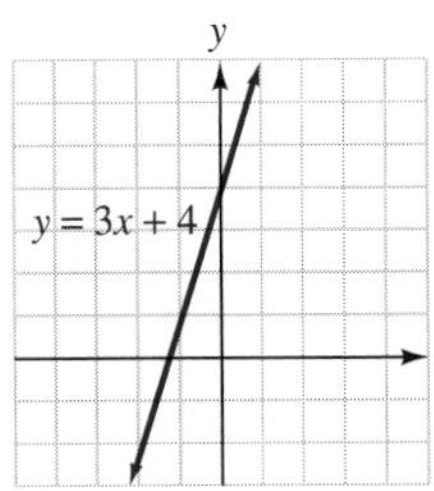

4. $x = 4 - 2y$

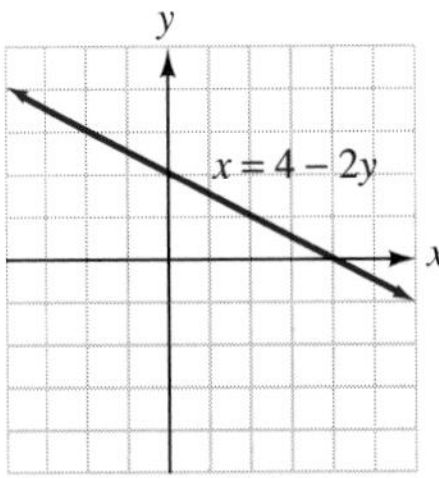

5. $y = 4$

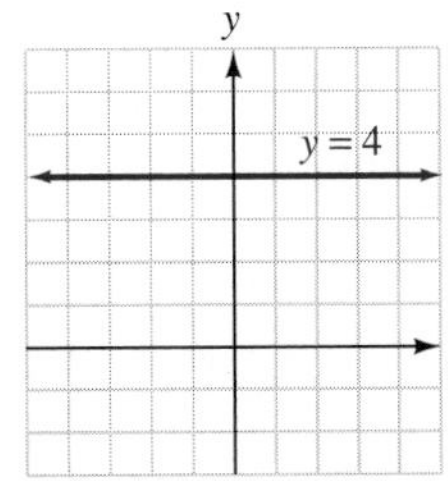

6. $x = -2$

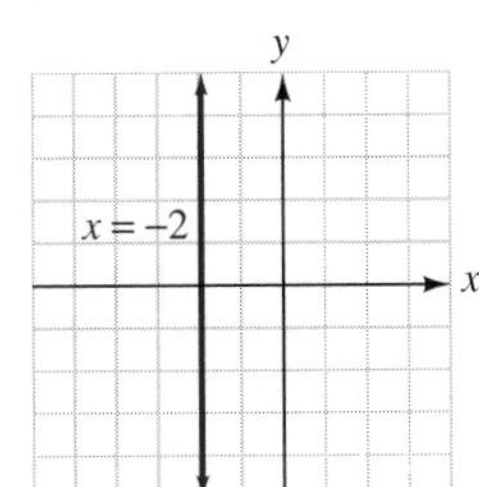

7. $2(x + 4) = x + 4$

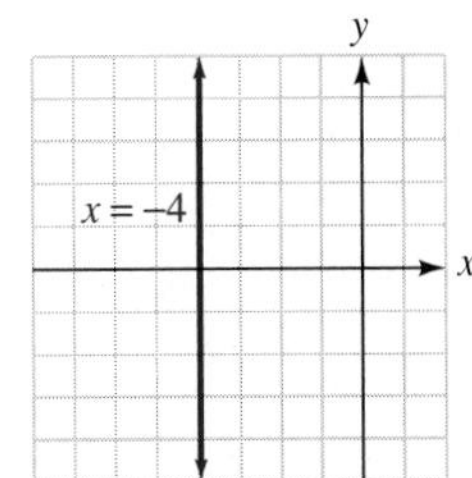

8. $3y = 2(y - 1)$

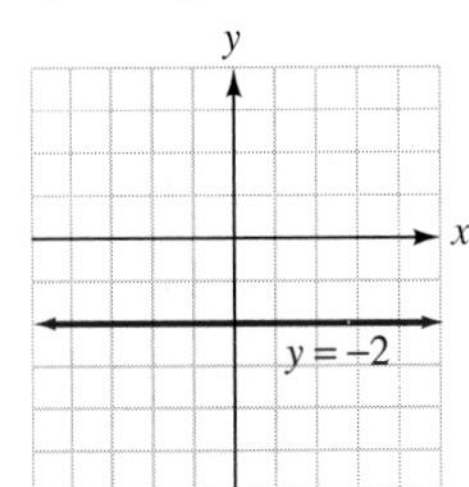

In Review Exercises 9–12, find the midpoint of each segment PQ.

9. $P(2, 6)$ and $Q(2, 12)$ $(2, 9)$

10. $P(8, -2)$ and $Q(-6, -2)$ $(1, -2)$

11. $P(2, -6)$ and $Q(5, 10)$ $\left(\frac{7}{2}, 2\right)$

12. $P(-3, -7)$ and $Q(10, -6)$ $\left(\frac{7}{2}, -\frac{13}{2}\right)$

In Review Exercises 13–16, find the slope of the line passing through points P and Q.

13. $P(2, 5)$ and $Q(5, 8)$ 1

14. $P(-3, -2)$ and $Q(6, 12)$ $\frac{14}{9}$

15. $P(-3, 4)$ and $Q(-5, -6)$ 5

16. $P(5, -4)$ and $Q(-6, -9)$ $\frac{5}{11}$

In Review Exercises 17–20, find the slope of the graph of each equation, if one exists.

17. $2x - 3y = 18$ $\frac{2}{3}$

18. $2x + y = 8$ -2

19. $-2(x - 3) = 10$ undefined

20. $3y + 1 = 7$ 0

In Review Exercises 21–24, tell whether the lines with the given slopes are parallel, perpendicular, or neither.

21. $m_1 = 4, m_2 = -\dfrac{1}{4}$ perpendicular

22. $m_1 = 0.5, m_2 = \dfrac{1}{2}$ parallel

23. $m_1 = 0.5, m_2 = -\dfrac{1}{2}$ neither

24. $m_1 = 5, m_2 = -0.2$ perpendicular

25. Use the point–slope form to write the equation of a line with a slope of $-\frac{3}{2}$ that passes through $P(-2, 5)$. $y = -\frac{3}{2}x + 2$

26. Use the slope–intercept form to write the equation of a line with a slope of $-\frac{3}{2}$ that passes through $P(-2, 5)$. $y = -\frac{3}{2}x + 2$

In Review Exercises 27–30, write the equation of the line with the given properties. Write each answer in general form.

27. Slope of 3; passing through $P(-8, 5)$. $3x - y = -29$

28. Passing through $(-2, 4)$ and $(6, -9)$. $13x + 8y = 6$

29. Passing through $(-3, -5)$; parallel to the graph of $3x - 2y = 7$. $3x - 2y = 1$

30. Passing through $(-3, -5)$; perpendicular to the graph of $3x - 2y = 7$. $2x + 3y = -21$

In Review Exercises 31–34, graph each equation.

31. $y = x^2 - 3$

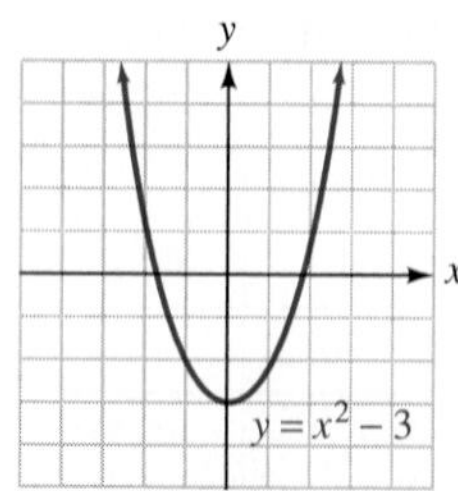

32. $y = |x| - 4$

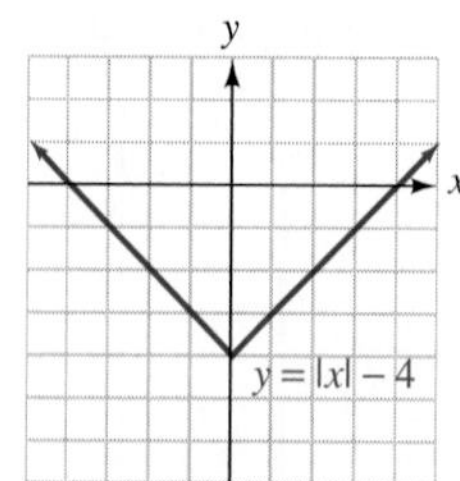

33. $xy = 10$

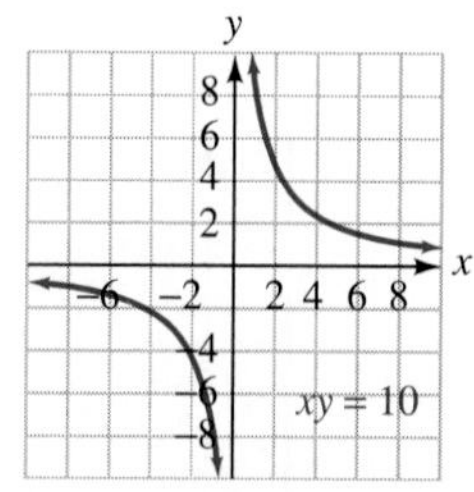

34. $y = x^2 - 6x$

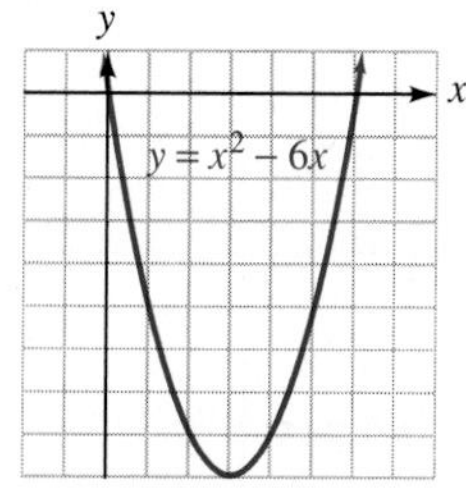

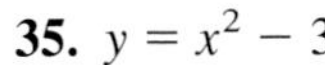

In Review Exercises 35–38, use a graphing calculator to graph each equation. Compare the results to Exercises 31–34.

35. $y = x^2 - 3$

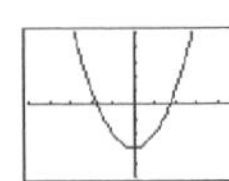

36. $y = |x| - 4$

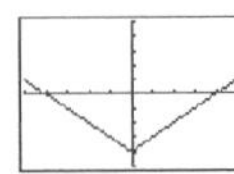

37. $xy = 10$

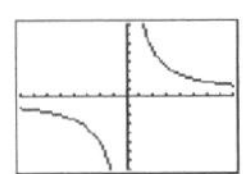

38. $y = x^2 - 6x$

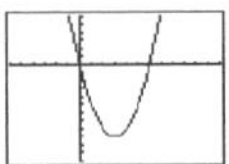

In Review Exercises 39–42, graph each inequality.

39. $2x + 3y > 6$

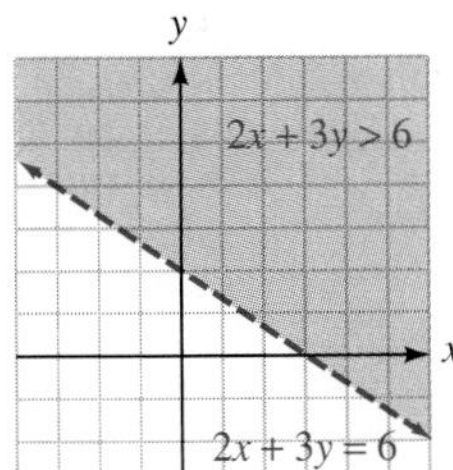

40. $y \leq 4 - x$

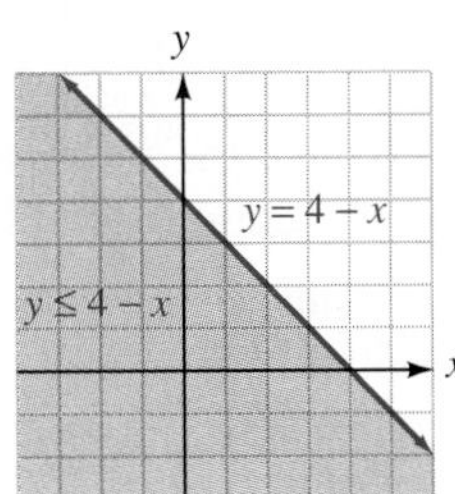

41. $-2 < x < 4$

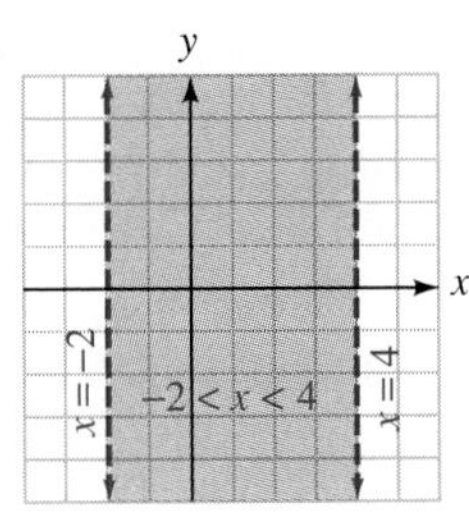

42. $y \leq -2$ or $y > 1$

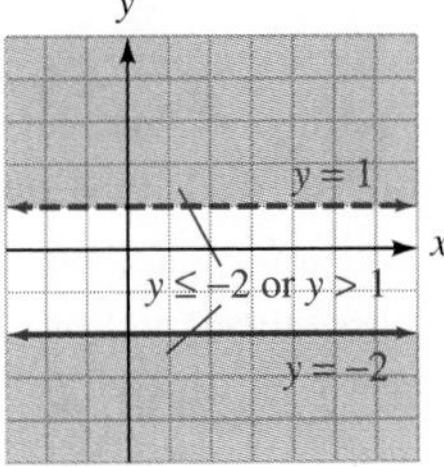

In Review Exercises 43–46, tell whether the given equation or inequality determines y to be a function of x.

43. $y > 6x - 4$ no

44. $y = 4 - x$ yes

45. $y = \dfrac{x + 2}{3}$ yes

46. $y = \dfrac{x^2 - 1}{2}$ yes

In Review Exercises 47–52, tell whether each equation defines y to be a function of x. Find the domain and range of each relation.

47. $y = 4x - 1$
function; domain is $\Re$; range is $\Re$

48. $x = 3y - 10$
function; domain is $\Re$; range is $\Re$

49. $y = 3x^2 + 1$ a function;
domain is $\Re$; range is $\{y \mid y \geq 1\}$

50. $y = \dfrac{4}{2 - x}$ function;
domain is $\Re$ but 2; range is $\Re$ but 0

51. $x = \dfrac{y + 3}{2}$
function; domain is $\Re$; range is $\Re$

52. $y^2 = 4x$
not a function; domain is $\{x \mid x \geq 0\}$; range is $\Re$

In Review Exercises 53–56, use the vertical line test to determine whether each graph represents a function.

53.

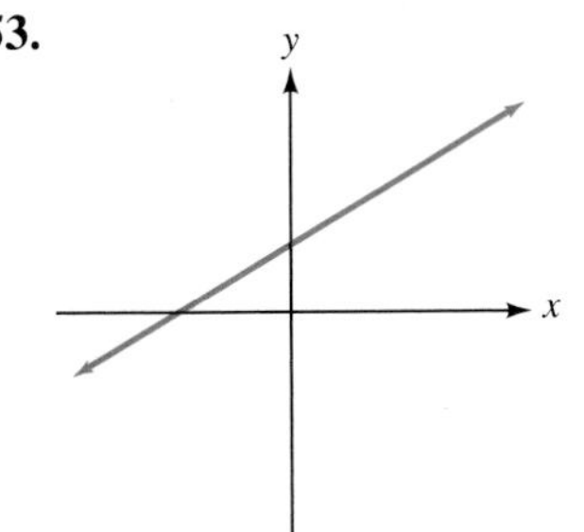

a function

54.

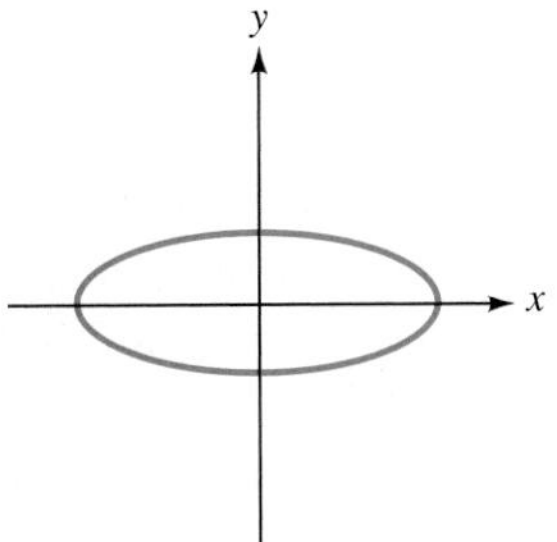

not a function

55.

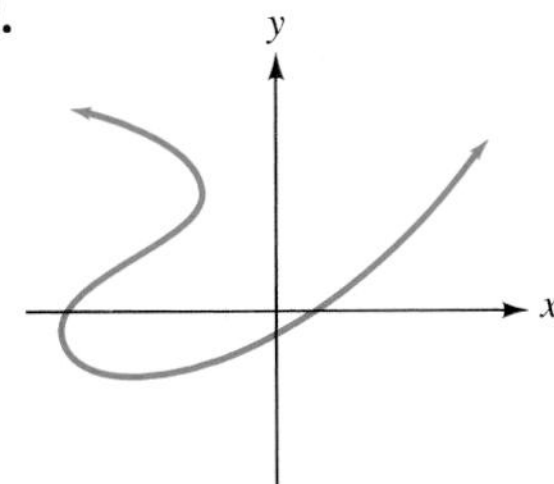

not a function

56.

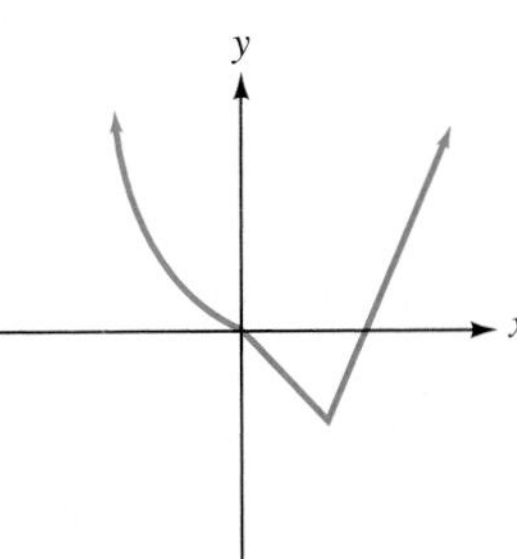

a function

In Review Exercises 57–60, assume that $f(x) = 3x + 2$ and $g(x) = x^2 - 4$. Find each value.

57. $f(-3)$ -7

58. $g(8)$ 60

59. $g(-2)$ 0

60. $f(5)$ 17

In Review Exercises 61–64, tell which equations define linear functions.

61. $y = 3x + 2$ yes

62. $y = \dfrac{x+5}{4}$ yes

63. $4x - 3y = 12$ yes

64. $y = x^2 - 25$ no

In Review Exercises 65–66, solve each proportion.

65. $\dfrac{x+1}{8} = \dfrac{4x-2}{24}$ 5

66. $\dfrac{1}{x+6} = \dfrac{x+10}{12}$ $-12, -4$

67. Assume that x varies directly with y. If $x = 12$ when $y = 2$, find the value of x when $y = 12$. 72

68. Assume the x varies inversely with y. If $x = 24$ when $y = 3$, find the value of y when $x = 12$. 6

69. Assume that x varies jointly with y and z. Find the constant of variation if $x = 24$ when $y = 3$ and $z = 4$. 2

70. Assume that x varies directly with t and inversely with y. Find the constant of variation if $x = 2$ when $t = 8$ and $y = 64$. 16

Chapter 6 Test

1. Graph the equation $2x - 5y = 10$.

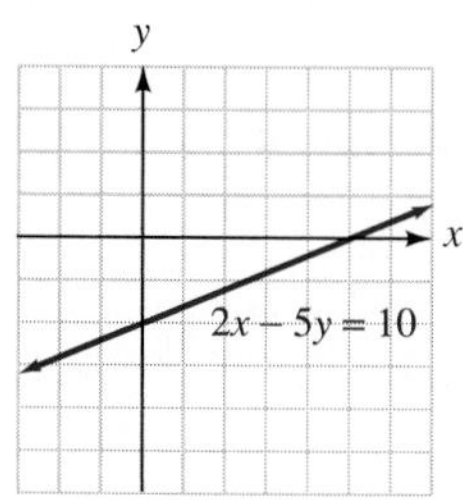

2. Find the x- and y-intercepts of the graph of $y = \frac{x-3}{5}$.
x-intercept $(3, 0)$, y-intercept $(0, -\frac{3}{5})$

In Problems 3–4, $P(-5, -6)$ and $Q(5, -2)$.

3. Find the midpoint of segment PQ. $(0, -4)$

4. Find the slope of the line PQ. $\frac{2}{5}$

5. Find the slope of the graph of $x = \frac{3y-8}{2}$. $\frac{2}{3}$

6. Find the slope of the graph of $y = 12$. 0

7. Write the equation of the line with slope of $\frac{2}{3}$ that passes through $P(4, -5)$. Give the answer in slope–intercept form. $y = \frac{2}{3}x - \frac{23}{3}$

8. Write the equation of the line that passes through $P(-2, 6)$ and $Q(-4, -10)$. Give the answer in general form. $8x - y = -22$

9. Find the slope and the y-intercept of the graph of $-2(x - 3) = 3(2y + 5)$.
$m = -\frac{1}{3}$, $(0, -\frac{3}{2})$

10. Determine whether the graphs of $4x - y = 12$ and $y = \frac{1}{4}x + 3$ are parallel, perpendicular, or neither.
neither

11. Determine whether the graphs of $y = -\frac{2}{3}x + 4$ and $2y = 3x - 3$ are parallel, perpendicular, or neither.
perpendicular

12. Write the equation of the line that passes through the origin and is parallel to the graph of $y = \frac{3}{2}x - 7$.
$y = \frac{3}{2}x$

13. Write the equation of the line that passes through $P(-3, 6)$ and is perpendicular to the graph of $y = -\frac{2}{3}x - 7$. $y = \frac{3}{2}x + \frac{21}{2}$

14. Graph $y = -x^2 + 3$.

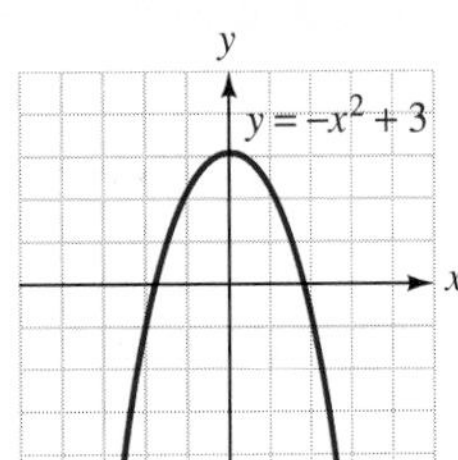

15. Graph $3x + 2y \geq 6$.

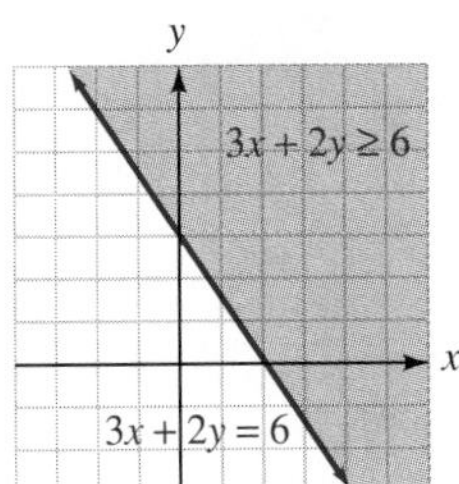

16. Graph $-2 \leq y < 5$.

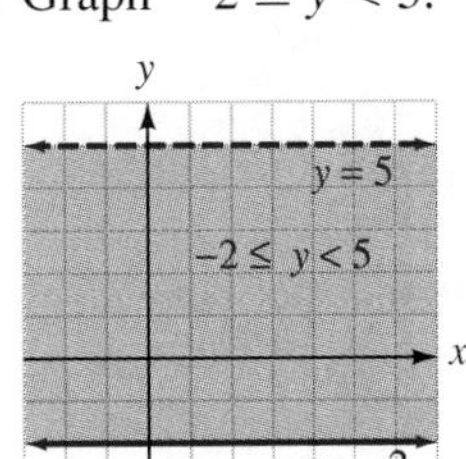

17. Does $|y| = x$ define y to be a function of x?
no

18. Find the domain and range of the function $y = |x|$.
domain is $\mathfrak{R}$; range is $\{y | y \geq 0\}$

In Problems 19–22, $f(x) = 3x + 1$ and $g(x) = x^2 - 2$. Find each value.

19. $f(3)$ 10

20. $g(0)$ -2

21. $f(a)$ $3a + 1$

22. $g(-x)$ $x^2 - 2$

23. Does $y(x + 3) + 4 = x(y - 2)$ define a linear function? yes

24. Solve the proportion: $\frac{3}{x - 2} = \frac{x + 3}{2x}$ 6, -1

25. The power (in watts) dissipated as heat in a resistor varies directly with the square of the voltage and inversely with the resistance. If k is the constant of variation, express this relationship with an equation.
$P = \frac{kv^2}{r}$

26. x varies directly with y. If $x = 30$ when $y = 4$, find x when $y = 9$.
$\frac{135}{2}$

27. V varies inversely with t. If $V = 55$ when $t = 20$, find t when $V = 75$. $\frac{44}{3}$

28. Does the graph define a function?
yes

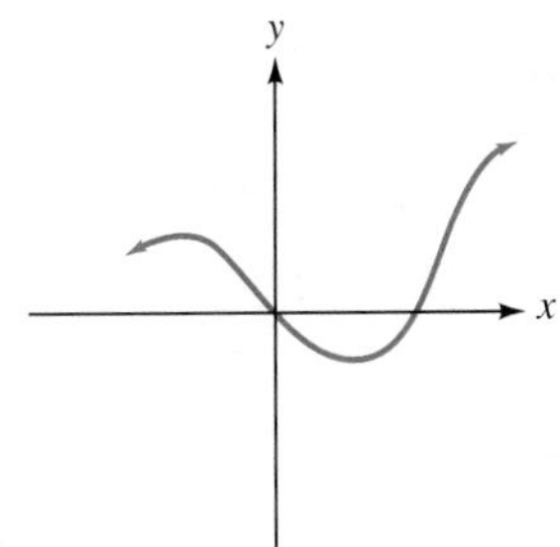

Cumulative Review Exercises

If you teach graphing early, skip these exercises.

In Exercises 1–12, factor each expression.

1. $3r^2s^3 - 6rs^4$ $3rs^3(r - 2s)$
2. $5(x - y) - a(x - y)$ $(x - y)(5 - a)$
3. $xu + yv + xv + yu$ $(x + y)(u + v)$
4. $81x^4 - 16y^4$ $(9x^2 + 4y^2)(3x + 2y)(3x - 2y)$
5. $8x^3 - 27y^6$ $(2x - 3y^2)(4x^2 + 6xy^2 + 9y^4)$
6. $6x^2 + 5x - 6$ $(2x + 3)(3x - 2)$
7. $9x^2 - 30x + 25$ $(3x - 5)^2$
8. $15x^2 - x - 6$ $(5x + 3)(3x - 2)$
9. $27a^3 + 8b^3$ $(3a + 2b)(9a^2 - 6ab + 4b^2)$
10. $6x^2 + x - 35$ $(2x + 5)(3x - 7)$
11. $x^2 + 10x + 25 - y^4$ $(x + 5 + y^2)(x + 5 - y^2)$
12. $x^4 + 8x^2 + 36$ $(x^2 - 2x + 6)(x^2 + 2x + 6)$

In Exercises 13–14, solve each equation by factoring.

13. $x^3 - 4x = 0$ 0, 2, −2
14. $6x^2 + 7 = -23x$ $-\frac{1}{3}, -\frac{7}{2}$

In Exercises 15–18, simplify each expression.

15. $\dfrac{2x^2y + xy - 6y}{3x^2y + 5xy - 2y}$ $\frac{2x - 3}{3x - 1}$
16. $\dfrac{x^2 - 4}{x^2 + 9x + 20} \div \dfrac{x^2 + 5x + 6}{x^2 + 4x - 5} \cdot \dfrac{x^2 + 3x - 4}{(x - 1)^2}$ $\frac{x - 2}{x + 3}$
17. $\dfrac{2}{x + y} + \dfrac{3}{x - y} - \dfrac{x - 3y}{x^2 - y^2}$ $\frac{4}{x - y}$
18. $\dfrac{\frac{a}{b} + b}{a - \frac{b}{a}}$ $\dfrac{a^2 + ab^2}{a^2b - b^2}$

In Exercises 19–20, solve each equation.

19. $\dfrac{5x - 3}{x + 2} = \dfrac{5x + 3}{x - 2}$ 0
20. $\dfrac{3}{x - 2} + \dfrac{x^2}{(x + 3)(x - 2)} = \dfrac{x + 4}{x + 3}$ −17
21. Graph $2x - 3y = 6$ and tell whether it defines a function. yes
22. Find the midpoint of a line segment with endpoints at $P(-2, 5)$ and $Q(8, -9)$. $(3, -2)$
23. Find the slope of a line passing through $P(-2, 5)$ and $Q(8, -9)$. $-\frac{7}{5}$
24. Write the equation of the line passing through $P(-2, 5)$ and $Q(8, -9)$. $y = -\frac{7}{5}x + \frac{11}{5}$

In Exercises 25–26, tell whether the lines represented by the equations are parallel or perpendicular.

25. $3x + 2y = 12$, $2x - 3y = 5$ perpendicular
26. $3x = y + 4$, $y = 3(x - 4) - 1$ parallel
27. Write the equation of the line passing through $P(-2, 3)$ and parallel to the graph of $3x + y = 8$. $y = -3x - 3$
28. Write the equation of the line passing through $P(-2, 3)$ and perpendicular to the graph of $3x + y = 8$. $y = \frac{1}{3}x + \frac{11}{3}$

In Exercises 29–30, graph each inequality.

29. $2x - 3y \leq 12$

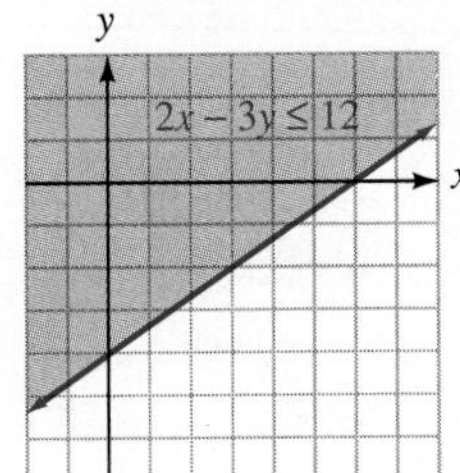

30. $3 > x \geq -2$

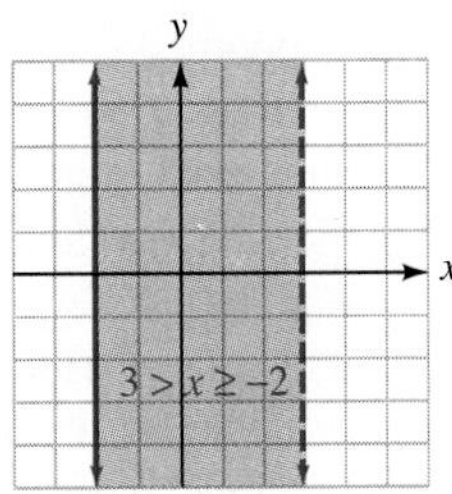

In Exercises 31–34, $f(x) = 3x^2 + 2$ *and* $g(x) = 2x - 1$. *Evaluate each expression.*

31. $f(-1)$ 5

32. $g(0)$ -1

33. $g(t)$ $2t - 1$

34. $f(-r)$ $3r^2 + 2$

35. Solve the proportion $\frac{x+3}{2x} = \frac{3x}{6x+5}$. $-\frac{15}{23}$

36. Express as a formula: *y varies directly with the product of x and y, but inversely with r.* $y = \frac{kxy}{r}$

37. The volume of a cylindrical tank varies jointly with the height of the tank and the square of the radius of its circular base. If the volume is 4π cubic feet when $h = 4$ feet and $r = 1$ foot, find the height when $V = 8\pi$ cubic feet and $r = 2$ feet. 2 ft

38. Does the following graph represent a function? no

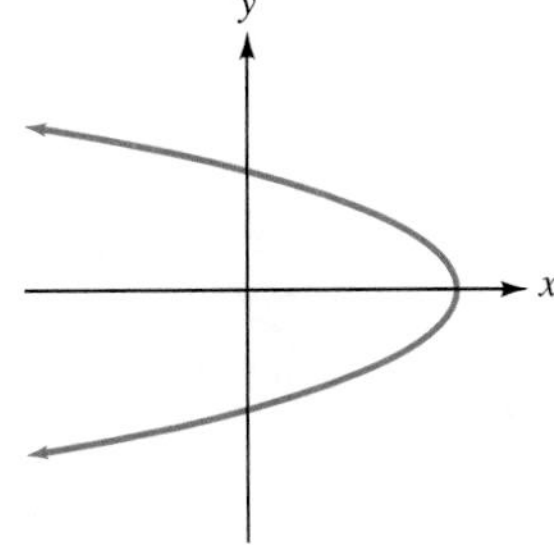

CAREERS & MATHEMATICS

Economist

Economists study the way a society uses resources such as land, labor, raw materials, and machinery to provide goods and services. They analyze the results of their research to determine the costs and benefits of making, distributing, and using resources in a particular way. Some economists are theoreticians who use mathematical models to explain the causes of recession and inflation. Most economists, however, are concerned with practical applications of economic policy in a particular area.

Qualifications

Economists must thoroughly understand economic theory, mathematical methods of economic analysis, and basic statistical procedures. Training in computer science is highly recommended.

Job Outlook

Employment of economists is expected to grow faster than the average for all occupations during the next decade. Opportunities should be best for economists in business and industry, research organizations, and consulting firms.

Example Application

An electronics firm manufactures tape recorders, receiving \$120 for each unit it makes. If x represents the number of recorders produced, then the income received is determined by the *revenue function*, given by the linear equation

$$R(x) = 120x$$

The manufacturer has determined that the *fixed costs* for advertising, insurance, utilities, and so on, are \$12,000 per month, and the *variable cost* for materials is \$57.50 for each machine produced. Thus, the *cost function* is given by the linear equation

$$\begin{aligned} C(x) &= \textit{variable cost} + \textit{fixed costs} \\ &= 57.50x + 12{,}000 \end{aligned}$$

The company's profit is the amount by which revenue exceeds costs. It is determined by the *profit function*, given by the equation

$$\begin{aligned} \textit{Profit} &= \textit{revenue} - \textit{costs} \\ P(x) &= R(x) - C(x) \\ &= 120x - (57.50x + 12{,}000) \\ &= 62.50x - 12{,}000 \end{aligned}$$

If $P(x) > 0$, the company is making money. If $P(x) < 0$, it is operating at a loss. How many recorders must the company manufacture to break even?

Solution Graph the profit function $y = P(x) = 62.50x - 12{,}000$. The break-even point is that value of x that gives a profit of zero. It is the x-intercept of the graph of the profit function. (See Illustration 1.) To find it, set $P(x)$ equal to 0 and solve for x.

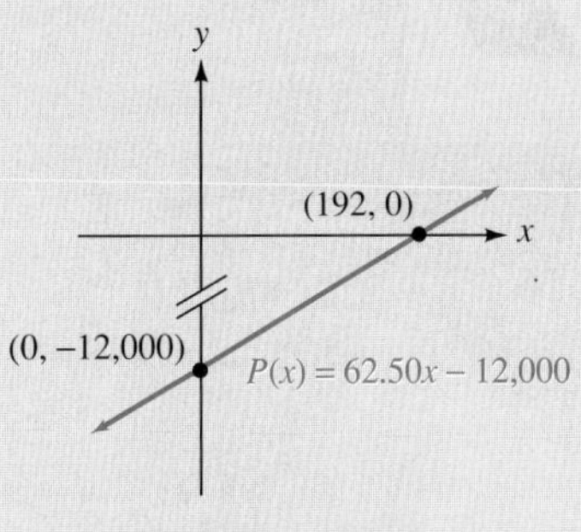

ILLUSTRATION 1

$$\begin{aligned} P(x) &= 0 \\ 62.50x - 12{,}000 &= 0 \\ 62.50x &= 12{,}000 && \text{Add 12,000 to both sides.} \\ x &= 192 && \text{Divide both sides by 62.50.} \end{aligned}$$

The company must manufacture and sell 192 tape recorders each month to break even.

EXERCISES

1. Find the revenue and the cost of manufacturing 192 units, and verify that the revenue and the cost are equal.
2. Determine the company's profit if it manufactures 150 units each month.
3. Determine the company's profit if it manufactures 400 units each month.
4. How many units must be manufactured each month to produce a total profit of $47,375?

(*Answers:* **1.** $23,040 revenue and cost **2.** $2625 loss **3.** $13,000 profit **4.** 90 units)

7 Rational Exponents and Radicals

In previous chapters, we squared and cubed numbers and raised them to other powers. We now discuss how to reverse this process and find roots of numbers.

A number b is called the ***n*th root of *a*** if the nth power of b is equal to a. For example,

4 is a square root of 16, because $4^2 = 16$

5 is a cube root of 125, because $5^3 = 125$

3 is a fourth root of 81, because $3^4 = 81$

The following statement is the general definition of the nth root of a number.

nth Root If n is a natural number, b is an nth root of a whenever

$$b^n = a$$

Either rational (fractional) exponents or radicals are used to denote nth roots of numbers.

7.1 Rational Exponents

■ Exponential Expressions with Nonnegative Bases ■ Exponential Expressions with Negative Bases ■ Fractional Exponents with Numerators Other Than 1 ■ Exponential Expressions with Variable Bases

In Chapter 2, we discussed the following rules of integral exponents.

Rules of Exponents If there are no divisions by 0, then for all integers m and n

1. $x^m x^n = x^{m+n}$ **2.** $(x^m)^n = x^{mn}$ **3.** $(xy)^n = x^n y^n$ **4.** $\left(\frac{x}{y}\right)^n = \frac{x^n}{y^n}$

5. $x^0 = 1\ (x \neq 0)$ **6.** $x^{-n} = \frac{1}{x^n}$ **7.** $\frac{x^m}{x^n} = x^{m-n}$ **8.** $\left(\frac{x}{y}\right)^{-n} = \left(\frac{y}{x}\right)^n$

We now extend these rules of exponents to include rational (fractional) exponents.

■ Exponential Expressions with Nonnegative Bases

Historical Note: The first recorded use of fractional exponents was in the work of Nicole Oresme (1323–1382).

If we apply Rule 2 of exponents to an expression such as $(16^{1/2})^2$, we have

$$\begin{aligned}(16^{1/2})^2 &= 16^{1/2 \cdot 2}\\ &= 16^1\\ &= 16\end{aligned}$$

Carl Friedrich Gauss (1777–1855)
Many people consider Gauss to be the greatest mathematician of all time. He made contributions in the areas of number theory, solutions of equations, geometry of curved surfaces, and statistics. For his efforts, he earned the title "Prince of the Mathematicians."

Thus, $16^{1/2}$ is a number whose square is 16. Although $4^2 = 16$ and $(-4)^2 = 16$, we define $16^{1/2}$ to be 4, which is the positive number whose square is 16:

$16^{1/2} = 4$ Read as "The square root of 16 is 4."

If we apply Rule 2 to the expression $(27^{1/3})^3$, we have

$$\begin{aligned}(27^{1/3})^3 &= 27^{1/3 \cdot 3} \\ &= 27^1 \\ &= 27\end{aligned}$$

Thus, $27^{1/3}$ is a number whose cube is 27. Since the cube of 3 is 27,

$27^{1/3} = 3$ Read $27^{1/3}$ as "The cube root of 27."

Similarly,

$81^{1/4} = 3$	because	$3^4 = 81$	Read $81^{1/4}$ as "the fourth root of 81."
$32^{1/5} = 2$	because	$2^5 = 32$	Read $32^{1/5}$ as "the fifth root of 32."
$729^{1/6} = 3$	because	$3^6 = 729$	Read $729^{1/6}$ as "the sixth root of 729."
$0^{1/7} = 0$	because	$0^7 = 0$	Read $0^{1/7}$ as "the seventh root of 0."

WARNING! By definition, 0^0 is undefined, and 0 raised to a negative exponent is undefined. For example,

$$0^{-2} = \frac{1}{0^2}$$

is undefined, because division by 0 is undefined.

(a, c, f, g) **EXAMPLE 1**

a. $9^{1/2} = 3$ because $3^2 = 9$

b. $\left(\frac{16}{9}\right)^{1/2} = \frac{4}{3}$ because $\left(\frac{4}{3}\right)^2 = \frac{16}{9}$

c. $64^{1/3} = 4$ because $4^3 = 64$

d. $16^{1/4} = 2$ because $2^4 = 16$

e. $\left(\frac{1}{32}\right)^{1/5} = \frac{1}{2}$ because $\left(\frac{1}{2}\right)^5 = \frac{1}{32}$

f. $0^{1/8} = 0$ because $0^8 = 0$

g. $-32^{1/5} = -(32^{1/5}) = -(2) = -2$ ■

Exponential Expressions with Negative Bases

If the base of an exponential expression is negative and the denominator of its fractional exponent is odd, the expression is a real number. For example,

$(-8)^{1/3} = -2$ because $(-2)^3 = -8$

$(-32)^{1/5} = -2$ because $(-2)^5 = -32$

$\left(-\frac{27}{64}\right)^{1/3} = -\frac{3}{4}$ because $\left(-\frac{3}{4}\right)^3 = -\frac{27}{64}$

However, if the base of an exponential expression is negative and the denominator of the fractional exponent is even, the expression is not a real number. For example,

$(-25)^{1/2}$ is not a real number, because no real number has a square of -25.

$(-16)^{1/4}$ is not a real number, because no real number has a fourth power of -16.

$(-64)^{1/6}$ is not a real number, because no real number has a sixth power of -64.

The previous three expressions are called **imaginary numbers,** a topic in the next chapter.

(a, c) **EXAMPLE 2**

a. $(-64)^{1/3} = -4$, because $(-4)^3 = -64$.

b. $\left(-\frac{3125}{32}\right)^{1/5} = -\frac{5}{2}$, because $\left(-\frac{5}{2}\right)^5 = -\frac{3125}{32}$.

c. $\left(-\frac{1}{32}\right)^{1/4}$ is an imaginary number, because no real number has a fourth power of $-\frac{1}{32}$. ■

In general, the following is true.

Rational Exponents If n is a natural number and a is a real number, then

	***n* is odd**	***n* is even**
If $a > 0$	$(a^{1/n})^n = a$	$(a^{1/n})^n = a$
If $a = 0$	$a^{1/n} = 0$	$a^{1/n} = 0$
If $a < 0$	$(a^{1/n})^n = a$	$a^{1/n}$ is an imaginary number.

Fractional Exponents with Numerators Other Than 1

The definition of $a^{1/n}$ can be extended to include rational exponents whose numerators are not 1. For example, $4^{3/2}$ can be written as

$$(4^{1/2})^3 \quad \text{or} \quad (4^3)^{1/2}$$

because of a power rule of exponents. In general, we have the following rule.

Definition of $a^{m/n}$ If m and n are positive integers, the fraction $\frac{m}{n}$ has been simplified to lowest terms, and $a^{1/n}$ is a real number, then

$$a^{m/n} = (a^{1/n})^m = (a^m)^{1/n}$$

Because of the previous statement, we can interpret the expression $a^{m/n}$ in two ways:

1. $a^{m/n}$ means the mth power of the nth root of a.
2. $a^{m/n}$ means the nth root of the mth power of a.

For example, $16^{3/4}$ can be interpreted as

$$16^{3/4} = (16^{1/4})^3 = (2)^3 = 8 \quad \text{or} \quad 16^{3/4} = (16^3)^{1/4} = (4096)^{1/4} = 8$$

As this example suggests, it is usually easier to find the root of the base first in order to avoid large numbers.

To be consistent with the definition of negative exponents, we define $a^{-m/n}$ as follows.

Definition of $a^{-m/n}$ If m and n are positive integers, the fraction $\frac{m}{n}$ has been simplified to lowest terms, and $a^{1/n}$ is a real number, then

$$a^{-m/n} = \frac{1}{a^{m/n}} \quad \text{and} \quad \frac{1}{a^{-m/n}} = a^{m/n} \qquad (a \neq 0)$$

(a, d)

EXAMPLE 3

a. $$64^{-2/3} = \frac{1}{64^{2/3}} = \frac{1}{(64^{1/3})^2} = \frac{1}{4^2} = \frac{1}{16}$$

b. $$16^{-3/2} = \frac{1}{16^{3/2}} = \frac{1}{(16^{1/2})^3} = \frac{1}{4^3} = \frac{1}{64}$$

c. $$(-32)^{-2/5} = \frac{1}{(-32)^{2/5}} = \frac{1}{[(-32)^{1/5}]^2} = \frac{1}{(-2)^2} = \frac{1}{4}$$

d. $(-16)^{-3/4}$ is an imaginary number, since $(-16)^{1/4}$ is not a real number. ■

In the following example, each rule for exponents is illustrated for expressions containing rational exponents.

EXAMPLE 4 Assume that all variables represent positive numbers and write all answers without using negative exponents.

a. $a^m a^n = a^{m+n}$ $\qquad 7^{2/5} \cdot 7^{1/5} = 7^{3/5}$

b. $(a^m)^n = a^{mn}$ $\qquad (4^{1/3})^5 = 4^{(1/3)5} = 4^{5/3}$

c. $(ab)^n = a^n b^n$ $\qquad (5 \cdot 4)^{4/3} = 5^{4/3} \cdot 4^{4/3}$

d. $\dfrac{a^m}{a^n} = a^{m-n}$ $\qquad \dfrac{8^{3/7}}{8^{2/7}} = 8^{3/7-2/7} = 8^{1/7}$

e. $\left(\dfrac{a}{b}\right)^n = \dfrac{a^n}{b^n}$ $\qquad \left(\dfrac{2}{3}\right)^{2/5} = \dfrac{2^{2/5}}{3^{2/5}}$

f. $a^{-n} = \dfrac{1}{a^n}$ $\qquad 5^{-2/3} = \dfrac{1}{5^{2/3}}$

g. $\dfrac{1}{a^{-n}} = a^n$ $\qquad \dfrac{1}{4^{-2/3}} = 4^{2/3}$

h. $a^0 = 1$ $\qquad (2.5)^0 = 1$ ■

Exponential Expressions with Variable Bases

 (a, e)

EXAMPLE 5 Assume that all variables represent positive numbers. Write all answers without using negative exponents.

a. $$(25x)^{1/2} = 25^{1/2}x^{1/2} = 5x^{1/2}$$

b. $$\frac{(a^{2/3}b^{1/2})^6}{(y^2)^3} = \frac{(a^{2/3})^6(b^{1/2})^6}{y^6} = \frac{a^{12/3}b^{6/2}}{y^6} = \frac{a^4b^3}{y^6}$$

c. $$\frac{a^{5/3}a^{1/4}}{a^{1/6}} = a^{5/3+1/4-1/6} = a^{20/12+3/12-2/12} = a^{21/12} = a^{7/4}$$

d. $$\frac{(12ab^2)^{1/2}(12ab)^{1/2}}{ab^{3/2}} = \frac{12^{1/2}a^{1/2}b \cdot 12^{1/2}a^{1/2}b^{1/2}}{ab^{3/2}} = \frac{12^{1/2}12^{1/2}a^{1/2}a^{1/2}bb^{1/2}}{ab^{3/2}} = \frac{12^{1/2+1/2}a^{1/2+1/2}b^{1+1/2}}{ab^{3/2}} = \frac{12ab^{3/2}}{ab^{3/2}}$$

$= 12$ Divide out $ab^{3/2}$.

e. $\dfrac{a^{x/2}a^{x/3}}{a^{x/4}} = a^{x/2+x/3-x/4}$

$= a^{6x/12+4x/12-3x/12}$

$= a^{7x/12}$ ■

(b, d) **EXAMPLE 6** Assume that all variables represent positive numbers and do the operations. Write all answers without using negative exponents.

a. $a^{4/5}(a^{1/5} + a^{5/4}) = a^{4/5}a^{1/5} + a^{4/5}a^{5/4}$ Use the distributive property.

$= a^{4/5+1/5} + a^{4/5+5/4}$ Use the rule $a^m a^n = a^{m+n}$.

$= a^{5/5} + a^{16/20+25/20}$ Add the exponents.

$= a + a^{41/20}$ Simplify.

WARNING! Note that $a + a^{41/20} \neq a^{1+41/20}$.

b. $x^{1/2}(x^{-1/2} + x^{1/2}) = x^{1/2}x^{-1/2} + x^{1/2}x^{1/2}$ Use the distributive property.

$= x^{1/2-1/2} + x^{1/2+1/2}$ Use the rule $a^m a^n = a^{m+n}$.

$= x^0 + x^1$ Simplify.

$= 1 + x$

c. $(x^{2/3} + 1)(x^{2/3} - 1) = x^{4/3} - x^{2/3} + x^{2/3} - 1$ Use the FOIL method.

$= x^{4/3} - 1$

d. $(x^{1/2} + y^{1/2})^2 = (x^{1/2} + y^{1/2})(x^{1/2} + y^{1/2})$

$= x + 2x^{1/2}y^{1/2} + y$ Use the FOIL method. ■

ORALS *Simplify each expression.*

1. $49^{1/2}$ 7
2. $9^{1/2}$ 3
3. $27^{1/3}$ 3
4. $1^{1/4}$ 1
5. $4^{3/2}$ 8
6. $8^{2/3}$ 4
7. $\left(\dfrac{1}{4}\right)^{1/2}$ $\frac{1}{2}$
8. $\left(\dfrac{1}{4}\right)^{-1/2}$ 2
9. $(8x^3)^{1/3}$ $2x$
10. $(16x^8)^{1/4}$ $2x^2$

EXERCISE 7.1

In Exercises 1–44, simplify each expression if possible. Write answers without negative exponents.

1. $4^{1/2}$ 2
2. $25^{1/2}$ 5
3. $8^{1/3}$ 2
4. $125^{1/3}$ 5

5. $16^{1/4}$ 2 **6.** $625^{1/4}$ 5 **7.** $32^{1/5}$ 2 **8.** $0^{1/5}$ 0

9. $\left(\frac{1}{4}\right)^{1/2}$ $\frac{1}{2}$ **10.** $\left(\frac{1}{16}\right)^{1/2}$ $\frac{1}{4}$ **11.** $\left(\frac{1}{8}\right)^{1/3}$ $\frac{1}{2}$ **12.** $\left(\frac{1}{16}\right)^{1/4}$ $\frac{1}{2}$

13. $-16^{1/4}$ -2 **14.** $-125^{1/3}$ -5 **15.** $(-27)^{1/3}$ -3 **16.** $(-125)^{1/3}$ -5

17. $(-64)^{1/2}$ imaginary **18.** $(-216)^{1/2}$ imaginary **19.** $0^{1/3}$ 0 **20.** $25^{3/2}$ 125

21. $36^{3/2}$ 216 **22.** $27^{2/3}$ 9 **23.** $81^{3/4}$ 27 **24.** $100^{3/2}$ 1000

25. $144^{3/2}$ 1728 **26.** $1000^{2/3}$ 100 **27.** $\left(\frac{1}{8}\right)^{2/3}$ $\frac{1}{4}$ **28.** $\left(\frac{4}{9}\right)^{3/2}$ $\frac{8}{27}$

29. $\left(\frac{8}{27}\right)^{2/3}$ $\frac{4}{9}$ **30.** $\left(\frac{27}{64}\right)^{2/3}$ $\frac{9}{16}$ **31.** $4^{-3/2}$ $\frac{1}{8}$ **32.** $25^{-5/2}$ $\frac{1}{3125}$

33. $16^{-3/2}$ $\frac{1}{64}$ **34.** $81^{-3/2}$ $\frac{1}{729}$ **35.** $(-27)^{-2/3}$ $\frac{1}{9}$ **36.** $(-8)^{-2/3}$ $\frac{1}{4}$

37. $(-32)^{-2/5}$ $\frac{1}{4}$ **38.** $16^{-5/2}$ $\frac{1}{1024}$ **39.** $\left(\frac{1}{4}\right)^{-3/2}$ 8 **40.** $\left(\frac{4}{25}\right)^{-3/2}$ $\frac{125}{8}$

41. $\left(\frac{27}{8}\right)^{-4/3}$ $\frac{16}{81}$ **42.** $\left(\frac{25}{49}\right)^{-3/2}$ $\frac{343}{125}$ **43.** $\left(-\frac{8}{27}\right)^{-1/3}$ $-\frac{3}{2}$ **44.** $\left(\frac{16}{81}\right)^{-3/4}$ $\frac{27}{8}$

In Exercises 45–56, do the operations. Write answers without negative exponents.

45. $5^{3/7}5^{2/7}$ $5^{5/7}$ **46.** $4^{2/5}4^{2/5}$ $4^{4/5}$ **47.** $(4^{1/5})^3$ $4^{3/5}$ **48.** $(3^{1/3})^5$ $3^{5/3}$

49. $\frac{9^{4/5}}{9^{3/5}}$ $9^{1/5}$ **50.** $\frac{7^{2/3}}{7^{1/2}}$ $7^{1/6}$ **51.** $\frac{7^{1/2}}{7^0}$ $7^{1/2}$ **52.** $5^{1/3}5^{-5/3}$ $\frac{1}{5^{4/3}}$

53. $6^{-2/3}6^{-4/3}$ $\frac{1}{36}$ **54.** $\frac{3^{4/3}3^{1/3}}{3^{2/3}}$ 3 **55.** $\frac{2^{5/6}2^{1/3}}{2^{1/2}}$ $2^{2/3}$ **56.** $\frac{5^{1/3}5^{1/2}}{5^{1/3}}$ $5^{1/2}$

In Exercises 57–82, do the operations. Write answers without negative exponents. Assume that variables represent positive numbers.

57. $(8x)^{1/3}$ $2x^{1/3}$ **58.** $(32x^3)^{1/5}$ $2x^{3/5}$ **59.** $(27x^2y)^{2/3}$ $9x^{4/3}y^{2/3}$ **60.** $(81x^2y)^{3/4}$ $27x^{3/2}y^{3/4}$

61. $(25x^2y)^{3/2}$ $125x^3y^{3/2}$ **62.** $(125x^2y)^{4/3}$ $625x^{8/3}y^{4/3}$ **63.** $(4xy^3)^{-1/2}$ $\frac{1}{2x^{1/2}y^{3/2}}$ **64.** $(8x^2y)^{-1/3}$ $\frac{1}{2x^{2/3}y^{1/3}}$

65. $(27x^{3/4}y^{3/2})^{1/3}$ $3x^{1/4}y^{1/2}$ **66.** $(16x^{2/3}y^{2/5})^{1/2}$ $4x^{1/3}y^{1/5}$

67. $\frac{(4x^3y)^{1/2}}{(9xy)^{1/2}}$ $\frac{2x}{3}$ **68.** $\frac{(27x^3y)^{1/3}}{(8xy^2)^{2/3}}$ $\frac{3x^{1/3}}{4y}$ **69.** $\frac{(8xy^2)^{1/2}(2xy)^{1/2}}{xy^{1/2}}$ $4y$ **70.** $\frac{(4x^2y)^{1/3}(2xy^2)^{1/3}}{(xy^2)^{2/3}}$ $\frac{2x^{1/3}}{y^{1/3}}$

71. $(27x^{-3})^{-1/3}$ $\frac{1}{3}x$ **72.** $(16a^{-2})^{-1/2}$ $\frac{1}{4}a$ **73.** $(8x^{-3/2})^{2/3}$ $\frac{4}{x}$ **74.** $(a^{-1/4}c^3)^{-4/3}$ $\frac{a^{1/3}}{c^4}$

75. $x^{a/3}x^{a/2}$ $x^{5a/6}$ **76.** $x^{a/4}x^{a/3}$ $x^{7a/12}$ **77.** $\frac{x^{a/4}}{x^{a/3}}$ $\frac{1}{x^{a/12}}$ **78.** $\frac{x^{a/2}}{x^{a/3}}$ $x^{a/6}$

79. $\frac{x^{a/5}x^{2a/5}}{x^{4a/5}}$ $\frac{1}{x^{a/5}}$ **80.** $\frac{x^{1/b}x^{2/b}}{x^{4/b}}$ $\frac{1}{x^{1/b}}$ **81.** $\left(\frac{x^{2/a}x^{3/a}}{x^{1/b}}\right)^{ab}$ x^{5b-a} **82.** $\left(\frac{x^by^c}{z^b}\right)^{c/b}$ $\frac{x^cy^{c^2/b}}{z^c}$

In Exercises 83–108, do the multiplications. Write answers without using negative exponents. Assume that variables represent positive numbers.

83. $y^{1/3}(y^{2/3} + y^{5/3})$ $y + y^2$

84. $y^{2/5}(y^{-2/5} + y^{3/5})$ $1 + y$

85. $a^{2/3}(a^{4/3} + a^{-2/3})$ $a^2 + 1$

86. $a^{-2/3}(a^{5/3} + a^{2/3})$ $a + 1$

87. $x^{3/5}(x^{7/5} - x^{2/5} + 1)$ $x^2 - x + x^{3/5}$

88. $x^{4/3}(x^{2/3} + 3x^{5/3} - 4)$ $x^2 + 3x^3 - 4x^{4/3}$

89. $z^3y^{1/2}(z^{-3} + 1 - y^{-1/2})$ $y^{1/2} + z^3y^{1/2} - z^3$

90. $z^{-2/3}x^{1/3}(z^{5/3}x^{2/3} - z^{-1/3}x^{-1/3})$ $xz - \frac{1}{z}$

91. $(x^{1/2} + 2)(x^{1/2} - 2)$ $x - 4$

92. $(x^{1/2} + y^{1/2})(x^{1/2} - y^{1/2})$ $x - y$

93. $(x^{2/3} - x)(x^{2/3} + x)$ $x^{4/3} - x^2$

94. $(x^{1/2} + y^{1/2})(x^{-1/2} + y^{-1/2})$ $2 + \dfrac{x^{1/2}}{y^{1/2}} + \dfrac{y^{1/2}}{x^{1/2}}$

95. $(x^{-1/2} + y^{1/2})(x^{1/2} + y^{-1/2})$ $2 + x^{1/2}y^{1/2} + \dfrac{1}{x^{1/2}y^{1/2}}$

96. $(x^{1/3} + x^2)(x^{1/3} - x^2)$ $x^{2/3} - x^4$

97. $(x^{2/3} + y^{2/3})^2$ $x^{4/3} + 2x^{2/3}y^{2/3} + y^{4/3}$

98. $(a^{1/2} - b^{2/3})^2$ $a - 2a^{1/2}b^{2/3} + b^{4/3}$

99. $(a^{3/2} - b^{3/2})^2$ $a^3 - 2a^{3/2}b^{3/2} + b^3$

100. $(x^{-1/2} - x^{1/2})^2$ $\frac{1}{x} - 2 + x$

101. $(x^{1/3} + 1)(x^{2/3} - x^{1/3} + 1)$ $x + 1$

102. $(x^{1/3} - 2)(x^{2/3} + 2x^{1/3} + 4)$ $x - 8$

103. $\left[(3 - x^{1/3})(9 + x^{2/3} + 3x^{1/3})\right]^2$ $729 - 54x + x^2$

104. $\left[x^{-1/3}(x^{-2/3} + x^{1/3})\right]^2$ $\dfrac{1}{x^2} + \dfrac{2}{x} + 1$

105. $(x^{1/3} + y^{1/3})^2$ $x^{2/3} + 2x^{1/3}y^{1/3} + y^{2/3}$

106. $(a^{1/4} + b^{1/4})^2$ $a^{1/2} + 2a^{1/4}b^{1/4} + b^{1/2}$

107. $(p^{2/3} - q^{2/3})^2$ $p^{4/3} - 2p^{2/3}q^{2/3} + q^{4/3}$

108. $(c^{1/3} - d^{1/3})^2$ $c^{2/3} - 2c^{1/3}d^{1/3} + d^{2/3}$

Writing Exercises ■ *Write a paragraph using your own words.*

1. Explain how you would decide whether $a^{1/n}$ represents an imaginary number.

2. The expression $(a^{1/2} + b^{1/2})^2$ is not equal to $a + b$. Explain.

Something to Think About ■

1. The fraction $\frac{2}{4}$ is equal to $\frac{1}{2}$. Is $16^{2/4}$ equal to $16^{1/2}$? Explain.

2. How would you evaluate an expression with a mixed-number exponent? For example, what is $8^{1\frac{1}{3}}$? What is $25^{2\frac{1}{2}}$? Discuss.

Review Exercises ■ *Solve each inequality.*

1. $5x - 4 < 11$ $x < 3$

2. $2(3t - 5) \geq 8$ $t \geq 3$

3. $\dfrac{4}{5}(r - 3) > \dfrac{2}{3}(r + 2)$ $r > 28$

4. $-4 < 2x - 4 \leq 8$ $0 < x \leq 6$

5. How much water must be added to 5 pints of a 20% alcohol solution to dilute it to a 15% solution? $1\frac{2}{3}$ pints

6. A grocer bought some boxes of apples for \$70. However, 4 boxes were spoiled. The grocer sold the remaining boxes at a profit of \$2 each. How many boxes did the grocer sell if she managed to break even? 10

7.2 Radical Expressions

■ Radicals with Positive Radicands ■ Radicals with Negative Radicands ■ Converting between Radicals and Rational Exponents ■ Unrestricted Variables ■ Properties of Radicals

We have defined the symbol $a^{1/n}$ to be the nth root of a. We now introduce another symbol, called a **radical,** for the nth root of a.

Radicals If n is a natural number greater than 1 and if $a^{1/n}$ is a real number, then

$$\sqrt[n]{a} = a^{1/n}$$

Historical Note: Christoff Rudolff (1500–1545?) was the first to use the $\sqrt{\ }$ symbol.

In the **radical expression** $\sqrt[n]{a}$, a is called the **radicand**, n is called the **index** (or **order**) of the radical, and the symbol $\sqrt{\ }$ is called a **radical sign**.

If the order of a radical is 2, the radical is called a **square root**, and we do not write the index. Thus,

$$\sqrt{a} \quad \text{means} \quad \sqrt[2]{a}$$

Radicals with Positive Radicands

We now restate the definition for the nth root of a *positive number*, using radical notation.

Definition of $\sqrt[n]{a}$ If n is a natural number greater than 1 and $a > 0$, then $\sqrt[n]{a}$ is the positive number whose nth power is a. In symbols,

$$\left(\sqrt[n]{a}\right)^n = a$$

The positive number $\sqrt[n]{a}$ is called the **principal nth root** of a.

If 2 is substituted for n in the equation $\left(\sqrt[n]{a}\right)^n = a$, we have

$$\left(\sqrt[2]{a}\right)^2 = \left(\sqrt{a}\right)^2 = \sqrt{a}\,\sqrt{a} = a$$

Thus, if a can be factored into two equal factors, then either of those factors is a square root of a.

Definition of $\sqrt[n]{0}$ $\sqrt[n]{0} = 0$

(a, c, d, e) **EXAMPLE 1**

a. $\sqrt{36} = 6$ because $6^2 = 36$ — Read $\sqrt{36}$ as "the square root of 36."

b. $\sqrt[3]{8} = 2$ because $2^3 = 8$ — Read $\sqrt[3]{8}$ as "the cube root of 8."

c. $\sqrt[4]{81} = 3$ because $3^4 = 81$ — Read $\sqrt[4]{81}$ as "the fourth root of 81."

d. $\sqrt[3]{\dfrac{8}{27}} = \dfrac{2}{3}$ because $\left(\dfrac{2}{3}\right)^3 = \dfrac{8}{27}$ — Read $\sqrt[3]{\dfrac{8}{27}}$ as "the cube root of $\dfrac{8}{27}$."

e. $-\sqrt[5]{32} = -\left(\sqrt[5]{32}\right) = -2$ — Read $\sqrt[5]{32}$ as "the fifth root of 32."

f. $\sqrt[11]{0} = 0$ because $0^{11} = 0$ — Read $\sqrt[11]{0}$ as "the eleventh root of 0." ■

Historical Note: The Babylonians (2000 B.C.) approximated $\sqrt{2}$ to be $\frac{17}{12}$.

Radicals with Negative Radicands

We can find the nth root of a negative number when the index of the radical is odd. For example,

$$\sqrt[3]{-8} = -2 \qquad \text{because} \qquad (-2)^3 = -8$$

However, if the index is even, then the nth root of a negative number is an imaginary number. For example,

$\sqrt[4]{-81}$ is an imaginary number, because no real number raised to the 4th power is -81.

The following chart summarizes the definitions concerning the radical expression $\sqrt[n]{a}$.

Radicals If n is a natural number greater than 1 and a is a real number, then

	n is odd	n is even
If $a > 0$	$\left(\sqrt[n]{a}\right)^n = a$	$\left(\sqrt[n]{a}\right)^n = a$
If $a = 0$	$\left(\sqrt[n]{a}\right)^n = 0$	$\left(\sqrt[n]{a}\right)^n = 0$
If $a < 0$	$\left(\sqrt[n]{a}\right)^n = a$	$\sqrt[n]{a}$ is an imaginary number.

Converting between Radicals and Rational Exponents

In the previous section, we saw that $x^{m/n} = (x^m)^{1/n} = (x^{1/n})^m$. The same fact stated in radical notation is

1. $x^{m/n} = \sqrt[n]{x^m} = \left(\sqrt[n]{x}\right)^m$ **provided $\sqrt[n]{x}$ is a real number**

Thus, the nth root of x^m is the same as the mth power of the nth root of x.

Because of Equation 1, all exponential expressions containing fractional exponents can be written in radical notation.

(a, c, e) **EXAMPLE 2** Use the equation $x^{m/n} = \sqrt[n]{x^m}$ to write each exponential expression as a radical.

a. $5^{1/2} = \sqrt{5}$ **b.** $8^{2/3} = \sqrt[3]{8^2}$

c. $5x^{3/2} = 5\sqrt{x^3}$ **d.** $(5a)^{3/4} = \sqrt[4]{(5a)^3} = \sqrt[4]{125a^3}$

e. $y^{-3/2} = \dfrac{1}{y^{3/2}} = \dfrac{1}{\sqrt{y^3}}$ **f.** $(x^2 - y^2)^{1/2} = \sqrt{x^2 - y^2}$ ■

WARNING! In part **f**, the expression $\sqrt{x^2 - y^2}$ does not equal $\sqrt{x^2} - \sqrt{y^2}$. To see why, we substitute 5 for x and 4 for y and get different answers:

$$\begin{aligned} \sqrt{x^2 - y^2} &= \sqrt{5^2 - 4^2} \\ &= \sqrt{25 - 16} \\ &= \sqrt{9} \\ &= 3 \end{aligned} \qquad \text{but} \qquad \begin{aligned} \sqrt{x^2} - \sqrt{y^2} &= \sqrt{5^2} - \sqrt{4^2} \\ &= \sqrt{25} - \sqrt{16} \\ &= 5 - 4 \\ &= 1 \end{aligned}$$

Thus, $\sqrt{x^2 - y^2} \neq \sqrt{x^2} - \sqrt{y^2}$.

In Example 3, we show that radical expressions can be written as exponential expressions with fractional exponents.

(b, c, e) **EXAMPLE 3** Use the equation $x^{m/n} = \sqrt[n]{x^m}$ to write each radical as an exponential expression.

a. $\sqrt{5} = 5^{1/2}$ **b.** $\sqrt[3]{7^4} = 7^{4/3}$

c. $\sqrt[5]{6^{-3}} = 6^{-3/5}$ **d.** $\sqrt{5a} = (5a)^{1/2}$

e. $\dfrac{1}{\sqrt{z^5}} = \dfrac{1}{z^{5/2}} = z^{-5/2}$ **f.** $\sqrt{x^2 + y^2} = (x^2 + y^2)^{1/2}$ ■

 (a) **EXAMPLE 4** Use Equation 1 to simplify **a.** $\sqrt[3]{8^2}$ and **b.** $\sqrt{9^3}$ in two ways.

Solution **a.** $\sqrt[3]{8^2} = \sqrt[3]{64} = 4$ or $\sqrt[3]{8^2} = \left(\sqrt[3]{8}\right)^2 = 2^2 = 4$

b. $\sqrt{9^3} = \sqrt{729} = 27$ or $\sqrt{9^3} = \left(\sqrt{9}\right)^3 = 3^3 = 27$ ■

Unrestricted Variables

All positive numbers have two square roots, one positive and one negative. For example, a square root of 100 is any number whose square is 100. There are two such numbers; 10 and -10. However, the radical $\sqrt{100}$ represents only the positive square root of 100. Thus,

$$\sqrt{100} = 10$$

The negative square root of 100 is represented by $-\sqrt{100}$. Thus,

$$-\sqrt{100} = -10$$

A similar situation occurs when the index n is any even natural number. The symbol $\sqrt[4]{16}$, for example, means the positive fourth root of 16. Thus,

$$\sqrt[4]{16} = 2 \qquad \text{Because } 2^4 = 16.$$

The symbol $-\sqrt[4]{16}$ represents the negative fourth root of 16.

$$-\sqrt[4]{16} = -2$$

In general, a^2 has two square roots: a and $-a$. One of these numbers is positive and the other is negative. If a might be negative, we must enclose it within absolute value symbols to guarantee that $\sqrt{a^2}$ is positive. Thus, if a is unrestricted,

$$\sqrt{a^2} = |a|$$

For example, if $x = -3$, we have

$$\begin{aligned}\sqrt{(-3)^2} &= |-3| \\ &= 3\end{aligned}$$

When the index n of a radical is an odd natural number, there is only one nth root of a, and we do not need to use absolute value symbols. For example,

$$\sqrt[3]{8} = 2 \qquad \text{and} \qquad \sqrt[3]{-8} = -2$$

In general, we have the following rules.

Definition of $\sqrt[n]{a^n}$ If n is an even natural number, then $\sqrt[n]{a^n} = |a|$.
If n is an odd natural number, then $\sqrt[n]{a^n} = a$.

(a, d, e) **EXAMPLE 5**

a. $\sqrt{x^2} = |x|$ Because $|x|^2 = x^2$ and $|x| \geq 0$.

b. $\sqrt[3]{x^3} = x$ Because $x^3 = x^3$.

c. $\sqrt[4]{a^4} = |a|$ Because $|a|^4 = a^4$ and $|a| \geq 0$.

d. $\sqrt[5]{-32x^5} = -2x$ Because $(-2x)^5 = -32x^5$.

e. $\sqrt{x^4} = x^2$ Because $(x^2)^2 = x^4$. No absolute value signs are needed, since $x^2 \geq 0$. ∎

Properties of Radicals

Many properties of exponents have counterparts in radical notation. For example, because $a^{1/n}b^{1/n} = (ab)^{1/n}$, we have

$$\sqrt[n]{a}\sqrt[n]{b} = \sqrt[n]{ab}$$

As long as all radicals represent real numbers, *the product of the nth roots of two numbers is equal to the nth root of their product.*

Property 1 of Radicals If $\sqrt[n]{a}$ and $\sqrt[n]{b}$ are real numbers, then

$$\sqrt[n]{ab} = \sqrt[n]{a}\sqrt[n]{b}$$

WARNING! Property 1 of radicals applies to the nth root of the product of two numbers. There is no such property for sums. For example,

$$\sqrt{9+4} \neq \sqrt{9} + \sqrt{4}$$

since

$$\sqrt{9+4} = \sqrt{13} \qquad \text{but} \qquad \sqrt{9} + \sqrt{4} = 3 + 2 = 5$$

Numbers like 1, 4, 9, 16, 25, and 36 that are squares of integers are called **perfect squares**. Numbers like 1, 8, 27, 64, 125, and 216 that are cubes of integers are called **perfect cubes**.

There are also perfect fourth powers, perfect fifth powers, and so on. We can use the concept of perfect powers and Property 1 of radicals to simplify many radical expressions.

 (a, c) **EXAMPLE 6** Simplify **a.** $\sqrt{12}$, **b.** $\sqrt{98}$, and **c.** $\sqrt[3]{54}$.

Another Approach: Use a factoring tree.

a.

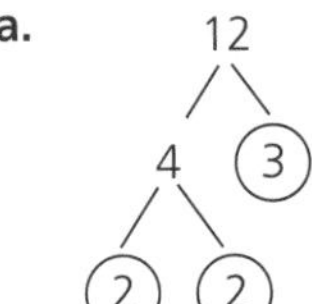

b. 98 → 2, 49; 49 → 7, 7

Solution **a.** We factor 12 so that one factor is the largest perfect square that divides 12. Since this largest perfect square is 4, we write 12 as $4 \cdot 3$, use Property 1 of radicals, and simplify.

$$\begin{aligned}\sqrt{12} &= \sqrt{4 \cdot 3}\\ &= \sqrt{4}\sqrt{3}\\ &= 2\sqrt{3}\end{aligned}$$

b. The largest perfect square factor of 98 is 49. Thus,

$$\begin{aligned}\sqrt{98} &= \sqrt{49 \cdot 2}\\ &= \sqrt{49}\sqrt{2}\\ &= 7\sqrt{2}\end{aligned}$$

c.

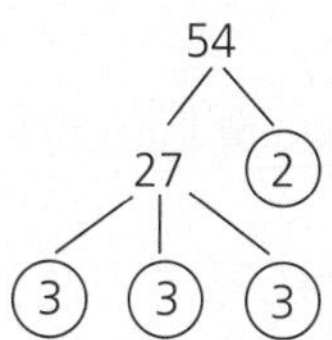

In parts a and b, each factor pair produces a single number outside the radical. In part c, each factor triple produces a single number outside the radical.

c. The largest perfect cube factor of 54 is 27. Thus,

$$\begin{aligned}\sqrt[3]{54} &= \sqrt[3]{27 \cdot 2} \\ &= \sqrt[3]{27}\sqrt[3]{2} \\ &= 3\sqrt[3]{2}\end{aligned}$$

■

Another property involves the quotient of two radicals. Because

$$\frac{a^{1/n}}{b^{1/n}} = \left(\frac{a}{b}\right)^{1/n}$$

it follows that

$$\frac{\sqrt[n]{a}}{\sqrt[n]{b}} = \sqrt[n]{\frac{a}{b}} \qquad (b \neq 0)$$

As long as all radicals represent real numbers, *the quotient of the nth roots of two numbers is equal to the nth root of their quotient.*

Property 2 of Radicals If $\sqrt[n]{a}$ and $\sqrt[n]{b}$ are real numbers and $b \neq 0$, then

$$\sqrt[n]{\frac{a}{b}} = \frac{\sqrt[n]{a}}{\sqrt[n]{b}}$$

 (b)

EXAMPLE 7 Simplify **a.** $\dfrac{\sqrt{72}}{\sqrt{2}}$ and **b.** $\sqrt[3]{\dfrac{9}{27}}$.

Solution **a.** We can write the quotient of the square roots as the square root of the quotient and simplify.

$$\begin{aligned}\frac{\sqrt{72}}{\sqrt{2}} &= \sqrt{\frac{72}{2}} && \text{Use Property 2 of radicals.} \\ &= \sqrt{36} \\ &= 6\end{aligned}$$

b. We can write the cube root of the quotient as the quotient of two cube roots.

$$\begin{aligned}\sqrt[3]{\frac{9}{27}} &= \frac{\sqrt[3]{9}}{\sqrt[3]{27}} && \text{Use Property 2 of radicals.} \\ &= \frac{\sqrt[3]{9}}{3}\end{aligned}$$

■

We will use Properties 1 and 2 of radicals to simplify many radical expressions. A radical expression is said to be in simplest form only if each of the following statements is true.

Simplified Form of a Radical

A radical expression is in simplest form when

1. No radicals appear in the denominator of a fraction.
2. The radicand contains no fractions or negative numbers.
3. Each factor in the radicand appears to a power that is less than the index of the radical.

(a, d)

EXAMPLE 8 Simplify each of the following expressions (all variables represent positive numbers):

a. $\sqrt{128a^5}$, **b.** $\sqrt[3]{24x^5}$, **c.** $\dfrac{\sqrt{45xy^2}}{\sqrt{5x}}$, and **d.** $\dfrac{\sqrt[3]{-432x^5}}{\sqrt[3]{8x}}$.

Solution **a.** We write $128a^5$ as $64a^4 \cdot 2a$ and use Property 1 of radicals.

$$\sqrt{128a^5} = \sqrt{64a^4 \cdot 2a} \qquad 64a^4 \text{ is the largest perfect square that divides } 128a^5.$$
$$= \sqrt{64a^4}\sqrt{2a} \qquad \text{Use Property 1 of radicals.}$$
$$= 8a^2\sqrt{2a} \qquad \sqrt{64a^4} = 8a^2.$$

b. We write $24x^5$ as $8x^3 \cdot 3x^2$ and use Property 1 of radicals.

$$\sqrt[3]{24x^5} = \sqrt[3]{8x^3 \cdot 3x^2} \qquad 8x^3 \text{ is the largest perfect cube that divides } 24x^5.$$
$$= \sqrt[3]{8x^3}\sqrt[3]{3x^2} \qquad \text{Use Property 1 of radicals.}$$
$$= 2x\sqrt[3]{3x^2} \qquad \sqrt{8x^3} = 2x.$$

c. We can write the quotient of the square roots as the square root of a quotient.

$$\frac{\sqrt{45xy^2}}{\sqrt{5x}} = \sqrt{\frac{45xy^2}{5x}} \qquad \text{Use Property 2 of radicals.}$$
$$= \sqrt{9y^2} \qquad \text{Simplify the fraction.}$$
$$= 3y$$

d. We can write the quotient of the cube roots as the cube root of a quotient.

$$\frac{\sqrt[3]{-432x^5}}{\sqrt[3]{8x}} = \sqrt[3]{\frac{-432x^5}{8x}} \qquad \text{Use Property 2 of radicals.}$$
$$= \sqrt[3]{-54x^4} \qquad \text{Simplify the fraction.}$$
$$= \sqrt[3]{-27x^3 \cdot 2x}$$
$$= \sqrt[3]{-27x^3}\sqrt[3]{2x}$$
$$= -3x\sqrt[3]{2x}$$

■

To simplify more complicated radicals, we can use the prime factorization of the radicand to find its perfect square factors. For example, to simplify the radical $\sqrt{3168x^5y^7}$, we first find the prime factorization of $3168x^5y^7$.

$$3168x^5y^7 = 2^5 \cdot 3^2 \cdot 11 \cdot x^5 \cdot y^7$$

Then we have

$$\begin{aligned}\sqrt{3168x^5y^7} &= \sqrt{2^4 \cdot 3^2 \cdot x^4 \cdot y^6 \cdot 2 \cdot 11 \cdot x \cdot y} \\ &= \sqrt{2^4 \cdot 3^2 \cdot x^4 \cdot y^6}\sqrt{2 \cdot 11 \cdot x \cdot y} \\ &= 2^2 \cdot 3x^2y^3\sqrt{22xy} \\ &= 12x^2y^3\sqrt{22xy}\end{aligned}$$

EXAMPLE 9 Simplify $\sqrt{x^2 + 6x + 9}$. Assume that x is positive.

Solution After factoring the radicand, the radical can be written in the form $\sqrt{(x + 3)^2}$. Because $x + 3$ is positive, it follows that $\sqrt{(x + 3)^2} = x + 3$. Thus,

$$\begin{aligned}\sqrt{x^2 + 6x + 9} &= \sqrt{(x + 3)^2} \\ &= x + 3\end{aligned}$$

■

ORALS *Simplify each radical. (Assume $x \geq 0$.)*

1. $\sqrt{9}$ 3
2. $\sqrt{16}$ 4
3. $\sqrt[3]{8}$ 2
4. $\sqrt[5]{32}$ 2
5. $\sqrt{64x^2}$ $8x$
6. $\sqrt[3]{-8x^3}$ $-2x$
7. $\dfrac{\sqrt{8}}{\sqrt{2}}$ 2
8. $\dfrac{\sqrt{27}}{\sqrt{3}}$ 3

EXERCISE 7.2

In Exercises 1–24, simplify each radical.

1. $\sqrt{121}$ 11
2. $\sqrt{144}$ 12
3. $-\sqrt{64}$ -8
4. $-\sqrt{1}$ -1
5. $\sqrt[3]{1}$ 1
6. $\sqrt[3]{-8}$ -2
7. $\sqrt[3]{-125}$ -5
8. $\sqrt[3]{512}$ 8
9. $\sqrt[4]{81}$ 3
10. $\sqrt[6]{64}$ 2
11. $-\sqrt[5]{243}$ -3
12. $-\sqrt[4]{625}$ -5
13. $\sqrt[5]{-32}$ -2
14. $\sqrt[3]{-512}$ -8
15. $\sqrt{9^3}$ 27
16. $\sqrt[3]{8^4}$ 16
17. $\sqrt[4]{16^3}$ 8
18. $\sqrt[3]{(-27)^2}$ 9
19. $\sqrt[5]{-32^2}$ -4
20. $\sqrt[4]{\left(\frac{1}{16}\right)^3}$ $\frac{1}{8}$
21. $\sqrt[3]{\left(-\frac{1}{8}\right)^2}$ $\frac{1}{4}$
22. $\sqrt[3]{\left(\frac{8}{27}\right)^2}$ $\frac{4}{9}$
23. $\sqrt[5]{243^{-2}}$ $\frac{1}{9}$
24. $\sqrt[5]{32^{-3}}$ $\frac{1}{8}$

In Exercises 25–36, change each exponential expression to a radical. ***Do not simplify.***

25. $7^{1/3}$ $\sqrt[3]{7}$ **26.** $6^{3/4}$ $\sqrt[4]{6^3}$ **27.** $5a^{3/2}$ $5\sqrt{a^3}$ **28.** $(5a)^{3/2}$ $\sqrt{125a^3}$

29. $x^{-2/5}$ $\dfrac{1}{\sqrt[5]{x^2}}$ **30.** $(ab)^{-1/4}$ $\dfrac{1}{\sqrt[4]{ab}}$ **31.** $(8p)^{3/8}$ $\sqrt[8]{512p^3}$ **32.** $8p^{3/8}$ $8\sqrt[8]{p^3}$

33. $(x^2 + y^2)^{1/2}$ $\sqrt{x^2 + y^2}$ **34.** $(x^3 + y^3)^{1/2}$ $\sqrt{x^3 + y^3}$

35. $(x^2 - y^2)^{3/4}$ $\sqrt[4]{(x^2 - y^2)^3}$ **36.** $(x^3 - y^3)^{-2/3}$ $\dfrac{1}{\sqrt[3]{(x^3 - y^3)^2}}$

In Exercises 37–48, change each radical to an exponential expression.

37. $\sqrt{11}$ $11^{1/2}$ **38.** $\sqrt[3]{12}$ $12^{1/3}$ **39.** $\sqrt[5]{5^2}$ $5^{2/5}$ **40.** $\sqrt[6]{7^5}$ $7^{5/6}$

41. $\sqrt{3a}$ $(3a)^{1/2}$ **42.** $3\sqrt{a}$ $3a^{1/2}$ **43.** $\dfrac{1}{\sqrt[3]{4z}}$ $(4z)^{-1/3}$ **44.** $\dfrac{1}{\sqrt{7^3a^3}}$ $(7a)^{-3/2}$

45. $\dfrac{2}{\sqrt[3]{x^4}}$ $2x^{-4/3}$ **46.** $\dfrac{3}{\sqrt[3]{2q}}$ $3(2q)^{-1/3}$ **47.** $\sqrt{a^2 - b^2}$ $(a^2 - b^2)^{1/2}$ **48.** $\sqrt[3]{x^2 + y^2}$ $(x^2 + y^2)^{1/3}$

In Exercises 49–60, simplify each radical. Use absolute value symbols where necessary.

49. $\sqrt{4x^2}$ $2|x|$ **50.** $\sqrt{9a^2}$ $3|a|$ **51.** $\sqrt[3]{8a^3}$ $2a$ **52.** $\sqrt[3]{27x^6}$ $3x^2$

53. $\sqrt[4]{x^4}$ $|x|$ **54.** $\sqrt[4]{x^8}$ x^2 **55.** $\sqrt[4]{x^{12}}$ $|x^3|$ **56.** $\sqrt{x^8}$ x^4

57. $\sqrt[5]{-x^5}$ $-x$ **58.** $\sqrt[3]{-x^6}$ $-x^2$ **59.** $\sqrt[3]{-27a^6}$ $-3a^2$ **60.** $\sqrt[5]{-32x^5}$ $-2x$

In Exercises 61–112, simplify each expression. All variables represent positive numbers.

61. $\sqrt{5}\sqrt{5}$ 5 **62.** $\sqrt{11}\sqrt{11}$ 11 **63.** $\sqrt[3]{2}\sqrt[3]{2}\sqrt[3]{2}$ 2 **64.** $\sqrt[3]{4}\sqrt[3]{4}\sqrt[3]{4}$ 4

65. $\sqrt{t}\sqrt{t}$ t **66.** $-\sqrt{z}\sqrt{z}$ $-z$ **67.** $\sqrt{20}$ $2\sqrt{5}$ **68.** $\sqrt{8}$ $2\sqrt{2}$

69. $\sqrt{24}$ $2\sqrt{6}$ **70.** $\sqrt{50}$ $5\sqrt{2}$ **71.** $\sqrt{200}$ $10\sqrt{2}$ **72.** $\sqrt{250}$ $5\sqrt{10}$

73. $-\sqrt{50x^2}$ $-5x\sqrt{2}$ **74.** $-\sqrt{75a^2}$ $-5a\sqrt{3}$ **75.** $\sqrt{32b}$ $4\sqrt{2b}$ **76.** $-\sqrt{80c}$ $-4\sqrt{5c}$

77. $-\sqrt{112a^3}$ $-4a\sqrt{7a}$ **78.** $\sqrt{147a^5}$ $7a^2\sqrt{3a}$ **79.** $\sqrt{175a^2b^3}$ $5ab\sqrt{7b}$ **80.** $-\sqrt{128a^3b^5}$ $-8ab^2\sqrt{2ab}$

81. $-\sqrt{300xy}$ $-10\sqrt{3xy}$ **82.** $\sqrt{200x^2y}$ $10x\sqrt{2y}$ **83.** $\sqrt[3]{81}$ $3\sqrt[3]{3}$ **84.** $\sqrt[3]{-72}$ $-2\sqrt[3]{9}$

85. $\sqrt[3]{-80}$ $-2\sqrt[3]{10}$ **86.** $\sqrt[3]{270}$ $3\sqrt[3]{10}$ **87.** $\sqrt[3]{-54x^6}$ $-3x^2\sqrt[3]{2}$ **88.** $-\sqrt[3]{-81a^3}$ $3a\sqrt[3]{3}$

89. $-\sqrt[3]{16x^{12}y^3}$ $-2x^4y\sqrt[3]{2}$ **90.** $\sqrt[3]{40a^3b^6}$ $2ab^2\sqrt[3]{5}$ **91.** $\sqrt[4]{32x^{12}y^4}$ $2x^3y\sqrt[4]{2}$ **92.** $-\sqrt[4]{243x^{20}y^8}$ $-3x^5y^2\sqrt[4]{3}$

93. $-\sqrt[5]{64x^{10}y^5}$ $-2x^2y\sqrt[5]{2}$ **94.** $\sqrt[5]{486a^{25}b^{20}}$ $3a^5b^4\sqrt[5]{2}$ **95.** $-\dfrac{\sqrt{500}}{\sqrt{5}}$ -10 **96.** $\dfrac{\sqrt{128}}{\sqrt{2}}$ 8

97. $\dfrac{\sqrt{98}}{\sqrt{2}}$ 7 **98.** $-\dfrac{\sqrt{75}}{\sqrt{3}}$ -5 **99.** $\dfrac{\sqrt{180ab^4}}{\sqrt{5ab^2}}$ $6b$ **100.** $\dfrac{\sqrt{112xy^2}}{\sqrt{7xy}}$ $4\sqrt{y}$

101. $\dfrac{\sqrt{128x^7}}{\sqrt{8x^2}}$ $4x^2\sqrt{x}$

102. $\dfrac{\sqrt{245a^7}}{\sqrt{5a}}$ $7a^3$

103. $\dfrac{\sqrt[3]{48}}{\sqrt[3]{6}}$ 2

104. $\dfrac{\sqrt[3]{64}}{\sqrt[3]{8}}$ 2

105. $\dfrac{\sqrt[3]{189a^4}}{\sqrt[3]{7a}}$ $3a$

106. $\dfrac{\sqrt[3]{243x^2}}{\sqrt[3]{9x}}$ $3\sqrt[3]{x}$

107. $\sqrt{\dfrac{7}{9}}$ $\dfrac{\sqrt{7}}{3}$

108. $\sqrt{\dfrac{3}{4}}$ $\dfrac{\sqrt{3}}{2}$

109. $\sqrt{\dfrac{3x}{48x^3}}$ $\frac{1}{4x}$

110. $\sqrt{\dfrac{2a^3}{128a^5}}$ $\frac{1}{8a}$

111. $\sqrt{\dfrac{10abc^2}{98a^3b^5}}$ $\dfrac{c\sqrt{5}}{7ab^2}$

112. $\sqrt{\dfrac{14tu^2}{128t^3}}$ $\dfrac{u\sqrt{7}}{8t}$

In Exercises 113–124, simplify each expression. All variables represent positive numbers.

113. $\sqrt[4]{x^5y^3z^8}$ $xz^2\sqrt[4]{xy^3}$

114. $\sqrt[4]{x^7y^5z^{13}}$ $xyz^3\sqrt[4]{x^3yz}$

115. $\sqrt[3]{-16a^5b^3c^2}$ $-2ab\sqrt[3]{2a^2c^2}$

116. $\sqrt[3]{-54a^{11}b^{13}c^{16}}$ $-3a^3b^4c^5\sqrt[3]{2a^2bc}$

117. $\sqrt[5]{-64a^6b^6c^{11}}$ $-2abc^2\sqrt[5]{2abc}$

118. $\sqrt[5]{486x^7y^8z^9}$ $3xyz\sqrt[5]{2x^2y^3z^4}$

119. $\sqrt{x^2 + 2x + 1}$ $x + 1$

120. $\sqrt{x^2 + 4x + 4}$ $x + 2$

121. $\sqrt{x^2 + 8x + 16}$ $x + 4$

122. $\sqrt{x^2 + 6x + 9}$ $x + 3$

123. $\sqrt{x^4 + 6x^2 + 9}$ $x^2 + 3$

124. $\sqrt{x^4 + 10x^2 + 25}$ $x^2 + 5$

Writing Exercises ■ *Write a paragraph using your own words.*

1. If x is any real number, then $\sqrt{x^2} = x$ is not correct. Explain.

2. If x is any real number, then $\sqrt[3]{x^3} = |x|$ is not correct. Explain.

Something to Think About ■ **1.** Explain why $\sqrt[4]{9} = \sqrt{3}$.

2. Simplify $\sqrt[6]{36}$. $\sqrt[3]{6}$

Review Exercises ■ *Simplify each fraction.*

1. $\dfrac{x^2 + 7x + 12}{x^2 - 16}$ $\dfrac{x + 3}{x - 4}$

2. $\dfrac{a^3 - b^3}{b^2 - a^2}$ $\dfrac{-(a^2 + ab + b^2)}{a + b}$

Do the operations.

3. $\dfrac{x^2 - x - 6}{x^2 - 2x - 3} \cdot \dfrac{x^2 - 1}{x^2 + x - 2}$ 1

4. $\dfrac{x^2 - 3x - 4}{x^2 - 5x + 6} \div \dfrac{x^2 - 2x - 3}{x^2 - x - 2}$ $\dfrac{(x - 4)(x + 1)}{(x - 3)^2}$

5. $\dfrac{3}{m + 1} + \dfrac{3m}{m - 1}$ $\dfrac{3(m^2 + 2m - 1)}{(m + 1)(m - 1)}$

6. $\dfrac{2x + 3}{3x - 1} - \dfrac{x - 4}{2x + 1}$ $\dfrac{x^2 + 21x - 1}{(3x - 1)(2x + 1)}$

7.3 Adding and Subtracting Radical Expressions

Radical expressions with the same index and the same radicand are called **like** or **similar radicals.** For example, $3\sqrt{2}$ and $2\sqrt{2}$ are like radicals. However,

$3\sqrt{5}$ and $4\sqrt{2}$ are not like radicals, because the radicands are different.

$3\sqrt{5}$ and $2\sqrt[3]{5}$ are not like radicals, because the indexes are different.

Because of the distributive property, we can combine like radicals. To simplify the expression $3\sqrt{2} + 2\sqrt{2}$, we use the distributive property to factor out $\sqrt{2}$ and simplify.

Teaching Tip: Just as 3 sevenths + 2 sevenths = 5 sevenths, we have 3 square roots of 2 + 2 square roots of 2 = 5 square roots of 2

$$\begin{aligned} 3\sqrt{2} + 2\sqrt{2} &= (3 + 2)\sqrt{2} \\ &= 5\sqrt{2} \end{aligned}$$

Radical expressions with the same index but different radicands can often be written as like radicals. For example, to simplify the expression $\sqrt{27} - \sqrt{12}$, we simplify both radicals and then combine the like radicals.

$$\begin{aligned} \sqrt{27} - \sqrt{12} &= \sqrt{9 \cdot 3} - \sqrt{4 \cdot 3} \\ &= \sqrt{9}\sqrt{3} - \sqrt{4}\sqrt{3} \\ &= 3\sqrt{3} - 2\sqrt{3} \\ &= (3 - 2)\sqrt{3} \\ &= \sqrt{3} \end{aligned}$$

As the previous examples suggest, we can use the following rule to add or subtract radicals.

Adding and Subtracting Radicals To add or subtract radicals, simplify each radical and combine all like radicals. To combine like radicals, add their coefficients and keep the common radical.

EXAMPLE 1 Simplify $2\sqrt{12} - 3\sqrt{48} + 3\sqrt{3}$.

Solution We simplify each radical separately and combine like radicals.

$$\begin{aligned} 2\sqrt{12} - 3\sqrt{48} + 3\sqrt{3} &= 2\sqrt{4 \cdot 3} - 3\sqrt{16 \cdot 3} + 3\sqrt{3} \\ &= 2\sqrt{4}\sqrt{3} - 3\sqrt{16}\sqrt{3} + 3\sqrt{3} \\ &= 2(2)\sqrt{3} - 3(4)\sqrt{3} + 3\sqrt{3} \\ &= 4\sqrt{3} - 12\sqrt{3} + 3\sqrt{3} \\ &= (4 - 12 + 3)\sqrt{3} \\ &= -5\sqrt{3} \end{aligned}$$

■

EXAMPLE 2 Simplify $\sqrt[3]{16} - \sqrt[3]{54} + \sqrt[3]{24}$.

Solution We simplify each radical separately and proceed as follows:

$$\begin{aligned}\sqrt[3]{16} - \sqrt[3]{54} + \sqrt[3]{24} &= \sqrt[3]{8 \cdot 2} - \sqrt[3]{27 \cdot 2} + \sqrt[3]{8 \cdot 3}\\ &= \sqrt[3]{8}\sqrt[3]{2} - \sqrt[3]{27}\sqrt[3]{2} + \sqrt[3]{8}\sqrt[3]{3}\\ &= 2\sqrt[3]{2} - 3\sqrt[3]{2} + 2\sqrt[3]{3}\\ &= -\sqrt[3]{2} + 2\sqrt[3]{3}\end{aligned}$$

WARNING! We cannot combine $-\sqrt[3]{2}$ and $2\sqrt[3]{3}$, because the radicals have different radicands. ■

EXAMPLE 3 Simplify $\sqrt[3]{16x^4} + \sqrt[3]{54x^4} - \sqrt[3]{-128x^4}$ $(x > 0)$.

Solution We simplify each radical expression separately, factor out $\sqrt[3]{2x}$, and simplify.

$$\begin{aligned}\sqrt[3]{16x^4} + \sqrt[3]{54x^4} - \sqrt[3]{-128x^4} &= \sqrt[3]{8x^3 \cdot 2x} + \sqrt[3]{27x^3 \cdot 2x} - \sqrt[3]{-64x^3 \cdot 2x}\\ &= \sqrt[3]{8x^3}\sqrt[3]{2x} + \sqrt[3]{27x^3}\sqrt[3]{2x} - \sqrt[3]{-64x^3}\sqrt[3]{2x}\\ &= 2x\sqrt[3]{2x} + 3x\sqrt[3]{2x} + 4x\sqrt[3]{2x}\\ &= (2x + 3x + 4x)\sqrt[3]{2x}\\ &= 9x\sqrt[3]{2x}\end{aligned}$$

■

EXAMPLE 4 Simplify $3\sqrt[3]{24} - 8\sqrt[3]{81} - 2\sqrt[3]{\dfrac{3}{27}}$.

Solution We simplify each radical expression separately and combine like radicals.

$$\begin{aligned}3\sqrt[3]{24} - 8\sqrt[3]{81} - 2\sqrt[3]{\frac{3}{27}} &= 3\sqrt[3]{8 \cdot 3} - 8\sqrt[3]{27 \cdot 3} - 2\frac{\sqrt[3]{3}}{\sqrt[3]{27}}\\ &= 3\sqrt[3]{8}\sqrt[3]{3} - 8\sqrt[3]{27}\sqrt[3]{3} - 2\frac{\sqrt[3]{3}}{3}\\ &= 3(2)\sqrt[3]{3} - 8(3)\sqrt[3]{3} - \frac{2}{3}\sqrt[3]{3}\\ &= 6\sqrt[3]{3} - 24\sqrt[3]{3} - \frac{2}{3}\sqrt[3]{3}\\ &= \frac{18}{3}\sqrt[3]{3} - \frac{72}{3}\sqrt[3]{3} - \frac{2}{3}\sqrt[3]{3}\\ &= \left(\frac{18}{3} - \frac{72}{3} - \frac{2}{3}\right)\sqrt[3]{3}\\ &= -\frac{56}{3}\sqrt[3]{3}\end{aligned}$$

■

ORALS *Combine like radicals.*

1. $3\sqrt{3} + 4\sqrt{3}$ $7\sqrt{3}$

2. $5\sqrt{7} - 2\sqrt{7}$ $3\sqrt{7}$

3. $2\sqrt[3]{9} + 3\sqrt[3]{9}$ $5\sqrt[3]{9}$

4. $10\sqrt[4]{5} - 2\sqrt[4]{5}$ $8\sqrt[4]{5}$

Simplify.

5. $\sqrt{18}$ $3\sqrt{2}$

6. $\sqrt[3]{16}$ $2\sqrt[3]{2}$

EXERCISE 7.3

In Exercises 1–54, simplify and combine like radicals. All variables represent positive numbers.

1. $4\sqrt{2} + 6\sqrt{2}$ $10\sqrt{2}$

2. $6\sqrt{5} + 3\sqrt{5}$ $9\sqrt{5}$

3. $8\sqrt[5]{7} - 7\sqrt[5]{7}$ $\sqrt[5]{7}$

4. $10\sqrt[6]{12} - \sqrt[6]{12}$ $9\sqrt[6]{12}$

5. $8\sqrt{x} + 6\sqrt{x}$ $14\sqrt{x}$

6. $10\sqrt{xy} - 2\sqrt{xy}$ $8\sqrt{xy}$

7. $\sqrt{3} + \sqrt{27}$ $4\sqrt{3}$

8. $\sqrt{8} + \sqrt{32}$ $6\sqrt{2}$

9. $\sqrt{2} - \sqrt{8}$ $-\sqrt{2}$

10. $\sqrt{20} - \sqrt{125}$ $-3\sqrt{5}$

11. $\sqrt{98} - \sqrt{50}$ $2\sqrt{2}$

12. $\sqrt{72} - \sqrt{200}$ $-4\sqrt{2}$

13. $3\sqrt{24} + \sqrt{54}$ $9\sqrt{6}$

14. $\sqrt{18} + 2\sqrt{50}$ $13\sqrt{2}$

15. $\sqrt[3]{24} + \sqrt[3]{3}$ $3\sqrt[3]{3}$

16. $\sqrt[3]{16} + \sqrt[3]{128}$ $6\sqrt[3]{2}$

17. $\sqrt[3]{32} - \sqrt[3]{108}$ $-\sqrt[3]{4}$

18. $\sqrt[3]{80} - \sqrt[3]{10{,}000}$ $-8\sqrt[3]{10}$

19. $2\sqrt[3]{125} - 5\sqrt[3]{64}$ -10

20. $3\sqrt[3]{27} + 12\sqrt[3]{216}$ 81

21. $3\sqrt[3]{-54} + 8\sqrt[3]{-128}$ $-41\sqrt[3]{2}$

22. $5\sqrt[3]{-81} - 7\sqrt[3]{-375}$ $20\sqrt[3]{3}$

23. $14\sqrt[4]{32} - 15\sqrt[4]{162}$ $-17\sqrt[4]{2}$

24. $23\sqrt[4]{768} + \sqrt[4]{48}$ $94\sqrt[4]{3}$

25. $3\sqrt[4]{512} + 2\sqrt[4]{32}$ $16\sqrt[4]{2}$

26. $4\sqrt[4]{243} - \sqrt[4]{48}$ $10\sqrt[4]{3}$

27. $\sqrt{98} - \sqrt{50} - \sqrt{72}$ $-4\sqrt{2}$

28. $\sqrt{20} + \sqrt{125} - \sqrt{80}$ $3\sqrt{5}$

29. $\sqrt{18} + \sqrt{300} - \sqrt{243}$ $3\sqrt{2} + \sqrt{3}$

30. $\sqrt{80} - \sqrt{128} + \sqrt{288}$ $4\sqrt{5} + 4\sqrt{2}$

31. $2\sqrt[3]{16} - \sqrt[3]{54} - 3\sqrt[3]{128}$ $-11\sqrt[3]{2}$

32. $\sqrt[4]{48} - \sqrt[4]{243} - \sqrt[4]{768}$ $-5\sqrt[4]{3}$

33. $\sqrt{25y^2z} - \sqrt{16y^2z}$ $y\sqrt{z}$

34. $\sqrt{25yz^2} + \sqrt{9yz^2}$ $8z\sqrt{y}$

35. $\sqrt{36xy^2} + \sqrt{49xy^2}$ $13y\sqrt{x}$

36. $3\sqrt{2x} - \sqrt{8x}$ $\sqrt{2x}$

37. $2\sqrt[3]{64a} + 2\sqrt[3]{8a}$ $12\sqrt[3]{a}$

38. $3\sqrt[4]{x^4y} - 2\sqrt[4]{x^4y}$ $x\sqrt[4]{y}$

39. $\sqrt{y^5} - \sqrt{9y^5} - \sqrt{25y^5}$ $-7y^2\sqrt{y}$

40. $\sqrt{8y^7} + \sqrt{32y^7} - \sqrt{2y^7}$ $5y^3\sqrt{2y}$

41. $\sqrt[5]{x^6y^2} + \sqrt[5]{32x^6y^2} + \sqrt[5]{x^6y^2}$ $4x\sqrt[5]{xy^2}$

42. $\sqrt[3]{xy^4} + \sqrt[3]{8xy^4} - \sqrt[3]{27xy^4}$ 0

43. $\sqrt{x^2 + 2x + 1} + \sqrt{x^2 + 2x + 1}$ $2x + 2$

44. $\sqrt{4x^2 + 12x + 9} + \sqrt{9x^2 + 6x + 1}$ $5x + 4$

45. $\sqrt{3x^2 + 6x + 3} + \sqrt{3x^2}$ $(2x + 1)\sqrt{3}$

46. $\sqrt{5x^2 + 10x + 5} - \sqrt{5x^2 + 20x + 20}$ $-\sqrt{5}$

47. $\sqrt{18} - \sqrt{12} + \dfrac{\sqrt{12}}{6} - \dfrac{\sqrt{8}}{4}$ $\quad \frac{5}{2}\sqrt{2} - \frac{5}{3}\sqrt{3}$

48. $\sqrt{80} + \sqrt{28} + \dfrac{\sqrt{20}}{5} + \dfrac{\sqrt{28}}{7}$ $\quad \dfrac{22\sqrt{5}}{5} + \dfrac{16\sqrt{7}}{7}$

49. $\dfrac{\sqrt{48a}}{3a} - \dfrac{\sqrt{108a^3}}{9a^2} + \dfrac{\sqrt{12a}}{6a}$ $\quad \dfrac{\sqrt{3a}}{a}$

50. $\dfrac{\sqrt{200y}}{4y} + \dfrac{\sqrt{98y}}{2y} - \dfrac{\sqrt{288y^3}}{6y^2}$ $\quad \dfrac{4\sqrt{2y}}{y}$

51. $\dfrac{\sqrt{108x^3}}{4x^2} + \dfrac{\sqrt{75x}}{x} - \dfrac{\sqrt{147x}}{2x}$ $\quad \dfrac{3\sqrt{3x}}{x}$

52. $\dfrac{\sqrt{50a^5}}{10a^4} + \dfrac{5\sqrt{2a^3}}{2a^3} - \dfrac{\sqrt{200a}}{10a^2}$ $\quad \dfrac{2\sqrt{2a}}{a^2}$

53. $\dfrac{3\sqrt{28x^5y}}{4x} + \dfrac{2x\sqrt{7xy^3}}{3y} - \dfrac{\sqrt{112x^5y^3}}{3xy}$ $\quad \dfrac{5x\sqrt{7xy}}{6}$

54. $\dfrac{y\sqrt{18x^5}}{2x^2} + \dfrac{7\sqrt{2xy^4}}{4y} - \dfrac{\sqrt{18xy^2}}{12}$ $\quad 3y\sqrt{2x}$

Writing Exercises ■ *Write a paragraph using your own words.*

1. Explain how to recognize like radicals.

2. Explain how to combine like radicals.

Something to Think About ■

1. A rectangle is $\sqrt{8}$ inches wide and $\sqrt{50}$ inches long. Find the perimeter. $14\sqrt{2}$ in.

2. Find the sum: $\sqrt{3} + \sqrt{3^2} + \sqrt{3^3} + \sqrt{3^4} + \sqrt{3^5}$ $\quad 12 + 13\sqrt{3}$

Review Exercises ■ *Do each operation.*

1. $3x^2y^3(-5x^3y^{-4})$ $\quad \dfrac{-15x^5}{y}$

2. $-2a^2b^{-2}(4a^{-2}b^4 - 2a^2b + 3a^3b^2)$ $\quad -8b^2 + \dfrac{4a^4}{b} - 6a^5$

3. $(3t + 2)^2$ $\quad 9t^2 + 12t + 4$

4. $(5r - 3s)(5r + 2s)$ $\quad 25r^2 - 5rs - 6s^2$

5. $2p - 5\overline{)6p^2 - 7p - 25}$ $\quad 3p + 4 + \dfrac{-5}{2p - 5}$

6. $3m + n\overline{)6m^3 - m^2n + 2mn^2 + n^3}$ $\quad 2m^2 - mn + n^2$

7.4 Multiplying and Dividing Radical Expressions

■ Multiplying a Monomial by a Monomial ■ Multiplying a Polynomial by a Monomial ■ Multiplying a Polynomial by a Polynomial ■ Rationalizing Denominators

Radical expressions can be multiplied and divided.

■ Multiplying a Monomial by a Monomial

EXAMPLE 1 Multiply $3\sqrt{6}$ by $2\sqrt{3}$.

Solution We use the commutative and associative properties of multiplication to multiply the coefficients and the radicals separately. We then simplify any radicals in the product, if possible.

$$\begin{aligned} 3\sqrt{6}\cdot 2\sqrt{3} &= 3(2)\sqrt{6}\sqrt{3} && \text{Use the commutative property of multiplication.} \\ &= 6\sqrt{18} && \sqrt{6}\sqrt{3} = \sqrt{6\cdot 3} = \sqrt{18}. \\ &= 6\sqrt{9}\sqrt{2} && \sqrt{18} = \sqrt{9\cdot 2} = \sqrt{9}\sqrt{2}. \\ &= 6(3)\sqrt{2} && \sqrt{9} = 3. \\ &= 18\sqrt{2} \end{aligned}$$

■

Multiplying a Polynomial by a Monomial

To multiply a polynomial by a monomial, we use the distributive property to remove parentheses and then simplify each term.

EXAMPLE 2 Simplify $3\sqrt{3}\left(4\sqrt{8} - 5\sqrt{27}\right)$.

Solution

$$\begin{aligned} 3\sqrt{3}\left(4\sqrt{8} - 5\sqrt{27}\right) &= 3\sqrt{3}\cdot 4\sqrt{8} - 3\sqrt{3}\cdot 5\sqrt{27} \\ &= 12\sqrt{24} - 15\sqrt{81} \\ &= 12\sqrt{4}\sqrt{6} - 15(9) \\ &= 12(2)\sqrt{6} - 135 \\ &= 24\sqrt{6} - 135 \end{aligned}$$

■

Multiplying a Polynomial by a Polynomial

To multiply a binomial by a binomial, we use the FOIL method.

EXAMPLE 3 Multiply $\left(\sqrt{7} + \sqrt{2}\right)\left(\sqrt{7} - 3\sqrt{2}\right)$.

Solution

$$\begin{aligned} \left(\sqrt{7} + \sqrt{2}\right)\left(\sqrt{7} - 3\sqrt{2}\right) &= \left(\sqrt{7}\right)^2 - 3\sqrt{7}\sqrt{2} + \sqrt{2}\sqrt{7} - 3\sqrt{2}\sqrt{2} \\ &= 7 - 3\sqrt{14} + \sqrt{14} - 3(2) \\ &= 7 - 2\sqrt{14} - 6 \\ &= 1 - 2\sqrt{14} \end{aligned}$$

■

EXAMPLE 4 Multiply $\left(\sqrt{3x} - \sqrt{5}\right)\left(\sqrt{2x} + \sqrt{10}\right)$.

Solution

$$\begin{aligned} \left(\sqrt{3x} - \sqrt{5}\right)\left(\sqrt{2x} + \sqrt{10}\right) &= \sqrt{3}\sqrt{2}x^2 + \sqrt{3}\sqrt{10}x - \sqrt{5}\sqrt{2}x - \sqrt{5}\sqrt{10} \\ &= \sqrt{6}x^2 + \sqrt{30}x - \sqrt{10}x - \sqrt{50} \\ &= \sqrt{6}x^2 + \sqrt{30}x - \sqrt{10}x - \sqrt{25}\sqrt{2} \\ &= \sqrt{6}x^2 + \sqrt{30}x - \sqrt{10}x - 5\sqrt{2} \end{aligned}$$

■

WARNING! It is important to draw radical signs carefully so that they completely cover a radicand, but no more than the radicand. To avoid confusion, we often write an expression such as $\sqrt{3x}$ in the form $x\sqrt{3}$.

Rationalizing Denominators

To divide radical expressions, we use a process called **rationalizing the denominator** to replace the divisor with a rational number. For example, to divide $\sqrt{70}$ by $\sqrt{3}$, we first write the division as the fraction

$$\frac{\sqrt{70}}{\sqrt{3}}$$

To eliminate the radical in the denominator of the fraction, we multiply both the numerator and the denominator by a number that will give a perfect square under the radical sign in the denominator. Because $3 \cdot 3 = 9$ is a perfect square, $\sqrt{3}$ is such a number.

$$\begin{aligned}\frac{\sqrt{70}}{\sqrt{3}} &= \frac{\sqrt{70} \cdot \sqrt{3}}{\sqrt{3} \cdot \sqrt{3}}\\ &= \frac{\sqrt{210}}{3}\end{aligned}$$

Since there is no radical in the denominator and $\sqrt{210}$ cannot be simplified, the expression $\sqrt{210}/3$ is in simplest form, and the division is complete.

EXAMPLE 5 Rationalize the denominator to simplify $\sqrt{\frac{20}{7}}$.

Solution We first write the square root of the quotient as the quotient of two square roots:

$$\sqrt{\frac{20}{7}} = \frac{\sqrt{20}}{\sqrt{7}}$$

We then rationalize the denominator by multiplying both the numerator and the denominator by $\sqrt{7}$:

$$\begin{aligned}\frac{\sqrt{20}}{\sqrt{7}} &= \frac{\sqrt{20} \cdot \sqrt{7}}{\sqrt{7} \cdot \sqrt{7}}\\ &= \frac{\sqrt{140}}{7}\end{aligned}$$

Another Approach: You may want to simplify $\sqrt{\frac{20}{7}}$ as follows:

$$\begin{aligned}\sqrt{\frac{20}{7}} &= \sqrt{\frac{20 \cdot 7}{7 \cdot 7}}\\ &= \sqrt{\frac{4 \cdot 5 \cdot 7}{49}}\\ &= \frac{\sqrt{4}\sqrt{35}}{\sqrt{49}}\\ &= \frac{2\sqrt{35}}{7}\end{aligned}$$

The radical in the numerator of $\sqrt{140}/7$ can be simplified and the fraction can be written as follows:

$$\frac{\sqrt{140}}{7} = \frac{\sqrt{4 \cdot 35}}{7} = \frac{\sqrt{4} \cdot \sqrt{35}}{7} = \frac{2\sqrt{35}}{7}$$

■

EXAMPLE 6 Rationalize the denominator of $\dfrac{1}{\sqrt[3]{2}}$.

Solution To rationalize the denominator, we multiply both the numerator and the denominator by a number that will result in a perfect cube under the radical sign. Because $2 \cdot 4 = 8$ is a perfect cube, $\sqrt[3]{4}$ is such a number.

Another Approach:

$$\frac{1}{\sqrt[3]{2}} = \frac{1}{\sqrt[3]{2}} \cdot \frac{\sqrt[3]{2^2}}{\sqrt[3]{2^2}} = \frac{\sqrt[3]{4}}{\sqrt[3]{2^3}} = \frac{\sqrt[3]{4}}{2}$$

$$\frac{1}{\sqrt[3]{2}} = \frac{1 \cdot \sqrt[3]{4}}{\sqrt[3]{2} \cdot \sqrt[3]{4}} = \frac{\sqrt[3]{4}}{\sqrt[3]{8}} = \frac{\sqrt[3]{4}}{2}$$

■

EXAMPLE 7 Rationalize the denominator of $\dfrac{\sqrt[3]{5}}{\sqrt[3]{18}}$.

Solution To rationalize the denominator, we multiply both the numerator and the denominator by a number that will result in a perfect cube under the radical sign in the denominator.

Since 216 is the smallest perfect cube that is divisible by 18 ($216 \div 18 = 12$), multiplying both the numerator and the denominator by $\sqrt[3]{12}$ will give the smallest possible perfect cube under the radical in the denominator.

Another Approach: Since $18 = 2 \cdot 3^2$, we need factors of $2^2 \cdot 3$ to get $2^3 \cdot 3^3$.

$$\frac{\sqrt[3]{5}}{\sqrt[3]{18}} = \frac{\sqrt[3]{5}\sqrt[3]{2^2 \cdot 3}}{\sqrt[3]{2 \cdot 3^2}\sqrt[3]{2^2 \cdot 3}} = \frac{\sqrt[3]{5}\sqrt[3]{12}}{\sqrt[3]{2^3 \cdot 3^3}}$$

etc.

$$\frac{\sqrt[3]{5}}{\sqrt[3]{18}} = \frac{\sqrt[3]{5} \cdot \sqrt[3]{12}}{\sqrt[3]{18} \cdot \sqrt[3]{12}} = \frac{\sqrt[3]{60}}{\sqrt[3]{216}} = \frac{\sqrt[3]{60}}{6}$$

■

EXAMPLE 8 Rationalize the denominator of $\dfrac{\sqrt{5xy^2}}{\sqrt{xy^3}}$. ($x$ and y are positive numbers.)

Solution 1

$$\frac{\sqrt{5xy^2}}{\sqrt{xy^3}} = \sqrt{\frac{5xy^2}{xy^3}} = \sqrt{\frac{5}{y}} = \frac{\sqrt{5}}{\sqrt{y}} = \frac{\sqrt{5}\sqrt{y}}{\sqrt{y}\sqrt{y}} = \frac{\sqrt{5y}}{y}$$

Solution 2

$$\frac{\sqrt{5xy^2}}{\sqrt{xy^3}} = \sqrt{\frac{5xy^2}{xy^3}} = \sqrt{\frac{5}{y}} = \sqrt{\frac{5y}{yy}} = \frac{\sqrt{5y}}{\sqrt{y^2}} = \frac{\sqrt{5y}}{y}$$

■

We now consider rationalizing denominators that are binomials. Suppose we wish to rationalize the denominator of the fraction

$$\frac{1}{\sqrt{2}+1}$$

It is not sufficient to multiply the numerator and the denominator by $\sqrt{2}$. (Try it and discover why.) Instead we shall multiply both the numerator and the denominator by $\sqrt{2}-1$. This binomial, called the **conjugate** of $\sqrt{2}+1$, is the same binomial as the denominator, but with the opposite sign between its terms.

$$\begin{aligned}\frac{1}{\sqrt{2}+1} &= \frac{1(\sqrt{2}-1)}{(\sqrt{2}+1)(\sqrt{2}-1)} \\ &= \frac{\sqrt{2}-1}{(\sqrt{2})^2-1} \qquad (\sqrt{2}+1)(\sqrt{2}-1) = (\sqrt{2})^2 - 1. \\ &= \frac{\sqrt{2}-1}{2-1} \\ &= \frac{\sqrt{2}-1}{1} \\ &= \sqrt{2}-1\end{aligned}$$

Conjugate Binomials **Conjugate binomials** are binomials that are the same except for the sign between their terms.

The conjugate of $a + b$ is $a - b$, and the conjugate of $a - b$ is $a + b$.

EXAMPLE 9 Rationalize the denominator of $\dfrac{\sqrt{x}-\sqrt{3}}{\sqrt{x}-\sqrt{2}}$ $(x > 0)$.

Solution We multiply both the numerator and denominator of the fraction by $\sqrt{x}+\sqrt{2}$, which is the conjugate of $\sqrt{x}-\sqrt{2}$. We then can simplify.

$$\frac{\sqrt{x}-\sqrt{3}}{\sqrt{x}-\sqrt{2}} = \frac{(\sqrt{x}-\sqrt{3})(\sqrt{x}+\sqrt{2})}{(\sqrt{x}-\sqrt{2})(\sqrt{x}+\sqrt{2})}$$

$$= \frac{x+\sqrt{2x}-\sqrt{3x}-\sqrt{6}}{x-2}$$

■

EXAMPLE 10 Rationalize the denominator of $\dfrac{9x^2}{\sqrt{3x}(\sqrt{5}-\sqrt{2})}$ $(x > 0)$.

Solution We multiply both the numerator and denominator of the fraction by $\sqrt{3x}(\sqrt{5}+\sqrt{2})$ and simplify.

$$\frac{9x^2}{\sqrt{3x}(\sqrt{5}-\sqrt{2})} = \frac{9x^2 \cdot \sqrt{3x}(\sqrt{5}+\sqrt{2})}{\sqrt{3x}(\sqrt{5}-\sqrt{2}) \cdot \sqrt{3x}(\sqrt{5}+\sqrt{2})}$$

$$= \frac{9x^2\sqrt{3x}(\sqrt{5}+\sqrt{2})}{\sqrt{3x}\sqrt{3x}(\sqrt{5}-\sqrt{2})(\sqrt{5}+\sqrt{2})}$$

$$= \frac{9x^2\sqrt{3x}(\sqrt{5}+\sqrt{2})}{3x(5-2)}$$

$$= \frac{9x^2\sqrt{3x}(\sqrt{5}+\sqrt{2})}{9x}$$

$$= x\sqrt{3x}(\sqrt{5}+\sqrt{2})$$

■

ORALS *Multiply and simplify.*

1. $\sqrt{3}\sqrt{3}$ $\quad 3$
2. $\sqrt[3]{2}\sqrt[3]{2}\sqrt[3]{2}$ $\quad 2$
3. $\sqrt{3}\sqrt{9}$ $\quad 3\sqrt{3}$
4. $\sqrt{a^3b}\sqrt{ab}$ $\quad a^2|b|$
5. $3\sqrt{2}(\sqrt{2}+1)$ $\quad 6+3\sqrt{2}$
6. $(\sqrt{2}+1)(\sqrt{2}-1)$ $\quad 1$
7. $\dfrac{1}{\sqrt{2}}$ $\quad \dfrac{\sqrt{2}}{2}$
8. $\dfrac{1}{\sqrt{3}-1}$ $\quad \dfrac{\sqrt{3}+1}{2}$

EXERCISE 7.4

In Exercises 1–24, do each multiplication and simplify, if possible. All variables represent positive numbers.

1. $\sqrt{2}\sqrt{8}$ 4
2. $\sqrt{3}\sqrt{27}$ 9
3. $\sqrt{5}\sqrt{10}$ $5\sqrt{2}$
4. $\sqrt{7}\sqrt{35}$ $7\sqrt{5}$
5. $\sqrt{3}\sqrt{6}$ $3\sqrt{2}$
6. $\sqrt{11}\sqrt{33}$ $11\sqrt{3}$
7. $\sqrt[3]{5}\sqrt[3]{25}$ 5
8. $\sqrt[3]{7}\sqrt[3]{49}$ 7
9. $\sqrt[3]{9}\sqrt[3]{3}$ 3
10. $\sqrt[3]{16}\sqrt[3]{4}$ 4
11. $\sqrt[3]{2}\sqrt[3]{12}$ $2\sqrt[3]{3}$
12. $\sqrt[3]{3}\sqrt[3]{18}$ $3\sqrt[3]{2}$
13. $\sqrt{ab^3}\sqrt{ab}$ ab^2
14. $\sqrt{8x}\sqrt{2x^3y}$ $4x^2\sqrt{y}$
15. $\sqrt{5ab}\sqrt{5a}$ $5a\sqrt{b}$
16. $\sqrt{15rs^2}\sqrt{10r}$ $5rs\sqrt{6}$
17. $\sqrt[3]{5r^2s}\sqrt[3]{2r}$ $r\sqrt[3]{10s}$
18. $\sqrt[3]{3xy^2}\sqrt[3]{9x^3}$ $3x\sqrt[3]{xy^2}$
19. $\sqrt[3]{a^5b}\sqrt[3]{16ab^5}$ $2a^2b^2\sqrt[3]{2}$
20. $\sqrt[3]{3x^4y}\sqrt[3]{18x}$ $3x\sqrt[3]{2x^2y}$
21. $\sqrt{x(x+3)}\sqrt{x^3(x+3)}$ $x^2(x+3)$
22. $\sqrt{y^2(x+y)}\sqrt{(x+y)^3}$ $y(x+y)^2$
23. $\sqrt[3]{6x^2(y+z)^2}\sqrt[3]{18x(y+z)}$ $3x(y+z)\sqrt[3]{4}$
24. $\sqrt[3]{9x^2y(z+1)^2}\sqrt[3]{6xy^2(z+1)}$ $3xy(z+1)\sqrt[3]{2}$

In Exercises 25–44, do each multiplication and simplify. All variables represent positive numbers.

25. $3\sqrt{5}(4-\sqrt{5})$ $12\sqrt{5}-15$
26. $2\sqrt{7}(3\sqrt{7}-1)$ $42-2\sqrt{7}$
27. $3\sqrt{2}(4\sqrt{3}+2\sqrt{7})$ $12\sqrt{6}+6\sqrt{14}$
28. $-\sqrt{3}(\sqrt{7}-\sqrt{5})$ $-\sqrt{21}+\sqrt{15}$
29. $-2\sqrt{5x}(4\sqrt{2x}-3\sqrt{3})$ $-8x\sqrt{10}+6\sqrt{15x}$
30. $3\sqrt{7t}(2\sqrt{7t}+3\sqrt{3t^2})$ $42t+9t\sqrt{21t}$
31. $(\sqrt{2}+1)(\sqrt{2}-3)$ $-1-2\sqrt{2}$
32. $(2\sqrt{3}+1)(\sqrt{3}-1)$ $5-\sqrt{3}$
33. $(4\sqrt{3x}+3)(2\sqrt{3x}-5)$ $24x-14\sqrt{3x}-15$
34. $(7\sqrt{2y}+2)(3\sqrt{2y}-5)$ $42y-29\sqrt{2y}-10$
35. $(\sqrt{5z}+\sqrt{3})(\sqrt{5z}+\sqrt{3})$ $5z+2\sqrt{15z}+3$
36. $(\sqrt{3p}-\sqrt{2})(\sqrt{3p}+\sqrt{2})$ $3p-2$
37. $(\sqrt{3x}-\sqrt{2y})(\sqrt{3x}+\sqrt{2y})$ $3x-2y$
38. $(\sqrt{3m}+\sqrt{2n})(\sqrt{3m}+\sqrt{2n})$ $3m+2\sqrt{6mn}+2n$
39. $(2\sqrt{3a}-\sqrt{5b})(\sqrt{3a}+3\sqrt{5b})$ $6a+5\sqrt{15ab}-15b$
40. $(5\sqrt{2p}-\sqrt{3q})(2\sqrt{2p}+2\sqrt{3q})$ $20p+8\sqrt{6pq}-6q$
41. $(3\sqrt{2r}-2)^2$ $18r-12\sqrt{2r}+4$
42. $(2\sqrt{3t}+5)^2$ $12t+20\sqrt{3t}+25$
43. $-2\sqrt{3x}(\sqrt{3x}+\sqrt{3})^2$ $-6x\sqrt{3x}-12x\sqrt{3}-6\sqrt{3x}$
44. $\sqrt{2}(2\sqrt{5x}+3\sqrt{3})^2$ $20x\sqrt{2}+12\sqrt{30x}+27\sqrt{2}$

In Exercises 45–68, rationalize each denominator. All variables represent positive numbers.

45. $\sqrt{\dfrac{1}{7}}$ $\dfrac{\sqrt{7}}{7}$
46. $\sqrt{\dfrac{5}{3}}$ $\dfrac{\sqrt{15}}{3}$
47. $\sqrt{\dfrac{2}{3}}$ $\dfrac{\sqrt{6}}{3}$
48. $\sqrt{\dfrac{3}{2}}$ $\dfrac{\sqrt{6}}{2}$
49. $\dfrac{\sqrt{5}}{\sqrt{8}}$ $\dfrac{\sqrt{10}}{4}$
50. $\dfrac{\sqrt{3}}{\sqrt{50}}$ $\dfrac{\sqrt{6}}{10}$
51. $\dfrac{\sqrt{8}}{\sqrt{2}}$ 2
52. $\dfrac{\sqrt{27}}{\sqrt{3}}$ 3
53. $\dfrac{1}{\sqrt[3]{2}}$ $\dfrac{\sqrt[3]{4}}{2}$
54. $\dfrac{2}{\sqrt[3]{6}}$ $\dfrac{\sqrt[3]{36}}{3}$
55. $\dfrac{3}{\sqrt[3]{9}}$ $\sqrt[3]{3}$
56. $\dfrac{2}{\sqrt[3]{a}}$ $\dfrac{2\sqrt[3]{a^2}}{a}$
57. $\dfrac{\sqrt[3]{2}}{\sqrt[3]{9}}$ $\dfrac{\sqrt[3]{6}}{3}$
58. $\dfrac{\sqrt[3]{9}}{\sqrt[3]{54}}$ $\dfrac{\sqrt[3]{36}}{6}$
59. $\dfrac{\sqrt{8x^2y}}{\sqrt{xy}}$ $2\sqrt{2x}$
60. $\dfrac{\sqrt{9xy}}{\sqrt{3x^2y}}$ $\dfrac{\sqrt{3x}}{x}$

61. $\dfrac{\sqrt{10xy^2}}{\sqrt{2xy^3}}$ $\dfrac{\sqrt{5y}}{y}$

62. $\dfrac{\sqrt{5ab^2c}}{\sqrt{10abc}}$ $\dfrac{\sqrt{2b}}{2}$

63. $\dfrac{\sqrt[3]{4a^2}}{\sqrt[3]{2ab}}$ $\dfrac{\sqrt[3]{2ab^2}}{b}$

64. $\dfrac{\sqrt[3]{9x}}{\sqrt[3]{3xy}}$ $\dfrac{\sqrt[3]{3y^2}}{y}$

65. $\dfrac{1}{\sqrt[4]{3}}$ $\dfrac{\sqrt[4]{27}}{3}$

66. $\dfrac{1}{\sqrt[5]{2}}$ $\dfrac{\sqrt[5]{16}}{2}$

67. $\dfrac{4}{\sqrt[5]{16}}$ $2\sqrt[5]{2}$

68. $\dfrac{4}{\sqrt[4]{32}}$ $\sqrt[4]{8}$

In Exercises 69–96, do each division by rationalizing the denominator and simplifying. All variables represent positive numbers.

69. $\dfrac{1}{\sqrt{2}-1}$ $\sqrt{2}+1$

70. $\dfrac{2}{\sqrt{3}-1}$ $\sqrt{3}+1$

71. $\dfrac{-6}{\sqrt{5}+4}$ $\dfrac{6(\sqrt{5}-4)}{11}$

72. $\dfrac{-10}{\sqrt{5}-1}$ $\dfrac{-5(\sqrt{5}+1)}{2}$

73. $\dfrac{2}{\sqrt{3}+1}$ $\sqrt{3}-1$

74. $\dfrac{2}{\sqrt{5}+1}$ $\dfrac{\sqrt{5}-1}{2}$

75. $\dfrac{25}{\sqrt{6}+1}$ $5(\sqrt{6}-1)$

76. $\dfrac{50}{\sqrt{7}+1}$ $\dfrac{25\sqrt{7}-25}{3}$

77. $\dfrac{\sqrt{2}}{\sqrt{5}+3}$ $\dfrac{3\sqrt{2}-\sqrt{10}}{4}$

78. $\dfrac{\sqrt{3}}{\sqrt{3}-2}$ $-3-2\sqrt{3}$

79. $\dfrac{\sqrt{7}}{2-\sqrt{5}}$ $-2\sqrt{7}-\sqrt{35}$

80. $\dfrac{\sqrt{11}}{3+\sqrt{7}}$ $\dfrac{3\sqrt{11}-\sqrt{77}}{2}$

81. $\dfrac{2}{\sqrt{7}-\sqrt{5}}$ $\sqrt{7}+\sqrt{5}$

82. $\dfrac{5}{\sqrt{7}-\sqrt{2}}$ $\sqrt{7}+\sqrt{2}$

83. $\dfrac{20}{\sqrt{3}+1}$ $10(\sqrt{3}-1)$

84. $\dfrac{36}{\sqrt{5}+2}$ $36(\sqrt{5}-2)$

85. $\dfrac{\sqrt{3}+1}{\sqrt{3}-1}$ $2+\sqrt{3}$

86. $\dfrac{\sqrt{2}-1}{\sqrt{2}+1}$ $3-2\sqrt{2}$

87. $\dfrac{\sqrt{7}-\sqrt{2}}{\sqrt{2}+\sqrt{7}}$ $\dfrac{9-2\sqrt{14}}{5}$

88. $\dfrac{\sqrt{3}+\sqrt{2}}{\sqrt{3}-\sqrt{2}}$ $5+2\sqrt{6}$

89. $\dfrac{2}{\sqrt{x}+1}$ $\dfrac{2(\sqrt{x}-1)}{x-1}$

90. $\dfrac{3}{\sqrt{x}-2}$ $\dfrac{3\sqrt{x}+6}{x-4}$

91. $\dfrac{x}{\sqrt{x}-4}$ $\dfrac{x(\sqrt{x}+4)}{x-16}$

92. $\dfrac{2x}{\sqrt{x}+1}$ $\dfrac{2x\sqrt{x}-2x}{x-1}$

93. $\dfrac{2z-1}{\sqrt{2z}-1}$ $\sqrt{2z}+1$

94. $\dfrac{3t-1}{\sqrt{3t}+1}$ $\sqrt{3t}-1$

95. $\dfrac{\sqrt{x}-\sqrt{y}}{\sqrt{x}+\sqrt{y}}$ $\dfrac{x-2\sqrt{xy}+y}{x-y}$

96. $\dfrac{\sqrt{x}+\sqrt{y}}{\sqrt{x}-\sqrt{y}}$ $\dfrac{x+2\sqrt{xy}+y}{x-y}$

In Exercises 97–102, simplify each radical expression and combine like radicals when possible.

97. $\sqrt{8}+\dfrac{4}{\sqrt{2}}-\sqrt{32}$ 0

98. $\sqrt{12}+\dfrac{6}{\sqrt{3}}-3\sqrt{243}$ $-23\sqrt{3}$

99. $\sqrt{16x}-\dfrac{x}{\sqrt{x}}$ $3\sqrt{x}$

100. $\sqrt{25a^3}+\dfrac{\sqrt{a^4}}{a}$ $5a\sqrt{a}+a$

101. $\dfrac{1}{\sqrt{x}}+\dfrac{\sqrt{x}}{x}+\dfrac{x}{\sqrt{x^3}}$ $\dfrac{3\sqrt{x}}{x}$

102. $\sqrt{\dfrac{1}{xy}}+\dfrac{\sqrt{x}}{x\sqrt{y}}+\dfrac{\sqrt{y}}{y\sqrt{x}}$ $\dfrac{3\sqrt{xy}}{xy}$

Writing Exercises ■ *Write a paragraph using your own words.*

1. Explain how to simplify a fraction with the denominator $\sqrt[3]{3}$.
2. Explain how to simplify a fraction with the denominator $\sqrt[3]{9}$.

Something to Think About ■

1. Change the numerator of $\dfrac{\sqrt{x}-3}{4}$ to a rational number. $\dfrac{x-9}{4(\sqrt{x}+3)}$
2. Rationalize the numerator of $\dfrac{2\sqrt{3a}+4}{\sqrt{3a}-1}$. $\dfrac{6a-8}{3a-3\sqrt{3a}+2}$

Review Exercises ■ *Solve each equation.*

1. $\dfrac{2}{3-a}=1$ 1
2. $5(s-4)=-5(s-4)$ 4
3. $\dfrac{8}{b-2}+\dfrac{3}{2-b}=-\dfrac{1}{b}$ $\frac{1}{3}$
4. $\dfrac{2}{x-2}+\dfrac{1}{x+1}=\dfrac{1}{(x+1)(x-2)}$ $\frac{1}{3}$

7.5 Radical Equations

■ Equations Containing One Radical ■ Equations Containing Two Radicals ■ Equations Containing Three Radicals

To solve equations that contain radicals, we will use the following theorem, called the **power rule.**

Power Rule If a, b, and n are real numbers and $a = b$, then

$$a^n = b^n$$

If we raise both sides of an equation to the same power, the resulting equation may or may not be equivalent to the original equation. For example, if we square both sides of the equation

Teaching Tip: Explain that the power rule is not true in the other direction. $a^n = b^n$ does not guarantee that $a = b$.

1. $x = 3$ With a solution set of $\{3\}$

we obtain the equation

2. $x^2 = 9$ With a solution set of $\{3, -3\}$

Equations 1 and 2 are not equivalent because they have different solution sets, and the solution -3 in Equation 2 does not satisfy Equation 1. Because raising both sides of an equation to the same power can produce an equation with roots that don't check in the original equation, we must check each suspected solution in the original equation.

Equations Containing One Radical

EXAMPLE 1 Solve the equation $\sqrt{x+3} = 4$.

Solution To eliminate the radical, we apply the power rule, squaring both sides of the equation, and proceed as follows:

$$\sqrt{x+3} = 4$$

$$\left(\sqrt{x+3}\right)^2 = (4)^2 \qquad \text{Square both sides.}$$

$$x + 3 = 16$$

$$x = 13 \qquad \text{Subtract 3 from both sides.}$$

We must check the apparent solution of 13 to see if it satisfies the original equation.

Check:

$$\sqrt{x+3} = 4$$

$$\sqrt{13+3} \stackrel{?}{=} 4$$

$$\sqrt{16} \stackrel{?}{=} 4$$

$$4 = 4$$

Since 13 satisfies the original equation, it is the solution. ■

To solve an equation with radicals, we follow these steps.

Steps for Solving an Equation with Radicals

1. Isolate one radical expression on one side of the equation.
2. Raise both sides of the equation to the power that is the same as the index of the radical.
3. Solve the resulting equation. If it still contains a radical, go back to Step 1.
4. Check the solutions to eliminate all extraneous roots.

EXAMPLE 2 Solve the equation $\sqrt{3x+1} + 1 = x$.

Solution We first subtract 1 from both sides to isolate the radical. Then, we square both sides of the equation to eliminate the radical.

$$\sqrt{3x+1} + 1 = x$$

$$\sqrt{3x+1} = x - 1 \qquad \text{Subtract 1 from both sides.}$$

$$\left(\sqrt{3x+1}\right)^2 = (x-1)^2 \qquad \text{Square both sides to eliminate the square root.}$$

$$3x + 1 = x^2 - 2x + 1 \qquad \text{Remove parentheses.}$$

$$0 = x^2 - 5x \qquad \text{Subtract } 3x \text{ and 1 from both sides.}$$

$$0 = x(x-5) \qquad \text{Factor } x^2 - 5x.$$

Teaching Tip: Students often forget the middle term when squaring a binomial on the right-hand side. To avoid this, have students write

$$\left(\sqrt{3x+1}\right)^2 = (x-1)(x-1)$$

$$x = 0 \quad \text{or} \quad x - 5 = 0 \qquad \text{Set each factor equal to 0.}$$
$$x = 0 \quad | \quad x = 5$$
(Extraneous.)

We must check each apparent solution to see if it is extraneous.

Check:

$\sqrt{3x+1} + 1 = x$	$\sqrt{3x+1} + 1 = x$
$\sqrt{3(0)+1} + 1 \stackrel{?}{=} 0$	$\sqrt{3(5)+1} + 1 \stackrel{?}{=} 5$
$\sqrt{1} + 1 \stackrel{?}{=} 0$	$\sqrt{16} + 1 \stackrel{?}{=} 5$
$2 \neq 0$	$5 = 5$

Since the apparent solution 0 does not check, it is extraneous and must be discarded. The solution set of the original equation is $\{5\}$. ■

EXAMPLE 3 Solve the equation $\sqrt[3]{x^3 + 7} = x + 1$.

Solution To eliminate the radical, we cube both sides of the equation and proceed as follows:

Teaching Tip: Point out that $(x + 1)^3 = (x + 1)(x + 1)(x + 1)$

$$\sqrt[3]{x^3 + 7} = x + 1$$
$$\left(\sqrt[3]{x^3 + 7}\right)^3 = (x + 1)^3 \qquad \text{Cube both sides to eliminate the cube root.}$$
$$x^3 + 7 = x^3 + 3x^2 + 3x + 1$$
$$0 = 3x^2 + 3x - 6 \qquad \text{Subtract } x^3 \text{ and 7 from both sides.}$$
$$0 = x^2 + x - 2 \qquad \text{Divide both sides by 3.}$$
$$0 = (x + 2)(x - 1) \qquad \text{Factor } x^2 + x - 2.$$

$$x + 2 = 0 \quad \text{or} \quad x - 1 = 0 \qquad \text{Set each factor equal to 0.}$$
$$x = -2 \quad | \quad x = 1$$

We check each apparent solution to see if it is extraneous.

Check:

$\sqrt[3]{x^3 + 7} = x + 1$	$\sqrt[3]{x^3 + 7} = x + 1$
$\sqrt[3]{(-2)^3 + 7} \stackrel{?}{=} -2 + 1$	$\sqrt[3]{1^3 + 7} \stackrel{?}{=} 1 + 1$
$\sqrt[3]{-8 + 7} \stackrel{?}{=} -1$	$\sqrt[3]{8} \stackrel{?}{=} 2$
$\sqrt[3]{-1} \stackrel{?}{=} -1$	$2 = 2$
$-1 = -1$	

Both solutions satisfy the original equation. ■

Equations Containing Two Radicals

When more than one radical appears in an equation, it is often necessary to apply the power rule more than once.

EXAMPLE 4 Solve the equation $\sqrt{x} + \sqrt{x+2} = 2$.

Solution To remove the radicals, we must square both sides of the equation. This is easier to do if one radical is on each side of the equation. So we subtract $\sqrt{x}$ from both sides to isolate one radical on one side of the equation.

$$\sqrt{x} + \sqrt{x+2} = 2$$

$$\sqrt{x+2} = 2 - \sqrt{x} \qquad \text{Subtract } \sqrt{x} \text{ from both sides.}$$

$$\left(\sqrt{x+2}\right)^2 = \left(2 - \sqrt{x}\right)^2 \qquad \text{Square both sides to eliminate the square root.}$$

$$x + 2 = 4 - 4\sqrt{x} + x \qquad \begin{aligned}\left(2 - \sqrt{x}\right)\left(2 - \sqrt{x}\right) &= 4 - 2\sqrt{x} - 2\sqrt{x} + x \\ &= 4 - 4\sqrt{x} + x\end{aligned}$$

$$2 = 4 - 4\sqrt{x} \qquad \text{Subtract } x \text{ from both sides.}$$

$$-2 = -4\sqrt{x} \qquad \text{Subtract 4 from both sides.}$$

$$\frac{1}{2} = \sqrt{x} \qquad \text{Divide both sides by } -4.$$

$$\frac{1}{4} = x \qquad \text{Square both sides.}$$

Check:

$$\sqrt{x} + \sqrt{x+2} = 2$$

$$\sqrt{\frac{1}{4}} + \sqrt{\frac{1}{4} + 2} \stackrel{?}{=} 2$$

$$\frac{1}{2} + \sqrt{\frac{9}{4}} \stackrel{?}{=} 2$$

$$\frac{1}{2} + \frac{3}{2} \stackrel{?}{=} 2$$

$$2 = 2$$

The solution checks. ■

Equations Containing Three Radicals

EXAMPLE 5 Solve the equation $\sqrt{x+2} + \sqrt{2x} = \sqrt{18 - x}$.

Solution In this case, it is impossible to isolate one radical on each side of the equation, so we begin by squaring both sides. Then we proceed as follows.

$$\sqrt{x+2} + \sqrt{2x} = \sqrt{18-x}$$

$$\left(\sqrt{x+2} + \sqrt{2x}\right)^2 = \left(\sqrt{18-x}\right)^2$$ Square both sides to eliminate one square root.

$$x + 2 + 2\sqrt{x+2}\sqrt{2x} + 2x = 18 - x$$

$$2\sqrt{x+2}\sqrt{2x} = 16 - 4x$$ Subtract $3x$ and 2 from both sides.

$$\sqrt{x+2}\sqrt{2x} = 8 - 2x$$ Divide both sides by 2.

$$\left(\sqrt{x+2}\sqrt{2x}\right)^2 = (8-2x)^2$$ Square both sides to eliminate the square roots.

$$(x+2)2x = 64 - 32x + 4x^2$$

$$2x^2 + 4x = 64 - 32x + 4x^2$$

$$0 = 2x^2 - 36x + 64$$ Subtract $2x^2$ and $4x$ from both sides.

$$0 = x^2 - 18x + 32$$ Divide both sides by 2.

$$0 = (x-16)(x-2)$$ Factor $x^2 - 18x + 32$.

$$x - 16 = 0 \quad \text{or} \quad x - 2 = 0$$ Set each factor equal to 0.

$$x = 16 \quad | \quad x = 2$$

(Extraneous.)

Verify that 2 satisfies the equation, but 16 does not. The solution set is $\{2\}$. ■

ORALS *Solve each equation.*

1. $\sqrt{x+2} = 3$ 7
2. $\sqrt{x-2} = 1$ 3
3. $\sqrt[3]{x+1} = 1$ 0
4. $\sqrt[3]{x-1} = 2$ 9
5. $\sqrt[4]{x-1} = 2$ 17
6. $\sqrt[5]{x+1} = 2$ 31

EXERCISE 7.5

In Exercises 1–54, solve each equation. Write all solutions and cross out those that are extraneous.

1. $\sqrt{5x-6} = 2$ 2
2. $\sqrt{7x-10} = 12$ 22
3. $\sqrt{6x+1} + 2 = 7$ 4
4. $\sqrt{6x+13} - 2 = 5$ 6
5. $2\sqrt{(4x+1)} = \sqrt{x+4}$ 0
6. $\sqrt{3(x+4)} = \sqrt{5x-12}$ 12
7. $\sqrt[3]{7n-1} = 3$ 4
8. $\sqrt[3]{12m+4} = 4$ 5
9. $\sqrt[4]{10p+1} = \sqrt[4]{11p-7}$ 8
10. $\sqrt[4]{10y+2} = 2\sqrt[4]{2}$ 3
11. $x = \dfrac{\sqrt{12x-5}}{2}$ $\frac{5}{2}, \frac{1}{2}$
12. $x = \dfrac{\sqrt{16x-12}}{2}$ 1, 3
13. $\sqrt{x+2} = \sqrt{4-x}$ 1
14. $\sqrt{6-x} = \sqrt{2x+3}$ 1

15. $2\sqrt{x} = \sqrt{5x - 16}$ 16

16. $3\sqrt{x} = \sqrt{3x + 12}$ 2

17. $r + 2 = \sqrt{5r + 34}$ 6, ~~-5~~

18. $s + 3 = 2\sqrt{5 - s}$ 1, ~~-11~~

19. $\sqrt{-5x + 24} = 6 - x$ 4, 3

20. $\sqrt{-x + 2} = x - 2$ 2, ~~1~~

21. $\sqrt{y + 2} = 4 - y$ 2, ~~7~~

22. $\sqrt{22y + 86} = y + 9$ 5, -1

23. $\sqrt{x}\sqrt{x + 16} = 15$ 9, ~~-25~~

24. $\sqrt{x}\sqrt{x + 6} = 4$ 2, ~~-8~~

25. $\sqrt[3]{x^3 - 7} = x - 1$ 2, -1

26. $\sqrt[3]{x^3 + 56} - 2 = x$ 2, -4

27. $\sqrt[4]{x^4 + 4x^2 - 4} = x$ 1, ~~-1~~

28. $\sqrt[4]{8x - 8} + 2 = 0$ ~~3~~, no solutions

29. $\sqrt[4]{12t + 4} + 2 = 0$ ~~1~~, no solutions

30. $u = \sqrt[4]{u^4 - 6u^2 + 24}$ 2, ~~-2~~

31. $\sqrt{2y + 1} = 1 - 2\sqrt{y}$ 0, ~~4~~

32. $\sqrt{u} + 3 = \sqrt{u - 3}$ ~~4~~, no solutions

33. $\sqrt{y + 7} + 3 = \sqrt{y + 4}$ ~~-3~~, no solutions

34. $1 + \sqrt{z} = \sqrt{z + 3}$ 1

35. $\sqrt{v} + \sqrt{3} = \sqrt{v + 3}$ 0

36. $\sqrt{x} + 2 = \sqrt{x + 4}$ 0

37. $2 + \sqrt{u} = \sqrt{2u + 7}$ 1, 9

38. $5r + 4 = \sqrt{5r + 20} + 4r$ 1, -4

39. $\sqrt{6t + 1} - 3\sqrt{t} = -1$ 4, ~~0~~

40. $\sqrt{4s + 1} - \sqrt{6s} = -1$ 6, ~~0~~

41. $\sqrt{2x + 5} + \sqrt{x + 2} = 5$ 2, ~~142~~

42. $\sqrt{2x + 5} + \sqrt{2x + 1} + 4 = 0$ ~~$\frac{5}{8}$~~, no solutions

43. $\sqrt{z - 1} + \sqrt{z + 2} = 3$ 2

44. $\sqrt{16v + 1} + \sqrt{8v + 1} = 12$ 3, ~~105~~

45. $\sqrt{x - 5} - \sqrt{x + 3} = 4$ ~~6~~, no solutions

46. $\sqrt{x + 8} - \sqrt{x - 4} = -2$ ~~8~~, no solutions

47. $\sqrt{x + 1} + \sqrt{3x} = \sqrt{5x + 1}$ 0, ~~$-\frac{12}{11}$~~

48. $\sqrt{3x} - \sqrt{x + 1} = \sqrt{x - 2}$ 3, ~~-1~~

49. $\sqrt{\sqrt{a} + \sqrt{a + 8}} = 2$ 1

50. $\sqrt{\sqrt{2y} - \sqrt{y - 1}} = 1$ 2

51. $\dfrac{6}{\sqrt{x + 5}} = \sqrt{x}$ 4, ~~-9~~

52. $\dfrac{\sqrt{2x}}{\sqrt{x + 2}} = \sqrt{x - 1}$ 2, ~~-1~~

53. $\sqrt{x + 2} + \sqrt{2x - 3} = \sqrt{11 - x}$ 2, ~~$\frac{21}{2}$~~

54. $\sqrt{8 - x} - \sqrt{3x - 8} = \sqrt{x - 4}$ 4, ~~$\frac{68}{13}$~~

Writing Exercises ■ *Write a paragraph using your own words.*

1. If both sides of an equation are raised to the same power, the resulting equation might not be equivalent to the original equation. Explain.

2. Explain why you must check each apparent solution of a radical equation.

Something to Think About ■

1. Solve $\sqrt[3]{2x} = \sqrt{x}$. 0, 4
(*Hint:* Square and then cube both sides.)

2. Solve $\sqrt[4]{x} = \sqrt{\dfrac{x}{4}}$. 0, 16

Review Exercises ■ *If* $P(x) = 3x^2 - 4x + 2$, *find each quantity.*

1. $P(2)$ 6

2. $P(-3)$ 41

3. $P(2)$ 6

4. $P(9)$ 209

7.6 Applications of Radicals

■ The Pythagorean Theorem ■ The Distance Formula ■ Other Applications

One of the important results from geometry is the Pythagorean theorem.

The Pythagorean Theorem

If we know the lengths of two legs of a right triangle, we can always find the length of the **hypotenuse** (the side opposite the 90° angle) by using the **Pythagorean theorem.**

Pythagorean Theorem If a and b are the lengths of two legs of a right triangle and c is the length of the hypotenuse, then

$$a^2 + b^2 = c^2$$

Pierre de Fermat (1601–1665)
Pierre de Fermat shares the honor with Descartes for discovering analytic geometry, and with Pascal for developing the theory of probability. But to Fermat alone goes credit for founding number theory. He is probably most famous for a theorem called *Fermat's last theorem*. It states that if n represents a number greater than 2, there are no whole numbers, a, b, and c that satisfy the equation $a^n + b^n = c^n$.

In words, the Pythagorean theorem says:

In any right triangle, the square of the hypotenuse is equal to the sum of squares of the other two sides.

Suppose the right triangle shown in Figure 7-1 has legs of length 3 and 4 units. To find the length of the hypotenuse, we use the Pythagorean theorem.

$$\begin{aligned} a^2 + b^2 &= c^2 \\ 3^2 + 4^2 &= c^2 \\ 9 + 16 &= c^2 \\ 25 &= c^2 \\ \sqrt{25} &= \sqrt{c^2} \\ 5 &= c \end{aligned}$$

FIGURE 7-1

The length of the hypotenuse is 5 units.

PERSPECTIVE

Don't Buck the System

Pythagoras was a teacher. Although it was unusual for schools at that time, his classes were coeducational. According to some legends, Pythagoras married one of his students. He and his followers formed a secret society with two rules: membership was for life, and members could not reveal the secrets they knew.

Much of their teaching was good mathematics, but some ideas were strange. To them, numbers were sacred. Because beans were used as counters to represent numbers, Pythagoreans refused to eat beans. They also believed that the *only* numbers were the whole numbers. To them, fractions were not numbers; $\frac{2}{3}$ was just a way of comparing the whole numbers 2 and 3. They believed that whole numbers were the building blocks of the universe, just as atoms are to us. The basic Pythagorean doctrine was "All things are number," and they meant *whole* number.

The Pythagorean theorem was an important discovery of the Pythagorean school, yet it caused some division in the ranks. The right triangle in Illustration 1 has two legs of length 1. By the Pythagorean theorem, the length of the hypotenuse is $\sqrt{2}$. One of their own group, Hippasus of Metapontum, discovered that $\sqrt{2}$ is an irrational number: there are *no* whole numbers *a* and *b* that make the fraction $\frac{a}{b}$ exactly equal to $\sqrt{2}$. This discovery was not appreciated by the other Pythagoreans. How could everything in the universe be described with whole numbers, when the side of this simple triangle couldn't? The Pythagoreans had a choice. Either revise and expand their beliefs, or cling to the old. According to legend, the group was at sea at the time of the discovery. Rather than upset the system, they threw Hippasus overboard.

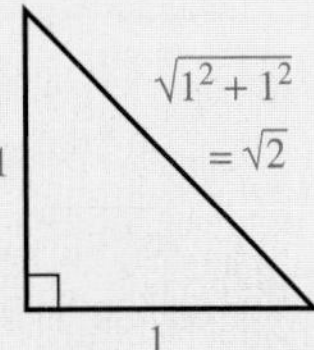

ILLUSTRATION 1

EXAMPLE 1 **Fighting fires** To fight a forest fire, the forestry department plans to clear a rectangular fire break around the fire, as shown in Figure 7-2. Crews are equipped with mobile communications with a 3000-yard range. Can crews at points A and B remain in radio contact?

Solution Points A, B, and C form a right triangle. To find the distance c from point A to point B, we can use the Pythagorean theorem, substituting 2400 for a and 1000 for b and solving for c.

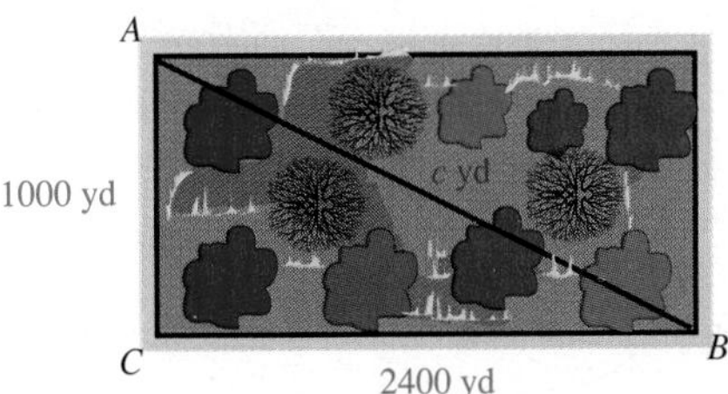

FIGURE 7-2

$$a^2 + b^2 = c^2$$
$$\mathbf{2400}^2 + \mathbf{1000}^2 = c^2$$
$$5{,}760{,}000 + 1{,}000{,}000 = c^2 \quad \text{Use a calculator.}$$
$$6{,}760{,}000 = c^2$$
$$\sqrt{6{,}760{,}000} = \sqrt{c^2} \quad \text{Take the square root of both sides.}$$
$$2600 = c \quad \text{Use a calculator to find the square root.}$$

The two fire crews are 2600 yards apart. Because this distance is less than the range of the radios, the crews can communicate. ■

The Distance Formula

Now that we have radicals, we can develop the **distance formula.** With it, we can find the distance between any two points that are graphed on a rectangular coordinate system.

To find the distance d between points $P(x_1, y_1)$ and $Q(x_2, y_2)$ shown in Figure 7-3, we construct the right triangle PRQ. The distance between P and R is $|x_2 - x_1|$, and the distance between R and Q is $|y_2 - y_1|$. We apply the Pythagorean theorem to the right triangle PRQ to get

$$[d(PQ)]^2 = |x_2 - x_1|^2 + |y_2 - y_1|^2 \quad \text{Read } d(PQ) \text{ as "the distance between } P \text{ and } Q\text{."}$$
$$= (x_2 - x_1)^2 + (y_2 - y_1)^2 \quad \text{Because } |x_2 - x_1|^2 = (x_2 - x_1)^2 \text{ and } |y_2 - y_1|^2 = (y_2 - y_1)^2.$$

or

1. $d(PQ) = \sqrt{(x_2 - x_1)^2 + (y_2 - y_1)^2}$

Equation 1 is the distance formula. Because it is one of the most important formulas in mathematics, take the time to memorize it.

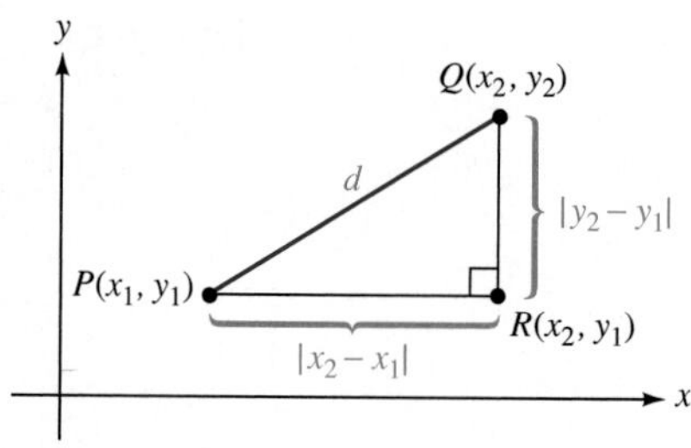

FIGURE 7-3

Distance Formula The distance between two points $P(x_1, y_1)$ and $Q(x_2, y_2)$ is given by the formula

$$d(PQ) = \sqrt{(x_2 - x_1)^2 + (y_2 - y_1)^2}$$

EXAMPLE 2 Find the distance between points $P(-2, 3)$ and $Q(4, -5)$.

Solution To find the distance, we can use the distance formula by substituting 4 for x_2, -2 for x_1, -5 for y_2, and 3 for y_1.

$$\begin{aligned} d(PQ) &= \sqrt{(x_2 - x_1)^2 + (y_2 - y_1)^2} \\ &= \sqrt{[4 - (-2)]^2 + (-5 - 3)^2} \\ &= \sqrt{(4 + 2)^2 + (-5 - 3)^2} \\ &= \sqrt{6^2 + (-8)^2} \\ &= \sqrt{36 + 64} \\ &= \sqrt{100} \\ &= 10 \end{aligned}$$

The distance between points P and Q is 10 units. ■

EXAMPLE 3 **Building a freeway** In a certain city, streets run north and south and avenues run east and west. Streets are 750 feet apart, and so are avenues. The city plans to construct a straight freeway from the intersection of 21st Street and 4th Avenue to the intersection of 111th Street and 60th Avenue. How long will the freeway be?

Solution We can represent the roads of the city by the coordinate system in Figure 7-4, where the units on each axis represent 750 feet. We represent the end of the freeway at 21st Street and 4th Avenue by the point $(x_1, y_1) = (21, 4)$. The other end is $(x_2, y_2) = (111, 60)$.

We use the distance formula to find the length of the freeway.

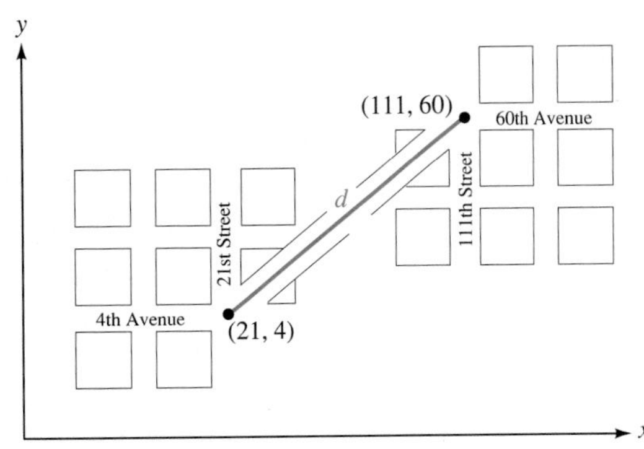

FIGURE 7-4

$$d = \sqrt{(x_2 - x_1)^2 + (y_2 - y_1)^2}$$
$$d = \sqrt{(111 - 21)^2 + (60 - 4)^2}$$
$$= \sqrt{8100 + 3136}$$
$$= \sqrt{11{,}236}$$
$$= 106 \qquad \text{Use a calculator to find the square root.}$$

Because each unit represents 750 feet, the length of the freeway is 79,500 feet, or about 15 miles. ■

Other Applications

EXAMPLE 4 **Bowling** The velocity, v, of an object after it has fallen d feet is given by the equation $v^2 = 64d$. An inexperienced bowler lofts the ball 4 feet. With what velocity does it strike the alley?

Solution We find the velocity by substituting 4 for d in the equation $v^2 = 64d$ and solving for v.

$$v^2 = 64d$$
$$v^2 = 64(4)$$
$$v^2 = 256$$
$$v = \sqrt{256} \qquad \text{Take the square root of both sides. Only the positive square root is meaningful.}$$
$$= 16$$

The ball strikes the alley at a velocity of 16 feet per second. ■

EXAMPLE 5 **Horizon distance** The higher a lookout tower is built, the farther an observer can see. That distance d, (called the **horizon distance,** measured in miles) is related to the height, h, of the observer (measured in feet) by the equation

$$d = 1.4\sqrt{h}$$

A forest fire watch tower is to be built on a hill overlooking a national park. Observers must be able to see to the edge of the forest, 25 miles away. How high above the terrain must the tower be built?

Solution Because the horizon distance is 25 miles, we substitute 25 for d in the equation and solve for h.

$$d = 1.4\sqrt{h}$$
$$25 = 1.4\sqrt{h}$$
$$25^2 = \left(1.4\sqrt{h}\right)^2 \qquad \text{Square both sides.}$$
$$625 = 1.96h$$

$$\frac{625}{1.96} = h \qquad \text{Divide both sides by 1.96.}$$

$$319 \approx h \qquad \text{Read} \approx \text{as "approximately equal to."}$$

The observers must be approximately 319 feet above the surrounding forest. ■

ORALS *Evaluate each expression.*

1. $\sqrt{25}$ 5
2. $\sqrt{100}$ 10
3. $\sqrt{169}$ 13
4. $\sqrt{3^2 + 4^2}$ 5
5. $\sqrt{8^2 + 6^2}$ 10
6. $\sqrt{5^2 + 12^2}$ 13
7. $\sqrt{5^2 - 3^2}$ 4
8. $\sqrt{5^2 - 4^2}$ 3
9. $\sqrt{169 - 12^2}$ 5

EXERCISE 7.6

In Exercises 1–4, the lengths of two sides of the right triangle ABC shown in Illustration 1 are given. Find the length of the missing side.

1. $a = 6$ ft and $b = 8$ ft 10 ft
2. $a = 10$ cm and $c = 26$ cm 24 cm
3. $b = 18$ m and $c = 82$ m 80 m
4. $a = 14$ in. and $c = 50$ in. 48 in.

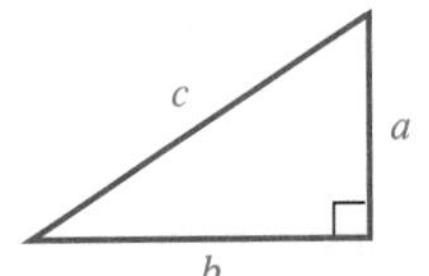

ILLUSTRATION 1

5. **Sailing** Refer to the sailboat in Illustration 2. How long must a rope be to reach from the top of the mast to the bow? 13 ft

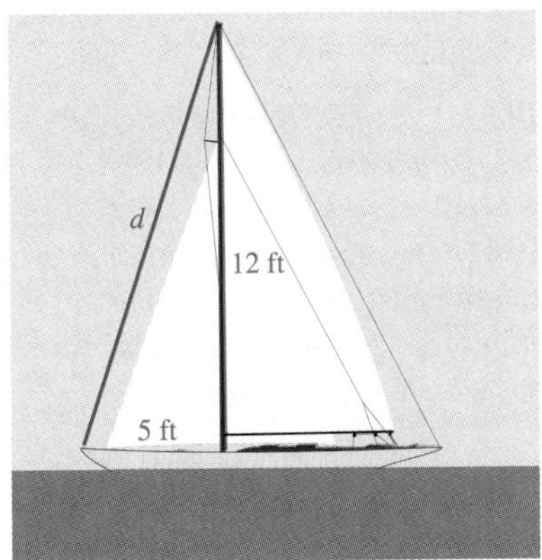

ILLUSTRATION 2

6. **Carpentry** The gable end of the roof shown in Illustration 3 is divided in half by vertical brace. How long is the brace? 8 ft

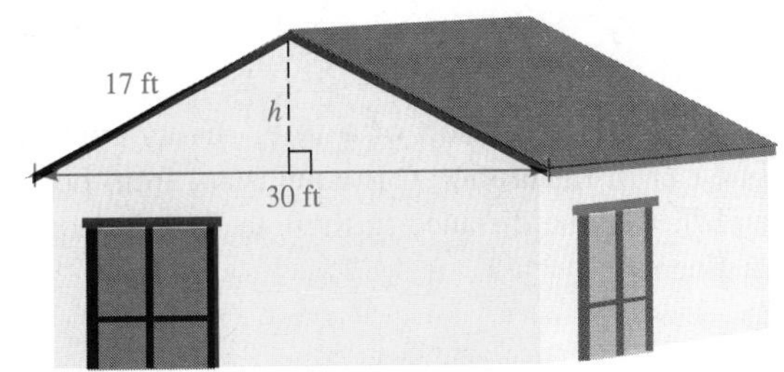

ILLUSTRATION 3

7. **Baseball** The baseball diamond shown in Illustration 4 is a square, 90 feet on a side, and the pitcher's mound is 60 feet, 6 inches from home plate. How far is the mound from second base? 66.8 ft

8. **Baseball** A shortstop fields a grounder at a point one-third of the way from second base to third base. How far will the shortstop have to throw the ball to make an out at first base? 94.9 ft

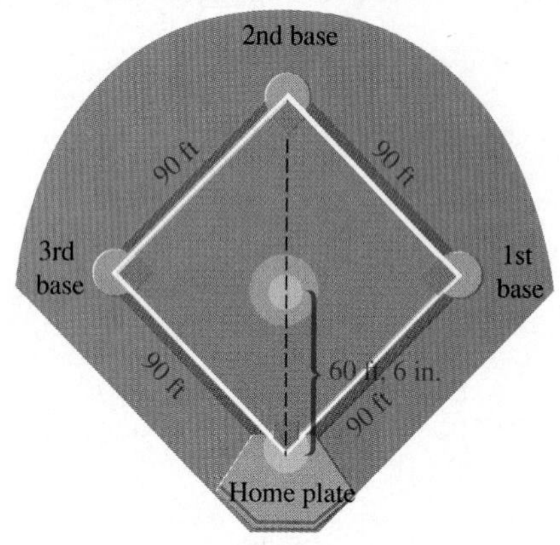

ILLUSTRATION 4

9. **Geometry** Find the length of the diagonal of one of the faces of the cube shown in Illustration 5. 9.9 cm

10. **Geometry** Find the length of the diagonal of the cube shown in Illustration 5. 12.1 cm

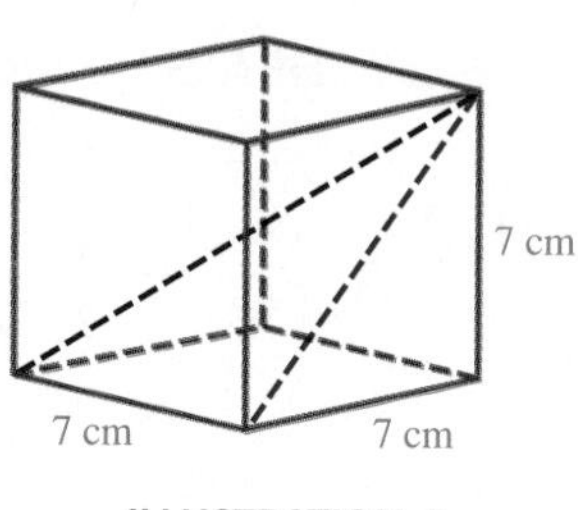

ILLUSTRATION 5

In Exercises 11–20, find the distance between P and Q.

11. $Q(0, 0)$, $P(3, -4)$ 5
12. $Q(0, 0)$, $P(-6, 8)$ 10
13. $P(2, 4)$, $Q(5, 8)$ 5
14. $P(5, 9)$, $Q(8, 13)$ 5
15. $P(-2, -8)$, $Q(3, 4)$ 13
16. $P(-5, -2)$, $Q(7, 3)$ 13
17. $P(6, 8)$, $Q(12, 16)$ 10
18. $P(10, 4)$, $Q(2, -2)$ 10
19. $Q(-3, 5)$, $P(-5, -5)$ $2\sqrt{26}$
20. $Q(2, -3)$, $P(4, -8)$ $\sqrt{29}$

21. **Geometry** Show that a triangle with vertices at $(-2, 4)$, $(2, 8)$, and $(6, 4)$ is isosceles.

22. **Geometry** Show that a triangle with vertices at $(-2, 13)$, $(-8, 9)$, and $(-2, 5)$ is isosceles.

23. **Finding the equation of a line** Every point on the line CD in Illustration 6 is equidistant from points A and B. Use the distance formula to find the equation of line CD. $x = 7$

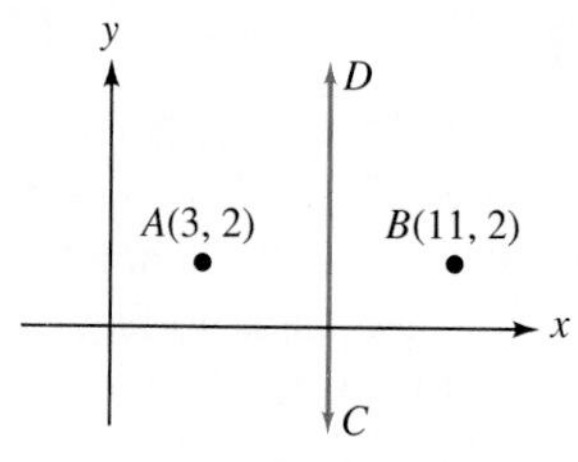

ILLUSTRATION 6

24. **Geometry** Show that a triangle with vertices at $(2, 3)$, $(-3, 4)$, and $(1, -2)$ is a right triangle. (*Hint:* If the Pythagorean relation holds, then the triangle is a right triangle.)

25. **Geometry** Find the coordinates of the two points on the x-axis that are $\sqrt{5}$ units from the point $(5, 1)$. $(7, 0)$ and $(3, 0)$

26. **Geometry** The square in Illustration 7 has an area of 18 square units, and its diagonals lie on the x- and y-axes. Find the coordinates of each corner of the square. $(3, 0)$, $(0, 3)$, $(-3, 0)$, $(0, -3)$

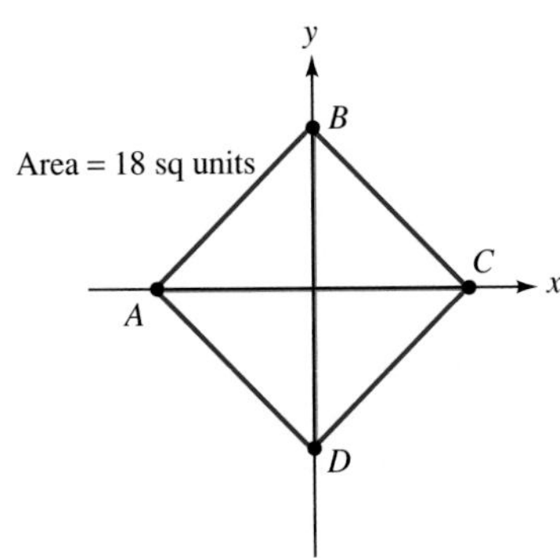

ILLUSTRATION 7

27. Packing a shotgun The diagonal d of a rectangular box with dimensions $a \times b \times c$ is given by the formula $d = \sqrt{a^2 + b^2 + c^2}$. Can a hunter fit a 32-inch shotgun in the shipping carton in Illustration 8? not quite

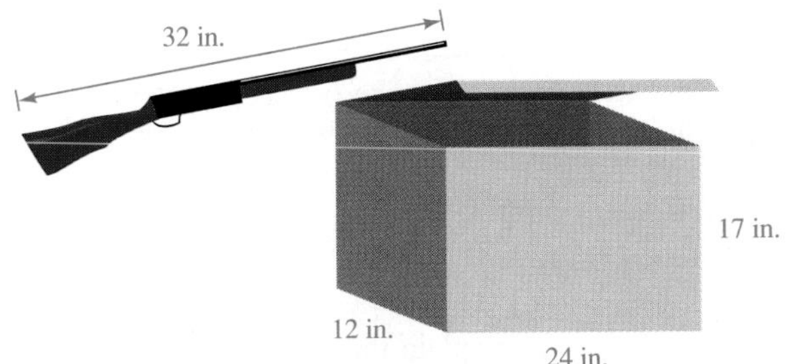

ILLUSTRATION 8

28. Shipping packages A delivery service won't accept a package for shipping if any dimension exceeds 21 inches. An archeologist wants to ship a 36-inch femur. Will it fit in a 3-inch tall box that has a square base, 21 inches on each side? no

29. Shipping packages Can the archeologist in Exercise 28 ship the femur in a cubical box 21 inches on an edge? yes

30. Reach of a ladder The base of the 37-foot ladder in Illustration 9 is 9 feet from the wall. Will the top reach a window ledge that is 35 feet above the ground? yes

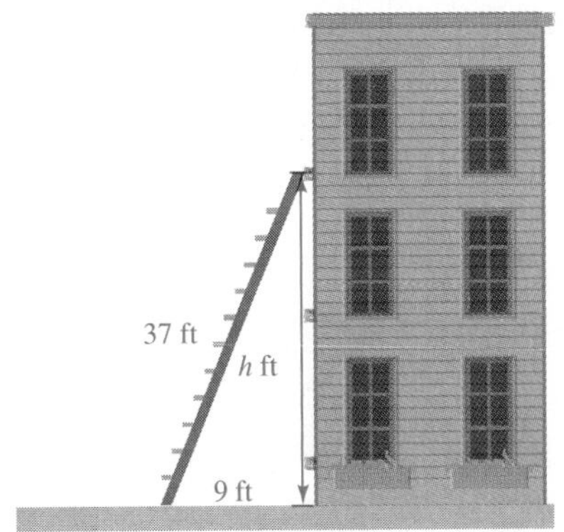

ILLUSTRATION 9

31. Telephone service The telephone cable in Illustration 10 runs from A to B to C to D. How much cable is required to run from A to D directly? 173 yd

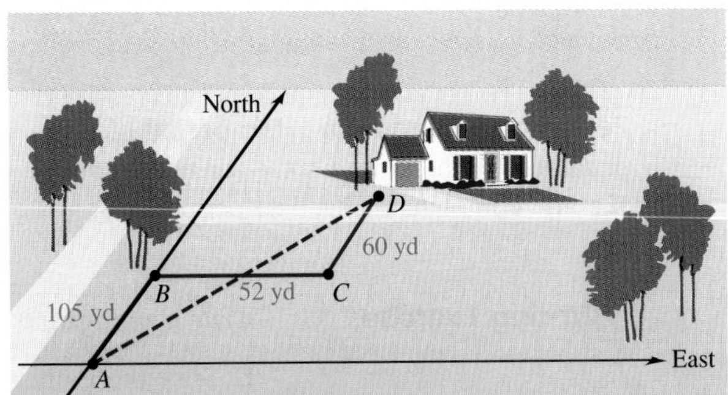

ILLUSTRATION 10

32. Electric service The power company routes its lines as in Illustration 11. How much wire could be saved by going directly from A to E? 90 yd

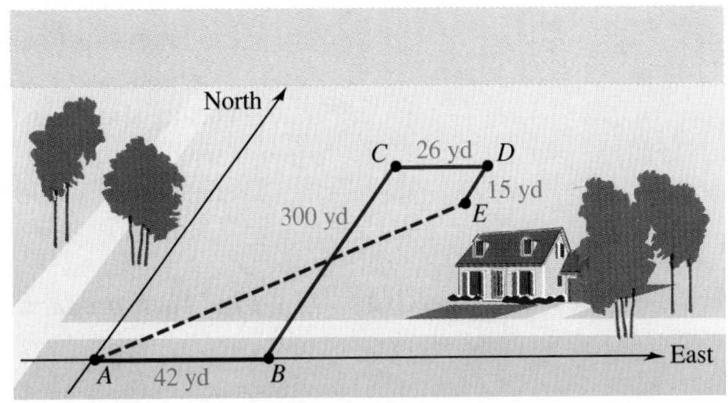

ILLUSTRATION 11

33. **Supporting a weight** A weight placed on the tight wire in Illustration 12 pulls the center down 1 foot. By how much is the wire stretched? Round the answer to the nearest hundredth of a foot. 0.05 ft

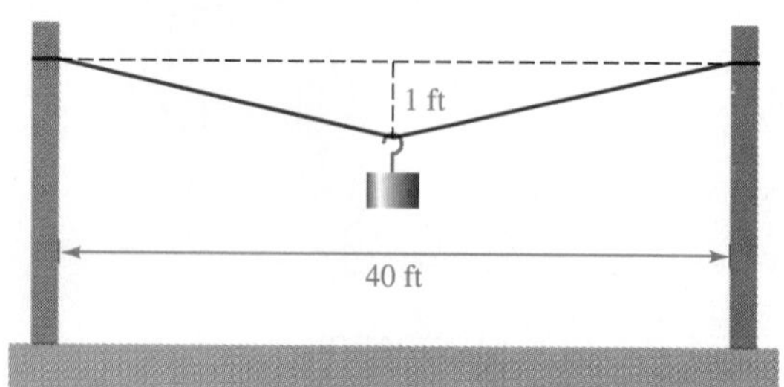

ILLUSTRATION 12

34. **Geometry** The side, s, of a square with area A square feet is given by the formula $s = \sqrt{A}$. Find the perimeter of a square with an area of 49 square feet. 28 ft

35. **Period of a pendulum** The **period** of a pendulum is the time it takes for the pendulum to make one complete swing. The period t (in seconds) is related to the length L (in feet) of the pendulum by the equation $t = 1.11\sqrt{L}$. What is the period of a pendulum that is 3 feet long? 1.92 sec

36. **Length of a pendulum** Find the length of a pendulum with a period of 2.94 seconds. (Refer to Exercise 35.) 7 ft

37. **Volume of a cube** The total surface area, A, of a cube is related to its volume, V, by the formula $A = 6\sqrt[3]{V^2}$. Find the volume of a cube with a surface area of 24 square centimeters. 8 cm^3

38. **Area of many cubes** A grain of table salt is a cube with a volume of approximately 6×10^{-6} cubic inches, and there are about 1.5 million grains of salt in one cup. Find the total surface area of the salt in one cup. (See Exercise 37.) almost 3000 in.^{2}, or 20 ft^2

39. **View from a submarine** A submarine's periscope extends 4.7 feet above the surface. How far is the horizon? (See Example 5.) 3 mi

40. **View from a submarine** How far out of the water must a submarine periscope extend to provide a 4-mile horizon? 8.2 ft

Writing Exercises

Write a paragraph using your own words.

1. State the Pythagorean theorem.
2. Explain the distance formula.

Something to Think About

1. The formula

$$I = \frac{704w}{h^2}$$

where w is weight in pounds and h is height in inches, can be used to estimate body mass index I. The scale, shown in Table 1, can be used to judge a person's risk of heart attack. Find your risk factor.

Mass Index	Risk
20–26	normal
27–29	higher risk
30 and above	very high risk

TABLE 1

2. A girl, weighing 104 pounds, has a body mass index of 25. How tall is she? 54.1 in.

Review Exercises

Find each product.

1. $(4x + 2)(3x - 5)$ $12x^2 - 14x - 10$
2. $(3y - 5)(2y + 3)$ $6y^2 - y - 15$
3. $(5t + 4s)(3t - 2s)$ $15t^2 + 2ts - 8s^2$
4. $(4r - 3)(2r^2 + 3r - 4)$ $8r^3 + 6r^2 - 25r + 12$

7.7 Radicals with Different Orders (Optional)

Teaching Tip:

$\sqrt[4]{4} = \sqrt{\sqrt{4}}$

With a calculator, have students verify that

4 [√] [√] = 2 [√]

Fractional exponents enable us to simplify many radical expressions. For example, to simplify $\sqrt[4]{4}$, we write the expression in rational exponent form and proceed as follows:

$$\begin{aligned}\sqrt[4]{4} &= \sqrt[4]{2^2}\\ &= (2^2)^{1/4}\\ &= 2^{2 \cdot 1/4}\\ &= 2^{1/2}\\ &= \sqrt{2}\end{aligned}$$

(a, b)

EXAMPLE 1 Simplify **a.** $\sqrt[6]{25}$, **b.** $\sqrt[12]{x^4}$, **c.** $\sqrt[20]{x^6y^2}$, and **d.** $\sqrt[9]{8x^6}$. In all cases, $x > 0$ and $y > 0$.

Solution

a. $$\begin{aligned}\sqrt[6]{25} &= \sqrt[6]{5^2}\\ &= 5^{2/6}\\ &= 5^{1/3}\\ &= \sqrt[3]{5}\end{aligned}$$

b. $$\begin{aligned}\sqrt[12]{x^4} &= x^{4/12}\\ &= x^{1/3}\\ &= \sqrt[3]{x}\end{aligned}$$

c. $$\begin{aligned}\sqrt[20]{x^6y^2} &= (x^6y^2)^{1/20}\\ &= x^{6/20}y^{2/20}\\ &= x^{3/10}y^{1/10}\\ &= (x^3y)^{1/10}\\ &= \sqrt[10]{x^3y}\end{aligned}$$

d. $$\begin{aligned}\sqrt[9]{8x^6} &= \sqrt[9]{2^3x^6}\\ &= (2^3x^6)^{1/9}\\ &= 2^{3/9}x^{6/9}\\ &= 2^{1/3}x^{2/3}\\ &= (2x^2)^{1/3}\\ &= \sqrt[3]{2x^2}\end{aligned}$$ ■

By using rational exponents, we can add, subtract, multiply, and divide many radicals with different orders (radicals with different indexes). For example, to add the radicals $\sqrt[4]{25}$ and $\sqrt{20}$, we proceed as follows:

$$\begin{aligned}\sqrt[4]{25} + \sqrt{20} &= \sqrt[4]{5^2} + \sqrt{4 \cdot 5}\\ &= (5^2)^{1/4} + \sqrt{4}\sqrt{5}\\ &= 5^{2/4} + 2\sqrt{5}\\ &= 5^{1/2} + 2\sqrt{5}\\ &= \sqrt{5} + 2\sqrt{5}\\ &= 3\sqrt{5}\end{aligned}$$

To multiply $\sqrt{2}$ by $\sqrt[3]{5}$, we first write each number as a sixth root. To do so, we write each radical as an expression with a rational exponent, change each exponent to a fraction with a denominator of 6, and change back to radical notation:

$$\begin{aligned}\sqrt{2} &= 2^{1/2} \\ &= 2^{3/6} \\ &= (2^3)^{1/6} \\ &= \sqrt[6]{2^3} \\ &= \sqrt[6]{8}\end{aligned} \qquad \begin{aligned}\sqrt[3]{5} &= 5^{1/3} \\ &= 5^{2/6} \\ &= (5^2)^{1/6} \\ &= \sqrt[6]{5^2} \\ &= \sqrt[6]{25}\end{aligned}$$

We can then multiply the sixth roots:

$$\begin{aligned}\sqrt{2}\sqrt[3]{5} &= \sqrt[6]{8}\sqrt[6]{25} \\ &= \sqrt[6]{8 \cdot 25} \\ &= \sqrt[6]{200}\end{aligned}$$

EXAMPLE 2 Write $\dfrac{\sqrt{3}}{\sqrt[3]{4}}$ as an expression containing a single radical.

Solution We first eliminate the radical in the denominator by multiplying the numerator and denominator by $\sqrt[3]{2}$ and then proceed as follows:

$$\begin{aligned}
\frac{\sqrt{3}}{\sqrt[3]{4}} &= \frac{\sqrt{3}\sqrt[3]{2}}{\sqrt[3]{4}\sqrt[3]{2}} && \text{Multiply numerator and denominator by } \sqrt[3]{2}. \\
&= \frac{\sqrt{3}\sqrt[3]{2}}{\sqrt[3]{8}} && \text{Simplify.} \\
&= \frac{\sqrt{3}\sqrt[3]{2}}{2} && \sqrt[3]{8} = 2. \\
&= \frac{3^{1/2}2^{1/3}}{2} && \text{Write the radicals in the numerator in fractional-exponent form.} \\
&= \frac{3^{3/6}2^{2/6}}{2} && \tfrac{1}{2} = \tfrac{3}{6}, \text{ and } \tfrac{1}{3} = \tfrac{2}{6}. \\
&= \frac{\sqrt[6]{3^3}\sqrt[6]{2^2}}{2} && \text{Write the expression with fractional exponents in radical form.} \\
&= \frac{\sqrt[6]{3^3 \cdot 2^2}}{2} && \text{Multiply the radicals.} \\
&= \frac{\sqrt[6]{27 \cdot 4}}{2} && 3^3 = 27, \text{ and } 2^2 = 4. \\
&= \frac{\sqrt[6]{108}}{2} && 27 \cdot 4 = 108.
\end{aligned}$$

■

EXAMPLE 3 Write $\sqrt{8x^3y}\sqrt[3]{9xy^4}$ as a single radical ($x > 0$, $y > 0$).

Solution We first remove all perfect square factors from the radicand within the square root and all perfect cube factors from the radicand within the cube root:

$$\begin{aligned}\sqrt{8x^3y}\sqrt[3]{9xy^4} &= \sqrt{4 \cdot x^2 \cdot 2xy}\sqrt[3]{y^3 \cdot 9xy} \\ &= 2x\sqrt{2xy}\, y\sqrt[3]{9xy} \\ &= 2xy\sqrt{2xy}\sqrt[3]{9xy}\end{aligned}$$

We now write each radical as an expression with fractional exponents with a common denominator:

$$\begin{aligned}\sqrt{8x^3y}\sqrt[3]{9xy^4} &= 2xy(2xy)^{1/2}(9xy)^{1/3} \\ &= 2xy(2xy)^{3/6}(9xy)^{2/6}\end{aligned}$$

Finally, we change the expression back to radical form and find the product of the sixth roots:

$$\begin{aligned}\sqrt{8x^3y}\sqrt[3]{9xy^4} &= 2xy\sqrt[6]{(2xy)^3}\,\sqrt[6]{(9xy)^2} \\ &= 2xy\sqrt[6]{8x^3y^3 \cdot 81x^2y^2} \\ &= 2xy\sqrt[6]{648x^5y^5}\end{aligned}$$

■

EXAMPLE 4 Write $\sqrt[3]{3x}\sqrt[5]{2x^2}$ as a single radical.

Solution We change both radicals to expressions with fractional exponents with the same denominator, change back to radical form, and find the product of the fifteenth roots:

$$\begin{aligned}\sqrt[3]{3x}\sqrt[5]{2x^2} &= (3x)^{1/3}(2x^2)^{1/5} \\ &= (3x)^{5/15}(2x^2)^{3/15} \\ &= \sqrt[15]{(3x)^5}\,\sqrt[15]{(2x^2)^3} \\ &= \sqrt[15]{243x^5}\,\sqrt[15]{8x^6} \\ &= \sqrt[15]{243 \cdot 8x^5x^6} \\ &= \sqrt[15]{1944x^{11}}\end{aligned}$$

■

ORALS *Simplify each radical.*

1. $\sqrt[4]{25}$ $\sqrt{5}$

2. $\sqrt[9]{8}$ $\sqrt[3]{2}$

Give each radical as a sixth root.

3. $\sqrt[3]{2}$ $\sqrt[6]{4}$

4. $\sqrt{3}$ $\sqrt[6]{27}$

EXERCISE 7.7

In Exercises 1–16, simplify each radical. All variables are positive numbers.

1. $\sqrt[4]{9}$ $\sqrt{3}$
2. $\sqrt[4]{64}$ $2\sqrt{2}$
3. $\sqrt[18]{x^9}$ $\sqrt{x}$
4. $\sqrt[12]{x^6}$ $\sqrt{x}$
5. $\sqrt[4]{x^8}$ x^2
6. $\sqrt[5]{y^{10}}$ y^2
7. $\sqrt[6]{y^{18}}$ y^3
8. $\sqrt[8]{y^{32}}$ y^4
9. $\sqrt[4]{16x^{12}}$ $2x^3$
10. $\sqrt[6]{27x^3}$ $\sqrt{3x}$
11. $\sqrt[8]{81x^4}$ $\sqrt{3x}$
12. $\sqrt[10]{243x^{10}}$ $x\sqrt{3}$
13. $\sqrt[9]{-8x^9}$ $-x\sqrt[3]{2}$
14. $\sqrt[9]{-27x^{18}}$ $-x^2\sqrt[3]{3}$
15. $\sqrt[3]{-64a^{12}}$ $-4a^4$
16. $\sqrt[3]{-125a^3b^6}$ $-5ab^2$

In Exercises 17–26, simplify all radicals and combine terms, if possible. All variables are positive numbers.

17. $\sqrt{12} + \sqrt[4]{9}$ $3\sqrt{3}$
18. $\sqrt{8} - \sqrt[4]{4}$ $\sqrt{2}$
19. $\sqrt[4]{64} - \sqrt{32}$ $-2\sqrt{2}$
20. $\sqrt[4]{144} + \sqrt{12}$ $4\sqrt{3}$
21. $\sqrt[4]{x^6} - \sqrt[6]{x^9}$ 0
22. $\sqrt[3]{a^4} + \sqrt[6]{a^8}$ $2a\sqrt[3]{a}$
23. $\sqrt[6]{x^6y^2} + 2x\sqrt[3]{y}$ $3x\sqrt[3]{y}$
24. $\sqrt[3]{a^2b^3} + \sqrt[9]{a^6b^9}$ $2b\sqrt[3]{a^2}$
25. $5\sqrt[5]{x^6y^{11}z^{16}} + 12\sqrt[10]{x^{12}y^{22}z^{32}}$ $17xy^2z^3\sqrt[5]{xyz}$
26. $13\sqrt[6]{a^8b^6c^{12}} - 4\sqrt[3]{a^4b^3c^6}$ $9abc^2\sqrt[3]{a}$

In Exercises 27–42, write each radical as another radical with the indicated order. All variables are positive numbers.

27. $\sqrt{3}$, sixth order $\sqrt[6]{27}$
28. $\sqrt{2}$, sixth order $\sqrt[6]{8}$
29. $\sqrt{5}$, eighth order $\sqrt[8]{625}$
30. $\sqrt{7}$, eighth order $\sqrt[8]{2401}$
31. $\sqrt[3]{15}$, sixth order $\sqrt[6]{225}$
32. $\sqrt[3]{9}$, sixth order $\sqrt[6]{81}$
33. $\sqrt[3]{4}$, ninth order $\sqrt[9]{64}$
34. $\sqrt[3]{7}$, ninth order $\sqrt[9]{343}$
35. $\sqrt[6]{3^3}$, second order $\sqrt{3}$
36. $\sqrt[8]{7^4}$, fourth order $\sqrt[4]{49}$
37. $\sqrt[9]{5^3}$, third order $\sqrt[3]{5}$
38. $\sqrt[9]{7^6}$, third order $\sqrt[3]{49}$
39. $\sqrt{2xy}$, sixth order $\sqrt[6]{8x^3y^3}$
40. $\sqrt{3x^3y}$, sixth order $x\sqrt[6]{27x^3y^3}$
41. $\sqrt[3]{4x^2y}$, sixth order $\sqrt[6]{16x^4y^2}$
42. $\sqrt[3]{9xy^2}$, sixth order $\sqrt[6]{81x^2y^4}$

In Exercises 43–66, write each expression as a single radical. All variables are positive numbers.

43. $\sqrt{5}\sqrt[3]{7}$ $\sqrt[6]{6125}$
44. $\sqrt[3]{9}\sqrt{2}$ $\sqrt[6]{648}$
45. $\sqrt[3]{3}\sqrt{5}$ $\sqrt[6]{1125}$
46. $\sqrt[3]{9}\sqrt{5}$ $\sqrt[6]{10{,}125}$
47. $\sqrt[3]{2}\sqrt{2}$ $\sqrt[6]{32}$
48. $\sqrt{5}\sqrt[3]{5}$ $\sqrt[6]{3125}$
49. $\sqrt[3]{2}\sqrt[5]{8}$ $\sqrt[15]{16{,}384}$
50. $\sqrt[5]{3}\sqrt[3]{2}$ $\sqrt[15]{864}$
51. $\sqrt{5x}\sqrt[3]{xy}$ $\sqrt[6]{125x^5y^2}$
52. $\sqrt[3]{4x}\sqrt{xy}$ $\sqrt[6]{16x^5y^3}$
53. $\sqrt[3]{25x}\sqrt[5]{3y^2}$ $\sqrt[15]{5^{10}3^3x^5y^6}$
54. $\sqrt[3]{3x}\sqrt[5]{2x^2}$ $\sqrt[15]{1944x^{11}}$
55. $\sqrt[7]{xy^3}\sqrt[5]{x^3y^4}$ $y\sqrt[35]{x^{26}y^8}$
56. $\sqrt[9]{x^2y}\sqrt[3]{xy^2}$ $\sqrt[9]{x^5y^7}$
57. $\dfrac{\sqrt[3]{3}}{\sqrt{2}}$ $\dfrac{\sqrt[6]{72}}{2}$
58. $\dfrac{\sqrt{2}}{\sqrt[3]{3}}$ $\dfrac{\sqrt[6]{648}}{3}$
59. $\dfrac{\sqrt[5]{3}}{\sqrt[3]{4}}$ $\dfrac{\sqrt[15]{864}}{2}$
60. $\dfrac{\sqrt[5]{4}}{\sqrt[3]{4}}$ $\dfrac{\sqrt[15]{2048}}{2}$

61. $\dfrac{\sqrt[5]{3}}{\sqrt{3}}$ $\dfrac{\sqrt[10]{2187}}{3}$

62. $\dfrac{\sqrt{3}}{\sqrt[5]{3}}$ $\sqrt[10]{27}$

63. $\dfrac{\sqrt[7]{4x}}{\sqrt{2x}}$ $\dfrac{\sqrt[14]{2048x^9}}{2x}$

64. $\dfrac{\sqrt[9]{xy}}{\sqrt{xy}}$ $\dfrac{\sqrt[18]{x^{11}y^{11}}}{xy}$

65. $\dfrac{\sqrt[5]{3x}}{\sqrt[3]{3}}$ $\dfrac{\sqrt[15]{3^{13}x^3}}{3}$

66. $\dfrac{\sqrt[7]{3x}}{\sqrt[3]{3}}$ $\dfrac{\sqrt[21]{3^{17}x^3}}{3}$

Writing Exercises *Write a paragraph using your own words.*

1. Explain how to simplify $\sqrt[6]{9}$.

2. Explain how to simplify $\dfrac{1}{\sqrt[4]{4}}$.

Something to Think About

1. Find $\sqrt[4]{\sqrt{3}}$. $\sqrt[8]{3}$

2. Find $\sqrt[3]{\sqrt{5}}$. $\sqrt[6]{5}$

Review Exercises *Solve each equation.*

1. $|3a - 2| = 10$ $4, -\frac{8}{3}$

2. $\left|\dfrac{2x + 1}{3}\right| = 3$ $4, -5$

Solve each inequality.

3. $|3a - 2| < 7$ $\left(-\frac{5}{3}, 3\right)$

4. $\left|\dfrac{2n + 3}{3}\right| \geq 5$ $(-\infty, -9] \cup [6, \infty)$

PROJECT Square Corners

By the Pythagorean theorem, any triangle with sides a, b, and c for which $a^2 + b^2 = c^2$ must be a right triangle. For example, a triangle with sides of 3, 4, and 5 feet is a right triangle, because $3^2 + 4^2 = 5^2$ $(9 + 16 = 25)$.

Carpenters and builders make use of this fact to determine whether the walls of a room or the corner of a foundation are really perpendicular (meet at a right angle). To decide whether the walls in Figure 7-5 are square, a carpenter would make a mark 3 feet from the corner on one wall, and 4 feet from the corner on the other wall. If the distance between marks is exactly 5 feet, the corner is square.

- Use this technique to determine whether the walls of a few rooms are square. Try it in some rooms of recent construction, and in some rooms of an older building. Are craftsmen more careful now, or then?
- If you were cutting the corners for a countertop that needed to fit the corner of a kitchen, what single measurement would help you get the correct angle? Explain.

(continued)

PROJECT ■ Square Corners *(continued)*

- Devise a technique using the Pythagorean theorem to determine whether the outside corner of a building is square. Use it to check the outside corner of your house or a building on your campus.

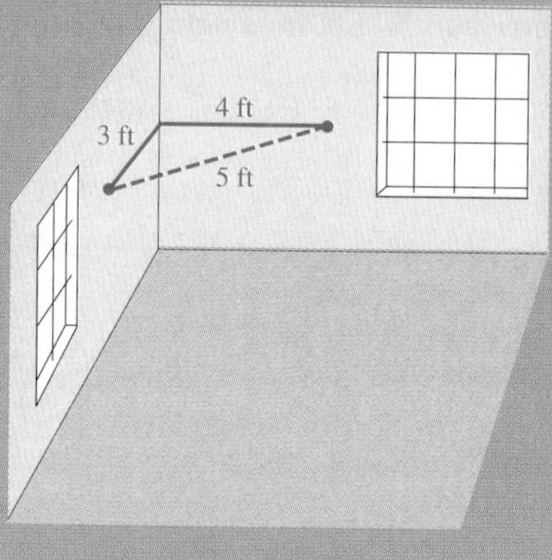

FIGURE 7-5

Chapter Summary

KEY WORDS

conjugate binomials (7.4)
cube root (7.2)
index of a radical (7.2)
like radicals (7.3)
power rule (7.5)
principle *n*th root (7.2)
radical expression (7.2)
radical sign (7.2)
radicand (7.2)
rationalizing the denominator (7.4)
similar radicals (7.3)
square root (7.2)

KEY IDEAS

(7.1) If n is a natural number and a is a real number, then

	n is odd	n is even
If $a > 0$	$(a^{1/n})^n = a$	$(a^{1/n})^n = a$
If $a = 0$	$a^{1/n} = 0$	$a^{1/n} = 0$
If $a < 0$	$(a^{1/n})^n = a$	$a^{1/n}$ is an imaginary number.

If m and n are positive integers, $\frac{m}{n}$ is in simplest form, and $a^{1/n}$ is a real number,

$$a^{m/n} = (a^{1/n})^m = (a^m)^{1/n}$$

$$a^{-m/n} = \frac{1}{a^{m/n}} \text{ and } \frac{1}{a^{-m/n}} = a^{m/n} \text{ (provided } a \neq 0)$$

(7.2) If n is a natural number greater than 1 and a is a real number, then

	n is odd	n is even
If $a > 0$	$(\sqrt[n]{a})^n = a$	$(\sqrt[n]{a})^n = a$
If $a = 0$	$(\sqrt[n]{a})^n = 0$	$(\sqrt[n]{a})^n = 0$
If $a < 0$	$(\sqrt[n]{a})^n = a$	$\sqrt[n]{a}$ is an imaginary number.

If n is an even natural number, then $\sqrt[n]{a^n} = |a|$.

If n is an odd natural number, then $\sqrt[n]{a^n} = a$.

Properties of radicals: If all radicals are real numbers,

$$\sqrt[n]{ab} = \sqrt[n]{a}\sqrt[n]{b}$$

and if $b \neq 0$,

$$\sqrt[n]{\frac{a}{b}} = \frac{\sqrt[n]{a}}{\sqrt[n]{b}}$$

(7.3) Like radicals can be combined by addition and subtraction: $3\sqrt{2} + 5\sqrt{2} = 8\sqrt{2}$.

Radicals that are not similar can often be converted to radicals that are similar and then be combined:

$$\sqrt{2} + \sqrt{8} = \sqrt{2} + \sqrt{4}\sqrt{2} = \sqrt{2} + 2\sqrt{2} = 3\sqrt{2}$$

(7.4) If two radicals have the same index, they can be multiplied: $\sqrt{3x}\sqrt{6x} = \sqrt{18x^2} = 3x\sqrt{2}$, provided that $x \geq 0$.

Radicals can be removed from the denominator of a fraction by rationalizing the denominator.

To rationalize the binomial denominator of a fraction, multiply both the numerator and the denominator by the conjugate of the binomial in the denominator.

(7.5) **The power rule:** If $a = b$, then $a^n = b^n$.

Raising both sides of an equation to the same power can lead to extraneous solutions. Be sure to check all suspected solutions.

(7.6) **The Pythagorean theorem:** If a and b are the lengths of the legs of a right triangle and c is the length of the hypotenuse, then $a^2 + b^2 = c^2$.

The distance formula: $d = \sqrt{(x_2 - x_1)^2 + (y_2 - y_1)^2}$

(7.7 **Optional**) Radicals with different orders can be changed to radicals with the same order.

Chapter 7 Review Exercises

In Review Exercises 1–16, do each operation and simplify the result, if possible. All variables represent positive numbers.

1. $25^{1/2}$ 5
2. $-36^{1/2}$ -6
3. $9^{3/2}$ 27
4. $16^{3/2}$ 64
5. $(-8)^{1/3}$ -2
6. $-8^{2/3}$ -4
7. $8^{-2/3}$ $\frac{1}{4}$
8. $8^{-1/3}$ $\frac{1}{2}$
9. $-49^{5/2}$ $-16{,}807$
10. $\dfrac{1}{25^{7/2}}$ $\frac{1}{78{,}125}$
11. $\left(\dfrac{1}{4}\right)^{-3/2}$ 8
12. $\left(\dfrac{4}{9}\right)^{-3/2}$ $\frac{27}{8}$
13. $(27x^3y)^{1/3}$ $3xy^{1/3}$
14. $(81x^4y^2)^{1/4}$ $3xy^{1/2}$
15. $(25x^4y^4)^{3/2}$ $125x^6y^6$
16. $(27u^6v^3)^{-2/3}$ $\dfrac{1}{3u^{4/3}v^2}$

In Review Exercises 17–20, do the multiplications. Write answers without negative exponents. All variables represent positive numbers.

17. $u^{1/2}(u^{1/2} - u^{-1/2})$ $u - 1$
18. $v^{2/3}(v^{1/3} + v^{4/3})$ $v + v^2$
19. $(x^{1/2} + y^{1/2})^2$ $x + 2x^{1/2}y^{1/2} + y$
20. $(a^{2/3} + b^{2/3})(a^{2/3} - b^{2/3})$ $a^{4/3} - b^{4/3}$

In Review Exercises 21–28, simplify each radical.

21. $\sqrt{49}$ 7 **22.** $-\sqrt{121}$ -11 **23.** $-\sqrt{36}$ -6 **24.** $\sqrt{225}$ 15

25. $\sqrt[3]{-27}$ -3 **26.** $-\sqrt[3]{216}$ -6 **27.** $\sqrt[4]{625}$ 5 **28.** $\sqrt[5]{-32}$ -2

In Review Exercises 29–36, simplify each radical. Use absolute value symbols where necessary.

29. $\sqrt{240}$ $4\sqrt{15}$ **30.** $\sqrt[3]{54}$ $3\sqrt[3]{2}$ **31.** $\sqrt[4]{32}$ $2\sqrt[4]{2}$ **32.** $\sqrt[5]{96}$ $2\sqrt[5]{3}$

33. $\sqrt{8x^2}$ $2|x|\sqrt{2}$ **34.** $\sqrt{18x^4y^2}$ $3x^2|y|\sqrt{2}$ **35.** $\sqrt[3]{16x^5y^4}$ $2xy\sqrt[3]{2x^2y}$ **36.** $\sqrt[3]{54x^7y^3}$ $3x^2y\sqrt[3]{2x}$

In Review Exercises 37–42, simplify each radical expression. All variables represent positive numbers.

37. $\sqrt{5}\sqrt{5}$ 5 **38.** $\sqrt{6}\sqrt{216}$ 36 **39.** $\sqrt{9x}\sqrt{x}$ $3x$ **40.** $\sqrt[3]{3}\sqrt[3]{9}$ 3

41. $-\sqrt[3]{2x^2}\sqrt[3]{4x}$ $-2x$ **42.** $-\sqrt[4]{256x^5y^{11}}\sqrt[4]{625x^8y^2}$ $-20x^3y^3\sqrt[4]{xy}$

In Review Exercises 43–44, simplify each radical expression. All variables represent positive numbers.

43. $\sqrt{x^2+6x+9}$ $x+3$ **44.** $\sqrt[3]{(x+1)(x^2+2x+1)}$ $x+1$

In Review Exercises 45–52, simplify and combine like radicals. All variables represent positive numbers.

45. $\sqrt{2}+\sqrt{8}$ $3\sqrt{2}$ **46.** $\sqrt{20}-\sqrt{5}$ $\sqrt{5}$ **47.** $2\sqrt[3]{3}-\sqrt[3]{24}$ 0 **48.** $\sqrt[4]{32}+2\sqrt[4]{162}$ $8\sqrt[4]{2}$

49. $2\sqrt{8}+2\sqrt{200}+\sqrt{50}$ $29\sqrt{2}$ **50.** $3\sqrt{27}-2\sqrt{3}+5\sqrt{75}$ $32\sqrt{3}$

51. $\sqrt[3]{54}-3\sqrt[3]{16}+4\sqrt[3]{128}$ $13\sqrt[3]{2}$ **52.** $2\sqrt[4]{32}+4\sqrt[4]{162}-5\sqrt[4]{512}$ $-4\sqrt[4]{2}$

In Review Exercises 53–60, simplify each radical expression. All variables represents positive numbers.

53. $\sqrt{2}\sqrt{8}$ 4 **54.** $\sqrt{2}(\sqrt{2}+3)$ $2+3\sqrt{2}$ **55.** $\sqrt{5}(\sqrt{2}-1)$ $\sqrt{10}-\sqrt{5}$ **56.** $\sqrt{3}(\sqrt{3}+\sqrt{2})$ $3+\sqrt{6}$

57. $(\sqrt{2}+1)(\sqrt{2}-1)$ 1 **58.** $(\sqrt{3}+\sqrt{2})(\sqrt{3}+\sqrt{2})$ $5+2\sqrt{6}$

59. $(\sqrt{x}+\sqrt{y})(\sqrt{x}-\sqrt{y})$ $x-y$ **60.** $(2\sqrt{u}+3)(3\sqrt{u}-4)$ $6u-12+\sqrt{u}$

In Review Exercises 61–64, rationalize each denominator.

61. $\dfrac{1}{\sqrt{3}}$ $\dfrac{\sqrt{3}}{3}$ **62.** $\dfrac{\sqrt{3}}{\sqrt{5}}$ $\dfrac{\sqrt{15}}{5}$ **63.** $\dfrac{x}{\sqrt{xy}}$ $\dfrac{\sqrt{xy}}{y}$ **64.** $\dfrac{\sqrt[3]{uv}}{\sqrt[3]{u^5v^7}}$ $\dfrac{\sqrt[3]{u^2}}{u^2v^2}$

In Review Exercises 65–68, rationalize each denominator and simplify. All variables represent positive numbers.

65. $\dfrac{2}{\sqrt{2}-1}$ $2(\sqrt{2}+1)$ **66.** $\dfrac{\sqrt{2}}{\sqrt{3}-1}$ $\dfrac{\sqrt{6}+\sqrt{2}}{2}$ **67.** $\dfrac{2x-32}{\sqrt{x}+4}$ $2(\sqrt{x}-4)$ **68.** $\dfrac{\sqrt{a}+1}{\sqrt{a}-1}$ $\dfrac{a+2\sqrt{a}+1}{a-1}$

In Review Exercises 69–70, simplify each radical expression and combine terms. All variables represent positive numbers.

69. $\sqrt{18a}-\dfrac{2a}{\sqrt{2a}}+\sqrt{50a}$ $7\sqrt{2a}$ **70.** $\sqrt[3]{16x}+\sqrt[3]{54x}-\dfrac{2x}{\sqrt[3]{4x^2}}$ $4\sqrt[3]{2x}$

In Review Exercises 71–76, solve each radical equation.

71. $\sqrt{y-5} = \sqrt{2y+17}$ ~~−22~~, no solution

72. $u = \sqrt{25u - 144}$ 16, 9

73. $r = \sqrt{12r - 27}$ 3, 9

74. $\sqrt{z+1} + \sqrt{z} = 2$ $\frac{9}{16}$

75. $\sqrt{2x+5} - \sqrt{2x} = 1$ 2

76. $\sqrt[3]{x^3 + 8} = x + 2$ 0, −2

77. Sailing A technique called **tacking** allows a sailboat to make progress into the wind. A sailboat follows the course in Illustration 1. Find d, the distance the boat advances into the wind. 88 yd

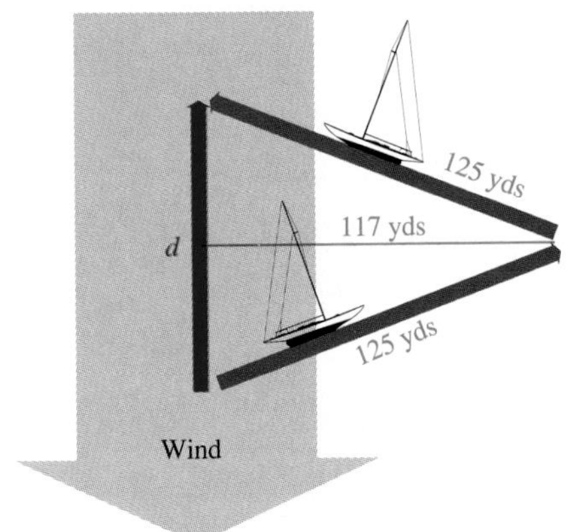

ILLUSTRATION 1

78. Communications Some campers 3900 yards from a highway are chatting with truckers on a citizen's band transceiver with an 8900-yard range. Over what length of highway can these conversations take place? (See Illustration 2.) 16,000 yd, or about 9 mi

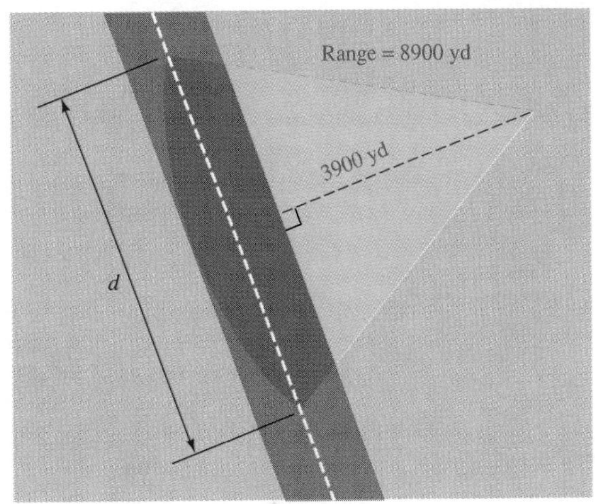

ILLUSTRATION 2

Review Exercises 79–82 come from an optional section. Simplify each radical. All variables are positive numbers.

79. $\sqrt[4]{49}$ $\sqrt{7}$

80. $\sqrt[8]{81}$ $\sqrt{3}$

81. $\sqrt[6]{8}$ $\sqrt{2}$

82. $\sqrt[6]{27}$ $\sqrt{3}$

Review Exercises 83–86 come from an optional section. Write each expression as a single radical.

83. $\sqrt{5}\sqrt[3]{2}$ $\sqrt[6]{500}$

84. $\sqrt{2}\sqrt[3]{5}$ $\sqrt[6]{200}$

85. $\dfrac{\sqrt{5}}{\sqrt[3]{2}}$ $\dfrac{\sqrt[6]{2000}}{2}$

86. $\dfrac{\sqrt{2}}{\sqrt[3]{3}}$ $\dfrac{\sqrt[6]{648}}{3}$

Chapter 7 Test

In Problems 1–6, simplify each expression. Write answers without negative exponents. All variables represent positive numbers.

1. $16^{1/4}$ 2

2. $27^{2/3}$ 9

3. $36^{-3/2}$ $\frac{1}{216}$

4. $\left(-\dfrac{8}{27}\right)^{-2/3}$ $\frac{9}{4}$

5. $\dfrac{2^{5/3}2^{1/6}}{2^{1/2}}$ $2^{4/3}$

6. $\dfrac{(8x^3y)^{1/2}(8xy^5)^{1/2}}{(x^3y^6)^{1/3}}$ $8xy$

In Problems 7–10, simplify each expression. All variables represent positive numbers.

7. $\sqrt{48}$ $4\sqrt{3}$

8. $\sqrt{250x^3y^5}$ $5xy^2\sqrt{10xy}$

9. $\dfrac{\sqrt[3]{24x^{15}y^4}}{\sqrt[3]{y}}$ $2x^5y\sqrt[3]{3}$

10. $\sqrt{\dfrac{3a^5}{48a^7}}$ $\frac{1}{4a}$

In Problems 11–12, simplify each expression. Use absolute values when needed.

11. $\sqrt{8x^3}$ $2|x|\sqrt{2x}$

12. $\sqrt{18x^4y^9}$ $3x^2y^4\sqrt{2y}$

In Problems 13–16, simplify and combine like radicals. All variables represent positive numbers.

13. $3\sqrt{3} + \sqrt{12} - \sqrt{27}$ $2\sqrt{3}$

14. $2\sqrt[3]{40} - \sqrt[3]{5000} + 4\sqrt[3]{625}$ $14\sqrt[3]{5}$

15. $2\sqrt{48y^5} - 3y\sqrt{12y^3}$ $2y^2\sqrt{3y}$

16. $\sqrt[4]{768z^5} + z\sqrt[4]{48z}$ $6z\sqrt[4]{3z}$

In Problems 17–18, do each operation and simplify, if possible. All variables represent positive numbers.

17. $-2\sqrt{xy}\left(3\sqrt{x} + \sqrt{xy^3}\right)$ $-6x\sqrt{y} - 2xy^2$

18. $\left(3\sqrt{2} + \sqrt{3}\right)\left(2\sqrt{2} - 3\sqrt{3}\right)$ $3 - 7\sqrt{6}$

In Problems 19–22, rationalize each denominator.

19. $\dfrac{1}{\sqrt{5}}$ $\dfrac{\sqrt{5}}{5}$

20. $\dfrac{\sqrt{3}}{\sqrt{8}}$ $\dfrac{\sqrt{6}}{4}$

21. $\dfrac{6}{\sqrt[3]{9}}$ $2\sqrt[3]{3}$

22. $\dfrac{\sqrt{18ab}}{\sqrt{6a^2b}}$ $\dfrac{\sqrt{3a}}{a}$

In Problems 23–24, rationalize each denominator and simplify, if possible. All variables represent positive numbers.

23. $\dfrac{-\sqrt{2}}{\sqrt{5} + 3}$ $\dfrac{\sqrt{2}\left(\sqrt{5} - 3\right)}{4}$

24. $\dfrac{3t - 1}{\sqrt{3t} - 1}$ $\sqrt{3t} + 1$

In Problems 25–26, simplify each radical expression and combine like radicals.

25. $\sqrt{50} - \dfrac{2}{\sqrt{2}}$ $4\sqrt{2}$

26. $\sqrt[3]{16a} + \dfrac{2a}{\sqrt[3]{4a^2}}$ $3\sqrt[3]{2a}$

In Problems 27–28, solve and check each equation.

27. $\sqrt[3]{6n + 4} - 4 = 0$ 10

28. $1 - \sqrt{u} = \sqrt{u - 3}$ 4 is extraneous

29. **Shipping crates** The diagonal brace on the shipping crate in Illustration 1 is 53 inches. Find the height, h, of the crate. 28 in.

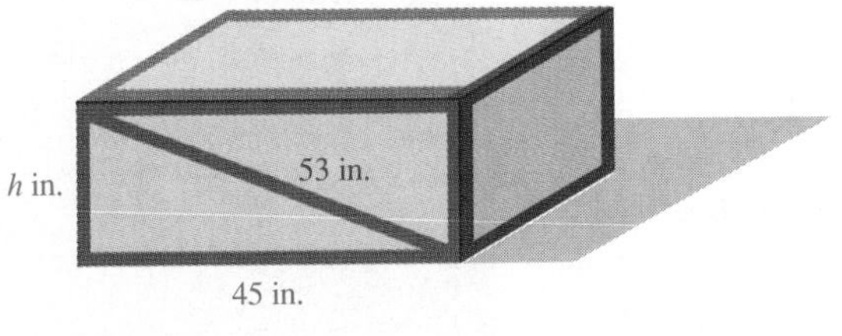

ILLUSTRATION 1

30. **Pendulum** The 2-meter pendulum in Illustration 2 rises 0.1 meters at the extremes of its swing. Find the width, w, of the swing. 1.25 m

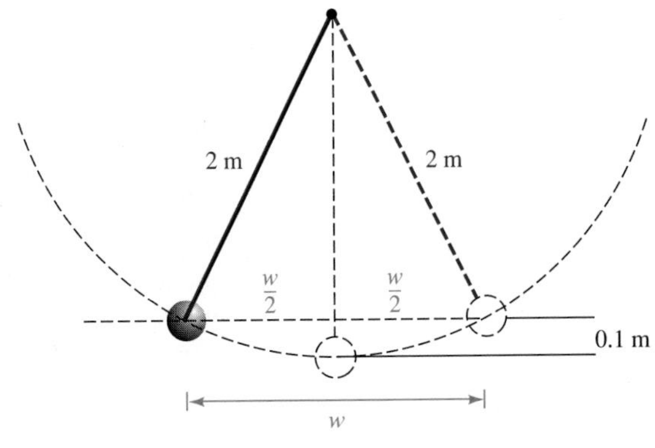

ILLUSTRATION 2

Problems 31–32 come from an optional section.

31. Simplify $\sqrt[4]{64}$. $2\sqrt{2}$

32. Write as a single radical: $\dfrac{\sqrt[4]{3}}{\sqrt{3}}$ $\dfrac{\sqrt[4]{27}}{3}$

CAREERS & MATHEMATICS

Photographer

Photographers use cameras and film to portray people, objects, places, and events. Some specialize in scientific, medical, or engineering photography and provide illustrations and documentation for publications and research reports. Others specialize in portrait, fashion, or industrial photography and provide the pictures for catalogs and other publications. Photojournalists capture newsworthy events, people, and places, and their work is seen in newspapers and magazines, as well as on television.

Qualifications Employers want applicants with a broad technical understanding of photography, as well as imagination, creativity, and a good sense of timing. Some knowledge of mathematics, physics, and chemistry is helpful for understanding lenses, films, lighting, and development processes.

Job Outlook Job opportunities will be excellent through the year 2005. Business and industry will need photographers to provide visual aids for meetings and reports, sales campaigns, and public relations work. Law enforcement agencies and scientific and medical research organizations will also require photographers with appropriate technical skills.

Example Application In each photographic lens, there is an adjustable circular opening called the **aperture,** which controls the amount of light that passes through the lens. Various lenses—wide-angle, close-up, and telephoto—are distinguished by their **focal length.** The diameter, d, of the aperture and the focal length, f, of the lens determine the ***f*-number** of the lens by the formula

$$f\text{-number} = \frac{f}{d}$$

Thus, a lens with a focal length of 12 centimeters and an aperture with a diameter of 6 centimeters has an f-number of $\frac{12}{6}$, or 2. It would be an $f/2$ lens.

Find the f-number of this lens if the *area* of its aperture is cut in half, so as to admit only half as much light.

Solution First determine the area of a circle with diameter d. Then substitute 6 for d to find the area of the aperture.

$$A = \pi r^2 \qquad \text{The formula for the area of a circle.}$$

$$A = \pi\left(\frac{d}{2}\right)^2 \qquad \text{Substitute } \tfrac{d}{2} \text{ for } r.$$

$$A = \pi\left(\frac{6}{2}\right)^2 \qquad \text{Substitute 6 for } d.$$

$$A = 9\pi$$

To find the diameter of a circle with area equal to one-half of 9π, substitute $\frac{9\pi}{2}$ for A in the formula for the area of a circle and solve for d.

$$A = \pi r^2 \qquad \text{The formula for the area of a circle.}$$

$$\frac{9\pi}{2} = \pi\left(\frac{d}{2}\right)^2 \qquad \text{Substitute } \tfrac{9\pi}{2} \text{ for } A \text{ and } \tfrac{d}{2} \text{ for } r.$$

$$\frac{9\pi}{2} = \frac{\pi d^2}{4}$$

$$18 = d^2 \qquad \text{Multiply both sides by 4.}$$

$$3\sqrt{2} = d \qquad \sqrt{18} = \sqrt{9}\sqrt{2} = 3\sqrt{2}.$$

If the diameter of the aperture were reduced from 6 centimeters to $3\sqrt{2}$ centimeters, the area (and the light admitted) would be cut in half. The f-number of the lens is now found by letting $f = 12$ (as before) and $d = 3\sqrt{2}$ in the formula for an f-number.

$$f\text{-number} = \frac{f}{d} = \frac{12}{3\sqrt{2}} = 2\sqrt{2} \approx 2.8$$

An $f/2.8$ lens admits one-half the light admitted by an $f/2$ lens.

If the light were cut in half again, the f-number would be $2.8\sqrt{2}$, or 4. The next f-number, representing another halving of the light admitted, would be $4\sqrt{2}$, or 5.6. These numbers,

$$f/2, \quad f/2.8, \quad f/4, \quad f/5.6, \quad f/8, \quad f/11, \quad f/16$$

are called ***f*-stops.** They are well known to all professional photographers.

EXERCISES

1. What would be the f-number of a lens with focal length of 20 cm and aperture with diameter of 5 cm?
2. If the focal length of the lens of Exercise 1 were doubled and the aperture held constant, what would be the f-number?
3. What diameter would give a lens with focal length 55 mm an f-number of $f/3.5$?
4. The **speed** of a lens is the square of the reciprocal of the f-number. How many times faster is an $f/2$ lens than an $f/4.5$ lens?

(*Answers:* **1.** $f/4$ **2.** $f/8$ **3.** 15.7 mm **4.** 5 times faster)

8 Quadratic Equations

So far we have discussed how to solve linear equations and certain quadratic equations in which the quadratic expression was factorable. In this chapter, we will discuss more general methods for solving quadratic equations and will consider their graphs.

8.1 Completing the Square and the Quadratic Formula

■ Completing the Square ■ Solving Equations by Completing the Square
■ The Quadratic Formula ■ Problem Solving

Historical Note: The Babylonians could solve certain quadratic equations in 2000 B.C.

A quadratic equation is an equation of the form $ax^2 + bx + c = 0$, where a, b, and c are real numbers and $a \neq 0$. In Chapter 4, we discussed how to solve certain quadratic equations by using factoring. For example, to solve $6x^2 - 7x - 3 = 0$, we proceed as follows:

$$6x^2 - 7x - 3 = 0$$

$$(2x - 3)(3x + 1) = 0 \qquad \text{Factor the quadratic trinomial.}$$

$$2x - 3 = 0 \quad \text{or} \quad 3x + 1 = 0$$

$$x = \frac{3}{2} \quad \Big| \quad x = -\frac{1}{3}$$

Both solutions satisfy the original equation.

Many quadratic expressions do not factor easily. For example, it would be difficult to factor the left-hand side of the equation $2x^2 + 4x + 1 = 0$, because it cannot be factored by using only integers.

To develop methods for solving all quadratic equations, we first consider the equation $x^2 = c$. If $c > 0$, the two real solutions of $x^2 = c$ can be found as follows:

$$x^2 = c$$

$$x^2 - c = 0 \qquad \text{Subtract } c \text{ from both sides.}$$

$$x^2 - (\sqrt{c})^2 = 0 \qquad c = (\sqrt{c})^2.$$

$$(x + \sqrt{c})(x - \sqrt{c}) = 0 \qquad \text{Factor the difference of two squares.}$$

$$x + \sqrt{c} = 0 \quad \text{or} \quad x - \sqrt{c} = 0 \qquad \text{Set each factor equal to 0.}$$

$$x = -\sqrt{c} \quad \Big| \quad x = \sqrt{c}$$

Thus, the solutions of the equation $x^2 = c$ are $x = \sqrt{c}$ and $x = -\sqrt{c}$.

Square Root Property If $c > 0$, the equation $x^2 = c$ has two real solutions. They are

$$x = \sqrt{c} \quad \text{and} \quad x = -\sqrt{c}$$

EXAMPLE 1 Solve the equation $x^2 - 12 = 0$.

Solution We can write the equation as $x^2 = 12$ and use the square root property.

$$x^2 - 12 = 0$$

$$x^2 = 12 \qquad \text{Add 12 to both sides.}$$

$$x = \sqrt{12} \quad \text{or} \quad x = -\sqrt{12} \qquad \text{Use the square root property.}$$

$$x = 2\sqrt{3} \quad | \quad x = -2\sqrt{3} \qquad \text{Simplify each radical.}$$

Verify that both solutions satisfy the equation. ■

EXAMPLE 2 Solve the equation $(x - 3)^2 = 16$.

Solution We use the square root property.

$$(x - 3)^2 = 16$$

$$x - 3 = \sqrt{16} \quad \text{or} \quad x - 3 = -\sqrt{16}$$

$$x - 3 = 4 \quad | \quad x - 3 = -4$$

$$x = 3 + 4 \quad | \quad x = 3 - 4$$

$$x = 7 \quad | \quad x = -1$$

Verify that both solutions satisfy the equation. ■

Completing the Square

Teaching Tip: Tell the students that the process of completing the square will be important when graphing conic sections and in calculus.

All quadratic equations can be solved by **completing the square**. This method is based on the special products

$$x^2 + 2ax + a^2 = (x + a)^2$$

and

$$x^2 - 2ax + a^2 = (x - a)^2$$

The trinomials $x^2 + 2ax + a^2$ and $x^2 - 2ax + a^2$ are both perfect square trinomials, because both factor as the square of a binomial. In each case, the coefficient of the first term is 1, and if we take one-half of the coefficient of x and square it, we obtain the third term.

$$\left[\frac{1}{2}(2a)\right]^2 = a^2$$

$$\left[\frac{1}{2}(-2a)\right]^2 = (-a)^2 = a^2$$

To make a binomial such as $x^2 + 10x$ a perfect square trinomial, we find one-half of 10 to get 5, square 5 to get 25, and add 25 to $x^2 + 10x$.

$$x^2 + 10x + \left[\frac{1}{2}(10)\right]^2 = x^2 + 10x + (5)^2$$
$$= x^2 + 10x + 25$$

Note that $x^2 + 10x + 25 = (x + 5)^2$.

To make the binomial $x^2 - 6x$ a perfect square trinomial, we find one-half of -6 to get -3, square -3 to get 9, and add 9 to $x^2 - 6x$.

$$x^2 - 6x + \left[\frac{1}{2}(-6)\right]^2 = x^2 - 6x + (-3)^2$$
$$= x^2 - 6x + 9$$

Note that $x^2 - 6x + 9 = (x - 3)^2$.

Solving Equations by Completing the Square

EXAMPLE 3 Use completing the square to solve the equation $x^2 + 8x + 7 = 0$.

Solution In this example, the coefficient of x^2 is 1. We start by adding -7 to both sides to get the constant on the right-hand side of the equal sign:

$$x^2 + 8x + 7 = 0$$
$$x^2 + 8x = -7$$

To complete the square, we must add a number k to both sides of the equation so that $x^2 + 8x + k$ is a perfect square trinomial. To find k, we take one-half of the coefficient of x $\left(\frac{1}{2} \cdot 8 = 4\right)$, square it $(4^2 = 16)$, and add 16 to both sides of the equation.

1. $x^2 + 8x + 16 = 16 - 7$

The left-hand side of Equation 1 is a perfect square trinomial, because $x^2 + 8x + 16 = (x + 4)^2$. We can now factor the left-hand side and combine terms on the right-hand side to obtain

$$(x + 4)^2 = 9$$

We can solve this equation by using the square root property.

$$(x + 4)^2 = 9$$

$x + 4 = \sqrt{9}$ or $x + 4 = -\sqrt{9}$ — Use the square root property.

$x + 4 = 3$ | $x + 4 = -3$

$x = -1$ | $x = -7$

Verify that both solutions satisfy the equation. ■

EXAMPLE 4 Solve $6x^2 + 5x - 6 = 0$.

Solution In this example, the coefficient of x^2 is not 1. To make it 1, we divide both sides of the equation by 6. We then proceed as follows:

$$6x^2 + 5x - 6 = 0$$

$$\frac{6x^2}{6} + \frac{5}{6}x - \frac{6}{6} = \frac{0}{6} \quad \text{Divide both sides by 6.}$$

$$x^2 + \frac{5}{6}x - 1 = 0 \quad \text{Simplify.}$$

$$x^2 + \frac{5}{6}x = 1 \quad \text{Add 1 to both sides.}$$

To complete the square on x, we take one-half of the coefficient of x $\left(\frac{1}{2} \cdot \frac{5}{6} = \frac{5}{12}\right)$, square $\frac{5}{12}$ to get $\frac{25}{144}$ and add $\frac{25}{144}$ to both sides of the equation:

$$x^2 + \frac{5}{6}x + \frac{25}{144} = 1 + \frac{25}{144}$$

$$\left(x + \frac{5}{12}\right)^2 = \frac{169}{144} \quad \text{Factor and add the fractions.}$$

We can now use the square root property:

$$x + \frac{5}{12} = \sqrt{\frac{169}{144}} \quad \text{or} \quad x + \frac{5}{12} = -\sqrt{\frac{169}{144}}$$

$$x + \frac{5}{12} = \frac{13}{12} \quad\quad x + \frac{5}{12} = -\frac{13}{12}$$

$$x = -\frac{5}{12} + \frac{13}{12} \quad\quad x = -\frac{5}{12} - \frac{13}{12}$$

$$x = \frac{8}{12} \quad\quad x = -\frac{18}{12}$$

$$x = \frac{2}{3} \quad\quad x = -\frac{3}{2}$$

Verify that both solutions satisfy the original equation. ■

The method used in Example 4 suggests a list of steps to follow when using completing the square to solve quadratic equations of the form $ax^2 + bx + c = 0$.

Completing the Square

1. Make sure that the coefficient of x^2 is 1. If it is not, make it 1 by dividing both sides of the equation by the coefficient of x^2.
2. If necessary, add a number to both sides of the equation to get the constant term on the right-hand side of the equal sign.
3. Complete the square:
   a. Identify the coefficient of x.
   b. Find one-half of the coefficient of x and square it.
   c. Add that square to both sides of the equation.
4. Factor the trinomial square and combine terms.
5. Solve the resulting equation by using the square root property.

EXAMPLE 5 Solve $2x^2 + 4x + 1 = 0$.

Solution We use completing the square.

$$2x^2 + 4x + 1 = 0$$

$$x^2 + 2x + \frac{1}{2} = \frac{0}{2}$$ Divide both sides by 2 to make the coefficient of x^2 equal to 1.

$$x^2 + 2x = -\frac{1}{2}$$ Subtract $\frac{1}{2}$ from both sides.

$$x^2 + 2x + 1 = 1 - \frac{1}{2}$$ Square half the coefficient of x and add it to both sides.

$$(x + 1)^2 = \frac{1}{2}$$ Factor and combine terms.

$$x + 1 = \sqrt{\frac{1}{2}} \quad \text{or} \quad x + 1 = -\sqrt{\frac{1}{2}}$$

$$x + 1 = \frac{\sqrt{2}}{2} \qquad x + 1 = -\frac{\sqrt{2}}{2}$$

$$x = -1 + \frac{\sqrt{2}}{2} \qquad x = -1 - \frac{\sqrt{2}}{2}$$

$$x = \frac{-2 + \sqrt{2}}{2} \qquad x = \frac{-2 - \sqrt{2}}{2}$$

Both values check. ■

The Quadratic Formula

Although all quadratic equations can be solved by completing the square, the method is often tedious. The **quadratic formula** will give the solutions of a qua-

Alternate Approach: You may want to develop the quadratic formula this way:

$ax^2 + bx + c = 0$

$ax^2 + bx = -c$

$4a^2x^2 + 4abx = -4ac$

$4a^2x^2 + 4abx + b^2 = b^2 - 4ac$

$(2ax + b)^2 = b^2 - 4ac$

$2ax + b = \pm\sqrt{b^2 - 4ac}$

$2ax = -b \pm \sqrt{b^2 - 4ac}$

$x = \dfrac{-b \pm\sqrt{b^2 - 4ac}}{2a}$

dratic equation with much less effort. To develop this formula, we will complete the square.

A quadratic equation is an equation that can be written in the form $ax^2 + bx + c = 0$, with $a \neq 0$. If we solve this general form by completing the square, we are solving every possible quadratic equation at one time. The solution of the general quadratic equation results in the quadratic formula.

To solve the equation $ax^2 + bx + c = 0$, with $a \neq 0$, we proceed as follows:

$$ax^2 + bx + c = 0$$

$$\frac{ax^2}{a} + \frac{bx}{a} + \frac{c}{a} = \frac{0}{a}$$ Since $a \neq 0$, we can divide both sides by a.

$$x^2 + \frac{bx}{a} = -\frac{c}{a}$$ Simplify and subtract $\frac{c}{a}$ from both sides.

$$x^2 + \frac{b}{a}x + \left(\frac{b}{2a}\right)^2 = \left(\frac{b}{2a}\right)^2 - \frac{c}{a}$$ Complete the square on x and add $\left(\frac{b}{2a}\right)^2$ to both sides.

$$x^2 + \frac{b}{a}x + \frac{b^2}{4a^2} = \frac{b^2}{4a^2} - \frac{4ac}{4aa}$$ Remove parentheses and get a common denominator on the right-hand side.

2. $$\left(x + \frac{b}{2a}\right)^2 = \frac{b^2 - 4ac}{4a^2}$$ Factor the left-hand side and add the fractions on the right-hand side.

We can solve Equation 2 by using the square root property.

$$x + \frac{b}{2a} = \sqrt{\frac{b^2 - 4ac}{4a^2}} \quad \text{or} \quad x + \frac{b}{2a} = -\sqrt{\frac{b^2 - 4ac}{4a^2}}$$

$$x + \frac{b}{2a} = \frac{\sqrt{b^2 - 4ac}}{2a} \qquad x + \frac{b}{2a} = -\frac{\sqrt{b^2 - 4ac}}{2a}$$

$$x = -\frac{b}{2a} + \frac{\sqrt{b^2 - 4ac}}{2a} \qquad x = -\frac{b}{2a} - \frac{\sqrt{b^2 - 4ac}}{2a}$$

$$= \frac{-b + \sqrt{b^2 - 4ac}}{2a} \qquad = \frac{-b - \sqrt{b^2 - 4ac}}{2a}$$

These two values of x are the solutions to the equation $ax^2 + bx + c = 0$, with $a \neq 0$. They are usually written as a single expression called the **quadratic formula**. Read the symbol $\pm$ as "plus or minus."

Quadratic Formula If $a \neq 0$, the solutions of $ax^2 + bx + c = 0$ are given by the formula

$$x = \frac{-b \pm \sqrt{b^2 - 4ac}}{2a}$$

WARNING! Be sure to draw the fraction bar under both parts of the numerator, and be sure to draw the radical sign exactly over $b^2 - 4ac$. Do not write the quadratic formula as

$$x = -b \pm \frac{\sqrt{b^2 - 4ac}}{2a}$$

EXAMPLE 6 Solve the equation $2x^2 - 3x - 5 = 0$.

Solution In this equation, $a = 2$, $b = -3$, and $c = -5$. We substitute these values into the quadratic formula and simplify:

$$\begin{aligned} x &= \frac{-b \pm \sqrt{b^2 - 4ac}}{2a} \\ &= \frac{-(-3) \pm \sqrt{(-3)^2 - 4(2)(-5)}}{2(2)} \\ &= \frac{3 \pm \sqrt{9 + 40}}{4} \\ &= \frac{3 \pm \sqrt{49}}{4} \\ &= \frac{3 \pm 7}{4} \end{aligned}$$

$$\begin{array}{ccc|c} x = \frac{3+7}{4} & \text{or} & x = \frac{3-7}{4} \\ x = \frac{10}{4} & & x = \frac{-4}{4} \\ x = \frac{5}{2} & & x = -1 \end{array}$$

Verify that both solutions satisfy the original equation. ■

EXAMPLE 7 Solve the equation $2x^2 + 4x + 1 = 0$.

Solution In this equation, $a = 2$, $b = 4$, and $c = 1$. We substitute these values into the quadratic formula and simplify:

$$\begin{aligned} x &= \frac{-b \pm \sqrt{b^2 - 4ac}}{2a} \\ &= \frac{-4 \pm \sqrt{4^2 - 4(2)(1)}}{2(2)} \end{aligned}$$

$$= \frac{-4 \pm \sqrt{16 - 8}}{4}$$

$$= \frac{-4 \pm \sqrt{8}}{4}$$

$$= \frac{-4 \pm 2\sqrt{2}}{4}$$

$$= \frac{-2 \pm \sqrt{2}}{2} \qquad \frac{-4 \pm 2\sqrt{2}}{4} = \frac{2(-2 \pm \sqrt{2})}{4} = \frac{-2 \pm \sqrt{2}}{2}.$$

$$x = \frac{-2 + \sqrt{2}}{2} \quad \text{or} \quad x = \frac{-2 - \sqrt{2}}{2}$$

■

Problem Solving

EXAMPLE 8 **Dimensions of a rectangle** A rectangle is 12 centimeters longer than it is wide, and its area is 253 square centimeters. Find the dimensions of the rectangle.

Solution We can let w represent the width of the rectangle.
Then $w + 12$ represents its length.

Since the area is found by multiplying the length and the width, and since the area is given to be 253 square centimeters, we can form the equation

$$w(w + 12) = 253$$

We can solve this equation as follows:

$$w(w + 12) = 253$$

$$w^2 + 12w = 253 \qquad \text{Use the distributive property to remove parentheses.}$$

$$w^2 + 12w - 253 = 0 \qquad \text{Subtract 253 from both sides.}$$

Solution by factoring

$$(w - 11)(w + 23) = 0$$

$$w - 11 = 0 \quad \text{or} \quad w + 23 = 0$$

$$w = 11 \quad | \quad w = -23$$

Solution by formula

$$w = \frac{-12 \pm \sqrt{12^2 - 4(1)(-253)}}{2(1)}$$

$$= \frac{-12 \pm \sqrt{144 + 1012}}{2}$$

$$= \frac{-12 \pm \sqrt{1156}}{2}$$

$$= \frac{-12 \pm 34}{2}$$

$$w = 11 \quad \text{or} \quad w = -23$$

Since a rectangle cannot have a negative width, the solution -23 must be discarded. Thus, the dimensions of the rectangle are 11 centimeters by $(11 + 12)$ centimeters, or 11 centimeters by 23 centimeters.

Check: 23 is 12 more than 11, and the area of a rectangle with dimensions of 23 centimeters by 11 centimeters is 253 square centimeters. ■

ORALS *Solve each equation.*

1. $x^2 = 49$ ± 7

2. $x^2 = 10$ $\pm\sqrt{10}$

Find the number that must be added to the binomial to make it a perfect square trinomial.

3. $x^2 + 4x$ 4

4. $x^2 - 6x$ 9

5. $x^2 - 3x$ $\frac{9}{4}$

6. $x^2 + 5x$ $\frac{25}{4}$

Identify a, b, and c in each quadratic equation.

7. $3x^2 - 4x + 7 = 0$ 3, −4, 7

8. $-2x^2 + x = 5$ −2, 1, −5

EXERCISE 8.1

In Exercises 1–12, use factoring to solve each equation.

1. $6x^2 + 12x = 0$ 0, −2

2. $5x^2 + 11x = 0$ $0, -\frac{11}{5}$

3. $2y^2 - 50 = 0$ 5, −5

4. $4y^2 - 64 = 0$ 4, −4

5. $r^2 + 6r + 8 = 0$ −2, −4

6. $x^2 + 9x + 20 = 0$ −5, −4

7. $x^2 - 7x + 6 = 0$ 6, 1

8. $t^2 - 5t + 6 = 0$ 3, 2

9. $2z^2 - 5z + 2 = 0$ $2, \frac{1}{2}$

10. $2x^2 - x - 1 = 0$ $1, -\frac{1}{2}$

11. $6s^2 + 11s - 10 = 0$ $\frac{2}{3}, -\frac{5}{2}$

12. $3x^2 + 10x - 8 = 0$ $-4, \frac{2}{3}$

In Exercises 13–24, use the square root property to solve each equation.

13. $x^2 = 36$ 6, −6

14. $x^2 = 144$ 12, −12

15. $z^2 = 5$ $\sqrt{5}, -\sqrt{5}$

16. $u^2 = 24$ $2\sqrt{6}, -2\sqrt{6}$

17. $3x^2 - 16 = 0$ $\pm\frac{4\sqrt{3}}{3}$

18. $5x^2 - 49 = 0$ $\pm\frac{7\sqrt{5}}{5}$

19. $(x + 1)^2 = 1$ 0, −2

20. $(x - 1)^2 = 4$ 3, −1

21. $(s - 7)^2 - 9 = 0$ 4, 10

22. $(t + 4)^2 = 16$ 0, −8

23. $(x + 5)^2 - 3 = 0$ $-5 \pm \sqrt{3}$

24. $(x + 3)^2 - 7 = 0$ $-3 \pm \sqrt{7}$

In Exercises 25–38, use completing the square to solve each equation.

25. $x^2 + 2x - 8 = 0$ 2, −4

26. $x^2 + 6x + 5 = 0$ −1, −5

27. $x^2 - 6x + 8 = 0$ 2, 4

28. $x^2 + 8x + 15 = 0$ −3, −5

29. $x^2 + 5x + 4 = 0$ −1, −4

30. $x^2 - 11x + 30 = 0$ 5, 6

31. $2x^2 - x - 1 = 0$ $1, -\frac{1}{2}$

32. $2x^2 - 5x + 2 = 0$ $2, \frac{1}{2}$

33. $6x^2 + 11x + 3 = 0$ $-\frac{1}{3}, -\frac{3}{2}$

34. $6x^2 + x - 2 = 0$ $\frac{1}{2}, -\frac{2}{3}$

35. $8r^2 + 6r = 9$ $\frac{3}{4}, -\frac{3}{2}$

36. $3w^2 - 11w = -10$ $2, \frac{5}{3}$

37. $\frac{7x + 1}{5} = -x^2$ $\frac{-7 \pm \sqrt{29}}{10}$

38. $\frac{3x^2}{8} = \frac{1}{8} - x$ $\frac{-4 \pm \sqrt{19}}{3}$

In Exercises 39–50, use the quadratic formula to solve each equation.

39. $x^2 + 3x + 2 = 0$ $-1, -2$

40. $x^2 - 3x + 2 = 0$ $1, 2$

41. $x^2 + 12x = -36$ $-6, -6$

42. $y^2 - 18y = -81$ $9, 9$

43. $5x^2 + 5x + 1 = 0$ $\dfrac{-5 \pm \sqrt{5}}{10}$

44. $4w^2 + 6w + 1 = 0$ $\dfrac{-3 \pm \sqrt{5}}{4}$

45. $8u = -4u^2 - 3$ $-\frac{3}{2}, -\frac{1}{2}$

46. $4t + 3 = 4t^2$ $\frac{3}{2}, -\frac{1}{2}$

47. $16y^2 + 8y - 3 = 0$ $\frac{1}{4}, -\frac{3}{4}$

48. $16x^2 + 16x + 3 = 0$ $-\frac{1}{4}, -\frac{3}{4}$

49. $\dfrac{x^2}{2} + \dfrac{5}{2}x = -1$ $\dfrac{-5 \pm \sqrt{17}}{2}$

50. $-3x = \dfrac{x^2}{2} + 2$ $-3 \pm \sqrt{5}$

51. Integer problem The product of two consecutive positive even integers is 288. Find the integers. 16, 18

52. Integer problem The product of two consecutive odd negative integers is 143. Find the integers. $-13, -11$

53. Integer problem The sum of the squares of two consecutive positive integers is 85. Find the integers. 6, 7

54. Integer problem The sum of the squares of three consecutive positive integers is 77. Find the integers. 4, 5, 6

55. Dimensions of a rectangle A rectangle is 4 feet longer than it is wide, and its area is 96 square feet. Find its dimensions. 8 ft by 12 ft

56. Dimensions of a rectangle One side of a rectangle is 3 times as long as another. The area of the rectangle is 147 square meters. Find its dimensions. 7 m by 21 m

57. Side of a square The area of a square is numerically equal to its perimeter. Find the length of one side of the square. 4 units

58. Perimeter of a rectangle A rectangle is 2 inches longer than it is wide. Numerically, its area exceeds its perimeter by 11. Find the perimeter. 24 in.

59. Base of a triangle The height of a triangle is 5 centimeters longer than three times its base. Find the base of the triangle if its area is 6 square centimeters. $\frac{4}{3}$ cm

60. Height of a triangle The height of a triangle is 4 meters longer than twice its base. Find the height of the triangle if its area is 15 square meters. 10 m

61. Finding rates A woman drives her snowmobile 150 miles at the rate of r miles per hour. She could have gone the same distance in 2 hours less time if she had increased her speed by 20 miles per hour. Find r. 30 mph

62. Finding rates Jeff bicycles 160 miles at the rate of r miles per hour. The same trip would have taken 2 hours longer if he had decreased his speed by 4 miles per hour. Find r. 20 mph

63. Pricing concert tickets Tickets to a rock concert cost \$4, and the projected attendance is 300 persons. It is further projected that for every 10¢ increase in ticket price, the average attendance will decrease by 5. At what ticket price will the nightly receipts be \$1248? \$4.80 or \$5.20

64. Setting bus fares A bus company has 3000 passengers daily, paying a 25¢ fare. For each nickel increase in fare, the company estimates that it will lose 80 passengers. What increase in fare will produce \$994 in daily revenue? 10¢

65. Computing revenue The *Gazette's* profit is \$20 per year for each of its 3000 subscribers. Management estimates that the profit per subscriber will increase by 1¢ for each additional subscriber over the current 3000. How many subscribers will bring a total profit of \$120,000? 4000

66. Finding interest rates A woman invests \$1000 in a mutual fund for which interest is compounded annually at a rate r. After one year, she deposits an additional \$2000. After two years the balance in the account is

$$\$1000(1 + r)^2 + \$2000(1 + r)$$

If this amount is \$3368.10, find r. 9%

67. **Framing a picture** The frame surrounding the picture in Illustration 1 has a constant width, and the area of the frame equals the area of the picture. How wide is the frame? 2.26 in.

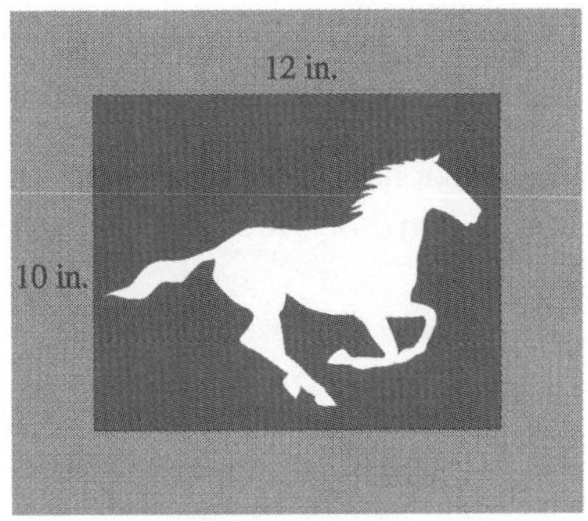

ILLUSTRATION 1

68. **Metal fabrication** A box with no top is to be made by cutting a 2-inch square from each corner of the square sheet of metal shown in Illustration 2. After bending up the sides, the volume of the box is to be 200 cubic inches. How large should the piece of metal be? 14 in. × 14 in.

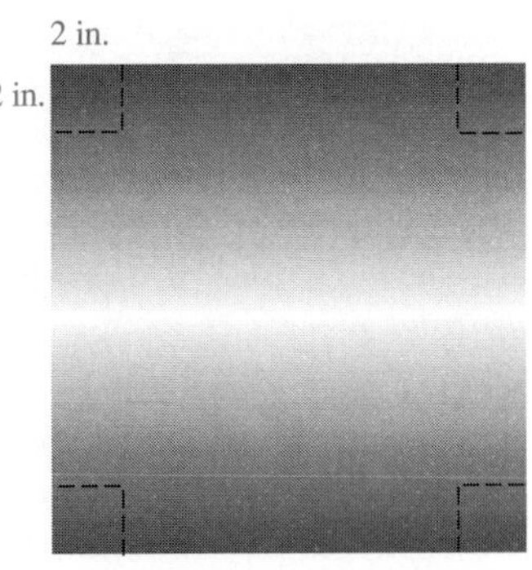

ILLUSTRATION 2

69. Find a quadratic equation with solutions of 3 and 5. $x^2 - 8x + 15 = 0$

70. Find a quadratic equation with solutions of -4 and 6. $x^2 - 2x - 24 = 0$

71. Find a third-degree equation with solutions of 2, 3, and -4. $x^3 - x^2 - 14x + 24 = 0$

72. Find a fourth-degree equation with solutions of 3, -3, 4, -4. $x^4 - 25x^2 + 144 = 0$

Writing Exercises *Write a paragraph using your own words.*

1. Explain how to complete the square.
2. Tell why a cannot be 0 in the quadratic equation $ax^2 + bx + c = 0$.

Something to Think About

1. What number must be added to $x^2 + \sqrt{3}x$ to make it a perfect square? $\frac{3}{4}$
2. Solve $x^2 + \sqrt{3}x - \frac{1}{4} = 0$ by completing the square. $-\frac{\sqrt{3}}{2} \pm 1$

Review Exercises *Solve each equation or inequality.*

1. $\frac{t+9}{2} + \frac{t+2}{5} = \frac{8}{5} + 4t$ 1
2. $\frac{1-5x}{2x} + 4 = \frac{x+3}{x}$ 5
3. $3(t-3) + 3t - 5 \le 2(t+1) + t - 4$ $t \le 4$
4. $-2(y+4) - 2y + 8 \ge 3(2y-3)$ $y \le \frac{9}{10}$

Solve for the indicated variable.

5. $Ax + By = C$ for B $B = \frac{-Ax + C}{y}$
6. $R = \frac{kL}{d^2}$ for L $L = \frac{Rd^2}{k}$

8.2 Complex Numbers

■ Powers of *i* ■ Simplifying Imaginary Numbers ■ Complex Numbers
■ Arithmetic of Complex Numbers ■ Rationalizing the Denominator
■ Absolute Value of a Complex Number

So far, all of our work with quadratic equations has involved real numbers only. However, the solutions to many quadratic equations are not real numbers, as the following example will show.

EXAMPLE 1 Solve the quadratic equation $x^2 + x + 1 = 0$.

Solution Because the quadratic trinomial is prime and cannot be factored, we will use the quadratic formula, with $a = 1$, $b = 1$, and $c = 1$:

$$\begin{aligned} x &= \frac{-b \pm \sqrt{b^2 - 4ac}}{2a} \\ &= \frac{-1 \pm \sqrt{1^2 - 4(1)(1)}}{2(1)} \\ &= \frac{-1 \pm \sqrt{1 - 4}}{2} \\ &= \frac{-1 \pm \sqrt{-3}}{2} \end{aligned}$$

$$x = \frac{-1 + \sqrt{-3}}{2} \quad \text{or} \quad x = \frac{-1 - \sqrt{-3}}{2}$$

■

Each solution in Example 1 contains the number $\sqrt{-3}$. Since the square of no real number is -3, $\sqrt{-3}$ cannot be a real number. For years, people believed that numbers like

$$\sqrt{-1}, \quad \sqrt{-3}, \quad \sqrt{-4}, \quad \text{and} \quad \sqrt{-9}$$

were nonsense. The great mathematician Sir Isaac Newton (1642–1727) called them impossible. In the 17th century, René Descartes (1596–1650) called them **imaginary numbers.**

Historical Note: The term *real number* is also attributed to Descartes.

Today, we no longer think of imaginary numbers as being fictitious. In fact, imaginary numbers have many important uses, such as describing the behavior of alternating current in electronics.

The imaginary number $\sqrt{-1}$ is often denoted by the letter i:

$$i = \sqrt{-1}$$

Teaching Tip: Technical math books often use j to represent $\sqrt{-1}$. This is to distinguish it from the letter i, which is used to represent current.

Because i represents the square root of -1, it follows that

$$i^2 = -1$$

Powers of *i*

The powers of i produce an interesting pattern:

$i^1 = \sqrt{-1} = i$	$i^5 = i^4 i = 1i = i$
$i^2 = \sqrt{-1}\sqrt{-1} = -1$	$i^6 = i^4 i^2 = 1(-1) = -1$
$i^3 = i^2 i = -1i = -i$	$i^7 = i^4 i^3 = 1(-i) = -i$
$i^4 = i^2 i^2 = (-1)(-1) = 1$	$i^8 = i^4 i^4 = (1)(1) = 1$

The pattern continues $i, -1, -i, 1, \ldots$.

EXAMPLE 2 Simplify i^{29}.

Solution We note that 29 divided by 4 gives a quotient of 7 and a remainder of 1. Thus, $29 = 4 \cdot 7 + 1$, and because $i^4 = 1$, we have

$$\begin{aligned} i^{29} &= i^{4 \cdot 7+1} \\ &= (i^4)^7 \cdot i^1 \\ &= 1^7 \cdot i \\ &= i \end{aligned}$$

■

PERSPECTIVE

From Ridicule to Respect

The Pythagoreans (ca. 500 B.C.) understood the universe as a harmony of whole numbers. They did not classify fractions as numbers, and were upset that $\sqrt{2}$ was not the ratio of whole numbers. For 2000 years, little progress was made in the understanding of the various kinds of numbers.

The father of algebra, François Vieta (1540–1603), understood the whole numbers, fractions, and certain irrational numbers. But he was unable to accept negative numbers, and certainly not imaginary numbers.

Girolamo Cardano (1501–1576) was more daring, even though negative and imaginary numbers puzzled him. He attacked the problem *Divide 10 into two parts whose product is 40.* "Of course, that is impossible," said Cardano. "Nevertheless, we will operate." (Cardano had some medical training!) He found solutions: the two numbers $5 + i\sqrt{15}$ and $5 - i\sqrt{15}$. (Can you show that their sum is 10 and their product is 40?) Cardano was impressed. "These numbers," he said, "are truly sophisticated." Unfortunately, he decided that this kind of arithmetic would be useless. Cardano was ahead of his time, but he didn't appreciate the value of his discovery.

René Descartes (1596–1650) thought these numbers to be nothing more than figments of his imagination, so he called them *imaginary numbers.* Leonhard Euler (1707–1783) used the letter i for $\sqrt{-1}$; Augustin Cauchy (1789–1857) used the term *conjugate*; and Karl Gauss (1777–1855) first used the word *complex.*

Today, we accept complex numbers without question, and use them freely in science, economics, medicine, and industry. But it took many centuries and the work of many mathematicians to make them respectable.

The results of Example 2 illustrate the following theorem.

Powers of i If n is a natural number that leaves a remainder of r when divided by 4, then

$$i^n = i^r$$

EXAMPLE 3 Simplify i^{55}.

Solution We divide 55 by 4 and get a remainder of 3. Therefore,

$$i^{55} = i^3$$

and $i^3 = -1$. ■

Simplifying Imaginary Numbers

If we assume that multiplication of imaginary numbers is commutative and associative, then

$$\begin{aligned}(2i)^2 &= 2^2i^2 \\ &= 4(-1) \\ &= -4\end{aligned}$$

Because $(2i)^2 = -4$, it follows that $2i$ is a square root of -4, and we can write

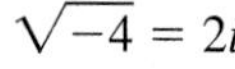

$$\sqrt{-4} = 2i$$

This result can also be obtained by using the multiplication property of radicals:

$$\sqrt{-4} = \sqrt{4(-1)} = \sqrt{4}\sqrt{-1} = 2i$$

In like fashion, we can use the multiplication property of radicals to simplify any imaginary number. For example,

$$\sqrt{-25} = \sqrt{25(-1)} = \sqrt{25}\sqrt{-1} = 5i$$

$$\sqrt{-\frac{1}{9}} = \sqrt{\frac{1}{9}(-1)} = \sqrt{\frac{1}{9}}\sqrt{-1} = \frac{1}{3}i$$

$$\sqrt{\frac{-100}{49}} = \sqrt{\frac{100}{49}(-1)} = \frac{\sqrt{100}}{\sqrt{49}}\sqrt{-1} = \frac{10}{7}i$$

Dr. Bernoit B. Mandelbrot has used mathematics to study unpredictable and irregular events in the physical, behavioral, and biological sciences. His study uses properties of the complex numbers to produce intricate images like this fractal.

WARNING! If a and b are both negative, then $\sqrt{ab} \neq \sqrt{a}\sqrt{b}$. For example,

$$\sqrt{(-16)(-4)} = \sqrt{64}$$
$$= 8$$

but

$$\sqrt{-16}\sqrt{-4} = (4i)(2i)$$
$$= 8i^2$$
$$= 8(-1)$$
$$= -8$$

Complex Numbers

The imaginary numbers are a subset of a set of numbers called the **complex numbers.**

Complex Numbers A **complex number** is any number that can be written in the form $a + bi$, where a and b are real numbers and $i = \sqrt{-1}$.

In the complex number $a + bi$, a is called the **real part,** and b is called the **imaginary part.**

If $b = 0$, the complex number $a + bi$ is a real number. If $b \neq 0$ and $a = 0$, the complex number $0 + bi$ (or just bi) is an imaginary number.

Any imaginary number can be expressed in bi form. For example,

$$\sqrt{-1} = i$$
$$\sqrt{-9} = \sqrt{9(-1)} = \sqrt{9}\sqrt{-1} = 3i$$
$$\sqrt{-3} = \sqrt{3(-1)} = \sqrt{3}\sqrt{-1} = \sqrt{3}i$$

WARNING! The expression $\sqrt{3}i$ is often written as $i\sqrt{3}$ to make it clear that i is not part of the radicand. Do not confuse $\sqrt{3}i$ with $\sqrt{3i}$.

The relationship between the real numbers, the imaginary numbers, and the complex numbers is shown in Figure 8-1.

Complex Numbers $a + bi$

Real Numbers $a + 0i$	**Imaginary Numbers $0 + bi$ $(b \neq 0)$**
$3, \frac{7}{3}, \pi, 125.345$	$4i, -12i, \sqrt{-4}$

$$4 + 7i,\ 5 - 16i,\ \frac{1}{32 - 12i},\ 15 + \sqrt{-25}$$

FIGURE 8-1

Equality of Complex Numbers The complex numbers $a + bi$ and $c + di$ are equal if and only if

$$a = c \quad \text{and} \quad b = d$$

Because of the previous definition, complex numbers are equal if and only if their real parts are equal and their imaginary parts are equal.

(c) **EXAMPLE 4**

a. $2 + 3i = \sqrt{4} + \frac{6}{2}i$ because $2 = \sqrt{4}$ and $3 = \frac{6}{2}$.

b. $4 - 5i = \frac{12}{3} - \sqrt{25}i$ because $4 = \frac{12}{3}$ and $-5 = -\sqrt{25}$.

c. $x + yi = 4 + 7i$ if and only if $x = 4$ and $y = 7$. ■

Arithmetic of Complex Numbers

Addition of Complex Numbers Complex numbers are added as if they were binomials:

$$(a + bi) + (c + di) = (a + c) + (b + d)i$$

(a, c) **EXAMPLE 5**

a. $(8 + 4i) + (12 + 8i) = 8 + 4i + 12 + 8i$
$= 20 + 12i$

b. $(7 - 4i) + (9 + 2i) = 7 - 4i + 9 + 2i$
$= 16 - 2i$

c. $(-6 + i) - (3 - 4i) = -6 + i - 3 + 4i$
$= -9 + 5i$

d. $(2 - 4i) - (-4 + 3i) = 2 - 4i + 4 - 3i$
$= 6 - 7i$ ■

To multiply a complex number by an imaginary number, we use the distributive property to remove parentheses and then simplify. For example,

$$\begin{aligned} -5i(4 - 8i) &= -5i(4) - (-5i)8i && \text{Use the distributive property.} \\ &= -20i + 40i^2 && \text{Simplify.} \\ &= -20i + 40(-1) && i^2 = -1. \\ &= -40 - 20i \end{aligned}$$

WARNING! When working with complex numbers, always be sure to substitute -1 for i^2 and simplify until the answer is in $a + bi$ form.

To multiply two complex numbers, we use the following definition.

Multiplying Complex Numbers Complex numbers are multiplied as if they were binomials, with $i^2 = -1$:

$$(a + bi)(c + di) = ac + adi + bci + bdi^2 = (ac - bd) + (ad + bc)i$$

(a, b)

EXAMPLE 6

a. $(2 + 3i)(3 - 2i) = 6 - 4i + 9i - 6i^2$ Use the FOIL method.
$= 6 + 5i + 6$ $i^2 = -1$.
$= 12 + 5i$

b. $(3 + i)(1 + 2i) = 3 + 6i + i + 2i^2$ Use the FOIL method.
$= 3 + 7i - 2$ $i^2 = -1$.
$= 1 + 7i$

c. $(-4 + 2i)(2 + i) = -8 - 4i + 4i + 2i^2$ Use the FOIL method.
$= -8 - 2$ $i^2 = -1$.
$= -10$ ■

The next two examples show how to write complex numbers in $a + bi$ form. When writing answers, it is acceptable to use $a - bi$ as a substitute for $a + (-b)i$.

EXAMPLE 7

a. $7 = 7 + 0i$

b. $3i = 0 + 3i$

c. $4 - \sqrt{-16} = 4 - \sqrt{16(-1)} = 4 - \sqrt{16}\sqrt{-1} = 4 - 4i$

d. $5 + \sqrt{-11} = 5 + \sqrt{11(-1)} = 5 + \sqrt{11}\sqrt{-1} = 5 + \sqrt{11}i$ ■

EXAMPLE 8

a. $2i^2 + 4i^3 = 2(-1) + 4(-i) = -2 - 4i$

b. $\dfrac{3}{2i} = \dfrac{3}{2i} \cdot \dfrac{i}{i} = \dfrac{3i}{2i^2} = \dfrac{3i}{2(-1)} = \dfrac{3i}{-2} = 0 - \dfrac{3}{2}i$

c. $-\frac{5}{i} = -\frac{5}{i} \cdot \frac{i^3}{i^3}$

$= -\frac{5i^3}{i^4}$

$= -\frac{5(-i)}{1}$

$= 5i$

$= 0 + 5i$

d. $\frac{6}{i^3} = \frac{6i}{i^3 i}$

$= \frac{6i}{i^4}$

$= \frac{6i}{1}$

$= 0 + 6i$

■

Complex Conjugates The complex numbers $a + bi$ and $a - bi$ are called **complex conjugates** of each other.

Note: Since $a + bi$ and $a - bi$ are complex conjugates, $bi + a$ and $bi - a$ are not. If they were, $a + bi$ would have two complex conjugates:
$a - bi$ and $bi - a$

For example,

$3 + 4i$ and $3 - 4i$ are complex conjugates.

$5 - 7i$ and $5 + 7i$ are complex conjugates.

$8 + 17i$ and $8 - 17i$ are complex conjugates.

EXAMPLE 9 Find the product of $3 + i$ and its complex conjugate.

Solution The complex conjugate of $3 + i$ is $3 - i$. We can find the product as follows:

$(3 + i)(3 - i) = 9 - 3i + 3i - i^2$ Use the FOIL method.

$= 9 - i^2$ Combine terms.

$= 9 - (-1)$ $i^2 = -1$.

$= 10$

■

The product of the complex number $a + bi$ and its complex conjugate $a - bi$ is always a real number, as the following work shows:

$(a + bi)(a - bi) = a^2 - abi + abi - b^2i^2$ Use the FOIL method.

$= a^2 - b^2(-1)$ $i^2 = -1$.

$= a^2 + b^2$

Thus,

$$(a + bi)(a - bi) = a^2 + b^2$$

Rationalizing the Denominator

We must rationalize the denominators to write complex numbers such as

$$\frac{1}{3 + i}, \quad \frac{3 - i}{2 + i}, \quad \text{and} \quad \frac{4 + 4i}{2 + 2i}$$

in $a + bi$ form. To do so, we multiply both the numerator and denominator by the complex conjugate of the denominator.

EXAMPLE 10 Write $\dfrac{1}{3+i}$ in $a + bi$ form.

Solution Because the product of $3 + i$ and its conjugate is a real number, we can rationalize the denominator by multiplying both the numerator and the denominator of the fraction by the complex conjugate of the denominator.

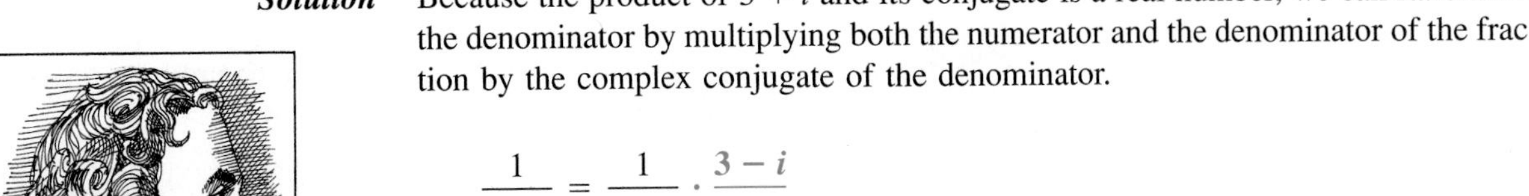

Abraham De Moivre
(1667–1754)
De Moivre is best known for a theorem used to find complex roots of certain equations. He also made several contributions to the field of probability.

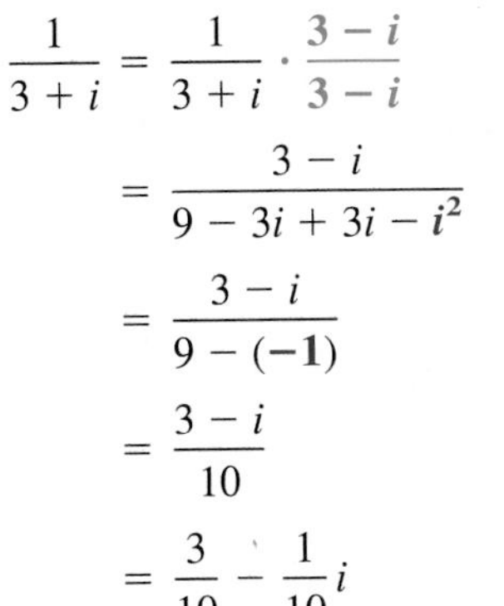

$$\frac{1}{3+i} = \frac{1}{3+i} \cdot \frac{3-i}{3-i}$$

$$= \frac{3-i}{9-3i+3i-i^2} \quad \text{Multiply the numerators and multiply the denominators.}$$

$$= \frac{3-i}{9-(-1)} \quad i^2 = -1.$$

$$= \frac{3-i}{10}$$

$$= \frac{3}{10} - \frac{1}{10}i$$

■

EXAMPLE 11 Write $\dfrac{3-i}{2+i}$ in $a + bi$ form.

Solution We can rationalize the denominator by multiplying both the numerator and the denominator of the fraction by the complex conjugate of the denominator.

$$\frac{3-i}{2+i} = \frac{3-i}{2+i} \cdot \frac{2-i}{2-i}$$

$$= \frac{6-3i-2i+i^2}{4-2i+2i-i^2} \quad \text{Multiply the numerators and multiply the denominators.}$$

$$= \frac{5-5i}{4-(-1)} \quad i^2 = -1.$$

$$= \frac{5(1-i)}{5} \quad \text{Factor out 5 in the numerator.}$$

$$= 1 - i \quad \text{Simplify.}$$

■

EXAMPLE 12 Write $\dfrac{4+\sqrt{-16}}{2+\sqrt{-4}}$ in $a + bi$ form.

Solution

$$\frac{4+\sqrt{-16}}{2+\sqrt{-4}} = \frac{4+4i}{2+2i}$$ Write each number in $a + bi$ form.

$$= \frac{2(2+2i)}{2+2i}$$ Factor out 2 in the numerator and simplify.

$$= 2 + 0i$$

■

WARNING! To avoid mistakes, always put complex numbers in $a + bi$ form before doing any complex number arithmetic.

Absolute Value of a Complex Number

Absolute Value of a Complex Number

The **absolute value** of the complex number $a + bi$ is $\sqrt{a^2 + b^2}$. In symbols,

$$|a + bi| = \sqrt{a^2 + b^2}$$

(c) **EXAMPLE 13**

a. $|3 + 4i| = \sqrt{3^2 + 4^2}$
$= \sqrt{9 + 16}$
$= \sqrt{25}$
$= 5$

b. $|3 - 4i| = \sqrt{3^2 + (-4)^2}$
$= \sqrt{9 + 16}$
$= \sqrt{25}$
$= 5$

Teaching Tip: Point out how part **d** extends the earlier definition of absolute value.

c. $|-5 - 12i| = \sqrt{(-5)^2 + (-12)^2}$
$= \sqrt{25 + 144}$
$= \sqrt{169}$
$= 13$

d. $|a + 0i| = \sqrt{a^2 + 0^2}$
$= \sqrt{a^2}$
$= |a|$

Each of these examples illustrates that the absolute value of a complex number is always a nonnegative real number. ■

ORALS

Simplify each power of i.

1. i^3 $-i$ **2.** i^2 -1 **3.** i^4 1 **4.** i^5 i

Write each imaginary number in bi form.

5. $\sqrt{-49}$ $7i$ **6.** $\sqrt{-64}$ $8i$ **7.** $\sqrt{-100}$ $10i$ **8.** $\sqrt{-81}$ $9i$

Find each absolute value.

9. $|3 + 4i|$ 5 **10.** $|5 - 12i|$ 13

EXERCISE 8.2

In Exercises 1–12, solve each equation. Write all roots in bi or a + bi form.

1. $x^2 + 9 = 0$ $\pm 3i$
2. $x^2 + 16 = 0$ $\pm 4i$
3. $3x^2 = -16$ $\pm\frac{4\sqrt{3}}{3}i$
4. $2x^2 = -25$ $\pm\frac{5\sqrt{2}}{2}i$
5. $x^2 + 2x + 2 = 0$ $-1 \pm i$
6. $x^2 + 3x + 3 = 0$ $-\frac{3}{2} \pm \frac{\sqrt{3}}{2}i$
7. $2x^2 + x + 1 = 0$ $-\frac{1}{4} \pm \frac{\sqrt{7}}{4}i$
8. $3x^2 + 2x + 1 = 0$ $-\frac{1}{3} \pm \frac{\sqrt{2}}{3}i$
9. $3x^2 - 4x = -2$ $\frac{2}{3} \pm \frac{\sqrt{2}}{3}i$
10. $2x^2 + 3x = -3$ $-\frac{3}{4} \pm \frac{\sqrt{15}}{4}i$
11. $3x^2 - 2x = -3$ $\frac{1}{3} \pm \frac{2\sqrt{2}}{3}i$
12. $5x^2 = 2x - 1$ $\frac{1}{5} \pm \frac{2}{5}i$

In Exercises 13–20, simplify each expression.

13. i^{21} i
14. i^{19} $-i$
15. i^{27} $-i$
16. i^{22} -1
17. i^{100} 1
18. i^{42} -1
19. i^{97} i
20. i^{200} 1

In Exercises 21–26, tell whether the complex numbers are equal.

21. $3 + 7i, \sqrt{9} + (5 + 2)i$ yes
22. $\sqrt{4} + \sqrt{25}i, 2 - (-5)i$ yes
23. $8 + 5i, 2^3 + \sqrt{25}i^3$ no
24. $4 - 7i, -4i^2 + 7i^3$ yes
25. $\sqrt{4} + \sqrt{-4}, 2 - 2i$ no
26. $\sqrt{-9} - i, 4i$ no

In Exercises 27–88, do the operations. Write all answers in a + bi form.

27. $(3 + 4i) + (5 - 6i)$ $8 - 2i$
28. $(5 + 3i) - (6 - 9i)$ $-1 + 12i$
29. $(7 - 3i) - (4 + 2i)$ $3 - 5i$
30. $(8 + 3i) + (-7 - 2i)$ $1 + i$
31. $(8 + 5i) + (7 + 2i)$ $15 + 7i$
32. $(-7 + 9i) - (-2 - 8i)$ $-5 + 17i$
33. $(1 + i) - 2i + (5 - 7i)$ $6 - 8i$
34. $(-9 + i) - 5i + (2 + 7i)$ $-7 + 3i$
35. $(5 + 3i) - (3 - 5i) + \sqrt{-1}$ $2 + 9i$
36. $(8 + 7i) - (-7 - \sqrt{-64}) + (3 - i)$ $18 + 14i$
37. $(-8 - \sqrt{3}i) - (7 - 3\sqrt{3}i)$ $-15 + 2\sqrt{3}i$
38. $(2 + 2\sqrt{2}i) + (-3 - \sqrt{2}i)$ $-1 + \sqrt{2}i$
39. $3i(2 - i)$ $3 + 6i$
40. $-4i(3 + 4i)$ $16 - 12i$
41. $-5i(5 - 5i)$ $-25 - 25i$
42. $2i(7 + 2i)$ $-4 + 14i$
43. $(2 + i)(3 - i)$ $7 + i$
44. $(4 - i)(2 + i)$ $9 + 2i$
45. $(2 - 4i)(3 + 2i)$ $14 - 8i$
46. $(3 - 2i)(4 - 3i)$ $6 - 17i$
47. $(2 + \sqrt{2}i)(3 - \sqrt{2}i)$ $8 + \sqrt{2}i$
48. $(5 + \sqrt{3}i)(2 - \sqrt{3}i)$ $13 - 3\sqrt{3}i$
49. $(8 - \sqrt{-1})(-2 - \sqrt{-16})$ $-20 - 30i$
50. $(-1 + \sqrt{-4})(2 + \sqrt{-9})$ $-8 + i$
51. $(2 + i)^2$ $3 + 4i$
52. $(3 - 2i)^2$ $5 - 12i$
53. $(2 + 3i)^2$ $-5 + 12i$
54. $(1 - 3i)^2$ $-8 - 6i$
55. $i(5 + i)(3 - 2i)$ $7 + 17i$
56. $i(-3 - 2i)(1 - 2i)$ $-4 - 7i$
57. $(2 + i)(2 - i)(1 + i)$ $5 + 5i$
58. $(3 + 2i)(3 - 2i)(i + 1)$ $13 + 13i$
59. $(3 + i)[(3 - 2i) + (2 + i)]$ $16 + 2i$
60. $(2 - 3i)[(5 - 2i) - (2i + 1)]$ $-4 - 20i$

61. $\dfrac{1}{i}$ $0 - i$

62. $\dfrac{1}{i^3}$ $0 + i$

63. $\dfrac{4}{5i^3}$ $0 + \frac{4}{5}i$

64. $\dfrac{3}{2i}$ $0 - \frac{3}{2}i$

65. $\dfrac{3i}{8\sqrt{-9}}$ $\frac{1}{8} + 0i$

66. $\dfrac{5i^3}{2\sqrt{-4}}$ $-\frac{5}{4} + 0i$

67. $\dfrac{-3}{5i^5}$ $0 + \frac{3}{5}i$

68. $\dfrac{-4}{6i^7}$ $0 - \frac{2}{3}i$

69. $\dfrac{5}{2 - i}$ $2 + i$

70. $\dfrac{26}{3 - 2i}$ $6 + 4i$

71. $\dfrac{13i}{5 + i}$ $\frac{1}{2} + \frac{5}{2}i$

72. $\dfrac{2i}{5 + 3i}$ $\frac{3}{17} + \frac{5}{17}i$

73. $\dfrac{-12}{7 - \sqrt{-1}}$ $-\frac{42}{25} - \frac{6}{25}i$

74. $\dfrac{4}{3 + \sqrt{-1}}$ $\frac{6}{5} - \frac{2}{5}i$

75. $\dfrac{5i}{6 + 2i}$ $\frac{1}{4} + \frac{3}{4}i$

76. $\dfrac{-4i}{2 - 6i}$ $\frac{3}{5} - \frac{1}{5}i$

77. $\dfrac{3 - 2i}{3 + 2i}$ $\frac{5}{13} - \frac{12}{13}i$

78. $\dfrac{2 + 3i}{2 - 3i}$ $-\frac{5}{13} + \frac{12}{13}i$

79. $\dfrac{3 + 2i}{3 + i}$ $\frac{11}{10} + \frac{3}{10}i$

80. $\dfrac{2 - 5i}{2 + 5i}$ $-\frac{21}{29} - \frac{20}{29}i$

81. $\dfrac{\sqrt{5} - \sqrt{3}i}{\sqrt{5} + \sqrt{3}i}$ $\dfrac{1}{4} - \dfrac{\sqrt{15}}{4}i$

82. $\dfrac{\sqrt{3} + \sqrt{2}i}{\sqrt{3} - \sqrt{2}i}$ $\dfrac{1}{5} + \dfrac{2\sqrt{6}}{5}i$

83. $\left(\dfrac{i}{3 + 2i}\right)^2$ $-\frac{5}{169} + \frac{12}{169}i$

84. $\left(\dfrac{5 + i}{2 + i}\right)^2$ $\frac{112}{25} - \frac{66}{25}i$

85. $\dfrac{i(3 - i)}{3 + i}$ $\frac{3}{5} + \frac{4}{5}i$

86. $\dfrac{5 + 3i}{i(3 - 5i)}$ 1

87. $\dfrac{(2 - 5i) - (5 - 2i)}{5 - i}$ $-\frac{6}{13} - \frac{9}{13}i$

88. $\dfrac{5i}{(5 + 2i) + (2 + i)}$ $\frac{15}{58} + \frac{35}{58}i$

In Exercises 89–96, find each value.

89. $|6 + 8i|$ 10

90. $|12 + 5i|$ 13

91. $|12 - 5i|$ 13

92. $|3 - 4i|$ 5

93. $|5 + 7i|$ $\sqrt{74}$

94. $|6 - 5i|$ $\sqrt{61}$

95. $\left|\dfrac{3}{5} - \dfrac{4}{5}i\right|$ 1

96. $\left|\dfrac{5}{13} + \dfrac{12}{13}i\right|$ 1

97. Show that $1 - 5i$ is a solution of $x^2 - 2x + 26 = 0$.

98. Show that $3 - 2i$ is a solution of $x^2 - 6x + 13 = 0$.

99. Show that i is a solution of $x^4 - 3x^2 - 4 = 0$.

100. Show that $2 + i$ is *not* a solution of $x^2 + x + 1 = 0$.

Writing Exercises ■ *Write a paragraph using your own words.*

1. Tell how to decide whether two complex numbers are equal.
2. Define the complex conjugate of a complex number.

Something to Think About ■

1. Rationalize the numerator of $\dfrac{3 - i}{2}$. $\dfrac{5}{3 + i}$
2. Rationalize the numerator of $\dfrac{2 + 3i}{2 - 3i}$. $\dfrac{13}{-5 - 12i}$

Review Exercises ■ *Do each operation.*

1. $\dfrac{x^2 - x - 6}{9 - x^2} \cdot \dfrac{x^2 + x - 6}{x^2 - 4}$ -1
2. $\dfrac{3x + 4}{x - 2} + \dfrac{x - 4}{x + 2}$ $\dfrac{4x^2 + 4x + 16}{(x + 2)(x - 2)}$

3. **Wind speed** A plane that can fly 200 miles per hour in still air makes a 330-mile flight with a tail wind and returns, flying into the same wind. Find the speed of the wind if the total flying time is $3\frac{1}{3}$ hours. 20 mph

4. **Finding rates** A student drove a distance of 105 miles. On his return trip, his average speed was 5 miles per hour faster, and he saved 30 minutes of driving time. Find his faster rate.
35 mph

8.3 More on Quadratic Equations

■ Equations That Can Be Put into Quadratic Form ■ Solutions of a Quadratic Equation

It is possible to predict what type of numbers will be solutions of a given quadratic equation without solving the equation. To see how, we suppose that the coefficients a, b, and c in the quadratic equation $ax^2 + bx + c = 0$ are real numbers. Then the solutions of the equation are given by the quadratic formula

$$x = \frac{-b \pm \sqrt{b^2 - 4ac}}{2a} \qquad (a \neq 0)$$

If $b^2 - 4ac \geq 0$, the solutions are real numbers. If $b^2 - 4ac < 0$, the solutions are nonreal complex numbers. Thus, the value of $b^2 - 4ac$, called the **discriminant,** determines the nature of the roots of a quadratic equation.

The Discriminant If a, b, and c are real numbers and

If $b^2 - 4ac$ is . . .	the solutions are . . .
positive,	real numbers and unequal.
0,	real numbers and equal.
negative,	nonreal complex numbers and complex conjugates.

If a, b, and c are rational numbers and

If $b^2 - 4ac$ is . . .	the solutions are . . .
a perfect square,	rational numbers and unequal.
positive and not a perfect square,	irrational numbers and unequal.

EXAMPLE 1 Determine the type of solutions for the quadratic equation $x^2 + x + 1 = 0$.

Solution We calculate the discriminant:

$$b^2 - 4ac = 1^2 - 4(1)(1) \qquad a = 1,\ b = 1,\ \text{and } c = 1.$$
$$= -3$$

Since $b^2 - 4ac < 0$, the solutions are nonreal complex numbers. ■

EXAMPLE 2 Determine the type of solutions for the quadratic equation $3x^2 + 5x + 2 = 0$.

Solution We calculate the discriminant:

$$b^2 - 4ac = 5^2 - 4(3)(2) \qquad a = 3,\ b = 5,\ \text{and } c = 2.$$
$$= 25 - 24$$
$$= 1$$

Since $b^2 - 4ac$ is a perfect square, the solutions are rational and unequal. ■

EXAMPLE 3 What value of k will make the solutions of the equation $kx^2 - 12x + 9 = 0$ equal?

Solution We calculate the discriminant:

$$b^2 - 4ac = (-12)^2 - 4(k)(9) \qquad a = k,\ b = -12,\ \text{and } c = 9.$$
$$= 144 - 36k$$

Since the solutions are to be equal, we let $-36k + 144 = 0$ and solve for k.

$$-36k + 144 = 0$$
$$-36k = -144 \qquad \text{Subtract 144 from both sides.}$$
$$k = 4 \qquad \text{Divide both sides by } -36.$$

If $k = 4$, the solutions will be equal. Verify this by solving $4x^2 - 12x + 9 = 0$ and showing that the solutions are equal. ■

Equations That Can Be Put into Quadratic Form

Many nonquadratic equations can be put into quadratic form and then solved with the techniques used for solving quadratic equations. For example, to solve $x^4 - 5x^2 + 4 = 0$, we can proceed as follows:

$$x^4 - 5x^2 + 4 = 0$$
$$(x^2)^2 - 5(x^2) + 4 = 0$$
$$y^2 - 5y + 4 = 0 \qquad \text{Let } y = x^2.$$
$$(y - 4)(y - 1) = 0 \qquad \text{Factor } y^2 - 5y + 4.$$

$$y - 4 = 0 \quad \text{or} \quad y - 1 = 0$$
$$y = 4 \quad | \quad y = 1$$

Since $x^2 = y$, it follows that $x^2 = 4$ or $x^2 = 1$. Thus,

$$x^2 = 4 \qquad \text{or} \qquad x^2 = 1$$
$$x = 2 \quad \text{or} \quad x = -2 \quad | \quad x = 1 \quad \text{or} \quad x = -1$$

This equation has the four solutions 1, -1, 2, and -2. Verify that each one satisfies the original equation.

This equation could be solved directly by factoring.

EXAMPLE 4 Solve the equation $x - 7x^{1/2} + 12 = 0$.

Solution This equation is not a quadratic equation. However, if y^2 is substituted for x and y is substituted for $x^{1/2}$, the equation

$$x - 7x^{1/2} + 12 = 0$$

becomes a quadratic equation that can be solved by factoring:

$$y^2 - 7y + 12 = 0 \qquad \text{Substitute } y^2 \text{ for } x \text{ and } y \text{ for } x^{1/2}.$$
$$(y - 3)(y - 4) = 0 \qquad \text{Factor } y^2 - 7y + 12 = 0.$$

$$y - 3 = 0 \quad \text{or} \quad y - 4 = 0 \qquad \text{Set each factor equal to 0.}$$
$$y = 3 \quad | \quad y = 4$$

Because $x = y^2$, it follows that

$$x = 3^2 \quad \text{or} \quad x = 4^2$$
$$x = 9 \quad | \quad x = 16$$

Verify that both solutions satisfy the original equation. ■

EXAMPLE 5 Solve the equation $\dfrac{24}{x} + \dfrac{12}{x+1} = 11$.

Solution Since the denominator of a fraction cannot be 0, x cannot be 0 or -1. If either 0 or -1 appears as a suspected solution, it is extraneous and must be discarded.

We can solve the equation as follows:

$$\frac{24}{x} + \frac{12}{x+1} = 11$$

$$x(x+1)\left(\frac{24}{x} + \frac{12}{x+1}\right) = x(x+1)11 \qquad \text{Multiply both sides by } x(x+1).$$

$$24(x+1) + 12x = (x^2 + x)11 \qquad \text{Simplify.}$$

$$24x + 24 + 12x = 11x^2 + 11x \qquad \text{Use the distributive property to remove parentheses.}$$

$$36x + 24 = 11x^2 + 11x \qquad \text{Combine terms.}$$

$$0 = 11x^2 - 25x - 24 \qquad \text{Subtract } 36x \text{ and 24 from both sides.}$$

$$0 = (11x + 8)(x - 3) \qquad \text{Factor } 11x^2 - 25x - 24.$$

$$11x + 8 = 0 \quad \text{or} \quad x - 3 = 0 \qquad \text{Set each factor equal to 0.}$$

$$x = -\frac{8}{11} \quad \Big| \quad x = 3$$

Verify that both $-\frac{8}{11}$ and 3 satisfy the original equation. ■

EXAMPLE 6 Solve the formula $s = 16t^2 - 32$ for t.

Solution We proceed as follows:

$$s = 16t^2 - 32$$

$$s + 32 = 16t^2 \qquad \text{Add 32 to both sides.}$$

$$\frac{s+32}{16} = t^2 \qquad \text{Divide both sides by 16.}$$

$$t^2 = \frac{s+32}{16} \qquad \text{Apply the symmetric property of equality.}$$

$$t = \pm\sqrt{\frac{s+32}{16}} \qquad \text{Apply the square root property.}$$

$$= \frac{\pm\sqrt{s+32}}{\sqrt{16}}$$

$$= \frac{\pm\sqrt{s+32}}{4}$$

Thus, $t = \frac{\pm\sqrt{s+32}}{4}$. ■

Solutions of a Quadratic Equation

The following theorem gives an interesting property of the solutions of a quadratic equation.

Solutions of a Quadratic Equation If r_1 and r_2 are the solutions of the quadratic equation $ax^2 + bx + c = 0$, with $a \neq 0$, then

$$r_1 + r_2 = -\frac{b}{a} \quad \text{and} \quad r_1 r_2 = \frac{c}{a}$$

Proof To prove this theorem, we note that the solutions to the equation are given by the quadratic formula

$$r_1 = \frac{-b + \sqrt{b^2 - 4ac}}{2a} \quad \text{and} \quad r_2 = \frac{-b - \sqrt{b^2 - 4ac}}{2a}$$

Thus,

$$\begin{aligned} r_1 + r_2 &= \frac{-b + \sqrt{b^2 - 4ac}}{2a} + \frac{-b - \sqrt{b^2 - 4ac}}{2a} \\ &= \frac{-b + \sqrt{b^2 - 4ac} - b - \sqrt{b^2 - 4ac}}{2a} \\ &= -\frac{2b}{2a} \\ &= -\frac{b}{a} \end{aligned}$$

Keep the denominator and add the numerators.

and

$$\begin{aligned} r_1 r_2 &= \frac{-b + \sqrt{b^2 - 4ac}}{2a} \cdot \frac{-b - \sqrt{b^2 - 4ac}}{2a} \\ &= \frac{b^2 - (b^2 - 4ac)}{4a^2} \\ &= \frac{b^2 - b^2 + 4ac}{4a^2} \\ &= \frac{4ac}{4a^2} \\ &= \frac{c}{a} \end{aligned}$$

Multiply the numerators and multiply the denominators.

□

It can also be shown that if

$$r_1 + r_2 = -\frac{b}{a} \quad \text{and} \quad r_1 r_2 = \frac{c}{a}$$

then r_1 and r_2 are solutions of $ax^2 + bx + c = 0$. We can use this fact to check the solutions of quadratic equations.

EXAMPLE 7 Show that $\frac{3}{2}$ and $-\frac{1}{3}$ are solutions of the quadratic equation $6x^2 - 7x - 3 = 0$.

Solution Since $a = 6$, $b = -7$, and $c = -3$, we have

$$-\frac{b}{a} = -\frac{-7}{6} = \frac{7}{6} \quad \text{and} \quad \frac{c}{a} = \frac{-3}{6} = -\frac{1}{2}$$

Since $\frac{3}{2} + \left(-\frac{1}{3}\right) = \frac{7}{6}$ and $\left(\frac{3}{2}\right)\left(-\frac{1}{3}\right) = -\frac{1}{2}$, these numbers are solutions. ■

ORALS *Find $b^2 - 4ac$ when*

1. $a = 1, b = 1, c = 1$ -3

2. $a = 2, b = 1, c = 1$ -7

Determine the type of solutions for

3. $x^2 - 4x + 1 = 0$ irrational and unequal

4. $8x^2 - x + 2 = 0$ complex conjugates

Are the following numbers solutions of $x^2 - 7x + 6 = 0$?

5. 1, 5 no

6. 1, 6 yes

EXERCISE 8.3

In Exercises 1–8, use the discriminant to determine what type of solutions exist for each quadratic equation. ***Do not solve the equation.***

1. $4x^2 - 4x + 1 = 0$ rational, equal

2. $6x^2 - 5x - 6 = 0$ rational, unequal

3. $5x^2 + x + 2 = 0$ complex conjugates

4. $3x^2 + 10x - 2 = 0$ irrational, unequal

5. $2x^2 = 4x - 1$ irrational, unequal

6. $9x^2 = 12x - 4$ rational, equal

7. $x(2x - 3) = 20$ rational, unequal

8. $x(x - 3) = -10$ complex conjugates

In Exercises 9–16, find the value(s) of k that will make the solutions of each given quadratic equation equal.

9. $x^2 + kx + 9 = 0$ 6, −6

10. $kx^2 - 12x + 4 = 0$ 9

11. $9x^2 + 4 = -kx$ 12, −12

12. $9x^2 - kx + 25 = 0$ 30, −30

13. $(k - 1)x^2 + (k - 1)x + 1 = 0$ 5

14. $(k + 3)x^2 + 2kx + 4 = 0$ 6, −2

15. $(k + 4)x^2 + 2kx + 9 = 0$ 12, −3

16. $(k + 15)x^2 + (k - 30)x + 4 = 0$ 66, 10

17. Use the discriminant to determine whether the solutions of $1492x^2 + 1776x - 1984 = 0$ are real numbers. yes

18. Use the discriminant to determine whether the solutions of $1776x^2 - 1492x + 1984 = 0$ are real numbers. no

19. Determine k such that the solutions of $3x^2 + 4x = k$ are nonreal complex numbers. $k < -\frac{4}{3}$

20. Determine k such that the solutions of $kx^2 - 4x = 7$ are nonreal complex numbers. $k < -\frac{4}{7}$

In Exercises 21–48, solve each equation.

21. $x^4 - 17x^2 + 16 = 0$ $1, -1, 4, -4$

22. $x^4 - 10x^2 + 9 = 0$ $3, -3, 1, -1$

23. $x^4 - 3x^2 = -2$ $1, -1, \sqrt{2}, -\sqrt{2}$

24. $x^4 - 29x^2 = -100$ $5, -5, 2, -2$

25. $x^4 = 6x^2 - 5$ $1, -1, \sqrt{5}, -\sqrt{5}$

26. $x^4 = 8x^2 - 7$ $1, -1, \sqrt{7}, -\sqrt{7}$

27. $2x^4 - 10x^2 = -8$ $1, -1, 2, -2$

28. $2x^4 + 24 = 26x^2$ $1, -1, 2\sqrt{3}, -2\sqrt{3}$

29. $2x + x^{1/2} - 3 = 0$ 1

30. $2x - x^{1/2} - 1 = 0$ 1

31. $3x + 5x^{1/2} + 2 = 0$ no solution

32. $3x - 4x^{1/2} + 1 = 0$ $1, \frac{1}{9}$

33. $x^{2/3} + 5x^{1/3} + 6 = 0$ $-8, -27$

34. $x^{2/3} - 7x^{1/3} + 12 = 0$ $64, 27$

35. $x^{2/3} - 2x^{1/3} - 3 = 0$ $-1, 27$

36. $x^{2/3} + 4x^{1/3} - 5 = 0$ $-125, 1$

37. $x + 5 + \dfrac{4}{x} = 0$ $-1, -4$

38. $x - 4 + \dfrac{3}{x} = 0$ $3, 1$

39. $x + 1 = \dfrac{20}{x}$ $4, -5$

40. $x + \dfrac{15}{x} = 8$ $3, 5$

41. $\dfrac{1}{x-1} + \dfrac{3}{x+1} = 2$ $0, 2$

42. $\dfrac{6}{x-2} - \dfrac{12}{x-1} = -1$ $4, 5$

43. $\dfrac{1}{x+2} + \dfrac{24}{x+3} = 13$ $-1, -\frac{27}{13}$

44. $\dfrac{3}{x} + \dfrac{4}{x+1} = 2$ $3, -\frac{1}{2}$

45. $x^{-4} - 2x^{-2} + 1 = 0$ $1, 1, -1, -1$

46. $4x^{-4} + 1 = 5x^{-2}$ $1, -1, 2, -2$

47. $x + \dfrac{2}{x-2} = 0$ $1 \pm i$

48. $x + \dfrac{x+5}{x-3} = 0$ $1 \pm 2i$

In Exercises 49–56, solve each equation for the indicated variable.

49. $x^2 + y^2 = r^2$ for x $x = \pm\sqrt{r^2 - y^2}$

50. $x^2 + y^2 = r^2$ for y $y = \pm\sqrt{r^2 - x^2}$

51. $I = \dfrac{k}{d^2}$ for d $d = \pm\sqrt{\dfrac{k}{I}} = \pm\dfrac{\sqrt{kI}}{I}$

52. $V = \dfrac{1}{3}\pi r^2 h$ for r $r = \pm\sqrt{\dfrac{3V}{\pi h}} = \pm\dfrac{\sqrt{3\pi V h}}{\pi h}$

53. $xy^2 + 3xy + 7 = 0$ for y $y = \dfrac{-3x \pm \sqrt{9x^2 - 28x}}{2x}$

54. $kx = ay - x^2$ for x $x = \dfrac{-k \pm \sqrt{k^2 + 4ay}}{2}$

55. $\sigma = \sqrt{\dfrac{\Sigma x^2}{N} - \mu^2}$ for μ^2 $\mu^2 = \dfrac{\Sigma x^2}{N} - \sigma^2$

56. $\sigma = \sqrt{\dfrac{\Sigma x^2}{N} - \mu^2}$ for N $N = \dfrac{\Sigma x^2}{\sigma^2 + \mu^2}$

In Exercises 57–64, solve each quadratic equation and verify that the sum of the solutions is $-\frac{b}{a}$ and that the product of the solutions is $\frac{c}{a}$.

57. $12x^2 - 5x - 2 = 0$ $\frac{2}{3}, -\frac{1}{4}$

58. $8x^2 - 2x - 3 = 0$ $\frac{3}{4}, -\frac{1}{2}$

59. $2x^2 + 5x + 1 = 0$ $\dfrac{-5 \pm \sqrt{17}}{4}$

60. $3x^2 + 9x + 1 = 0$ $\dfrac{-9 \pm \sqrt{69}}{6}$

61. $3x^2 - 2x + 4 = 0$ $\frac{1 \pm i\sqrt{11}}{3}$

62. $2x^2 - x + 4 = 0$ $\frac{1}{4} \pm \frac{\sqrt{31}}{4}i$

63. $x^2 + 2x + 5 = 0$ $-1 \pm 2i$

64. $x^2 - 4x + 13 = 0$ $2 \pm 3i$

Writing Exercises ■ *Write a paragraph using your own words.*

1. Describe how to predict what type of solutions the equation $3x^2 - 4x + 5 = 0$ will have.

2. How is the discriminant related to the quadratic formula?

Something to Think About ■

1. Can a quadratic equation with integer coefficients have one real and one complex solution? Why?
no

2. Can a quadratic equation with complex coefficients have one real and one complex solution? Why?
yes

Review Exercises ■ *Solve each equation.*

1. $\frac{1}{4} + \frac{1}{t} = \frac{1}{2t}$ -2

2. $\frac{p-3}{3p} + \frac{1}{2p} = \frac{1}{4}$ 6

3. Find the slope of the line passing through $P(-2, -4)$ and $Q(3, 5)$.
$\frac{9}{5}$

4. Write the equation of the line passing through $P(-2, -4)$ and $Q(3, 5)$ in general form.
$9x - 5y = 2$

8.4 Graphs of Quadratic Functions

■ Vertex and Axis of Symmetry of a Parabola ■ Graphing Calculators ■ Problem Solving ■ Formula for the Vertex of a Parabola

We have defined a linear function as a correspondence between x and y determined by a first-degree equation that can be written in the form $y = mx + b$. We now discuss another function called a **quadratic function.**

Quadratic Function A **quadratic function** is a function defined by a second-degree polynomial equation of the form

$$y = ax^2 + bx + c$$

where a, b, and c are real numbers and $a \neq 0$.

To graph the quadratic function determined by $y = x^2 - 3$, for example, we find several ordered pairs that satisfy the equation, plot each point, and join them with a smooth curve to get the **parabola** shown in Figure 8-2. Note that the graph passes the vertical line test.

$y = x^2 - 3$

x	y
-3	6
-2	1
-1	-2
0	-3
1	-2
2	1
3	6

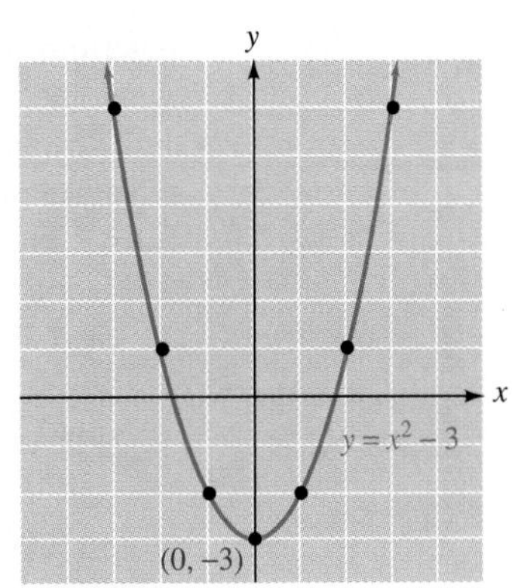

FIGURE 8-2

EXAMPLE 1 Graph the quadratic function determined by $y = -x^2 + 2x + 1$.

Solution We plot several points whose coordinates satisfy the equation and join them with a smooth curve to obtain the parabola shown in Figure 8-3.

$y = -x^2 + 2x + 1$

x	y
-1	-2
0	1
1	2
2	1
3	-2

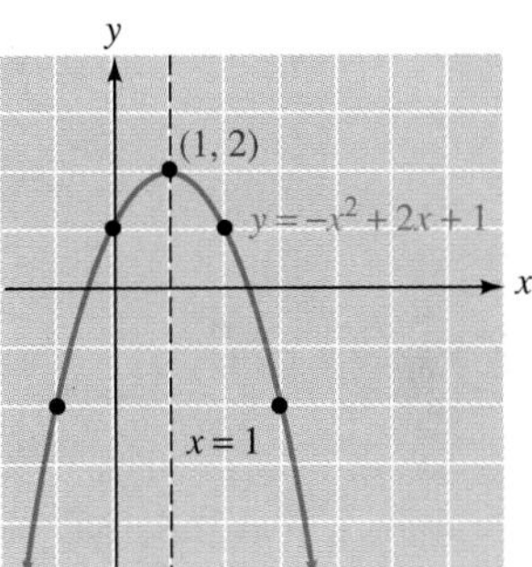

FIGURE 8-3

■

Vertex and Axis of Symmetry of a Parabola

The graphs of all functions determined by equations of the form $y = ax^2 + bx + c$, where $a \neq 0$, are parabolas. They open upward when $a > 0$ (as in Figure 8-2) and downward when $a < 0$ (as in Figure 8-3).

The bottom point of a parabola that opens upward, or the top point of a parabola that opens downward, is called the **vertex** of the parabola. The vertex of the parabola shown in Figure 8-2 is the point $(0, -3)$. The vertex of the parabola in Figure 8-3 is the point $(1, 2)$.

The vertical line that passes through the vertex of a parabola is called an **axis of symmetry,** because it divides the parabola into two congruent halves. The axis of symmetry in Figure 8-2 is the y-axis. The axis of symmetry in Figure 8-3 is the line $x = 1$.

If a, h, and k are constants and $a \neq 0$, the equation

$$y = a(x - h)^2 + k$$

determines a quadratic function, because it takes on the form $y = ax^2 + bx + c$ when the right-hand side is expanded and simplified.

The graph of the equation $y = a(x - h)^2 + k$ is a parabola that opens upward when $a > 0$, or downward when $a < 0$. This form of the quadratic equation is useful because it displays the coordinates of the vertex of its parabolic graph, as the following discussion will show.

If $a > 0$, then the graph of

$$y = a(x - h)^2 + k$$

is a parabola opening upward, as in Figure 8-4. The vertex of the parabola is the point on the graph that has the least possible value for its y-coordinate. Since $a > 0$, the least possible y-coordinate will occur when the nonnegative quantity $a(x - h)^2$ on the right-hand side of the equal sign is 0. This occurs when $x = h$. When $x = h$, then $a(x - h)^2 = 0$, and $y = k$. Thus, the vertex of the parabola is at the point with coordinates (h, k). A similar argument holds when $a < 0$.

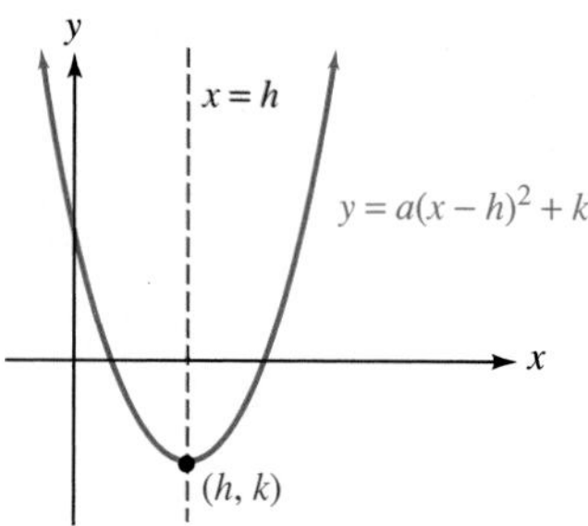

FIGURE 8-4

The previous discussion leads to the following theorem.

Vertex and Axis of Symmetry of a Parabola The graph of the equation

$$y = a(x - h)^2 + k$$

where $a \neq 0$, is a parabola with vertex at (h, k).

The parabola opens upward when $a > 0$ and downward when $a < 0$. The axis of symmetry is the line $x = h$.

EXAMPLE 2 Find the vertex and axis of symmetry of the parabola determined by the quadratic function $y = 2x^2 + 6x - 3$ and graph the parabola.

Solution To write the equation in the form $y = a(x - h)^2 + k$, we complete the square on the right-hand side of the equation:

$$y = 2x^2 + 6x - 3$$

$$= 2(x^2 + 3x) - 3$$ Factor out 2 from $2x^2 + 6x$ to make the coefficient of x^2 equal to 1.

$$= 2\left(x^2 + 3x + \frac{9}{4} - \frac{9}{4}\right) - 3$$ Add and subtract $\frac{9}{4}$ to complete the square.

$$= 2\left(x^2 + 3x + \frac{9}{4}\right) - 3 - \frac{9}{2}$$ $2\left(-\frac{9}{4}\right) = -\frac{9}{2}$

$$= 2\left(x + \frac{3}{2}\right)^2 - \frac{15}{2}$$ Factor $x^2 + 3x + \frac{9}{4}$ and add -3 and $-\frac{9}{2}$.

$$= 2\left[x - \left(-\frac{3}{2}\right)\right]^2 + \left(-\frac{15}{2}\right)$$ To write the equation in the form $y = a(x - h)^2 + k$, write $x + \frac{3}{2}$ as $x - \left(-\frac{3}{2}\right)$ and $-\frac{15}{2}$ as $+\left(-\frac{15}{2}\right)$.

Thus, the vertex is the point with coordinates $\left(-\frac{3}{2}, -\frac{15}{2}\right)$, as shown in the graph in Figure 8-5. The parabola has the line $x = -\frac{3}{2}$ as the axis of symmetry.

$y = 2x^2 + 6x - 3$

x	y
−4	5
−3	−3
−2	−7
−1	−7
0	−3
1	5
2	17

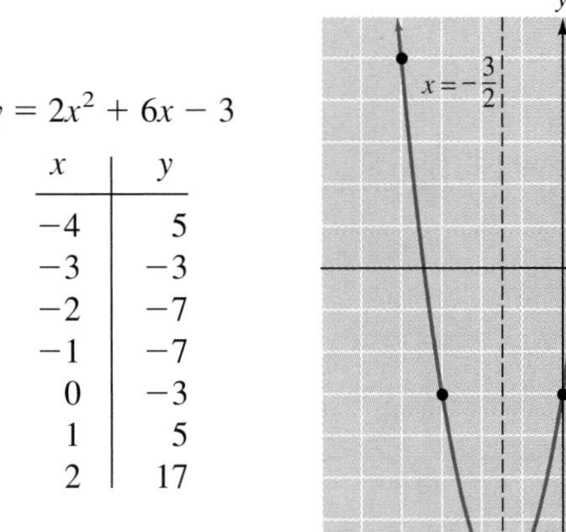

FIGURE 8-5

■

Graphing Calculators

It is easy to approximate the coordinates of the vertex of a parabola with a graphing calculator. For example, to find the vertex of the parabola determined by $y = 0.7x^2 + 2x - 3.5$, we use the standard RANGE settings to graph the parabola, as in Figure 8-6(a). We then press the TRACE key and move the cursor to the lowest point on the graph, as in Figure 8-6(b). From the calculator, we can read that the coordinates of the vertex are approximately

$$x = -1.368421 \qquad \text{and} \qquad y = -4.926039$$

For better results, we can use the ZOOM key.

It is also easy to solve a quadratic equation with a graphing calculator.

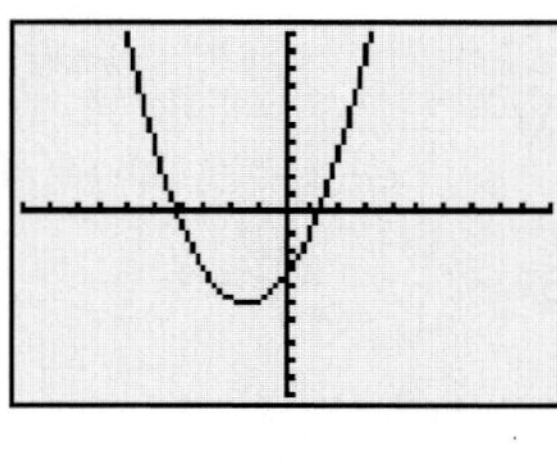

(a)

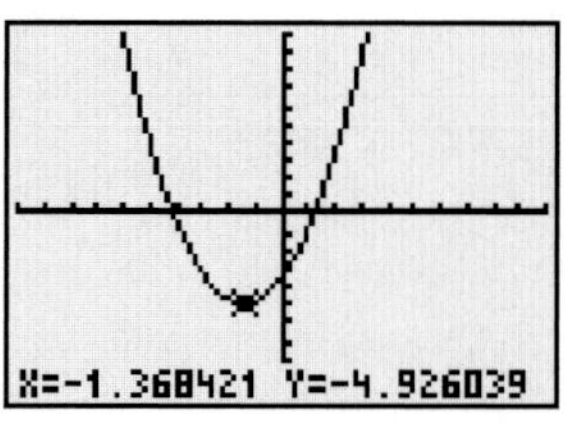

(b)

FIGURE 8-6

EXAMPLE 3 Solve the quadratic equation $0.7x^2 + 2x - 3.5 = 0$.

Solution The solution of $0.7x^2 + 2x - 3.5 = 0$ are the numbers x that will make $y = 0$ in the quadratic function $y = 0.7x^2 + 2x - 3.5$. To find these numbers, we can graph the quadratic function and read the x-intercepts from the graph.

As before, we use the standard RANGE settings and graph the quadratic function, as in Figure 8-7(a). We then press the TRACE key and move the cursor to each x-intercept, as in Figures 8-7(b) and 8-7(c). From the calculator, we can read the value of the x-coordinate of each x-intercept. The solutions to the quadratic equation are approximately

$$x = -4.105263 \qquad \text{and} \qquad x = 1.1578947$$

For better results, we can use the ZOOM key.

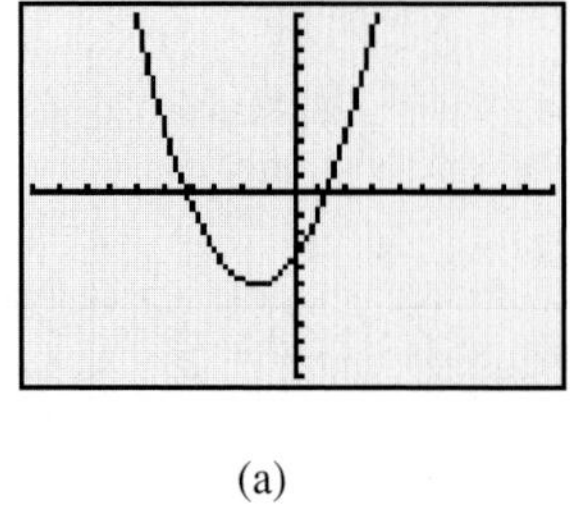

(a)

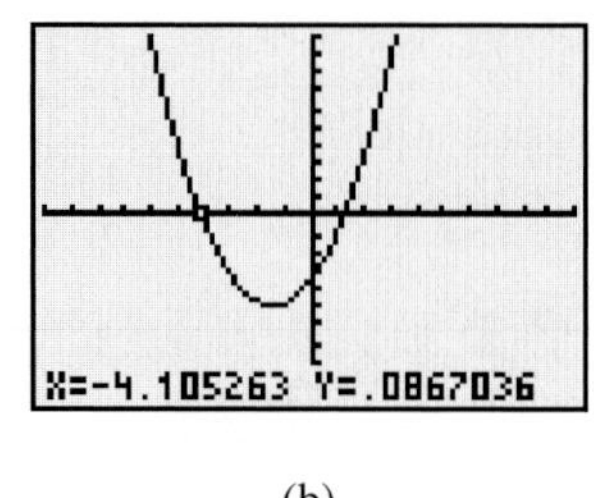

(b)

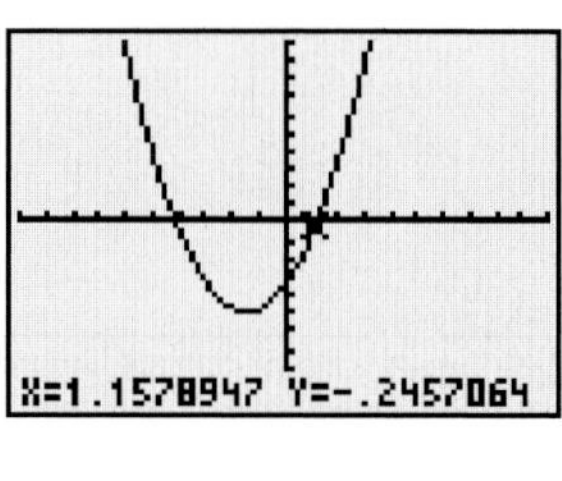

(c)

FIGURE 8-7

Problem Solving

EXAMPLE 4 **Ballistics** Suppose that the ball shown in Figure 8-8 is thrown straight up with a velocity of 128 feet per second. The quadratic function $s = 128t - 16t^2$ gives the relation between s and t, where s represents the number of feet the ball is above the ground and t represents the time measured in seconds. How high does the ball travel?

Solution The graph of the equation $s = -16t^2 + 128t$ is a parabola. Since the coefficient of t^2 is negative, the parabola opens downward, and the maximum height attained by the ball is given by the s-coordinate of the vertex of the parabola. We can find the coordinates of the vertex by completing the square:

$$
\begin{aligned}
s &= -16t^2 + 128t \\
&= -16(t^2 - 8t) && \text{Factor out } -16. \\
&= -16(t^2 - 8t + 16 - 16) && \text{Add and subtract 16.} \\
&= -16(t^2 - 8t + 16) + 256 && (-16)(-16) = 256. \\
&= -16(t - 4)^2 + 256 && \text{Factor } t^2 - 8t + 16.
\end{aligned}
$$

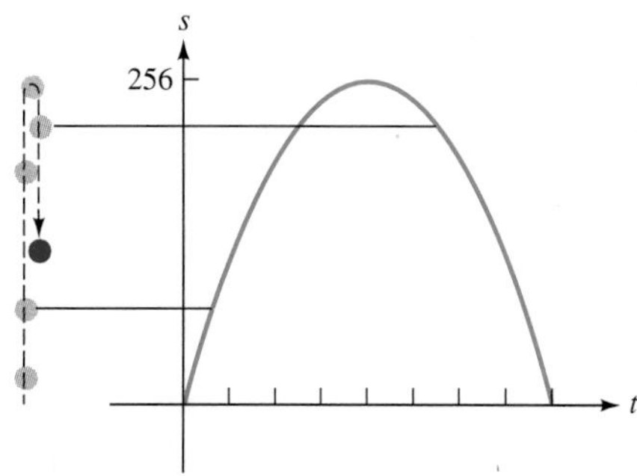

FIGURE 8-8

Thus, the coordinates of the vertex are (4, 256). Since $t = 4$ and $s = 256$ are the coordinates of the vertex, the ball reaches a maximum height of 256 feet in 4 seconds.

Since a parabola is symmetric, it takes 4 more seconds for the ball to return to earth. The total time of flight is 8 seconds.

WARNING! Although this quadratic function describes the height of the ball in relation to time, it does not describe the path traveled by the ball. ■

EXAMPLE 5 **Maximizing area** A man wants to build the rectangular pen shown in Figure 8-9 to house his dog. To save fencing, he intends to use one side of his garage. Find the maximum area that he can enclose with 80 feet of fencing.

Solution We can let the width of the area be represented by w. Then the length is represented by $80 - 2w$.

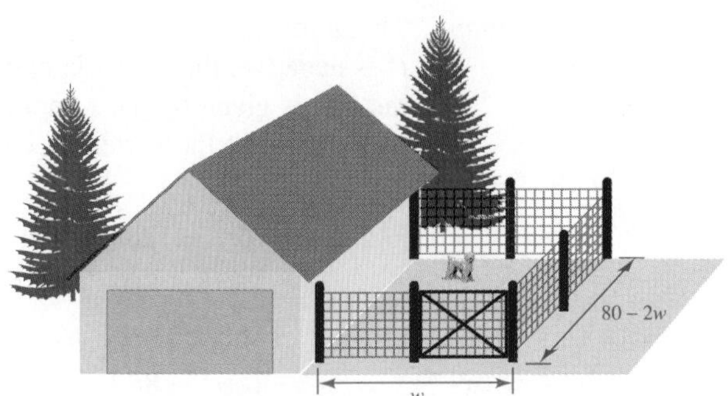

FIGURE 8-9

We can find the maximum value of A as follows:

$A = (80 - 2w)w$	$A = lw$.
$= 80w - 2w^2$	Use the distributive property to remove parentheses.
$= -2(w^2 - 40w)$	Factor out -2.
$= \mathbf{-2}(w^2 - 40w + \mathbf{400 - 400})$	Subtract and add 400.
$= -2(w^2 - 40w + 400) + 800$	
$= -2(w - 20)^2 + 800$	Factor $w^2 - 40w + 400$.

Thus, the coordinates of the vertex of the graph of the quadratic function are (20, 800), and the maximum area is 800 square feet. ■

Formula for the Vertex of a Parabola

We can show that the vertex of the parabola determined by the equation $y = ax^2 + bx + c$ is the point with coordinates

$$\left(-\frac{b}{2a}, c - \frac{b^2}{4a}\right)$$

To do so, we complete the square to write the equation $y = ax^2 + bx + c$ in the form $y = a(x - h)^2 + k$ to find the coordinates of the vertex.

$y = ax^2 + bx + c$	
$= a\left(x^2 + \frac{b}{a}x\right) + c$	Factor out a.
$= a\left(x^2 + \frac{b}{a}x + \frac{b^2}{4a^2} - \frac{b^2}{4a^2}\right) + c$	Add and subtract $\frac{b^2}{4a^2}$.
$= a\left(x^2 + \frac{b}{a}x + \frac{b^2}{4a^2}\right) + c - \frac{b^2}{4a}$	$a\left(-\frac{b^2}{4a^2}\right) = -\frac{b^2}{4a}$.

$$= a\left(x + \frac{b}{2a}\right)^2 + c - \frac{b^2}{4a} \qquad \text{Factor the trinomial.}$$

$$= a\left[x - \left(-\frac{b}{2a}\right)\right]^2 + \left(c - \frac{b^2}{4a}\right)$$

Thus, the vertex is the point with coordinates of $\left(-\frac{b}{2a}, c - \frac{b^2}{4a}\right)$.

EXAMPLE 6 **Maximizing the revenue** The revenue R received for selling x stereos is given by the equation

$$R = -\frac{x^2}{1000} + 10x$$

Find the number of stereos that must be sold to obtain the maximum revenue. Find the maximum revenue.

Solution The graph of the equation is a parabola that opens downward. Thus, the maximum value of R will be the R-coordinate of the vertex. The value of x at the vertex is

$$\begin{aligned} x &= -\frac{b}{2a} \\ &= \frac{-10}{2\left(-\frac{1}{1000}\right)} \\ &= \frac{-10}{-\frac{1}{500}} \\ &= (-10)(-500) \\ &= 5000 \end{aligned}$$

The value of R at the vertex is

$$\begin{aligned} R &= c - \frac{b^2}{4a} \\ &= 0 - \frac{10^2}{4\left(-\frac{1}{1000}\right)} \\ &= -\frac{100}{-\frac{1}{250}} \\ &= (-100)(-250) \\ &= 25{,}000 \end{aligned}$$

If the company sells 5000 stereos, it will obtain the maximum revenue of \$25,000. ■

ORALS *Tell whether the graph of each equation opens up or down.*

1. $y = -3x^2 + x - 5$ down
2. $y = 4x^2 + 2x - 3$ up
3. $y = 2(x - 3)^2 - 1$ up
4. $y = -3(x + 2)^2 + 2$ down

Find the vertex of the parabola determined by each equation.

5. $y = 2(x - 3)^2 - 1$ $(3, -1)$
6. $y = -3(x + 2)^2 + 2$ $(-2, 2)$

EXERCISE 8.4

In Exercises 1–12, graph each function determined by the given quadratic equation.

1. $y = x^2$

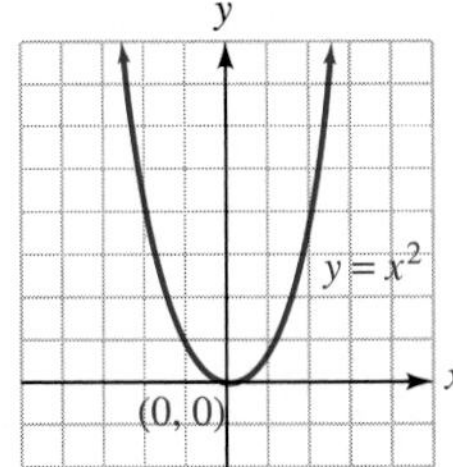

2. $y = -x^2$

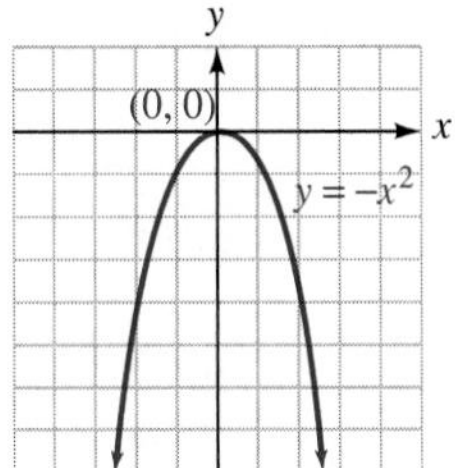

3. $y = x^2 + 2$

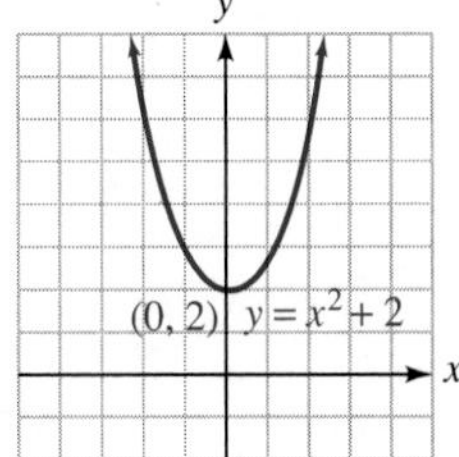

4. $y = x^2 - 3$

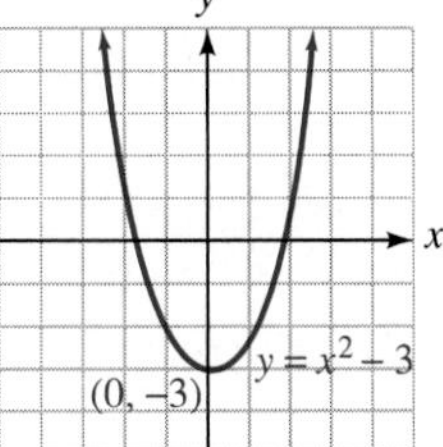

5. $y = -(x - 2)^2$

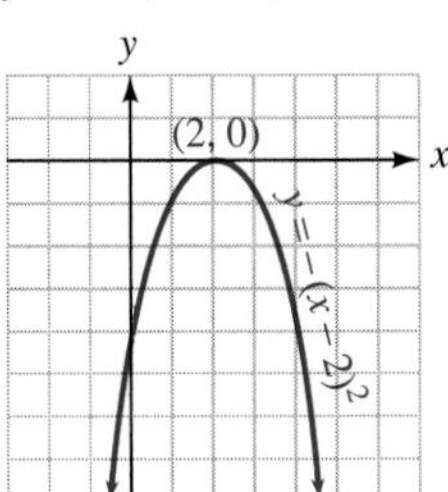

6. $y = (x + 2)^2$

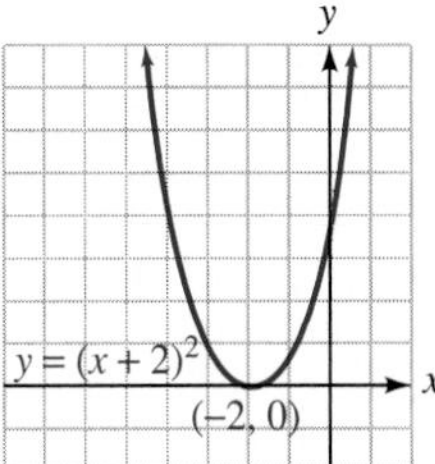

7. $y = -3x^2 + 2x$

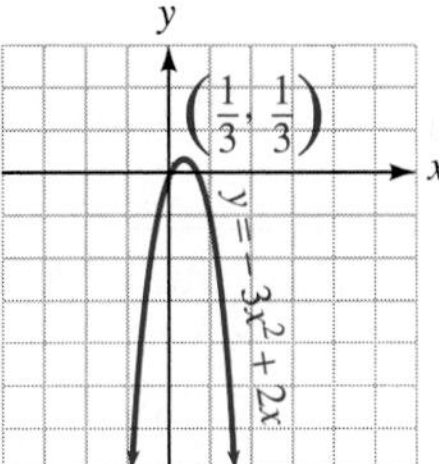

8. $y = 5x + x^2$

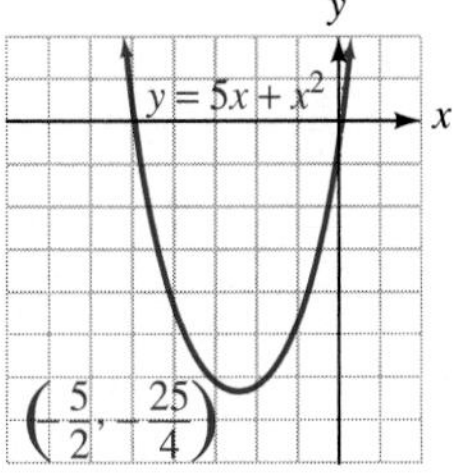

9. $y = x^2 + x - 6$

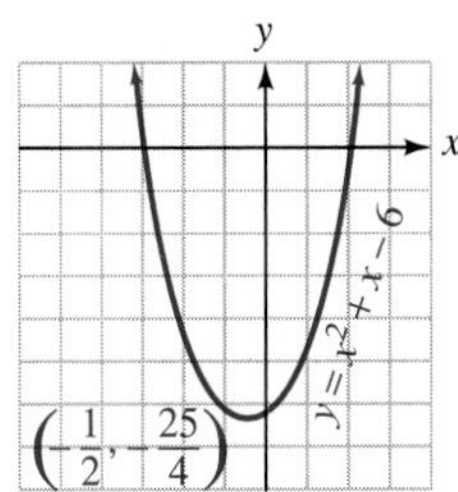

10. $y = x^2 - x - 6$

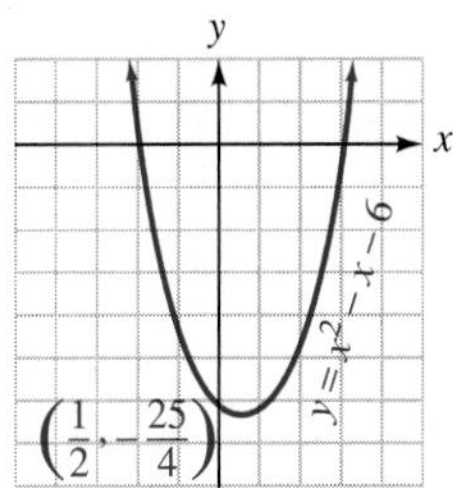

11. $y = 12x^2 + 6x - 6$

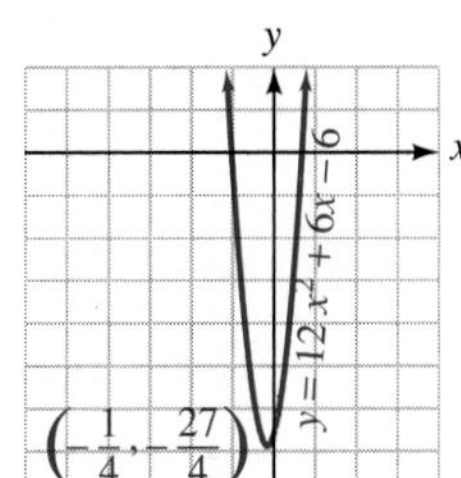

12. $y = -2x^2 + 4x + 3$

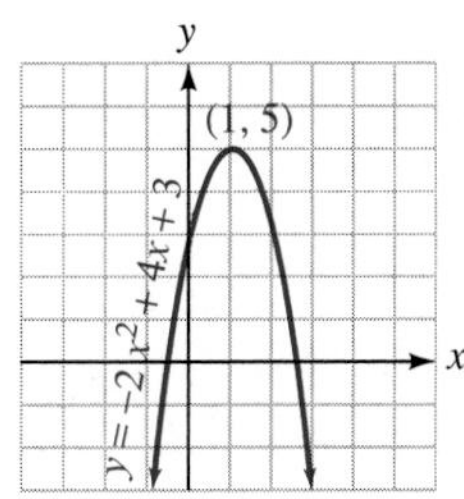

In Exercises 13–24, find the coordinates of the vertex and the axis of symmetry of the graph of each equation. If necessary, complete the square on x to write the equation in the form $y = a(x - h)^2 + k$. ***Do not graph the equation.***

13. $y = (x - 1)^2 + 2$ $(1, 2), x = 1$

14. $y = 2(x - 2)^2 - 1$ $(2, -1), x = 2$

15. $y = 2(x + 3)^2 - 4$ $(-3, -4), x = -3$

16. $y = -3(x + 1)^2 + 3$ $(-1, 3), x = -1$

17. $y = -3x^2$ $(0, 0), x = 0$

18. $y = 3x^2 - 3$ $(0, -3), x = 0$

19. $y = 2x^2 - 4x$ $(1, -2), x = 1$

20. $y = 3x^2 + 6x$ $(-1, -3), x = -1$

21. $y = -4x^2 + 16x + 5$ $(2, 21), x = 2$

22. $y = 5x^2 + 20x + 25$ $(-2, 5), x = -2$

23. $y - 7 = 6x^2 - 5x$ $\left(\frac{5}{12}, \frac{143}{24}\right), x = \frac{5}{12}$

24. $y - 2 = 3x^2 + 4x$ $\left(-\frac{2}{3}, \frac{2}{3}\right), x = -\frac{2}{3}$

25. The equation $y - 2 = (x - 5)^2$ represents a quadratic function whose graph is a parabola. Find its vertex. $(5, 2)$

26. Show that the graph of $y = ax^2$, where $a \neq 0$, is a parabola with vertex at the origin.

In Exercises 27–30, use a graphing calculator to find the coordinates of the vertex of the graph of each quadratic function. Round to the nearest hundredth.

27. $y = 2x^2 - x + 1$ $(0.25, 0.88)$

28. $y = x^2 + 5x - 6$ $(-2.50, -12.25)$

29. $y = 7 + x - x^2$ $(0.50, 7.25)$

30. $y = 2x^2 - 3x + 2$ $(0.75, 0.88)$

In Exercises 31–34, use a graphing calculator to solve each equation.

31. $x^2 + x - 6 = 0$ $-3, 2$

32. $2x^2 - 5x - 3 = 0$ $-0.5, 3$

33. $0.5x^2 - 0.7x - 3 = 0$ $-1.85, 3.25$

34. $2x^2 - 0.5x - 2 = 0$ $-.88, 1.13$

35. Integer problem The sum of two integers is 50, and their product is maximum. Find the integers. 25, 25

36. Integer problem The sum of two integers is 10, and the sum of their squares is minimum. Find the integers. 5, 5

37. Ballistics If a ball is thrown straight up with an initial velocity of 48 feet per second, its height s after t seconds is given by the equation $s = 48t - 16t^2$. Find the maximum height attained by the ball and the time it takes for the ball to return to earth. 36 ft, 3 sec

38. Ballistics From the top of the building in Illustration 1, a ball is thrown straight up with an initial velocity of 32 feet per second. The equation $s = -16t^2 + 32t + 48$ gives the height s of the ball t seconds after it was thrown. Find the maximum height reached by the ball and the time it takes for the ball to hit the ground. 64 ft, 3 sec

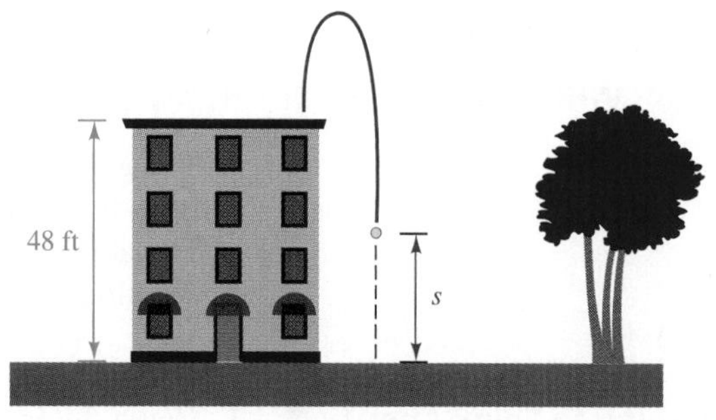

ILLUSTRATION 1

39. Maximizing area Find the dimensions of the rectangle of maximum area that can be constructed with 200 feet of fencing. Find the maximum area. 50 ft by 50 ft, 2500 ft^2

40. Fencing a field A farmer wants to fence in three sides of a rectangular field (shown in Illustration 2) with 1000 feet of fencing. The other side of the rectangle will be a river. If the enclosed area is to be maximum, find the dimensions of the field.
250 ft by 500 ft

ILLUSTRATION 2

41. Maximizing revenue The revenue received for selling x radios is given by the formula

$$R = -\frac{x^2}{728} + 9x$$

How many radios must be sold to obtain the maximum revenue? Find the maximum revenue.
3276, $14,742

42. Maximizing revenue The revenue received for selling x stereos is given by the formula

$$R = -\frac{x^2}{5} + 80x - 1000$$

How many stereos must be sold to obtain the maximum revenue? Find the maximum revenue.
200, $7000

43. Maximizing revenue When priced at $30 each, a toy has annual sales of 4000 units. The manufacturer estimates that each $1 increase in cost will decrease sales by 100 units. Find the unit price that will maximize total revenue. (*Hint:* Total revenue = price · the number of units sold.) $35

44. Maximizing revenue When priced at $57, one type of camera has annual sales of 525 units. For each $1 the camera is reduced in price, management expects to sell an additional 75 cameras. Find the unit price that will maximize total revenue. (*Hint:* Total revenue = price · the number of units sold.) $32

Writing Exercises ■ *Write a paragraph using your own words.*

1. The graph of $y = ax^2 + bx + c$ $(a \neq 0)$ passes the vertical line test. Explain why this shows that the equation defines a function.
2. The graph of $x = y^2 - 2y$ is a parabola. Explain why its graph does not represent a function.

Something to Think About ■

1. Graph the function $y = x^2 + k$ for several values of k. What do you notice?
2. Graph the function $y = (x - k)^2$ for several values of k. What do you notice?

Review Exercises ■ *Factor each expression.*

1. $x^2y - y^3$
   $y(x + y)(x - y)$
2. $8x^3t^2 + 10x^2t^2 - 3xt^2$
   $xt^2(4x - 1)(2x + 3)$
3. $8x^6 + 27y^9$
   $(2x^2 + 3y^3)(4x^4 - 6x^2y^3 + 9y^6)$
4. $q^2 - p^2 - 2p - 1$
   $(q + p + 1)(q - p - 1)$

5. Madison and St. Louis are 385 miles apart. One train leaves Madison and heads toward St. Louis at the rate of 30 miles per hour. Three hours later, a second train leaves Madison, also bound for St. Louis. If the second train travels at the rate of 55 miles per hour, in how many hours will the faster train overtake the slower train? $3\frac{3}{5}$ hr

6. A woman invests \$25,000, some at 7% annual interest and the rest at 8%. If the annual income from both investments is \$1900, how much is invested at the higher rate? \$15,000

8.5 Quadratic and Other Nonlinear Inequalities

■ Solving Quadratic Inequalities ■ Solving Other Inequalities ■ Graphing Calculators ■ Graphs of Nonlinear Inequalities in Two Variables

■ Solving Quadratic Inequalities

We begin this section by solving quadratic inequalities such as $x^2 + x - 6 < 0$. To solve this inequality, we must find the numbers x that make the inequality true. To find these values, we must factor the trinomial to obtain

$$(x + 3)(x - 2) < 0$$

Since the product of $x + 3$ and $x - 2$ is less than 0, their values must be opposite in sign. To find the intervals where this is true, we keep track of their signs by constructing the chart in Figure 8-10. The chart shows that

- The binomial $x - 2 = 0$ when $x = 2$, the binomial is positive when $x > 2$, and the binomial is negative when $x < 2$.
- The binomial $x + 3 = 0$ when $x = -3$, the binomial is positive when $x > -3$, and the binomial is negative when $x < -3$.

Alternate Approach: You may want to solve the equation $x^2 + x - 6 = 0$ to establish the critical points of -3 and 2 on the number line. Then you can use test points in each of the three resulting regions to find the solution.

Thus, the only place where the values of the binomial are opposite in sign is in the interval $(-3, 2)$. Therefore,

$$-3 < x < 2$$

The graph of the solution set is shown on the number line in Figure 8-10.

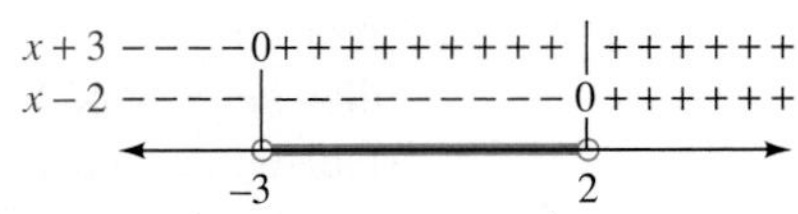

FIGURE 8-10

EXAMPLE 1 Solve the inequality $x^2 + 2x - 3 \geq 0$.

Solution We factor the quadratic trinomial to get $(x - 1)(x + 3)$ and construct a sign chart as in Figure 8-11.

- The binomial $x - 1$ is 0 when $x = 1$, is positive when $x > 1$, and is negative when $x < 1$.
- The binomial $x + 3$ is 0 when $x = -3$, is positive when $x > -3$, and is negative when $x < -3$.

The product of $x - 1$ and $x + 3$ will be greater than 0 when the signs of the binomial factors are the same. This occurs in the intervals $(-\infty, -3)$ or $(1, \infty)$. The numbers -3 and 1 are also included because they make the product equal to 0. Thus, the solution set is

$$(-\infty, -3] \cup [1, \infty) \qquad \text{or} \qquad x \leq -3 \text{ or } x \geq 1$$

The graph of the solution set is shown on the number line in Figure 8-11.

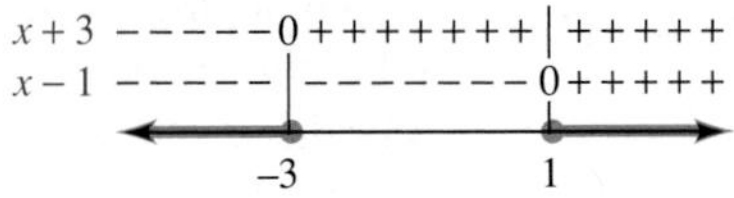

FIGURE 8-11

■

Solving Other Inequalities

A sign chart is useful for solving many inequalities that are neither linear nor quadratic.

EXAMPLE 2 Solve the inequality $\frac{1}{x} < 6$.

Solution We subtract 6 from both sides to make the right-hand side equal to 0. We then find a common denominator and add the fractions:

Teaching Tip: Emphasize that we must get 0 on the right-hand side.

$$\frac{1}{x} < 6$$

$$\frac{1}{x} - 6 < 0 \qquad \text{Subtract 6 from both sides.}$$

$$\frac{1}{x} - \frac{6x}{x} < 0 \qquad \text{Get a common denominator.}$$

$$\frac{1 - 6x}{x} < 0 \qquad \text{Keep the denominator and subtract the numerators.}$$

We now make a sign chart as in Figure 8-12.

- The denominator x is 0 when $x = 0$, is positive when $x > 0$, and is negative when $x < 0$.
- The numerator $1 - 6x$ is 0 when $x = \frac{1}{6}$, is positive when $x < \frac{1}{6}$, and is negative when $x > \frac{1}{6}$.

The fraction $\frac{1-6x}{x}$ will be less than 0 when the numerator and denominator are opposite in sign. This occurs in the union of two intervals:

$$(-\infty, 0) \cup \left(\frac{1}{6}, \infty\right) \qquad \text{which is} \qquad x < 0 \text{ or } x > \frac{1}{6}$$

The graph of this solution is shown in Figure 8-12.

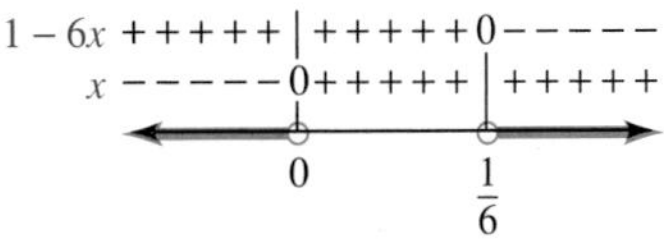

FIGURE 8-12

■

WARNING! Multiplying both sides of the inequality $\frac{1}{x} < 6$ by x is a three-case situation:

- If $x > 0$, then $1 < 6x$.
- If $x = 0$, then the fraction $\frac{1}{x}$ is undefined.
- If $x < 0$, then $1 > 6x$.

If you multiply both sides by x and solve the linear inequality $1 < 6x$, you are only considering one case and will get only part of the answer.

EXAMPLE 3 Solve the inequality $\dfrac{x^2 - 3x + 2}{x - 3} \geq 0$.

Solution We write the fraction with the numerator in factored form.

$$\frac{(x-2)(x-1)}{x-3} \geq 0$$

To keep track of the signs of the three binomials, we construct the sign chart shown in Figure 8-13. The fraction will be positive in the intervals where all factors are positive, or where two factors are negative. The numbers 1 and 2 are included because they make the numerator (and thus the fraction) equal to 0. The number 3 is not included because it gives a 0 in the denominator.

The solution is the interval $[1, 2] \cup (3, \infty)$. The graph appears in Figure 8-13.

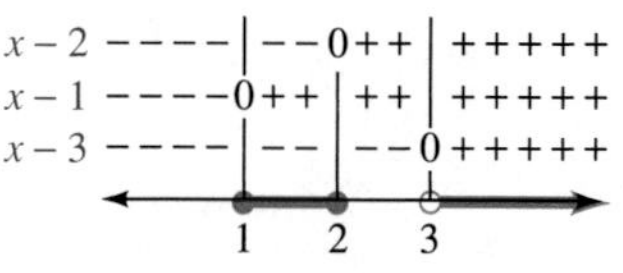

FIGURE 8-13

EXAMPLE 4 Solve the inequality $\frac{3}{x-1} < \frac{2}{x}$.

Solution We subtract $\frac{2}{x}$ from both sides to get 0 on the right-hand side and proceed as follows:

$$\frac{3}{x-1} < \frac{2}{x}$$

$$\frac{3}{x-1} - \frac{2}{x} < 0 \qquad \text{Subtract } \tfrac{2}{x} \text{ from both sides.}$$

Teaching Tip: Emphasize that we must get 0 on the right-hand side.

$$\frac{3x}{(x-1)x} - \frac{2(x-1)}{x(x-1)} < 0 \qquad \text{Get a common denominator.}$$

$$\frac{3x - 2x + 2}{x(x-1)} < 0 \qquad \text{Keep the denominator and subtract the numerators.}$$

$$\frac{x+2}{x(x-1)} < 0 \qquad \text{Combine terms.}$$

We can keep track of the signs of the three polynomials with the sign chart shown in Figure 8-14. The fraction will be negative in the intervals with either one or three negative factors. The numbers 0 and 1 are not included because they give a 0 in the denominator, and the number -2 is not included because it does not satisfy the inequality.

The solution is the interval $(-\infty, -2) \cup (0, 1)$, as shown in Figure 8-14.

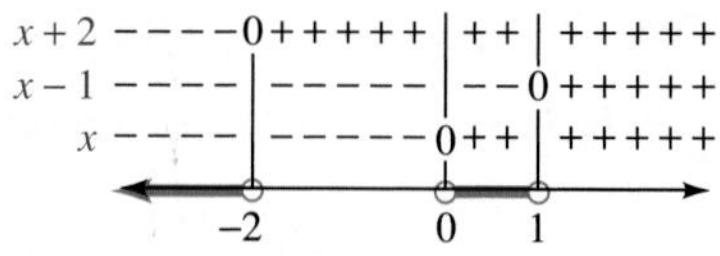

FIGURE 8-14

Graphing Calculators

We can use graphing calculators to solve inequalities.

EXAMPLE 5 Solve the inequality $x^2 + 2x - 3 \geq 0$.

Solution We have already solved this inequality in Example 1.

To solve the inequality with a graphing calculator, we use the standard RANGE settings and graph the quadratic function $y = x^2 + 2x - 3$, as in Figure 8-15. The solution of the inequality will be those numbers x for which the graph of $y = x^2 + 2x - 3$ lies above or on the x-axis. We can use the TRACE key to find that this interval is $(-\infty, -3] \cup [1, \infty)$.

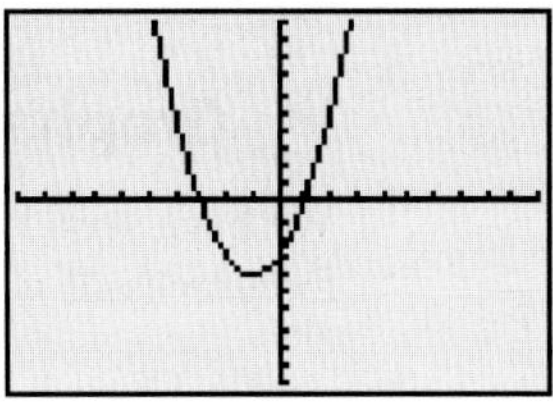

FIGURE 8-15 ■

EXAMPLE 6 Solve the inequality $\dfrac{3}{x-1} < \dfrac{2}{x}$.

Solution We have already solved this inequality in Example 4. As in Example 4, we first write the inequality in the form

$$\frac{x+2}{x(x-1)} < 0$$

To solve the inequality with a graphing calculator, we use the standard RANGE settings and graph the function $y = \frac{x+2}{x(x-1)}$, as in Figure 8-16(a). The solution of the inequality will be those numbers x for which the graph lies below the x-axis.

We can use the TRACE key to find that the graph is below the x-axis when x is less than -2. However, we cannot see the graph in the interval $0 < x < 1$. To view the graph in this interval, we change the RANGE settings to

$$\text{Xmin} = -1, \qquad \text{Xmax} = 2, \qquad \text{Ymin} = -25, \qquad \text{Ymax} = 10$$

and graph the function to get the graph shown in Figure 8-16(b).

We now see that the graph is below the x-axis in the interval $(0, 1)$. Thus, the solution to the inequality is the union of two intervals:

$$(-\infty, -2) \cup (0, 1)$$

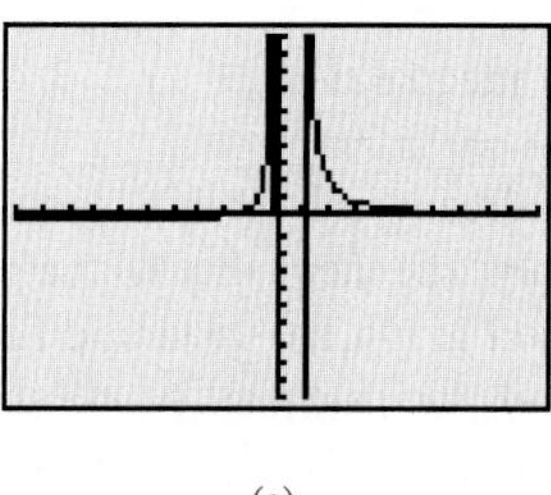

(a)

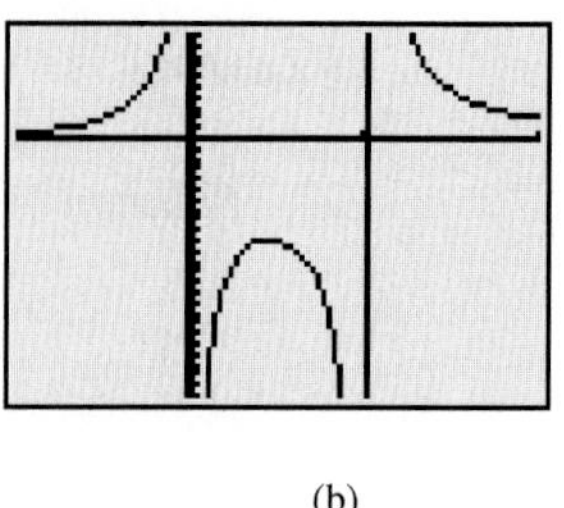

(b)

FIGURE 8-16

Graphs of Nonlinear Inequalities in Two Variables

We have seen how to graph linear inequalities in two variables. We now consider the graphs of nonlinear inequalities in two variables.

EXAMPLE 7 Graph the inequality $y < -x^2 + 4$.

Solution The graph of $y = -x^2 + 4$ is the parabolic boundary separating the region representing $y < -x^2 + 4$ and the region representing $y > -x^2 + 4$.

We graph the quadratic function $y = -x^2 + 4$ as a broken parabola, because equality is not permitted. Since the coordinates of the origin satisfy the inequality $y < -x^2 + 4$, the point (0, 0) is in the graph. The complete graph is shown in Figure 8-17.

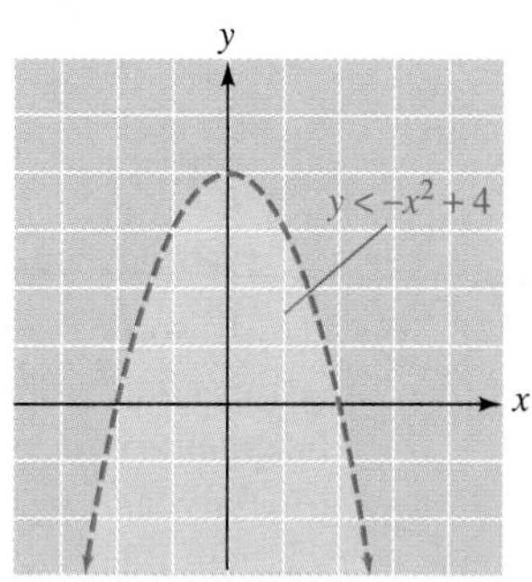

FIGURE 8-17

EXAMPLE 8 Graph the inequality $x \leq |y|$.

Solution We first graph $x = |y|$ as in Figure 8-18(a), using a solid line because equality is permitted. Because the origin is on the graph, we cannot use the origin as a test point. However, another point, such as (1, 0), will do. We substitute 1 for x and 0 for y into the inequality to get

$$x \leq |y|$$
$$1 \leq |0|$$
$$1 \leq 0$$

Since $1 \leq 0$ is a false statement, the point (1, 0) does not satisfy the inequality and is not part of the graph. Thus, the graph of $x \leq |y|$ is to the left of the boundary.

The complete graph is shown in Figure 8-18(b).

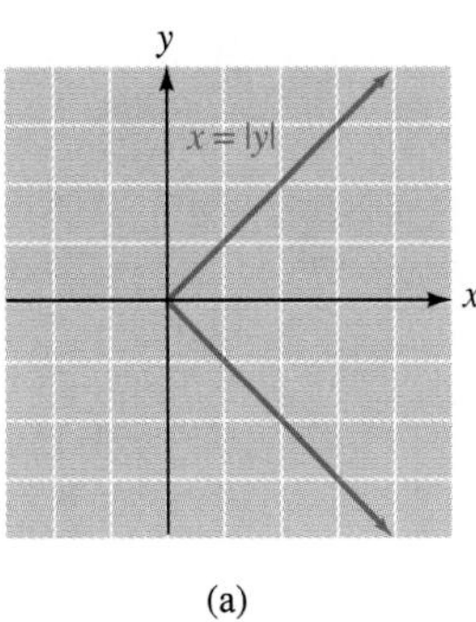

(a)

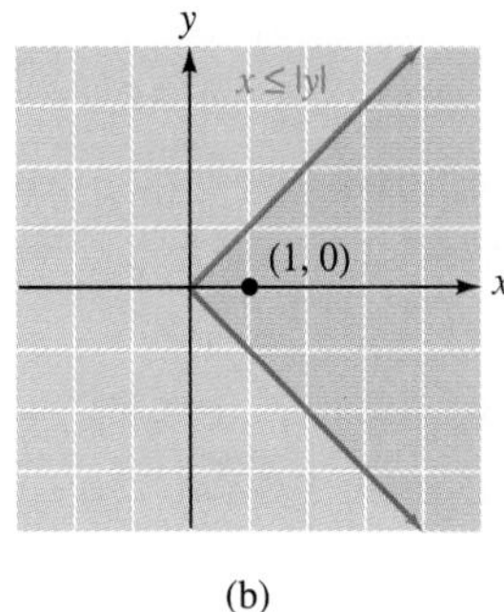

(b)

FIGURE 8-18

ORALS *Find the interval where $x - 2$ is*

1. 0 $x = 2$ **2.** positive $x > 2$ **3.** negative $x < 2$

Find the interval where $x + 3$ is

4. 0 $x = -3$ **5.** positive $x > -3$ **6.** negative $x < -3$

Multiply both sides of the inequality $\frac{1}{x} < 2$ by x when x is

7. positive $1 < 2x$ **8.** negative $1 > 2x$

EXERCISE 8.5

In Exercises 1–40, solve each inequality. Give each result in interval notation and graph the solution set.

1. $x^2 - 5x + 4 < 0$
(1, 4)
1 4

2. $x^2 - 3x - 4 > 0$
$(-\infty, -1) \cup (4, \infty)$
−1 4

3. $x^2 - 8x + 15 > 0$
$(-\infty, 3) \cup (5, \infty)$
3 5

4. $x^2 + 2x - 8 < 0$
(−4, 2)
−4 2

5. $x^2 + x - 12 \leq 0$
[−4, 3]

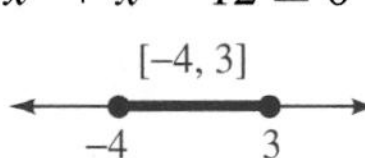

6. $x^2 + 7x + 12 \geq 0$
$(-\infty, -4] \cup [-3, \infty)$
−4 −3

7. $x^2 + 2x \geq 15$
$(-\infty, -5] \cup [3, \infty)$
−5 3

8. $x^2 - 8x \leq -15$
[3, 5]
3 5

9. $x^2 + 8x < -16$
no solutions

10. $x^2 - 6x > -9$
$(-\infty, 3) \cup (3, \infty)$

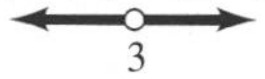

11. $x^2 \geq 9$
$(-\infty, -3] \cup [3, \infty)$

12. $x^2 \geq 16$
$(-\infty, -4] \cup [4, \infty)$

13. $2x^2 - 50 < 0$
$(-5, 5)$

14. $3x^2 - 243 < 0$
$(-9, 9)$

15. $\frac{1}{x} < 2$
$(-\infty, 0) \cup (1/2, \infty)$

16. $\frac{1}{x} > 3$
$(0, 1/3)$

17. $\frac{4}{x} \geq 2$
$(0, 2]$

18. $-\frac{6}{x} < 12$
$(-\infty, -1/2) \cup (0, \infty)$

19. $-\frac{5}{x} < 3$
$(-\infty, -5/3) \cup (0, \infty)$

20. $\frac{4}{x} \geq 8$
$(0, 1/2]$

21. $\frac{x^2 - x - 12}{x - 1} < 0$
$(-\infty, -3) \cup (1, 4)$

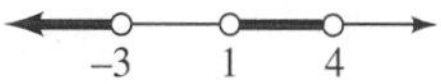

22. $\frac{x^2 + x - 6}{x - 4} \geq 0$
$[-3, 2] \cup (4, \infty)$

23. $\frac{x^2 + x - 20}{x + 2} \geq 0$
$[-5, -2) \cup [4, \infty)$

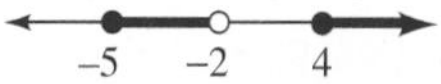

24. $\frac{x^2 - 10x + 25}{x + 5} < 0$
$(-\infty, -5)$

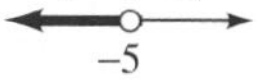

25. $\frac{2x^2 + x - 6}{x - 3} \leq 0$
$(-\infty, -2] \cup [3/2, 3)$

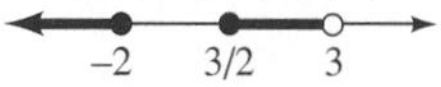

26. $\frac{2x^2 - 5x + 2}{x + 2} > 0$
$(-2, 1/2) \cup (2, \infty)$

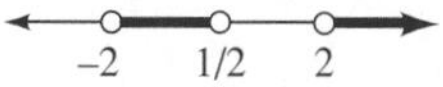

27. $\frac{6x^2 - 5x + 1}{2x + 1} > 0$
$(-1/2, 1/3) \cup (1/2, \infty)$

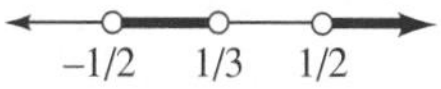

28. $\frac{6x^2 + 11x + 3}{3x - 1} < 0$
$(-\infty, -3/2) \cup (-1/3, 1/3)$

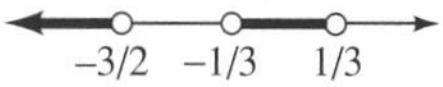

29. $\frac{3}{x - 2} < \frac{4}{x}$
$(0, 2) \cup (8, \infty)$

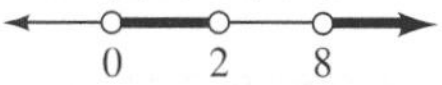

30. $\frac{-6}{x + 1} \geq \frac{1}{x}$
$(-\infty, -1) \cup [-1/7, 0)$

31. $\frac{-5}{x + 2} \geq \frac{4}{2 - x}$
$(-\infty, -2) \cup (2, 18]$

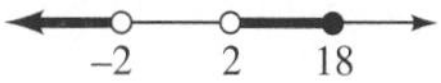

32. $\frac{-6}{x - 3} < \frac{5}{3 - x}$
$(3, \infty)$

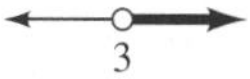

33. $\frac{7}{x - 3} \geq \frac{2}{x + 4}$
$[-34/5, -4) \cup (3, \infty)$

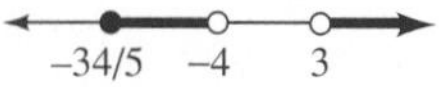

34. $\frac{-5}{x - 4} < \frac{3}{x + 1}$
$(-1, 7/8) \cup (4, \infty)$

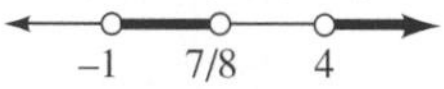

35. $\frac{x}{x + 4} \leq \frac{1}{x + 1}$
$(-4, -2] \cup (-1, 2]$

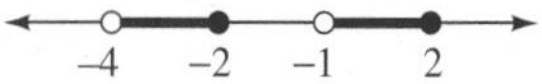

36. $\frac{x}{x + 9} \geq \frac{1}{x + 1}$
$(-\infty, -9) \cup [-3, -1) \cup [3, \infty)$

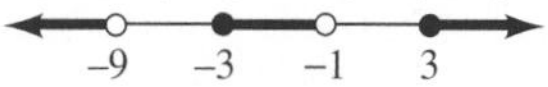

37. $\frac{x}{x + 16} > \frac{1}{x + 1}$
$(-\infty, -16) \cup (-4, -1) \cup (4, \infty)$

38. $\frac{x}{x + 25} < \frac{1}{x + 1}$
$(-25, -5) \cup (-1, 5)$

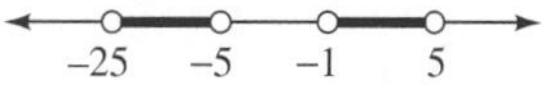

39. $(x + 2)^2 > 0$
$(-\infty, -2) \cup (-2, \infty)$

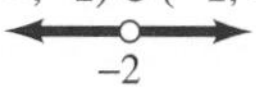

40. $(x - 3)^2 < 0$ no solutions

In Exercises 41–44, use a graphing calculator to solve each inequality. Give the answer in interval notation.

41. $x^2 - 2x - 3 < 0$
$(-1, 3)$

42. $x^2 + x - 6 > 0$
$(-\infty, -3) \cup (2, \infty)$

43. $\dfrac{x+3}{x-2} > 0$
$(-\infty, -3) \cup (2, \infty)$

44. $\dfrac{3}{x} < 2$
$(-\infty, 0) \cup \left(\frac{3}{2}, \infty\right)$

In Exercises 45–56, graph each inequality.

45. $y < x^2 + 1$

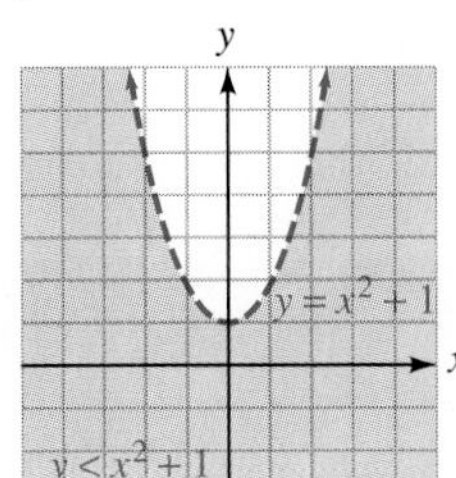

46. $y > x^2 - 3$

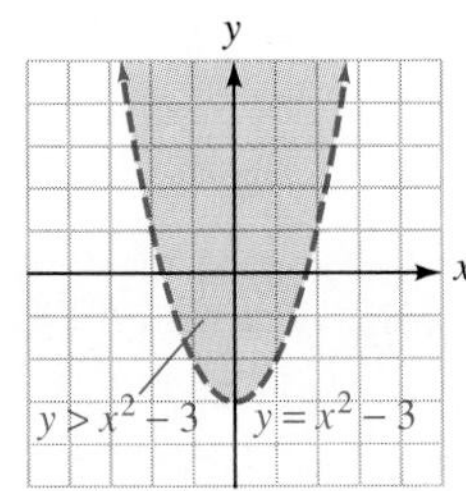

47. $y \le x^2 + 5x + 6$

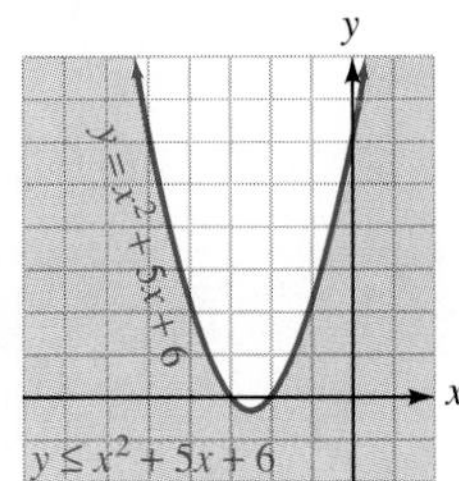

48. $y \ge x^2 + 5x + 4$

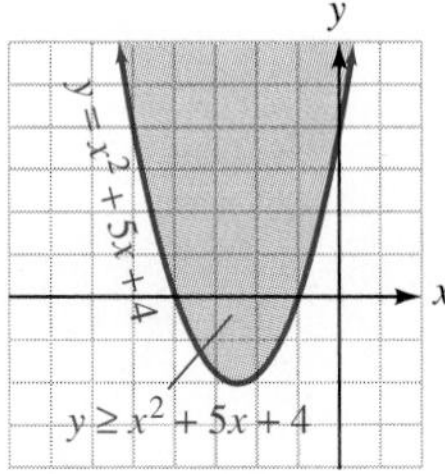

49. $x \ge y^2 - 3$

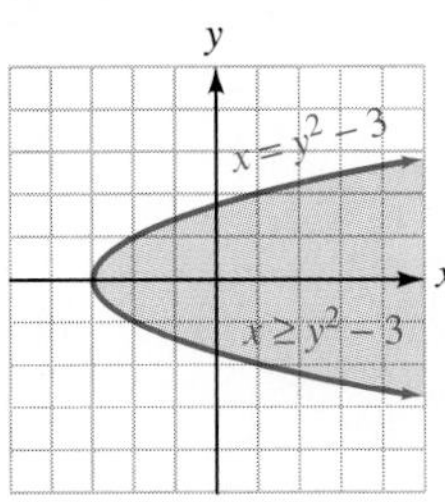

50. $x \le y^2 + 1$

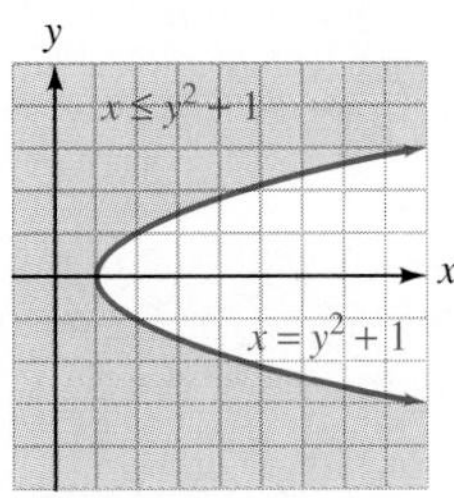

51. $-x^2 - y + 6 > -x$

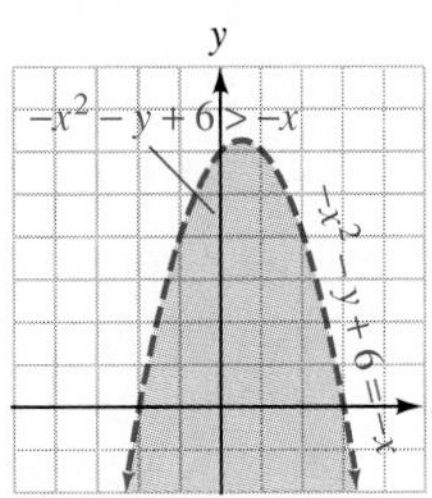

52. $y > (x + 3)(x - 2)$

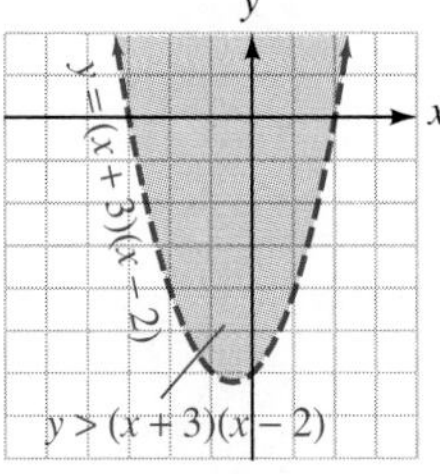

53. $y < |x + 4|$

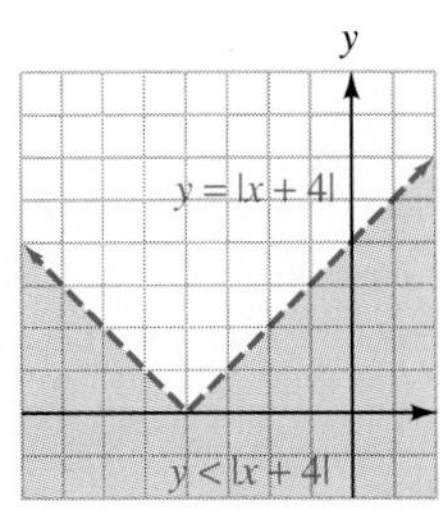

54. $y \ge |x - 3|$

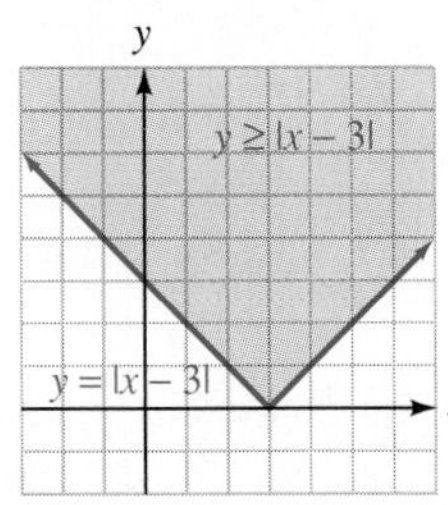

55. $y \le -|x| + 2$

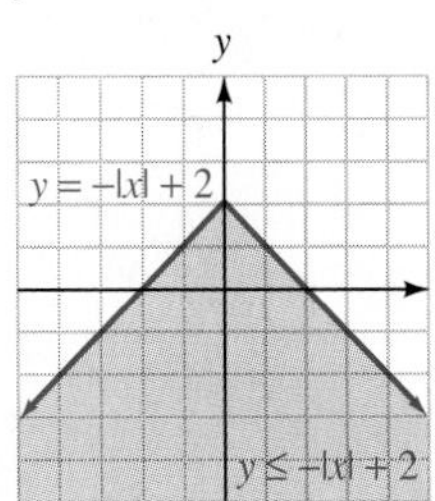

56. $y > |x| - 2$

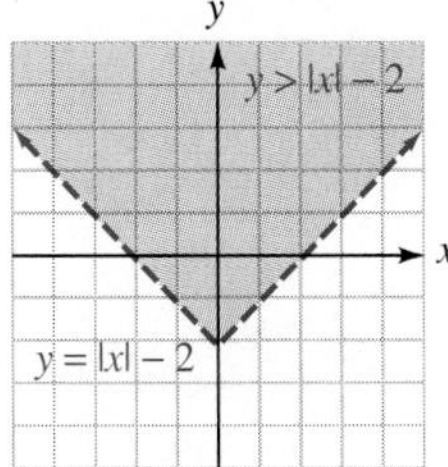

Writing Exercises ■ *Write a paragraph using your own words.*

1. Explain why $(x - 4)(x + 5)$ will be positive only when the signs of $x - 4$ and $x + 5$ are the same.

2. Tell how to find the graph of $y \ge x^2$.

Something to Think About ■

1. Under what conditions will the fraction $\frac{(x-1)(x+4)}{(x+2)(x+1)}$ be positive?
when 4 factors are negative, 2 factors are negative, or no factors are negative

2. Under what conditions will the fraction $\frac{(x-1)(x+4)}{(x+2)(x+1)}$ be negative?
when 3 factors are negative or one factor is negative

Review Exercises *Write each expression as an equation.*

1. x varies directly with y. $x = ky$
2. y varies inversely with t. $y = \frac{k}{t}$
3. t varies jointly with x and y. $t = kxy$
4. d varies directly with t but inversely with u^2. $d = \dfrac{kt}{u^2}$

Find the slope of the graph of each equation.

5. $y = 3x - 4$ 3
6. $\dfrac{2x - y}{5} = 8$ 2

PROJECT How Tall Is the Tree?

You don't need a measuring tape to find the height of a tall tree. All you need is a stone, a stopwatch, and a friend. The method uses the sum of the solutions of a quadratic equation.

If a stone is thrown upward with a velocity v, then its height, h, above the ground after t seconds is given by

$$h = -16t^2 + vt$$

A stone thrown straight upward next to a tree will pass the top of the tree at two different times: first on the way up, and later on the way down. Call these two times T_1 and T_2. If H is the height of the tree, then T_1 and T_2 are the two solutions of the equation

$$H = -16t^2 + vt$$

or the equation

$$16t^2 - vt + H = 0$$

- The product of the roots of this quadratic equation is T_1T_2. Find the product of the roots a different way (see Section 8.3). Form an equation and solve for H.
- Find a tree (or a flagpole) far from people, cars, and windows. Station a friend some distance away, with a stopwatch. Next to the tree, throw the stone straight upward (and get out of the way!). Have your friend use the watch's lap timer to find the two times when the stone passes the top of the tree. Use your formula to compute the height of the tree. Because the stone isn't released from ground level, consider adding an appropriate "head start" distance. Do the experiment several times and record the results.
- Wait for a sunny day, and use the lengths of the shadows of the tree and a yardstick to compute the tree's height by similar triangles (see Section 5.9). Write a brief report showing your derivation of the formula and discussing the accuracy and convenience of these two methods.

Chapter Summary

KEY WORDS

- axis of symmetry (8.4)
- completing the square (8.1)
- complex conjugates (8.2)
- complex numbers (8.2)
- discriminant (8.3)
- imaginary numbers (8.2)
- parabola (8.4)
- quadratic function (8.4)
- vertex of a parabola (8.4)
- vertical line test (8.4)

KEY IDEAS

(8.1) The **square root property:** If $c > 0$, the equation $x^2 = c$ has two real solutions. They are

$$x = \sqrt{c} \quad \text{and} \quad x = -\sqrt{c}$$

The **quadratic formula:** $x = \dfrac{-b \pm \sqrt{b^2 - 4ac}}{2a} \quad (a \neq 0)$

(8.2) **Properties of complex numbers:** If a, b, c, and d are real numbers and $i^2 = -1$, then

$$a + bi = c + di \quad \text{if and only if} \quad a = c \text{ and } b = d$$
$$(a + bi) + (c + di) = (a + c) + (b + d)i$$
$$(a + bi)(c + di) = (ac - bd) + (ad + bc)i$$
$$|a + bi| = \sqrt{a^2 + b^2}$$

(8.3) The **discriminant:** For real numbers a, b, and c

If $b^2 - 4ac > 0$, the solutions of $ax^2 + bx + c = 0$ are unequal real numbers.

If $b^2 - 4ac = 0$, the solutions of $ax^2 + bx + c = 0$ are equal real numbers.

If $b^2 - 4ac < 0$, the solutions of $ax^2 + bx + c = 0$ are complex conjugates.

If r_1 and r_2 are solutions of $ax^2 + bx + c = 0$, then

$$r_1 + r_2 = -\frac{b}{a} \quad \text{and} \quad r_1 r_2 = \frac{c}{a}$$

(8.4) If $a \neq 0$, the graph of the equation $y = a(x - h)^2 + k$ is a parabola with vertex at (h, k). It opens upward when $a > 0$ and downward when $a < 0$.

If every vertical line that intersects a graph does so exactly once, the graph represents a function.

(8.5) To graph a quadratic inequality in one variable, make a sign chart.

To solve inequalities with rational expressions, get 0 on the right-hand side, add the fractions, and then factor the numerator and denominator. Then use a sign chart.

To graph an inequality such as $y > -x^2 + 4$, first graph the equation $y = -x^2 + 4$. Then determine the region that represents the graph of $y > -x^2 + 4$.

Chapter 8 Review Exercises

In Review Exercises 1–4, solve each equation by factoring or by using the square root property.

1. $12x^2 + x - 6 = 0$ $\frac{2}{3}, -\frac{3}{4}$
2. $6x^2 + 17x + 5 = 0$ $-\frac{1}{3}, -\frac{5}{2}$
3. $15x^2 + 2x - 8 = 0$ $\frac{2}{3}, -\frac{4}{5}$
4. $(x + 2)^2 = 36$ $4, -8$

In Review Exercises 5–6, solve each equation by completing the square.

5. $x^2 + 6x + 8 = 0$ $-4, -2$
6. $2x^2 - 9x + 7 = 0$ $\frac{7}{2}, 1$

In Review Exercises 7–10, solve each equation by using the quadratic formula.

7. $x^2 - 8x - 9 = 0$ $9, -1$

8. $x^2 - 10x = 0$ $0, 10$

9. $2x^2 + 13x - 7 = 0$ $\frac{1}{2}, -7$

10. $3x^2 - 20x - 7 = 0$ $7, -\frac{1}{3}$

In Review Exercises 11–28, do the operations and give all answers in a + bi form.

11. $(5 + 4i) + (7 - 12i)$ $12 - 8i$

12. $(-6 - 40i) - (-8 + 28i)$ $2 - 68i$

13. $\left(-32 + \sqrt{-144}\right) - \left(64 + \sqrt{-81}\right)$ $-96 + 3i$

14. $\left(-8 + \sqrt{-8}\right) + \left(6 - \sqrt{-32}\right)$ $-2 - 2\sqrt{2}i$

15. $(2 - 7i)(-3 + 4i)$ $22 + 29i$

16. $(-5 + 6i)(2 + i)$ $-16 + 7i$

17. $\left(5 - \sqrt{-27}\right)\left(-6 + \sqrt{-12}\right)$ $-12 + 28\sqrt{3}i$

18. $\left(2 + \sqrt{-128}\right)\left(3 - \sqrt{-98}\right)$ $118 + 10\sqrt{2}i$

19. $\dfrac{3}{4i}$ $0 - \frac{3}{4}i$

20. $\dfrac{-2}{5i^3}$ $0 - \frac{2}{5}i$

21. $\dfrac{6}{2 + i}$ $\frac{12}{5} - \frac{6}{5}i$

22. $\dfrac{7}{3 - i}$ $\frac{21}{10} + \frac{7}{10}i$

23. $\dfrac{4 + i}{4 - i}$ $\frac{15}{17} + \frac{8}{17}i$

24. $\dfrac{3 - i}{3 + i}$ $\frac{4}{5} - \frac{3}{5}i$

25. $\dfrac{3}{5 + \sqrt{-4}}$ $\frac{15}{29} - \frac{6}{29}i$

26. $\dfrac{2}{3 - \sqrt{-9}}$ $\frac{1}{3} + \frac{1}{3}i$

27. $|9 + 12i|$ $15 + 0i$

28. $|24 - 10i|$ $26 + 0i$

In Review Exercises 29–30, use the discriminant to determine what type of solutions exist for each equation.

29. $3x^2 + 4x - 3 = 0$ irrational, unequal

30. $4x^2 - 5x + 7 = 0$ complex conjugates

31. Find the values of k that will make the solutions of $(k - 8)x^2 + (k + 16)x = -49$ equal. 12, 152

32. Find the values of k such that the solutions of $3x^2 + 4x = k + 1$ are real numbers. $k \geq -\frac{7}{3}$

33. Dimensions of a rectangle A rectangle is 2 centimeters longer than it is wide. If both the length and width are doubled, its area is increased by 72 square centimeters. Find the dimensions of the original rectangle. 4 cm by 6 cm

34. Dimensions of a rectangle A rectangle is 1 foot longer than it is wide. If the length is tripled and the width is doubled, its area is increased by 30 square feet. Find the dimensions of the original rectangle. 2 ft by 3 ft

In Review Exercises 35–38, solve each equation.

35. $x - 13x^{1/2} + 12 = 0$ $1, 144$

36. $a^{2/3} + a^{1/3} - 6 = 0$ $8, -27$

37. $\dfrac{1}{x + 1} - \dfrac{1}{x} = -\dfrac{1}{x + 1}$ 1

38. $\dfrac{6}{x + 2} + \dfrac{6}{x + 1} = 5$ $1, -\frac{8}{5}$

39. Find the sum of the solutions of the equation $3x^2 - 14x + 3 = 0$. $\frac{14}{3}$

40. Find the product of the solutions of the equation $3x^2 - 14x + 3 = 0$. 1

In Review Exercises 41–44, graph each equation and give the coordinates of the vertex of the resulting parabola.

41. $y = 2x^2 - 3$

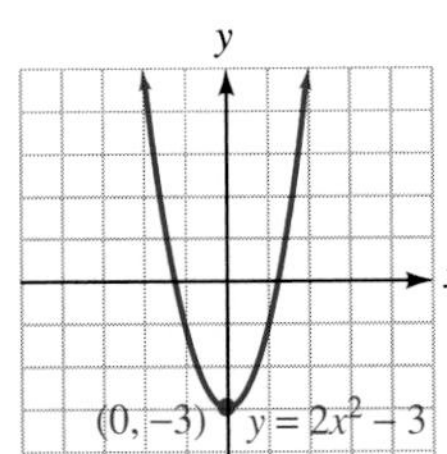

42. $y = -2x^2 - 1$

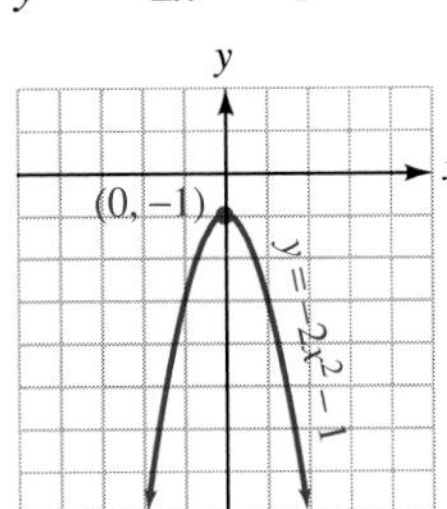

43. $y = -4(x - 2)^2 + 1$

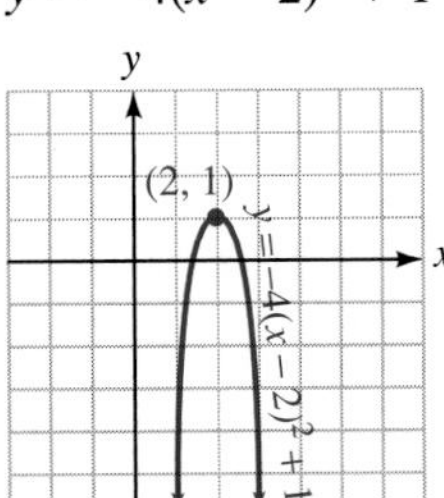

44. $y = 5(x + 1)^2 - 6$

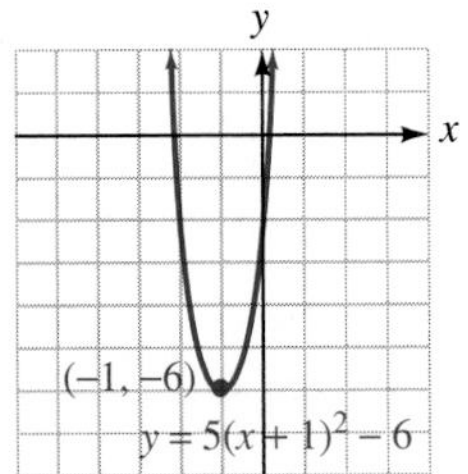

In Review Exercises 45–48, solve each inequality. Give each result in interval notation and graph the solution set.

45. $x^2 + 2x - 35 > 0$

$(-\infty, -7) \cup (5, \infty)$

−7 5

46. $x^2 + 7x - 18 < 0$

$(-9, 2)$

−9 2

47. $\dfrac{3}{x} \leq 5$

$(-\infty, 0) \cup [3/5, \infty)$

0 3/5

48. $\dfrac{2x^2 - x - 28}{x - 1} > 0$

$(-7/2, 1) \cup (4, \infty)$

−7/2 1 4

In Review Exercises 49–52, use a graphing calculator to solve each inequality. Compare the results with Review Exercises 45–48.

49. $x^2 + 2x - 35 > 0$

50. $x^2 + 7x - 18 < 0$

51. $\dfrac{3}{x} \leq 5$

52. $\dfrac{2x^2 - x - 28}{x - 1} > 0$

In Review Exercises 53–54, graph each inequality.

53. $y < \dfrac{1}{2}x^2 - 1$

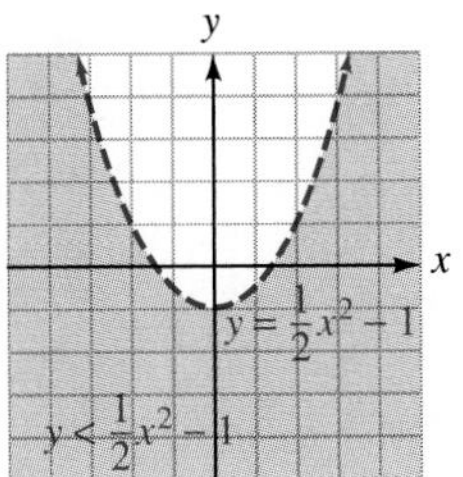

54. $y \geq -|x|$

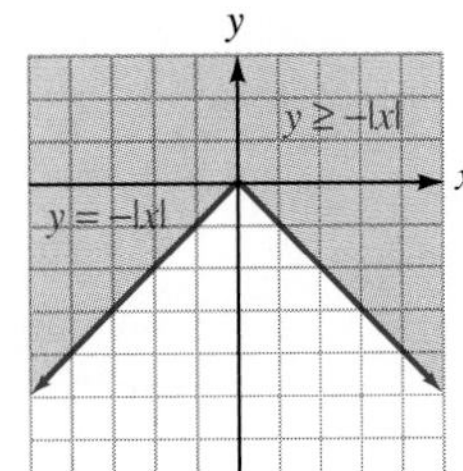

Chapter 8 Test

In Problems 1–2, solve each equation.

1. $x^2 + 3x - 18 = 0$ 3, −6

2. $x(6x + 19) = -15$ $-\frac{3}{2}, -\frac{5}{3}$

In Problems 3–4, determine what number must be added to each binomial to make a perfect square.

3. $x^2 + 24x$ 144

4. $x^2 - 50x$ 625

In Problems 5–6, solve each equation.

5. $x^2 + 4x + 1 = 0$ $-2 \pm \sqrt{3}$

6. $x^2 - 5x - 3 = 0$ $\dfrac{5 \pm \sqrt{37}}{2}$

In Problems 7–12, do the operations. Give all answers in $a + bi$ form.

7. $(2 + 4i) + (-3 + 7i)$ $-1 + 11i$

8. $\left(3 - \sqrt{-9}\right) - \left(-1 + \sqrt{-16}\right)$ $4 - 7i$

9. $2i(3 - 4i)$ $8 + 6i$

10. $(3 + 2i)(-4 - i)$ $-10 - 11i$

11. $\dfrac{1}{i\sqrt{2}}$ $0 - \dfrac{\sqrt{2}}{2}i$

12. $\dfrac{2 + i}{3 - i}$ $\frac{1}{2} + \frac{1}{2}i$

13. Determine whether the solutions of $3x^2 + 5x + 17 = 0$ are real or nonreal. nonreal

14. For what value(s) of k are the solutions of $4x^2 - 2kx + k - 1 = 0$ equal? 2

15. One leg of a right triangle is 14 inches longer than the other, and the hypotenuse is 26 inches. Find the length of the shortest leg. 10 in.

16. Solve the equation $2y - 3y^{1/2} + 1 = 0$. 1, $\frac{1}{4}$

17. Graph the equation $y = \dfrac{1}{2}x^2 + 5$ and give the coordinates of its vertex.

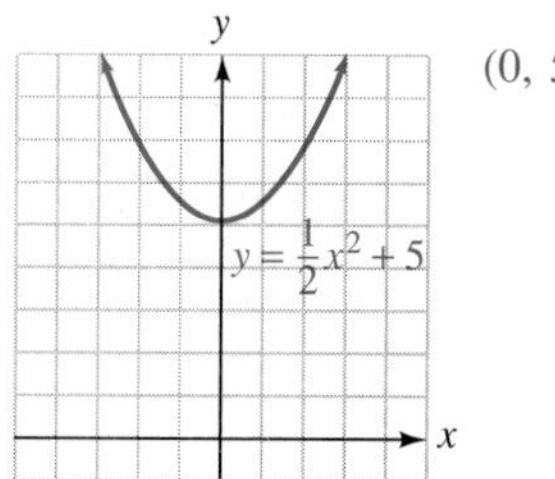

(0, 5)

18. Graph the inequality $y \le -x^2 + 3$.

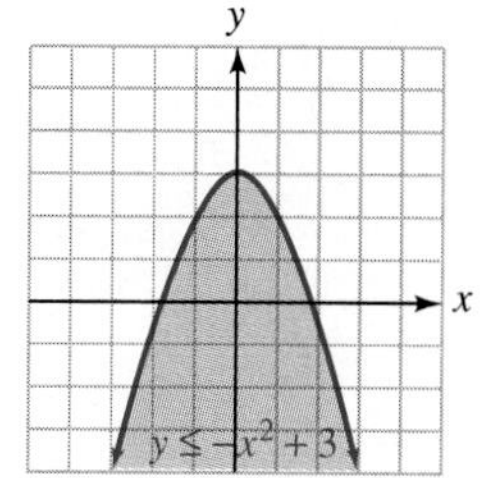

In Problems 19–20, solve each inequality and graph the solution set.

19. $x^2 - 2x - 8 > 0$

$(-\infty, -2) \cup (4, \infty)$

−2 4

20. $\dfrac{x - 2}{x + 3} \leq 0$

$(-3, 2]$

−3 2

CAREERS & MATHEMATICS

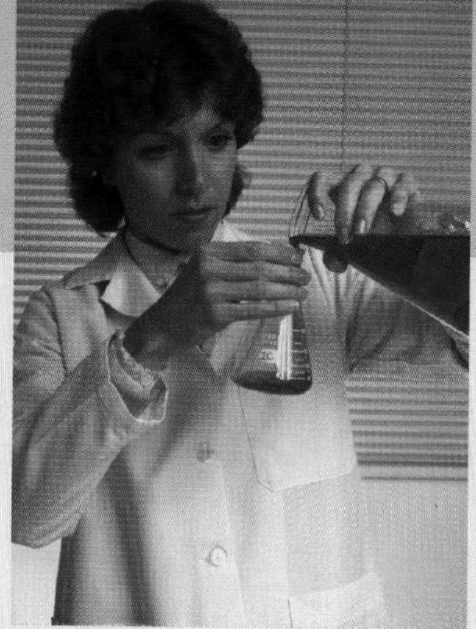

Chemist

Chemists search for new knowledge about substances and put it to practical use. Over half of all chemists work in research and development. In basic research, chemists investigate the properties, composition, and structure of matter and the laws that govern the combination of elements and the reactions of substances. Their research has resulted in the development of a tremendous variety of synthetic materials, of ingredients that have improved other substances, and of processes that help save energy and reduce pollution. In applied research and development, they create new products or improve existing ones, often using knowledge gained from basic research.

Qualifications A bachelor's degree in chemistry or a related discipline is sufficient for many beginning jobs as a chemist. However, graduate training is required for most research jobs, and most college teaching jobs require a Ph.D. degree.

Job Outlook The employment of chemists is expected to grow about as fast as the average for all occupations through the year 2005.

The majority of job openings are expected to be in private industry, primarily in the development of new products. In addition, industrial companies and government agencies will need chemists to help solve problems related to energy shortages, pollution control, and health care.

Example Application A certain weak acid (0.1 M concentration) will break down into free cations (the hydrogen ion, H^+) and anions (A^-). When this acid dissociates, the following equilibrium equation is established.

1. $$\frac{[H^+][A^-]}{[HA]} = 4 \times 10^{-4}$$

where $[H^+]$, the hydrogen ion concentration, is equal to $[A^-]$, the anion concentration. [HA] is the concentration of the undissociated acid itself. Find $[H^+]$ at equilibrium.

Solution Let x be the concentration of H^+. Then x is also $[A^-]$. From chemistry, it turns out that the concentration [HA] of the undissociated acid is $0.1 - x$. Substituting these concentrations into Equation 1 gives the equation

$$\frac{x^2}{0.1 - x} = 4 \times 10^{-4}$$

which can be solved as follows:

$$\frac{x^2}{0.1 - x} = 0.0004 \quad \text{Because } 4 \times 10^{-4} = 0.0004.$$

$$x^2 = 0.0004(0.1 - x) \quad \text{Multiply both sides by } 0.1 - x.$$

$$x^2 = 0.00004 - 0.0004x \quad \text{Remove parentheses.}$$

$$x^2 + 0.0004x - 0.00004 = 0 \quad \text{Add } 0.0004x - 0.00004 \text{ to both sides.}$$

Substitute into the quadratic formula with $a = 1$, $b = 0.0004$, and $c = -0.00004$. We only need to consider the positive root.

$$x = \frac{-b + \sqrt{b^2 - 4ac}}{2a}$$

$$x = \frac{\mathbf{-0.0004} + \sqrt{(\mathbf{0.0004})^2 - 4(\mathbf{1})(\mathbf{-0.00004})}}{2(\mathbf{1})}$$

$$x = \frac{-0.0004 + 0.01266}{2}$$

$$x = 0.00613$$

At equilibrium, $[H^+]$ is approximately 6.13×10^{-3} M.

EXERCISES *Use a scientific calculator to solve these problems.*

1. A saturated solution of hydrogen sulfide (concentration 0.1 M) dissociates into cation H^+ and anion HS^-. When this solution dissociates, the following equilibrium equation is established.

$$\frac{[H^+][HS^-]}{[HHS]} = 1.0 \times 10^{-7}$$

Find $[H^+]$.

2. An HS^- anion dissociates into cation H^+ and anion S^{--} with equilibrium equation

$$\frac{[H^+][S^{--}]}{[HS^-]} = 1.3 \times 10^{-13}$$

Assume the concentration of HS^- to be 1×10^{-4} M. Find $[H^+]$.

3. Suppose the concentration of the acid of the example were 0.2 M. Find the hydrogen ion concentration, $[H^+]$, at equilibrium.

4. Show that the equation of the example has only one meaningful solution by showing that its other solution is negative.

(*Answers:* **1.** 9.995×10^{-5} **2.** 3.605×10^{-9} **3.** 8.75×10^{-3})

9 Systems of Equations and Inequalities

We have considered linear equations with two variables (usually x and y). We found that there was an unlimited number of pairs (x, y) that satisfied each equation, and that we could graph each equation on a rectangular coordinate system.

We now consider certain **systems of linear equations:** either two equations (each with two variables) or three equations (each with three variables).

9.1 Solution by Graphing

■ Inconsistent Systems ■ Dependent Equations ■ Graphing Calculators

In the pair of equations

$$\begin{cases} x + 2y = 4 \\ 2x - y = 3 \end{cases}$$

there are infinitely many pairs (x, y) that satisfy the first equation and infinitely many pairs (x, y) that satisfy the second equation. However, there is only one pair (x, y) that satisfies both equations at the same time. The process of finding the pair (x, y) that satisfies both of these equations simultaneously is called **solving the system of simultaneous equations.**

We follow these steps to solve a system of two equations in two variables by graphing.

Steps to Follow When Using the Graphing Method

1. On a single set of coordinate axes, graph each equation.
2. Find the coordinates of the point (or points) where the graphs intersect. These coordinates give the solution of the system.
3. If the graphs have no point in common, the system has no solution.
4. Check the solution in both of the original equations.

EXAMPLE 1 Solve the system $\begin{cases} x + 2y = 4 \\ 2x - y = 3 \end{cases}$.

Solution We graph both equations on a single set of coordinate axes, as shown in Figure 9-1.

Although an infinite number of pairs (x, y) satisfy $x + 2y = 4$, and an infinite number of pairs (x, y) satisfy $2x - y = 3$, only the coordinates of the point where the graphs intersect satisfy both equations simultaneously. Since the intersection point has coordinates of $(2, 1)$, the solution is the pair $(2, 1)$ or $x = 2$ and $y = 1$.

To check the solution, we substitute 2 for x and 1 for y in each equation and verify that $(2, 1)$ satisfies each equation.

$x + 2y = 4$

x	y
4	0
0	2
−2	3

$2x - y = 3$

x	y
1	−1
0	−3
−1	−5

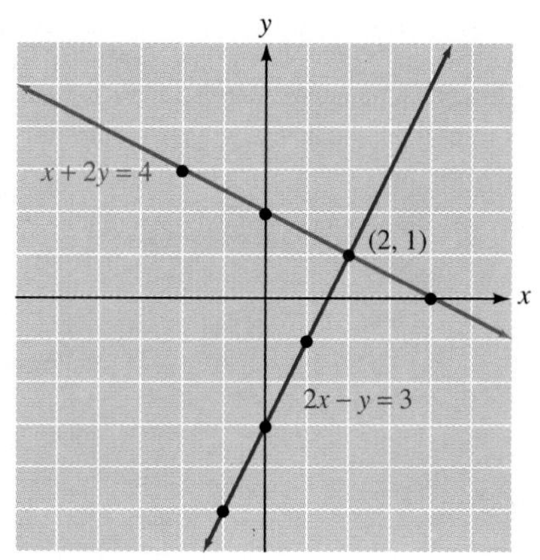

FIGURE 9-1

Inconsistent Systems

When a system of equations (as in Example 1) has a solution, the system is called a **consistent system of equations.** A system with no solution is called an **inconsistent system.**

EXAMPLE 2 Solve the system $\begin{cases} 2x + 3y = 6 \\ 4x + 6y = 24 \end{cases}$.

Solution We graph both equations on the same set of coordinate axes, as shown in Figure 9-2. In this example, the graphs are parallel, because the slopes of the two lines are equal. We can see that the slope of each line is $-\frac{2}{3}$ by writing each equation in slope–intercept form.

Since the graphs are parallel lines, the lines do not intersect, and the system does not have a solution. It is an inconsistent system.

$2x + 3y = 6$

x	y
3	0
0	2
−3	4

$4x + 6y = 24$

x	y
6	0
0	4
−3	6

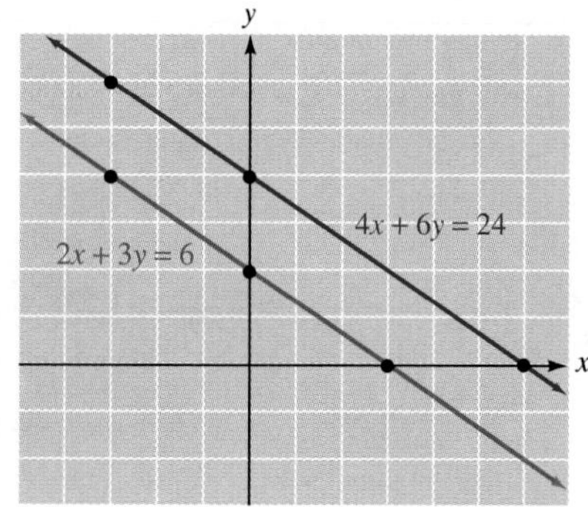

FIGURE 9-2

Dependent Equations

When the equations of a system have different graphs (as in Examples 1 and 2), the equations are called **independent equations.** Two equations with the same graph are called **dependent equations.**

EXAMPLE 3 Solve the system $\begin{cases} 2y - x = 4 \\ 2x + 8 = 4y \end{cases}$.

Solution We graph each equation on the same set of coordinate axes, as shown in Figure 9-3. Since the graphs coincide, the system has infinitely many solutions. Any pair (x, y) that satisfies one equation satisfies the other also.

From the tables of ordered pairs shown in Figure 9-3, we see that $(-4, 0)$ and $(0, 2)$ are solutions. We can find many more solutions by finding additional pairs (x, y) that satisfy either equation.

Because the two equations have the same graph, they are dependent equations.

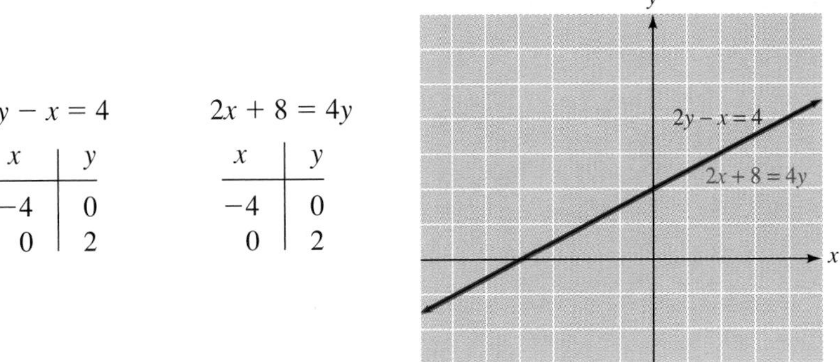

$2y - x = 4$

x	y
-4	0
0	2

$2x + 8 = 4y$

x	y
-4	0
0	2

FIGURE 9-3

We summarize the possibilities that can occur when graphing two linear equations, each with two variables.

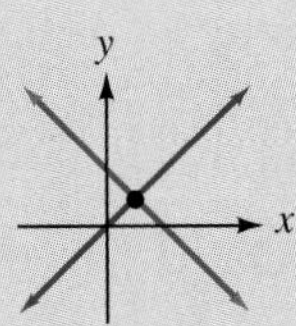

If the lines are different and intersect, the equations are independent and the system is consistent. *One solution exists.*

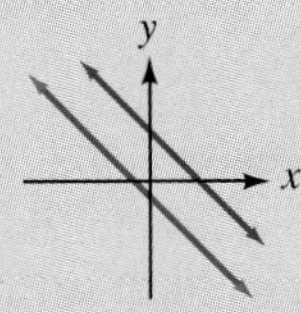

If the lines are different and parallel, the equations are independent and the system is inconsistent. *No solution exists.*

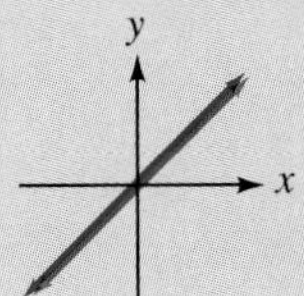

If the lines coincide, the equations are dependent and the system is consistent. *An infinite number of solutions exist.*

Two systems are called **equivalent systems** when they have the same solution set. In the next example, we solve a more difficult system by changing it into a simpler equivalent system.

EXAMPLE 4 Solve the system $\begin{cases} \frac{3}{2}x - y = \frac{5}{2} \\ x + \frac{1}{2}y = 4 \end{cases}$.

Solution We multiply both sides of $\frac{3}{2}x - y = \frac{5}{2}$ by 2 to clear the fractions and obtain the equation $3x - 2y = 5$. We multiply both sides of $x + \frac{1}{2}y = 4$ by 2 to clear the fractions and obtain the equation $2x + y = 8$.

The new system

$$\begin{cases} 3x - 2y = 5 \\ 2x + y = 8 \end{cases}$$

is equivalent to the original system, but it is easier to solve since it has no fractions. If we graph each equation in the new system, as in Figure 9-4, we see that the coordinates of the point where the two lines intersect are (3, 2). Verify that $x = 3$ and $y = 2$ satisfy each equation in the original system.

$3x - 2y = 5$

x	y
0	$-\frac{5}{2}$
$\frac{5}{3}$	0

$2x + y = 8$

x	y
4	0
1	6

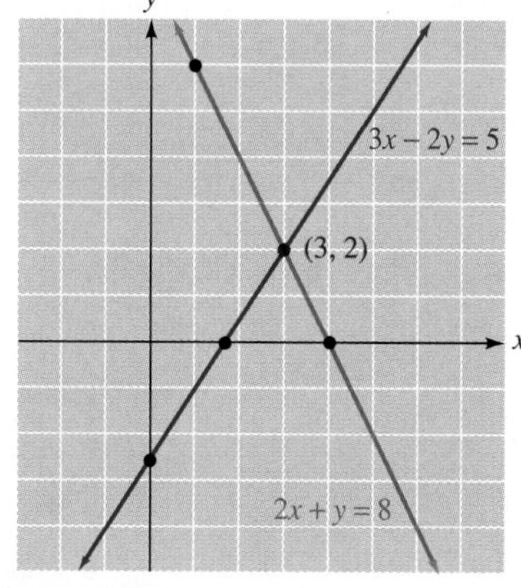

FIGURE 9-4

Graphing Calculators

The graphing method has limitations. First, the method is limited to equations with two variables. Systems with three or more variables cannot be solved graphically. Second, it is often difficult to find exact solutions graphically. However, the TRACE and ZOOM capabilities of a graphing calculator enable us to get very good approximations of such solutions.

EXAMPLE 5 Use a graphing calculator to solve the system $\begin{cases} 3x + 2y = 12 \\ 2x - 3y = 12 \end{cases}$.

Solution To enter the equations into a graphing calculator, we must first solve them for y to get the following equivalent system:

$$\begin{cases} y = -\frac{3}{2}x + 6 \\ y = \frac{2}{3}x - 4 \end{cases}$$

If we use a graphing calculator with standard RANGE settings, the graphs of the equations will look like those in Figure 9-5(a). If we zoom in on the intersection point of the two lines as in Figure 9-5(b) and use the TRACE function, we find that an approximate solution is $x = 4.6052632$ and $y = -.9078947$. To get better results, we can do additional zooms. Verify that the exact solution is $x = \frac{60}{13}$ and $y = -\frac{12}{13}$. ■

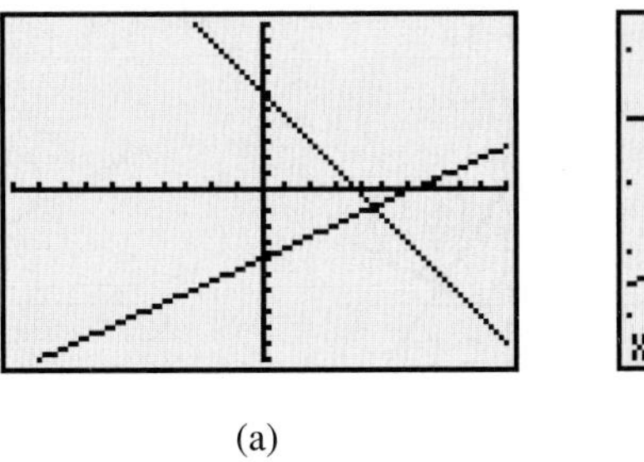

(a)

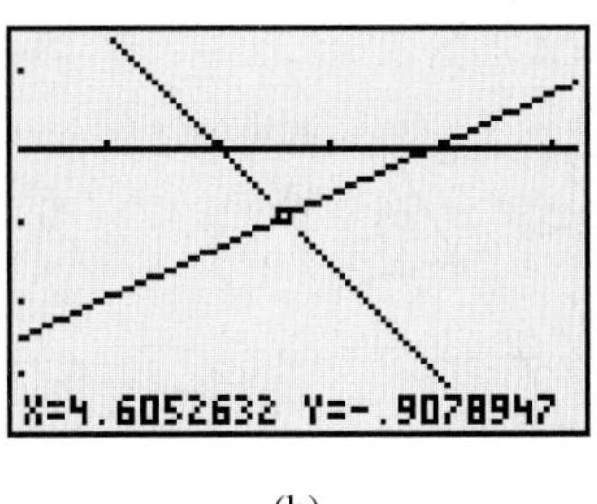

(b)

FIGURE 9-5

ORALS *Tell whether the following systems will have one solution, no solution, or an infinite number of solutions.*

1. $\begin{cases} y = 2x \\ y = 2x + 5 \end{cases}$ no solution

2. $\begin{cases} y = 2x \\ y = x + x \end{cases}$ infinite number

3. $\begin{cases} y = 2x \\ y = -2x \end{cases}$ one solution

4. $\begin{cases} y = 2x + 1 \\ 2x = y \end{cases}$ no solution

EXERCISE 9.1

In Exercises 1–4, tell whether the ordered pair is a solution of the system of equations.

1. $(1, 2)$; $\begin{cases} y = 2x \\ y = \dfrac{1}{2}x + \dfrac{3}{2} \end{cases}$ yes

2. $(-1, 2)$; $\begin{cases} y = 3x + 5 \\ y = x + 4 \end{cases}$ no

3. $(2, -3)$; $\begin{cases} y = \dfrac{1}{2}x - 2 \\ 3x + 2y = 0 \end{cases}$ no

4. $(-4, 3)$; $\begin{cases} 4x - y = -19 \\ 3x + 2y = -6 \end{cases}$ yes

In Exercises 5–24, solve each system by graphing, if possible. Check your graphs with a graphing calculator.

5. $\begin{cases} x + y = 6 \\ x - y = 2 \end{cases}$

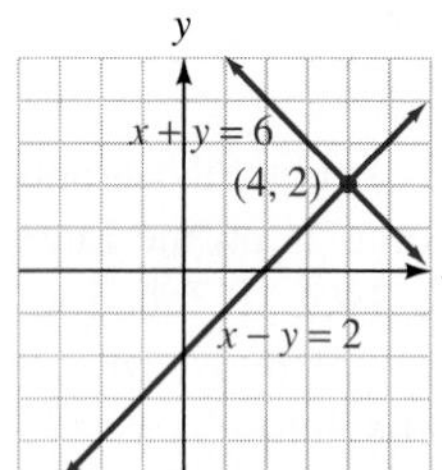

6. $\begin{cases} x - y = 4 \\ 2x + y = 5 \end{cases}$

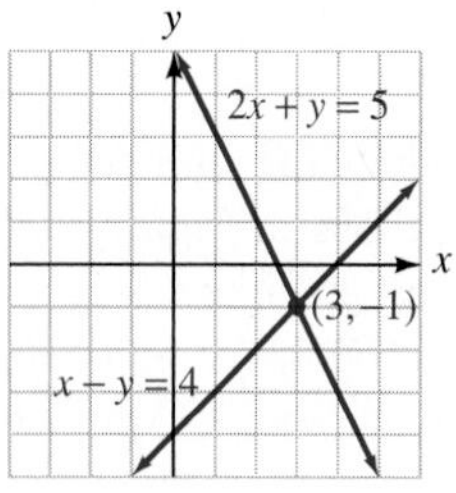

7. $\begin{cases} 2x + y = 1 \\ x - 2y = -7 \end{cases}$

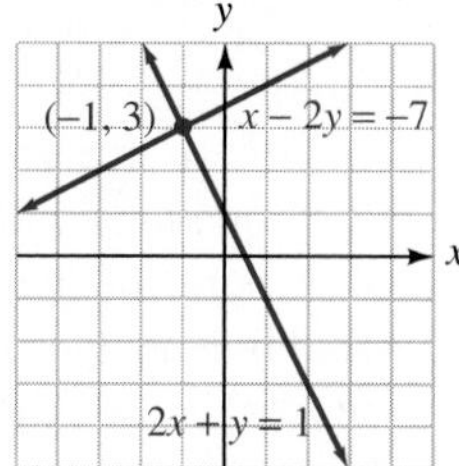

8. $\begin{cases} 3x - y = -3 \\ 2x + y = -7 \end{cases}$

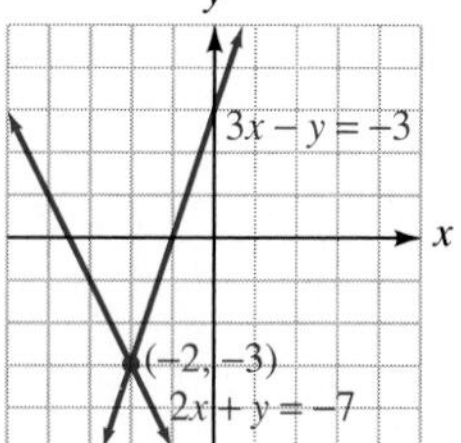

9. $\begin{cases} x = 13 - 4y \\ 3x = 4 + 2y \end{cases}$

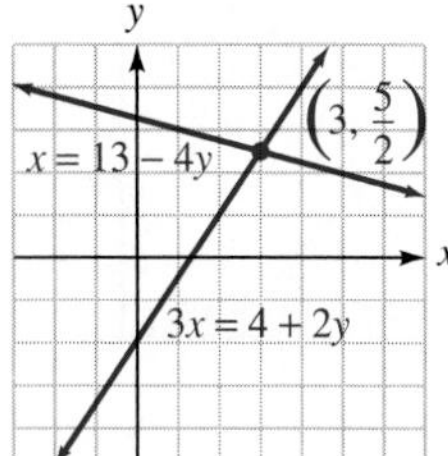

10. $\begin{cases} 3x = 7 - 2y \\ 2x = 2 + 4y \end{cases}$

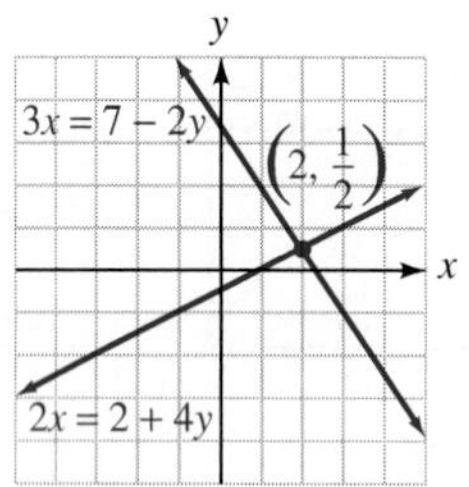

11. $\begin{cases} x = 3 - 2y \\ 2x + 4y = 6 \end{cases}$

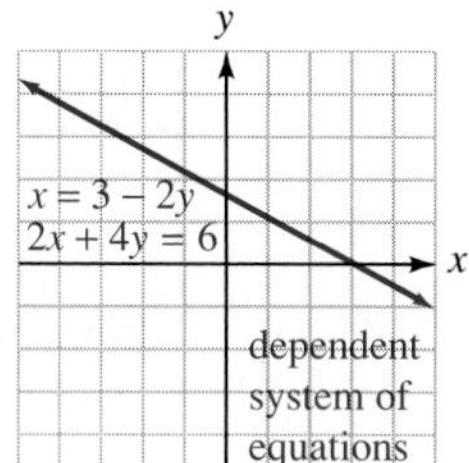

12. $\begin{cases} 3x = 5 - 2y \\ 3x + 2y = 7 \end{cases}$

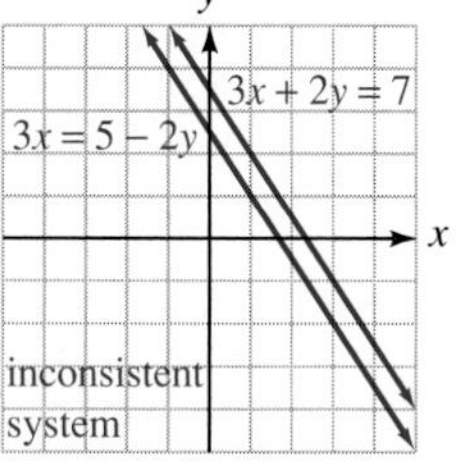

13. $\begin{cases} x = 2 \\ y = \dfrac{4 - x}{2} \end{cases}$

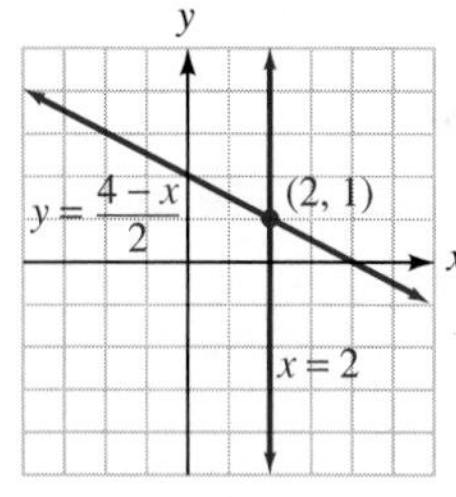

14. $\begin{cases} y = -2 \\ x = \dfrac{4 + 3y}{2} \end{cases}$

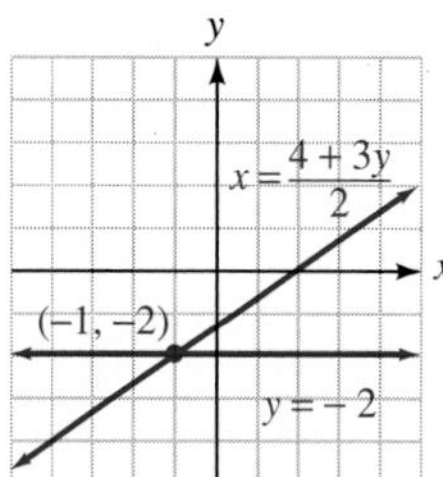

15. $\begin{cases} y = 3 \\ x = 2 \end{cases}$

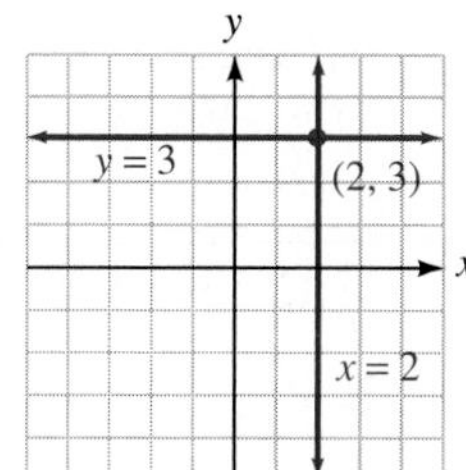

16. $\begin{cases} 2x + 3y = -15 \\ 2x + y = -9 \end{cases}$

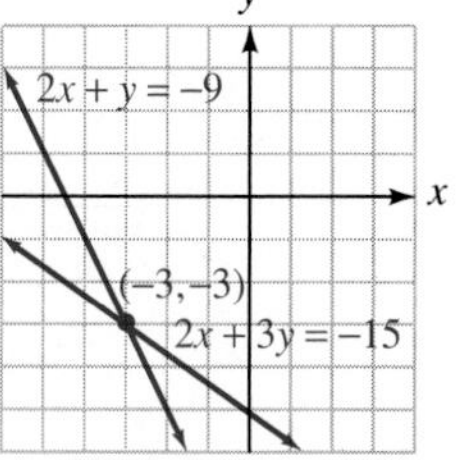

17. $\begin{cases} x = \dfrac{11 - 2y}{3} \\ y = \dfrac{11 - 6x}{4} \end{cases}$

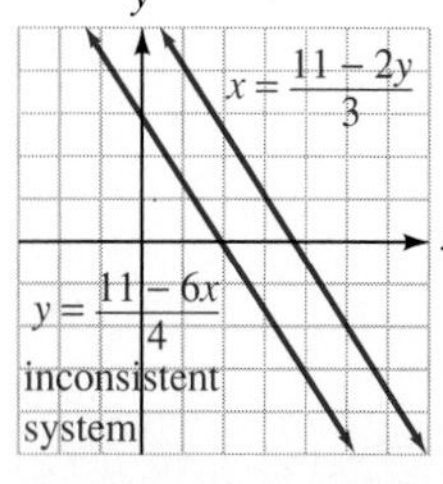

18. $\begin{cases} x = \dfrac{1 - 3y}{4} \\ y = \dfrac{12 + 3x}{2} \end{cases}$

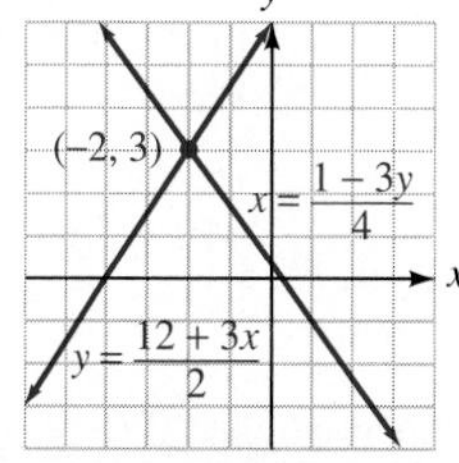

19. $\begin{cases} \dfrac{5}{2}x + y = \dfrac{1}{2} \\ 2x - \dfrac{3}{2}y = 5 \end{cases}$

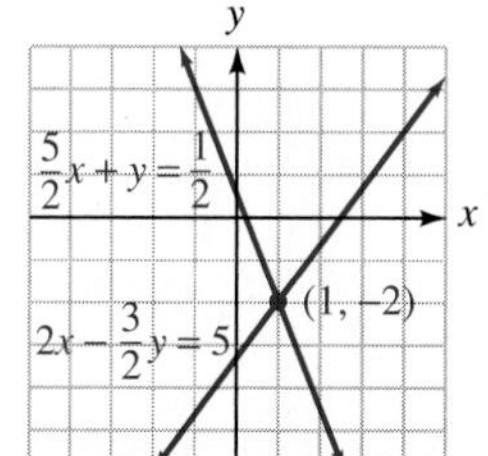

20. $\begin{cases} \dfrac{5}{2}x + 3y = 6 \\ y = \dfrac{24 - 10x}{12} \end{cases}$

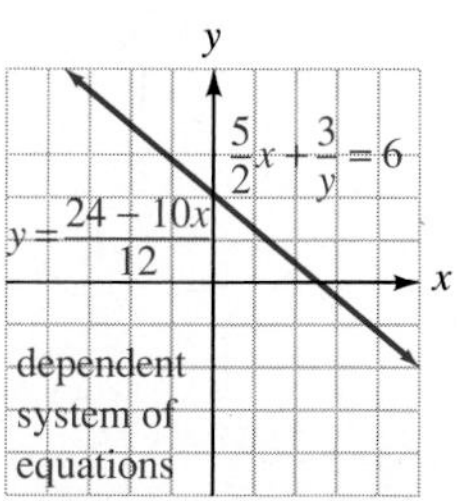

21. $\begin{cases} x = \dfrac{5y - 4}{2} \\ x - \dfrac{5}{3}y + \dfrac{1}{3} = 0 \end{cases}$

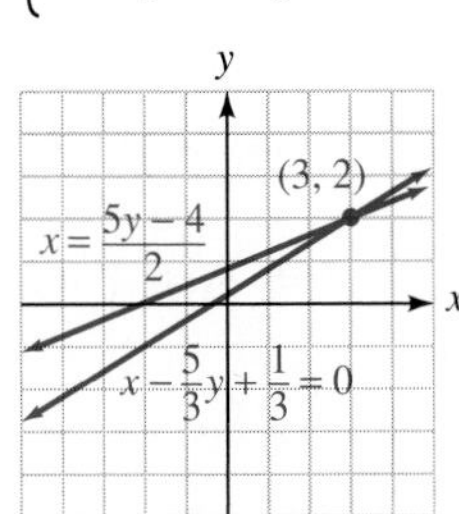

22. $\begin{cases} 2x = 5y - 11 \\ 3x = 2y \end{cases}$

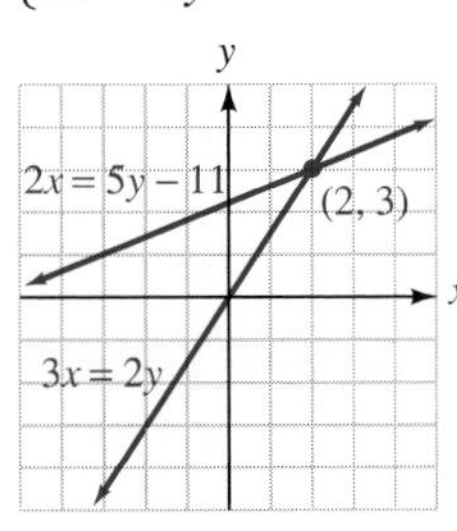

23. $\begin{cases} x = -\dfrac{3}{2}y \\ x = \dfrac{3}{2}y - 2 \end{cases}$

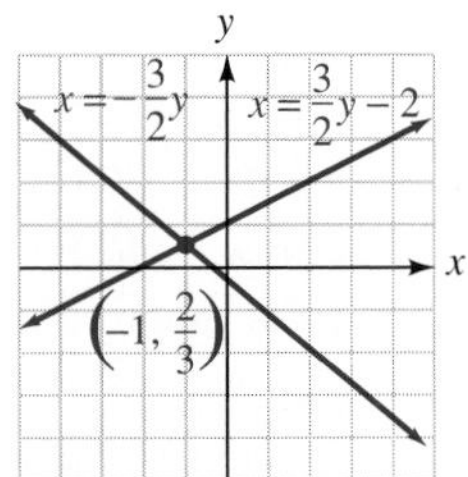

24. $\begin{cases} x = \dfrac{3y - 1}{4} \\ y = \dfrac{4 - 8x}{3} \end{cases}$

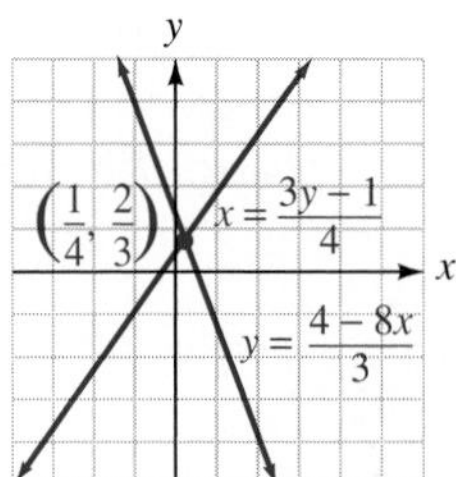

In Exercises 25–28, use a graphing calculator to solve each system. Give answers to the nearest hundredth.

25. $\begin{cases} y = 3.2x - 1.5 \\ y = -2.7x - 3.7 \end{cases}$ $(-0.37, -2.69)$

26. $\begin{cases} y = -0.45x + 5 \\ y = 5.55x - 13.7 \end{cases}$ $(3.12, 3.60)$

27. $\begin{cases} 1.7x + 2.3y = 3.2 \\ y = 0.25x + 8.95 \end{cases}$ $(-7.64, 7.04)$

28. $\begin{cases} 2.75x = 12.9y - 3.79 \\ 7.1x - y = 35.76 \end{cases}$ $(5.24, 1.41)$

Writing Exercises ■ *Write a paragraph using your own words.*

1. Explain how to solve a system of two equations in two variables.

2. Can a system of two equations in two variables have exactly two solutions? Why or why not?

Something to Think About ■

1. Form an independent system of equations with a solution of $(-5, 2)$.

One possible answer is $\begin{cases} x + y = -3 \\ x - y = -7 \end{cases}$.

2. Form a dependent system of equations with a solution of $(-5, 2)$.

One possible answer is $\begin{cases} y = 3x + 17 \\ 2y = 6x + 34 \end{cases}$.

Review Exercises ■ *Write each number in scientific notation.*

1. 93,000,000 — 9.3×10^7

2. 0.0000000236 — 2.36×10^{-8}

3. 345×10^2 — 3.45×10^4

4. 752×10^{-5} — 7.52×10^{-3}

9.2 Solution by Substitution and Addition

■ The Substitution Method ■ The Addition Method ■ An Inconsistent System ■ A System with Infinitely Many Solutions ■ Problem Solving

The graphing method is a nice way to visualize the process of solving systems of equations. However, it cannot be used to solve systems of higher order, such as three equations, each with three variables. In this section, we will discuss algebraic methods of solving systems of equations.

The Substitution Method

We use the following steps to solve a system of two equations, each with two variables, by substitution.

Steps to Follow When Using the Substitution Method

1. If necessary, solve one equation for one of its variables, preferably a variable with a coefficient of 1.
2. Substitute the resulting expression for the variable obtained in Step 1 into the other equation and solve that equation.
3. Find the value of the other variable by substituting the value of the variable found in Step 2 in any equation containing both variables.
4. Check the solution in both of the original equations.

EXAMPLE 1 Use the substitution method to solve the system $\begin{cases} 4x + y = 13 \\ -2x + 3y = -17 \end{cases}$.

Solution We solve the first equation for y, because y has a coefficient of 1 and no fractions are introduced.

$$4x + y = 13$$

1. $y = -4x + 13$ Subtract $4x$ from both sides.

We then substitute $-4x + 13$ for y in the second equation of the system and solve for x.

$$\begin{aligned} -2x + 3y &= -17 \\ -2x + 3(-4x + 13) &= -17 && \text{Substitute } -4x + 13 \text{ for } y. \\ -2x - 12x + 39 &= -17 && \text{Use the distributive property to remove parentheses.} \\ -14x &= -56 && \text{Combine terms and subtract 39 from both sides.} \\ x &= 4 && \text{Divide both sides by } -14. \end{aligned}$$

To find y, we substitute 4 for x in Equation 1 and simplify:

$$\begin{aligned} y &= -4x + 13 \\ &= -4(4) + 13 && \text{Substitute 4 for } x. \\ &= -3 \end{aligned}$$

The solution is $x = 4$ and $y = -3$, or just $(4, -3)$. The graphs of these two equations would intersect at the point $(4, -3)$. Verify that this solution satisfies each equation in the given system. ■

EXAMPLE 2 Use the substitution method to solve the system $\begin{cases} \frac{4}{3}x + \frac{1}{2}y = -\frac{2}{3} \\ \frac{1}{2}x + \frac{2}{3}y = \frac{5}{3} \end{cases}$.

Solution First we find an equivalent system without fractions by multiplying each side of each equation by 6 to obtain the system

$$\begin{array}{ll} \textbf{2.} & \left\{\begin{array}{l} 8x + 3y = -4 \\ 3x + 4y = 10 \end{array}\right. \\ \textbf{3.} & \end{array}$$

Because no variable in either equation has a coefficient of 1, it is impossible to avoid fractions when solving for a variable. We can solve Equation 3 for x:

$$3x + 4y = 10$$

$$3x = -4y + 10 \qquad \text{Subtract } 4y \text{ from both sides.}$$

4. $$x = -\frac{4}{3}y + \frac{10}{3} \qquad \text{Divide both sides by 3.}$$

We then substitute $-\frac{4}{3}y + \frac{10}{3}$ for x in Equation 2 and solve for y.

$$\begin{aligned} 8x + 3y &= -4 \\ 8\left(-\frac{4}{3}y + \frac{10}{3}\right) + 3y &= -4 && \text{Substitute } -\tfrac{4}{3}y + \tfrac{10}{3} \text{ for } x. \\ \frac{8}{3}(-4y + 10) + 3y &= -4 && \text{Factor out } \tfrac{1}{3}. \\ 8(10 - 4y) + 9y &= -12 && \text{Multiply both sides by 3.} \\ 80 - 32y + 9y &= -12 && \text{Use the distributive property to remove parentheses.} \\ -23y &= -92 && \text{Combine terms and subtract 80 from both sides.} \\ y &= 4 && \text{Divide both sides by } -23. \end{aligned}$$

We can find x by substituting 4 for y in Equation 4 and simplifying:

$$\begin{aligned} x &= -\frac{4}{3}y + \frac{10}{3} \\ &= -\frac{4}{3}(4) + \frac{10}{3} && \text{Substitute 4 for } y. \\ &= -\frac{6}{3} && -\frac{16}{3} + \frac{10}{3} = -\frac{6}{3}. \\ &= -2 \end{aligned}$$

The solution is the pair $(-2, 4)$. Verify that this solution satisfies each equation in the original system. ■

The Addition Method

In the addition method, we combine the equations of the system in a way that will eliminate terms involving one of the variables.

Steps to Follow When Using the Addition Method

1. Write both equations of the system in general form.
2. Multiply the terms of one or both of the equations by constants chosen to make the coefficients of x (or y) differ only in sign.
3. Add the equations and solve the equation that results, if possible.
4. Substitute the value obtained in Step 3 into either of the original equations and solve for the remaining variable.
5. The results obtained in Steps 3 and 4 give the solution of the system.
6. Check the solution in both of the original equations.

EXAMPLE 3 Use the addition method to solve the system $\begin{cases} 4x + y = 13 \\ -2x + 3y = -17 \end{cases}$.

Solution This is the system discussed in Example 1. To solve it by addition, we multiply the second equation by 2 to obtain the system

$$\textbf{5.} \quad \begin{cases} 4x + y = 13 \\ -4x + 6y = -34 \end{cases}$$

When these equations are added, the terms involving x drop out, and we get

$$7y = -21$$
$$y = -3 \qquad \text{Divide both sides by 7.}$$

To find x, we substitute -3 for y in either of the original equations and solve for x. If we use Equation 5, we have

$$4x + y = 13$$
$$4x + (-3) = 13 \qquad \text{Substitute } -3 \text{ for } y.$$
$$4x = 16 \qquad \text{Add 3 to both sides.}$$
$$x = 4$$

The solution is $x = 4$ and $y = -3$, or just $(4, -3)$. ■

Teaching Tip: To find x, it would be just as easy to eliminate y. Just multiply the first equation by -3 and add.

$$\begin{aligned} -12x - 3y &= -39 \\ -2x + 3y &= -17 \\ \hline -14x \quad &= -56 \\ x &= 4 \end{aligned}$$

EXAMPLE 4 Use the addition method to solve the system $\begin{cases} \frac{4}{3}x + \frac{1}{2}y = -\frac{2}{3} \\ \frac{1}{2}x + \frac{2}{3}y = \frac{5}{3} \end{cases}$.

Solution This system is the system discussed in Example 2. To solve it by addition, we find an equivalent system with no fractions by multiplying both sides of each equation by 6 to obtain

$$\begin{array}{ll} \textbf{6.} & \left\{ \begin{array}{l} 8x + 3y = -4 \\ 3x + 4y = 10 \end{array} \right. \\ \textbf{7.} & \end{array}$$

To make the y terms drop out when we add the equations, we multiply both sides of Equation 6 by 4 and both sides of Equation 7 by -3 to get

$$\begin{cases} 32x + 12y = -16 \\ -9x - 12y = -30 \end{cases}$$

When these equations are added, the y terms drop out, and we get

$$23x = -46$$

$$x = -2 \qquad \text{Divide both sides by 23.}$$

To find y, we substitute -2 for x in either Equation 6 or Equation 7. If we substitute -2 for x in Equation 7, we get

$$\begin{aligned} 3x + 4y &= 10 \\ 3(-2) + 4y &= 10 && \text{Substitute } -2 \text{ for } x. \\ -6 + 4y &= 10 && \text{Simplify.} \\ 4y &= 16 && \text{Add 6 to both sides.} \\ y &= 4 && \text{Divide both sides by 4.} \end{aligned}$$

The solution is $(-2, 4)$. ■

An Inconsistent System

EXAMPLE 5 Solve the system $\begin{cases} y = 2x + 4 \\ 8x - 4y = 7 \end{cases}$.

Solution Because the first equation is already solved for y, we use the substitution method.

$$\begin{aligned} 8x - 4y &= 7 \\ 8x - 4(2x + 4) &= 7 && \text{Substitute } 2x + 4 \text{ for } y. \end{aligned}$$

We then solve this equation for x:

$$\begin{aligned} 8x - 8x - 16 &= 7 && \text{Use the distributive property to remove parentheses.} \\ -16 &= 7 && \text{Combine terms.} \end{aligned}$$

This impossible result shows that the equations in the system are independent and that the system is inconsistent. Since the system has no solution, the graphs of the equations in the system would be parallel. ■

A System with Infinitely Many Solutions

EXAMPLE 6 Solve the system $\begin{cases} 4x + 6y = 12 \\ -2x - 3y = -6 \end{cases}$.

Solution Since the equations are written in general form, we use the addition method and multiply both sides of the second equation by 2 to get

$$\begin{cases} 4x + 6y = 12 \\ -4x - 6y = -12 \end{cases}$$

After adding the left-hand sides and the right-hand sides, we get

$$0x + 0y = 0$$
$$0 = 0$$

Here, both the x and y terms drop out. The true statement $0 = 0$ shows that the equations in this system are dependent and that the system is consistent.

Note that the equations of the system are equivalent because, when the second equation is multiplied by -2, it becomes the first equation.

The line graphs of these equations would coincide. Any ordered pair that satisfies one of the equations also satisfies the other. ■

Problem Solving

EXAMPLE 7 **Retail sales** Hi-Fi Electronics advertises two types of car radios, one selling for \$67 and the other for \$100. If the receipts from the sale of 36 radios totaled \$2940, how many of each type were sold?

Solution We can let x represent the number of radios sold for \$67 and let y represent the number of radios sold for \$100. Then the receipts for the sale of the less expensive radio are \$$67x$, and the receipts for the sale of the more expensive radios are \$$100y$. The information in the problem gives the following two equations:

The number of less expensive radios	+	the number of more expensive radios	=	the total number of radios.
x	+	y	=	36

The value of the less expensive radios	+	the value of the more expensive radios	=	the total receipts.
$67x$	+	$100y$	=	2940

We can solve the following system for x and y to find out how many of each type were sold:

$$\begin{cases} x + y = 36 & \textbf{8.} \\ 67x + 100y = 2940 & \textbf{9.} \end{cases}$$

We multiply both sides of Equation 8 by -100, add the resulting equation to Equation 9, and solve for x:

$$\begin{aligned} -100x - 100y &= -3600 \\ 67x + 100y &= 2940 \\ \hline -33x &= -660 \\ x &= 20 \end{aligned}$$

Divide both sides by -33.

To find y, we substitute 20 for x in Equation 8 and solve for y:

$$x + y = 36$$

$$20 + y = 36 \qquad \text{Substitute 20 for } x.$$

$$y = 16 \qquad \text{Subtract 20 from both sides.}$$

The store sold 20 of the less expensive radios and 16 of the more expensive radios.

Check: If 20 of one type were sold and 16 of the other type were sold, a total of 36 radios were sold.

Since the value of the less expensive radios is 20($67) = $1340 and the value of the more expensive radios is 16($100) = $1600, the total receipts are $2940. ■

EXAMPLE 8 **Mixing solutions** How many ounces of a 10% saline solution and how many ounces of a 20% saline solution must be mixed together to obtain 50 ounces of a 15% saline solution?

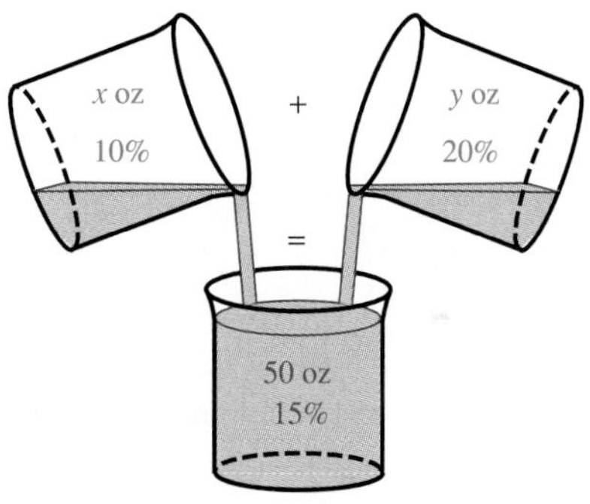

FIGURE 9-6

Solution We can let x represent the number of ounces of the 10% solution and let y represent the number of ounces of the 20% solution that are to be mixed. (See Figure 9-6.) The information given in the problem gives the following two equations:

The number of ounces of 10% solution	+	the number of ounces of 20% solution	=	the total number of ounces in the mixture.
x	+	y	=	50

The salt in the 10% solution	+	the salt in the 20% solution	=	the salt in the mixture.
$0.10x$	+	$0.20y$	=	$0.15(50)$

We can solve the following system for x and y to find how many ounces of each are needed:

10. / **11.**
$$\begin{cases} x + y = 50 \\ 0.10x + 0.20y = 7.5 \end{cases}$$

To solve this system by substitution, we can solve Equation 10 for y

$$x + y = 50$$

12. $y = 50 - x$ Subtract x from both sides.

and then substitute $50 - x$ for y in Equation 11.

$$0.10x + 0.20y = 7.5$$

$0.10x + 0.20(50 - x) = 7.5$ Substitute $50 - x$ for y.

$10x + 20(50 - x) = 750$ Multiply both sides by 100.

$10x + 1000 - 20x = 750$ Use the distributive property to remove parentheses.

$-10x = -250$ Combine terms and subtract 1000 from both sides.

$x = 25$ Divide both sides by -10.

To find y, we can substitute 25 for x in Equation 12:

$y = 50 - x$

$= 50 - 25$ Substitute 25 for x.

$= 25$

To obtain 50 ounces of a 15% solution, we must mix 25 ounces each of the 10% and 20% solutions. ■

ORALS *Solve each system for x.*

1. $\begin{cases} y = 2x \\ x + y = 6 \end{cases}$ 2

2. $\begin{cases} y = -x \\ 2x + y = 4 \end{cases}$ 4

3. $\begin{cases} x - y = 6 \\ x + y = 2 \end{cases}$ 4

4. $\begin{cases} x + y = 4 \\ 2x - y = 5 \end{cases}$ 3

EXERCISE 9.2

In Exercises 1–12, solve each system by substitution, if possible.

1. $\begin{cases} y = x \\ x + y = 4 \end{cases}$
(2, 2)

2. $\begin{cases} y = x + 2 \\ x + 2y = 16 \end{cases}$
(4, 6)

3. $\begin{cases} x - y = 2 \\ 2x + y = 13 \end{cases}$
(5, 3)

4. $\begin{cases} x - y = -4 \\ 3x - 2y = -5 \end{cases}$
(3, 7)

5. $\begin{cases} x + 2y = 6 \\ 3x - y = -10 \end{cases}$ $(-2, 4)$

6. $\begin{cases} 2x - y = -21 \\ 4x + 5y = 7 \end{cases}$ $(-7, 7)$

7. $\begin{cases} 3x = 2y - 4 \\ 6x - 4y = -4 \end{cases}$ no solution

8. $\begin{cases} 8x = 4y + 10 \\ 4x - 2y = 5 \end{cases}$ dependent equations

9. $\begin{cases} 3x - 4y = 9 \\ x + 2y = 8 \end{cases}$ $\left(5, \frac{3}{2}\right)$

10. $\begin{cases} 3x - 2y = -10 \\ 6x + 5y = 25 \end{cases}$ $(0, 5)$

11. $\begin{cases} 2x + 2y = -1 \\ 3x + 4y = 0 \end{cases}$ $\left(-2, \frac{3}{2}\right)$

12. $\begin{cases} 5x + 3y = -7 \\ 3x - 3y = 7 \end{cases}$ $\left(0, -\frac{7}{3}\right)$

In Exercises 13–24, solve each system by addition, if possible.

13. $\begin{cases} x - y = 3 \\ x + y = 7 \end{cases}$ $(5, 2)$

14. $\begin{cases} x + y = 1 \\ x - y = 7 \end{cases}$ $(4, -3)$

15. $\begin{cases} 2x + y = -10 \\ 2x - y = -6 \end{cases}$ $(-4, -2)$

16. $\begin{cases} x + 2y = -9 \\ x - 2y = -1 \end{cases}$ $(-5, -2)$

17. $\begin{cases} 2x + 3y = 8 \\ 3x - 2y = -1 \end{cases}$ $(1, 2)$

18. $\begin{cases} 5x - 2y = 19 \\ 3x + 4y = 1 \end{cases}$ $(3, -2)$

19. $\begin{cases} 4x + 9y = 8 \\ 2x - 6y = -3 \end{cases}$ $\left(\frac{1}{2}, \frac{2}{3}\right)$

20. $\begin{cases} 4x + 6y = 5 \\ 8x - 9y = 3 \end{cases}$ $\left(\frac{3}{4}, \frac{1}{3}\right)$

21. $\begin{cases} 8x - 4y = 16 \\ 2x - 4 = y \end{cases}$ dependent equations

22. $\begin{cases} 2y - 3x = -13 \\ 3x - 17 = 4y \end{cases}$ $(3, -2)$

23. $\begin{cases} x = \frac{3}{2}y + 5 \\ 2x - 3y = 8 \end{cases}$ no solution

24. $\begin{cases} x = \frac{2}{3}y \\ y = 4x + 5 \end{cases}$ $(-2, -3)$

In Exercises 25–32, solve each system by any method.

25. $\begin{cases} \frac{x}{2} + \frac{y}{2} = 6 \\ \frac{x}{2} - \frac{y}{2} = -2 \end{cases}$ $(4, 8)$

26. $\begin{cases} \frac{x}{2} - \frac{y}{3} = -4 \\ \frac{x}{2} + \frac{y}{9} = 0 \end{cases}$ $(-2, 9)$

27. $\begin{cases} \frac{3}{4}x + \frac{2}{3}y = 7 \\ \frac{3}{5}x - \frac{1}{2}y = 18 \end{cases}$ $(20, -12)$

28. $\begin{cases} \frac{2}{3}x - \frac{1}{4}y = -8 \\ \frac{1}{2}x - \frac{3}{8}y = -9 \end{cases}$ $(-6, 16)$

29. $\begin{cases} \frac{3x}{2} - \frac{2y}{3} = 0 \\ \frac{3x}{4} + \frac{4y}{3} = \frac{5}{2} \end{cases}$ $\left(\frac{2}{3}, \frac{3}{2}\right)$

30. $\begin{cases} \frac{3x}{5} + \frac{5y}{3} = 2 \\ \frac{6x}{5} - \frac{5y}{3} = 1 \end{cases}$ $\left(\frac{5}{3}, \frac{3}{5}\right)$

31. $\begin{cases} \frac{2}{5}x - \frac{1}{6}y = \frac{7}{10} \\ \frac{3}{4}x - \frac{2}{3}y = \frac{19}{8} \end{cases}$ $\left(\frac{1}{2}, -3\right)$

32. $\begin{cases} \frac{5}{6}x + \frac{2}{3}y = \frac{7}{6} \\ \frac{10}{7}x - \frac{4}{9}y = \frac{17}{21} \end{cases}$ $\left(\frac{4}{5}, \frac{3}{4}\right)$

In Exercises 33–36, solve each system for x and y. Solve for $\frac{1}{x}$ and $\frac{1}{y}$ first.

33. $\begin{cases} \frac{1}{x} + \frac{1}{y} = \frac{5}{6} \\ \frac{1}{x} - \frac{1}{y} = \frac{1}{6} \end{cases}$ $(2, 3)$

34. $\begin{cases} \frac{1}{x} + \frac{1}{y} = \frac{9}{20} \\ \frac{1}{x} - \frac{1}{y} = \frac{1}{20} \end{cases}$ $(4, 5)$

35. $\begin{cases} \frac{1}{x} + \frac{2}{y} = -1 \\ \frac{2}{x} - \frac{1}{y} = -7 \end{cases}$ $\left(-\frac{1}{3}, 1\right)$

36. $\begin{cases} \frac{3}{x} - \frac{2}{y} = -30 \\ \frac{2}{x} - \frac{3}{y} = -30 \end{cases}$ $\left(-\frac{1}{6}, \frac{1}{6}\right)$

In Exercises 37–54, use two variables to solve each problem.

37. **Integer problem** The sum of two integers is 49. If their difference is 7, find the integers. 28, 21

38. **Integer problem** The sum of two integers is −7. The first integer minus the second is 3. Find the integers. −2, −5

39. **Merchandising** A pair of shoes and a sweater cost \$98. If the sweater cost \$16 more than the shoes, how much did the sweater cost? \$57

40. **Merchandising** A sporting goods salesperson sells 2 fishing reels and 5 rods for \$270. The next day, the salesperson sells 4 reels and 2 rods for \$220. How much does each cost? \$35 reel, \$40 rod

41. **Electronics** Two resistors in the voltage divider circuit in Illustration 1 have a total resistance of 1375 ohms. To provide the required voltage, R_1 must be 125 ohms greater than R_2. Find both resistances.
625 and 750 ohms

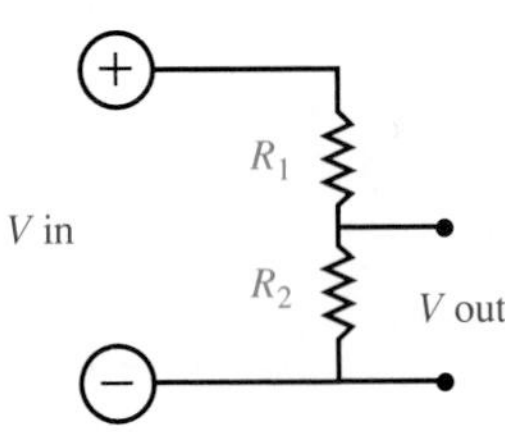

ILLUSTRATION 1

42. **Stowing baggage** A small aircraft can carry 950 pounds of baggage, distributed between two storage compartments. On one flight, the plane is fully loaded, with 150 pounds more baggage in one compartment than the other. How much is stowed in each compartment? 400, 550 lb

43. **Geometry problem** The rectangular field in Illustration 2 is surrounded by 72 meters of fencing. If the field is partitioned as shown, a total of 88 meters of fencing is required. Find the dimensions of the field. 16 m by 20 m

ILLUSTRATION 2

44. **Geometry problem** In a right triangle, one acute angle is 15° greater than two times the other acute angle. Find the difference between the angles. 40°

45. **Investment income** Part of \$8000 was invested at 10% interest and the rest at 12%. If the annual income from these investments was \$900, how much was invested at each rate?
\$3000 at 10%, \$5000 at 12%

46. **Investment income** Part of \$12,000 was invested at 6% interest and the rest at 7.5%. If the annual income from these investments was \$810, how much was invested at each rate?
\$6000 at 6%, \$6000 at 7.5%

47. **Mixing a solution** How many ounces of the two alcohol solutions in Illustration 3 must be mixed to obtain 100 ounces of a 12.2% solution?
40 oz of 8% solution, 60 oz of 15% solution

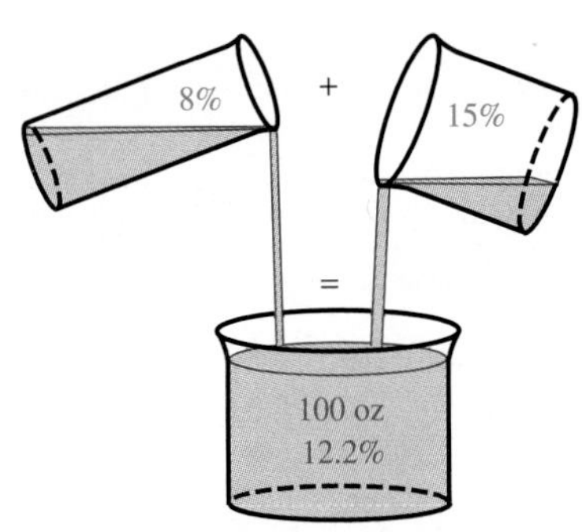

ILLUSTRATION 3

48. **Mixing nuts** How many pounds of nuts that cost \$2 per pound and nuts that cost \$4 per pound must be mixed to obtain 60 pounds of nuts that are worth \$3 per pound? 30 lb of each

49. **Travel** A car travels 50 miles in the same time that a plane travels 180 miles. The speed of the plane is 143 miles per hour faster than the speed of the car. Find the speed of the car. 55 mph

50. **Making bicycles** A bicycle manufacturer builds racing bikes and mountain bikes, with the per unit manufacturing costs shown in Illustration 4. The company has budgeted \$15,900 for labor and \$13,075 for materials. How many bicycles of each type can be built?
85 racing bikes, 120 mountain bikes

Model	Cost of materials	Cost of labor
racing	\$55	\$60
mountain	\$70	\$90

ILLUSTRATION 4

51. **Feeding cattle** A farmer keeps some animals on a strict diet. Each animal is to receive 15 grams of protein and 7.5 grams of carbohydrates. The farmer uses two food mixes with the nutrients shown in Illustration 5.

Mix	Protein	Carbohydrates
A	12%	9%
B	15%	5%

ILLUSTRATION 5

How many grams of each mix should the farmer use to provide the correct nutrients for each animal? 50 gm of A, 60 gm of B

52. **Renting apartments** The manager of an apartment complex is also a tenant. He pays only three-quarters of the rent that each of the remaining 5 tenants pays. Each month, the landlord collects a total of $2070 from the 6 occupants. How much rent does the manager pay? $270

53. **Curve fitting** The equation of the parabola in Illustration 6 is of the form $y = ax^2 + bx + c$. Find the equation by determining a, b, and c. (*Hint:* The equation is satisfied by the coordinates of each point.) $y = x^2 - 4x$

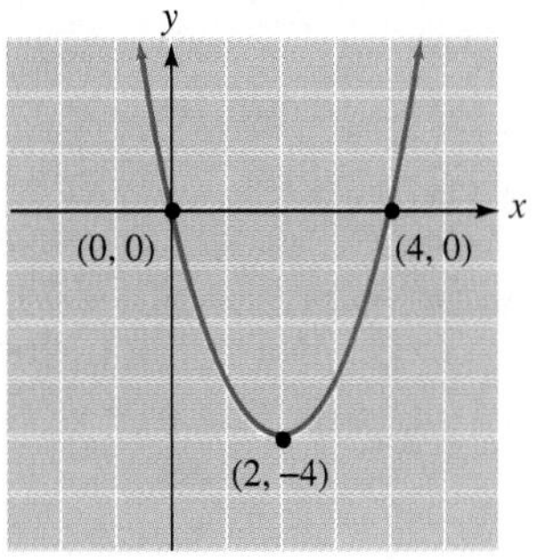

ILLUSTRATION 6

54. **Comparing salary plans** A sales clerk can choose from two salary options:
   1. a straight 7% commission
   2. $150 + 2% commission

How much would the clerk have to sell for each plan to produce the same monthly paycheck? $3000

Writing Exercises ■ *Write a paragraph using your own words.*

1. Tell which method you would use to solve the following system. Why?
$$\begin{cases} y = 3x + 1 \\ 3x + 2y = 12 \end{cases}$$

2. Tell which method you would use to solve the following system. Why?
$$\begin{cases} 2x + 4y = 9 \\ 3x - 5y = 20 \end{cases}$$

Something to Think About ■

1. Under what conditions will a system of two equations in two variables be inconsistent?
2. Under what conditions will the equations of a system of two equations in two variables be dependent?

Review Exercises ■ *Simplify each expression. Write all answers without using negative exponents.*

1. $(a^2a^3)^2(a^4a^2)^2$ a^{22}
2. $\left(\dfrac{a^2b^3c^4d}{ab^2c^3d^4}\right)^{-3}$ $\dfrac{d^9}{a^3b^3c^3}$
3. $\left(\dfrac{-3x^3y^4}{x^{-5}y^3}\right)^{-4}$ $\dfrac{1}{81x^{32}y^4}$
4. $\dfrac{3t^0 - 4t^0 + 5}{5t^0 + 2t^0}$ $\dfrac{4}{7}$

9.3 Solution of Three Equations in Three Variables

■ Problem Solving

In this section, we extend the definition of a linear equation to include any equation of the form $ax + by + cz = d$. The solution of a system of three linear equations with three variables is an ordered triple of numbers. For example, the solution of the system

$$\begin{cases} 2x + 3y + 4z = 20 \\ 3x + 4y + 2z = 17 \\ 3x + 2y + 3z = 16 \end{cases}$$

is the ordered triple (1, 2, 3). Verify that the numbers in the triple satisfy each of the three equations in the system.

The graph of an equation of the form $ax + by + cz = d$ is a flat surface called a **plane.** A system of three linear equations in three variables is consistent or inconsistent, depending on how the three planes corresponding to the three equations intersect. Figure 9-7 illustrates some of the possibilities.

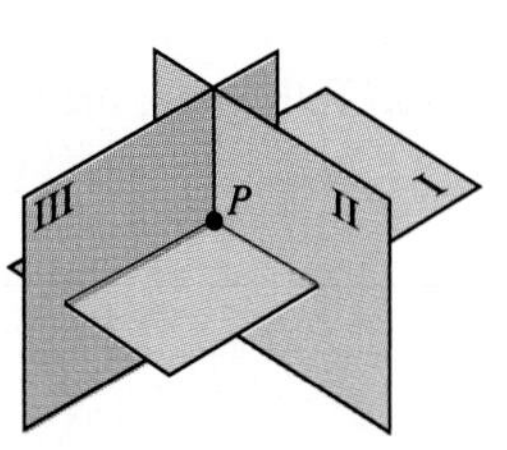

The three planes intersect at a single point P: One solution

(a)

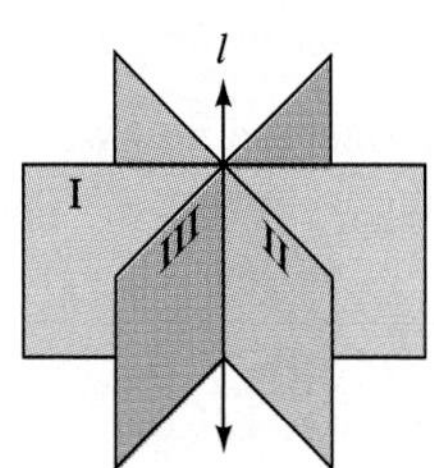

The three planes have a line l in common: An infinite number of solutions

(b)

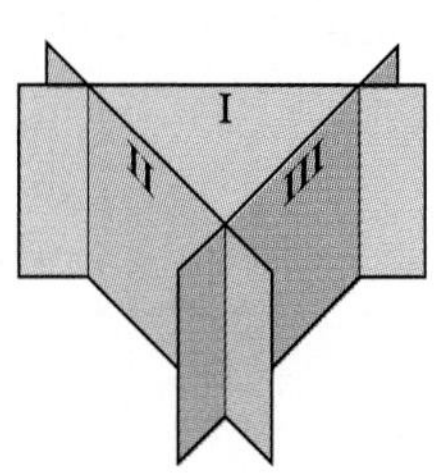

The three planes have no point in common: No solutions

(c)

FIGURE 9-7

To solve a system of three linear equations in three variables, we follow these steps.

Steps to Follow When Solving Three Equations in Three Variables

1. Pick any two equations and eliminate a variable.
2. Pick a different pair of equations and eliminate the same variable.
3. Solve the resulting pair of two equations in two variables.
4. To find the value of the third variable, substitute the values of the two variables found in Step 3 into any equation containing all three variables and solve the equation.
5. Check the solution in all three of the original equations.

Example 1 discusses a consistent system of three equations in three variables. Example 2 discusses an inconsistent system.

EXAMPLE 1 Solve the system $\begin{cases} 2x + y + 4z = 12 \\ x + 2y + 2z = 9. \\ 3x - 3y - 2z = 1 \end{cases}$

Solution We are given the system

$$\begin{array}{ll} \textbf{1.} & \\ \textbf{2.} & \\ \textbf{3.} & \end{array}\begin{cases} 2x + y + 4z = 12 \\ x + 2y + 2z = 9 \\ 3x - 3y - 2z = 1 \end{cases}$$

If we pick Equations 2 and 3 and add them, the variable z is eliminated:

$$\begin{array}{ll} \textbf{2.} & x + 2y + 2z = 9 \\ \textbf{3.} & \underline{3x - 3y - 2z = 1} \\ \textbf{4.} & 4x - y = 10 \end{array}$$

We now pick a different pair of equations (Equations 1 and 3) and eliminate z again. If each side of Equation 3 is multiplied by 2 and the resulting equation is added to Equation 1, z is again eliminated:

$$\begin{array}{ll} \textbf{1.} & 2x + y + 4z = 12 \\ & \underline{6x - 6y - 4z = 2} \\ \textbf{5.} & 8x - 5y = 14 \end{array}$$

Equations 4 and 5 form a system of two equations in two variables:

$$\begin{array}{ll} \textbf{4.} & \\ \textbf{5.} & \end{array}\begin{cases} 4x - y = 10 \\ 8x - 5y = 14 \end{cases}$$

To solve this system, we multiply Equation 4 by -5 and add the resulting equation to Equation 5 to eliminate y:

$$\begin{array}{rrcl} & -20x + 5y &=& -50 \\ \textbf{5.} & 8x - 5y &=& 14 \\ \hline & -12x \quad\quad &=& -36 \\ & x &=& 3 \end{array} \qquad \text{Divide both sides by } -12.$$

To find y, we substitute 3 for x in any equation containing x and y (such as Equation 5) and solve for y:

$$\begin{array}{rrcll} \textbf{5.} & 8x - 5y &=& 14 & \\ & 8(3) - 5y &=& 14 & \text{Substitute 3 for } x. \\ & 24 - 5y &=& 14 & \text{Simplify.} \\ & -5y &=& -10 & \text{Subtract 24 from both sides.} \\ & y &=& 2 & \text{Divide both sides by } -5. \end{array}$$

To find z, we substitute 3 for x and 2 for y in an equation containing x, y, and z (such as Equation 1) and solve for z:

$$\begin{array}{rrcll} \textbf{1.} & 2x + y + 4z &=& 12 & \\ & 2(3) + 2 + 4z &=& 12 & \text{Substitute 3 for } x \text{ and 2 for } y. \\ & 8 + 4z &=& 12 & \text{Simplify.} \\ & 4z &=& 4 & \text{Subtract 8 from both sides.} \\ & z &=& 1 & \text{Divide both sides by 4.} \end{array}$$

The solution of the system is $(x, y, z) = (3, 2, 1)$. Verify that these values satisfy each equation in the original system. ■

EXAMPLE 2 Solve the system $\begin{cases} 2x + y - 3z = -3 \\ 3x - 2y + 4z = 2. \\ 4x + 2y - 6z = -7 \end{cases}$

Solution We are given the system of equations

$$\begin{array}{l} \textbf{1.} \\ \textbf{2.} \\ \textbf{3.} \end{array} \begin{cases} 2x + y - 3z = -3 \\ 3x - 2y + 4z = 2 \\ 4x + 2y - 6z = -7 \end{cases}$$

We can multiply Equation 1 by 2 and add the resulting equation to Equation 2 to eliminate y:

$$\begin{array}{rrcl} & 4x + 2y - 6z &=& -6 \\ \textbf{2.} & 3x - 2y + 4z &=& 2 \\ \hline \textbf{4.} & 7x \quad\quad - 2z &=& -4 \end{array}$$

We now add Equations 2 and 3 to eliminate y again:

$$\begin{array}{lrcl} \textbf{2.} & 3x - 2y + 4z &=& 2 \\ \textbf{3.} & \underline{4x + 2y - 6z} &=& \underline{-7} \\ \textbf{5.} & 7x \qquad - 2z &=& -5 \end{array}$$

Equations 4 and 5 form the system

$$\begin{array}{l} \textbf{4.} \\ \textbf{5.} \end{array} \begin{cases} 7x - 2z = -4 \\ 7x - 2z = -5 \end{cases}$$

Since $7x - 2z$ cannot equal both -4 and -5, this system is inconsistent. Thus, it has no solution. ■

Problem Solving

EXAMPLE 3 **Integer problem** The sum of three integers is 2. The third integer is 2 greater than the second and 17 greater than the first. Find the integers.

Solution Let a, b, and c represent the integers. Because their sum is 2, we know that

$$a + b + c = 2$$

Because the third integer is 2 greater than the second and 17 greater than the first, we know that

$$c - b = 2$$
$$c - a = 17$$

Put these three equations together to form a system of three equations in three variables:

$$\begin{array}{l} \textbf{1.} \\ \textbf{2.} \\ \textbf{3.} \end{array} \begin{cases} a + b + c = 2 \\ \quad\;\; - b + c = 2 \\ -a \qquad + c = 17 \end{cases}$$

We add Equations 1 and 2 to get Equation 4:

$$\textbf{4.} \quad a + 2c = 4$$

Because Equations 3 and 4 have only the variables a and c, Equations 3 and 4 form a system of two equations in two variables.

$$\begin{array}{l} \textbf{3.} \\ \textbf{4.} \end{array} \begin{cases} -a + \;\; c = 17 \\ \;\;\, a + 2c = 4 \end{cases}$$

We add Equations 3 and 4 to get the equation

$$3c = 21$$
$$c = 7$$

We substitute 7 for c in Equation 4 to find a:

4. $a + 2c = 4$

$a + 2(7) = 4$

$a + 14 = 4$ Simplify.

$a = -10$ Add -14 to both sides.

We substitute 7 for c in Equation 2 to find b:

2. $-b + c = 2$

$-b + 7 = 2$

$-b = -5$ Subtract 7 from both sides.

$b = 5$ Divide both sides by -1.

Thus, the integers are -10, 5, and 7. These integers have a sum of 2; 7 is 2 greater than 5; and 7 is 17 greater than -10. ■

EXAMPLE 4 **Manufacturing** A company manufactures three types of hammers—good, better, and best. The cost of manufacturing each type is \$4, \$6, and \$7, respectively, and the hammers sell for \$6, \$9, and \$12. Each day, the cost for manufacturing 100 hammers is \$520, and the daily revenue from their sales is \$810. How many of each type are manufactured?

Solution We can let x represent the number of good hammers manufactured, y represent the number of better hammers manufactured, and z represent the number of best hammers manufactured. Then

The number of good hammers	+	the number of better hammers	+	the number of best hammers	=	the total number of hammers.
x	+	y	+	z	=	100

The cost of good hammers	+	the cost of better hammers	+	the cost of best hammers	=	the total cost.
$4x$	+	$6y$	+	$7z$	=	520

The revenue from good hammers	+	the revenue from better hammers	+	the revenue from best hammers	=	the total revenue.
$6x$	+	$9y$	+	$12z$	=	810

These three equations give the system

$$\begin{array}{l} \textbf{1.} \\ \textbf{2.} \\ \textbf{3.} \end{array} \begin{cases} x + y + z = 100 \\ 4x + 6y + 7z = 520 \\ 6x + 9y + 12z = 810 \end{cases}$$

If we multiply Equation 1 by -7 and add the result to Equation 2, we get

$$\begin{aligned} -7x - 7y - 7z &= -700 \\ 4x + 6y + 7z &= 520 \\ \hline \textbf{4.} \quad -3x - y \quad &= -180 \end{aligned}$$

We can eliminate z again if we multiply Equation 1 by -12 and add the result to Equation 3.

$$\begin{aligned} -12x - 12y - 12z &= -1200 \\ 6x + 9y + 12z &= 810 \\ \hline \textbf{5.} \quad -6x - 3y \quad &= -390 \end{aligned}$$

If we multiply Equation 4 by -3 and add it to Equation 5, we get

$$\begin{aligned} 9x + 3y &= 540 \\ -6x - 3y &= -390 \\ \hline 3x \quad &= 150 \\ x &= 50 \quad \text{Divide both sides by 3.} \end{aligned}$$

To find y, we substitute 50 for x in Equation 4:

$$\begin{aligned} \textbf{4.} \quad -3x - y &= -180 \\ -3(\mathbf{50}) - y &= -180 \quad \text{Substitute 50 for } x. \\ -150 - y &= -180 \\ -y &= -30 \quad \text{Add 150 to both sides.} \\ y &= 30 \quad \text{Divide both sides by } -1. \end{aligned}$$

To find z, we substitute 50 for x and 30 for y in Equation 1:

$$\begin{aligned} \textbf{1.} \quad x + y + z &= 100 \\ \mathbf{50} + \mathbf{30} + z &= 100 \\ z &= 20 \quad \text{Subtract 80 from both sides.} \end{aligned}$$

The company manufactures 50 good hammers, 30 better hammers, and 20 best hammers each day. ■

ORALS *Is the given triple a solution of the system?*

1. (1, 1, 1), $\begin{cases} 2x + y - 3z = 0 \\ 3x - 2y + 4z = 5 \\ 4x + 2y - 6z = 0 \end{cases}$ yes

2. (2, 0, 1), $\begin{cases} 3x + 2y - z = 5 \\ 2x - 3y + 2z = 4 \\ 4x - 2y + 3z = 10 \end{cases}$ no

EXERCISE 9.3

In Exercises 1–12, solve each system.

1. $\begin{cases} x + y + z = 4 \\ 2x + y - z = 1 \\ 2x - 3y + z = 1 \end{cases}$ (1, 1, 2)

2. $\begin{cases} x + y + z = 4 \\ x - y + z = 2 \\ x - y - z = 0 \end{cases}$ (2, 1, 1)

3. $\begin{cases} 2x + 2y + 3z = 10 \\ 3x + y - z = 0 \\ x + y + 2z = 6 \end{cases}$ (0, 2, 2)

4. $\begin{cases} x - y + z = 4 \\ x + 2y - z = -1 \\ x + y - 3z = -2 \end{cases}$ (2, −1, 1)

5. $\begin{cases} x + y + 2z = 7 \\ x + 2y + z = 8 \\ 2x + y + z = 9 \end{cases}$ (3, 2, 1)

6. $\begin{cases} x + 2y + 2z = 10 \\ 2x + y + 2z = 9 \\ 2x + 2y + z = 11 \end{cases}$ (2, 3, 1)

7. $\begin{cases} 2x + y - z = 1 \\ x + 2y + 2z = 2 \\ 4x + 5y + 3z = 3 \end{cases}$ no solution

8. $\begin{cases} 4x + 3z = 4 \\ 2y - 6z = -1 \\ 8x + 4y + 3z = 9 \end{cases}$ $\left(\frac{3}{4}, \frac{1}{2}, \frac{1}{3}\right)$

9. $\begin{cases} 2x + 3y + 4z - 6 = 0 \\ 2x - 3y - 4z + 4 = 0 \\ 4x + 6y + 8z - 12 = 0 \end{cases}$ dependent equations

10. $\begin{cases} x - 3y + 4z - 2 = 0 \\ 2x + y + 2z - 3 = 0 \\ 4x - 5y + 10z - 7 = 0 \end{cases}$ dependent equations

11. $\begin{cases} x + \frac{1}{3}y + z = 13 \\ \frac{1}{2}x - y + \frac{1}{3}z = -2 \\ x + \frac{1}{2}y - \frac{1}{3}z = 2 \end{cases}$ (2, 6, 9)

12. $\begin{cases} x - \frac{1}{5}y - z = 9 \\ \frac{1}{4}x + \frac{1}{5}y - \frac{1}{2}z = 5 \\ 2x + y + \frac{1}{6}z = 12 \end{cases}$ (4, 5, −6)

In Exercises 13–24, solve each problem.

13. **Integer problem** The sum of three integers is 18. The third integer is four times the second, and the second integer is 6 more than the first. Find the integers. −2, 4, 16

14. **Integer problem** The sum of three integers is 48. If the first integer is doubled, the sum is 60. If the second integer is doubled, the sum is 63. Find the integers. 12, 15, 21

15. **Geometry problem** The sum of the angles in any triangle is 180°. In triangle *ABC*, angle *A* is 100° less that the sum of angles *B* and *C*, and angle *C* is 40° less than twice angle *B*. Find each angle.
$A = 40°, B = 60°, C = 80°$

16. **Geometry problem** The sum of the angles of any four-sided figure is 360°. In the quadrilateral shown in Illustration 1, angle A = angle B, angle C is 20° greater than angle A, and angle $D = 40°$. Find the angles. 100°, 100°, 120°, 40°

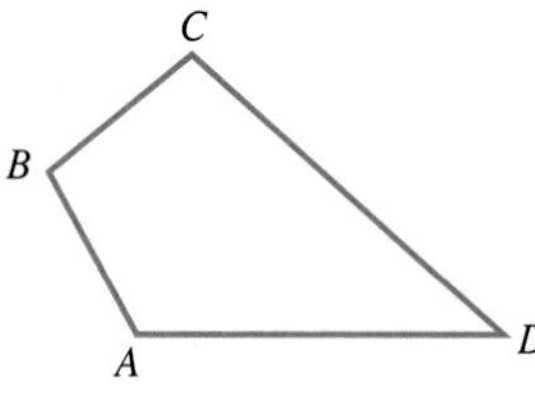

ILLUSTRATION 1

17. **Nutritional planning** A unit of food contains 1 gram of fat, 1 gram of carbohydrate, and 2 grams of protein. A second contains 2 grams of fat, 1 gram of carbohydrate, and 1 gram of protein. A third contains 2 grams of fat, 1 gram of carbohydrate, and 2 grams of protein. How many units of each must be used to provide exactly 11 grams of fat, 6 grams of carbohydrate, and 10 grams of protein?
1 unit, 2 units, 3 units

18. **Nutritional planning** A unit of food contains 2 grams of fat, 1 gram of carbohydrate, and 2 grams of protein. A second contains 3 grams of fat, 2 grams of carbohydrate, and 1 gram of protein. A third contains 1 gram of fat, 1 gram of carbohydrate, and 2 grams of protein. How many units of each must be used to provide exactly 14 grams of fat, 9 grams of carbohydrate, and 9 grams of protein?
2 units, 3 units, 1 unit

19. **Making statues** An artist makes three types of ceramic statues at a monthly cost of $650 for 180 statues. The manufacturing costs for the three types are $5, $4, and $3. If the statues sell for $20, $12, and $9, respectively, how many of each type should be made to produce $2100 in monthly revenue?
30 expensive, 50 middle-priced, 100 inexpensive

20. **Manufacturing footballs** A factory manufactures three types of footballs at a monthly cost of $2425 for 1125 footballs. The manufacturing costs for the three types of footballs are $4, $3, and $2. These footballs sell for $16, $12, and $10, respectively. How many of each type are manufactured if the monthly profit is $9275? (*Hint:* Profit = income − cost.) 50 expensive, 75 middle-priced, 1000 cheap

21. **Concert tickets** Tickets for a concert cost $5, $3, and $2. Twice as many $5 tickets were sold as $2 tickets. The receipts for 750 tickets were $2625. How many of each price ticket were sold?
250 $5 tickets, 375 $3 tickets, 125 $2 tickets

22. **Mixing nuts** The owner of a candy store wants to mix some peanuts worth $3 per pound, some cashews worth $9 per pound, and some Brazil nuts worth $9 per pound to get 50 pounds of a mixture that will sell for $6 per pound. She used 15 fewer pounds of cashews than peanuts. How many pounds of each did she use? 25 lb peanuts, 10 lb cashews, 15 lb Brazil nuts

23. **Chainsaw sculpting** A northwoods sculptor carves three types of statues with a chainsaw. The time required for carving, sanding, and painting a totem pole, a bear, and a deer are shown in the table. How many of each should be produced to use all available labor hours? 3 poles, 2 bears, 4 deer

	Totem pole	Bear	Deer	Time available
Carving	2 hours	2 hours	1 hour	14 hours
Sanding	1 hour	2 hours	2 hours	15 hours
Painting	3 hours	2 hours	2 hours	21 hours

24. **Making clothing** A clothing manufacturer makes coats, shirts, and slacks. The time required for cutting, sewing, and packaging each item are shown in the table. How many of each should be made to use all available labor hours? 120 coats, 200 shirts, 150 slacks

	Coat	Shirt	Slacks	Time available
Cutting	20 minutes	15 minutes	10 minutes	115 hours
Sewing	60 minutes	30 minutes	24 minutes	280 hours
Packaging	5 minutes	12 minutes	6 minutes	65 hours

Writing Exercises ■ *Write a paragraph using your own words.*

1. What makes a system of three equations in three variables inconsistent?

2. What makes the equations of a system of three equations in three variables dependent?

Something to Think About

1. Solve the system

$$\begin{cases} x + y + z + w = 3 \\ x - y - z - w = -1 \\ x + y - z - w = 1 \\ x + y - z + w = 3 \end{cases} \quad (1, 1, 0, 1)$$

2. Solve the system

$$\begin{cases} 2x + y + z + w = 3 \\ x - 2y - z + w = -3 \\ x - y - 2z - w = -3 \\ x + y - z + 2w = 4 \end{cases} \quad (0, 2, 0, 1)$$

Review Exercises

Solve each equation.

Note: If you teach graphing early, skip these review exercises.

1. $6x^2 + 11x + 4 = 0$ $\quad -\frac{4}{3}, -\frac{1}{2}$

2. $\left|\frac{2x - 4}{5}\right| = 20$ $\quad 52, -48$

3. $|3x - 4| = |12 - x|$ $\quad 4, -4$

4. $\frac{3x - 2}{x - 3} = 5$ $\quad \frac{13}{2}$

9.4 Determinants

■ Determinants ■ Cramer's Rule

Another method of solving systems of equations involves a rectangular array of numbers called a **matrix** and a related concept called a **determinant.**

Matrix A matrix is any rectangular array of numbers.

Some examples of matrices are

$$A = \begin{bmatrix} 1 & 2 & 3 \\ 4 & 5 & 6 \end{bmatrix} \quad B = \begin{bmatrix} 1 & 2 \\ 3 & 4 \\ 5 & 6 \end{bmatrix} \quad C = \begin{bmatrix} 2 & 4 & 6 \\ 8 & 10 & 12 \\ 14 & 16 & 18 \end{bmatrix}$$

The numbers in each matrix are called its **elements.** Because matrix A has two rows and three columns, it is called a 2×3 matrix (read "2 by 3" matrix). Matrix B is a 3×2 matrix, because the matrix has three rows and two columns. Matrix C is a 3×3 matrix (three rows and three columns).

Any matrix with the same number of rows and columns is called a **square matrix.** Matrix C is an example of a square matrix.

Determinants

A determinant is a number that is associated with a square matrix. For any square matrix A, the symbol $|A|$ represents the determinant of A.

Value of a 2 × 2 Determinant If a, b, c, and d are numbers, the determinant of $A = \begin{bmatrix} a & b \\ c & d \end{bmatrix}$ is

$$\begin{vmatrix} a & b \\ c & d \end{vmatrix} = ad - bc$$

The determinant of a 2 × 2 matrix is the number that is equal to the product of the numbers on the major diagonal

$$\begin{vmatrix} a & b \\ c & d \end{vmatrix}$$

minus the product of the numbers on the other diagonal

$$\begin{vmatrix} a & b \\ c & d \end{vmatrix}$$

EXAMPLE 1 Evaluate the determinants **a.** $\begin{vmatrix} 3 & 2 \\ 6 & 9 \end{vmatrix}$ and **b.** $\begin{vmatrix} -5 & \frac{1}{2} \\ -1 & 0 \end{vmatrix}$.

Solution **a.**
$$\begin{vmatrix} 3 & 2 \\ 6 & 9 \end{vmatrix} = 3(9) - 2(6) = 27 - 12 = 15$$

b.
$$\begin{vmatrix} -5 & \frac{1}{2} \\ -1 & 0 \end{vmatrix} = -5(0) - \frac{1}{2}(-1) = 0 + \frac{1}{2} = \frac{1}{2}$$

■

A 3 × 3 determinant is evaluated by expanding by **minors.**

Value of a 3 × 3 Determinant

$$\begin{vmatrix} a_1 & b_1 & c_1 \\ a_2 & b_2 & c_2 \\ a_3 & b_3 & c_3 \end{vmatrix} = a_1 \underbrace{\begin{vmatrix} b_2 & c_2 \\ b_3 & c_3 \end{vmatrix}}_{\text{Minor of } a_1} - b_1 \underbrace{\begin{vmatrix} a_2 & c_2 \\ a_3 & c_3 \end{vmatrix}}_{\text{Minor of } b_1} + c_1 \underbrace{\begin{vmatrix} a_2 & b_2 \\ a_3 & b_3 \end{vmatrix}}_{\text{Minor of } c_1}$$

To find the minor of a_1, we cross out the elements of the determinant that are in the same row and column as a_1:

$$\begin{vmatrix} a_1 & b_1 & c_1 \\ a_2 & b_2 & c_2 \\ a_3 & b_3 & c_3 \end{vmatrix}$$

The minor of a_1 is $\begin{vmatrix} b_2 & c_2 \\ b_3 & c_3 \end{vmatrix}$.

To find the minor of b_1, we cross out the elements of the determinant that are in the same row and column as b_1:

$$\begin{vmatrix} \not{a_1} & \not{b_1} & \not{c_1} \\ a_2 & \not{b_2} & c_2 \\ a_3 & \not{b_3} & c_3 \end{vmatrix} \qquad \text{The minor of } b_1 \text{ is } \begin{vmatrix} a_2 & c_2 \\ a_3 & c_3 \end{vmatrix}.$$

To find the minor of c_1, we cross out the elements of the determinant that are in the same row and column as c_1:

$$\begin{vmatrix} \not{a_1} & \not{b_1} & \not{c_1} \\ a_2 & b_2 & \not{c_2} \\ a_3 & b_3 & \not{c_3} \end{vmatrix} \qquad \text{The minor of } c_1 \text{ is } \begin{vmatrix} a_2 & b_2 \\ a_3 & b_3 \end{vmatrix}.$$

EXAMPLE 2 Evaluate the determinant $\begin{vmatrix} 1 & 3 & -2 \\ 2 & 1 & 3 \\ 1 & 2 & 3 \end{vmatrix}$.

Solution

Minor of 1 ↓ Minor of 3 ↓ Minor of −2 ↓

$$\begin{vmatrix} 1 & 3 & -2 \\ 2 & 1 & 3 \\ 1 & 2 & 3 \end{vmatrix} = 1\begin{vmatrix} 1 & 3 \\ 2 & 3 \end{vmatrix} - 3\begin{vmatrix} 2 & 3 \\ 1 & 3 \end{vmatrix} + (-2)\begin{vmatrix} 2 & 1 \\ 1 & 2 \end{vmatrix}$$

$$= 1(3 - 6) - 3(6 - 3) - 2(4 - 1)$$

$$= -3 - 9 - 6$$

$$= -18$$

■

We can evaluate a 3×3 determinant by expanding it along any row or column. To determine the signs between the terms of the expansion of a 3×3 determinant, we use the following array of signs.

Array of Signs for a 3 × 3 Determinant

$$\begin{matrix} + & - & + \\ - & + & - \\ + & - & + \end{matrix}$$

EXAMPLE 3 Evaluate the determinant $\begin{vmatrix} 1 & 3 & -2 \\ 2 & 1 & 3 \\ 1 & 2 & 3 \end{vmatrix}$ by expanding on the middle column.

Solution This is the determinant of Example 2. To expand it along the middle column, we use the signs of the middle column of the array of signs:

$$\begin{vmatrix} 1 & 3 & -2 \\ 2 & 1 & 3 \\ 1 & 2 & 3 \end{vmatrix} = -3\underbrace{\begin{vmatrix} 2 & 3 \\ 1 & 3 \end{vmatrix}}_{\text{Minor of 3}} + 1\underbrace{\begin{vmatrix} 1 & -2 \\ 1 & 3 \end{vmatrix}}_{\text{Minor of 1}} - 2\underbrace{\begin{vmatrix} 1 & -2 \\ 2 & 3 \end{vmatrix}}_{\text{Minor of 2}}$$

$$= -3(6-3) + 1[3-(-2)] - 2[3-(-4)]$$
$$= -3(3) + 1(5) - 2(7)$$
$$= -9 + 5 - 14$$
$$= -18$$

As expected, we get the same value as in Example 2. ■

Cramer's Rule

Gabriel Cramer (1704–1752)
Although other mathematicians had worked with determinants, it was the work of Cramer that popularized them.

The method of using determinants to solve systems of equations is called **Cramer's rule,** named after the 18th-century mathematician Gabriel Cramer. To develop Cramer's rule, we consider the system

$$\begin{cases} ax + by = e \\ cx + dy = f \end{cases}$$

where x and y are variables and a, b, c, d, e, and f are constants.

If we multiply both sides of the first equation by d and multiply both sides of the second equation by $-b$, we can add the equations and eliminate y:

$$\begin{array}{rl} adx + bdy &= ed \\ -bcx - bdy &= -bf \\ \hline adx - bcx \qquad &= ed - bf \end{array}$$

To solve for x, we factor out x on the left-hand side and divide each side by $ad - bc$:

$$(ad - bc)x = ed - bf$$
$$x = \frac{ed - bf}{ad - bc} \qquad (ad - bc \neq 0)$$

We can find y in a similar manner. After eliminating the variable x, we get

$$y = \frac{af - ec}{ad - bc} \qquad (ad - bc \neq 0)$$

Determinants provide an easy way of remembering these formulas. Note that the denominator for both x and y is

$$\begin{vmatrix} a & b \\ c & d \end{vmatrix} = ad - bc$$

PERSPECTIVE

Hand-Held Matrices

Not many years ago, hand-held calculators supported only simple mathematical operations, and calculations with matrices and determinants required more powerful computers. However, matrices and determinants are used so often in mathematics and science, and the technology has improved so much, that several scientific calculators now support matrix arithmetic.

For example, on the TI-80 series of calculators made by Texas Instruments, the matrix

$$A = \begin{bmatrix} 2 & 3 \\ -4 & 1 \end{bmatrix}$$

is easily entered, row by row, as prompted by the screen in Illustration 1.

```
[A] 2×2
1,1=2
1,2=3
2,1=-4
2,2=1█
```

ILLUSTRATION 1

Another matrix,

$$B = \begin{bmatrix} 3 & -2 & 2 \\ 1 & 0 & -3 \\ 2 & -1 & 1 \end{bmatrix}$$

is entered similarly.

The screens in Illustration 2 and 3 show that the determinant of A is 14, and the determinant of B is 3.

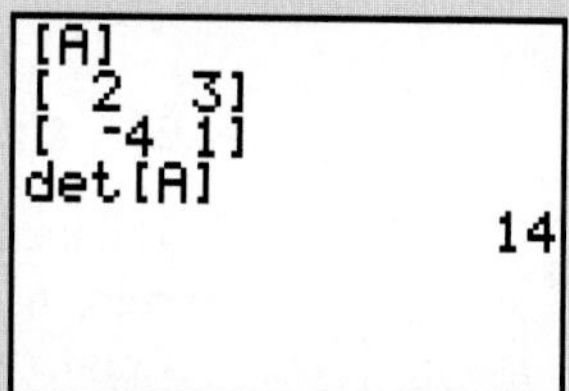

ILLUSTRATION 2

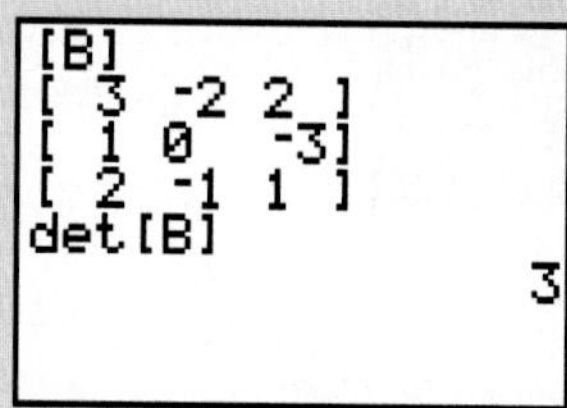

ILLUSTRATION 3

The numerators can be expressed as determinants also:

$$x = \frac{ed - bf}{ad - bc} = \frac{\begin{vmatrix} e & b \\ f & d \end{vmatrix}}{\begin{vmatrix} a & b \\ c & d \end{vmatrix}} \quad \text{and} \quad y = \frac{af - ec}{ad - bc} = \frac{\begin{vmatrix} a & e \\ c & f \end{vmatrix}}{\begin{vmatrix} a & b \\ c & d \end{vmatrix}}$$

If we compare these formulas with the original system

$$\begin{cases} ax + by = e \\ cx + dy = f \end{cases}$$

we note that in the expressions for x and y above, the denominator determinant is formed by using the coefficients a, b, c, and d of the variables in the equations.

The numerator determinants are the same as the denominator determinant, except that the column of coefficients of the variable for which we are solving is replaced with the column of constants e and f.

Cramer's Rule for Two Equations in Two Variables

The solution of the system $\begin{cases} ax + by = e \\ cx + dy = f \end{cases}$ is given by

$$x = \frac{D_x}{D} = \frac{\begin{vmatrix} e & b \\ f & d \end{vmatrix}}{\begin{vmatrix} a & b \\ c & d \end{vmatrix}} \quad \text{and} \quad y = \frac{D_y}{D} = \frac{\begin{vmatrix} a & e \\ c & f \end{vmatrix}}{\begin{vmatrix} a & b \\ c & d \end{vmatrix}}$$

If every determinant is 0, the system is consistent but the equations are dependent.

If $D = 0$ and D_x or D_y is nonzero, the system is inconsistent.

EXAMPLE 4 Use Cramer's rule to solve the system $\begin{cases} 4x - 3y = 6 \\ -2x + 5y = 4 \end{cases}$.

Solution The value of x is the quotient of two determinants. The denominator determinant is made up of the coefficients of x and y:

$$D = \begin{vmatrix} 4 & -3 \\ -2 & 5 \end{vmatrix}$$

To solve for x, we form the numerator determinant from the denominator determinant by replacing its first column (the coefficients of x) with the column of constants (6 and 4).

To solve for y, we form the numerator determinant from the denominator determinant by replacing the second column (the coefficients of y) with the column of constants (6 and 4).

To find the values of x and y, we evaluate each determinant:

$$x = \frac{\begin{vmatrix} 6 & -3 \\ 4 & 5 \end{vmatrix}}{\begin{vmatrix} 4 & -3 \\ -2 & 5 \end{vmatrix}} = \frac{6(5) - (-3)(4)}{4(5) - (-3)(-2)} = \frac{30 + 12}{20 - 6} = \frac{42}{14} = 3$$

$$y = \frac{\begin{vmatrix} 4 & 6 \\ -2 & 4 \end{vmatrix}}{\begin{vmatrix} 4 & -3 \\ -2 & 5 \end{vmatrix}} = \frac{4(4) - 6(-2)}{4(5) - (-3)(-2)} = \frac{16 + 12}{20 - 6} = \frac{28}{14} = 2$$

The solution to this system is (3, 2). Verify that $x = 3$ and $y = 2$ satisfy each equation in the given system. ■

EXAMPLE 5 Use Cramer's rule to solve the system $\begin{cases} 7x = 8 - 4y \\ 2y = 3 - \frac{7}{2}x \end{cases}$.

Solution We multiply both sides of the second equation by 2 to eliminate the fraction and write the system in the form

$$\begin{cases} 7x + 4y = 8 \\ 7x + 4y = 6 \end{cases}$$

When we attempt to use Cramer's rule to solve this system, we find that the determinant in the denominator is 0:

$$\begin{vmatrix} 7 & 4 \\ 7 & 4 \end{vmatrix} = 7(4) - 4(7) = 0$$

Since the two equations are different (independent) and because the determinant in the denominator is 0, the system is inconsistent. It has no solutions. ■

Cramer's Rule for Three Equations in Three Variables

The solution of the system $\begin{cases} ax + by + cz = j \\ dx + ey + fz = k \\ gx + hy + iz = l \end{cases}$ is given by

$$x = \frac{D_x}{D}, \quad y = \frac{D_y}{D}, \quad \text{and} \quad z = \frac{D_z}{D}$$

where

$$D = \begin{vmatrix} a & b & c \\ d & e & f \\ g & h & i \end{vmatrix} \qquad D_x = \begin{vmatrix} j & b & c \\ k & e & f \\ l & h & i \end{vmatrix}$$

$$D_y = \begin{vmatrix} a & j & c \\ d & k & f \\ g & l & i \end{vmatrix} \qquad D_z = \begin{vmatrix} a & b & j \\ d & e & k \\ g & h & l \end{vmatrix}$$

If every determinant is 0, the system is consistent but the equations are dependent.

If $D = 0$ and D_x or D_y or D_z is nonzero, the system is inconsistent.

EXAMPLE 6 Use Cramer's rule to solve the system $\begin{cases} 2x + y + 4z = 12 \\ x + 2y + 2z = 9 \\ 3x - 3y - 2z = 1 \end{cases}$.

Solution The denominator determinant is the determinant formed by the coefficients of the variables. The numerator determinants are formed by replacing the coefficients of the variable being solved for by the column of constants. We form the quotients for x, y, and z and evaluate the determinants:

$$x = \frac{\begin{vmatrix} 12 & 1 & 4 \\ 9 & 2 & 2 \\ 1 & -3 & -2 \end{vmatrix}}{\begin{vmatrix} 2 & 1 & 4 \\ 1 & 2 & 2 \\ 3 & -3 & -2 \end{vmatrix}} = \frac{12\begin{vmatrix} 2 & 2 \\ -3 & -2 \end{vmatrix} - 1\begin{vmatrix} 9 & 2 \\ 1 & -2 \end{vmatrix} + 4\begin{vmatrix} 9 & 2 \\ 1 & -3 \end{vmatrix}}{2\begin{vmatrix} 2 & 2 \\ -3 & -2 \end{vmatrix} - 1\begin{vmatrix} 1 & 2 \\ 3 & -2 \end{vmatrix} + 4\begin{vmatrix} 1 & 2 \\ 3 & -3 \end{vmatrix}} = \frac{12(2) - (-20) + 4(-29)}{2(2) - (-8) + 4(-9)} = \frac{-72}{-24} = 3$$

$$y = \frac{\begin{vmatrix} 2 & 12 & 4 \\ 1 & 9 & 2 \\ 3 & 1 & -2 \end{vmatrix}}{\begin{vmatrix} 2 & 1 & 4 \\ 1 & 2 & 2 \\ 3 & -3 & -2 \end{vmatrix}} = \frac{2\begin{vmatrix} 9 & 2 \\ 1 & -2 \end{vmatrix} - 12\begin{vmatrix} 1 & 2 \\ 3 & -2 \end{vmatrix} + 4\begin{vmatrix} 1 & 9 \\ 3 & 1 \end{vmatrix}}{-24} = \frac{2(-20) - 12(-8) + 4(-26)}{-24} = \frac{-48}{-24} = 2$$

$$z = \frac{\begin{vmatrix} 2 & 1 & 12 \\ 1 & 2 & 9 \\ 3 & -3 & 1 \end{vmatrix}}{\begin{vmatrix} 2 & 1 & 4 \\ 1 & 2 & 2 \\ 3 & -3 & -2 \end{vmatrix}} = \frac{2\begin{vmatrix} 2 & 9 \\ -3 & 1 \end{vmatrix} - 1\begin{vmatrix} 1 & 9 \\ 3 & 1 \end{vmatrix} + 12\begin{vmatrix} 1 & 2 \\ 3 & -3 \end{vmatrix}}{-24} = \frac{2(29) - (-26) + 12(-9)}{-24} = \frac{-24}{-24} = 1$$

The solution to this system is (3, 2, 1). ■

ORALS *Evaluate each determinant.*

1. $\begin{vmatrix} 2 & 1 \\ 1 & 1 \end{vmatrix}$ 1 **2.** $\begin{vmatrix} 0 & 2 \\ 1 & 1 \end{vmatrix}$ −2 **3.** $\begin{vmatrix} 0 & 1 \\ 0 & 1 \end{vmatrix}$ 0

When using Cramer's rule to solve the system $\begin{cases} x + 2y = 5 \\ 2x - y = 4 \end{cases}$,

4. Find the denominator determinant for x. $\begin{vmatrix} 1 & 2 \\ 2 & -1 \end{vmatrix}$

5. Find the numerator determinant for x. $\begin{vmatrix} 5 & 2 \\ 4 & -1 \end{vmatrix}$

6. Find the numerator determinant for y. $\begin{vmatrix} 1 & 5 \\ 2 & 4 \end{vmatrix}$

EXERCISE 9.4

In Exercises 1–18, evaluate each determinant.

1. $\begin{vmatrix} 2 & 3 \\ -2 & 1 \end{vmatrix}$ 8

2. $\begin{vmatrix} 3 & -2 \\ -2 & 4 \end{vmatrix}$ 8

3. $\begin{vmatrix} -1 & 2 \\ 3 & -4 \end{vmatrix}$ −2

4. $\begin{vmatrix} -1 & -2 \\ -3 & -4 \end{vmatrix}$ -2

5. $\begin{vmatrix} x & y \\ y & x \end{vmatrix}$ $x^2 - y^2$

6. $\begin{vmatrix} x+y & x-y \\ x-y & x+y \end{vmatrix}$ $4xy$

7. $\begin{vmatrix} 1 & 0 & 1 \\ 0 & 1 & 0 \\ 1 & 1 & 1 \end{vmatrix}$ 0

8. $\begin{vmatrix} 1 & 2 & 0 \\ 0 & 1 & 2 \\ 0 & 0 & 1 \end{vmatrix}$ 1

9. $\begin{vmatrix} -1 & 2 & 1 \\ 2 & 1 & -3 \\ 1 & 1 & 1 \end{vmatrix}$ -13

10. $\begin{vmatrix} 1 & 2 & 3 \\ 1 & 2 & 3 \\ 1 & 2 & 3 \end{vmatrix}$ 0

11. $\begin{vmatrix} 1 & -2 & 3 \\ -2 & 1 & 1 \\ -3 & -2 & 1 \end{vmatrix}$ 26

12. $\begin{vmatrix} 1 & 1 & 2 \\ 2 & 1 & -2 \\ 3 & 1 & 3 \end{vmatrix}$ -9

13. $\begin{vmatrix} 1 & 2 & 3 \\ 4 & 5 & 6 \\ 7 & 8 & 9 \end{vmatrix}$ 0

14. $\begin{vmatrix} 1 & 4 & 7 \\ 2 & 5 & 8 \\ 3 & 6 & 9 \end{vmatrix}$ 0

15. $\begin{vmatrix} a & 2a & -a \\ 2 & -1 & 3 \\ 1 & 2 & -3 \end{vmatrix}$ $10a$

16. $\begin{vmatrix} 1 & 2b & -3 \\ 2 & -b & 2 \\ 1 & 3b & 1 \end{vmatrix}$ $-28b$

17. $\begin{vmatrix} 1 & a & b \\ 1 & 2a & 2b \\ 1 & 3a & 3b \end{vmatrix}$ 0

18. $\begin{vmatrix} a & b & c \\ 0 & b & c \\ 0 & 0 & c \end{vmatrix}$ abc

In Exercises 19–44, use Cramer's rule to solve each system of equations, if possible.

19. $\begin{cases} x + y = 6 \\ x - y = 2 \end{cases}$ $(4, 2)$

20. $\begin{cases} x - y = 4 \\ 2x + y = 5 \end{cases}$ $(3, -1)$

21. $\begin{cases} 2x + y = 1 \\ x - 2y = -7 \end{cases}$ $(-1, 3)$

22. $\begin{cases} 3x - y = -3 \\ 2x + y = -7 \end{cases}$ $(-2, -3)$

23. $\begin{cases} 2x + 3y = 0 \\ 4x - 6y = -4 \end{cases}$ $\left(-\frac{1}{2}, \frac{1}{3}\right)$

24. $\begin{cases} 4x - 3y = -1 \\ 8x + 3y = 4 \end{cases}$ $\left(\frac{1}{4}, \frac{2}{3}\right)$

25. $\begin{cases} y = \dfrac{-2x + 1}{3} \\ 3x - 2y = 8 \end{cases}$ $(2, -1)$

26. $\begin{cases} 2x + 3y = -1 \\ x = \dfrac{y - 9}{4} \end{cases}$ $(-2, 1)$

27. $\begin{cases} y = \dfrac{11 - 3x}{2} \\ x = \dfrac{11 - 4y}{6} \end{cases}$ no solution

28. $\begin{cases} x = \dfrac{12 - 6y}{5} \\ y = \dfrac{24 - 10x}{12} \end{cases}$ dependent equations

29. $\begin{cases} x = \dfrac{5y - 4}{2} \\ y = \dfrac{3x - 1}{5} \end{cases}$ $\left(5, \frac{14}{5}\right)$

30. $\begin{cases} y = \dfrac{1 - 5x}{2} \\ x = \dfrac{3y + 10}{4} \end{cases}$ $(1, -2)$

31. $\begin{cases} x + y + z = 4 \\ x + y - z = 0 \\ x - y + z = 2 \end{cases}$ $(1, 1, 2)$

32. $\begin{cases} x + y + z = 4 \\ x - y + z = 2 \\ x - y - z = 0 \end{cases}$ $(2, 1, 1)$

33. $\begin{cases} x + y + 2z = 7 \\ x + 2y + z = 8 \\ 2x + y + z = 9 \end{cases}$ $(3, 2, 1)$

34. $\begin{cases} x + 2y + 2z = 10 \\ 2x + y + 2z = 9 \\ 2x + 2y + z = 1 \end{cases}$ $(-2, -1, 7)$

35. $\begin{cases} 2x + y - z = 1 \\ x + 2y + 2z = 2 \\ 4x + 5y + 3z = 3 \end{cases}$ no solution

36. $\begin{cases} 4x + 3z = 4 \\ 2y - 6z = -1 \\ 8x + 4y + 3z = 9 \end{cases}$ $\left(\frac{3}{4}, \frac{1}{2}, \frac{1}{3}\right)$

37. $\begin{cases} 2x + y + z = 5 \\ x - 2y + 3z = 10 \\ x + y - 4z = -3 \end{cases}$ $(3, -2, 1)$

38. $\begin{cases} 3x + 2y - z = -8 \\ 2x - y + 7z = 10 \\ 2x + 2y - 3z = -10 \end{cases}$ $(-2, 0, 2)$

39. $\begin{cases} 2x + 3y + 4z = 6 \\ 2x - 3y - 4z = -4 \\ 4x + 6y + 8z = 12 \end{cases}$ dependent equations

40. $\begin{cases} x - 3y + 4z - 2 = 0 \\ 2x + y + 2z - 3 = 0 \\ 4x - 5y + 10z - 7 = 0 \end{cases}$
dependent equations

41. $\begin{cases} x + y = 1 \\ \frac{1}{2}y + z = \frac{5}{2} \\ x - z = -3 \end{cases}$
$(-2, 3, 1)$

42. $\begin{cases} 3x + 4y + 14z = 7 \\ -\frac{1}{2}x - y + 2z = \frac{3}{2} \\ x + \frac{3}{2}y + \frac{5}{2}z = 1 \end{cases}$
dependent equations

43. $\begin{cases} 2x - y + 4z + 2 = 0 \\ 5x + 8y + 7z = -8 \\ x + 3y + z + 3 = 0 \end{cases}$ no solution

44. $\begin{cases} x + 2y + 2z + 3 = 0 \\ x + \frac{1}{2}y + z - \frac{1}{2} = 0 \\ 2x + 2y + z + 1 = 0 \end{cases}$ $\left(\frac{9}{5}, -\frac{11}{5}, -\frac{1}{5}\right)$

In Exercises 45–48, evaluate each determinant and solve the resulting equation.

45. $\begin{vmatrix} x & 1 \\ 3 & 2 \end{vmatrix} = 1$ 2

46. $\begin{vmatrix} x & -x \\ 2 & -3 \end{vmatrix} = -5$ 5

47. $\begin{vmatrix} x & -2 \\ 3 & 1 \end{vmatrix} = \begin{vmatrix} 4 & 2 \\ x & 3 \end{vmatrix}$ 2

48. $\begin{vmatrix} x & 3 \\ x & 2 \end{vmatrix} = \begin{vmatrix} 3 & 2 \\ 1 & 1 \end{vmatrix}$ -1

Writing Exercises ■ *Write a paragraph using your own words.*

1. Tell how to find the minor of a determinant.
2. Tell how to find x when using Cramer's rule.

Something to Think About ■

1. Evaluate
$$\begin{vmatrix} 1 & 3 & 5 \\ 2 & 1 & 5 \\ 3 & 0 & 9 \end{vmatrix} + \begin{vmatrix} 1 & 3 & 5 \\ 3 & 0 & 9 \\ 2 & 1 & 5 \end{vmatrix}$$ 0
(*Hint:* You don't need to evaluate either determinant.)

2. If $\begin{vmatrix} a & b & c \\ 1 & 3 & 7 \\ 2 & 8 & 9 \end{vmatrix} = \begin{vmatrix} 1 & 3 & 7 \\ a & b & c \\ 2 & 8 & 9 \end{vmatrix}$,
find the value of $\begin{vmatrix} 2 & 8 & 9 \\ 1 & 3 & 7 \\ a & b & c \end{vmatrix}$. 0

Determinants with more than 3 rows and 3 columns can be evaluated by expanding them by minors. The sign array for a 4 × 4 determinant is

$$\begin{matrix} + & - & + & - \\ - & + & - & + \\ + & - & + & - \\ - & + & - & + \end{matrix}$$

Evaluate each determinant.

3. $\begin{vmatrix} 1 & 0 & 2 & 1 \\ 2 & 1 & 1 & 3 \\ 1 & 1 & 1 & 1 \\ 2 & 1 & 1 & 1 \end{vmatrix}$ -4

4. $\begin{vmatrix} 1 & 2 & -1 & 1 \\ -2 & 1 & 3 & -1 \\ 0 & 1 & 1 & 2 \\ 2 & 0 & 3 & 1 \end{vmatrix}$ -53

Review Exercises ■ *Simplify each fraction.*

Note: If you teach graphing early, skip these review exercises.

1. $\dfrac{x^2 - 4}{x^2 - 4x + 4}$ $\quad \dfrac{x + 2}{x - 2}$

2. $\dfrac{x^3 + y^3}{x^2 - y^2}$ $\quad \dfrac{x^2 - xy + y^2}{x - y}$

3. $\dfrac{2m^2 + 3mn - 5n^2}{m^3 - n^3}$ $\quad \dfrac{2m + 5n}{m^2 + mn + n^2}$

4. $\dfrac{2ac - 2ad + 2bc - 2bd}{4a^2 + 6ab + 2b^2}$ $\quad \dfrac{c - d}{2a + b}$

9.5 Matrices

■ Gaussian Elimination ■ Systems with More Equations Than Variables
■ Systems with More Variables Than Equations

Arthur Cayley (1821–1895)
Cayley taught mathematics at Cambridge University. When he refused to take religious vows, he was fired and became a lawyer. After 14 years, he returned to mathematics and to Cambridge. Cayley was a major force in developing the theory of matrices.

To show how to use matrices to solve systems of linear equations, we consider the following system

$$\begin{cases} x - 2y - z = 6 \\ 2x + 2y - z = 1 \\ -x - y + 2z = 1 \end{cases}$$

which can be represented by the following matrix, called an **augmented matrix:**

$$\left[\begin{array}{rrr|r} 1 & -2 & -1 & 6 \\ 2 & 2 & -1 & 1 \\ -1 & -1 & 2 & 1 \end{array}\right]$$

The 3×3 matrix to the left of the dashed line, called the **coefficient matrix,** is determined by the coefficients of x, y, and z in the equations of the system. The 3×1 matrix to the right of the dashed line is determined by the constants in the equations. Each row of the augmented matrix represents exactly one equation of the system:

$$\left[\begin{array}{rrr|r} 1 & -2 & -1 & 6 \\ 2 & 2 & -1 & 1 \\ -1 & -1 & 2 & 1 \end{array}\right] \begin{array}{c} \leftrightarrow \\ \leftrightarrow \\ \leftrightarrow \end{array} \begin{cases} x - 2y - z = 6 \\ 2x + 2y - z = 1 \\ -x - y + 2z = 1 \end{cases}$$

■ Gaussian Elimination

To solve a system by **Gaussian elimination,** we transform the augmented matrix into the following matrix that has all 0s below its main diagonal, formed by elements a, e, and h.

$$\left[\begin{array}{ccc|c} a & b & c & d \\ 0 & e & f & g \\ 0 & 0 & h & i \end{array}\right] \qquad (a, b, c, \ldots, i \text{ are real numbers.})$$

We can often write a matrix in this form, called **triangular form,** by using three operations called **elementary row operations.**

Elementary Row Operations

1. Any two rows of a matrix can be interchanged.
2. Any row of a matrix can be multiplied by a nonzero constant.
3. Any row of a matrix can be changed by adding a constant multiple of another row to it.

- A type 1 row operation corresponds to interchanging two equations of the system.
- A type 2 row operation corresponds to multiplying both sides of an equation by a nonzero constant.
- A type 3 row operation corresponds to adding a multiple of one equation to another.

None of these operations will change the solution of the given system of equations.

After we have written the matrix in triangular form, we can solve the corresponding system of equations by a substitution process, as shown in Example 1.

EXAMPLE 1 Solve the system $\begin{cases} x - 2y - z = 6 \\ 2x + 2y - z = 1. \\ -x - y + 2z = 1 \end{cases}$

Solution We can represent the system with the following augmented matrix:

$$\left[\begin{array}{ccc|c} 1 & -2 & -1 & 6 \\ 2 & 2 & -1 & 1 \\ -1 & -1 & 2 & 1 \end{array}\right]$$

To get 0's under the 1 in the first column, we use a type 3 row operation twice:

Multiply row 1 by -2 and add to row 2. Multiply row 1 by 1 and add to row 3.

$$\left[\begin{array}{ccc|c} 1 & -2 & -1 & 6 \\ 2 & 2 & -1 & 1 \\ -1 & -1 & 2 & 1 \end{array}\right] \approx \left[\begin{array}{ccc|c} 1 & -2 & -1 & 6 \\ 0 & 6 & 1 & -11 \\ -1 & -1 & 2 & 1 \end{array}\right] \approx \left[\begin{array}{ccc|c} 1 & -2 & -1 & 6 \\ 0 & 6 & 1 & -11 \\ 0 & -3 & 1 & 7 \end{array}\right]$$

The symbol "$\approx$" is read as "is row equivalent to." Each of the above matrices represents a system of equations, and they are all equivalent.

To get a 0 under the 6 in the second column of the last matrix, we use another type 3 row operation:

Multiply row 2 by $\frac{1}{2}$ and add to row 3.

$$\left[\begin{array}{rrr|r} 1 & -2 & -1 & 6 \\ 0 & 6 & 1 & -11 \\ 0 & -3 & 1 & 7 \end{array}\right] \approx \left[\begin{array}{rrr|r} 1 & -2 & -1 & 6 \\ 0 & 6 & 1 & -11 \\ 0 & 0 & \frac{3}{2} & \frac{3}{2} \end{array}\right]$$

Finally, we use a type 2 row operation:

Multiply row 3 by $\frac{2}{3}$.

$$\left[\begin{array}{rrr|r} 1 & -2 & -1 & 6 \\ 0 & 6 & 1 & -11 \\ 0 & 0 & \frac{3}{2} & \frac{3}{2} \end{array}\right] \approx \left[\begin{array}{rrr|r} 1 & -2 & -1 & 6 \\ 0 & 6 & 1 & -11 \\ 0 & 0 & 1 & 1 \end{array}\right]$$

The final matrix represents the system of equations

$$\begin{array}{l} \textbf{1.} \\ \textbf{2.} \\ \textbf{3.} \end{array} \begin{cases} x - 2y - z = 6 \\ 0x + 6y + z = -11 \\ 0x + 0y + z = 1 \end{cases}$$

From Equation 3, we can read that $z = 1$. To find y, we substitute 1 for z in Equation 2 and solve for y:

2. $6y + z = -11$

$6y + 1 = -11$ Substitute 1 for z.

$6y = -12$ Subtract 1 from both sides.

$y = -2$ Divide both sides by 6.

Thus, $y = -2$. To find x, we substitute 1 for z and -2 for y in Equation 1 and solve for x:

1. $x - 2y - z = 6$

$x - 2(-2) - 1 = 6$ Substitute 1 for z and -2 for y.

$x + 4 - 1 = 6$ $(-2)(-2) = +4$.

$x + 3 = 6$ Simplify.

$x = 3$ Subtract 3 from both sides.

Thus, $x = 3$. The solution to the given system is $(3, -2, 1)$. Verify that this triple satisfies each equation of the original system. ■

Systems with More Equations Than Variables

We can use matrices to solve systems with more equations than variables.

EXAMPLE 2 Solve the system $\begin{cases} x + y = -1 \\ 2x - y = 7. \\ -x + 2y = -8 \end{cases}$

Solution This system, with 3 equations and 2 variables, can be represented by a 3×3 augmented matrix:

$$\left[\begin{array}{rr|r} 1 & 1 & -1 \\ 2 & -1 & 7 \\ -1 & 2 & -8 \end{array}\right]$$

To get 0's under the 1 in the first column, we do a type 3 row operation twice:

Multiply row 1 by -2 and add to row 2. Multiply row 1 by 1 and add to row 3.

$$\left[\begin{array}{rr|r} 1 & 1 & -1 \\ 2 & -1 & 7 \\ -1 & 2 & -8 \end{array}\right] \approx \left[\begin{array}{rr|r} 1 & 1 & -1 \\ 0 & -3 & 9 \\ -1 & 2 & -8 \end{array}\right] \approx \left[\begin{array}{rr|r} 1 & 1 & -1 \\ 0 & -3 & 9 \\ 0 & 3 & -9 \end{array}\right]$$

We can do other row operations to get

Add row 2 to to row 3. Multiply row 2 by $-\frac{1}{3}$.

$$\left[\begin{array}{rr|r} 1 & 1 & -1 \\ 0 & -3 & 9 \\ 0 & 3 & -9 \end{array}\right] \approx \left[\begin{array}{rr|r} 1 & 1 & -1 \\ 0 & -3 & 9 \\ 0 & 0 & 0 \end{array}\right] \approx \left[\begin{array}{rr|r} 1 & 1 & -1 \\ 0 & 1 & -3 \\ 0 & 0 & 0 \end{array}\right]$$

The final matrix represents the system

$$\begin{cases} x + y = -1 \\ 0x + y = -3 \\ 0x + 0y = 0 \end{cases}$$

The third equation can be discarded, because $0x + 0y = 0$ for all x and y. From the second equation, we can read that $y = -3$. To find x, we substitute -3 for y in the first equation and solve for x:

$$x + y = -1$$

$$x - \mathbf{3} = -1 \quad \text{Substitute } -3 \text{ for } y.$$

$$x = 2 \quad \text{Add 3 to both sides.}$$

The solution is $(2, -3)$. Verify that this solution satisfies all three equations of the original system. ■

If the last row of the final matrix in Example 2 had been of the form $0x + 0y = k$, where $k \neq 0$, the system would have no solution. No values of x and y could make the expression $0x + 0y$ equal to a nonzero constant k.

Systems with More Variables Than Equations

We can also solve many systems with more variables than equations.

EXAMPLE 3 Solve the system $\begin{cases} x + y - 2z = -1 \\ 2x - y + z = -3 \end{cases}$.

Solution This system has 2 equations and 3 variables. We can start to solve it by doing a type 3 row operation to get a 0 under the 1 in the first column.

Multiply row 1 by -2 and add to row 2.

$$\left[\begin{array}{ccc|c} 1 & 1 & -2 & -1 \\ 2 & -1 & 1 & -3 \end{array}\right] \approx \left[\begin{array}{ccc|c} 1 & 1 & -2 & -1 \\ 0 & -3 & 5 & -1 \end{array}\right]$$

We then do a type 2 row operation:

Multiply row 2 by $-\frac{1}{3}$.

$$\left[\begin{array}{ccc|c} 1 & 1 & -2 & -1 \\ 0 & -3 & 5 & -1 \end{array}\right] \approx \left[\begin{array}{ccc|c} 1 & 1 & -2 & -1 \\ 0 & 1 & -\frac{5}{3} & \frac{1}{3} \end{array}\right]$$

The final matrix represents the system

$$\begin{cases} x + y - 2z = -1 \\ \quad\;\; y - \dfrac{5}{3}z = \dfrac{1}{3} \end{cases}$$

We add $\frac{5}{3}z$ to both sides of the second equation to obtain

$$y = \frac{1}{3} + \frac{5}{3}z$$

We substitute $\frac{1}{3} + \frac{5}{3}z$ for y in the first equation and simplify to get

$$x + y - 2z = -1$$

$$x + \frac{1}{3} + \frac{5}{3}z - 2z = -1 \qquad \text{Substitute } \tfrac{1}{3} + \tfrac{5}{3}z \text{ for } y.$$

$$x + \frac{1}{3} - \frac{1}{3}z = -1 \qquad \tfrac{5}{3}z - 2z = \tfrac{5}{3}z - \tfrac{6}{3}z = -\tfrac{1}{3}z.$$

$$x - \frac{1}{3}z = -\frac{4}{3} \qquad \text{Subtract } \tfrac{1}{3} \text{ from both sides.}$$

$$x = -\frac{4}{3} + \frac{1}{3}z \qquad \text{Add } \tfrac{1}{3}z \text{ to both sides.}$$

A solution to this system must have the form

$$\left(-\frac{4}{3}+\frac{1}{3}z,\quad \frac{1}{3}+\frac{5}{3}z,\quad z\right)$$

for all values of z. This system has an infinite number of solutions, a different one for each value of z. For example,

- If $z = 0$, the corresponding solution is $\left(-\frac{4}{3}, \frac{1}{3}, 0\right)$.
- If $z = 1$, the corresponding solution is $(-1, 2, 1)$.

Verify that both of these solutions satisfy each equation of the given system. ■

ORALS *Consider the system* $\begin{cases} 3x + 2y = 8 \\ 4x - 3y = 6 \end{cases}$.

1. Find the coefficient matrix. $\begin{bmatrix} 3 & 2 \\ 4 & -3 \end{bmatrix}$

2. Find the augmented matrix. $\left[\begin{array}{cc|c} 3 & 2 & 8 \\ 4 & -3 & 6 \end{array}\right]$

Tell if each matrix is in triangular form.

3. $\begin{bmatrix} 4 & 1 & 5 \\ 0 & 2 & 7 \\ 0 & 0 & 4 \end{bmatrix}$ yes

4. $\begin{bmatrix} 8 & 5 & 2 \\ 0 & 4 & 5 \\ 0 & 7 & 0 \end{bmatrix}$ no

EXERCISE 9.5

In Exercises 1–12, use matrices to solve each system of equations. Each system has one solution.

1. $\begin{cases} x + y = 2 \\ x - y = 0 \end{cases}$ (1, 1)

2. $\begin{cases} x + y = 3 \\ x - y = -1 \end{cases}$ (1, 2)

3. $\begin{cases} x + 2y = -4 \\ 2x + y = 1 \end{cases}$ (2, −3)

4. $\begin{cases} 2x - 3y = 16 \\ -4x + y = -22 \end{cases}$ (5, −2)

5. $\begin{cases} 3x + 4y = -12 \\ 9x - 2y = 6 \end{cases}$ (0, −3)

6. $\begin{cases} 5x - 4y = 10 \\ x - 7y = 2 \end{cases}$ (2, 0)

7. $\begin{cases} x + y + z = 6 \\ x + 2y + z = 8 \\ x + y + 2z = 9 \end{cases}$ (1, 2, 3)

8. $\begin{cases} x - y + z = 2 \\ x + 2y - z = 6 \\ 2x - y - z = 3 \end{cases}$ (3, 2, 1)

9. $\begin{cases} 2x + y + 3z = 3 \\ -2x - y + z = 5 \\ 4x - 2y + 2z = 2 \end{cases}$ (−1, −1, 2)

10. $\begin{cases} 3x + 2y + z = 8 \\ 6x - y + 2z = 16 \\ -9x + y - z = -20 \end{cases}$ (2, 0, 2)

11. $\begin{cases} 3x - 2y + 4z = 4 \\ x + y + z = 3 \\ 6x - 2y - 3z = 10 \end{cases}$ (2, 1, 0)

12. $\begin{cases} 2x + 3y - z = -8 \\ x - y - z = -2 \\ -4x + 3y + z = 6 \end{cases}$ (−2, −1, 1)

In Exercises 13–20, use matrices to solve each system of equations. If a system has no solution, so indicate.

13. $\begin{cases} x + y = 3 \\ 3x - y = 1 \\ 2x + y = 4 \end{cases}$ (1, 2)

14. $\begin{cases} x - y = -5 \\ 2x + 3y = 5 \\ x + y = 1 \end{cases}$ (−2, 3)

15. $\begin{cases} 2x - y = 4 \\ x + 3y = 2 \\ -x - 4y = -2 \end{cases}$ (2, 0)

16. $\begin{cases} 3x - 2y = 5 \\ x + 2y = 7 \\ -3x - y = -11 \end{cases}$ (3, 2)

17. $\begin{cases} 2x + y = 7 \\ x - y = 2 \\ -x + 3y = -2 \end{cases}$ no solution

18. $\begin{cases} 3x - y = 2 \\ -6x + 3y = 0 \\ -x + 2y = -4 \end{cases}$ no solution

19. $\begin{cases} x + 3y = 7 \\ x + y = 3 \\ 3x + y = 5 \end{cases}$ (1, 2)

20. $\begin{cases} x + y = 3 \\ x - 2y = -3 \\ x - y = 1 \end{cases}$ no solution

In Exercises 21–24, use matrices to solve each system of equations.

21. $\begin{cases} x + 2y + 3z = -2 \\ -x - y - 2z = 4 \end{cases}$ $(-6 - z, 2 - z, z)$

22. $\begin{cases} 2x - 4y + 3z = 6 \\ -4x + 6y + 4z = -6 \end{cases}$ $\left(\frac{17}{2}z - 3, 5z - 3, z\right)$

23. $\begin{cases} x - y = 1 \\ y + z = 1 \\ x + z = 2 \end{cases}$ $(2 - z, 1 - z, z)$

24. $\begin{cases} x + z = 1 \\ x + y = 2 \\ 2x + y + z = 3 \end{cases}$ $(1 - z, 1 + z, z)$

Writing Exercises ■ *Write a paragraph using your own words.*

1. Explain how to check the solution of a system of equations.
2. Explain how to perform a type 3 row operation.

Something to Think About ■

1. If the system represented by $\left[\begin{array}{ccc|c} 1 & 1 & 0 & 1 \\ 0 & 0 & 1 & 2 \\ 0 & 0 & 0 & k \end{array}\right]$ has no solution, what do you know about k? $k \neq 0$
2. Is it possible for a system with fewer equations than variables to have no solution? Illustrate. yes; $\begin{cases} x + y + z = 1 \\ x + y + z = 2 \end{cases}$

Review Exercises ■ *Solve each equation.*

Note: If you teach graphing early, skip these review exercises.

1. $x^4 - 50x^2 + 49 = 0$ 1, −1, 7, −7
2. $n\left(30 - \frac{13}{n}\right) = \frac{10}{n}$ $\frac{5}{6}, -\frac{2}{5}$
3. $\frac{w + 4}{2w} = \frac{4w - 2}{4}$ 2, −1
4. $x^2 + 2x + 6 = 0$ $-1 \pm i\sqrt{5}$

9.6 Systems of Inequalities

We have previously considered graphs of inequalities containing two variables. We now consider the graphs of systems of such inequalities.

EXAMPLE 1 Graph the solution set of the system $\begin{cases} x + y \le 1 \\ 2x - y > 2 \end{cases}$.

Solution On the same set of coordinate axes, we graph each inequality, as in Figure 9-8.

The graph of the inequality $x + y \le 1$ includes the line graph of the equation $x + y = 1$ and all points below it. Since the boundary line is included, it is drawn as a solid line.

The graph of the inequality $2x - y > 2$ contains only those points below the graph of the equation $2x - y = 2$. Since the boundary line is not included, it is drawn as a broken line.

The area where the half planes intersect represents the simultaneous solutions of the given system of inequalities, because any point in that region has coordinates that will satisfy both inequalities.

Teaching Tip: Emphasize that the solution is the intersection of the two graphs.

$x + y = 1$

x	y
0	1
1	0

$2x - y = 2$

x	y
0	−2
1	0

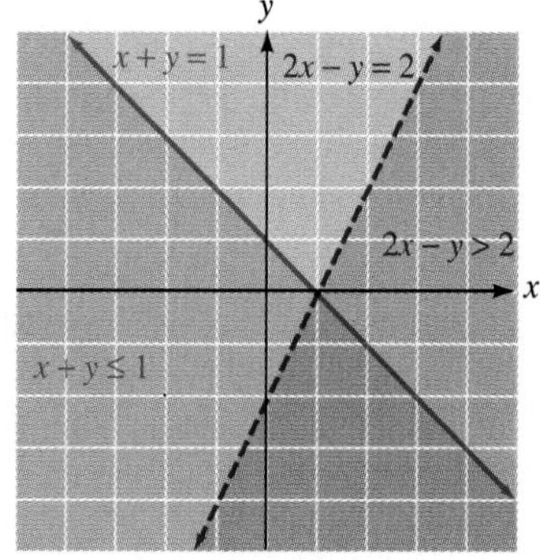

FIGURE 9-8

EXAMPLE 2 Graph the solution set of the system $\begin{cases} y < x^2 \\ y > \dfrac{x^2}{4} - 2 \end{cases}$.

Solution The graph of $y = x^2$ is the parabola shown in Figure 9-9, which opens upward and has its vertex at the origin. The points with coordinates that satisfy the inequality $y < x^2$ are those points below the parabola.

The graph of $y = \dfrac{x^2}{4} - 2$ is a parabola opening upward, with vertex at $(0, -2)$. However, this time the points with coordinates that satisfy the inequality are those points above the parabola. Thus, the graph of the solution set of the system is the area between the parabolas.

$y = x^2$

x	y
0	0
1	1
-1	1
2	4
-2	4

$y = \frac{x^2}{4} - 2$

x	y
0	-2
2	-1
-2	-1
4	2
-4	2

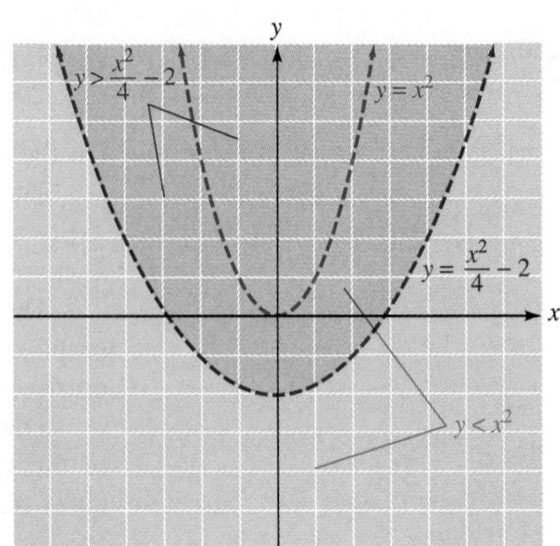

FIGURE 9-9

EXAMPLE 3 Graph the solution set of the system $\begin{cases} x \geq 1 \\ y \geq x \\ 4x + 5y < 20 \end{cases}$.

Solution The graph of the solution set of the inequality $x \geq 1$ includes those points on the graph of the equation $x = 1$ and to the right, as in Figure 9-10(a).

The graph of the solution set of the inequality $y \geq x$ includes those points on the graph of the equation $y = x$ and above it, as in Figure 9-10(b).

The graph of the solution set of the inequality $4x + 5y < 20$ includes those points below the line graph of the equation $4x + 5y = 20$, as in Figure 9-10(c).

If these three graphs are merged onto a single set of coordinate axes, the graph of the original system of inequalities includes those points within the shaded triangle, together with the points on the sides of the triangle drawn as solid lines. See Figure 9-10(d).

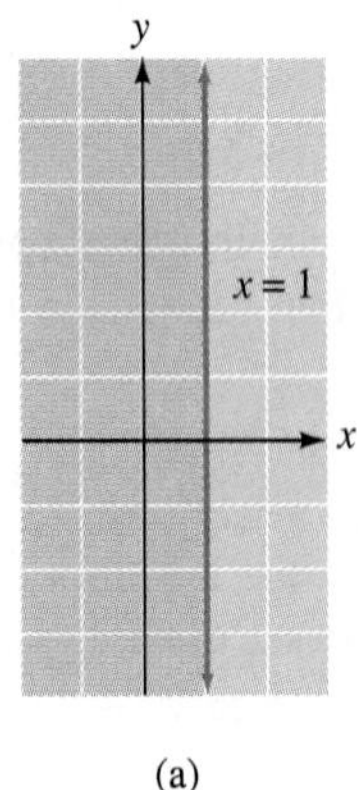

(a)

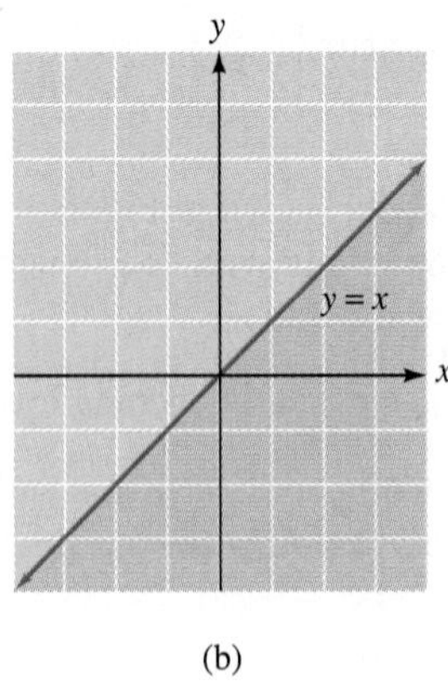

(b)

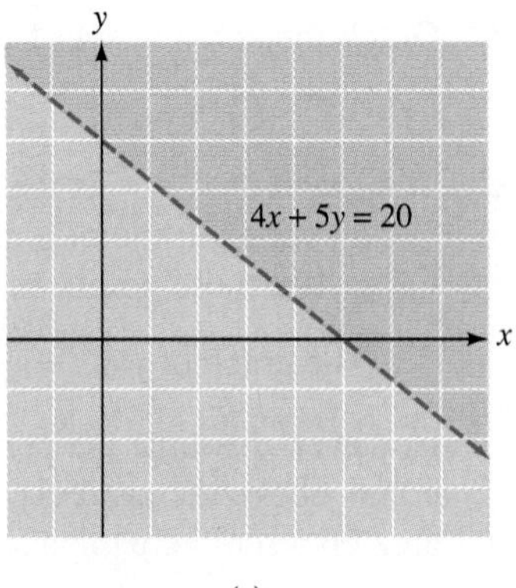

(c)

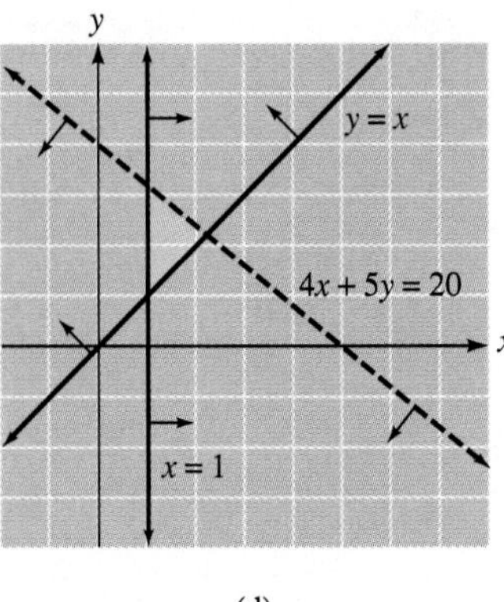

(d)

FIGURE 9-10

EXERCISE 9.6

In Exercises 1–16, graph the solution set of each system of inequalities.

1. $\begin{cases} y < 3x + 2 \\ y < -2x + 3 \end{cases}$

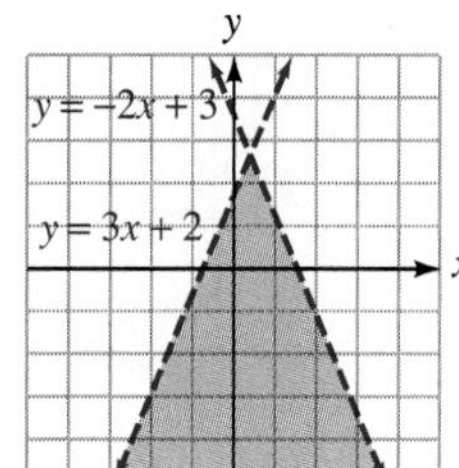

2. $\begin{cases} y \le x - 2 \\ y \ge 2x + 1 \end{cases}$

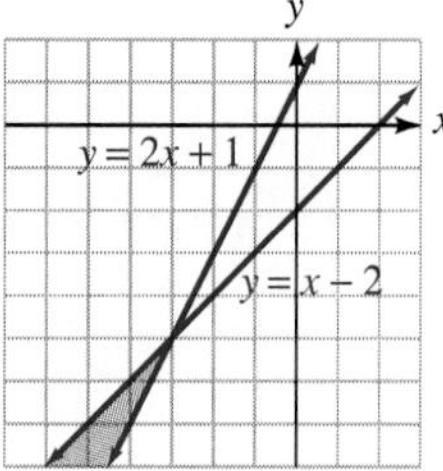

3. $\begin{cases} 3x + 2y > 6 \\ x + 3y \le 2 \end{cases}$

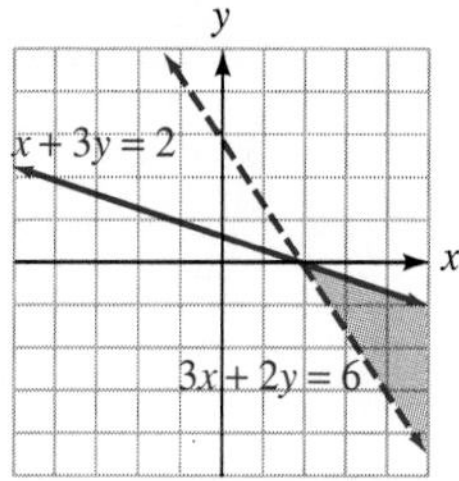

4. $\begin{cases} x + y < 2 \\ x + y \le 1 \end{cases}$

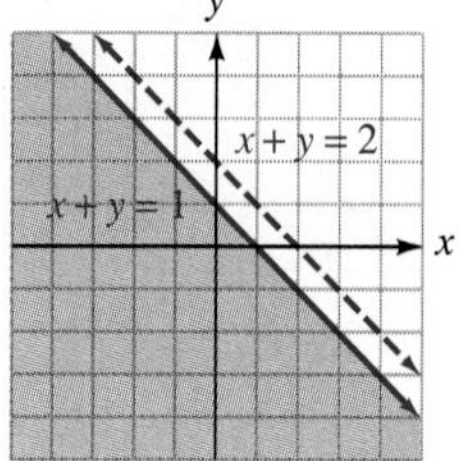

5. $\begin{cases} 3x + y \le 1 \\ -x + 2y \ge 9 \end{cases}$

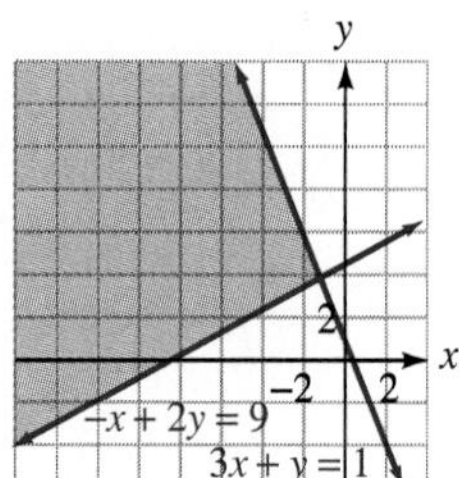

6. $\begin{cases} x + 2y < 3 \\ 2x + 4y < 8 \end{cases}$

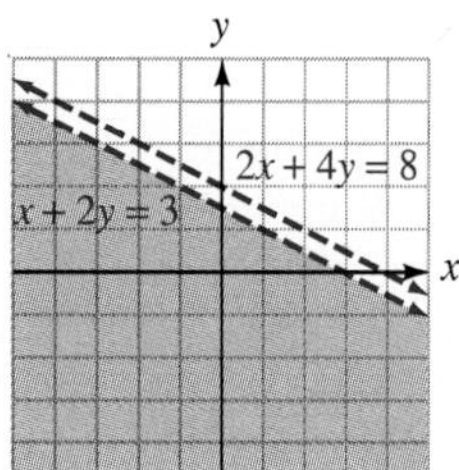

7. $\begin{cases} 2x - y > 4 \\ y < -x^2 + 2 \end{cases}$

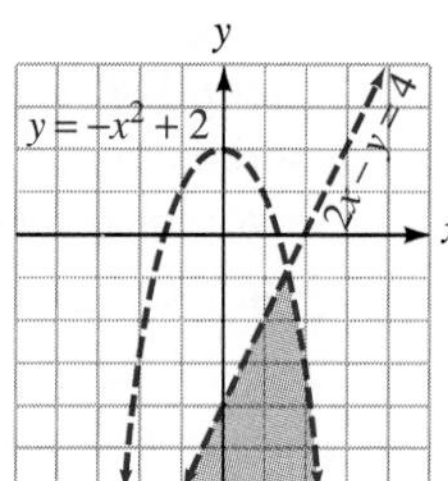

8. $\begin{cases} x \le y^2 \\ y \ge x \end{cases}$

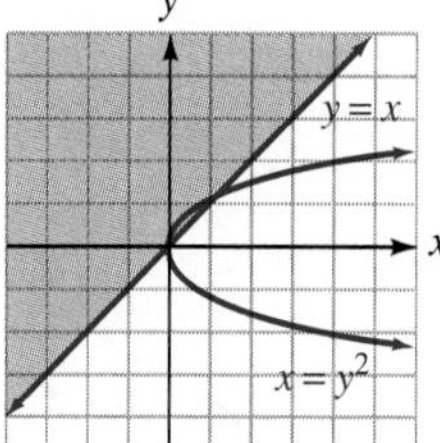

9. $\begin{cases} y > x^2 - 4 \\ y < -x^2 + 4 \end{cases}$

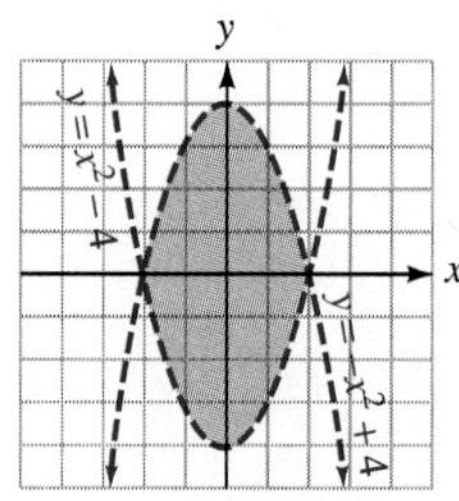

10. $\begin{cases} x \ge y^2 \\ y \ge x^2 \end{cases}$

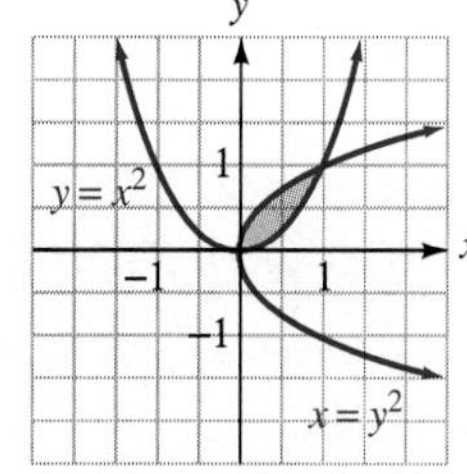

11. $\begin{cases} 2x + 3y \le 6 \\ 3x + y \le 1 \\ x \le 0 \end{cases}$

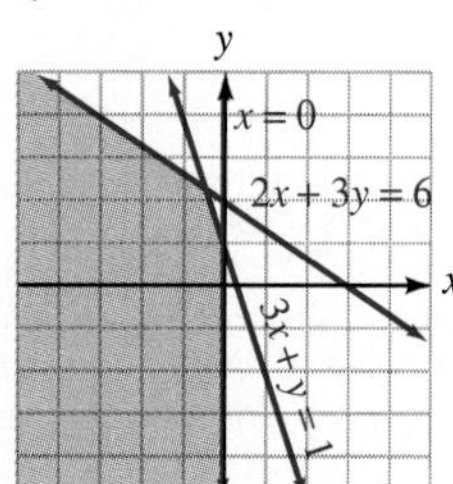

12. $\begin{cases} 2x + y \le 2 \\ y \ge x \\ x \ge 0 \end{cases}$

13. $\begin{cases} x - y < 4 \\ y \le 0 \\ x \ge 0 \end{cases}$

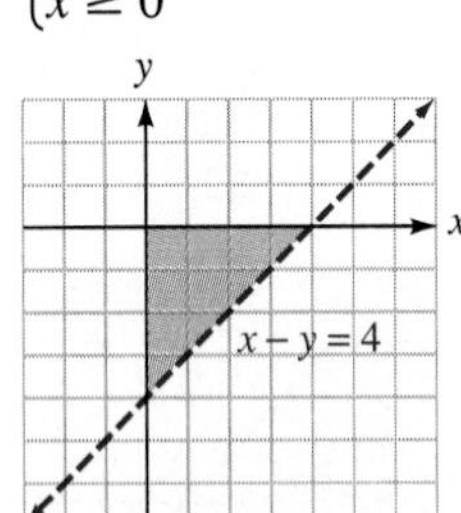

14. $\begin{cases} xy \le 1 \\ x \ge 0 \\ y \ge 0 \end{cases}$

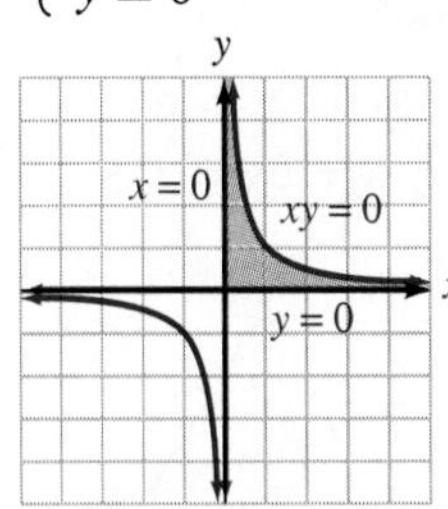

15. $\begin{cases} x \ge 0 \\ y \ge 0 \\ 9x + 3y \le 18 \\ 3x + 6y \le 18 \end{cases}$

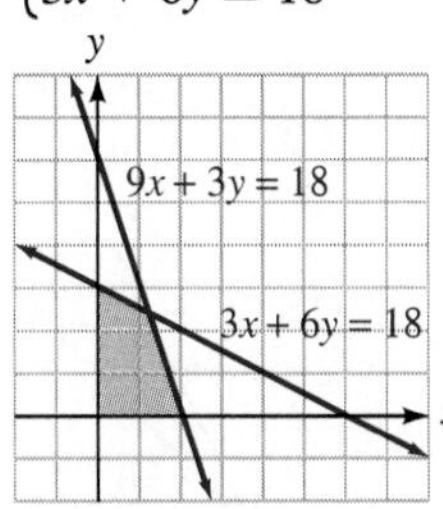

16. $\begin{cases} x + y \ge 1 \\ x - y \le 1 \\ x - y \ge 0 \\ x \le 2 \end{cases}$

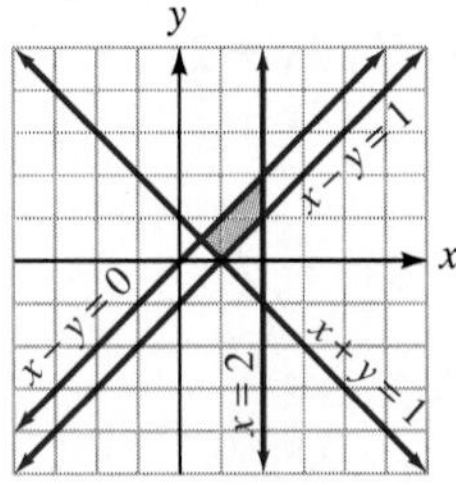

Writing Exercises ■ *Write a paragraph using your own words.*

1. When graphing a system of linear inequalities, explain how to decide which region to shade.

2. Explain how a system of two linear inequalities might have no solution.

Something to Think About ■

1. The solution of a system of inequalities in two variables is **bounded** if it is possible to draw a circle around it. Can the solution of two linear inequalities be bounded? no

2. The solution of $\begin{cases} y \ge |x| \\ y \le k \end{cases}$ has an area of 25. Find k. 5

Review Exercises ■ *Solve each formula for the given variable.*

Note: If you teach graphing early, skip these review exercises.

1. $A = p + prt$ for r $\quad r = \dfrac{A - p}{pt}$

2. $A = p + prt$ for p $\quad p = \dfrac{A}{1 + rt}$

3. $\dfrac{1}{r} = \dfrac{1}{r_1} + \dfrac{1}{r_2}$ for r $\quad r = \dfrac{r_1 r_2}{r_2 + r_1}$

4. $\dfrac{1}{r} = \dfrac{1}{r_1} + \dfrac{1}{r_2}$ for r_1 $\quad r_1 = \dfrac{r r_2}{r_2 - r}$

Solve each equation.

5. $3(x + 2) - 6(x - 5) - 1 = 2x - 6$ $\quad \frac{41}{5}$

6. $4x^2 + 45 = 29x$ $\quad 5, \frac{9}{4}$

PROJECT ■ Lines, Triangles, and Determinants

In this project, you will explore two of the many uses of determinants. In the first, you will discover that the equation of a line can be written as a determinant equation. If you are given the coordinates of two fixed points, you can use a determinant to write the equation of the line passing through them.

- The equation of the line passing through the points $P(2, 3)$ and $Q(-1, 4)$ is

$$\begin{vmatrix} x & y & 1 \\ 2 & 3 & 1 \\ -1 & 4 & 1 \end{vmatrix} = 0$$

  Verify this by expanding the determinant and graphing the resulting equation.
- In general, the **two-point form** of the equation of the line passing through the points $P(x_1, y_1)$ and $Q(x_2, y_2)$ is

$$\begin{vmatrix} x & y & 1 \\ x_1 & y_1 & 1 \\ x_2 & y_2 & 1 \end{vmatrix} = 0$$

  Find the equation of the line passing through $P(-4, 5)$ and $Q(1, -3)$.
- Does this equation still work if the x-coordinates of the two points are equal? (For then the line would be vertical and therefore have no defined slope.)

As a second application, the formula for the area of a triangle can be written as a determinant.

- The vertices of the triangle in Illustration 1 are $A(-3, -2)$, $B(4, -2)$, and $C(4, 4)$. Clearly, ABC is a right triangle, and it is easy to find its area by the formula $A = \frac{1}{2}bh$. Show that the area is also given by

$$A = \frac{1}{2}\begin{vmatrix} -3 & -2 & 1 \\ 4 & -2 & 1 \\ 4 & 4 & 1 \end{vmatrix}$$

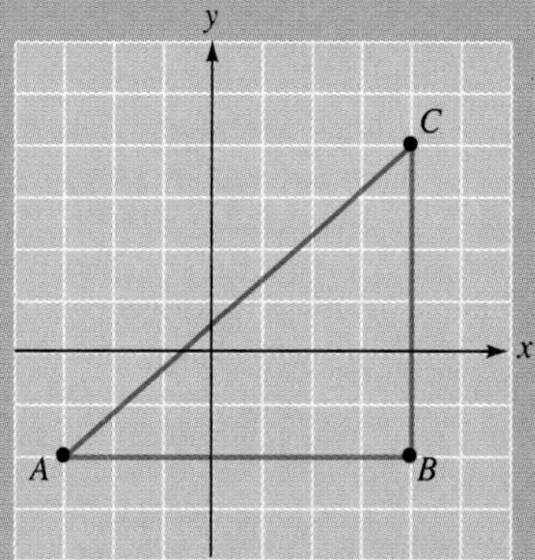

ILLUSTRATION 1

The formula works for any triangle, not just right triangles. The area of the triangle with vertices $A(x_1, y_1)$, $B(x_2, y_2)$, and $C(x_3, y_3)$ is

$$A = \frac{1}{2}\begin{vmatrix} x_1 & y_1 & 1 \\ x_2 & y_2 & 1 \\ x_3 & y_3 & 1 \end{vmatrix}$$

Chapter Summary

KEY WORDS

augmented matrix (9.5)
coefficient matrix (9.5)
consistent system of equations (9.1)
Cramer's rule (9.4)
dependent equations (9.1)
determinant (9.4)
elementary row operations (9.5)
element of a matrix (9.4)
equivalent systems (9.1)
Gaussian elimination (9.5)
inconsistent system of equations (9.1)
independent equations (9.1)
matrix (9.4, 9.5)
minor (9.4)
square matrix (9.4)
systems of equations (9.1)
triangular form of a matrix (9.5)

KEY IDEAS

(9.1) Systems of two linear equations in two variables can be solved by graphing.

If a system of equations has at least one solution, the system is a **consistent system.** Otherwise, the system is an **inconsistent system.**

If the graphs of the equations of a system are distinct, the equations are **independent equations.** Otherwise, the equations are **dependent equations.**

(9.2) Systems of two linear equations in two variables can be solved by the **substitution method** or the **addition method.**

(9.3) A system of three linear equations in three variables can be solved by using the addition method.

(9.4) A **matrix** is any rectangular array of numbers.

A **determinant of a square matrix** is a real number:

$$\begin{vmatrix} a & b \\ c & d \end{vmatrix} = ad - bc$$

$$\begin{vmatrix} a_1 & b_1 & c_1 \\ a_2 & b_2 & c_2 \\ a_3 & b_3 & c_3 \end{vmatrix}$$

$$= a_1 \begin{vmatrix} b_2 & c_2 \\ b_3 & c_3 \end{vmatrix} - b_1 \begin{vmatrix} a_2 & c_2 \\ a_3 & c_3 \end{vmatrix} + c_1 \begin{vmatrix} a_2 & b_2 \\ a_3 & b_3 \end{vmatrix}$$

Many systems of linear equations can be solved by using **Cramer's rule.**

(9.5) Many systems of linear equations can be solved by using matrices and the method of **Gaussian elimination.**

(9.6) Systems of inequalities can be solved by graphing.

Chapter 9 Review Exercises

In Review Exercises 1–4, solve each system of equations by the graphing method.

1. $\begin{cases} 2x + y = 11 \\ -x + 2y = 7 \end{cases}$

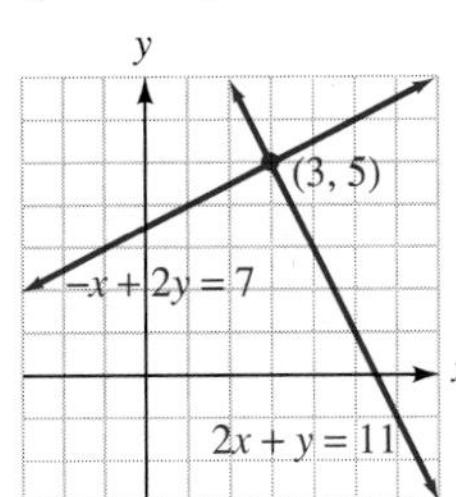

2. $\begin{cases} 3x + 2y = 0 \\ 2x - 3y = -13 \end{cases}$

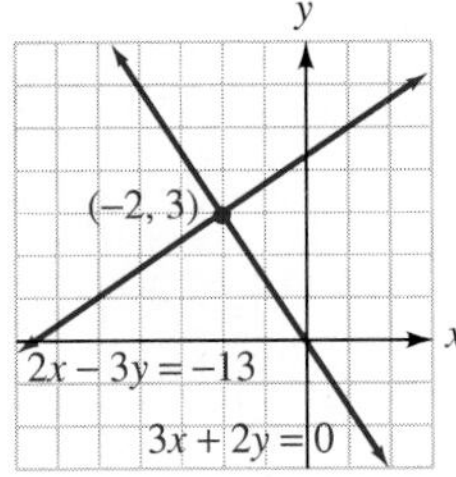

3. $\begin{cases} \frac{1}{2}x + \frac{1}{3}y = 2 \\ y = 6 - \frac{3}{2}x \end{cases}$

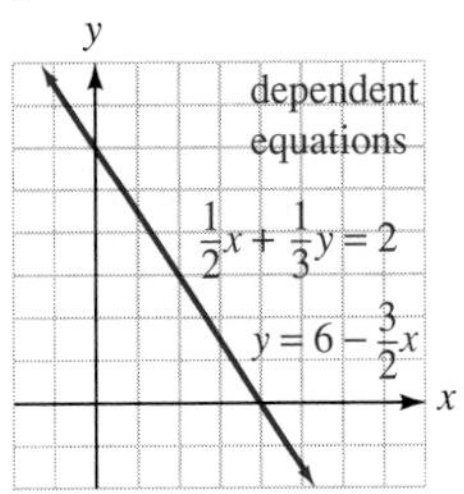

4. $\begin{cases} \frac{1}{3}x - \frac{1}{2}y = 1 \\ 6x - 9y = 2 \end{cases}$

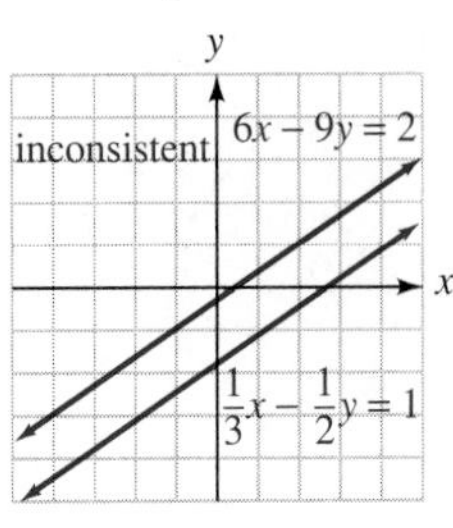

In Review Exercises 5–8, solve each system of equations by substitution.

5. $\begin{cases} y = x + 4 \\ 2x + 3y = 7 \end{cases}$ $(-1, 3)$

6. $\begin{cases} y = 2x + 5 \\ 3x - 5y = -4 \end{cases}$ $(-3, -1)$

7. $\begin{cases} x + 2y = 11 \\ 2x - y = 2 \end{cases}$ $(3, 4)$

8. $\begin{cases} 2x + 3y = -2 \\ 3x + 5y = -2 \end{cases}$ $(-4, 2)$

In Review Exercises 9–14, solve each system of equations by addition.

9. $\begin{cases} x + y = -2 \\ 2x + 3y = -3 \end{cases}$ $(-3, 1)$

10. $\begin{cases} 3x + 2y = 1 \\ 2x - 3y = 5 \end{cases}$ $(1, -1)$

11. $\begin{cases} x + \frac{1}{2}y = 7 \\ -2x = 3y - 6 \end{cases}$ $(9, -4)$

12. $\begin{cases} y = \dfrac{x - 3}{2} \\ x = \dfrac{2y + 7}{2} \end{cases}$ $(4, \frac{1}{2})$

13. $\begin{cases} x + y + z = 6 \\ x - y - z = -4 \\ -x + y - z = -2 \end{cases}$ $(1, 2, 3)$

14. $\begin{cases} 2x + 3y + z = -5 \\ -x + 2y - z = -6 \\ 3x + y + 2z = 4 \end{cases}$ inconsistent system

In Review Exercises 15–18, evaluate each determinant.

15. $\begin{vmatrix} 2 & 3 \\ -4 & 3 \end{vmatrix}$ 18

16. $\begin{vmatrix} -3 & -4 \\ 5 & -6 \end{vmatrix}$ 38

17. $\begin{vmatrix} -1 & 2 & -1 \\ 2 & -1 & 3 \\ 1 & -2 & 2 \end{vmatrix}$ -3

18. $\begin{vmatrix} 3 & -2 & 2 \\ 1 & -2 & -2 \\ 2 & 1 & -1 \end{vmatrix}$ 28

In Review Exercises 19–22, use Cramer's rule to solve each system of equations.

19. $\begin{cases} 3x + 4y = 10 \\ 2x - 3y = 1 \end{cases}$
$(2, 1)$

20. $\begin{cases} 2x - 5y = -17 \\ 3x + 2y = 3 \end{cases}$
$(-1, 3)$

21. $\begin{cases} x + 2y + z = 0 \\ 2x + y + z = 3 \\ x + y + 2z = 5 \end{cases}$
$(1, -2, 3)$

22. $\begin{cases} 2x + 3y + z = 2 \\ x + 3y + 2z = 7 \\ x - y - z = -7 \end{cases}$
$(-3, 2, 2)$

In Review Exercises 23–24, solve each system of equations by using matrices.

23. $\begin{cases} x + 2y = 4 \\ 2x - y = 3 \end{cases}$ (2, 1)

24. $\begin{cases} x + y + z = 6 \\ 2x - y + z = 1 \\ 4x + y - z = 5 \end{cases}$ (1, 3, 2)

In Review Exercises 25–26, graph the solution set of each system of inequalities.

25. $\begin{cases} y \geq x + 1 \\ 3x + 2y < 6 \end{cases}$

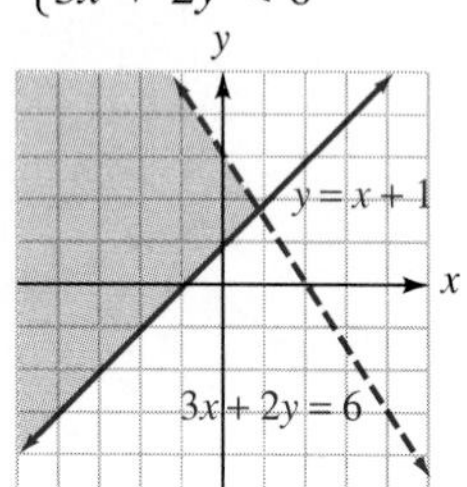

26. $\begin{cases} y \geq x^2 - 4 \\ y < x + 3 \end{cases}$

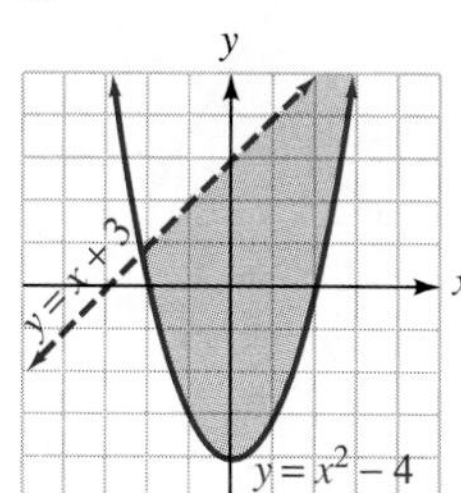

Chapter 9 Test

1. Solve $\begin{cases} 2x + y = 5 \\ y = 2x - 3 \end{cases}$ by graphing.

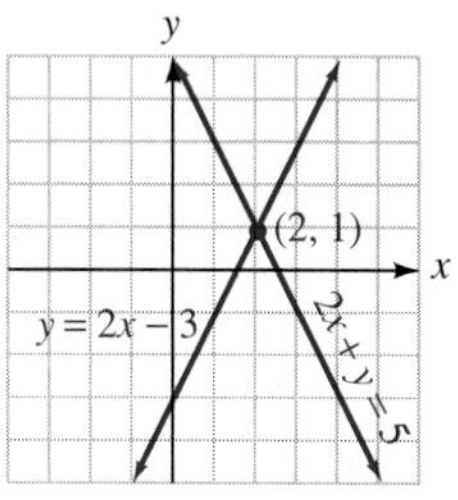

2. Use substitution to solve $\begin{cases} 2x - 4y = 14 \\ x = -2y + 7 \end{cases}$. (7, 0)

3. Use addition to solve $\begin{cases} 2x + 3y = -5 \\ 3x - 2y = 12 \end{cases}$. (2, −3)

4. Use any method to solve $\begin{cases} \dfrac{x}{2} - \dfrac{y}{4} = -4 \\ x + y = -2 \end{cases}$. (−6, 4)

In Problems 5–6, consider the system $\begin{cases} 3(x + y) = x - 3 \\ -y = \dfrac{2x + 3}{3} \end{cases}$.

5. Are the equations of the system dependent or independent? dependent

6. Is the system consistent or inconsistent? consistent

In Problems 7–10, evaluate each determinant.

7. $\begin{vmatrix} 2 & -3 \\ 4 & 5 \end{vmatrix}$ 24

8. $\begin{vmatrix} -3 & -4 \\ -2 & 3 \end{vmatrix}$ −17

9. $\begin{vmatrix} 1 & 2 & 0 \\ 2 & 0 & 3 \\ 1 & -2 & 2 \end{vmatrix}$ 4

10. $\begin{vmatrix} 2 & -1 & 1 \\ 3 & 1 & 0 \\ 0 & 1 & 2 \end{vmatrix}$ 13

In Problems 11–14, consider the system $\begin{cases} x - y = -6 \\ 3x + y = -6 \end{cases}$, *which is to be solved with Cramer's rule.*

11. When solving for x, what is the numerator determinant? **(Don't evaluate it.)** $\begin{vmatrix} -6 & -1 \\ -6 & 1 \end{vmatrix}$

12. When solving for y, what is the denominator determinant? **(Don't evaluate it.)** $\begin{vmatrix} 1 & -1 \\ 3 & 1 \end{vmatrix}$

13. Solve the system for x. -3

14. Solve the system for y. 3

In Problems 15–18, consider the system $\begin{cases} x + y + z = 4 \\ x + y - z = 6 \\ 2x - 3y + z = -1 \end{cases}$.

15. Solve for x. 3

16. Solve for y. 2

17. Solve for z. -1

18. Write the augmented matrix that represents the system. $\left[\begin{array}{ccc|c} 1 & 1 & 1 & 4 \\ 1 & 1 & -1 & 6 \\ 2 & -3 & 1 & -1 \end{array}\right]$

In Problems 19–20, use graphing to solve each system.

19. $\begin{cases} 2x - 3y \geq 6 \\ y \leq -x + 1 \end{cases}$

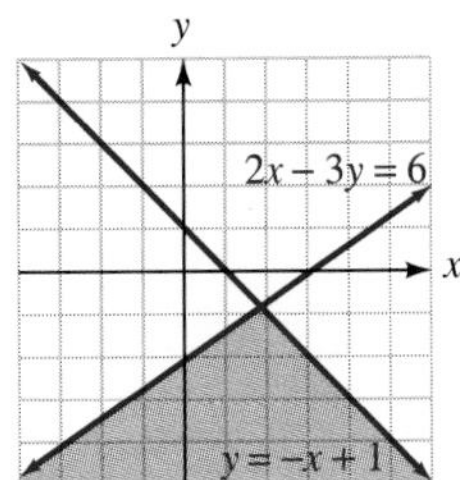

20. $\begin{cases} y \geq x^2 \\ y < x + 3 \end{cases}$

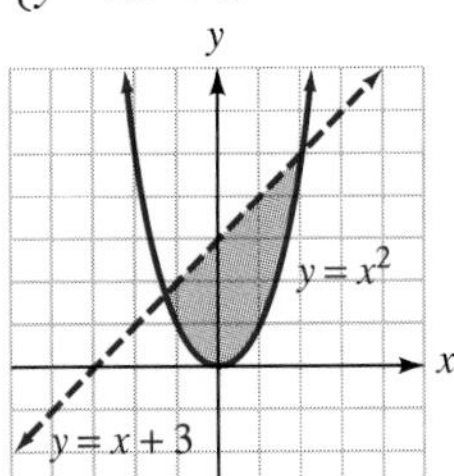

Cumulative Review Exercises

Note: If you teach graphing early, skip these review exercises.

In Exercises 1–16, simplify each expression. Assume that all variables represent positive numbers, and write answers without using negative exponents.

1. $64^{2/3}$ 16

2. $8^{-1/3}$ $\frac{1}{2}$

3. $\frac{y^{2/3}y^{5/3}}{y^{1/3}}$ y^2

4. $\frac{x^{5/3}x^{1/2}}{x^{3/4}}$ $x^{17/12}$

5. $(x^{2/3} - x^{1/3})(x^{2/3} + x^{1/3})$ $x^{4/3} - x^{2/3}$

6. $(x^{-1/2} + x^{1/2})^2$ $\frac{1}{x} + 2 + x$

7. $\sqrt[3]{-27x^3}$ $-3x$

8. $\sqrt{48t^3}$ $4t\sqrt{3t}$

9. $\sqrt[3]{\frac{128x^4}{2x}}$ $4x$

10. $\sqrt{x^2 + 6x + 9}$ $x + 3$

11. $\sqrt{50} - \sqrt{8} + \sqrt{32}$ $7\sqrt{2}$

12. $-3\sqrt[4]{32} - 2\sqrt[4]{162} + 5\sqrt[4]{48}$ $-12\sqrt[4]{2} + 10\sqrt[4]{3}$

13. $3\sqrt{2}(2\sqrt{3} - 4\sqrt{12})$ $-18\sqrt{6}$

14. $\frac{5}{\sqrt[3]{x}}$ $\frac{5\sqrt[3]{x^2}}{x}$

15. $\dfrac{\sqrt{x}+2}{\sqrt{x}-1}$ $\dfrac{x+3\sqrt{x}+2}{x-1}$

16. $\sqrt[6]{x^3y^3}$ $\sqrt{xy}$
(This problem is from an optional section.)

In Exercises 17–18, solve each equation.

17. $5\sqrt{x+2} = x+8$ 2, 7

18. $\sqrt{x} + \sqrt{x+2} = 2$ $\frac{1}{4}$

19. Use the method of completing the square to solve the equation $2x^2 + x - 3 = 0$. 1, $-\frac{3}{2}$

20. Use the quadratic formula to solve the equation $3x^2 + 4x - 1 = 0$. $\dfrac{-2 \pm \sqrt{7}}{3}$

In Exercises 21–28, write each complex number in a + bi form.

21. $(3+5i) + (4-3i)$ $7+2i$

22. $(7-4i) - (12+3i)$ $-5-7i$

23. $(2-3i)(2+3i)$ 13

24. $(3+i)(3-3i)$ $12-6i$

25. $(3-2i) - (4+i)^2$ $-12-10i$

26. $\dfrac{5}{3-i}$ $\frac{3}{2}+\frac{1}{2}i$

27. $|3+2i|$ $\sqrt{13}$

28. $|5-6i|$ $\sqrt{61}$

29. For what values of k will the solutions of $2x^2 + 4x = k$ be equal? -2

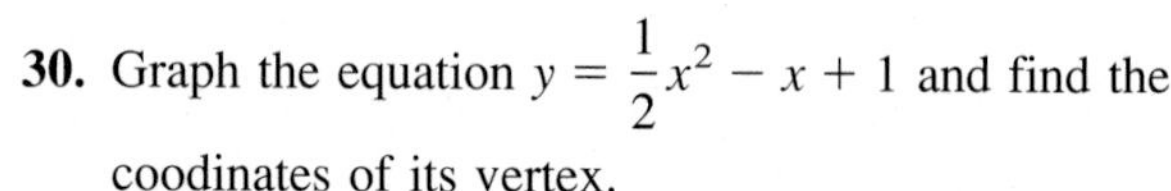
30. Graph the equation $y = \dfrac{1}{2}x^2 - x + 1$ and find the coodinates of its vertex.

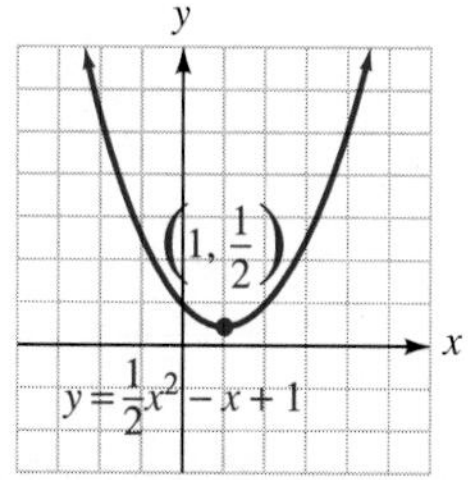

31. Solve $x^2 - x - 6 > 0$.
$(-\infty, -2) \cup (3, \infty)$
−2 3

32. $x^2 - x - 6 \leq 0$.

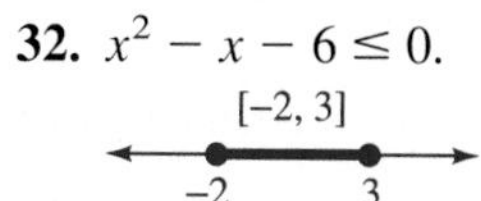

33. Use graphing to solve $\begin{cases} 2x + y = 5 \\ x - 2y = 0 \end{cases}$. (2, 1)

34. Use substitution to solve $\begin{cases} 3x + y = 4 \\ 2x - 3y = -1 \end{cases}$. (1, 1)

35. Use addition to solve $\begin{cases} x + 2y = -2 \\ 2x - y = 6 \end{cases}$. (2, −2)

36. Use any method to solve $\begin{cases} \dfrac{x}{10} + \dfrac{y}{5} = \dfrac{1}{2} \\ \dfrac{x}{2} - \dfrac{y}{5} = \dfrac{13}{10} \end{cases}$. (3, 1)

37. Evaluate: $\begin{vmatrix} 3 & -2 \\ 1 & -1 \end{vmatrix}$ -1

38. Use Cramer's rule and solve for y only: $\begin{cases} 4x - 3y = -1 \\ 3x + 4y = -7 \end{cases}$ -1

39. Solve: $\begin{cases} x + y + z = 1 \\ 2x - y - z = -4 \\ x - 2y + z = 4 \end{cases}$ $(-1, -1, 3)$

40. Solve for z only: $\begin{cases} x + 2y + 3z = 6 \\ 3x + 2y + z = 6 \\ 2x + 3y + z = 6 \end{cases}$ 1

41. Solve $\begin{cases} 3x - 2y < 6 \\ y < -x + 2 \end{cases}$.

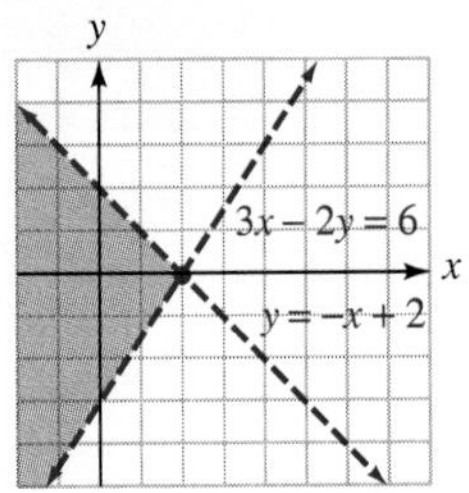

42. Solve $\begin{cases} y < x + 2 \\ 3x + y \leq 6 \end{cases}$.

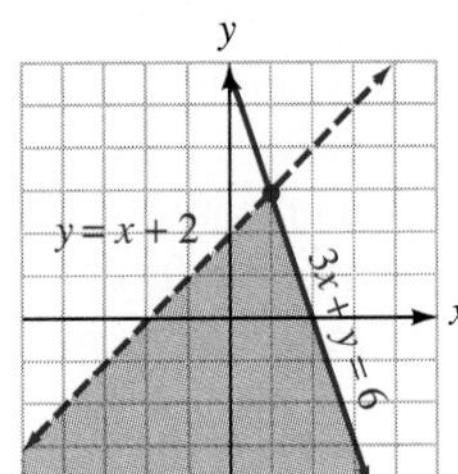

CAREERS & MATHEMATICS

Electrical/Electronic Engineer

Electrical engineers design, develop, test, and supervise the manufacture of electrical and electronic equipment. Electrical engineers who work with electronic equipment are often called **electronic engineers.**

Electrical engineers generally specialize in a major area, such as power-distributing equipment, integrated circuits, computers, electrical equipment manufacturing, or communications. Besides manufacturing and research, development, and design, many are employed in administration and management, technical sales, or teaching.

Qualifications A bachelor's degree in engineering is generally acceptable for beginning engineering jobs. College graduates with a degree in natural science or mathematics also may qualify for some jobs.

Engineers should be able to work as part of a team and should have creativity, an analytical mind, and a capacity for detail. In addition, engineers should be able to express themselves well—both orally and in writing.

Job Outlook Employment of electrical engineers is expected to increase faster than the average for all occupations through the year 2005. Although increased demand for computers, communications equipment, and electronics is expected to be the major contributor to this growth, demand for electrical and electronic consumer goods, along with increased research and development in new types of power generation, should create additional jobs.

Example Application In a radio, an inductor and a capacitor are used in a resonant circuit to select a desired station at frequency f and reject all others. The inductance L and the capacitance C determine the inductive reactance X_L and the capacitive reactance X_C of that circuit, where

$$X_L = 2\pi fL \qquad \text{and} \qquad X_C = \frac{1}{2\pi fC}$$

The radio station selected will be at the frequency f for which the inductive reactance and capacitive reactance are equal. To find that frequency, solve for f in terms of L and C.

Solution Since X_L and X_C are to be equal, we can solve the system of equations

$$\begin{cases} X_L = 2\pi fL \\ X_C = \dfrac{1}{2\pi fC} \\ X_L = X_C \end{cases}$$

by substitution to express f in terms of L and C.

$$X_L = X_C$$

$$2\pi f L = \frac{1}{2\pi f C}$$

$$(2\pi f L)(2\pi f C) = 1 \qquad \text{Multiply both sides by } 2\pi f C.$$

$$4\pi^2 f^2 LC = 1 \qquad \text{Simplify.}$$

$$f^2 = \frac{1}{4\pi^2 LC} \qquad \text{Divide both sides by } 4\pi^2 LC.$$

$$f = \sqrt{\frac{1}{4\pi^2 LC}} \qquad \text{Take the positive square root of both sides.}$$

$$f = \frac{1}{2\pi\sqrt{LC}} \qquad \text{Simplify.}$$

EXERCISES

1. At what frequency will a 0.0001 farad capacitor and a 0.005 henry inductor resonate? (*Hint:* $C = 0.0001$ and $L = 0.005$.)
2. If the inductor and the capacitor in Exercise 1 were both doubled, what would the resonant frequency become?
3. At what frequency will a 0.0008 farad capacitor and a 0.002 henry inductor resonate?
4. If the inductor of Exercise 3 were doubled and the capacitance reduced by one-half, what would the resonant frequency be?

(*Answers:* **1.** 225 hertz **2.** 113 hertz **3.** 126 hertz **4.** 126 hertz)

10

More Functions and Conic Sections

We have seen that the graphs of linear functions are straight lines, and that the graphs of quadratic functions are parabolas. In this chapter, we will discuss the graphs of other functions and discuss some special graphs, called **conic sections.**

10.1 Polynomial and Other Functions

■ Graphs of Polynomial Functions ■ Symmetries of Graphs ■ Absolute Value Functions ■ Square Root Functions ■ Piecewise-Defined Functions ■ Step Functions

The graphs of higher-degree polynomial functions are curves, often with more than one x-intercept.

■ Graphs of Polynomial Functions

Polynomial Function A **polynomial function in one variable** (say, x) is defined by an equation of the form $y = P(x)$, where $P(x)$ is a polynomial in the variable x.

The **degree of the polynomial function** $y = P(x)$ is the degree of $P(x)$.

EXAMPLE 1 Graph the polynomial function $y = f(x) = x^4 - 5x^2 + 4$.

Solution We plot several points (x, y) whose coordinates satisfy the equation, as in Figure 10-1(a), and then draw the graph by joining the points with a smooth curve.

To draw the graph with a graphing calculator set for the standard RANGE settings, we enter the equation and press the GRAPH key to obtain the graph shown in Figure 10-1(b). With the TRACE key, we can find the coordinates of its lowest point.

From the graph in Figure 10-1(b) we can see that the domain is the interval $(-\infty, \infty)$, and the range is $[-2.25, \infty)$.

$y = x^4 - 5x^2 + 4$

x	y
0	4
± 1	0
$\pm\frac{3}{2}$	$-\frac{35}{16}$
± 2	0
± 3	40

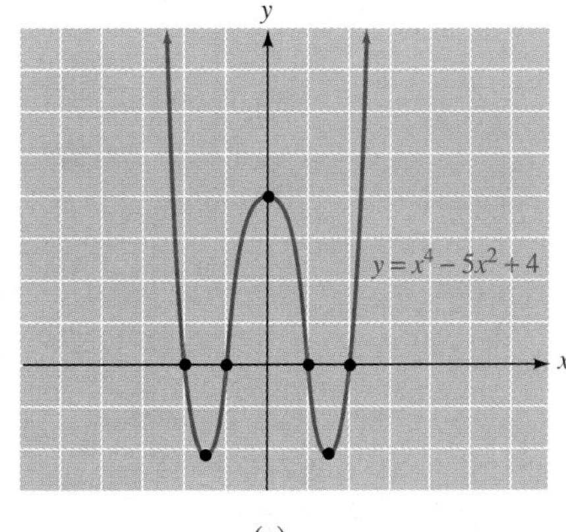

(a)

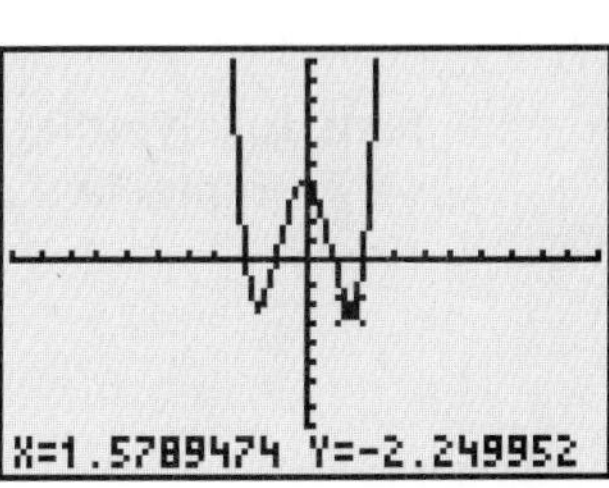

(b)

FIGURE 10-1

■

The x-intercepts of the graph in Example 1 are the points $(-2, 0)$, $(-1, 0)$, $(1, 0)$, and $(2, 0)$, and the y-intercept is the point $(0, 4)$. We note that if we folded the graph along the y-axis, the left-hand side would exactly match the right-hand side. For this reason, we say that the graph in Example 1 is symmetric about the y-axis.

It is often easier to draw graphs of polynomial functions if we first find the x- and y-intercepts and find any symmetries of the graph.

Symmetries of Graphs

y-axis symmetry: If the point $(-x, y)$ lies on a graph whenever the point (x, y) does, as in Figure 10-2(a), we say that the graph is **symmetric about the y-axis.**

Symmetry about the origin: If the point $(-x, -y)$ lies on the graph whenever the point (x, y) does, as in Figure 10-2(b), we say that the graph is **symmetric about the origin.**

x-axis symmetry: If the point $(x, -y)$ lies on the graph whenever the point (x, y) does, as in Figure 10-2(c), we say that the graph is **symmetric about the x-axis.**

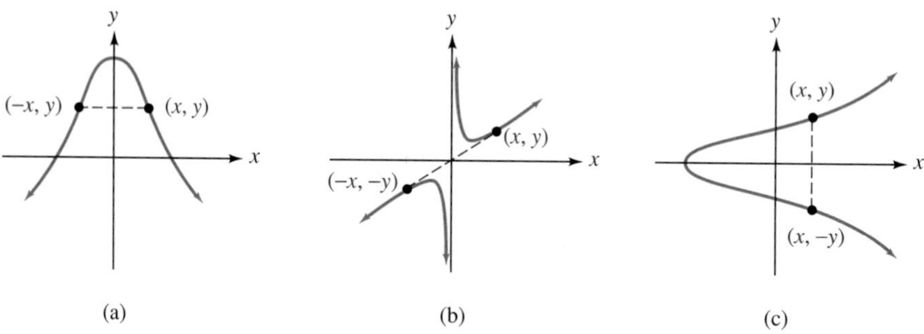

FIGURE 10-2

Tests for Symmetry for Graphs in x and y

- To test a graph for y-axis symmetry, replace x with $-x$. If the new equation is equivalent to the original equation, the graph is symmetric about the y-axis. Symmetry about the y-axis will occur whenever x appears with only even exponents.
- To test a graph for symmetry about the origin, replace x with $-x$ and y with $-y$. If the resulting equation is equivalent to the original equation, the graph is symmetric about the origin.
- To test a graph for x-axis symmetry, replace y with $-y$. If the resulting equation is equivalent to the original equation, the graph is symmetric about the x-axis.

EXAMPLE 2 Find the intercepts and the symmetries of the graph of $y = f(x) = x^3 - 9x$. Then graph the function.

Solution **x-intercepts:** To find the x-intercepts, we let $y = 0$ and solve for x:

$$y = x^3 - 9x$$

$$0 = x^3 - 9x \quad \text{Substitute 0 for } y.$$

$$0 = x(x^2 - 9) \quad \text{Factor out } x.$$

$$0 = x(x + 3)(x - 3) \quad \text{Factor } x^2 - 9.$$

$$x = 0 \quad \text{or} \quad x + 3 = 0 \quad \text{or} \quad x - 3 = 0 \quad \text{Set each factor equal to 0.}$$

$$| \quad x = -3 \quad | \quad x = 3$$

Since the x-coordinates of the x-intercepts are 0, -3, and 3, the graph intersects the x-axis at (0, 0), $(-3, 0)$, and (3, 0).

y-intercepts: To find the y-intercepts, we let $x = 0$ and solve for y.

$$y = x^3 - 9x$$

$$y = 0^3 - 9(0) \quad \text{Substitute 0 for } x.$$

$$y = 0$$

Since the y-coordinate of the y-intercept is 0, the graph intersects the y-axis at (0, 0).

Symmetry: We test for symmetry about the y-axis by replacing x with $-x$, simplifying, and comparing the result to the original equation.

1. $y = x^3 - 9x$ The original equation.

$y = (-x)^3 - 9(-x)$ Replace x with $-x$.

2. $y = -x^3 + 9x$ Simplify.

Because Equation 2 is not equivalent to Equation 1, the graph is not symmetric about the y-axis.

We test for symmetry about the origin by replacing x and y with $-x$ and $-y$, respectively, and comparing the result to the original equation.

1. $y = x^3 - 9x$ The original equation.

$-y = (-x)^3 - 9(-x)$ Replace x with $-x$, and y with $-y$.

$-y = -x^3 + 9x$ Simplify.

3. $y = x^3 - 9x$ Multiply both sides by -1.

Because Equation 3 is equivalent to Equation 1, the graph is symmetric about the origin.

We test for symmetry about the x-axis by replacing y with $-y$ and comparing the result to the original equation.

1. $y = x^3 - 9x$ The original equation.

$-y = x^3 - 9x$ Replace y with $-y$.

4. $y = -x^3 + 9x$ Multiply both sides by -1.

Because Equation 4 is not equivalent to Equation 1, the graph is not symmetric about the x-axis.

To graph the equation, we plot the x-intercepts of $(-3, 0)$, $(0, 0)$, and $(3, 0)$ and the y-intercept of $(0, 0)$. We also plot other points for positive values of x and use the symmetry about the origin to draw the rest of the graph, as in Figure 10-3(a). (Note that the scale on the x-axis is different from the scale on the y-axis.)

To draw the graph with a graphing calculator set for the standard RANGE settings, we enter the equation and press the GRAPH key to obtain the graph shown in Figure 10-3(b).

From the graph, we can see that the domain is the interval $(-\infty, \infty)$, and the range is the interval $(-\infty, \infty)$.

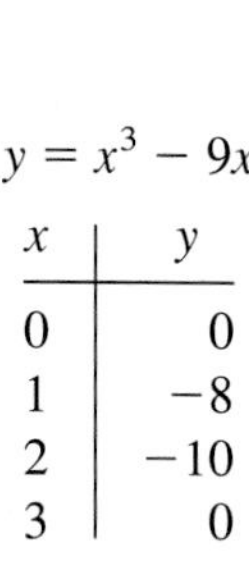

$y = x^3 - 9x$

x	y
0	0
1	−8
2	−10
3	0

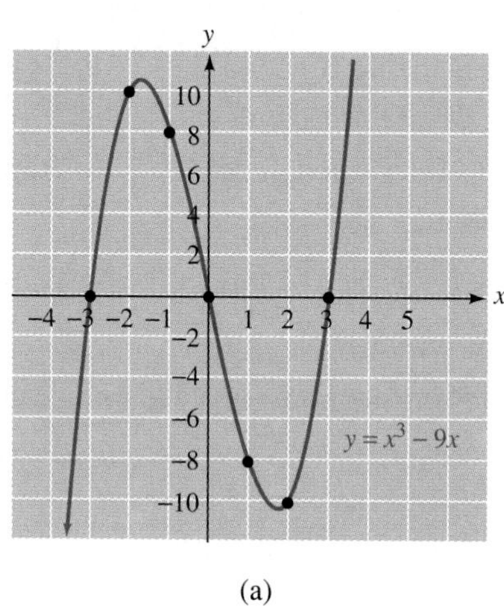

(a)

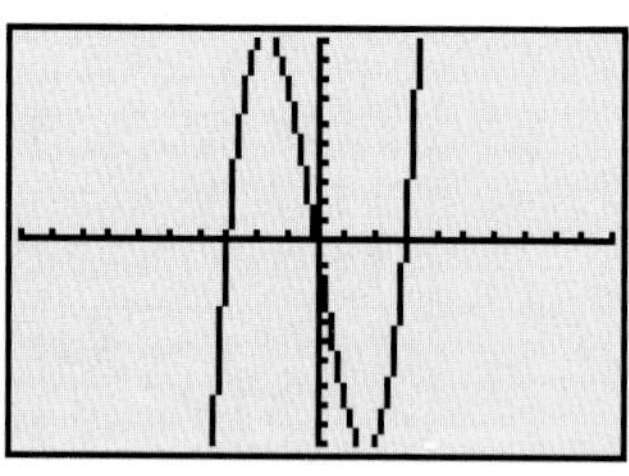

(b)

FIGURE 10-3

Absolute Value Functions

Absolute Value Function An **absolute value function** is any function defined by an equation of the form $y = |u| + k$, where u is an algebraic expression in x and k is a constant.

EXAMPLE 3 Graph the function $y = f(x) = |x| - 2$.

Solution **x-intercepts:** To find the x-intercepts, we let $y = 0$ and solve for x:

$$y = |x| - 2$$
$$0 = |x| - 2$$
$$2 = |x|$$
$$x = -2 \quad \text{or} \quad x = 2$$

Since -2 and 2 are solutions, the points $(-2, 0)$ and $(2, 0)$ are the x-intercepts, and the graph passes through $(-2, 0)$ and $(2, 0)$.

y-intercepts: To find the y-intercepts, we let $x = 0$ and solve for y:

$$y = |x| - 2$$
$$y = |0| - 2$$

Since $y = -2$, $(0, -2)$ is the y-intercept, the graph passes through the point $(0, -2)$.

Symmetry: To test for y-axis symmetry, we replace x with $-x$.

5. $y = |x| - 2$ — The original equation.

$y = |-x| - 2$ — Replace x with $-x$.

6. $y = |x| - 2$ — $|-x| = |x|$.

Since Equation 6 is equivalent to Equation 5, the graph is symmetric about the y-axis. The graph has no other symmetries.

We plot the x- and y-intercepts and several other points (x, y), and use the y-axis symmetry to obtain the graph shown in Figure 10-4(a).

To draw the graph with a graphing calculator set for the standard RANGE settings, we enter the equation and press the GRAPH key to obtain the graph shown in Figure 10-4(b).

From the graph, we see that the domain is the interval $(-\infty, \infty)$, and the range is the interval $[-2, \infty)$.

$y = |x| - 2$

x	y
0	-2
1	-1
2	0
3	1
4	2

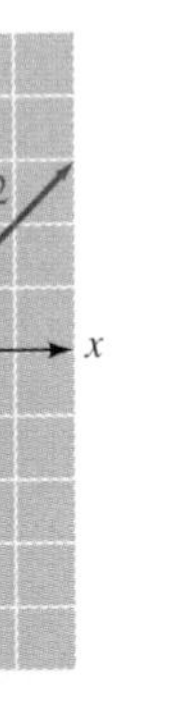

(a)

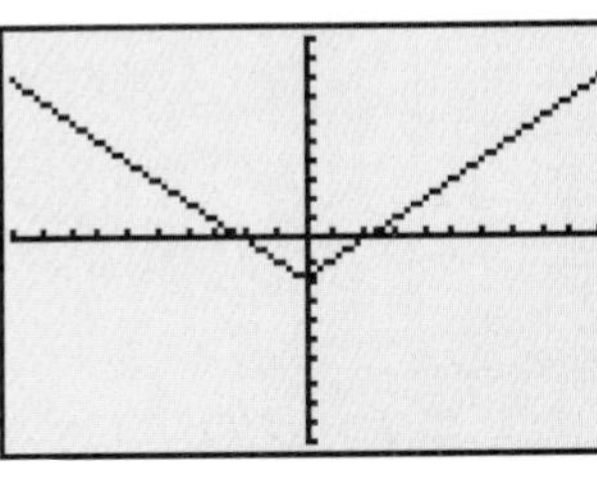

(b)

FIGURE 10-4

Square Root Functions

Square Root Function A **square root function** is any function defined by an equation of the form $y = \sqrt{u} + k$ $(u \geq 0)$, where u is an algebraic expression in x and k is a constant.

EXAMPLE 4 Graph the function $y = f(x) = \sqrt{x + 4}$.

Solution **x-intercepts:** To find the x-intercepts, we let $y = 0$ and solve for x:

$$\begin{aligned} y &= \sqrt{x+4} \\ 0 &= \sqrt{x+4} \\ (0)^2 &= \left(\sqrt{x+4}\right)^2 && \text{Square both sides.} \\ 0 &= x + 4 \\ x &= -4 \end{aligned}$$

Since -4 is a solution, $(-4, 0)$ is the x-intercept, and the graph passes through $(-4, 0)$.

y-intercepts: To find the y-intercepts, we let $x = 0$ and solve for y:

$$\begin{aligned} y &= \sqrt{x+4} \\ y &= \sqrt{0+4} \\ y &= \sqrt{4} \\ y &= 2 \end{aligned}$$

Since $y = 2$, $(0, 2)$ is the y-intercept and the graph passes through the point $(0, 2)$.

Symmetry: This graph has no symmetries.

We plot the x- and y-intercepts and several other pairs (x, y) to obtain the graph shown in Figure 10-5(a).

To draw the graph with a graphing calculator set for the standard RANGE settings, we enter the equation and press the GRAPH key to obtain the graph shown in Figure 10-5(b).

From the graph, we can see that the domain is the interval $[-4, \infty)$, and the range is the interval $[0, \infty)$.

$y = \sqrt{x + 4}$

x	y
-4	0
-3	1
0	2
5	3
12	4

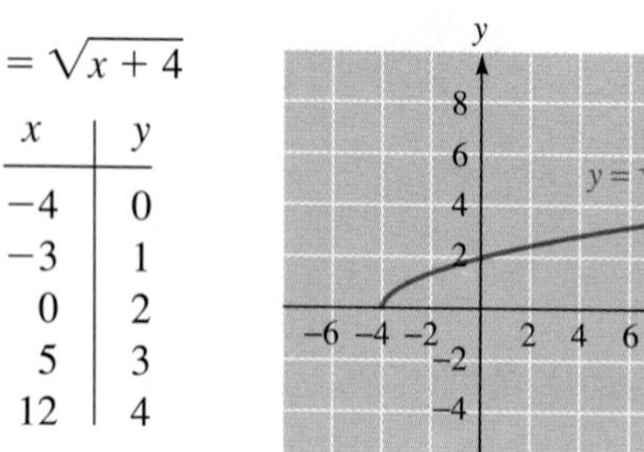

(a)

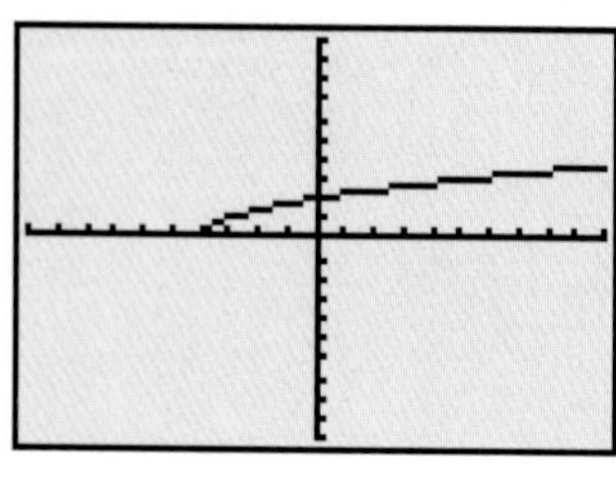

(b)

FIGURE 10-5 ■

Piecewise-Defined Functions

Some functions are defined by using different equations for different parts of their domains. Such functions are called **piecewise-defined functions.**

EXAMPLE 5 Graph the piecewise-defined function given by

$$y = \begin{cases} x^2 & \text{when } x \le 0 \\ x & \text{when } 0 < x < 2 \\ -1 & \text{when } x \ge 2 \end{cases}$$

Solution For each number x, we decide which of these three equations will be used to find the corresponding value of y:

- For numbers $x \le 0$, y is determined by the equation $y = x^2$, and the graph is the left half of a parabola. (See Figure 10-6.)
- For numbers x between 0 and 2, y is determined by the equation $y = x$, and the graph is part of a line.
- For numbers $x \ge 2$, y is the constant -1, and the graph is part of a horizontal line.

The use of solid and open circles in the graph indicates that $y = -1$ when $x = 2$.

The domain of this function is the interval $(-\infty, \infty)$, because every number x gives a single value y. The range is the interval $[0, \infty) \cup \{-1\}$.

If $x \le 0$
$y = x^2$

x	y
0	0
-1	1
-2	4
-3	9

If $0 < x < 2$
$y = x$

x	y
$\frac{1}{2}$	$\frac{1}{2}$
1	1
$\frac{3}{2}$	$\frac{3}{2}$

If $x \ge 2$
$y = -1$

x	y
2	-1
3	-1
5	-1

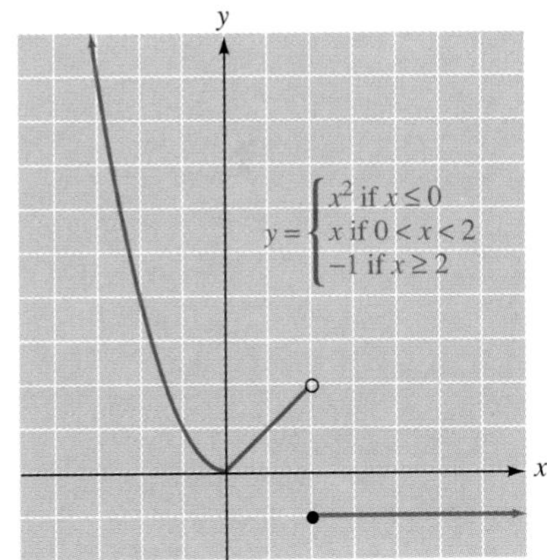

FIGURE 10-6

Step Functions

The **greatest integer function** is important in computer applications. This function is determined by the equation

$y = [\![x]\!]$ Read as "y equals the greatest integer in x."

where the value of y that corresponds to x is the greatest integer that is less than or equal to x. For example,

$$[\![4.7]\!] = 4, \qquad \left[\!\!\left[2\frac{1}{2}\right]\!\!\right] = 2, \qquad [\![\pi]\!] = 3, \qquad [\![-3.7]\!] = -4, \qquad [\![-5.7]\!] = -6$$

EXAMPLE 6 Graph $y = [\![x]\!]$.

Solution We list several intervals and the corresponding values of the greatest integer function:

$[0, 1)$	$y = [\![x]\!] = 0$	For numbers from 0 to 1, not including 1, the greatest integer in the interval is 0.
$[1, 2)$	$y = [\![x]\!] = 1$	For numbers from 1 to 2, not including 2, the greatest integer in the interval is 1.
$[2, 3)$	$y = [\![x]\!] = 2$	For numbers from 2 to 3, not including 3, the greatest integer in the interval is 2.

In each interval, the values of y are constant, but they jump by 1 at integer values of x. The graph is shown in Figure 10-7. From the graph, we see that the domain is $(-\infty, \infty)$, and the range is the set of integers $\{. . . , -3, -2, -1, 0, 1, 2, 3, . . .\}$.

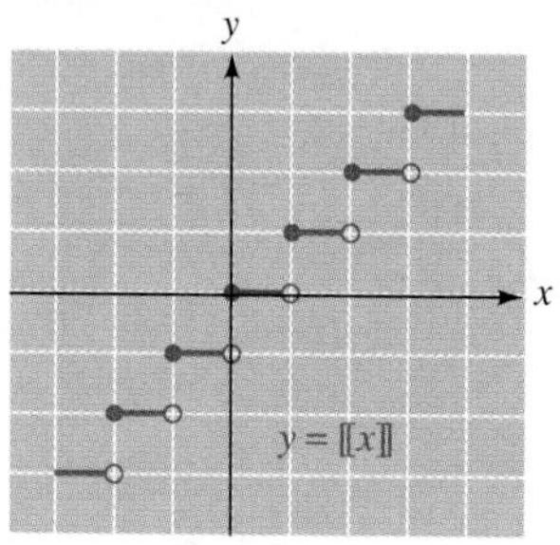

FIGURE 10-7

■

Since the greatest integer function is made up of a series of horizontal line segments, it is an example of a group of functions called **step functions.**

EXAMPLE 7 **Printing costs** To print stationery, a printer charges \$10 for setup charges, plus \$20 for each box. The printer counts any portion of a box as a full box. Graph this step function.

Solution If we order stationery and cancel before it is printed, the cost will be \$10. Thus, the ordered pair (0, 10) will be on the graph.

If we purchase 1 box, the cost will be \$10 for setup plus \$20 for printing, for a total cost of \$30. Thus, the ordered pair (1, 30) will be on the graph.

The cost of $1\frac{1}{2}$ boxes will be the same as the cost of 2 boxes, or \$50. Thus, the ordered pairs (1.5, 50) and (2, 50) will be on the graph.

The complete graph is shown in Figure 10-8.

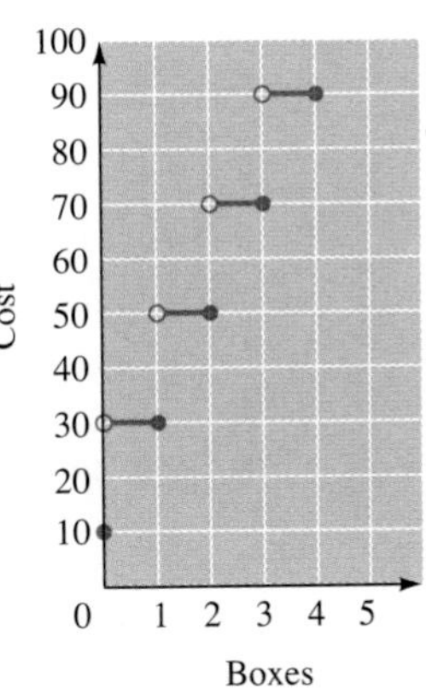

FIGURE 10-8

■

ORALS *Find the symmetries of the graph of each equation.*

1. $y = x^2$ y-axis
2. $y^2 = x$ x-axis
3. $y = x$ origin
4. $y = |x|$ y-axis
5. $y = \sqrt{x}$ none
6. $|y| = x$ x-axis

Give the domain and range of the relation determined by each graph.

7.

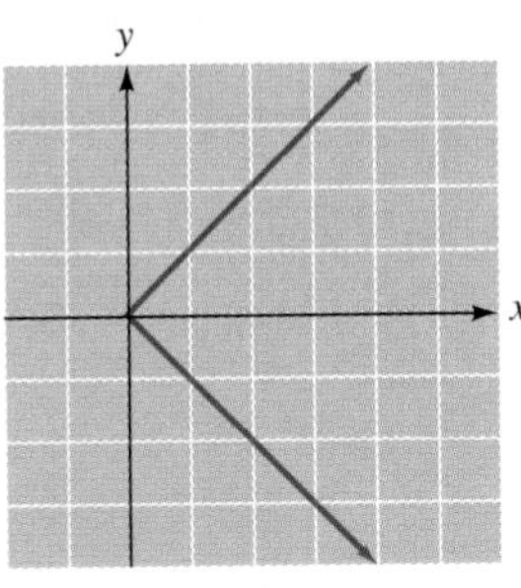

$[0, \infty), (-\infty, \infty)$

8.

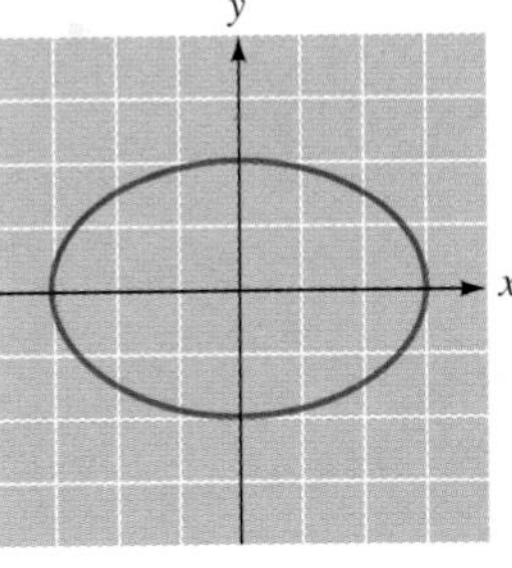

$[-3, 3], [-2, 2]$

EXERCISE 10.1

In Exercises 1–12, find the symmetries of the graph of each relation. **Do not draw the graph.**

1. $y = x^2 - 1$ y-axis
2. $y = x^3$ origin
3. $y = x^5$ origin
4. $y = x^4$ y-axis
5. $y = -x^2 + 2$ y-axis
6. $y = x^3 + 1$ none
7. $y = x^2 - x$ none
8. $y^2 = x + 7$ x-axis
9. $y = -|x + 2|$ none
10. $y = |x| - 3$ y-axis
11. $|y| = x$ x-axis
12. $y = 2\sqrt{x}$ none

In Exercises 13–28, graph each function and give its domain and range. Check your graph with a graphing calculator.

13. $y = x^4 - 4$

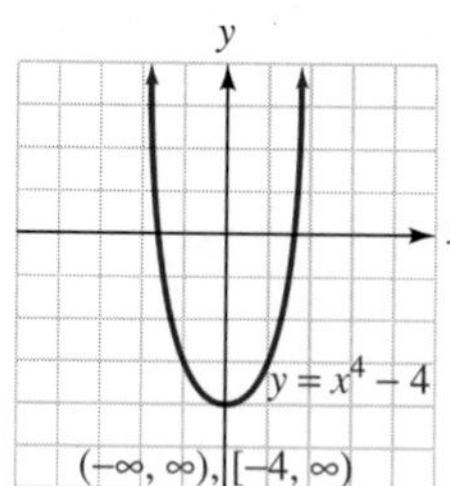

14. $y = \frac{1}{2}x^4 - 1$

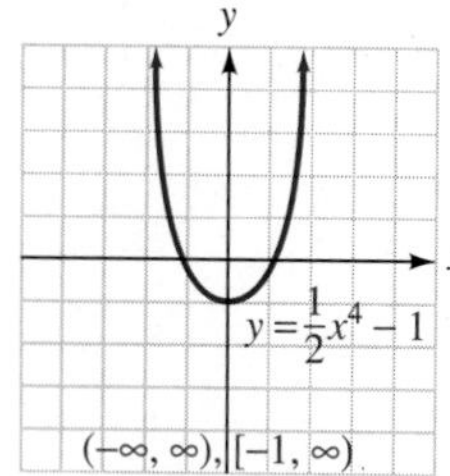

15. $y = -x^3$

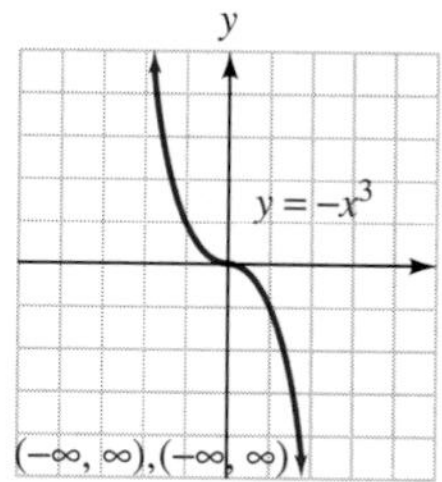

16. $y = x^3 + 2$

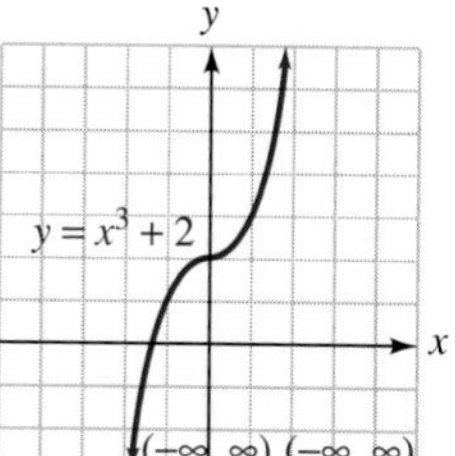

17. $y = x^4 + x^2$

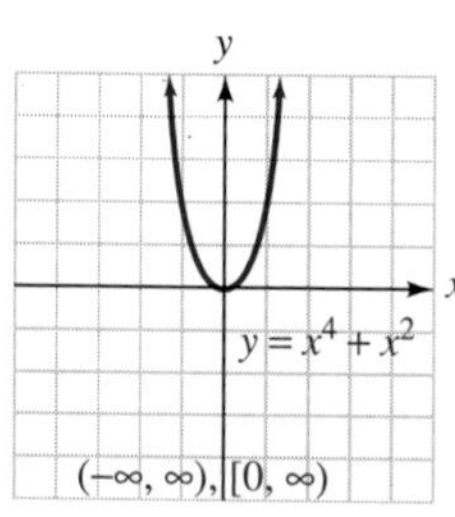

18. $y = 3 - x^4$

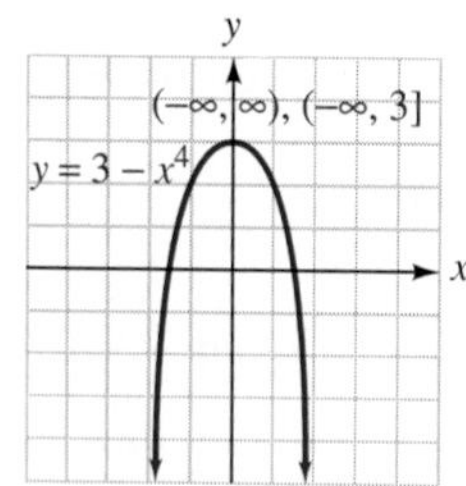

19. $y = x^3 - x$

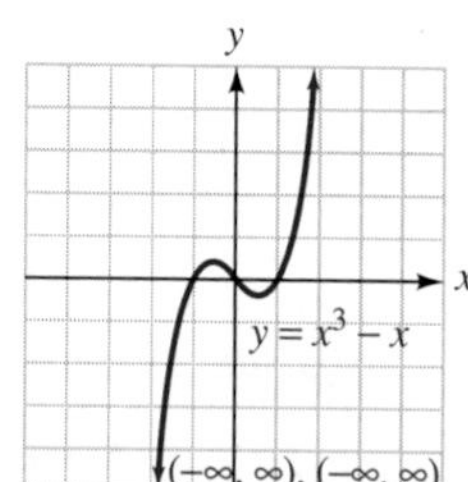

20. $y = x^3 + x$

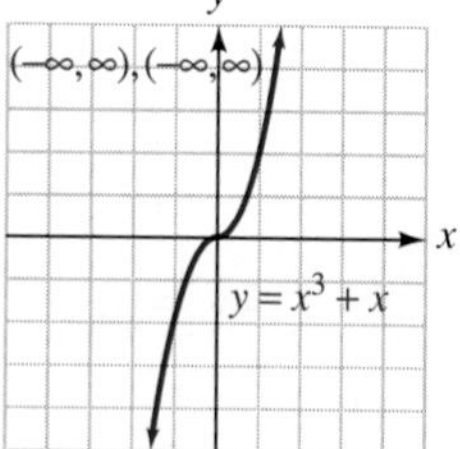

21. $y = \frac{1}{2}|x| - 1$

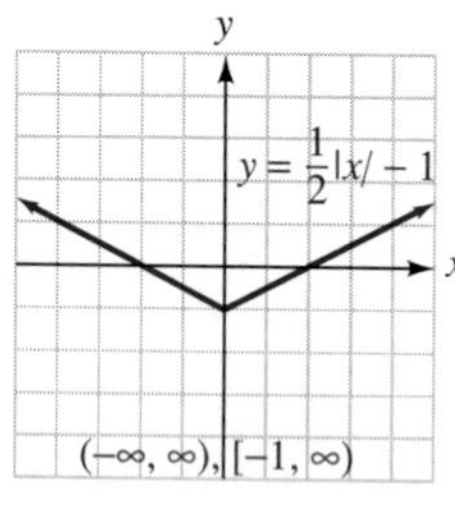

22. $y = -|x| + 1$

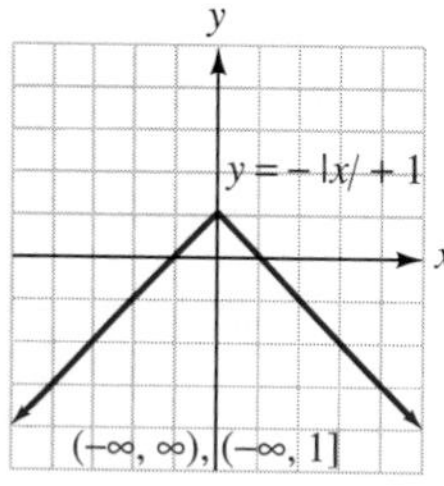

23. $y = -|x + 2|$

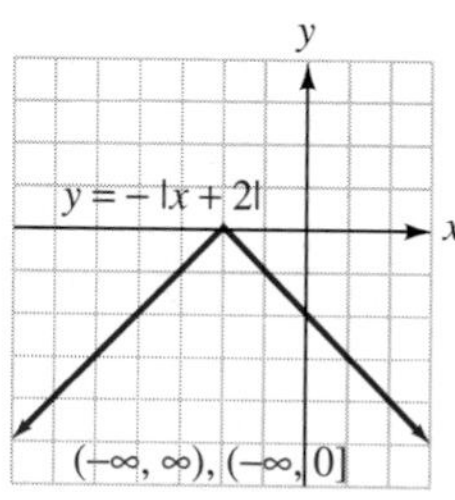

24. $y = |x - 2|$

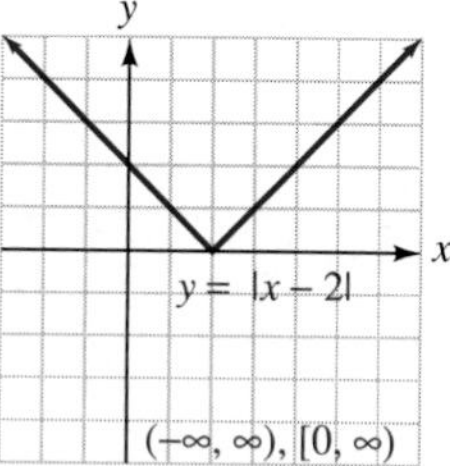

25. $y = \sqrt{x + 1}$

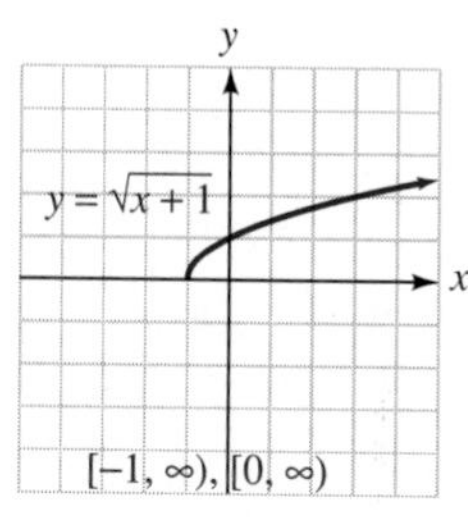

26. $y = -\sqrt{x + 4}$

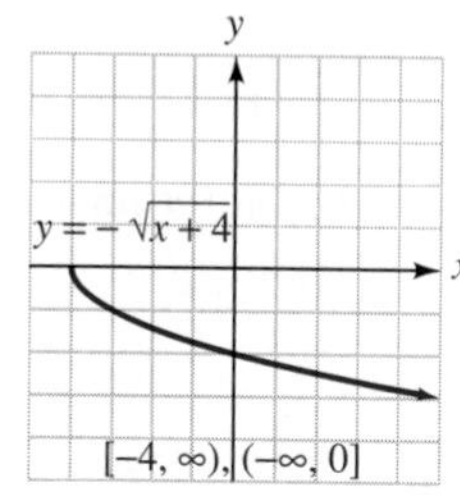

27. $y = -\sqrt{x} + 1$

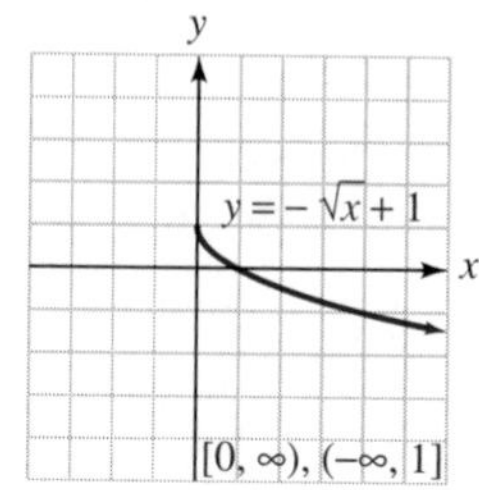

28. $y = \sqrt{x} - 2$

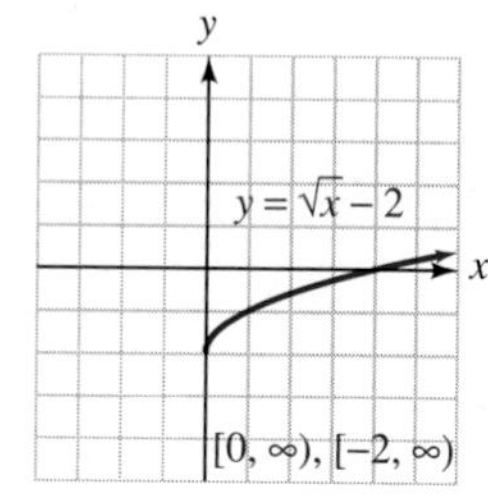

In Exercises 29–34, graph each function.

29. $y = \begin{cases} -1 & \text{if } x \leq 0 \\ x & \text{if } x > 0 \end{cases}$

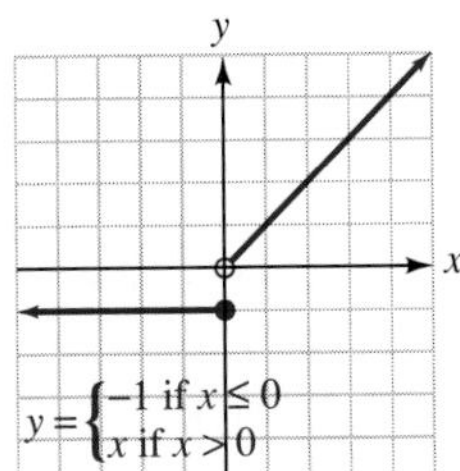

30. $y = \begin{cases} -2 & \text{if } x \leq 0 \\ x^2 & \text{if } x > 0 \end{cases}$

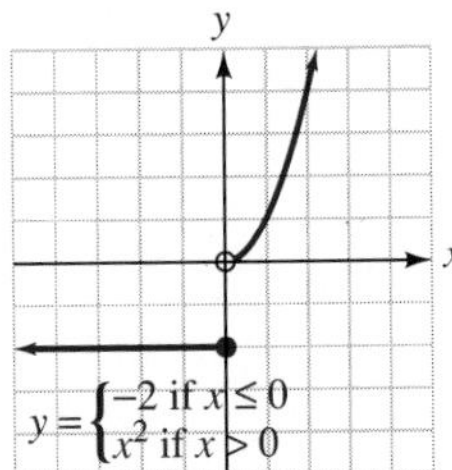

31. $y = \begin{cases} -x & \text{if } x \leq 0 \\ x & \text{if } 0 < x < 2 \\ -x & \text{if } x \geq 2 \end{cases}$

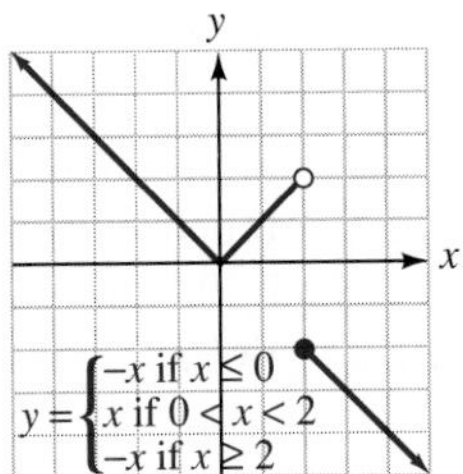

32. $y = \begin{cases} -x & \text{if } x < 0 \\ x^2 & \text{if } 0 \leq x \leq 1 \\ 1 & \text{if } x > 1 \end{cases}$

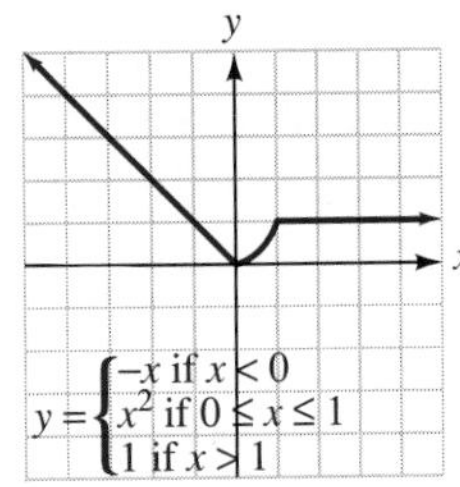

33. $y = -[\![x]\!]$

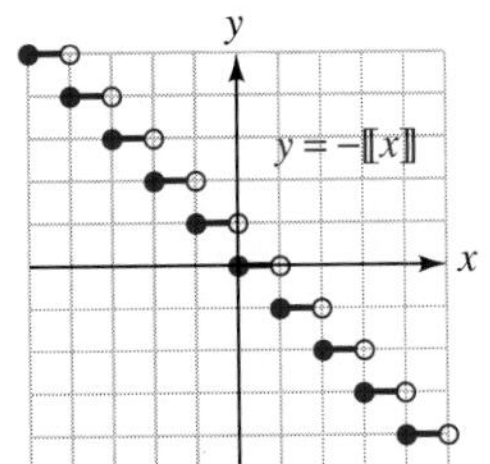

34. $y = [\![x]\!] + 2$

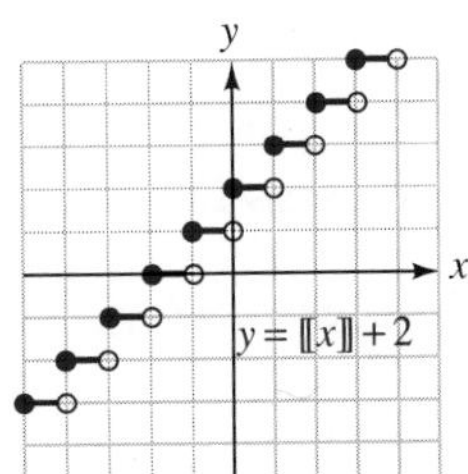

35. Signum function Computer programmers use a function, denoted by $y = \text{sgn } x$, that is defined in the following way:

$$y = \begin{cases} -1 & \text{if } x < 0 \\ 0 & \text{if } x = 0 \\ 1 & \text{if } x > 0 \end{cases}$$

Graph this function.

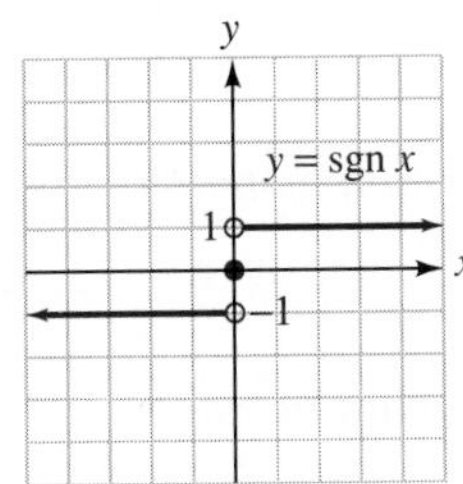

36. Heaviside unit step function This function, used in calculus, is defined by

$$y = \begin{cases} 1 & \text{if } x > 0 \\ 0 & \text{if } x < 0 \end{cases}$$

Graph this function.

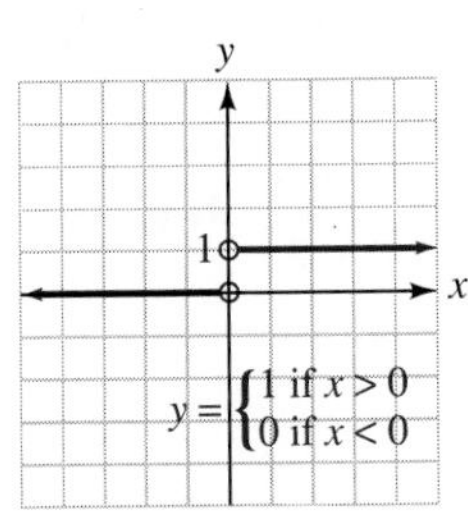

37. Renting a jetski A marina charges \$20 to rent a jetski for 1 hour, plus \$5 for every extra hour (or portion of an hour). Graph the ordered pairs (h, c), where h represents the number of hours and c represents the cost. Find the cost if the ski is used for 2.5 hours. \$30

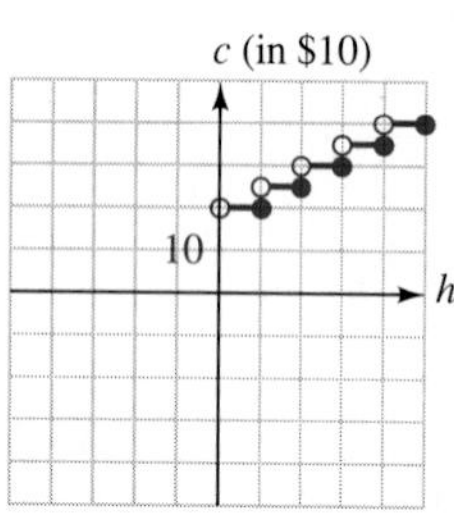

38. Riding in a taxi A cab company charges \$3 for a trip up to 1 mile, and \$2 for every extra mile (or portion of a mile). Graph the ordered pairs (m, c), where m represents the number of miles traveled and c represents the cost. Find the cost to ride $10\frac{1}{4}$ miles. \$23

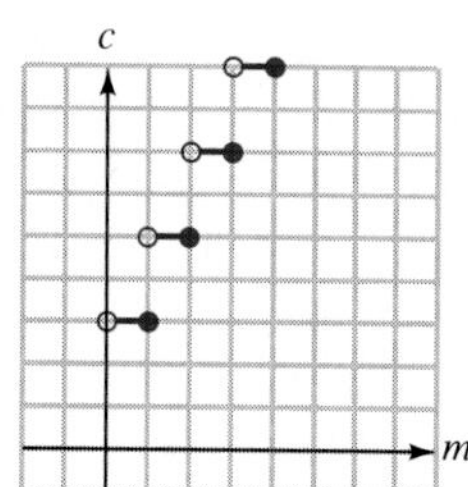

39. Information access Computer access to one international data network A costs \$10 per day plus \$8 per hour or fraction of an hour. Network B charges \$15 per day, but only \$6 per hour or fraction. For each network, graph the ordered pairs (t, C), where t represents connect time and C represents the total cost. Find the minimal daily usage at which it would be more economical to use network B.
After 2 hours, network B is cheaper.

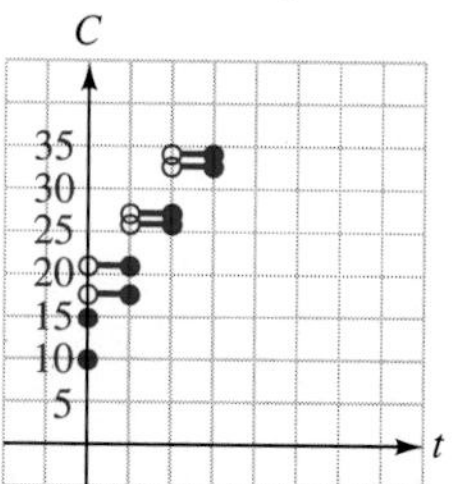

40. Royalties A publisher has agreed to pay the author of a popular novel 7% royalties on sales of the first 50,000 copies, 10% on sales of the next 50,000, and 15% on sales thereafter. If the book sells for \$10, express royalty income, I, as a function of s, the number of books sold, and graph the function. (*Hint:* When sales are into the second 50,000 copies, how much was earned on the first 50,000?)

$$I = \begin{cases} 0.7s & \text{if } 0 \leq s \leq 50{,}000 \\ s - 15{,}000 & \text{if } 50{,}000 < s \leq 100{,}000 \\ 1.5s - 65{,}000 & \text{if } s > 100{,}000 \end{cases}$$

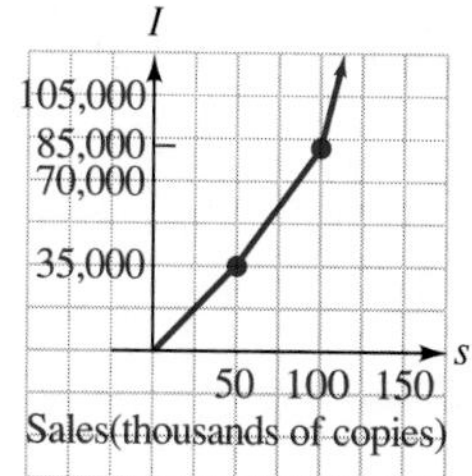

Writing Exercises

Write a paragraph using your own words.

1. Use a graphing calculator to graph $y = x^2 + k$ for several values of k. What do you discover?

2. Use a graphing calculator to graph $y = (x - k)^2$ for several values of k. What do you discover?

Something to Think About

1. How do the graphs of $y = [\![|x|]\!]$ and $y = |[\![x]\!]|$ differ?

2. How do the graphs of $y = -[\![x]\!]$ and $y = [\![-x]\!]$ differ?

Review Exercises

Write each number in scientific notation.

1. 9,300,000 9.3×10^6

2. 0.00023 2.3×10^{-4}

Write each number in standard notation.

3. 6.7×10^5 670,000

4. 6.89×10^{-6} 0.00000689

10.2 Algebra and Composition of Functions

■ Algebra of Functions ■ Composition of Functions ■ Identity Function
■ Problem Solving

■ Algebra of Functions

Functions can be added, subtracted, multiplied, and divided to form new functions.

Operations on Functions

If the ranges of functions f and g are subsets of the real numbers, then

The **sum** of f and g, denoted as $\boldsymbol{f + g}$, is defined by

$$(f + g)(x) = f(x) + g(x)$$

The **difference** of f and g, denoted as $\boldsymbol{f - g}$, is defined by

$$(f - g)(x) = f(x) - g(x)$$

The **product** of f and g, denoted as $\boldsymbol{f \cdot g}$, is defined by

$$(f \cdot g)(x) = f(x)g(x)$$

The **quotient** of f and g, denoted as $\boldsymbol{f/g}$, is defined by

$$(f/g)(x) = \frac{f(x)}{g(x)} \qquad (g(x) \neq 0)$$

The domain of each of these functions is the set of real numbers x that are in the domain of both f and g. In the case of the quotient, there is the further restriction that $g(x) \neq 0$.

 (a)

EXAMPLE 1 Let $f(x) = 2x^2 + 1$ and $g(x) = 5x - 3$. Find each of the following functions and its domain: **a.** $f + g$ and **b.** $f - g$.

Solution **a.**

$$\begin{aligned}(f + g)(x) &= \boldsymbol{f(x) + g(x)} \\ &= \boldsymbol{(2x^2 + 1) + (5x - 3)} \\ &= 2x^2 + 5x - 2\end{aligned}$$

The domain of $f + g$ is the set of real numbers that are in the domain of both f and g. Since the domain of both f and g is the set of real numbers, the domain of $f + g$ is also the set of real numbers.

b.

$$\begin{aligned}(f - g)(x) &= \boldsymbol{f(x) - g(x)} \\ &= \boldsymbol{(2x^2 + 1) - (5x - 3)} \\ &= 2x^2 + 1 - 5x + 3 && \text{Remove parentheses.} \\ &= 2x^2 - 5x + 4 && \text{Combine like terms.}\end{aligned}$$

Since the domain of both f and g is the set of real numbers, the domain of $f - g$ is also the set of real numbers. ■

 (b)

EXAMPLE 2 Let $f(x) = 2x^2 + 1$ and $g(x) = 5x - 3$. Find each of the following functions and its domain: **a.** $f \cdot g$ and **b.** f/g.

Solution **a.**

$$\begin{aligned}(f \cdot g)(x) &= f(x)g(x) \\ &= (2x^2 + 1)(5x - 3) \\ &= 10x^3 - 6x^2 + 5x - 3\end{aligned}$$

The domain $f \cdot g$ is the set of real numbers that are in the domain of both f and g. Since the domain of both f and g is the set of real numbers, the domain of $f \cdot g$ is also the set of real numbers.

b.

$$\begin{aligned}(f/g)(x) &= \frac{f(x)}{g(x)} \\ &= \frac{2x^2 + 1}{5x - 3}\end{aligned}$$

Since the denominator of the fraction cannot be 0, $x \neq \frac{3}{5}$. Thus, the domain of f/g is the set of real numbers, except $\frac{3}{5}$. ■

Évariste Galois
(1811–1832)
Galois was a French revolutionary and mathematician who was imprisoned twice for his republican sympathies. By the age of 17, he had made important contributions to the fields of number theory, functions, and theory of equations. At the age of 21, he was killed in a duel by a political opponent.

Composition of Functions

Often one quantity is a function of a second quantity that depends, in turn, on a third quantity. For example, the cost of a vacation is a function of the gasoline consumed. The cost of gas, in turn, is a function of the number of miles driven. Such chains of dependence can be analyzed mathematically as **compositions of functions.**

Suppose that $y = f(x)$ and $y = g(x)$ define two functions. Any number x in the domain of g will produce the corresponding value $g(x)$ in the range of g. If $g(x)$ is in the domain of function f, then $g(x)$ can be substituted into f, and a corresponding value $f(\mathbf{g(x)})$ will be determined. This two-step process defines a new function, called a **composite function,** denoted by $f \circ g$.

Composite Function The **composite function $f \circ g$** is defined by

$$(f \circ g)(x) = f(g(x))$$

For example, if $f(x) = 4x$ and $g(x) = 3x + 2$, then

$$\begin{aligned}(f \circ g)(x) &= f(\mathbf{g(x)}) \\ &= f(\mathbf{3x + 2}) \\ &= 4(\mathbf{3x + 2}) \\ &= 12x + 8\end{aligned}$$

or

$$\begin{aligned}(g \circ f)(x) &= g(\mathbf{f(x)}) \\ &= g(\mathbf{4x}) \\ &= 3(\mathbf{4x}) + 2 \\ &= 12x + 2\end{aligned}$$

WARNING! Note that in the previous example, $(f \circ g)(x) \neq (g \circ f)(x)$. This shows that the composition of functions is not commutative.

We have seen that a function can be represented by a machine: We put in a number from the domain, and a number from the range comes out. For example, if we put the number 2 into the machine shown in Figure 10-9(a), the number $f(2) = 8$ comes out. In general, if we put x into the machine shown in Figure 10-9(b), the value $f(x)$ comes out.

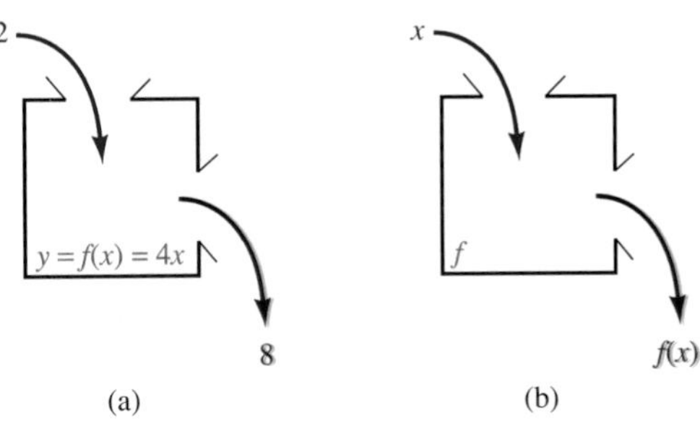

FIGURE 10-9

The function machines shown in Figure 10-10 illustrate the composition $f \circ g$. When we put a number x into the function g, $g(x)$ comes out. The value $g(x)$ immediately falls into function f, which transforms $g(x)$ into $f(g(x))$. If functions g and f are thought of as two-stage internal workings of a single function machine, that single machine would be named $f \circ g$.

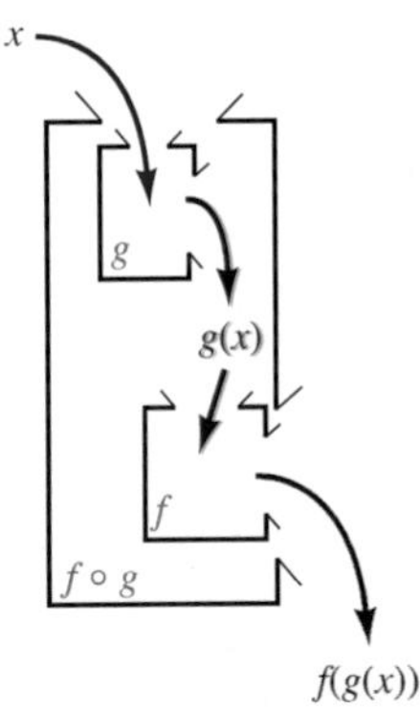

FIGURE 10-10

To be in the domain of the composite function $f \circ g$, a number x has to be in the domain of g. Also, the output of g must be in the domain of f. Thus, the domain of $f \circ g$ consists of those numbers x that are in the domain of g, and for which $g(x)$ is in the domain of f.

(a, c)

EXAMPLE 3 Let $f(x) = 2x + 1$ and $g(x) = x - 4$. Find **a.** $(f \circ g)(9)$, **b.** $(g \circ f)(-2)$, and **c.** $(f \circ g)(x)$.

Solution **a.** $(f \circ g)(9)$ means $f(g(9))$. In Figure 10-11(a), function g receives the number 9, subtracts 4, and releases the number $g(x) = 5$. The 5 falls into the f function, which doubles 5 and adds 1. The final result, 11, is the output of the composite function $f \circ g$:

$$(f \circ g)(9) = f(g(9)) = f(5) = 2(5) + 1 = 11$$

b. $(g \circ f)(-2)$ means $g(f(-2))$. In Figure 10-11(b), function f receives the number -2, doubles it and adds 1, and releases -3 into the g function. Function g subtracts 4 from -3 and releases a final result of -7. Thus,

$$(g \circ f)(-2) = g(f(-2)) = g(-3) = -3 - 4 = -7$$

c. $(f \circ g)(x)$ means $f(g(x))$. Refer to Figure 10-11(c) to see how the input value x is processed in two stages to produce the final output value, $2x - 7$:

$$(f \circ g)(x) = f(g(x)) = f(x - 4) = 2(x - 4) + 1 = 2x - 7$$

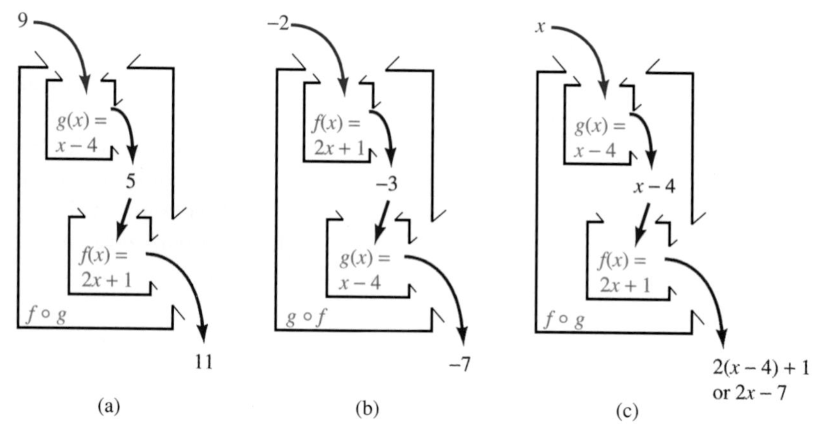

FIGURE 10-11

Identity Function

A special function, called the **identity function,** is defined by the equation $I(x) = x$. Under this function, the value that corresponds to any real number x is x itself. If f is any function, the composition of f with the identity function is just the function f:

$$(f \circ I)(x) = (I \circ f)(x) = f(x)$$

EXAMPLE 4 Let f be any function and I be the identity function, $I(x) = x$. Show that **a.** $(f \circ I)(x) = f(x)$ and **b.** $(I \circ f)(x) = f(x)$.

Solution **a.** $(f \circ I)(x)$ means $f(I(x))$. Because $I(x) = x$, we have

$$(f \circ I)(x) = f(I(x)) = f(x)$$

b. $(I \circ f)(x)$ means $I(f(x))$. Because I passes any number through unchanged, we have $I(f(x)) = f(x)$ and

$$(I \circ f)(x) = I(f(x)) = f(x)$$

Problem Solving

EXAMPLE 5 **Temperature change** A laboratory sample is removed from a cooler at a temperature of 15° Fahrenheit. Technicians are warming the sample at a controlled rate of 3° F per hour. Express the sample's Celsius temperature as a function of the time, t, since it was removed from refrigeration.

Solution The temperature of the sample is 15° F when $t = 0$. Because it warms 3° F per hour after that, its Fahrenheit temperature is given by the function

$$F(t) = 3t + 15$$

The Celsius temperature, C, is a function of the Fahrenheit temperature, F, given by the equation

$$C(F) = \frac{5}{9}(F - 32)$$

To express the sample's Celsius temperature as a function of time, we find the composition function

$$\begin{aligned}(C \circ F)(t) &= C(F(t))\\ &= \frac{5}{9}(F(t) - 32)\\ &= \frac{5}{9}[(3t + 15) - 32]\\ &= \frac{5}{9}(3t - 17) \qquad \text{Remove brackets and combine terms.}\end{aligned}$$

■

ORALS *If $f(x) = 2x$, $g(x) = 3x$, and $h(x) = 4x$, find*

1. $f + g$ $5x$ **2.** $h - g$ x **3.** $f \cdot h$ $8x^2$
4. g/f $\frac{3}{2}$ **5.** h/f 2 **6.** $g \cdot h$ $12x^2$
7. $f \circ h$ $8x$ **8.** $f \circ g$ $6x$ **9.** $g \circ h$ $12x$

EXERCISE 10.2

In Exercises 1–8, $f(x) = 3x$ and $g(x) = 4x$. Find each function and its domain.

1. $f + g$ $7x, (-\infty, \infty)$ **2.** $f - g$ $-x, (-\infty, \infty)$ **3.** $f \cdot g$ $12x^2, (-\infty, \infty)$ **4.** f/g $\frac{3}{4}, (-\infty, 0) \cup (0, \infty)$
5. $g - f$ $x, (-\infty, \infty)$ **6.** $g + f$ $7x, (-\infty, \infty)$ **7.** g/f $\frac{4}{3}, (-\infty, 0) \cup (0, \infty)$ **8.** $g \cdot f$ $12x^2, (-\infty, \infty)$

In Exercises 9–16, $f(x) = 2x + 1$ and $g(x) = x - 3$. Find each function and its domain.

9. $f + g$ $3x - 2, (-\infty, \infty)$ **10.** $f - g$ $x + 4, (-\infty, \infty)$ **11.** $f \cdot g$ $2x^2 - 5x - 3, (-\infty, \infty)$ **12.** f/g $\frac{2x + 1}{x - 3}, (-\infty, 3) \cup (3, \infty)$

13. $g - f$
$-x - 4, (-\infty, \infty)$

14. $g + f$
$3x - 2, (-\infty, \infty)$

15. g/f
$\dfrac{x-3}{2x+1}, \left(-\infty, -\frac{1}{2}\right) \cup \left(-\frac{1}{2}, \infty\right)$

16. $g \cdot f$
$2x^2 - 5x - 3, (-\infty, \infty)$

In Exercises 17–20, $f(x) = 3x - 2$ and $g(x) = 2x^2 + 1$. Find each function and its domain.

17. $f - g$
$-2x^2 + 3x - 3, (-\infty, \infty)$

18. $f + g$
$2x^2 + 3x - 1, (-\infty, \infty)$

19. f/g
$(3x - 2)/(2x^2 + 1), (-\infty, \infty)$

20. $f \cdot g$
$6x^3 - 4x^2 + 3x - 2, (-\infty, \infty)$

In Exercises 21–24, $f(x) = x^2 - 1$ and $g(x) = x^2 - 4$. Find each function and its domain.

21. $f - g$
$3, (-\infty, \infty)$

22. $f + g$
$2x^2 - 5, (-\infty, \infty)$

23. g/f $(x^2 - 4)/(x^2 - 1)$,
$(-\infty, -1) \cup (-1, 1) \cup (1, \infty)$

24. $g \cdot f$
$x^4 - 5x^2 + 4, (-\infty, \infty)$

In Exercises 25–36, $f(x) = 2x + 1$ and $g(x) = x^2 - 1$. Find each value.

25. $(f \circ g)(2)$ 7

26. $(g \circ f)(2)$ 24

27. $(g \circ f)(-3)$ 24

28. $(f \circ g)(-3)$ 17

29. $(f \circ g)(0)$ -1

30. $(g \circ f)(0)$ 0

31. $(f \circ g)\left(\dfrac{1}{2}\right)$ $-\frac{1}{2}$

32. $(g \circ f)\left(\dfrac{1}{3}\right)$ $\frac{16}{9}$

33. $(f \circ g)(x)$ $2x^2 - 1$

34. $(g \circ f)(x)$ $4x^2 + 4x$

35. $(f \circ g)(2x)$ $16x^2 + 8x$

36. $(g \circ f)(2x)$ $8x^2 - 1$

In Exercises 37–44, $f(x) = 3x - 2$ and $g(x) = x^2 + x$. Find each value.

37. $(f \circ g)(4)$ 58

38. $(g \circ f)(4)$ 110

39. $(g \circ f)(-3)$ 110

40. $(f \circ g)(-3)$ 16

41. $(g \circ f)(0)$ 2

42. $(f \circ g)(0)$ -2

43. $(f \circ g)(x)$ $3x^2 + 3x - 2$

44. $(g \circ f)(x)$ $9x^2 - 9x + 2$

45. If $f(x) = x + 1$ and $g(x) = 2x - 5$, show that $(f \circ g)(x) \neq (g \circ f)(x)$.

46. If $f(x) = x^2 + 1$ and $g(x) = 3x^2 - 2$, show that $(f \circ g)(x) \neq (g \circ f)(x)$.

47. If $f(x) = x^2 + 2x - 3$, find $f(a)$, $f(h)$, and $f(a + h)$. Then show that $f(a + h) \neq f(a) + f(h)$.

48. If $g(x) = 2x^2 + 10$, find $g(a)$, $g(h)$, and $g(a + h)$. Then show that $g(a + h) \neq g(a) + g(h)$.

49. If $f(x) = x^2 + 2$, find $\dfrac{f(x+h) - f(x)}{h}$. $2x + h$

50. If $f(x) = x^3 - 1$, find $\dfrac{f(x+h) - f(x)}{h}$. $3x^2 + 3xh + h^2$

51. **Alloys** A molten alloy must be cooled slowly to control crystalization. When removed from the furnace, its temperature is 2700° Fahrenheit, and it will be cooled at 200° per hour. Express the Celsius temperature as a function of the number of hours, t, since cooling began.
$C(t) = \frac{5}{9}(2668 - 200t)$

52. **Weather forecasting** A high pressure area promises increasingly warmer weather for the next 48 hours. The temperature is now 34° Celsius and is expected to rise 1° every 6 hours. Express the Fahrenheit temperature as a function of the number of hours from now. (*Hint:* $F = \frac{9}{5}C + 32$.)
$F(t) = \frac{9}{5}(34 + \frac{1}{6}t) + 32$

Writing Exercises ■ *Write a paragraph using your own words.*

1. Explain how to find the domain of f/g.

2. Explain how to find the domain of $(f \circ g)(x)$.

Something to Think About

1. Is composition of functions associative? Choose functions f, g, and h, and determine if $[f \circ (g \circ h)](x) = [(f \circ g) \circ h](x)$.

2. Let $f(x) = [\![x]\!]$. What is $(f \circ f)(x)$? $[\![[\![x]\!]]\!] = [\![x]\!]$

Review Exercises

Simplify each expression.

1. $\dfrac{3x^2 + x - 14}{4 - x^2}$ $-\dfrac{3x + 7}{x + 2}$

2. $\dfrac{2x^3 + 14x^2}{3 + 2x - x^2} \cdot \dfrac{x^2 - 3x}{x}$ $-\dfrac{2x^2(x + 7)}{x + 1}$

3. $\dfrac{8 + 2x - x^2}{12 + x - 3x^2} \div \dfrac{3x^2 + 5x - 2}{3x - 1}$ $\dfrac{x - 4}{3x^2 - x - 12}$

4. $\dfrac{x - 1}{1 + \dfrac{x}{x - 2}}$ $\dfrac{x - 2}{2}$

10.3 Inverses of Functions

■ Inverses of Functions ■ One-to-One Functions ■ The Horizontal Line Test

The linear function defined by the equation $C = \frac{5}{9}(F - 32)$ shows how to convert degrees Fahrenheit to degrees Celsius. If we substitute a Fahrenheit reading into the formula, a Celsius reading comes out. For example, if we substitute 41° for F in the formula, we obtain a Celsius reading of 5°:

$$C = \frac{5}{9}(F - 32)$$
$$= \frac{5}{9}(\mathbf{41} - 32)$$
$$= \frac{5}{9}(9)$$
$$= 5$$

If we want to find a Fahrenheit reading from a Celsius reading, we need a formula into which we can substitute a Celsius reading and have a Fahrenheit reading come out. Such a formula is $F = \frac{9}{5}C + 32$, which takes the Celsius reading of 5° and turns it back into a Fahrenheit reading of 41°.

$$F = \frac{9}{5}C + 32$$
$$= \frac{9}{5}(\mathbf{5}) + 32$$
$$= 41$$

The functions defined by these two formulas do opposite things. The first turns 41° into 5°, and the second turns 5° back into 41°. Such functions are called **inverse functions.**

Inverses of Functions

If R is a function determined by the set of ordered pairs

$$R = \{(1, 10), (2, 20), (3, 30)\}$$

then R turns the number 1 into 10, turns 2 into 20, and turns 3 into 30. Since the inverse of R, denoted as R^{-1}, must turn 10 back into 1, turn 20 back into 2, and turn 30 back into 3, it must be the set of ordered pairs

$$R^{-1} = \{(10, 1), (20, 2), (30, 3)\}$$

To form the inverse of the function R, we simply interchange the coordinates of each ordered pair that determines R.

WARNING! The -1 in the notation R^{-1} is not an exponent. It refers to the inverse of R. The symbol R^{-1} is read as "the inverse relation of R" or just "R inverse."

The domain of R and the range of R^{-1} is $\{1, 2, 3\}$. The range of R and the domain of R^{-1} is $\{10, 20, 30\}$.

The function R can be illustrated by the diagram in Figure 10-12(a). The function R is represented by arrows drawn from each number in the domain to each corresponding value of y in the range.

R^{-1} is illustrated in Figure 10-12(b). In this figure, the direction of the arrows has been reversed. Because each arrow in the figure points to exactly one value of y, R^{-1} is also a function.

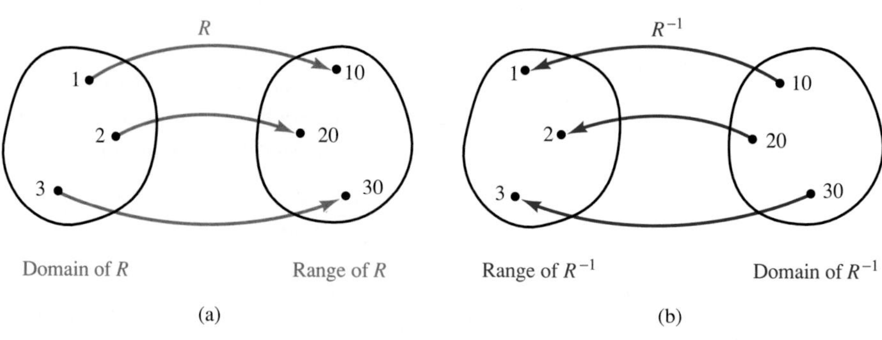

FIGURE 10-12

Inverse Relations If R is any function and R^{-1} is the relation obtained from R by interchanging the components of each ordered pair of R, then R^{-1} is called the **inverse relation of R.**

The domain of R^{-1} is the range of R, and the range of R^{-1} is the domain of R.

Recall that any linear equation in x and y determines a linear function. For example, $y = 4x + 2$ determines the linear function R consisting of infinitely many ordered pairs (x, y). If $x = 0$, the corresponding value of y is

$$4(\mathbf{0}) + 2$$

or 2. Thus, the pair (0, 2) is an element of R. A partial listing of ordered pairs in R is

$$R = \left\{(0, 2), (1, 6), \left(\frac{1}{4}, 3\right), (-3, -10), \ldots\right\}$$

To form a partial listing of R^{-1}, we interchange the x- and y-coordinates of the elements of R to obtain

$$R^{-1} = \left\{(2, 0), (6, 1), \left(3, \frac{1}{4}\right), (-10, -3), \ldots\right\}$$

To generate other elements of R^{-1} conveniently, and to decide whether R^{-1} is a function, we need an equation. To find an equation for R^{-1}, we proceed as in Example 1.

EXAMPLE 1 Find R^{-1}, the inverse relation of $y = 4x + 2$, and tell whether the inverse is a function.

Solution To find the inverse relation of $y = 4x + 2$, we interchange the variables x and y to obtain

1. $x = 4y + 2$

To decide whether the inverse relation is a function, we solve Equation 1 for y:

$$x = 4y + 2$$

$$x - 2 = 4y \qquad \text{Subtract 2 from both sides.}$$

2. $$y = \frac{x - 2}{4} \qquad \text{Divide both sides by 4 and use the symmetric property.}$$

Because each number x that is substituted into Equation 2 gives a single value y, the inverse relation R^{-1} is a function. ■

In Example 1, the inverse relation of the function $y = 4x + 2$ was found to be the function $y = \frac{x-2}{4}$. In function notation, this inverse function can be denoted as

$$f^{-1}(x) = \frac{x-2}{4} \qquad \text{Read as "}f\text{ inverse of }x\text{ is }\tfrac{x-2}{4}\text{."}$$

To see an important relationship between a function and its inverse, we substitute some number x, such as $x = 3$, into the function $f(x) = 4x + 2$ of Example 1. The corresponding value of y produced is

$$\begin{aligned} y &= f(\mathbf{3}) \\ &= 4(\mathbf{3}) + 2 \\ &= 14 \end{aligned}$$

If we substitute 14 into the inverse function, f^{-1}, the corresponding value of y that is produced is

$$\begin{aligned} y &= f^{-1}(\mathbf{14}) \\ &= \frac{\mathbf{14} - 2}{4} \\ &= 3 \end{aligned}$$

Thus, the function f turns 3 into 14, and the inverse function f^{-1} turns 14 back into 3. In general, the composition of a function and its inverse is the identity function.

To prove that $f(x) = 4x + 2$ and $f^{-1}(x) = \frac{x-2}{4}$ are inverse functions, we must show that their composition (in both directions) is the identity function:

$$\begin{aligned} (f \circ f^{-1})(x) &= f(\mathbf{f^{-1}(x)}) \\ &= f\left(\frac{\mathbf{x-2}}{\mathbf{4}}\right) \\ &= 4\left(\frac{\mathbf{x-2}}{\mathbf{4}}\right) + 2 \\ &= x - 2 + 2 \\ &= x \end{aligned}$$

$$\begin{aligned} (f^{-1} \circ f)(x) &= f^{-1}(\mathbf{f(x)}) \\ &= f^{-1}(\mathbf{4x + 2}) \\ &= \frac{\mathbf{4x + 2} - 2}{4} \\ &= \frac{4x}{4} \\ &= x \end{aligned}$$

Thus, $(f \circ f^{-1})(x) = (f^{-1} \circ f)(x) = x$, which is the identity function $I(x)$.

Steps for Finding the Inverse of a Relation

1. If the relation is given as a set of ordered pairs, interchange the x- and y-values to obtain the inverse relation.
2. If the relation is given as an equation,
   a. Interchange the variables x and y.
   b. Solve the resulting equation for y, if possible.
   c. This equation is $y = f^{-1}(x)$, which defines the inverse relation.

EXAMPLE 2 The set of all pairs (x, y) determined by the equation $3x + 2y = 6$ is a function. Find its inverse function and graph both functions on a single coordinate system.

Solution To find the inverse function of $3x + 2y = 6$, we interchange the variables x and y to obtain

$$3y + 2x = 6$$

and then solve the equation for y.

$$3y + 2x = 6$$

$$3y = -2x + 6 \qquad \text{Subtract } 2x \text{ from both sides.}$$

$$y = -\frac{2}{3}x + 2 \qquad \text{Divide both sides by 3.}$$

Thus, $y = f^{-1}(x) = -\dfrac{2}{3}x + 2$.

The graphs of the equations $3x + 2y = 6$ and $y = f^{-1}(x) = -\frac{2}{3}x + 2$ appear in Figure 10-13.

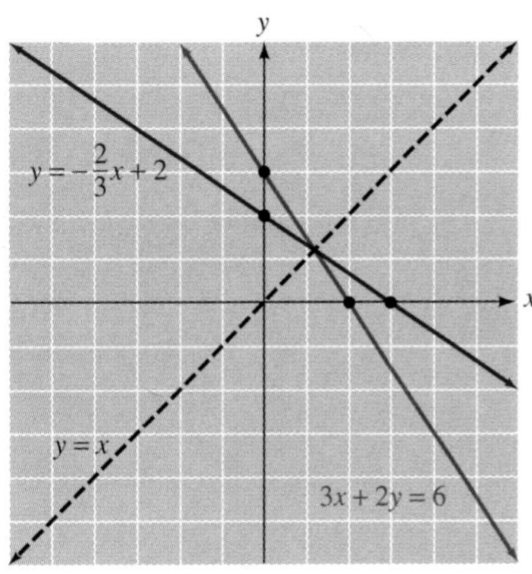

FIGURE 10-13

In Example 2, the graphs of the equations $3x + 2y = 6$ and $y = f^{-1}(x) = -\frac{2}{3}x + 2$ are symmetric about the line $y = x$. This is always the case, because when the coordinates (a, b) satisfy an equation, the coordinates (b, a) will satisfy its inverse.

In each example so far, the inverse of a function has been another function. This is not always true, as the following example will show.

EXAMPLE 3 Find the inverse relation of the function determined by $y = x^2$.

Solution

$$y = x^2$$

$$x = y^2 \qquad \text{Interchange } x \text{ and } y.$$

$$y = \pm\sqrt{x} \qquad \text{Use the symmetric property of equality and the square root property.}$$

Because each number x in the domain determines two values of y, the inverse relation is not a function. When the inverse relation $y = \pm\sqrt{x}$ is graphed as in Figure 10-14, we see that the graph does not pass the vertical line test.

The graph of $y = x^2$ is also shown in the figure. As expected, the graphs of $y = x^2$ and $y = \pm\sqrt{x}$ are symmetric about the line $y = x$.

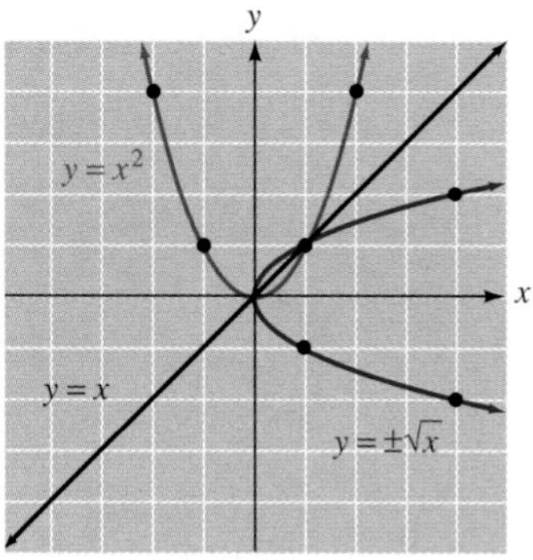

FIGURE 10-14

One-to-One Functions

We have seen that a function f can be illustrated with a diagram like the one shown in Figure 10-15(a), where the function f is represented by arrows drawn from each number x in the domain to its corresponding value y in the range.

We can visualize the inverse of f by reversing the direction of the arrows in Figure 10-15(a) to produce the diagram shown in Figure 10-15(b). In both parts of Figure 10-15, each arrow points to a single value y. Thus, both diagrams represent functions, and each function is the inverse of the other.

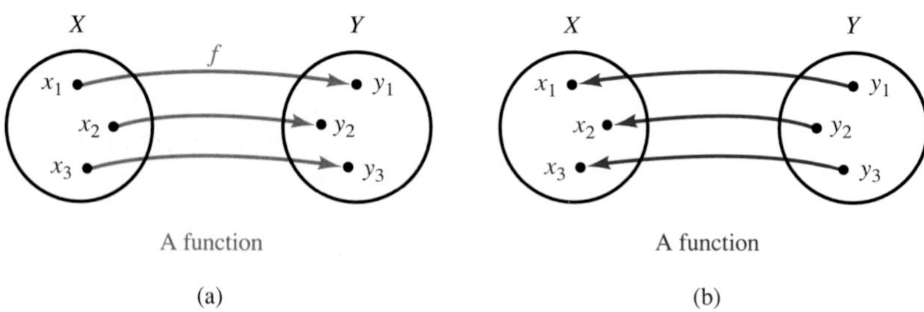

FIGURE 10-15

The correspondence shown in Figure 10-16(a) is a function from set X to set Y. However, when the arrows are reversed, as in Figure 10-16(b), the correspondence is not a function from set Y to set X, because to a single number y there correspond several values of x.

The diagrams in Figures 10-15 and 10-16 suggest these facts:

- If each y corresponds to a single number x, as in Figure 10-15(a), then the correspondence represents a function when the arrows are reversed.
- However, if one y corresponds to several numbers x, as in Figure 10-16(a), then the inverse is not a function.

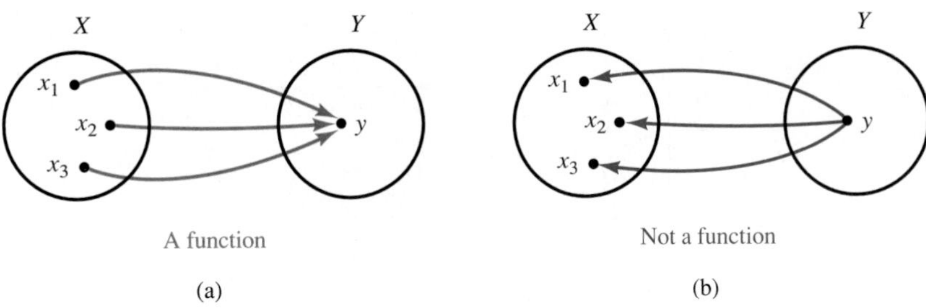

FIGURE 10-16

One-to-One Function A function is called **one-to-one** if and only if each value of y in the range corresponds to only one number x in the domain.

Because of the definition, the inverse of any one-to-one function is also a function.

The Horizontal Line Test

The **horizontal line test** can be used to decide whether a graph of a function represents a one-to-one function. If any horizontal line that intersects the graph of a function does so only once, the function is one-to-one. Otherwise, the function is not one-to-one.

EXAMPLE 4 Use the horizontal line test to decide whether the functions defined by **a.** $y = x^2 - 4$ and **b.** $y = x^3$ are one-to-one.

Solution **a.** Graph the equation as in Figure 10-17(a). Because many horizontal lines that intersect the graph do so twice, some values of y correspond to two numbers x. Thus, the function is not one-to-one.

b. Graph the equation as in Figure 10-17(b). Because each horizontal line that intersects the graph does so exactly once, each value of y corresponds to only one number x. Thus, the function defined by $y = x^3$ is one-to-one.

$y = x^2 - 4$

x	y
-3	5
-2	0
-1	-3
0	-4
1	-3
2	0
3	5

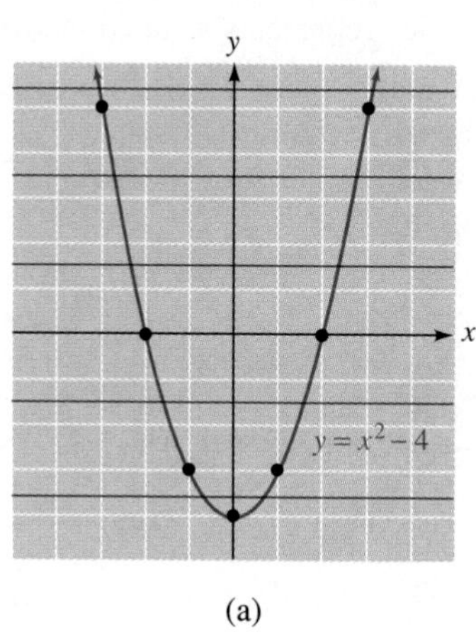

(a)

$y = x^3$

x	y
-2	-8
-1	-1
0	0
1	1
2	8

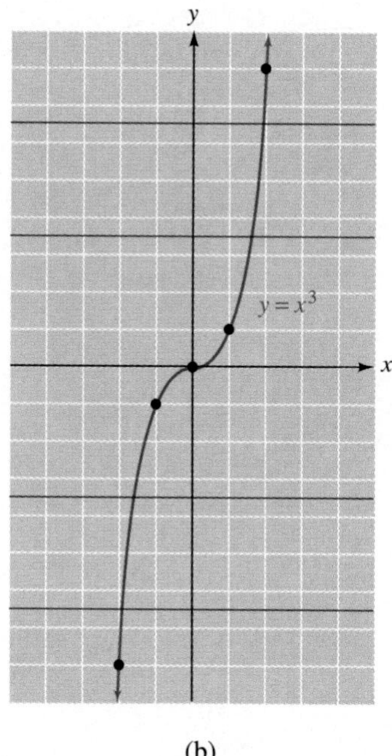

(b)

FIGURE 10-17 ■

WARNING! Make sure to use the vertical line test to determine if the graph of a relation is a function. If it is, use the horizontal line test to determine if the function is one-to-one.

EXAMPLE 5 Express the inverse of the function $y = f(x) = x^3$ in $y = f^{-1}(x)$ form.

Solution To find the inverse of the function, we proceed as follows:

$$y = x^3$$

$$x = y^3 \quad \text{Interchange the variables } x \text{ and } y.$$

$$\sqrt[3]{x} = y \quad \text{Take the cube root of both sides.}$$

We note that to each number x there corresponds a single cube root. Thus, the equation $y = \sqrt[3]{x}$ represents a function. In $f^{-1}(x)$ notation, we have

$$y = f^{-1}(x) = \sqrt[3]{x}$$ ■

If a function is not one-to-one, it is often possible to make it one-to-one by restricting its domain.

EXAMPLE 6 Find the inverse of the function defined by $y = x^2$ with $x \geq 0$. Then tell whether the inverse relation is a function. Graph the function and its inverse on a single set of coordinate axes.

Solution The inverse of the function $y = x^2$ with $x \geq 0$ is

$$x = y^2 \quad \text{with} \quad y \geq 0 \qquad \text{Interchange the variables } x \text{ and } y.$$

This equation can be written in the form

$$y = \pm\sqrt{x} \quad \text{with} \quad y \geq 0$$

Since $y \geq 0$, each number x gives only one value of y: $y = \sqrt{x}$. Thus, the inverse relation is a function.

The graphs of the two functions appear in Figure 10-18. The line $y = x$ is included so that we can see that the graphs are symmetric about the line $y = x$.

$y = x^2$ and $x \geq 0$

x	y
0	0
1	1
2	4
3	9

$x = y^2$ and $y \geq 0$

x	y
0	0
1	1
4	2
9	3

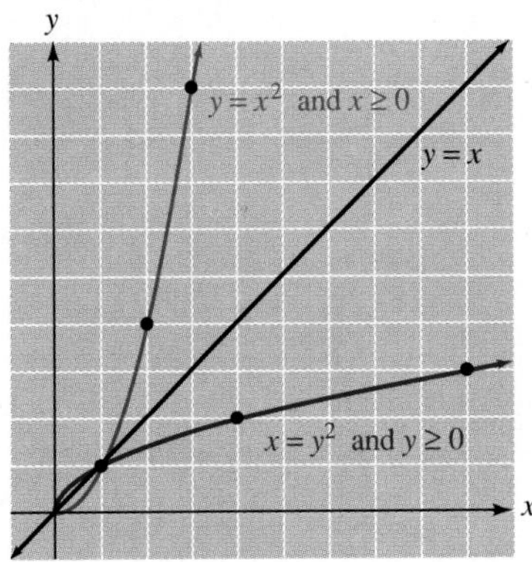

FIGURE 10-18

ORALS *Find the inverse relation of each set of ordered pairs.*

1. $\{(1, 2), (2, 3), (5, 10)\}$
   $\{(2, 1), (3, 2), (10, 5)\}$
2. $\{(1, 1), (2, 8), (4, 64)\}$
   $\{(1, 1), (8, 2), (64, 4)\}$

Find the inverse function of each linear function.

3. $y = \frac{1}{2}x$ $\quad f^{-1}(x) = 2x$
4. $y = 2x$ $\quad f^{-1}(x) = \frac{1}{2}x$

Tell whether each function is one-to-one.

5. $y = x^2 - 2$ no
6. $y = x^5$ yes

EXERCISE 10.3

In Exercises 1–6, find the inverse relation of each set of ordered pairs (x, y) and tell whether the inverse relation is a function.

1. $\{(3, 2), (2, 1), (1, 0)\}$
   $\{(2, 3), (1, 2), (0, 1)\}$, yes
2. $\{(4, 1), (5, 1), (6, 1), (7, 1)\}$
   $\{(1, 4), (1, 5), (1, 6), (1, 7)\}$, no
3. $\{(1, 2), (2, 3), (1, 3), (1, 5)\}$
   $\{(2, 1), (3, 2), (3, 1), (5, 1)\}$, no

4. $\{(-1, -1), (0, 0), (1, 1), (2, 2)\}$
$\{(-1, -1), (0, 0), (1, 1), (2, 2)\}$, yes

5. $\{(1, 1), (2, 4), (3, 9), (4, 16)\}$
$\{(1, 1), (4, 2), (9, 3), (16, 4)\}$, yes

6. $\{(1, 1), (2, 1), (3, 1), (4, 1)\}$
$\{(1, 1), (1, 2), (1, 3), (1, 4)\}$, no

In Exercises 7–14, find the inverse of the relation determined by the given equation and tell whether that inverse relation is a function. If the inverse relation is a function, express it in the form $y = f^{-1}(x)$.

7. $y = 3x + 1$
$y = f^{-1}(x) = \frac{x-1}{3}$

8. $y + 1 = 5x$
$y = f^{-1}(x) = \frac{x+1}{5}$

9. $x + 4 = 5y$
$y = f^{-1}(x) = 5x - 4$

10. $x = 3y + 1$
$y = f^{-1}(x) = 3x + 1$

11. $y = \dfrac{x-4}{5}$
$y = f^{-1}(x) = 5x + 4$

12. $y = \dfrac{2x+6}{3}$
$y = f^{-1}(x) = \frac{3x-6}{2}$

13. $4x - 5y = 20$
$y = f^{-1}(x) = \frac{5x+20}{4}$

14. $3x + 5y = 15$
$y = f^{-1}(x) = \frac{15-5x}{3}$

In Exercises 15–24, find the inverse of each linear function. Then graph both the function and its inverse on a single coordinate system. Find the equation of the line of symmetry.

15. $y = 4x + 3$

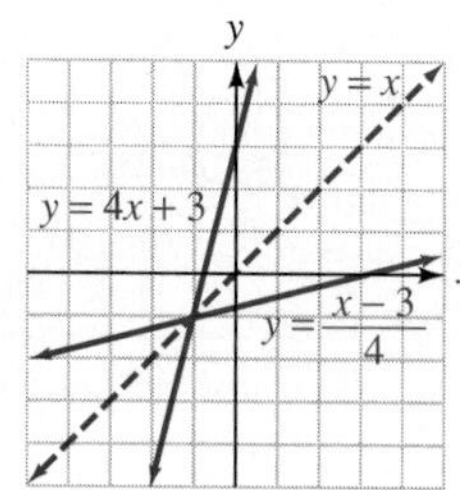

16. $x = 3y - 1$

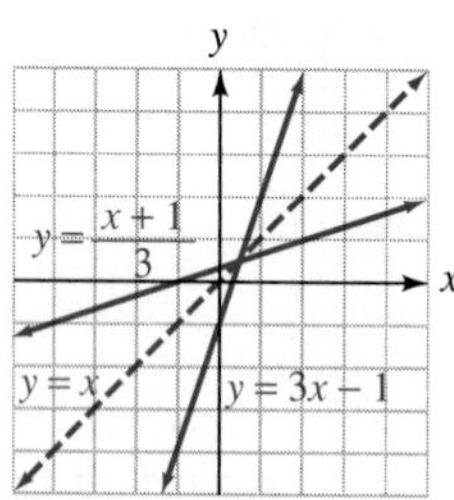

17. $x = \dfrac{y-2}{3}$

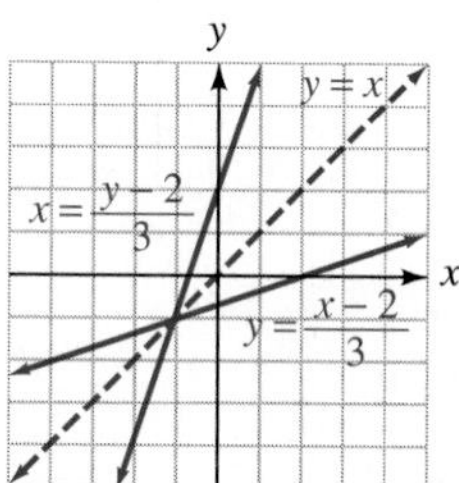

18. $y = \dfrac{x+3}{4}$

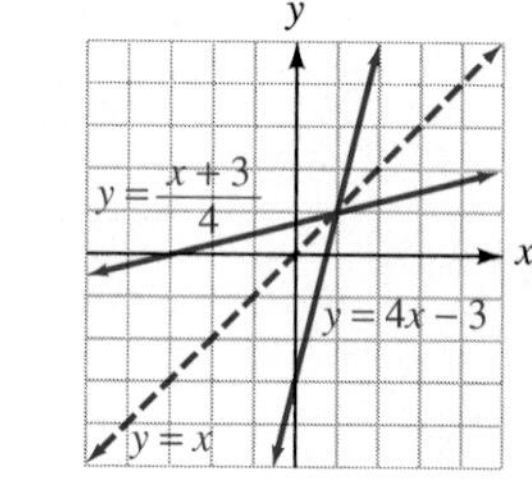

19. $3x - y = 5$

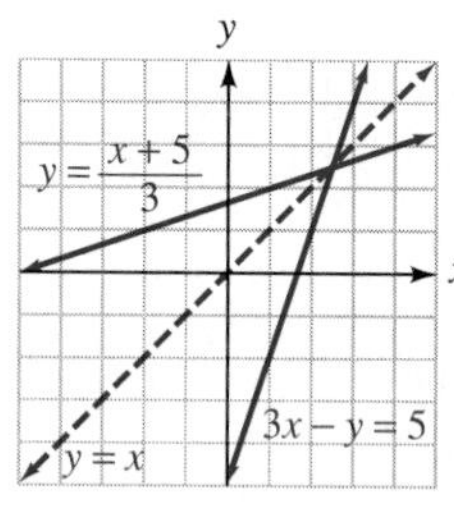

20. $2x + 3y = 9$

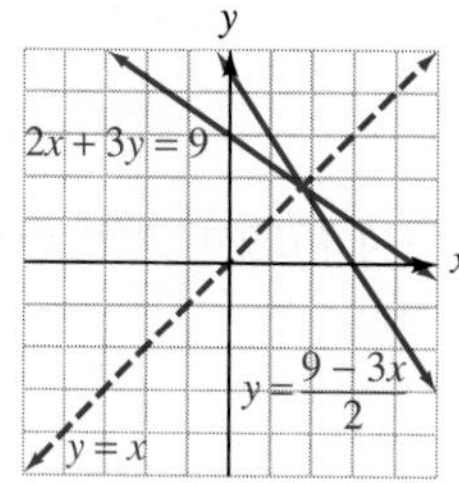

21. $3(x + y) = 2x + 4$

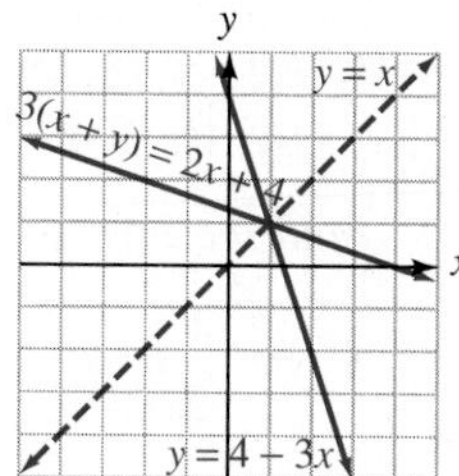

22. $-4(y - 1) + x = 2$

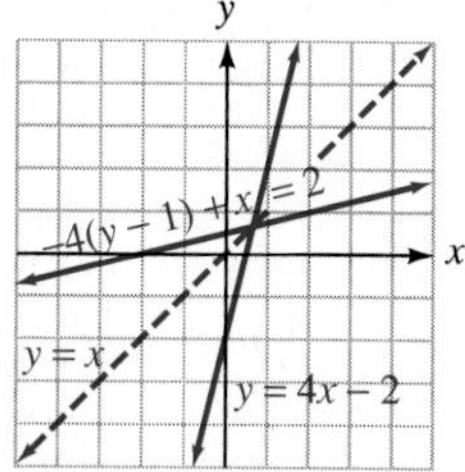

23. $3x = 2(1 - y)$

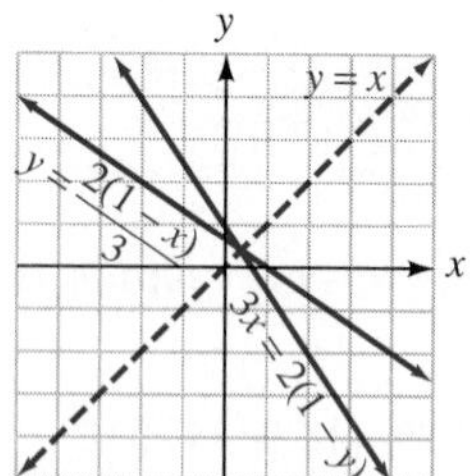

24. $2\left(y + \dfrac{3}{2}\right) = -x$

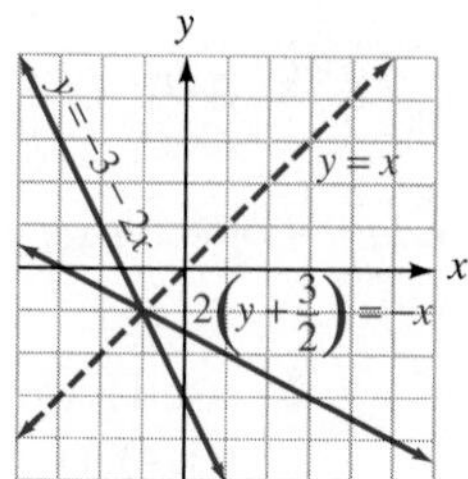

In Exercises 25–34, find the inverse of the relation determined by each equation. Tell whether the inverse relation is a function.

25. $y = x^2 + 4$
$y = \pm\sqrt{x - 4}$; not a function

26. $x = y^2 - 2$
$y = x^2 - 2$; a function

27. $x = y^2 - 4$
$y = x^2 - 4$; a function

28. $y = x^2 + 5$
$y = \pm\sqrt{x - 5}$; not a function

29. $y = x^3$
$y = \sqrt[3]{x}$; a function

30. $xy = 4$
$y = \frac{4}{x}$; a function

31. $y = \pm\sqrt{x}$
$y = x^2$; a function

32. $y = \sqrt[3]{x}$
$y = x^3$; a function

33. $x = \sqrt{y}$ $y = \sqrt{x}$; a function

34. $4y^2 = x - 3$ $y = 4x^2 + 3$; a function

In Exercises 35–36, show that the inverse of the function determined by each equation is also a function. Express it using $f^{-1}(x)$ notation.

35. $y = 2x^3 - 3$ $y = f^{-1}(x) = \sqrt[3]{\dfrac{x+3}{2}}$

36. $y = \dfrac{3}{x^3} - 1$ $y = f^{-1}(x) = \sqrt[3]{\dfrac{3}{x+1}}$

In Exercises 37–40, graph each equation and its inverse on one set of coordinate axes. Find the axis of symmetry.

37. $y = x^2 + 1$

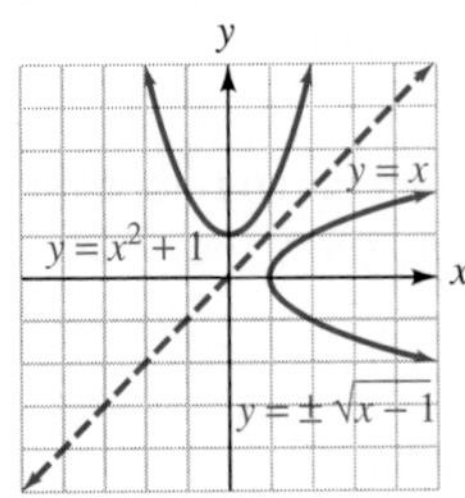

38. $y = \dfrac{1}{4}x^2 - 3$

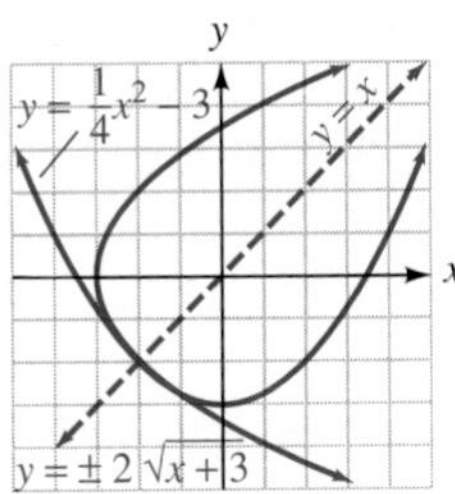

39. $y = \sqrt{x}$

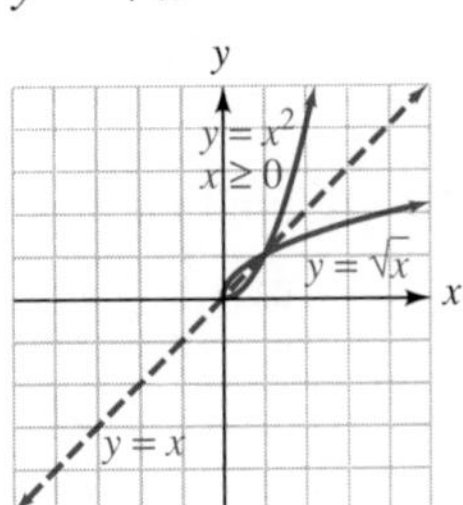

40. $y = |x|$

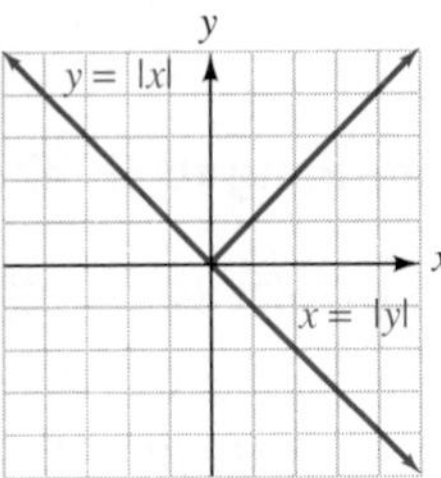

In Exercises 41–50, graph the function represented by each equation. Use the horizontal line test to decide whether each function is one-to-one.

41. $y = 3x + 2$

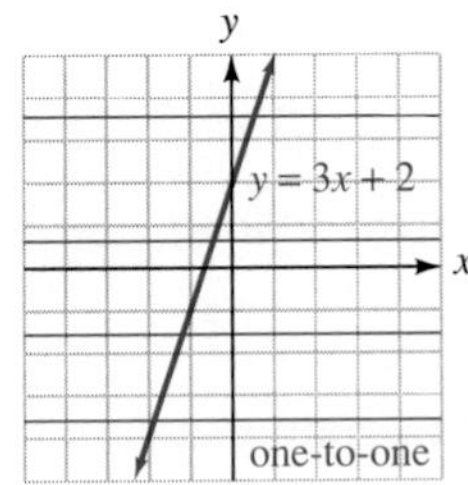

42. $y = 5 - 3x$

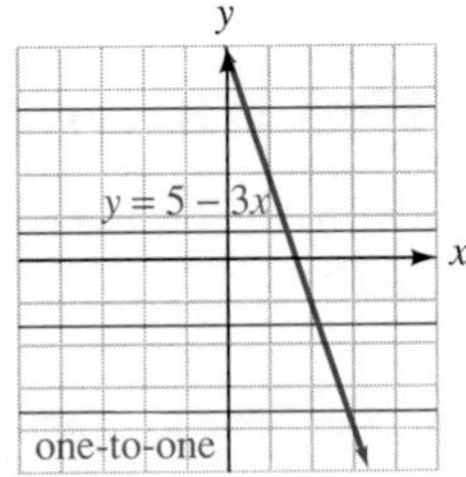

43. $y = \dfrac{x+5}{2}$

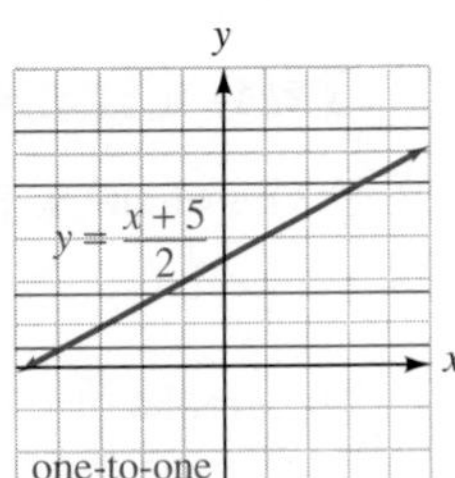

44. $y = \dfrac{5-x}{2}$

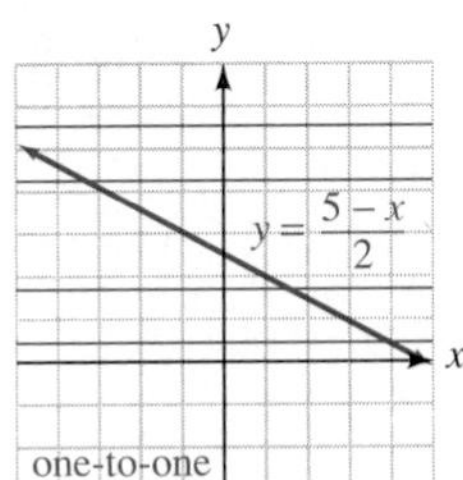

45. $y = 3x^2 + 2$

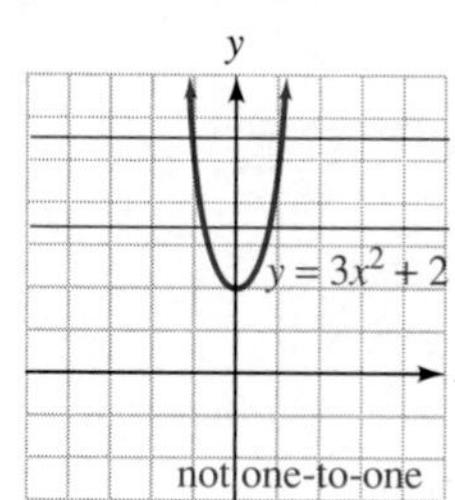

46. $y = 5 - x^2$

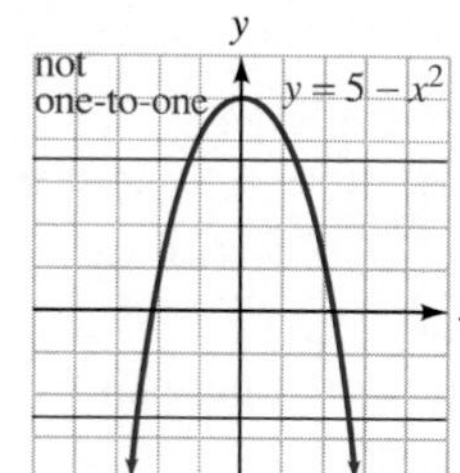

47. $y = \sqrt[3]{x}$

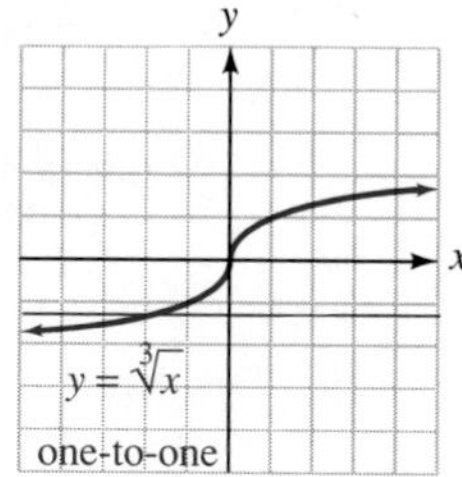

48. $y = \sqrt{x}$

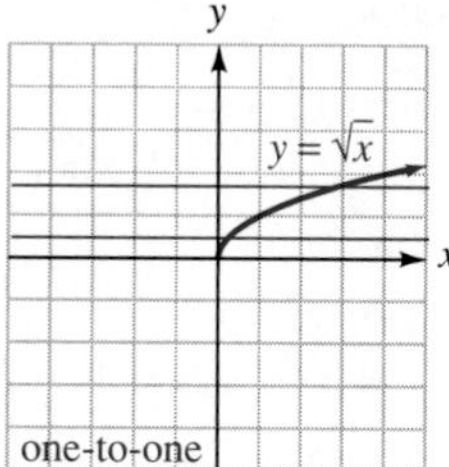

49. $y = x^3 - x$

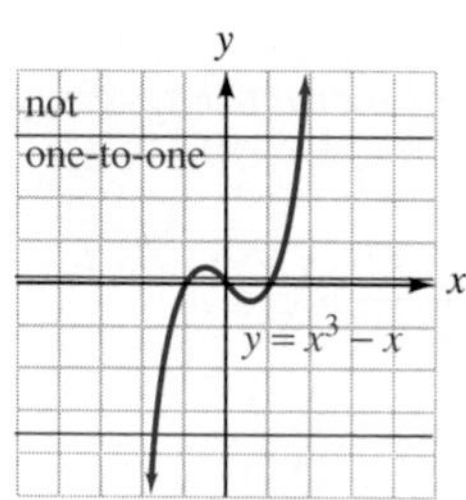

50. $y = -x^4 + x^2$

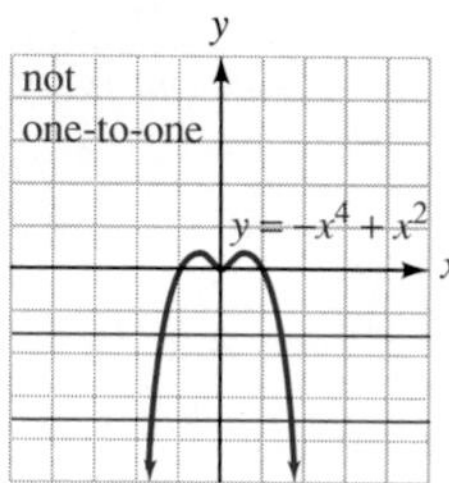

Writing Exercises ■ *Write a paragraph using your own words.*

1. Explain the purpose of the vertical line test.

2. Explain the purpose of the horizontal line test.

Something to Think About ■ **1.** Find the inverse of $y = f(x) = \dfrac{x+1}{x-1}$. $f^{-1}(x) = \dfrac{x+1}{x-1}$

2. Using the function and its inverse of Question 1, show that $(f \circ f^{-1})(x) = x$.

Review Exercises ■ *Write each complex number in a + bi form or find each value.*

1. $3 - \sqrt{-64}$ $3 - 8i$

2. $(2 - 3i) + (4 + 5i)$ $6 + 2i$

3. $(3 + 4i)(2 - 3i)$ $18 - i$

4. $\dfrac{6 + 7i}{3 - 4i}$ $-\dfrac{2}{5} + \dfrac{9}{5}i$

5. $|6 - 8i|$ 10

6. $\left|\dfrac{2 + i}{3 - i}\right|$ $\dfrac{\sqrt{2}}{2}$

10.4 The Circle and the Parabola

■ The Circle ■ Problem Solving ■ The Parabola ■ Problem Solving

The graphs of second-degree equations in x and y represent figures that have interested people since the time of the ancient Greeks. The equations of these graphs were studied carefully in the 17th century, when René Descartes (1596–1650) and Blaise Pascal (1623–1662) began investigating them.

Descartes discovered that the graphs of second-degree equations fall into one of several categories: a pair of lines, a point, a circle, a parabola, an ellipse, a hyperbola, or no graph at all. Because all of these graphs can be formed by the intersection of a plane and a right-circular cone, they are called **conic sections.** See Figure 10-19.

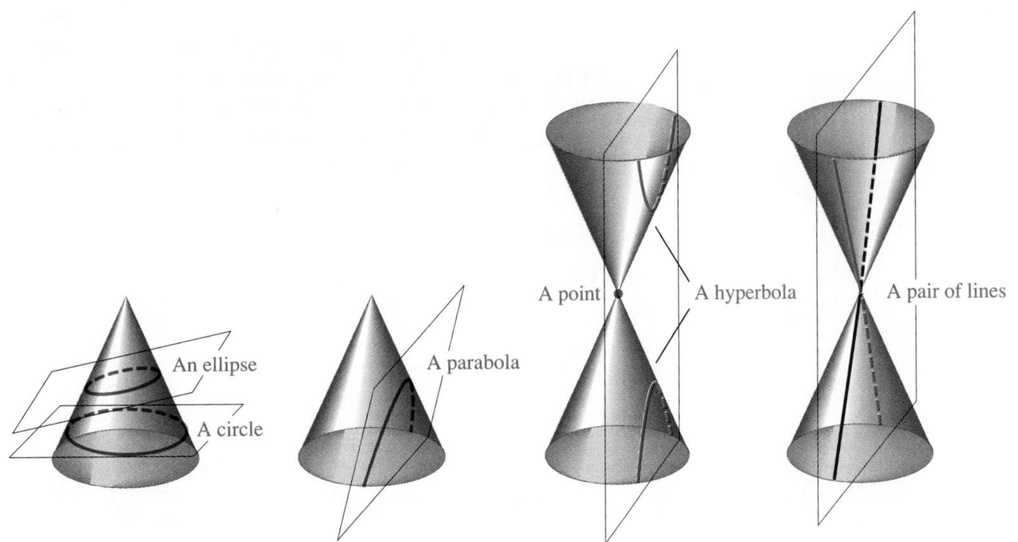

FIGURE 10-19

Conic sections have many applications. For example, a parabola can be rotated to generate a dish-shaped surface called a **paraboloid.** Any light or sound placed at a certain point, called the *focus* of the paraboloid, is reflected outward in parallel paths. This property makes parabolic surfaces ideal for flashlight and headlight reflectors.

Using the same property in reverse makes parabolic surfaces good antennas, because signals captured by such an antenna are concentrated at the focus. A parabolic mirror is capable of concentrating the rays of the sun at a single point and thereby generating tremendous heat. This fact is used in the design of certain solar furnaces. Any object that is thrown upward and outward travels in a parabolic path.

In architecture, many arches are parabolic in shape because of their strength, and the cable that supports a suspension bridge hangs in the form of a parabola.

Ellipses have optical and acoustical properties that are useful in architecture and engineering. For example, many arches are portions of an ellipse because the shape is pleasing to the eye. Gears are often cut into elliptical shapes, to provide nonuniform motion. The planets and some comets have elliptical orbits.

Hyperbolas serve as the basis of a navigational system known as LORAN (LOng RAnge Navigation). They are also used to find the source of a distress signal, are the basis for the design of hypoid gears, and describe the orbits of some comets.

The Circle

The Circle A **circle** is the set of all points in a plane that are a fixed distance from a point called its **center.**

The fixed distance is called the **radius** of the circle.

To develop the general equation of a circle, we must write the equation of a circle with a radius of r and with a center at some point $C(h, k)$, as in Figure 10-20. This task is equivalent to finding all points $P(x, y)$ such that the length of line segment CP is r. We can use the distance formula to find r.

$$r = \sqrt{(x - h)^2 + (y - k)^2}$$

We then square both sides to obtain

1. $r^2 = (x - h)^2 + (y - k)^2$

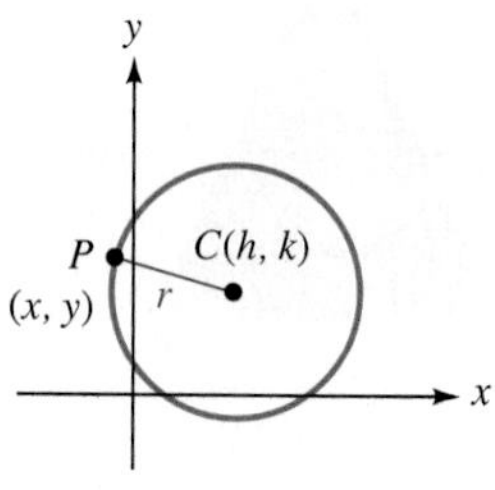

FIGURE 10-20

Equation 1 is called the **standard form of the equation of a circle** with a radius of r and center at the point with coordinates (h, k).

Standard Equation of a Circle with Center at (h, k) Any equation that can be written in the form

$$(x - h)^2 + (y - k)^2 = r^2$$

has a graph that is a circle with radius r and center at point (h, k).

If $r = 0$, the graph reduces to a single point called a **point circle.** If $r < 0$, then a circle does not exist. If both h and k are 0, then the center of the circle is the origin.

Standard Equation of a Circle with Center at (0, 0) Any equation that can be written in the form

$$x^2 + y^2 = r^2$$

has a graph that is a circle with radius r and center at the origin.

EXAMPLE 1 Graph the equation $x^2 + y^2 = 25$.

Solution Because this equation can be written in the form $x^2 + y^2 = r^2$, its graph is a circle with center at the origin. Since $r^2 = 25 = 5^2$, the circle has a radius of 5. The graph appears in Figure 10-21.

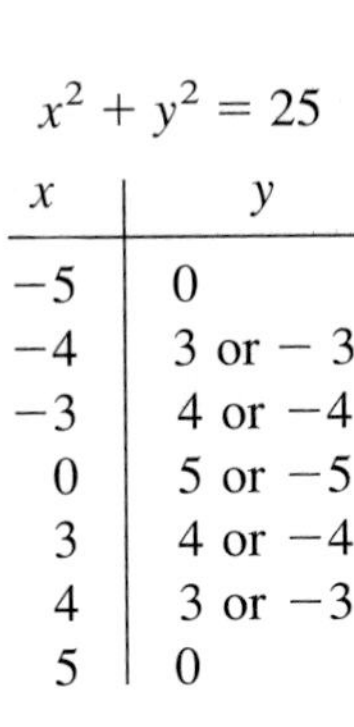

$x^2 + y^2 = 25$

x	y
-5	0
-4	3 or -3
-3	4 or -4
0	5 or -5
3	4 or -4
4	3 or -3
5	0

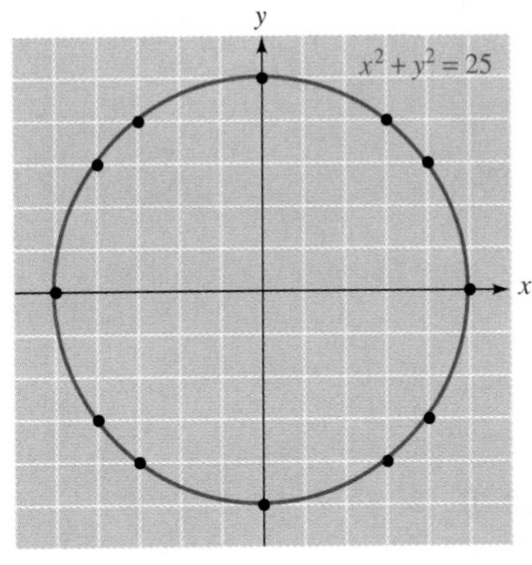

FIGURE 10-21

■

EXAMPLE 2 Find the equation of the circle with radius 5 and center at $C(3, 2)$.

Solution We substitute 5 for r, 3 for h, and 2 for k in standard form and simplify.

$$(x - \mathbf{h})^2 + (y - \mathbf{k})^2 = \mathbf{r}^2$$
$$(x - \mathbf{3})^2 + (y - \mathbf{2})^2 = \mathbf{5}^2$$
$$x^2 - 6x + 9 + y^2 - 4y + 4 = 25$$
$$x^2 + y^2 - 6x - 4y - 12 = 0$$

The equation of the circle is $x^2 + y^2 - 6x - 4y - 12 = 0$. ■

EXAMPLE 3 Graph the circle $x^2 + y^2 - 4x + 2y = 20$.

Solution Because the equation is not in standard form, the coordinates of the center and the length of the radius are not obvious. To put the equation in standard form, we complete the square on both x and y as follows:

$$x^2 + y^2 - 4x + 2y = 20$$
$$x^2 - 4x + y^2 + 2y = 20$$
$$x^2 - 4x + \mathbf{4} + y^2 + 2y + \mathbf{1} = 20 + \mathbf{4} + \mathbf{1}$$ Add 4 and 1 to both sides to complete the squares.

$$(x - 2)^2 + (y + 1)^2 = 25$$ Factor $x^2 - 4x + 4$ and $y^2 + 2y + 1$.

$$(x - 2)^2 + [y - (-1)]^2 = 5^2$$

The radius of the circle is 5, and the coordinates of its center are $h = 2$ and $k = -1$. We plot the center of the circle and draw a circle with a radius of 5 units, as shown in Figure 10-22.

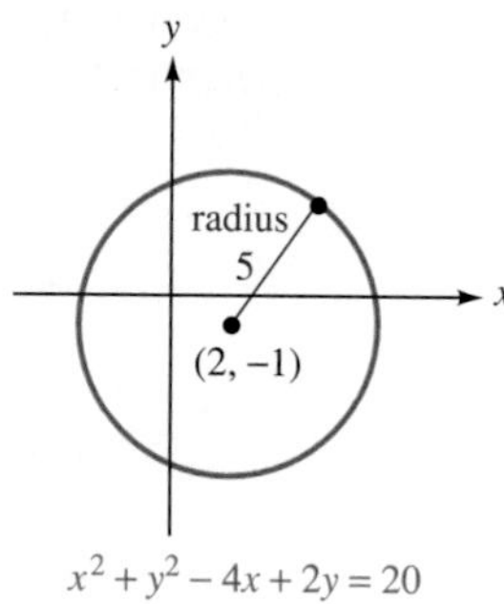

FIGURE 10-22

■

PERSPECTIVE

Focus on Conics

Satellite dishes and flashlight reflectors are familiar examples of a conic's ability to reflect a beam of light or to concentrate incoming satellite signals at one point. That property is shown in Illustration 1.

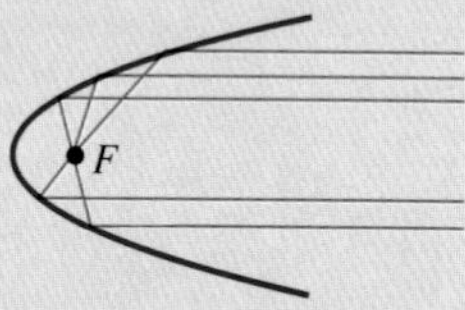

ILLUSTRATION 1

An ellipse has two foci, the points labeled F in Illustration 2. Any light or signal that starts at one focus will be reflected to the other. This property is the basis of whispering galleries, where a person standing at one focus can clearly hear another person speaking at the other focus.

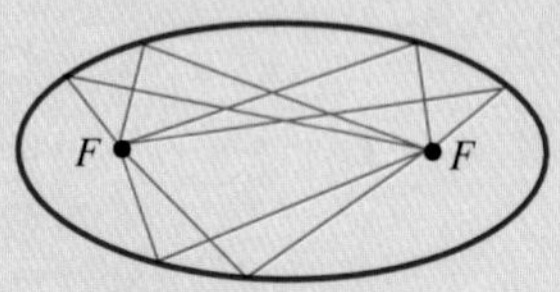

ILLUSTRATION 2

The focal property of the ellipse is also used in **lithotripsy,** a medical procedure for treating kidney stones. The patient is placed in an elliptical tank of water with the kidney stone at one focus. Shock waves from a small controlled explosion at the other focus are concentrated on the stone, pulverizing it.

The hyperbola also has two foci, the two points labeled *F* in Illustration 3. As in the ellipse, light aimed at one focus is reflected toward the other. Hyperbolic mirrors are used in some reflecting telescopes.

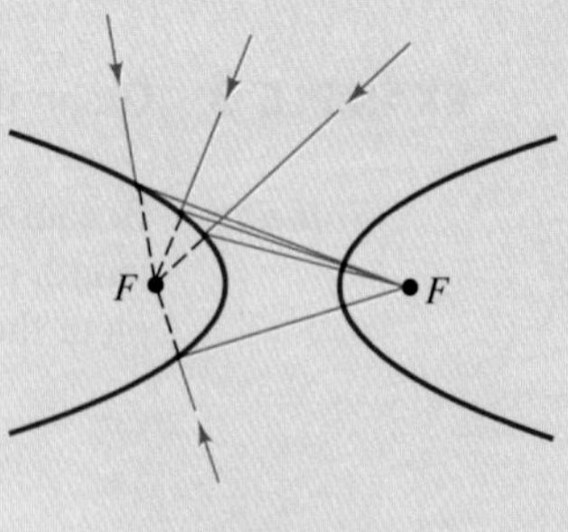

ILLUSTRATION 3

Problem Solving

EXAMPLE 4 **Radio translators** The effective broadcast area of a television station is bounded by the circle $x^2 + y^2 = 3600$, where x and y are measured in miles. A translator station picks up the signal and retransmits it from the center of a circular area bounded by $(x + 30)^2 + (y - 40)^2 = 1600$. Find the location of the translator and the greatest distance from the main transmitter that the signal can be received.

Solution The coverage of the television station is bounded by $x^2 + y^2 = 60^2$, a circle centered at the origin with a radius of 60 miles, as shown in Figure 10-23. Because the translator is at the center of the circle $(x + 30)^2 + (y - 40)^2 = 1600$, it is located at $(-30, 40)$, a point 30 miles west and 40 miles north of the television station. The radius of the translator's coverage is $\sqrt{1600}$, or 40 miles.

As shown in Figure 10-23, the greatest distance of reception is the sum of A, the distance of the translator from the television station, and 40 miles, the radius of the translator's coverage.

To find A, we use the distance formula to find the distance between $(x_1, y_1) = (-30, 40)$ and the origin, $(0, 0)$.

$$\begin{aligned} A &= \sqrt{(x_1 - x_2)^2 + (y_1 - y_2)^2} \\ A &= \sqrt{(\mathbf{-30} - \mathbf{0})^2 + (\mathbf{40} - \mathbf{0})^2} \\ &= \sqrt{(-30)^2 + 40^2} \\ &= \sqrt{2500} \\ &= 50 \end{aligned}$$

Thus, the translator is located 50 miles from the television station, and it broadcasts the signal an additional 40 miles. The greatest reception distance is 50 + 40, or 90 miles.

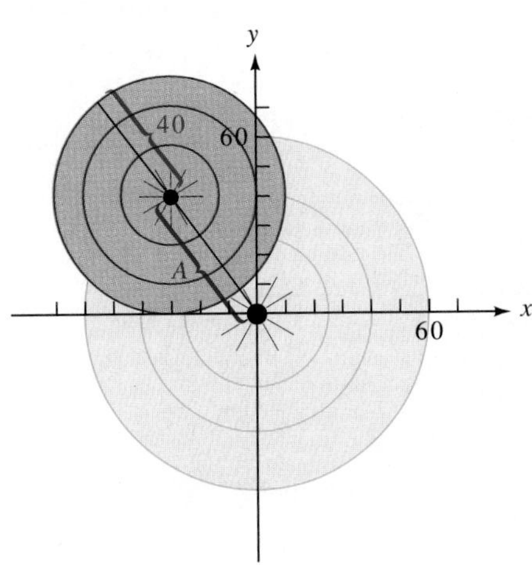

FIGURE 10-23

■

The Parabola

We have seen that equations of the form $y = a(x - h)^2 + k$, with $a \neq 0$, represent parabolas with vertex at the point (h, k). They open upward when $a > 0$ and downward when $a < 0$.

Equations of the form $x = a(y - k)^2 + h$ also represent parabolas with vertex at the point (h, k). However, they open to the right when $a > 0$ and to the left when $a < 0$. Parabolas that open to the right or left do not represent functions, because their graphs do not pass the vertical line test.

Several types of parabolas are summarized in the following chart. (In all cases, $a > 0$.)

Equations of Parabolas ($a > 0$)

Parabola opening	Vertex at origin	Vertex at (h, k)
Up	$y = ax^2$	$y = a(x - h)^2 + k$
Down	$y = -ax^2$	$y = -a(x - h)^2 + k$
Right	$x = ay^2$	$x = a(y - k)^2 + h$
Left	$x = -ay^2$	$x = -a(y - k)^2 + h$

EXAMPLE 5 Graph the equations **a.** $x = \frac{1}{2}y^2$ and **b.** $x = -2(y - 2)^2 + 3$.

Solution **a.** We make a table of ordered pairs, plot each pair, and draw the parabola as in Figure 10-24(a). Because the equation is of the form $x = ay^2$ with $a > 0$, the parabola opens to the right and has its vertex at the origin.

b. We make a table of ordered pairs, plot each pair, and draw the parabolas as in Figure 10-24(b). Because the equation is of the form $x = -a(y - k)^2 + h$, the parabola opens to the left and has its vertex at the point with coordinates (3, 2).

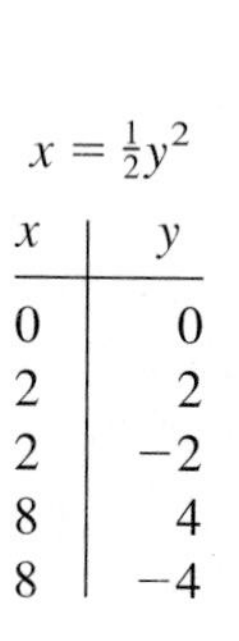

$x = \frac{1}{2}y^2$

x	y
0	0
2	2
2	−2
8	4
8	−4

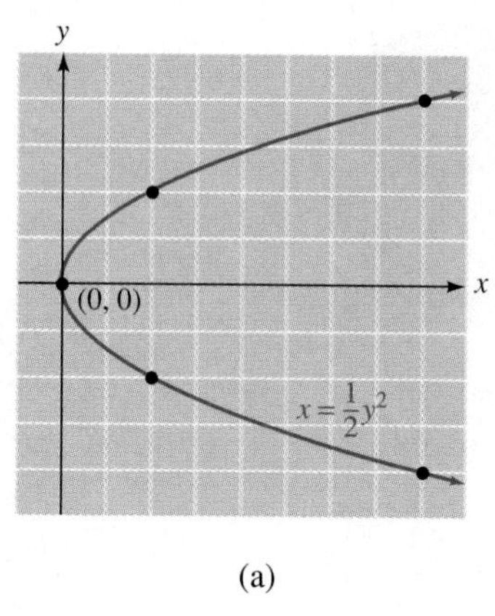

(a)

$x = -2(y - 2)^2 + 3$

x	y
−5	0
1	1
3	2
1	3
−5	4

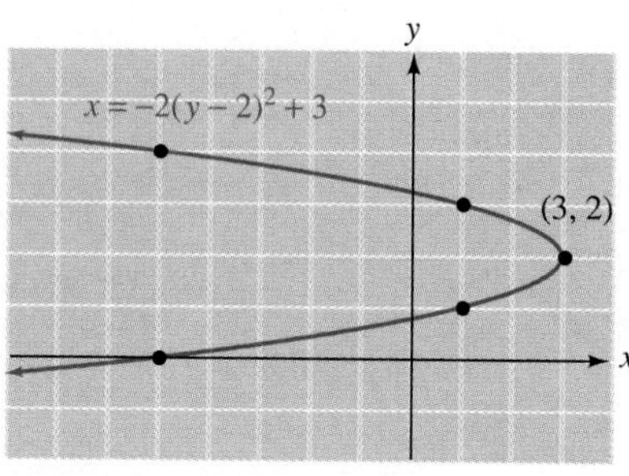

(b)

FIGURE 10-24

EXAMPLE 6 Graph the equation $y = -2x^2 + 12x - 15$.

Solution Because the equation is not in standard form, the coordinates of its vertex are not obvious. To put the equation into standard form, we complete the square on x.

$$y = -2x^2 + 12x - 15$$
$$y = -2(x^2 - 6x \quad) - 15 \qquad \text{Factor out } -2 \text{ from } -2x^2 + 12x.$$
$$y = -2(x^2 - 6x + \mathbf{9} - \mathbf{9}) - 15 \qquad \text{Add and subtract 9.}$$
$$y = -2(x^2 - 6x + 9) + 18 - 15$$
$$y = -2(x - 3)^2 + 3$$

Because the equation is in the form $y = -a(x - h)^2 + k$, we can see that the parabola opens downward and has its vertex at (3, 3). The graph of the equation appears in Figure 10-25.

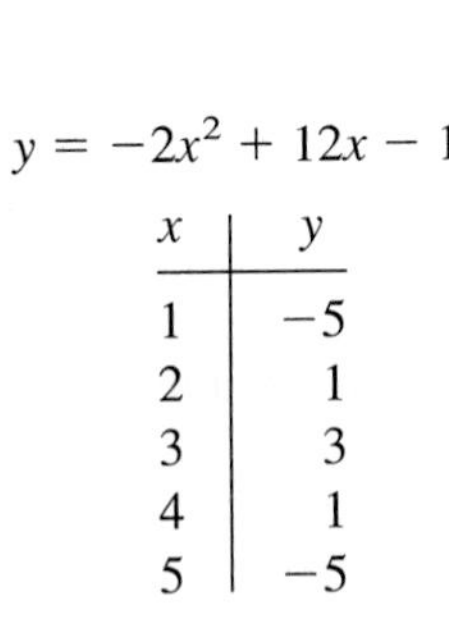
$y = -2x^2 + 12x - 15$

x	y
1	−5
2	1
3	3
4	1
5	−5

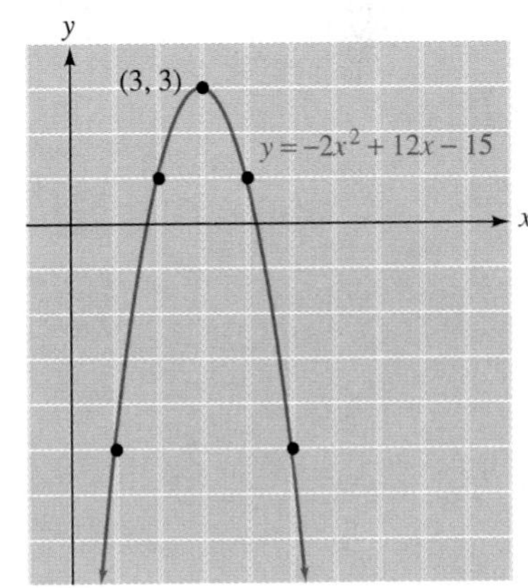

FIGURE 10-25

■

Problem Solving

EXAMPLE 7 **Gateway Arch** The shape of the Gateway Arch in Saint Louis is approximately a parabola, as shown in Figure 10-26(a). How high is the arch 100 feet from its foundation?

Teaching Tip The shape is an inverted catenary.

Solution We place the parabola in a coordinate system as in Figure 10-26(b), with ground level on the x-axis and the vertex of the parabola at the point (0, 630). The equation of this downward-opening parabola has the form

$$y = -a(x - h)^2 + k$$
$$y = -a(x - \mathbf{0})^2 + \mathbf{630} \qquad \text{Substitute } h = 0 \text{ and } y = 630.$$
$$y = -ax^2 + 630 \qquad \text{Simplify.}$$

Because the Gateway Arch is 630 feet wide at its base, the parabola passes through the point $\left(\frac{630}{2}, 0\right)$, or (315, 0). To find a in the equation of the parabola, we substitute $x = 315$ and $y = 0$ and proceed as follows:

$$y = -ax^2 + 630$$

$$\mathbf{0} = -a \cdot \mathbf{315}^2 + 630$$

$$\frac{-630}{315^2} = -a \qquad \text{Subtract 630 from both sides and divide both sides by } 315^2.$$

$$\frac{2}{315} = a \qquad \text{Multiply both sides by } -1 \text{ and simplify.}$$

Thus, the equation of the parabola that approximates the shape of the Gateway Arch is

$$y = -\frac{2}{315}x^2 + 630$$

To find the height of the arch at a point 100 feet from its foundation, we substitute 315 − 100, or 215, for x into the equation of the parabola and solve for y.

$$y = -\frac{2}{315}x^2 + 630$$

$$y = -\frac{2}{315}(\mathbf{215})^2 + 630$$

$$\approx 336.5 \qquad \text{Use a calculator.}$$

At a point 100 feet from the foundation, the height of the arch is 336.5 feet.

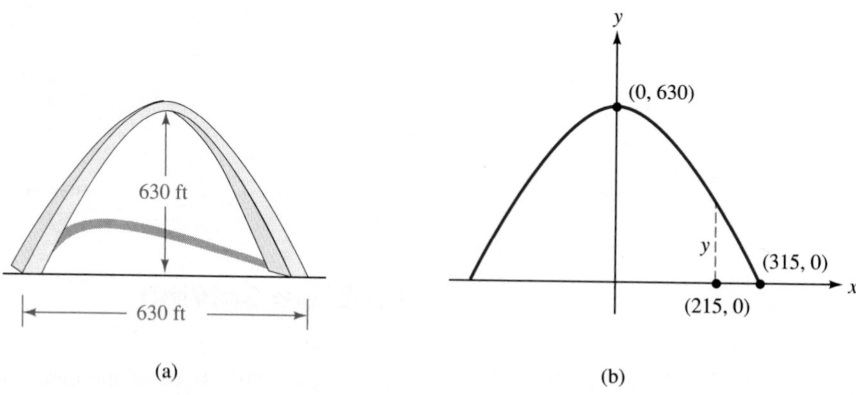

FIGURE 10-26

ORALS *Find the center and the radius of each circle.*

1. $x^2 + y^2 = 144$ (0, 0), 12
2. $x^2 + y^2 = 121$ (0, 0), 11
3. $(x - 2)^2 + y^2 = 16$ (2, 0), 4
4. $x^2 + (y + 1)^2 = 9$ (0, −1), 3

Tell whether each parabola opens up or down or left or right.

5. $y = -3x^2 - 2$ down
6. $y = 7x^2 - 5$ up
7. $x = -3y^2$ left
8. $x = (y - 3)^2$ right

EXERCISE 10.4

In Exercises 1–10, graph each equation.

1. $x^2 + y^2 = 9$

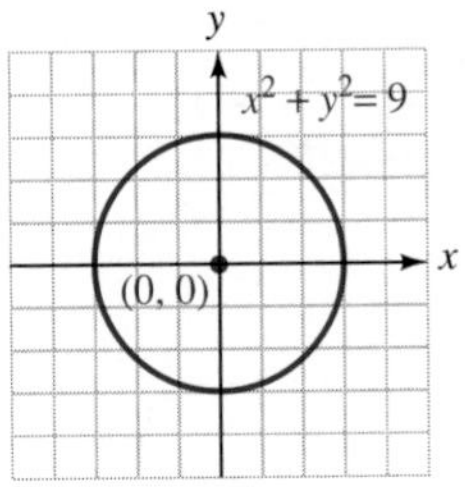

2. $x^2 + y^2 = 16$

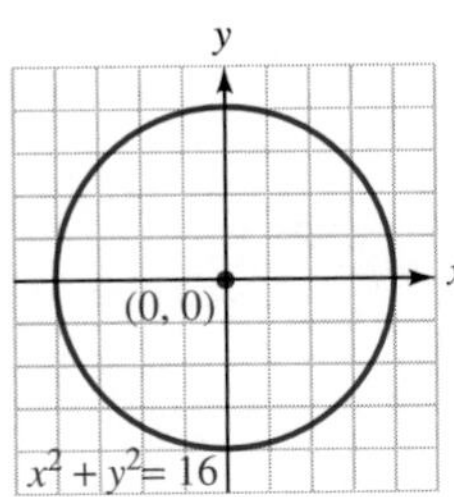

3. $(x - 2)^2 + y^2 = 9$

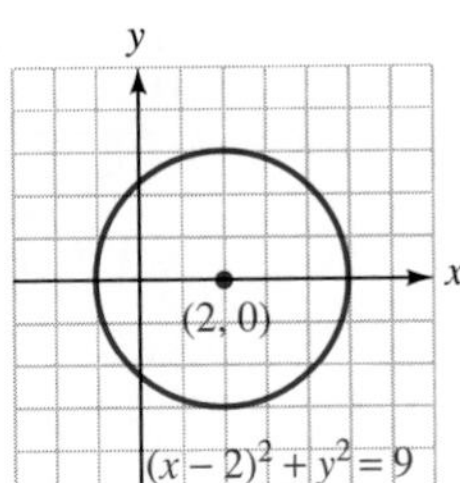

4. $x^2 + (y - 3)^2 = 4$

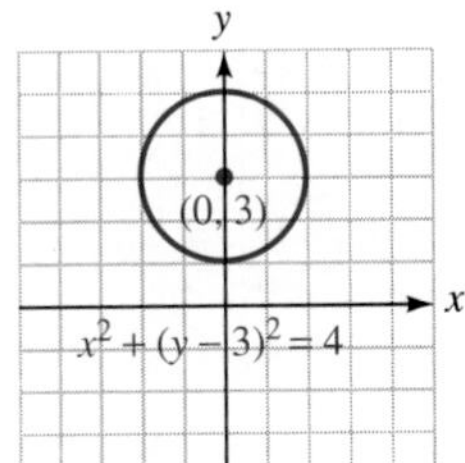

5. $(x - 2)^2 + (y - 4)^2 = 4$

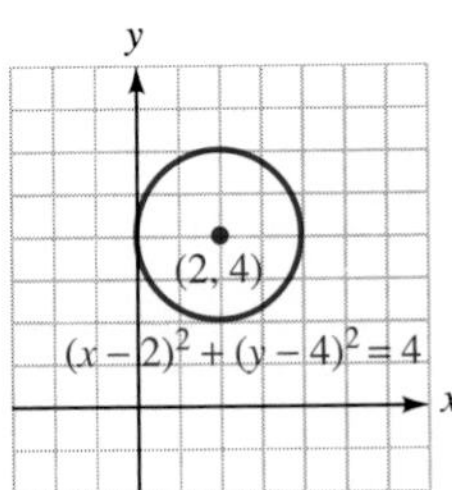

6. $(x - 3)^2 + (y - 2)^2 = 4$

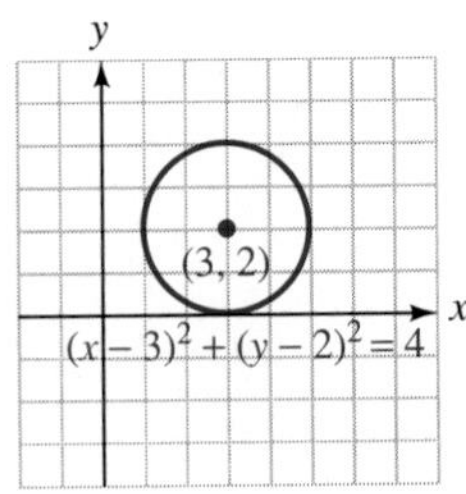

7. $(x + 3)^2 + (y - 1)^2 = 16$

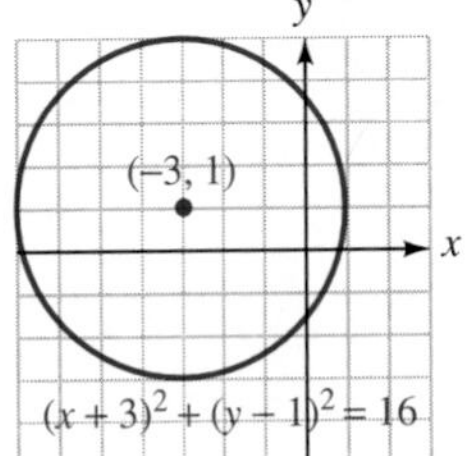

8. $(x - 1)^2 + (y + 4)^2 = 9$

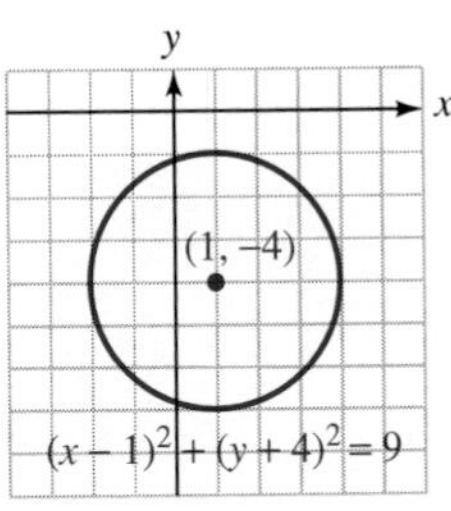

9. $x^2 + (y + 3)^2 = 1$

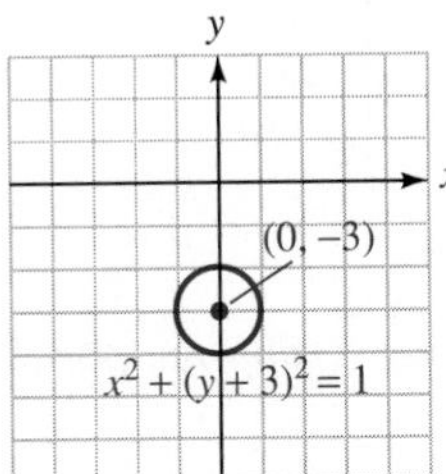

10. $(x + 4)^2 + y^2 = 1$

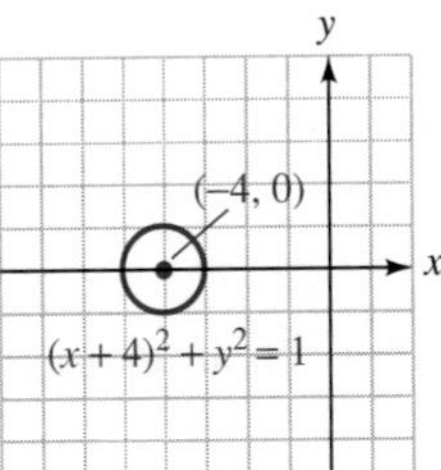

In Exercises 11–18, write the equation of the circle with the following properties.

11. Center at the origin; radius of 1 $x^2 + y^2 = 1$

12. Center at the origin; radius of 4 $x^2 + y^2 = 16$

13. Center at (6, 8); radius of 5 $(x - 6)^2 + (y - 8)^2 = 25$

14. Center at (5, 3); radius of 2 $(x - 5)^2 + (y - 3)^2 = 4$

15. Center at (−2, 6); radius of 12 $(x + 2)^2 + (y - 6)^2 = 144$

16. Center at (5, −4); radius of 6 $(x - 5)^2 + (y + 4)^2 = 36$

17. Center at the origin; diameter of $2\sqrt{2}$ $x^2 + y^2 = 2$

18. Center at the origin; diameter of $4\sqrt{3}$ $x^2 + y^2 = 12$

In Exercises 19–26, graph each circle. Give the coordinates of the center.

19. $x^2 + y^2 + 2x - 8 = 0$

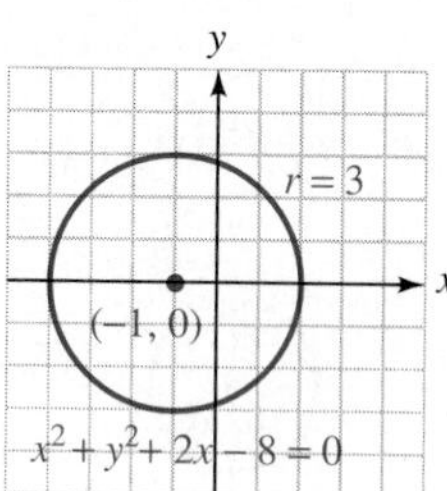

20. $x^2 + y^2 - 4y = 12$

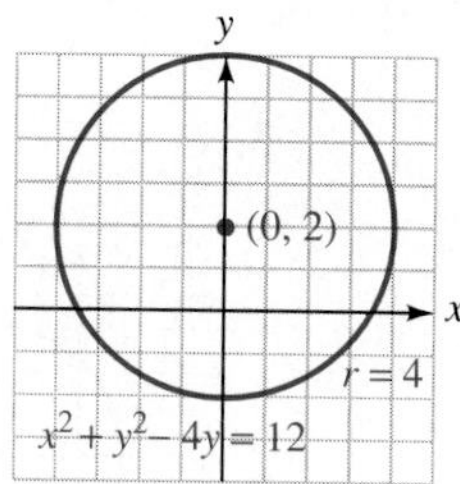

21. $9x^2 + 9y^2 - 12y = 5$

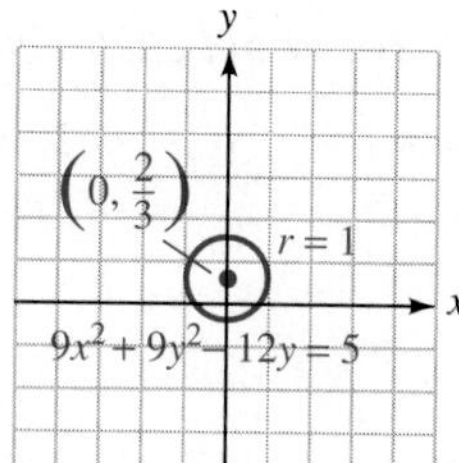

22. $4x^2 + 4y^2 + 4y = 15$

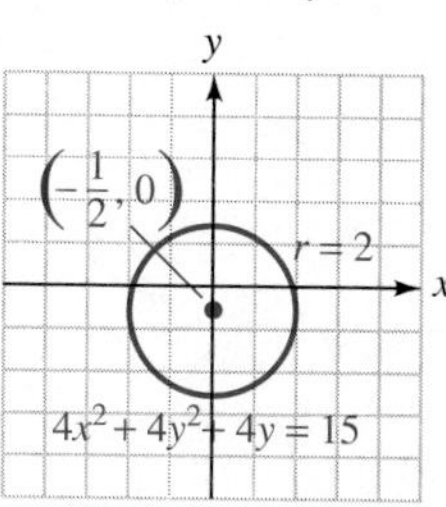

23. $x^2 + y^2 - 2x + 4y = -1$

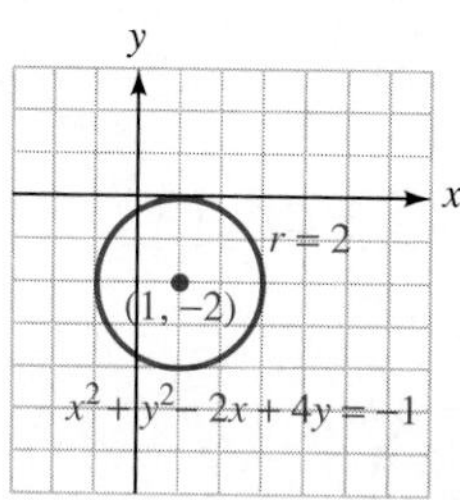

24. $x^2 + y^2 + 4x + 2y = 4$

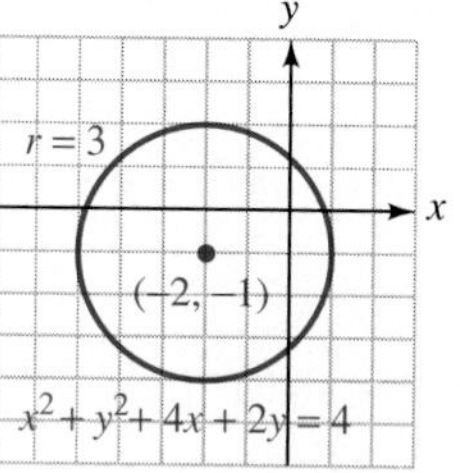

25. $x^2 + y^2 + 6x - 4y = -12$

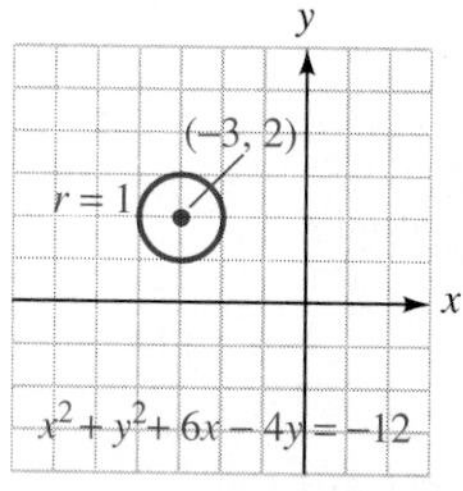

26. $x^2 + y^2 + 8x + 2y = -13$

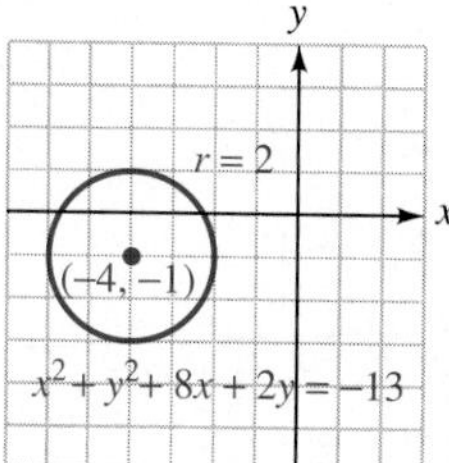

In Exercises 27–40, find the vertex of each parabola. Then graph the parabola.

27. $x = y^2$

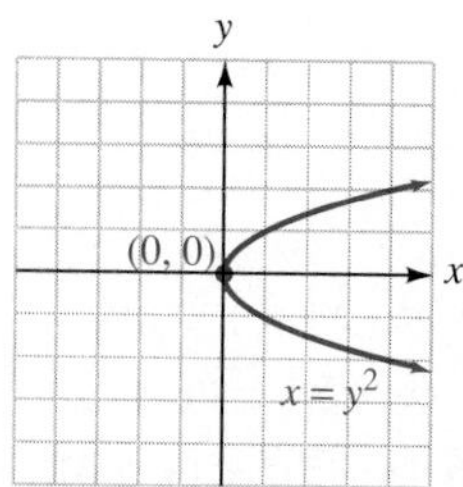

28. $x = -y^2 + 1$

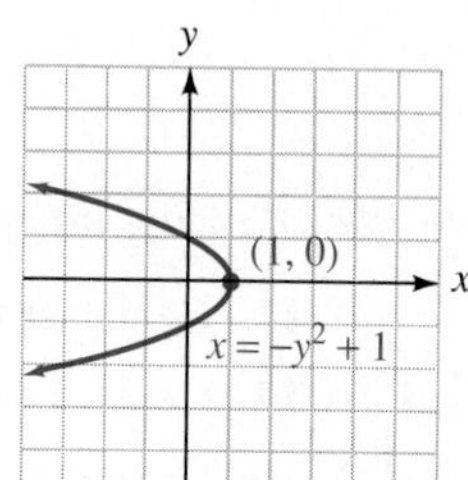

29. $x = -\frac{1}{4}y^2$

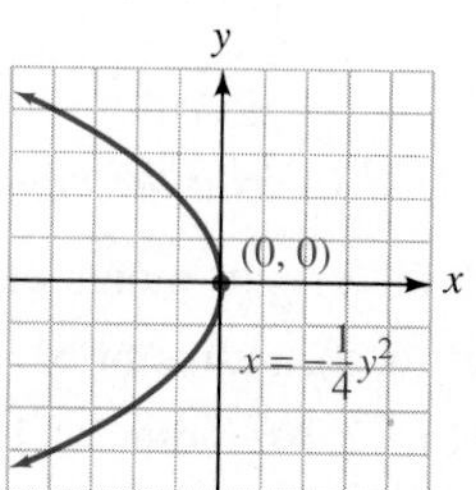

30. $x = 4y^2$

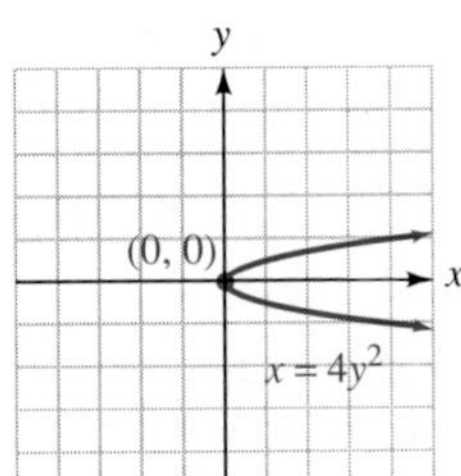

31. $y = x^2 + 4x + 5$

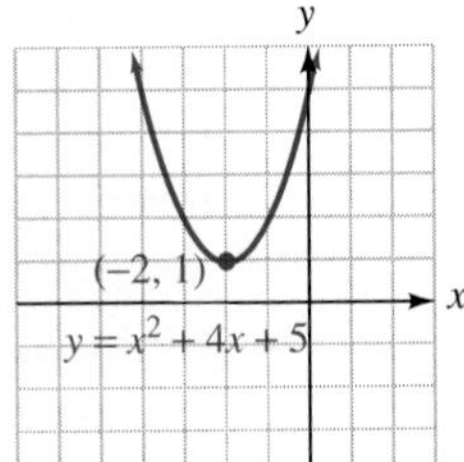

32. $y = -x^2 - 2x + 3$

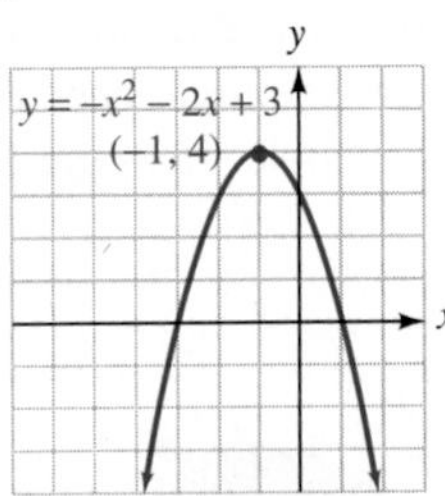

33. $y = -x^2 - x + 1$

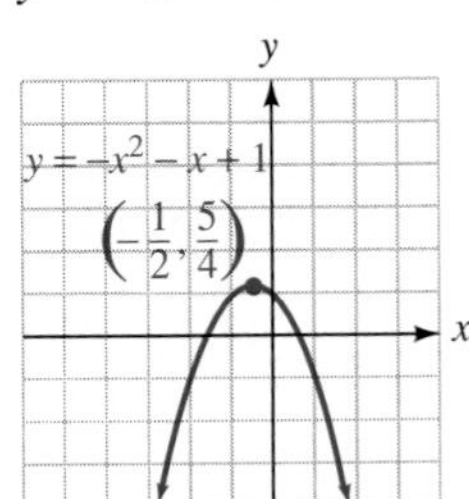

34. $x = \frac{1}{2}y^2 + 2y$

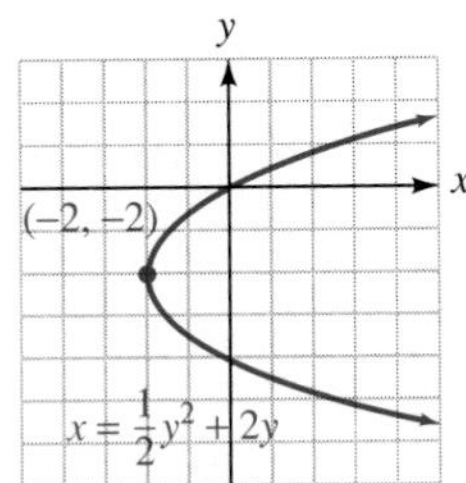

35. $y^2 + 4x - 6y = -1$

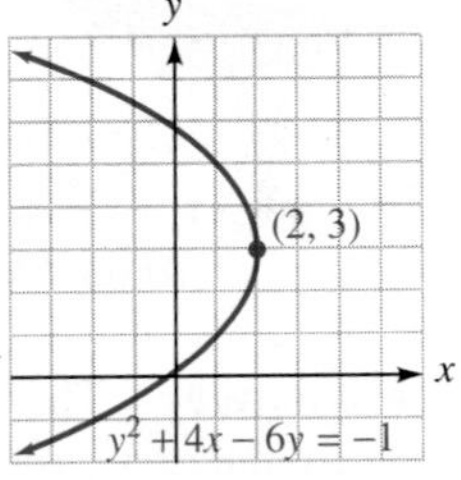

36. $x^2 - 2y - 2x = -7$

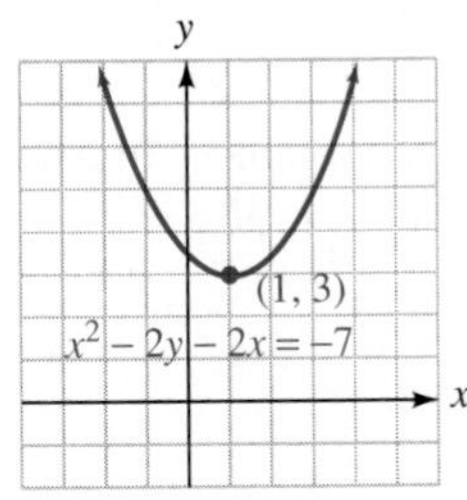

37. $y = 2(x - 1)^2 + 3$

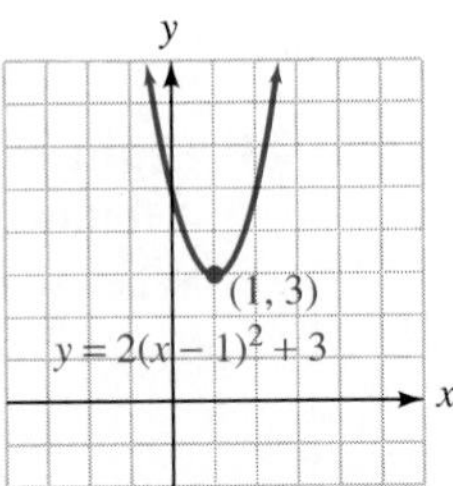

38. $y = -2(x + 1)^2 + 2$

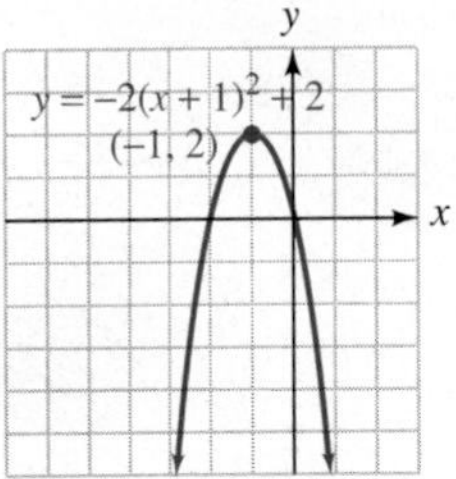

39. $x = -3(y + 2)^2 - 2$

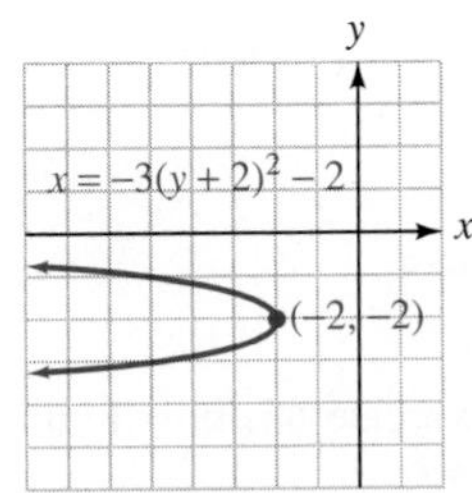

40. $x = 2(y - 3)^2 - 4$

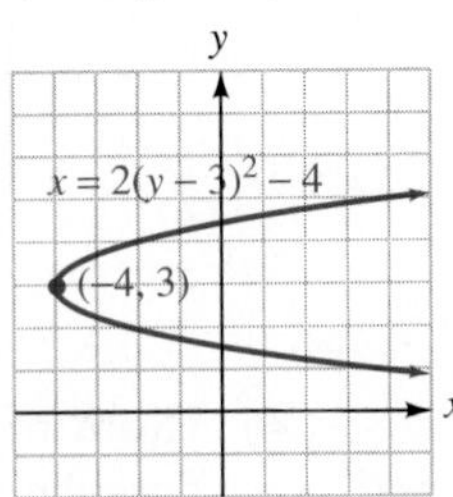

41. **Meshing gears** For design purposes, the large gear in Illustration 1 is the circle $x^2 + y^2 = 16$. The smaller gear is a circle centered at $(7, 0)$ and tangent to the larger circle. Find the equation of the smaller gear. $(x - 7)^2 + y^2 = 9$

ILLUSTRATION 1

42. **Width of a walkway** The walkway in Illustration 2 is bounded by the two circles $x^2 + y^2 = 2500$ and $(x - 10)^2 + y^2 = 900$, measured in feet. Find the largest and the smallest width of the walkway. 30 ft and 10 ft

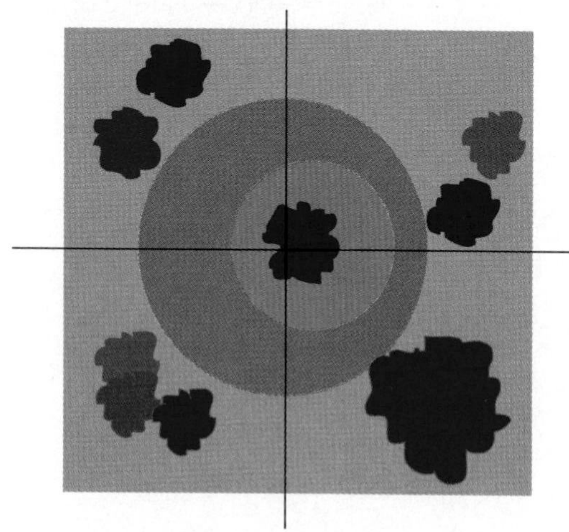

ILLUSTRATION 2

43. **Broadcast ranges** Radio stations applying for licensing may not use the same frequency if their broadcast areas overlap. One station's coverage is bounded by $x^2 + y^2 - 8x - 20y + 16 = 0$, and the other's by $x^2 + y^2 + 2x + 4y - 11 = 0$. May they be licensed for the same frequency? no

44. **Highway curves** Highway design engineers want to join two sections of highway with a curve that is one-quarter of a circle, as in Illustration 3. The equation of the circle is $x^2 + y^2 - 16x - 20y + 155 = 0$, where distances are measured in kilometers. Find the locations (relative to the center of town at the origin) of the intersections of the highway with State and with Main. 11 km, 13 km

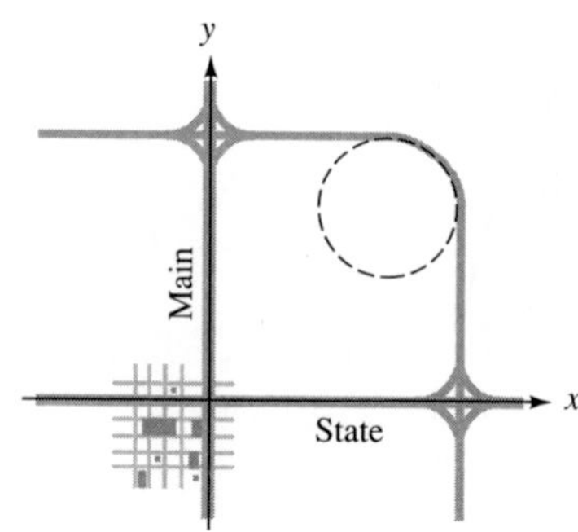

ILLUSTRATION 3

45. **Flight of a projectile** The cannonball in Illustration 4 follows the parabolic trajectory $y = 30x - x^2$. Where does it land? 30 ft away

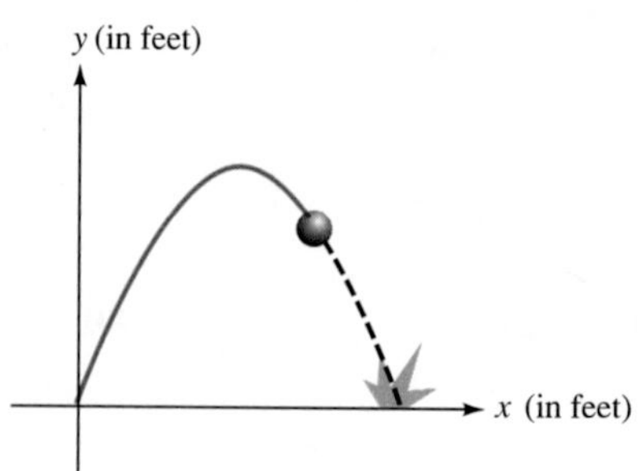

ILLUSTRATION 4

46. **Flight of a projectile** In Exercise 45, how high does the cannonball get? 225 ft

47. Orbit of a comet If the orbit of the comet shown in Illustration 5 is given by the equation $2y^2 - 9x = 18$, how far is it from the sun at the vertex of the orbit? Distances are in astronomical units (AU). 2 AU

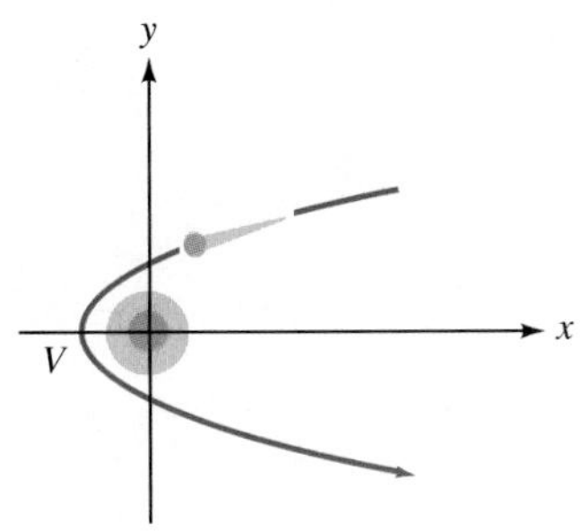

ILLUSTRATION 5

48. Satellite antenna The cross section of the satellite antenna in Illustration 6 is a parabola given by the equation $y = \frac{1}{16}x^2$, with distances measured in feet. If the dish is 8 feet wide, how deep is it? 1 ft

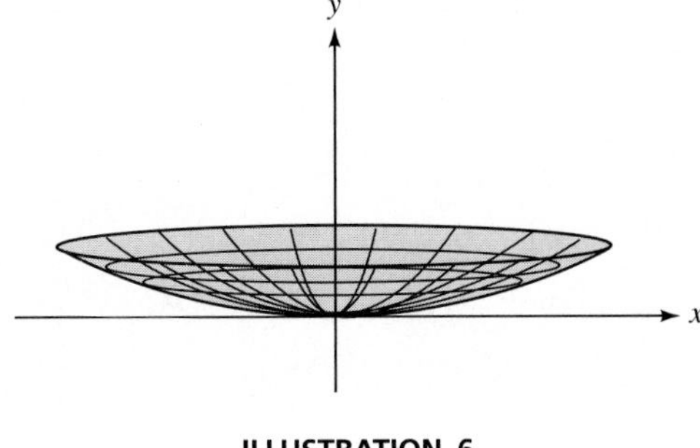

ILLUSTRATION 6

Writing Exercises ■ *Write a paragraph using your own words.*

1. Explain how to decide from its equation whether the graph of a parabola opens up, down, right, or left.
2. From the equation of a circle, explain how to determine the radius and the coordinates of the center.

Something to Think About ■

1. From the values of a, h, and k, explain how to determine the number of x-intercepts of the graph of $y = a(x - h)^2 + k$.
2. Under what conditions will the graph of $x = a(y - k)^2 + h$ have no y-intercepts?

Review Exercises ■ *Solve each equation.*

1. $|3x - 4| = 11$ $5, -\frac{7}{3}$
2. $\left|\frac{4 - 3x}{5}\right| = 12$ $\frac{64}{3}, -\frac{56}{3}$
3. $|3x + 4| = |5x - 2|$ $3, -\frac{1}{4}$
4. $|6 - 4x| = |x + 2|$ $\frac{4}{5}, \frac{8}{3}$

10.5 The Ellipse and the Hyperbola

■ The Ellipse ■ Problem Solving ■ The Hyperbola ■ Problem Solving

■ The Ellipse

The Ellipse An **ellipse** is the set of all points P in the plane the sum of whose distances from two fixed points is a constant. See Figure 10-27, in which $d_1 + d_2$ is a constant.

(continued)

The Ellipse *(continued)* Each of the two points is called a **focus.** Midway between the foci is the **center** of the ellipse.

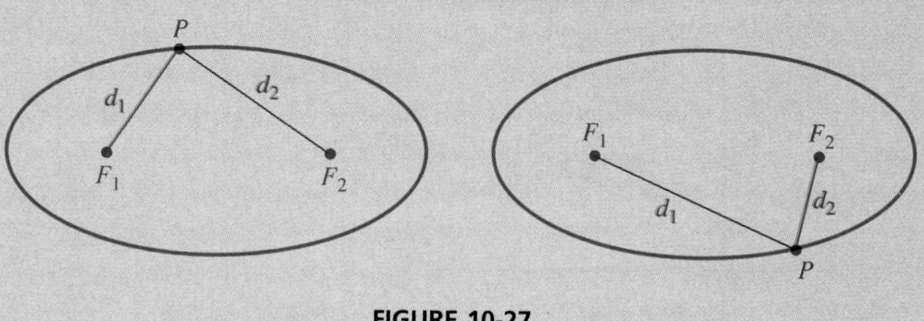

FIGURE 10-27

Because of the previous definition, we can construct an ellipse by placing two thumbtacks fairly close together, as in Figure 10-28. We then tie each end of a piece of string to a thumbtack, catch the loop with the point of a pencil, and while keeping the string taut, draw the ellipse.

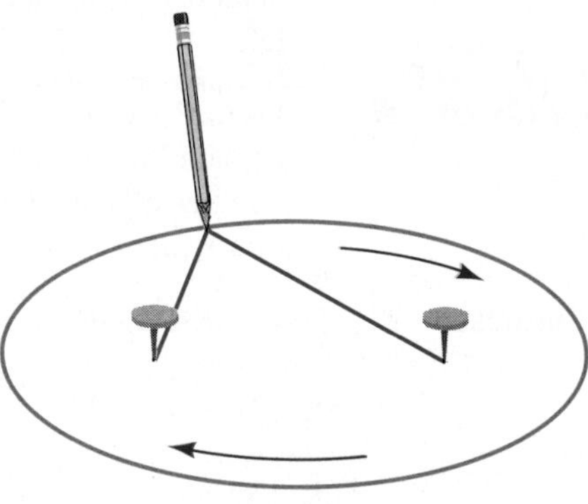

FIGURE 10-28

The graph of the equation

$$\frac{x^2}{36} + \frac{y^2}{9} = 1$$

is an ellipse. To graph the ellipse, we make a table of ordered pairs, plot each pair, and join them with a smooth curve, as shown in Figure 10-29.

The center of the ellipse is at the origin; it intersects the x-axis at points $(6, 0)$ and $(-6, 0)$ and the y-axis at the points $(0, 3)$ and $(0, -3)$.

$$\frac{x^2}{36} + \frac{y^2}{9} = 1$$

x	y
-6	0
-4	$\sqrt{5}$ or $-\sqrt{5}$
-2	$2\sqrt{2}$ or $-2\sqrt{2}$
0	3 or -3
2	$2\sqrt{2}$ or $-2\sqrt{2}$
4	$\sqrt{5}$ or $-\sqrt{5}$
6	0

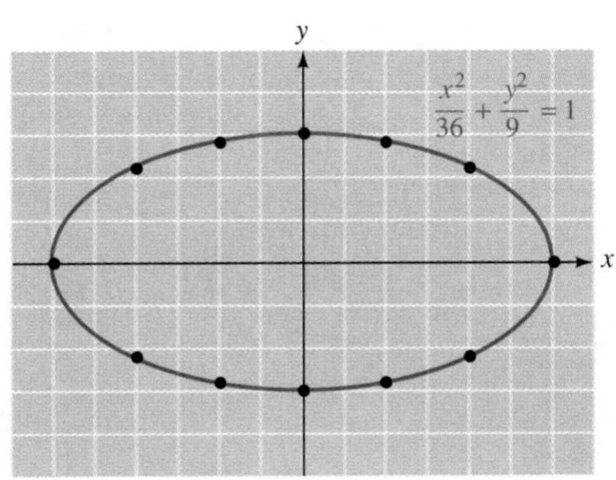

FIGURE 10-29

The previous discussion illustrates this general theorem.

Equations of an Ellipse Centered at the Origin

Any equation that can be written in the form

$$\frac{x^2}{a^2} + \frac{y^2}{b^2} = 1 \qquad (a > b)$$

has a graph that is an **ellipse** centered at the origin, as in Figure 10-30(a). The x-intercepts are the **vertices** $V(a, 0)$ and $V'(-a, 0)$. (Read V' as "V prime.") The y-intercepts are the points $(0, b)$ and $(0, -b)$.

Any equation that can be written in the form

$$\frac{x^2}{b^2} + \frac{y^2}{a^2} = 1 \qquad (a > b)$$

has a graph that is also an ellipse centered at the origin, as in Figure 10-30(b). The y-intercepts are the vertices $V(0, a)$ and $V'(0, -a)$. The x-intercepts are the points $(b, 0)$ and $(-b, 0)$.

The point midway between the vertices is the **center** of the ellipse.

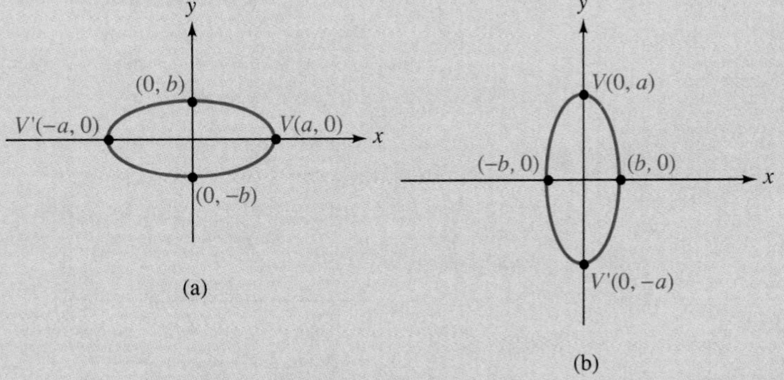

FIGURE 10-30

The following theorem gives the equation of an ellipse with center at a point with coordinates (h, k).

Equations of an Ellipse Centered at (h, k)

Any equation that can be written in the form

$$\frac{(x-h)^2}{a^2} + \frac{(y-k)^2}{b^2} = 1 \quad \text{or} \quad \frac{(x-h)^2}{b^2} + \frac{(y-k)^2}{a^2} = 1$$

(with $a > b$) is an ellipse with center at the point (h, k).

EXAMPLE 1 Graph the ellipse $\frac{(x-2)^2}{16} + \frac{(y+3)^2}{25} = 1$.

Solution We write the equation in the form

$$\frac{(x-2)^2}{16} + \frac{[y-(-3)]^2}{25} = 1$$

to see that the center of the ellipse is at the point $(2, -3)$. To find some points on the ellipse, we let $x = 2$ and solve for y.

$$\frac{(x-2)^2}{16} + \frac{(y+3)^2}{25} = 1$$

$$\frac{(2-2)^2}{16} + \frac{(y+3)^2}{25} = 1$$

$$0 + \frac{(y+3)^2}{25} = 1$$

$$(y+3)^2 = 25$$

$y + 3 = 5$ or $y + 3 = -5$ Use the square root theorem.

$y = 2$ | $y = -8$

Thus, the points with coordinates of $(2, 2)$ and $(2, -8)$ lie on the graph. These points are the vertices of the ellipse. Now we let $y = -3$ and solve for x.

$$\frac{(x-2)^2}{16} + \frac{(y+3)^2}{25} = 1$$

$$\frac{(x-2)^2}{16} + \frac{(-3+3)^2}{25} = 1$$

$$(x-2)^2 = 16$$

$$x - 2 = 4 \quad \text{or} \quad x - 2 = -4$$
$$x = 6 \quad \bigg| \quad x = -2$$

Thus, the points with coordinates of $(6, -3)$ and $(-2, -3)$ lie on the graph. The graph appears in Figure 10-31.

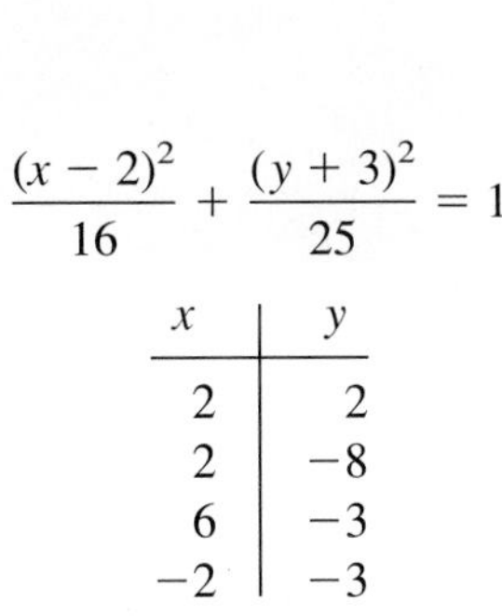

$$\frac{(x-2)^2}{16} + \frac{(y+3)^2}{25} = 1$$

x	y
2	2
2	−8
6	−3
−2	−3

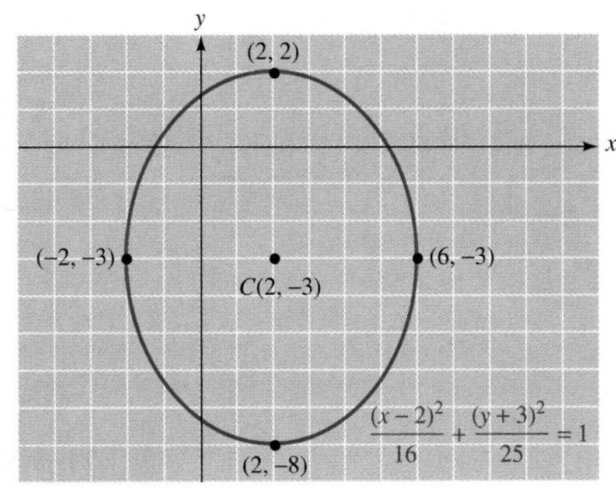

FIGURE 10-31

■

■ Problem Solving

EXAMPLE 2 **Landscape design** A landscape architect is designing an elliptical pool that will fit in the center of a 20-by-30-foot rectangular garden, leaving at least 5 feet of space on all sides. Find the equation of the ellipse.

Solution We place the rectangular garden in a coordinate system, as in Figure 10-32. To maintain 5 feet of clearance at the ends of the ellipse, the vertices must be the points $V(10, 0)$ and $V'(-10, 0)$. Similarly, the y-intercepts are the points $(0, 5)$ and $(0, -5)$.

The equation of the ellipse has the form

$$\frac{x^2}{a^2} + \frac{y^2}{b^2} = 1$$

with $a = 10$ and $b = 5$. Thus, the equation of the boundary of the pool is

$$\frac{x^2}{100} + \frac{y^2}{25} = 1$$

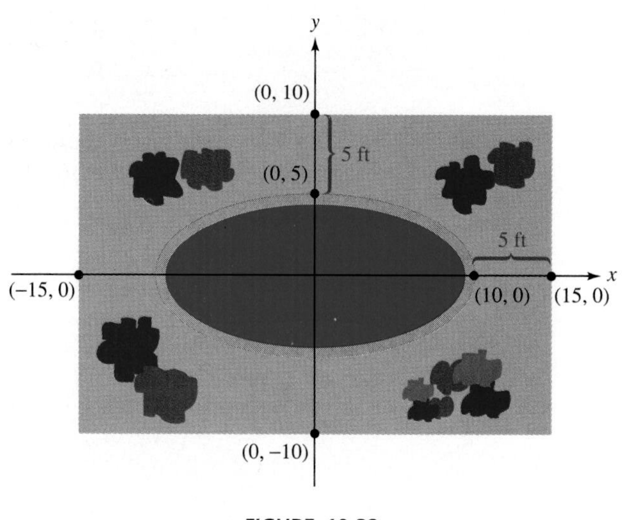

FIGURE 10-32

The Hyperbola

The Hyperbola A **hyperbola** is the set of all points P in the plane for which the difference of the distances of each point on the hyperbola from two fixed points is a constant. See Figure 10-33, in which $d_1 - d_2$ is a constant.

Each of the two points is called a **focus.** Midway between the foci is the **center** of the hyperbola.

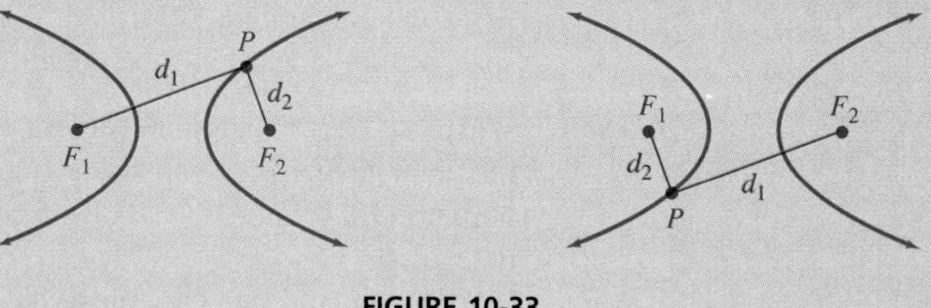

FIGURE 10-33

The graph of the equation

$$\frac{x^2}{25} - \frac{y^2}{9} = 1$$

is a hyperbola. To graph this hyperbola, we make a table of ordered pairs, plot each pair, and join the points with a smooth curve as in Figure 10-34.

$$\frac{x^2}{25} - \frac{y^2}{9} = 1$$

x	y
-7	$\pm \frac{6\sqrt{6}}{5} \approx \pm 2.9$
-6	$\pm \frac{3\sqrt{11}}{5} \approx \pm 2.0$
-5	0
5	0
6	$\pm \frac{3\sqrt{11}}{5} \approx \pm 2.0$
7	$\pm \frac{6\sqrt{6}}{5} \approx \pm 2.9$

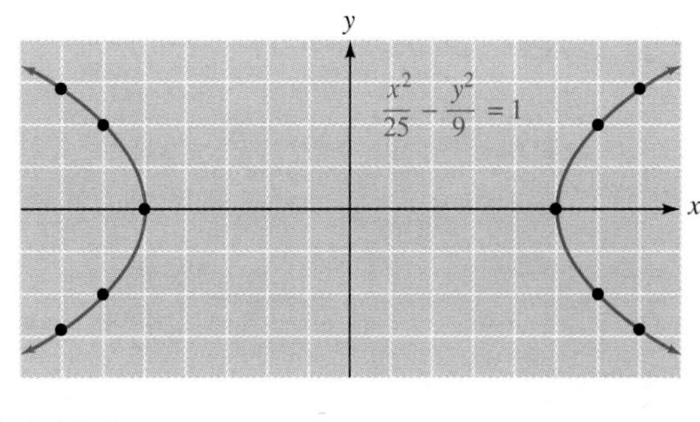

FIGURE 10-34

This graph is centered at the origin and intersects the x-axis at $(5, 0)$ and $(-5, 0)$. We also note that the graph does not intersect the y-axis.

Although it is possible to draw any hyperbola by plotting many points and joining them with a smooth curve, there is an easier way. For example, if we want to graph the hyperbola with an equation of

$$\frac{x^2}{a^2} - \frac{y^2}{b^2} = 1$$

we first look at the x- and y-intercepts. To find the x-intercepts, we let $y = 0$ and solve for x:

$$\frac{x^2}{a^2} - \frac{0^2}{b^2} = 1$$
$$x^2 = a^2$$
$$x = \pm a$$

Thus, the hyperbola crosses the x-axis at the points $V(a, 0)$ and $V'(-a, 0)$, called the **vertices** of the hyperbola. (See Figure 10-35.)

To attempt to find the y-intercepts, we let $x = 0$ and solve for y:

$$\frac{0^2}{a^2} - \frac{y^2}{b^2} = 1$$
$$y^2 = -b^2$$
$$y = \pm\sqrt{-b^2}$$

Since $b^2 > 0$, $\sqrt{-b^2}$ is an imaginary number. This means that the hyperbola does not cross the y-axis.

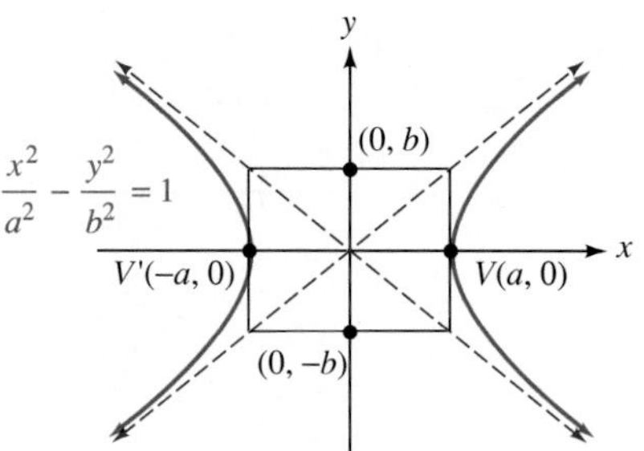

FIGURE 10-35

If we construct a rectangle whose sides pass horizontally through $\pm b$ on the y-axis and vertically through $\pm a$ on the x-axis, the extended diagonals of the rectangle will be a useful aid in drawing the graph. As points on the branches of the hyperbola move further away from the origin, they get closer to these extended diagonals. The extended diagonals of the rectangle, called the **fundamental rectangle,** are called **asymptotes.**

The previous discussion illustrates the following theorem.

Equation of a Hyperbola Centered at the Origin

Any equation that can be written in the form

$$\frac{x^2}{a^2} - \frac{y^2}{b^2} = 1$$

has a graph that is a **hyperbola** centered at the origin, as in Figure 10-36. The x-intercepts are the **vertices** $V(a, 0)$ and $V'(-a, 0)$. There are no y-intercepts.

The **asymptotes** of the hyperbola are the extended diagonals of the rectangle in the figure.

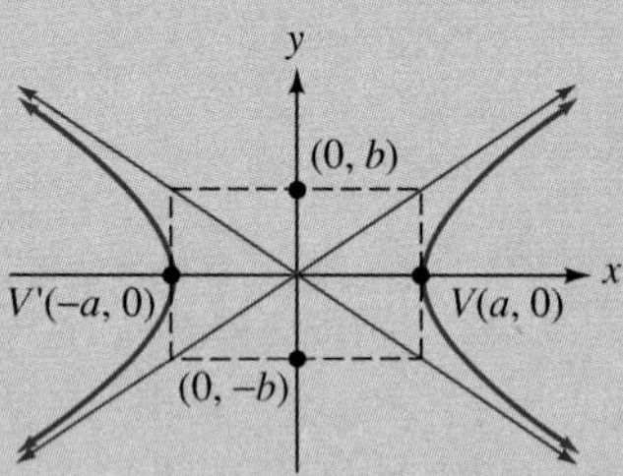

FIGURE 10-36

The branches of the hyperbola in previous discussions open to the left and to the right. It is possible for hyperbolas to have different orientations with respect to the x- and y-axes. For example, the branches of a hyperbola can open upward and downward. In that case, the following theorem applies.

Equation of a Hyperbola Centered at the Origin

Any equation that can be written in the form

$$\frac{y^2}{a^2} - \frac{x^2}{b^2} = 1$$

has a graph that is a **hyperbola** centered at the origin, as in Figure 10-37. The y-intercepts are the **vertices** $V(0, a)$ and $V'(0, -a)$. There are no x-intercepts.

The **asymptotes** of the hyperbola are the extended diagonals of the rectangle in the figure.

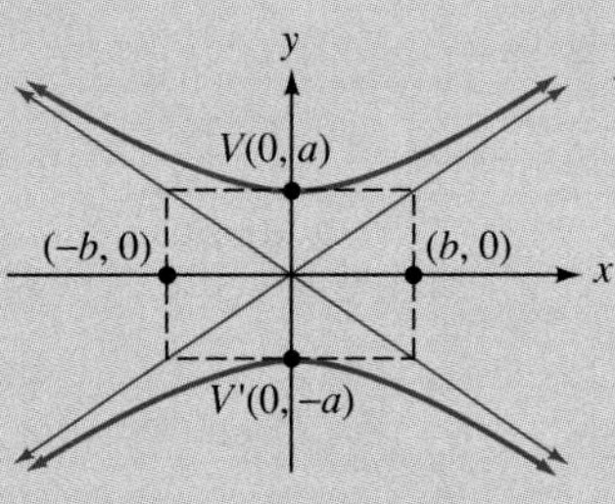

FIGURE 10-37

EXAMPLE 3 Graph the equation $9y^2 - 4x^2 = 36$.

Solution To write the equation in standard form, we divide both sides by 36 to obtain

$$\frac{9y^2}{36} - \frac{4x^2}{36} = 1$$

$$\frac{y^2}{4} - \frac{x^2}{9} = 1 \quad \text{Simplify each fraction.}$$

We then find the y-intercepts by letting $x = 0$ and solving for y:

$$\frac{y^2}{4} - \frac{0^2}{9} = 1$$

$$y^2 = 4$$

Thus, $y = \pm 2$, and the vertices of the hyperbola are $V(0, 2)$ and $V'(0, -2)$. See Figure 10-38.

Since $b = \pm\sqrt{9} = \pm 3$, we use the points $(3, 0)$ and $(-3, 0)$ on the x-axis to help draw the fundamental rectangle. We then draw its extended diagonals and sketch the hyperbola.

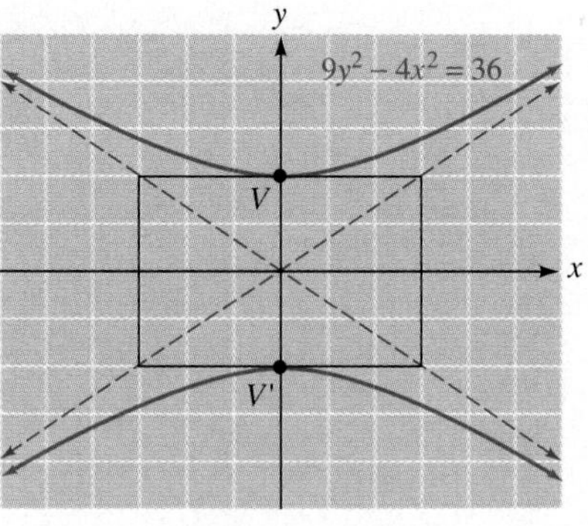

FIGURE 10-38

If a hyperbola is centered at a point with coordinates (h, k), the following theorem applies.

Equations of Hyperbolas Centered at (h, k)

Any equation that can be written in the form

$$\frac{(x-h)^2}{a^2} - \frac{(y-k)^2}{b^2} = 1$$

is a hyperbola with center at (h, k) that opens left and right.

Any equation of the form

$$\frac{(y-k)^2}{a^2} - \frac{(x-h)^2}{b^2} = 1$$

is a hyperbola with center at (h, k) that opens up and down.

EXAMPLE 4 Graph the hyperbola $\dfrac{(x-3)^2}{16} - \dfrac{(y+1)^2}{4} = 1.$

Solution We write the equation in the form

$$\frac{(x-3)^2}{16} - \frac{[y-(-1)]^2}{4} = 1$$

to see that the hyperbola is of the form that opens left and right and is centered at the point $(h, k) = (3, -1)$. Because this hyperbola has its center at $(3, -1)$, its vertices are located at $a = 4$ units to the right and left of the center, at $(7, -1)$ and $(-1, -1)$. Since $b = 2$, we can locate the points 2 units above and 2 units below the center. With these points, we can draw the fundamental rectangle along with its extended diagonals. We can then sketch the hyperbola as shown in Figure 10-39.

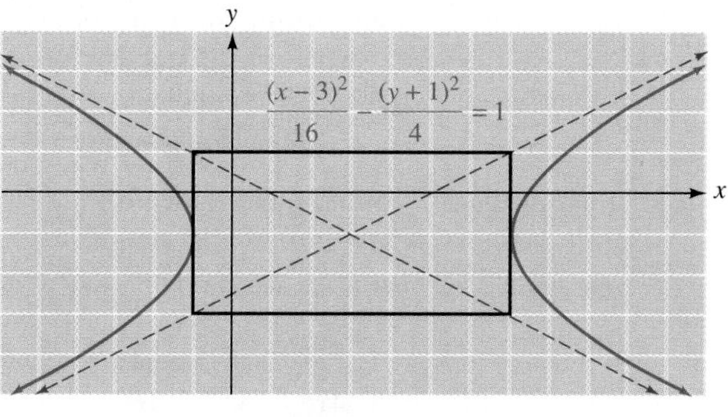

FIGURE 10-39 ■

There is a special type of hyperbola (also centered at the origin) that does not intersect either the x- or the y-axis. These hyperbolas have equations of the form $xy = k$, where $k \neq 0$.

EXAMPLE 5 Graph the equation $xy = -8$.

Solution We make a table of ordered pairs, plot each pair, and join the points with a smooth curve to obtain the hyperbola in Figure 10-40.

$xy = -8$

x	y
1	−8
2	−4
4	−2
8	−1
−1	8
−2	4
−4	2
−8	1

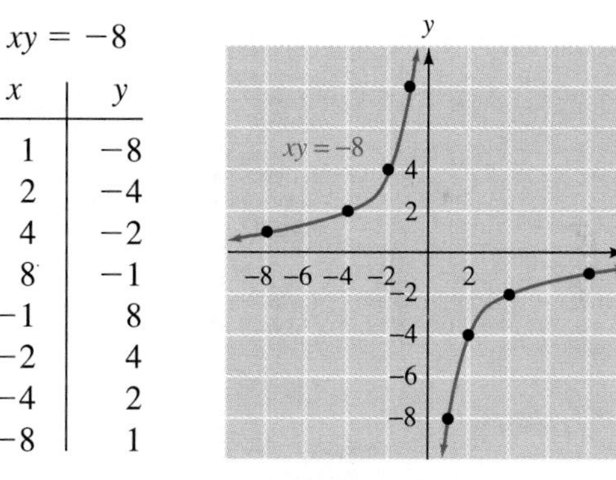

FIGURE 10-40 ■

The result in Example 5 suggests the following theorem.

Theorem Any equation of the form $xy = k$, where $k \neq 0$, has a graph that is a **hyperbola,** which does not intersect either the x- or the y-axis.

Problem Solving

EXAMPLE 6 **Atomic structure** In an experiment that led to the discovery of the atomic structure of matter, Lord Rutherford (1871–1937) shot high-energy alpha particles toward a thin sheet of gold. Because many were reflected, Rutherford showed the existence of the nucleus of a gold atom. The alpha particle in Figure 10-41 is repelled by the nucleus at the origin, and travels along the hyperbolic path given by $4x^2 - y^2 = 16$. How close does the particle come to the nucleus?

Solution To find the distance from the nucleus at the origin, we must find the coordinates of the vertex V. To do so, we write the equation of the particle's path in standard form:

$$4x^2 - y^2 = 16$$

$$\frac{4x^2}{\mathbf{16}} - \frac{y^2}{\mathbf{16}} = \frac{16}{\mathbf{16}} \quad \text{Divide both sides by 16.}$$

$$\frac{x^2}{4} - \frac{y^2}{16} = 1 \quad \text{Simplify.}$$

$$\frac{x^2}{2^2} - \frac{y^2}{4^2} = 1 \quad \text{Write 4 as } 2^2 \text{ and 16 as } 4^2.$$

This equation is in the form $\frac{x^2}{a^2} - \frac{y^2}{b^2} = 1$, with $a = 2$. Thus, the vertex of the path is (2, 0). The particle is never closer than 2 units from the nucleus.

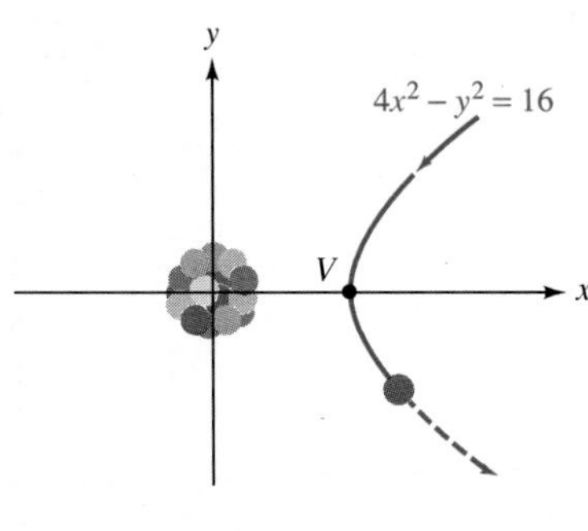

FIGURE 10-41

■

ORALS *Find the x- and y-intercepts of each ellipse.*

1. $\frac{x^2}{9} + \frac{y^2}{16} = 1$ $(\pm 3, 0), (0, \pm 4)$

2. $\frac{x^2}{25} + \frac{y^2}{36} = 1$ $(\pm 5, 0), (0, \pm 6)$

Find the center of each ellipse.

3. $\frac{(x-2)^2}{9} + \frac{y^2}{16} = 1$ $(2, 0)$

4. $\frac{x^2}{25} + \frac{(y+1)^2}{36} = 1$ $(0, -1)$

Find the x- or y-intercepts of each hyperbola.

5. $\frac{x^2}{9} - \frac{y^2}{16} = 1$ $(\pm 3, 0)$

6. $\frac{y^2}{25} - \frac{x^2}{36} = 1$ $(0, \pm 5)$

EXERCISE 10.5

In Exercises 1–10, graph each equation.

1. $\dfrac{x^2}{4} + \dfrac{y^2}{9} = 1$

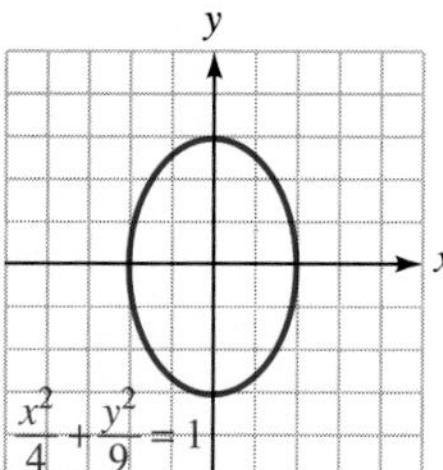

2. $x^2 + \dfrac{y^2}{9} = 1$

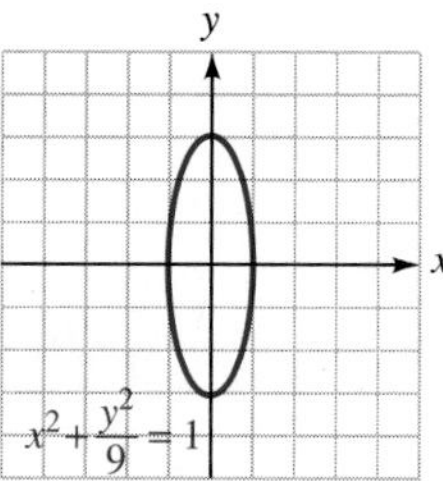

3. $x^2 + 9y^2 = 9$

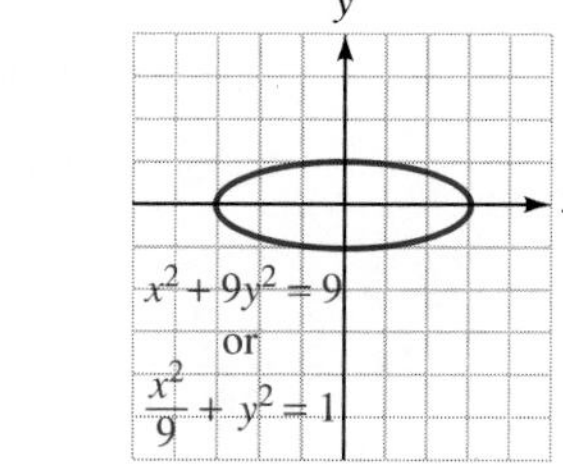

4. $25x^2 + 9y^2 = 225$

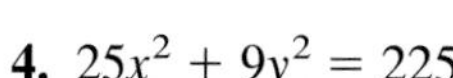

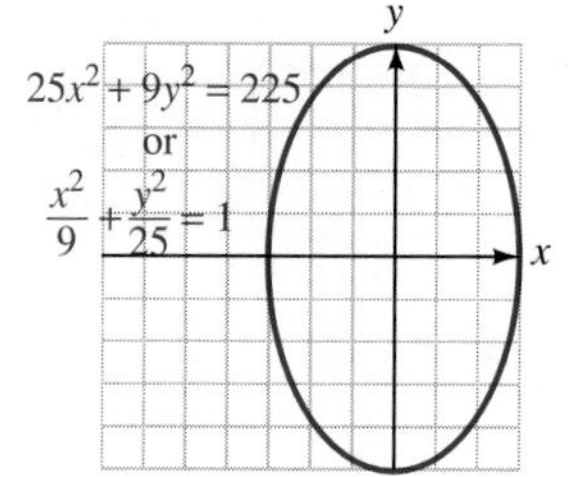

5. $16x^2 + 4y^2 = 64$

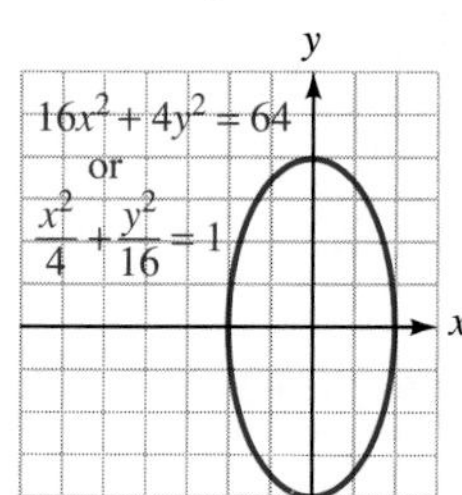

6. $4x^2 + 9y^2 = 36$

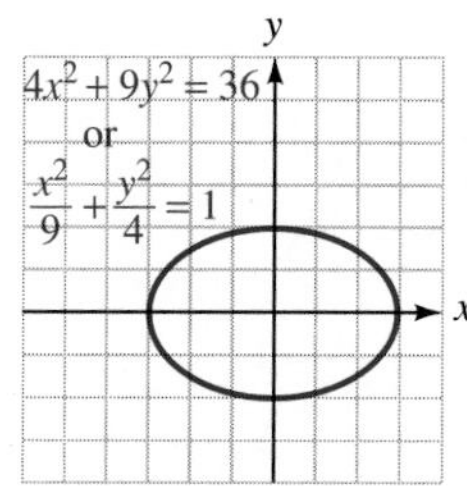

7. $\dfrac{(x-2)^2}{9} + \dfrac{(y-1)^2}{4} = 1$

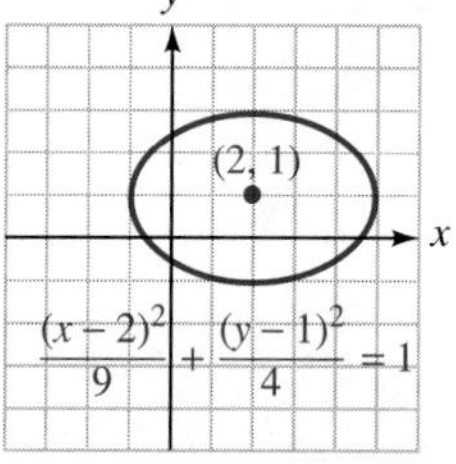

8. $\dfrac{(x-1)^2}{9} + \dfrac{(y-3)^2}{4} = 1$

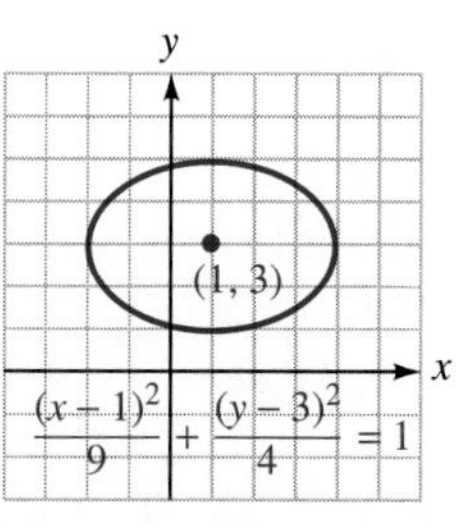

9. $(x+1)^2 + 4(y+2)^2 = 4$

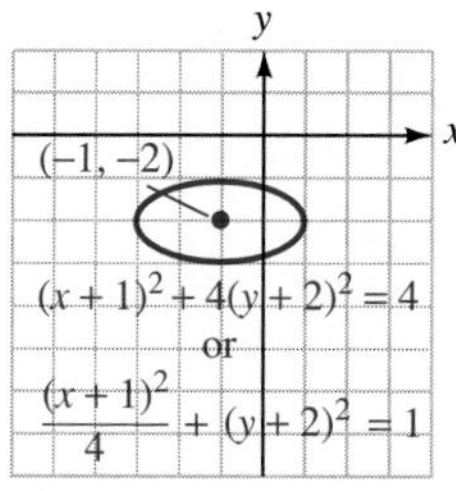

10. $25(x+1)^2 + 9y^2 = 225$

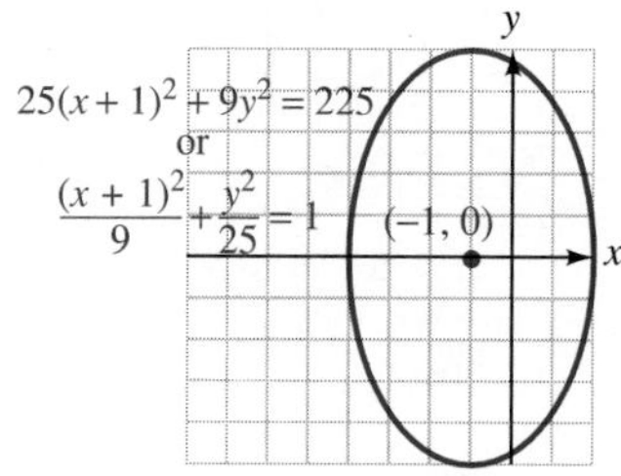

In Exercises 11–22, graph each hyperbola.

11. $\dfrac{x^2}{9} - \dfrac{y^2}{4} = 1$

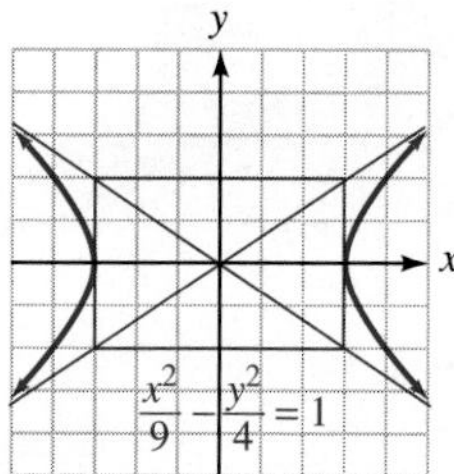

12. $\dfrac{x^2}{4} - \dfrac{y^2}{4} = 1$

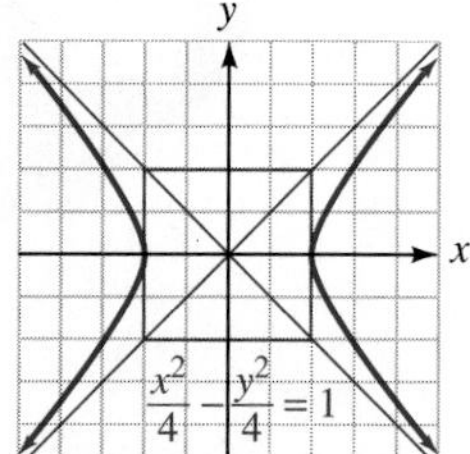

13. $\dfrac{y^2}{4} - \dfrac{x^2}{9} = 1$

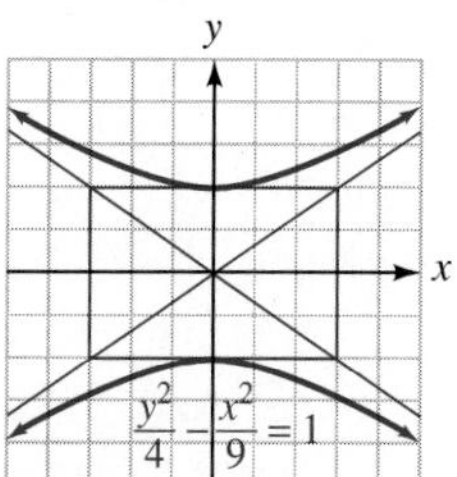

14. $\dfrac{y^2}{4} - \dfrac{x^2}{64} = 1$

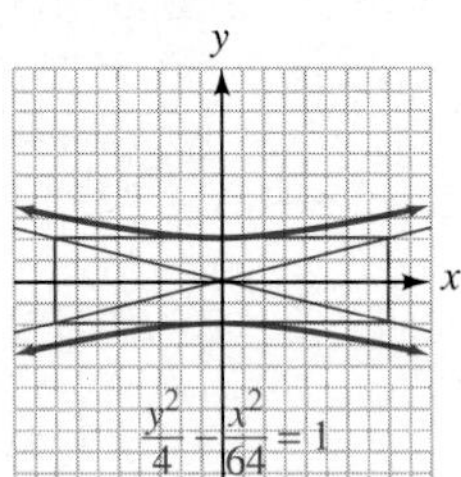

15. $25x^2 - y^2 = 25$

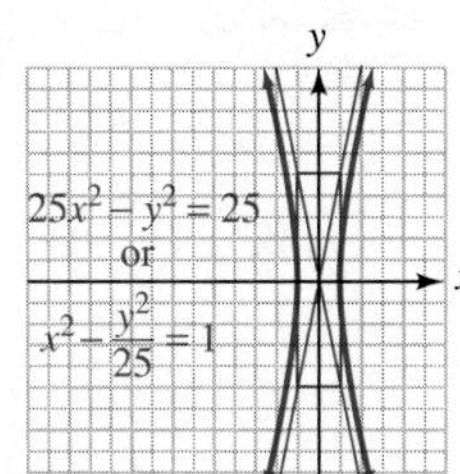

16. $9x^2 - 4y^2 = 36$

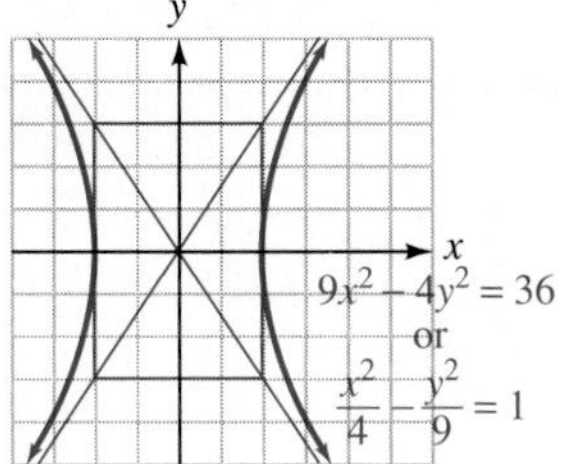

17. $\dfrac{(x-2)^2}{9} - \dfrac{y^2}{16} = 1$

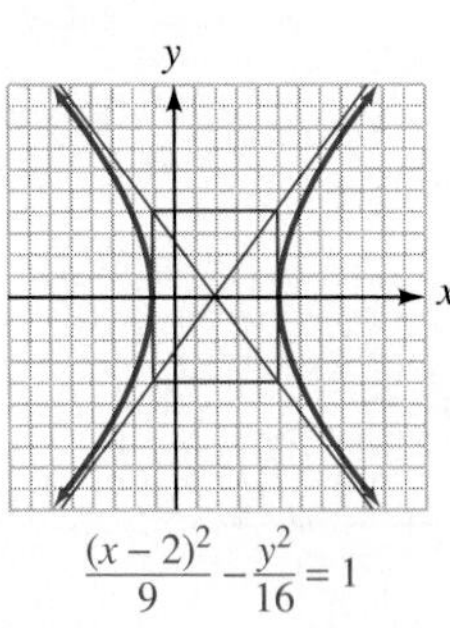

18. $\dfrac{(x+2)^2}{16} - \dfrac{(y-3)^2}{25} = 1$

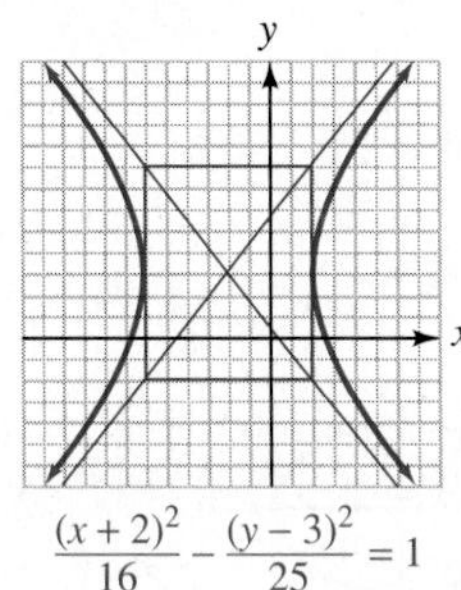

19. $4(x+3)^2 - (y-1)^2 = 4$

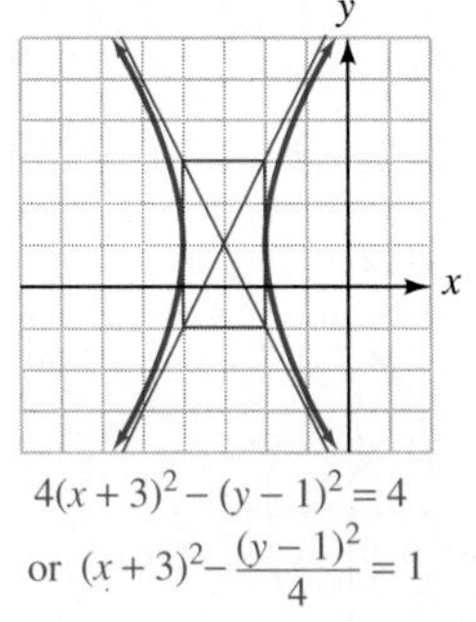

20. $(x+5)^2 - 16y^2 = 16$

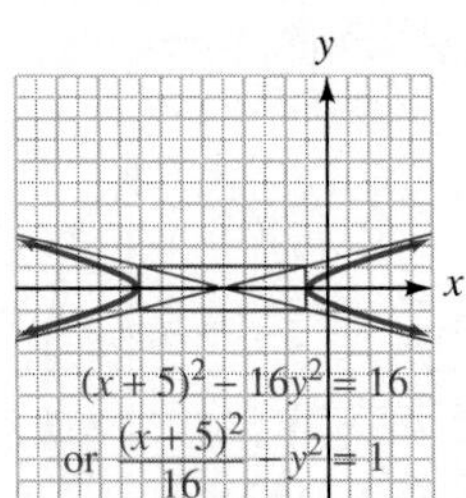

21. $xy = 8$

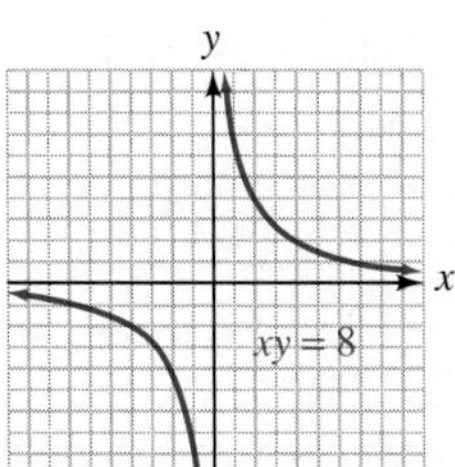

22. $xy = -10$

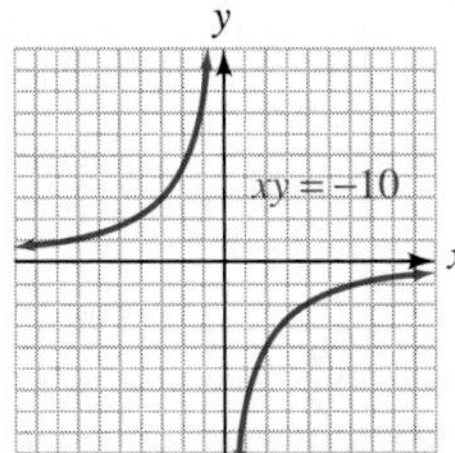

23. Designing an underpass The arch of the underpass in Illustration 1 is a part of an ellipse. Find the equation of the arch. $\dfrac{x^2}{400} + \dfrac{y^2}{100} = 1$

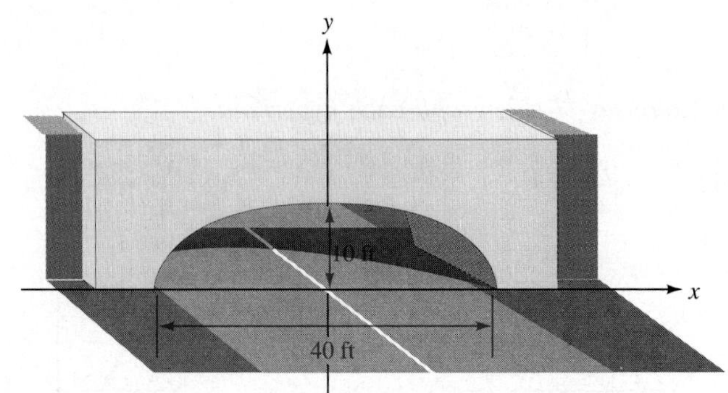

ILLUSTRATION 1

24. Calculating clearance Find the height of the elliptical arch in Exercise 23 at a point 10 feet from the center of the roadway. $5\sqrt{3}$ ft

25. Area of an ellipse The area A of the ellipse $\frac{x^2}{a^2} + \frac{y^2}{b^2} = 1$ is given by $A = \pi ab$. Find the area of the ellipse $9x^2 + 16y^2 = 144$. 12π

26. Area of a track The elliptical track in Illustration 2 is bounded by the ellipses

$$4x^2 + 9y^2 = 576 \text{ and } 9x^2 + 25y^2 = 900$$

Find the area of the track. (See Exercise 25.) 36π

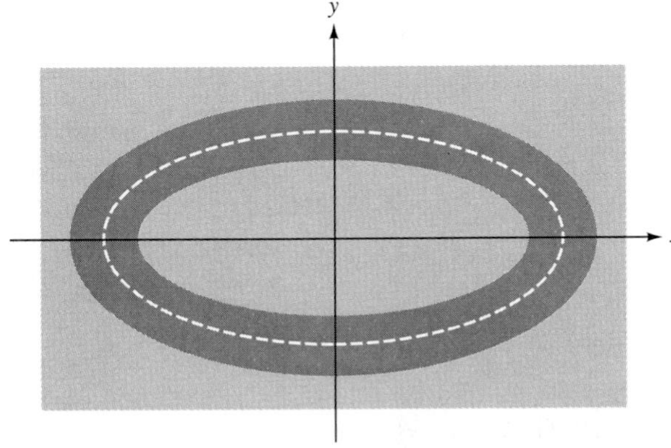

ILLUSTRATION 2

27. Alpha particles The particle in Illustration 3 approaches the nucleus at the origin along the path $9y^2 - x^2 = 81$. How close does the particle come to the nucleus? 3 units

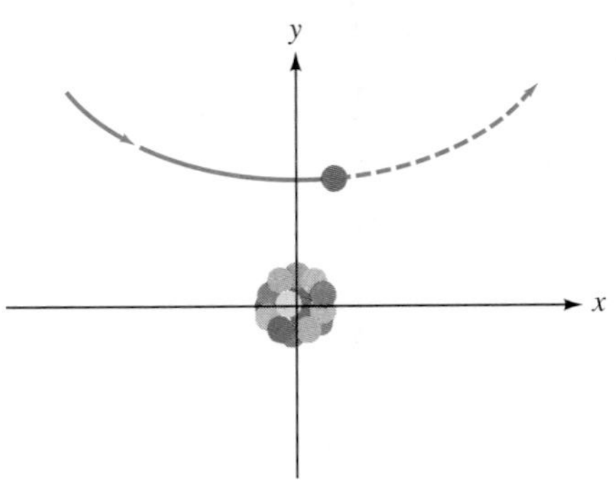

ILLUSTRATION 3

28. LORAN By determining the difference of the distances between the ship in Illustration 4 and two radio transmitters, the LORAN system places the ship on the hyperbola $x^2 - 4y^2 = 576$. If the ship is also 5 miles out to sea, find its coordinates.
$x = 26$ mi, $y = 5$ mi

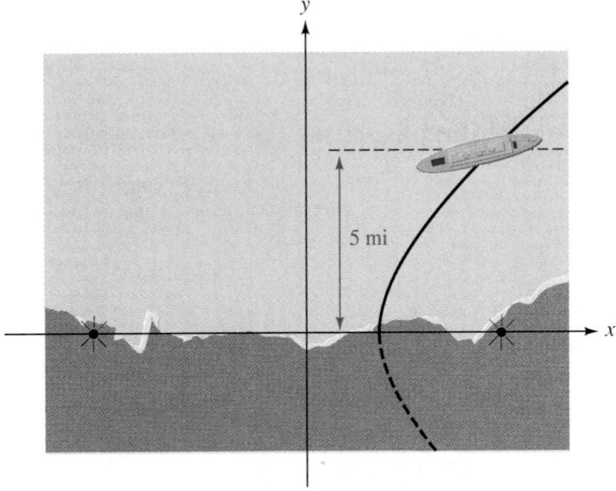

ILLUSTRATION 4

29. Sonic boom The position of the sonic boom caused by the faster-than-sound aircraft in Illustration 5 is the hyperbola $y^2 - x^2 = 25$ in the coordinate system shown. How wide is the hyperbola 5 units from its vertex? $4\sqrt{3}$ mi

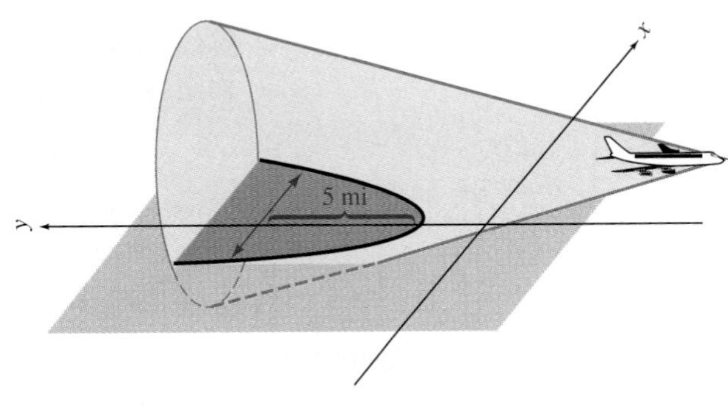

ILLUSTRATION 5

30. Electrostatic repulsion Two similarly charged particles are shot together for an almost head-on collision, as in Illustration 6. They repel each other and travel the two branches of the hyperbola given by $x^2 - 4y^2 = 4$. How close do they get? 4 units

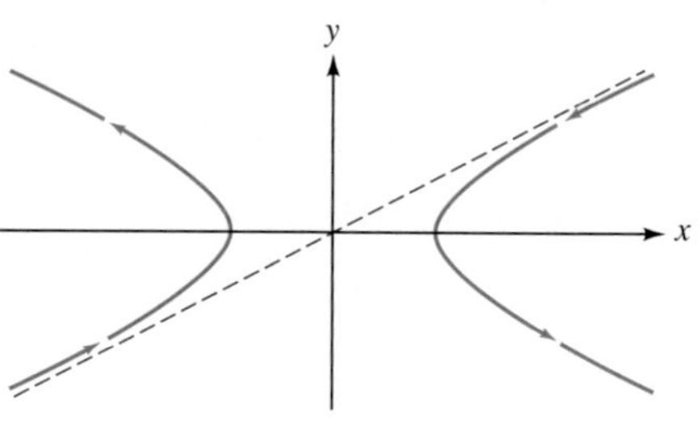

ILLUSTRATION 6

Writing Exercises *Write a paragraph using your own words.*

1. Explain how to find the x- and the y-intercepts of the graph of the ellipse $\frac{x^2}{a^2} + \frac{y^2}{b^2} = 1$.
2. Explain why the graph of the hyperbola $\frac{x^2}{a^2} - \frac{y^2}{b^2} = 1$ has no y-intercept.

Something to Think About

1. What happens to the graph of $\frac{x^2}{a^2} + \frac{y^2}{b^2} = 1$ if $a = b$?
2. The hyperbolas $x^2 - y^2 = 1$ and $y^2 - x^2 = 1$ are called **conjugate** hyperbolas. Graph both on the same axes. What do they have in common?

Review Exercises *Find each product.*

1. $3x^{-2}y^2(4x^2 + 3y^{-2})$ $\quad 12y^2 + \frac{9}{x^2}$
2. $(2a^{-2} - b^{-2})(2a^{-2} + b^{-2})$ $\quad \frac{4}{a^4} - \frac{1}{b^4}$

Write each expression without using negative exponents.

3. $\frac{x^{-2} + y^{-2}}{x^{-2} - y^{-2}}$ $\quad \frac{y^2 + x^2}{y^2 - x^2}$
4. $\frac{2x^{-3} - 2y^{-3}}{4x^{-3} + 4y^{-3}}$ $\quad \frac{y^3 - x^3}{2(y^3 + x^3)}$

10.6 Solving Simultaneous Second-Degree Equations

We now discuss ways to solve systems of two equations in two variables where at least one of the equations is of second degree.

EXAMPLE 1 Solve the system $\begin{cases} x^2 + y^2 = 25 \\ 2x + y = 10 \end{cases}$ by graphing.

Solution The graph of the equation $x^2 + y^2 = 25$ is a circle with center at the origin and radius of 5. The graph of the equation $2x + y = 10$ is a straight line. Depending on whether the line is a secant (intersecting the circle at two points) or a tangent (intersecting the circle at one point) or does not intersect the circle at all, there are two, one, or no solutions to the system, respectively.

After graphing the circle and the line, as shown in Figure 10-42, we see that there are two intersection points, $P(3, 4)$ and $P'(5, 0)$. Thus, the solutions to the system of equations are

$$\begin{cases} x = 3 \\ y = 4 \end{cases} \quad \text{and} \quad \begin{cases} x = 5 \\ y = 0 \end{cases}$$

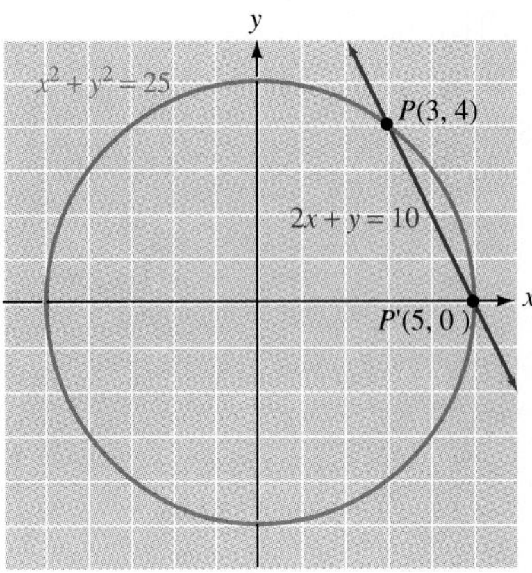

FIGURE 10-42

Verify that these are exact solutions. ■

Graphical methods of solving systems of equations usually give only approximate solutions. A second method, using algebra, can be used to find exact solutions.

EXAMPLE 2 Solve the system $\begin{cases} x^2 + y^2 = 25 \\ 2x + y = 10 \end{cases}$ algebraically.

Solution This system contains one equation of second degree and another of first degree. We can solve this type of system by substitution. Solving the linear equation for y gives

$$2x + y = 10$$

1. $$y = -2x + 10$$

We substitute $-2x + 10$ for y in the second-degree equation and solve the resulting quadratic equation for x:

$$x^2 + y^2 = 25$$

$$x^2 + (-2x + 10)^2 = 25$$

$$x^2 + 4x^2 - 40x + 100 = 25 \qquad (-2x + 10)(-2x + 10) = 4x^2 - 40x + 100.$$

$$5x^2 - 40x + 75 = 0 \qquad \text{Subtract 25 from both sides and combine terms.}$$

$$x^2 - 8x + 15 = 0 \qquad \text{Divide both sides by 5.}$$

$$(x - 5)(x - 3) = 0 \qquad \text{Factor } x^2 - 8x + 15.$$

$$x - 5 = 0 \quad \text{or} \quad x - 3 = 0 \qquad \text{Set each factor equal to 0.}$$

$$x = 5 \quad | \quad x = 3$$

If we substitute 5 for x in Equation 1, we get $y = 0$. If we substitute 3 for x in Equation 1, we get $y = 4$. The two solutions are

$$\begin{cases} x = 5 \\ y = 0 \end{cases} \quad \text{or} \quad \begin{cases} x = 3 \\ y = 4 \end{cases}$$

■

EXAMPLE 3 Solve the system $\begin{cases} 4x^2 + 9y^2 = 5 \\ y = x^2 \end{cases}$ algebraically.

Solution We can solve this system by substitution.

$$\begin{aligned} 4x^2 + 9y^2 &= 5 \\ 4y + 9y^2 &= 5 && \text{Substitute } y \text{ for } x^2. \\ 9y^2 + 4y - 5 &= 0 && \text{Subtract 5 from both sides.} \\ (9y - 5)(y + 1) &= 0 && \text{Factor } 9y^2 + 4y - 5. \end{aligned}$$

$$9y - 5 = 0 \quad \text{or} \quad y + 1 = 0 \qquad \text{Set each factor equal to 0.}$$

$$y = \frac{5}{9} \quad \Big| \quad y = -1$$

Since $y = x^2$, the values of x are found by solving the equations

$$x^2 = \frac{5}{9} \quad \text{and} \quad x^2 = -1$$

Because the equation $x^2 = -1$ has no real solutions, this possibility is discarded. The solutions of the equation $x^2 = \frac{5}{9}$ are

$$x = \frac{\sqrt{5}}{3} \quad \text{or} \quad x = \frac{-\sqrt{5}}{3}$$

Thus, the solutions of the system are

$$\left(\frac{\sqrt{5}}{3}, \frac{5}{9}\right) \quad \text{and} \quad \left(\frac{-\sqrt{5}}{3}, \frac{5}{9}\right)$$

■

EXAMPLE 4 Solve the system $\begin{cases} 3x^2 + 2y^2 = 36 \\ 4x^2 - y^2 = 4 \end{cases}$ algebraically.

Solution In this system, both equations are of second degree and in the form $ax^2 + by^2 = c$. We can solve systems like this by addition.

We can copy the first equation and multiply the second equation by 2 to obtain the equivalent system of equations:

$$\begin{cases} 3x^2 + 2y^2 = 36 \\ 8x^2 - 2y^2 = 8 \end{cases}$$

We add the equations to eliminate y and solve the resulting equation for x:

$$11x^2 = 44$$
$$x^2 = 4$$
$$x = 2 \quad \text{or} \quad x = -2$$

To find y, we substitute 2 for x and then -2 for x in the first equation and proceed as follows:

For $x = 2$	**For $x = -2$**
$3x^2 + 2y^2 = 36$	$3x^2 + 2y^2 = 36$
$3(2)^2 + 2y^2 = 36$	$3(-2)^2 + 2y^2 = 36$
$12 + 2y^2 = 36$	$12 + 2y^2 = 36$
$2y^2 = 24$	$2y^2 = 24$
$y^2 = 12$	$y^2 = 12$
$y = +\sqrt{12}$ or $y = -\sqrt{12}$	$y = +\sqrt{12}$ or $y = -\sqrt{12}$
$y = 2\sqrt{3}$ | $y = -2\sqrt{3}$	$y = 2\sqrt{3}$ | $y = -2\sqrt{3}$

The four solutions of this system are

$$(2, 2\sqrt{3}), \quad (2, -2\sqrt{3}), \quad (-2, 2\sqrt{3}), \quad \text{and} \quad (-2, -2\sqrt{3})$$ ■

ORALS *Give the possible number of solutions of a system when the graphs of the equations are*

1. a line and a parabola.
2. a line and a hyperbola.
3. a circle and a parabola.
4. a circle and a hyperbola.

EXERCISE 10.6

In Exercises 1–10, solve each system of equations by graphing.

1. $\begin{cases} 8x^2 + 32y^2 = 256 \\ x = 2y \end{cases}$

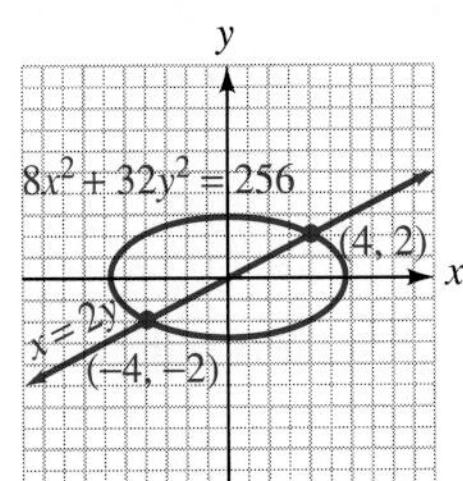

2. $\begin{cases} x^2 + y^2 = 2 \\ x + y = 2 \end{cases}$

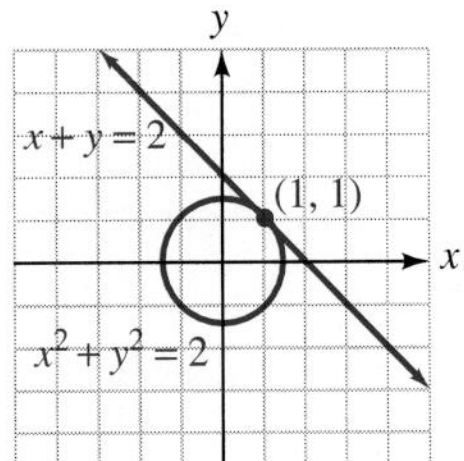

3. $\begin{cases} x^2 + y^2 = 10 \\ y = 3x^2 \end{cases}$

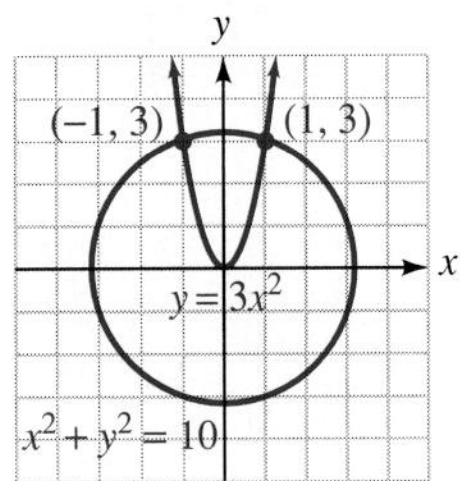

4. $\begin{cases} x^2 + y^2 = 5 \\ x + y = 3 \end{cases}$

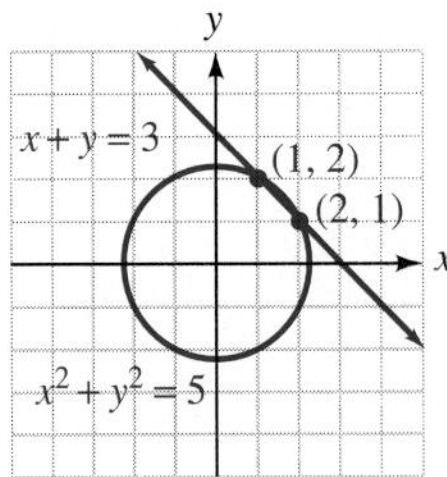

5. $\begin{cases} x^2 + y^2 = 25 \\ 12x^2 + 64y^2 = 768 \end{cases}$

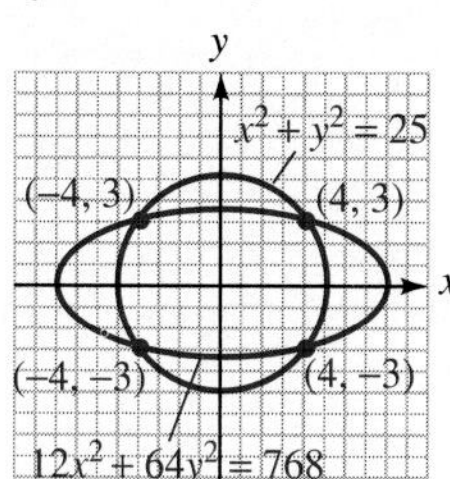

6. $\begin{cases} x^2 + y^2 = 13 \\ y = x^2 - 1 \end{cases}$

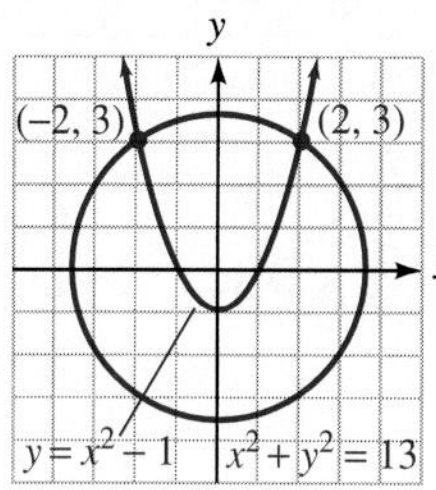

7. $\begin{cases} x^2 - 13 = -y^2 \\ y = 2x - 4 \end{cases}$

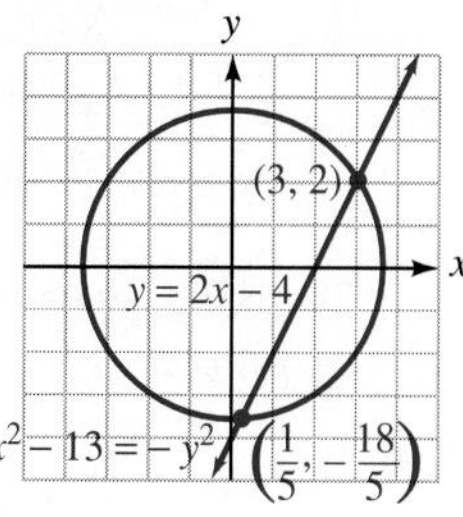

8. $\begin{cases} x^2 + y^2 = 20 \\ y = x^2 \end{cases}$

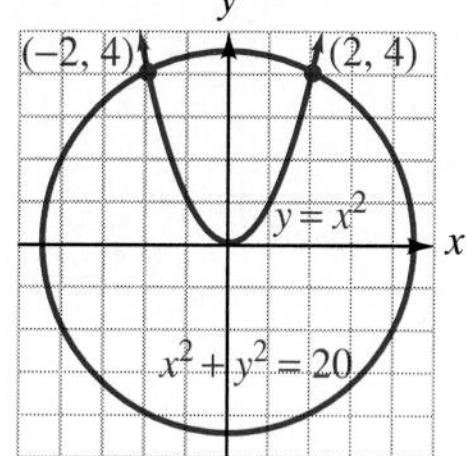

9. $\begin{cases} x^2 - 6x - y = -5 \\ x^2 - 6x + y = -5 \end{cases}$

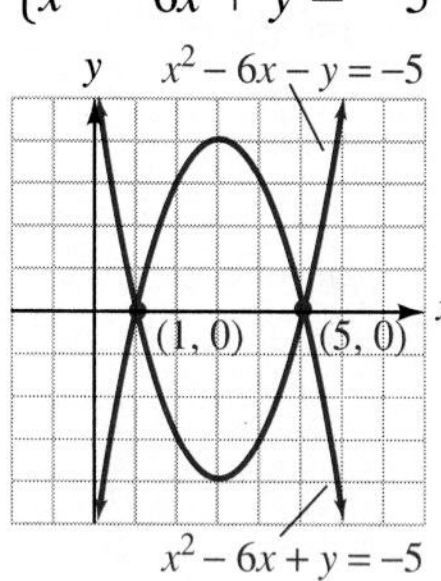

10. $\begin{cases} x^2 - y^2 = -5 \\ 3x^2 + 2y^2 = 30 \end{cases}$

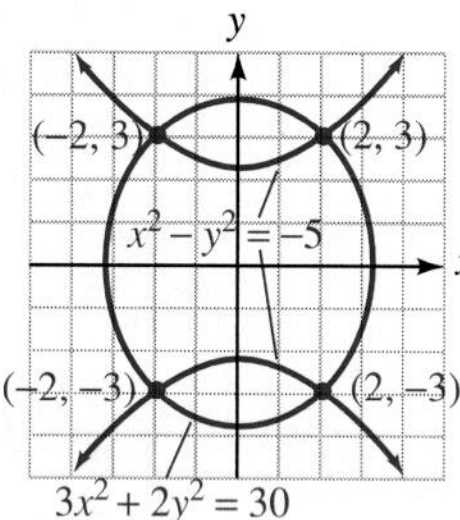

In Exercises 11–36, solve each system of equations algebraically for real values of x and y.

11. $\begin{cases} 25x^2 + 9y^2 = 225 \\ 5x + 3y = 15 \end{cases}$
(3, 0), (0, 5)

12. $\begin{cases} x^2 + y^2 = 20 \\ y = x^2 \end{cases}$
(2, 4), (−2, 4)

13. $\begin{cases} x^2 + y^2 = 2 \\ x + y = 2 \end{cases}$
(1, 1)

14. $\begin{cases} x^2 + y^2 = 36 \\ 49x^2 + 36y^2 = 1764 \end{cases}$
(6, 0), (−6, 0)

15. $\begin{cases} x^2 + y^2 = 5 \\ x + y = 3 \end{cases}$
(1, 2), (2, 1)

16. $\begin{cases} x^2 - x - y = 2 \\ 4x - 3y = 0 \end{cases}$
$\left(-\frac{2}{3}, -\frac{8}{9}\right)$, (3, 4)

17. $\begin{cases} x^2 + y^2 = 13 \\ y = x^2 - 1 \end{cases}$
(−2, 3), (2, 3)

18. $\begin{cases} x^2 + y^2 = 25 \\ 2x^2 - 3y^2 = 5 \end{cases}$
(4, 3), (−4, 3), (4, −3), (−4, −3)

19. $\begin{cases} x^2 + y^2 = 30 \\ y = x^2 \end{cases}$
$(\sqrt{5}, 5), (-\sqrt{5}, 5)$

20. $\begin{cases} 9x^2 - 7y^2 = 81 \\ x^2 + y^2 = 9 \end{cases}$
(3, 0), (−3, 0)

21. $\begin{cases} x^2 + y^2 = 13 \\ x^2 - y^2 = 5 \end{cases}$
(3, 2), (3, −2), (−3, 2), (−3, −2)

22. $\begin{cases} 2x^2 + y^2 = 6 \\ x^2 - y^2 = 3 \end{cases}$
$(\sqrt{3}, 0), (-\sqrt{3}, 0)$

23. $\begin{cases} x^2 + y^2 = 20 \\ x^2 - y^2 = -12 \end{cases}$
(2, 4), (2, −4), (−2, 4), (−2, −4)

24. $\begin{cases} xy = -\dfrac{9}{2} \\ 3x + 2y = 6 \end{cases}$
$\left(-1, \frac{9}{2}\right), \left(3, -\frac{3}{2}\right)$

25. $\begin{cases} y^2 = 40 - x^2 \\ y = x^2 - 10 \end{cases}$
$(-\sqrt{15}, 5), (\sqrt{15}, 5)$, (−2, −6), (2, −6)

26. $\begin{cases} x^2 - 6x - y = -5 \\ x^2 - 6x + y = -5 \end{cases}$
(5, 0), (1, 0)

27. $\begin{cases} y = x^2 - 4 \\ x^2 - y^2 = -16 \end{cases}$
(0, −4), (−3, 5), (3, 5)

28. $\begin{cases} 6x^2 + 8y^2 = 182 \\ 8x^2 - 3y^2 = 24 \end{cases}$
(3, 4), (3, −4), (−3, 4), (−3, −4)

29. $\begin{cases} x^2 - y^2 = -5 \\ 3x^2 + 2y^2 = 30 \end{cases}$
(−2, 3), (2, 3), (−2, −3), (2, −3)

30. $\begin{cases} \dfrac{1}{x} + \dfrac{1}{y} = 5 \\ \dfrac{1}{x} - \dfrac{1}{y} = -3 \end{cases}$
$\left(1, \frac{1}{4}\right)$

31. $\begin{cases} \dfrac{1}{x} + \dfrac{2}{y} = 1 \\ \dfrac{2}{x} - \dfrac{1}{y} = \dfrac{1}{3} \end{cases}$ (3, 3)

32. $\begin{cases} \dfrac{1}{x} + \dfrac{3}{y} = 4 \\ \dfrac{2}{x} - \dfrac{1}{y} = 7 \end{cases}$ $\left(\frac{7}{25}, 7\right)$

33. $\begin{cases} 3y^2 = xy \\ 2x^2 + xy - 84 = 0 \end{cases}$
$(6, 2), (-6, -2), (\sqrt{42}, 0), (-\sqrt{42}, 0)$

34. $\begin{cases} x^2 + y^2 = 10 \\ 2x^2 - 3y^2 = 5 \end{cases}$
$(\sqrt{7}, \sqrt{3}), (\sqrt{7}, -\sqrt{3}), (-\sqrt{7}, \sqrt{3}), (-\sqrt{7}, -\sqrt{3})$

35. $\begin{cases} xy = \dfrac{1}{6} \\ y + x = 5xy; \end{cases}$ $\left(\frac{1}{2}, \frac{1}{3}\right), \left(\frac{1}{3}, \frac{1}{2}\right)$

36. $\begin{cases} xy = \dfrac{1}{12} \\ y + x = 7xy \end{cases}$ $\left(\frac{1}{4}, \frac{1}{3}\right), \left(\frac{1}{3}, \frac{1}{4}\right)$

37. **Geometry problem** The area of a rectangle is 63 square centimeters, and its perimeter is 32 centimeters. Find the dimensions of the rectangle. 7 by 9 cm

38. **Integer problem** The product of two integers is 32, and their sum is 12. Find the integers. 4 and 8

39. **Number problem** The sum of the squares of two numbers is 221, and the sum of the numbers is 212 less. Find the numbers. 14 and −5

40. **Investing money** Grant receives \$225 annual income from one investment. Jeff invested \$500 more than Grant, but at an annual rate of 1% less. Jeff's annual income is \$240. What is the amount and rate of Grant's investment? \$2500 at 9%

41. **Investing money** Carol receives \$67.50 annual income from one investment. John invested \$150 more than Carol at an annual rate of $1\frac{1}{2}$% more. John's annual income is \$94.50. What is the amount and rate of Carol's investment? (*Hint:* There are two answers.) either \$750 at 9% or \$900 at 7.5%

42. **Artillery** The shell fired from the base of the hill in Illustration 1 follows the parabolic path $y = -\frac{1}{6}x^2 + 2x$, with distances measured in miles. The hill has a slope of $\frac{1}{3}$. How far from the gun is the point of impact? (*Hint:* Find the coordinates of the point and the distance.) $\frac{10}{3}\sqrt{10}$ mi

43. **Driving rates** Jim drove 306 miles. Jim's brother made the same trip at a speed of 17 miles per hour slower than Jim did and required an extra $1\frac{1}{2}$ hours. What was Jim's rate and time? 68 mph, 4.5 hr

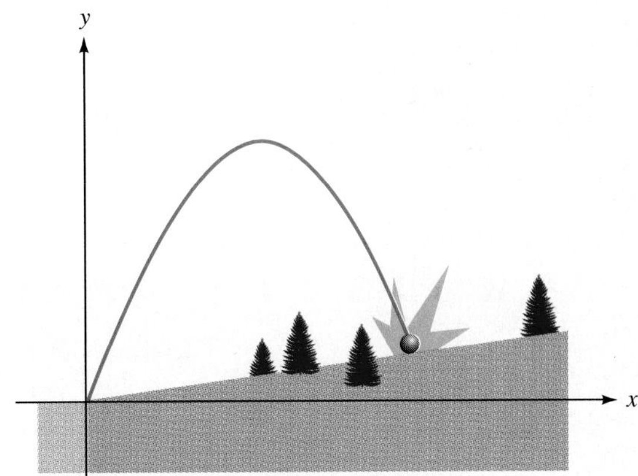

ILLUSTRATION 1

44. **Fencing a pasture** A rectangular pasture is to be fenced in along a riverbank, as shown in Illustration 2. If 260 feet of fencing is to enclose an area of 8000 square feet, find the dimensions of the pasture. 80 ft by 100 ft

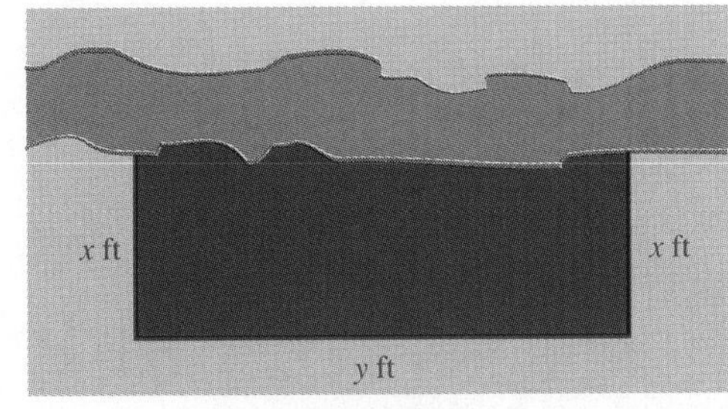

ILLUSTRATION 2

Writing Exercises *Write a paragraph using your own words.*

1. Describe the benefits of the graphical method for solving a system of equations.

2. Describe the drawbacks of the graphical method.

Something to Think About

1. The graphs of the two equations of a system are parabolas. How many solutions might the system have?

2. The graphs of the two equations of a system are a line and a hyperbola. How many solutions might the system have?

Review Exercises *Simplify each radical expression. Assume that all variables represent positive numbers.*

1. $\sqrt{200x^2} - 3\sqrt{98x^2}$ $\quad -11x\sqrt{2}$

2. $a\sqrt{112a} - 5\sqrt{175a^3}$ $\quad -21a\sqrt{7a}$

3. $\dfrac{3t\sqrt{2t} - 2\sqrt{2t^3}}{\sqrt{18t} - \sqrt{2t}}$ $\quad \dfrac{t}{2}$

4. $\sqrt[3]{\dfrac{x}{4}} + \sqrt[3]{\dfrac{x}{32}} - \sqrt[3]{\dfrac{x}{500}}$ $\quad \dfrac{13\sqrt[3]{2x}}{20}$

PROJECT Graphing Conics

The graphs of many conics do not pass the vertical line test, so their equations do not represent functions. Such conics are not easy to graph on a graphing calculator, because graphing calculators are most useful for graphing functions. For example, the circle $(x - 1)^2 + (y - 2)^2 = 4$ does not represent a function. To graph it with a graphing calculator, we have to split its equation into two functions and graph each one separately. We begin by solving the equation for y.

$$(x - 1)^2 + (y - 2)^2 = 4$$

$$(y - 2)^2 = 4 - (x - 1)^2 \qquad \text{Subtract } (x-1)^2 \text{ from both sides.}$$

$$y - 2 = \pm\sqrt{4 - (x - 1)^2} \qquad \text{Take the square root of both sides. There are two possibilities.}$$

$$y = 2 \pm \sqrt{4 - (x - 1)^2} \qquad \text{Add 2 to both sides.}$$

This equation defines two functions. The graphs of the two functions

$$Y_1 = 2 + \sqrt{4 - (x - 1)^2}$$
$$Y_2 = 2 - \sqrt{4 - (x - 1)^2}$$

produce the circle in Illustration 1.

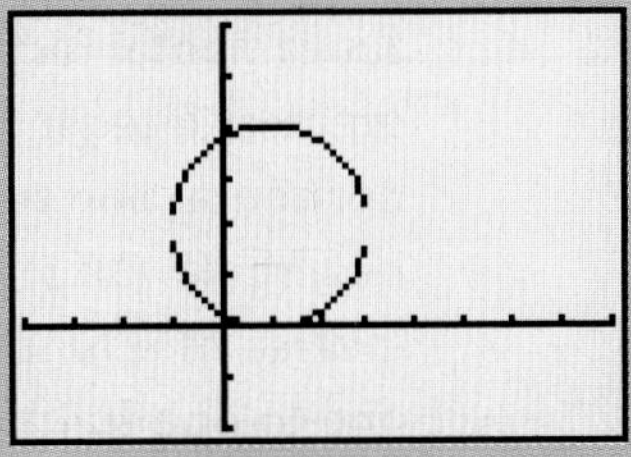

ILLUSTRATION 1

Similarly, the equation of the hyperbola $x^2 - y^2 = 1$ is graphed in two parts:

$$Y_1 = \sqrt{x^2 - 1} \quad \text{and} \quad Y_2 = -\sqrt{x^2 - 1}$$

The graph appears in Illustration 2.

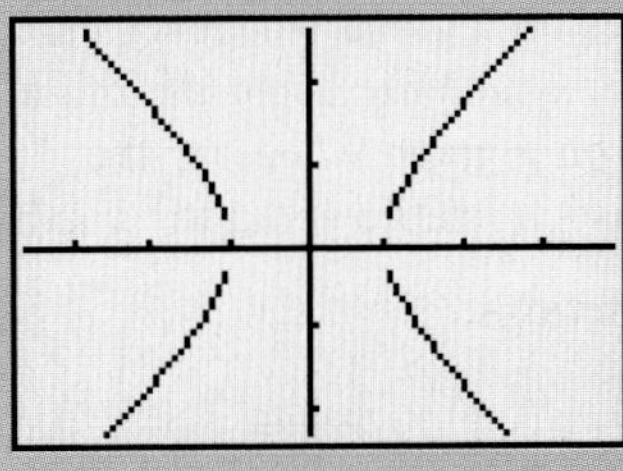

ILLUSTRATION 2

- Break the equation of the parabola $x = \frac{y^2}{4}$ into two functions, and use a graphing calculator to graph both.
- Graph the ellipse $(x - 3)^2 + y^2 = 9$.
- Graph the hyperbola $y^2 - x^2 = 1$.
- On the same axes that you graphed the previous hyperbola, also graph the hyperbola of Illustration 2. What do you notice about their graphs? The two hyperbolas are called **conjugate hyperbolas.**

Chapter 10 Summary

KEY WORDS

absolute value function (10.1)
asymptotes (10.5)
circle (10.4)
composition of functions (10.2)
conic sections (10.4)
degree of a polynomial function (10.1)
difference of two functions (10.2)

ellipse (10.5)
focus (10.5)
fundamental rectangle (10.5)
greatest integer function (10.1)
horizontal line test (10.3)
hyperbola (10.5)
inverse of a function (10.3)
one-to-one function (10.3)
parabola (10.4)
paraboloid (10.4)
piecewise-defined functions (10.1)
polynomial function (10.1)
product of two functions (10.2)
quotient of two functions (10.2)
square root function (10.1)
step functions (10.1)
sum of two functions (10.2)
symmetry of graphs (10.1)

KEY IDEAS

(10.1) **Symmetries of graphs:**

If the point $(-x, y)$ lies on a graph whenever the point (x, y) does, the graph is symmetric about the y-axis.

If the point $(-x, -y)$ lies on a graph whenever the point (x, y) does, the graph is symmetric about the origin.

If the point $(x, -y)$ lies on a graph whenever the point (x, y) does, the graph is symmetric about the x-axis.

(10.2) **Operations with functions:**

$$(f+g)(x) = f(x) + g(x)$$

$$(f-g)(x) = f(x) - g(x)$$

$$(f \cdot g)(x) = f(x)g(x)$$

$$(f/g)(x) = \frac{f(x)}{g(x)} \quad (g(x) \neq 0)$$

$$(f \circ g)(x) = f(g(x))$$

(10.3) To find the inverse relation of a function, interchange the positions of variables x and y. If possible, solve for y.

The horizontal line test: If any horizontal line that intersects the graph of a function does so only once, the function is one-to-one.

(10.4) **Equations of a circle:**

$(x-h)^2 + (y-k)^2 = r^2$ center (h, k), radius r

$x^2 + y^2 = r^2$ center $(0, 0)$, radius r

Equations of parabolas:

Parabola opening	Vertex at origin	Vertex at (h, k)
Up	$y = ax^2$	$y = a(x-h)^2 + k$
Down	$y = -ax^2$	$y = -a(x-h)^2 + k$
Right	$x = ay^2$	$x = a(y-k)^2 + h$
Left	$x = -ay^2$	$x = -a(y-k)^2 + h$

(10.5) **Equations of an ellipse:**

$\frac{x^2}{a^2} + \frac{y^2}{b^2} = 1$ center $(0, 0)$

$\frac{(x-h)^2}{a^2} + \frac{(y-k)^2}{b^2} = 1$ center (h, k)

Equations of a hyperbola:

$\frac{x^2}{a^2} - \frac{y^2}{b^2} = 1$ or $\frac{y^2}{a^2} - \frac{x^2}{b^2} = 1$ center $(0, 0)$

$$\left.\begin{aligned} \frac{(x-h)^2}{a^2} - \frac{(y-k)^2}{b^2} &= 1 \\ \frac{(y-k)^2}{a^2} - \frac{(x-h)^2}{b^2} &= 1 \end{aligned}\right\} \text{center } (h, k)$$

(10.6) Good estimates for solutions to systems of simultaneous second-degree equations can be found by graphing.

Exact solutions to systems of simultaneous second-degree equations can be found with algebraic techniques.

Chapter 10 Review Exercises

In Review Exercises 1–4, find the symmetries of the graph of each equation.

1. $y = x^2 - 5$ y-axis
2. $|y| = x$ x-axis
3. $y = x^3 - x$ origin
4. $y = \sqrt{x} + 3$ none

In Review Exercises 5–8, graph each function and give the domain.

5. $y = x^3 - 2$

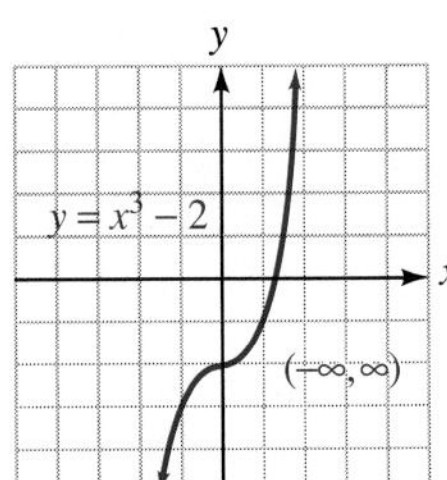

6. $y = x^4 - 3x^2 + 2$

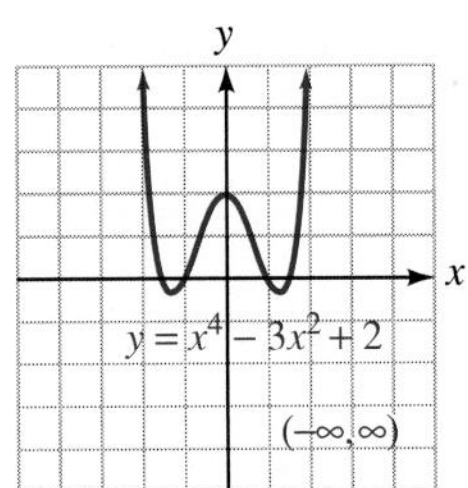

7. $y = -\frac{1}{2}|x|$

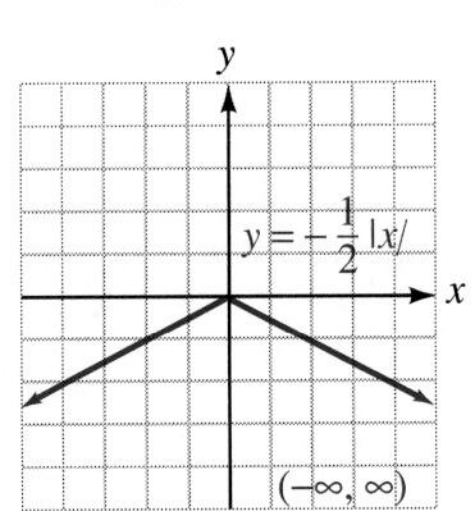

8. $y = \frac{1}{2}\sqrt{x}$

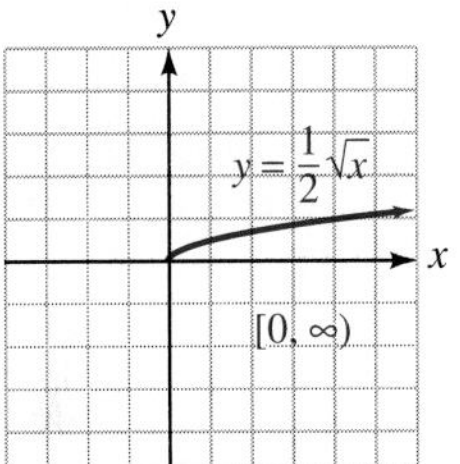

In Review Exercises 9–10, graph each function.

9. $y = \begin{cases} x & \text{if } x \leq 1 \\ -x^2 & \text{if } x > 1 \end{cases}$

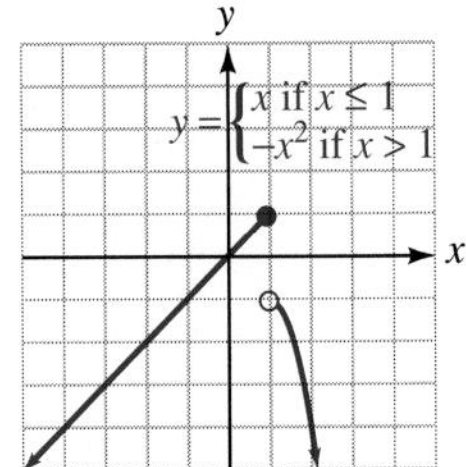

10. $y = 2[\![x]\!]$

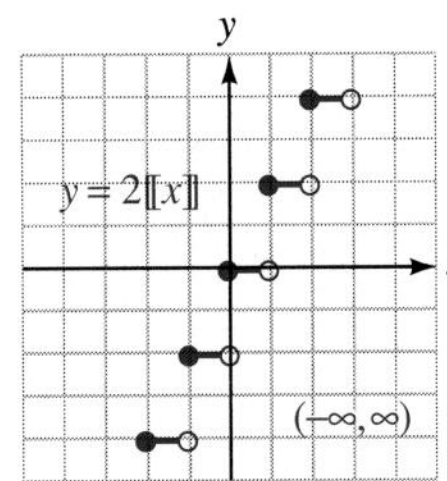

In Review Exercises 11–18, f(x) = 2x and g(x) = x + 1. Find each function.

11. $f + g$ $(f + g)(x) = 3x + 1$
12. $f - g$ $(f - g)(x) = x - 1$
13. $f \cdot g$ $(f \cdot g)(x) = 2x^2 + 2x$
14. f/g $(f/g)(x) = \frac{2x}{x+1}$
15. $(f \circ g)(2)$ 6
16. $(g \circ f)(-1)$ -1
17. $(f \circ g)(x)$ $2(x + 1)$
18. $(g \circ f)(x)$ $2x + 1$

In Review Exercises 19–22, find the inverse of each function.

19. $y = 6x - 3$
$y = \frac{x + 3}{6}$

20. $y = 4x + 5$
$y = \frac{x - 5}{4}$

21. $y = 2x^2 - 1$ $(x \geq 0)$
$y = \sqrt{\frac{x + 1}{2}}$

22. $y = |x|$
$x = |y|$

In Review Exercises 23–24, graph each function and use the horizontal line test to decide whether the function is one-to-one.

23. $y = 2(x - 3)$

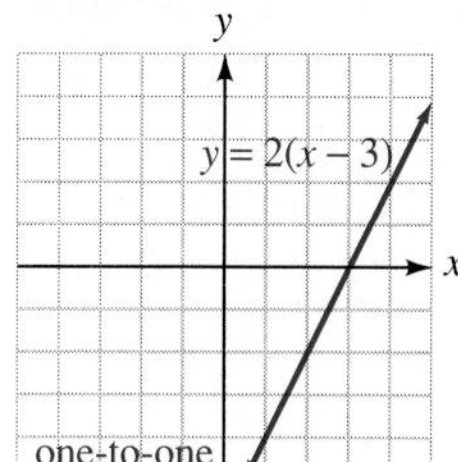

24. $y = x^2 - 4$

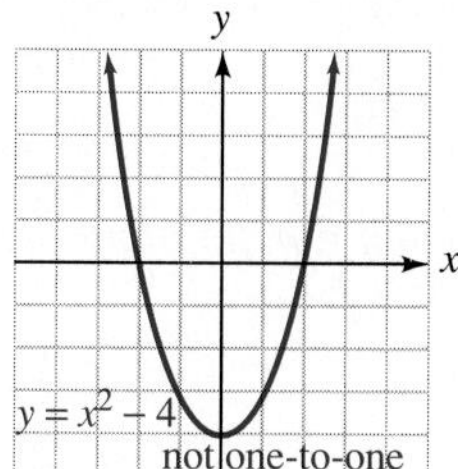

In Review Exercises 25–34, graph each equation.

25. $x^2 + y^2 = 16$

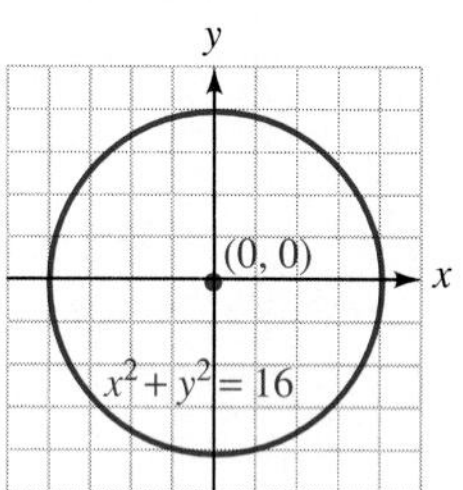

26. $(x - 3)^2 + (y + 2)^2 = 4$

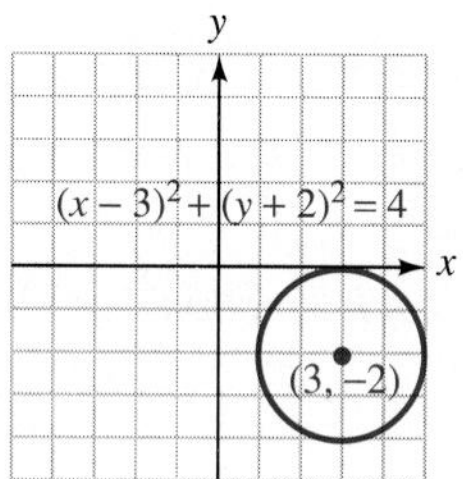

27. $x = -3(y - 2)^2 + 5$

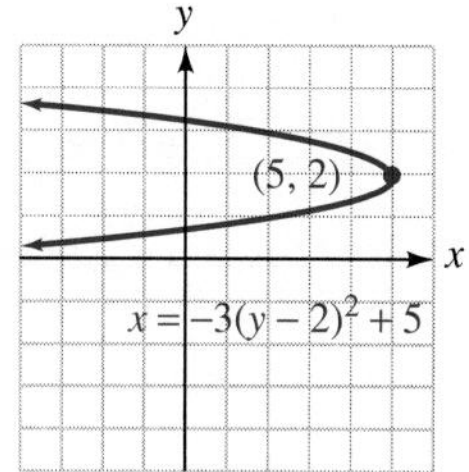

28. $x = 2(y + 1)^2 - 2$

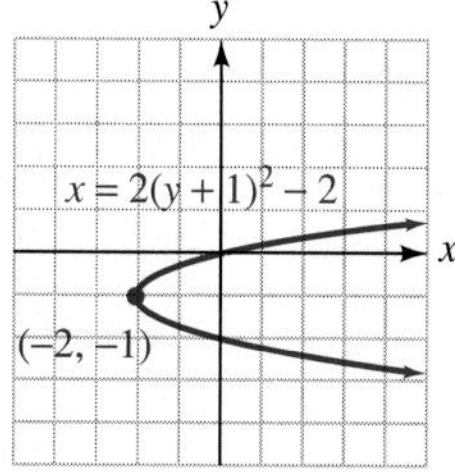

29. $9x^2 + 16y^2 = 144$

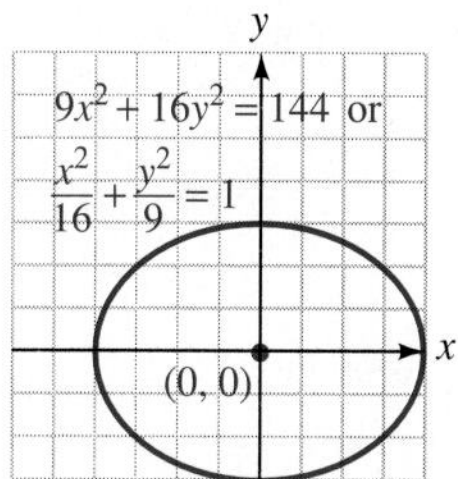

30. $\dfrac{(x - 2)^2}{4} + \dfrac{(y - 1)^2}{9} = 1$

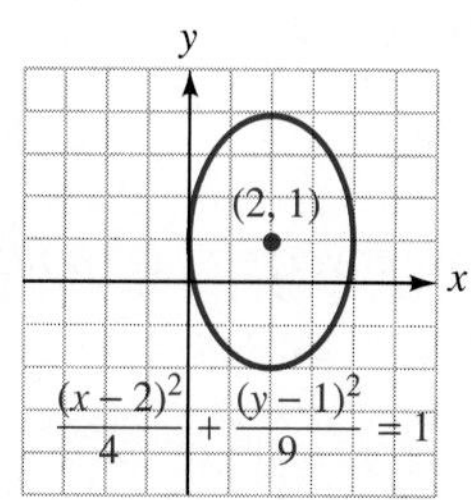

31. $xy = 9$

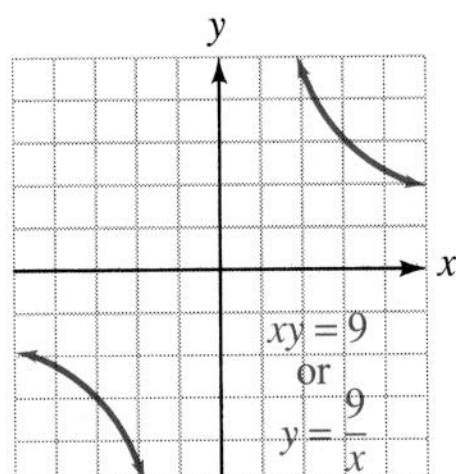

32. $9x^2 - y^2 = -9$

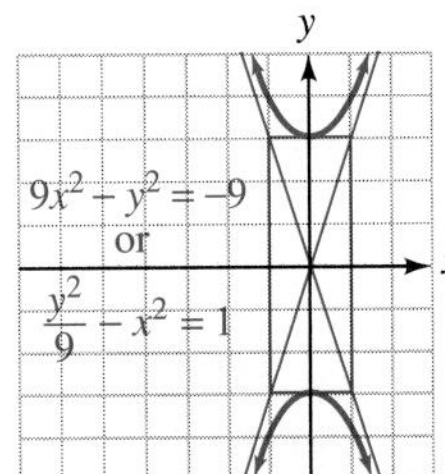

33. $y(x + y) = (y + 2)(y - 2)$

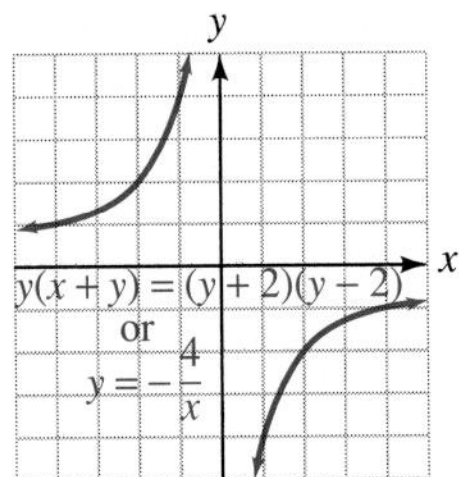

34. $x(x - y) = (x + 1)(x - 1)$

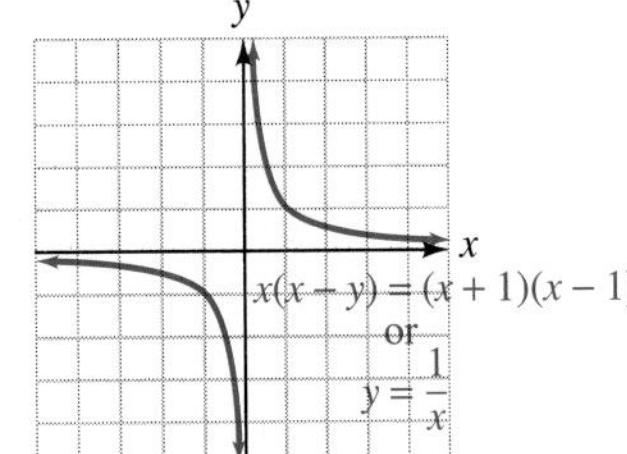

In Review Exercises 35–36, solve each system.

35. $\begin{cases} 3x^2 + y^2 = 52 \\ x^2 - y^2 = 12 \end{cases}$ (4, 2), (4, −2), (−4, 2), (−4, −2)

36. $\begin{cases} \dfrac{x^2}{16} + \dfrac{y^2}{12} = 1 \\ x^2 - \dfrac{y^2}{3} = 1 \end{cases}$ (2, 3), (2, −3), (−2, 3), (−2, −3)

Chapter 10 Test

In Problems 1–2, find the symmetries of the graph of each equation.

1. $y = x^3 - x$ origin

2. $y = x^8 - 5x^4$ y-axis

In Problems 3–6, graph each function.

3. $y = -x^4 - 5x^2 + 4$

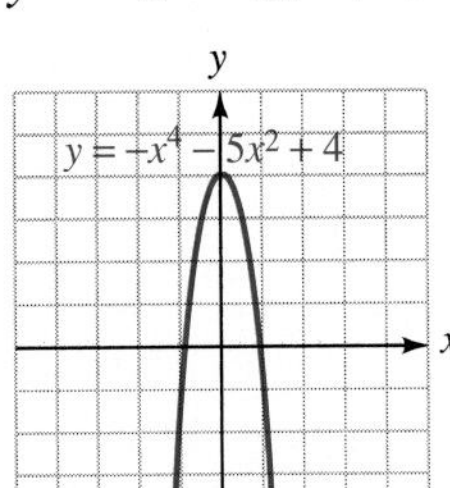

4. $y = -\sqrt{x} + 3$

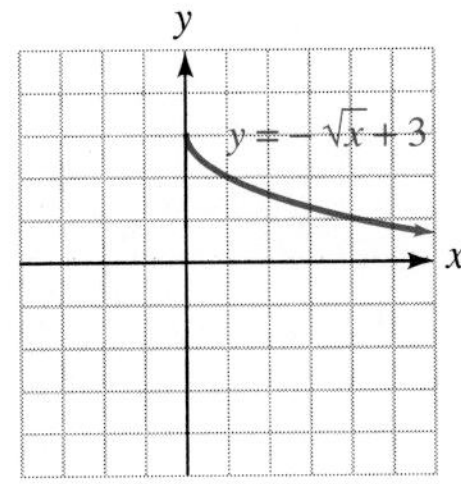

5. $y = \begin{cases} x^2 - 1 \text{ if } x \leq 0 \\ -x^2 + 1 \text{ if } x > 0 \end{cases}$

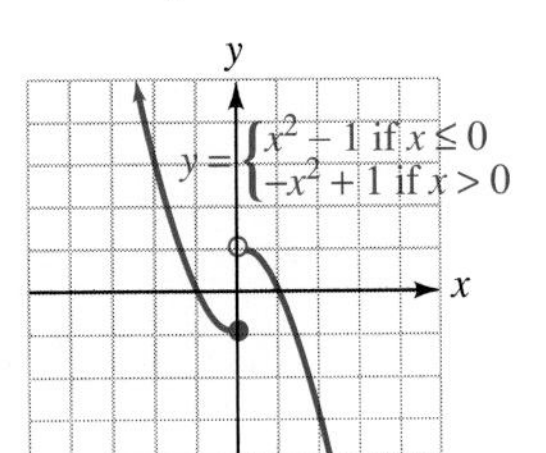

6. $y = [\![x]\!] - 1$

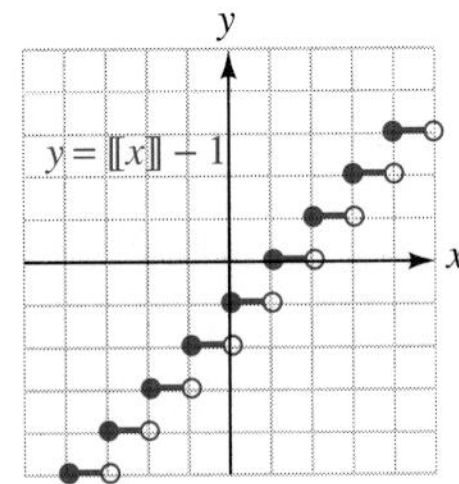

In Problems 7–10, $f(x) = 4x$ and $g(x) = x - 1$. Find each function.

7. $g + f$ $(g + f)(x) = 5x - 1$ **8.** $f - g$ $(f - g)(x) = 3x + 1$ **9.** $g \cdot f$ $(g \cdot f)(x) = 4x^2 - 4x$ **10.** g/f $(g/f)(x) = \frac{x-1}{4x}$

In Problems 11–14, $f(x) = 4x$ and $g(x) = x - 1$. Find each value.

11. $(g \circ f)(1)$ 3 **12.** $(f \circ g)(0)$ -4 **13.** $(f \circ g)(-1)$ -8 **14.** $(g \circ f)(-2)$ -9

In Problems 15–16, $f(x) = 4x$ and $g(x) = x - 1$. Find each function.

15. $(f \circ g)(x)$ $4(x - 1)$

16. $(g \circ f)(x)$ $4x - 1$

In Problems 17–18, find the inverse of each function.

17. $3x + 2y = 12$ $y = \dfrac{12 - 2x}{3}$

18. $y = 3x^2 + 4$ $(x \le 0)$ $y = -\sqrt{\dfrac{x - 4}{3}}$

19. Graph the function $y = \frac{1}{4}x^2 - 3$ and use the horizontal line test to decide whether the function is one-to-one.

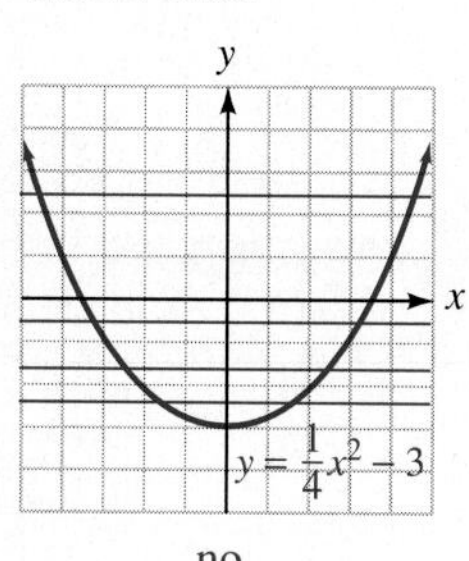

no

20. Find the center and the radius of the circle $(x - 2)^2 + (y + 3)^2 = 4$ and graph it.

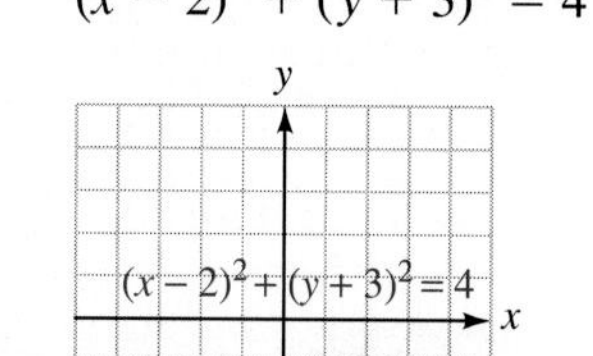

In Problems 21–22, graph each equation.

21. $9x^2 + 4y^2 = 36$

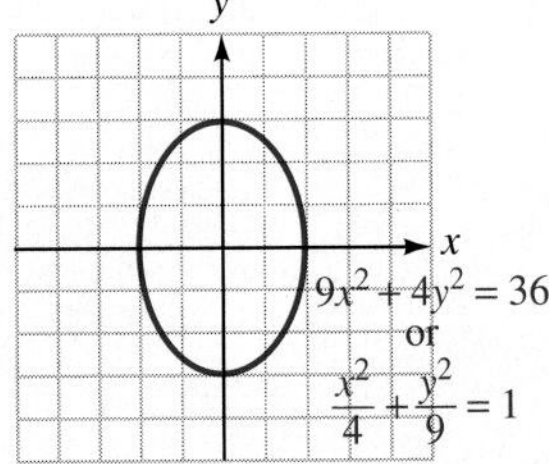

22. $\dfrac{(x-2)^2}{9} - y^2 = 1$

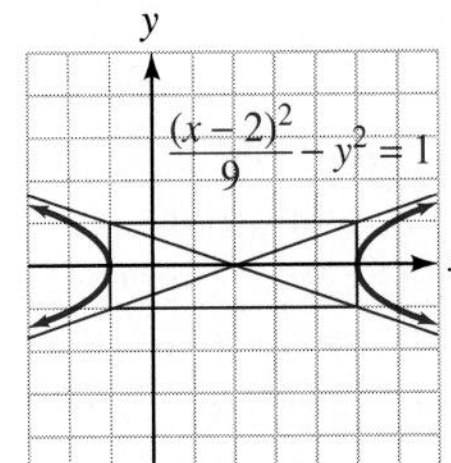

In Problems 23–24, solve each system.

23. $\begin{cases} 2x - y = -2 \\ x^2 + y^2 = 16 + 4y \end{cases}$ (2, 6), (−2, −2)

24. $\begin{cases} x^2 + y^2 = 25 \\ 4x^2 - 9y = 0 \end{cases}$ (3, 4), (−3, 4)

CAREERS & MATHEMATICS

Secondary School Teacher

Secondary school teachers help students delve more deeply into subjects introduced in elementary school and learn more about the world and about themselves. Teachers plan and evaluate lessons, prepare tests, grade papers, fill out report cards, oversee study halls and homerooms, supervise extracurricular activities, and meet with parents and school staff. In recent years, teachers have become more involved in curriculum design, such as choosing textbooks and evaluating teaching methods.

Qualifications

All of the 50 states and the District of Columbia require public secondary school teachers to be certified. Requirements for regular certificates vary by state. However, all states require a bachelor's degree and completion of an approved teacher-training program, including a prescribed number of subjects and educational credits and supervised practice teaching in a secondary school.

Job Outlook

Employment of secondary school teachers is expected to increase faster than the average for all occupations through the year 2005. Job openings for secondary school teachers are expected to increase substantially by the end of the decade as the large number of teachers now in their 40s and 50s reach retirement age.

Example Application

An art teacher wants to paint a mural with an elliptical background that is 10 feet wide and 6 feet high. To construct the ellipse, she uses some algebra and the definition of an ellipse.

By that definition, the total distance from one focus, F, out to any point on the ellipse and back to the other focus, F', is always constant. That fact provides an easy way to draw an ellipse: tie each end of a piece of string to a thumbtack, as in Illustration 1. Catch the loop with the point of a pencil and keep the string taut. Because the length of the string doesn't change as she slides the pencil around, she will draw an ellipse.

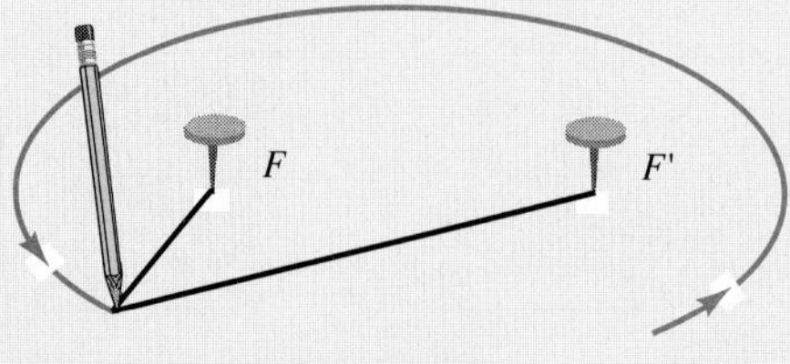

ILLUSTRATION 1

To draw a 10-foot-wide-by-6-foot-high ellipse, the teacher must find the length of the string and the distance between the thumbtacks. She would let a represent the distance from center to vertex V, as shown in Illustration 2, and let c be the distance between the center and either focus. When the pencil is at vertex V, the length of the string is $c + a + (a - c)$, or just $2a$. Because $2a$ is the 10-foot width of the ellipse, the art teacher will use a 10-foot-long string.

The distance $2a$ is constant for *any* point on the ellipse, including point B in Illustration 3.

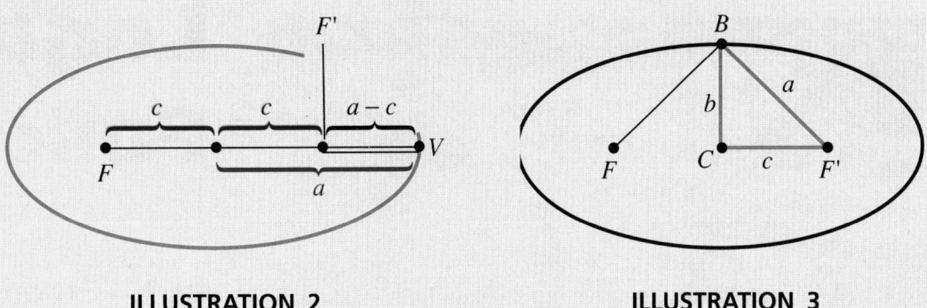

ILLUSTRATION 2 **ILLUSTRATION 3**

From the right triangle CBF' and the Pythagorean theorem, the teacher can find c:

$$a^2 = b^2 + c^2 \quad \text{or} \quad c = \sqrt{a^2 - b^2}$$

Because distance b is one-half of the height of the ellipse, $b = 3$. Because $2a = 10$, $a = 5$. She substitutes $a = 5$ and $b = 3$ into the formula to find c:

$$\begin{aligned} c &= \sqrt{5^2 - 3^2} \\ &= \sqrt{25 - 9} \\ &= \sqrt{16} \\ &= 4 \end{aligned}$$

Since the distance between the center and either focus is 4 feet, the distance between the thumbtacks must be 8 feet. The teacher can draw her ellipse by tying a 10-foot string between thumbtacks 8 feet apart.

EXERCISES

1. Using a 10-inch string and thumbtacks 8 inches apart, draw a miniature of the teacher's ellipse. Is it the expected size?
2. Draw an ellipse that is 26 inches long and 10 inches wide. How long was the string, and how far apart were the thumbtacks?
3. As the ellipse gets longer relative to its width, what happens to the distance between the thumbtacks?
4. If the length of the ellipse were equal to its width, how far apart would the thumbtacks be? What is another name for such an ellipse?

(*Answers:* **2.** 26-in. string; 24 in. between tacks **3.** The distance increases. **4.** zero; a circle)

11 Exponential and Logarithmic Functions

In this chapter, we will discuss two functions that are important in many applications of mathematics. **Exponential functions** are used to compute compound interest, to model population growth, and to find radioactive decay.

Logarithmic functions are used to measure the acidity of solutions, drug dosage, the gain of an amplifier, the intensity of earthquakes, and safe noise levels in factories.

11.1 Exponential Functions

■ Irrational Exponents ■ Graphing Exponential Functions ■ Applications of Exponential Functions ■ Watching Money Grow ■ Base-*e* Exponential Functions

■ Irrational Exponents

We have considered exponential expressions such as $2^{1/2}$ and $3^{3/4}$, which are of the form b^x where x is a rational number. To give meaning to the exponential expression b^x where x is an irrational number, we consider the expression

$3^{\sqrt{2}}$ where $\sqrt{2}$ is the irrational number 1.414213562373 . . .

Because $1 < \sqrt{2} < 2$, it can be shown that

$$3^1 < 3^{\sqrt{2}} < 3^2$$

and because $1.4 < \sqrt{2} < 1.5$, it can be shown that

$$3^{1.4} < 3^{\sqrt{2}} < 3^{1.5}$$

The value of $3^{\sqrt{2}}$ is bounded by two numbers involving only rational powers of 3, as shown in the following list. As the exponent on each side of the list gets closer to $\sqrt{2}$, the value of $3^{\sqrt{2}}$ gets squeezed into a smaller and smaller interval:

$$3^1 = \mathbf{3} < \mathbf{3^{\sqrt{2}}} < \mathbf{9} = 3^2$$
$$3^{1.4} \approx \mathbf{4.656} < \mathbf{3^{\sqrt{2}}} < \mathbf{5.196} \approx 3^{1.5}$$
$$3^{1.41} \approx \mathbf{4.7070} < \mathbf{3^{\sqrt{2}}} < \mathbf{4.7590} \approx 3^{1.42}$$
$$3^{1.414} \approx \mathbf{4.727695} < \mathbf{3^{\sqrt{2}}} < \mathbf{4.732892} \approx 3^{1.415}$$

There is exactly one real number that is greater than any of the increasing numbers on the left side of the previous list and less than all of the decreasing numbers on the right side. We define this number to be $3^{\sqrt{2}}$.

To find an approximation for $3^{\sqrt{2}}$, we can press these keys on a scientific calculator:

3 [y^x] 2 [$\sqrt{x}$] [=]   You may have to press [2nd] before [$\sqrt{x}$].

The display will show 4.728804388. Thus, to four decimal places, $3^{\sqrt{2}} = 4.7288$.

For the keys to press on a graphing calculator, consult the owner's manual.

Classroom Activity: Discuss what happens when $b = 1$ in the equation $y = b^x$ and tell why a base of 1 is excluded in the definition.

In general, if $b > 0$ and x is any real number, the exponential expression b^x represents a single positive number. It can be shown that all of the familiar properties of exponents hold for irrational exponents as well.

If $b > 0$ and $b \neq 1$, the function defined by the equation $y = f(x) = b^x$ is called an **exponential function.**

Exponential Function with Base *b*

An **exponential function with base *b*** is defined by the equation

$$y = b^x \qquad \text{where } b > 0,\ b \neq 1, \text{ and } x \text{ is a real number}$$

The **domain** of an exponential function is the set of real numbers, and the **range** is the set of positive real numbers.

Graphing Exponential Functions

Because the domain and range of the exponential function $y = f(x) = b^x$ are sets of real numbers, we can graph exponential functions on a rectangular coordinate system.

EXAMPLE 1 Graph the exponential functions **a.** $y = 2^x$ and **b.** $y = 4^x$.

Solution

a. To graph the exponential function $y = 2^x$, we find several points (x, y) whose coordinates satisfy the equation, plot those points, and join them with a smooth curve. The graph is shown in Figure 11-1(a). Check the graph with a graphing calculator.

b. To graph the exponential function $y = 4^x$, we proceed as in part **a.** The graph is shown in Figure 11-1(b). Check the graph with a graphing calculator.

By looking at the graphs, we see that the domain of each function is the interval $(-\infty, \infty)$, and that the range is the interval $(0, \infty)$. ■

In each graph in Example 1, the values of y increase as the values of x increase. Thus, the functions defined by $y = 2^x$ and $y = 4^x$ are called **increasing functions.** We note that each graph approaches the x-axis as x gets smaller, and each graph passes through the point $(0, 1)$. The graph of $y = 2^x$ passes through the point $(1, 2)$, and the graph of $y = 4^x$ passes through the point $(1, 4)$.

$y = 2^x$

x	y
-1	$\frac{1}{2}$
0	1
1	2
2	4
3	8

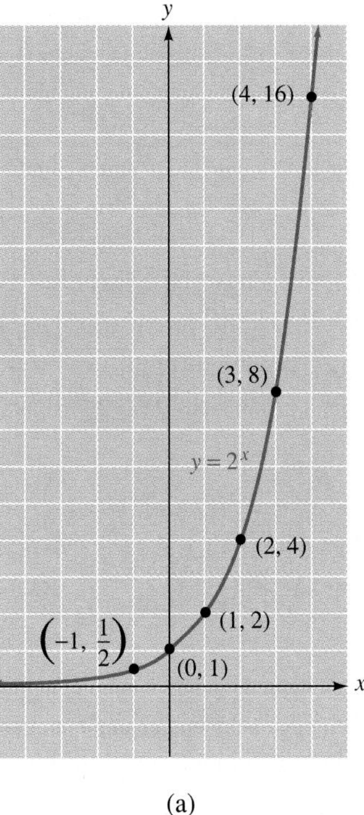

(a)

$y = 4^x$

x	y
-1	$\frac{1}{4}$
0	1
1	4
2	16

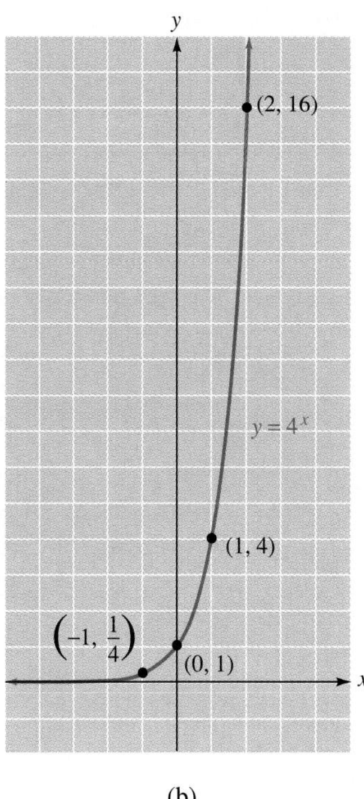

(b)

FIGURE 11-1

EXAMPLE 2 Graph the exponential functions $y = \left(\frac{1}{2}\right)^x$ and $y = \left(\frac{1}{4}\right)^x$.

Solution We find and plot pairs (x, y) that satisfy each equation. The graph of $y = \left(\frac{1}{2}\right)^x$ appears in Figure 11-2(a), and the graph of $y = \left(\frac{1}{4}\right)^x$ appears in Figure 11-2(b). Check each graph with a graphing calculator.

$y = \left(\frac{1}{2}\right)^x$

x	y
-1	2
0	1
1	$\frac{1}{2}$

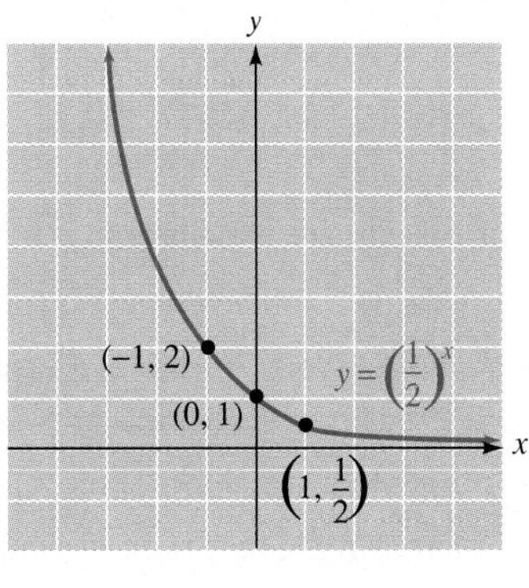

(a)

$y = \left(\frac{1}{4}\right)^x$

x	y
-1	4
0	1
1	$\frac{1}{4}$

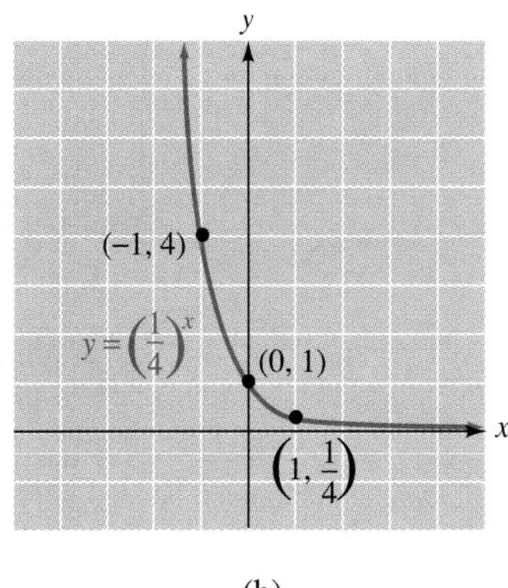

(b)

FIGURE 11-2

By looking at the graphs, we see that the domain of each function is the interval $(-\infty, \infty)$, and the range is the interval $(0, \infty)$. ■

In each graph in Example 2, the values of y decrease as the values of x increase. Thus, the functions defined by $y = \left(\frac{1}{2}\right)^x$ and $y = \left(\frac{1}{4}\right)^x$ are called **decreasing functions.**

Each graph approaches the x-axis as x gets larger, and each graph passes through the point $(0, 1)$. The graph of $y = \left(\frac{1}{2}\right)^x$ passes through the point $\left(1, \frac{1}{2}\right)$ and the graph of $y = \left(\frac{1}{4}\right)^x$ passes through the point $\left(1, \frac{1}{4}\right)$.

Examples 1 and 2 illustrate these facts.

Points on the Graph of $y = b^x$ The graph of the exponential function defined by $y = f(x) = b^x$ passes through the points $(0, 1)$ and $(1, b)$.

An exponential function with base b is either increasing (for $b > 1$) or decreasing ($0 < b < 1$). Since distinct real numbers x will determine distinct values b^x, exponential functions are one-to-one.

Exponential Function An exponential function defined by

$$y = f(x) = b^x \qquad \text{where } b > 0 \text{ and } b \neq 1$$

is one-to-one. Thus,

1. If $b^r = b^s$, then $r = s$.

2. If $r \neq s$, then $b^r \neq b^s$.

EXAMPLE 3 On the same set of coordinate axes, graph **a.** $y = \left(\dfrac{2}{3}\right)^x$ and **b.** $y = \left(\dfrac{3}{2}\right)^x$.

Solution **a.** The graph of $y = \left(\frac{2}{3}\right)^x$ passes through the points $(0, 1)$ and $\left(1, \frac{2}{3}\right)$. Since $\frac{2}{3} < 1$, the function is decreasing. The graph appears in Figure 11-3.

b. The graph of $y = \left(\frac{3}{2}\right)^x$ passes through the points $(0, 1)$ and $\left(1, \frac{3}{2}\right)$. Since $\frac{3}{2} > 1$, the function is increasing. The graph also appears in Figure 11-3.

Check each graph with a graphing calculator. Since both graphs pass the horizontal line test, each function is one-to-one.

$y = \left(\frac{2}{3}\right)^x$

x	y
-2	$\frac{9}{4}$
-1	$\frac{3}{2}$
0	1
1	$\frac{2}{3}$
2	$\frac{4}{9}$

$y = \left(\frac{3}{2}\right)^x$

x	y
-2	$\frac{4}{9}$
-1	$\frac{2}{3}$
0	1
1	$\frac{3}{2}$
2	$\frac{9}{4}$

FIGURE 11-3

Applications of Exponential Functions

A mathematical description of an observed event is called a **model** of that event. Many events that change with time are modeled by functions defined by equations of the form

$$y = f(t) = ab^{kt} \qquad \text{Remember that } ab^{kt} \text{ means } a(b^{kt}).$$

where a, b, and k are constants and t represents time. If the function is an increasing function, then y is said to **grow exponentially.** If the function is a decreasing function, then y is said to **decay exponentially.**

One example of exponential growth is **compound interest.** When interest is allowed to accumulate in a savings account, the balance will grow exponentially according to the following model.

Formula for Compound Interest If $\$A_0$ is deposited in an account, and interest is paid k times a year at an annual rate r, the amount $\$A$ in the account after t years is given by

$$A = A_0\left(1 + \frac{r}{k}\right)^{kt}$$

EXAMPLE 4 **Saving for college** To save for college, parents invest \$12,000 for their newborn child in a mutual fund that should average 10% annual interest. If the quarterly dividends are reinvested, how much will be available in 18 years?

Solution We substitute 12,000 for A_0, 0.10 for r, and 18 for t into the formula for compound interest and find A. Since interest is paid quarterly, $k = 4$.

$$A = A_0\left(1 + \frac{r}{k}\right)^{kt}$$

$$A = \mathbf{12{,}000}\left(1 + \frac{\mathbf{0.10}}{\mathbf{4}}\right)^{\mathbf{4(18)}}$$

$$= 12{,}000(1 + 0.025)^{72}$$

$$= 12{,}000(1.025)^{72}$$

$$= 71{,}006.74$$

Use a calculator, and press these keys: 1.025 y^x 72 = × 12,000 =.

In 18 years, the account will contain more than \$71,000. ■

In business applications, the initial amount of money deposited is called the **present value** (*PV*). The amount to which the money will grow is called the **future value** (*FV*). The interest rate used for each compounding period is the **periodic interest rate** (i), and the number of times interest is compounded is the **number of compounding periods** (n). Using these definitions, an alternate formula for the compound interest formula is as follows.

Formula for Compound Interest

$$FV = PV(1 + i)^n$$

This alternate formula appears on business calculators. To use this formula to solve Example 4, we proceed as follows:

$$FV = \mathbf{PV}(1 + \mathbf{i})^{\mathbf{n}}$$

$$FV = \mathbf{12{,}000}(1 + \mathbf{0.025})^{\mathbf{72}}$$

$$= 71{,}006.74$$

$i = \frac{0.10}{4} = 0.025$ and $n = 4(18) = 72$.

Watching Money Grow

EXAMPLE 5 If \$1 is deposited in an account earning 6% annual interest, compounded monthly, estimate how much will be in the account in 100 years.

Solution We can substitute 1 for A_0, 0.06 for r, and 12 for k into the formula

$$A = A_0\left(1 + \frac{r}{k}\right)^{kt}$$

and simplify to get

$$A = (1.005)^{12t}$$

Teaching Tip: Although the variables in the example are *A* and *t*, the graphing calculator still uses Y and X.

We now use a graphing calculator to graph $A = (1.005)^{12t}$ in the viewing window, using $0 \leq t \leq 120$ and $0 \leq A \leq 400$ to obtain the graph shown in Figure 11-4. We can then use the TRACE and ZOOM features to watch \$1 grow to approximately \$397 in 100 years.

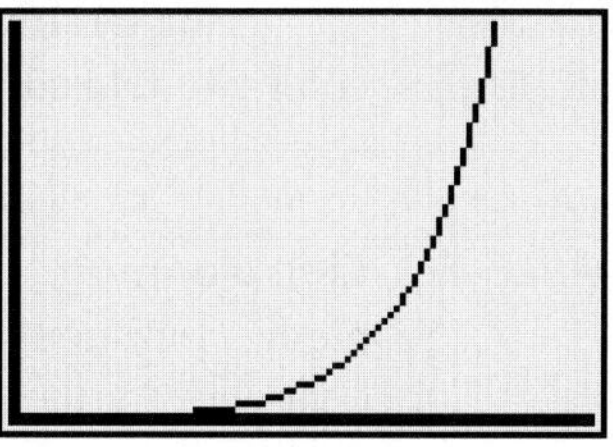

FIGURE 11-4

Base-e Exponential Functions

When banks compound interest quarterly, the account balance increases four times each year. In other situations, quantities grow continuously. For example, the population of a town does not increase abruptly every quarter but grows steadily all year long.

Exponential expressions can predict the future value FV of a quantity that increases or decreases continuously from its present value PV. These formulas have the following form.

Formula for Exponential Growth

$$FV = PVe^{rt}$$

where r is the rate of increase, t is the length of time, and $e = 2.718281828459 \ldots$.

Leonhard Euler (1707–1783)
Euler first used the letter i to represent $\sqrt{-1}$, the letter e for the base of natural logarithms, and the symbol Σ for summation. Euler was one of the most prolific mathematicians of all time, contributing to almost all areas of mathematics. Much of his work was accomplished after he became blind.

If time is measured in years, then r is called the **annual growth rate.** If r is negative, the "growth" represents a decrease.

Banks often compound interest more frequently than quarterly. Monthly and daily compounding is common, but it is possible to compound interest *continuously.* To compute the future value to which a present value of \$12,000 will grow if invested for 18 years at 10% annual interest compounded continuously, we substitute 12,000 for PV, 0.10 for r, and 18 for t in the formula for exponential growth:

$$FV = \mathbf{PV}e^{\mathbf{rt}}$$

$$FV = \mathbf{12{,}000}e^{\mathbf{0.10(18)}}$$

$$= 12{,}000e^{1.8}$$

$$= 72{,}595.77$$

Use a calculator and press these keys: 1.8 e^x × 12,000 =.

After 18 years, the account will contain \$72,595.77. This is over \$1589 more than the result in Example 4, where interest was compounded quarterly.

If your calculator does not have an e^x key, try pressing INV $\ln x$, or consult your owner's manual.

In the **Malthusian model for population growth,** the future population of a colony is related to the present population by the formula for exponential growth.

EXAMPLE 6 **City planning** The population of a city is currently 15,000, but changing economic conditions are causing the population to decrease 2% each year. If this trend continues by the Malthusian model, find the population in 30 years.

Solution Since the population is decreasing 2% each year, the annual growth rate is −2%, or −0.02. We can substitute −0.02 for r, 30 for t, and 15,000 for PV in the formula for exponential growth and find FV.

$$FV = PVe^{rt}$$
$$FV = \mathbf{15{,}000}e^{\mathbf{-0.02(30)}}$$
$$= 15{,}000e^{-0.6}$$
$$= 8232.17$$

In 30 years, city planners expect a population of approximately 8232 persons. ■

ORALS *If $x = 2$, evaluate each expression.*

1. 2^x 4 **2.** 5^x 25 **3.** $2(3^x)$ 18 **4.** 3^{x-1} 3

If $x = -2$, evaluate each expression.

5. 2^x $\frac{1}{4}$ **6.** 5^x $\frac{1}{25}$ **7.** $2(3^x)$ $\frac{2}{9}$ **8.** 3^{x-1} $\frac{1}{27}$

EXERCISE 11.1

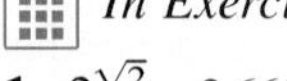

In Exercises 1–4, find each value to four decimal places.

1. $2^{\sqrt{2}}$ 2.6651 **2.** $7^{\sqrt{2}}$ 15.6729 **3.** $5^{\sqrt{5}}$ 36.5548 **4.** $6^{\sqrt{3}}$ 22.2740

In Exercises 5–12, graph each exponential function. Check your work with a graphing calculator.

5. $y = 3^x$

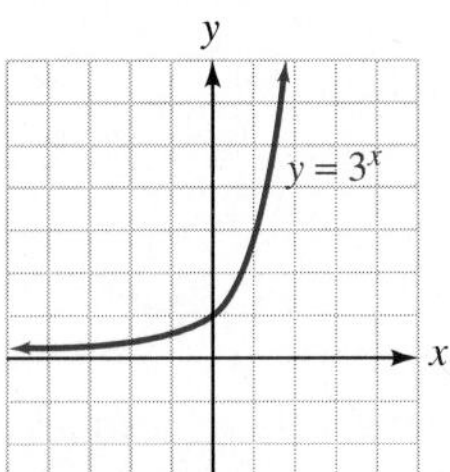

6. $y = 4^x$

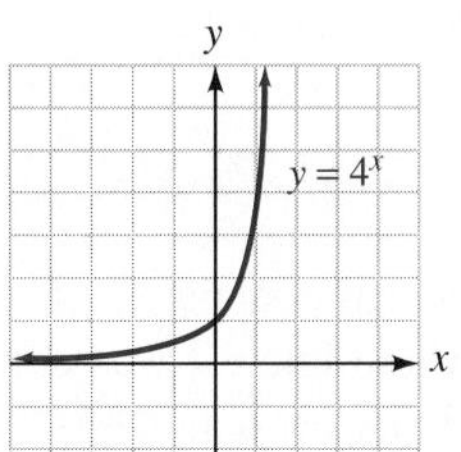

7. $y = 5^x$

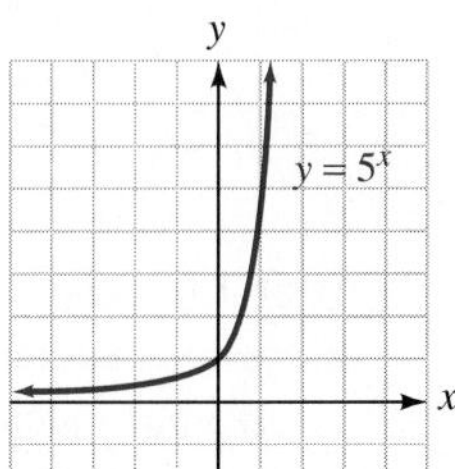

8. $y = 6^x$

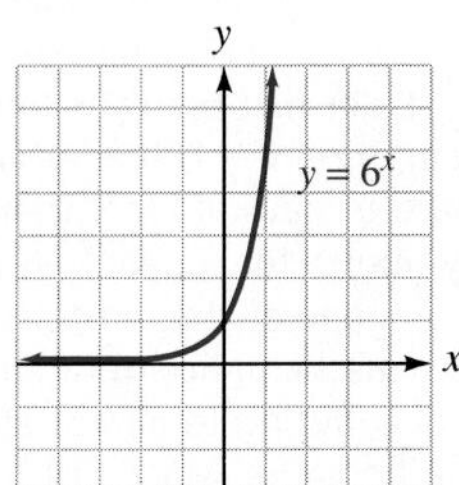

9. $y = \left(\frac{1}{3}\right)^x$

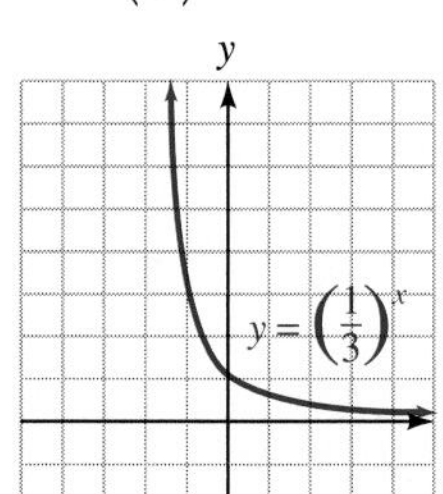

10. $y = \left(\frac{1}{4}\right)^x$

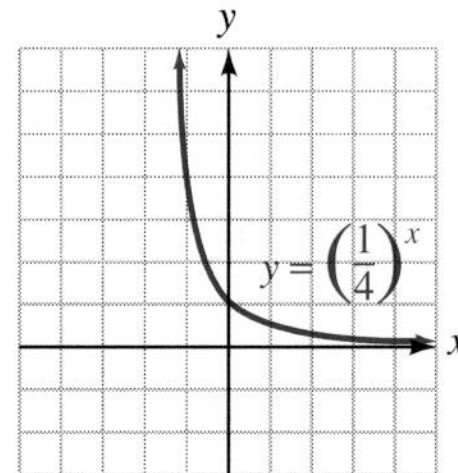

11. $y = \left(\frac{1}{5}\right)^x$

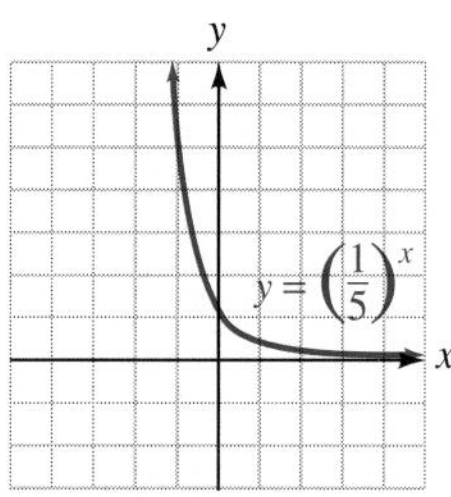

12. $y = \left(\frac{1}{6}\right)^x$

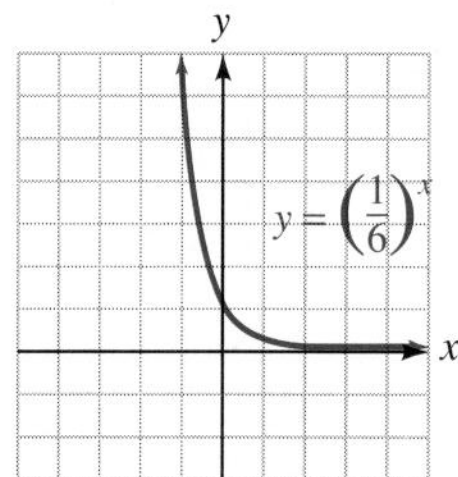

In Exercises 13–20, graph each function. Check your work with a graphing calculator.

13. $y = 2 + 2^x$

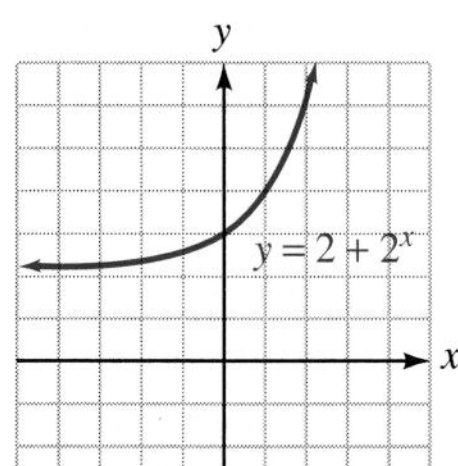

14. $y = -2 + 3^x$

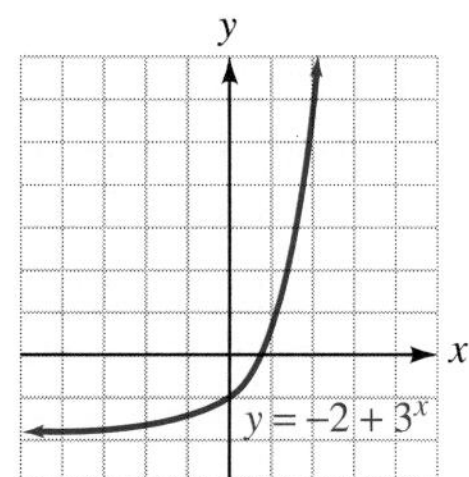

15. $y = 3(2^x)$

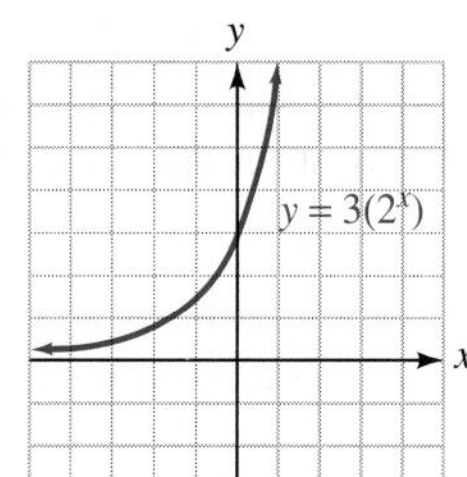

16. $y = 2(3^x)$

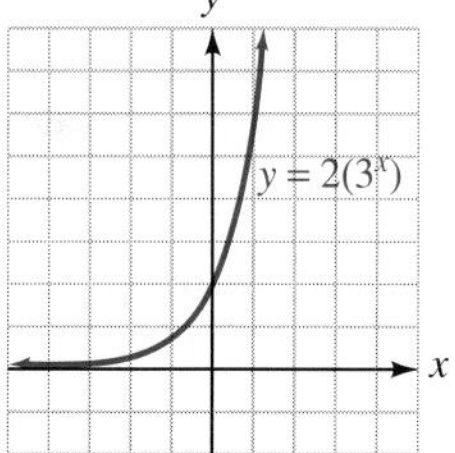

17. $y = 2^{-x}$

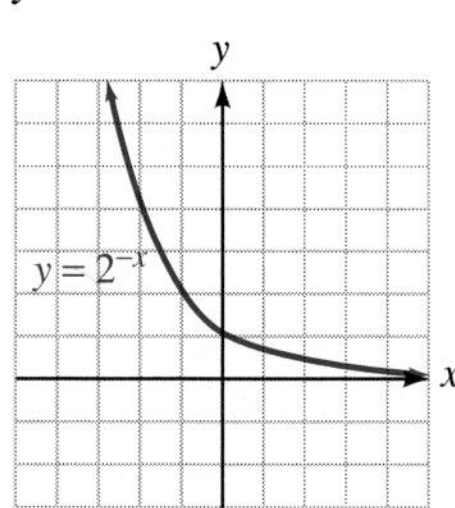

18. $y = 3^{-x}$

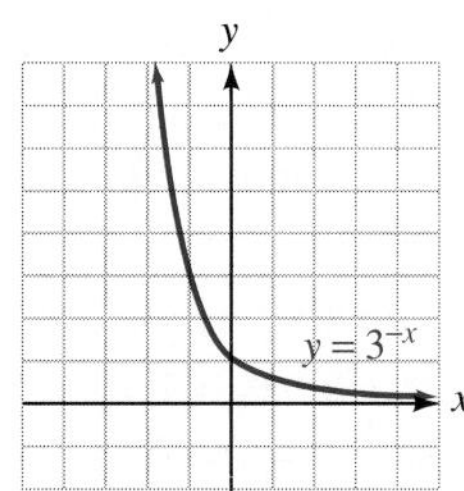

19. $y = 2^{x+1}$

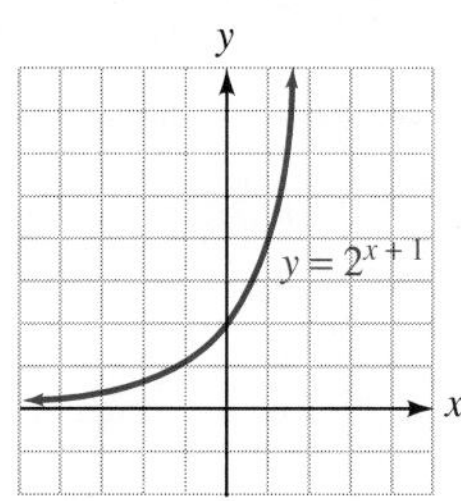

20. $y = 4^{x-1}$

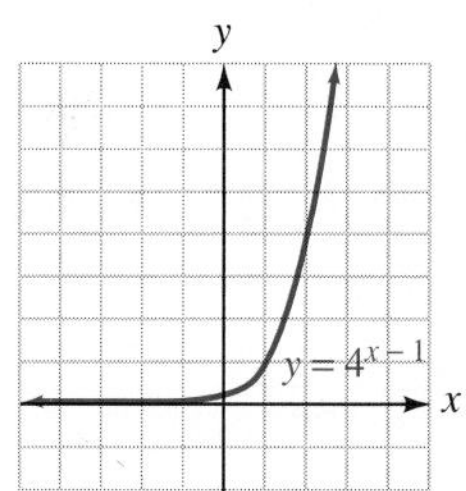

In Exercises 21–28, find the value of b, if any, that would cause the graph of $y = b^x$ to look like the graph indicated.

21.

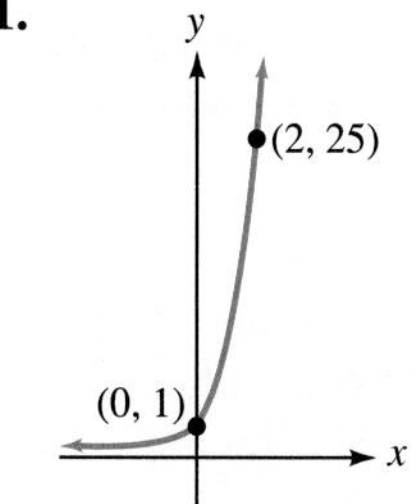

5

22.

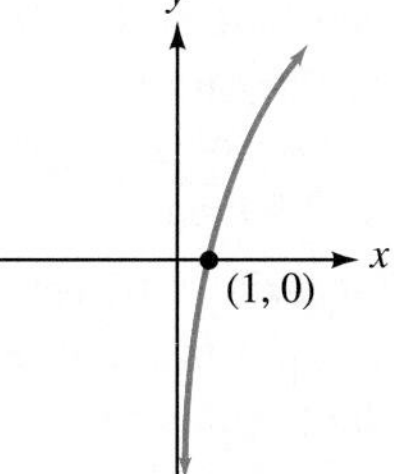

no value of b

23.

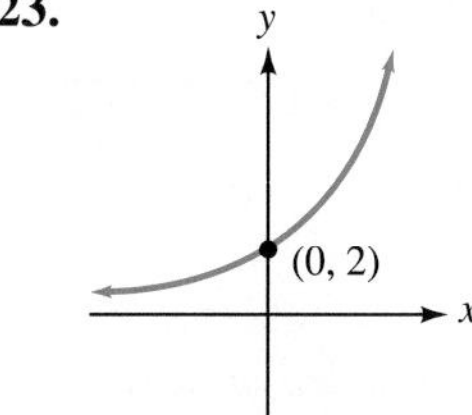

no value of b

24.

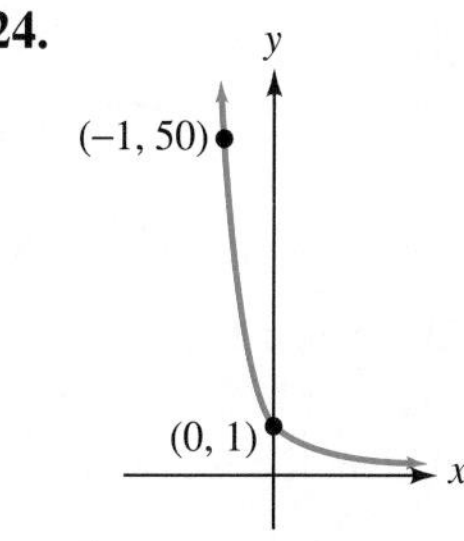

$\frac{1}{50}$

25.

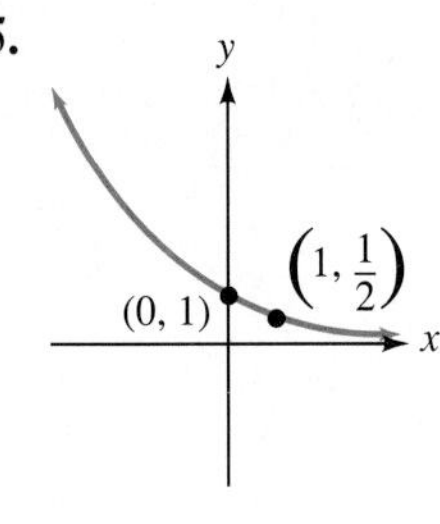

$\frac{1}{2}$

26.

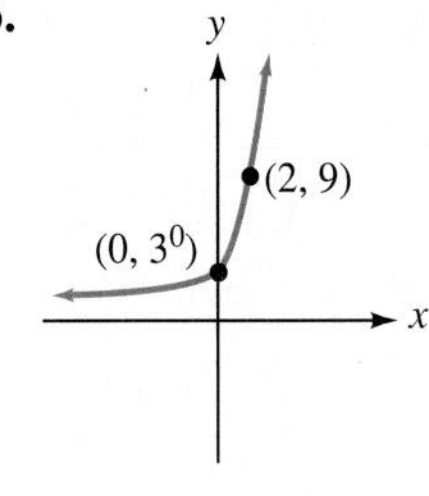

3

27.

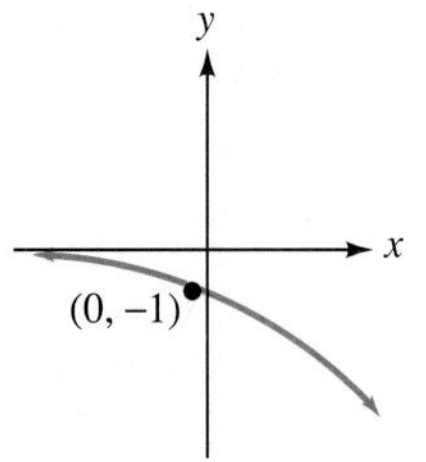

no value of b

28.

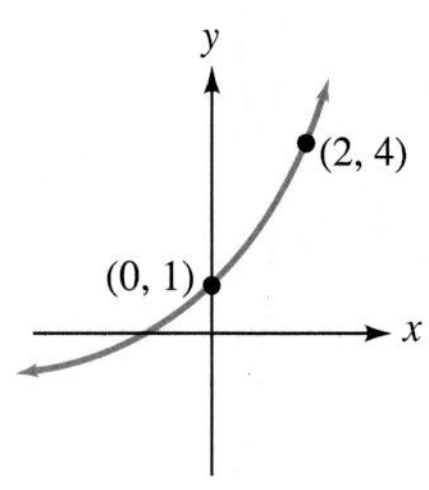

no value of b

In Exercises 29–34, assume there are no deposits or withdrawals.

29. **Compound interest** An initial deposit of \$10,000 earns 8% interest, compounded quarterly. How much will be in the account after 10 years? \$22,080.40

30. **Compound interest** An initial deposit of \$10,000 earns 8% interest, compounded monthly. How much will be in the account in after 10 years? \$22,196.40

31. **Comparing interest rates** How much more interest could \$1000 earn in 5 years, compounded quarterly, if the annual interest rate were $5\frac{1}{2}\%$ instead of 5%? \$32.03

32. **Comparing savings plans** Which institution in Illustration 1 provides the better investment? Fidelity

Fidelity Savings & Loan
earn 5.25% compounded monthly

Union Trust
Money Market account paying 5.35% compounded annually

ILLUSTRATION 1

33. **Compound interest** If \$1 had been invested on July 4, 1776 at 5% interest, compounded annually, what will it be worth on July 4, 2076? \$2,273,996.13

34. **Frequency of compounding** \$10,000 is invested in each of two accounts, both paying 6% annual interest. In the first account, interest compounds quarterly, and in the second account, interest compounds daily. Find the difference between the accounts after 20 years. \$291.26

35. **Fish population** A population of fish is growing according to the Malthusian model. How many fish will there be in 10 years if the annual growth rate is 3% and the initial population is 2700 fish? 3645

36. **Town population** The population of a town is 1350. The town is expected to grow according to the Malthusian model, with an annual growth rate of 6%. Find the population of the town in 20 years. 4482

37. **World population** The population of the world is approximately 5 billion. If the population is growing according to the Malthusian model with an annual growth rate of 1.8%, what will be the population of the world in 30 years? 8.58 billion

38. **World population** The population of the world is approximately 5 billion. If the population is growing according to the Malthusian model with an annual growth rate of 1.8%, what will be the population of the world in 60 years? 14.72 billion

39. **Bacteria cultures** A colony of 6 million bacteria is growing in a culture medium. (See Illustration 2.) The population P after t hours is given by the

formula $P = (6 \times 10^6)(2.3)^t$. Find the population after 4 hours. 1.679046×10^8

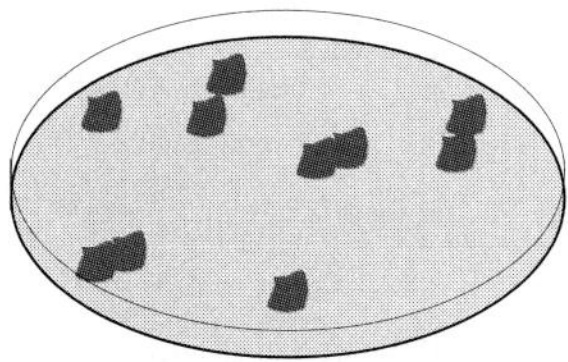

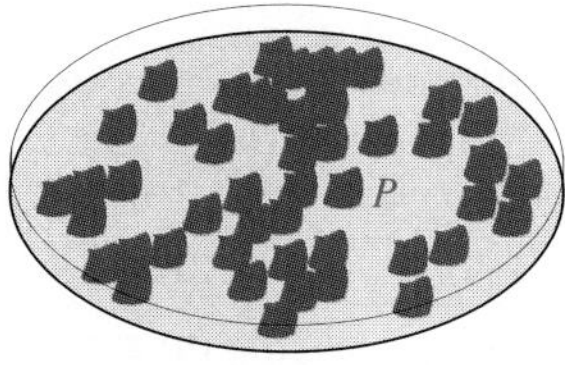

12:00 PM 4:00 PM

ILLUSTRATION 2

40. **Discharging a battery** The charge remaining in a battery decreases as the battery discharges. The charge C after t days is given by the formula $C = (3 \times 10^{-4})(0.7)^t$. Find the charge after 5 days. 5.0421×10^{-5} coulombs

41. **Radioactive decay** A radioactive material decays according to the formula $A = A_0\left(\frac{2}{3}\right)^t$, where A_0 is the initial amount present and t is measured in years. Find the amount present in 5 years. $\frac{32}{243}A_0$

42. **Town population** The population of North Rivers is decreasing exponentially according to the formula $P = 3745(0.93)^t$, where t is measured in years from the present date. Find the population in 6 years, 9 months. 2295

43. **Depreciation** A camping trailer originally purchased for \$4570 is losing value at the rate of 6% per year. Find its value when it is $6\frac{1}{2}$ years old. \$3094.15

44. **Depreciation** A boat purchased for \$7500 has been decreasing in value at the rate of 2% each year. It is now 8 years, 3 months old. Find its value. \$6359.20

Writing Exercises *Write a paragraph using your own words.*

1. If world population is increasing exponentially, why is there cause for concern?
2. How do the graphs of $y = b^x$ differ when $b > 1$ and $0 < b < 1$?

Something to Think About

1. In the definition of the exponential function, b could not equal 0. Why not?
2. In the definition of the exponential function, b could not be negative. Why not?

Review Exercises *Simplify each radical expression. Assume that variables represent positive numbers.*

1. $\sqrt{240x^5}$ $4x^2\sqrt{15x}$
2. $\sqrt[3]{-125x^5y^4}$ $-5xy\sqrt[3]{x^2y}$
3. $4\sqrt{48y^3} - 3y\sqrt{12y}$ $10y\sqrt{3y}$
4. $\sqrt[4]{48z^5} + \sqrt[4]{768z^5}$ $6z\sqrt[4]{3z}$

11.2 Logarithmic Functions

■ Common Logarithms ■ Natural Logarithms ■ Graphing Logarithmic Functions

Since an exponential function defined by $y = b^x$ is one-to-one, it has an inverse function that is defined by the equation $x = b^y$. To express this inverse function in the form $y = f^{-1}(x)$, we must solve the equation $x = b^y$ for y. To do this, we need the following definition.

Logarithmic Function If $b > 0$ and $b \neq 1$, a **logarithmic function with base b** is defined by

$$y = \log_b x \qquad \text{if and only if} \qquad x = b^y$$

The **domain of the logarithmic function** is the set of positive real numbers. The **range of the logarithmic function** is the set of real numbers.

Since the function $y = \log_b x$ is the inverse of the one-to-one exponential function $y = b^x$, the logarithmic function is also a one-to-one function.

WARNING! Because the domain of the logarithmic function is the set of positive numbers, it is impossible to find the logarithm of 0 or the logarithm of a negative number.

The previous definition guarantees that any pair (x, y) that satisfies the equation $y = \log_b x$ also satisfies the equation $x = b^y$. Thus,

$\log_7 1 = 0$ because $1 = 7^0$

$\log_5 25 = 2$ because $25 = 5^2$

$\log_5 \frac{1}{25} = -2$ because $\frac{1}{25} = 5^{-2}$

$\log_{16} 4 = \frac{1}{2}$ because $4 = 16^{1/2}$

$\log_2 \frac{1}{8} = -3$ because $\frac{1}{8} = 2^{-3}$

$\log_b x = y$ because $x = b^y$

In each of the previous examples, the logarithm of a number is an exponent, and furthermore,

$\log_b x$ is the exponent to which b is raised to get x.

To say this with an equation, we write

$$b^{\log_b x} = x$$

Alternate Approach: Since $b^y = x$ is equivalent to $y = \log_b x$, then

$$b^{\log_b x} = x$$

by substitution.

EXAMPLE 1 Find y in each equation: **a.** $\log_6 1 = y$, **b.** $\log_3 27 = y$, and **c.** $\log_5 \frac{1}{5} = y$.

Solution **a.** We can change the equation $\log_6 1 = y$ into the equivalent exponential equation $1 = 6^y$. Since $1 = 6^0$, it follows that $y = 0$. Thus,

$$\log_6 1 = 0$$

b. $\log_3 27 = y$ is equivalent to $27 = 3^y$. Since $27 = 3^3$, it follows that $y = 3$. Thus,

$$\log_3 27 = 3$$

c. $\log_5 \frac{1}{5} = y$ is equivalent to $\frac{1}{5} = 5^y$. Since $\frac{1}{5} = 5^{-1}$, it follows that $y = -1$. Thus,

$$\log_5 \frac{1}{5} = -1$$

■

(b) **EXAMPLE 2** Find the value of x in each equation: **a.** $\log_3 81 = x$, **b.** $\log_x 125 = 3$, and **c.** $\log_4 x = 3$.

Solution **a.** $\log_3 81 = x$ is equivalent to $3^x = 81$. Because $3^4 = 81$, it follows that $x = 4$.

b. $\log_x 125 = 3$ is equivalent to $x^3 = 125$. Because $5^3 = 125$, it follows that $x = 5$.

c. $\log_4 x = 3$ is equivalent to $4^3 = x$. Because $4^3 = 64$, it follows that $x = 64$.

■

(b) **EXAMPLE 3** Find the value of x in each equation: **a.** $\log_{1/3} x = 2$, **b.** $\log_{1/3} x = -2$, and **c.** $\log_{1/3} \frac{1}{27} = x$.

Solution **a.** $\log_{1/3} x = 2$ is equivalent to $\left(\frac{1}{3}\right)^2 = x$. Thus, $x = \frac{1}{9}$.

b. $\log_{1/3} x = -2$ is equivalent to $\left(\frac{1}{3}\right)^{-2} = x$. Thus, $x = \left(\frac{1}{3}\right)^{-2} = 3^2 = 9$

c. $\log_{1/3} \frac{1}{27} = x$ is equivalent to $\left(\frac{1}{3}\right)^x = \frac{1}{27}$. Because $\left(\frac{1}{3}\right)^3 = \frac{1}{27}$, it follows that $x = 3$.

■

Common Logarithms

Historical Note: Henry Briggs (1561–1631) published base-10 logarithm tables that were computed to 14 decimal places.

For computational purposes, base-10 logarithms are convenient. For this reason, base-10 logarithms are called **common logarithms.** When the base b is not indicated in the notation $\log x$, we assume that $b = 10$:

$$\log x \quad \text{means} \quad \log_{10} x$$

Because base-10 logarithms appear often in mathematics and science, it is helpful to become familiar with the following base-10 logarithms:

$$\log \frac{1}{100} = -2 \qquad \text{because} \qquad 10^{-2} = \frac{1}{100}$$

$$\log \frac{1}{10} = -1 \qquad \text{because} \qquad 10^{-1} = \frac{1}{10}$$

$$\log 1 = 0 \qquad \text{because} \qquad 10^0 = 1$$

$$\log 10 = 1 \qquad \text{because} \qquad 10^1 = 10$$

$$\log 100 = 2 \qquad \text{because} \qquad 10^2 = 100$$

$$\log 1000 = 3 \qquad \text{because} \qquad 10^3 = 1000$$

In general, we have

$$\log 10^x = x$$

Base-10 logarithms are easy to find with a calculator. For example, to find log 6.35, we press these keys on a scientific calculator:

6.35 LOG   You may have to press a 2nd function key before LOG.

The display will read 0.802773725. Thus, to four decimal places,

$$\log 6.35 = 0.8028$$

For the keys to press on a graphing calculator, consult the owner's manual.

EXAMPLE 4 Find x in the equation $\log x = 0.3568$.

Solution The equation $\log x = 0.3568$ is equivalent to the equation $10^{0.3568} = x$. To find x with a scientific calculator, we press these keys:

10 y^x .3568 =

The display will read 2.274049951. Thus, to four decimal places,

$$x = 2.2740$$

If your calculator has a 10^x key, enter .3568 and press it to get the same result. (You may have to press a 2nd function key first.) For the keys to press on a graphing calculator, consult the owner's manual. ■

John Napier
(1550–1617)
Napier is famous for his work with natural logarithms. In fact, natural logarithms are often called *Napierian logarithms.* He also invented a device, called *Napier's rods,* that did multiplications mechanically. His device was a forerunner of modern-day computers.

Natural Logarithms

Because the number e appears often in mathematical models of natural events, base-e logarithms are called **natural logarithms.** Natural logarithms are usually denoted by the symbol $\ln x$, rather than $\log_e x$:

$$\ln x \quad \text{means} \quad \log_e x$$

We can also find natural logarithms with a calculator. For example, to find ln 6.35 with a scientific calculator, we press these keys:

6.35 LN   (You may have to press a 2nd function key first.)

The display will read 1.848454813. Thus, to four decimal places,

$$\ln 6.35 = 1.8485$$

For the keys to press on a graphing calculator, consult the owner's manual.

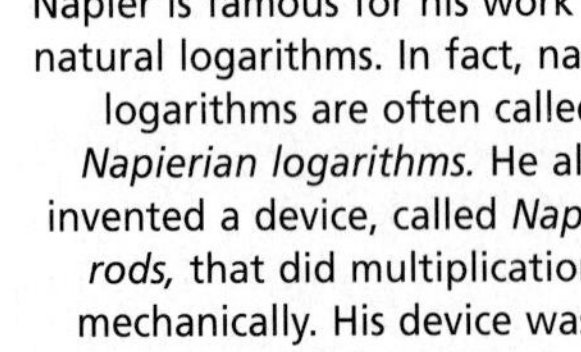

EXAMPLE 5 Find x in each equation: **a.** $\ln x = 3.441$ and **b.** $\ln x = \log 6.7$.

Solution **a.** The equation $\ln x = 3.441$ is equivalent to the equation $e^{3.441} = x$. To use a scientific calculator to find x, we press these keys:

3.441 e^x (You may have to press a 2nd function key first.)

The display will read 31.21816073. Thus, to four decimal places,

$$x = 31.2182$$

For the keys to press on a graphing calculator, consult the owner's manual.

b. The equation $\ln x = \log 6.7$ is equivalent to the equation $e^{\log 6.7} = x$. To use a scientific calculator to find x, we press these keys:

6.7 LOG e^x (You may have to press a 2nd function key first.)

The display will read 2.284334655. Thus, to four decimal places,

$$x = 2.2843$$

For the keys to press on a graphing calculator, consult the owner's manual. ■

Graphing Logarithmic Functions

To graph the logarithmic function $y = \log_2 x$, we find and plot several points with coordinates (x, y) that satisfy the equation $x = 2^y$. After joining them with a smooth curve, we have the graph that is shown in Figure 11-5(a). Check the graph with a graphing calculator.

To graph $y = \log_{1/2} x$, we find and plot several points with coordinates (x, y) that satisfy the equation $x = \left(\frac{1}{2}\right)^y$. After joining them with a smooth curve, we have the graph shown in Figure 11-5(b). Check the graph with a graphing calculator.

We can see from the graphs that the domain of each function is the interval $(0, \infty)$, and the range is the interval $(-\infty, \infty)$.

$y = \log_2 x$

x	y
$\frac{1}{4}$	-2
$\frac{1}{2}$	-1
1	0
2	1
4	2
8	3

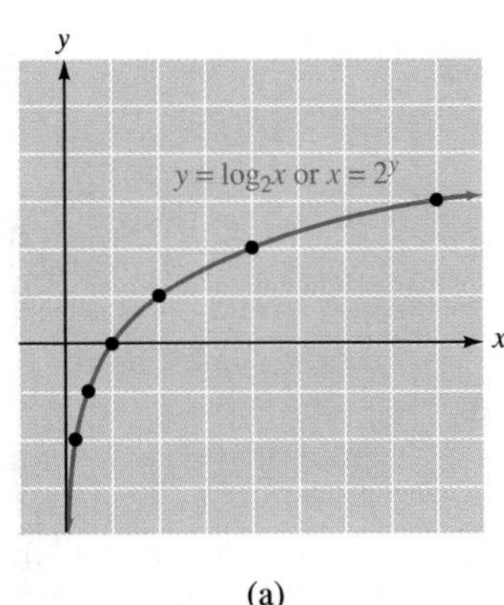

(a)

$y = \log_{1/2} x$

x	y
8	-3
4	-2
2	-1
1	0
$\frac{1}{2}$	1
$\frac{1}{4}$	2

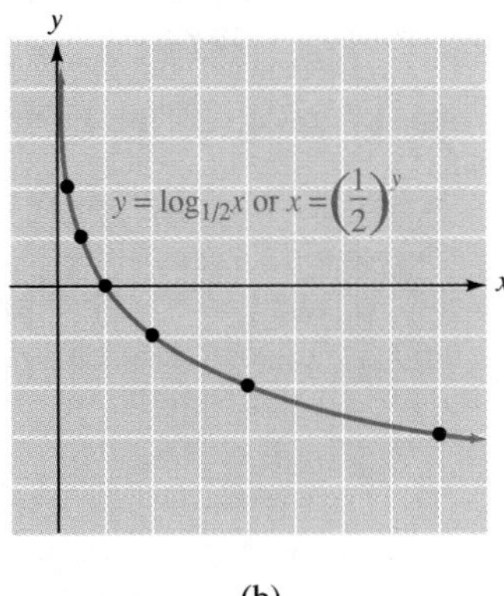

(b)

FIGURE 11-5

 (a)

EXAMPLE 6 Graph the functions **a.** $y = \log x$ and **b.** $y = \ln x$.

Solution **a.** The equation $y = \log x$ is equivalent to the equation $x = 10^y$. To get the graph of $y = \log x$, we can plot points that satisfy the equation $x = 10^y$ and join them with a smooth curve, as in Figure 11-6(a). Check the graph with a graphing calculator.

b. The equation $y = \ln x$ is equivalent to the equation $x = e^y$. To get the graph of $y = \ln x$, we can plot points that satisfy the equation $x = e^y$ and join them with a smooth curve, as in Figure 11-6(b). Check the graph with a graphing calculator.

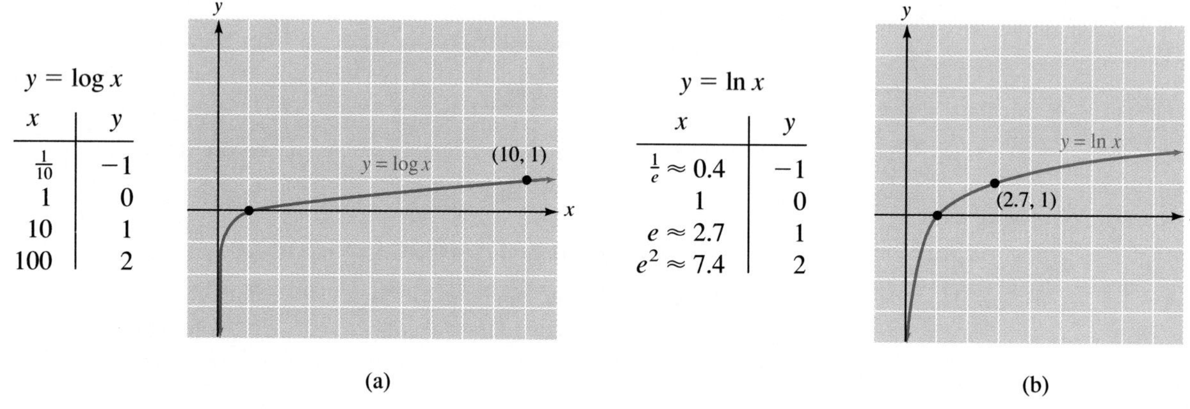

$y = \log x$

x	y
$\frac{1}{10}$	-1
1	0
10	1
100	2

$y = \ln x$

x	y
$\frac{1}{e} \approx 0.4$	-1
1	0
$e \approx 2.7$	1
$e^2 \approx 7.4$	2

(a) (b)

FIGURE 11-6

We can see from the graphs that the domain of each function is the interval $(0, \infty)$, and the range is the interval $(-\infty, \infty)$. ■

PERSPECTIVE

Another Graph Paper

Because an exponential function increases rapidly, it is difficult to fit the graph into a reasonable space. For example, between $x = 0$ and $x = 2$, the exponential function $y = 10^x$ changes from a low of 1 to a high of 100.

On the **semi-logarithmic graph paper** in Illustration 1, the vertical scale is marked so that the distance between 1 and 2 is the logarithm of 2, the distance between 1 and 3 is log 3, and so on. On this logarithmic scale, the distance between 1 and 2 represents a doubling of y. That same distance, between 2 and 4, also represents a doubling, as does that between 4 and 8, and between 8 and 16, and so on. On a logarithmic y-axis, equal vertical distances represent increasing y by the same *factor.*

Illustration 2 shows the graphs of three exponential functions in the first quadrant of a Cartesian coordinate system. All of them quickly run off of the grid. Illustration 3 shows the graphs of the same exponential functions on a semilog graph. Note how easily the coordinate system accommodates large values of y. Note also that the graph of an exponential function on a semi-log coordinate system is a straight line.

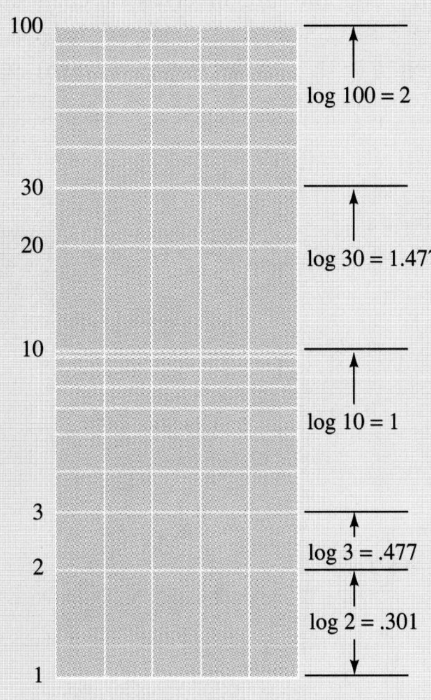

ILLUSTRATION 1

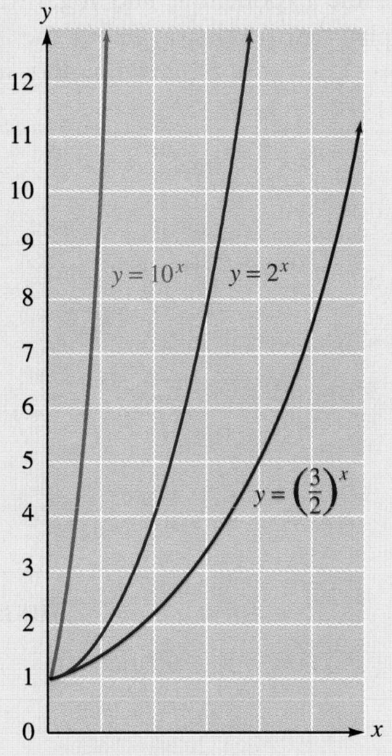

ILLUSTRATION 2

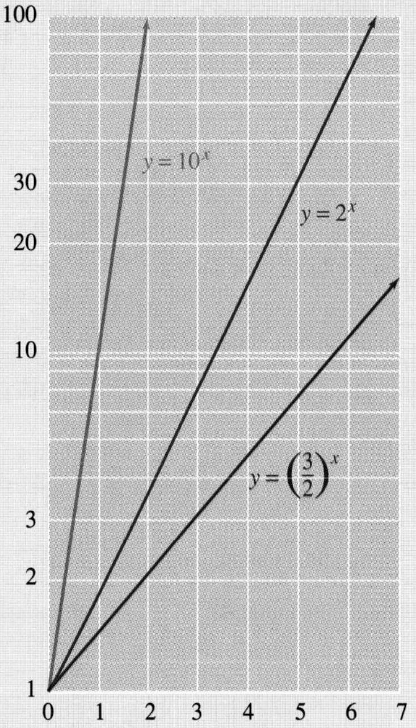

ILLUSTRATION 3

The previous examples suggest that the graphs of all logarithmic functions are similar to those shown in Figure 11-7. If $b > 1$, the logarithmic function is increasing as in Figure 11-7(a). If $0 < b < 1$, the logarithmic function is decreasing, as in Figure 11-7(b).

Each graph of $y = \log_b x$ passes through the points $(1, 0)$ and $(b, 1)$ and has the y-axis as an asymptote.

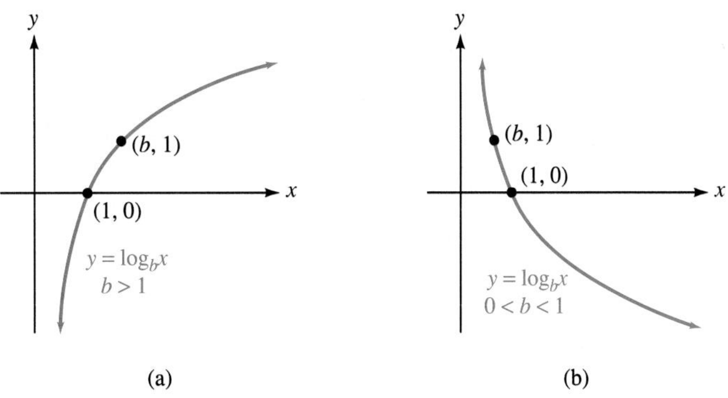

FIGURE 11-7

Since the exponential and logarithmic functions are inverses of each other, their graphs are symmetric about the line $y = x$. The graphs of $y = \log_b x$ and $y = b^x$ are shown in Figure 11-8(a) when $b > 0$, and in Figure 11-8(b) when $0 < b < 1$.

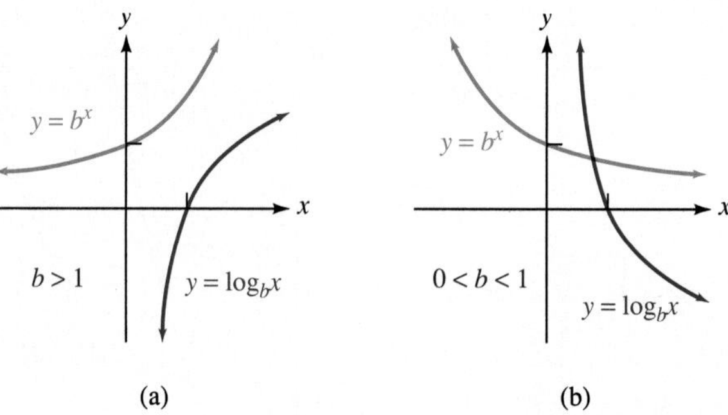

FIGURE 11-8

ORALS *Find the value of x in each equation.*

1. $\log_2 8 = x$ 3
2. $\log_3 9 = x$ 2
3. $\log_x 125 = 3$ 5
4. $\log_x 8 = 3$ 2
5. $\log_4 16 = x$ 2
6. $\log_x 32 = 5$ 2
7. $\log_{1/2} x = 2$ $\frac{1}{4}$
8. $\log_9 3 = x$ $\frac{1}{2}$
9. $\log_x \frac{1}{4} = -2$ 2

EXERCISE 11.2

In Exercises 1–8, write each equation in exponential form.

1. $\log_3 81 = 4$ $3^4 = 81$
2. $\log_7 7 = 1$ $7^1 = 7$
3. $\log_{1/2} \frac{1}{8} = 3$ $\left(\frac{1}{2}\right)^3 = \frac{1}{8}$
4. $\log_{1/5} 1 = 0$ $\left(\frac{1}{5}\right)^0 = 1$
5. $\log_4 \frac{1}{64} = -3$ $4^{-3} = \frac{1}{64}$
6. $\log_6 \frac{1}{36} = -2$ $6^{-2} = \frac{1}{36}$
7. $\log_{1/2} \frac{1}{8} = 3$ $\left(\frac{1}{2}\right)^3 = \frac{1}{8}$
8. $\log_4 \frac{1}{4} = -1$ $4^{-1} = \frac{1}{4}$

In Exercises 9–16, write each equation in logarithmic form.

9. $8^2 = 64$ $\log_8 64 = 2$
10. $10^3 = 1000$ $\log 1000 = 3$
11. $4^{-2} = \frac{1}{16}$ $\log_4 \frac{1}{16} = -2$
12. $3^{-4} = \frac{1}{81}$ $\log_3 \frac{1}{81} = -4$
13. $\left(\frac{1}{2}\right)^{-5} = 32$ $\log_{1/2} 32 = -5$
14. $\left(\frac{1}{3}\right)^{-3} = 27$ $\log_{1/3} 27 = -3$
15. $x^y = z$ $\log_x z = y$
16. $m^n = p$ $\log_m p = n$

In Exercises 17–56, find each value of x.

17. $\log_2 8 = x$ 3
18. $\log_3 9 = x$ 2
19. $\log_4 64 = x$ 3
20. $\log_6 216 = x$ 3
21. $\log_{1/2} \frac{1}{8} = x$ 3
22. $\log_{1/3} \frac{1}{81} = x$ 4
23. $\log_9 3 = x$ $\frac{1}{2}$
24. $\log_{125} 5 = x$ $\frac{1}{3}$
25. $\log_{1/2} 8 = x$ -3
26. $\log_{1/2} 16 = x$ -4
27. $\log_8 x = 2$ 64
28. $\log_7 x = 0$ 1
29. $\log_7 x = 1$ 7
30. $\log_2 x = 3$ 8
31. $\log_{25} x = \frac{1}{2}$ 5
32. $\log_4 x = \frac{1}{2}$ 2
33. $\log_5 x = -2$ $\frac{1}{25}$
34. $\log_3 x = -4$ $\frac{1}{81}$
35. $\log_{36} x = -\frac{1}{2}$ $\frac{1}{6}$
36. $\log_{27} x = -\frac{1}{3}$ $\frac{1}{3}$
37. $\log_{100} \frac{1}{1000} = x$ $-\frac{3}{2}$
38. $\log_{5/2} \frac{4}{25} = x$ -2
39. $\log_{27} 9 = x$ $\frac{2}{3}$
40. $\log_{12} x = 0$ 1
41. $\log_x 5^3 = 3$ 5
42. $\log_x 5 = 1$ 5
43. $\log_x \frac{9}{4} = 2$ $\frac{3}{2}$
44. $\log_x \frac{\sqrt{3}}{3} = \frac{1}{2}$ $\frac{1}{3}$
45. $\log_x \frac{1}{64} = -3$ 4
46. $\log_x \frac{1}{100} = -2$ 10
47. $\log_{2\sqrt{2}} x = 2$ 8
48. $\log_4 8 = x$ $\frac{3}{2}$
49. $2^{\log_2 5} = x$ 5
50. $3^{\log_3 4} = x$ 4
51. $x^{\log_4 6} = 6$ 4
52. $x^{\log_3 8} = 8$ 3
53. $\log 10^3 = x$ 3
54. $\log 10^{-2} = x$ -2
55. $10^{\log x} = 100$ 100
56. $10^{\log x} = \frac{1}{10}$ $\frac{1}{10}$

In Exercises 57–68, use a calculator to find each value, if possible. Give answers to four decimal places.

57. log 3.25 0.5119
58. log 0.57 −0.2441
59. log 0.00467 −2.3307
60. log 375.876 2.5750
61. ln 0.93 −0.0726
62. ln 7.39 2.0001
63. ln 37.896 3.6348
64. ln 0.00465 −5.3709
65. log (ln 1.7) −0.2752
66. ln (log 9.8) −0.0088
67. ln (log 0.1) impossible
68. log (ln 0.01) impossible

In Exercises 69–76, use a calculator to find each value of y, if possible. Give answers to two decimal places.

69. $\log y = 1.4023$ 25.25
70. $\ln y = 2.6490$ 14.14
71. $\ln y = 4.24$ 69.41
72. $\log y = 0.926$ 8.43
73. $\log y = -3.71$ 0.00
74. $\ln y = -0.28$ 0.76
75. $\log y = \ln 8$ 120.07
76. $\ln y = \log 7$ 2.33

In Exercises 77–80, graph the function defined by each equation.

77. $y = \log_3 x$

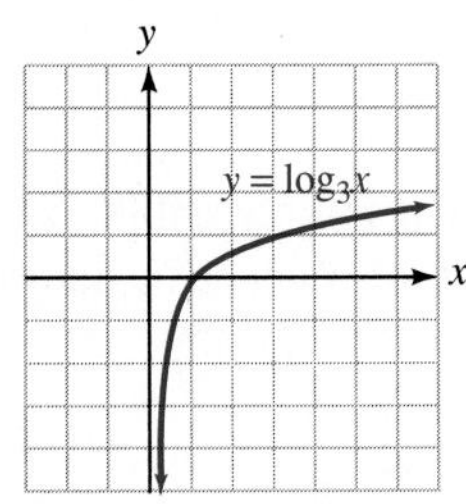

78. $y = \log_{1/3} x$

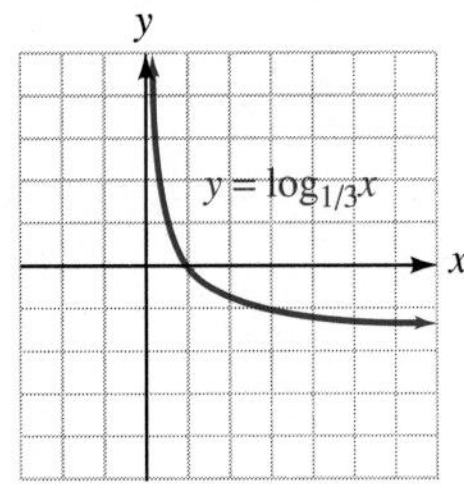

79. $y = \log_{1/2} x$

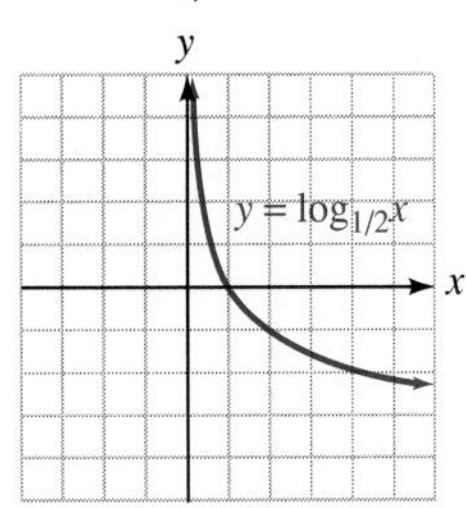

80. $y = \log_4 x$

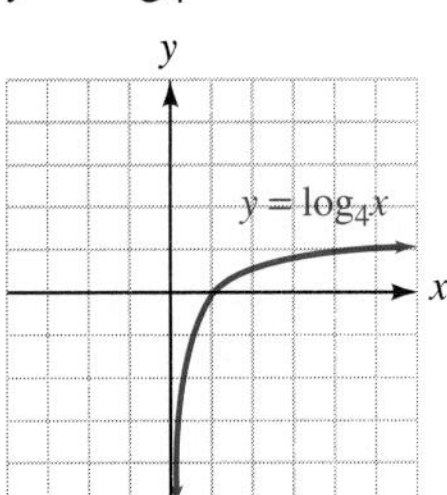

In Exercises 81–84, graph each pair of inverse functions.

81. $y = 2^x$, $y = \log_2 x$

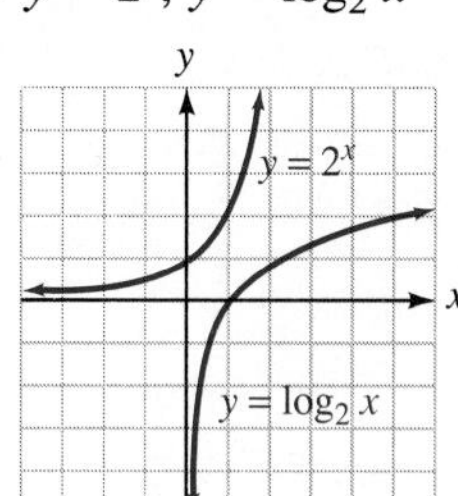

82. $y = \left(\frac{1}{2}\right)^x$, $y = \log_{1/2} x$

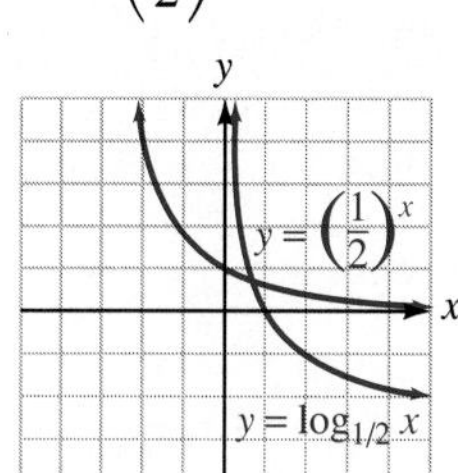

83. $y = \left(\frac{1}{4}\right)^x$, $y = \log_{1/4} x$

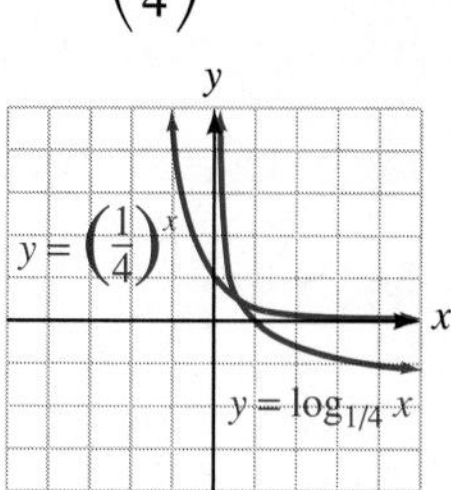

84. $y = 4^x$, $y = \log_4 x$

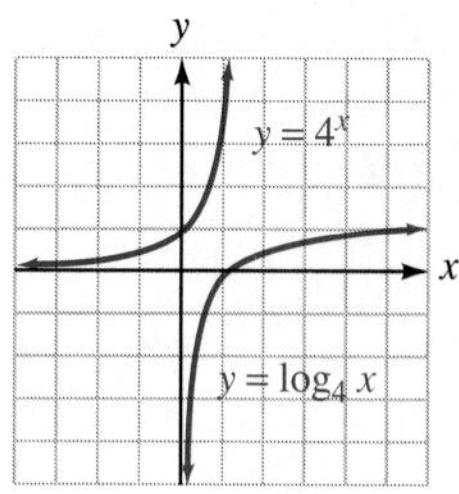

In Exercises 85–92, find the value of b, if any, that would cause the graph of $y = \log_b x$ to look like the graph indicated.

85.

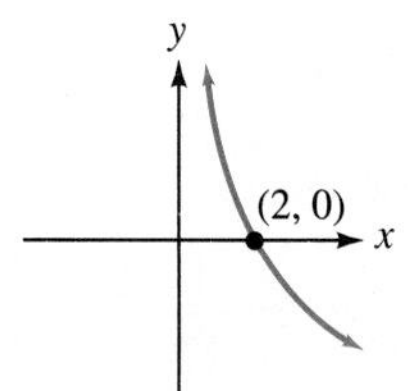

no value of b

86.

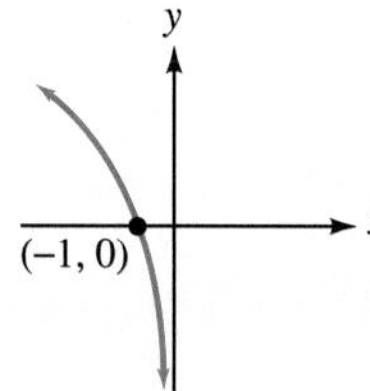

no value of b

87.

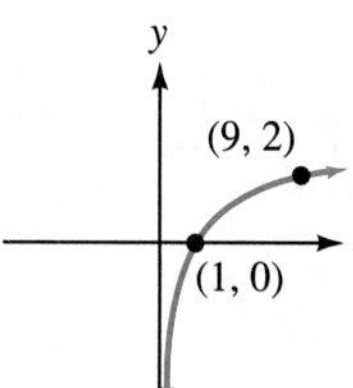

3

88.

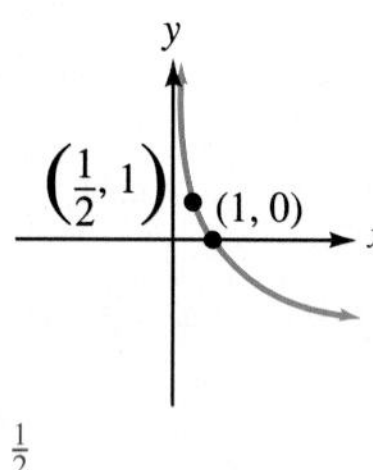

$\frac{1}{2}$

89.

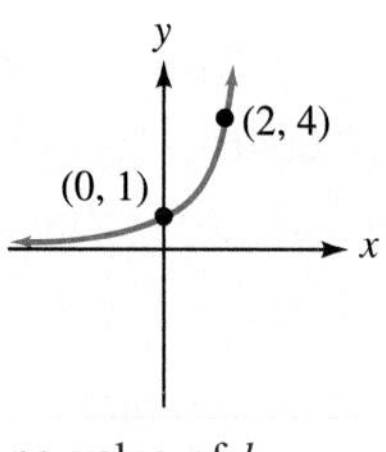

no value of b

90.

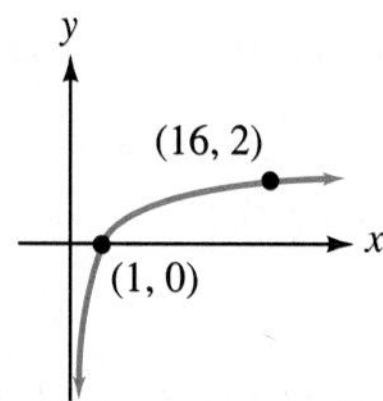

4

91.

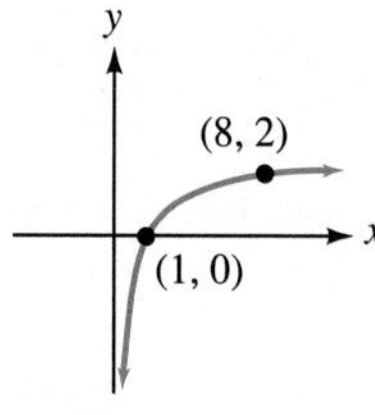

$2\sqrt{2}$

92.

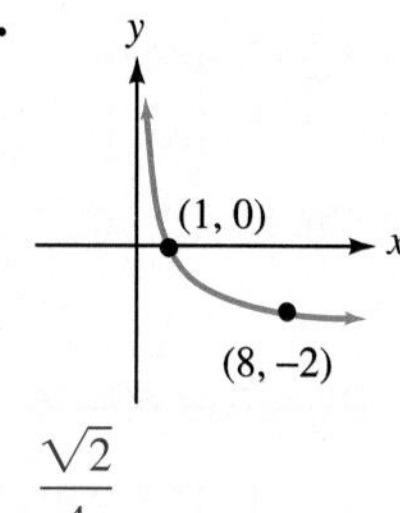

$\frac{\sqrt{2}}{4}$

In Exercises 93–96, use a graphing calculator to do each experiment. Describe what you find.

93. Graph $y = k + \log x$ for many values of k.

94. Graph $y = \log (x + k)$ for many values of k.

95. Graph $y = k \log x$ for many values of k.

96. Graph $y = \log kx$ for many values of k.

Writing Exercises ■ *Write a paragraph using your own words.*

1. Describe the appearance of the graph of $y = \log_b x$ when $0 < b < 1$ and when $b > 1$.

2. Explain why it is impossible to find the logarithm of a negative number.

Something to Think About

1. Find the value of $\left(1 + \frac{1}{p}\right)^p$ for $p = 1, 10, 100, 1000$, and 10,000. What do you find?

2. Add the first 3 terms, the first 4 terms, the first 5 terms, and the first 6 terms of the list

$$1, 1, \frac{1}{2}, \frac{1}{2 \cdot 3}, \frac{1}{2 \cdot 3 \cdot 4}, \frac{1}{2 \cdot 3 \cdot 4 \cdot 5}$$

What do you find?

Review Exercises

Solve each equation.

1. $\sqrt[3]{6x + 4} = 4$
   10
2. $\sqrt{3x - 4} = \sqrt{-7x + 2}$
   $\frac{3}{5}$ is extraneous
3. $\sqrt{a + 1} - 1 = 3a$
   0; $-\frac{5}{9}$ is extraneous
4. $3 - \sqrt{t - 3} = \sqrt{t}$
   4

11.3 Properties of Logarithms

- The Change-of-Base Formula

Since logarithms are exponents, it is not surprising that the properties of exponents have counterparts in logarithmic notation.

Properties of Logarithms

If b is a positive number and $b \neq 1$, then

1. $\log_b 1 = 0$
2. $\log_b b = 1$
3. $\log_b b^x = x$
4. $b^{\log_b x} = x \quad (x > 0)$

Properties 1 through 4 follow directly from the definition of logarithms:

1. $\log_b 1 = 0$ because $b^0 = 1$.
2. $\log_b b = 1$ because $b^1 = b$.
3. $\log_b b^x = x$ because $b^x = b^x$.
4. $b^{\log_b x} = x$ because $\log_b x$ is the exponent to which b is raised to get x.

EXAMPLE 1 Simplify each expression: **a.** $\log_5 1$, **b.** $\log_3 3$, **c.** $\log_7 7^3$, and **d.** $b^{\log_b 7}$.

Solution **a.** By Property 1, $\log_5 1 = 0$ because $5^0 = 1$.

b. By Property 2, $\log_3 3 = 1$ because $3^1 = 3$.

c. By Property 3, $\log_7 7^3 = 3$ because $7^3 = 7^3$.

d. By Property 4, $b^{\log_b 7} = 7$ because $\log_b 7$ is the power to which b is raised to get 7. ■

Properties 3 and 4 show that the compositions of the exponential and logarithmic functions (in both directions) are the identity function. This is expected, because the exponential and logarithmic functions are inverse functions.

More Properties of Logarithms

If M, N, and b are positive numbers and $b \neq 1$, then

5. $\log_b MN = \log_b M + \log_b N$ **6.** $\log_b \frac{M}{N} = \log_b M - \log_b N$

7. $\log_b M^p = p \log_b M$ **8.** If $\log_b x = \log_b y$, then $x = y$.

Proof To prove Property 5, we let $x = \log_b M$ and $y = \log_b N$. Because of the definition of logarithms, these equations can be written in the form

$$M = b^x \quad \text{and} \quad N = b^y$$

Then $MN = b^x b^y$ and the properties of exponents give

$$MN = b^{x+y}$$

Using the definition of logarithms gives

$$\log_b MN = x + y$$

Substituting the values of x and y completes the proof:

$$\log_b MN = \log_b M + \log_b N$$ □

WARNING! Property 5 of logarithms asserts that the logarithm of the product of two numbers is equal to the sum of their logarithms. The logarithm of a sum or a difference usually does not simplify. In general,

$$\log_b (M + N) \neq \log_b M + \log_b N \quad \text{and} \quad \log_b (M - N) \neq \log_b M - \log_b N$$

Proof To prove Property 6, we again let $x = \log_b M$ and $y = \log_b N$. These equations can be written as

$$M = b^x \quad \text{and} \quad N = b^y$$

Then $\frac{M}{N} = \frac{b^x}{b^y}$ and the properties of exponents give

$$\frac{M}{N} = b^{x-y}$$

Using the definition of logarithms gives

$$\log_b \frac{M}{N} = x - y$$

Substituting the values of x and y completes the proof:

$$\log_b \frac{M}{N} = \log_b M - \log_b N$$

□

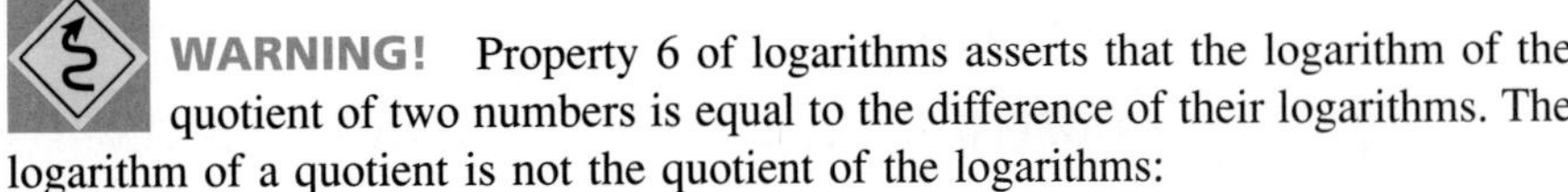

WARNING! Property 6 of logarithms asserts that the logarithm of the quotient of two numbers is equal to the difference of their logarithms. The logarithm of a quotient is not the quotient of the logarithms:

$$\log_b \frac{M}{N} \neq \frac{\log_b M}{\log_b N}$$

Proof To prove Property 7, we let $x = \log_b M$, write the expression in exponential form, and raise both sides to the pth power:

$$\begin{aligned} M &= b^x \\ (M)^p &= (b^x)^p \\ M^p &= b^{px} \end{aligned}$$

Using the definition of logarithms gives

$$\log_b M^p = px$$

Substituting the value for x completes the proof:

$$\log_b M^p = p\log_b M$$

□

Property 8 follows from the fact that the logarithmic function is a one-to-one function.

EXAMPLE 2 Use a calculator to verify Property 6 of logarithms by showing that

$$\ln \frac{25.37}{37.25} = \ln 25.37 - \ln 37.25$$

Solution We can find the left- and right-hand sides of the equation separately and compare the results. To use a scientific calculator to find $\ln \frac{25.37}{37.25}$, we press these keys:

25.37 [÷] 37.25 [=] [LN]

The display will read -0.384084571.

To find ln 25.37 − ln 37.25, we press these keys:

25.37 LN − 37.25 LN =

The display will read −0.384084571. Since the left- and right-hand sides are equal, the equation is verified. For the keys to press on a graphing calculator, consult the owner's manual. ■

Properties 5, 6, and 7 can be used to expand or condense logarithmic expressions.

(a, c)

EXAMPLE 3 Assume that x, y, z, and b are positive numbers ($b \neq 1$). Use the properties of logarithms to write each expression in terms of the logarithms of x, y, and z. **a.** $\log_b xyz$, **b.** $\log_b \dfrac{x}{yz}$, and **c.** $\log_b \dfrac{y^2\sqrt{z}}{x}$

Solution **a.**

$$\log_b xyz = \log_b (xy)z \qquad \text{Use the associative property of multiplication.}$$

$$= \log_b (xy) + \log_b z \qquad \text{Use Property 5 of logarithms.}$$

$$= \log_b x + \log_b y + \log_b z \qquad \text{Use Property 5 of logarithms.}$$

b.

$$\log_b \frac{x}{yz} = \log_b x - \log_b (yz) \qquad \text{Use Property 6 of logarithms.}$$

$$= \log_b x - (\log_b y + \log_b z) \qquad \text{Use Property 5 of logarithms.}$$

$$= \log_b x - \log_b y - \log_b z \qquad \text{Remove parentheses.}$$

c.

$$\log_b \frac{y^2\sqrt{x}}{z} = \log_b (y^2\sqrt{x}) - \log_b z \qquad \text{Use Property 6 of logarithms.}$$

$$= \log_b y^2 + \log_b \sqrt{x} - \log_b z \qquad \text{Use Property 5 of logarithms.}$$

$$= \log_b y^2 + \log_b x^{1/2} - \log_b z \qquad \text{Write } \sqrt{x} \text{ as } x^{1/2}.$$

$$= 2\log_b y + \frac{1}{2}\log_b x - \log_b z \qquad \text{Use Property 7 of logarithms twice.}$$

■

(b)

EXAMPLE 4 Assume that x, y, z, and b are positive numbers ($b \neq 1$). Use the properties of logarithms to write each expression as the logarithm of a single quantity. **a.** $2\log_b x + \frac{1}{3}\log_b y$ and **b.** $\frac{1}{2}\log_b (x-2) - \log_b y + 3\log_b z$

Solution **a.**

$$2\log_b x + \frac{1}{3}\log_b y = \log_b x^2 + \log_b y^{1/3} \qquad \text{Use Property 7 of logarithms.}$$

$$= \log_b (x^2y^{1/3}) \qquad \text{Use Property 5 of logarithms.}$$

$$= \log_b (x^2\sqrt[3]{y}) \qquad \text{Change } y^{1/3} \text{ to } \sqrt[3]{y}.$$

b. $\frac{1}{2}\log_b (x-2) - \log_b y + 3\log_b z$

$= \log_b (x-2)^{1/2} - \log_b y + \log_b z^3$ Use Property 7 of logarithms.

$= \log_b \dfrac{(x-2)^{1/2}}{y} + \log_b z^3$ Use Property 6 of logarithms.

$= \log_b \dfrac{z^3\sqrt{x-2}}{y}$ Use Property 5 of logarithms. ■

(b, e)

EXAMPLE 5 Given that $\log 2 \approx 0.3010$ and $\log 3 \approx 0.4771$, find approximations for **a.** $\log 6$, **b.** $\log 9$, **c.** $\log 5$, **d.** $\log \sqrt{5}$, and **e.** $\log 1.5$.

Solution **a.** $\log 6 = \log (2 \cdot 3)$ Factor 6 as $2 \cdot 3$.

$= \log 2 + \log 3$ Use Property 5 of logarithms.

$\approx 0.3010 + 0.4771$ Substitute the value of each logarithm.

≈ 0.7781

b. $\log 9 = \log (3^2)$ Write 9 as 3^2.

$= 2 \log 3$ Use Property 7 of logarithms.

$= 2(0.4771)$ Substitute the value of the logarithm of 3.

$= 0.9542$

c. $\log 5 = \log \left(\dfrac{10}{2}\right)$ Write 5 as $\frac{10}{2}$.

$= \log 10 - \log 2$ Use Property 6 of logarithms.

$= 1 - \log 2$ Use Property 2 of logarithms.

$= 1 - 0.3010$ Substitute the value of the logarithm of 2.

$= 0.6990$

d. $\log \sqrt{5} = \log (5^{1/2})$ Write $\sqrt{5}$ as $5^{1/2}$.

$= \frac{1}{2}\log 5$ Use Property 7 of logarithms.

$= \frac{1}{2}(0.6990)$ Use the answer from Part c.

$= 0.3495$

e. $\log 1.5 = \log\left(\dfrac{3}{2}\right)$ Write 1.5 as $\frac{3}{2}$.

$= \log 3 - \log 2$ Use Property 6 of logarithms.

$= 0.4771 - 0.3010$ Substitute the value of the logarithm of 2.

$= 0.1761$ ■

The Change-of-Base Formula

If we know the base-a logarithm of a number, we can find its logarithm to some other base b by using the **change-of-base formula.**

Change-of-Base Formula If a, b, and x are positive real numbers, then

$$\log_b x = \frac{\log_a x}{\log_a b} \quad (a \neq 1, b \neq 1)$$

Proof We begin with the equation $\log_b x = y$ and proceed as follows:

1. $\log_b x = y$

$b^y = x$ Change the equation into logarithmic form.

$\log_a b^y = \log_a x$ Take the base-a logarithm of both sides.

$y \log_a b = \log_a x$ Use Property 7 of logarithms.

$y = \dfrac{\log_a x}{\log_a b}$ Divide both sides by $\log_a b$.

$\log_b x = \dfrac{\log_a x}{\log_a b}$ Refer to Equation 1 and substitute $\log_b x$ for y. □

If we know logarithms to some base a (for example, $a = 10$), we can find the logarithm of x to a new base b. We simply divide the base-a logarithm of x by the base-a logarithm of b.

EXAMPLE 6 Use the change-of-base formula to find $\log_3 5$.

Solution We substitute 3 for b, 10 for a, and 5 for x into the change-of-base formula:

$$\log_b x = \frac{\log_a x}{\log_a b}$$

$$\log_3 5 = \frac{\log_{10} 5}{\log_{10} 3}$$

$$\approx 1.464973521$$

Thus, to four decimal places, $\log_3 5 = 1.4650$. ■

ORALS *Find the value of x.*

1. $\log_3 9 = x$ 2 **2.** $\log_x 5 = 1$ 5 **3.** $\log_7 x = 49$ 7^{49} **4.** $\log_2 x = -2$ $\frac{1}{4}$

5. $\log_4 x = \frac{1}{2}$ 2 **6.** $\log_x 4 = 2$ 2 **7.** $\log_3 \frac{1}{9} = x$ -2 **8.** $\log_{\sqrt{3}} x = 2$ 3

EXERCISE 11.3

In Exercises 1–8, simplify each expression.

1. $\log_7 1$ 0
2. $\log_5 5$ 1
3. $\log_3 3^7$ 7
4. $5^{\log_5 8}$ 8
5. $8^{\log_8 10}$ 10
6. $\log_4 4^2$ 2
7. $\log_9 9$ 1
8. $\log_3 1$ 0

In Exercises 9–14, use a calculator to verify each equation.

9. $\log [(2.5)(3.7)] = \log 2.5 + \log 3.7$

10. $\ln \frac{11.3}{6.1} = \ln 11.3 - \ln 6.1$

11. $\ln (2.25)^4 = 4 \ln 2.25$

12. $\log 45.37 = \frac{\ln 45.37}{\ln 10}$

13. $\log \sqrt{24.3} = \frac{1}{2} \log 24.3$

14. $\ln 8.75 = \frac{\log 8.75}{\log e}$

In Exercises 15–26, assume that x, y, and z are positive numbers. Use the properties of logarithms to write each expression in terms of the logarithms of x, y, and z.

15. $\log_b xyz$ $\log_b x + \log_b y + \log_b z$

16. $\log_b 4xz$ $\log_b 4 + \log_b x + \log_b z$

17. $\log_b \frac{2x}{y}$ $\log_b 2 + \log_b x - \log_b y$

18. $\log_b \frac{x}{yz}$ $\log_b x - \log_b y - \log_b z$

19. $\log_b x^3y^2$ $3 \log_b x + 2 \log_b y$

20. $\log_b xy^2z^3$ $\log_b x + 2 \log_b y + 3 \log_b z$

21. $\log_b (xy)^{1/2}$ $\frac{1}{2}(\log_b x + \log_b y)$

22. $\log_b x^3y^{1/2}$ $3 \log_b x + \frac{1}{2} \log_b y$

23. $\log_b x\sqrt{z}$ $\log_b x + \frac{1}{2} \log_b z$

24. $\log_b \sqrt{xy}$ $\frac{1}{2}(\log_b x + \log_b y)$

25. $\log_b \frac{\sqrt[3]{x}}{\sqrt[4]{yz}}$ $\frac{1}{3} \log_b x - \frac{1}{4} \log_b y - \frac{1}{4} \log_b z$

26. $\log_b \sqrt[4]{\frac{x^3y^2}{z^4}}$ $\frac{3}{4} \log_b x + \frac{1}{2} \log_b y - \log_b z$

In Exercises 27–34, assume that x, y, and z are positive numbers. Use the properties of logarithms to write each expression as the logarithm of a single quantity.

27. $\log_b (x + 1) - \log_b x$ $\log_b \frac{x+1}{x}$

28. $\log_b x + \log_b (x + 2) - \log_b 8$ $\log_b \frac{x(x+2)}{8}$

29. $2 \log_b x + \frac{1}{2} \log_b y$ $\log_b x^2\sqrt{y}$

30. $-2 \log_b x - 3 \log_b y + \log_b z$ $\log_b \frac{z}{x^2y^3}$

31. $-3 \log_b x - 2 \log_b y + \frac{1}{2} \log_b z$ $\log_b \frac{\sqrt{z}}{x^3y^2}$

32. $3 \log_b (x + 1) - 2 \log_b (x + 2) + \log_b x$ $\log_b \frac{x(x+1)^3}{(x+2)^2}$

33. $\log_b \left(\frac{x}{z} + x\right) - \log_b \left(\frac{y}{z} + y\right)$ $\log_b \frac{\frac{x}{z} + x}{\frac{y}{z} + y} = \log_b \frac{x}{y}$

34. $\log_b (xy + y^2) - \log_b (xz + yz) + \log_b z$ $\log_b y$

In Exercises 35–54, tell whether the given statement is true. All variables are positive, and $b \neq 1$.

35. $\log_b 0 = 1$ false

36. $\log_b (x + y) \neq \log_b x + \log_b y$ true

37. $\log_b xy = (\log_b x)(\log_b y)$ false

38. $\log_b ab = \log_b a + 1$ true

39. $\log_7 7^7 = 7$ true

40. $7^{\log_7 7} = 7$ true

41. $\dfrac{\log_b A}{\log_b B} = \log_b A - \log_b B$ false

42. $\log_b (A - B) = \dfrac{\log_b A}{\log_b B}$ false

43. $3 \log_b \sqrt[3]{a} = \log_b a$ true

44. $\dfrac{1}{3} \log_b a^3 = \log_b a$ true

45. $\log_b \dfrac{1}{a} = -\log_b a$ true

46. $\log_b 2 = \log_2 b$ false

47. If $\log_a b = c$, then $\log_b a = c$. false

48. If $\log_a b = c$, then $\log_b a = \dfrac{1}{c}$. true

49. $\log_b (-x) = -\log_b x$ false

50. If $\log_b a = c$, then $\log_b a^p = pc$. true

51. $\log_b \dfrac{1}{5} = -\log_b 5$ true

52. $\log_{4/3} y = -\log_{3/4} y$ true

53. $\log_b y + \log_{1/b} y = 0$ true

54. $\log_{10} 10^3 = 3(10^{\log_{10} 3})$ false

In Exercises 55–66, assume that $\log 4 = 0.6021$, $\log 7 = 0.8451$, *and* $\log 9 = 0.9542$. *Use these values and the properties of logarithms to find each value.*

55. $\log 28$ 1.4472

56. $\log \dfrac{7}{4}$ 0.2430

57. $\log 2.25$ 0.3521

58. $\log 36$ 1.5563

59. $\log \dfrac{63}{4}$ 1.1972

60. $\log \dfrac{4}{63}$ −1.1972

61. $\log 252$ 2.4014

62. $\log 49$ 1.6902

63. $\log 112$ 2.0493

64. $\log 324$ 2.5105

65. $\log \dfrac{144}{49}$ 0.4682

66. $\log \dfrac{324}{63}$ 0.7112

In Exercises 67–74, use the change-of-base formula to find each logarithm to four decimal places.

67. $\log_3 7$ 1.7712

68. $\log_7 3$ 0.5646

69. $\log_{1/3} 3$ −1.0000

70. $\log_{1/2} 6$ −2.5850

71. $\log_3 8$ 1.8928

72. $\log_5 10$ 1.4307

73. $\log_{\sqrt{2}} \sqrt{5}$ 2.3219

74. $\log_\pi e$ 0.8736

Writing Exercises

Write a paragraph using your own words.

1. Explain why $\ln(\log 0.9)$ is undefined.
2. Explain why $\log_b (\ln 1)$ is undefined.
3. State Properties 5–7 of logarithms in words.
4. State the change-of-base formula in words.

Something to Think About

1. Show that $\ln(e^x) = x$.
2. If $\log_b 3x = 1 + \log_b x$, find b. 3
3. Show that $\log_{b^2} x = \dfrac{1}{2} \log_b x$.
4. Show that $e^{x \ln a} = a^x$.

Review Exercises

Consider the line that passes through $P(-2, 3)$ *and* $Q(4, -4)$.

1. Find the slope of line PQ. $-\frac{7}{6}$
2. Find the distance PQ. $\sqrt{85}$
3. Find the midpoint of segment PQ. $\left(1, -\frac{1}{2}\right)$
4. Write the equation of line PQ. $y = -\frac{7}{6}x + \frac{2}{3}$

11.4 Applications of Logarithms

■ Applications of Base-10 Logarithms ■ Applications of Base-e Logarithms

In this section, we consider many applications of logarithmic functions.

■ Applications of Base-10 Logarithms

Chemistry In chemistry, common logarithms are used to express the acidity of solutions. The more acidic a solution, the greater the concentration of hydrogen ions. This concentration is indicated indirectly by the **pH scale,** or **hydrogen ion index.** The pH of a solution is defined by the following equation.

pH of a Solution If $[H^+]$ is the hydrogen ion concentration in gram-ions per liter, then

$$pH = -\log[H^+]$$

EXAMPLE 1 Find the pH of pure water, which has a hydrogen ion concentration of 10^{-7} gram-ions per liter.

Solution Since the hydrogen ion concentration of pure water is 10^{-7} gram-ions per liter, its pH is

$$\begin{aligned} pH &= -\log 10^{-7} \\ &= -(-7)\log 10 && \text{Use Property 7 of logarithms.} \\ &= -(-7)\cdot 1 && \log 10 = 1. \\ &= 7 \end{aligned}$$

■

EXAMPLE 2 Find the hydrogen ion concentration of seawater if its pH is 8.5.

Solution To find its hydrogen ion concentration, we solve the following equation for $[H^+]$.

$$\begin{aligned} 8.5 &= -\log [H^+] \\ -8.5 &= \log [H^+] && \text{Multiply both sides by } -1. \\ [H^+] &= 10^{-8.5} && \text{Change the equation to exponential form.} \end{aligned}$$

We can use a calculator to find that

$$[H^+] \approx 3.2 \times 10^{-9} \text{ gram-ions per liter.}$$

■

Electrical engineering Common logarithms are used in electrical engineering to express the voltage gain (or loss) of an electronic device such as an amplifier. The unit of gain (or loss), called the **decibel,** is defined by the following logarithmic relation.

Decibel Voltage Gain If E_O is the output voltage of a device and E_I is the input voltage, the decibel voltage gain is given by

$$\text{db gain} = 20 \log \frac{E_O}{E_I}$$

EXAMPLE 3 If the input to an amplifier is 0.5 volt and the output is 40 volts, find the decibel voltage gain of the amplifier.

Solution We can find the decibel voltage gain by substituting 0.5 for E_I and 40 for E_O into the formula for db gain:

$$\begin{aligned} \text{db gain} &= 20 \log \frac{E_O}{E_I} \\ \text{db gain} &= 20 \log \frac{40}{0.5} \\ &= 20 \log 80 \\ &\approx 38 && \text{Use a calculator.} \end{aligned}$$

The amplifier provides a 38-decibel voltage gain. ■

Geology In seismology, common logarithms are used to measure the intensity of earthquakes on the **Richter scale.** The intensity of an earthquake is given by the following logarithmic function.

Richter Scale If A is the amplitude of an earthquake (the earth movement, measured in micrometers), and P is the period (the time of one oscillation of the earth's surface measured in seconds), then the intensity R on the Richter scale is

$$R = \log \frac{A}{P}$$

EXAMPLE 4 Find the measure on the Richter scale of an earthquake with an amplitude of 10,000 micrometers (1 centimeter) and a period of 0.1 second.

Solution We substitute 10,000 for A and 0.1 for P in the Richter scale formula and simplify:

$$\begin{aligned} R &= \log \frac{A}{P} \\ R &= \log \frac{10{,}000}{0.1} \\ &= \log 100{,}000 \\ &= \log 10^5 \\ &= 5 \log 10 && \text{Use Property 7 of logarithms.} \\ &= 5 && \text{Use Property 2 of logarithms.} \end{aligned}$$

The earthquake measures 5 on the Richter scale. ■

Applications of Base-e Logarithms

Population growth If a population grows exponentially at a certain annual rate, the time required for the population to double is called the **doubling time** and is given by the following formula.

Formula for Doubling Time If r is the annual rate and t is time required for a population to double, then

$$t = \frac{\ln 2}{r}$$

EXAMPLE 5 The population of the earth is growing at the rate of approximately 2% per year. If this rate continues, how long will it take the population to double?

Solution Because the population is growing at the rate of 2% per year, we substitute 0.02 for r into the formula for doubling time and simplify.

$$t = \frac{\ln 2}{r}$$

$$t = \frac{\ln 2}{0.02}$$

$$\approx 34.65735903$$

At the current rate of growth, the population of the earth will double in about 35 years. ■

Economics The formula for doubling time can also be used to find the length of time it will take money to double when interest is compounded continuously.

EXAMPLE 6 How long will it take \$1000 to double if it is invested at an annual rate of 8% compounded continuously?

Solution We substitute 0.08 for r and simplify:

Teaching Tip: Because $\ln 2 \approx 0.7$, this rule is called the rule of 70: Divide 70 by the rate, expressed as a percent, to estimate the doubling time.

$$t = \frac{\ln 2}{r}$$

$$t = \frac{\ln 2}{0.08}$$

$$\approx 8.664339757 \quad \text{Use a calculator.}$$

It will take approximately $8\frac{2}{3}$ years for the money to double. ■

Physiology In physiology, experiments suggest that the relationship between the loudness and the intensity of sound is a logarithmic one known as the **Weber–Fechner law.**

Weber–Fechner Law If L is the apparent loudness of a sound, I is the actual intensity, and k is a constant, then

$$L = k \ln I$$

EXAMPLE 7 Find the increase in actual intensity that will cause the apparent loudness of a sound to double.

Solution If the original loudness L_0 is caused by an actual intensity I_0, then

1. $L_0 = k \ln I_0$

To double the apparent loudness, we multiply both sides of Equation 1 by 2 and use Property 7 of logarithms:

$$\begin{aligned} 2L_0 &= 2k \ln I_0 \\ &= k \ln (I_0)^2 \end{aligned}$$

Thus, to double the apparent loudness of a sound, the intensity must be squared. ■

ORALS *Give the following formulas.*

1. pH $\text{pH} = -\log [\text{H}^+]$

2. db gain $20 \log \dfrac{E_O}{E_I}$

3. Richter scale $R = \log \frac{A}{P}$

4. Doubling time $t = \frac{\ln 2}{r}$

5. Weber–Fechner law $L = k \ln I$

EXERCISE 11.4

 *In Exercises 1–28, solve each problem.*

1. pH of a solution Find the pH of a solution with a hydrogen ion concentration of 1.7×10^{-5} gram-ions per liter. 4.77

2. Hydrogen ion concentration of calcium hydroxide Find the hydrogen ion concentration of a saturated solution of calcium hydroxide whose pH is 13.2. 6.31×10^{-14} gram-ions per liter

3. Aquariums To test for safe pH levels in a fresh-water aquarium, a test strip is compared with the scale shown in Illustration 1. Find the corresponding safe range in the hydrogen ion concentration from 2.5119×10^{-8} to 1.585×10^{-7}

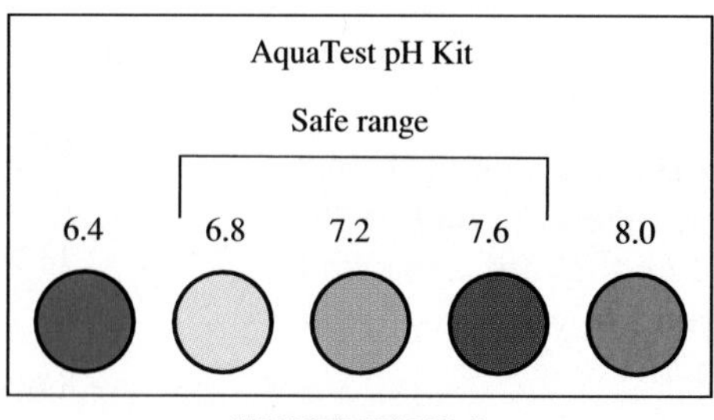

ILLUSTRATION 1

4. pH of sour pickles The hydrogen concentration of sour pickles is 6.31×10^{-4}. Find the pH. 3.2

5. **Finding input voltage** The db gain of an amplifier is 29. Find the input voltage when the output voltage is 20 volts. 0.71 V

6. **Finding output voltage** The db gain of an amplifier is 35. Find the output voltage when the input voltage is 0.05 volt. 2.8 V

7. **db gain of an amplifier** Find the db gain of the amplifer shown in Illustration 2. 49.5 db

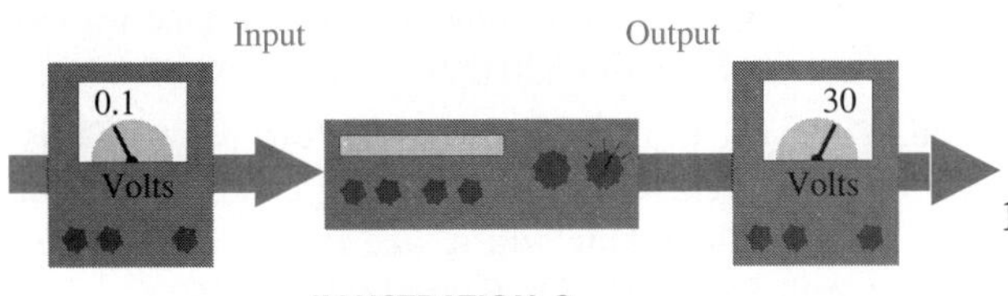

ILLUSTRATION 2

8. **db gain of an amplifier** An amplifier produces an output of 80 volts when driven by an input of 0.12 volts. Find the amplifier's db gain. 56.5 db

9. **Earthquakes** An earthquake has amplitude of 5000 micrometers and a period of 0.2 second. Find its measure on the Richter scale. 4.4

10. **Earthquakes** Find the period of an earthquake with amplitude of 80,000 micrometers that measures 6 on the Richter scale. 0.08 sec

11. **Earthquakes** An earthquake with a period of $\frac{1}{4}$ second measures 4 on the Richter scale. Find its amplitude. 2500 μm

12. **Earthquakes** By what factor must the amplitude of an earthquake change to increase its severity by 1 point on the Richter scale? Assume that the period remains constant. factor of 10

13. **Population growth** How long will it take the population of River City to double? (See Illustration 3.) 5.8 yr

River City

A growing community

- 6 parks • 12% annual growth
- 10 churches • low crime rate

ILLUSTRATION 3

14. **Doubling money** How long will it take \$1000 to double if it is invested at an annual rate of 5% compounded continuously? 13.9 yr

15. **Population growth** A population growing at an annual rate r will triple in a time t given by the formula

$$t = \frac{\ln 3}{r}$$

How long will it take the population of a town growing at the rate of 12% per year to triple? 9.2 yr

16. **Tripling money** Find the length of time for \$25,000 to triple if invested at 6% annual interest, compounded continuously. 18.3 yr

17. **Change in loudness** If the intensity of a sound is doubled, find the apparent change in loudness. It will increase by $k \ln 2$.

18. **Change in intensity** If the intensity of a sound is tripled, find the apparent change in loudness. It will increase by $k \ln 3$.

19. **Change in loudness** What change in intensity of sound will cause an apparent tripling of the loudness? The intensity must be cubed.

20. **Change in intensity** What increase in the intensity of a sound will cause the apparent loudness to be multiplied by 4? The intensity must be raised to the 4th power.

21. **Charging a battery** The time (in hours) required to charge a certain battery to a level C (expressed as a fraction of the battery's full charge) is

$$t = -3.7 \ln(1 - C)$$

How long would it take to bring the battery to 80% of its full charge? almost 6 hours

22. **Power output** If P_O is the output of an amplifier (in watts) and P_I is the power input, then the decibel voltage gain is

$$\text{db voltage gain} = 10 \log \frac{P_O}{P_I}$$

One brand of amplifier produces a 70 watt output with a 0.035 watt input. Find the db voltage gain. 33 db

23. **Depreciation** In business, equipment is often depreciated using the double declining-balance method. In this method, a piece of equipment with a life expectancy of N years, costing $\$C$, will depreciate to a value of $\$V$ in n years, where n is given by the formula

$$n = \frac{\log V - \log C}{\log\left(1 - \frac{2}{N}\right)}$$

A computer that cost \$37,000 has a life expectancy of 5 years. If it has depreciated to a value of \$8000, how old is it? 3 yr old

24. **Depreciation** A typewriter worth \$470 when new had a life expectancy of 12 years. If it is now worth \$189, how old is it? 5 yr old

25. **Time for money to double** If $\$P$ is invested at the end of each year in an annuity earning annual interest at rate r, then the amount in the account will be $\$A$ after n years, where

$$n = \frac{\log\left[\frac{Ar}{P} + 1\right]}{\log(1 + r)}$$

If \$1000 is invested each year in an annuity earning 12% annual interest, how long will it take for the account be worth \$20,000? 10.8 yr

26. **Time for money to grow tenfold** If \$5000 is invested each year in an annuity earning 8% annual interest, how long will it take for the account be worth \$50,000? (See Exercise 25.) 7.6 yr

27. Use the formula $P = P_0e^{rt}$ to verify that P will be twice P_0 when $t = \frac{\ln 2}{r}$.

28. Use the formula $P = P_0e^{rt}$ to verify that P will be three times as large as P_0 when $t = \frac{\ln 3}{r}$.

Writing Exercises *Write a paragraph using your own words.*

1. Explain why an earthquake measuring 7 on the Richter scale is much worse than an earthquake measuring 6.

2. The time it takes money to double at an annual rate of r, compounded continuously, is given by the formula $(\ln 2)/r$. Explain why money doubles more quickly the higher the rate.

Something to Think About

1. Find a formula to find how long it will take money to quadruple.

2. Graph the logistic function $y = \dfrac{1}{1 + e^{-2x}}$ and discuss its graph.

Review Exercises *Let $f(x) = 3x - 2$ and $g(x) = x^2 + 3$. Find each function or value.*

1. $f + g$ $x^2 + 3x + 1$

2. $f - g$ $-x^2 + 3x - 5$

3. $f \cdot g$ $3x^3 - 2x^2 + 9x - 6$

4. g/f $\dfrac{x^2 + 3}{3x - 2}$

5. $(g \circ f)(-2)$ 67

6. $(f \circ g)(x)$ $3x^2 + 7$

11.5 Exponential and Logarithmic Equations

■ Solving Exponential Equations ■ Solving Logarithmic Equations ■ Applications of Exponential and Logarithmic Equations ■ Solving Equations with a Graphing Calculator

An **exponential equation** is an equation that contains a variable in one of its exponents. Some examples of exponential equations are

$$3^x = 5, \qquad 6^{x-3} = 2^x, \qquad \text{and} \qquad 3^{2x+1} - 10(3^x) + 3 = 0$$

A **logarithmic equation** is an equation with a logarithmic expression that contains a variable. Some examples of logarithmic equations are

$$\log 2x = 25, \qquad \ln x - \ln (x - 12) = 24, \qquad \text{and} \qquad \log x = \log \frac{1}{x} + 4$$

In this section, we will learn how to solve many of these equations.

■ Solving Exponential Equations

EXAMPLE 1 Solve the exponential equation $3^x = 5$.

Solution Since logarithms of equal numbers are equal, we can take the common logarithm of each side of the equation. Property 7 of logarithms then provides a way of moving the variable x from its position as an exponent to a position as a coefficient.

$$3^x = 5$$

$$\log 3^x = \log 5 \qquad \text{Take the common logarithm of each side.}$$

$$x \log 3 = \log 5 \qquad \text{Use Property 7 of logarithms.}$$

1. $$x = \frac{\log 5}{\log 3} \qquad \text{Divide both sides by log 3.}$$

$$\approx 1.464973521 \qquad \text{Use a calculator.}$$

To four decimal places, $x = 1.4650$. ■

WARNING! A careless reading of Equation 1 leads to a common error. The right-hand side of Equation 1 calls for a division, not a subtraction.

$$\frac{\log 5}{\log 3} \quad \text{means} \quad (\log 5) \div (\log 3)$$

It is the expression $\log \frac{5}{3}$ that means $\log 5 - \log 3$.

EXAMPLE 2 Solve the exponential equation $6^{x-3} = 2^x$.

Solution

$6^{x-3} = 2^x$	
$\log 6^{x-3} = \log 2^x$	Take the common logarithm of each side.
$(x - 3) \log 6 = x \log 2$	Use Property 7 of logarithms.
$x \log 6 - 3 \log 6 = x \log 2$	Use the distributive property.
$x \log 6 - x \log 2 = 3 \log 6$	Add 3 log 6 and subtract x log 2 from both sides.
$x(\log 6 - \log 2) = 3 \log 6$	Factor out x on the left-hand side.
$x = \dfrac{3 \log 6}{\log 6 - \log 2}$	Divide both sides by log 6 − log 2.
$x \approx 4.892789261$	Use a calculator.

■

EXAMPLE 3 Solve the exponential equation $2^{x^2+2x} = \frac{1}{2}$.

Solution Since $\frac{1}{2} = 2^{-1}$, we can write the equation in the form

$$2^{x^2+2x} = 2^{-1}$$

Since equal quantities with equal bases have equal exponents, we have

$x^2 + 2x = -1$	
$x^2 + 2x + 1 = 0$	Add 1 to both sides.
$(x + 1)(x + 1) = 0$	Factor the trinomial.
$x + 1 = 0$ or $x + 1 = 0$	Set each factor equal to 0.
$x = -1$ | $x = -1$	

Verify that -1 satisfies the equation. ■

Solving Logarithmic Equations

EXAMPLE 4 Solve the logarithmic equation $\log x + \log (x - 3) = 1$.

Solution

$\log x + \log (x - 3) = 1$	
$\log x(x - 3) = 1$	Use Property 5 of logarithms.
$x(x - 3) = 10^1$	Use the definition of logarithms to change the equation to exponential form.
$x^2 - 3x - 10 = 0$	Remove parentheses and subtract 10 from both sides.
$(x + 2)(x - 5) = 0$	Factor the trinomial.
$x + 2 = 0$ or $x - 5 = 0$	Set each factor equal to 0.
$x = -2$ | $x = 5$	

Check: The number -2 is not a solution, because it does not satisfy the equation (a negative number does not have a logarithm). We will check the remaining number, 5.

$$\log x + \log (x - 3) = 1$$
$$\log 5 + \log (5 - 3) \stackrel{?}{=} 1 \qquad \text{Substitute 5 for } x.$$
$$\log 5 + \log 2 \stackrel{?}{=} 1$$
$$\log 10 \stackrel{?}{=} 1 \qquad \text{Use Property 5 of logarithms.}$$
$$1 = 1 \qquad \log 10 = 1.$$

Since 5 satisfies the equation, it is a solution. ■

EXAMPLE 5 Solve the logarithmic equation $\log_b (3x + 2) - \log_b (2x - 3) = 0$.

Solution

$$\log_b (3x + 2) - \log_b (2x - 3) = 0$$
$$\log_b (3x + 2) = \log_b (2x - 3) \qquad \text{Add } \log_b (2x - 3) \text{ to both sides.}$$
$$3x + 2 = 2x - 3 \qquad \text{If } \log_b r = \log_b s, \text{ then } r = s.$$
$$x = -5 \qquad \text{Subtract } 2x \text{ and 2 from both sides.}$$

Check:

$$\log_b (3x + 2) - \log_b (2x - 3) = 0$$
$$\log_b [3(-5) + 2] - \log_b [2(-5) - 3] \stackrel{?}{=} 0$$
$$\log_b (-13) - \log_b (-13) \stackrel{?}{=} 0$$

Since the logarithm of a negative number does not exist, -5 is an extraneous solution and must be discarded. This equation has no roots. ■

EXAMPLE 6 Solve the logarithmic equation $\dfrac{\log (5x - 6)}{\log x} = 2$.

Solution We can multiply both sides of the equation by $\log x$ to get

$$\log (5x - 6) = 2 \log x$$

and apply Property 7 of logarithms to get

$$\log (5x - 6) = \log x^2$$

By Property 8 of logarithms, $5x - 6 = x^2$, because they have equal logarithms. Thus,

$$5x - 6 = x^2$$
$$0 = x^2 - 5x + 6$$
$$0 = (x - 3)(x - 2)$$

$$x - 3 = 0 \quad \text{or} \quad x - 2 = 0$$
$$x = 3 \quad | \quad x = 2$$

Verify that both 2 and 3 satisfy the equation. ■

Applications of Exponential and Logarithmic Equations

Carbon-14 dating Experiments have determined the time it takes for half of a sample of a given radioactive material to decompose. This time is a constant, called the material's **half-life.**

When a living organism dies, the oxygen/carbon dioxide cycle common to all living things ceases, and carbon-14, a radioactive isotope with a half-life of 5700 years, is no longer absorbed. By measuring the amount of carbon-14 present in an ancient object, archeologists can estimate the object's age by using the radioactive decay formula.

Radioactive Decay Formula If A is the amount of radioactive material present at time t, A_0 was the amount present at $t = 0$, and h is the material's half-life, then

$$A = A_0 2^{-t/h}$$

EXAMPLE 7 How old is a wooden statue that contains only one-third of its original carbon-14 content?

Solution To find the time t when $A = \frac{1}{3} A_0$, we substitute $\frac{A_0}{3}$ for A and 5700 for h into the radioactive decay formula and solve for t:

$$A = A_0 2^{-t/h}$$

$$\frac{A_0}{3} = A_0 2^{-t/5700}$$

$$A_0 = 3A_0 2^{-t/5700} \quad \text{Multiply both sides by 3.}$$

$$1 = 3(2^{-t/5700}) \quad \text{Divide both sides by } A_0.$$

$$\log 1 = \log 3(2^{-t/5700}) \quad \text{Take the common logarithm of each side.}$$

$$0 = \log 3 + \log 2^{-t/5700} \quad \text{Use Properties 1 and 5 of logarithms.}$$

$$-\log 3 = -\frac{t}{5700} \log 2 \quad \text{Subtract log 3 from both sides and use Property 7 of logarithms.}$$

$$5700\left(\frac{\log 3}{\log 2}\right) = t \quad \text{Multiply both sides by } -\tfrac{5700}{\log 2}.$$

$$t \approx 9034.286254 \quad \text{Use a calculator.}$$

The wooden statue is approximately 9000 years old. ■

Population growth When there is sufficient food and space, populations of living organisms tend to increase exponentially according to the Malthusian growth model.

Malthusian Growth Model If P is the population at some time t, P_0 is the initial population at $t = 0$, and k depends on the rate of growth, then

$$P = P_0e^{kt}$$

EXAMPLE 8 The bacteria in a laboratory culture increased from an initial population of 500 to 1500 in 3 hours. How long will it take for the population to reach 10,000?

Solution We substitute 500 for P_0, 1500 for P, and 3 for t and simplify to find k:

$$P = P_0e^{kt}$$

$$1500 = 500(e^{k3}) \quad \text{Substitute 1500 for } P\text{, 500 for } P_0\text{, and 3 for } t.$$

$$3 = e^{3k} \quad \text{Divide both sides by 500.}$$

$$\ln 3 = \ln e^{3k} \quad \text{Take the natural logarithm of both sides.}$$

$$\ln 3 = 3k \quad \ln e^{3k} = 3k \ln e = 3k.$$

$$k = \frac{\ln 3}{3} \quad \text{Divide both sides by 3.}$$

To find when the population will reach 10,000, we substitute 10,000 for P, 500 for P_0, and $\frac{\ln 3}{3}$ for k in the equation $P = P_0e^{kt}$ and solve for t:

$$P = P_0e^{kt}$$

$$10{,}000 = 500e^{[(\ln 3)/3]t}$$

$$20 = e^{[(\ln 3)/3]t} \quad \text{Divide both sides by 500.}$$

$$\left(\frac{\ln 3}{3}\right)t = \ln 20 \quad \text{Change the equation to logarithmic form.}$$

$$t = \frac{3 \ln 20}{\ln 3} \quad \text{Multiply both sides by } \tfrac{3}{\ln 3}.$$

$$\approx 8.180499084 \quad \text{Use a calculator.}$$

The culture will reach 10,000 bacteria in a little more than 8 hours. ■

Solving Equations with a Graphing Calculator

EXAMPLE 9 Use a graphing calculator to solve the equation $\log x + \log (x - 3) = 1$.

Solution We can subtract 1 from both sides of the equation to get

$$\log x + \log (x - 3) - 1 = 0$$

and graph the corresponding function

$$y = \log x + \log (x - 3) - 1$$

in the viewing window, using $0 \leq x \leq 20$ and $-2 \leq y \leq 2$, to obtain the graph shown in Figure 11-9.

Since the root of the equation is given by the x-intercept of the graph, we can find the root by zooming in on the value of the x-intercept. The root is $x = 5$.

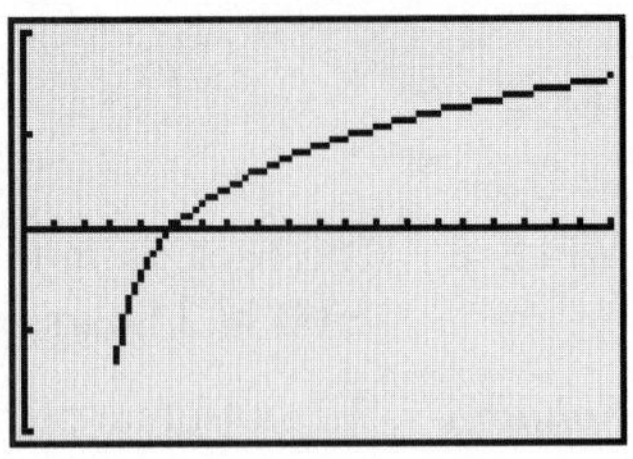

FIGURE 11-9

ORALS *Solve each equation for x. Do not simplify answers.*

1. $3^x = 5$ $x = \frac{\log 5}{\log 3}$
2. $5^x = 3$ $x = \frac{\log 3}{\log 5}$
3. $2^{-x} = 7$ $x = -\frac{\log 7}{\log 2}$
4. $6^{-x} = 2$ $x = -\frac{\log 2}{\log 6}$
5. $\log 2x = \log (x + 2)$ $x = 2$
6. $\log 2x = 0$ $x = \frac{1}{2}$
7. $\log x^4 = 4$ $x = 10$
8. $\log \sqrt{x} = \frac{1}{2}$ $x = 10$

EXERCISE 11.5

In Exercises 1–20, solve each exponential equation. Give all answers to four decimal places.

1. $4^x = 5$ 1.1610
2. $7^x = 12$ 1.2770
3. $13^{x-1} = 2$ 1.2702
4. $5^{x+1} = 3$ −0.3174
5. $2^{x+1} = 3^x$ 1.7095
6. $5^{x-3} = 3^{2x}$ −8.2144
7. $2^x = 3^x$ 0
8. $3^{2x} = 4^x$ 0
9. $7^{x^2} = 10$ ± 1.0878
10. $8^{x^2} = 11$ ± 1.0738
11. $8^{x^2} = 9^x$ 0, 1.0566
12. $5^{x^2} = 2^{5x}$ 2.1534
13. $2^{x^2-2x} = 8$ 3, −1
14. $3^{x^2-3x} = 81$ 4, −1
15. $3^{x^2+4x} = \frac{1}{81}$ −2, −2
16. $7^{x^2+3x} = \frac{1}{49}$ −2, −1
17. $4^{x+2} - 4^x = 15$ (*Hint:* $4^{x+2} = 4^x4^2$.) 0
18. $3^{x+3} + 3^x = 84$ (*Hint:* $3^{x+3} = 3^x3^3$.) 1
19. $2(3^x) = 6^{2x}$ 0.2789
20. $2(3^{x+1}) = 3(2^{x-1})$ −3.4190

In Exercises 21–50, solve each logarithmic equation.

21. $\log 2x = \log 4$ 2
22. $\log 3x = \log 9$ 3
23. $\log (3x + 1) = \log (x + 7)$ 3
24. $\log (x^2 + 4x) = \log (x^2 + 16)$ 4
25. $\log (2x - 3) - \log (x + 4) = 0$ 7
26. $\log (3x + 5) - \log (2x + 6) = 0$ 1
27. $\log \frac{4x + 1}{2x + 9} = 0$ 4
28. $\log \frac{5x + 2}{2(x + 7)} = 0$ 4
29. $\log x^2 = 2$ 10, −10
30. $\log x^3 = 3$ 10
31. $\log x + \log (x - 48) = 2$ 50
32. $\log x + \log (x + 9) = 1$ 1
33. $\log x + \log (x - 15) = 2$ 20
34. $\log x + \log (x + 21) = 2$ 4

35. $\log(x + 90) = 3 - \log x$ 10

36. $\log(x - 90) = 3 - \log x$ 100

37. $\log(x - 6) - \log(x - 2) = \log\frac{5}{x}$ 10

38. $\log(2x - 3) - \log(x - 1) = 0$ 2

39. $\log x^2 = (\log x)^2$ 1, 100

40. $\log(\log x) = 1$ 10^{10}

41. $\dfrac{\log(3x - 4)}{\log x} = 2$ no solution

42. $\dfrac{\log(8x - 7)}{\log x} = 2$ 7

43. $\dfrac{\log(5x + 6)}{2} = \log x$ 6

44. $\frac{1}{2}\log(4x + 5) = \log x$ 5

45. $\log_3 x = \log_3\left(\frac{1}{x}\right) + 4$ 9

46. $\log_5(7 + x) + \log_5(8 - x) - \log_5 2 = 2$ 3, −2

47. $2\log_2 x = 3 + \log_2(x - 2)$ 4

48. $2\log_3 x - \log_3(x - 4) = 2 + \log_3 2$ 6, 12

49. $\log(7y + 1) = 2\log(y + 3) - \log 2$ 1, 7

50. $2\log(y + 2) = \log(y + 2) - \log 12$ $-\frac{23}{12}$

In Exercises 51–68, solve each problem.

51. Tritium decay The half-life of tritium is 12.4 years. How long will it take for 25% of a sample of tritium to decompose? 5.1 yr

52. Radioactive decay In two years, 20% of a radioactive element decays. Find its half-life. 6.2 yr

53. Thorium decay An isotope of thorium, ^{227}Th, has a half-life of 18.4 days. How long will it take 80% of the sample to decompose? 42.7 days

54. Lead decay An isotope of lead, ^{201}Pb, has a half-life of 8.4 hours. How many hours ago was there 30% more of the substance? 3.2 hr

55. Carbon-14 dating The bone fragment shown in Illustration 1 contains 60% of the carbon-14 that it is assumed to have had initially. How old is the bone? 4200 yr

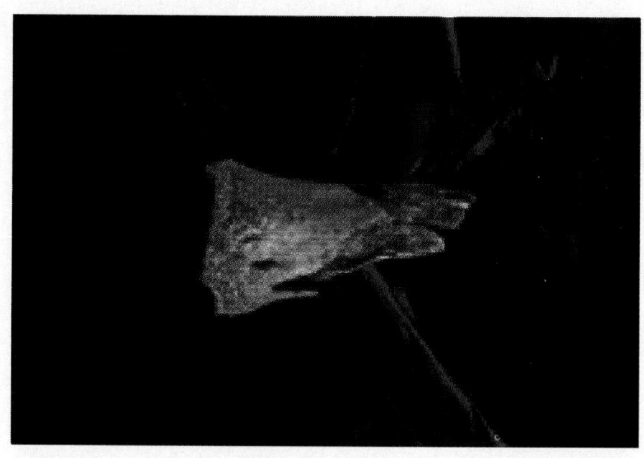

ILLUSTRATION 1

56. Carbon-14 dating Only 10% of the carbon-14 in a small wooden bowl remains. How old is the bowl? 19,000 yr

57. Compound interest If \$500 is deposited in an account paying 8.5% annual interest, compounded semiannually, how long will it take for the account to increase to \$800? 5.6 yr

58. Continuous compound interest In Exercise 57, how long will it take if the interest is compounded continuously? 5.5 yr

59. Compound interest If \$1300 is deposited in a savings account paying 9% interest, compounded quarterly, how long will it take the account to increase to \$2100? 5.4 yr

60. Compound interest A sum of \$5000 deposited in an account grows to \$7000 in 5 years. Assuming continuous compounding, what interest rate is being paid? 6.73%

61. Comparing investments Which of these investment opportunities is best: 4.9% compounded annually, 4.8% compounded quarterly, or 4.7% compounded monthly? 4.9 annually

62. Bacterial growth A bacteria culture grows according to the formula

$$P = P_0 a^t$$

If it takes 5 days for the culture to triple in size, how long will it take to double in size? 3.2 days

63. **Rodent control** The rodent population in a city is currently estimated at 30,000. If it is expected to double every 5 years, when will the population reach 1 million? 25.3 yr

64. **Population growth** The population of a city is expected to triple every 15 years. When can the city planners expect the population to be double the present census of 140 persons? 9.5 yr

65. **Bacteria culture** A bacteria culture doubles in size every 24 hours. By how much will it have increased in 36 hours? 2.828 times larger

66. **Oceanography** The intensity I of a light a distance x meters beneath the surface of a lake deceases exponentially. From Illustration 2, find the depth at which the intensity will be 20%. 27 m

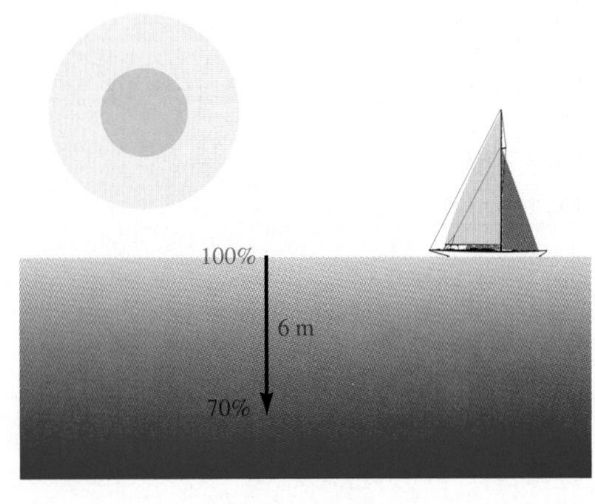

ILLUSTRATION 2

67. **Newton's law of cooling** Water initially at 100°C is left to cool in a room of temperature 60°C. After 3 minutes, the water temperature is 90°. If the water temperature T is a function of time t given by $T = 60 + 40e^{kt}$, find k. $\frac{\ln 0.75}{3}$

68. **Newton's law of cooling** Refer to Exercise 67 and find the time for the water temperature to reach 70°C. 14.5 min

In Exercises 69–72, use a graphing calculator to solve each equation.

69. $\log x + \log (x - 15) = 2$ 20

70. $\log x + \log (x + 3) = 1$ 2

71. $2^{x+1} = 7$ 1.8

72. $\ln (2x + 5) - \ln 3 = \ln (x - 1)$ 8

Writing Exercises *Write a paragraph using your own words.*

1. Explain how to solve the equation $2^x = 7$.

2. Explain how to solve the equation $x^2 = 7$.

Something to Think About

1. Without solving the following equation, find the values of x that cannot be a solution:

$$\log (x - 3) - \log (x^2 + 2) = 0$$

2. Solve the equation $x^{\log x} = 10{,}000$. 100, $\frac{1}{100}$

Review Exercises *Solve each equation.*

1. $5x^2 - 25x = 0$ 0, 5

2. $4y^2 - 25 = 0$ $\frac{5}{2}, -\frac{5}{2}$

3. $3p^2 + 10p = 8$ $\frac{2}{3}, -4$

4. $4t^2 + 1 = -6t$ $\frac{-3 \pm \sqrt{5}}{4}$

PROJECT ■ Napier's Bones

Many 17th-century discoveries in mathematics and science required extensive calculations, and there were no personal computers to help. John Napier (1550–1617) devised several methods to make calculating easier. One device, called **Napier's bones,** consists of rods with numbers printed on the four faces. Illustration 1 shows the faces for the digits 2 and 9. Along the length of each rod are the multiples of the digits, each written in a square with the tens and units digits separated by a diagonal line.

In Illustration 2, we use Napier's bones to multiply 1357 by 253. We place the bones with digits 1, 3, 5, and 7 next to each other, in order. We use the digits of the multiplier 253 to mark rows 2, 5, and 3. Each marked row determines a product: By adding the digits in a diagonal, row 2 determines the digits 2, (6 + 1), (0 + 1), 4, or 2714. This number is added to the others as shown in the illustration: 1357 × 253 = 343,321.

Napier also devised bones for finding square roots and cube roots. The development of logarithms provided another aid to tedious calculation. John Napier, always interested in easing these burdens, was also the inventor of logarithms.

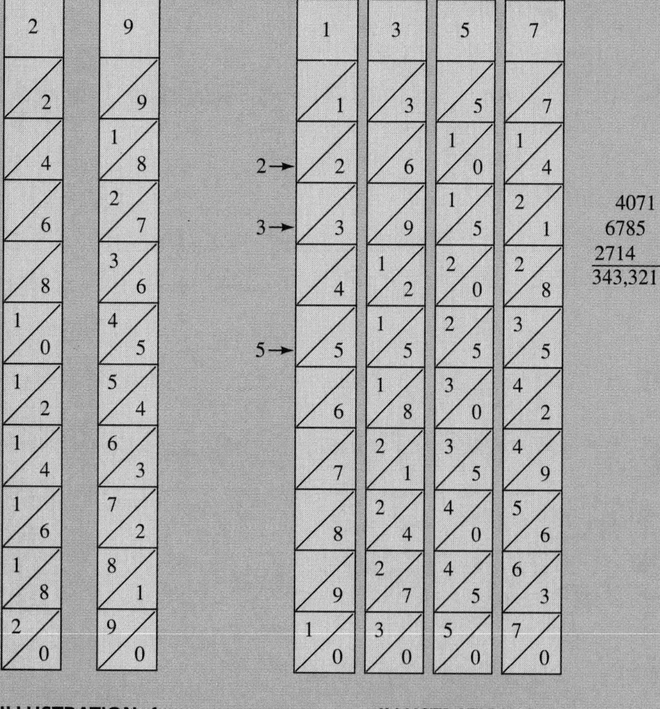

ILLUSTRATION 1   ILLUSTRATION 2

1. Construct a set of Napier's bones.
2. Use Napier's bones to find the products 2315 × 467 and 2359 × 1478.

Chapter Summary

KEY WORDS

change-of-base formula (11.3)
common logarithm (11.2)
compound interest (11.1)
continuous compound interest (11.1)
decreasing functions (11.1)
domain of an exponential function (11.1)
domain of a logarithmic function (11.2)
e (11.1)
exponential decay (11.1)
exponential equation (11.5)
exponential function (11.1)
exponential growth (11.1)
half-life (11.5)
increasing function (11.1)
logarithmic equation (11.5)
logarithmic function (11.2)
Malthusian population growth (11.1, 11.5)
natural logarithm (11.2)
radioactive decay (11.5)
range of an exponential function (11.1)
range of a logarithmic function (11.2)

KEY IDEAS

(11.1) The exponential function $y = b^x$, where $b > 0$, $b \neq 1$, and x is a real number, is one-to-one. Its domain is the set of real numbers, and its range is the set of positive numbers.

$e = 2.718281828\ldots$.

The exponential function $y = e^x$ is one-to-one. Its domain is the set of real numbers, and its range is the set of positive numbers.

(11.2) The logarithmic function $y = \log_b x$, where $b > 0$, $b \neq 1$, and x is a positive number, is one-to-one. Its domain is the set of positive numbers, and its range is the set of real numbers.

$y = \log_b x$ is equivalent to $x = b^y$.

Logarithms of negative numbers do not exist.

The functions defined by $y = \log_b x$ and $y = b^x$ are inverse functions.

Common logarithms are base-10 logarithms.

Natural logarithms are base-e logarithms.

(11.3) **Properties of logarithms:** If b is a positive number and $b \neq 1$, then

1. $\log_b 1 = 0$

2. $\log_b b = 1$

3. $\log_b b^x = x$

4. $b^{\log_b x} = x$

5. $\log_b MN = \log_b M + \log_b N$

6. $\log_b \dfrac{M}{N} = \log_b M - \log_b N$

7. $\log_b M^p = p \log_b M$

8. If $\log_b x = \log_b y$, then $x = y$.

The change-of-base formula: $\log_b y = \dfrac{\log_a y}{\log_a b}$

(11.4–11.5) Many applications lead to exponential and logarithmic equations.

Chapter 11 Review Exercises

In Review Exercises 1–4, graph the function defined by each equation.

1. $y = \left(\frac{6}{5}\right)^x$

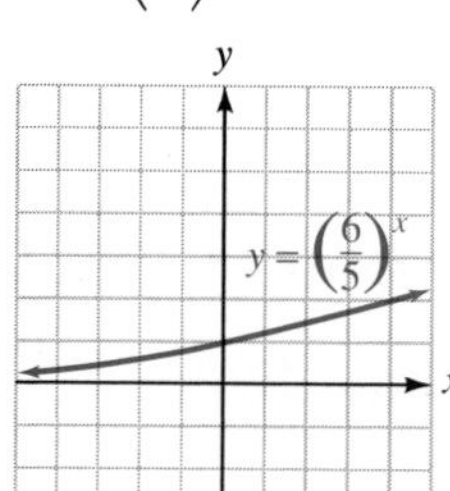

2. $y = e^x$

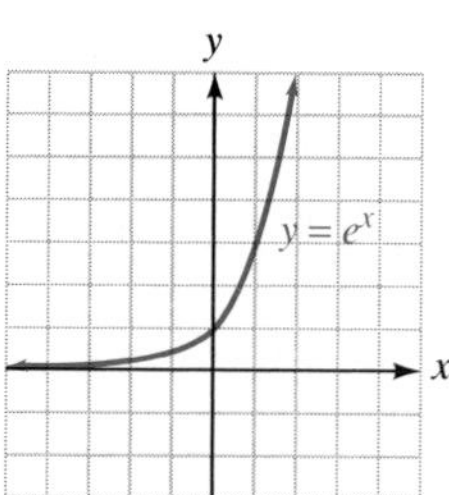

3. $y = \log x$

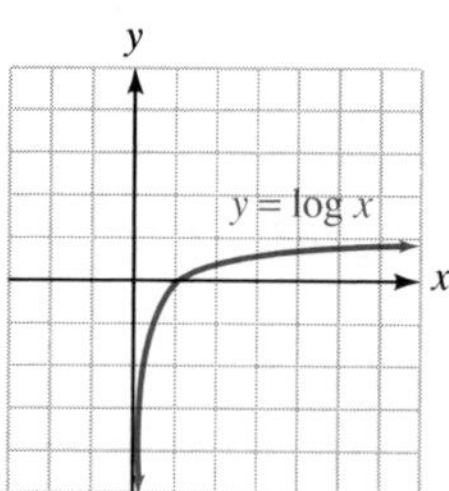

4. $y = \ln x$

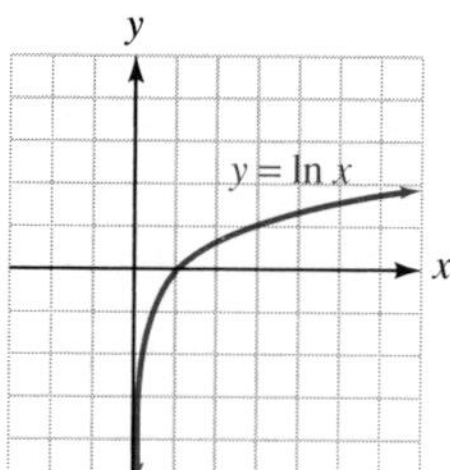

In Review Exercises 5–8, graph each pair of equations on one set of coordinate axes.

5. $y = \left(\frac{1}{3}\right)^x$ and $y = \log_{1/3} x$

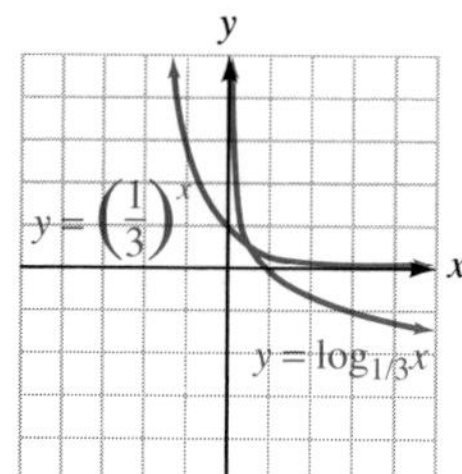

6. $y = \left(\frac{2}{5}\right)^x$ and $y = \log_{2/5} x$

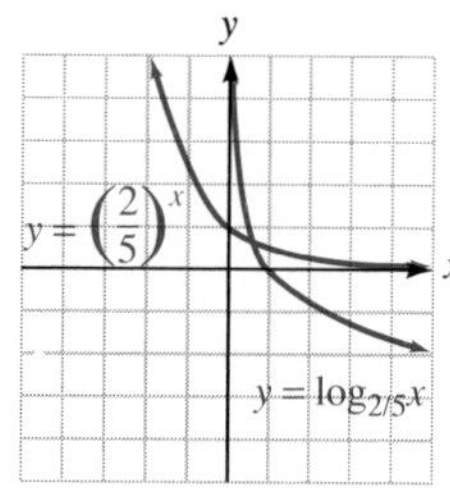

7. $y = 4^x$ and $y = \log_4 x$

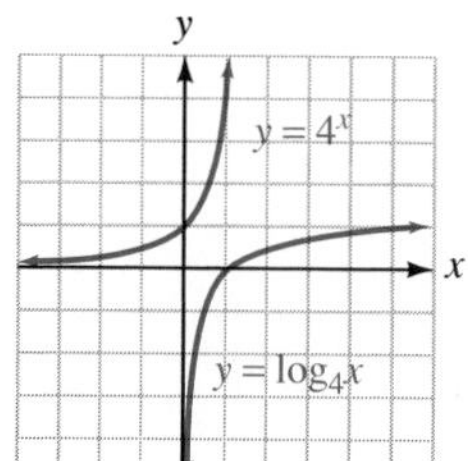

8. $y = 3^x$ and $y = \log_3 x$

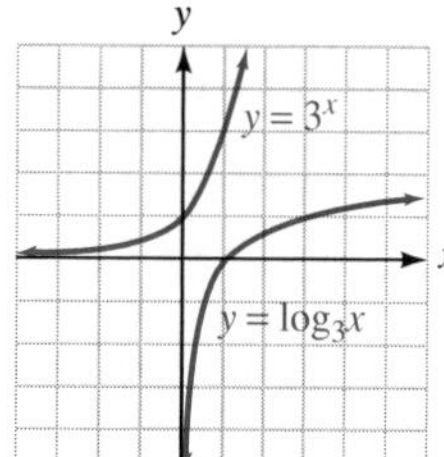

In Review Exercises 9–20, find each value.

9. $\log_3 9$ 2

10. $\log_9 \frac{1}{3}$ $-\frac{1}{2}$

11. $\log_\pi 1$ 0

12. $\log_5 0.04$ -2

13. $\log_a \sqrt{a}$ $\frac{1}{2}$

14. $\log_a \sqrt[3]{a}$ $\frac{1}{3}$

15. $\ln e^4$ 4

16. $\ln 1$ 0

17. $10^{\log_{10} 7}$ 7

18. $e^{\log_e 3}$ 3

19. $\log_b b^4$ 4

20. $\ln e^9$ 9

In Review Exercises 21–44, find x.

21. $\log_2 x = 3$ 8

22. $\log_3 x = -2$ $\frac{1}{9}$

23. $\log_x 9 = 2$ 3

24. $\log_x 0.125 = -3$ 2

25. $\log_7 7 = x$ 1

26. $\log_3 \sqrt{3} = x$ $\frac{1}{2}$

27. $\log_8 \sqrt{2} = x$ $\frac{1}{6}$

28. $\log_6 36 = x$ 2

29. $\log_{1/3} 9 = x$ -2

30. $\log_{1/2} 1 = x$ 0

31. $\log_x 3 = \dfrac{1}{3}$ 27

32. $\log_x 25 = -2$ $\frac{1}{5}$

33. $\log_2 x = 5$ 32

34. $\log_{\sqrt{3}} x = 4$ 9

35. $\log_{\sqrt{3}} x = 6$ 27

36. $\log_{0.1} 10 = x$ -1

37. $\log_x 2 = -\dfrac{1}{3}$ $\frac{1}{8}$

38. $\log_x 32 = 5$ 2

39. $\log_{0.25} x = -1$ 4

40. $\log_{0.125} x = -\dfrac{1}{3}$ 2

41. $\log_{\sqrt{2}} 32 = x$ 10

42. $\log_{\sqrt{5}} x = -4$ $\frac{1}{25}$

43. $\log_{\sqrt{3}} 9\sqrt{3} = x$ 5

44. $\log_{\sqrt{5}} 5\sqrt{5} = x$ 3

In Review Exercises 45–46, write each expression in terms of the logarithms of x, y, and z.

45. $\log_b \dfrac{x^2y^3}{z^4}$ $2 \log_b x + 3 \log_b y - 4 \log_b z$

46. $\log_b \sqrt{\dfrac{x}{yz^2}}$ $\frac{1}{2} \log_b x - \frac{1}{2} \log_b y - \log_b z$

In Review Exercises 47–48, write each expression as the logarithm of a single quantity.

47. $3 \log_b x - 5 \log_b y + 7 \log_b z$ $\log_b \dfrac{x^3z^7}{y^5}$

48. $\dfrac{1}{2} \log_b x + 3 \log_b y - 7 \log_b z$ $\log_b \dfrac{y^3\sqrt{x}}{z^7}$

In Review Exercises 49–52, assume that $\log a = 0.6$, $\log b = 0.36$, *and* $\log c = 2.4$. *Find each value.*

49. $\log abc$ 3.36

50. $\log a^2b$ 1.56

51. $\log \dfrac{ac}{b}$ 2.64

52. $\log \dfrac{a^2}{c^3b^2}$ -6.72

In Review Exercises 53–66, solve for x, where possible.

53. $3^x = 7$ $\frac{\log 7}{\log 3}$

54. $5^{x+2} = 625$ 2

55. $2^x = 3^{x-1}$ $\frac{\log 3}{\log 3 - \log 2}$

56. $2^{x^2+4x} = \dfrac{1}{8}$ $-3, -1$

57. $\log x + \log (29 - x) = 2$ 25, 4

58. $\log_2 x + \log_2 (x - 2) = 3$ 4

59. $\log_2 (x + 2) + \log_2 (x - 1) = 2$ 2

60. $\dfrac{\log (7x - 12)}{\log x} = 2$ 4, 3

61. $\log x + \log (x - 5) = \log 6$ 6

62. $\log 3 - \log (x - 1) = -1$ 31

63. $e^{x \ln 2} = 9$ $\frac{\ln 9}{\ln 2}$

64. $\ln x = \ln (x - 1)$ no solution

65. $\ln x = \ln (x - 1) + 1$ $\dfrac{e}{e - 1}$

66. $\ln x = \log_{10} x$ 1
(*Hint:* Use the change-of-base formula.)

In Review Exercises 67–70, solve each problem.

67. **Carbon-14 dating** A wooden statue excavated in Egypt has a carbon-14 content that is two-thirds of that found in living wood. If the half-life of carbon-14 is 5700 years, how old is the statue? about 3300 yr

68. **pH of grapefruit** The pH of grapefruit juice is approximately 3.1. Find its hydrogen ion concentration. about 7.94×10^{-4} gram-ions/liter

69. **Radioactive decay** One-third of a radioactive material decays in 20 years. Find its half-life. about 34.2 yr

70. **Formula for pH** Some chemistry textbooks define the pH of a solution with the formula

$$\text{pH} = \log_{10} \frac{1}{[\text{H}^+]}$$

Show that this definition is equivalent to the one given in this book.

Chapter 11 Test

In Questions 1–2, graph each function.

1. $y = 2^x + 1$

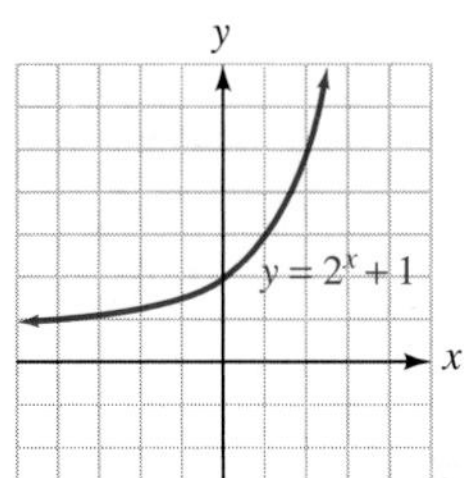

2. $y = 2^{-x}$

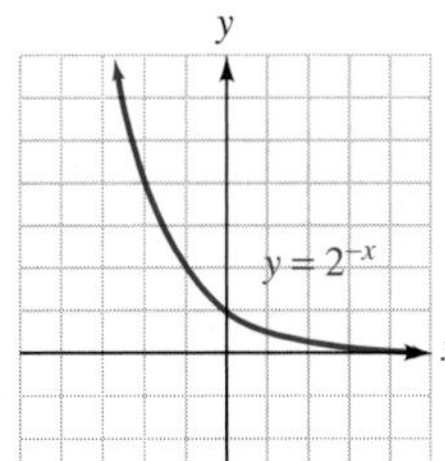

In Questions 3–4, solve each equation.

3. A radioactive material decays according to the formula $A = A_0(2)^{-t}$. How much of a 3-gram sample will be left in 6 years? $\frac{3}{64}$ gm

4. An initial deposit of \$1000 earns 6% interest, compounded twice a year. How much will be in the account in one year? \$1060.90

5. Graph the function $y = e^x$.

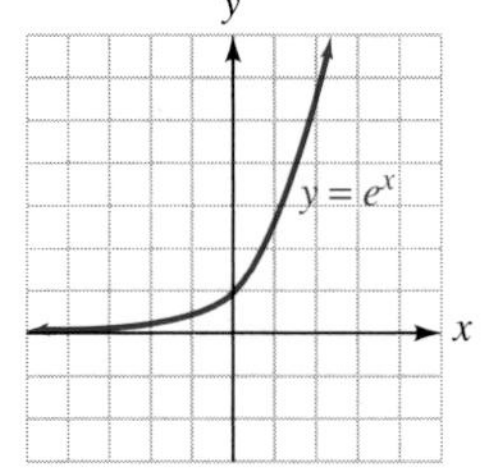

6. An account contains \$2000 and has been earning 8% interest, compounded continuously. How much will be in the account in 10 years? \$4451.08

In Questions 7–12, find x.

7. $\log_4 16 = x$ 2

8. $\log_x 81 = 4$ 3

9. $\log_3 x = -3$ $\frac{1}{27}$

10. $\log_x 100 = 2$ 10

11. $\log_{3/2} \frac{9}{4} = x$ 2

12. $\log_{2/3} x = -3$ $\frac{27}{8}$

In Questions 13–14, graph each function.

13. $y = -\log_3 x$

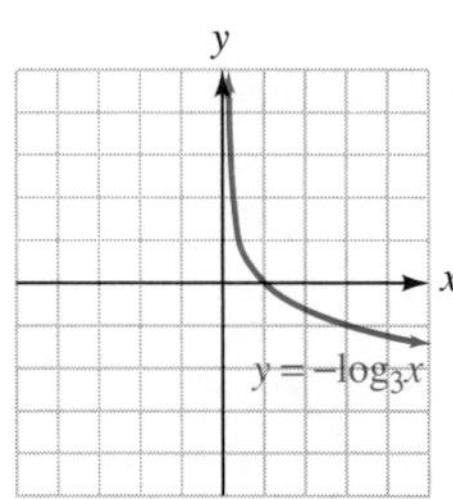

14. $y = \ln x$

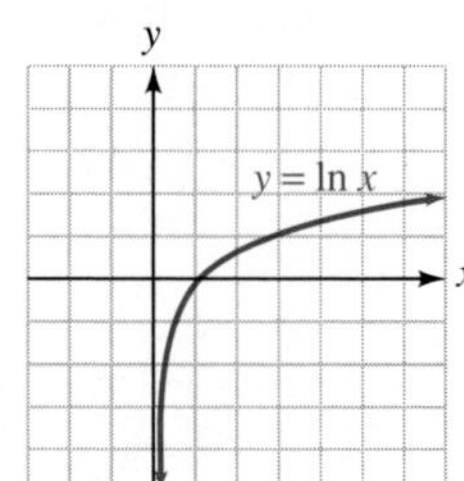

In Questions 15–16, write each expression in terms of the logarithms of a, b, and c.

15. $\log a^2bc^3$ $2 \log a + \log b + 3 \log c$

16. $\ln \sqrt{\frac{a}{b^2c}}$ $\frac{1}{2}(\ln a - 2 \ln b - \ln c)$

In Questions 17–18, write each expression as a logarithm of a single quantity.

17. $\frac{1}{2} \log (a + 2) + \log b - 2 \log c$ $\log \frac{b\sqrt{a+2}}{c^2}$

18. $\frac{1}{3}(\log a - 2 \log b) - \log c$ $\log \frac{\sqrt[3]{a}}{c\sqrt[3]{b^2}}$

In Questions 19–20, assume that $\log 2 = 0.3010$ *and* $\log 3 = 0.4771$. *Find each value.*

19. $\log 24$ 1.3801

20. $\log \frac{8}{3}$ 0.4259

In Questions 21–22, use the change-of-base formula to find each logarithm. Do not attempt to simplify the answer.

21. $\log_7 3$ $\frac{\log 3}{\log 7}$ or $\frac{\ln 3}{\ln 7}$

22. $\log_\pi e$ $\frac{\log e}{\log \pi}$ or $\frac{1}{\ln \pi}$

In Questions 23–26, tell whether each statement is true.

23. $\log_a ab = 1 + \log_a b$ true

24. $\frac{\log a}{\log b} = \log a - \log b$ false

25. $\log a^{-3} = \frac{1}{3 \log a}$ false

26. $\ln (-x) = -\ln x$ false

27. Find the pH of a solution with a hydrogen ion concentration of 3.7×10^{-7}. (*Hint:* pH = $-\log [H^+]$.) 6.4

28. Find the db gain of an amplifier when $E_O = 60$ volts and $E_I = 0.3$ volt. (*Hint:* db gain = $20 \log (E_O/E_I)$.) 46

In Questions 29–30, solve each equation. Do not simplify the logarithms.

29. $5^x = 3$ $\frac{\log 3}{\log 5}$

30. $3^{x-1} = 100^x$ $\frac{\log 3}{(\log 3) - 2}$

In Questions 31–32, solve each equation.

31. $\log (5x + 2) = \log (2x + 5)$ 1

32. $\log x + \log (x - 9) = 1$ 10

CAREERS & MATHEMATICS

Medical Laboratory Worker

Medical laboratory workers, also called clinical laboratory workers, include three levels of personnel: medical technologists, technicians, and assistants. They perform laboratory tests on specimens taken from patients by other health professionals, such as physicians.

Qualifications

The minimum educational requirement for a beginning job as a medical technologist is 4 years of college, including completion of a training program in medical technology.

Medical laboratory technicians acquire their training by enrolling in accredited 2-year programs offered by community and junior colleges, colleges, and universities. Some are trained in the armed forces.

Most lab assistants are either trained on the job or complete a one-year training program in a hospital or community college.

Job Outlook

Employment of medical laboratory workers is expected to expand at a rate about average for all occupations through the year 2005, as physicians continue to use laboratory tests in routine physical checkups and in the diagnosis and treatment of disease.

Example Application

During bacterial reproduction, the time required for a population to double is called the **generation time.** If B bacteria are inoculated into a medium, then after the generation time of the organism has elapsed, there are $2B$ cells. After another generation, there are $2(2B)$ or $4B$ cells, and so on. After n generations, the number of cells present will be

$$b = B \cdot 2^n$$

To find the number of generations that have passed while the population grows from B bacteria to b bacteria, we solve for n.

$$\log b = \log(B \cdot 2^n) \quad \text{Take the common logarithm of both sides.}$$

$$\log b = \log B + n \log 2 \quad \text{Apply Properties 5 and 7 of logarithms.}$$

$$\log b - \log B = n \log 2 \quad \text{Add } -\log B \text{ to both sides.}$$

$$n = \frac{\log b - \log B}{\log 2} \quad \text{Divide both sides by } \log 2.$$

If we substitute 0.301 for log 2 and simplify, we obtain the formula that gives the number of generations that have passed while the population grew from B bacteria to b bacteria.

$$n = \frac{\log b - \log B}{0.301}$$

$$n = \frac{1}{0.301} \log \frac{b}{B}$$

1. $$n = 3.3 \log \frac{b}{B}$$

The generation time G is given by the formula

$$G = \frac{t}{n} = \frac{t}{3.3 \log \frac{b}{B}}$$

where t is the length of the time of growth.

If a medium is inoculated with a bacterial culture that contains 1000 cells per milliliter, how many generations will have passed by the time the culture has grown to a population of 1 million cells per milliliter? What is the generation time if the culture is 10 hours old?

Solution To find the number of generations that have passed, substitute 1000 for B and 1,000,000 for b into Equation 1 and solve for n.

$$n = 3.3 \log \frac{1{,}000{,}000}{1000}$$

$n = 3.3 \log 1000$ Simplify.

$n = 3.3(3)$ $\log 1000 = 3$.

$n = 9.9$ Simplify.

Approximately 10 generations have passed.

To find the generation time, divide 10 hours by 9.9.

$$G = \frac{10}{9.9} \approx 1$$

The generation time is approximately 1 hour.

EXERCISES

1. If a medium is inoculated with a bacterial culture containing 500 cells per milliliter, how many generations have passed by the time the culture contains 5×10^6 cells per milliliter?
2. What is the generation time in Exercise 1 if the time period is 26.5 hours?
3. It takes 12 hours for a culture containing 300 cells per milliliter to grow to contain 7×10^6 cells per milliliter. What is the generation time?
4. Show that in any bacterial culture, the per-milliliter cell count will grow by a factor of k in $3.3 \log k$ generations.

(*Answers:* **1.** 13.2 **2.** 2 hours **3.** 0.83 hour)

12 Miscellaneous Topics

In this chapter, we introduce several topics that have applications in advanced mathematics and in certain occupational areas. The binomial theorem, permutations, and combinations are used in statistics. Arithmetic sequences and geometric sequences are used in the mathematics of finance.

12.1 The Binomial Theorem

■ Pascal's Triangle ■ Factorial Notation ■ The Binomial Theorem

We have discussed how to raise binomials to positive integral powers. For example, we know that

$$(a+b)^2 = a^2 + 2ab + b^2$$

and that

$$\begin{aligned}(a+b)^3 &= (a+b)(a+b)^2\\ &= (a+b)(a^2+2ab+b^2)\\ &= a^3 + 2a^2b + ab^2 + ba^2 + 2ab^2 + b^3\\ &= a^3 + 3a^2b + 3ab^2 + b^3\end{aligned}$$

To show how to raise binomials to positive integral powers without doing the actual multiplications, we consider the following binomial expansions:

$$\begin{aligned}(a+b)^0 &= 1\\ (a+b)^1 &= a+b\\ (a+b)^2 &= a^2+2ab+b^2\\ (a+b)^3 &= a^3+3a^2b+3ab^2+b^3\\ (a+b)^4 &= a^4+4a^3b+6a^2b^2+4ab^3+b^4\\ (a+b)^5 &= a^5+5a^4b+10a^3b^2+10a^2b^3+5ab^4+b^5\\ (a+b)^6 &= a^6+6a^5b+15a^4b^2+20a^3b^3+15a^2b^4+6ab^5+b^6\end{aligned}$$

Several patterns appear in these expansions:

1. Each expansion has one more term than the power of the binomial.
2. The degree of each term in each expansion is equal to the exponent of the binomial that is being expanded.
3. The first term in each expansion is a, raised to the power of the binomial.
4. The exponents of a decrease by 1 in each successive term. The exponents of b, beginning with $b^0 = 1$ in the first term, increase by 1 in each successive term. Thus, the variables have the pattern

$$a^n,\ a^{n-1}b,\ a^{n-2}b^2,\ \ldots,\ ab^{n-1},\ b^n$$

Blaise Pascal
(1623–1662)
Pascal was torn between the fields of religion and mathematics. Each surfaced at times in his life to dominate his interest. In mathematics, Pascal made contributions to the study of conic sections, probability, and differential calculus. At the age of 19, he invented a calculating machine. He is best known for a triangular array of numbers that bears his name.

Pascal's Triangle

To see another pattern, we write the coefficients of each binomial expansion in the following triangular array:

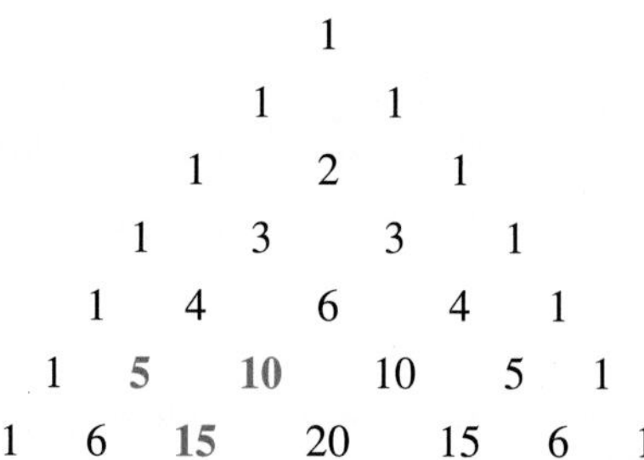

In this array, called **Pascal's triangle,** each entry between the 1's is the sum of the closest pair of numbers in the line immediately above it. For example, the first 15 in the bottom row is the sum of the 5 and 10 immediately above it. Pascal's triangle continues with the same pattern forever. The next two lines are

1 7 21 35 35 21 7 1
1 8 28 56 70 56 28 8 1

EXAMPLE 1 Expand $(x + y)^5$.

Solution The first term in the expansion is x^5, and the exponents of x decrease by 1 in each successive term. A y first appears in the second term, and the exponents of y increase by 1 in each successive term, concluding when the term y^5 is reached. Thus, the variables in the expansion are

$$x^5, \quad x^4y, \quad x^3y^2, \quad x^2y^3, \quad xy^4, \quad y^5$$

The coefficients of these variables are given in Pascal's triangle in the row whose second entry is 5:

1 5 10 10 5 1

Putting these two pieces of information together gives the required expansion:

$$(x + y)^5 = x^5 + 5x^4y + 10x^3y^2 + 10x^2y^3 + 5xy^4 + y^5$$ ■

EXAMPLE 2 Expand $(u - v)^4$.

Solution We note that the expression $(u - v)^4$ can be written in the form $[u + (-v)]^4$. The variables in this expansion are

$$u^4, \quad u^3(-v), \quad u^2(-v)^2, \quad u(-v)^3, \quad (-v)^4$$

and the coefficients are given in Pascal's triangle in the row whose second entry is 4:

$$1 \quad 4 \quad 6 \quad 4 \quad 1$$

Hence, the required expansion is

$$\begin{aligned}(u - v)^4 &= u^4 + 4u^3(-v) + 6u^2(-v)^2 + 4u(-v)^3 + (-v)^4 \\ &= u^4 - 4u^3v + 6u^2v^2 - 4uv^3 + v^4\end{aligned}$$

■

Factorial Notation

Although Pascal's triangle gives the coefficients of the terms in a binomial expansion, it is not the best way to expand a binomial. To develop another way to expand a binomial, we introduce **factorial notation.**

Factorial Notation If n is a natural number, the symbol $n!$ (read as ***n* factorial** or as **factorial *n***) is defined as

$$n! = n(n - 1)(n - 2)(n - 3) \cdot \cdot \cdot (3)(2)(1)$$

EXAMPLE 3 Find **a.** $2!$, **b.** $5!$, **c.** $-9!$, and **d.** $(n - 2)!$.

Solution

a. $2! = 2 \cdot 1 = 2$

b. $5! = 5 \cdot 4 \cdot 3 \cdot 2 \cdot 1 = 120$

c. $-9! = -9 \cdot 8 \cdot 7 \cdot 6 \cdot 5 \cdot 4 \cdot 3 \cdot 2 \cdot 1 = -362{,}880$

d. $(n - 2)! = (n - 2)(n - 3)(n - 4) \cdot \cdot \cdot \cdot \cdot 3 \cdot 2 \cdot 1$

WARNING! According to the previous definition, part **d** is meaningful only if $n - 2$ is a natural number. ■

We define zero factorial as follows.

Zero Factorial $0! = 1$

Classroom Activity: Have students discover the largest n for which $n!$ can be found with a calculator.

We note that

$$5 \cdot 4! = 5 \cdot 4 \cdot 3 \cdot 2 \cdot 1 = 5!$$
$$7 \cdot 6! = 7 \cdot 6 \cdot 5 \cdot 4 \cdot 3 \cdot 2 \cdot 1 = 7!$$
$$10 \cdot 9! = 10 \cdot 9 \cdot 8 \cdot 7 \cdot 6 \cdot 5 \cdot 4 \cdot 3 \cdot 2 \cdot 1 = 10!$$

These examples suggest the following theorem.

Theorem If n is a positive integer, then $n(n-1)! = n!$.

The Binomial Theorem

We now state the binomial theorem.

The Binomial Theorem If n is any positive integer, then

$$(a+b)^n = a^n + \frac{n!}{1!(n-1)!}a^{n-1}b + \frac{n!}{2!(n-2)!}a^{n-2}b^2 + \frac{n!}{3!(n-3)!}a^{n-3}b^3 + \cdots + \frac{n!}{r!(n-r)!}a^{n-r}b^r + \cdots + b^n$$

In the binomial theorem, the exponents of the variables follow the familiar pattern:

- The sum of the exponents of a and b in each term is n,
- the exponents of a decrease, and
- the exponents of b increase.

Only the method of finding the coefficients is different. Except for the first and last terms, the numerator of each coefficient is $n!$. If the exponent of b in a particular term is r, the denominator of the coefficient of that term is $r!(n-r)!$.

EXAMPLE 4 Use the binomial theorem to expand $(a+b)^3$.

Solution We can substitute directly into the binomial theorem and simplify:

$$\begin{aligned}(a+b)^3 &= a^3 + \frac{3!}{1!(3-1)!}a^2b + \frac{3!}{2!(3-2)!}ab^2 + b^3 \\ &= a^3 + \frac{3!}{1!2!}a^2b + \frac{3!}{2!1!}ab^2 + b^3 \\ &= a^3 + \frac{3\cdot 2\cdot 1}{1\cdot 2\cdot 1}a^2b + \frac{3\cdot 2\cdot 1}{2\cdot 1\cdot 1}ab^2 + b^3 \\ &= a^3 + 3a^2b + 3ab^2 + b^3\end{aligned}$$

■

EXAMPLE 5 Use the binomial theorem to expand $(x - y)^4$.

Solution We can write $(x - y)^4$ in the form $[x + (-y)]^4$, substitute directly into the binomial theorem, and simplify:

$$\begin{aligned}(x - y)^4 &= [x + (-y)]^4 \\ &= x^4 + \frac{4!}{1!(4-1)!}x^3(-y) + \frac{4!}{2!(4-2)!}x^2(-y)^2 + \frac{4!}{3!(4-3)!}x(-y)^3 + (-y)^4 \\ &= x^4 - \frac{4 \cdot 3!}{1!3!}x^3y + \frac{4 \cdot 3 \cdot 2!}{2!2!}x^2y^2 - \frac{4 \cdot 3!}{3!1!}xy^3 + y^4 \\ &= x^4 - 4x^3y + 6x^2y^2 - 4xy^3 + y^4\end{aligned}$$

■

EXAMPLE 6 Use the binomial theorem to expand $(3u - 2v)^4$.

Solution We write $(3u - 2v)^4$ in the form $[3u + (-2v)]^4$ and let $a = 3u$ and $b = -2v$. Then we can use the binomial theorem to expand $(a + b)^4$.

$$\begin{aligned}(a + b)^4 &= a^4 + \frac{4!}{1!(4-1)!}a^3b + \frac{4!}{2!(4-2)!}a^2b^2 + \frac{4!}{3!(4-3)!}ab^3 + b^4 \\ &= a^4 + 4a^3b + 6a^2b^2 + 4ab^3 + b^4\end{aligned}$$

Now we can substitute $3u$ for a and $-2v$ for b and simplify:

$$\begin{aligned}(3u - 2v)^4 &= (3u)^4 + 4(3u)^3(-2v) + 6(3u)^2(-2v)^2 + 4(3u)(-2v)^3 + (-2v)^4 \\ &= 81u^4 - 216u^3v + 216u^2v^2 - 96uv^3 + 16v^4\end{aligned}$$

■

ORALS *Find each value.*

1. 1! 1 **2.** 4! 24 **3.** 0! 1 **4.** 5! 120

Expand each binomial.

5. $(m + n)^2$ $m^2 + 2mn + n^2$ **6.** $(m - n)^2$ $m^2 - 2mn + n^2$

7. $(p + 2q)^2$ $p^2 + 4pq + 4q^2$ **8.** $(2p - q)^2$ $4p^2 - 4pq + q^2$

EXERCISE 12.1

In Exercises 1–20, evaluate each expression.

1. 3! 6 **2.** 7! 5040 **3.** −5! −120 **4.** −6! −720

5. 3! + 4! 30 **6.** 2!(3!) 12 **7.** 3!(4!) 144 **8.** 4! + 4! 48

9. 8(7!) 40,320 **10.** 4!(5) 120 **11.** $\frac{9!}{11!}$ $\frac{1}{110}$ **12.** $\frac{13!}{10!}$ 1716

13. $\frac{49!}{47!}$ 2352 **14.** $\frac{101!}{100!}$ 101 **15.** $\frac{5!}{3!(5-3)!}$ 10 **16.** $\frac{6!}{4!(6-4)!}$ 15

17. $\dfrac{7!}{5!(7-5)!}$ 21

18. $\dfrac{8!}{6!(8-6)!}$ 28

19. $\dfrac{5!(8-5)!}{4!7!}$ $\frac{1}{168}$

20. $\dfrac{6!7!}{(8-3)!(7-4)!}$ 5040

In Exercises 21–36, use the binomial theorem to expand each expression.

21. $(x+y)^2$ $x^2+2xy+y^2$

22. $(x+y)^4$ $x^4+4x^3y+6x^2y^2+4xy^3+y^4$

23. $(x-y)^4$ $x^4-4x^3y+6x^2y^2-4xy^3+y^4$

24. $(x-y)^3$ $x^3-3x^2y+3xy^2-y^3$

25. $(2x+y)^3$ $8x^3+12x^2y+6xy^2+y^3$

26. $(x+2y)^3$ $x^3+6x^2y+12xy^2+8y^3$

27. $(x-2y)^3$ $x^3-6x^2y+12xy^2-8y^3$

28. $(2x-y)^3$ $8x^3-12x^2y+6xy^2-y^3$

29. $(2x+3y)^3$ $8x^3+36x^2y+54xy^2+27y^3$

30. $(3x-2y)^3$ $27x^3-54x^2y+36xy^2-8y^3$

31. $\left(\dfrac{x}{2}-\dfrac{y}{3}\right)^3$ $\dfrac{x^3}{8}-\dfrac{x^2y}{4}+\dfrac{xy^2}{6}-\dfrac{y^3}{27}$

32. $\left(\dfrac{x}{3}+\dfrac{y}{2}\right)^3$ $\dfrac{x^3}{27}+\dfrac{x^2y}{6}+\dfrac{xy^2}{4}+\dfrac{y^3}{8}$

33. $(3+2y)^4$ $81+216y+216y^2+96y^3+16y^4$

34. $(2x+3)^4$ $16x^4+96x^3+216x^2+216x+81$

35. $\left(\dfrac{x}{3}-\dfrac{y}{2}\right)^4$ $\dfrac{x^4}{81}-\dfrac{2x^3y}{27}+\dfrac{x^2y^2}{6}-\dfrac{xy^3}{6}+\dfrac{y^4}{16}$

36. $\left(\dfrac{x}{2}+\dfrac{y}{3}\right)^4$ $\dfrac{x^4}{16}+\dfrac{x^3y}{6}+\dfrac{x^2y^2}{6}+\dfrac{2xy^3}{27}+\dfrac{y^4}{81}$

37. Without referring to the text, write the first ten rows of Pascal's triangle.

38. Find the sum of the numbers in each row of the first ten rows of Pascal's triangle. What is the pattern?
1, 2, 4, 8, 16, 32, 64, 128, 256, 512; the numbers are consecutive powers of 2

39. Find the sum of the numbers in the designated diagonal rows of Pascal's triangle shown in Illustration 1. What is the pattern?
1, 1, 2, 3, 5, 8, 13, . . .; beginning with 2, each number is the sum of the previous two numbers

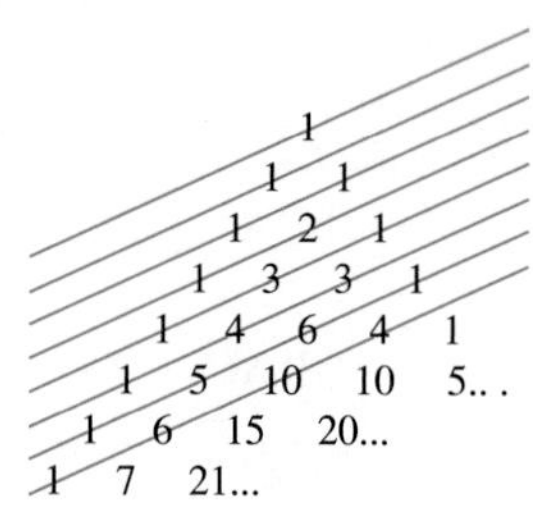

ILLUSTRATION 1

Writing Exercises ■ *Write a paragraph using your own words.*

1. Tell how to construct Pascal's triangle.

2. Tell how to find the variables in the expansion of $(r+s)^4$.

Something to Think About ■

1. If we apply the pattern of the coefficients to the coefficient of the first term in a binomial expansion, the coefficient would be $\frac{n!}{0!(n-0)!}$. Show that this expression is 1.

2. If we apply the pattern of the coefficients to the coefficient of the last term in a binomial expansion, the coefficient would be $\frac{n!}{n!(n-n)!}$. Show that this expression is 1.

Review Exercises ■ *Find each value of x.*

1. $\log_4 16 = x$ 2

2. $\log_x 49 = 2$ 7

3. $\log_{25} x = \dfrac{1}{2}$ 5

4. $\log_{1/2} \dfrac{1}{8} = x$ 3

12.2 The *n*th Term of a Binomial Expansion

To find the fourth term of the expansion of $(a + b)^9$, we could raise the binomial $a + b$ to the ninth power and look at the fourth term. However, this task would be very tedious. By using the binomial theorem, we can construct the fourth term without finding the complete expansion of $(a + b)^9$.

EXAMPLE 1 Find the fourth term in the expansion of $(a + b)^9$.

Solution Since b^1 appears in the second term, b^2 appears in the third term, and so on, the exponent of b in the fourth term is 3. Since the exponent of b added to the exponent of a must equal 9, the exponent of a must be 6. Thus, the variables of the fourth term are

Teaching Tip: Point out that the exponent of the fourth term is 3 because the exponents on *b* lag 1 behind the number of the term.

a^6b^3 The sum of the exponents must be 9.

Because of the binomial theorem, the coefficient of the variables must be

$$\frac{n!}{r!(n-r)!} = \frac{9!}{3!(9-3)!}$$

Thus, the complete fourth term is

$$\frac{9!}{3!(9-3)!}a^6b^3 = \frac{9 \cdot 8 \cdot 7 \cdot 6!}{3 \cdot 2 \cdot 1 \cdot 6!}a^6b^3$$
$$= 84a^6b^3$$

■

EXAMPLE 2 Find the sixth term in the expansion of $(x - y)^7$.

Solution We find the sixth term of $[x + (-y)]^7$. In the sixth term, the exponent of $(-y)$ is 5. Thus, the variables in the sixth term are

$x^2(-y)^5$ The sum of the exponents must be 7.

The coefficient of these variables is

$$\frac{n!}{r!(n-r)!} = \frac{7!}{5!(7-5)!}$$

The complete sixth term is

$$\frac{7!}{5!(7-5)!}x^2(-y)^5 = -\frac{7 \cdot 6 \cdot 5!}{5! \cdot 2 \cdot 1}x^2y^5$$
$$= -21x^2y^5$$

■

EXAMPLE 3 Find the fourth term of the expansion of $(2x - 3y)^6$.

Solution We can let $a = 2x$ and $b = -3y$ and find the fourth term of the expansion of $(a + b)^6$:

$$\frac{6!}{3!(6-3)!}a^3b^3 = \frac{6 \cdot 5 \cdot 4 \cdot 3!}{3 \cdot 2 \cdot 1 \cdot 3!}a^3b^3$$
$$= 20a^3b^3$$

We can now substitute $2x$ for a and $-3y$ for b and simplify:

$$20a^3b^3 = 20(2x)^3(-3y)^3$$
$$= -4320x^3y^3$$

The fourth term in the expansion of $(2x - 3y)^6$ is $-4320x^3y^3$. ■

ORALS *In the expansion of $(x + y)^8$, find the exponent of y in the*

1. third term 2 **2.** fourth term 3 **3.** seventh term 6

In the expansion of $(x + y)^8$, find the exponent of x in the

4. third term 6 **5.** fourth term 5 **6.** seventh term 2

In the expansion of $(x + y)^8$, find the coefficient of the

7. first term 1 **8.** second term 8 **9.** last term 1

EXERCISE 12.2

In Exercises 1–30, use the binomial theorem to find the required term of each expansion.

1. $(a + b)^3$; second term $3a^2b$

2. $(a + b)^3$; third term $3ab^2$

3. $(x - y)^4$; fourth term $-4xy^3$

4. $(x - y)^5$; second term $-5x^4y$

5. $(x + y)^6$; fifth term $15x^2y^4$

6. $(x + y)^7$; fifth term $35x^3y^4$

7. $(x - y)^8$; third term $28x^6y^2$

8. $(x - y)^9$; seventh term $84x^3y^6$

9. $(x + 3)^5$; third term $90x^3$

10. $(x - 2)^4$; second term $-8x^3$

11. $(4x + y)^5$; third term $640x^3y^2$

12. $(x + 4y)^5$; fourth term $640x^2y^3$

13. $(x - 3y)^4$; second term $-12x^3y$

14. $(3x - y)^5$; third term $270x^3y^2$

15. $(2x - 5)^7$; fourth term $-70{,}000x^4$

16. $(2x + 3)^6$; sixth term $2916x$

17. $(2x - 3y)^5$; fifth term $810xy^4$

18. $(3x - 2y)^4$; second term $-216x^3y$

19. $(\sqrt{2}x + \sqrt{3}y)^6$; third term $180x^4y^2$

20. $(\sqrt{3}x + \sqrt{2}y)^5$; second term $45\sqrt{2}x^4y$

21. $\left(\frac{x}{2} - \frac{y}{3}\right)^4$; second term $-\frac{1}{6}x^3y$

22. $\left(\frac{x}{3} + \frac{y}{2}\right)^5$; fourth term
$\frac{5x^2y^3}{36}$

23. $(a + b)^n$; fourth term
$\frac{n!}{3!(n-3)!} a^{n-3}b^3$

24. $(a + b)^n$; third term
$\frac{n!}{2!(n-2)!} a^{n-2}b^2$

25. $(a - b)^n$; fifth term
$\frac{n!}{4!(n-4)!} a^{n-4}b^4$

26. $(a - b)^n$; sixth term
$-\frac{n!}{5!(n-5)!} a^{n-5}b^5$

27. $(a + b)^n$; rth term
$\frac{n!}{(r-1)!(n-r+1)!} a^{n-r+1}b^{r-1}$

28. $(a + b)^n$; $(r + 1)$th term
$\frac{n!}{r!(n-r)!} a^{n-r}b^r$

29. $(2a - 3b)^n$; fifth term
$\frac{81(2^{n-4})n!}{4!(n-4)!} a^{n-4}b^4$

30. $(3a - 2b)^n$; fourth term
$-\frac{8(3^{n-3})n!}{3!(n-3)!} a^{n-3}b^3$

Writing Exercises ■ *Write a paragraph using your own words.*

1. Tell how to find the coefficients in the expansion of $(x + y)^5$.

2. Explain why the signs alternate in the expansion of $(x - y)^9$.

Something to Think About ■

1. Find the constant term in the expansion of $\left(x + \frac{1}{x}\right)^{10}$. 252

2. Find the coefficient of a^5 in the expansion of $\left(a - \frac{1}{a}\right)^9$. 36

Review Exercises ■ *Solve each system of equations.*

1. $\begin{cases} 3x + 2y = 12 \\ 2x - y = 1 \end{cases}$ (2, 3)

2. $\begin{cases} a + b + c = 6 \\ 2a + b + 3c = 11 \\ 3a - b - c = 6 \end{cases}$ (3, 2, 1)

Evaluate each determinant.

3. $\begin{vmatrix} 2 & -3 \\ 4 & -2 \end{vmatrix}$ 8

4. $\begin{vmatrix} 1 & 2 & 3 \\ 4 & 5 & 0 \\ -1 & -2 & 1 \end{vmatrix}$ −12

12.3 Arithmetic Sequences

■ Arithmetic Sequences ■ Arithmetic Means ■ The Sum of the First n Terms of an Arithmetic Sequence ■ Summation Notation

A **sequence** is a function whose domain is the set of natural numbers. For example, the function $f(n) = 3n + 2$, where n is a natural number, is a sequence. Because a sequence is a function whose domain is the set of natural numbers, it is easy to

Leonardo Fibonacci
(late 12th and early 13th cent.)
Fibonacci, an Italian mathematician, is also known as Leonardo da Pisa. In his work *Liber abaci,* he advocated the adoption of Arabic numerals, the numerals that we use today. He is best known for a sequence of numbers that bears his name.

write its values as a list. If the natural numbers are substituted for n, the function $f(n) = 3n + 2$ generates the list

$$5, 8, 11, 14, 17, \ldots$$

It is common to call the list, as well as the function, a sequence. Each number in the list is called a **term** of the sequence. Other examples of sequences are

a. $1^3, 2^3, 3^3, 4^3, \ldots$ **b.** $4, 8, 12, 16, \ldots$

c. $2, 3, 5, 7, 11, \ldots$ **d.** $1, 1, 2, 3, 5, 8, 13, 21, \ldots$

The sequence in part **a** is the ordered list of the cubes of the natural numbers. The sequence in part **b** is the ordered list of the positive multiples of 4. The sequence in part **c** is the ordered list of the prime numbers. The sequence in part **d** is called the **Fibonacci sequence,** after the 12th-century mathematician Leonardo of Pisa—also known as Fibonacci. Beginning with the 2, each term of the Fibonacci sequence is the sum of the two preceding terms.

Arithmetic Sequences

One important type of sequence is the **arithmetic sequence.**

Arithmetic Sequence An **arithmetic sequence** is a sequence of the form

$$a, a + d, a + 2d, a + 3d, \ldots, a + (n - 1)d, \ldots$$

where a is the **first term,** $a + (n - 1)d$ is the ***n*th term,** and d is the **common difference.**

We note that the second term of an arithmetic sequence has an addend of $1d$, the third term has an addend of $2d$, the fourth term has an addend of $3d$, and the nth term has an addend of $(n - 1)d$. We also note that the difference between any two consecutive terms in an arithmetic sequence is d.

EXAMPLE 1 An arithmetic sequence has a first term of 5 and a common difference of 4.

a. Write the first six terms of the sequence.

b. Write the 25th term of the sequence.

Solution **a.** Because the first term is $a = 5$ and the common difference is $d = 4$, the first six terms are

$$5, \quad 5 + 4, \quad 5 + 2(4), \quad 5 + 3(4), \quad 5 + 4(4), \quad 5 + 5(4)$$

or

$$5, 9, 13, 17, 21, 25$$

b. The nth term is $a + (n - 1)d$. Because we want the 25th term, we let $n = 25$:

$$\begin{aligned} n\text{th term} &= a + (n - 1)d \\ 25\text{th term} &= 5 + (25 - 1)4 && \text{Remember that } a = 5 \text{ and } d = 4. \\ &= 5 + 24(4) \\ &= 5 + 96 \\ &= 101 \end{aligned}$$

■

EXAMPLE 2 The first three terms of an arithmetic sequence are 3, 8, and 13. Find **a.** the 67th term and **b.** the 100th term.

Solution We first find d, the common difference. It is the difference between successive terms:

$$d = 8 - 3 = 13 - 8 = 5$$

a. We substitute 3 for a, 67 for n, and 5 for d in the formula for the nth term and simplify:

$$\begin{aligned} n\text{th term} &= a + (n - 1)d \\ 67\text{th term} &= 3 + (67 - 1)5 \\ &= 3 + 66(5) \\ &= 333 \end{aligned}$$

b. We substitute 3 for a, 100 for n, and 5 for d in the formula for the nth term and simplify:

$$\begin{aligned} n\text{th term} &= a + (n - 1)d \\ 100\text{th term} &= 3 + (100 - 1)5 \\ &= 3 + 99(5) \\ &= 498 \end{aligned}$$

■

EXAMPLE 3 The first term of an arithmetic sequence is 12, and the 50th term is 3099. Write the first six terms of the sequence.

Solution The key is to find the common difference. Because the 50th term of this sequence is 3099, we can let $n = 50$ and solve the following equation for d:

$$\begin{aligned} 50\text{th term} &= a + (n - 1)d \\ 3099 &= 12 + (50 - 1)d \\ 3099 &= 12 + 49d && \text{Simplify.} \\ 3087 &= 49d && \text{Subtract 12 from both sides.} \\ 63 &= d && \text{Divide both sides by 49.} \end{aligned}$$

The first term of the sequence is 12, and the common difference is 63. Thus, its first six terms are

12, 75, 138, 201, 264, 327

■

Arithmetic Means

If numbers are inserted between two numbers a and b to form an arithmetic sequence, the inserted numbers are called **arithmetic means** between a and b.

If a single number is inserted between the numbers a and b, that number is called **the arithmetic mean** between a and b.

EXAMPLE 4 Insert two arithmetic means between 6 and 27.

Solution In this example, the first term is $a = 6$, and the fourth term (or the last term) is $l = 27$. We must find the common difference such that the terms

$$6, \mathbf{6 + d}, \mathbf{6 + 2d}, 27$$

form an arithmetic sequence. To find d, we can substitute 6 for a and 4 for n into the formula for the nth term:

$$\begin{aligned} n\text{th term} &= a + (n - 1)d \\ 4\text{th term} &= 6 + (4 - 1)d \\ 27 &= 6 + 3d && \text{Simplify.} \\ 21 &= 3d && \text{Subtract 6 from both sides.} \\ 7 &= d && \text{Divide both sides by 3.} \end{aligned}$$

The two arithmetic means between 6 and 27 are

$$\begin{aligned} 6 + d &= 6 + 7 \\ &= 13 \end{aligned} \quad \text{or} \quad \begin{aligned} 6 + 2d &= 6 + 2(7) \\ &= 6 + 14 \\ &= 20 \end{aligned}$$

The numbers 6, 13, 20, and 27 are the first four terms of an arithmetic sequence. ■

The Sum of the First *n* Terms of an Arithmetic Sequence

There is a formula that gives the sum of the first n terms of an arithmetic sequence. To develop this formula, we let S_n represent the sum of the first n terms of an arithmetic sequence:

$$S_n = a + [a + d] + [a + 2d] + \cdots + [a + (n - 1)d]$$

We write the same sum again, but in reverse order:

$$S_n = [a + (n - 1)d] + [a + (n - 2)d] + [a + (n - 3)d] + \cdots + a$$

We add these two equations together, term by term, to get

$$2S_n = [2a + (n - 1)d] + [2a + (n - 1)d] + [2a + (n - 1)d] + \cdots + [2a + (n - 1)d]$$

Because there are n equal terms on the right-hand side of the preceding equation, we can write

$$2S_n = n[2a + (n-1)d]$$

$$2S_n = n[\mathbf{a + a} + (n-1)d] \quad 2a = a + a.$$

$$2S_n = n[a + l] \quad \text{Substitute } l \text{ for } a + (n-1)d, \text{ because } a + (n-1)d \text{ in the last term of the sequence.}$$

$$S_n = \frac{n(a+l)}{2} \quad \text{Divide both sides by 2.}$$

This reasoning establishes the following theorem.

Sum of the First n Terms of an Arithmetic Sequence

The sum of the first n terms of an arithmetic sequence is given by the formula

$$S_n = \frac{n(a+l)}{2} \qquad \text{with } l = a + (n-1)d$$

where a is the first term, l is the last (or nth) term, and n is the number of terms in the sequence.

EXAMPLE 5 Find the sum of the first 40 terms of the arithmetic sequence 4, 10, 16, ...

Solution In this example, we let $a = 4$, $n = 40$, $d = 6$, and $l = 4 + (40 - 1)6 = 238$ and substitute these values into the formula for S_n:

$$\begin{aligned} S_n &= \frac{n(a+l)}{2} \\ &= \frac{\mathbf{40}(\mathbf{4} + \mathbf{238})}{2} \\ &= 20(242) \\ &= 4840 \end{aligned}$$

The sum of the first 40 terms is 4840. ■

Summation Notation

There is a shorthand notation for indicating the sum of a finite (ending) number of consecutive terms in a sequence. This notation, called **summation notation,** involves the Greek letter Σ (sigma). The expression

$$\sum_{k=2}^{5} 3k \quad \text{Read as "the summation of } 3k \text{ as } k \text{ runs from 2 to 5."}$$

designates the sum of all terms obtained if we successively substitute the numbers 2, 3, 4, and 5 for k, called the **index of the summation.** Thus, we have

$$\begin{aligned}\sum_{k=2}^{5} 3k &= 3(2) + 3(3) + 3(4) + 3(5)\\ &= 6 + 9 + 12 + 15\\ &= 42\end{aligned}$$

(a)

EXAMPLE 6 Find each sum: **a.** $\sum_{k=3}^{5} (2k + 1)$, **b.** $\sum_{k=2}^{5} k^2$, and **c.** $\sum_{k=1}^{3} (3k^2 + 3)$.

Solution **a.**
$$\begin{aligned}\sum_{k=3}^{5} (2k + 1) &= [2(3) + 1] + [2(4) + 1] + [2(5) + 1]\\ &= 7 + 9 + 11\\ &= 27\end{aligned}$$

b.
$$\begin{aligned}\sum_{k=2}^{5} k^2 &= 2^2 + 3^2 + 4^2 + 5^2\\ &= 4 + 9 + 16 + 25\\ &= 54\end{aligned}$$

c.
$$\begin{aligned}\sum_{k=1}^{3} (3k^2 + 3) &= [3(1)^2 + 3] + [3(2)^2 + 3] + [3(3)^2 + 3]\\ &= 6 + 15 + 30\\ &= 51\end{aligned}$$

■

ORALS *Find the next term in each arithmetic sequence.*

1. 2, 6, 10, . . . 14

2. 10, 7, 4, . . . 1

Find the common difference in each arithmetic sequence.

3. −2, 3, 8, . . . 5

4. 5, −1, −7, . . . −6

Find each sum.

5. $\sum_{k=1}^{2} k$ 3

6. $\sum_{k=2}^{3} k$ 5

EXERCISE 12.3

In Exercises 1–14, write the first five terms of each arithmetic sequence with the given properties.

1. $a = 3, d = 2$ 3, 5, 7, 9, 11

2. $a = -2, d = 3$ −2, 1, 4, 7, 10

3. $a = -5, d = -3$ −5, −8, −11, −14, −17

4. $a = 8, d = -5$ 8, 3, −2, −7, −12

5. $a = 5$, fifth term is 29 5, 11, 17, 23, 29

6. $a = 4$, sixth term is 39 4, 11, 18, 25, 32

7. $a = -4$, sixth term is -39 −4, −11, −18, −25, −32

8. $a = -5$, fifth term is -37 −5, −13, −21, −29, −37

9. $d = 7$, sixth term is -83 −118, −111, −104, −97, −90

10. $d = 3$, seventh term is 12 −6, −3, 0, 3, 6

11. $d = -3$, seventh term is 16 34, 31, 28, 25, 22

12. $d = -5$, seventh term is -12 18, 13, 8, 3, −2

13. The 19th term is 131, and the 20th term is 138. 5, 12, 19, 26, 33

14. The 16th term is 70, and the 18th term is 78. 10, 14, 18, 22, 26

15. Find the 30th term of the arithmetic sequence with $a = 7$ and $d = 12$. 355

16. Find the 55th term of the arithmetic sequence with $a = -5$ and $d = 4$. 211

17. Find the 37th term of the arithmetic sequence with a second term of -4 and a third term of -9. −179

18. Find the 40th term of the arithmetic sequence with a second term of 6 and a fourth term of 16. 196

19. Find the first term of the arithmetic sequence with a common difference of 11 and whose 27th term is 263. −23

20. Find the common difference of the arithmetic sequence with a first term of -164 if its 36th term is -24. 4

21. Find the common difference of the arithmetic sequence with a first term of 40 if its 44th term is 556. 12

22. Find the first term of the arithmetic sequence with a common difference of -5 and whose 23rd term is -625. −515

23. Insert three arithmetic means between 2 and 11. $\frac{17}{4}, \frac{13}{2}, \frac{35}{4}$

24. Insert four arithmetic means between 5 and 25. 9, 13, 17, 21

25. Insert four arithmetic means between 10 and 20. 12, 14, 16, 18

26. Insert three arithmetic means between 20 and 30. $\frac{45}{2}, 25, \frac{55}{2}$

27. Find the arithmetic mean between 10 and 19. $\frac{29}{2}$

28. Find the arithmetic mean between 5 and 23. 14

29. Find the arithmetic mean between -4.5 and 7. $\frac{5}{4}$

30. Find the arithmetic mean between -6.3 and -5.2. −5.75

In Exercises 31–38, find the sum of the first n terms of each arithmetic sequence.

31. 1, 4, 7, . . . , $n = 30$ 1335

32. 2, 6, 10, . . .; $n = 28$ 1568

33. −5, −1, 3, . . .; $n = 17$ 459

34. −7, −1, 5, . . .; $n = 15$ 525

35. Second term is 7, third term is 12; $n = 12$ 354

36. Second term is 5, fourth term is 9; $n = 16$ 288

37. $f(n) = 2n + 1$, nth term is 31; n is a natural number 255

38. $f(n) = 4n + 3$, nth term is 23; n is a natural number 75

39. Find the sum of the first 50 natural numbers. 1275

40. Find the sum of the first 100 natural numbers. 5050

41. Find the sum of the first 50 odd natural numbers. 2500

42. Find the sum of the first 50 even natural numbers. 2550

43. **Saving money** Fred puts \$60 into a safety deposit box. After each succeeding month, he puts \$50 more in the safety deposit box. Write the first six terms of an arithmetic sequence that gives the monthly amounts in his savings, and find his savings after 10 years. 60, 110, 160, 210, 260, 310; \$6060

44. **Installment loan** Freda borrowed \$10,000, interest-free, from her mother. Freda agreed to pay back the loan in monthly installments of \$275. Write the first six terms of an arithmetic sequence that shows the balance due after each month, and find the balance due after 17 months.
9725, 9450, 9175, 8900, 8625, 8350; \$5325

45. **Designing a patio** Each row of bricks in the triangular patio in Illustration 1 is to have one more brick than the previous row, ending with the longest row of 150 bricks. How many bricks will be needed?
11,325

ILLUSTRATION 1

46. **Falling object** The equation $s = 16t^2$ represents the distance s in feet that an object will fall in t seconds. After 1 second, the object has fallen 16 feet. After 2 seconds, the object has fallen 64 feet, and so on. Find the distance that the object will fall during the second and third seconds. 48 ft; 80 ft

47. **Falling object** Refer to Exercise 46. How far will the object fall during the 12th second? 368 ft

48. **Interior angles** The sums of the angles of several polygons are given in Illustration 2. Assuming that the pattern continues, find the sum of the interior angles of an octagon (8 sides) and dodecagon (12 sides). 1080°, 1800°

Figure	Number of sides	Sum of angles
Triangle	3	180°
Quadralateral	4	360°
Pentagon	5	540°
Hexagon	6	720°

ILLUSTRATION 2

49. Show that the arithmetic mean between a and b is the average of a and b: $\frac{a+b}{2}$.

50. Show that the sum of the two arithmetic means between a and b is $a + b$.

In Exercises 51–56, find each sum.

51. $\sum_{k=1}^{4} 6k$ 60

52. $\sum_{k=2}^{5} 3k$ 42

53. $\sum_{k=3}^{4} (k^2 + 3)$ 31

54. $\sum_{k=2}^{6} (k^2 + 1)$ 95

55. $\sum_{k=4}^{4} (2k + 4)$ 12

56. $\sum_{k=3}^{5} (3k^2 - 7)$ 129

57. Show that $\sum_{k=1}^{5} 5k = 5 \sum_{k=1}^{5} k$.

58. Show that $\sum_{k=3}^{6} (k^2 + 3k) = \sum_{k=3}^{6} k^2 + \sum_{k=3}^{6} 3k$.

59. Show that $\sum_{k=1}^{n} 3 = 3n$. (*Hint:* Consider 3 to be $3k^0$.)

60. Show that $\sum_{k=1}^{3} \frac{k^2}{k} \neq \frac{\sum_{k=1}^{3} k^2}{\sum_{k=1}^{3} k}$.

Writing Exercises ■ *Write a paragraph using your own words.*

1. Define an arithmetic sequence.
2. Develop the formula for finding the sum of the first n terms of an arithmetic sequence.

Something to Think About ■

1. Write the first 6 terms of the arithmetic sequence given by

$$\sum_{n=1}^{n}\left(\frac{1}{2}n+1\right)$$ $\frac{3}{2}, 2, \frac{5}{2}, 3, \frac{7}{2}, 4$

2. Find the sum of the first 6 terms of the sequence given in Problem 1. 16.5

Review Exercises ■ *Do the operations and simplify if possible.*

1. $3(2x^2 - 4x + 7) + 4(3x^2 + 5x - 6)$ $18x^2 + 8x - 3$
2. $(2p + q)(3p^2 + 4pq - 3q^2)$ $6p^3 + 11p^2q - 2pq^2 - 3q^3$
3. $\dfrac{3a+4}{a-2} + \dfrac{3a-4}{a+2}$ $\dfrac{6a^2+16}{(a+2)(a-2)}$
4. $2t - 3\overline{)8t^4 - 12t^3 + 8t^2 - 16t + 6}$ $4t^3 + 4t - 2$

12.4 Geometric Sequences

■ Geometric Sequences ■ Geometric Means ■ The Sum of the First n Terms of a Geometric Sequence ■ Population Growth

Another important sequence is called a **geometric sequence.**

■ Geometric Sequences

Geometric Sequence A **geometric sequence** is a sequence of the form

$$a, ar, ar^2, ar^3, \ldots, ar^{n-1}, \ldots$$

where a is the **first term,** ar^{n-1} is the ***n*th term,** and r is the **common ratio.**

We note that the second term of a geometric sequence has a factor of r^1, the third term has a factor of r^2, the fourth term has a factor of r^3, and the nth term has a factor of r^{n-1}. We also note that the quotient obtained when any term is divided by the previous term is r.

EXAMPLE 1 A geometric sequence has a first term of 5 and a common ratio of 3.

a. Write the first five terms of the sequence.

b. Write the ninth term of the sequence.

Solution **a.** Because the first term is $a = 5$ and the common ratio is $r = 3$, the first five terms are

$$5, \quad 5(3), \quad 5(3^2), \quad 5(3^3), \quad 5(3^4)$$

or

$$5, 15, 45, 135, 405$$

b. The nth term is ar^{n-1} where $a = 5$ and $r = 3$. Because we want the ninth term, we let $n = 9$:

$$\begin{aligned} n\text{th term} &= ar^{n-1} \\ 9\text{th term} &= 5(3)^{9-1} \\ &= 5(3)^8 \\ &= 5(6561) \\ &= 32{,}805 \end{aligned}$$

■

EXAMPLE 2 The first three terms of a geometric sequence are 16, 4, and 1. Find the seventh term of the sequence.

Solution We substitute 16 for a, $\frac{1}{4}$ for r, and 7 for n in the formula for the nth term and simplify:

$$\begin{aligned} n\text{th term} &= ar^{n-1} \\ 7\text{th term} &= 16\left(\frac{1}{4}\right)^{7-1} \qquad r = \tfrac{4}{16} = \tfrac{1}{4}. \\ &= 16\left(\frac{1}{4}\right)^{6} \\ &= 16\left(\frac{1}{4096}\right) \\ &= \frac{1}{256} \end{aligned}$$

■

Geometric Means

If numbers are inserted between two numbers a and b to form a geometric sequence, the inserted numbers are called **geometric means** between a and b.

If a single number is inserted between the numbers a and b, that number is called a **geometric mean** between a and b.

EXAMPLE 3 Insert two geometric means between 7 and 1512.

Solution In this example, the first term is $a = 7$, and the fourth term (or last term) is $l = 1512$. To find the common ratio r such that the terms

$$7, \quad 7r, \quad 7r^2, \quad 1512$$

form a geometric sequence, we substitute 4 for n and 7 for a into the formula for the nth term of a geometric sequence and solve for r.

$$\begin{aligned} n\text{th term} &= ar^{n-1} \\ 4\text{th term} &= 7r^{4-1} \\ 1512 &= 7r^3 \\ 216 &= r^3 && \text{Divide both sides by 7.} \\ 6 &= r && \text{Take the cube root of both sides.} \end{aligned}$$

The two geometric means between 7 and 1512 are

$$7r = 7(6) = 42$$

and

$$7r^2 = 7(6)^2 = 7(36) = 252$$

The numbers 7, 42, 252, and 1512 are the first four terms of a geometric sequence. ■

EXAMPLE 4 Find a geometric mean between 2 and 20.

Solution We want to find the middle term of the three-termed geometric sequence

$$2, \quad 2r, \quad 20$$

with $a = 2$, $l = 20$, and $n = 3$. To find r, we substitute these values into the formula for the nth term of a geometric sequence:

$$\begin{aligned} n\text{th term} &= ar^{n-1} \\ 3\text{rd term} &= 2r^{3-1} \\ 20 &= 2r^2 \\ 10 &= r^2 && \text{Divide both sides by 2.} \\ \pm\sqrt{10} &= r && \text{Take the square root of both sides.} \end{aligned}$$

Teaching Tip: $-2\sqrt{10}$ is called a geometric mean *between* 2 and 20 even though $2 < -2\sqrt{10} < 20$ is false.

Because r can be either $\sqrt{10}$ or $-\sqrt{10}$, there are two values for a geometric mean. They are

$$2r = 2\sqrt{10}$$

and

$$2r = -2\sqrt{10}$$

The numbers 2, $2\sqrt{10}$, 20 and 2, $-2\sqrt{10}$, 20 both form geometric sequences. The common ratio of the first sequence is $\sqrt{10}$, and the common ratio of the second sequence is $-\sqrt{10}$. ■

The Sum of the First n Terms of a Geometric Sequence

There is a formula that gives the sum of the first n terms of a geometric sequence. To develop this formula, we let S_n represent the sum of the first n terms of a geometric sequence.

1. $S_n = a + ar + ar^2 + ar^3 + \cdots + ar^{n-1}$

We multiply both sides of Equation 1 by r to get

2. $S_n r = \quad ar + ar^2 + ar^3 + \cdots + ar^{n-1} + ar^n$

We now subtract Equation 2 from Equation 1 and solve for S_n:

$$S_n - S_n r = a - ar^n$$

$$S_n(1 - r) = a - ar^n \qquad \text{Factor out } S_n \text{ from the left side.}$$

$$S_n = \frac{a - ar^n}{1 - r} \qquad \text{Divide both sides by } 1 - r.$$

This reasoning establishes the following theorem.

Sum of the First n Terms of a Geometric Sequence

The sum of the first n terms of a geometric sequence is given by the formula

$$S_n = \frac{a - ar^n}{1 - r} \qquad (r \neq 1)$$

where S_n is the sum, a is the first term, r is the common ratio, and n is the number of terms.

EXAMPLE 5 Find the sum of the first six terms of the geometric sequence 250, 50, 10, ...

Solution In this sequence, $a = 250$, $r = \frac{1}{5}$, and $n = 6$. We substitute these values into the formula for the sum of the first n terms of a geometric sequence and simplify.

$$S_n = \frac{a - ar^n}{1 - r} = \frac{250 - 250\left(\frac{1}{5}\right)^6}{1 - \frac{1}{5}}$$

$$= \frac{250 - 250\left(\frac{1}{15{,}625}\right)}{\frac{4}{5}}$$

$$= \frac{5}{4}\left(250 - \frac{250}{15{,}625}\right)$$

$$= \frac{5}{4}\left(\frac{3{,}906{,}000}{15{,}625}\right)$$

$$= 312.48$$

The sum of the first six terms is 312.48. ■

Population Growth

EXAMPLE 6 The mayor of Eagle River (population 1500) predicts a growth rate of 4% each year for the next ten years. Find the population of Eagle River ten years from now.

Solution Let P_0 be the initial population of Eagle River. After 1 year, there will be a different population P_1. The initial population (P_0) plus the growth (the product of P_0 and the rate of growth, r) will equal this new population, P_1:

$$P_1 = P_0 + P_0 r = P_0(1 + r)$$

The population after 2 years will be P_2, and

$$\begin{aligned} P_2 &= P_1 + P_1 r \\ &= \mathbf{P_1}(1 + r) && \text{Factor out } P_1. \\ &= \mathbf{P_0(1 + r)}(1 + r) && \text{Remember that } P_1 = P_0(1 + r). \\ &= P_0(1 + r)^2 \end{aligned}$$

The population after 3 years will be P_3, and

$$\begin{aligned} P_3 &= P_2 + P_2 r \\ &= \mathbf{P_2}(1 + r) && \text{Factor out } P_2. \\ &= \mathbf{P_0(1 + r)^2}(1 + r) && \text{Remember that } P_2 = P_0(1 + r)^2 \\ &= P_0(1 + r)^3 \end{aligned}$$

The yearly population figures

$$P_0, \quad P_1, \quad P_2, \quad P_3, \ldots$$

or

$$P_0 \quad P_0(1 + r), \quad P_0(1 + r)^2, \quad P_0(1 + r)^3, \ldots$$

Teaching Tip: You may want to relate this to the formula
$$A = A_0\left(1 + \frac{r}{k}\right)^{kt}$$

form a geometric sequence with a first term of P_0 and a common ratio of $1 + r$. The population of Eagle River after 10 years is P_{10}, which is the 11th term of this sequence:

$$\begin{aligned} n\text{th term} &= ar^{n-1} \\ P_{10} = 11\text{th term} &= P_0(1 + r)^{10} \\ &= 1500(1 + 0.04)^{10} \\ &= 1500(1.04)^{10} \\ &\approx 2220 \qquad \text{Use a calculator.} \end{aligned}$$

The estimated population ten years from now is 2220 people. ■

ORALS *Find the next term in each geometric sequence.*

1. 1, 3, 9, . . . 27

2. $1, \frac{1}{3}, \frac{1}{9}, \ldots$ $\frac{1}{27}$

Find the common ratio in each geometric sequence.

3. 0.2, 0.5, 1.25, . . . 2.5

4. $\sqrt{3}, 3, 3\sqrt{3}, \ldots$ $\sqrt{3}$

Find x in each geometric sequence.

5. 2, x, 18, 54, . . . 6

6. $3, x, \frac{1}{3}, \frac{1}{9}, \ldots$ 1

EXERCISE 12.4

In Exercises 1–14, write the first five terms of each geometric sequence with the given properties.

1. $a = 3, r = 2$ 3, 6, 12, 24, 48

2. $a = -2, r = 2$ −2, −4, −8, −16, −32

3. $a = -5, r = \frac{1}{5}$ $-5, -1, -\frac{1}{5}, -\frac{1}{25}, -\frac{1}{125}$

4. $a = 8, r = \frac{1}{2}$ $8, 4, 2, 1, \frac{1}{2}$

5. $a = 2, r > 0$, third term is 32 2, 8, 32, 128, 512

6. $a = 3$, fourth term is 24 3, 6, 12, 24, 48

7. $a = -3$, fourth term is −192
−3, −12, −48, −192, −768

8. $a = 2, r < 0$, third term is 50
2, −10, 50, −250, 1250

9. $a = -64, r < 0$, fifth term is −4
−64, 32, −16, 8, −4

10. $a = -64, r > 0$, fifth term is −4
−64, −32, −16, −8, −4

11. $a = -64$, sixth term is −2 −64, −32, −16, −8, −4

12. $a = -81$, sixth term is $\frac{1}{3}$ −81, 27, −9, 3, −1

13. The second term is 10, and the third term is 50.
2, 10, 50, 250, 1250

14. The third term is −27, and the fourth term is 81.
−3, 9, −27, 81, −243

15. Find the tenth term of the geometric sequence with $a = 7$ and $r = 2$. 3584

16. Find the 12th term of the geometric sequence with $a = 64$ and $r = \frac{1}{2}$. $\frac{1}{32}$

17. Find the first term of the geometric sequence with a common ratio of -3 and an eighth term of -81. $\frac{1}{27}$

18. Find the first term of the geometric sequence with a common ratio of 2 and a tenth term of 384. $\frac{3}{4}$

19. Find the common ratio of the geometric sequence with a first term of -8 and a sixth term of -1944. 3

20. Find the common ratio of the geometric sequence with a first term of 12 and a sixth term of $\frac{3}{8}$. $\frac{1}{2}$

21. Insert three positive geometric means between 2 and 162. 6, 18, 54

22. Insert four geometric means between 3 and 96. 6, 12, 24, 48

23. Insert four geometric means between -4 and $-12{,}500$. $-20, -100, -500, -2500$

24. Insert three geometric means (two positive and one negative) between -64 and -1024. $128, -256, 512$

25. Find the negative geometric mean between 2 and 128. -16

26. Find the positive geometric mean between 3 and 243. 27

27. Find the positive geometric mean between 10 and 20. $10\sqrt{2}$

28. Find the negative geometric mean between 5 and 15. $-5\sqrt{3}$

29. Find a geometric mean, if possible, between -50 and 10. No geometric mean exists.

30. Find a negative geometric mean, if possible, between -25 and -5. $-5\sqrt{5}$

In Exercises 31–42, find the sum of the first n terms of each geometric sequence.

31. $2, 6, 18, \ldots; n = 6$ 728

32. $2, -6, 18, \ldots; n = 6$ -364

33. $2, -6, 18, \ldots; n = 5$ 122

34. $2, 6, 18, \ldots; n = 5$ 242

35. $3, -6, 12, \ldots; n = 8$ -255

36. $3, 6, 12, \ldots; n = 8$ 765

37. $3, 6, 12, \ldots; n = 7$ 381

38. $3, -6, 12, \ldots; n = 7$ 129

39. The second term is 1 and the third term is $\frac{1}{5}$; $n = 4$. $\frac{156}{25}$

40. The second term is 1 and the third term is 4; $n = 5$. $\frac{341}{4}$

41. The third term is -2 and the fourth term is 1; $n = 6$. $-\frac{21}{4}$

42. The third term is -3 and the fourth term is 1; $n = 5$. $-\frac{61}{3}$

In Exercises 43–50, use a calculator to solve each problem.

43. Population growth The population of Union is predicted to increase by 6% each year. What will be the population of Union 5 years from now if its current population is 500? about 669

44. Population decline The population of Hicksville is decreasing by 10% each year. If its current population is 98, what will be the population 8 years from now? 42

45. Declining savings John has $10,000 in a safety deposit box. Each year he spends 12% of what is left in the box. How much will be in the box after 15 years? $1469.74

46. Savings growth Sally has $5000 in an investment account earning 12% annual interest. How much will be in her account 10 years from now? (Assume that Sally makes no deposits or withdrawals.) $15,529.24

47. House appreciation A house appreciates by 6% each year. If the house is worth $70,000 today, how much will it be worth 12 years from now? $140,853.75

48. Motorboat depreciation A motorboat that cost $5000 when new depreciates at a rate of 9% per year. How much will the boat be worth in 5 years? $3120.16

49. Inscribed squares Each inscribed square in Illustration 1 joins the midpoints of the sides of the next larger square. The area of the first square, the largest, is 1. Find the area of the 12th square. $\left(\frac{1}{2}\right)^{11} \approx 0.0005$

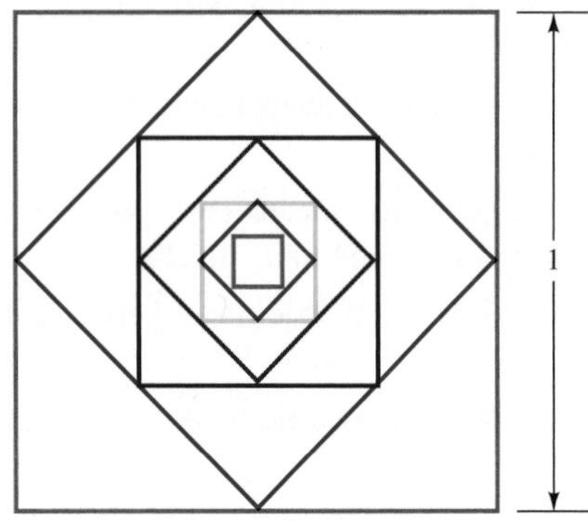

ILLUSTRATION 1

50. Genealogy The family tree in Illustration 2 spans three generations and lists seven people. How many names would be listed in a family tree that spans ten generations? $2^{10} - 1 = 1023$

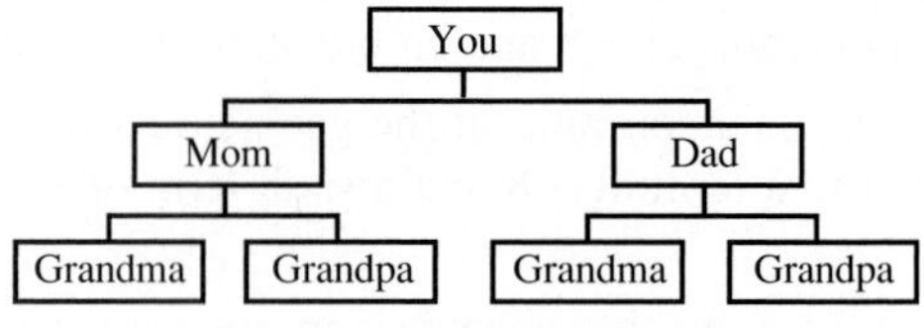

ILLUSTRATION 2

51. Show that the formula for the sum of the first n terms of a geometric sequence can be written in the form

$$S_n = \frac{a - lr}{1 - r} \qquad \text{where } l = ar^{n-1}$$

52. Show that the formula for the sum of the first n terms of a geometric sequence can be written in the form

$$S_n = \frac{a(1 - r^n)}{1 - r}$$

Writing Exercises ■ *Write a paragraph using your own words.*

1. Define a geometric sequence.
2. Develop the formula for finding the sum of the first n terms of a geometric sequence.

Something to Think About ■

1. Show that the arithmetic mean between a and b is $\frac{a+b}{2}$.
2. Show that the geometric mean between a and b is $\sqrt{ab}$.
3. If $a > b > 0$, which is larger: the arithmetic mean between a and b or the geometric mean between a and b? arithmetic mean
4. Is there a geometric mean between $-a$ and a? no

Review Exercises ■ *Solve each inequality.*

1. $x^2 - 5x - 6 \leq 0$ $[-1, 6]$
2. $a^2 - 7a + 12 \geq 0$ $(-\infty, 3] \cup [4, \infty)$
3. $\dfrac{x - 4}{x + 3} > 0$ $(-\infty, -3) \cup (4, \infty)$
4. $\dfrac{t^2 + t - 20}{t + 2} < 0$ $(-\infty, -5) \cup (-2, 4)$

12.5 Infinite Geometric Sequences

■ The Sum of an Infinite Geometric Sequence

An **infinite geometric sequence** is a geometric sequence with an infinite number of terms. Two examples of infinite geometric sequences are

$$2, 6, 18, 54, 162, \ldots \qquad (r = 3)$$

$$\frac{3}{2}, \frac{3}{4}, \frac{3}{8}, \frac{3}{16}, \frac{3}{32}, \ldots \qquad \left(r = \frac{1}{2}\right)$$

■ The Sum of an Infinite Geometric Sequence

Under certain conditions, we can find the sum of all the terms of an infinite geometric sequence. To define this sum, we consider the geometric sequence

$$a, ar, ar^2, ar^3, \ldots, ar^{n-1}, \ldots$$

- The first **partial sum,** S_1, of the sequence is $S_1 = a$.
- The second partial sum, S_2, of the sequence is $S_2 = a + ar$.
- The third partial sum, S_3, of the sequence is $S_3 = a + ar + ar^2$.
- The nth partial sum, S_n, of the sequence is $S_n = a + ar + ar^2 + \cdots + ar^{n-1}$.

If the nth partial sum, S_n, approaches some number S as n approaches infinity, then S is called the **sum of the infinite geometric sequence.**

To develop a formula for finding the sum of all the terms in an infinite geometric sequence, we consider the formula that gives the sum of the first n terms of the sequence.

$$S_n = \frac{a - ar^n}{1 - r} \qquad (r \neq 1)$$

If $|r| < 1$ and a is constant, then the term ar^n in the above formula approaches 0 as n becomes very large. For example,

Teaching Tip: S is sometimes expressed as S_∞ or $\sum_{k=1}^{\infty} a_k$.

$$a\left(\frac{1}{4}\right)^1 = \frac{1}{4}a, \qquad a\left(\frac{1}{4}\right)^2 = \frac{1}{16}a, \qquad a\left(\frac{1}{4}\right)^3 = \frac{1}{64}a$$

and so on. Thus, when n is very large, the value of ar^n is negligible, and the term ar^n in the above formula can be ignored. This reasoning justifies the following theorem.

Sum of an Infinite Geometric Sequence If a is the first term and r is the common ratio of an infinite geometric sequence, and if $|r| < 1$, then the sum of the terms of the sequence is given by the formula

$$S = \frac{a}{1 - r}$$

EXAMPLE 1 Find the sum of the terms of the infinite geometric sequence 125, 25, 5, . . .

Solution In this geometric sequence, $a = 125$ and $r = \frac{1}{5}$. Because $|r| = \left|\frac{1}{5}\right| = \frac{1}{5} < 1$, we can find the sum of all the terms of the sequence. We do this by substituting 125 for a and $\frac{1}{5}$ for r in the formula $S = \frac{a}{1 - r}$ and simplifying:

$$S = \frac{a}{1 - r} = \frac{\mathbf{125}}{1 - \frac{\mathbf{1}}{\mathbf{5}}} = \frac{125}{\frac{4}{5}} = \frac{5}{4}(125) = \frac{625}{4}$$

The sum of all the terms of the sequence 125, 25, 5, . . . is $\frac{625}{4}$. ■

EXAMPLE 2 Find the sum of the infinite geometric sequence $64, -4, \frac{1}{4}, \ldots$.

Solution In this geometric sequence, $a = 64$ and $r = -\frac{1}{16}$. Because $|r| = \left|-\frac{1}{16}\right| = \frac{1}{16} < 1$, we can find the sum of all the terms of the sequence. We substitute 64 for a and $-\frac{1}{16}$ for r in the formula $S = \frac{a}{1 - r}$ and simplify:

$$S = \frac{a}{1 - r} = \frac{\mathbf{64}}{1 - \left(-\frac{\mathbf{1}}{\mathbf{16}}\right)} = \frac{64}{\frac{17}{16}} = \frac{16}{17}(64) = \frac{1024}{17}$$

The sum of all the terms of the geometric sequence $64, -4, \frac{1}{4}, \ldots$ is $\frac{1024}{17}$. ■

EXAMPLE 3 Change $0.\overline{8}$ to a common fraction.

Solution The decimal $0.\overline{8}$ can be written as the sum of an infinite geometric sequence.

$$0.\overline{8} = 0.888\ldots = \frac{8}{10} + \frac{8}{100} + \frac{8}{1000} + \cdots$$

where $a = \frac{8}{10}$ and $r = \frac{1}{10}$. Because $|r| = \left|\frac{1}{10}\right| = \frac{1}{10} < 1$, we can find the sum as follows:

$$S = \frac{a}{1-r} = \frac{\frac{8}{10}}{1 - \frac{1}{10}} = \frac{\frac{8}{10}}{\frac{9}{10}} = \frac{8}{9}$$

Thus, $0.\overline{8} = \frac{8}{9}$. Long division will verify that $\frac{8}{9} = 0.888\ldots$. ■

EXAMPLE 4 Change $0.\overline{25}$ to a common fraction.

Solution The decimal $0.\overline{25}$ can be written as the sum of an infinite geometric sequence

$$0.\overline{25} = 0.252525\ldots = \frac{25}{100} + \frac{25}{10{,}000} + \frac{25}{1{,}000{,}000} + \cdots$$

where $a = \frac{25}{100}$ and $r = \frac{1}{100}$. Because $|r| = \left|\frac{1}{100}\right| = \frac{1}{100} < 1$, we can find the sum as follows:

$$S = \frac{a}{1-r} = \frac{\frac{25}{100}}{1 - \frac{1}{100}} = \frac{\frac{25}{100}}{\frac{99}{100}} = \frac{25}{99}$$

Thus, $0.\overline{25} = \frac{25}{99}$. Long division will verify that this is true. ■

ORALS *Find the common ratio in each infinite geometric sequence.*

1. $\frac{1}{64}, \frac{1}{8}, 1, \ldots$ 8

2. $1, \frac{1}{8}, \frac{1}{64}, \ldots$ $\frac{1}{8}$

3. $\frac{2}{3}, \frac{1}{3}, \frac{1}{6}, \ldots$ $\frac{1}{2}$

4. $64, 8, 1, \ldots$ $\frac{1}{8}$

Find the sum of the terms in each infinite geometric sequence.

5. $18, 6, 2, \ldots$ 27

6. $12, 3, \frac{3}{4}, \ldots$ 16

EXERCISE 12.5

In Exercises 1–12, find the sum of each infinite geometric sequence, if possible.

1. $8, 4, 2, \ldots$ 16

2. $12, 6, 3, \ldots$ 24

3. $54, 18, 6, \ldots$ 81

4. $45, 15, 5, \ldots$ $\frac{135}{2}$

5. $12, -6, 3, \ldots$ 8

6. $8, -4, 2, \ldots$ $\frac{16}{3}$

7. $-45, 15, -5, \ldots$ $-\frac{135}{4}$

8. $-54, 18, -6, \ldots$ $-\frac{81}{2}$

9. $\frac{9}{2}, 6, 8, \ldots$ no sum

10. $-112, -28, -7, \ldots$ $-\frac{448}{3}$

11. $-\frac{27}{2}, -9, -6, \ldots$ $-\frac{81}{2}$

12. $\frac{18}{25}, \frac{6}{5}, 2, \ldots$ no sum

In Exercises 13–20, change each decimal to a common fraction. Then check the answer by division.

13. $0.\overline{1}$ $\frac{1}{9}$

14. $0.\overline{2}$ $\frac{2}{9}$

15. $-0.\overline{3}$ $-\frac{1}{3}$

16. $-0.\overline{4}$ $-\frac{4}{9}$

17. $0.\overline{12}$ $\frac{4}{33}$

18. $0.\overline{21}$ $\frac{7}{33}$

19. $0.\overline{75}$ $\frac{25}{33}$

20. $0.\overline{57}$ $\frac{19}{33}$

21. Bouncing ball On each bounce, the rubber ball in Illustration 1 rebounds to a height one-half of that from which it fell. Find the total distance the ball travels. 30 m

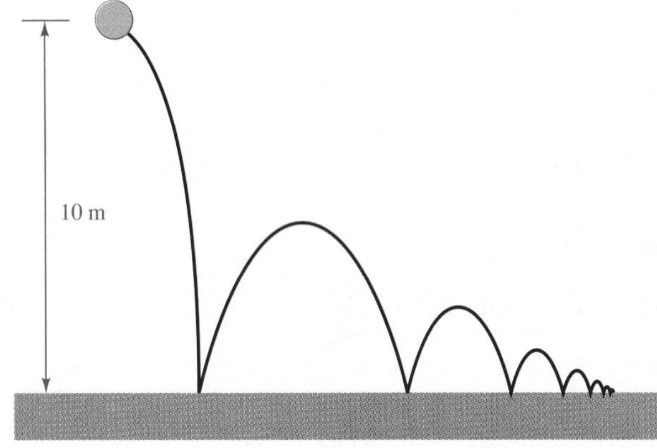

ILLUSTRATION 1

22. Bouncing ball A golf ball is dropped from a height of 12 feet. On each bounce it returns to a height two-thirds of that from which it fell. Find the total distance the ball travels. 60 ft

23. Controling moths To reduce the population of a destructive moth, biologists release 1000 sterilized male moths into the environment each day. If 80% of these moths alive one day survive until the next, then after a long time the population of sterile males is the sum of the infinite geometric series

$$1000 + 1000(0.8) + 1000(0.8)^2 + 1000(0.8)^3 + \cdots$$

Find the long-term population. 5000

24. Controling moths If mild weather increases the day-to-day survival rate of the sterile male moths in Exercise 23 to 90%, find the long-term population. 10,000

25. Show that $0.\overline{9} = 1$.

26. Show that $1.\overline{9} = 2$.

27. Does $0.999999 = 1$? Explain. no; $0.999999 = \frac{999,999}{1,000,000} < 1$

28. If $f(x) = 1 + x + x^2 + x^3 + x^4 + \cdots$, find $f\left(\frac{1}{2}\right)$ and $f\left(-\frac{1}{2}\right)$. $f\left(\frac{1}{2}\right) = 2, f\left(-\frac{1}{2}\right) = \frac{2}{3}$

Writing Exercises ■ *Write a paragraph using your own words.*

1. Why must the common ratio be less than 1 before an infinite geometric sequence can have a sum?

2. If its common difference is not 0, can an infinite arithmetic sequence have a sum?

Something to Think About ■

1. An infinite geometric sequence has a sum of 5 and a first term of 1. Find the common ratio. $\frac{4}{5}$

2. An infinite geometric sequence has a common ratio of $-\frac{2}{3}$ and a sum of 9. Find the first term. 15

Review Exercises ■ *Determine whether each equation determines y to be a function of x.*

1. $y = 3x^3 - 4$ yes **2.** $xy = 12$ yes **3.** $3x = y^2 + 4$ no **4.** $x = |y|$ no

12.6 Permutations and Combinations

■ The Multiplication Principle for Events ■ Permutations ■ Combinations ■ Alternative Form of the Binomial Theorem

■ The Multiplication Principle for Events

A student goes to the cafeteria for lunch. He has a choice of three different sandwiches (hamburger, hot dog, or ham and cheese) and four different beverages (cola, root beer, orange, or milk). How many different lunches can he choose?

The student has three choices of sandwich, and for any one of these choices, he has four choices of drink. The different options are shown in the *tree diagram* in Figure 12-1. The tree diagram shows that there is a total of 12 different lunches to choose from. One possibility is a hamburger with a cola, and another is a hot dog with milk.

Teaching Tip: Point out that the student has three choices of sandwich and four choices of beverage. Thus, he can choose from

$3 \cdot 4 = 12$

different meals.

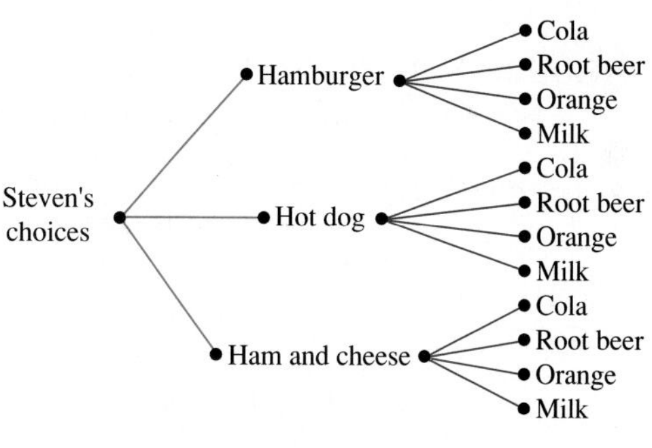

FIGURE 12-1

A situation that can have several different outcomes—such as choosing a sandwich—is called an **event.** Choosing a sandwich and choosing a beverage can be thought of as two events. The preceding example illustrates the **multiplication principle for events.**

Multiplication Principle for Events Let E_1 and E_2 be two events. If E_1 can be done in a_1 ways, and if—after E_1 has occurred—E_2 can be done in a_2 ways, then the event "E_1 followed by E_2" can be done in $a_1 \cdot a_2$ ways.

EXAMPLE 1 Before studying for an exam, Heidi plans to watch the early evening news and then a situation comedy on television. If she has a choice of four news broadcasts and two comedies, in how many ways can she choose to watch television?

Solution Let E_1 be the event "watching the news" and E_2 be the event "watching a comedy." Because there are four ways to accomplish E_1 and two ways to accomplish E_2, the number of choices that Heidi has is $4 \cdot 2 = 8$. ■

The multiplication principle can be extended to any number of events. In Example 2, we use it to compute the number of ways that we can arrange objects in a row.

EXAMPLE 2 In how many ways can we arrange five books on a shelf?

Solution We can fill the first space with any of the 5 books, the second space with any of the remaining 4 books, the third space with any of the remaining 3 books, the fourth space with any of the remaining 2 books, and the fifth space with the remaining 1 (or last) book. According to the multiplication principle for events, the number of ways that the books can be arranged is

$$5 \cdot 4 \cdot 3 \cdot 2 \cdot 1 = 120$$ ■

EXAMPLE 3 If a sailor has six flags, each of a different color, to hang on a flagpole, how many different signals can the sailor send by using four flags?

Solution The sailor must find the number of arrangements of 4 flags when there are 6 flags to choose from. The sailor can hang any one of the 6 flags in the top position, any one of the remaining 5 flags in the second position, any one of the remaining 4 flags in the third position, and any one of the remaining 3 flags in the lowest position. According to the multiplication principle for events, the total number of signals that can be sent is

$$6 \cdot 5 \cdot 4 \cdot 3 = 360$$ ■

Permutations

Teaching Tip: Emphasize that the order in which the flags hang determines the message. A permutation describes the number of arrangements—order is important.

When computing the number of possible arrangements of objects such as books on a shelf or flags on a pole, we are finding the number of **permutations** of those objects. In Example 2, we found that the number of permutations of five books, using all five of them, is 120. In Example 3, we found that the number of permutations of six flags, using four of them, is 360.

The symbol $P(n, r)$, read as "the number of permutations of n things r at a time," is often used to express permutation problems. In Example 2, we found that $P(5, 5) = 120$. In Example 3, we found that $P(6, 4) = 360$.

EXAMPLE 4 If Sarah has seven flags, each of a different color, to hang on a flagpole, how many different signals can she send by using three flags?

Solution Sarah must find $P(7, 3)$ (the number of permutations of 7 things 3 at a time). In the top position, Sarah can hang any of the 7 flags; in the middle position, any one of the remaining 6 flags; and in the bottom position, any one of the remaining 5 flags. According to the multiplication principle for events,

$$\begin{aligned} P(7, 3) &= 7 \cdot 6 \cdot 5 \\ &= 210 \end{aligned}$$

Sarah can send 210 signals using only three of the available seven flags. ■

Although it is correct to write $P(7, 3) = 7 \cdot 6 \cdot 5$, there is an advantage in changing the form of this answer to obtain a formula for computing $P(7, 3)$:

$$\begin{aligned} P(7, 3) &= 7 \cdot 6 \cdot 5 \\ &= \frac{7 \cdot 6 \cdot 5 \cdot \mathbf{4} \cdot \mathbf{3} \cdot \mathbf{2} \cdot \mathbf{1}}{\mathbf{4} \cdot \mathbf{3} \cdot \mathbf{2} \cdot \mathbf{1}} && \text{Multiply both the numerator and denominator by } 4 \cdot 3 \cdot 2 \cdot 1. \\ &= \frac{7!}{4!} \\ &= \frac{7!}{(7-3)!} \end{aligned}$$

The generalization of this idea gives the following formula.

Formula for Finding $P(n, r)$ The number of permutations of n things r at a time is given by the formula

$$P(n, r) = \frac{n!}{(n-r)!}$$

EXAMPLE 5 Compute **a.** $P(8, 2)$, **b.** $P(7, 5)$, **c.** $P(n, n)$, and **d.** $P(n, 0)$

Solution **a.**

$$\begin{aligned} P(8, 2) &= \frac{8!}{(8-2)!} \\ &= \frac{8 \cdot 7 \cdot 6!}{6!} \\ &= 8 \cdot 7 \\ &= 56 \end{aligned}$$

b.

$$\begin{aligned} P(7, 5) &= \frac{7!}{(7-5)!} \\ &= \frac{7 \cdot 6 \cdot 5 \cdot 4 \cdot 3 \cdot 2!}{2!} \\ &= 7 \cdot 6 \cdot 5 \cdot 4 \cdot 3 \\ &= 2520 \end{aligned}$$

c. $P(n, n) = \dfrac{n!}{(n-n)!} = \dfrac{n!}{0!} = \dfrac{n!}{1} = n!$

d. $P(n, 0) = \dfrac{n!}{(n-0)!} = \dfrac{n!}{n!} = 1$

Parts **c** and **d** of Example 5 establish the following formulas.

Formulas for Finding $P(n, n)$ and $P(n, 0)$

The number of permutations of n things n at a time and n things 0 at a time are given by the formula

$$P(n, n) = n! \qquad \text{and} \qquad P(n, 0) = 1$$

 (a)

EXAMPLE 6

a. In how many ways can a television executive arrange the Saturday night lineup of 6 programs if there are 15 programs to choose from?

b. If there are only 6 programs to choose from?

Solution

a. To find the number of permutations of 15 programs 6 at a time, we use the formula $P(n, r) = \frac{n!}{(n-r)!}$ with $n = 15$ and $r = 6$.

$$P(15, 6) = \frac{15!}{(15-6)!} = \frac{15 \cdot 14 \cdot 13 \cdot 12 \cdot 11 \cdot 10 \cdot 9!}{9!}$$
$$= 15 \cdot 14 \cdot 13 \cdot 12 \cdot 11 \cdot 10$$
$$= 3{,}603{,}600$$

b. To find the number of permutations of 6 programs 6 at a time, we use the formula $P(n, n) = n!$ with $n = 6$.

$$P(6, 6) = 6! = 720$$

Norbert Wiener (1894–1964)
A child prodigy, Norbert Wiener received his Ph.D. from Harvard University at the age of 19. As a professor of Mathematics at M.I.T., Wiener analyzed the nature of information and communication and created a new field called *cybernetics.* Without this study, modern computers would not exist.

Combinations

Suppose that Raul must read 4 books from a reading list of 10 books. The order in which he reads them is not important. For the moment, however, let's assume that order is important and find the number of permutations of 10 things 4 at a time:

$$P(10, 4) = \frac{10!}{(10-4)!}$$
$$= \frac{10 \cdot 9 \cdot 8 \cdot 7 \cdot 6!}{6!}$$
$$= 10 \cdot 9 \cdot 8 \cdot 7$$
$$= 5040$$

If order is important, there are 5040 ways of choosing 4 books when there are 10 books to choose from.

However, because the order in which Raul reads the books does not matter, the previous result of 5040 is too big. Since there are 4! ways of ordering the 4 books that are chosen, the result of 5040 is exactly 4! (or 24) times too big. Thus, the number of choices that Raul has is the number of permutations of 10 things 4 at a time, divided by 24:

$$\frac{P(10, 4)}{24} = \frac{5040}{24}$$
$$= 210$$

Raul has 210 ways of choosing 4 books to read from the list of 10 books.

In situations where order is *not* important, we are interested in **combinations,** not permutations. The symbols $C(n, r)$ and $\binom{n}{r}$ both mean the number of combinations of n things r at a time.

If a selection of r books is chosen from a total of n books, the number of possible selections is $C(n, r)$, and there are $r!$ arrangements of the r books in each selection. If we consider the selected books as an ordered grouping, the number of orderings is $P(n, r)$. Thus, we have

1. $r! \cdot C(n, r) = P(n, r)$

We can divide both sides of Equation 1 by $r!$ to get the formula for finding $C(n, r)$:

$$C(n, r) = \binom{n}{r} = \frac{P(n, r)}{r!} = \frac{n!}{r!(n - r)!}$$

Formula for Finding C(*n*, *r*) The number of combinations of n things r at a time is given by

$$C(n, r) = \frac{n!}{r!(n - r)!}$$

EXAMPLE 7 Compute **a.** $C(8, 5)$, **b.** $\binom{7}{2}$, **c.** $C(n, n)$, and **d.** $C(n, 0)$

Solution **a.** $C(8, 5) = \frac{8!}{5!(8 - 5)!}$
$$= \frac{8 \cdot 7 \cdot 6 \cdot 5!}{5!3!}$$
$$= 8 \cdot 7$$
$$= 56$$

b. $\binom{7}{2} = \frac{7!}{2!(7 - 2)!}$
$$= \frac{7 \cdot 6 \cdot 5!}{2 \cdot 1 \cdot 5!}$$
$$= 21$$

c. $$C(n, n) = \frac{n!}{n!(n-n)!} = \frac{1}{0!} = \frac{1}{1} = 1$$

d. $$C(n, 0) = \frac{n!}{0!(n-0)!} = \frac{n!}{0!n!} = \frac{1}{0!} = \frac{1}{1} = 1$$

The symbol $C(n, 0)$ indicates that we choose 0 things from the available n things. ■

Parts **c** and **d** of Example 7 establish the following formulas.

Formulas for Finding $C(n, n)$ and $C(n, 0)$ The number of combinations of n things n at a time is 1. The number of combinations of n things 0 at a time is 1.

$$C(n, n) = 1 \quad \text{and} \quad C(n, 0) = 1$$

PERSPECTIVE

Winning the Lottery

Gambling is an occasional diversion for some and an obsession for others. Whether it is horse racing, slot machines, or state lotteries, the lure of instant riches is hard to resist.

Many states conduct lotteries, and the systems vary. One scheme is typical: For $1, you have two chances to match 6 numbers chosen from 55 numbers and win a grand prize of about $5 million. How likely are you to win? Is it worth $1 to play the game?

To match the six numbers chosen from 55, you must choose the one winning combination out of C(55, 6) possibilities:

$$C(n, r) = \frac{n!}{r!(n-r)!}$$

$$C(55, 6) = \frac{55!}{6!(55-6)!}$$

$$= \frac{55 \cdot 54 \cdot 53 \cdot 52 \cdot 51 \cdot 50}{6 \cdot 5 \cdot 4 \cdot 3 \cdot 2}$$

$$= 28{,}989{,}675$$

In this game, you have two chances in about 29 million of winning $5 million. Over the long haul, you will win $\frac{2}{29{,}000{,}000}$ of the time, so each ticket is worth $\frac{2}{29{,}000{,}000}$ of $5,000,000, or about 35¢, if you don't have to share the prize with another winner. For every dollar spent to play the game, you can expect to throw away 65¢. This state lottery is a poor bet. Casinos pay better than 50¢ on the dollar, with some slot machines returning 90¢. "You can't win if you're not in!" is the claim of the lottery promoters. A better claim would be "You won't regret if you don't bet!"

EXAMPLE 8 If 15 students want to pick a committee of 4 students to plan a party, how many different committees are possible?

Solution Since the ordering of people on each possible committee is unimportant, we find the number of combinations of 15 people 4 at a time:

$$\begin{aligned} C(15, 4) &= \frac{15!}{4!(15-4)!} \\ &= \frac{15 \cdot 14 \cdot 13 \cdot 12 \cdot 11!}{4 \cdot 3 \cdot 2 \cdot 1 \cdot 11!} \\ &= \frac{15 \cdot 14 \cdot 13 \cdot 12}{4 \cdot 3 \cdot 2 \cdot 1} \\ &= 1365 \end{aligned}$$

There are 1365 possible committees. ■

EXAMPLE 9 A committee in Congress consists of 10 Democrats and 8 Republicans. In how many ways can a subcommittee be chosen if it is to contain 5 Democrats and 4 Republicans?

Solution There are $C(10, 5)$ ways of choosing the 5 Democrats and $C(8, 4)$ ways of choosing the 4 Republicans. By the multiplication principle for events, there are $C(10, 5) \cdot C(8, 4)$ ways of choosing the subcommittee:

$$\begin{aligned} C(10, 5) \cdot C(8, 4) &= \frac{10!}{5!(10-5)!} \cdot \frac{8!}{4!(8-4)!} \\ &= \frac{10 \cdot 9 \cdot 8 \cdot 7 \cdot 6 \cdot 5!}{120 \cdot 5!} \cdot \frac{8 \cdot 7 \cdot 6 \cdot 5 \cdot 4!}{24 \cdot 4!} \\ &= \frac{10 \cdot 9 \cdot 8 \cdot 7 \cdot 6}{120} \cdot \frac{8 \cdot 7 \cdot 6 \cdot 5}{24} \\ &= 17{,}640 \end{aligned}$$

There are 17,640 possible subcommittees. ■

Alternative Form of the Binomial Theorem

We have seen that the expansion of $(x + y)^3$ is

$$(x + y)^3 = \mathbf{1}x^3 + \mathbf{3}x^2y + \mathbf{3}xy^2 + \mathbf{1}y^3$$

and that

$$\binom{3}{0} = 1, \quad \binom{3}{1} = 3, \quad \binom{3}{2} = 3, \quad \text{and} \quad \binom{3}{3} = 1$$

Putting these facts together gives the following way of writing the expansion of $(x + y)^3$:

$$(x + y)^3 = \binom{3}{0}x^3 + \binom{3}{1}x^2y + \binom{3}{2}xy^2 + \binom{3}{3}y^3$$

Likewise, we have

$$(x + y)^4 = \binom{4}{0}x^4 + \binom{4}{1}x^3y + \binom{4}{2}x^2y^2 + \binom{4}{3}xy^3 + \binom{4}{4}y^4$$

The generalization of this idea allows us to state the binomial theorem in an alternative form, using combinatorial notation.

Alternative Form of the Binomial Theorem

If n is any positive integer, then

$$(a + b)^n = \binom{n}{0}a^n + \binom{n}{1}a^{n-1}b + \binom{n}{2}a^{n-2}b^2 + \cdot \cdot \cdot + \binom{n}{r}a^{n-r}b^r + \cdot \cdot \cdot + \binom{n}{n}b^n$$

EXAMPLE 10 Use the alternative form of the binomial theorem to expand $(x + y)^6$.

Solution

$$(x + y)^6 = \binom{6}{0}x^6 + \binom{6}{1}x^5y + \binom{6}{2}x^4y^2 + \binom{6}{3}x^3y^3 + \binom{6}{4}x^2y^4 + \binom{6}{5}xy^5 + \binom{6}{6}y^6$$
$$= x^6 + 6x^5y + 15x^4y^2 + 20x^3y^3 + 15x^2y^4 + 6xy^5 + y^6$$

■

EXAMPLE 11 Use the alternative form of the binomial theorem to expand $(2x - y)^3$.

Solution

$$(2x - y)^3 = [2x + (-y)]^3$$
$$= \binom{3}{0}(2x)^3 + \binom{3}{1}(2x)^2(-y) + \binom{3}{2}(2x)(-y)^2 + \binom{3}{3}(-y)^3$$
$$= 1(2x)^3 + 3(4x^2)(-y) + 3(2x)(-y)^2 + (-y)^3$$
$$= 8x^3 - 12x^2y + 6xy^2 - y^3$$

■

ORALS

1. If there are three books and five records, in how many ways can you pick one book and one record? 15
2. In how many ways can five soldiers stand in line? 120
3. Find $P(3, 1)$ 3
4. Find $P(3, 3)$ 6
5. Find $C(3, 0)$ 1
6. Find $C(3, 3)$ 1

EXERCISE 12.6

1. **Arranging an evening** Kristy intends to go out to dinner and then see a movie. In how many ways can she arrange her evening if she has a choice of five movies and seven restaurants? 35
2. **Travel choices** Paula has five ways to travel from New York to Chicago, three ways to travel from Chicago to Denver, and four ways to travel from Denver to Los Angeles. How many choices are available to Paula if she travels from New York to Los Angeles? 60
3. **Making license plates** How many six-digit license plates can be manufactured? (Note that there are ten choices—0, 1, 2, 3, 4, 5, 6, 7, 8, 9—for each digit.) 1,000,000
4. **Making license plates** How many six-digit license plates can be manufactured if no digit can be repeated? 151,200
5. **Making license plates** How many six-digit license plates can be manufactured if no license number can begin with 0 and if no digit can be repeated? 136,080
6. **Making license plates** How many license plates can be manufactured with two letters followed by four digits? 6,760,000
7. **Phone numbers** How many 7-digit phone numbers are available in area code 815 if no phone number can begin with 0 or 1? 8,000,000
8. **Phone numbers** How many 10-digit phone numbers are available if area codes 000 and 911 cannot be used and if no local number can begin with 0 or 1? 7.984×10^9

In Exercises 9–18, evaluate each expression.

9. $P(5, 5)$ 120
10. $P(4, 4)$ 24
11. $P(5, 3)$ 60
12. $P(3, 2)$ 6
13. $P(2, 2) \cdot P(3, 3)$ 12
14. $P(3, 2) \cdot P(3, 3)$ 36
15. $\dfrac{P(5, 3)}{P(4, 2)}$ 5
16. $\dfrac{P(6, 2)}{P(5, 4)}$ $\frac{1}{4}$
17. $\dfrac{P(6, 2) \cdot P(7, 3)}{P(5, 1)}$ 1260
18. $\dfrac{P(8, 3)}{P(5, 3) \cdot P(4, 3)}$ $\frac{7}{30}$

19. **Lining up** In how many ways can six girls be placed in a line? 720
20. **Arranging books** In how many ways can seven books be placed on a shelf? 5040
21. **Arranging books** In how many ways can four novels and five biographies be arranged on a shelf if the novels are placed first? 2880
22. **Making a ballot** In how many ways can six candidates for mayor and four candidates for the county board be arranged on a ballot if all of the candidates for mayor must be placed first? 17,280
23. **Combination locks** How many permutations does a combination lock have if each combination has three numbers, no two numbers of any combination are equal, and the lock has 25 numbers? 13,800
24. **Combination locks** How many permutations does a combination lock have if each combination has three numbers, no two numbers of any combination are equal, and the lock has 50 numbers? 117,600
25. **Arranging appointments** The receptionist at a dental office has only three appointment times available before next Tuesday, and there are ten patients with a toothache. In how many ways can the receptionist fill those appointments? 720
26. **Computers** In many computers, a *word* consists of 32 *bits*—a string of thirty-two 1's and 0's. How many different words are possible? 4,294,967,296

27. **Palindromes** A palindrome is any word, such as *madam* or *radar,* that reads the same backward and forward. How many five-digit numerical palindromes (like 13531) are there? (*Hint:* A leading zero would be dropped.) 900

28. **Call letters** The call letters of a U.S. commercial radio station have either three or four letters, and the first is either a *W* or a *K*. How many radio stations could this system support? 36,504

In Exercises 29–40, evaluate each expression.

29. $C(5, 3)$ 10

30. $C(5, 4)$ 5

31. $\binom{6}{3}$ 20

32. $\binom{6}{4}$ 15

33. $\binom{5}{4}\binom{5}{3}$ 50

34. $\binom{6}{5}\binom{6}{4}$ 90

35. $\dfrac{C(38, 37)}{C(19, 18)}$ 2

36. $\dfrac{C(25, 23)}{C(40, 39)}$ $\frac{15}{2}$

37. $C(12, 0)C(12, 12)$ 1

38. $\dfrac{C(8, 0)}{C(8, 1)}$ $\frac{1}{8}$

39. $C(n, 2)$ $\frac{n!}{2!(n-2)!}$

40. $C(n, 3)$ $\frac{n!}{3!(n-3)!}$

41. **Planning a picnic** A class of 14 students wants to pick a committee of 3 students to plan a picnic. How many committees are possible? 364

42. **Choosing books** Jeffrey must read 3 books from a reading list of 15 books. How many choices does he have? 455

43. **Forming a committee** The number of three-person committees that can be formed from a group of persons is ten. How many persons are in the group? 5

44. **Forming a committee** The number of three-person committees that can be formed from a group of persons is 20. How many persons are in the group? 6

45. **Winning a lottery** In one state lottery, anyone who picks the correct six numbers (in any order) wins. With the numbers 0 through 99 available, how many choices are possible? 1,192,052,400

46. **Taking a test** The instructions on a test read, *Answer any ten of the following fifteen questions. Then choose one of the remaining questions for homework, and turn in its solution tomorrow.* In how many ways can the questions be chosen? 15,015

47. **Forming a committee** In how many ways can we select a committee of two boys and two girls from a group containing three boys and four girls? 18

48. **Forming a committee** In how many ways can we select a committee of three boys and two girls from a group containing five boys and three girls? 30

49. **Choosing clothes** In how many ways can we select 2 shirts and 3 neckties from a group of 12 shirts and 10 neckties? 7920

50. **Choosing clothes** In how many ways can we select five dresses and two coats from a wardrobe containing nine dresses and three coats? 378

In Exercises 51–56, use the alternative form of the binomial theorem to expand each expression.

51. $(x + y)^4$ $x^4 + 4x^3y + 6x^2y^2 + 4xy^3 + y^4$

52. $(x - y)^2$ $x^2 - 2xy + y^2$

53. $(2x + y)^3$ $8x^3 + 12x^2y + 6xy^2 + y^3$

54. $(2x + 1)^4$ $16x^4 + 32x^3 + 24x^2 + 8x + 1$

55. $(3x - 2)^4$ $81x^4 - 216x^3 + 216x^2 - 96x + 16$

56. $(3 - x^2)^3$ $27 - 27x^2 + 9x^4 - x^6$

In Exercises 57–60, find the indicated term of the binomial expansion.

57. $(x - 5y)^5$; fourth term $-1250x^2y^3$

58. $(2x - y)^5$; third term $80x^3y^2$

59. $(x^2 - y^3)^4$; second term $-4x^6y^3$

60. $(x^3 - y^2)^4$; fourth term $-4x^3y^6$

Writing Exercises *Write a paragraph using your own words.*

1. State the multiplication principle for events.
2. Explain why *permutation lock* would be a better name for a combination lock.

Something to Think About

1. How many ways could five people stand in line if two people insist on standing together? 48
2. How many ways could five people stand in line if two people refuse to stand next to each other? 72

Review Exercises *Solve each equation.*

1. $|2x - 3| = 9$ 6, −3
2. $2x^2 - x = 15$ 3, $-\frac{5}{2}$
3. $\frac{3}{x - 5} = \frac{8}{x}$ 8
4. $\frac{3}{x} = \frac{x - 2}{8}$ 6, −4

PROJECT The Fertile Triangle

Pascal's triangle contains a wealth of interesting patterns. You have seen two in Exercises 38 and 39 of Section 12.1. Here are a few more.

- Find the hockey-stick pattern in the numbers in Illustration 1. What would be the missing number in the rightmost hockey stick? Does this pattern work for larger hockey sticks? Experiment.

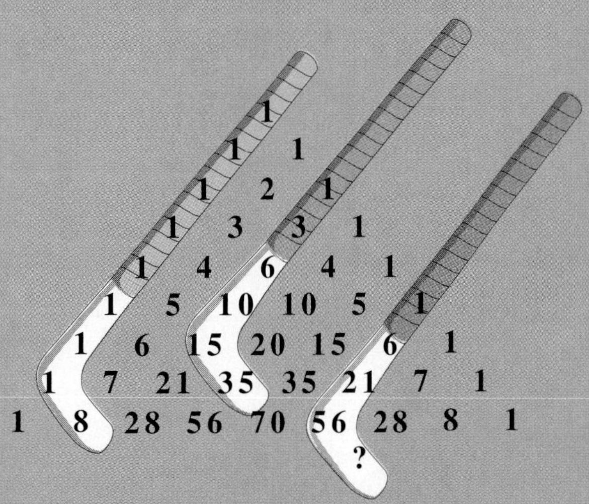

ILLUSTRATION 1

(continued)

PROJECT ■ The Fertile Triangle *(continued)*

- In Illustration 2, find the pattern in the sums of increasingly larger portions of Pascal's triangle. Find the sum of all of the numbers up to and including the row that begins 1 10 45 . . . .

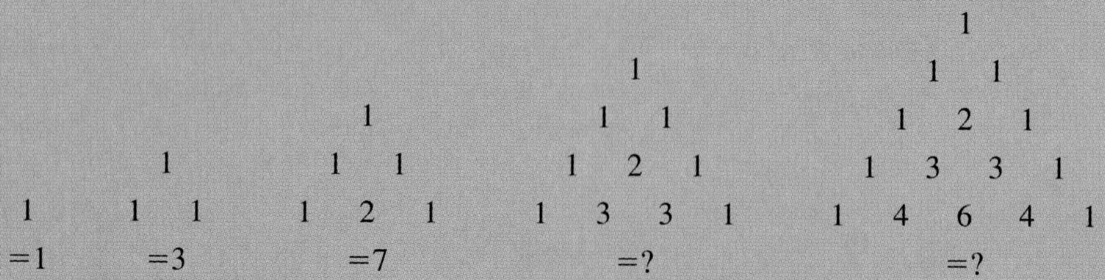

ILLUSTRATION 2

- In Illustration 3, find the pattern in the sums of the squares of the numbers in each row of the triangle. What is the sum of the squares of the numbers in the row that begins 1 10 45 . . .? (*Hint:* Calculate $P(2, 1)$, $P(4, 2)$, $P(6, 3)$, ... Do these numbers appear elsewhere in the triangle?)

$$1^2 = 1$$
$$1^2 + 1^2 = 2$$
$$1^2 + 2^2 + 1^2 = 6$$
$$1^2 + 3^2 + 3^2 + 1^2 = 20$$
$$1^2 + 4^2 + 6^2 + 4^2 + 1^2 = 70$$
$$1^2 + 5^2 + 10^2 + 10^2 + 5^2 + 1^2 = ?$$
$$1^2 + 6^2 + 15^2 + 20^2 + 15^2 + 6^2 + 1^2 = ?$$

ILLUSTRATION 3

- In 1653, Pascal described the triangle in *Treatise on the Arithmetic Triangle,* writing, "I have left out many more properties than I have included. It is remarkable how fertile in properties this triangle is. *Everyone can try his hand.*" Accept Pascal's invitation. Find some of the triangle's patterns for yourself and share your discoveries with your class. Illustration 4 is an idea to get you started.

1
1 1
1 2 1
1 3 3 1
1 4 6 4 1
1 5 10 10 5 1
1 6 15 20 15 6 1
1 7 21 35 35 21 7 1
1 8 28 56 70 56 28 8 1

ILLUSTRATION 4

Chapter Summary

KEY WORDS

arithmetic mean (12.3)
arithmetic sequence (12.3)
binomial theorem (12.1)
combinations (12.6)
event (12.6)
factorial notation (12.1)
Fibonacci sequence (12.3)
geometric means (12.4)
geometric sequence (12.4)
index of summation (12.3)
infinite geometric sequence (12.5)
multiplication principle for events (12.6)
Pascal's triangle (12.1)
permutations (12.6)
sequence (12.3)
summation notation (12.3)

KEY IDEAS

(12.1) The symbol $n!$ (***n* factorial**) is defined as

$$n! = n(n-1)(n-2)(n-3) \cdot \cdots \cdot 3 \cdot 2 \cdot 1$$

where n is a natural number.

$0! = 1$

$n(n-1)! = n!$, provided that n is a natural number.

The binomial theorem:

$$(a+b)^n = a^n + \frac{n!}{1!(n-1)!}a^{n-1}b + \frac{n!}{2!(n-2)!}a^{n-2}b^2 + \frac{n!}{3!(n-3)!}a^{n-3}b^3 + \cdots + b^n$$

(12.2) The binomial theorem can be used to find the nth term of a binomial expansion.

(12.3) An **arithmetic sequence** is a sequence of the form

$$a, a+d, a+2d, a+3d, \ldots, a+(n-1)d$$

where a is the first term, $a+(n-1)d$ is the nth term, and d is the common difference.

If numbers are inserted between two given numbers a and b to form an arithmetic sequence, the inserted numbers are **arithmetic means** between a and b.

The sum of the first n terms of an arithmetic sequence is given by the formula

$$S_n = \frac{n(a+l)}{2} \quad \text{with } l = a+(n-1)d$$

where a is the first term, l is the last (or nth) term, and n is the number of terms in the sequence.

Summation notation:

$$\sum_{k=1}^{n} f(k) = f(1) + f(2) + f(3) + \cdots + f(n)$$

(12.4) A **geometric sequence** is a sequence of the form

$$a, ar, ar^2, ar^3, \ldots, ar^{n-1}$$

where a is the first term, ar^{n-1} is the nth term, and r is the common ratio.

If numbers are inserted between two numbers a and b to form a geometric sequence, the inserted numbers are **geometric means** between a and b.

The sum of the first n terms of a geometric sequence is given by the formula

$$S_n = \frac{a - ar^n}{1-r} \quad (r \neq 1)$$

(continued)

where S_n is the sum, a is the first term, r is the common ratio, and n is the number of terms in the sequence.

(12.5) If r is the common ratio of an infinite geometric sequence, and if $|r| < 1$, then the sum of the terms of the infinite geometric sequence is given by the formula

$$S = \frac{a}{1 - r}$$

where a is the first term and r is the common ratio.

(12.6) The **multiplication principle for events:** If E_1 and E_2 are two events, and if E_1 can be done in a_1 ways and E_2 can be done in a_2 ways, then the event "E_1 followed by E_2" can be done in $a_1 \cdot a_2$ ways.

The formula for computing the number of permutations of n things r at a time is

$$P(n, r) = \frac{n!}{(n - r)!}$$

The number of permutations of n things n and n things 0 at a time are

$$P(n, n) = n! \quad \text{and} \quad P(n, 0) = 1$$

The formula for computing the number of combinations of n things r at a time is

$$C(n, r) = \binom{n}{r} = \frac{n!}{r!(n - r)!}$$

The number of combinations of n things n at a time is

$$C(n, n) = \binom{n}{n} = 1$$

The number of combinations of n things 0 at a time is

$$C(n, 0) = \binom{n}{0} = 1$$

Chapter 12 Review Exercises

In Review Exercises 1–4, evaluate each expression.

1. $(4!)(3!)$ 144

2. $\frac{5!}{3!}$ 20

3. $\frac{6!}{2!(6 - 2)!}$ 15

4. $\frac{12!}{3!(12 - 3)!}$ 220

In Review Exercises 5–8, use the binomial theorem to find each expansion.

5. $(x + y)^5$ $x^5 + 5x^4y + 10x^3y^2 + 10x^2y^3 + 5xy^4 + y^5$

6. $(x - y)^4$ $x^4 - 4x^3y + 6x^2y^2 - 4xy^3 + y^4$

7. $(4x - y)^3$ $64x^3 - 48x^2y + 12xy^2 - y^3$

8. $(x + 4y)^3$ $x^3 + 12x^2y + 48xy^2 + 64y^3$

In Review Exercises 9–12, find the required term in each expansion.

9. $(x + y)^4$; third term $6x^2y^2$

10. $(x - y)^5$; fourth term $-10x^2y^3$

11. $(3x - 4y)^3$; second term $-108x^2y$

12. $(4x + 3y)^4$; third term $864x^2y^2$

13. Write the first five terms of the arithmetic sequence whose ninth term is 242 and whose seventh term is 212. 122, 137, 152, 167, 182

14. Find two arithmetic means between 8 and 25. $\frac{41}{3}, \frac{58}{3}$

15. Find the sum of the first 20 terms of the sequence 11, 18, 25, ... 1550

16. Find the sum of the first ten terms of the sequence 9, $6\frac{1}{2}$, 4, ... $-\frac{45}{2}$

17. Write the first five terms of the geometric sequence whose fourth term is 3 and whose fifth term is $\frac{3}{2}$. 24, 12, 6, 3, $\frac{3}{2}$

18. Find two geometric means between -6 and 384. 24, -96

19. Find the sum of the first eight terms of the sequence $\frac{1}{8}, -\frac{1}{4}, \frac{1}{2}, \ldots$. $-\frac{85}{8}$

20. Find the sum of the first seven terms of the sequence 162, 54, 18, ... $\frac{2186}{9}$

21. Find the sum of the infinite geometric sequence 25, 20, 16, ... 125

22. Change the decimal $0.\overline{05}$ to a common fraction. $\frac{5}{99}$

23. Car depreciation A $5000 car depreciates at the rate of 20% of the previous year's value. How much is the car worth after five years? $1638.40

24. Stock appreciation The value of Mia's stock portfolio is expected to appreciate at the rate of 18% per year. How much will the portfolio be worth in ten years if its current value is $25,700? $134,509.57

25. Planting corn A farmer planted 300 acres in corn this year. He intends to plant an additional 75 acres in corn in each successive year until he has 1200 acres in corn. In how many years will that be? 12 yr

26. Falling object If an object is in free fall, the sequence 16, 48, 80, . . . represents the distance (in feet) that object falls during the first second, during the second second, during the third second, and so on. How far will the object fall during the first ten seconds? 1600 ft

In Review Exercises 27–40, evaluate each expression.

27. $P(7, 7)$ 5040

28. $P(7, 0)$ 1

29. $P(8, 6)$ 20,160

30. $\dfrac{P(9, 6)}{P(10, 7)}$ $\frac{1}{10}$

31. $C(7, 7)$ 1

32. $C(7, 0)$ 1

33. $\dbinom{8}{6}$ 28

34. $\dbinom{9}{6}$ 84

35. $C(6, 3) \cdot C(7, 3)$ 700

36. $\dfrac{C(7, 3)}{C(6, 3)}$ $\frac{7}{4}$

37. $\displaystyle\sum_{k=4}^{6} \frac{1}{2}k$ $\frac{15}{2}$

38. $\displaystyle\sum_{k=2}^{5} 7k^2$ 378

39. $\displaystyle\sum_{k=1}^{4} (3k - 4)$ 14

40. $\displaystyle\sum_{k=10}^{10} 36k$ 360

41. Lining up In how many ways can five persons be arranged in a line? 120

42. Lining up In how many ways can three boys and five girls be arranged in a line if the girls are placed ahead of the boys? 720

43. Choosing people In how many ways can we pick three persons from a group of ten persons? 120

44. Forming a committee In how many ways can we pick a committee of two Democrats and two Republicans from a group containing five Democrats and six Republicans? 150

Chapter 12 Test

1. Evaluate $\frac{7!}{4!}$. 210

2. Evaluate 0!. 1

3. Find the second term in the expansion of $(x - y)^5$. $-5x^4y$

4. Find the third term in the expansion of $(x + 2y)^4$. $24x^2y^2$

5. Find the tenth term of an arithmetic sequence whose first three terms are 3, 10, and 17. 66

6. Find the sum of the first 12 terms of the sequence $-2, 3, 8, \ldots$. 306

7. Find two arithmetic means between 2 and 98. 34, 66

8. Evaluate $\sum_{k=1}^{3} (2k - 3)$. 3

9. Find the seventh term of the geometric sequence whose first three terms are $-\frac{1}{9}$, $-\frac{1}{3}$, and -1. -81

10. Find the sum of the first six terms of the sequence $\frac{1}{27}, \frac{1}{9}, \frac{1}{3}, \ldots$. $\frac{364}{27}$

11. Find two geometric means between 3 and 648. 18, 108

12. Find the sum of all of the terms of the infinite geometric sequence $9, 3, 1, \ldots$. $\frac{27}{2}$

In Problems 13–20, find the value of each expression.

13. $P(5, 4)$ 120

14. $P(8, 8)$ 40,320

15. $C(6, 4)$ 15

16. $C(8, 3)$ 56

17. $C(6, 0) \cdot P(6, 5)$ 720

18. $P(8, 7) \cdot C(8, 7)$ 322,560

19. $\frac{P(6, 4)}{C(6, 4)}$ 24

20. $\frac{C(9, 6)}{P(6, 4)}$ $\frac{7}{30}$

21. In how many ways can we pick 3 persons from a group of 7 persons? 35

22. From a group of 5 men and 4 women, how many 3-person committees can be chosen that will include exactly 2 women? 30

Cumulative Review Exercises

In Exercises 1–2, find the symmetries of the graph of each function.

1. $y = x^4 + x^2 - 5$ y-axis

2. $y = x^3 - x$ origin

In Exercises 3–4, graph each function and give its domain and range. Check your work with a graphing calculator.

3. $y = \frac{1}{2}x^2 - 2$

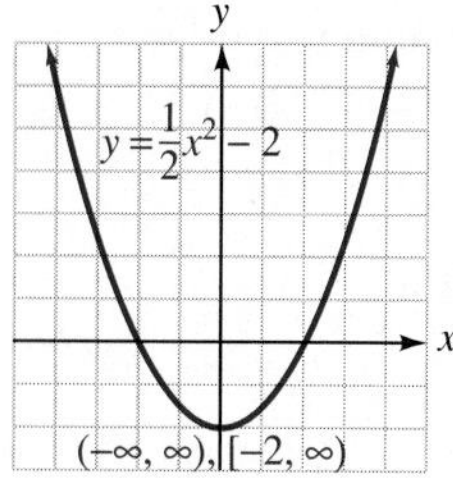

4. $y = (x - 2)^3$

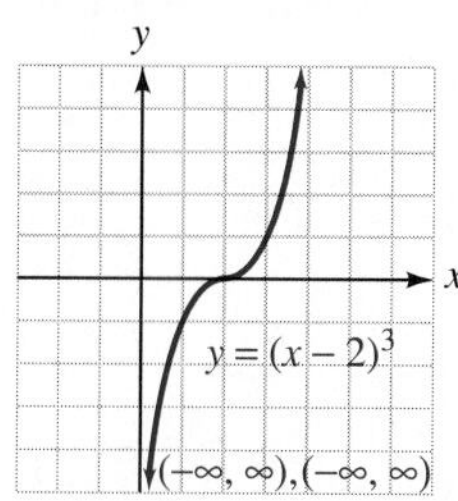

In Exercises 5–6, $f(x) = 3x$ and $g(x) = x^2 - 1$.

5. Find $(f \cdot g)(x)$. $3x^3 - 3x$

6. Find $(f \circ g)(x)$. $3x^2 - 3$

In Exercises 7–8, find the inverse relation of each function and tell whether the inverse is a function.

7. $y = f(x) = \dfrac{x + 3}{3}$ $f^{-1}(x) = 3x - 3$, yes

8. $y = f(x) = x^2 - 3$ $y = \pm\sqrt{x + 3}$, no

In Exercises 9–10, graph each equation.

9. $x^2 + (y + 1)^2 = 9$

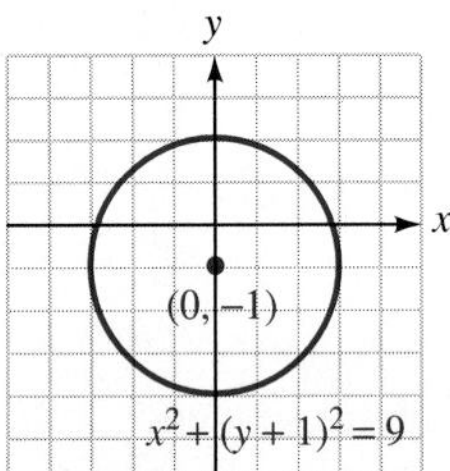

10. $x^2 - 9(y + 1)^2 = 9$

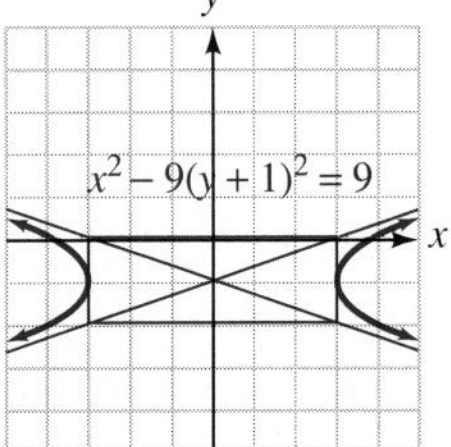

11. Graph $y = \left(\dfrac{1}{2}\right)^x$.

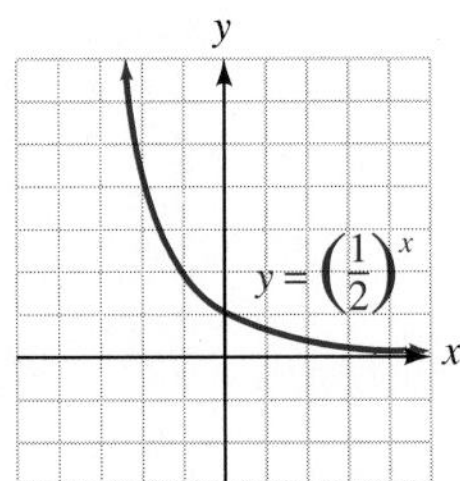

12. Write $y = \log_2 x$ as an exponential equation. $2^y = x$

In Exercises 13–16, find x.

13. $\log_x 25 = 2$ 5

14. $\log_5 125 = x$ 3

15. $\log_3 x = -3$ $\frac{1}{27}$

16. $\log_5 x = 0$ 1

17. Find the inverse of $y = \log_2 x$. $y = 2^x$

18. Simplify $\log 10^x$. x

In Exercises 19–22, $\log 7 = 0.8451$ *and* $\log 14 = 1.1461$. *Evaluate each expression without using a calculator.*

19. $\log 98$ 1.9912

20. $\log 2$ 0.301

21. $\log 49$ 1.6902

22. $\log \frac{7}{5}$ (*Hint:* $\log 10 = 1$.) 0.1461

23. Solve: $2^{x+2} = 3^x$. $\frac{2 \log 2}{\log 3 - \log 2}$

24. Solve: $2 \log 5 + \log x - \log 4 = 2$. 16

In Exercises 25–26, use a calculator.

25. Boat depreciation How much will a \$9000 boat be worth after 9 years if it depreciates 12% per year? \$2848.31

26. Find $\log_6 8$. 1.16056

27. Evaluate $\frac{6!7!}{5!}$. 30,240

28. Use the binomial theorem to expand $(3a - b)^4$. $81a^4 - 108a^3b + 54a^2b^2 - 12ab^3 + b^4$

29. Find the seventh term in the expansion of $(2x - y)^8$. $112x^2y^6$

30. Find the 20th term of an arithmetic sequence with a first term of -11 and a common difference of 6. 103

31. Find the sum of the first 20 terms of an arithmetic sequence with a first term of 6 and a common difference of 3. 690

32. Insert two arithmetic means between -3 and 30. 8 and 19

33. Evaluate $\sum_{k=1}^{3} 3k^2$. 42

34. Evaluate $\sum_{k=3}^{5} (2k + 1)$. 27

35. Find the seventh term of a geometric sequence with a first term of $\frac{1}{27}$ and a common ratio of 3. 27

36. Find the sum of the first ten terms of the sequence $\frac{1}{64}, \frac{1}{32}, \frac{1}{16}, \ldots$. $\frac{1023}{64}$

37. Insert two geometric means between -3 and 192. 12, -48

38. Find the sum of all the terms of the sequence $9, 3, 1, \ldots$. $\frac{27}{2}$

39. Evaluate $P(9, 3)$. 504

40. Evaluate $C(7, 4)$. 35

41. Evaluate $\dfrac{C(8, 4)C(8, 0)}{P(6, 2)}$. $\frac{7}{3}$

42. If $n > 1$, which is smaller: $P(n, n)$ or $C(n, n)$? $C(n\ n)$

43. Lining up In how many ways can seven people stand in a line? 5040

44. Forming a committee In how many ways can a committee of three people be chosen from a group containing nine people? 84

CAREERS & MATHEMATICS

Actuary

Why do young persons pay more for automobile insurance than older persons? How much should an insurance policy cost? How much should an organization contribute each year to its pension fund? Answers to these and similar questions are provided by actuaries, who design insurance and pension plans and follow their performance to make sure that the plans are maintained on a sound financial basis.

Qualifications A good educational background for a beginning job in a large life insurance or casualty insurance company is a bachelor's degree in mathematics or statistics; a degree in actuarial science is even better. Companies prefer well-rounded individuals with a liberal arts background, including social science and communications courses, as well as a good technical background.

Job Outlook Shifts in the age distribution of the population will result in large increases in the number of people with established careers and family responsibilities. This group will account for increases in sales of insurance and pension plans. The outlook for employment is very good through the year 2005. Job opportunities will be best for new college graduates who have passed at least two actuarial examinations while still in school and have a strong mathematical and statistical background.

Example Application 1 An **annuity** is a sequence of equal payments made periodically over a length of time. The sum of the payments and the interest earned during the **term** of the annuity is called the **amount** of the annuity.

After a sales clerk works six months, her employer will begin an annuity for her and will contribute $500 semiannually to a fund that pays 8% annual interest. After she has been employed two years, what will be the amount of her annuity?

Solution Because the payments are to be made semiannually, there will be four payments of $500, each earning a rate of 4% per six-month period. These payments will occur at the end of 6 months, 12 months, 18 months, and 24 months. The first payment, to be made after 6 months, will earn interest for three interest periods. Thus, the amount of the first payment after two years is $\$500(1.04)^3$. The amounts of each of the four payments after two years are shown in Table 1.

Payment (at the end of period)	Amount of payment at the end of 2 years
1	$\$500(1.04)^3 = \$\ 562.43$
2	$\$500(1.04)^2 = \$\ 540.80$
3	$\$500(1.04)^1 = \$\ 520.00$
4	$\$500 = \$\ 500.00$
	$A_n = \$2123.23$

TABLE 1

The amount of the annuity is the sum of the amounts of the individual payments. This sum is \$2123.23.

Example Application 2

The **present value** of the above annuity is the lump-sum principal that must be invested on the first day of her employment at 8% annual interest, compounded semiannually, to grow to \$2123.23 at the end of two years. Find the present value of the annuity.

Solution A sum of $\$500(1.04)^{-1}$ invested at 4% interest per six-month period will grow to be the first payment of \$500, to be made after 6 months. A sum of $\$500(1.04)^{-2}$ will grow to be the second payment of \$500, to be made after 12 months. The present values of each of the four payments are shown in Table 2.

Payment	Present value of each payment
1	$\$500(1.04)^{-1} = \$\ 480.77$
2	$\$500(1.04)^{-2} = \$\ 462.28$
3	$\$500(1.04)^{-3} = \$\ 444.50$
4	$\$500(1.04)^{-4} = \$\ 427.40$
	Present value = \$1814.95

TABLE 2

The present value of the annuity is \$1814.95. Verify that \$1814.95 will grow to \$2123.23 in two years at 8% annual interest, compounded semiannually.

EXERCISES

1. Find the amount of an annuity if \$1000 is paid semiannually for two years at 6% annual interest. Assume that the first of the four payments is made immediately.
2. Find the present value of the annuity of Exercise 1.
3. Note that the amounts for the payments in the Table 1 form a geometric progression. Verify the answer for the first example by using the formula for the sum of a geometric progression.
4. Note that the present values for the payments in the Table 2 form a geometric progression. Verify the answer for the second example by using the formula for the sum of a geometric progression.

(*Answers:* **1.** \$4309.14 **2.** \$3828.61)

APPENDIX I SAMPLE FINAL EXAMINATION

1. How many prime numbers are there in the interval from 40 to 50?
(**a.**) 3 **b.** 4 **c.** 5 **d.** none of the above

2. The commutative property of multiplication is written symbolically as
(**a.**) $ab = ba$ **b.** $(ab)c = a(bc)$ **c.** If $a = b$, then $b = a$ **d.** none of the above

3. If $a = 3$, $b = -2$, and $c = 6$, the value of $\frac{c - ab}{bc}$ is
a. 1 **b.** 0 (**c.**) -1 **d.** none of the above

4. Evaluate $\frac{1}{2} + \frac{3}{4} \div \frac{5}{6}$.
a. $\frac{3}{2}$ **b.** $\frac{7}{2}$ **c.** $\frac{9}{8}$ (**d.**) none of the above

5. The expression $\left(\frac{a^2}{a^5}\right)^{-5}$ can be written as
a. 15 **b.** a^{-15} (**c.**) a^{15} **d.** none of the above

6. Write 0.0000234 in scientific notation.
(**a.**) 2.34×10^{-5} **b.** 2.34×10^5 **c.** $234. \times 10^{-7}$
d. none of the above

7. If $P(x) = 2x^2 - x - 1$, find $P(-1)$.
a. 4 **b.** 0 (**c.**) 2 **d.** none of the above

8. Simplify: $(3x + 2) - (2x - 1) + (x - 3)$.
a. $2x - 2$ (**b.**) $2x$ **c.** 2 **d.** none of the above

9. Multiply: $(3x - 2)(2x + 3)$.
(**a.**) $6x^2 + 5x - 6$ **b.** $6x^2 - 5x - 6$ **c.** $6x^2 + 5x + 6$
d. none of the above

10. Divide: $2x + 1\overline{)2x^2 - 3x - 2}$
a. $x + 2$ **b.** $x - 1$ (**c.**) $x - 2$ **d.** none of the above

11. Solve for x: $5x - 3 = -2x + 10$.
a. $\frac{11}{7}$ **b.** $\frac{13}{3}$ **c.** 1 (**d.**) none of the above

12. The sum of two consecutive odd integers is 44. The product of the integers is
a. an even integer **b.** 463 (**c.**) 483 **d.** none of the above

13. Solve for x: $2ax - a = b + x$.
a. $\frac{a + b}{2a}$ (**b.**) $\frac{a + b}{2a - 1}$ **c.** $a + b - 2a$
d. $-a + b + 1$

14. The sum of the solutions of $|2x + 5| = 13$ is
a. 4 **b.** 8 **c.** 12 (**d.**) none of the above

15. Solve for x: $-2x + 5 > 9$.
a. $x > 7$ **b.** $x < 7$ (**c.**) $x < -2$ **d.** $x > -2$

16. Solve for x: $|2x - 5| \le 9$.
a. $2 \le x \le 7$ **b.** $-2 \ge x$ and $x \le 7$ **c.** $x \le 7$
(**d.**) $-2 \le x \le 7$

17. Factor completely: $3ax^2 + 6a^2x$.
a. $3(ax^2 + 2a^2x)$ **b.** $3a(x^2 + 2ax)$ **c.** $3x(ax + 2a^2)$
(**d.**) none of the above

18. The sum of the prime factors of $x^4 - 16$ is
a. $2x^2$ (**b.**) $x^2 + 2x + 4$ **c.** $4x + 4$ **d.** none of the above

19. The sum of the factors of $8x^2 - 2x - 3$ is
(**a.**) $6x - 2$ **b.** $8x - 3$ **c.** $6x + 2$ **d.** none of the above

20. One of the factors of $27a^3 + 8$ is
a. $3a - 2$ **b.** $9a^2 + 12a + 4$ **c.** $9a^2 + 6a + 4$
(**d.**) none of the above

21. The smallest solution of the equation $6x^2 - 5x - 6 = 0$ is
a. $-\frac{3}{2}$ **b.** $\frac{2}{3}$ (**c.**) $-\frac{2}{3}$ **d.** none of the above

22. Simplify: $\frac{x^2 + 5x + 6}{x^2 - 9}$.
a. $-\frac{2}{3}$ (**b.**) $\frac{x + 2}{x - 3}$ **c.** $\frac{-5x + 6}{9}$ **d.** none of the above

23. Simplify: $\dfrac{3x+6}{x+3} - \dfrac{x^2-4}{x^2+x-6}$.
a. $\dfrac{3(x+2)(x+2)}{(x+3)(x+3)}$ **b.** 3 **c.** $\dfrac{1}{3}$ (**d.**) none of these

24. The numerator of the sum $\dfrac{y}{x+y} + \dfrac{x}{x-y}$ is
(**a.**) $x^2 + 2xy - y^2$ **b.** $y + x$ **c.** $y - x$ **d.** none of the above

25. Simplify: $\dfrac{\frac{1}{x}+\frac{1}{y}}{\frac{1}{y}}$
a. $\dfrac{1}{xy}$ **b.** 1 (**c.**) $\dfrac{y+x}{x}$ **d.** none of the above

26. Solve for y: $\dfrac{2}{y+1} = \dfrac{1}{y+1} - \dfrac{1}{3}$.
a. 4 **b.** 3 (**c.**) -4 **d.** none of the above

27. The sum of the x- and y-intercepts of the graph of $2x + 3y = 6$ is
a. $-\dfrac{2}{3}$ **b.** 0 (**c.**) 5 **d.** none of the above

28. The slope of the line passing through $(3, -2)$ and $(5, -1)$ is
a. $-\dfrac{1}{2}$ **b.** 2 **c.** -2 (**d.**) $\dfrac{1}{2}$

29. The graphs of the equations $\begin{cases} 2x - 3y = 4 \\ 3x + 2y = 1 \end{cases}$
a. are parallel (**b.**) are perpendicular **c.** do not intersect **d.** are the same line

30. The equation of the line passing through $(-2, 5)$ and $(6, 7)$ is
a. $y = -\dfrac{1}{4}x - \dfrac{11}{2}$ (**b.**) $y = \dfrac{1}{4}x + \dfrac{11}{2}$
c. $y = \dfrac{1}{4}x - \dfrac{11}{2}$ **d.** $y = -\dfrac{1}{4}x + \dfrac{11}{2}$

31. If $g(x) = x^2 - 3$, then $g(t + 1)$ is
a. $t^2 - 2$ **b.** -2 **c.** $t - 2$ (**d.**) $t^2 + 2t - 2$

32. Assume that d varies directly with t. Find the constant of variation if $d = 12$ when $t = 3$.
a. 36 (**b.**) 4 **c.** -36 **d.** none of the above

33. The expression $x^{a/2}x^{a/5}$ can be expressed as
a. $x^{a/7}$ (**b.**) $x^{7a/10}$ **c.** $x^{a/10}$ **d.** $x^{a/5}$

34. The product of $(x^{1/2} + 2)$ and $(x^{-1/2} - 2)$ is
a. -3 **b.** $-3 + 2x$ (**c.**) $-3 - 2x^{1/2} + 2x^{-1/2}$
d. none of the above

35. Completely simplify $\sqrt{112a^3}$ $(a \geq 0)$.
a. $a\sqrt{112a}$ **b.** $4a^2\sqrt{7a}$ **c.** $4\sqrt{7a^2}$ (**d.**) none of the above

36. Simplify and combine terms: $\sqrt{50} - \sqrt{98} + \sqrt{128}$.
a. $-2\sqrt{2} + \sqrt{128}$ (**b.**) $6\sqrt{2}$ **c.** $20\sqrt{2}$
d. $-4\sqrt{2}$

37. Rationalize the denominator and simplify: $\dfrac{3}{2-\sqrt{3}}$.
a. $-6 - \sqrt{3}$ (**b.**) $6 + 3\sqrt{3}$ **c.** $2(2 + \sqrt{3})$
d. $6(2 + \sqrt{3})$

38. The distance between the points $(-2, 3)$ and $(6, -8)$ is
(**a.**) $\sqrt{185}$ **b.** $\sqrt{41}$ **c.** $\sqrt{57}$ **d.** none of the above

39. Solve for x: $\sqrt{x+7} - 2x = -1$.
(**a.**) 2 **b.** $-\dfrac{3}{4}$ **c.** solutions are extraneous
d. $2, -\dfrac{3}{4}$

40. The graph of $y > 3x + 2$ contains no points in
a. quadrant I **b.** quadrant II **c.** quadrant III
(**d.**) quadrant IV

41. The quadratic formula is
a. $x = \dfrac{b \pm \sqrt{b^2 - 4ac}}{2a}$ **b.** $x = \dfrac{-b \pm \sqrt{b^2 - 4ac}}{a}$
c. $x = \dfrac{-b \pm \sqrt{b - 4ac}}{a}$ (**d.**) $x = \dfrac{-b \pm \sqrt{b^2 - 4ac}}{2a}$

42. Write the complex number $(2 + 3i)^2$ in $a + bi$ form.
(**a.**) $-5 + 12i$ **b.** -5 **c.** $13 + 12i$ **d.** 13

43. Write the complex number $\dfrac{i}{3+i}$ in $a + bi$ form.
a. $\dfrac{1}{3} + 0i$ (**b.**) $\dfrac{1}{10} + \dfrac{3}{10}i$ **c.** $\dfrac{1}{8} + \dfrac{3}{8}i$ **d.** none of the above

44. The vertex of the parabola determined by $y = 2x^2 + 4x - 3$ is at point
a. $(0, -3)$ **b.** $(2, 13)$ (**c.**) $(-1, -5)$ **d.** $(1, 3)$

45. Solve for x: $\frac{2}{x} < 3$.

a. $0 < x < \frac{2}{3}$ **b.** $x > \frac{2}{3}$ **(c.)** $x < 0$ or $x > \frac{2}{3}$

d. $x < 0$ and $x > \frac{2}{3}$

46. If $f(x) = 2x^2 + 1$, then $f(3) =$

a. 7 **(b.)** 19 **c.** 17 **d.** 37

47. The inverse function of $y = 3x + 2$ is

a. $y = \frac{x+2}{3}$ **b.** $x = 3y - 2$ **(c.)** $y = \frac{x-2}{3}$

d. $x = \frac{y-2}{3}$

48. The equation of a circle with center at $(-2, 4)$ and radius of 4 units is

(a.) $(x+2)^2 + (y-4)^2 = 16$
b. $(x+2)^2 + (y-4)^2 = 4$
c. $(x+2)^2 + (y-4)^2 = 2$
d. $(x-2)^2 + (y+4)^2 = 16$

49. The sum of the solutions of the system

$$\begin{cases} \frac{4}{x} + \frac{2}{y} = 2 \\ \frac{2}{x} - \frac{3}{y} = -1 \end{cases}$$ is

(a.) 6 **b.** -6 **c.** 2 **d.** -4

50. The y value of the solution of the system $\begin{cases} 4x + 6y = 5 \\ 8x - 9y = 3 \end{cases}$ is

a. $\frac{3}{4}$ **(b.)** $\frac{1}{3}$ **c.** $\frac{1}{2}$ **d.** $\frac{2}{3}$

51. The value of the determinant $\begin{vmatrix} 2 & -3 \\ 4 & 4 \end{vmatrix}$ is

a. 0 **(b.)** 20 **c.** -4 **d.** -20

52. The value of z in the system $\begin{cases} x + y + z = 4 \\ 2x + y + z = 6 \\ 3x + y + 2z = 8 \end{cases}$ is

(a.) 0 **b.** 1 **c.** 2 **d.** 3

53. $\log_a N = x$ means

(a.) $a^x = N$ **b.** $a^N = x$ **c.** $x^N = a$ **d.** $N^a = x$

54. $\log_2 \frac{1}{32} =$

a. 5 **b.** $\frac{1}{5}$ **(c.)** -5 **d.** $-\frac{1}{5}$

55. $\log 7 + \log 5 =$

a. $\log 12$ **b.** $\log \frac{7}{12}$ **c.** $\log 2$ **(d.)** $\log 35$

56. $b^{\log_b x} =$

a. b **(b.)** x **c.** 10 **d.** 0

57. Solve for y: $\log y + \log(y+3) = 1$.

a. $-5, 2$ **(b.)** 2 **c.** 5 **d.** none of the above

58. The coefficient of the third term in the expansion of $(a+b)^6$ is

a. 30 **b.** 2! **(c.)** 15 **d.** 120

59. Compute $P(7, 3)$.

a. 35 **(b.)** 210 **c.** 21 **d.** none of the above

60. Compute $C(7, 3)$.

(a.) 35 **b.** 210 **c.** $7! \cdot 3!$ **d.** none of the above

61. Find the 100th term of 2, 5, 8, 11, ...

a. 301 **(b.)** 299 **c.** 297 **d.** 295

62. Find the sum of all the terms of $1, \frac{1}{3}, \frac{1}{9}, \frac{1}{27}, \ldots$.

a. 2 **b.** $\frac{2}{3}$ **(c.)** $\frac{3}{2}$ **d.** $\frac{5}{3}$

APPENDIX II TABLES

TABLE A Powers and Roots

n	n^2	$\sqrt{n}$	n^3	$\sqrt[3]{n}$	n	n^2	$\sqrt{n}$	n^3	$\sqrt[3]{n}$
1	1	1.000	1	1.000	51	2,601	7.141	132,651	3.708
2	4	1.414	8	1.260	52	2,704	7.211	140,608	3.733
3	9	1.732	27	1.442	53	2,809	7.280	148,877	3.756
4	16	2.000	64	1.587	54	2,916	7.348	157,464	3.780
5	25	2.236	125	1.710	55	3,025	7.416	166,375	3.803
6	36	2.449	216	1.817	56	3,136	7.483	175,616	3.826
7	49	2.646	343	1.913	57	3,249	7.550	185,193	3.849
8	64	2.828	512	2.000	58	3,364	7.616	195,112	3.871
9	81	3.000	729	2.080	59	3,481	7.681	205,379	3.893
10	100	3.162	1,000	2.154	60	3,600	7.746	216,000	3.915
11	121	3.317	1,331	2.224	61	3,721	7.810	226,981	3.936
12	144	3.464	1,728	2.289	62	3,844	7.874	238,328	3.958
13	169	3.606	2,197	2.351	63	3,969	7.937	250,047	3.979
14	196	3.742	2,744	2.410	64	4,096	8.000	262,144	4.000
15	225	3.873	3,375	2.466	65	4,225	8.062	274,625	4.021
16	256	4.000	4,096	2.520	66	4,356	8.124	287,496	4.041
17	289	4.123	4,913	2.571	67	4,489	8.185	300,763	4.062
18	324	4.243	5,832	2.621	68	4,624	8.246	314,432	4.082
19	361	4.359	6,859	2.668	69	4,761	8.307	328,509	4.102
20	400	4.472	8,000	2.714	70	4,900	8.367	343,000	4.121
21	441	4.583	9,261	2.759	71	5,041	8.426	357,911	4.141
22	484	4.690	10,648	2.802	72	5,184	8.485	373,248	4.160
23	529	4.796	12,167	2.844	73	5,329	8.544	389,017	4.179
24	576	4.899	13,824	2.884	74	5,476	8.602	405,224	4.198
25	625	5.000	15,625	2.924	75	5,625	8.660	421,875	4.217
26	676	5.099	17,576	2.962	76	5,776	8.718	438,976	4.236
27	729	5.196	19,683	3.000	77	5,929	8.775	456,533	4.254
28	784	5.292	21,952	3.037	78	6,084	8.832	474,552	4.273
29	841	5.385	24,389	3.072	79	6,241	8.888	493,039	4.291
30	900	5.477	27,000	3.107	80	6,400	8.944	512,000	4.309
31	961	5.568	29,791	3.141	81	6,561	9.000	531,441	4.327
32	1,024	5.657	32,768	3.175	82	6,724	9.055	551,368	4.344
33	1,089	5.745	35,937	3.208	83	6,889	9.110	571,787	4.362
34	1,156	5.831	39,304	3.240	84	7,056	9.165	592,704	4.380
35	1,225	5.916	42,875	3.271	85	7,225	9.220	614,125	4.397
36	1,296	6.000	46,656	3.302	86	7,396	9.274	636,056	4.414
37	1,369	6.083	50,653	3.332	87	7,569	9.327	658,503	4.431
38	1,444	6.164	54,872	3.362	88	7,744	9.381	681,472	4.448
39	1,521	6.245	59,319	3.391	89	7,921	9.434	704,969	4.465
40	1,600	6.325	64,000	3.420	90	8,100	9.487	729,000	4.481
41	1,681	6.403	68,921	3.448	91	8,281	9.539	753,571	4.498
42	1,764	6.481	74,088	3.476	92	8,464	9.592	778,688	4.514
43	1,849	6.557	79,507	3.503	93	8,649	9.644	804,357	4.531
44	1,936	6.633	85,184	3.530	94	8,836	9.695	830,584	4.547
45	2,025	6.708	91,125	3.557	95	9,025	9.747	857,375	4.563
46	2,116	6.782	97,336	3.583	96	9,216	9.798	884,736	4.579
47	2,209	6.856	103,823	3.609	97	9,409	9.849	912,673	4.595
48	2,304	6.928	110,592	3.634	98	9,604	9.899	941,192	4.610
49	2,401	7.000	117,649	3.659	99	9,801	9.950	970,299	4.626
50	2,500	7.071	125,000	3.684	100	10,000	10.000	1,000,000	4.642

TABLE B Base-10 Logarithms

N	0	1	2	3	4	5	6	7	8	9
1.0	.0000	.0043	.0086	.0128	.0170	.0212	.0253	.0294	.0334	.0374
1.1	.0414	.0453	.0492	.0531	.0569	.0607	.0645	.0682	.0719	.0755
1.2	.0792	.0828	.0864	.0899	.0934	.0969	.1004	.1038	.1072	.1106
1.3	.1139	.1173	.1206	.1239	.1271	.1303	.1335	.1367	.1399	.1430
1.4	.1461	.1492	.1523	.1553	.1584	.1614	.1644	.1673	.1703	.1732
1.5	.1761	.1790	.1818	.1847	.1875	.1903	.1931	.1959	.1987	.2014
1.6	.2041	.2068	.2095	.2122	.2148	.2175	.2201	.2227	.2253	.2279
1.7	.2304	.2330	.2355	.2380	.2405	.2430	.2455	.2480	.2504	.2529
1.8	.2553	.2577	.2601	.2625	.2648	.2672	.2695	.2718	.2742	.2765
1.9	.2788	.2810	.2833	.2856	.2878	.2900	.2923	.2945	.2967	.2989
2.0	.3010	.3032	.3054	.3075	.3096	.3118	.3139	.3160	.3181	.3201
2.1	.3222	.3243	.3263	.3284	.3304	.3324	.3345	.3365	.3385	.3404
2.2	.3424	.3444	.3464	.3483	.3502	.3522	.3541	.3560	.3579	.3598
2.3	.3617	.3636	.3655	.3674	.3692	.3711	.3729	.3747	.3766	.3784
2.4	.3802	.3820	.3838	.3856	.3874	.3892	.3909	.3927	.3945	.3962
2.5	.3979	.3997	.4014	.4031	.4048	.4065	.4082	.4099	.4116	.4133
2.6	.4150	.4166	.4183	.4200	.4216	.4232	.4249	.4265	.4281	.4298
2.7	.4314	.4330	.4346	.4362	.4378	.4393	.4409	.4425	.4440	.4456
2.8	.4472	.4487	.4502	.4518	.4533	.4548	.4564	.4579	.4594	.4609
2.9	.4624	.4639	.4654	.4669	.4683	.4698	.4713	.4728	.4742	.4757
3.0	.4771	.4786	.4800	.4814	.4829	.4843	.4857	.4871	.4886	.4900
3.1	.4914	.4928	.4942	.4955	.4969	.4983	.4997	.5011	.5024	.5038
3.2	.5051	.5065	.5079	.5092	.5105	.5119	.5132	.5145	.5159	.5172
3.3	.5185	.5198	.5211	.5224	.5237	.5250	.5263	.5276	.5289	.5302
3.4	.5315	.5328	.5340	.5353	.5366	.5378	.5391	.5403	.5416	.5428
3.5	.5441	.5453	.5465	.5478	.5490	.5502	.5514	.5527	.5539	.5551
3.6	.5563	.5575	.5587	.5599	.5611	.5623	.5635	.5647	.5658	.5670
3.7	.5682	.5694	.5705	.5717	.5729	.5740	.5752	.5763	.5775	.5786
3.8	.5798	.5809	.5821	.5832	.5843	.5855	.5866	.5877	.5888	.5899
3.9	.5911	.5922	.5933	.5944	.5955	.5966	.5977	.5988	.5999	.6010
4.0	.6021	.6031	.6042	.6053	.6064	.6075	.6085	.6096	.6107	.6117
4.1	.6128	.6138	.6149	.6160	.6170	.6180	.6191	.6201	.6212	.6222
4.2	.6232	.6243	.6253	.6263	.6274	.6284	.6294	.6304	.6314	.6325
4.3	.6335	.6345	.6355	.6365	.6375	.6385	.6395	.6405	.6415	.6425
4.4	.6435	.6444	.6454	.6464	.6474	.6484	.6493	.6503	.6513	.6522
4.5	.6532	.6542	.6551	.6561	.6571	.6580	.6590	.6599	.6609	.6618
4.6	.6628	.6637	.6646	.6656	.6665	.6675	.6684	.6693	.6702	.6712
4.7	.6721	.6730	.6739	.6749	.6758	.6767	.6776	.6785	.6794	.6803
4.8	.6812	.6821	.6830	.6839	.6848	.6857	.6866	.6875	.6884	.6893
4.9	.6902	.6911	.6920	.6928	.6937	.6946	.6955	.6964	.6972	.6981
5.0	.6990	.6998	.7007	.7016	.7024	.7033	.7042	.7050	.7059	.7067
5.1	.7076	.7084	.7093	.7101	.7110	.7118	.7126	.7135	.7143	.7152
5.2	.7160	.7168	.7177	.7185	.7193	.7202	.7210	.7218	.7226	.7235
5.3	.7243	.7251	.7259	.7267	.7275	.7284	.7292	.7300	.7308	.7316
5.4	.7324	.7332	.7340	.7348	.7356	.7364	.7372	.7380	.7388	.7396

TABLE B *(continued)*

N	0	1	2	3	4	5	6	7	8	9
5.5	.7404	.7412	.7419	.7427	.7435	.7443	.7451	.7459	.7466	.7474
5.6	.7482	.7490	.7497	.7505	.7513	.7520	.7528	.7536	.7543	.7551
5.7	.7559	.7566	.7574	.7582	.7589	.7597	.7604	.7612	.7619	.7627
5.8	.7634	.7642	.7649	.7657	.7664	.7672	.7679	.7686	.7694	.7701
5.9	.7709	.7716	.7723	.7731	.7738	.7745	.7752	.7760	.7767	.7774
6.0	.7782	.7789	.7796	.7803	.7810	.7818	.7825	.7832	.7839	.7846
6.1	.7853	.7860	.7868	.7875	.7882	.7889	.7896	.7903	.7910	.7917
6.2	.7924	.7931	.7938	.7945	.7952	.7959	.7966	.7973	.7980	.7987
6.3	.7993	.8000	.8007	.8014	.8021	.8028	.8035	.8041	.8048	.8055
6.4	.8062	.8069	.8075	.8082	.8089	.8096	.8102	.8109	.8116	.8122
6.5	.8129	.8136	.8142	.8149	.8156	.8162	.8169	.8176	.8182	.8189
6.6	.8195	.8202	.8209	.8215	.8222	.8228	.8235	.8241	.8248	.8254
6.7	.8261	.8267	.8274	.8280	.8287	.8293	.8299	.8306	.8312	.8319
6.8	.8325	.8331	.8338	.8344	.8351	.8357	.8363	.8370	.8376	.8382
6.9	.8388	.8395	.8401	.8407	.8414	.8420	.8426	.8432	.8439	.8445
7.0	.8451	.8457	.8463	.8470	.8476	.8482	.8488	.8494	.8500	.8506
7.1	.8513	.8519	.8525	.8531	.8537	.8543	.8549	.8555	.8561	.8567
7.2	.8573	.8579	.8585	.8591	.8597	.8603	.8609	.8615	.8621	.8627
7.3	.8633	.8639	.8645	.8651	.8657	.8663	.8669	.8675	.8681	.8686
7.4	.8692	.8698	.8704	.8710	.8716	.8722	.8727	.8733	.8739	.8745
7.5	.8751	.8756	.8762	.8768	.8774	.8779	.8785	.8791	.8797	.8802
7.6	.8808	.8814	.8820	.8825	.8831	.8837	.8842	.8848	.8854	.8859
7.7	.8865	.8871	.8876	.8882	.8887	.8893	.8899	.8904	.8910	.8915
7.8	.8921	.8927	.8932	.8938	.8943	.8949	.8954	.8960	.8965	.8971
7.9	.8976	.8982	.8987	.8993	.8998	.9004	.9009	.9015	.9020	.9025
8.0	.9031	.9036	.9042	.9047	.9053	.9058	.9063	.9069	.9074	.9079
8.1	.9085	.9090	.9096	.9101	.9106	.9112	.9117	.9122	.9128	.9133
8.2	.9138	.9143	.9149	.9154	.9159	.9165	.9170	.9175	.9180	.9186
8.3	.9191	.9196	.9201	.9206	.9212	.9217	.9222	.9227	.9232	.9238
8.4	.9243	.9248	.9253	.9258	.9263	.9269	.9274	.9279	.9284	.9289
8.5	.9294	.9299	.9304	.9309	.9315	.9320	.9325	.9330	.9335	.9340
8.6	.9345	.9350	.9355	.9360	.9365	.9370	.9375	.9380	.9385	.9390
8.7	.9395	.9400	.9405	.9410	.9415	.9420	.9425	.9430	.9435	.9440
8.8	.9445	.9450	.9455	.9460	.9465	.9469	.9474	.9479	.9484	.9489
8.9	.9494	.9499	.9504	.9509	.9513	.9518	.9523	.9528	.9533	.9538
9.0	.9542	.9547	.9552	.9557	.9562	.9566	.9571	.9576	.9581	.9586
9.1	.9590	.9595	.9600	.9605	.9609	.9614	.9619	.9624	.9628	.9633
9.2	.9638	.9643	.9647	.9652	.9657	.9661	.9666	.9671	.9675	.9680
9.3	.9685	.9689	.9694	.9699	.9703	.9708	.9713	.9717	.9722	.9727
9.4	.9731	.9736	.9741	.9745	.9750	.9754	.9759	.9763	.9768	.9773
9.5	.9777	.9782	.9786	.9791	.9795	.9800	.9805	.9809	.9814	.9818
9.6	.9823	.9827	.9832	.9836	.9841	.9845	.9850	.9854	.9859	.9863
9.7	.9868	.9872	.9877	.9881	.9886	.9890	.9894	.9899	.9903	.9908
9.8	.9912	.9917	.9921	.9926	.9930	.9934	.9939	.9943	.9948	.9952
9.9	.9956	.9961	.9965	.9969	.9974	.9978	.9983	.9987	.9991	.9996

TABLE C Base-*e* Logarithms

N	0	1	2	3	4	5	6	7	8	9
1.0	.0000	.0100	.0198	.0296	.0392	.0488	.0583	.0677	.0770	.0862
1.1	.0953	.1044	.1133	.1222	.1310	.1398	.1484	.1570	.1655	.1740
1.2	.1823	.1906	.1989	.2070	.2151	.2231	.2311	.2390	.2469	.2546
1.3	.2624	.2700	.2776	.2852	.2927	.3001	.3075	.3148	.3221	.3293
1.4	.3365	.3436	.3507	.3577	.3646	.3716	.3784	.3853	.3920	.3988
1.5	.4055	.4121	.4187	.4253	.4318	.4383	.4447	.4511	.4574	.4637
1.6	.4700	.4762	.4824	.4886	.4947	.5008	.5068	.5128	.5188	.5247
1.7	.5306	.5365	.5423	.5481	.5539	.5596	.5653	.5710	.5766	.5822
1.8	.5878	.5933	.5988	.6043	.6098	.6152	.6206	.6259	.6313	.6366
1.9	.6419	.6471	.6523	.6575	.6627	.6678	.6729	.6780	.6831	.6881
2.0	.6931	.6981	.7031	.7080	.7129	.7178	.7227	.7275	.7324	.7372
2.1	.7419	.7467	.7514	.7561	.7608	.7655	.7701	.7747	.7793	.7839
2.2	.7885	.7930	.7975	.8020	.8065	.8109	.8154	.8198	.8242	.8286
2.3	.8329	.8372	.8416	.8459	.8502	.8544	.8587	.8629	.8671	.8713
2.4	.8755	.8796	.8838	.8879	.8920	.8961	.9002	.9042	.9083	.9123
2.5	.9163	.9203	.9243	.9282	.9322	.9361	.9400	.9439	.9478	.9517
2.6	.9555	.9594	.9632	.9670	.9708	.9746	.9783	.9821	.9858	.9895
2.7	.9933	.9969	1.0006	.0043	.0080	.0116	.0152	.0188	.0225	.0260
2.8	1.0296	.0332	.0367	.0403	.0438	.0473	.0508	.0543	.0578	.0613
2.9	.0647	.0682	.0716	.0750	.0784	.0818	.0852	.0886	.0919	.0953
3.0	1.0986	.1019	.1053	.1086	.1119	.1151	.1184	.1217	.1249	.1282
3.1	.1314	.1346	.1378	.1410	.1442	.1474	.1506	.1537	.1569	.1600
3.2	.1632	.1663	.1694	.1725	.1756	.1787	.1817	.1848	.1878	.1909
3.3	.1939	.1969	.2000	.2030	.2060	.2090	.2119	.2149	.2179	.2208
3.4	.2238	.2267	.2296	.2326	.2355	.2384	.2413	.2442	.2470	.2499
3.5	1.2528	.2556	.2585	.2613	.2641	.2669	.2698	.2726	.2754	.2782
3.6	.2809	.2837	.2865	.2892	.2920	.2947	.2975	.3002	.3029	.3056
3.7	.3083	.3110	.3137	.3164	.3191	.3218	.3244	.3271	.3297	.3324
3.8	.3350	.3376	.3403	.3429	.3455	.3481	.3507	.3533	.3558	.3584
3.9	.3610	.3635	.3661	.3686	.3712	.3737	.3762	.3788	.3813	.3838
4.0	1.3863	.3888	.3913	.3938	.3962	.3987	.4012	.4036	.4061	.4085
4.1	.4110	.4134	.4159	.4183	.4207	.4231	.4255	.4279	.4303	.4327
4.2	.4351	.4375	.4398	.4422	.4446	.4469	.4493	.4516	.4540	.4563
4.3	.4586	.4609	.4633	.4656	.4679	.4702	.4725	.4748	.4770	.4793
4.4	.4816	.4839	.4861	.4884	.4907	.4929	.4951	.4974	.4996	.5019
4.5	1.5041	.5063	.5085	.5107	.5129	.5151	.5173	.5195	.5217	.5239
4.6	.5261	.5282	.5304	.5326	.5347	.5369	.5390	.5412	.5433	.5454
4.7	.5476	.5497	.5518	.5539	.5560	.5581	.5602	.5623	.5644	.5665
4.8	.5686	.5707	.5728	.5748	.5769	.5790	.5810	.5831	.5851	.5872
4.9	.5892	.5913	.5933	.5953	.5974	.5994	.6014	.6034	.6054	.6074
5.0	1.6094	.6114	.6134	.6154	.6174	.6194	.6214	.6233	.6253	.6273
5.1	.6292	.6312	.6332	.6351	.6371	.6390	.6409	.6429	.6448	.6467
5.2	.6487	.6506	.6525	.6544	.6563	.6582	.6601	.6620	.6639	.6658
5.3	.6677	.6696	.6715	.6734	.6752	.6771	.6790	.6808	.6827	.6845
5.4	.6864	.6882	.6901	.6919	.6938	.6956	.6974	.6993	.7011	.7029

Use the properties of logarithms and ln 10 ≈ 2.3026 to find logarithms of numbers less than 1 or greater than 10.

TABLE C (continued)

N	0	1	2	3	4	5	6	7	8	9
5.5	1.7047	.7066	.7084	.7102	.7120	.7138	.7156	.7174	.7192	.7210
5.6	.7228	.7246	.7263	.7281	.7299	.7317	.7334	.7352	.7370	.7387
5.7	.7405	.7422	.7440	.7457	.7475	.7492	.7509	.7527	.7544	.7561
5.8	.7579	.7596	.7613	.7630	.7647	.7664	.7681	.7699	.7716	.7733
5.9	.7750	.7766	.7783	.7800	.7817	.7834	.7851	.7867	.7884	.7901
6.0	1.7918	.7934	.7951	.7967	.7984	.8001	.8017	.8034	.8050	.8066
6.1	.8083	.8099	.8116	.8132	.8148	.8165	.8181	.8197	.8213	.8229
6.2	.8245	.8262	.8278	.8294	.8310	.8326	.8342	.8358	.8374	.8390
6.3	.8405	.8421	.8437	.8453	.8469	.8485	.8500	.8516	.8532	.8547
6.4	.8563	.8579	.8594	.8610	.8625	.8641	.8656	.8672	.8687	.8703
6.5	1.8718	.8733	.8749	.8764	.8779	.8795	.8810	.8825	.8840	.8856
6.6	.8871	.8886	.8901	.8916	.8931	.8946	.8961	.8976	.8991	.9006
6.7	.9021	.9036	.9051	.9066	.9081	.9095	.9110	.9125	.9140	.9155
6.8	.9169	.9184	.9199	.9213	.9228	.9242	.9257	.9272	.9286	.9301
6.9	.9315	.9330	.9344	.9359	.9373	.9387	.9402	.9416	.9430	.9445
7.0	1.9459	.9473	.9488	.9502	.9516	.9530	.9544	.9559	.9573	.9587
7.1	.9601	.9615	.9629	.9643	.9657	.9671	.9685	.9699	.9713	.9727
7.2	.9741	.9755	.9769	.9782	.9796	.9810	.9824	.9838	.9851	.9865
7.3	.9879	.9892	.9906	.9920	.9933	.9947	.9961	.9974	.9988	2.0001
7.4	2.0015	.0028	.0042	.0055	.0069	.0082	.0096	.0109	.0122	.0136
7.5	2.0149	.0162	.0176	.0189	.0202	.0215	.0229	.0242	.0255	.0268
7.6	.0281	.0295	.0308	.0321	.0334	.0347	.0360	.0373	.0386	.0399
7.7	.0412	.0425	.0438	.0451	.0464	.0477	.0490	.0503	.0516	.0528
7.8	.0541	.0554	.0567	.0580	.0592	.0605	.0618	.0631	.0643	.0656
7.9	.0669	.0681	.0694	.0707	.0719	.0732	.0744	.0757	.0769	.0782
8.0	2.0794	.0807	.0819	.0832	.0844	.0857	.0869	.0882	.0894	.0906
8.1	.0919	.0931	.0943	.0956	.0968	.0980	.0992	.1005	.1017	.1029
8.2	.1041	.1054	.1066	.1078	.1090	.1102	.1114	.1126	.1138	.1150
8.3	.1163	.1175	.1187	.1199	.1211	.1223	.1235	.1247	.1258	.1270
8.4	.1282	.1294	.1306	.1318	.1330	.1342	.1353	.1365	.1377	.1389
8.5	2.1401	.1412	.1424	.1436	.1448	.1459	.1471	.1483	.1494	.1506
8.6	.1518	.1529	.1541	.1552	.1564	.1576	.1587	.1599	.1610	.1622
8.7	.1633	.1645	.1656	.1668	.1679	.1691	.1702	.1713	.1725	.1736
8.8	.1748	.1759	.1770	.1782	.1793	.1804	.1815	.1827	.1838	.1849
8.9	.1861	.1872	.1883	.1894	.1905	.1917	.1928	.1939	.1950	.1961
9.0	2.1972	.1983	.1994	.2006	.2017	.2028	.2039	.2050	.2061	.2072
9.1	.2083	.2094	.2105	.2116	.2127	.2138	.2148	.2159	.2170	.2181
9.2	.2192	.2203	.2214	.2225	.2235	.2246	.2257	.2268	.2279	.2289
9.3	.2300	.2311	.2322	.2332	.2343	.2354	.2364	.2375	.2386	.2396
9.4	.2407	.2418	.2428	.2439	.2450	.2460	.2471	.2481	.2492	.2502
9.5	2.2513	.2523	.2534	.2544	.2555	.2565	.2576	.2586	.2597	.2607
9.6	.2618	.2628	.2638	.2649	.2659	.2670	.2680	.2690	.2701	.2711
9.7	.2721	.2732	.2742	.2752	.2762	.2773	.2783	.2793	.2803	.2814
9.8	.2824	.2834	.2844	.2854	.2865	.2875	.2885	.2895	.2905	.2915
9.9	.2925	.2935	.2946	.2956	.2966	.2976	.2986	.2996	.3006	.3016

INDEX

INDEX OF APPLICATIONS

PROJECTS

CAREERS IN MATHEMATICS

APPLICATIONS

Boldface page numbers refer to Examples; lightface numbers refer to Exercises.

Business and Industry

Finance and Investing

Geometry

Home and Farm Management

Medicine and Health

Miscellaneous

Recreation, Entertainment, and Travel

Science, Technology, and Engineering